Sixth Scientific Conference "Renewable Energy and Its Applications" 2018

AA006038

Journal of Physics: Conference Series Volume 1032

Karbala, Iraq
21 – 22 February 2018

ISBN: 978-1-5108-6407-8
ISSN: 1742-6588

Printed from e-media with permission by:

Curran Associates, Inc.
57 Morehouse Lane
Red Hook, NY 12571

Some format issues inherent in the e-media version may also appear in this print version.

Printed by Curran Associates, Inc. (2018)

For permission requests, please contact the Institute of Physics
at the address below.

Institute of Physics
Dirac House, Temple Back
Bristol BS1 6BE UK

Phone: 44 1 17 929 7481
Fax: 44 1 17 920 0979

techtracking@iop.org

Additional copies of this publication are available from:

Curran Associates, Inc.
57 Morehouse Lane
Red Hook, NY 12571 USA
Phone: 845-758-0400
Fax: 845-758-2633
Email: curran@proceedings.com
Web: www.proceedings.com

Table of contents

Volume 1032

The Sixth Scientific Conference "Renewable Energy and its Applications"

21–22 February 2018, Karbala, Iraq

Accepted papers received: 16 May 2018
Published online: 4 June 2018

Preface

Papers

IOP Conf. Series: Journal of Physics: Conf. Series **1032** (2018) 011001 doi:10.1088/1742-6596/1032/1/011001

The Sixth Scientific Conference "Renewable Energy and its Applications"

21-22 February 2018, Karbala, Iraq

The sixth conference entitle: "Renewable Energy and its applications" had been held in the main building of College of Science, 56001 Karbala, IRAQ. The aim of it is to investigate the main challenges of current energy resources in the world, to identify the new energy resources and how to reduce the energy consumption by using the renewable and green technology concept, to create the scientific platform for renewable energy researcher and prepare the suitable environment to apply, discuss and share their knowledge for more enhancements in this field and activate the role of University to develop the society and the country.

IOP Conf. Series: Journal of Physics: Conf. Series **1032** (2018) 011002 doi:10.1088/1742-6596/1032/1/011002

Peer review statement

All papers published in this volume of *Journal of Physics: Conference Series* have been peer reviewed through processes administered by the proceedings Editors. Reviews were conducted by expert referees to the professional and scientific standards expected of a proceedings journal published by IOP Publishing.

IOP Conf. Series: Journal of Physics: Conf. Series **1032** (2018) 012001 doi:10.1088/1742-6596/1032/1/012001

Study of structural and optical properties of the ferrite films prepared by chemical pyrolysis spray technique

Gh. J. Alyasary[1],H. H. Almarmade[2] and D. h. Altiee[1]

[1]University of Al-Qadisiyah College of Education/Dep. of Physics.

[2] University of Sumer College of basic education.

Ghafran.altiee@gmail.com, sumer55noor@gmail.com

Abstract. Copper-zinc ferrite ($CuFe_2O_4$) films have been deposited successfully by spray pyrolysis technique. Ferrite films deposited of thickness (450 ± 15nm) at temperature (400 ± 10 °C) and annealing (600 °C) for range of (4h). The compound was doping with zinc at rates (x= 0,0.2,0.4). The crystalline structure of the prepared films was determined Tetragonal a spin Ferrite in peak (211), and when it was doping with zinc, it turned into cubic in peaks (311). The particle size was found to decrease by increasing the concentration of (Zn). The optical properties have been identified of films wavelengths of (300-900) nm, The results showed that the transmittance for all thin films increases as the wavelength increases then saturates at higher wavelengths.

1. Introduction

The Formula AB_2O_4 refers to spinal ferrites are very big class of oxides with very large application in industry from simple permanent magnets to microwave application ,magnetic recording ,gases sensors[1]. The relative control of the magnetizations of the layers in these devices necessitates to spin the magnetization of the layers this is due to by magnetically coupling this sheet to a hard one[2]. To deposition thin films of ferrite[3], scientist have find a number of methods such as plating ferrite, sol-gel method ,Chemical method[4] ,electro deposition method ,sputtering , RF sputtering and spin spray pyrolysis[5]. Microstructure of thin film is an important parameter in determining their magnetic[6]. There are various growth parameters that affect the microstructure of the films like substrate temperature during deposition [7], Thermal expansion coefficient and lattice mismatch between the target and substrate [8].

2. Experiments

$Zn_xCu_{1-x}Fe_2O_4$ thin films with (x=0.0,0.2,0.4) of thickness (450 ± 15nm) were prepared by spray pyrolysis technique on substrate of SiO_2/Si at (400±10°C) substrate cleaning plays an important aim in the deposition of thin films because for the chemical deposition as the contaminated surface provide nucleation sites facilitating growth resulting into non-homogeneous film with different orientation and impurities $Zn_xCu_{1-x}Fe_2O_4$ thin film were prepared by spray pyrolysis technique, Analytic reagent grade chemical ($Fe(C_2O_4).2H_2O$), ($ZnCl_2$) ($Cu(NO_3)_2.3H_2O$) were used as raw materials, solutions were dissolved separately in de- ionized water at the concentration of(0.0 2) M for Zn^{+2} final solution were prepared by mixing these three solution in 1:2 with the ration of (x=0.0,0.2,0.4).

$ZnCl_2 + 2H_2O \xrightarrow{heat} ZnO_{(S)} + Cl_{2(g)} + H_2O_{(L)} + H_{2(g)}$ .... (1)

$2Cu\ (NO_3)_2.3H_2O \xrightarrow{heat} 2CuO_{(s)} + 4\ NO_{2(g)} + O_{2(g)} + 3\ H_2O_{(L)}$ .... (2)

$Fe(C_2O_4) \xrightarrow{heat} FeO_{(S)} + CO_{2(g)} + CO_{(g)} \rightarrow 2Fe_2O_{(s)} + 1/2\ O_{2\ (g)} + heat \rightarrow \gamma\text{-}Fe_2O_3$ .... (3)

$(1\text{-}x)ZnO + (x)CuO + Fe_2O_3 \xrightarrow{heat} Zn_{(1\text{-}x)}Cu_xFe_2O_4$ .... (4)

3. Results and Discussion

3.1. Structure properties

From figure (1) X-ray diffraction results showed that all films were prepared by chemical pyrolysis spray method were spinal ferrite .The diffraction patterns corresponding after annealing reveal that the film crystallize in polycrystalline preferred orientation along (211) plan type of tetragonal When it was added (Zn^{+2}) it turned into (cubic) in peaks (311). The lattice constant was calculated for pure Cu Fe_2O_4 films. The results in Table (1) were found to have increase value than the deflection ratio. The average size was calculated with the increase in the ratio of fouling and according to the results listed in Table (1) because the radius of the ionic atom of the copper is smaller than the radius of the ionic atom of the zinc of the crystal and thus reduce average size. The lattice constant (a_o) calculated from equation [9,10]:

$$a_o = d\sqrt{h^2 + k^2 + l^2}$$

The average crystallite size (D) was calculated by using Scherrer's formula [11,12].

$$D = (0.9 \times \lambda)/ (\beta \times \cos\theta)$$

where, λ the wavelength of X-rays (1.5406 Å), β is the full-width at half-maximum in radian and θ the angle of diffraction. It is seen that as the Zn content increases the crystallite size decreases from 17.44 to 14.96 nm.

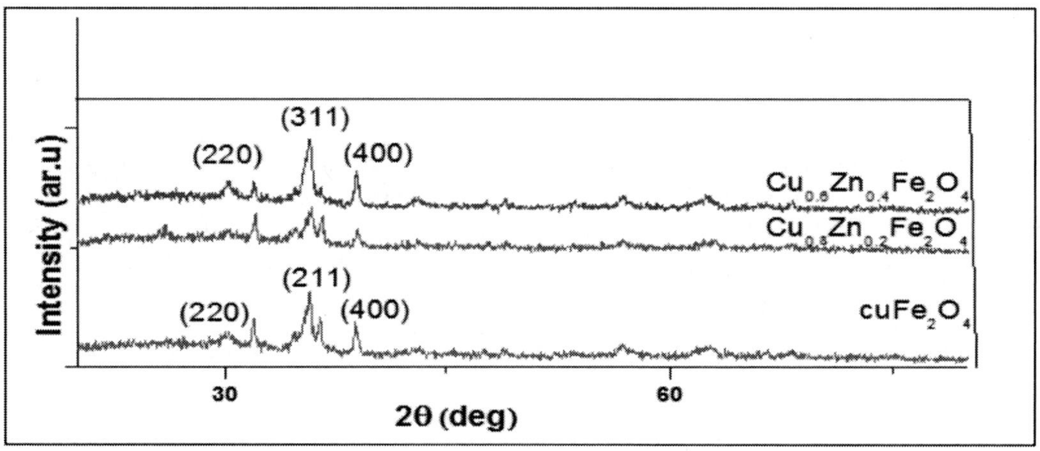

Figure 1. X-ray diffraction patterns.

IOP Conf. Series: Journal of Physics: Conf. Series **1032** (2018) 012001 doi:10.1088/1742-6596/1032/1/012001

Table 1. lattice constant (a), crystallite size(D_{ave}).

x-value	Lattice Parameter (A°)	D (nm)
0.0	a= 5.8005, C=8.5985	17.44
0.2	8.39	15.92
0.4	8.42	14.96

3.2. Optical properties

3.2.1. Transmission (T)

Figure (2) shows the relation between transmittance and wavelength in the range of (300 - 900) nm for Cu-Zinc Ferrite films. The transmittance for all thin films increases as the wavelength increases in the wavelength range of about (300-600) nm, and then saturates at higher wavelengths. This material is good in absorbance at wavelengths of (300-600) so it can be used in the work of reagents or solar cells. The spectrum shows high transmittance in the visible and infrared regions, and low in the ultraviolet region.

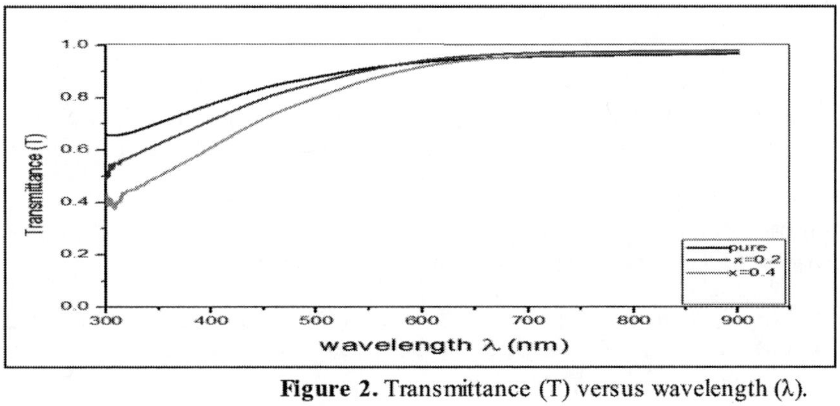

Figure 2. Transmittance (T) versus wavelength (λ).

3.2.2. Absorbance (A)

The variation of the absorbance spectrum with wavelength is opposite to the transmittance spectrum. The study of absorbance was in the range of (300– 900) nm. Figure (3) shows the relation between absorbance (A) and wavelength for Copper -Zinc Ferrite thin films. It is clear that absorbance decreases rabidily the wavelength increases in the wavelength range of about of (300-600) nm, and then decreases slowly at higher wavelengths.

It can be noticed that the absorbance decreases as the concentration of Zinc increases and this can be attributed to the localized conductivity levels introduced by Zn^{2+} cations.

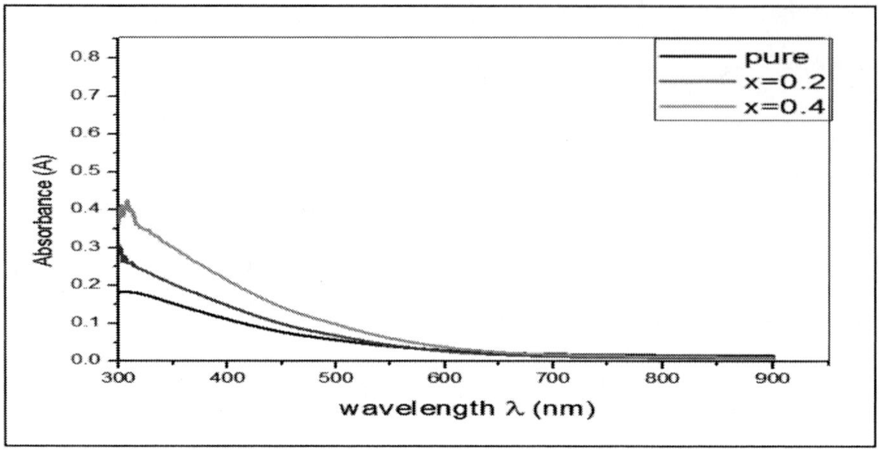

Figure3. Absorbance (A) versus wavelength (λ).

4. Conclusion

The results obtained from the study of the structural characteristics of $Zn_xCu_{1-x}Fe_2O_4$ films and in terms of (0,0.2,0.4) showed that the process of attribution to the variable of the nature of crystalline structure, The prepared films are Tetragonal type the preferred direction(211)When doping with (Zn^{+2}) Turned into cubic type the preferred direction (311). From the optical properties found increase in transmittance and decrease in absorbance which meant an improvement in crystalline structure.

Reference

[1] D. K. Pawara, S. M. Pawarb, P. S. Patil and S. S. Kolekara,"Synthesis of Nanocrystalline Nickel–Zinc Ferrite($Ni_{0.8}Zn_{0.2}Fe_2O_4$) Thin Films by Chemical Bath DepositionMethod", Journal of Alloys andCompounds, Vol. 509, pp. 3587-3591, (2011)

[2] Anuj Jain ,Ravi Kanl Baranwal ,Ajaya Bharti,Z.Vakil and C.S.Prajapati,"Study of Zn-Cu Ferrite nanoparticales for LPG sensing " ,Journal of the scientific world ,NO. ,PP. ,(2013).

[3] A.Manikandan ,J.Judith Vijaya,L.John Kenned and M.Bonoudina "Structural ,optical and magnetic properties of $Zn_{1-x}Cu_xFe_2O_4$ nanoparticales prepared by Microwave combusition method" Journal of molecular structure ,vol.1035,pp.338-340,(2013) .

[4] S. S. Kumbhar, M. A. Mahadik, V. S. Mohite, K. Y. Rajpure andC. H. Bhosale, "Synthesis and Characterization of SprayDeposited Nickel-Zinc Ferrite Thin Films", 4th International Conference on Advances in Energy Research, Vol. 54, pp. 599-605, (2014).

[5] P. J. Brown and j. B. Forsyth, "The crystal structure of solid", Arnold, (1973).

[6] M. Caglar, S. Ilican and Y. Caglar, "Influnce of substrate temperature on structural and electrical properties of ZnO films", Trakya Univ. J Sci, Vol. 7, No. 2, pp. (153-158), (2006).

[7] A. T. Raghavender, N. H. Hong, E. Chikoidze, Y. Dumont and M.Kurisu, "Effect of Zinc Doping on The Structural andMagnetic Properties of Nickel Ferrite Thin Films FabricatedPulsed Laser Deposition Technique", Journal of Magnetismand Magnetic Materials, Vol. 378, pp. 358-361, (2015).

[8] M.I. Khan, K.A. Bhatti, Rabia Qindeel," Characterizations of multilayer ZnO thin films deposited by sol-gel spincoating technique", Journal of results in physics, No.7, pp.651–655, (2017).

[9] Liao DL, Badour CA, Liao BQ. Preparation of nano sized TiO2/ZnO composite catalyst and its photocatalytic activity for degradation of methyl orange. JPhotochem Photobiol A Chem 2008;194:11.

[10] S.N. Klausen, P.A. Lindgard, K. Lefmann, F. Bodker, S. Moorup, Phys. Status Solidi A, 189 (2002) 1039.

[11] R. Laishram, C. Prakash, Journal of Magnetism and Magnetic Materials, 305 (2006) 35. 24. Cao Jun-gang, Li. Jian-jun, Duan Hai-feng, Lin Ying-jie, Chem. Res. Chinese Universities, 28 (2012) 590.

[12] P. Priyadharsini, A. Pradeep, P. S. Rao, G. Chandrasekaran, Mater. Chem. Phys., 116 (2009) 207.

IOP Conf. Series: Journal of Physics: Conf. Series **1032** (2018) 012002 doi:10.1088/1742-6596/1032/1/012002

Studying the Effect of Deposition Time on Optical Properties of CdS Thin Films

R. A. Al-Wardy

Department of clinical laboratory science/college of pharmacy/Al-mustansiriyah university

e-mail: dr.rusuladnan@uomustansiriyah.edu.iq

Abstract. We have studied in this work the effect of deposition time on the absorption spectra and the optical energy gap of CdS thin film. That prepared by chemical bath deposition method (CBD) on glass substrate at temperature of (80 ± 5)°C. The structural features were examined by X-ray diffraction (XRD) is wurtzite structure for all samples. Transmittance spectra have been recorded in order to determine absorbance, reflectivity, absorption coefficient and the optical constants such as refractive index, extinction coefficient, real and imaginary part of dielectric constant. It was seen that all the parameters under investigation affected by deposition time (1,2 3, and 4) hours.

Key words: chemical bath technique, optical constants, Cadmium sulfide

1. Introduction

Direct band gap thin film cadmium sulfide (CdS) has been doing the subject of intensive research because of its intermediate band gap, high absorption coefficient, reasonable conversion efficiency and stability [1]. CdS thin films are being widely investigated for applications in photo electrochemical cells, photo catalysis, optical switching and single electron transistors, light emitting diodes, laser materials, optical waveguides, electrochemical cells, gas sensors, and metal-Schottky barrier cells [2]. There are various techniques have been explored to deposition of CdS film, such as thermal evaporation, sputtering, molecular beam epitaxy, spray pyrolysis, chemical bath deposition, and close space sublimation (CSS), etc [3-8].

The chemical bath deposition (CBD) is a simple and inexpensive deposition technique most widely employed to obtain homogeneous, hard, adherent, transparent and stoichiometric CdS thin films [9]. Most CBD films reported are fairly transparent to very transparent, typically between 60 and 90% transparent in the sub-band gap region [10-12]. The deposition time of the films increases [13] or decrease [10] the optical transmission, according to deposition time of film. The absorption edge shifts towards higher energy region and increase in the band gap [11,13, 14]. Ali et al related the grain size dependence on the band gap for the films [10].

The aim of this work is studying the effect of various deposition time (1,2,3,and 4) hours on optical properties of CdS thin films prepared by CBD method. In the optical studies of CdS thin films showed a high absorbance in the area of visible region, this made the deposited film suitable in the designing of solar cells.

2. Experimental Work

CdS thin films were deposited on a cleaned glass substrate by the chemical bath deposition CBD method at various deposition times. First, placed 0.384gm of cadmium chloride ($CdCl_2$) as a source of cadmium in 240ml distilled water under vigorous stirring about 15 min, (NaOH) was added drop by drop in ($CdCl_2$) solution under stirring to maintain the pH value of the solution is (10.5) and a fuzzy solution was obtained, then placed 0.768gm of thiourea $CS(NH_2)_2$ as a source of sulfur in 240ml distilled water under vigorous stirring about 15 min.

The solution was mixed in 500 ml beaker with added distilled water to complete 500ml with continuous stirring by magnetic stirrer about 20 min, the solution temperature was kept constant at (80 ±5 °C), the solution color change from fuzzy to yellow. This equation illustrates the interaction[3]:

$$CdCl_2 + CS(NH_2)_2 + 2H_2O \rightarrow\downarrow CdS + \uparrow 2NH_4Cl + \uparrow CO_2$$

The glass substrate that used for deposition CdS was cleaned by methanol and acetone for 10 min by using an ultrasonic cleaner and then cleaned with de ionized water and dried. The clean glass slides were vertically immersed in the beaker contain the solution whose temperature were kept around (80 ±5 °C) during the growth. After the deposition time was change from (1,2,3and 4) hours, the samples were pulled out from the bath and washed in distilled water and finally dried in air. CdS films obtained were hard, and had very good adhesion to the glass substrate. Finally, all films of CdS annealing at (300°C) about 45 min. Figure 1 shows CdS thin film deposited on glass-substrate by CBD method with deposition time at 1hour. The measurements of CdS thin films were investigated using SP-3000NANO optima (UV-Vis spectrophotometer) in the wavelength region of (200 to 1100) nm to exam the optical transmission and absorbance, while XRD results of these films where measured by SHIMADZU-6000 (CuK_α with wavelength 1.54056Å), the 2θ around $(20-80)^\circ$.

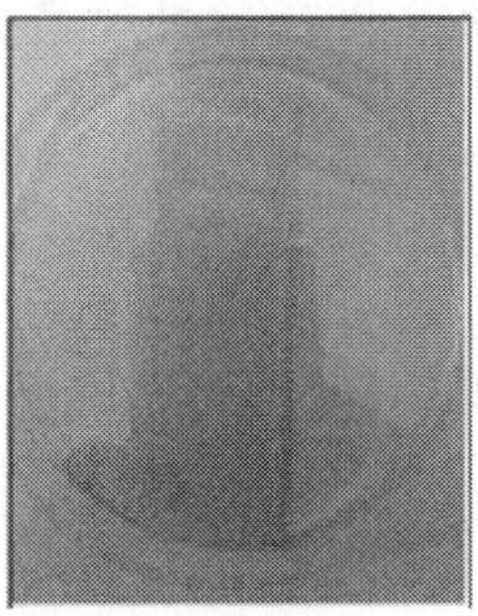

Figure 1. CdS thin film deposited on glass-substrate by CBD method with deposition time at 1hour.

3. Results and Discussion

3.1 Structural Properties by XRD

Figure 2 shows the crystalline structure that analyses by XRD of the CdS thin films are prepared by chemical bath deposition on the glass substrate with varying deposition times (1,2,3and 4) hours

IOP Conf. Series: Journal of Physics: Conf. Series **1032** (2018) 012002 doi:10.1088/1742-6596/1032/1/012002

respectively, all deposited films have polycrystalline with the hexagonal phase also all diffraction peaks of CdS samples at different deposition time correspond to (JCPDS no. 77-2306) this agreement with [15], it indicates that the very intense (002) plane with a preferential orientation (001) for all sample of wurtize CdS also two lattice constants a = 4.1409Å, c= 6.7198Å. The average crystal size (D) of CdS thin films about (22.25 nm) was derived from FWHM of peak corresponding to 2θ= 24.93 for sample that calculates using Scherer's equation [16]:

$$D = \frac{0.9\,\lambda}{\beta\, cos\theta} \quad \dots (1)$$

Where D is the average grain size, λ is of X-ray wavelength (0.15418 nm), β is the full-width at half maximum (FWHM) of the peak which has maximum intensity and θ is the Bragg's angle. this obvious that the films made up of nanocrystal particles this agreement with [17], the range of 2θ 20° to 80°. In fact, CdS has three forms of structure are cubic, hexagonal and rock salt, the structure more stable is the hexagonal phase due to the method of preparation cadmium sulfide in addition to be the structure more favorable for manufacturing of a solar cells [1] .

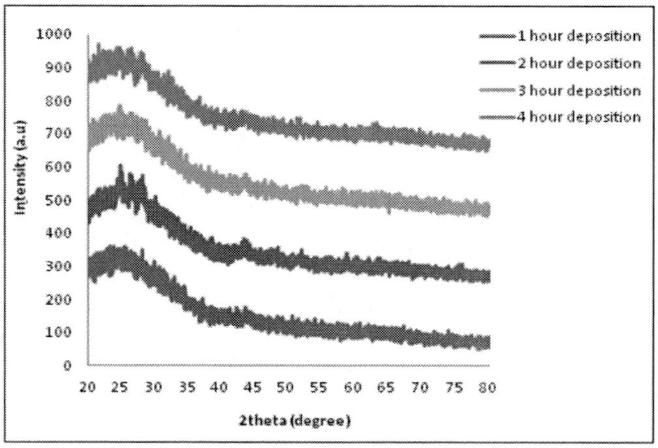

Figure 2: Thin films of CdS deposited on glass substrate by CBD at different time.

3.2 The Transmission and Absorption

Optical transmittance and absorbance spectra as a function of wavelength for CdS thin films at different deposition times (1,2,3 and 4) hours, are shown in figure 3. A maximum value transmittance at (77%) in the near-infrared region for CdS film at deposition time (1hour) while increased to (82, 83, 86)% for CdS film at deposition time (2,3,4) hours, respectively. The increase in transmittance in the near infra-red region shows decrease absorbance as the wavelength increases along the near infra-red region of the spectrum and height absorption in the visible light region for all samples of CdS this result is good agreement with research [13].

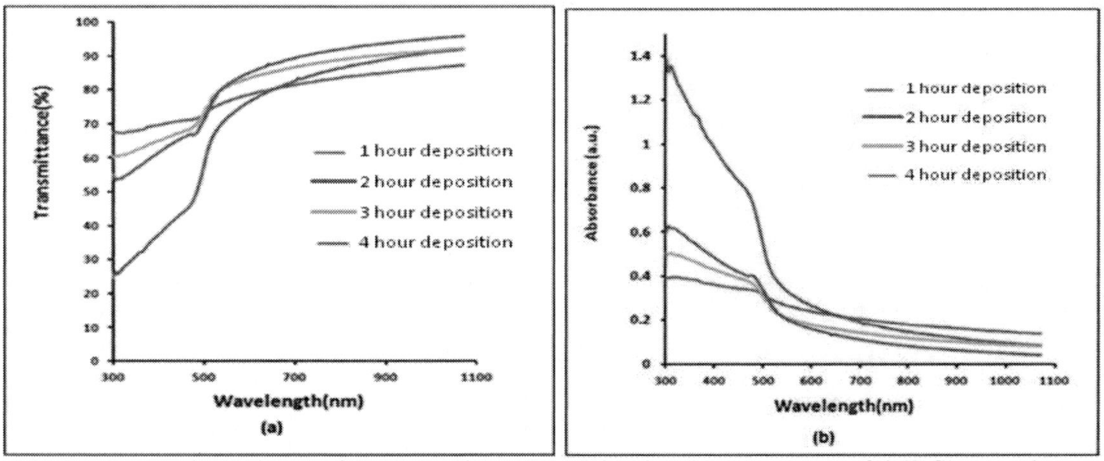

Figure 3. The dependence of (a) transmittance and (b) absorbance spectra of CdS films on the wavelength.

3.3 The Absorption Coefficient:

The absorption coefficient (α) associated with the strong absorption region of the film was calculated from absorbance (A) and the thin film thickness (t) as equation [18]:

$$\boldsymbol{\alpha = 2.303\frac{A}{t}} \quad ....(2)$$

Where $A = \log\frac{1}{T}$ is the absorbance and t is the thickness of the CdS film about100 nm was measured by gravimetric method.

The behavior of absorption coefficient (α) of CdS thin films for different deposition times (1,2,3 and 4) hours as a function of photon energy and wavelength is shown in figure 4. Absorption coefficient (α) decrease in the law photon energy because the probability of the electrical transfer between valence band and the conduction band is very rare and it will gradual increase in the absorption coefficient toward the high energy ($hv > 2eV$)for all the samples [19], as the deposition time increase, there is a blue shift of the absorption edge (i.e. shift towards short wavelengths). This result is a good agreement with result of literature [20].

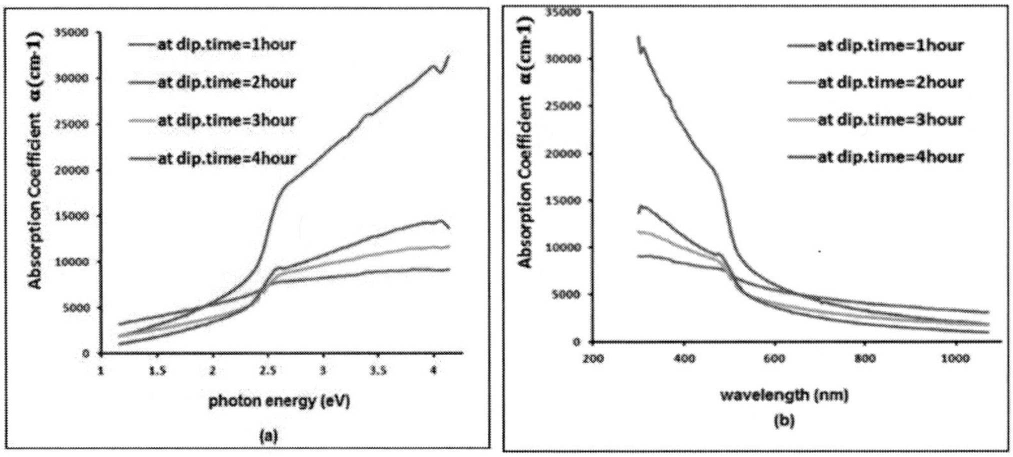

Figure 4. Optical absorption coefficient of CdS thin films prepared by CBD as a function of (a) photon energy (b) wavelength.

3.4 The Optical Band Gap

The absorption coefficient and optical band gap (E_g) are related by Tauc equation [21]:

$$\alpha = \left(hv - E_g\right)^2 \qquad ...(3)$$

Where hv is the incident photon energy, h Planck constant, v the frequency of incident photon, E_g the optical band gap.

The $(\alpha hv)^2$ versus (hv) plots for CdS films at different deposition times (1,2,3 and 4) hours are shown in figure 5. The optical band gap (E_g) value is determined by extrapolating the linear part of the plot $(\alpha hv)^2$ versus (hv) in the abscissa (x axis), which indicates a direct optical transition. The band gap value of the CdS film was found increases from (2.25, 2.4, 2.5 and 2.65) eV with increasing deposition times at (1,2,3 and 4) hours, respectively. This behavior is identical to the results of previous researches [10,11,13,14]. The increase in band gap value is due to crystal size decreases below a certain limiting size, associated with its exciton Bohr diameter. The spacing between levels in the bands becomes larger so that the energy structure changes from continuum band to discontinuous or discrete quantized levels and the band gap increases.

IOP Conf. Series: Journal of Physics: Conf. Series **1032** (2018) 012002 doi:10.1088/1742-6596/1032/1/012002

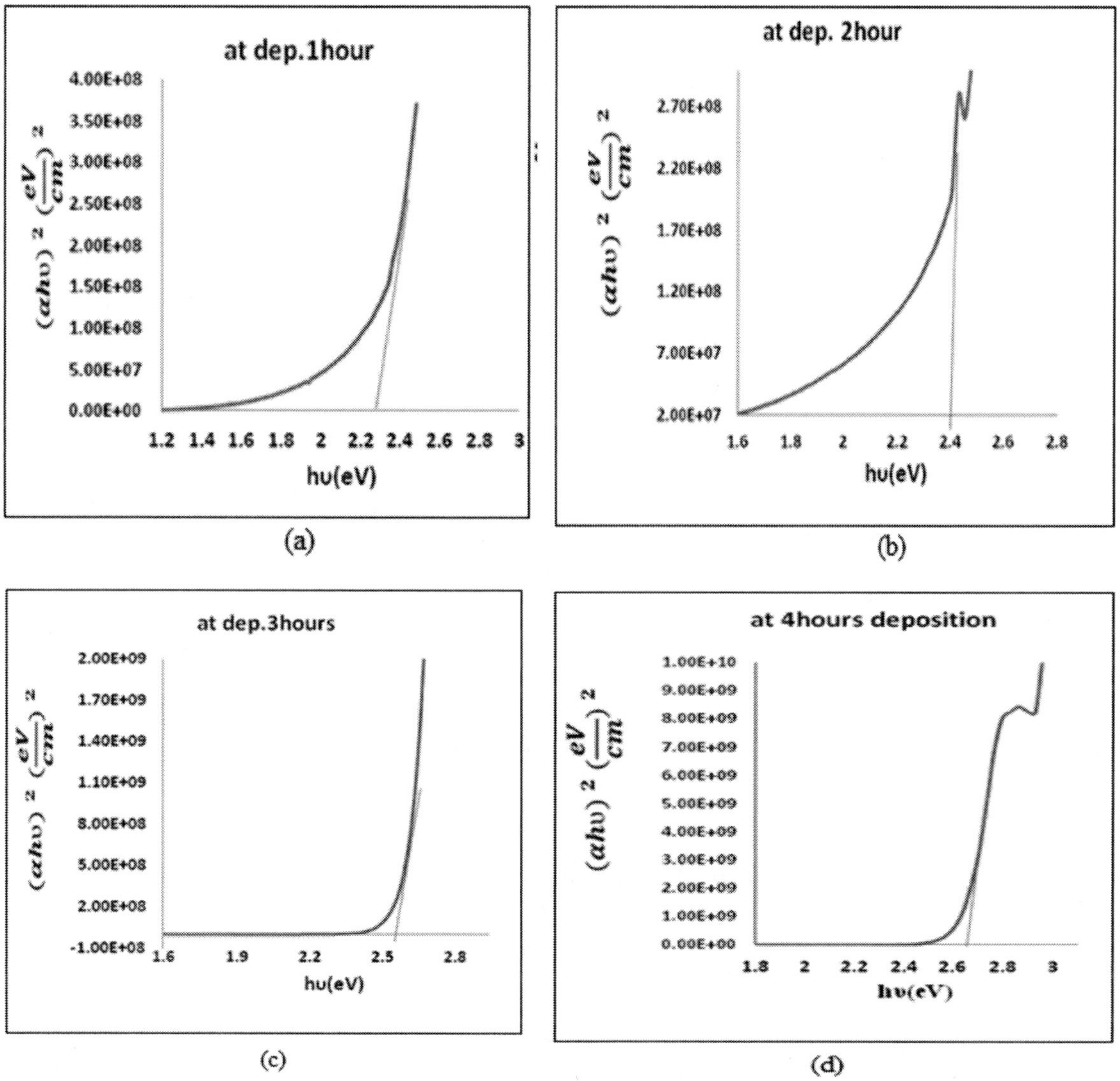

Figure 5. Variation of band gap for CdS thin films prepared by CBD at different deposition times .

3.5 The Extinction Coefficient

The extinction coefficient was evaluated using equation [22]:

$$K = \frac{\alpha\lambda}{4\pi} \quad \dots (4)$$

Figure 6 shows variation of the extinction coefficient with wavelength for CdS films with different deposition time. The extinction coefficient decreases as the wavelength increases, and its decreases as the deposition time increases. This decrease in K is due to the deposition time working on decrease the absorbance where K depends on the absorption coefficient (α) as which comes from decreasing the tails width of the localized states within the energy gap and then decreasing K according to equation 4.

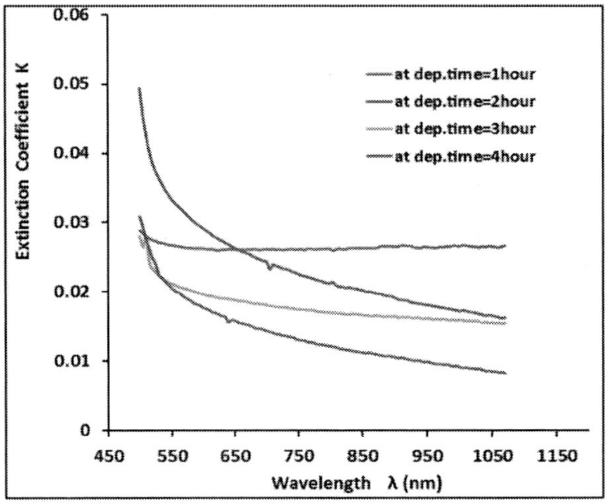

Figure 6. Extinction coefficient as a function of wavelength λ for CdS films at different deposition time.

3.6 The Refractive Index

The refraction index can be calculated from the relationship [23]:

$$n = \frac{1+R^{\frac{1}{2}}}{1-R^{\frac{1}{2}}} \qquad \dots.(5)$$

Where $R = 1 - T - A$ is the reflection of the films.

Figure 7 shows the variation of refractive index of CdS films with wavelength at different deposition time. The increase in the deposition time results in the overall decrease in the refractive index. This decrease in the refractive index indicates that the deposition time acts on the surface change of the film, leading to a change in the accurate structure of the film. Thus decreasing the packing density which increase the speed of light in thin film material, leads to decrease in the reflectance with the deposition time and decrease refractive index, according to equation 5.

IOP Conf. Series: Journal of Physics: Conf. Series **1032** (2018) 012002 doi:10.1088/1742-6596/1032/1/012002

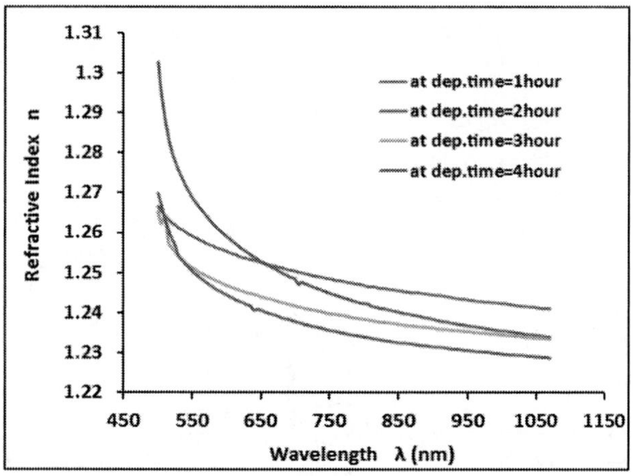

Figure 7. Refractive index as a function of wavelength λ for CdS films at different deposition time.

3.7 The Dielectric Constant

The real and imaginary parts of complex dielectric constant given by the following relations [24]:

$$\epsilon_1 = n^2 - K^2 \qquad \text{... (6)}$$

$$\epsilon_2 = 2nK \qquad \text{... (7)}$$

Figure 8 shows the variation of real and imaginary part of dielectric constant of the CdS thin films with wavelength at different deposition time (1,2,3 and 4) hours. The obtained results show that both values of real part ϵ_1 and imaginary part ϵ_2 of dielectric constant are decreased with increasing of wavelength for CdS thin films, and are decrease with increasing of deposition time. The behavior of ϵ_1 with the wavelength and the deposition time is the same as the behavior of the refractive index n to adopt ϵ_1 on the refractive index according to equation 6, While the behavior of ϵ_2 with the wavelength and the deposition time is the same as the extinction coefficient K, according to equation 7.

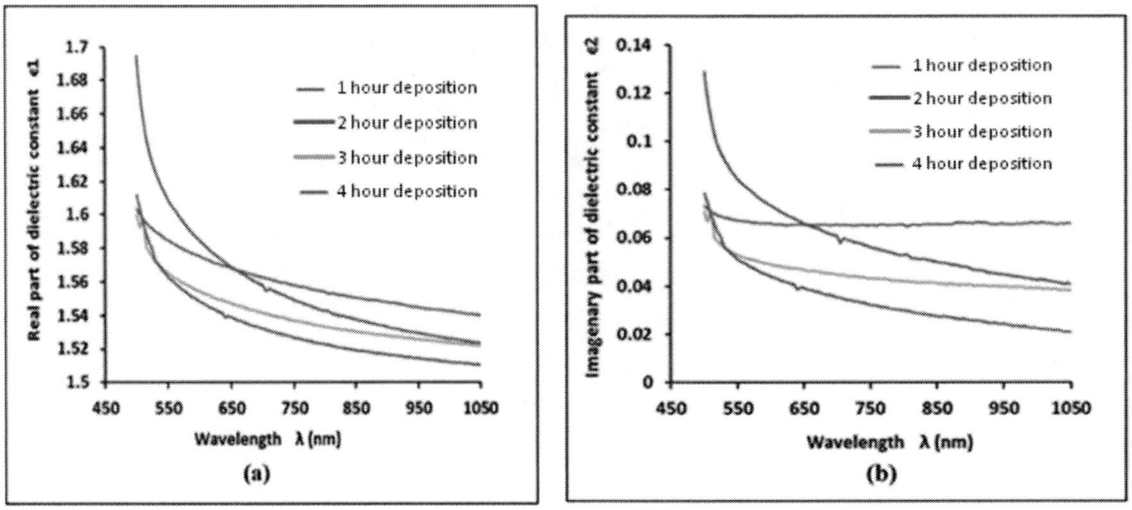

Figure 8. the Real and Imaginary Parts of Dielectric Constant as a function of wavelength λ for CdS films at different deposition time.

3.8 The Optical Conductivity

To calculate the optical conductivity used the equation [25]

$$\sigma_{optical} = \frac{\alpha\, n\, c}{4\,\pi} \qquad \dots (8)$$

Where $\sigma_{optical}$ is the optical conductance, c is the velocity of the radiation in the space, n is the refractive index and α is the absorption coefficient.

Figure 9, Shows the variation of optical conductivity as a function of photon energy for different deposition time of CdS thin films. From the figure, the optical conductivity of CdS thin films were increased with photon energy. This suggests that the increase in optical conductivity is due to electron exited by photon energy, and the optical conductivity of the films decreases with increasing deposition time.

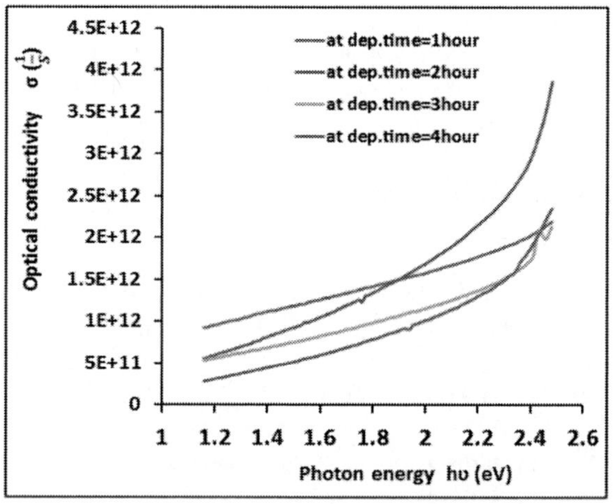

Figure 9. The Optical Conductivity of CdS thin films with different deposition time.

4.Conclusions

The CdS thin films with different deposition time (1,2,3 and 4) hours were prepared by chemical bath deposition method on glass substrate. The structure of the four films is hexagonal structure (Wurtize) according to XRD measurements. The optical properties showed that the transmittance for all films increased with the increasing in the wavelength range, and increases with the increase in the deposition time. The band gap increases when the deposition time increases and the band gap changed from 2.25 eV (for dep. time 1hour) to 2.65eV (for dep. Time 4hour), as a result of quantum confinement, which can be used in the application of thin film as window layer in solar cell fabrication.

Acknowledgment

I should like to grateful each of Assist prof. Sabah Ibrahim and lecture Hanan Abd Ali in physics department in college science/ Wassit university to help me to achieve this work.

References

[1] Feng K and Shenglin L 2017 *J. materials today communications* **13** 1-5 https://doi.org/10.1016/j.mtcomm.2017.08.003
[2] Lisco F, Kaminski P, Abbas A, Bass K, Bowers J, Claudio G, Losurdo M and Walls J 2015 *Thin Solid Films* **582** 323–327. http://doi.org/10.1016/j.tsf.2014.11.062
[3] Abdulla M, Raied K and Hani J 2011 *Adv. in Mate. Phys. and Chem.* **1** 64-69.
[4] Hussain K, Begum T, Mahmood Z, Syed I and Ahmed S 2014 *Int. J. Nanoscience and Nanoengineering* **1** 34-38.
[5] Romeo N, Bosio A, Canevari V and Seuret D 1987 *Solar Cells* **22** 23-27.
[6] Petillon S, Dinger A, Gru M, Hetterich M, Kazukauskas V, Klingshirn C, Liang J, Weise B and Wagner V 1999 *J. Cryst. Growth* **202** 453-456.

[7] Ashour A 2003 *Turk. J. Phys.* **27** 551-558.

[8] Calixto M, Albor-Aguilera M, Tufiño-Velázquez M, Contreras-Puente G and Morales-Acevedo A 2011 *Solar Cells – Thin-Film Technologies* ISBN 978-953-307-570-9.

[9] Al-Hussam A and Jassim S 2012 *Journal of the Association of Arab Universities for Basic and Applied Sciences* **11** 27–31.

Mousa A, Haider A and Al-Jawad S 2011 *Journal of Materials Science and Engineering* **5** 184-491.http://doi.org/10.17265/2161-6213/2011.02.011

[10]

[11] Al-Jawad S, Mousa A and Taha W 2009 *Um-Salama Science Journal* **6** 1.

[12] Al-Jawad S, Ahmed Z and Taha W 2009 *Iraq. J. App. Phys. Lett* **2** 2.

[13] Ezekoye B, Ighodalo K, Ezekoye V, Emeakaroha T, Ezema J and Offor P 2015 *International Journal of Physical Sciences* **10** 403-412.http://doi.org/10.5897/ijps2015.4360

[14] Rupom R, Matin R, Bashar M, Sultana M, Rahaman M, Gafur A, Hakim M, Hossain M, Bhuiyan M and Ahmed F 2016 *American International Journal of Research in Science, Technology, Engineering & Mathematics* **16** 109.

[15] Liu Y, Ma Y, Liu W, Shang Y, Zhu A, Tan P and Xiong X 2017 *Journal of Colloid and Interface Science* **S0021-9797** (17)31326-7 https://doi.org/10.1016/j.jcis.2017.11.030

[16] Cullity B, 1972 *Elements of X-ray Diffraction* (Addison-Wesley Reading) p 102

[17] Ouachtari F, Rmili A, Elidrissi S, Bouaoud A, Erguig H and Elies P 2011 *Journal of Modern Physics* **2** 1073-1082. https://doi.org/10.4236/jmp.2011.29131

[18] Hasnat A and Podder J 2012 *Advances in Materials Physics and Chemistry* **2**. http://doi.org/10.4236/ampo.2012.24034

[19] Saeed N 2011 *Journal of Al-Nahrain University* **14** 86-92

[20] Li X, Barnes T , Barnes T and Coutts T 2001 *Mat. Res. symp.***666** 18-23.

[21] Tauc T and Abeies F 1970 *optical properties of solids* IOP Publishing Ltd (North Holland, Amsterdam) p 903.

[22] Tauc J 1974 *Amorphouse and Liquid Semiconductor* (Plenums Press. New York and London).

[23] Pankove J 1975 *Optical Process in Semiconductor* (Prentice Hall Inc).

[24] Frank G, Kauer E and Kostlin H 1981 *Thin solid film* **77** 107.

[25] Abakaliki and Nigeria 2000 *The Pacific Journal of Science and Technology* **7**

IOP Conf. Series: Journal of Physics: Conf. Series **1032** (2018) 012003 doi:10.1088/1742-6596/1032/1/012003

Enhancement of MEH-PPV:CNT for H_2S gas sensor

I M Ibrahim[1] , A H Khalid[2] and M H A.Wahid[3]

[1]Department of Physics, Science College, University of Baghdad.
[2] Ministry of Science and Technology, Laser and Optoelectronic Center ,Iraq-Bagdad.
[3]Semiconductor Photonics & Integrated Lightwave Systems (SPILS), School of Microelectronic Engineering,University Malaysia Perlis, 02600 Arau, Perlis, Malaysia

Odourflower25@yahoo.com

Abstract. This work presents improvement of gas sensing capability by introducing CNTs in conductive polymer namely "MEH-PPV". Firstly, the organic conjugated polymer poly (2-methoxy-5- (2'-ethythexyloxy) - 1,4 -phenlenevinylene) was dissolved in chloroform solvent and multi wall carbon nanotube (MWCNTs) were softened in DMF solvent. The two solutions "MEH-PPV" and MWCNTs were mixed in a volume ratio of 0.95:0.05, respectively and spin-coated on Si substrate for realizing facile and low-cost sensors. FTIR results indicated that "MEH-PPV" is contacted to the surface of carbon nanotube by an ester bond. All films were examined by Atomic Force Microscope (AFM) which revealed the average roughness increment from 0.168 to 0.336 nm with increase the concentration of MWCNT. The "MEH-PPV:MWCNTs" based sensors also shown noticeable responses when the sensors exposure to H_2S gas at the concentration of 25ppm. The sensitivity enhanced by adding MWCNT to MEH-PPV was recorded to be 104.45% at 100°C where as the response and recovery time was ~ 42.9 s and ~36.8 s, respectively. Finally, results of "MEH-PPV:MWCNTs" nanocomposite thin films as active organic layer for H_2S gas sensors were measured.

1.Introduction

In the last year, it has been a great interest by researchers to develop the sensors and detectors for detecting chemical agents with quickly response time and low amounts of gases [1,2]. H_2S gas is one of the mischievous gas to the human healthy when the exposure time and the concentration be sufficient to cause death [3]. The development of commercial device has suffered from the following features power exhaustion, operating temperatures and high- cost [4–6]. In order to obtain portable and sensitive gas sensors we need to develop the sensing structure of the H_2S gas sensors.

Conductive polymers (CPs) such as "MEH-PPV" known as poly (2-methoxy -5 - (2'- ethylhexyloxy)-1,4 -phenylenevinylene), the movement of electrons is easy between chains therefore these polymers supports contact charger [7,8]. The backbone of the conjugated polymers created by a series of alternating single and double bonds[9]. The single and double bonds chemically are strong, represented by localized (σ) bond and less strongly localized (π)bond [10]. The overlap of (π) orbitals in the row (π) bonds allowed the electrons moving between atoms and to be delocalized electrons [7,9]. The preparation conditions such as solvent type, spin coating speed, concentration and molecular weight affect on optical parameters and film morphology[11]. Many reports specially in the organic material recorded sensors with a different sensitivities and responses by made a little alteration in preparation of the organic polymers[12].Gas sensors that are based on conducting polymers have many improved properties such as short response time and high sensitivity ; even at room temperature comparison with available sensors. Chemical and electrochemical methods can be easily used to synthesis the conducting polymers, and it is possible to modified the chain structure of the conducting polymers by copolymerization [13-15]. carbon nanotube (CNTs) have many properties such as high aspect ratio, flexibility and diameter in nanometer scale make the carbon nanotube attractive to mixed with polymer as nanocomposites material [16,17].

The high surface to volume ratio of carbon nanotube (CNTs) one of the most important property to be appropriate for gas sensing application. The large capacity for gas sensor provides by the large surface area is obtained from inside and outside walls in carbon nanotube. The interaction between carbon nanotube with some gasses result in alter the electrical properties of carbon nanotube even at room temperature. In addition, carbon nanotube that mixed with polymers to create nanocomposites that combine the advantage of carbon nanotube, thus produce a new active material with unique gas sensing behavior[18].

When the H_2S gas sensors exposure to the molecules gas, as a result the molecules adsorbs at the surface and modulates the conductivity. The relative alteration of the resistance is recognized as sensor response: where R_g and R_0 are the resistances of the sensor in the existence and the absence of the molecular gas, respectively. The sensor response is typically estimated five times and the results are the average of all sensors responses [19].

$$R(\%) = R_g - R_0 /R_0 \times 100 \quad (1)$$

The response time is recognized as the time required for the sensor to arrive (90%) of its "maximum value/minimum value" of conductance upon introduction of the "reducing gas /oxidizing gas". Similarly, the recovery time is recognized as the required time to recover within (10%) of the original baseline when the flow of reducing or gas is removed. The response time and the recovery time can be calculated as follows [20]:

$$\text{Response time} = /t\ \text{gas(on)} - t\ \text{gas(off)}\ / \times 0.9 \quad (2)$$

$$\text{Recovery time} = /t\ \text{gas(off)} - t\ \text{gas(recovery)}\ / \times 0.9 \quad (3)$$

This study focuses on the fabrication of MEH-PPV:CNTs blended thin films toward the application of organic gas sensing with various concentration of CNTs.

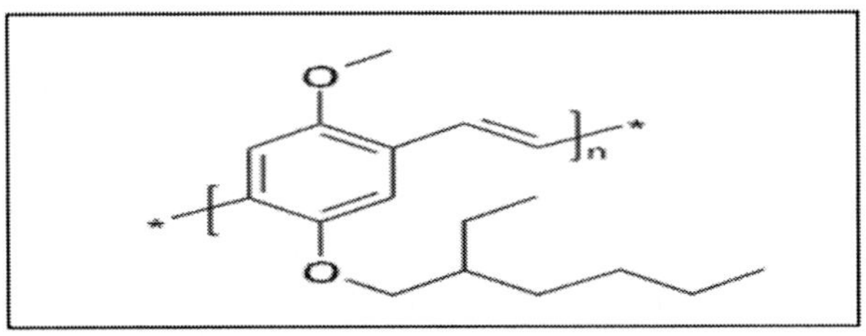

Figure 1. The structure of "MEH-PPV"

2.Experimental Procedure

2.1 Preparation of Multi Wall Carbon Nanotube (MWCNT) and MEH-PPV solutions

10 mg of multi wall carbon nanotube MWCNT was first dissolved in 10 ml of (N,N-dimethylfromamide DMF) to obtain the concentration 1 mg/ml. The solution was placed on the magnetic stirrer for 6 hours and ultrasonication for 8 hours in order to obtain homogeneity solution. This solution is diluted to the concentration of multi wall carbon nanotube MWCNTs 0.01 mg/ml and 0.1 mg/ml .
The concentration of polymer poly (2-methoxy-5-(2'-ethythexyloxy)-1,4-phenlene vinylene) MEH-PPV, (Sigma-Aldrich , Mn =40.000-70.000) was prepared by weighted of 10 mg of the polymer and dissolved in 10 ml of chloroform solvent ($CHCL_3$). The solution was put on the magnetic stirrer for half an hour to obtain "MEH-PPV" solution with concentration 1mg/ml and the powder dispersed well in solvent. Next the multi wall carbon nanotube MWCNT solutions were added to the "MEH-PPV" and sonicated in ultrasonic bath for 15 minutes without heat treatment. The ultrasonication process is perform to enable the multi wall carbon nanotube MWCNT and the "MEH-PPV" to be well-mixed at volume ratio of 0.05:0.95. The nanocomposites of "MEH-PPV:MWCNTs" were deposited on Si substrate by spin coating for 1 minute and 1500 rpm. Spin coating system ACE 200 was used in this investigation. Image of the nanocomposite thin films was monitored by using (Atomic Force Microscope SPM, model AA3000, and tip NSC35/AIBS) from Angstrom Advanced Incorporated USA. Fourier Transform Infrared FTIR test is used to observe the spectra of samples utilizing (a Thermo Scientific Nicolet iN10 FTIR Spectrometer). The data was recorded in the wave numbers range of 400 to 4000 cm^{-1}. Thermal stabilities of "MEH-PPV" powder was studied by Thermal Advantage Q500 V20.13 Build 39 from TA Instruments, USA. The testing system for the suggested gas sensor consists of cylindrical stainless steel chamber with diameter of (18 cm) and height of (8 cm). The volume of the chamber is about (2034 cm^3). It has an inlet for allowing the target gas to flow in and an air admittance valve to allow the flow of atmospheric air after evacuation. Multi pin feed through at the base of the chamber to allow the electrical connections to be established at the heater which consists K-type thermocouple and sensor electrodes. Figure 2 illustrated the thin films structure.

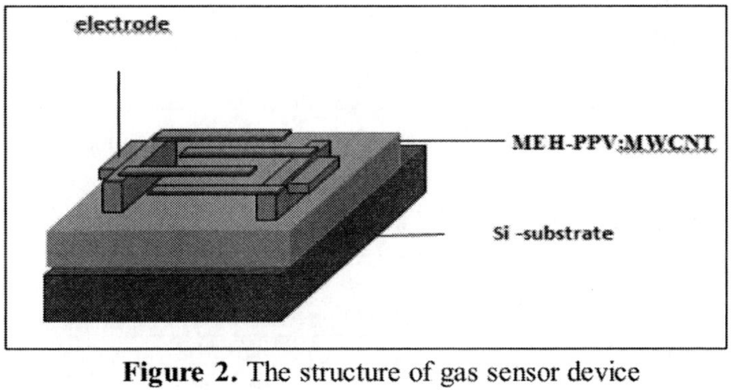

Figure 2. The structure of gas sensor device

3. Results and discussions

3.1. FTIR spectrum

Fourier transform infrared -(FTIR) spectra were obtained for both organic polymer "MEH-PPV" (PURE) and "MEH-PPV:MWCNTs" in the volume ratio 0.95:0.05 respectively. Figure 3 displays the FTIR spectrum of the isolated "MEH-PPV". Addition of multi wall carbon nanotube MWCNT in different concentrations (0.01 and 0.1mg/ml) is to make sure the existence of ester bond in the "MEH-PPV:MWCNT" structure . Absorption bands of "MEH-PPV" (pure) as well as "MEH-PPV:MWCNTs" are shown in Table 1. The bands are almost similar to reference [21] as comparison and verification.

"MEH-PPV:MWCNT" exhibits alike FTIR spectral pattern of pure "MEH-PPV" in one of the common bands (~ 1203 cm^{-1}phenyl C–O stretching),(~ 1462 cm^{-1}asymmetric alkyl CH_2),(~ 2870 $cm^{-1}CH_2$ stretching),(~ 2924 cm^{-1}CH stretching),and(~2954 cm^{-1} asymmetric stretching CH_3). In addition, (a C=C stretching band) at (~1,552 cm^{-1}) is due to graphene, and (ester C=O) and (C–O stretching bands) at (~1,735 cm^{-1}) and (~1,119 cm^{-1}) respectively. The FTIR bands promote that "MEH-PPV" is chemically contacted to the carbon nanotube surfaces through an ester bond in "MEH-PPV: MWCNT".

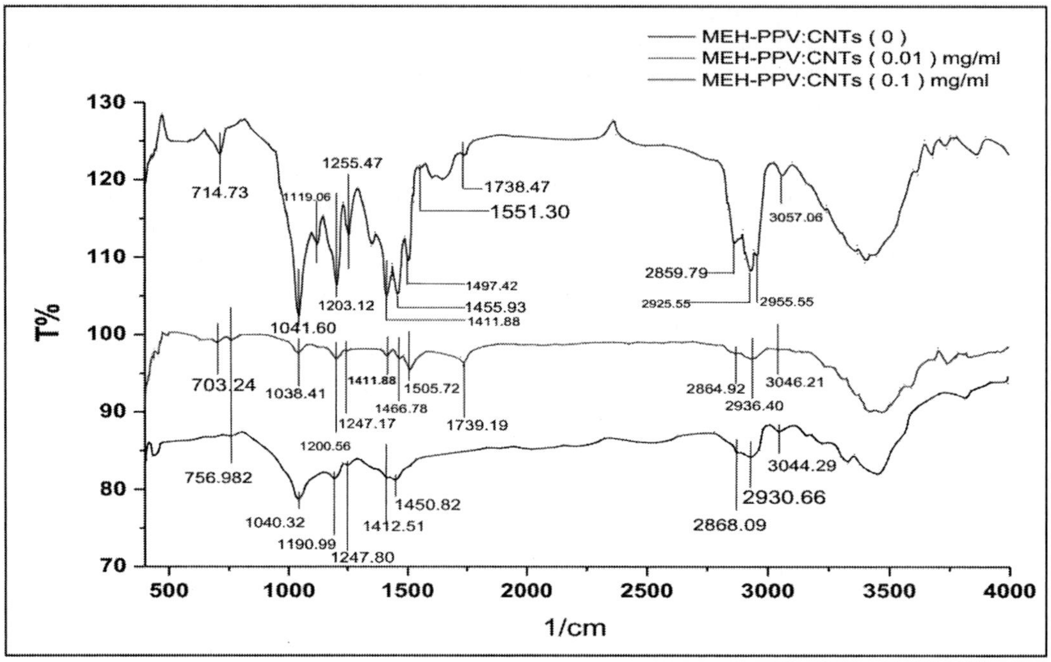

Figure 3. FTIR spectra of MEH-PPV:CNTs with two conc.of CNTs

Table 1. Absorption bands of pure MEH-PPV and MEH-PPV:MWCNT with different concentration

Bond type	MEH-PPV *(PURE)*	MEH-PPV:MWCNT (0.01) *mg/ml*	MEH-PPV:MWCNT *(0.1) mg/ml*
C-H monosubstituted aromatic bending	~756.98 cm^{-1}	~703.24 , 756.98 cm^{-1}	~714.73 cm^{-1}
C-O alky stretching.	~1040.32 cm^{-1}	~ 1038.41 cm^{-1}	~1041.60 cm^{-1}
C-O phenyl stretching	~ 1190.99 cm^{-1}	~ 1200.56 cm^{-1}	~ 1203.12 cm^{-1}
C -O ethene stretching	~1247.80 cm^{-1}	~1247.12 cm^{-1}	~1225.47 cm^{-1}
"Semi- circular stretch bands" of phenyl ring	~1412.51 cm^{-1}	~1411.88 ,1505.72 cm^{-1}	~ 1411.88,1497.42 cm^{-1}
CH2 asymmetric alkyl	~ 1450.82cm^{-1}	~1466.78 cm^{-1}	~1455.3cm^{-1}
C = O ester		~1736.19 cm^{-1}	~1738.74 cm^{-1}
CH2 stretching	~ 2868.09 cm^{-1}	~2864.90 cm^{-1}	~2859.79 cm^{-1}
CH stretching	~2930.66 cm^{-1}	~2936.40 cm^{-1}	~2925.55 cm^{-1}

C-H aromatic stretching	~3044.29 cm^{-1}	~3046.21 cm^{-1}	~ 3057.06cm^{-1}
CH3 asymmetric stretching			~2955.55 cm^{-1}
C-O stretch			~ 1119.06 cm^{-1}
C=C stretch			~1551.30 cm^{-1}

3.2. Thermal analysis -Thermogravimetric analysis (TGA)

Thermogravimetric analyzer (TGA) is a thermal analysis technique which measures the mass change of a sample under heating treatment. These mass changes are displayed on a thermo gram, which graphically displays the increasing or decreasing change in mass as a function of time or temperature.The TGA has been used as a tool for measuring the thermal stability of polymer films [22]. It was shown that analyzing MEH-PPV under nitrogen resulted in a two-step decomposition, where air yielded a multi-step. TGA analysis results for MEH-PPV are shown in Fig.4.

The analyzed MEH-PPV:MWCNT(20wt%) under air yielded a multi-step. It is observed that there was no distinctive weight loss until 270°C, which clearly indicated that MEH-PPV:MWCNT was stable up to 270°C and start to decompose after this temperature. It was also noted that the first decomposition of the sample took place between 270-370°C with the corresponding weight loss of about-17.5124 % at 366.009°C. The second decomposition started between 370-440°C with weight loss of -27.8667 % at 437.511 °C. The weight loss increased rapidly, later at 592.975°C the remaining residue was recorded around-35.8151% . Eventually, the weight loss profiles of the films can be attained.

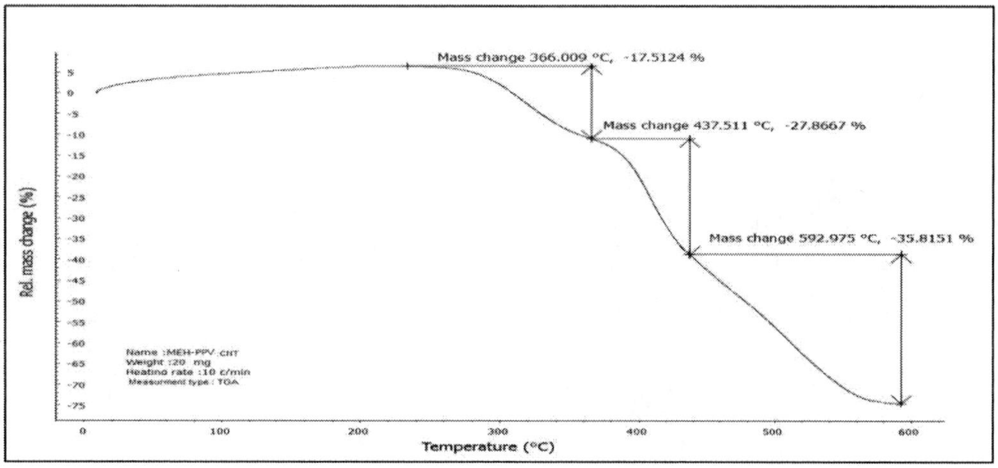

Figure 4. Thermal Weight Loss (TGA) Curve of MEH-PV:MWCNT Powder

3.3. Atomic Force Microscope analysis

Dispersion of multi wall carbon nanotube MWCNTs in polymer as nanocmposite material was calculated via analysis the surface morphology utilizing (AFM) images in the tapping mode on 2.05 × 2.05 μm^2 area, as illustrated in Fig.5. It was conducted for pure thin films of "MEH-PPV" and "MEH-PPV:MWCNTs" films in the volume ratio of 0.95:0.05 at different concentration of multi wall carbon nanotube MWCNT (0.01 and 0.1 mg/ml).

The best regularity sample which directly influence the H_2S gas sensing is the concentration of 0.1 mg/ml as shown in Fig. 5c. In Figs. 5b and 5c, the 2D micrograph that indicated by blue arrows, the MWCNT can be observed as a shadow bundle, which shows that a polymer layer has wrapped around the nanotubes. The presence of MWCNTs was proven within the film as revealed in 2D micrograph even at a low concentration 0.01 mg/ml. The values of average roughness of "MEH-PPV"(pure) is 0.168 nm as shown in Fig.5a. The roughness of "MEH-PPV:MWCNTs" at 0.01 mg/ml and "MEH-PPV:MWCNTs" at 0.1mg/ml are 0.221 and 0.336 nm, respectively. The increment roughness of the surface is due to the aggregation of CNTs, as the concentration of carbon nanotube increases within "MEH-PPV" polymer matrix. The AFM study present a great reliance on the nature of multi wall carbon nanotube MWCNTs in the polymer matrix.

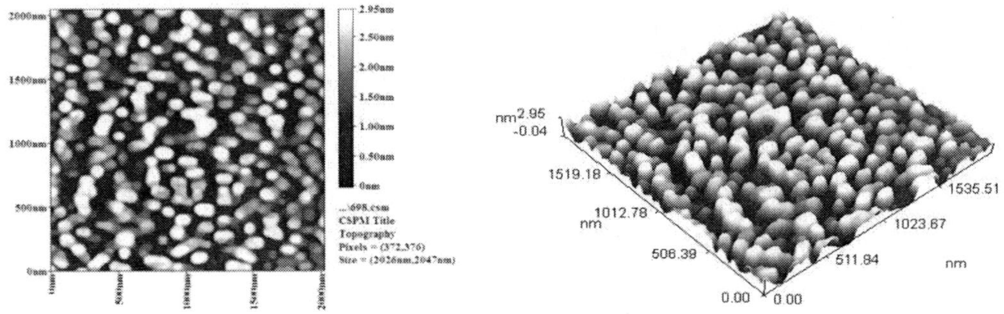

(a)

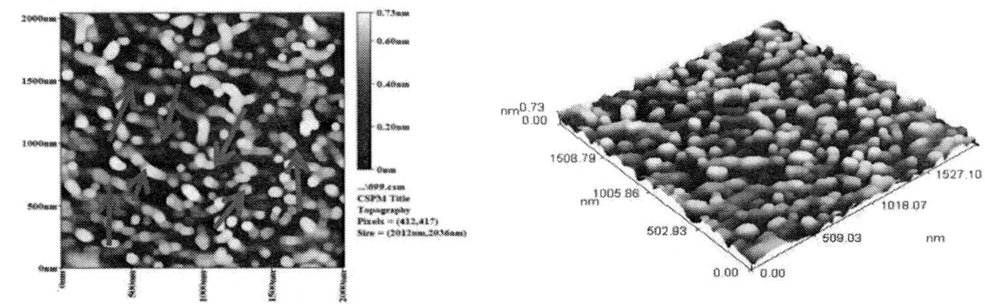

(b)

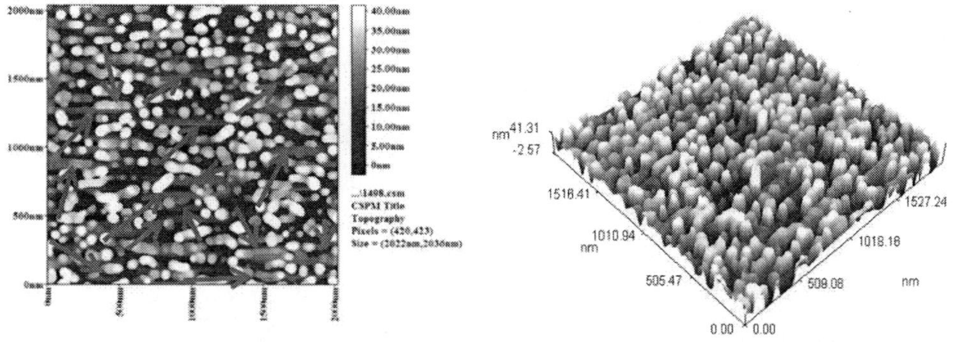

(c)

Figure 5. AFM images (tapping mode) 2D to the left side and 3D to the right side of " MEH-PPV " and "MEH-PPV:MWCNT" nanocomposite thin films ,(a) " MEH-PPV " (pure), (b) "MEH-PPV:MWCNTs" (0.01) mg/ml and (c)" MEH-PPV:MWCNTs" (0.1) mg/ml

3.4. Gas sensor measurements

Figure 6 presents the dynamics of the gas sensor response to the H_2S gas at 100°C. The results demonstrate the sensors capability to detect H_2S gas as low as 25 ppm. Figure 5a shows the sensitivity of the "MEH-PPV" (pure) sensor with a low sensitivity 7.8% while the sensitivity increases with increase the concentration of multi wall carbon nanotube MWCNT at 0.1mg/ml up to 104.45%, as displayed in Figure6c. The multi wall carbon nanotube MWCNT provides a massive specific surface area, which provide a massive number of sites for interaction with gases. The detection of different gases can be provided as possible as lowering the temperatures of preheating of the body sensor[23].

The sensing mechanism of nanocomposite films "MEH-PPV:MWCNTs" to molecular gas may be attributable to adsorption the moleculear gaseous at the surface of the films. The moleculear gaseous interaction with (π) electron network of the polymer that has entrenched nanoparticles results in capturing of electrons, depending upon the nature of reduction gaseous molecules, leading to the decrease of the resistance [24]. Reduced resistance or an increase conductivity is due to protonation reaction. H_2S dissociation then followed by protonating (H^+) the polymer, semiquinone radicals generates from the protonation reaction , these radicals are conjugated and therefore the conductivity an increase[25]. The response time and the recovery time are depicted in Fig.6 with red and green arrows at 42.9 s and 36.8 s, respectively. Table 2 summarizes the findings.

IOP Conf. Series: Journal of Physics: Conf. Series **1032** (2018) 012003 doi:10.1088/1742-6596/1032/1/012003

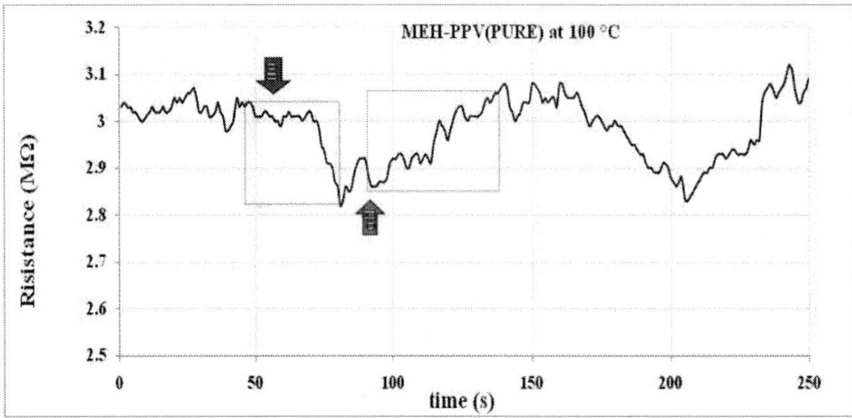

(a)

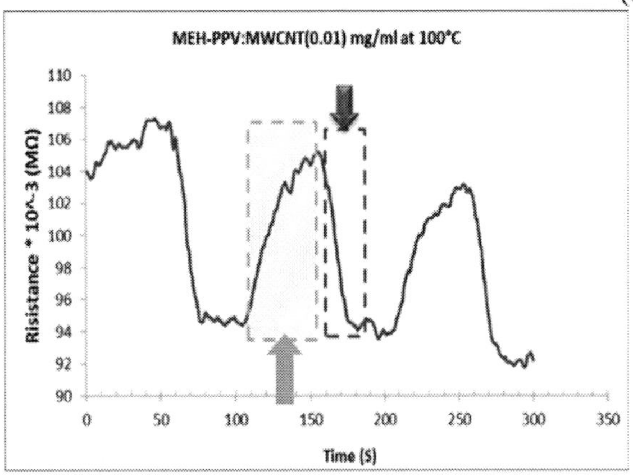

(b)

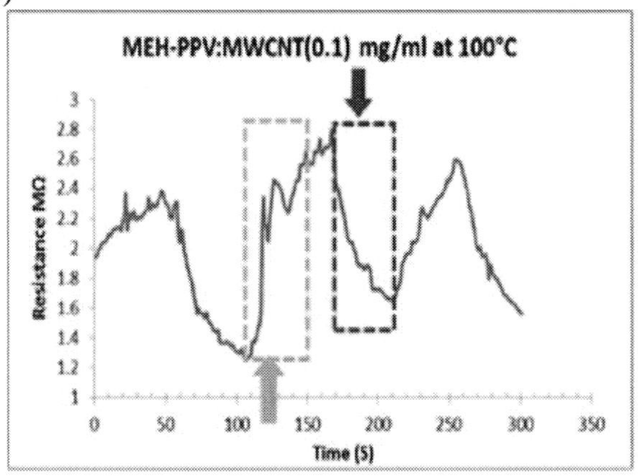

(c)

Figure 6. A The variation of resistance with time for MEH-PPV and MEH-PPV:MWCNT device as H_2S gas sensing,(a)MEH-PPV(pure),(b)MEH-PPV:MWCNT(0.01)mg/ml,(c)MEH-PPV:MWCNT(0.1)mg/ml

Table 2. The sensitivity of MEH-PPV and MEH-PPV:MWCNT device as H_2S gas sensing

Sample	Operating temp (°C)	Sensitivity (%)	Response time (s)	Recover time (s)
MEH-PPV (pure)	100	7.8	34.7	47.8
MEH-PPV:MWCNT(0.01)mg/ml	100	11.70	23.90	47.81
MEH-PPV:MWCNT(0.1)mg/ml	100	104.45	42.9	36.8

4. Conclusion

Nanocomposite thin films "MEH-PPV:MWCNTs" prepared by spin coating has successfully enhanced the sensitivity with addition multi wall carbon nanotube MWCNT in different concentrations to the organic polymer "MEH-PPV" at the same volume ratio 0.05 and at the temperature 100°C, the

IOP Conf. Series: Journal of Physics: Conf. Series **1032** (2018) 012003 doi:10.1088/1742-6596/1032/1/012003

enhancement sensitivity is 104.45%, with response time and recovery time are 42.9 Sec and 36.8 Sec respectively. Increment of the surface roughness with increase the concentration of multi wall carbon nanotube, the polymer wrapped around the CNT as shown in the AFM measurement.

References

[1] Dai Zh, Xu L, Duan G, Li T, Zhang H, Li Y, Wang Y and Cai W 2012 *J. Sci. Rep.* **3** 1669.

[2] Valentini L, Armentano I, Kenny J.M, Cantalini C, Lozzi L and Santucci S 2003 *J.ppl.Phys.Lett.* **82** 961 .

[3] Nasresfahani Sh, Doroodmand M, Sheikhi M, Ghasemi A 2011 *J. Nanoeng. Nanomanuf.* **1** 226–228.

[4] Verma M and Gupta V 2012 *J. Sens. Actuators B* **166** 378–385.

[5] Zhang R, Zhang X and Hu S 2013 *J. Sens. Actuators B* **177** 163–171.

[6] Shao v, Hoffmann M, Prades J, Zamania R, Arbiol J, Morante J, Varechkina E, Rumyantseva M, Gaskov A, Giebelhaus I, Fischer T, Mathur S and Hernández-Ramírez F 2013 *J. Sens. Actuators B* **181** 130–135.

[7] Ghasemi-Mobarakeh L, Prabhakaran MP, Morshed M, Nasr-Esfahani MH, Baharvand H and et al. 2011 *J. Tissue EngRegen Med* **5** e17-35.

[8] Suhail M, Ibrahim I, Ahmad Z, Abdullah S and Sulaiman K 2013 *J.IJAIEM* **2** 2319 – 4847.

[9] Balint R, Cassidy NJ and Cartmell SH 2014 *J. ActaBiomater* **10**: 2341-2353,2014.

[10] Ravichandran R, Sundarrajan S, Venugopal JR, Mukherjee S and Ramakrishna S 2010 *J R Soc* Interface 7 Suppl **5** S559-579.

[11] Suhail M and Ibrahim I 2013 *J. of APP. RESE.* **3** 2249-555X.

[12] Ahmad Z, Suhail M, Ibrahim I, Al-Rawi W, Sulaiman K, Zafar Q, Hamzah A, and Shaameri Z 2013 *J. Chin. Phys.B* **22** 100701.

[13] Dubbe A 2003 *J. Sens. Actuators B* **88** 138-148.

[14] Zakrzewska K 2001 *J. Thin Solid Films* **391** 229-238.

[15] Timmer B, Olthuis W and van den Berg A 2005 *J.Sens. Actuators B* **107** 666-677.

[16] Fournet P, Coleman J, Lahr B, Drury A, Blau W, Brien D and Hörhold H 2001 *J. of App. Phys.* **90** 969-975.

[17] Kymakis E and Amaratunga G 2001 *J. of App. Phys. Lett.* **80** 112- 114.

[18] Penza M, Rossi R, Alvisi M, Signore M, Cassano G, Dimaio D, Pentassuglia R, Piscopiello E, Serra E and Falconieri M. 2009 J. Thin Solid Films **517** 6211-6216.

[19] Asada M, Sheikhia M, Pourfathb M and Moradid M 2015 J. *Elsever Sensors and Actuators B* **210** 1–8.

[20] Patil D, Patil L, Jain G and Wagh M 2006 *J. Sensors Transducers* **74** 874–883.

[21] Prajongtat P, Suramitr S, Gleeson M, Mitsuke K ans Hannongbua S 2013 *J. Monatsh Chem* **144** 925–935.

[22] Cossiello RF, Akcelrud L and. Atvars TDZ 2005 *J. of the Brazilian Chemical Society* **16** 74-86.

[23] Aroutiounian V 2015 *J. of Physics*, **55** 319–329.

[24] Kaushik A, Kumar R, Arya S, Nair M, Malhotra B and Bhansali S 2015 J. Chem. Rev.**115** 4571–4606.

[25] Clarke E and Glew D 1971 J. Chem. **49** 691-698.

IOP Conf. Series: Journal of Physics: Conf. Series **1032** (2018) 012004 doi:10.1088/1742-6596/1032/1/012004

Design and Fabrication of Hydrogen Production System

Yaseen .H. Mahmood[1] and Rafa Y. J. Al-Salih[2]

[1,2]Department of Physics, College of Sciences, University of Tikrit, Iraq

D.yaseen.ph.sc@tu.edu.iq[1], Aburafall965@yahoo.com[2]

Abstract. A system of water electrolysis have been designed and fabricated locally by using simple home materials. Many types of water solutions and the effect of its concentrations on the electrical current have been studied. The effect of voltage and power on the quantity of gas production and the analysis quickness have been studied also. It is noticed that the current will increases as concentration increase. In addition, the productive gas quantity is increases as the voltage increases. The productive gas can be used in a welding processes, or as a hydrogen fuel for of electricity generators and cars engines.

Keywords: Hydrogen production, Hydrogen fuel, Water electrolysis, Water splitting, Renewable energy, Energy conversion.

1- INTRODUCTION

Hydrogen is the most important element in the universe. Many of planets and all stars are basically consist of hydrogen. But on the earth, the hydrogen is rarely found as a free gas. The atmosphere contains a very little amounts of it "0.07 percent". It is usually mixed with other natural gases in crustal reservoirs. However, it have been discovered in a large amounts in few wells. The principle of hydrogen energy production covered a whole array of methods, such as electrolysis, thermal photolysis, and thermo chemical cycles [1].Hydrogen energy one of most important source for Renewable energy. Many searches studied method of Hydrogen preparing. Because of that the Hydrogen is most spread in the universe, it is rarely found as a pure element in the nature [2]. But it is available as a compound. The water compound is most spread, therefore the water considered as an important and suitable for renewable energy. The hydrogen is used as a fuel for space crafts, water heating, and for cooking. It is more convenient to produce hydrogen by electrolysis using excess electricity and conversion efficient ranging from (80-95%) are achievable [3]. A theoretical and experimental studies was done about using solar thermal to produce hydrogen [4,5]. Theoretical studies has been done on vicinage effect in ions collisions with solid energy matter [6]. Hydrogen will be required to supplement, but eventually, replaced rapidly diminishing by hydrocarbon as a fuel resource for internal combustion engine [7]. Electronic capture by hydrogen projectiles in gazing scattering from Aluminum thin film with different thicknesses had been studied theoretically using perturbation method [8]. Characterization of water electrolysis cell and high frequency behavior have been evaluated in this work. The experimental electrolytic bath of the mentioned research for each one has been arranged to be differs from other setups in distance between the electrodes, electrode size, and electrolyte morality [9].

IOP Conf. Series: Journal of Physics: Conf. Series **1032** (2018) 012004 doi:10.1088/1742-6596/1032/1/012004

2- ESSIENTIAL CONCEPTS OF WATER ELECTROLYSIS

The water electrolysis is occurred when a DC current passing between two electrodes immersed in an electrolyte. Oxygen is formed at the anode and hydrogen at the cathode. The current passing between any two electrodes is the main parameter of hydrogen production . Well known "Michael Faraday's laws" of electrolysis illustrated that [10]:

1. "The mass of a substance changed during electrolysis at an electrode is directly proportional to the quantity of electric charge (Q) transferred".

2. "For a specific quantity of electric charge, the mass of the material altered at an electrode is directly proportional to the equivalent weight of the element. So, the equivalent weight of the substance is usually equal to its molar mass divided by the variation in oxidation state which it happens upon electrolysis. The electrodes should be resist corrosion, possess a good electric conductivity, bare a good catalytic properties, and have a suitable structural integrity. In addition, electrodes and the electrolyte should not be reacted with each other" [11]. By splitting of water, hydrogen can be produced using various processes; such as "water electrolysis, photo-electrolysis, photo-biological production and high-temperature water decomposition". Electrical water analysis one of the best method by passing the electrical current through a water solution. This method is used for many applications. It is easy to done with low cost, but it is not more efficiency. Electrolysis is the splitting of a water molecule with electricity to produce hydrogen and oxygen. The charge breaks the bond between the hydrogen and oxygen creating ions that form on two poles: the positively charged anode attracts the oxygen, and the negatively charged cathode attracts the hydrogen. One of The main disadvantage of this process is that the efficiency is relatively low, due to the indirectly generation of hydrogen via electricity. So, this process is economically feasible only in the countries which have superabundant hydro-generating quantity, or huge thermal resources. The total energy, which is necessary for water electrolysis is increasing with temperature slightly, however, the needed electrical energy decreases. When a high-temperature heat is available; a high-temperature electrolysis process might be preferable. This is important, because that the most of the produced electricity is usually based on fossil energy sources [12]. The normal chemical reaction of water electrolysis "without thermodynamic energy values" can be expressed as [13]:

"H_2O (liquid)$\rightarrow$ H_2(gas) $+\frac{1}{2}O_2$ (gas) .... (1)"

Separator or diaphragm is necessary to avoid recombination of the hydrogen with oxygen to preserve efficiency in the electrolysis technology, and it is also used in fuel cells [14]. Thermodynamically, in water electrolysis, thermal and electrical energy are converted into chemical energy "stored in hydrogen". The energy required to achieve the reaction is the enthalpy (ΔH) of water formation. The free energy of this reaction "called Gibbs free energy change ΔG", supplies the electrodes as an electrical energy [15]. The remainder is" thermal energy", which is the product of "process temperature T" multiplies with "entropy change ΔS", and if it is known that F is "the Faraday constant (96485.3365 C/mol)", enthalpy variation can be written as [16]:

"$\Delta H = \Delta G + T\Delta S = zF\,[T\,(\partial U_{rev}\,\partial T\,)\,p - U_{rev}]$ .... (2)"

where, "z is the number of moles of electrons transferred in the reaction", "U_{rev} is the reversible voltage", and "p is the prevailing pressure (in Pa)". The reversible voltage U_{rev} is the lowest required voltage required for the electrolysis to take place, and it is also known as "the equilibrium cell voltage", or the electromotive force. The electrical work applied by the electrolytic cell is equal to the free energy variation happened "at constant temperature, pressure and positive electromotive force" [14].

IOP Conf. Series: Journal of Physics: Conf. Series **1032** (2018) 012004 doi:10.1088/1742-6596/1032/1/012004

3- EXPERMENTAL WORK

The preferable way to produce hydrogen can be obtained by splitting water using electricity "electrolysis". This can be done by passing an electric current through the water. This will splits it into hydrogen and oxygen. In spite of the lowest efficiency of electrolysis with that of direct chemical path, but it offers virtually no toxic or pollution. In this study, the objectives of the work are illustrated below:

1- Metal piping: A metal pipe have been used to connect the plastic container with aid of Teflon tape to avoid gas leakage, as illustrated in Fig. 1 below.

Figure 1. Metal piping .

2- plastic containers: A plastic container (3 containers) which is always used in home for water purification have been used in this work; that it has been supplied with an O-ring to avoid gas leakage, and it is also easy to connection. See Fig. 2.

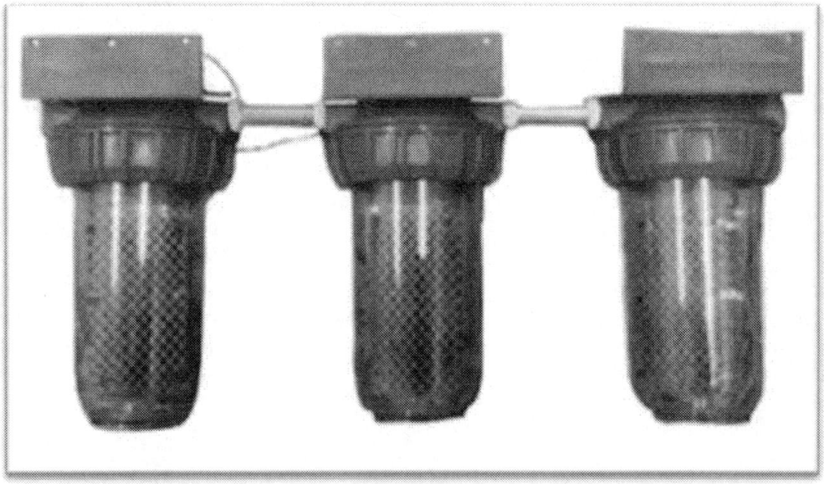

Figure 2. Plastic containers.

IOP Conf. Series: Journal of Physics: Conf. Series **1032** (2018) 012004 doi:10.1088/1742-6596/1032/1/012004

3- Plastic piping: Plastic piping have been used here to connected the output of containers with a storage tank.
4- Storage tank: A metal tank which is used to store cooling gas have been used as a tank for the system designed here.
5- Electrode: A rectangular shape of Aluminum sheets of (10 cm x 15 cm) dimensions have been designed as electrodes. Electrodes have been connected and separated by plastic screw using pieces of ceramic between electrodes to insulate them, where a three sheets connected to a positive electrode and the other three sheets connected to a negative by sandwich method using cupper wire to connected the electrodes with each other's and with the power supply, as demonstrated in Fig. 3.

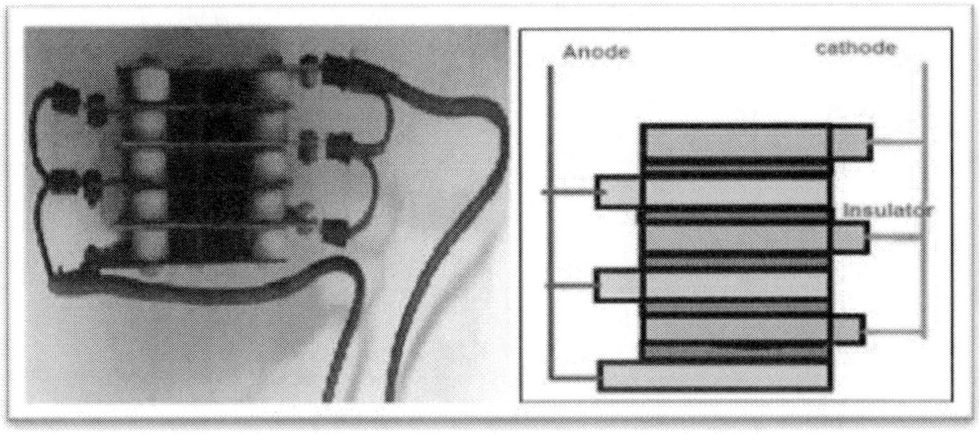

Figure 3. The electrodes assembling.

6- Power Supply: Variable voltage power supply of (30 V ,10 Amp.) has been used.

4-SYSTEM ASSEMBLING

Fig. 4 illustrates the diagram of four plastic containers connected with the terminals of electrodes with the power supply in parallel. A multimeter have been used to measures the current and the voltage.

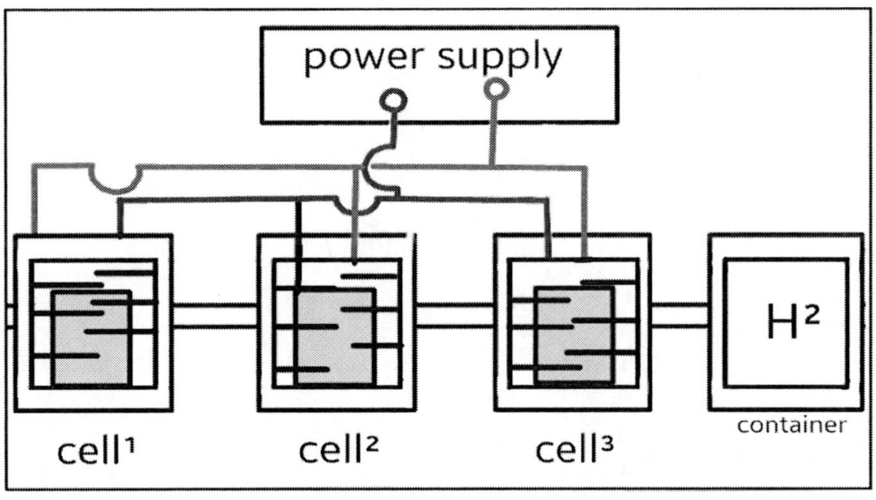

Fig. 4: Electrical Water analysis system diagram.

IOP Conf. Series: Journal of Physics: Conf. Series **1032** (2018) 012004 doi:10.1088/1742-6596/1032/1/012004

5-PERIMENT(1)

After filling the plastic containers with 8 liters of mineral water, and connecting the power supply of (12 Volt voltage & 2 Amp. Current) for 20 minutes, the gas increased in the storage tank. The test steps are repeated at 24 & 30 Volt. The results shown in Fig. 5.

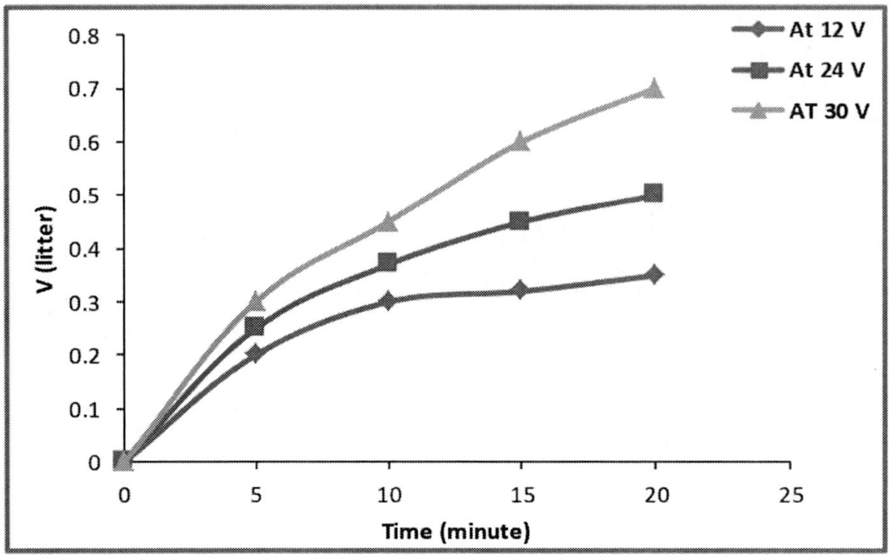

Fig. 5: Relation between volume of gas with time for mineral water.

EXPERIMENT (2)

This experiment have been done by repeating the same procedures in experiment (1), but a normal water (home water) is used, which consists of metal salts (K, Na,..). The results shown in Fig. 6.

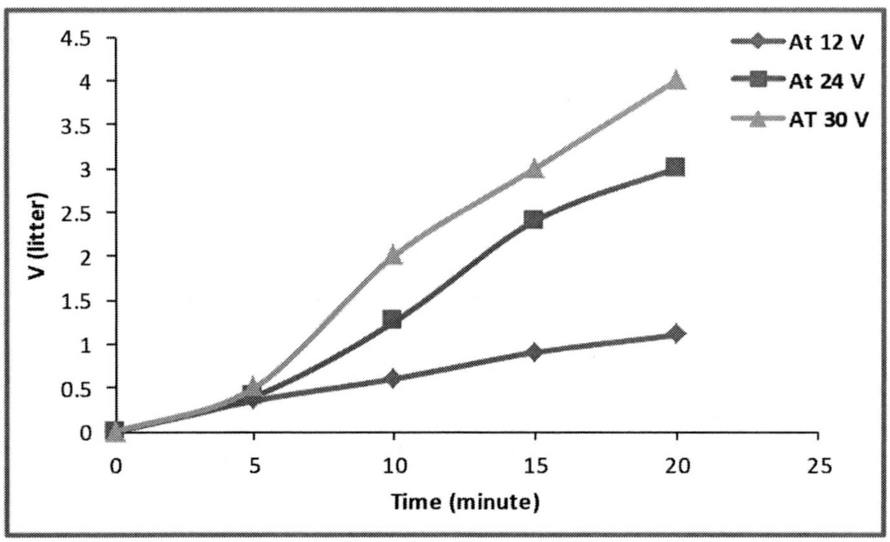

Fig. 6: Relation between volume of gas with time for normal water.

IOP Conf. Series: Journal of Physics: Conf. Series **1032** (2018) 012004 doi:10.1088/1742-6596/1032/1/012004

EXPERIMENT (3)

By adding 1 cm^3 Acid to each 1 liter of normal water used above, the same steps in the previous experiments have been followed in this experiment. The consequence is shown in Fig. 7 below.

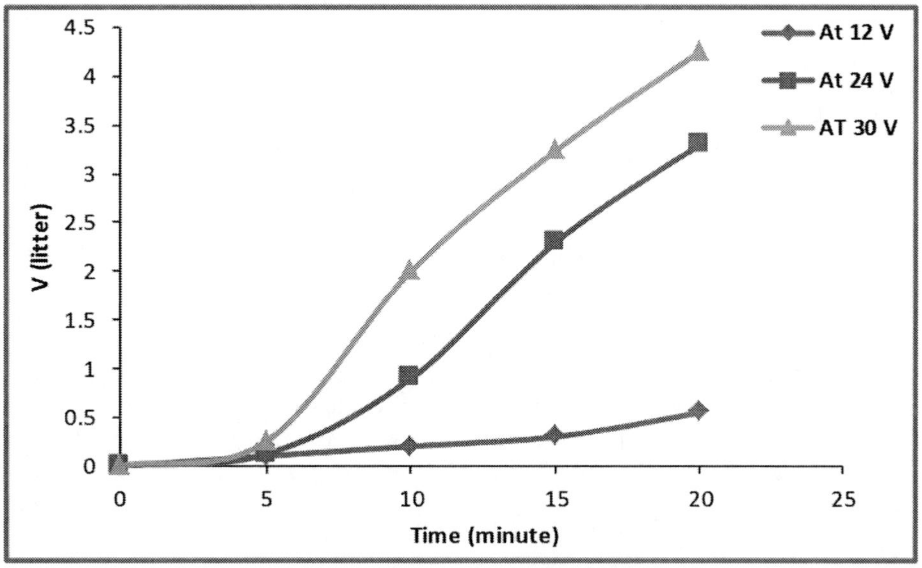

Fig. 7: Relation between volume of gas with time, by adding 1 cm^3 Acid to each 1liter of normal water.

6- RESULTS AND DISCUTION

The outcome of the three experiments mentioned above in this work, the electrical characteristics of the system designed here are shown in the following table 1.

Table 1:Electrical Voltage, Current, and Power for experimental work.

No.	Water type	V (volt)	I (Amp.)	P (Watt)
1	Mineral water	12	2	24
2	Mineral water	24	3.5	84
3	Mineral water	30	5.3	159
4	Normal water	12	2.4	28.8
5	Normal water	24	4.2	100.8
6	Normal water	30	6	180
7	Normal water with acid	12	2.6	31.2
8	Normal water with acid	24	5.3	127.2
9	Normal water with acid	30	7.5	225

From all the obtained data, the results of this work can be summarized in two points:

1- The electrical voltage and current: if the voltage, current passed in water solution with power more enough to heat the electrode, the molecules will be heated, moved, and vibrated. If the frequency of vibration coincides with the lattices vibration, resonance occurred to brakes the bonds of water to releasing hydrogen and oxygen in the solution. When the voltage and current increase, the quantity of released gas (oxygen and hydrogen) is also increases, because of the increasing in collisions between molecules due to the high temperature of electrodes [17].
2- Concentration result: Increasing in concentration of water solution will increase both of the electrical current and the reaction rate due to increase in collisions and vibrated resulting molecules break the bond to releasing gases (hydrogen and oxygen) and reduces the water solution resistance [18, 19, 20]. This continues until the solution will be saturated, whereupon, the water solution should be exchanges.

7- CONCLUSIONS

There are many advantages to use of H_2O as hydrogen source for applications, because it is easy and clean (no contaminations). In this paper, it is found that the suitable water solution should be chosen depending on the purpose of the work. If low power is needed, mineral water must be used with slow releasing of gas. But if high speed for releasing gas and high quantity at the same time is needed, high power (up to 30 Volt with high current 10 Amp.) must be used. So, the sea water can be used instead of normal water mixed with acid. The collected gas (enough quantity) can be used for wilding processes, or as an engine's fuel for electricity generators and cars.

ACKNOWLEDGMENT

We would like to express our sincere appreciation and gratitude to the Physics Department in the College of Sciences in Tikrit University for their kind support during the period of the research.

Our deeply grateful also to the Physics Department in the College of Sciences in Karbala University and to the supervisor committee of the 6th Karbala conference for sciences for their efforts in preparation of this conference.

REFERENCES

[1] Sherif, S 2003 principle of hydrogen energy product and storage and utilizati *(Journal of scientific and industrial research* on Vol. 62) Jan –Feb pp. 46-63

[2] Zeng , k. and Zhang 2010 Recent progress in alkaline water electrolysis for hydrogen production and applications (*Prog. Energy combust Science* Elsevier) 36(307)

[3] Ullber, O Morner S.O. 1997 TRNSYS simulation Models for solar – hydrogen system *(Solar energy* Vol.59) No.4-6,pp.271-279

[4] Nozik, NREL 2005 Basic research needs for solar energy utilization *report of the basic energy science workshop on solar energy utilization* Argonne national laboratory April 18-21

[5] Walsh F.C 2001 Electrochemical technology for environmental treatment and clean energy conversion *(Pure appl.chem.*,Vol.73) No.12 pp.1819-1837.

[6] Abd-Alazez, S 2011 vicinage effect in collision for hydrogen ions (*,Journal Almustansrya science* Vol.22) No.7 pp.53

[7] Al zubaidi, D 2013 study performance of NI engine fuelled with supplemented hydrogen to natural gas *(The Iraqi journal of machining and material engineering*,Vol. 13).No.2 PP.350-363

[8] Jassem, H.A. 2014 Nagative ion formation during scattering of H- Atom from AL (111) thin film : using perturbative *(Journal Thi –Gar*, Vol. 4) July.

[9] Mazloome, K 2013 analysis of frequency response of a water electrolysis cell (*,Int.J. Electrochem .Sci* Vol. 8). pp. 3731-3739

[10] Sundholm, G, Raitanen, E, Hakoila, E, Asplund, J, Lindström, M & Steinby, K1978, *Säh-kökemiallinen analyysi*, Teknillisten tieteiden akatemia, Helsinki.(In Finnish).

[11] Ursúa, A. , Gandia, L., Sanchis, P 2012 Hydrogen production from water electrolysis: current status and future trends (Proceedings of the IEEE, vol. 100) no. 2, pp. 410–426

[12] Hadi, R 2007 Theoretical Investigations of Vibrational – Rotational Energy of some Hydrogen Halide Molecules *(Um-Salama ScienceJournal*, Vol.4) (3)

[13] Koponen, J. 2015 Review of water electrolysis technology and design of renewable hydrogen production system", Master Thesis, Lappeenranta (University of Technology), pp. 15

[14] Decourt, B., Lajoie, B 2014 The hydrogen-based energy conversion FactBook", The SBC Energy Institute

[15] McAuliffe, C.1980 Hydrogen and Energy Macmillan Press Ltd.,(London)

[16] Bard, A. & Faulkner, L 1980 Electrochemical Methods: Fundamentals and Applications (New York John Wiley & Sons)

[17] Riis, T. 2006 Hydrogen production R&D: priorities and gaps (*International Energy Agency France by Stedi Media*) January

[18] Raissi, T.2002 Hydrogen Production Part II Hydrogen from Ammonia and Ammonia-Borane Complex forFuel Cell Applications *Proceedings of the* 2002 *U.S. DOE Hydrogen Program.* Review NREL/CP-610-32405,(University of Central Florida)

[19] Melnicki, R 2015 Sustained H2 Production Driven by Photosynthetic Water Splitting in 1a Unicellular Cyan bacterium",Downloaded from mbio.asm.orgon February, 15

[20] Rand, D 2006 Hydrogen product and storage ,properties and gap international energy Agency", Hydrogen Implanting agreement, Available from http://www.iea.org Text base/paper/ Hydrogen product, 2006.

IOP Conf. Series: Journal of Physics: Conf. Series **1032** (2018) 012005 doi:10.1088/1742-6596/1032/1/012005

Feasibility Study of a Solar Flat Plate Collector for Domestic and Commercial Applications under Iraq Climate

H. K. Jobair, O. I. Abdullah, Z. M. Atyia

University of Baghdad, College of Engineering, Energy Department, Baghdad-Aljadria 47024, Iraq

Husseinkhalaf82@gmail.com

Abstract. Experimental study had been accomplished to evaluate the performance of flat plate collector according to Iraq climate to find the practical benefit and to transform the idea into reality and to apply it for domestically and commercial field. The study had been achieved for a typical flat plate collector. The dimensions of the flat plate collector, solar intensity, ambient temperature and angles relating to this study have been practically applied and all of these information mentioned in this study. This study approved the feasibility of using the hot water as heating purposes for domestic and commercial applications instead of using the electricity for heating the water. This method will reduce the consumption of the electric power obtained from the national network. The reduction of consumption for the national network electric power is approved by comparing three models of houses that is using the electric power for heating water with the same houses models which are using a flat plate collector for heating purposes.

1. Introduction

Global warming, the greenhouse effect, climatic change, ozone layer depletion and acid rain are all accidental consequences of the extensive fossil fuel exploitation in almost all human activities. These phenomena are caused by the emission of carbon dioxide into the atmosphere.

All these phenomena and causes that are harmful to the environment and reflected on human life led to the opening of the mind of humans to how to deal with other types of energy-friendly environment and how to exploit them in a way that serves humanity and the lack of impact on the environment. Among the environmentally friendly types of energy, solar energy is the most important source of renewable energy sources due to the abundance of energy obtained from the sun. Since Iraq receives a large amount of solar energy, it is necessary to use this energy to serve the region rather than the traditional sources of energy that Iraq relies on to save energy for humans like oil. One of the most important applications that use solar energy as an energy source is the flat plate collector. Flat plate collector is the simplest and one of the cheapest means of collecting solar energy for use in system that require thermal energy at low temperature (< 100°C). Flat plate collector requires a minimum amount of material per square meter of collector and thus provides for the possibility of lower costs. Domestic hot water is the second-highest energy cost in the typical household in all over the world. In fact, for some homes it can be the highest energy expenditure. Solar water heating can reduce domestic water heating costs by as much as 70%. It is easy to install and almost maintenance free.

Yousef and Adam [1] investigated the effect of several parameters theoretically to the performance of a flat plate collector. Mass flow rate, flow channel depth and collector length on the system thermal performance and pressure drop through the collector with and without porous medium had been studied for a single and double flow rate of a flat plate collector. The results showed that at the same configuration and parameters for both single and double flow rate, the efficiency of the thermal system increases by 10-12% in double flow mode than single flow due to the increased of heat removal, and increase by 8% after using porous medium in the lower channel as a result of the increase of heat transfer area. The study also showed that the pressure drop increased and all collectors show improved efficiency when the collector operates at relatively high flow rates, and at relatively low collector temperature rise since the collector losses will be less in low temperature difference. Amrutkar et al. [2] evaluated the performance of FPC with different geometric absorber configuration. This study have been accomplished under laboratory conditions (as per IS standard available for the flat plate collector testing). The results of this study showed that with the same collector space higher thermal efficiency or higher water temperature can be obtained. Thus, cost of the FPC can be further bringing down by enhancing the collector efficiency. Chaji et al. [3] fabricated and tested a small flat plate solar collector (FPSC). The study had been achieved by using nanoparticle concentrations of TiO in water as base fluid. Different flow rates and concentration ratios was employed in this study. Ameen et al. [4] investigated experimental and numerical the heat transfer enhancement of flat plat collector (FPC) using three types of twisted tapes (single twisted tape (ST), double twisted tape (DT) and mixed twisted tape (SDT)) then compared with plain tube with twist ratios (TR=2). The study was considered under fully developed turbulent flow with solar radiation heat gain is changing with time. Experimental study showed that the DT are more efficient than ST and SDT and the temperature of outlet water from mixed twisted tape collector is higher than the other type of plain tube collector by 10°C. Numerical analysis of this study was based on finite volume numerical techniques to solve the governing partial differential equations in three dimensions, using ANSYS FLUENT commercial CFD software. The comparison between the experimental and numerical results shows a high agreement. Mankhi et al. [5] designed a number of solar systems that are fit for work in the atmosphere of the city of Basra. The study demonstrated the feasibility of obtaining temperatures in these systems may sometimes up to 90 degrees Celsius even in the coldest days. Basim et al. [6] investigated the experimentally and numerically the effect (ZnO-water) nano fluid on heat transfer flat plate solar collector. The experiment part is conducted in Karbala with the latitude of 32.6 0 N. The numerical part is applied by ANSYS 15software. a good agreement between the experimental and CFD results for outlet temperatures where the maximum error was (8.4%).

2. Solar Radiation in Iraq

The monthly average of solar radiation measured in Baghdad city, reported by NASA, is shown in figure 1. The annual averaged of energy received daily from the sun in Iraq ranges between 4.5 – 5.4 KWh/m^2. For horizontal surface, more insulation is available during summer rather than winter. If the surface is tilted, the available insolation can be increased due to focusing of the incident radiation to the optical area.

IOP Conf. Series: Journal of Physics: Conf. Series **1032** (2018) 012005 doi:10.1088/1742-6596/1032/1/012005

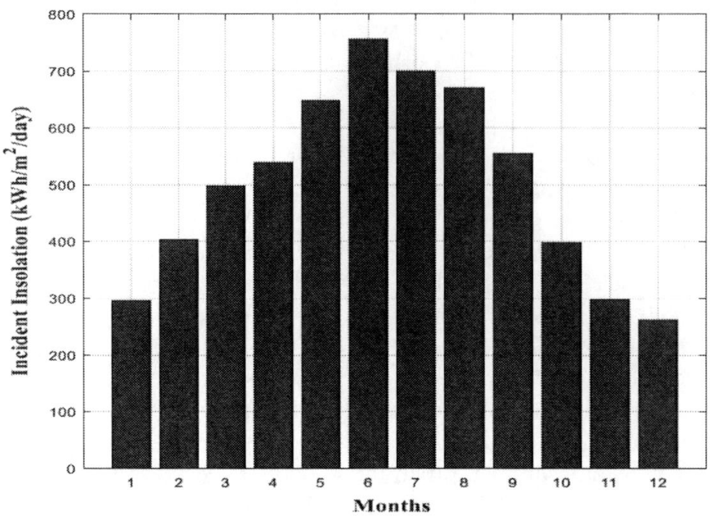

Figure 1. Monthly averaged insolation incident for year 2016 on a horizontal surface in Baghdad, adopted [7].

Because of geographical location and favourable climatic conditions, most parts of Iraq receive very high solar insolation about 2000 KWh/m^2/year. It is therefore necessary to go to the solar energy options to reduce the consumption of fossil fuel. Therefore, there is the possibility for converting a proportion of that solar energy into useful heat and electricity and subsequently solar thermal technology is commonly used to harness the sun's energy.

3. Collector Formulation

The first law of thermodynamics plays an important role in the performance calculations of the solar flat plate collector. Energy balance applied to a flat plate collector for computing the useful energy absorbed. For a simple flat plate collector computations, several assumptions which is acceptable for evaluating the energy gain. Steady state energy transfer condition was assumed and the energy exchanging occurred for a collector area Ac. The useful energy absorbed in flat plate collector is the difference between the absorbed solar radiation and thermal losses [8]:
Useful energy gain (Q_u) = energy absorbed by the collector – heat loss to the surroundings,

$$Q_u = A_c[S - U_L(T_i - T_a)] \tag{1}$$

where A_c represented Collector area in m^2, S is absorbed solar radiation in J/m^2, U_L is heat transfer loss coefficient in $J/m^2{}^\circ C$, T_i is the mean absorber plate temperature in °C and T_a is the ambient temperature in °C. Equation (1) is an extremely useful equation and applies to essentially all flat-plate collectors. To improve the performance of solar collector it is necessary either to reduce the overall energy loss coefficient or reduce area from which energy is lost. That is the maximum possible useful energy gain (heat transfer) in a solar collector occurs when the whole collector is at the inlet fluid temperature. Heat losses to the surroundings are then at a minimum. The above equation is correct and applicable if the actual useful energy gain of the collector is equal to the useful gain, which is mean that the heat removal factor F_R is unity. In equation 1, S represents the absorbed radiation and it is equal to:

IOP Conf. Series: Journal of Physics: Conf. Series **1032** (2018) 012005 doi:10.1088/1742-6596/1032/1/012005

$$S = I_b R_b (\tau\alpha)_b + I_d (\tau\alpha)_d \left(\frac{1+\cos\beta}{2}\right) + \rho_g (I_b + I_d)(\tau\alpha)_g \left(\frac{1-\cos\beta}{2}\right) \quad (2)$$

In equation 2, $(1 + \cos\beta/2), (1 - \cos\beta/2)$ are the view factors from the collector to the sky and from the collector to the ground, respectively. β represents a flat plate collector tilt angle from the horizontal. The subscripts $(b, d,$ and $g)$ represent beam, diffuse, and ground, respectively. $\tau\alpha$ is transmittance and absorption product. R_b is the ratio of beam radiation on the tilted surface to that on a horizontal surface at any time. ρ_g represents the ground reflectance. When measurements of incident solar radiation I_t are available, instead of equation (2), the following relation can be used [9]:

$$S = (\tau\alpha)_{av} I_t \quad (3)$$

$(\tau\alpha)_{av}$ can be obtained from,

$$(\tau\alpha)_{av} = 0.96(\tau\alpha)_b \quad (4)$$

The overall heat loss coefficient is a complicated function of the collector construction and its operating conditions, given by the following expression. In equation (1) U_L is the collector overall loss coefficient and it is equal to the sum of the top, bottom,and edge loss coefficients.

$$U_L = U_{top} + U_{bottom} + U_{edge} \quad (5)$$

Equation 5 represents the overall loss coefficient U_L. Energy diagram of typical flat collector is shown in figure 2. % 92 of the total sunshine reaches to the copper absorber, % 8 of the total sunshine is reflected from glass, %5 of the sunshine is emitted from the panel, %12 is lost through convection and conduction.

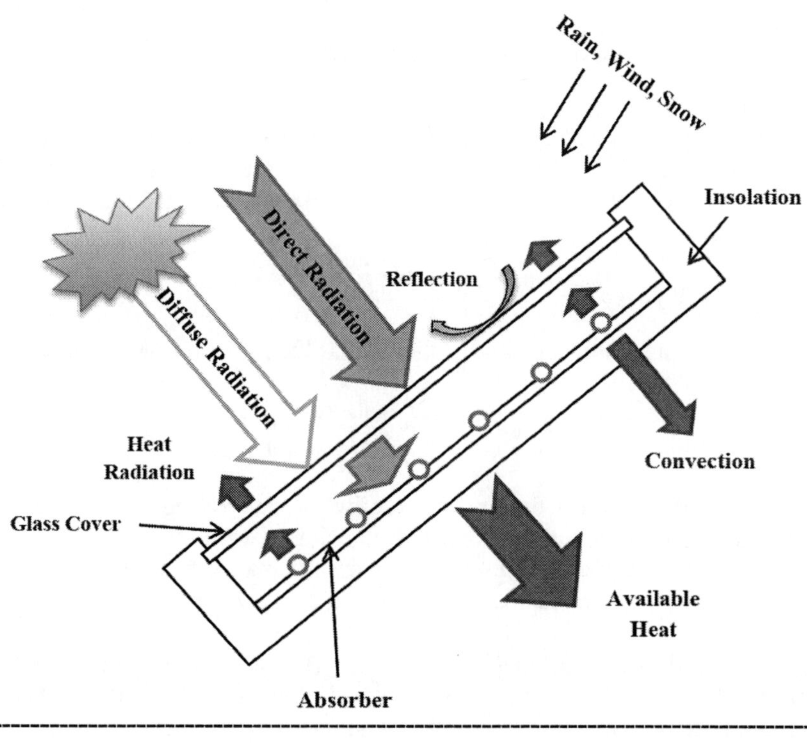

Figure 2. Energy diagram for a typical flat plate collector

The basic method of measuring collector performance is to expose the operating collector to solar radiation and measure the fluid inlet and outlet temperatures and the fluid flow rate.The useful gain is:

$$Q_u = A_c[S - U_L(T_i - T_a)] = \dot{m}C_p(T_o - T_i) \quad (6)$$

Where $\dot{m}$ is mass flow rate of working fluid, C_p is specific heat of working fluid and T_o,T_i represent outlet and inlet temperature of the working fluid, respectively. The instantaneous efficiency can be defined as

$$\eta = \frac{\dot{m}C_p(T_o - T_i)}{A_c I_t} \quad (7)$$

I_t represtnts the total incident radiation to the flat plate collector.

4. Orientation of Flat Plate Collector

For maximizing the performance of flat plate collector, they should be orientated and tilted properly. The collectors should be oriented directly toward the equator, facing south in the Northern Hemisphere and north in the Southern Hemisphere. The optimum tilt angle of the collector is equal to the latitude of the location, with angle variations of (10° to 15°) more or less, depending on the application [9]. If the application is solar cooling, then the optimum angle is latitude (-10°) so that the sun will be perpendicular to the collector during summertime, when the energy will be mostly required. If the application is space heating, then the optimal angle is latitude (+10°); whereas for annual hot water production, it is latitude (+5°), to have relatively better performance during wintertime, when hot water is mostly required.

$$H_G = H_b \frac{\sin(\alpha_a + \beta)}{\sin\alpha_a} + H_d \quad (8)$$

Where H_G is the monthly average of the global radiation in kWh/m^2, H_b, H_d are beam and diffuse monthly average solar radiation respectively in kWh/m^2. α_a is the monthly average altitude angle at noon. The total flux falling on the collector over the year is obtained by summing overall months,

$$H_G = \sum_{i=1}^{12} H_b \frac{\sin(\alpha_a + \beta)}{\sin\alpha_a} + H_d \quad (9)$$

For this annual flux to be a maximum, it can be shown that the optimum tilt angle β is given by:

$$\beta_{opt} = \tan^{-1}\left[\frac{\sum_{i=1}^{12} H_b \tan(\emptyset - \delta)}{\sum_{i=1}^{12} H_b}\right] \quad (10)$$

Where δ is declination angle which is the angle made by the line joining the center of the sun and the earth with its projection on the equatorial plane, $\emptyset$ is the latitude of the location (degree). The optimum tilt angle would be different if the nature of the energy demand is different. For example, for an application like space heating, demand may be high in the winter months of November, December, January and february. On the other hand, if solar energy were to be used for running an absrorption refrigeration plant, the duty would be highest in months like May, June, July and August have high ambient temperature. In such case, it would be obviously be desirable to use a tilt greater than the latitude for winter applications and the reverse for the summer applications.

Table 1. Solar angles of flat plate collector in Baghdad city for year 2016

Month	H_b kWh/m^2	δ Degree	$\emptyset - \delta$ Degree	$H_b \tan(\emptyset - \delta)$ kWh/m^2	$\beta_{opt.}$ Degree
Jan.	2.96	-20.7	54.63	4.078	54.02
Feb.	4.03	-12.3	45.63	4.119	45.62
Mar.	4.98	-1.8	35.13	3.503	35.12
Apr.	5.39	9.7	23.63	2.358	23.62
May	6.48	18.8	14.53	1.679	14.52
Jun.	7.56	23.0	10.33	1.377	10.32
Jul.	7.00	21.2	12.13	1.504	12.12
Aug.	6.71	13.7	19.63	2.393	30.23
Sep.	5.55	3.09	30.24	3.235	30.23
Oct.	3.98	-8.45	41.78	3.556	41.77
Nov.	2.98	-18.1	51.43	3.736	51.44
Dec.	2.62	-22.8	56.13	3.903	56.12
Σ	**60.24**			**35.441**	$\beta_{opt.}$ **Anuualy** **32.87**

5. Experiment Analysis of Flat Plate Collector

This section presents the details of the experimantal work using a typical flat plate collector installed in Baghdad city. Table 2 gives the dimensions of the flat plate collector and all the relevant properties. Table 3 lists the measured values of the incident solar intensity, inlet temperature, outlet temperature, mass flow rate and efficiency of flat plate collector according to Iraq climate on 4th of May 2016.

Table 2. Specification of flat plate collector

Length of the absorber plate, L [m]	2m	Thermal conductivity of insulating material	0.055 W/m.k
Width of the absorber plate, W [m]	1m	Density of insulating material	200 Kg/m^3
Plate thickness	0.005m	Material of collector tray	mild steel
Material of absorber plate	Copper	Thermal conductivity of collector tray	53.6W/m.k
Glass cover transmissivity	0.92	Density of collector tray	7833 Kg/m^3
Tilt angle of collector	33°	Specific heat of fluid	4200 J/Kg.k

Table 3. Experimental results of a flat plate collector

Time (hr)	I_t, (w/m^2)	T_i(C°)	T_o(C°)	$\dot{m}$ (kg/s)	η
10.00AM	900	16	29	0.0157	0.474
11:00AM	1000	21	39	0.0157	0.591
12:00 PM	1200	32	59	0.0157	0.739
1:00 PM	1150	40	62	0.0179	0.716
2:00 PM	1050	47	66	0.018	0.68

IOP Conf. Series: Journal of Physics: Conf. Series **1032** (2018) 012005 doi:10.1088/1742-6596/1032/1/012005

6. Results and Discussions

The discussion consists of two parts, the first part reveals a brief results of the measurements obtained from the experimental work, which is explained in figures 3 and 4. Figure 3 represents the variation of the flat plate collector efficiency with the local time. The value of the efficiency starts to increase during of time proceeding. It reaches the maximum value at the midday, then begins to fall. This is because of the decreasing in the incident solar radiation that is strikes the flat plate collector.

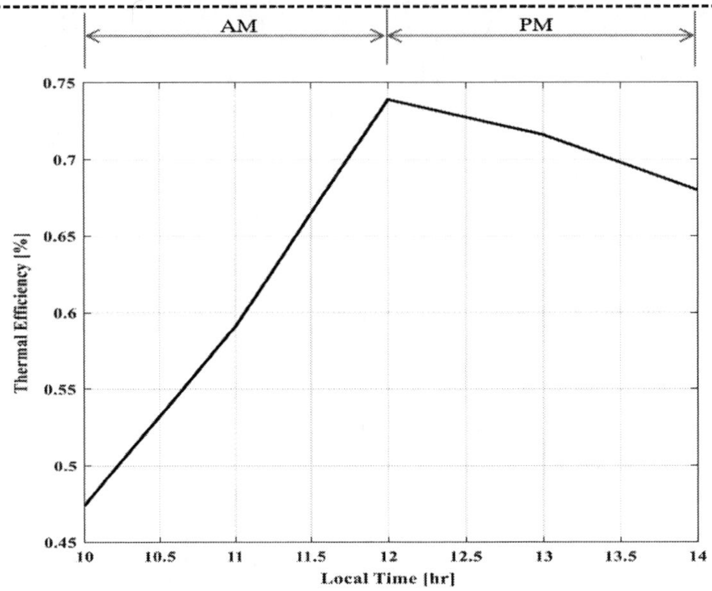

Figure 3. Thermal efficiency with local time

This decreasing in the incident solar radiation striking the flat plate collector appears in figure 4, which represents the variation of incident solar radiation with respect to time. The obtained heat gain in flat plate collector having the same pattern of incident flat plate collector, but in less values.

The second part of discussion presents the Statistics derived from the use of solar water heating instead of electrical heating to determine the obtained economic feasibility. Table 4 shows the fuel used based on economic surveys related to the availability of household fuel for various purposes.

The proper design and implementation of buildings, the use of efficient appliances and energy systems improve the performance of buildings. In addition, the rationalization of energy consumption and the improvement of the efficiency of use in the building sector depend on the procedures and actions of the users of these buildings and the management of energy in them, hence the importance of finding suitable opportunities for rationalizing energy consumption and improving the efficiency of its use in buildings. To find the actual benefit from the use of solar collectors and compensate them for devices used to heat water with electricity. Three models were taken for household consumption and how to distribute consumption according to the device used. The models were putted according to the classification of the Ministry of Electricity in Iraq [12].

In order to know the amount of electricity consumed to all of country. We must know the population of Iraq. The last census was recorded indicating that the population of Iraq almost (37 880 000) [13]. So the consumption is equal to approximately (37 880 000) $\times$ (0.0017×10^{6}) = (6.4×10^{10} MWh). By making some assumptions to find the percentage of consumption the electric power for heating water demands. The assumptions were based on dividing the population into families with different members. Each family lives in a separate house. Three types of families have been adopted. A family of five people, a family of six people, and the last assumption is a family of seven people. Iraq families

were divided into three groups as mentioned below. Each family assumed to be living in each three models (see Tables 5-7) of houses to find the value of electricity consumption by heating water.
Table 8 represented the statistic of consumption the electric power by heating water. This statistic is Acceptable to a high extent with an error percentage of (±2%).
Table 8 shows the quantity of electric power which was consumed to heat water in three selected models of houses in Iraq for one hour. The average heat gain from flat plate collector is roughly to (1301 w.h). Table 9 represents calculations about the reduction in consumption of electrical power from the national network.
From table 9, flat plate collector reduces the consumptions of electrical power of heating water from 100%. Which means there is no ever consumptions of electrical power to heating water, to 25% which represents the electrical power consumed to heat the water reduced by 25% when using flat plate collector for each house.

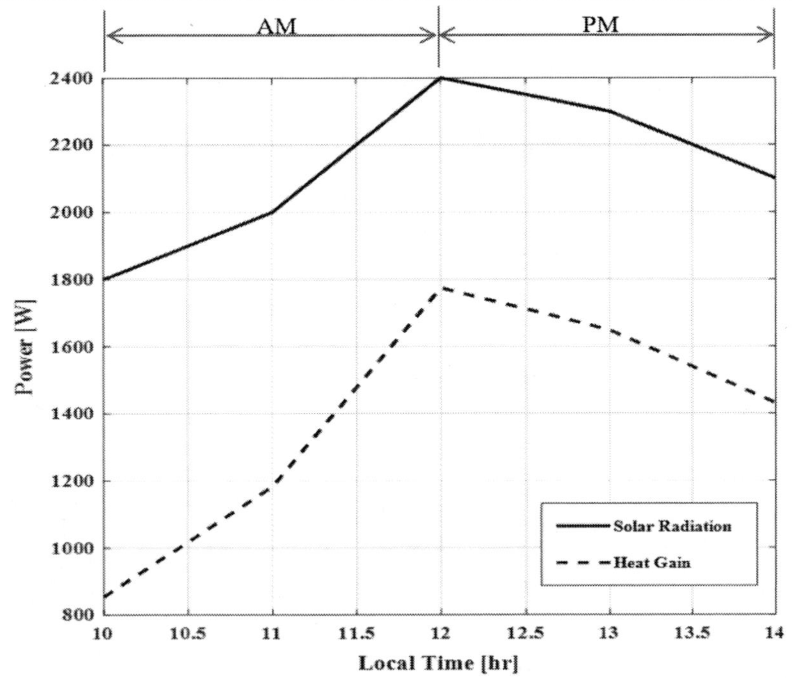

Figure 4. Variation of incident solar radiation on the collector and heat gain with local time

Table 4. Distribution of resources domestic consumption % [11]

Consumption Type	Electricity	Gas	Kerosene	Others	Total
Cooking	2.2	82.7	11.2	3.9	100
Lighting	98.4	0.1	1.5	-	100
Heating	13.9	0.9	81.0	4.2	100
Water heating	60.2	11.3	23.3	5.3	100

IOP Conf. Series: Journal of Physics: Conf. Series **1032** (2018) 012005 doi:10.1088/1742-6596/1032/1/012005

Table 5. Model (1) of calculation the electricity according to the domestic consumption

	Elec. Instruments	**Quantity**	**Elec. Power (W)**	**Average daily operating hours**	**Daily consumption (kWh)**	**Monthly consumption (kWh)**	**(%)**
1	Fluorescent lamp	5	40	8	1.6	48	9.6
2	Flash lamp	5	60	8	2.4	72	
3	Refrigerator	1	200	8	1.6	48	10.6
4	frozen	1	350	8	2.8	84	
5	TV	1	400	8	3.2	96	
6	Radio	1	60	2	0.12	3.6	
7	Washing machine without heating	1	200	1	0.2	6	10
8	Iron	1	1250	0.5	0.625	18.75	
9	Fan	5	100	10	5	150	26.4
10	Air cooler	2	300	10	6	180	7
11	**Water heater**	**1**	**3000**	**6**	**18**	**540**	**43.3**
		Total			41.545	1246.35	100

Table 6. Model (2) of calculation the electricity according to the domestic consumption

	Elec. Instruments	**Quantity**	**Elec. Power (W)**	**Average daily operating hours**	**Daily consumption (kWh)**	**Monthly consumption (kWh)**	**(%)**
1	Fluorescent lamp	6	40	8	1.92	57.6	6.6
2	Flash lamp	6	60	8	2.88	86.4	
3	Refrigerator	2	200	8	3.2	96	8.2
4	frozen	1	350	8	2.8	84	
5	TV	2	400	8	6.4	192	
6	Radio	1	60	2	0.12	3.6	
7	Washing machine	1	200	1	0.2	6	10.9
8	Iron	1	1250	1	1.250	37.5	
9	Fan	6	100	10	6	180	
10	Air cooler	2	300	10	6	180	49.5
11	Air Conditioner	1	3000	8	24.0	720	
12	**Water heater**	**1**	**3000**	**6**	**18**	**540**	**24.7**
		Total			72.77	2183.1	100

Table 7. Model (3) of calculation the electricity according to the domestic consumption

	Elec. Instruments	Quantity	Elec. Power (W)	Average daily operating hours	Daily consumption (kWh)	Monthly consumption (kWh)	(%)
1	Fluorescent lamp	8	40	8	2.56	76.8	5.2
2	Flash lamp	8	60	8	4.84	115.2	
3	Refrigerator	2	200	8	3.2	96	7.1
4	frozen	2	350	8	5.6	168	
5	TV	2	400	8	6.4	192	
6	Radio	1	60	2	0.12	3.6	
7	Washing machine	1	200	1	2.5	75	8.3
8	Iron	1	1250	1	1.25	37.5	
9	Fan	6	100	10	6	180	
10	Air cooler	1	300	8	2.4	72	64.9
11	Air Conditioner	3	3000	8	72	2160	
12	**Water heater**	**1**	**3000**	**6**	**18**	**540**	**14.5**
		Total			123.87	3716	100

Table 8. Statistic of consumption the electric power with typical houses models

	Number of houses	Power consumptions model (1)	Power consumptions model (2)	Power consumptions model (3)
Family of five people	7576000	3680.5	2099.5	1232.5
Family of six people	6313333	4416.6	2519.4	1479
Family of seven people	5411428	5152.7	2939.3	1725.5

Table 9. Percentege reduction in consumption of electrical national network

Number of houses	Reduction % in Elec. consumption, model (1)	Reduction % in Elec. consumption, model (2)	Reduction % in Elec. consumption, model (3)
7576000	35%	62%	100%
6313333	29%	52%	88%
5411428	25%	44%	75%

7. Conclusions

Experimental study supported by a numerical calculation had been accomplished to determine the economical feasibility of using flat plate collector technique. The study assumed there is a singular flat plate collector for each house in Iraq. The population were divided into three different models of houses, different in terms of electricity consumption and the consumption of electricity to heat the water. By using these assumptions, the electrical power consumptions appear to be decreased by a

(25% to 35%) from the total electric power consumptions in the case of using the electric power for heating water. The study had been achieved on 4th of May 2016. This month have the same properties for several months. This study could be achived in June, July,August, September and April. By using this techinque with these months, the consumption of electrical heating will be decresed by a large amount and this is helps to store electrical energy for using in another useful work.

8. References

[1] Yousef B and Adam N 2008 *Journal of Energy in Southern Africa* **19** **4** 32.

[2] Amrutkar S K, Ghodke S and Dr. Patil K N 2012 *IOSR Journal of Engineering (IOSRJEN)* **2** **2** 207.

[3] Chaji H, Ajabshirchi Y, Esmaeilzadeh E, Heris S Z, Hedayatizadeh M and Kahani M 2013 *Modern Applied Science* **7** **10** 60.

[4] Ameen B K, Alhadithi M B and Fadhil O T 2015 *Al-Nahrain University, College of Engineering Journal (NUCEJ)* **18** **2** 259.

[5] Mankhi K K, Al-Hashimi N H N and Al-Asadi J M 2014 *IOSR Journal of Engineering (IOSRJEN)* **04** **12** 15.

[6] Basim Z, Mohsen A and Shareef A S 2017 *ARPN Journal of Engineering and Applied Sciences* **12** **11** 3518.

[7] RETscreen Expert software

[8] Duffie J A and Beckman W A *Solar Engineering of Thermal Processes*, 4th ed. (New Jersey: Wiley).

[9] Kalogirou S A *Solar Energy Engineering Processes and Systems*.

[10] http://www.solcoast.com/go-solar-by-november-8th-and-save-big/

[11] Mohammed H H *Energy Statistics in Iraq, Directorate of Environment Statistics* Beirut 3-5 March 2009 Arabic

[12] https://moelc.gov.iq/index.php?name=Pages&op=page&pid=115

[13] Ministry of Planning Central Bureau of Statistics Iraq Figures and Indicators 2017 Arabic.

IOP Conf. Series: Journal of Physics: Conf. Series **1032** (2018) 012006 doi:10.1088/1742-6596/1032/1/012006

Experimental Study of Two Different Types of Solar Dish Characteristics and its Efficiency Based on Tikrit, Iraq Weather Conditions

Yaseen .H. Mahmood[1], Rafa Y. J. Al-Salih[2]

Department of Physics, College of Sciences, University of Tikrit, Iraq

D.yaseen.ph.sc@tu.edu.iq[1], Aburafal1965@yahoo.com[2]

Abstract. Two solar dish concentrators (SDC) with diameter of (2 m) have been designed for water heating purpose. This study applied on solar steam water heater (SSWH). Both of the dishes coated with thick layer of steel (galvanized), while its interior surface covered with a reflector layer. The first dish interior surface covered with a layer of reflectivity up to 80%, whereas the second one covered by a pieces of mirrors with a reflectivity of 95%. The receivers have been fixed in the focal position of both dishes. The measurement of temperature and solar power for both types have been documented and recorded. The optical energy for the 1st dish is 1909 W, the final energy is 1551 W, and the efficiency is 61 %, while the optical energy for the 2nd dish is 1607.68 W, the final energy is 1853 W, and the efficiency is 73 %. The obtained results affirmed that the materials used in this research are economic, with high quality to obtain solar concentrator, but the dish which is covered with pieces of mirrors is better, and it can be used for water heating.
Key words: Solar dish, Solar concentrator, Solar boiler, Solar thermal, Concentrator reflectance.

1- INTRODUCTION

In the last 50 years, a need of renewable energy have been appeared because of the contamination and the high cost of fuel . The solar thermal radiation is the main energy source in the world. many researchers worked on using the solar thermal energy to design a useful project; e.g. design and test solar trough with length 2.4 m and 1 m aperture for solar water heat, and it reached 82 ^{0}C without tracking [1]. To help soldier getting water in high places like Himalaya a solar dish with receiver has been used to reflect the solar radiation to this receiver an ice has been put in the receiver to melt it and get water and tank has been used to store the water, then the result analyzed for the system [2]." The Australian National University had been re-designed the big dish installation for mass production and commercialization [3]. The building had been designed and structured as an"SG4" big dish of (500 m^2) solar parabolic concentrator using direct-steam generation [4,5].By using a solar trough a 228 ^{0}C had been obtained by using metal reflector to produce electricity [6] and an "SG4" solar dish array with dual tracing used for electrical power [7]". Design and fabricate a

IOP Conf. Series: Journal of Physics: Conf. Series **1032** (2018) 012006 doi:10.1088/1742-6596/1032/1/012006

primary sample for a concentrator that was used as a water heater and also a Cu boiler was used and with efficiency of about 65.9% in April and May [8], making of a solar cooker by using empty tubes and a storage system to store the heat, and the system was tested in different condition in Japan to focus the radiation on boiler which are located at the focus of the concentrator [9]. This boiler absorbs and transports into thermal energy. This heat can be used directly for power generation.

The operation of solar thermal energy collector can be demonstrated as an energy balance between the absorbed solar radiation by the collector and the removed or lost thermal energy from the collector [10, 11] . Therefore, the main goal of this paper is to analyze and evaluate the effect of the location of the designed solar dish, and to study the reflectivity of concentrators, and its feasibility in the work location (Middle of Iraq).

2- THEORYTICAL DETAILS

1- Parabolic Geometry

"A parabola is the locus of a point that moves so that its distances from a fixed line and a fixed point are equal. This is shown in Fig.1, where the fixed line is called the *directrix* and the fixed point *F*, the focus. Note that the length *FR* equals the length *RD*. The line perpendicular to the directrix and passing through the focus *F* is called the axis of the parabola. The parabola intercepts its axis at a point *V* called " vertex", which is exactly midway between focus and directrix"[10].

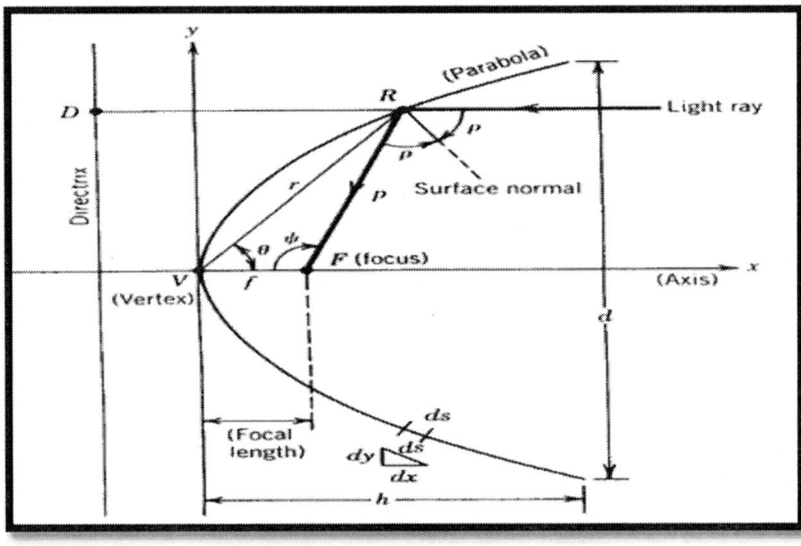

Fig.1: The parabola diagram.

2- Concentrator Geometry

The concentration ratio is usually used to describe the quantity of light energy concentration realized by a given collector. "Geometric Concentration Ratio" (CR_g) is described as the area of the collector aperture (A_a) divided by the receiver surface area (A_r), and it can be directly calculated according to the following equation:

$$CR_g = \frac{A_a}{A_r} \quad (1)$$

The distance from the aperture to the vertex (h) can be calculated in terms of the focal length (f) and the aperture diameter (d), as the following expression[10,11]:

$$h = \frac{d^2}{16f} \quad (2)$$

$$\frac{f}{d} = \frac{1}{4tan(\psi_{rim}/2)} \quad (3)$$

In a same manner, the rim angle (ψ_{rim}) can be expressed: as:

$$tan\,\psi_{rim} = \frac{1}{(d/8h)-(2h/d)} \quad (4)$$

3- Optical Energy Absorbed By The Reciever

"To calculate the heat transfer and optical energy (Q_{opt}) of the receiver; the following equation can be used [12,13]:

$$Q_{opt} = Aa\,p_{s.m}\,.\tau g.ar.S.Ia \quad (5)$$

where:

$p_{s.m}$: "Specular reflectance of concentrating foil".

τ_g: "Transmittance of any glass envelope which covering the receiver".

A_a: "Aperture of the collector".

S: "receiver shading factor" ("fraction of collector aperture, not shadow by the receiver").

I_a: "Insulation incident on the collector aperture".

a_r: "Absorbance of the of the receiver".

S, a_r, $p_{s.m}$, and τ_g: "constants". They are depending on the material type used, and on the geometric accuracy of the collector. These constants are nominally agglomerated into single constant term.

n_{opt}: "The optical efficiency of the collector".

The quantity of thermal energy generated by the solar collector is characterized by [12,14]:

"$Q_{out1}(v_1) = Q_{opt} - Q_{Loss1}(v_1)$ " (6)

4- Collector Efficiency

The collector efficiency ($\eta_{collector}$) and the receiver temperature ($T_{receiver}$) can be calculated using equation (6) above with another parameters, and expressed by [11]:

$$\eta_{collector}\,(v_1) = \frac{Q_{out-1}(v_1)}{Aa.Ia} \quad (7)$$

3- EXPERMENTAL WORK

1- Parabolic Dish

A concave dish "made of galvanized steel" and lined with a steel sheet have been chosen. The dimensions of the designed dish were given in table 1 below:

Table 1: Characteristics of the solar concentrator.

Diameter of opening of the parabola (d)	2 m
Surface collecting of the parabola	3.14 m^2
Depth of the parabola (h)	0.25 m
Focal distance (f)	1 m
Rim angle (ψ_{rim})	53°
f/d	0.5

Method of a given focus and directrix have been employed in the construction of the parabolic dish. The reflector plain mirror cut into small rectangular shapes and fixed by screws. This structures painted by steel paint and used as mirrors with reflectivity of (95%). In another dish, a miller paper have been covered the dish's metal as a reflective surface of (80% reflectivity). The reflective surfaces of the parabolic dishes used here are shown in Fig. 2.

Fig. 2: The designed parabolic dishes.

2- Reciever

The experimental receiver is made of stainless steel and coated by a thin layer of black paint; to decrease the reflection of the solar rays. It was located in the focal point of the parabola dish of (2 m) diameter; to study the radiation and temperature on a receiver's cavity. The geometry of the concentrator of this model has been done by applying equation (1) mentioned above. To obtain the optimum design of the receiver; heat losses of the receiver should be minimized. Convection and radiation heat transfers from the receiver to the surrounding should be investigated carefully, because of the significant difference between them. On the other hand, it is difficult to find simple method for predicate surrounding. The major concern for most of studies is the convection and radiation losses approximated by a simple equation when the operating temperature is low, however, the convection losses are also relatively low to high temperature. It is found that the magnitude and the direction of the winding can greatly affecting on the amount of the heat loss. The heat loss became higher in the case of wind parallel to the aperture plane than that of the case of the head–on wind.

In this work, the cylindrical receiver was chosen, because it has a small area facing the aperture window, which leads to minimize the reflection losses. Thus, this type of receiver design gives more chance to absorb the internal reflected photons. The receiver consists of a set of copper coil windings house within carbon steel conical cavity. All these parts were made locally. The dimensions of cylindrical cavity are (200 mm) depth and (170 mm) aperture diameter. The length of the coil of (4 m) consisting of (10 windings) of reduced diameter serially, with tube diameter of (12.5 mm). Fig. 3 shows the fabricated structure of the cylindrical receiver. The receiver design is explained in the next section.

Fig. 3: Experimental cylindrical receiver

The aperture diameter of the cylindrical receiver is chosen according to the diameter of the focus of the dish. Two thermocouples of k-type were used for measuring the inlet and the outlet temperature of the heat transfer fluid (HTF)[14].

4- EXPERMENTAL CALCULATIONS OF THE RECIEVER

The parameters of the receiver have been calculated using equations (5, 6,7) by using experimental values and heat transfer for the inlet temperature and the outlet temperature [15], and illustrated in the table 2 below:

IOP Conf. Series: Journal of Physics: Conf. Series **1032** (2018) 012006 doi:10.1088/1742-6596/1032/1/012006

Table 2: Experimental calculations of the receiver concentration.

Sample	a_r	$p_{s.m}$	τ_g	A_a (m^2)	I_a (w/m^2)	Q_{Loss}(w)	Q_{opt} (w)	Q_{final} (w)	$\eta_{collector}$
1	0.15	0.8	1	3.14	800	48.3	1607.68	1551.6	0.617
2	0.15	0.95	1	3.14	800	56	1909	1853	0.73

5- RESULTS AND DISCUTION

This study, demonstrated the relation between solar radiation and time in Tikrit, Iraq for latitude (34.616) and longitude (43.679) in one day at a middle of April and one day in the middle of May / 2016. Iraq is one of the 2nd favorable area solar application; due to the fact that it locates in the middle of solar radiation zone. Fig.4 illustrates the distribution of solar radiation according to local time. It can be seen that there is an increasing in the solar radiation from dawn to mid noon. After mid noon, it starts decreasing until sun set in May. Similarly, in April, with the important exception that decreasing in the solar radiation was dramatically observed at 10:10 AM; due to the passing of clouds at this period of time which causes this decreasing

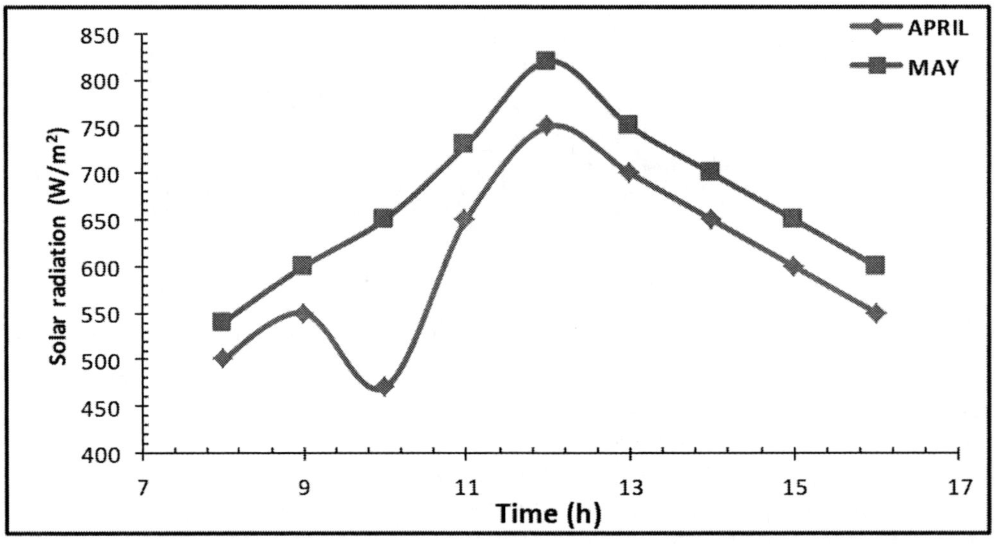

Fig. 4: Relation between solar radiation and time.

It is important to mention that the receiver used without water; as presented in Fig. 5. It is clear that the temperature increasing as the time increasing, due to the absorption of solar radiation by the receiver; so the heat increases.

IOP Conf. Series: Journal of Physics: Conf. Series **1032** (2018) 012006 doi:10.1088/1742-6596/1032/1/012006

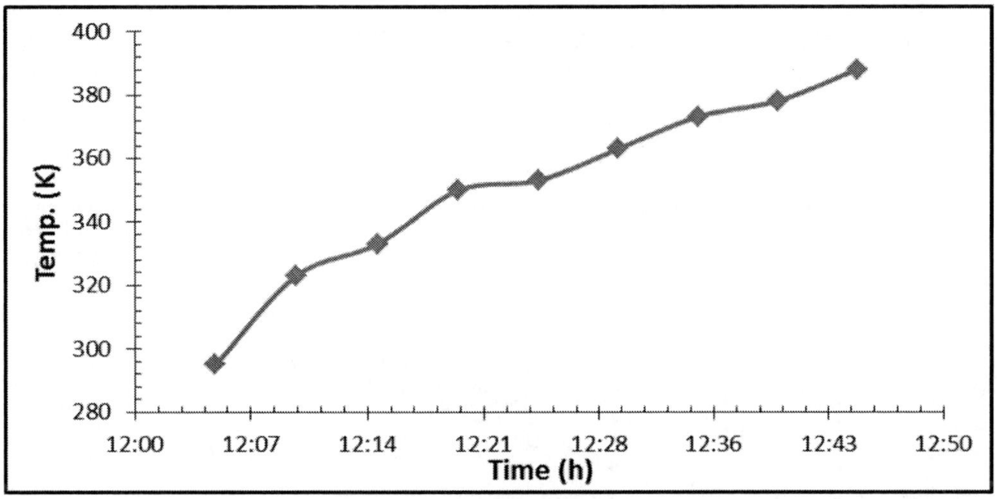

Fig. 5: Relation between temperature and time without water.

Fig. 6 shows the increasing of the temperature during the time for inlet, outlet, and ambient. In a clear day, where ambient temperature semi constant; the inlet, outlet temperature has been increased with time until the sun set. Practically a closed cycle between the storage tank and the boiler have been used. It is also noticed that the outlet temperature was greater than the inlet at 9:30 AM to 1:30 PM; because of the increasing in the solar radiation in this period of time [15].

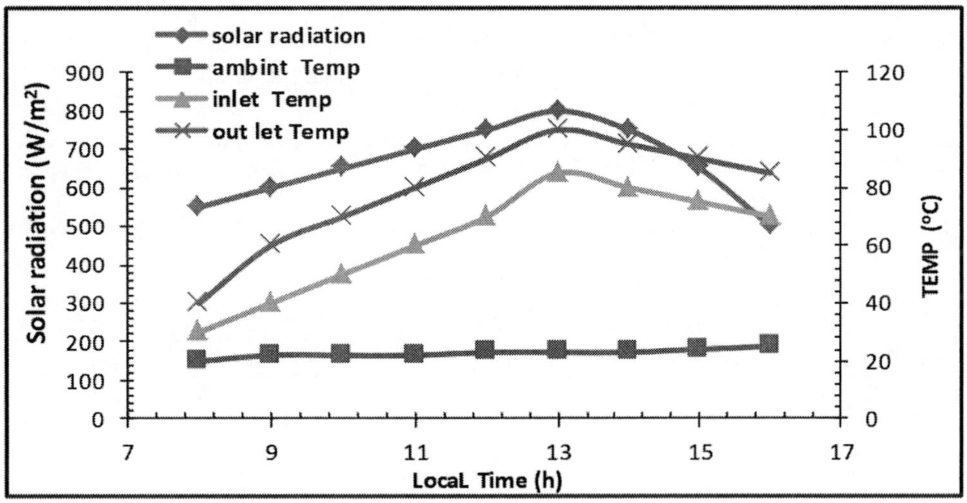

Fig. 6: Relation between temperature and time, with solar radiation

Moreover, in this work, the heat loss coefficient with time has been studied. When solar radiation increases, the stored heat increases also. So the heat radiation increases too, according to Steven- Boltzmann 4th power of source temperature. This fact is clearly appeared in Fig.7, where it is

IOP Conf. Series: Journal of Physics: Conf. Series **1032** (2018) 012006 doi:10.1088/1742-6596/1032/1/012006

clear that the curve of solar radiation of (800 w/m^2) is higher than that of (700 w/m^2). This figure shows increasing in heat loss coefficient with increasing of receiver heat. In addition, there is another factors which may be contribute in heat lose coefficient with mean temperature of receiver such as the wind speed, which is an important factor. The effect of ambient temperature is represented by the degradation in heat loss coefficient by radiation (h_r) of receiver at any specific temperature, when we left the winter months towards summer.

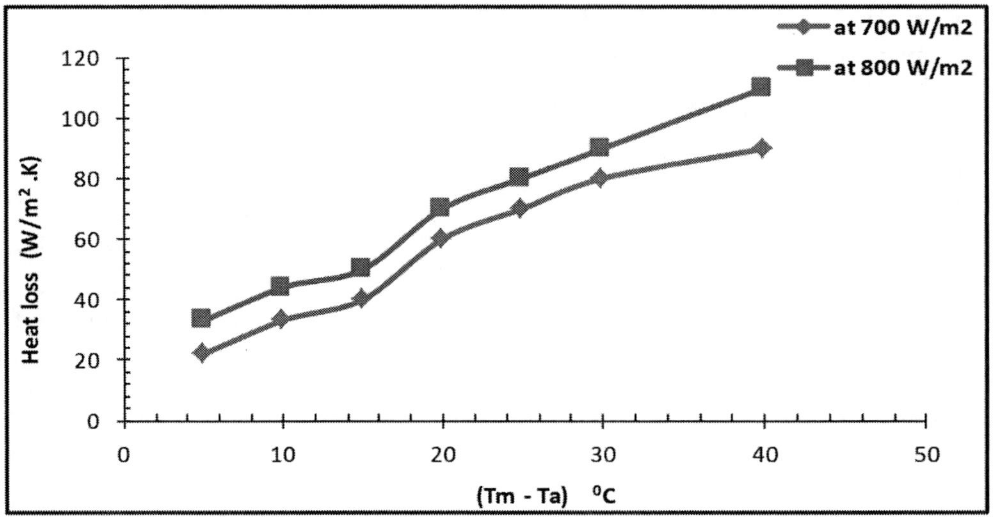

Fig. 7: Variation of heat loss coefficient of radiation with operating temperature of receiver.

Fig. 8 shows the variation of instantaneous efficiency with operating temperature (T_m - T_a) for receiver. Finally, the system efficiency is decreased with operating temperature.

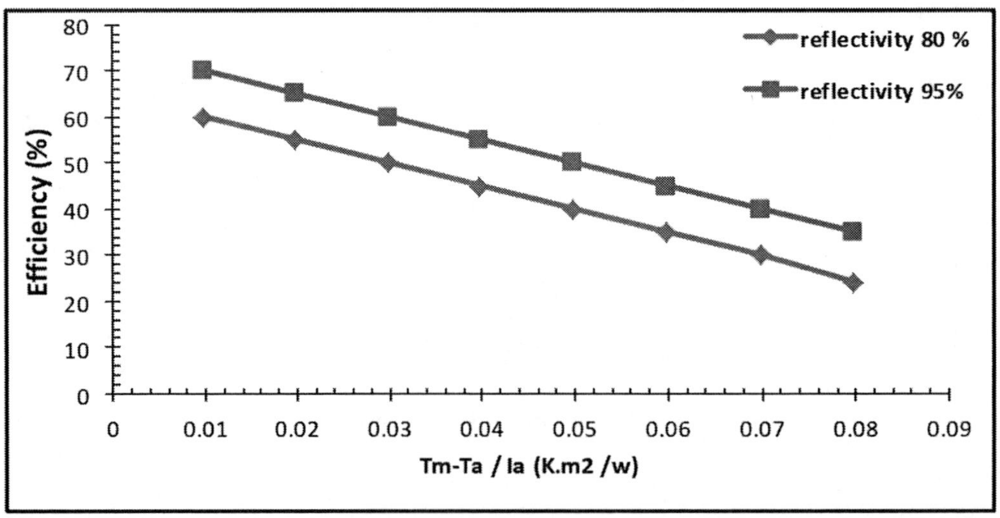

Fig. 8: Relation between efficiency of solar radiation with operating temperature.

6- CONCLUSIONS

Two solar concentrator designs have been belt and investigated. These designs were based on a reflector dish and a close receiver. Experimental measurements of solar flux, heat transfer, and temperature distribution on the receiver have been carried out. The obtained results described the awaited physical phenomenon correctly. The temperature in the receiver have been reached a value up to 350 °C. To improve the results a bigger solar concentrator and smaller receiver as possible should be done to get rid of the heat loss and to get the best useful energy by using best reflective mirrors to eliminate defects; such as spherical aberration and astigmatism. The optical energy, the final energy, and the efficiency for both types have been calculated. The optical energy for the 1st dish is 1909 W, the final energy is 1551 W, and the efficiency is 61 %, while the optical energy for the 2nd dish is 1607.68 W, the final energy is 1853 W, and the efficiency is 73 %. The obtained results affirms that the materials used in this research are economic and they having a high quality to obtain solar concentrator. The dish which is used mirrors in it; is better, and it can be used for water heating in the metals melting. From the outcome of this work, it can be concluded that the two dishes used are good for the solar thermal application, but the dish which is covered with mirrors is more efficient; because of its high reflectivity, and this method can be used as a solar water heat.

ACKNOWLEDGMENT

We would like to express our sincere appreciation and gratitude to the Physics Department in the College of Sciences in Tikrit University for their kind support during the period of the research.

Our deeply grateful also to the Physics Department in the College of Sciences in Karbala University and to the supervisor committee of the 6th Karbala conference for sciences for their efforts in preparation of this conference.

REFERENCES

[1] Eltahir Ahmed Mohamed 2013 Design and testing of a solar parabolic concentrating collector *(International Conference on Renewable Energies and Power Quality* (ICREPQ'13 Bilbao Spain) 20th to 22th ISSN 2172-038 X, No.11, March,

[2] Kailash Karunakaran 2014 Thermal Modeling of Snow to Water Conversion Device Using Solar Energy (International Journal of Mechanical And Production Engineering) ISSN: 2320-2092, Volume- 2, Issue- 2, Feb

[3] Lovegrove K., Burgess G, Pye J., *Solar Energy*, 85, (2011) pp.(620 – 626).

[4] Australian Radiation Protection and Nuclear Safety Agency (ARPANSA), "Radiation Protection Standard: Occupational exposure to ultraviolet radiation". *Radiation reaction Series* No. 12, December (2006).

[5] Burgess G., Shortis M., Kearton A 2009 Photogram try for dish concentrator construction Solar (09, 46th ANZSES Conference, Townsville) 29 Sept. – 02 Oct

[6] Parashar V., "Solar energy for domestic and small industries with help of parabolic trough solar concentrator" , International Journal of Engineering Research & Technology (IJERT) Vol. 2 Issue 10 IJERT ISSN: 2278-0181, October - 2013.

[7] Vinayak Sakhare V. N. Kapatkar, Experimental Analysis of Parabolic Solar Dish with Copper Helical coil Receiver" , International Journal of Innovative Research in Advanced Engineering (IJIRAE) ISSN: 2349-2163,Volume 1, Issue 8 (September

[8] Nilkanth Shinde 2012 Performance evaluation of parabolic dish type solar collector for industrial heating application (J. Energy Technology and Policy, Vol. 8) No. 1

[9] S. D. Sharma, T. Iwata, H. Kitano, and K. Sagara 2005 Thermal performance of a solar cooker based on an evacuated tube solar colletor with a PCM storage unit (Solar Energy, Vol. 78) , pp. 416-426

[10] William, B. S., and Michael G. 2001 Power from the Sun http://www.powerfromthesun.net/book.html.

[11] Newton, C. C.2007 A Concentration Solar Thermal SystemM.Sc. Thesis,(Florida State University)

[12] Duffie J. A. and Beckman W. A 2013Solar Engineering of Thermal Processes 4th Edition, (Wiley, New York).

[13] Karima Ghazouani, Safa Skouri 2017 Thermal study of solar parabolic concentrator (IOSR Journal of Mechanical and Civil Engineering) (IOSR-JMCE), -ISSN:2320–334X, pp. 118-123, www.iosrjournals.org(International Conference on Recent Innovations in Civil & Mechanical Engineering 118)

[14] Mahmood, Y. H. 2017 Design and fabrication of Solar dish array and study it characterization (Tikrit Journal of Pure Science, 22) (11), ISSN: 1813– 1662 (Print), E-ISSN: 2415– 1726 (On Line) 78.

[15] Mahmood, Y. H.2017 "Design and installation of (flower sun) a concentration dish and study its parameters(Tikrit Journal of Pure Science, 22) (11) (2017) ISSN: 1813 – 1662 (Print), E-ISSN: 2415 – 1726 (On Line).

IOP Conf. Series: Journal of Physics: Conf. Series **1032** (2018) 012007 doi:10.1088/1742-6596/1032/1/012007

Assessment of Solar Photovoltaic Potential in Iraq

Mohammed T. Hussain[1] and Emad.J..Mahdi[1]

[1]University of Baghdad, College of Sciences, Physics department (Baghdad – Iraq)

mohammedtake@gmail.com, emad.jaleel@yahoo.com

Abstract: This paper presents the comparison of global solar radiation data and PV potential measured at different sites in Iraq obtained from web (PV GIS data) by selecting seven locations which cover north middle and south of Iraq as follow (37°N - 43°E) , (33°N - 44°E) , (34°N - 43°E) , (29°N - 46°.17′E) , (33°N – 40° E) , (32°N – 44°.3′E) and (33°.4′N – 45°.3′E) respectively with the software program used empirical equation by visual basic net language to give a good agreement for assessment PV potential maps in Iraq using Arc GIS program version 9.3 .

1- Introduction

Photovoltaic solar projects can be generate electricity from solar radiation are foreseen as a technology option significantly contributing to the sustainable energy supply due to continuous technological development and cost reduction [1,2].

The geographical potential is critical for each renewable resource demarcate the locations where the energy - conversion can be deployed to preclude those zones reserved for other uses , land used typically residential cities, roads, and land cover such like lakes, rivers, beaches and their influence areas, and protected areas [2,3].

The technical potential should care about into account collection and conversion efficiencies for converting the resource power into electricity, these can be a function of geographical or climatic (e.g. temperature, irradiance or humidity) conditions [1,3].

2- Solar radiation resource assessment

Calculation of the hourly solar radiation falling off during the average day of each month is essential for calculating the solar collector performance throughout the year. The long-term models provide the mean hourly distribution of global solar radiation over the average day [2,4].

Satellite data provides information about solar radiation and meteorological conditions for locations where ground measurement data are not available [4,5,6,7].

The total irradiation H_{GIH} on a surface normal to the sun's rays is made up of normal direct irradiation H_{ND}, diffuse irradiation H_d, and reflected irradiation H_R [6,8,9 :

$$H_{GIH} = H_{ND} + H_d + H_R$$

IOP Conf. Series: Journal of Physics: Conf. Series **1032** (2018) 012007 doi:10.1088/1742-6596/1032/1/012007

3- Basics of solar potential calculation

The technical potential is the geographical potential multiplied by efficiency factors and performance ratios of the solar panels. This study's estimation of technical potential was based on the approach of any location [10-12].

Basically the amount of solar irradiation incident to the fraction of the land suitable for energy production was calculated and then the amount of energy that land could potentially generate if covered by PV panels was computed [10,13,14].

However, this approach has been modified in two ways, The power output of the PV array is given by equation [11, 15-17]:

$$P_k = \sum Y_{PV}\, D\left[1 + f\left(T_c - T_{c,STC}\right)\right] PV_n$$

Where: P_k . Annual PV array power output, Y_{PV}. PV panel power output , D.Degradation factor, f.Temperature loss coefficient,T_c.Operation temperature ,$T_{c,STC}$ .Temperature at standard test condition ,PV_n . Number of panels.

The term Y_{PV} which relation to the solar resource can be obtained from equation:

$$Y_{PV} = H_{GHI}\, A\, d_m\, \eta_{PV}$$

Where:H_{GHI} .Monthly average of solar radiation, A.Area of penal ,d_m. Numbers of days in month ,η_{PV}.PV conversion efficiency.

$$P_{sys} = P_k \,.\, L_{sys}$$

Where: P_{sys}.The power output of solar plant (kwh),L_{sys}.Product of all system losses.

4- Methodology

Is a source geographic information system (PVGIS)-based web application that provides solar radiation developers with tools and data for site selection and screening of potential PV solar energy plants. It provides data for solar radiation rates for ten years for the period 2001-2012 . The combination of solar radiation data obtained by PV GIG Web and software program prepared using empirical equation and visual basic language to manage process.

The estimation of the primary solar resource is based on the clear-sky (potential) global irradiation on a horizontal plane. The real global irradiation on a horizontal plane is estimated by multiplying the clear-sky values by the clear-sky index. These values are then used in the model for computation of global irradiation on a horizontal plane.

The empirical program used input parameter such like (latitude , longitude of the site ,elevation above sea level, the standard meridian for local time zone, solar constant, tilt angle, the date of day and month in

IOP Conf. Series: Journal of Physics: Conf. Series **1032** (2018) 012007 doi:10.1088/1742-6596/1032/1/012007

year for 200 sites in Iraq ,this study uses Arc GIS program version (9.3) for mapping plotting . To test the ability and accuracy of program data were selecting7 locations in northern, middle and southern Iraq (37°N - 43°E), (33°N - 44°E), (34°N - 43°E), (29°N - 46°.17′E), (33°N – 40°E), (32°N – 44°.3′E) and (33°.4′N – 45°.3′E). The program and enters the main parameters to the calculation process, each of the coordinates of the site (latitude and longitude), solar constant, altitude site, visibility and local solar time, as shown in figure (1) the flow chart of the first page (main function of program).

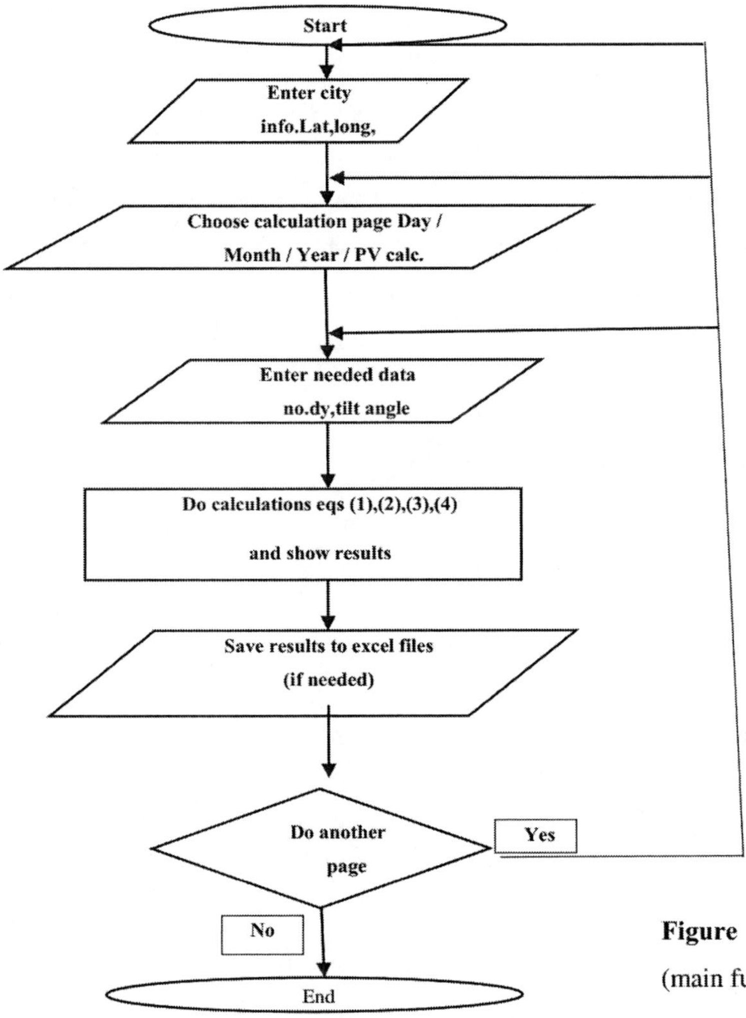

Figure 1. The flow chart of the first page (main function of program

IOP Conf. Series: Journal of Physics: Conf. Series **1032** (2018) 012007 doi:10.1088/1742-6596/1032/1/012007

5- Results and Discussion

The monthly values of the global solar radiation on horizontal and inclined surface were calculated for seven Sits by software program and camper with PVGIS data (Provided reliable and accurate solar, weather and solar electricity data that are used in the whole lifecycle of solar power plants, from prospection to development and operation). The comparison between them which gives a small difference between both data can be referred to the values of the assumed solar constant, daylight hours, and the accuracy of web data.

Figure (2) shows the comparison of monthly irradiation with time on horizontal plane in ($Wh/m^2/day$) for (37° N – 43°E) site in northern of Iraq. It has found that the maximum solar radiation is between June and July and there is a small difference between the two methods before the middle due to the process of measuring solar radiation at sunrise is influenced by the air mass of this site and weather conditions such like (cloud, fog and rain etc.) that consider as attenuation parameters which taken into account in PV GIS program.

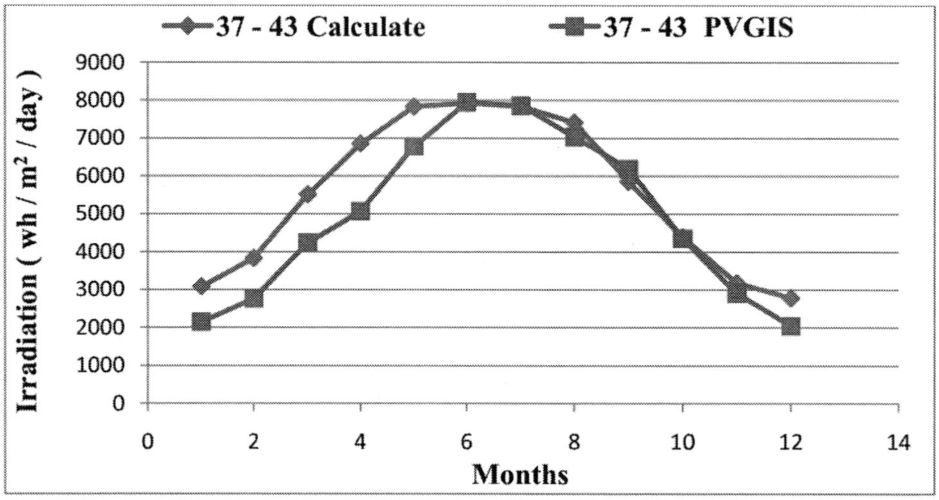

Figure 2. The comparison between the monthly irradiation with time on horizontal plane in ($Wh/m^2/day$) for (37°N – 43°E) site.

Figure (3) shows the comparison between the irradiation with time on tilt plane in ($Wh/m^2/day$) for (37°N – 43°E) site. It has observed the difference between the two curves in the values of solar radiation rates on tilt measurements.

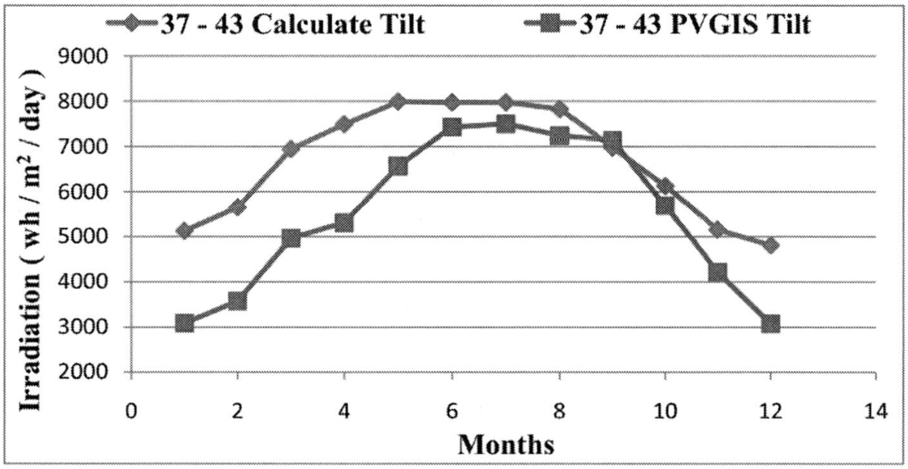

Figure 3. The comparison between the irradiation with time on tilt plane in (Wh/m^2/day) for (37° N – 43° E) site

Figure (4) shows the comparison of monthly irradiation with time on horizontal plane in (Wh/m^2/day) for (33° N – 44°E) site in central Iraq, we note that there is a close of between the two curves when sunrise and sunset for this site, There is little difference near midday this depends on the weather variables which measured by PVGIS web.

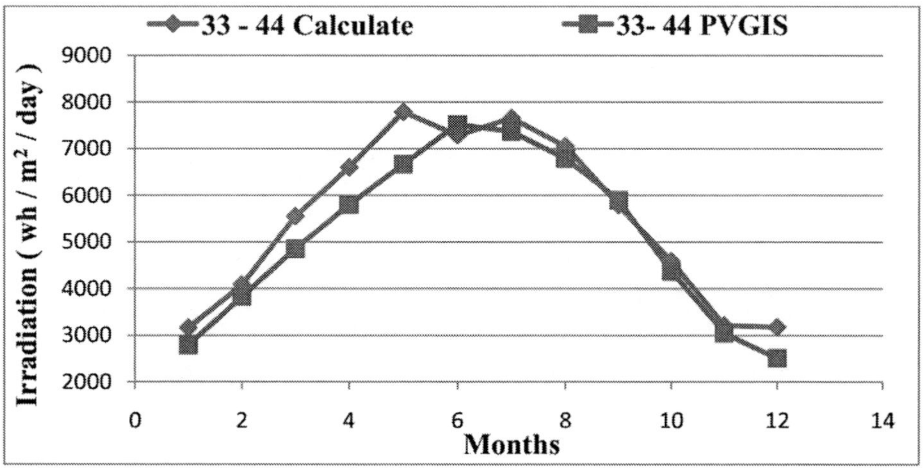

Figure 4. The comparison between the irradiation with time on horizontal plane in (Wh/m^2/day) for (33°N – 44°E) site

From the figure (5) show the comparison between the irradiation with time on tilt plane in (Wh/m^2/day) for (33°N – 44°E) site, The same behavior is observed in the difference between the two curves in the previous figure (4), but more in values, because of the tilt of the plane and weather conditions.

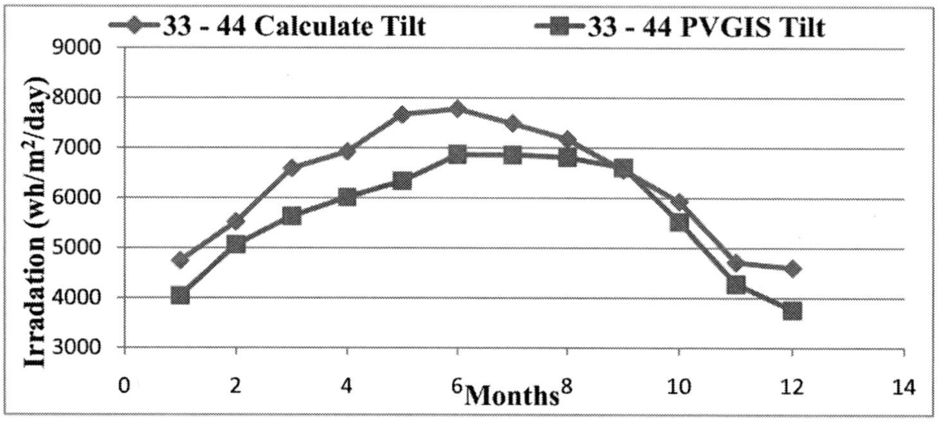

Figure 5. The comparison between the irradiation with time on tilt plane in (Wh/m^2/day) for (33°N – 44°E) site.

Figures (6),(7) show the comparison between the solar irradiation rates calculated by the program and the values recorded at the PVGIS web for (34°N – 43°E) site to horizontal and tilt plane , It is observed from this figure the same behavior of the previous figures.

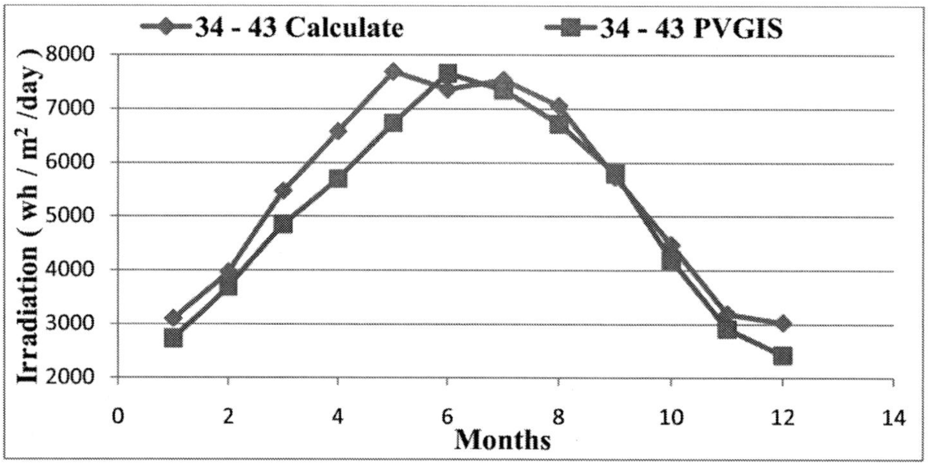

Figure 6. The comparison between the monthly irradiation on horizontal plane in (Wh/m^2/day) for (34°N – 43°E) site

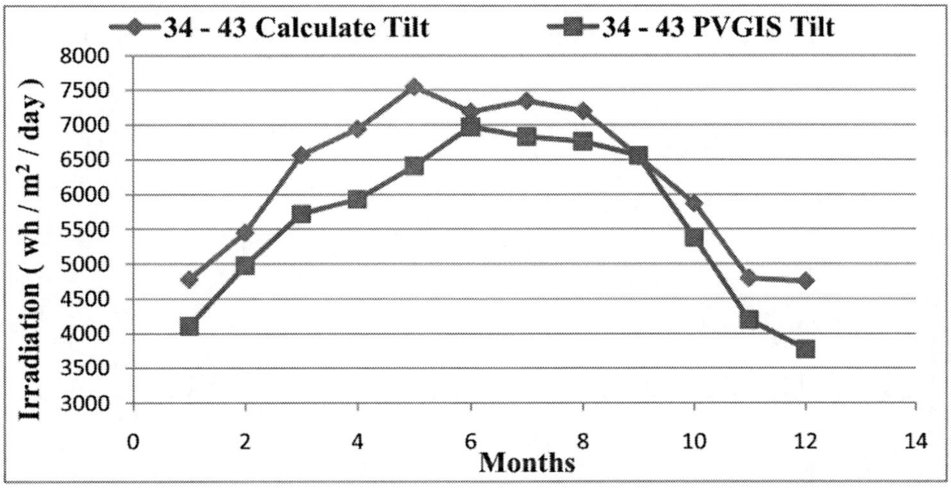

Figure 7. The comparison between the irradiation with on tilt plane in (Wh/m^2/day) for (34°N – 43°E) site

In the following two figures (8),(9) show a clear difference between the calculated and the record values of a site located in southern Iraq in all the daylight, this means that is a relative attenuation of light in this site.

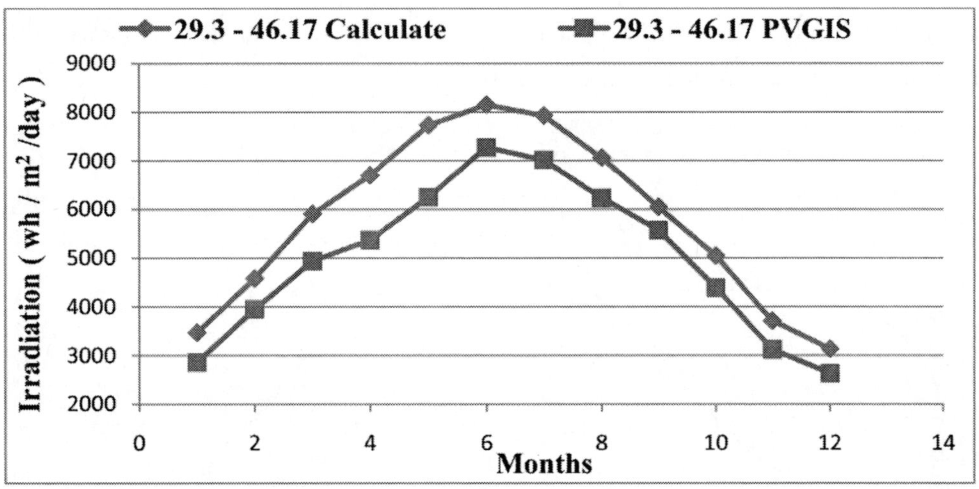

Figure 8. The comparison between the irradiation with time on horizontal plane in (Wh/m^2/day) for (29°N – 46°.17′E) site

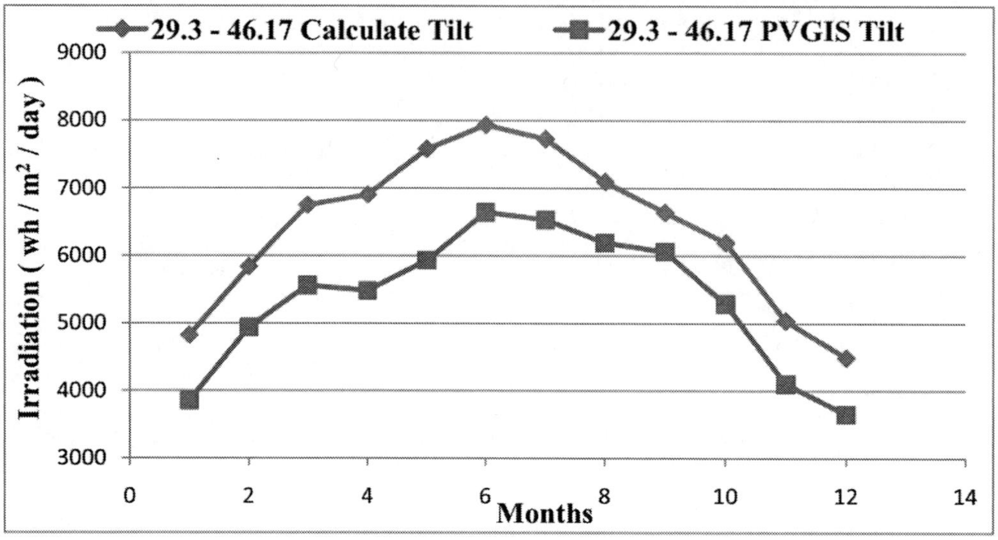

Figure 9. The comparison between the irradiation with time on tilt plane in (Wh/m^2/day) for (29°N – 46°.17′E) site

For a location in western Iraq (33°N – 40°E) site, it shows the close of to calculated and recorded values is illustrated by the figure (10) for horizontal plane and a little different in tilt plane as shown in figure (11), this means that attenuation in the light of this site is small.

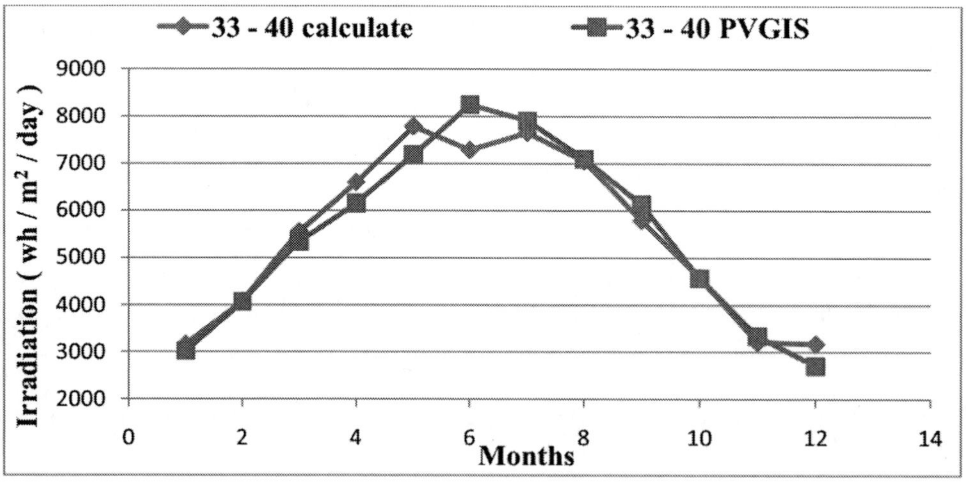

Figure 10. The comparison between the irradiation with time on horizontal plane in (Wh/m^2/day) for (33°N – 40°E) site

IOP Conf. Series: Journal of Physics: Conf. Series **1032** (2018) 012007 doi:10.1088/1742-6596/1032/1/012007

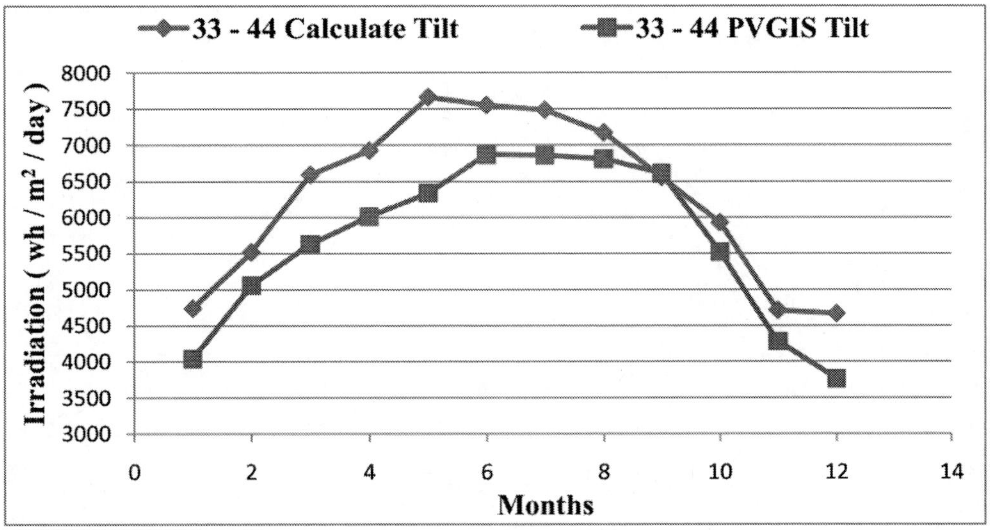

Figure 11. The comparison between the irradiation with time on tilt plane in (Wh/m^2/day) for (33°N – 40°E) site

The following figures (12),(13),(14),(15) are illustrate the same behavior, difference and close of previous figures, which are affected by the weather conditions of the site exposed to solar radiation attenuation for the length of the daylight .

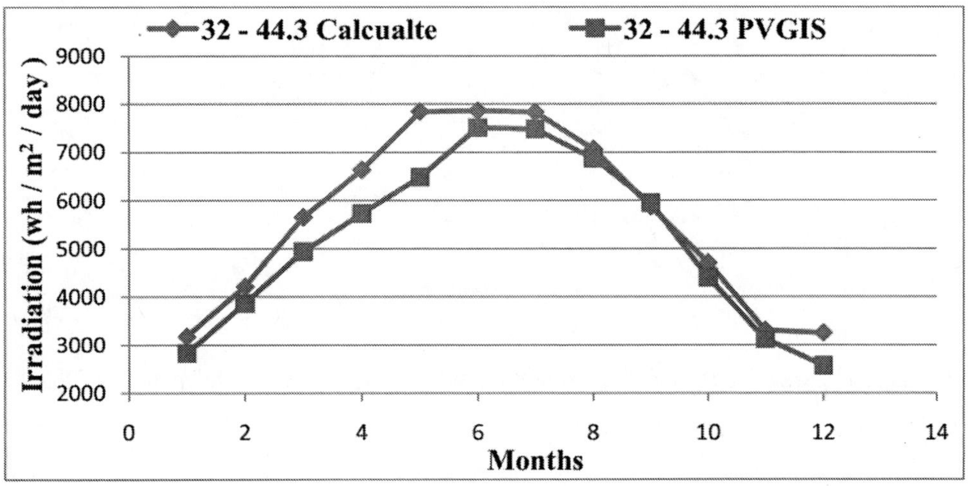

Figure 12. The comparison between the irradiation with time on horizontal plane in (Wh/m^2/day) for (32°N – 44°.3′E) site

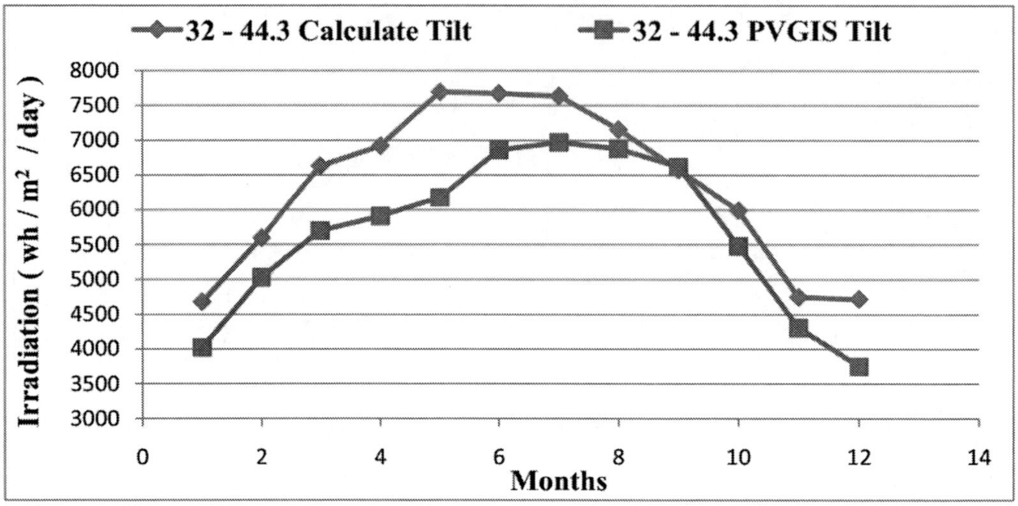

Figure 13. The comparison between the irradiation with time on tilt plane in (Wh/m^2/day) for (32°N – 44°.3′E) site

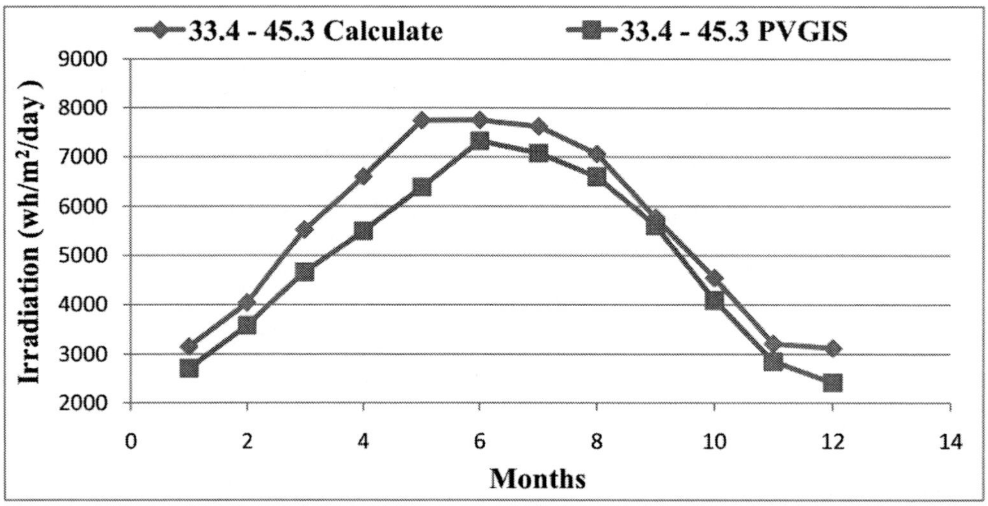

Figure 14. The comparison between the irradiation with time on horizontal plane in (Wh/m^2/day) for (33.4°N – 45°.3′E) site

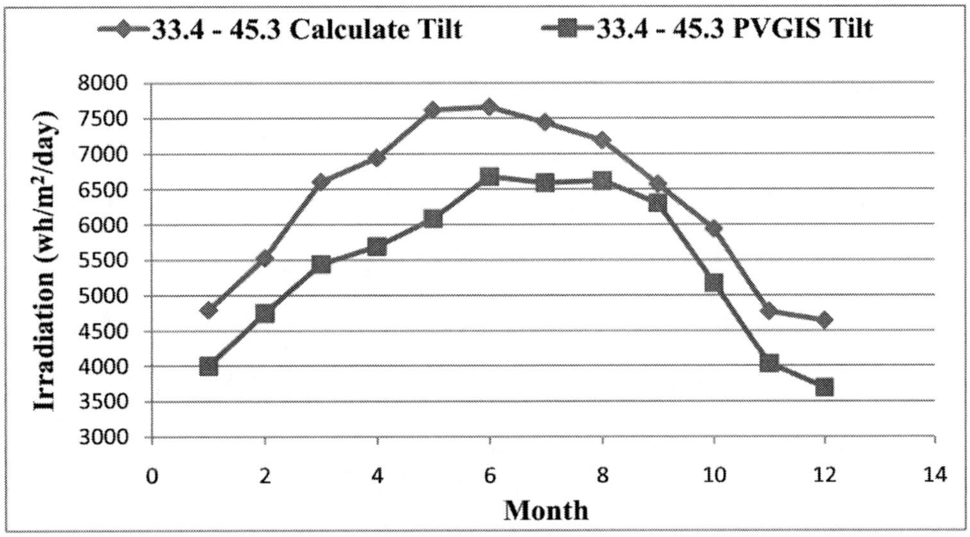

Figure 15. The comparison between the irradiation with time tilt plane in (Wh/m^2/day) for (33°.4′N – 45°.3′E) site

From the above results and the comparison between the software calculated values and measured data from PVGIS web. Led us to adopt the values to assessment the potential maps of Iraqi sites by choose about 200 site shows in figure (16) using Arc GIS program version 9.3 (is a computer-based systems that enable users to collect, store, and process, analyze and present spatial data), to representation Iraqi maps.

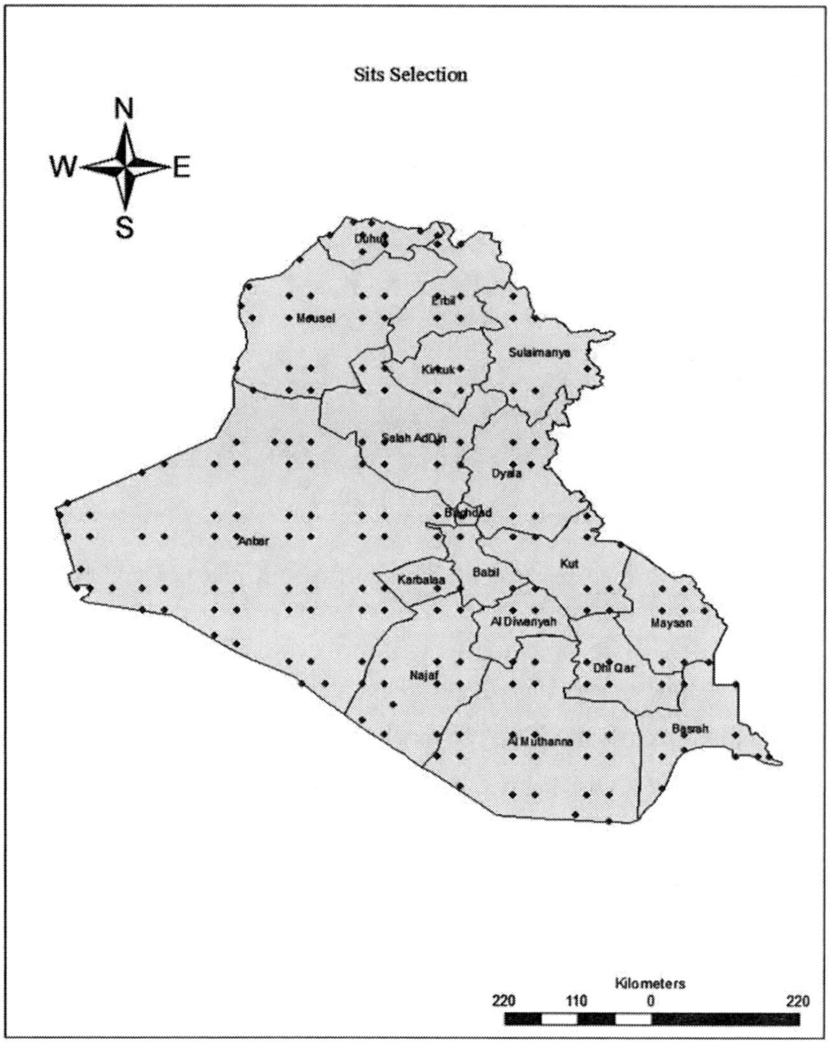

Figure 16. The selection sits for potential map

From PV GIS data for 200 sites, which simulate the solar radiation rates of each location depending on the weather conditions affecting that location , can be obtain on solar radiation potential of Iraqi map show in figure (17) the potential of irradiation on horizontal plane (kwh/ m^2/day), which ranged from (5.96 – 4.6)(kwh/m^2/day) .

IOP Conf. Series: Journal of Physics: Conf. Series **1032** (2018) 012007 doi:10.1088/1742-6596/1032/1/012007

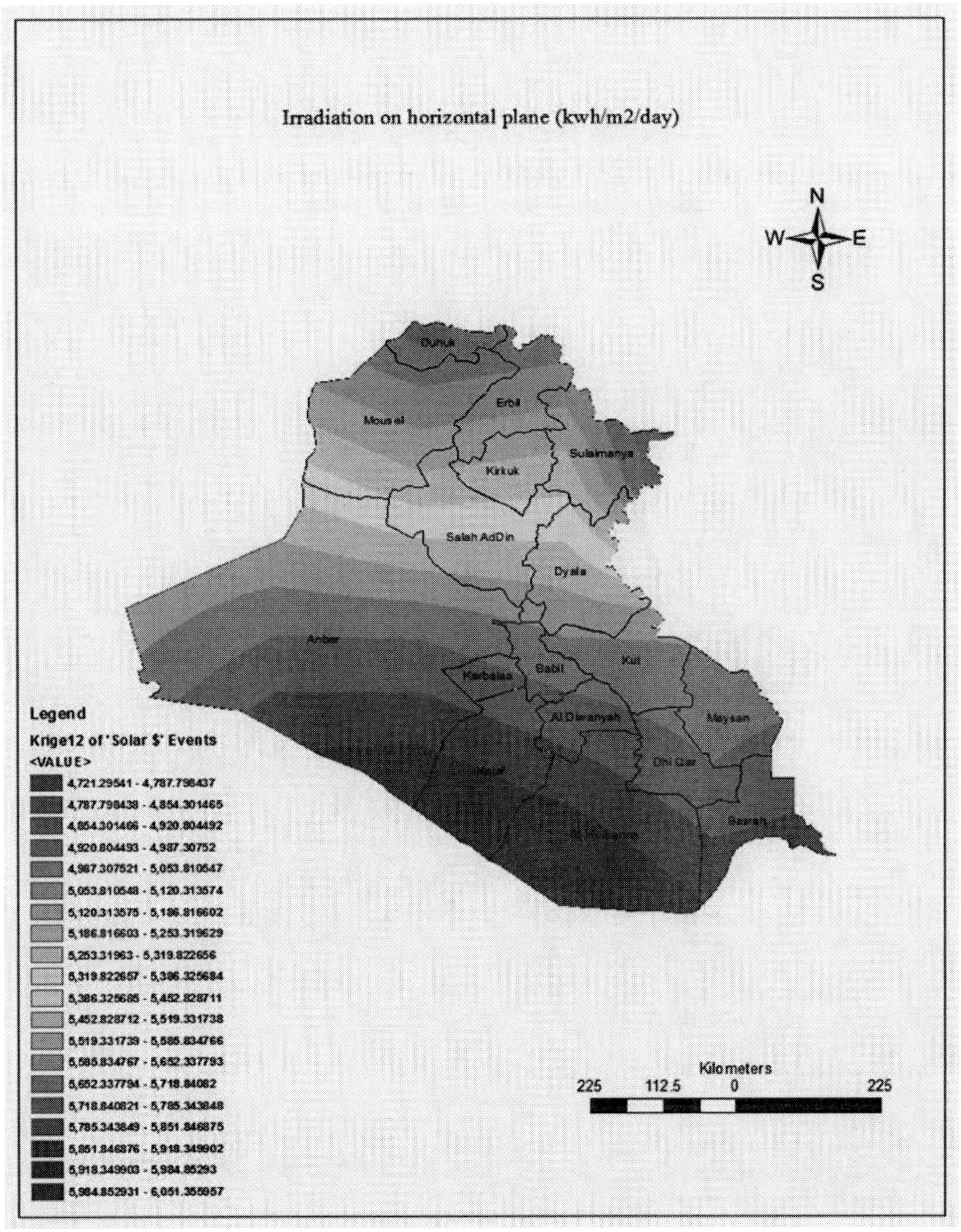

Figure 17. The potential of irradiation on horizontal plane (kwh/ m^2/day)

IOP Conf. Series: Journal of Physics: Conf. Series **1032** (2018) 012007 doi:10.1088/1742-6596/1032/1/012007

Figure (18) shows potential of irradiation on optimally inclined plane which ranged from (6.5- 5.07) (kwh/m^2/day).

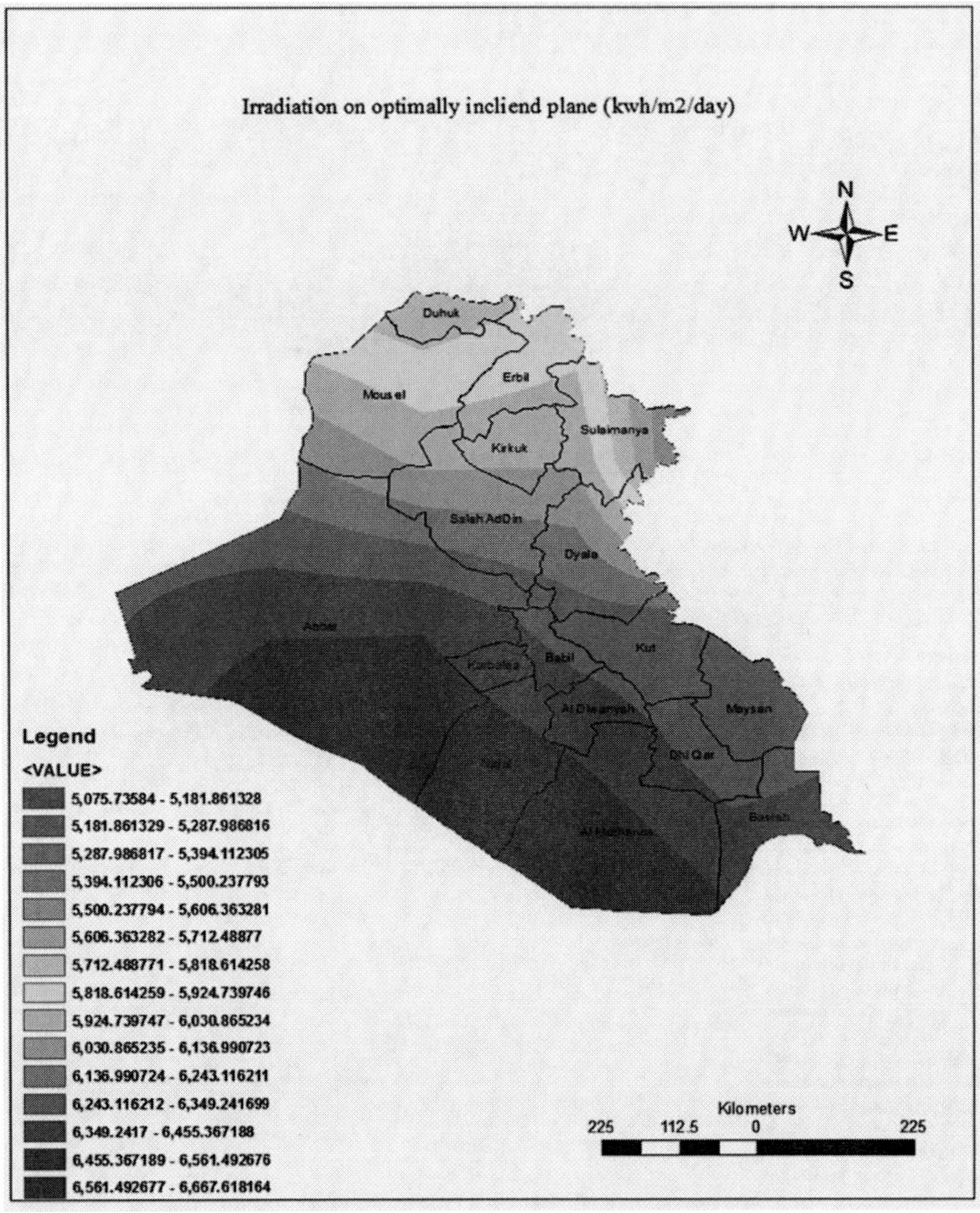

Figure 18. The potential of irradiation on optimally inclined plane (kwh/ m^2/day)

It has found that the monthly averages of irradiation on optimally inclined plane from data recorded in the PV GIS site show in figure (19) which ranged from (181.84 – 142.4) (kwh/m^2).

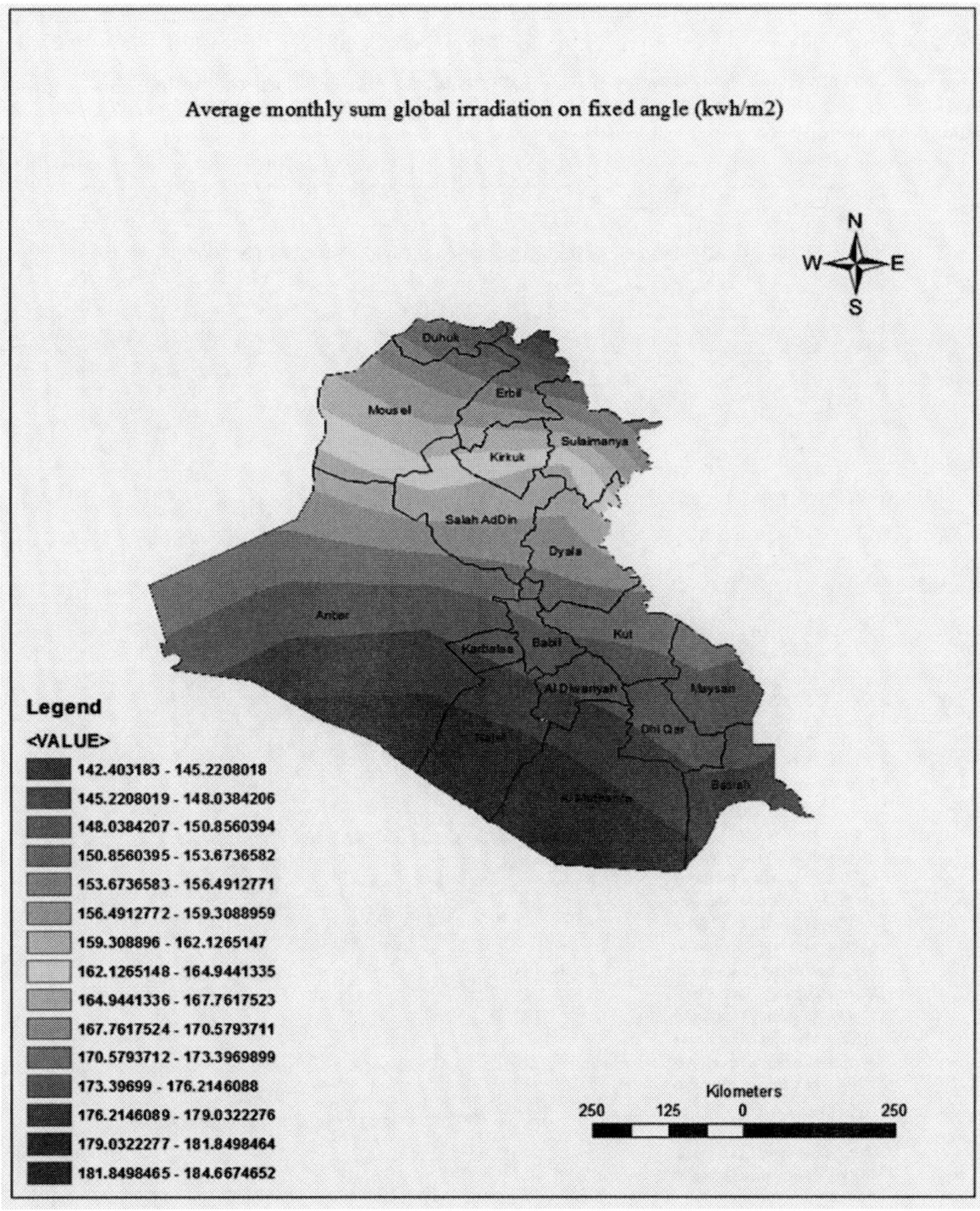

Figure 19. The average potential monthly of irradiation on optimally inclined plane (kwh /m^2)

IOP Conf. Series: Journal of Physics: Conf. Series **1032** (2018) 012007 doi:10.1088/1742-6596/1032/1/012007

Depending on the solar radiation data, it is possible to see the average of solar photovoltaic energy potential which produced by the mono crystalline silicon cell technology without any electric and dust losses to see the maximum production , because the value of any losses varies from location to another location and depending on the environmental conditions surrounding, figure (20) show the electricity potential in scale 1 MW, which ranged from (5476 - 4443) (kwh/ day), this value can be doubled as needed and available space .

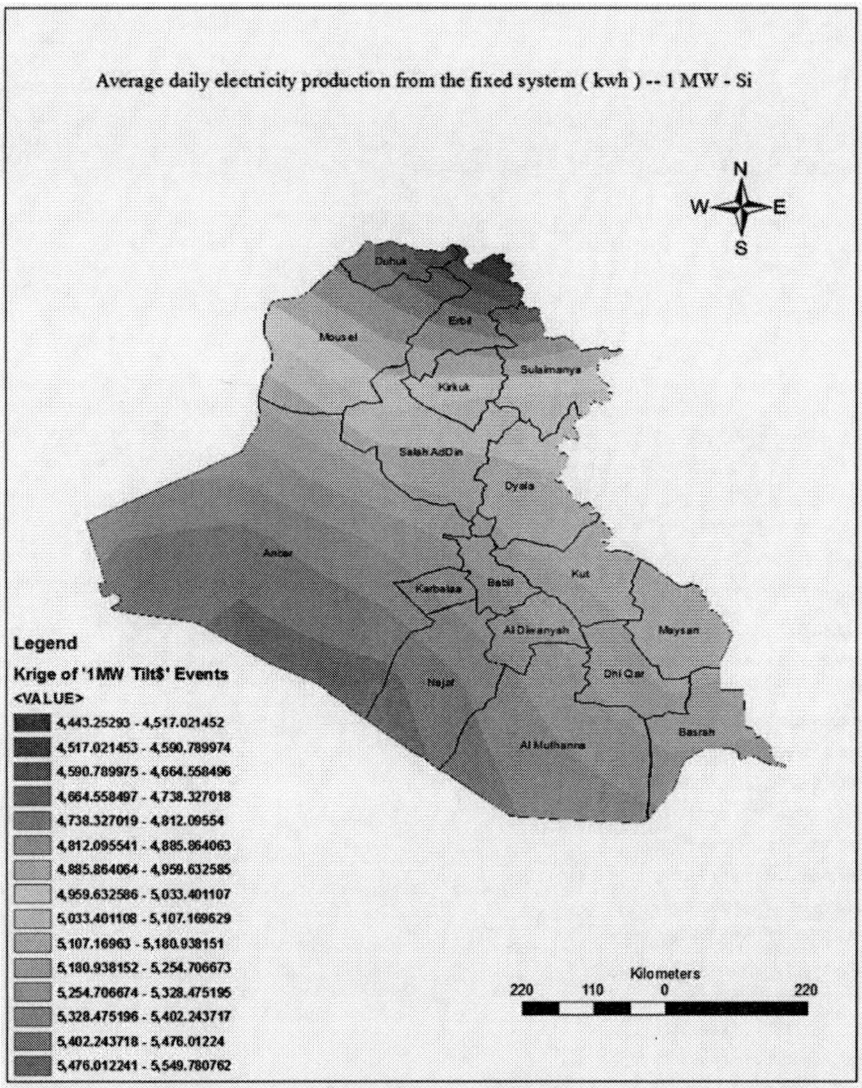

Figure 20. The daily of electricity on optimally inclined plane (kwh) – 1 MW – Si- mono-structure

IOP Conf. Series: Journal of Physics: Conf. Series **1032** (2018) 012007 doi:10.1088/1742-6596/1032/1/012007

Depending on the values shown in the above maps, could be identify Suggested sites for to construction of solar power plants show in figure (21) in scale 1Mw, which ranged from (160,523,0625-139,172,5469) (kwh/month), according to the construction of stations near power transmission lines, residential areas and the availability of manpower.

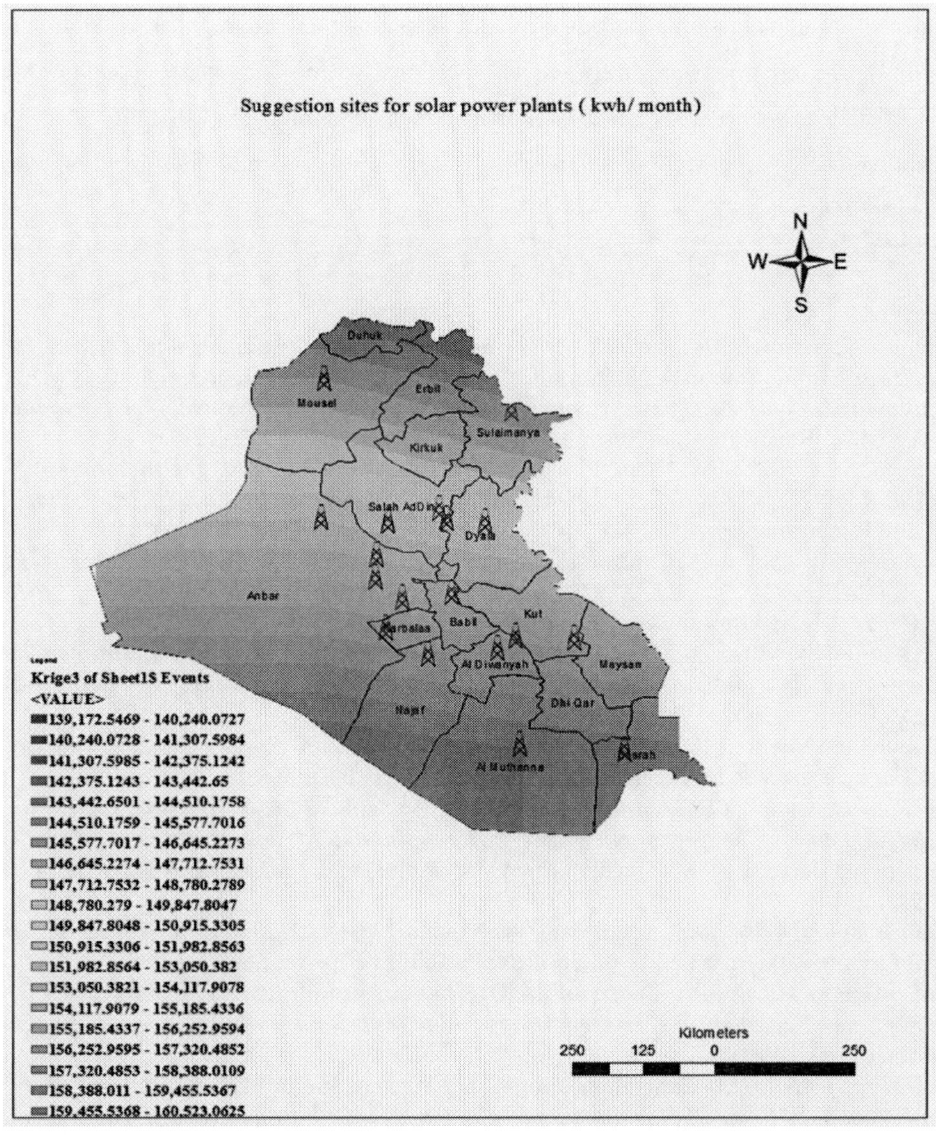

Figure 21. The suggestion Sites for solar power plants in scale 1Mw (kwh/month) on optimally inclined plane

IOP Conf. Series: Journal of Physics: Conf. Series **1032** (2018) 012007 doi:10.1088/1742-6596/1032/1/012007

6- Conclusion

Calculated the solar radiation values for the photovoltaic power rates of 200 sites by software program using empirical equation as visual basic net language to manage process assessment of PV potential , conclude that agreement between the software data and PVGIS data and the results show the most of these sites are eligible to build photovoltaic solar plants for their abundant solar radiation , but sites selection must be selected for construction solar power plants depending on geographical conditions (land cover and land used), the environment determinants, locations where appropriate conditions can be the construction of stations near power transmission lines, urban areas and the availability of manpower, to be sites for photovoltaic solar power plants in size (1) megawatts and multiples it as assessment of PV potential map for Iraq.

References:

[1] S. Ganguli, J. Singh 2010. *Estimating the Solar Photovoltaic Generation Potential and Possible Plant Capacity in Patiala.* International journal of applied engineering research .Dindigul v.1.No.2.

[2] M. A. Eltawil, Z. Zhao 2010. *Grid-Connected Photovoltaic Power Systems: Technical and Potential .* Problems -A review. Renewable and Sustainable Energy Reviews 14, 112–129.

[3] M. Luqman, S. R. Ahmad, S. Khan, U. Ahmad, A. Raza, F. Akmal 2015. *Estimation of Solar Energy Potential from Rooftop of Punjab Government Servants Cooperative Housing Society Lahore Using GIS.* Smart Grid and Renewable Energy,6, 128-139.

[4] F. Antonanzas-Torres, F. Canizares, O. Perpinan 2013. *Comparative Assessment of Global Irradiation from a Satellite Estimate Model (CM SAF) and On-Ground Measurements (SIAR):A Spanish case study.* Renewable and Sustainable Energy Reviews 21,248–261.

[5] Hamdy K. Elminir. V.T.ezslav Benda. JR.R.Tou.sek 2001.*Effects of Solar Irradiation Conditions and Other Factors on the Outdoor Performance of Photovoltaic Modules.* Journal of electrical engineering.Vol. 52, No. 5-6, 125 -133.

[6] C.Christopher Newton 2007.*A Concentrated Solar Thermal Energy System* .The Department of Mechanical Engineering.The Florida state university Famu - Fsu College of engineering. .

[7] A. Korfiati, C. Gkonos, F. Verones, A. Gaki, S. Grassi,R. Schenkalm S.Volkwien, M. Rantal, L. Hurni 2016.*Estimation of the Global Solar Energy Potential and Photovoltaic Cost with the Use of Open Data.* International Journal of Sustainable Energy Planning and Management Vol. 09. 17-30. Zurrich ,Switzeland.

[8] A. Belhamadia ,M. Mansor ,Mahmoud A. Younis 2014. *A Study on Wind and Solar Energy Potentials in Malaysia.* International Journal of Renewable Energy research . Vol.4, No.4.

[9] Saeb M. Besarati, Ricardo Vasquez Padilla, D. Yogi Goswami,Elias Stefanakos 2013. *The Potential of Harnessing Solar Radiation in Iran Generating Solar Maps and Viability Study of PV Power Plants.* Renewable Energy 53,193-199.

[10] G. Maximo Tiepolo, O. Canciglieri Junior, J. Urbanetz Junior 2013. *Analysis of the Electricity Generation Potential by Solar Photovoltaic Source in the State of Parana – Brazil.*The Italian Association of Chemical Engineering.Vol.35,ISBN 978-88-95608-26-6; ISSN 1974-9791. DOI:10.3303/CET1335100.

[11] L. A. Aguilar 2015. *Feasibility Study of Developing Large Scale Solar PV Project in Ghana: An Economical Analysis*. Department of Energy and Environment, Division of Electric Power Engineering, Chalmers University of Technology, Gothenburg, Sweden.
[12] Marcel S. Uri, Thomas A. Huld, Ewan D. Dunlop, Heinz A. Ossenbrink 2007. *Potential of Solar Electricity Generation in the European Union Member States and Candidate Countries*. Solar Energy 81 1295–1305.
[13] G. M. Tiepolo,J. U. Junior,O. C. Junior,T. Viana 2014. *Photovoltaic Generation Potential of Paraná State, Brazil – a Comparative Analysis with European Countries*. Energy Procedia 5 ,725 – 734.
[14] R. Khatri 2016. *Design and Assessment of Solar PV Plant for Girls Hostel (GARGI) of MNIT University, Jaipur city: A case study*. Energy Reports 2,89–98.
[15] T. V. Ramachandra 2007 . *Solar Energy Potential Assessment Using GIS*. Energy Education Science and Technology .V.(issue) 18(2), 101-114.
[16] M. Enjavi.Arsanjani,K. Hirbodi,M. Yaghoubi 2015. *Solar Energy Potential and Performance Assessment of CSP Plants in Different Areas of Iran*. International Conference on Concentrating Solar Power and Chemical Energy Systems, Energy Procedia 69,2039 – 2048.
[17] A. Singh , P. Baredar 2016.*Techno-Economic Assessment of a Solar PV, Fuel Cell, and Biomass Gasifier hybrid energy system*. Energy Reports 2 ,254–260.

IOP Conf. Series: Journal of Physics: Conf. Series **1032** (2018) 012008 doi:10.1088/1742-6596/1032/1/012008

Treatment of industries wastewater using solar light

A H Ali[1], A A Risn[1] and A Hussein[2]

[1]Chemical Engineering Department , College of Engineering, Al-Muthanna University, Al-Samawa, Iraq
[2]Civil Engineering Department, College of Engineering, Al-Muthanna University, Al-Samawa, Iraq
dr.ahmedha@mu.edu.iq, dr.ahmedha@gmail.com, a.hussein3@edu.salford.ac.uk

Abstract. The degradation of eosin yellow dye using ZnO as a photo catalysts in a slurry reactor exposed to solar light was studied by this paper . The optimum values of different parameters versus catalyst concentration, catalyst type and initial concentration of the substrate affecting the degradation rate of eosin yellow dye were obtained by varying catalyst concentration (255–1255 ppm), and initial dye concentration (25–100 ppm). Color disappearance as well as substrates with the depletion of total organic carbon content was monitored.
Keywords: Solar light ; eosin yellowish dye; Photo catalysis; Zinc oxide

1. Introduction

The degradation of organic pollutants in water by catalytic photovoltaic processes have been subjected to extensive research recently. The high photosensitivity, non-toxic nature, low cost and chemical stability of semiconductor oxides have recognized them as preferable components for catalytic photovoltaic processes. The semiconductor or photo catalysis is initiated by the surface trapping of photo-generated electrons (e-) and holes (h+), which induces interfacial electron-transfer reactions with adsorbed substrates. However, because of the wide band gap of ZnO and TiO_2 [1-3]. Solar Energy treatment is very important method used in this field because it is very cheap and available in most of world countries we can say it is future energy. The adoption of solar photocatalaytic reactors to treat textile processes wastewater is potentially viable in geographical areas receive heavy flucx of sunlight over the year [4]. And Iraq is a potential to introduce similar technology .Its reported that dyes absorb light highly within visible range, the summation of energy efficiency should be feasibly high in order to make the photo-assisted treatment of dye-polluted wastewater achievable [5,6]. Eosin yellowish was used as one of the most important dyes because of it is emotional-state lifetime which was considerably short. Moreover, It was visibly constant and in addition, It provides a helpful investigation for the both of removal of photo reactions and oxidation [7]. For the oxidation reaction about photo of Eosin yellowish in a slurry reactors and TiO_2 immobilized reactors, most studies had been made [8,9]. Estimating the execution of a fluidized bed photo catalytic reactor in a three-phase have been tested by previous studies and significant information were obtained. The photocatalytic degradation of eosin yellowish dye in aquatic solutions using two different photocatalysts had been examined. The major aims of the study were (a) effect of catalyst concentration; (b) effect of pH. (c) effect of initial dye concentration; (d) studies kinetic of reaction.

2. Experimental work

2.1. Materials

ZnO powder was used in this experiment and were gained from Merck Co. (Germany), eosin yellowish dye was bought from Merck (Germany) and was used without further treatment. Different solutions were prepared by using distilled water.

IOP Conf. Series: Journal of Physics: Conf. Series **1032** (2018) 012008 doi:10.1088/1742-6596/1032/1/012008

2.2. Instruments
Photo-chemical degradation was conducted by employing an especially designed insulation walled containers with volume of 500 ml, Continuous stirring while samples are exposed to solar light were achieved by magnetic stirrers. Temperature was maintained constant during the reaction by wetting the photo catalytic reaction vessel by water.

2.3. Irradiation experiments
Suspension of 0.25 l of the dye solution and photo catalyst was prepared and the mixture was exposed to irradiation. Experiments were conducted under solar light. The aqueous suspension was magnetically stirred over the experiment. At different time periods, an aliquot was obtained by using a syringe and so filtered through Millipore syringe filter paper of 0.45 μm.

2.4 The Measurements of Absorption
A Double Beam UV-1800 (Shmadzo spectrophotometer) was used to record absorption spectrum and then compared with distilled water as a reference liquid and the percentage rate of degradation was noticed in terms of change in intensity at λ_{max} of the dyes during irradiation time. The degradation efficiency (%) has been calculated as:

$$Degradation\ (\%) = \frac{Abs\ 0 - Abs\ t}{Abs\ 0} \times 100 \quad (1)$$

Where Degradation (%) is percentage of dye disappearance, Abs 0 is initial absorption of dye (at time = 0 min.) and Abs t is absorption of dye (at time = t min.)

Similar experiments have used different catalyst concentration (250–1250 ppm), to select the best concentration of catalyst ZnO at the same concentration of the dye with the solution (25 ppm).

3. Results and Discussion

3.1 Effect of Photo catalyst Concentration
The catalyst concentration effect on the degradation of Eosin yellowish (25 ppm) was examined by using commercial ZnO from 250 to 1250 ppm maintaining other parameters such as dye concentration, pH and temperature). Figure 1 showed that the degradation percentage is subject to increase in ZnO commercial concentration up to 0.75 g\l for ZnO. This change in the degredation percent could be explained by active sites available on the catalyst surface and the permeation of solar light into the suspension. The total active surface area is increased with any increasing the dosage of the catalyst. On other hand, due to suspension turbidity increase , there is a reduction in solar light permeation because of increased scattering effect and hence the photo activated volume of suspension decreases. Furthermore, at high catalyst loading, it was observed that maintaining the suspension homogenous was difficult due to particles agglomeration, which decreases the number of active sites [10-13]. The results showed the highest value of photo degradation for commercial ZnO (750 ppm) achieved after 90 minutes.

IOP Conf. Series: Journal of Physics: Conf. Series **1032** (2018) 012008 doi:10.1088/1742-6596/1032/1/012008

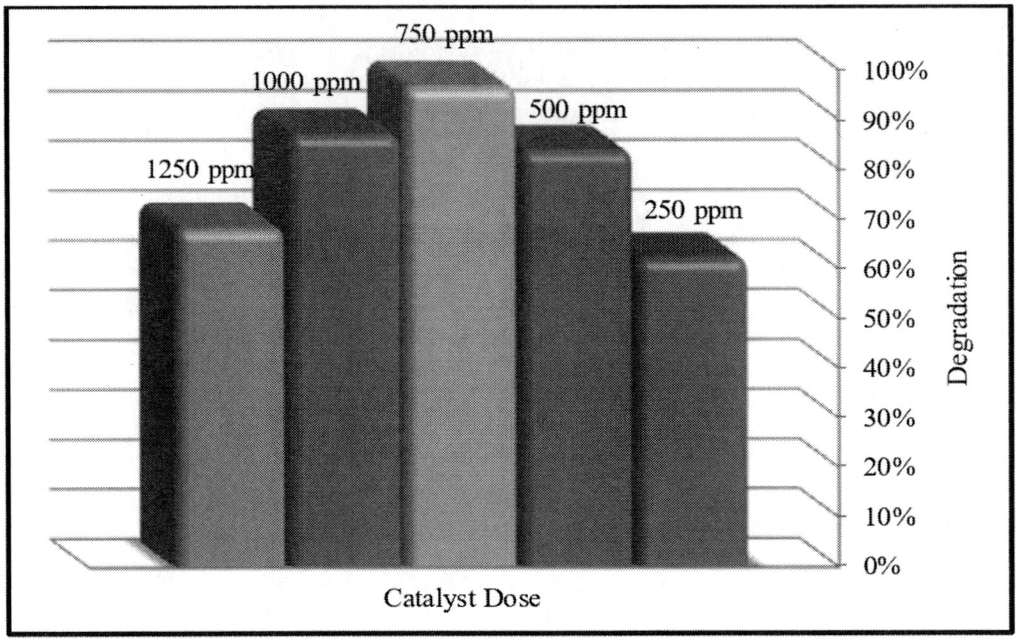

Figure 1. Effect the catalyst dose of ZnO on degradation of Eosin yellowish dye

3.2 pH Effective on Solution Treatment

The effluent on any type of wastewaters is produced at different pHs. So , study of pH is very important on photo degradation of Eosin yellowish dye. Experiments have been done at different values of pH varying from 3 to 11 for 25 ppm dye solution concentration and for the best concentration of the catalyst (750) ppm of ZnO. The photo degradation percentage against values of pH was showed in Figure 2. It is clearly increasing in pH up to 10 for ZnO because increasing in photo degradation activity [14]. The utilizing of ZnO as the catalyst is more appropriate at high pH values with the textile effluent; Textile industry processes are among the most environmentally unsustainable industrial processes [15-19]. Its complex to interpret the effect of pH on decolonization efficiency because many reactions can result in dye degradation such as "hydroxyl radical reaction, direct oxidation by the positive hole and direct reduction by the electron in the conducting band". The importance of each one depends upon the substrate nature and pH.

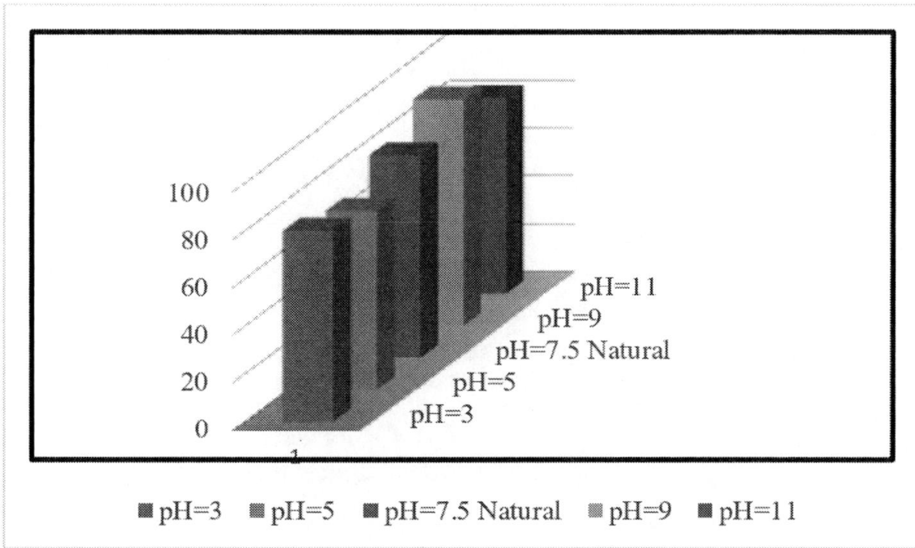

Figure 2. Effect the pH on degradation of Eosin yellowish dye

3.3 Dye Concentration effect

The photo degradation of Eosin yellowish was conducted by increasing the initial dye concentration from 25 ppm to 125 ppm. This step was conducted in order to find out the effect of initial concentration of the dye on the most effective catalyst kind and dose (commercial ZnO). Because the concentration of the dye was enhanced, the share of degradation reduced indicating that either to extend the catalyst concentration or time span for the entire reduction. Figure 3 depicts the time-dependent graphs of degradation of Eosin yellowish at different dye solutions (25–125 ppm). A 100% degradation took place within 90 & 150 minutes respectively when using dye solutions of 25 ppm and 150 ppm. However, when using 150 ppm, it took almost 300 minutes for accomplished degradation of the dye. The degradation percentage rate had further minimized when the dye concentration was increasing . The reason behind this behavior was the path length of the light photons entering the solution was reduced in high dye concentration. In other words , the photochemical reaction is tend to reduced but the number of photons absorbed by catalyst is high at low initial concentration of the dye concentration the amount of photon absorption by the catalyst in lower [19-25].

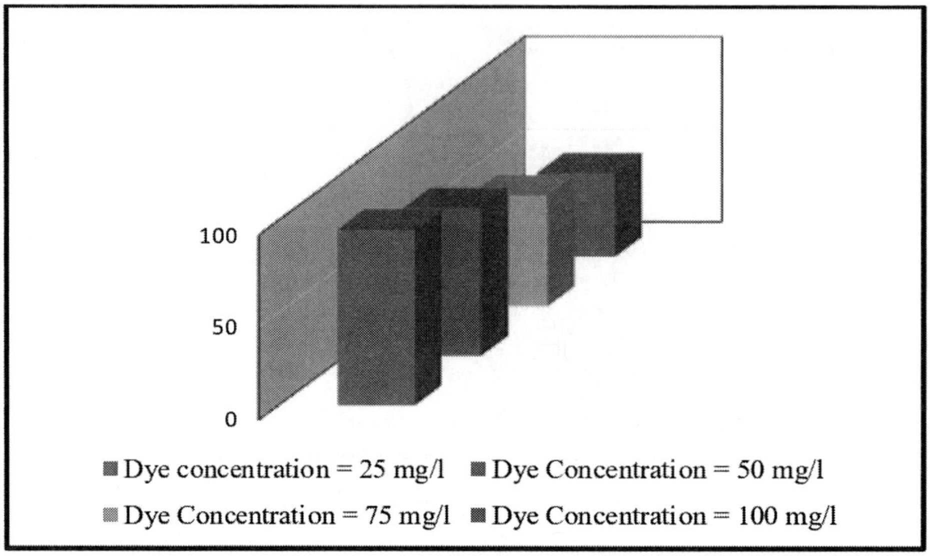

Figure 3. Effect the dye conc. (mg/l) on degradation of Eosin yellowish dye

3.4 Kinetic Study

Mechanism kinetic of dye degradation is shown by figure 4 (at optimum conditions). The results show that the disappeared of Eosin yellowish dye with zinc oxide with 1 g/l dose can be write by the following mathematical equation:-

$$\frac{Co}{C} = kt \qquad (2)$$

Where Co is feed dye concentration, C is dye concentration at time t (y=0.0422x) , k is equation constant and t is degradation time. The result data of the concentration with degradation time is gave a straight line and $R^2 = 0.9231$ (correlation factor) for the ZnO [26-28].

3.5 Mineralization studies

Since the decrease of chemically demanded oxygen (COD) in water reflects that degradation reactions or mineralization of the organic components have toke place. The removal percentage of COD was studied in case of initial concentration-25 ppm for dye samples by using optimized conditions of dose of catalyst of 0.75 g/l and a value of pH 9 as a function of irradiation time. It was found that after 20 minutes the removal COD percentage falls in the ranges of 85% and 90%. (see Figure 5). The percentage discoloring was higher than reduction of COD which could be explained by the formation of smaller colorless compounds. Therefore, it is quite evident that longer irradiation time is required in order to achieve complete mineralization of dyes.

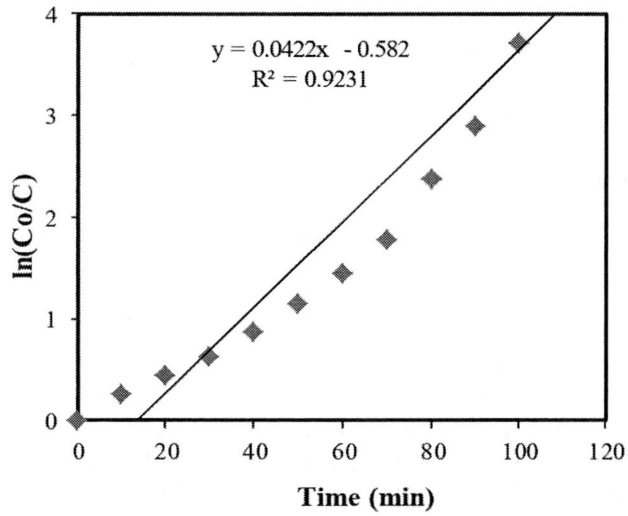

Figure 4. Kinetic study of the disappearance of Eosin yellowish dye

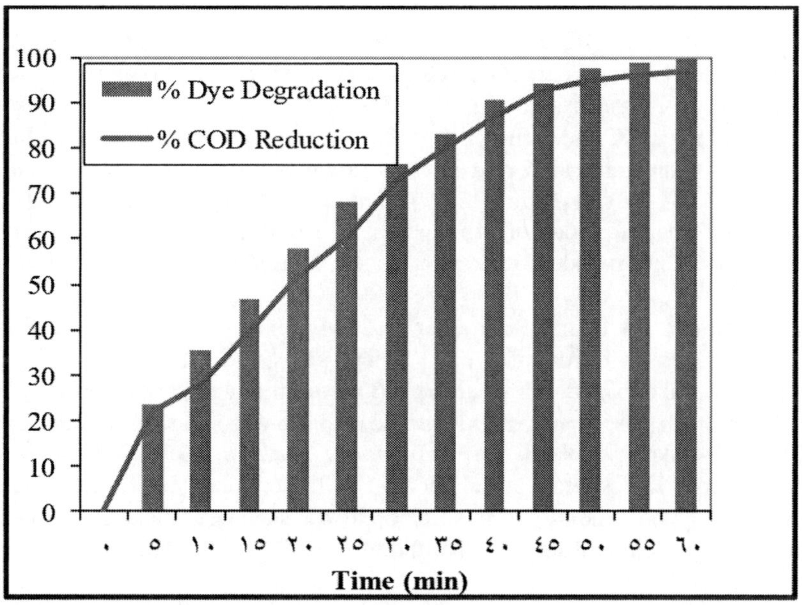

Figure 5. Comparison between percentage of discoloring and percentage COD reduction for eosin yellowish dye

IOP Conf. Series: Journal of Physics: Conf. Series **1032** (2018) 012008 doi:10.1088/1742-6596/1032/1/012008

3.6 Possibility of ZnO Reuse

The possibility of reusing ZnO for discoloring of Eosin yellowish dye was experimented. The solutions resulting from the photo catalytic discoloring of dye dried after filtering and washing the sample. The dried catalyst samples were used for the discoloring of dye solutions employing similar experimental conditions. It was found that considerable photo catalytic activity up to four cycles was shown by the sample. The activity of the sample tended to weakness to the extent of (8 – 10)% only after the forth trial. Calculation of Zn^{2+} ions in the solutions was done to determine the loss of Zn^{2+} ions within the solution as a result of dissolution of ZnO at alkaline pH (pH 9). The dissolution of ZnO was found to be negligible. Similar findings are reportable within the literature that no photo corrosion of ZnO takes place in alkaline conditions [27,28]. These results detect the stability of the prepared catalyst.

4. Conclusions

The photo catalytic degradation of aqueous solutions of Eosin yellowish dye has been examined by using a solar light-irradiated and ZnO catalyst. It has been found that the process leads to discoloring and, eventually to complete mineralization of the dye solution. Discoloring of the Eosin yellowish dye and finally complete consumption of the dye solution were achieved by the process. Mid and final products on the photo catalyst surface and the solution were monitored with various technologies that allowed the reaction pathways from adsorption of the dye molecules on the photo catalyst surface to the formation of the final product to be determined.

References

[1] Sakthivel S, Neppolian B, Shankar M V, Arabindoo B, Palanichamy M and Murugesan V 2003 Solar photocatalytic degradation of azo dye: comparison of photocatalytic efficiency of ZnO and TiO_2 *Sol. Energy Mater. Sol. Cells* **77** 65–82.

[2] Evgenidou Fytianos E K and oulios I P 2005 Semiconductor-sensitized photodegradation of dichlorvos in water using TiO2 and ZnO as catalysts *Appl. Catal. B: Environ.* **59** 81–89.

[3] Daneshvar N, Aber S, Seyed Dorraji M S, , Khataee A R and Rasoulifard M H 2005 Photocatalytic degradation of theinsecticide diazinon in the presence of prepared nanocrystalline ZnO powders under irradiation of UV-C light, *Sep. Purif. Technol.* **58** 91-98.

[4] Bousselmi L, Geissen S U and Schroeder H 2004 Textile wastewater treatment and reuse by solar catalysis: results from a pilot plant in Tunisia. *Water Sci. Technol.* **49** 331–37.

[5] Vinodgopal K, Wynkoop D and Kamat P 1996 Environmental photochemistry on semiconductor surfaces: photosensitized degradation of a textile azo dye, acid orange 7, on TiO2 particles using visible light. *Environ. Sci. Technol.* **30** 1660–66.

[6] Nasr C, Vinodgopal K, Fisher L, Hotchandani S, Chattopadhyay A K and Kamat P V 1996 Environmental photochemistry on semiconductor surfaces, Visible light induced degradation of a textile diazo dye, Naphthol Blue Black, on TiO2 nanoparticles. *J. Phys. Chem.* **100** 8436–42.

[7] Lathasree S, Nageswara R, Sivasankar B, Sadasivam V and Rengaraj K 2004 Heterogeneous photocatalytic mineralization of phenols in aqueous solutions. *J. Mol. Catal. A Chem.* **223** 101–105.

[8] Daneshvar N, Salari D and Khataee A R 2004 Photocatalytic degradation of azo dye acid red 14 in water on ZnO as an alternative catalyst to TiO2. *J. Photochem. Photobiol. A Chem.* **162** 317–22.

[9] Chen L C 2000 Effects of factors and interacted factors on the optimal decolorization process of Eosin yellowish by ozone. *Wat. Res.* **34** 974–82.

[10] Zhu C, Wang L, Kong L, Yang X, Wang L, Zheng S, Chen F, Maizhi F and Zong H 2000 Photocatalytic degradation of azo dyes by supported TiO2 and UV in aqueous solution. *Chemosphere*. **41** 303–309.

[11] Kansal S K, Ahmed Hassan Ali and Kapoor S 2010 Photocatalytic decolorization of biebrich scarlet dye in aqueous phase using different nanophotocatalysts. *Desalination* **259** 147–55

[12] Konstantinou I K and Albanis T A 2004 TiO2-assisted photocatalytic degradation of azo dyes in aqueous solution: kinetic and mechanistic investigation. *Appl. Catal. B: Environ.* **49** 1–14.

[13] Ali A H, Naser G F and Mohammed S A 2016 Photocatalytic Degradation of Methyl Orange Dye using Different Photocatalysts under Solar Light. *Internat. J. ChemTech Res.* **910.**

[14] Lizama C, Freer J, Baeza J and Mansilla H D 2002 Optimized photodegradation of reactive blue 19 on TiO2 and ZnO suspensions. *Catal. Today* **76** 235–246.

[15] Hussein A and Scholz M 2017 Dye wastewater treatment by vertical-flow constructed wetlands. *Ecol. engin.* **101** 28-38.

[16] Hussein A and Scholz M 2017 Treatment of artificial wastewater containing two azo textile dyes by vertical-flow constructed wetlands. *Environ. Sci. and Poll. Res.* 1-20.

[17] Hussein, A. 2017. *Azo textile dyes wastewater treatment with constructed wetlands: design and operation of experimental vertical-flow constructed wetlands applied for the treatment of azo textile dyes (with/without artificial wastewater)*(Doctoral dissertation, University of Salford).

[18] Hussein A M 2018 *constructed wetlands for treatment azo textile dyes wastewater* (Mauritius: LAP LAMBERT Academic publishing) p 293

[19] Ali. A H 2017 Preparation and Characterization of Nanoparticles TiO2 and Its Application for Methylene Blue Dye Degradation under Solar Light. *J. of Glob. Pharma Techno.* **9** 141-48

[20] Rabindranathan S, Suja D P and Yesodharan S 2003 Photocatalytic degradation of phosphamidon on semiconductor oxides, *J. Hazard. Mater. B.* **102** 217–29.

[21] Chan A H C, Porter J F, Barford J P and Chan C K 2002 Effect of thermal treatment on the photocatalytic activity of TiO2 coatings for photocatalytic oxidation of benzoic acid, *J. Mater. Res.* **17** 1758–65.

[22] Stylidi M Kondarides D I and Verykios X E 2003 Pathways of solar light-induced photocatalytic degradation of azo dyes in aqueous TiO2 suspentions, *Appl. Catal. B: Environ.* **40** 271–86.

[23] Sakthivel S, Neppolian B, Shankar M V, Arabindoo B, Palanichamy M and Murugesan V 2003 Solar photocatalytic degradation of azo dye comparison of photocatalytic efficiency of ZnO and TiO_2. *Sol. Energy Mater. Sol. cell.* **77** 65-82.

[24] Chakrabarti S and Dutta B K 2004 Photocatalytic degradation of model textile dyes in wastewater using ZnO as semiconductor catalyst, *J. Hazard. Mater.* **112** 269–78.

[25] Wang H, Xie C, Zhang W, Cai S, Cai Z, Yang Z and Gui Y 2007 Comparison of dye degradation efficiency using ZnO powders with various size scales, *J. Hazard. Mater.* **141** 645–52.

[26] Ali A H, Naser G F and Alkaim A F 2018 Treatment of Textile Industries Wastewater by Advance Oxidation Process. *J. Eng Appl. Sci.* **13** 1042-45.

[27] Daneshvar N, Salari D and Khataee A R 2003 Photocatalytic degradation of azo dye acid red 14 in water: investigation of the effect of operational parameters, *J. Photochem. Photobiol. A: Chem.* **157** 111–16.

[28] Shankar M V, Anandan S, Venkatachalam N, Arabindoo B and Murugesan V 2004 Novel thin-film reactor for photocatalytic degradation of pesticides in aqueous solutions, *J. Chem. Technol. Biotechnol.* **79** 1258–79.

IOP Conf. Series: Journal of Physics: Conf. Series **1032** (2018) 012009 doi:10.1088/1742-6596/1032/1/012009

Preparation of Nanoparticles Copper Oxide using an Atmospheric-Pressure Plasma Jet

Ahmed A Anber[1], Mohammed Sh Essa[2], Ghada A Kadhim[1] and Shaymaa S Hashim[1]

[1]Department of Physics, College of Science, University of Wasit, Wasit, Iraq
[2]Ministry of Science and Technology, Renewable Energy Directorate, Solar Energy Research Center, Baghdad, Iraq

ahmedkrm88@gmail.com

Abstract. In this work, nanoparticles copper oxide were prepared by using an atmospheric-pressure plasma jet in NaCl–NaOH–NaNO3 electrolytic system. Atmospheric-Pressure Plasma Jet was employed as the electrode (Cathode) while the anode was made of the copper strip. The formation and structure of the samples have been characterized via Fourier-transformation infrared (FTIR), X-ray diffraction patterns (XRD) and optical microscope. From the outcomes, the morphology was showed copper oxide nanocrystals by this technique is rely on the media of electrolytic and preparation time. It was cleared from (XRD), the addition of $CuCl_2$ with the compositions was led to increase appearance of copper oxide. The uniformity of copper oxide nanoparticles with the average grain size about 70,104 nm. These results encourage preparing these nanostructures for using in industrial applications.

Keywords Copper oxide, Atmospheric pressure plasma jet, electrolytic system, Nanostructures.

1. Introduction

Atmospheric pressure plasma contact with electrolytic media has been studied for different application which was considered as a nonthermal plasma [1]. The plasma electrode (atmospheric plasma jet electrochemical technique) is carry out at the gas-liquid media as the cathode to prepare nanoparticles of copper oxide. liquid reactions are take place by using plasma and the discharge system is represented as the system of electrolytic media. presently, plasma can be worked as an cathode in electrolytic media. The electrode of plasma supplies ions and electrons into the electrolyte system, and many reactions can be observed in electrolysis, also using different metal electrode works as the anode to immerse into electrolytic media [2]. First, its clarified the mechanism of copper oxide nanoparticles preparation using an $CuCl_2$ powder, which was not observed in the previous studies. Recently, nanoparticles were received interest for its applications and properties. Copper oxide is paramount material which has potential applications in solar energy conversion [3,4], sensors and catalysts [5,6–9]. In particularly, it has potential applications in photon catalytic degradation of organic pollutants under visible light [10]. Moreover, the characterization of copper oxide are cheap, low toxicity, readily available and good environmental acceptability [11–13]. Currently, nanostructures of Cu_2O with different forms was prepared by various approaches such as hydrothermal [14–15], electrochemical rout [10,16,17], chemical vapor deposition of precursors, solution synthesis method and there are no chemical methods [18,19,20]. Therefore, no report was researched on the preparation of copper oxide via atmospheric pressure plasma jet electrochemistry technology with AC power supply. The plasma jet have common interest from the plasma organization due to its characterization[22–24], atmospheric pressure stability and non–equilibrium thermodynamics [25,26,27]. These properties were made atmospheric pressure plasma important for many application, including gas treatment, nanofabrication, medicine, textiles and surface modification [28].

2. Experiment

Copper oxide nanoparticles were prepared by atmospheric plasma jet. A schematic diagram of the experimental set-up and a photograph of the discharge are shown in Figure 1. The system consist of a tube acted as the cathode (Inner diameter is 0.6 mm and length is 7 cm, made of stainless steel. It was located 3 cm onward from the anode. The copper strip was used as the anode (3 cm in width, 7 cm length), it was immersed in area is 2 cm^2 and the distance between the surface of liquid and the needle end was 3 mm. Argon was used as a discharge gas and joined with the syringe by using glass flowmeter to control the gas flow at 60 ml/min. The reaction was occurred in glass basin (width is 5.5 cm and length is 8.5 cm). The copper electrode have been polished and then washed by distilled water. After that the copper strip was immersed into electrolytic system. The electrolytic medium was consisted of 0.001 kg/L NaOH, 0.15 kg/L NaCl, 0.0013 kg/L $NaNO_3$ and 0.001 kg/L glucose (fructose) as stabilizers to prevent aggregation of particles with the distilled water. The second electrolyte was contained same previous component with 1.5 g/L $CuCl_2$. The discharge was ignited by applying of a high voltage of 3 kV with 40 KHz using AC power supply. The discharge current was ranged from 5 to 10 mA. The electrolytic system was exposure at different time, (10, 20 and 30 min) then, the color of electrolyte have been changed as shown in Figure 2. The sediment was centrifuged and dried at 50 °C for 3 h.

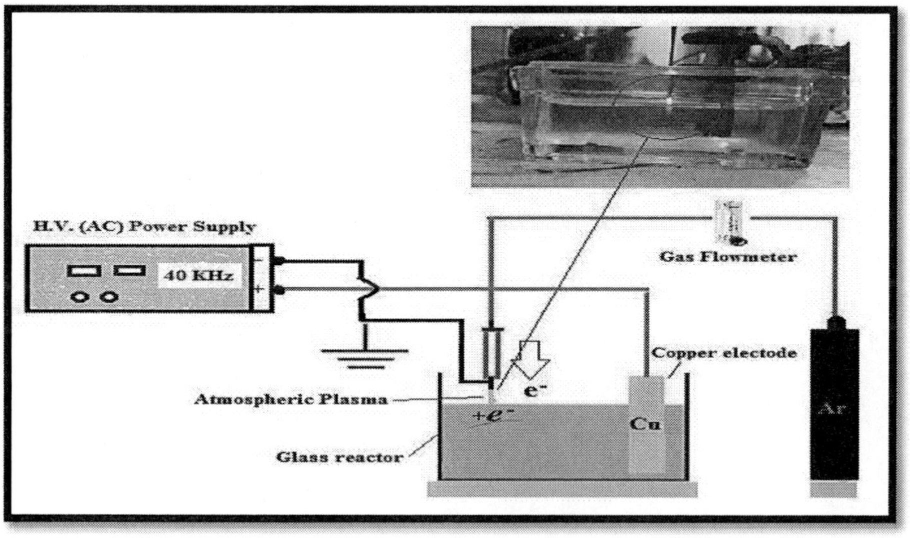

Figure 1. The schematic diagram of atmospheric plasma jet electrochemical synthesis of copper oxide nanoparticles.

IOP Conf. Series: Journal of Physics: Conf. Series **1032** (2018) 012009 doi:10.1088/1742-6596/1032/1/012009

Figure 2. (Color online) at different discharge duration: (a) 0, (b) 10, (c) 20 and (d) 30 min. The discharge current is 5 mA. The gap length is 3 mm.

3. Results and Discussion

Figure 3 shows the XRD patterns of the copper oxide nanoparticles and prepared with and without $CuCl_2$ on the electrolyte. It is clear that these powders are polycrystalline and include many crystal planes. Figure (3a) consist of four peaks belong to copper oxide nanocrystals was examined by the ICDD, USA (1979) JCPDS 1979, C 29-1133, (Joint Committee on Powder Standards). Moreover, by adding $CuCl_2$ powder with distilled water led to appear more peaks, it was indicated that the products prepared because of the existence of copper with chlorinate and contributing to produce copper oxide into electrolyte as shown in Figure (3b). This difference could clearly be observed from change the products colour as shown in Figure 4.

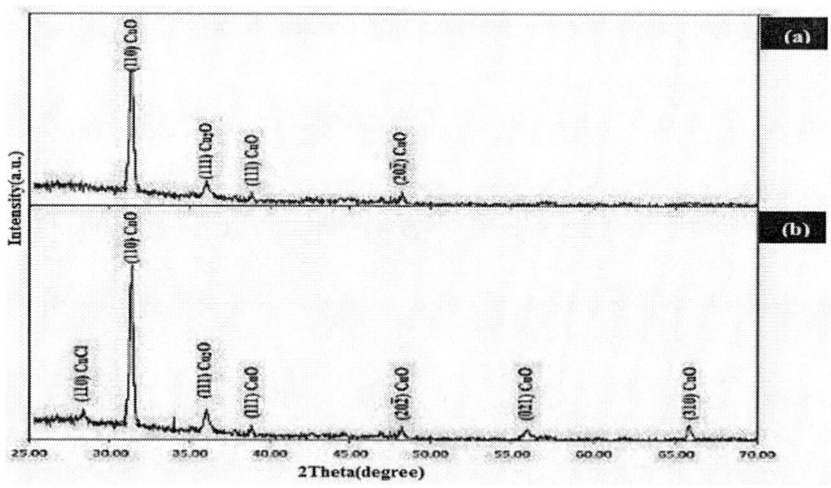

Figure 3. The XRD patterns of the products prepared by atmospheric pressure plasma jet electrochemical method in different compositions: (a) without addition $CuCl_2$ (b) with addition $CuCl_2$

IOP Conf. Series: Journal of Physics: Conf. Series **1032** (2018) 012009 doi:10.1088/1742-6596/1032/1/012009

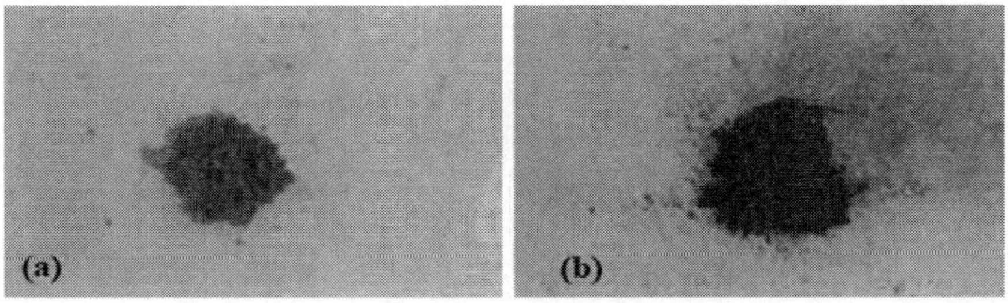

Figure 4. The pictures of products prepared by atmospheric-pressure plasma (a)Without adding $CuCl_2$ and (b) with adding $CuCl_2$.

FTIR was used to estimate the chemical bonding configurations in the sediments copper oxide nanoparticles. The recorded transmission spectra were in the range of 400–4000 cm^{-1}. Figures 5 a and 5 b, show that (Cu_2O) samples have a strong absorption band centered at around 416 cm^{-1} that can be identified with the stretching vibration mode of the Cu=O bond. One band of significant absorption can be seen around 3389 cm^{-1}, and can be attributed to the Cu-O stretching mode in CuO molecule. These broad bands indicate that the structure of copper oxide include nanocrystals. This shift can be interpreted by considering the oxygen incorporation into the network. The main characteristic peaks around 3000–3600 cm^{-1} (O–H bond) were observed and the broad absorption bands between 2800 and 4000 cm^{-1} belonging to O-H and Cu–O groups. There are two absorption peaks reveal the vibrational modes of copper oxide nanostructures in the range of 1400–1600 cm^{-1} in both figures. But in Figure (5b) The major peaks were observed from 800 $cm^{-1}$to 1200 cm^{-1} belonging to $CuCl_2$ group. A small shoulder at about 1615 cm^{-1} leaning against the water band about 1636 cm^{-1} (possibly due to copper carboxylate).

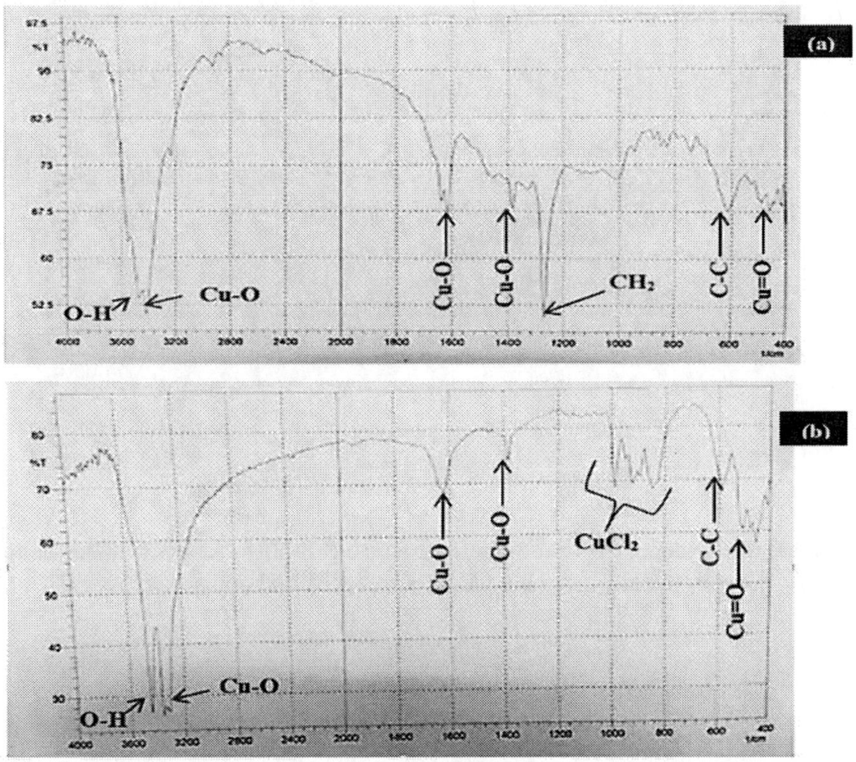

Figure 5. FTIR spectra of the copper oxide nanostructures prepared in this work at different compositions: (a) without addition $CuCl_2$ (b) with addition $CuCl_2$.

IOP Conf. Series: Journal of Physics: Conf. Series **1032** (2018) 012009 doi:10.1088/1742-6596/1032/1/012009

Optical micrographs for the prepared copper oxide nanoparticles are shown in Figure 6. The microstructure show two prevailing features; the first is no definite grain boundaries detected but white regions and green due to addition $CuCl_2$ in the matrix is shown in Figure 6 a. In the figure 6 b without addition $CuCl_2$ shows the second feature is that (dark regions) are of small population, confined and totally isolated. The lack of grain boundaries channels may lead to higher dielectric constants. By using Hand-Held model Skyray Instrument EDX-Pocket III/2012 was used to record the elements analysis, (from Sulfur to uranium) wide range, also to provides detection limits at the sub-part per million level. The concentrations was measured up to 100% easily can also simultaneously as shown in Figure 7.

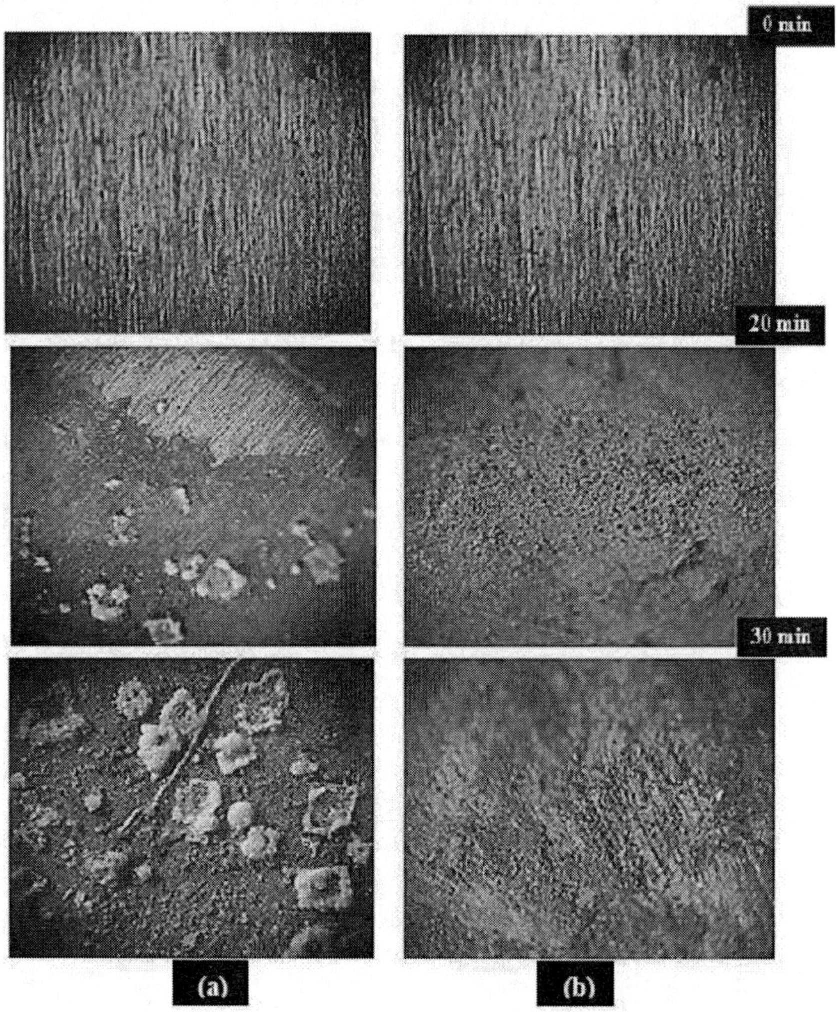

Figure 6. Optical microscope shows the microstructure (×800) for copper oxide nanoparticles on the copper sheet: (a) with addition $CuCl_2$ (b) without addition $CuCl_2$

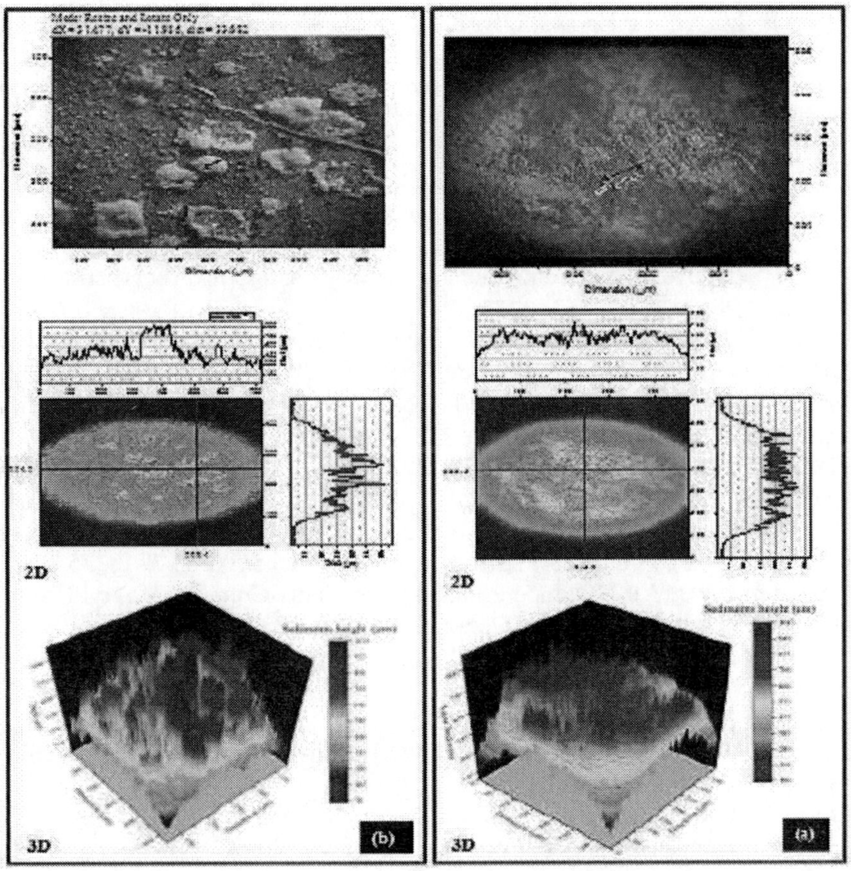

Figure 7. Optical microscope shows the image processing with upper and side-linked curve exhibited the surface profile of copper oxide nanoparticles on the copper sheet at 30 minute: (a) without addition $CuCl_2$ (b) with addition $CuCl_2$.

4. Conclusions

In concluding remarks, it was cleared that quality nanostructured copper oxide powder can be prepared by atmospheric-pressure plasma jet in the gas or liquid phase. However, the preparation nanoparticles via this technique is not common. To prove these results, its need intensive efforts for exploiting the atmospheric-pressure plasma jet. The prepared copper oxide nanoparticles have a widely application such as adsorption of organic pollutants. This result can be considered high and the work can be good attempt to prepare copper oxide by atmospheric-pressure plasma jet as a new method and employ nanostructures in such important application.

References

[1] P. Bruggeman, M. Kushner, B. Locke, J. Gardeniers, W. Graham, D. Graves, R. Hofman-Caris, D. Maric, J. Reid, E. Ceriani, D. Fernandez Rivas, J. Foster, S. Garrick, Y. Gorbanev, S. Hamaguchi, F. Iza, H. Jablonowski, E. Klimova, J. Kolb, F. Krcma, P. Lukes, Z. Machala, I. Marinov, D. Mariotti, S. Mededovic Thagard, D. Minakata, E. Neyts, J. Pawlat, Z. Lj Petrovic, R. Pflieger, S. Reuter, D. Schram, S. Schroter, M. Shiraiwa, B. Tarabova, P. Tsai, J. Verlet, T. von Woedtke, K. Wilson, K. Yasui, and G. Zvereva 2016 Plasma Sources Sci. Technol. **25** 053002.

[2] F. Tochikubo, Y. Shimokawa, N. Shirai, and S. Uchida, Jpn. 2014 J. Appl. Phys. **53** 126201.

[3] Wang, W. Z. *et al.* 2002 .Synthesis and Characterization of Cu_2O Nanowires by a Novel Reduction Route. Adv Mater **14** 67–69.

[4] Zhao, W. *et al.* 2010 Shape-controlled synthesis of Cu_2O microcrystals by electrochemical method. Appl Surf Sci **256** 2269–2275.

[5] Deng, S. *et al.* 2012Reduced Graphene Oxide Conjugated Cu_2O Nanowire Mesocrystals for High-Performance NO_2 Gas Sensor J Am Chem. Soc. **134** 4905–4917.

[6] E. de Jongh, P., Vanmaekelbergh, D. & J. Kelly, J. 1999 Cu_2O: a catalyst for the photochemical decomposition of water Chem. Commun. **12** 1069–1070 .

[7] White, B. *et al.* 2006 Complete CO Oxidation over Cu_2O Nanoparticles Supported on Silica Gel. Nano Lett. **6** 2095–2098.

[8] Kuo, C.-H. & Huang, M. H. Facile 2008 Synthesis of Cu_2O Nanocrystals with Systematic Shape Evolution from Cubic to Octahedral Structures. J Phys Chem C **112** 18355–18360 .

[9] Kim, J. Y. *et al.* 2009 Cu_2O Nanocube-Catalyzed Cross-Coupling of Aryl Halides with Phenols via Ullmann Coupling. Eur. J Inorg. Chem. **28** 4219–4223.

[10] Yuan, G., Zhu, J., Xie, F. & Chang, X. 2010 Shape-Controlled Synthesis of Cuprous Oxide Nanocrystals via the Electrochemical Route with H_2O-Polyol Mix-Solvent and Their Behaviors of Adsorption. J Nanosci Nanotechno **10** 5258–5264.

[11] Liu, R. *et al.* 2005 Epitaxial Electrodeposition of High-Aspect-Ratio Cu_2O(110) Nanostructures on InP(111). Chem Mater **17** 725–729.

[12] Li, J. *et al.* 2008 Patterning of Nanostructured Cuprous Oxide by Surfactant-Assisted Electrochemical Deposition. Cryst. Growth Des **8** 2652–2659.

[13] Bao, H. *et al.* 2010 Shape-Dependent Reducibility of Cuprous Oxide Nanocrystals. J Phys Chem C **114** 6676–6680.

[14] Xu, J. & Xue, D. 2007 Five branching growth patterns in the cubic crystal system: A direct observation of cuprous oxide microcrystals. Acta Mater **55** 2397–2406.

[15] Si, Y. & Xue, D. 2009 Hydrothermal fabrication of core-shell structured Cu_2O microspheres via an intermediate-template route. Mod Phys. Lett. B **23** 3851–3858.

[16] Kang, S. O. *et al.* 2009 Electrochemical growth and resistive switching of flat-surfaced and (111)-oriented Cu2O films. Appl. Phys. Lett. **95** 092108.

[17] Han, X. F., Han, K. H. & Tao, M. 2009 n-Type Cu_2O by Electrochemical Doping with Cl. Electrochem Solid St **12** H89–H91.

[18] Medina-Valtierra, J., Calixto, S. & Ruiz, F. 2004 Formation of copper oxide films on fiberglass by adsorption and reaction of cuprous ions. Thin Solid Films **460** 58–61.

[19] Yang, Z., Xu, J., Zhang, W., Liu, A. & Tang, S. 2007.Controlled synthesis of CuO nanostructures by a simple solution route. J Solid State Chem. **180** 1390–1396.

[20] Kumar, R. V.,Mastai, Y., Diamant, Y. & Gedanken, A. Sonochemical 2001 synthesis ofamorphous Cu and nanocrystalline Cu2O embedded in a polyaniline matrix.J Mater Chem. **11** 1209–1213.

[21] Becker, K. H., Schoenbach, K. H. & Eden, J. G. 2006 Microplasmas and applications. J Phys D: Appl Phys **39** R55.

IOP Conf. Series: Journal of Physics: Conf. Series **1032** (2018) 012009 doi:10.1088/1742-6596/1032/1/012009

[22] El-Habachi, A.& Schoenbach, K.H. 1998 Generation of intense excimer radiation from high-pressure hollow cathode discharges. Appl. Phys. Lett. **73** 885.
[23] Sankaran, R. M., Giapis, K. P., Moselhy, M. & Schoenbach, K. H. 2003 Argon excimer emission from high-pressure microdischarges in metal capillaries. Appl. Phys. Lett. **83** 4728.
[24] Park, S. J., Eden, J. G., Chen, J. & Liu, C. 2004 Microdischarge devices with 10 or 30 mm square silicon cathode cavities: pd scaling and production of the XeO excimer. Appl. Phys. Lett. **85** 4869.
[25] Park, S. J. & Eden, J. G. 2002 13–30 micron diameter microdischarge devices: Atomic ion and molecular emission at above atmospheric pressures. Appl. Phys. Lett. **81** 4127.
[26] Kurunczia, P.,Abramzona, N., Figusb,M.&Becker, K. 2004 Measurement of rotational temperatures in high-pressure microhollow cathode (MHC) and capillary plasma electrode (CPE) discharges. Acta Phys Slovaca **54** 115–124.
[27] Penache, C. *et al.* 2002 Characterization of a high-pressure microdischarge using diode laser atomic absorption spectroscopy. Plasma Sources Sci T **11** 476–483.
[28] Mariotti, D. & Sankaran, R.M. 2011 Perspectives on atmospheric-pressure plasmas for nanofabrication. J Phys D: Appl Phys **44** 174023.

Synthesis of ZnS Quantum Dots for QDs-LED hybrid device with different cathode materials

Ahmed F. Mohammed[1,*] **and Wasan R. Salah**[2]

[1]Ministry of Education, General Directorate of Education of Najaf, Iraq
[2]Department of Physics, College of Science, University of Baghdad, Baghdad, Iraq

Email: ahmedasker85@gmail.com (Corresponding author: Ahmed F. Mohammed)

Abstract. In this study, a chemical method was used to synthesise ZnS quantum dots (QDs) with low cost and simple process. A fixed molar ratio 1:1 of zinc chloride and sodium sulfide has been used for preparing two samples A and B. The X-ray diffraction (XRD) analyses prove the cubic phase of ZnS with an average particles size of (3-29) nm. According to UV-Vis. spectra analyses, a large blue shift over 32 and 48 nm was observed, while the band gaps energy (Eg) were 3.55 and 3.8 eV for A and B samples respectively. The emission spectrum of quantum dots carries more evidence on the presence of shallow at 373 and 367 nm originated from the interstitial Zn and traps at 428 and 439 nm, whose origin ascribed to defects zinc vacancy (VZn), while 541 nm was due to sulfide vacancy (VS). QDs-LED hybrid devices were fabricated using ITO/ PEDOT: PSS/ Poly-TPD/ ZnS-QDs/ layers with two types of cathode materials LiF and Al. Using LiF as cathode in addition to the Al gives good enhancement to light generation in compares with that of Al cathode only. The spectra 350-700 nm appeared a wide emission band according to the EL measurement.

1. Introduction

Semiconductor QDs are very important nanomaterials with unique physical, chemical, optical and electronic properties owing to the quantum confinement effect. These properties are the result of the confinement of electronic wave functions which can be manipulated by controlling the size of the QDs. To observe the confinement effects, the nano crystals size has to be in the range of nanometers a variety of low dimensional nanostructures such as zero-dimensional (0D) nanoparticles, one dimensional (1D) nanowire, nanotube and nanorods, two dimensional (2D) nanobelt and nanosheet are investigated extensively due to their novel and fascinating properties compared to their bulk counterparts. Nanoscale light sources have found numerous applications in biology, bioanalytics, and optoelectronics [1, 2]. (QDs) semiconductor particles have drawn considerable interest recently for their properties; large surface-to-volume ratio, quantum tunneling effect, increased activity, special electronic and optical properties in comparisem with bulk materials [3, 4]. These luminescents nanomaterials possess suitable characteristics for various applications in biological labeling and optoelectronic devices. Along with the direct use of QDs as efficient phosphors showing narrow emission, they can also be applied as host materials for accommodating dopant ions of interest [5]. Zinc sulfide (ZnS), as one of the most important members of the family of group II-VI wide-gap semiconductors with direct band gap for the bulk cubic as 3.66 eV. ZnS has been extensively investigated because of its potential applications in flat-panel display, light-emitting diodes (LEDs), infrared windows, electroluminescence, sensors, lasers and photocatalysis due to its diverse range of possible structures and morphologies, and superior chemical and thermal stabilities. The optical properties can be tunable by modulating the QDS size, chemical composition and defects concentration. It has been reported that QDs contain few hundred atoms and emit only one wavelength of light when they are excited. The color emitted is determined by the size of the quantum dots [6, 2].

Light emitting diodes (LEDs) based on colloidal QDs emitting layer are highly promising for the next generation of lighting and displays and come into being an undeniable competitor to organic LEDs due to their advantageous properties, such as the tunable color emission spectra, the narrow emission line width, and the cost-effective fabrication techniques [7]. The quantum dot light-emitting diodes (QD-LEDs), because of their unique potential properties, have attracted much attention and considered most practical candidates for flat panel displays and solid-state lighting sources, offering significant advantages over organic LEDs (OLEDs) [8].

Organic electronic devices required Ohmic contacts for most efficient performance. Metals-organic semiconductors contacts can be improved by placing charge injection layers between the semiconductor and the metal [9]. In this work, the structural and optical properties of synthesized ZnS QDs were studied. QDs-LED hybrid devices were fabricated using ITO/ PEDOT: PSS/ Poly-TPD/ ZnS-QDs layers with two types of cathode materials lithium florid (LiF) and aluminum (Al).

2. Experimental Procedure Chemical Materials

The chemical materials which is used in this research like zinc chloride ($ZnCl_2$, Merck, 99%), sodium sulfide (Na_2S, Merck, 99%) and ammonium hydroxide (NH_4OH, Merck, 99%) were used without further purification. The materials used for the QDs-LED fabrication are; Poly (3, 4 ethylenedioxythiophene): styrene sulfonic acid (PEDOT: PSS) were supplied from Ossila Limited. Poly [N, N'-bis(4-butylphenyl)-N, N'-bis(phenyl)-benzidine] (Poly-TPD) was purchased from American Dye Source co., while Lithium fluoride (LiF) of 99%) from Riedel-de Haen.

3. Synthesis of ZnS Quantum Dots Colloidal

A fixed molar ratio 1:1 of zinc chloride and sodium sulfide has been used for preparing the aqueous solution at room temperature to produce two samples A and B. An equimolar concentration of both $ZnCl_2$ and Na_2S were used to prepare QDs of ZnS. ($ZnCl_2$) solution has been prepared by dissolving 0.68 g of zinc chloride powder ($ZnCl_2$) in 50 ml of distilled water (d.w.). The solution is acidic with pH value of 6.1. In order to have a base solution with pH of 7.2, adding alkaline solution of ammonium hydroxide (NH_4OH) drop by drop to the previous ($ZnCl_2$) solution. Using alkaline solutions instead of the acidic solution because the rapid decay of the quantum dots dosen't occupy in the acidic range, as well as to reduce the aggregation [10]. (Na_2S) solution has been prepared by dissolving 0.39 g of sodium sulfide powder (Na_2S) in 50 ml of d.w. and gather in a three-neck flask. The flask put on a magnetic stirrer under a high stirring rate (700 r.p.m.) and flowing of argon gas continuously at room temperature. After which, the second solution was added drop by drop to the first solution, quit reaction for one hour until the mixture appears completely milky color, this indication of forming ZnS QDs. Stopped the flowing of argon gas and the solution was poured in a redundant amount of methanol. Two steps lead to diminishing the chemical reaction: first take amount of the deposited material to represent A sample. Secondly centrifuge the remaining residue and then wash with distilled water to have sample B. The reaction can represent as:

$$ZnCl_2 + Na_2S \xrightarrow{\Delta} ZnS\downarrow + 2NaCl$$

4. Assembling layers of Hybrid Devices

For fabrication process of the QD-LED device, the ITO conductive glass of resistivity of 12 Ω/ square as the substrate was used with dimensions of 25x25 mm^2 after cleaning by distilled water and then by isopropanol alcohol in the acoustic-wave ultrasonic and for a period of 10 minutes. All solutions were deposited layer by layer on the ITO glass by a spin-coating method. PEDOT: PSS as an anode buffer layer of 30 nm thickness was deposited on ITO glass at 6k r.p.m. for 30 s then annealed at 150°C for

IOP Conf. Series: Journal of Physics: Conf. Series **1032** (2018) 012010 doi:10.1088/1742-6596/1032/1/012010

15 minutes. After that, poly-TPD which represent a hole transport layer was prepared by dissolving 70 mg/ml in chlorobenzene solution then deposited a layer with thickness 135 nm at 2.5k r.p.m. for 30 s then annealed at 80°C for 30 minutes. Thereafter, ZnS QDs as an electron transport layer was prepared (250 mg/ml) in ethanol solution then deposited of 165 nm layer thickness at 2.5K r.p.m. for 30 s and then annealed at 60°C for 50 minutes. Finally, cathode layer was deposited once using Al material along with the first device, while for the second device using Al then LiF material. Al layer of 200 nm was deposited by a thermal evaporation method in high vacuum ~10^{-6} Torre with deposition rate ~11 Å/s, while LiF layer of 2 nm thickness was deposited by ion coater method at 3 mA/40s. The films thicknesses have been measured by the BLK-CXR-SR-25 spectrometer.

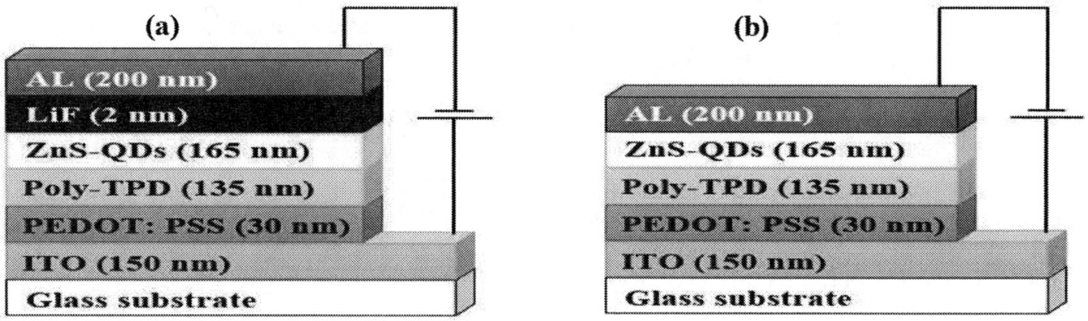

Figure 1. Schematic of QDs-LED with cathode layer materials of: (a) LiF with Al and (b) only Al.

5. Results and Discussion

5.1 Structural Measurements

The QDs of ZnS prepared by the chemical method which considers a low cost and simple process. The structural investigation of ZnS QDs was carried out and recorded using XRD - 6000 Labx, supplied by Shimadzu. Figure (2) shows the X-ray diffraction pattern of samples A and B respectively. Figure (2a) shows that sample A has scattering peak at (2θ) = 28.6°, 47.9° and 56.8° which corresponding to (111), (220) and (311) crystal planes of the zinc-blende (cubic) phase of ZnS. These results are according to card [JCPDS NO: 5-0566] while the peak at (2θ) = 31.7° referred to the appearance of sodium chloride in the prepared ZnS QDs. This is meaning that sample A was not pure ZnS QDs because of containing ions of reaction compounds that have not reacted due to a decrease in the base concentration of the solution these ions produced by adding hexamethylenetetramine [$(Ch_3)_6N_4$] [11].

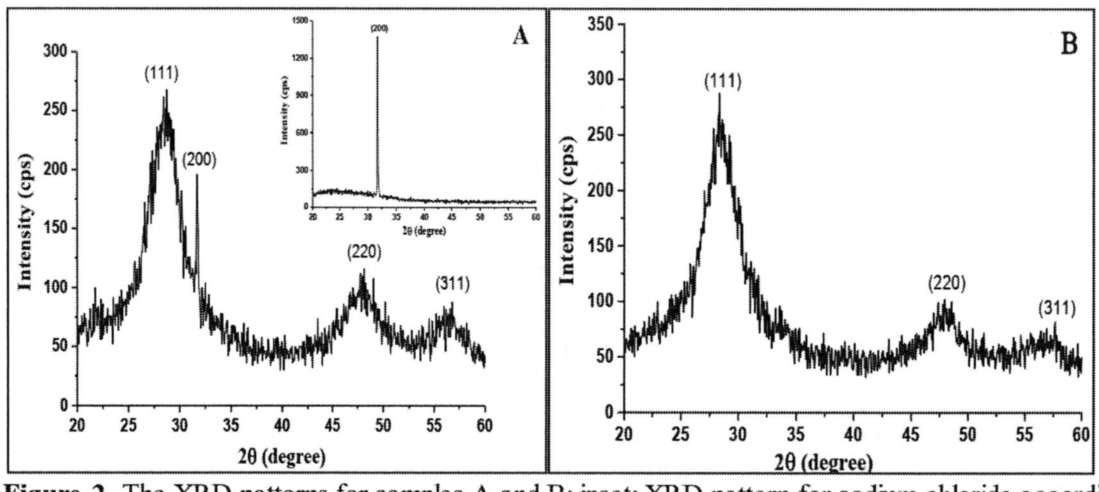

Figure 2. The XRD patterns for samples A and B: inset: XRD pattern for sodium chloride according to card number [JCPDS NO: 5-628].

Figure (2b) shows that sample B has characteristic diffraction peaks at (2θ) = 28.4°, 47.9° and 57.6° that appointed to (111), (220) and (311) reflection planes respectively of the face-centered cubic of ZnS. These results conformance according to card number [ICDD PDF 65-1691]. The absence of (NaCl) or any other peaks in sample B refer to the purity ZnS QDs as a final product. The average size (D) of ZnS QDs can be calculated according to Scherrer's equation [2].

$$D = \frac{0.9\,\lambda}{\beta cos\theta} \qquad (1)$$

where λ is the X-ray wavelength and equals to 0.15406 nm, β is the full width at half maximum. Table (1) summarized the XRD parameters and crystal size of the prepared ZnS QDs.

Table 1. XRD parameters and crystal size of the prepared ZnS QDs.

Samples	2θ *(degree)*	B *(degree)*	Crystal Size *(nm)*	(hkl)
A	28.6°	2.54	3.37	(111)
	47.9°	1.05	8.65	(220)
	56.8°	0.48	19.43	(311)
B	28.4°	2.23	3.82	(111)
	47.9°	0.54	16.77	(220)
	57.6°	0.31	29.91	(311)

It is clear that the crystal size for sample A is smaller than that for B. The reason may be attributed to the little period of staying sample A in alcohol, which led to has smaller crystal size.

5.2 Optical Measurements

The UV-Vis. spectra of ZnS QDs were measured using OPTIMA SP- 3000 UV –Vis. spectrophotometer and figure (3) shows these spectra. It is evident from the figure that the absorption edge λ_0 of the prepared ZnS QDs is observed at 318 and 302 nm for samples A and B respectively, which are very much shorter than 350 nm of the bulk ZnS [2] which indicating a clear blue shift. These blue shifts of the absorption edge can be attributed to the quantum confinement effect of the ZnS QDs. The band gap energy (E_g) can be estimated from the figure (4) using Tauc equation [2] by extrapolating the linear portions of $(\alpha h\nu)^2$ vs hν graph on the hν axis to α = 0.

$$(\alpha h\nu)^n = A\,(h\nu - Eg) \qquad (2)$$

where α is the absorption coefficient, hν is the incident photon energy and A is a constant depends on the electron–hole mobility. The energy gaps for direct transitions were found to be 3.55 and 3.8 eV for A and B samples respectively. Photoluminescence spectra were taken by SL 174 spectrofluorometer with an excitation wavelength of 300 nm and 280 nm respectively for samples A and B, figure (5) shows these spectra. Since QDs have a very high surface to the volume ratio, so they susceptible to various surface defects. The sources from PL are due to self-activated centers and then to crystal lattice vacancies such as sulfide vacancies according to [12]. The first emission peaks were detected at 373 and 367 nm for A and B samples respectively which originated from the interstitial Zn, while the decay time was 5.6 ns for both which consider quite fast such as in [13]. The second emission peaks which appeared at 428 and 439 nm for A and B samples, the origin were ascribed to defects zinc vacancy (V_{Zn}). The third peak at 541 nm for A sample was due to sulfide vacancy (V_S). The decay time was slow as 140 ns for the second peaks and 7.7 μs for third peak as in [14, 15].

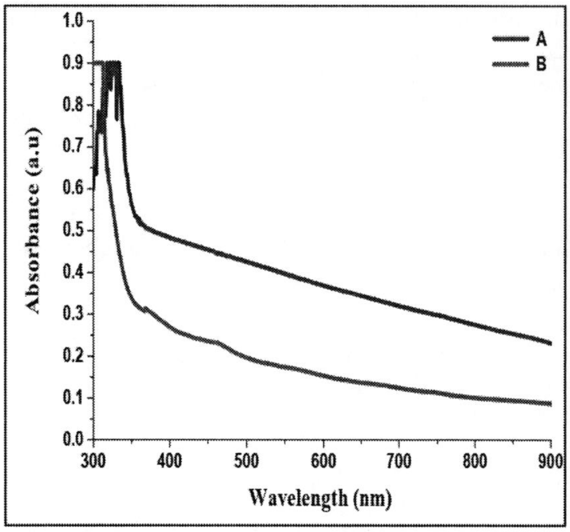

Figure 3. Absorption spectra of ZnS QDs samples A and B

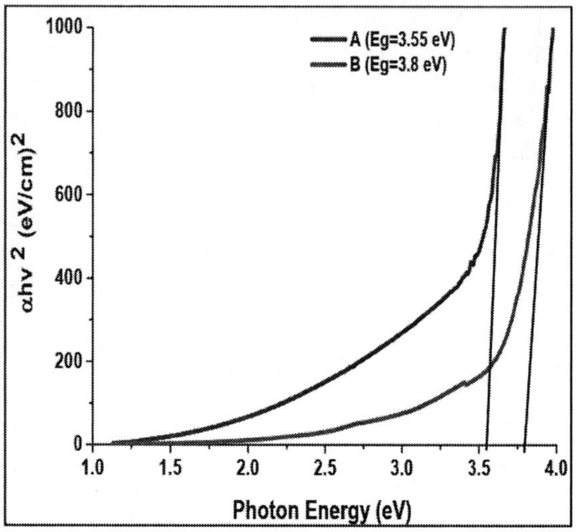

Figure 4. Energy gaps of ZnS QDs samples A and B.

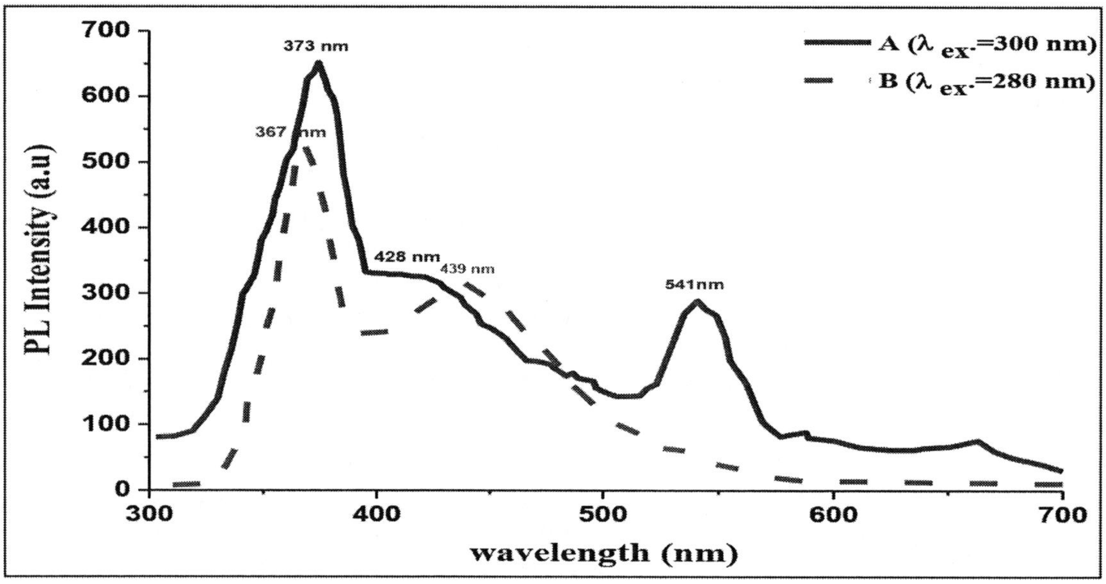

Figure 5. The PL spectra of the ZnS QDs of samples A and B.

5.3 Morphological Measurements

The topography of the film surface was recorded using CSPM AA3000 Atomic Force Microscope (AFM) supplied by Angstrom Company. Figure (6) shows the 3 dimensions AFM images and granularity distribution charts of ZnS QDs samples A and B. The average particles size according to the AFM images is about 2.88 and 36.44 nm for samples A and B respectively.

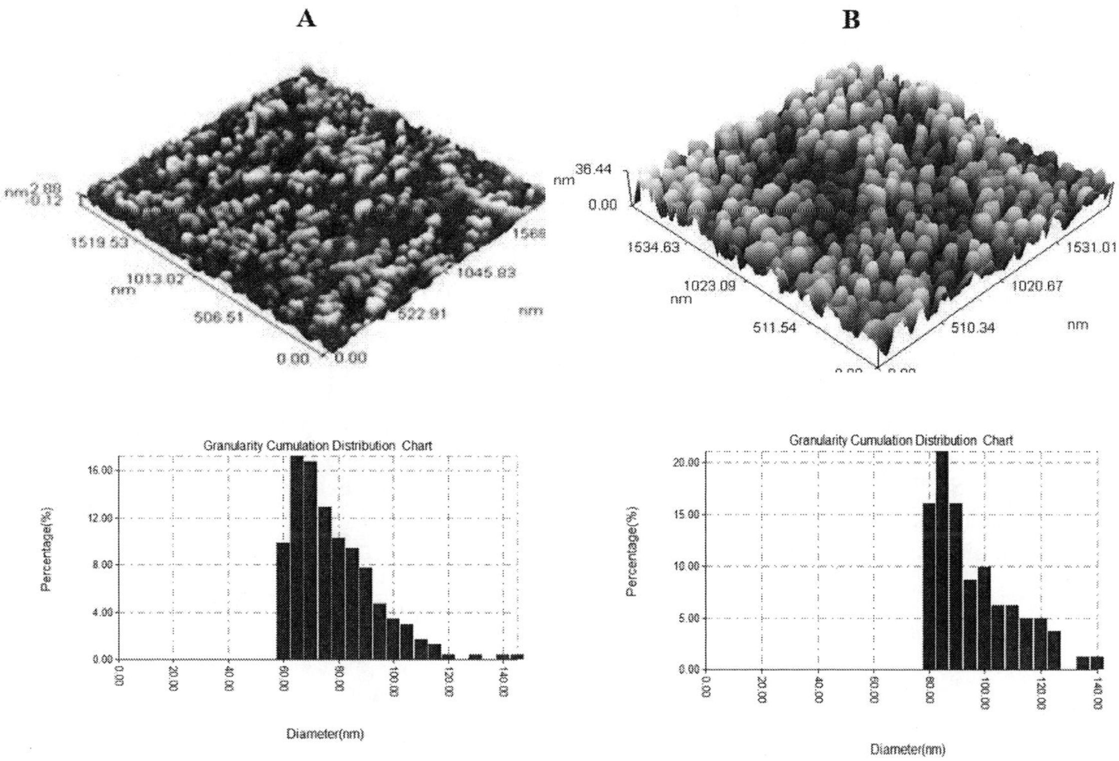

Figure 6. 3 dimensions AFM images and granularity distribution charts of ZnS QDs samples A and B.

Table (2) displays a summary of surface topography parameters of ZnS QDs. According to figure (6) sample B has low aggregation from sample A which is important for producing a homogeneous light from the QD-LED.

Table 2. AFM parameters of ZnS QDs.

Samples	Roughness Average Sa *(nm)*	Root Mean Square Sq *(nm)*	Ten Point High Sz *(nm)*	Average Diameter D *(nm)*
A	0.74	0.86	2.99	76.08
B	7.25	8.55	21.4	93.57

5.4 Hall Effect Measurements

Hall Effect was used to electrically characterize the prepared films. The measurements carried out using Ecopia HMS-3000 Hall measurement system. Table (3) summarizes Hall Effect results which revealed that both films n-type conductivity. Sample A has conductivity and mobility values larger than that of sample B, and this is may be attributed to existence of salts.

Table 3. Electrical parameters of ZnS QDs films.

Samples	Conductivity $(\Omega.cm)^{-1}$	Mobility (cm^2/Vs)	Hall coefficient (m^2/C)	Resistivity $(\Omega.cm)$
A	1.16×10^{-5}	2.56	-2.20×10^{5}	4.30×10^{4}
B	0.11×10^{-5}	1.22	-1.21×10^{5}	0.98×10^{4}

5.5 Current-Voltage Characteristics

Figure 7 shows the current-voltage measurements under forward bias voltage. By observing the curve, one can find that the two samples behave as diode behavior. This is called the exponential increase due to the increased saturation current between the hybrid junction, after the holes moved from PEDOT: PSS to the p-(Poly-TPD) towards depletion zone, At the same time, when the electrons moved from LiF to the n-(ZnS-QDs) towards the junction, reducing the potential barrier. Thin film LiF layers play a role in reducing the barrier for electron injection from Al layer into the organic semiconductors according to a change in the effective work function of the LiF/Al contact [16].

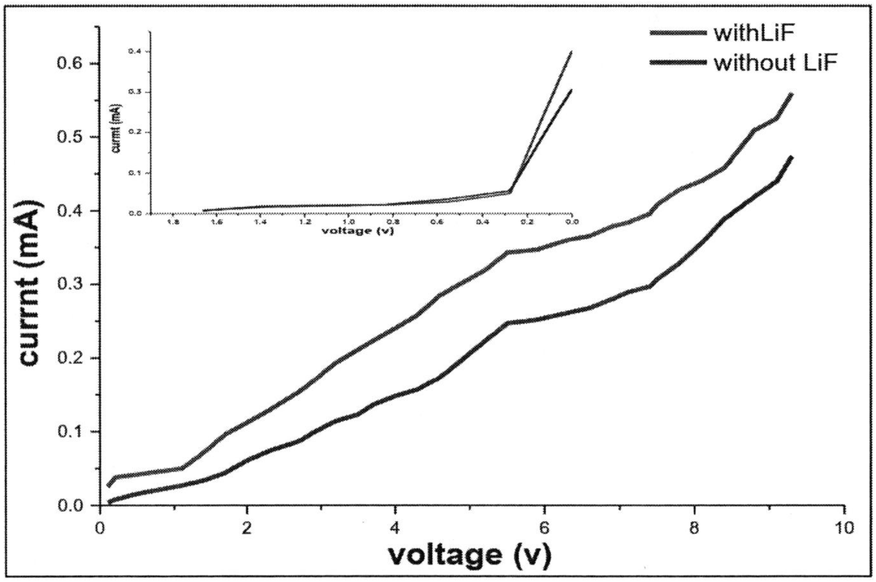

Figure 7. I-V forward characteristics of hybrid devices with and without using LiF as a cathode layer: inset behavior the devices from 5.5-9v.

According to the results of the current-voltage, thin film layer of LiF has provided the flow current largest in the device. However, the intensity of the light without using LiF is less than with that of LiF and this is what obvious in the electroluminescence measurements and photographic plat in both cases. Turn-on voltage of the devices are between (5.5 - 9 V), but at voltage greater than 9 V causes the device to collapse. Therefore, 8 V was chosen for electroluminescence measurements and device operating. The I-V characteristics of ITO/ PEDOT: PSS/ Poly-TPD/ ZnS-QDs/ LiF/ Al and ITO/ PEDOT: PSS/ Poly-TPD/ ZnS-QDs/ Al hybrid devices shown in figure (7).

5.6 Electroluminescence Measurements

The EL measurements for the QD hybrid devices shown in figure (8) were carried out using a photomultiplier detector at room temperature under forwarding bias voltage of 8 V.

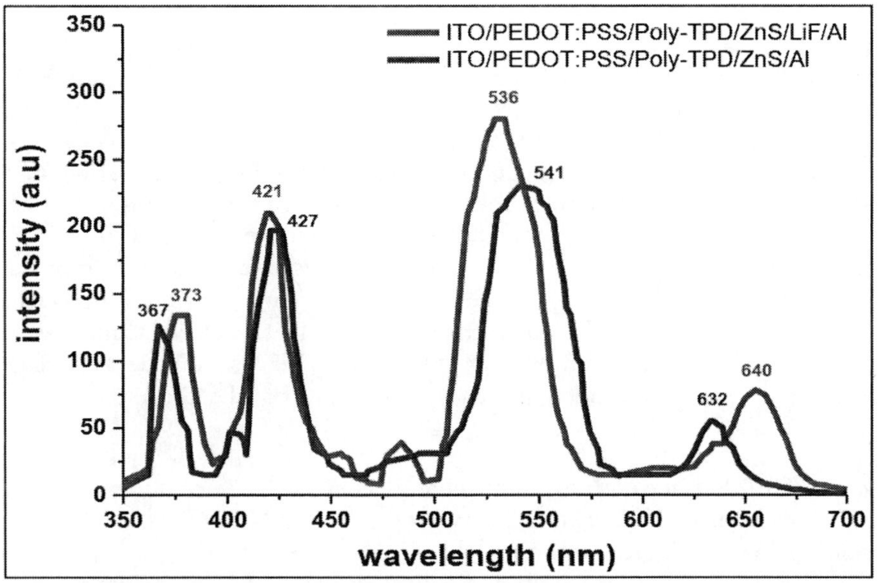

Figure 8. EL spectra of hybrid devices with and without using LiF as a cathode layer with forward bias voltage of 8 V.

The carrier transport mechanism in QDs-LED devices can summarized as follows; the Poly-TPD which consider a hole transporting material. Holes are injected from PEDOT: PSS, the buffer ITO anode, into the HOMO of the Poly-TPD matrix and transported to the valence band (V.B) or to the acceptor levels (V_{Zn} and V_S) of the ZnS QDs. Meanwhile, electrons are injected from the LiF (which consider as buffer to the Al) cathode into the conduction band (C.B). This is called the excitons from band edge recombination. The process of recombination of holes and electrons through defects in order to emitting different wavelength light called Shockley-Read Hall (SRH) recombination. [11]. According to figure (8), the device with LiF cathode has more intensity than the device without LiF. Actually the existence of LiF and PEDOT: PSS layers play a rule in producing a large difference in work function between the cathode and anode to be about 2.5 eV, which is very important in the manufacturing of diodes, which rapid the exciton production process and increase the resulted intensity also. Figure 9 shows a photographic plate of the light generation of ITO/PEDOT: PSS/Poly-TPD/ZnS/LiF/Al and ITO/PEDOT: PSS/Poly-TPD/ZnS/Al hybrid according to the emission lines of the EL spectrum at the 8 V. It is clear that the intensity of the device with LiF is higher and the color is closer to the white than the device without LiF which is closer to yellow light.

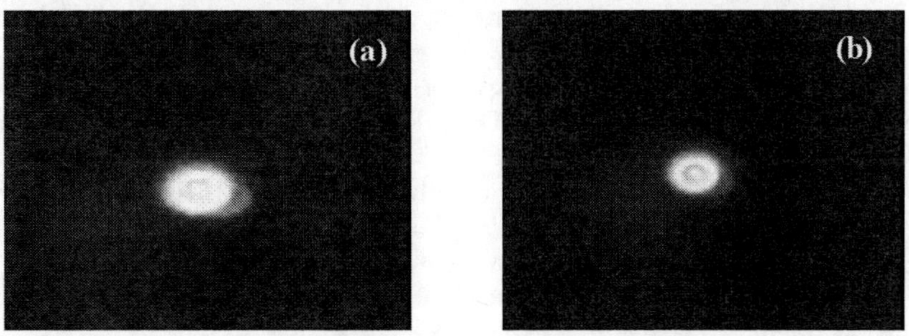

Figure 9. A photographic plate of the light generation of (a) ITO/PEDOT: PSS/Poly-TPD/ZnS/LiF/Al (b) ITO/PEDOT: PSS/Poly-TPD/ZnS/Al hybrid devices.

IOP Conf. Series: Journal of Physics: Conf. Series **1032** (2018) 012010 doi:10.1088/1742-6596/1032/1/012010

6. Conclusion

ZnS QDs were synthesized by a simple chemical method with low cost. XRD, UV-Vis., and PL measurements confirmed the presence of zinc and sulfide. The prepared ZnS QDs have a small average grains size of 3.37 and 3.82 nm. Optical measurements proved that the absorption edge was shifted largely towards blue region of the spectrum because of the confinement effect on the carriers. The PL spectrum of ZnS shows the defect related emission. The deep traps are attributed to the process of recombination at Zn^{2+} and S^{2-} vacancy levels. AFM images proved that the purified sample (B) has low aggregation which is important for producing a homogeneous light from the QD-LED. (I–V) characteristics indicate that for few voltages of (8V), the output current is good which consider acceptable for light generation process. Fabrication of QDs-LED devices from two layers' hole injection organic Poly-TPD and electron injection QDs using LiF as cathode in addition to the Al gives good enhancement to light generation in compares with that of Al cathode only.

7. References

[1] Dorokhin D V 2010 Surface engineered quantum dots in photoelectron chemistry and supramolecular assembly *Ph.D. Thesis University of Twente MESA⁺ Institute for Nanotechnology Enschede, The Netherlands.*

[2] Bodo B and Singha R 2016 Structural and optical properties of ZnS quantum dots synthesized by CBD method *International Journal of Scientific and Research Publications* **6(8)** 461-65.

[3] Nirmala Jothi N S, Joshi A G, Jerald Vijay R., Muthuvinayagam A and Sagayaraj P 2013 Investigation on one-pot hydrothermal synthesis, structural and optical properties of ZnS quantum dots *Materials Chemistry and Physics* **138** 186-91.

[4] Wu M, Wei Z, Zhao W, Wang X and Jiang J 2017 Optical and Magnetic Properties of Ni Doped ZnS Diluted Magnetic Semiconductors Synthesized by Hydrothermal Method *Journal of Nanomaterials* **2017 (1603450)**1-9.

[5] Thuy U T D, Maurice A, Liem N Q, Reiss P 2013 Europium doped In(Zn)P/ZnS colloidal quantum dots. *An international journal of inorganic chemistry* **42(35)** 12606-10.

[6] Zeng Z, Garoufalis C S, Baskoutas S and Bester G 2015 Excitonic optical properties of wurtzite ZnS quantum dots under pressure *Journal of Chemical Physics* **142(114305)** 1-5.

[7] Zhang X, Dai H, Zhao J, Wang S and Sun X 2016 All-solution processed composite hole transport layer for quantum dot light emitting diode *Thin Solid Films* **603** 187–92.

[8] Ji W, Tian Y, Zeng Q, Qu S, Zhang L, Jing P, Wang J and Zhao J 2014 Efficient quantum dot light-emitting diodes by controlling the carrier accumulation and exciton formation *ACS Applied Materials and Interfaces* **6(16)** 14001–7.

[9] Bory B F, Rocha P R F, Janssen R A J, Gomes H L, Leeuw D M D and Meskers S C J 2014 Lithium fluoride injection layers can form quasi-Ohmic contacts for both holes and electrons. *Applied Physics Letters* **105(123302)** 1-4.

[10] Saleh W R and Kadim A M 2017 Synthesis of ZnS QDs with Different Values of pH and White Light Generation from Fabrication of TPD: PMMA/ZnS Hybrid Devices. Nano Hybrids and Composites **14** 49–57.

[11] Ibrahim O A 2015 Electroluminescence of White Light from ZnO Nanostructures/ TPD Hybrid, *Ph.D. Thesis, Physics Department, College of Science, University of Baghdad.*

[12] Ni Y, Liu H, Wang F, Yin G, Hong J, Ma X and Xu Z 2004 Self-assembly of CuS nanoparticles to solid, hollow, spherical and tubular structures in a simple aqueous-phase reaction. *Applied Physics A: Materials Science and Processing* **79(8)** 2007–11.

[13] Yanagida T, Fujimoto Y and Yanagida S 2014 Optical and scintillation properties of pure ZnS crystal, *Proceedings of the 12th Asia Pacific Physics Conference* JPS Conf. Proc.**1(014031)** 1-4.

IOP Conf. Series: Journal of Physics: Conf. Series **1032** (2018) 012010 doi:10.1088/1742-6596/1032/1/012010

[14] Koda T and Shionoya S 1964 Nature of the self-activated blue luminescence center in cubic ZnS: Cl Single Crystals *Physical Review* **136(2A)** 541-55.

[15] . Georgobiani N, Kotlyarevskii M S, Mikhalenko V N and Shvetsov Y V 1981 Nature of luminescence in zinc sulfide with intrinsic-defect hole conductivity *Journal of Applied Spectroscopy* **35(4)** 1097–100.

[16] Willander M, Nur, Sadaf J R, Qadir M I, Zaman S, Zainelabdin A, Bano N and Hussain I 2010 Luminescence from zinc oxide nanostructures and polymers and their hybrid devices. *Materials* **3(4)** 2643–67.

Choosing of relevant low height for operation wind turbine over urban Baghdad city

Monim H Al-Jiboori and Mina S Qassim[1,2]

Department of atmospheric science, College of science, Mustansiriyah university, Iraq
[1]Department of astronomy and air activities, Directorate general of scientific welfare, Ministry of youth and sport, Iraq

monim.atmsc@uomustansiriyah.edu.iq

Abstract. Two different sets of data used to achieve this work were measured by fast- and slow-response devices, which located on the roof of building followed to College of Science, Mustansiriyah university. Wind speed data classified to calculate their frequencies and wind rose. Wind direction prevailing is northeast and most frequencies are at less than 3 m/s. To improve these speeds, the internal boundary layer (IBL) height establishing over urban Baghdad city was estimated to satisfy the airflow characteristics before entering the city. The values of IBL height were calculated according to air stability classifications: unstable, neutral and stable. These values are lower at stable, graduate increase at neutral, and extreme increase at unstable conditions. However, the better performance of wind turbine operation over the surface of Baghdad city is suggested to be at the low level of 78 m.

1. Introduction

All non-(and developed) countries depend on the existence of energy either directly or indirectly. A feasibility study of any wind energy project should certainly involve a study of the spatial, temporal, and directional variations of wind speed. In order to optimize wind energy conversion systems and maximize the energy extraction frequency distributions of variations are required [1]. Knowledge of the wind characteristics and variability in the lower of the atmosphere is important to, for example, exploitation of wind energy, planning of tall buildings, and monitoring of dispersion of pollutant substances. For example, in the field of energy production, these characteristics where wind turbine installations are planned are critical. A comprehensive review of the main issues related to the assessment and forecasting of the wind and wind energy has been shown by Koracin et al. [2]. The aim of this paper focuses on analyzing the wind speed and wind direction distributions, as well as estimating the IBL depth, which can be considered as lower height for installing wind turbine in urban boundary layer of Baghdad city.

2. The theoretical frame

Figure 1 shows that a complicated flow system developing above such an inhomogeneous surface. The reason is that over each surface a wind profile is generated dependent on the surface roughness. Due to the horizontal wind field, the different profiles are shifted downwind forming a layer of discontinuity, which is called an internal boundary layer (IBL). Therefore, neighboring areas can also affect the exchange of energy and matter above a different downwind area. The IBL is caused by different surface roughness and different source areas of surface temperature, moisture, etc. Consequently, IBL can also be found for neutral, stable, and unstable stratifications. Thus, the study of IBL is very important in the use of wind power [3].

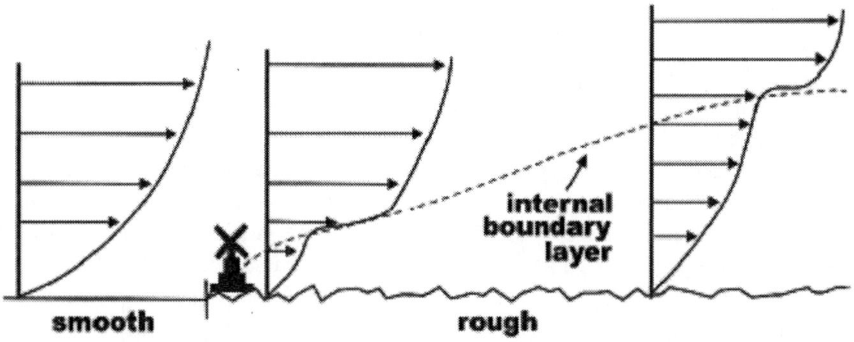

Figure 1. Growth of internal boundary layer height (H_{IBL}).

The distance (X) downwind from a change in surface features is called fetch. In other speech, it is the distance between the search area and the city influential limits in the wind flow. When making surface-layer measurements, it is desirable to locate the instrument mast far enough downwind of the border so that the depth of the IBL is greater than the height of the mast. Thus, with large fetch, the surface-layer measurements are necessary to observe characteristic of the local field in which the instruments are located. For this reason, many measurements are made within large uniform farm fields, meadows, land, or forests [3].

The height of the internal boundary layer, H_{IBL}, developing from rough-to-smooth transition can be estimated by the equation [4]:

$$\frac{H_{IBL}}{Z_0} = 0.3 \left(\frac{X}{Z_0}\right)^{\alpha} \quad (1)$$

where z_0: surface roughness length in (m). The exponent α is constant based on stability class: α~0.8 for neutral, ~0.6-0.7 for stable, and larger ~0.8-1.0 for unstable conditions [4].

Atmospheric stability has great importance in determining the wind flow and the logarithmic relationship of winds with altitude. It depends on both wind speed and lapse rate changes. Here the stability parameter was calculated by using local Obukhov length, Λ, follow as [5]:

$$\Lambda = -\frac{u_*^3 \bar{T}}{\kappa\, g\, \overline{w'T'}} \quad (2)$$

where $\overline{w'T'}$: sensible heat flux with w' (vertical fluctuation of wind $= w - \bar{w}$) and T' (The vertical temperature, $T' = T - \bar{T}$), $\bar{T}$: Average air temperature (in Kelivn), κ: von-kármán constant (~0.4), g: Accelerating gravity, u_*: Friction velocity ($u_* = \sqrt{|-\overline{u'w'}|}$) with $u' = u - \bar{u}$ is the longitude fluctuation of wind. The wind changes logarithmically with altitude, and the equation below is applied to describe the velocity of the wind at different altitudes [6]:

$$U(H_{IBL}) = U(H_{ref}) \left(\frac{H_{IBL}}{H_{ref}}\right)^{\beta} \quad (3)$$

where $U(H_{ref})$ is reference wind speed at height H_{ref}, and $U(H_{IBL})$ is the wind speed at height H_{IBL}. The power β is based on the experimental data and dependent on the surface roughness and air stability. The values of β are 0.15 for the unstable, 0.25 for neutral and 0.3 for the stable conditions [7]. The wind power density (WPD) available from moving is proportional to how much wind passes the wind turbine, i.e.,

$$WPD = \frac{1}{2}\, \rho\, A\, U^3 \quad (4)$$

where A is the airflow cross-section area and ρ is the air density (1.27 kg/m^3).

3. Site and data

IOP Conf. Series: Journal of Physics: Conf. Series **1032** (2018) 012011 doi:10.1088/1742-6596/1032/1/012011

In this work, two different sets of data were used: fast-response wind components (u, v, w) and temperature measured by 3D ultrasonic anemometer, and slow-response wind speed and its direction measured by the automatic weather station. These devices are located on the roof of building of atmospheric science department with 19 and 18 m height respectively above the ground level. For first set data, 30 days were available dispersed during six months of the year, 2016. They are sampled every second with frequency 1 Hz. The dates of these data were 4-5-6 Feb., 1-6-7-8-23 Mar., 8-9 May, 15-22-24-25 June, 5-7-8-9-10-13-14-15-22-23-27-28-30 July and 2-5-6 Aug. with total number 223 runs. The weather situations for these days and the certain observation hours (00-01, 05-06, 09-12, 15-18, 21-00) were clear sky. We took the sample length of a quarter of an hour.

Slow data were recorded for every hour of 2016, except for the last two months (Nov. and Dec.) because of no available and sometimes they observed for a quarter of an hour in some months. However, the total runs received are 15591, which we cannot get it from the first station (fast-response data). We used some programs such as Excel, Matlab and Origin version 9.3 to execute this work.

4. Results and discussion

4.1. Wind speed and it direction distributions

We could make a classification of the data taken from the automatic station by using the Matlab program to 10 intervals as shown in Table 1, and also used to know the average wind speed for each interval according to the classification to the initially identified. Although highest frequencies of wind are in the first three intervals, the wind speed with values higher than 3 m/s has good frequencies that can move turbine blades to produce electricity.

It can also see from table 1, most frequencies (about 75%) are for low values of wind velocity, while about 25% of high values of wind are available over the site. The wind power densities per unit area were also calculated from equation (4) for each mean wind velocity as reported in last column of table 1, in which the first three values of WPD are not convenient for generating electrical power, but they are accelerated up at increasing wind.

Table 1: The frequency of wind speed for ten intervals of wind speed.

U intervals (m/s)	$\bar{U}$ (m/s)	No. of runs	Percent freq. (%)	PWD (W/m^2)
0-0.9	0.6	4053	26.0	0.1
1-1.9	1.5	4159	26.6	2.1
2-2.9	2.4	3672	23.5	8.7
3-3.9	3.3	2270	14.6	22.6
4-4.9	4.2	1038	6.7	46.7
5-5.9	5.3	97	0.6	93.8
6-6.9	6.3	165	1.1	157.5
7-7.9	7.2	78	0.5	235.1
8-8.9	8.2	49	0.3	347.4
9-9.9	9.4	10	0.1	523.3
	Sum	15591	100.000	

To display information on the wind speed and its direction, they are combined in a diagram of a wind rose, which illustrates the distribution of wind in different directions and is useful in determining the prevailing trend of wind through the diagram and also mission in the design of wind farms and urban planning [8]. The chart was divided into 16 sectors. Wind rose defines the percentage of time that we get wind from a particular direction as well as repetition of wind direction. The wind velocity and its direction data taken from automatic weather station are plotted by Origin software to produce the wind rose as shown in Figure 2. The trend was calculated and found that the prevailing wind direction is northeast and east-north-east. The impact of high roughness elements surrounding the site has been changed the prevailing winds from northwest [9].

4.2. Wind power density at IBL height

To calculate fetch from the university site at eight directions according to the regions depending on the change of the region roughness including height of buildings and trees and density of buildings that work on the obstruction of the wind, it has been taken into account and by using Google earth map to surrounding the study site (see Figure 3), in which the numbers in (km) represents the fetches estimated from surface roughness change to the measurement site. Mean surface roughness length for Bab Al-Muadam area, ($\overline{z_0}$~1.2 m) was adapted as reported by Al-Jiboori and Al-Draji [10]. Before estimating IBL height, air stability was computed using equation (2) to determine the α value. Then the IBL height was calculated by equation (1). The results of height of IBL were found to be largest under unstable conditions as shown in table 2, while their values were lowest under stable conditions.

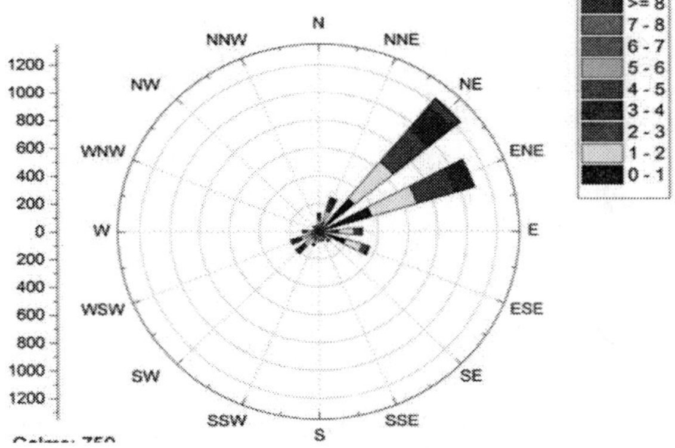

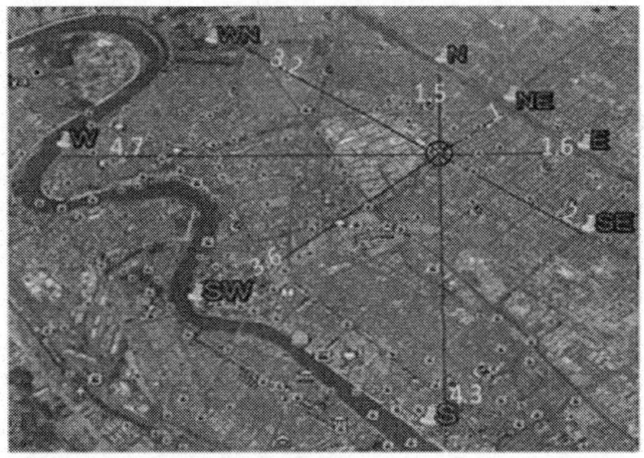

Figure 2. The wind rose diagram for study area.

Figure 3. Map of Baghdad city on which red lines and numbers are fetch values.

We estimated the wind velocity using equation (3) at IBL heights in different stability conditions (unstable, neutral and stable) and then found the average of wind speed for height as shown in table 2, where the initial height of the station (H_{ref}) is about 18 m and the second height (H_{IBL}) is estimated as illustrated in table 2, while the wind speed data (U_{ref}) obtained from the automatic weather station.

Table 2: The heights of IBL and wind speeds at them under different stabilities.

Dir. (°)	$\overline{U_{18}}$ (m/s)	Unstable class		Neutral class		Stable class		$\overline{U_{IBL}}$ (m/s)	WPD (W/m^2)
		H_{IBL} (m)	U (m/s)	H_{IBL} (m)	U (m/s)	H_{IBL} (m)	U (m/s)		
N	0.6	108	0.8	108	0.9	37	0.7	0.8	0.3
NE	1.5	153	2.1	**78**	2.16	28.4	1.7	2.0	5.0
E	2.4	223.7	3.5	113.8	3.8	38.6	3.0	3.4	24.8
SE	3.3	298	5.0	114.1	5.2	46	4.4	4.9	74.1
S	4.2	569	7.1	251	8.1	73.5	6.4	7.2	235.1
SW	5.3	484	8.7	217.7	9.9	65.5	7.8	8.8	429.3
W	6.3	1410	12.1	269	12.4	77.9	9.8	11.4	933.3
WN	7.2	436	11.6	198	13.1	68	10.7	11.8	1035.1

IOP Conf. Series: Journal of Physics: Conf. Series **1032** (2018) 012011 doi:10.1088/1742-6596/1032/1/012011

The average wind speed calculated at the IBL heights does not exceed 25 m/s, which cannot damage the wind turbine, which leads to a mall function or defect in the work of the turbine and therefore the speed is suitable for the installation of a wind turbine. The improvement of wind values at measurement site, surely, will be clear when calculating them at IBL height using equation (3) for all stability conditions as presented in table 2. For example, $\overline{U}$ =2.4 m/s for interval (2-2.9) m/s had been approximately 3.5 m/s for different stabilities. This means the wind velocity increased about 50 percent and could reach about 60 percent at upper remaining levels.

The improvement in WDP is also occurred in all mean wind velocities as shown in the last column of table 2, except the first two values. Also, it should notice that these values are better than those corresponding reported in table 1. The lowest IBL height that equals 78 m for neutral segment as average height is the best lowest for setting wind turbine for Mustansiriyah university. The neutral stratification can be considered as moderate state and also is not unstable nor stable stratification. It can be considered as transition state between in(-stability) classes.

5. Concluding remarks

Through data collection of wind speed and its direction for the year 2016 and its analysis in according to the different stability conditions that accompanied the data recording by using different devices mount on the roof of one of buildings belong to Mustansiriyah university. With changing roughness surface around the site, internal boundary layer height that is corresponding to estimate the minimum height for settling up the wind turbine over urban Baghdad city. The concluding remarks can be summarized as:

1. From wind speed frequencies, 75% of the wind speed (<3 m/s) are not encouraged to set up wind turbine due to the site is affected by tall buildings and trees.
2. Because high surface roughness, the prevailing wind direction was changed from north or northwest to be northeast.
3. To eliminate surface roughness effect, internal boundary layer depth was estimated, thus the suggestion height for best operation of wind turbine is at neutral condition (as an average) to generate wind energy at the university site is about 78 m.

References

[1] Shaw S 2012 Development and validation of the wind energy calculator for use as a module in the larger complimentary energy decision support tool project, M Sc thesis, University of Guelph, Canada, 146

[2] Koracin D, Belu R, Horvath K, Candillas B and Jiang J 2014 A review of challenges in assessment and forecasting of wind, energy resources, Croatian Meteor J **47** 13–33

[3] WMO 1981 *Meteorological aspects of the utilization of wind as an energy source* Technical Note **175** 180

[4] Foken T and Nappo C 2008 *Micrometeorology* (Springer-Verlag Berlin Heidelberg) p 360

[5] Stull R 2003 *An introduction to boundary layer meteorology* (Springer Science& Business Media) **13** p 669

[6] Al-Jiboori M 2010 Determining of neutral and unstable wind profile over Baghdad city, Iraqi J of Sci **51** 343-50

[7] Hanna S, Briggs G and Hosker R 1982 *Atmospheric diffusion*, Handbook no. NTIS DE81009890 (DOE/TIC-22800), Springfield, VA.

[8] Brower M 2012 *Wind resource assessment: A practical guide to developing a wind project* (John Wiley & Sons Inc.) p 296

[9] Manweel J, McGowan J and Rogres A, 2011 *Wind energy explained-theory, design and application* (John Wiley & Sons) p 689

[10] Al-Jiboori M and Al-Draji A 2010 Aerodynamic surface roughness length of Baghdad city, J of Al-Nahrain university **13** 96-102

Determination of Radioactivity Levels, Hazard, Cancer Risk and Radon Concentrations of Water and Sediment Samples in Al-Husseiniya River (Karbala, Iraq)

I. T. AL-Alawy[1,*] R. S. Mohammed[2] H. R. Fadhil[3] A. A. Hasan[4]

[1, 2] Al-Mustansiriyah University, Faculty of Science, Physics department, Baghdad – Iraq
[3] Directorate of liquidation of nuclear facilities and destroyed sites, Ministry of Science and Technology, Baghdad – Iraq
[4] Ministry of Education, Karbala - Iraq
[*]E-mail: profimantarik@gmail.com
drimantarik@uomustansiriyah.edu.iq
http://orcid.org/0000-0002-7692-0861

Abstract. Natural radioactivity U-235, Th-232, and K-40 of eight samples of water and sediments from Al-Husseiniya River have been investigated by using gamma spectrometer with High-Purity Germanium HPGe detector. Radon concentrations have been measured for these samples by using CR-39 detector. The average activity concentrations were 15.8±2.421 for U-238, 11.2±2.385 for Th-232, and 311±21.826 Bq/kg for K-40 in sediment samples; and 1.9±0.557 for U-238, 1.23±0.396 for Th-232, and 10.1±1.437Bq/L for K-40 in water samples respectively. These activities are compared with the worldwide limit. The average radium equivalent activity Ra_{eq} was 55.959±8.559Bq/kg in sediment. The average value of hazard index H was 0.344 in sediment samples. The average absorbed dose rate D_{out} was 27.511±4.095nGy/h and D_{in} was 51.997±7.639 nGy/h in sediment samples respectively. The average of total annual effective dose equivalent $AEDE$ for sediment was 288.815 $\mu Sv/y$. While the total of $ELCR$ was 1.011×10^{-3}. They are less than worldwide limits. Radon concentrations have been measured for the same samples. CR-39 detector was used for radon measurements. The average value in sediment was 6.00±0.759 Bq/kg, and 2.40±0.270 Bq/L in water samples. While the radium concentrations in sediment was 0.247±0.031Bq/kg. Both are less than the worldwide limit. The average value of E_A and C_d in water was found to be 1.783±0.202$mBq/m^2.h$ and 46.890±255.245Bq/L respectively. While the surface and mass exhalation rates of sediment sample was 35.283±4.458$mBq/m^2.h$ and 7.091±0.896$mBq/kg.h$. The total annual effective dose AED in water samples was 14.762 $\mu Sv/y$which are less than the worldwide value 1mSv/y. Therefore, water of Al-Husseiniya River is safe with regards to the concentration of radon and natural radioactivity.
Keywords: Natural radioactivity; Radon concentrations; Water and sediment; Cancer risk; HPGe and CR-39 detectors; Iraq

1. Introduction

Al-Husseiniya River is a branch of the Euphrates River located north-east of Karbala Governorate/Iraq. It is about 5km length and is one of the most important rivers that feed the orchards and used for drinking. To our knowledge, there seems to be no information about radioactivity level in Al-Husseiniya River water and sediments so far. For this reason, water and sediment samples were collected along Al-Husseiniya River. Al-Husseiniya River padding by sediments was taken from the sites of a military industrial facility belonging to the city of Karbala that may be contaminated with radioactive elements. Human exposure to natural radiation, from background radiation is responsible for large-scale population exposure to environmental radioactivity, and that it is present in different geological formations such as soil, rocks and plants, sand, water and air [1]. Humans also exposed to contamination of the food chain caused by direct deposition of radionuclides into plant leaves and root, absorption of contaminated soil, sediments, water, and direct ingestion of contaminated water [2]. The radiation dose received by human being from natural radiation sources is about 70% [3]. The presence of radioactive elements in water leads to the spread of contamination on the earth. The sediments were accumulated in river riverbeds, mostly on the bottom. These sediments include many natural radioactive isotopes, especially from the uranium chain, and often their concentration is higher than the average values in this area [4]. Long-term exposure to uranium, thorium and radium through inhalation has several health effects such as lung disease, anemia, mouth necrosis, nasal tumors, and leukemia, pancreas, and kidney cancers. Therefore, the distribution of radioactivity found in natural materials enables us to assess any possible radiation risk to humanity [1]. The worldwide average human exposure to natural sources is $2.4 mSv/y$, and the weight of water and food is about $0.3 mSv/y$ [3, 5]. The characteristic of Solid State Nuclear Detectors (SSND's) has made them an excellent tool for making inexpensive radiation measurements at the site. Plastic track detectors CR – 39 are more suitable for measuring alpha emitters in the environmental samples due to their high simplicity, sensitivity, rapidity and cheapen. A can technique is found to be the most appropriate and the most widely used technique because of its simplicity as well as being economical [6].

2. Materials and methods

2.1. Preparation of samples

The present study area of the Al-Husseiniya River in Karbala covered a total length of 5km, from which 8 consecutive sites (five sediment samples from different locations at each site as well as water), and thus 40 samples were collected. The samples were collected (60cm in depth) between May and June 2017. Sediment samples were placed in plastic bags, and marked with label. The samples were brushed and dried for 5days under the sun until all moisture was completely lost. The dried samples are grinded into a fine powder using a stainless steel ball mill and sieved through 1mm mesh. Each sample of soil grain with weights about 1000g. As well, water samples were filtered using filter paper to get rid of the suspended minutes with samples. Each sample was placed in plastic containers used to collect water samples with a capacity of 1 liter after washing them with HCl. Then tightly sealed and labeled. The studied area from which the samples were collected was shown in **Figure 1**.

Figure 1. Area of the study.

2.2. Calibration

2.2.1. Gamma-ray spectrometry. The Gamma-ray spectrometer used in this work is a High-Purity Germanium (HPGe) detector system, a coaxial cylindrical detector, manufactured by CANBERRA, its active volume is about 80 cm^3 and the size of the crystal is 3x3 inch with 65% relative efficiency. The shield, 12cm thick, was made of lead (99.9% purity) and the internal surface was degraded in atomic number with copper, aluminum and polyethylene. The energy, efficiency, resolution, detection limit and background measurement of high purity Germanium (HPGe) detector gamma spectrometry system have been done with standard source of known activity. **Figures 2 and 3** show the energy and efficiency calibration curves of (HPGe) detector using a mixed standard source with different energy ranging from 59.5 keV for 241Am to 1836.06 keV for 88Y.

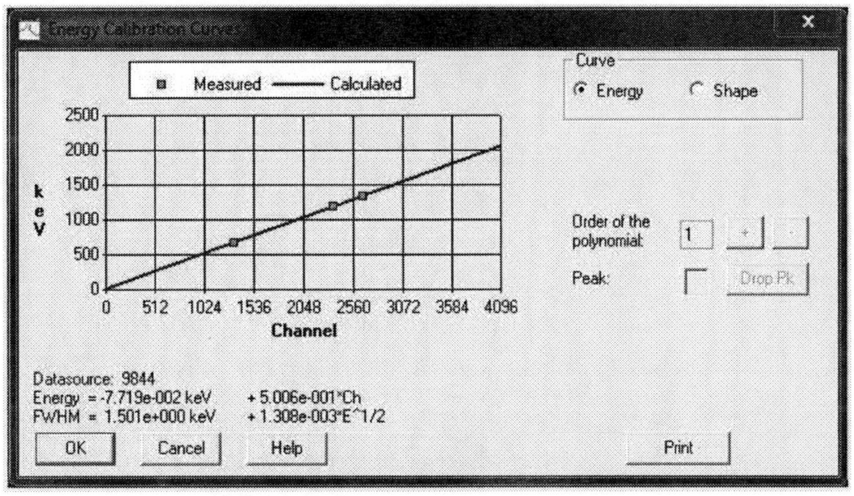

Figure 2. Energy calibration curve of HPGe detector.

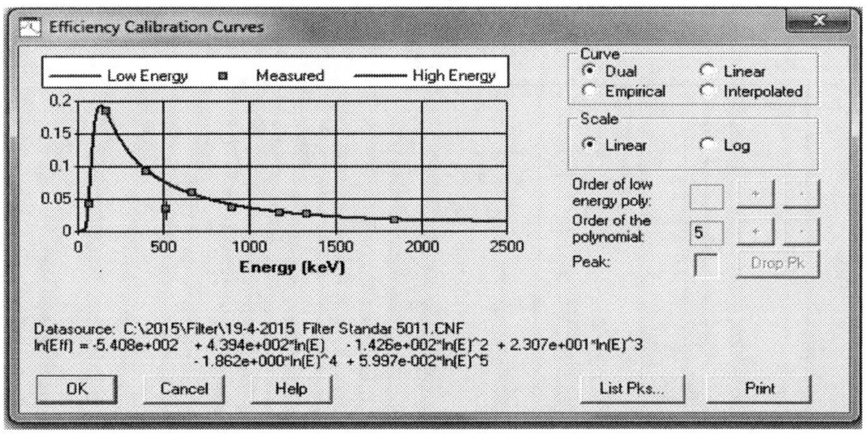

Figure 3. Efficiency calibration curve of HPGe detector.

2.2.2. CR-39 Detector. Radon concentrations in sediments and water have been determined by using the Can technique. A known amount of sample was placed at the bottom of the can which was then closed for one month in order to reach equilibrium between radon and radium members of the decay series. After one month, the CR-39 nuclear track detector, UK issued, of 500μm thickness and density of 1.36*gm/cm*^{3}, was cut into small pieces of *1cm×1cm* area. Then, they were fixed inside the cover of the can, as shown in **Figure 4.** After two months of exposure, detectors in aqueous solution 6.25*N* of sodium hydroxide (NaOH) were etched at 60°*C* for 5hr. The detectors were rinsed with distilled water and dried in the air. The track density was recorded using an optical microscope with 400X. The density of the tracks (ρ) was obtained in the samples according to the following relationship [7]:

$$Track\ density\ (\rho) = \frac{average\ number\ of\ total\ tracks\ (N_{ave})}{area\ of\ field\ view\ (A)} \tag{1}$$

Radon concentration was measured using the relationship [8]:

$$\frac{C_X}{\rho_X} = \frac{C_S}{\rho_S} \tag{2}$$

Where C_S , C_X: are radon exposure ($Bq.m^{-3}$) for standard and sample respectively. ρ_S ,ρ_X are track density, in($Track.mm^{-2}$), for standard and sample respectively.

and

$$C_X = C_S \frac{\rho_X}{\rho_S} \tag{3}$$

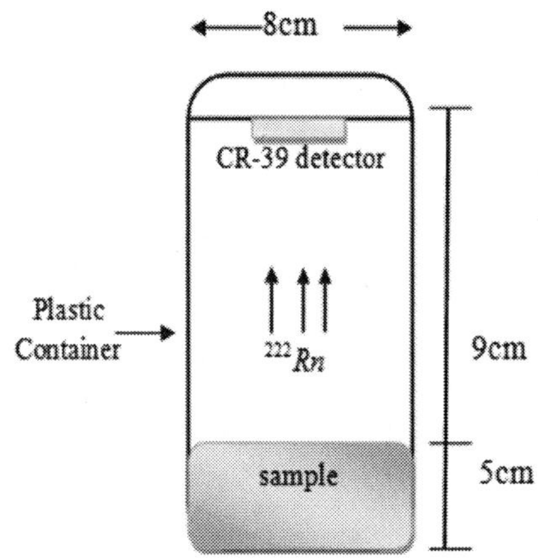

Figure 4. Schematic diagram of the sealed-can technique used for sediments and water samples.

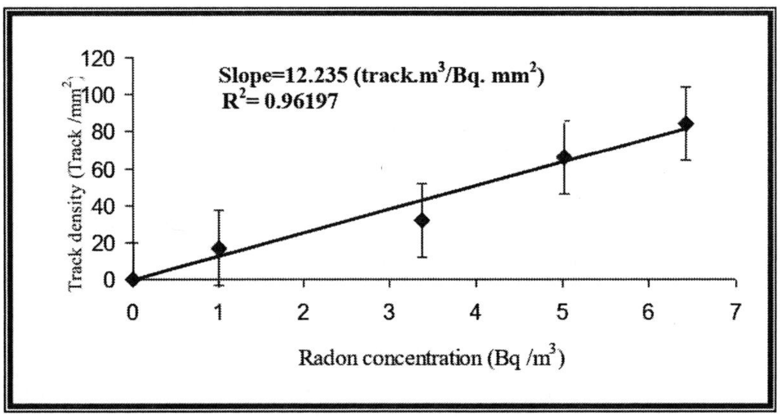

Figure 5. Relation between radon gas concentration and track density in soil standard samples.

IOP Conf. Series: Journal of Physics: Conf. Series **1032** (2018) 012012 doi:10.1088/1742-6596/1032/1/012012

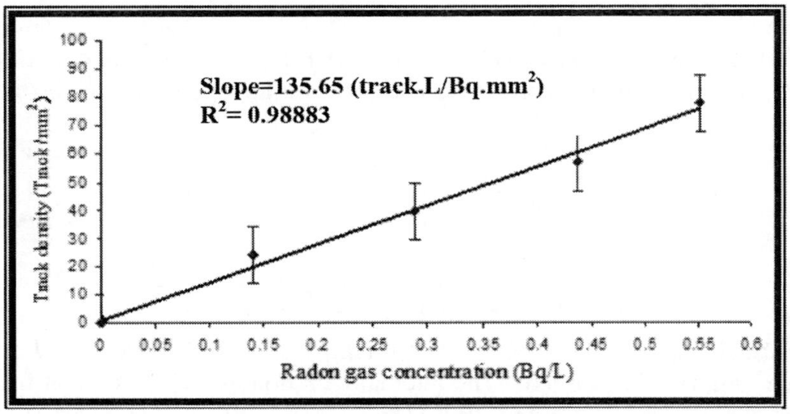

Figure 6. Relation between radon gas concentration and track density in water standard samples.

3. Parameters for sediments samples

3.1. Absorbed dose rate D

To convert the activity concentration of ^{238}U, ^{232}Th, and ^{40}K radionuclides into doses in $(nGyh^{-1}\ per\ Bqkg^{-1})$, UNSCEAR (1988) [9] has given the dose conversion factors as 0.427, 0.662 and 0.043 respectively. Using these factors, the **outdoor absorbed dose rate** is calculated using the following equation:

$$D_{out}(nGyh^{-1}) = 0.427\ A_U + 0.662A_{Th} + 0.043\ A_K \quad (4)$$

Where A_U, A_{Th} and $A_K$are the activity concentrations in $(Bqkg^{-1})$ of uranium, thorium and potassium respectively. The indoor gamma ray dose imparted by of ^{238}U, ^{232}Th, and ^{40}K radionuclides present indoor can be calculated by converting the absorbed dose rate to effective dose by using the conversion factors 0.92, 1.1 and 0.081 $nGyh^{-1}\ per\ Bqkg^{-1}$ respectively. By using these factors following, the following equation is used to calculate the **indoor dose rate,** given by UC European Commission (1999) [10]:

$$D_{in}(nGyh^{-1}) = 0.92\ A_U + 1.1A_{Th} + 0.081\ A_K \quad (5)$$

3.2. Radium equivalent activity Ra_{eq}

The absorbed activity dose is not provided as an exact indication of radiation hazard associated with sediments or any other materials because the concentration of U-238, Th-232 and K-40 are not uniformly distributed in sediments and most of environmental materials throughout the world. So that, for uniformity with respect to the exposure to radiation UNSCEAR (2000) [11] has defined the **radium equivalent activity** expressed by the following equation:

$$Ra_{eq}(Bq.kg^{-1}) = A_U + 1.43\ A_{Th} + 0.077\ A_K \quad (6)$$

Where A_U, A_{Th} and A_K are the activity concentration of U-238, Th-232 and K-40 in $(Bqkg^{-1})$ respectively. 1, 1.43 and 0.077 are the activity conversion rates of uranium, thorium and potassium respectively which outcome in same gamma dose rate at maximum permissible Ra_{eq} of 370 $Bqkg^{-1}$. Ra_{eq} is measured in $Bqkg^{-1}$ and its definition is based on the assumption that specific activity of 370 $Bqkg^{-1}$ for Ra-226 uniformly distributed in any environmental sample can perform in the annual effective dose of 1mSv at a height of 1 m above the ground level [12].

3.3. Hazard index H

Other radiological hazards are external (H_{ex}) and internal (H_{in}) hazard indices. H_{ex} is the radiation hazard due to external exposure to gamma ray. The **external hazard index** can be calculated from the following equation [11]:

$$H_{ex} = \frac{Ra_{eq}}{370\ Bq.kg^{-1}} \tag{7}$$

OR:

$$H_{ex} = \frac{A_U}{370\ Bq.kg^{-1}} + \frac{A_{Th}}{259\ Bq.kg^{-1}} + \frac{A_K}{4810\ Bq.kg^{-1}} \tag{8}$$

It is assumed that 370 $Bq.kg^{-1}$ of Ra-226, 259 $Bq.kg^{-1}$of Th-232, and 4810 $Bq.kg^{-1}$ of K-40, produce the same gamma dose rate [13]. The internal radiation exposure is quantified by the **internal hazard index** (H_{in}) given by UNSCEAR (2000) [11]:

$$H_{in} = \frac{A_U}{185\ Bq.kg^{-1}} + \frac{A_{Th}}{259\ Bq.kg^{-1}} + \frac{A_K}{4810\ Bq.kg^{-1}} \tag{9}$$

The internal exposure to radon Rn-222 and its decay products is controlled by internal hazard index (H_{in}) and for safe use UNSCEAR (2000) [11] provided that the value of the above indexes must be less than unity for the radiation hazard to be regarded as insignificant.

3.4. Annual effective dose equivalent (AEDE)

The **outdoor annual effective dose equivalent** ($AEDE_{out}$) was estimated to convert the outdoor absorbed dose in air to effective dose. While the **indoor annual effective dose equivalent** ($AEDE_{in}$) is estimated from indoor absorbed dose in air to convert it to the effective dose. UNSCEAR (2000) [11] reported the value 0.7 $SvGy^{-1}$ as conversion coefficient from absorbed dose in the air to the effective dose received by adults. While 0.2 and 0.8 represent the outdoor and indoor occupancy factors respectively. The annual effective dose equivalent can be calculated from the following equations as reported by UNSCEAR (2000) [11]:

$$AEDE_{out}(\mu Sv/y) = D_{out}(nGy/h) \times 8760(h/y) \times 0.20 \times 0.7(Sv/Gy) \times 10^{-3} \tag{10}$$

$$AEDE_{in}(\mu Sv/y) = D_{in}(nGy/h) \times 8760(h/y) \times 0.80 \times 0.7(Sv/Gy) \times 10^{-3} \tag{11}$$

3.5. Annual gonadal dose equivalent (AGDE)

The United Nations Scientific Committee on the Effects of Atomic Radiation UNSCEAR (1988) [9] has been interested active bone marrow and bone surface cells as organs. Therefore, the **annual gonadal dose equivalent** ($AGDE$) due to the specific activities of U-238, Th-232 and K-40 and conversion factors 3.09, 4.18 and 0.3144 $\mu Sv/y\ per\ Bqkg^{-1}$ respectively, was calculated using the following equation [14]:

$$AGDE(\mu Sv/y) = 3.09A_U + 4.18A_{Th} + 0.314A_K \tag{12}$$

3.6. Life-time cancer risk (ELCR)

The **excess life-time cancer risk** ($ELCR$) was estimated from annual effective dose equivalent using the equation [15]:

$$ELCR_{out} = AEDE_{out} \times DL \times RF \tag{13}$$

$$ELCR_{in} = AEDE_{in} \times DL \times RF \tag{14}$$

Where DL, and RF are the duration of life (70 years), and risk factor (0.05/Sv), respectively. Defined the risk factor as fatal cancer risk per Sievert is assigned a value of 0.05 by ICRP (2012) [16] for the public for random effects, for low-level radiations.

4. Parameters for water samples

4.1. Annual effective dose (AED)

The **annual effective dose** AED_{ing} due to the ingestion of radon was calculated for the individual consumer due to **intake of radon** from the consumption of drinking water in unit $(\mu Sv/y)$ according to the following equation [17,18]:

$$AED_{ing}(\mu Sv/y) = CR_n \, CR_w \, D_{cw} \tag{15}$$

Where CR_n is the concentration of radon in the water in (Bq/L) units, CR_w is the annual intake of drinking and is equal to 730 L/y for adults, and D_{cw} is the ingested dose conversion factor and is equal to 5×10^{-9} Sv/Bq [11].

The **annual effective dose** AED_{inh} due to the **inhalation of radon**, resulting from the radon concentration in indigenous water, was calculated according to the following expression UNSCEAR (2000) [11]:

$$AED_{inh}\,(\mu Sv/y) = C_{Rn-w} \; R_{a-w} \; EqF \; ExT \; 9nSv\,(kBq\,h\,L^{-1})^{-1} \tag{16}$$

Where C_{Rn-w} is the average radon concentration in water, in $kBq.L^{-1}$, R_{a-w} is the air–water concentration ratio (10^{-4}), EqF is the equilibrium factor between indoor radon and its short lived progeny (0.4), ExT is the exposure time to this concentration, in hours (assumed to be 7000 h per year), and $9nSv\,(kBq.h.L^{-1})^{-1}$ is the dose conversion factor.

4.2. Radium concentration C_{Ra}

The **radium concentration** of the sample (sediment) was calculated using the following relation [19]:

$$C_{Ra} = \frac{\rho h A}{K T_e M} \tag{17}$$

Where C_{Ra} is the effective radium content of sample (Bq/kg), ρ is the track density $(track/mm^2)$, M is the mass of the sample (kg), h is the distance between the detector and the top of the soil sample, A is the surface area from which radon is exhaled (m^2), t is the exposure time and K is the calibration constant which was determined to be: $K = 8.497 \times 10^{-3} track/mm^2$ h per Bq/m^3, and T_e is the effective exposure time, which is related to the actual exposure time t, by the following relation [20]:

$$T_e = t - 1/\lambda(1 - e^{-\lambda t}) \tag{18}$$

5. Determination of some radon parameters for water samples

5.1. Radon exhalation rate E

The **surface radon exhalation rate** E_A of the sample is defined as the flux of radon released from its surface. Since Rn-222 is one of the most important isotopes in the U-238 series and it is a short lived progenies, it can be regarded as a major indicator to lung dose. The radon exhalation rate in terms of area (surface exhalation rate) in units of $(Bq.m^{-2}.h^{-1})$ can be calculated according to the following equation [21,22,23]:

$$E_A = \frac{CV\lambda}{A[T+1/\lambda(e^{-\lambda T}-1)]} \tag{19}$$

Moreover, the **radon exhalation rate in terms of mass** E_M $(Bq.kg^{-1}.h^{-1})$ is determined by the following equation [22]:

$$E_M = \frac{CV\lambda}{M[T+1/\lambda(e^{-\lambda T}-1)]} \tag{20}$$

Where C is the integrated radon exposure as measured by CR-39 $(Bq.h/m^3)$ for sediment or $(Bq.h/L)$ for water, V is the volume of hollow air in cup (m^3) or (Liter), λ is the radon decay constant (h^{-1}),

A is the surface area of the sample from which radon is exhaled (m^2), M is the mass of the sample (kg), and T is the exposure time(h).

5.2. Dissolved radon concentration in water C_d

The **dissolved radon concentration C_d** in water was calculated in the unit (Bq/L) according to the following equation [22]:

$$C_d\ (Bq/L) = C_{Rn}\ \lambda\ h\ T\ /\ L \quad (21)$$

Where C_{Rn} is the concentration of radon in the air above the water sample (Bq/L), λ is the decay constant of radon gas and is equal to 0.181452day^{-1}, h is the high of detector from the surface of the water (9cm), T is the exposure time of the detector to water sample (60days), L is the height of the water inside the can (5cm).

The concentration of radon in the can air above the water samples was determined by measuring the tracks density on the detector according to the following relation [22]:

$$C_{Rn} = \frac{\rho}{KT} \quad (22)$$

Where C_{Rn} is the activity concentration of radon in the ingested water (Bq/L), λ is the decay constant of ^{222}Rn (h^{-1}), h is the distance from the surface of water sample in cup to the detector (m), T is the exposure time of the sample (h) and L is the depth of the sample (m).

6. Results and discussion

Gamma spectroscopy was used to determine the activities of U-238, Th-232, and K-40. The activity concentrations of Th-232 and U-238 were calculated assuming secular equilibrium with their decay products. The gamma ray transitions of energies 186.3 keV (Ra-226), 609.3 keV (Bi-214) and 351.9 keV (Pb-214) were used to determine the concentration of the U-238 series. The gamma-ray lines at 911.0 keV (Ac-228) and 583.3 keV (Tl-208) were used to determine the concentration of the Th-232 series. The 1460 keV gamma-ray transition was used to determine the concentration of K-40. Cs-137 does not exist in water and sediment naturally because it is producer of radioactivity repercussions.

Table 1 shows the activity concentrations of main gamma emitting radionuclides of U series, Th series, and K-40 in the water and sediment samples. The average concentrations of U-238, Th-232, and K-40 were found to be 15.8±2.421, 11.2±2.385, and 311.0±21.826 Bq/kg in sediment samples, respectively. The average concentrations of U-238, Th-232, and K-40 were found to be 1.9±0.557, 1.23±0.396 and 10.08±1.437 Bq/L in water samples, respectively.

The worldwide concentrations of the radionuclides U-238, Th-232, and K-40 have averages in sediment samples of 25, 25, and 373 Bq/kg respectively (UNSCEAR, 1988) **[9]**. Our results show that the mean activity concentrations of these radionuclides are lower than the worldwide concentrations according to the UNSCEAR report [24,25]. See **Figure 7** and **8**.

From these results, we can see that the concentration of U-238, Th-232 and K-40 in sediment samples was higher than that in water samples by 87.98%, 89.02% and 96.76% respectively. This is because the rapid movement of water and its constant change lead to the deposition of radionuclides, so their presence in the water is very low. **Table 2** shows the comparison of specific activity of U-238, Th-232, and K-40 in sediment and water samples with other references.

A comparison of radiological hazard indices Ra_{eq} , H_{ex} , H_{in} , D_{out} , D_{in} , $AEDE_{out}$, $AEDE_{in}$, $AGDE$, $ELCR_{out}$, $ELCR_{in}$, and total $ELCR$ is given in **Table 3.**

Table 3 presented the measured radium equivalent activity Ra_{eq} in sediment samples which were ranged from 43.804 to 62.847Bq/kg with an average value 55.959±8.559Bq/kg. The results are lower than the recommended limit 370 Bq/kg reported by UNSCEAR (1988, 2000) [9, 11], see **Figure 9**.

The calculated external hazard index H_{ex} ranged from 0.118 to 0.170 in sediment samples with an average value 0.151±0.023. While the calculated internal hazard index H_{in} ranged from 0.152 to 0.218

with an average value 0.193±0.029, which is lower than the worldwide limit ≤1 recommended by UNSCEAR (2000) [11]. These results are indicated in **Table 3** and **Figure 10.**

Table 1. Average activity concentration of U-238, Th-232, and K-40 in sediment and water samples in Al-Husseiniya River.

Sediment Samples (Bq/kg)				Water Samples (Bq/L)			
Sample Code*	U-238	Th-232	K-40	Sample Code*	U-238	Th-232	K-40
S-1	17.1	10.2	319.4	W-1	1.3	0.95	10.6
S-2	12.5	8.3	252.4	W-2	2.4	1.17	6.7
S-3	18.0	13.0	341.0	W-3	——	0.98	12.4
S-4	15.4	13.3	331.0	W-4	2.0	1.80	10.6
Min.	12.5	8.3	252.4	Min.	1.3	0.95	6.7
Max.	18.0	13.3	341.0	Max.	2.4	1.80	12.4
Ave.	15.8	11.2	311.0	Ave.	1.9	1.23	10.08
SD	±2.421	±2.385	±21.826	SD	±0.557	±0.396	±1.437
Worldwide [9]	25	25	370				

* Each sample code is an average of five samples collected from different locations at the same site.

Table 2. Activity concentration of U-238, Th-232 and K-40 in sediment and water samples.

Sample	Radio-Nuclide	Present Work	Ref. Range
Sediment (Bq/kg)	U-238	15.8±2.421	18.220±1.404 Tigris River in AL-Amara city [26] 16-113 Turkey (Fırtına River) [27] 2.00±0.077 mg/kg to 2.71±0.076 mg/kg Sediments in North central part of Nigeria [28] 4100 Beaches of Brazil [29]
	Th-232	11.2±2.385	13.792±1.302 Tigris River in AL-Amara city [26] 17-87 Turkey (Fırtına River) [27] 4.38±0.037mg/kg to 10.08±0.15mg/kg Sediments in North central part of Nigeria [28] 57000 Beaches of Brazil [29]
	K-40	311.0±21.826	317.343±16.997 Tigris River in AL-Amara city [26] 51-1605 Turkey (Fırtına River) [27] 0.0041±0.024% to 3.31±0.025% Sediments in north central part of Nigeria [28] 3000 Beaches of Brazil [29] 142±2.16 Madloom Region in Al-Najaf Al-Ashraf [30]
Water (Bq/L)	U-238	1.9±0.557	B.D.L. Tigris River in AL-Amara city [26] 0.0182 ± 0.0015 water at Johor bahru, Malaysia [31]
	Th-232	1.23±0.396	B.D.L. Tigris River in AL-Amara city [26] 0.0389 ± 0.0034 water at Johor bahru, Malaysia [31]
	K-40	10.08±1.437	6.818±1.999 Tigris River in AL-Amara city [26] 1.22 ± 0.05 water at Johor bahru, Malaysia [31]

Table 3 shows the calculated outdoor absorbed dose D_{out} due to the presence of U-238,Th-232 and K-40 in the sediments of Euphrates River that ranged from 21.685 to 30.955nGy/h with an average value of 27.511±4.095nGy/h which is lower than the worldwide limit of 59 nGy/h as presented by UNSCEAR (2000) [11]. The values of indoor absorbed dose D_{in} calculated during present study range from 41.074 to 58.481 nGy/h with an average of 51.997±7.639 nGy/h , which is lower than the worldwide limit of 84 nGy/h as recommended by UNSCEAR (2000) [11]. The total average absorbed dose 99.508 nGy/h is greater by 1.687 times of the outdoor limit and greater by 1.185 times of the indoor limit. See **Figure 11**.

The values of outdoor annual effective dose equivalent $AEDE_{out}$ for Euphrates River sediments ranges from 26.595 to 37.963 $\mu Sv/y$ with an average of 33.739±5.021 $\mu Sv/y$ which is lower than the worldwide average of 70 $\mu Sv/y$. The $AEDE_{in}$ calculated with a range from 201.493 to 286.884 $\mu Sv/y$ with an average value 255.076±37.476 $\mu Sv/y$ which is less than the worldwide limit of 450 $\mu Sv/y$ reported by UNSCEAR (2000) [11]. The total average annual effective dose equivalent was estimated to be 288.815 $\mu Sv/y$ which is 1.80 times lower than the worldwide limit of 520 $\mu Sv/y$ as predicted by UNSCEAR (2000) [11]. Spatial distribution of total annual effective dose equivalent in studied area is shown in **Figure12.** From **Table 3** it can be seen that the estimated values of annual effective dose equivalent are relatively on higher side.
The obtained values of $AGDE$ are listed in **Table 3**. The values of $AGDE$ varied from 152.573to 217.034$\mu Sv/y$ and the average value was found to be 193.122±28.395$\mu Sv/y$. The average value of AGDE was found to be 550.5 mSv/y in Rize, Turkey [32]. Also, the average value of $AGDE$ was found to be 2850 mSv/y in Kerala, India [33]. These two values of $AGDE$ are higher than our result. See **Figure 13.**
The excess lifetime cancer risk $ELCR$ for outdoor exposure calculated for sediment, given in **Table 3** varied between 0.093×10^{-3} and 0.133×10^{-3} with average value of 0.118×10^{-3} and standard deviation of 0.018×10^{-3}. This value was found to be less than the limit of 0.29×10^{-3} set by UNSCEAR (2000) [11]. For indoor exposure it is varied from 0.705×10^{-3} to 1.004×10^{-3} with an average of 0.893×10^{-3}. It is less than the limit 1.16×10^{-3} [11]. The total $ELCR$ ranges from 0.798×10^{-3} to 1.137×10^{-3} with an average value of 1.011×10^{-3}. The total $ELCR$ is 1.434 times less than the worldwide limit of 1.45×10^{-3}. The behavior of $ELCR$ in samples is shown in **Figure 14.**
The average value of $ELCR$ was found to be 3.21×10^{-3} in the sediment of Hunza, Gilgit and Indus Rivers from Northern Pakistan [34] which is higher than the world permissible value of 0.29×10^{-3} given by (UNSCEAR, 2000), while the calculated average value was 0.202×10^{-3} for surface sediment samples of Ponnaiyar River in India [35]. This average value of ELCR is less than the worldwide limit given by (UNSCEAR, 2000). Also the ELCR in sediment samples in the Eastern Black Sea, Turkey is 0.19×10^{-3} [36]. These last two values are lower than the world permissible value of 0.29×10^{-3}. However, the average ELCR value of present sediments is lower than the values obtained by Qureshi et al. (2014) [34] for sediment samples of Pakistan and higher than the average value obtained by Ramasamy et al. (2011) [35] and that obtained by Baltasa et al. (2018) [36] for sediment samples of India and Turkey respectively.
As is known the natural river water contains dissolved radon from the uranium series present in soil and rocks. Dissolved radon is easily released into the air when the water is used for showering, cleaning and other everyday purposes in homes. However, breathing radon released to air from water increases the risk of lung cancer. Some radon stays in the water; drinking water containing radon also presents a risk of developing internal organ cancers [37].
To be safe, several international health organizations determined the acceptable limit for radon concentrations in water. The USEPA (1991) has proposed that the allowed maximum contamination level (MCL) for radon concentration in water is 11 Bq/L [38], WHO (1993) defined the value 11.1 Bq/L [39], USEPA (1999) defined a value of 11.1 Bq/L for radon concentration in water [40], EPA (2000) defined a value of 10.0 Bq/L [41], UNSCEAR (2011) defined a value of 40 Bq/L [42], and the WHO (2012) defined a value of 100 Bq/L as an action limit [43]. While UNSCEAR (2012) [24] has proposed that the permissible limit of radon concentration in sediments is 30Bq/kg.
The purpose of this study was to investigate the radon concentrations in sediments and water of Euphrates River used for constructions, irrigation or drinking in some areas of Karbala governorate. Samples of sediments and water collected were examined to determine Rn-222 concentrations.
Table 4 presents the radon concentrations in sediments samples. The average value in sediment samples was 6±0.759 Bq/kg. This result showed that the radon concentration in sediment samples was below the permissible limit of radon exposure 30 Bq/kg, according to the UNSCEAR report [11], see **Figure 15**. The Current results in **Table 4** show that the average value of radon concentration in water samples was 2.40±0.270 Bq/L, see **Figure 16** This value is lower than the maximum worldwide

contaminant level 11.1 Bq/L set by WHO (1993) [38] and USEPA (1999) [40]. The total average radon concentration in sediment is greater 2.5 times that in water. It is clear that there is no significant difference in the concentration of radon in the waters of the Al-Husseiniya River, although this river carries mud, soil and various types of elements due to its passage in a military area. This should increase the probability of containing a large amount of radium element to appear in increasing the concentration of radon in the samples. The radon concentrations vary from 78.6±21.3 to 606.8±157.3 Bq/kg in the marine and coastal sediments of Khor-Abdullaha south of Iraq, northern west of the Arabian Gulf [44]. While the average radon concentration of Vakilabad river water in Mashhad City was 9.917±0.004Bq/L [45]. Also Radon concentration is 6.120(Bq/L) in tap water, 15.775(Bq/L) in drill well water, and 11.790(Bq/L) in ground water in Al-Hindiyah city of Karbala Governorate, Iraq [46].
The results of annual effective dose obtained are shown in **Table 4**. The estimated annual effective dose due to ingestion AED_{ing} ranged from 7.556 to 9.965 μSv/y in water samples. The average of total annual effective dose obtained in this study is 8.733±0.985 μSv/y. It was evident that the total annual effective dose resulting from radon in intake water are significantly lower than the permissible limit of 1 mSv/y provided by International Atomic Energy Agency, IAEA [47], Environmental Protection Agency (EPA, 2000) [41], United Nations Scientific Committee on the Effects of Radiation (UNSCEAR, 2012) [24], (ICRP, 1993) [48] and World Health Organization (WHO, 2012) [43]. The International Commission on Radiological Protection ICRP (1993) has recommended that the annual effective dose limit should be 1 mSv/y for each member of the public and 20 mSv/y for radiation workers. These dosage limits have been established on the utilization pattern approach assuming no threshold dose below which there would be no effect [48]. Thus, the water in Al-Husseiniya River is safe as far as radon concentration is concerned. Differences in the values of radon concentrations were observed in water samples. These differences can arise due to differences in the nature and content of the nuclei in these samples.
As can be observed from **Table 4**, the annual effective dose due to inhalation of radon AED_{inh} for the members of local population was ranged from 5.216 to 6.880 μSv/y . The calculated average dose of the four codes is 6.029±0.680μSv/y , which is not exceeded the 1 mSv/y exposure limits for the members of the public [11].
In general, from a radiological health point of view, these slightly elevated the effective doses due to ingestion of water from sources in the studied areas are not expected to pose any significant health concerns to the local population since there are other optional available water sources. Also, these results do not represent a large value when compared with extreme value ~100 mSv/y which does not constitute any harmful effects [11]. The total annual effective dose of Vakilabad river water in Mashhad City was 36.578 μSv/y [45].
The result for radium activity and radon exhalation rate in sediment samples in selected locations of Al-Husseiniya river, are presented in **Tables 4**. The radium activity in sediment samples were found to vary from 0.213 to 0.287$Bqkg^{-1}$. The radon exhalation rate in terms of area of Al-Husseiniya River sediments varies from 30.413 to 41.037$mBq/m^2.h$. Whereas the radon exhalation rate in sediment samples in terms of mass varies from 6.112 to 8.247$mBq/kg.h$. The values of radium concentration and radon exhalation rate are also found to be maximum in sample code S-3 and minimum in sample code S-2. The exhalation rate in terms of area of Al-Husseiniya River water varies from 1.542 to 2.037$mBq/m^2.h$. The values of exhalation rate in terms of area of Al-Husseiniya River are, generally, found to be more in sediment samples compared with water samples. The values of radium activity determined in sediment are less than the permissible value30$Bqkg^{-1}$, which is acceptable for safe use as recorded by UNSCEAR (2000) [11]. The dissolved radon concentration in water C_d varied from 40.509 to 53.562Bq/L with average value is 46.890±255.245 Bq/L respectively. These results are given in **Table 4**. It was noticed that annual effective dose AED, surface exhalation rate E_A, mass exhalation rate E_M and the dissolved radon concentration in water C_d, are increase with radon concentration. Thus, results show that the area is safe as far the health hazard effects are concerned.

IOP Conf. Series: Journal of Physics: Conf. Series **1032** (2018) 012012 doi:10.1088/1742-6596/1032/1/012012

Table 3. Measured values of Ra_{eq}, H_{ex}, H_{in}, D_y, $AEDE$, $AGDE$, and $ELCR$ for sediment samples from Al-Husseiniya River.

Sediment Sample Code	Ra_{eq} (Bq/kg)	H_{ex}	H_{in}	D_{out} (nGy/h)	D_{in} (nGy/h)	$AEDE_{out}$ $(\mu Sv/y)$	$AEDE_{in}$ $(\mu Sv/y)$	$AGDE$ $(\mu Sv/y)$	$ELCR_{out}$ $\times 10^{-3}$	$ELCR_{in}$ $\times 10^{-3}$	$ELCR$ $\times 10^{-3}$
S-1	56.280	0.152	0.198	27.790	52.823	34.082	259.129	195.767	0.119	0.907	1.026
S-2	43.804	0.118	0.152	21.685	41.074	26.595	201.493	152.573	0.093	0.705	0.798
S-3	62.847	0.170	0.218	30.955	58.481	37.963	286.884	217.034	0.133	1.004	1.137
S-4	60.906	0.162	0.203	29.613	55.609	36.317	272.796	207.115	0.127	0.954	1.081
Min.	43.804	0.118	0.152	21.685	41.074	26.595	201.493	152.573	0.093	0.705	0.798
Max.	62.847	0.170	0.218	30.955	58.481	37.963	286.884	217.034	0.133	1.004	1.137
Ave	55.959	0.151	0.193	27.511	51.997	33.739	255.076	193.122	0.118	0.893	1.011
SD	±8.559	±0.023	±0.029	±4.095	±7.639	±5.021	±37.476	±28.395	±0.018	±0.131	±0.149
Worldwide*	370*' **	≤1	≤1	59	84	70	450	1000	0.29	1.16	1.45

*UNSCEAR (2000) [11]
** UNSCEAR (1988) [9]

Table 4. Radon and radium concentrations, surface and mass exhalation rates of sediment samples. Radon concentration, Annual effective dose (AED), surface exhalation rate, and dissolved radon concentration of water samples, in Al-Husseiniya River.

Sediment Samples						
Sample Code	ρ (track/mm^2)	Radon concentration (Bq/kg)	Radium concentration (Bq/kg)	E_A $(mBq/m^2.h)$	E_M $(mBq/kg.h)$	
S-1	551.02	5.74	0.236	33.746	6.782	
S-2	496.60	5.17	0.213	30.413	6.112	
S-3	670.07	6.98	0.287	41.037	8.247	
S-4	586.74	6.11	0.251	35.934	7.221	
Min.	496.60	5.17	0.213	30.413	6.112	
Max.	670.07	6.98	0.287	41.037	8.247	
Ave.	576.11	6.00	0.247	35.283	7.091	
SD	±72.785	±0.759	±0.031	±4.458	±0.896	
Worldwide		30(Bq/kg) [11]				
Water Samples						
Sample Code	ρ (track/mm^2)	Radon concentration (Bq/L)	AED_{ing} $(\mu Sv/y)$	AED_{inh} $(\mu Sv/y)$	E_A $(mBq/m^2.h)$	C_d (Bq/L)
W-1	370.75	2.73	9.965	6.880	2.037	53.562
W-2	280.61	2.07	7.556	5.216	1.542	40.509
W-3	321.43	2.37	8.651	5.972	1.766	46.448
W-4	324.83	2.40	8.760	6.048	1.785	47.035
Min.	280.61	2.07	7.556	5.216	1.542	40.509
Max.	370.75	2.73	9.965	6.880	2.037	53.562
Ave.	324.41	2.40	8.733	6.029	1.783	46.890
SD	±36.855	±0.270	±0.985	±0.680	±0.202	±255.245
Worldwide		11.1 Bq/L [39,40]	0.1 mSv/y [47,49,50,51] 0.2-0.8 with average 0.3 mSv/y [11]	1 mSv/y [24,41,43,48] 0.2-10 with average 1.2 mSv/y [11]		

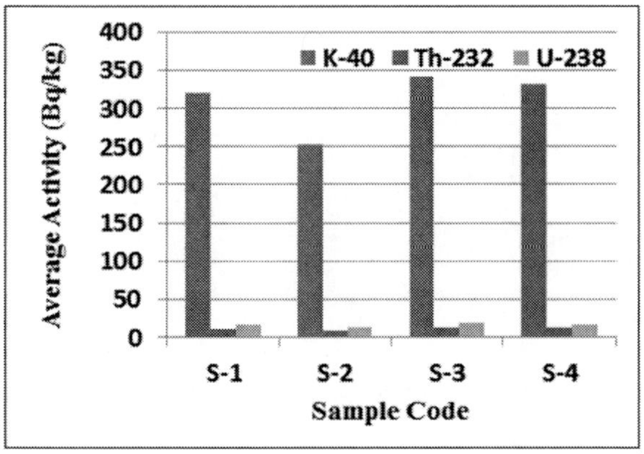

Figure 7. Activity of U-238, Th-232, and K-40 in sediment samples.

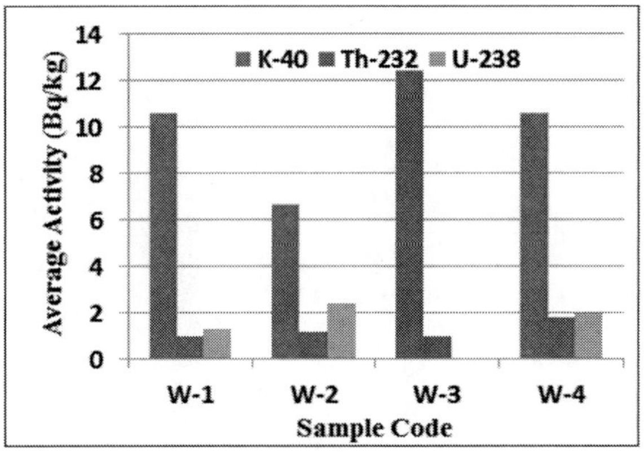

Figure 8. Activity of U-238, Th-232, and K-40 in water samples.

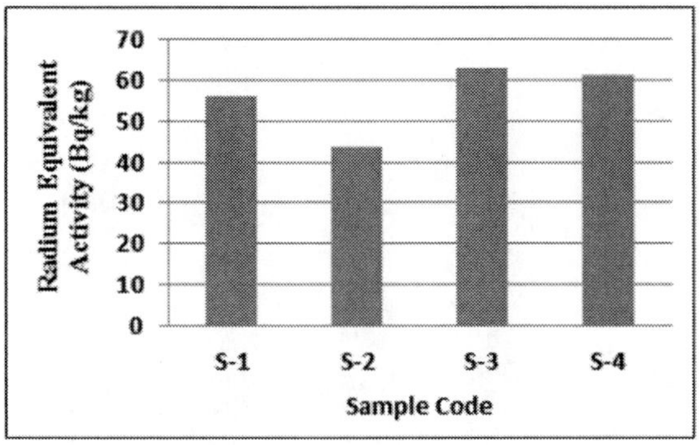

Figure 9. Radium equivalent activity of sediment samples.

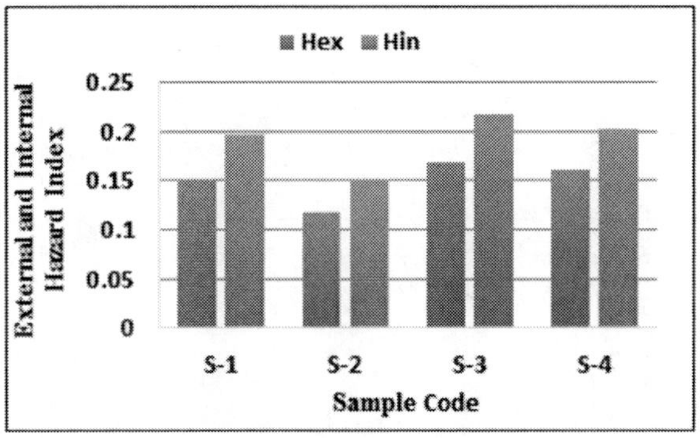

Figure 10. Hazard index of sediment samples.

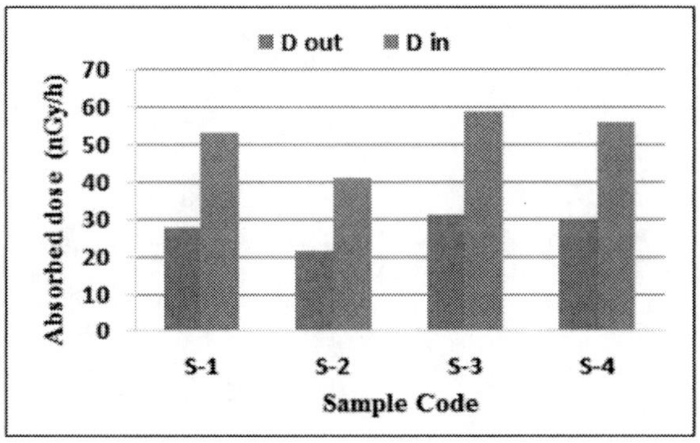

Figure 11. Absorbed dose of sediment samples.

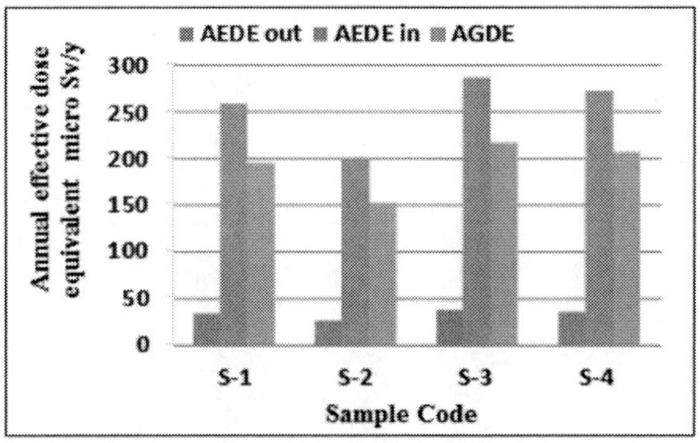

Figure 12. Annual effective dose equivalent of sediment

IOP Conf. Series: Journal of Physics: Conf. Series **1032** (2018) 012012 doi:10.1088/1742-6596/1032/1/012012

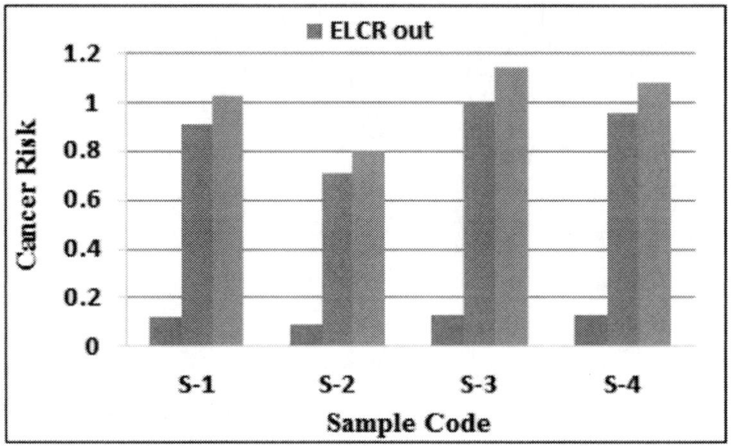

Figure 13. Excess lifetime cancer risk of sediment

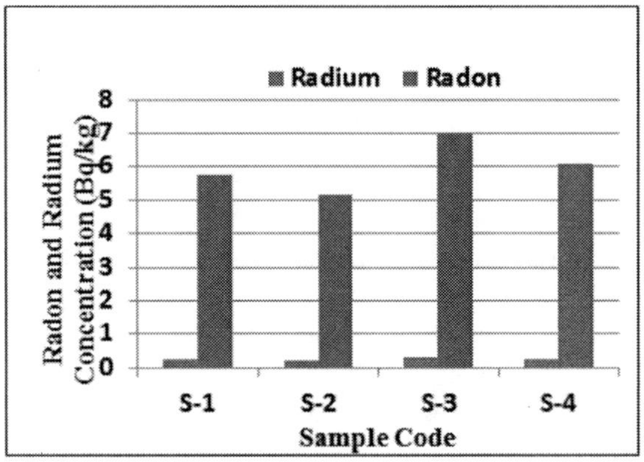

Figure 14. Radon and radium concentration in sediment samples.

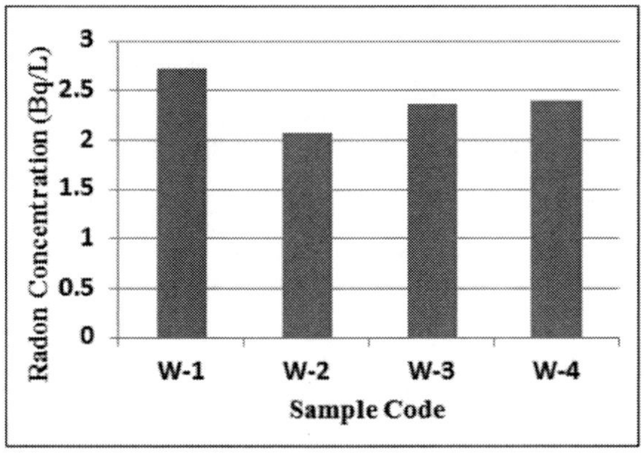

Figure 15. Radon concentration in water samples.

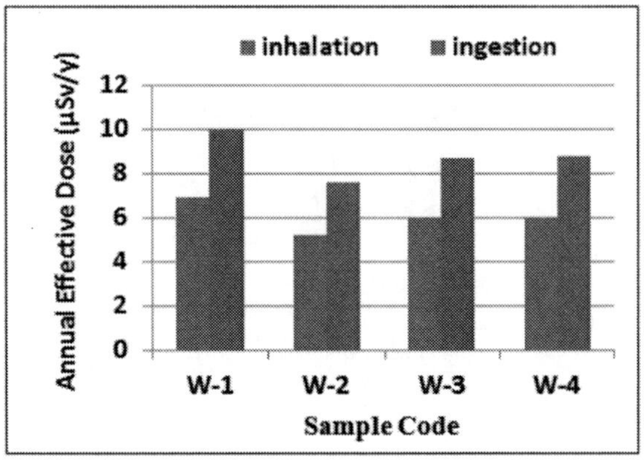

Figure 16. Annual effective dose for water samples.

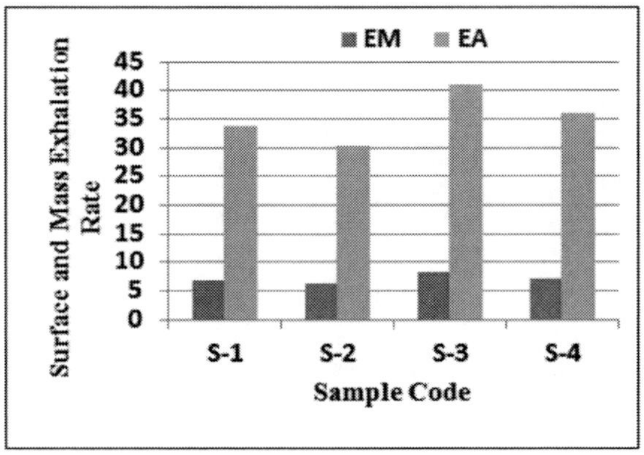

Figure 17. Surface and mass exhalation rate for sediment samples.

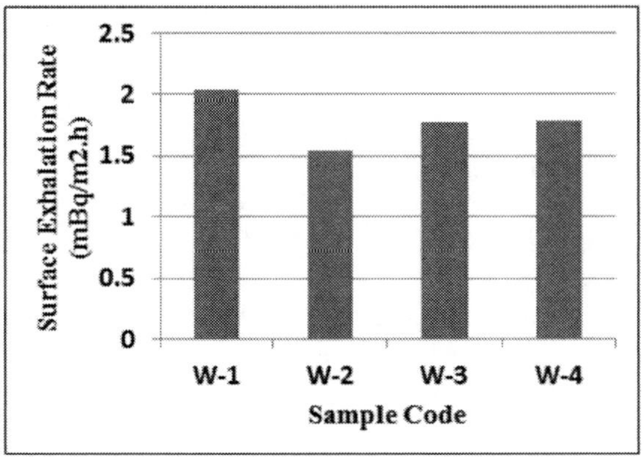

Figure 18. Surface exhalation rate for water samples.

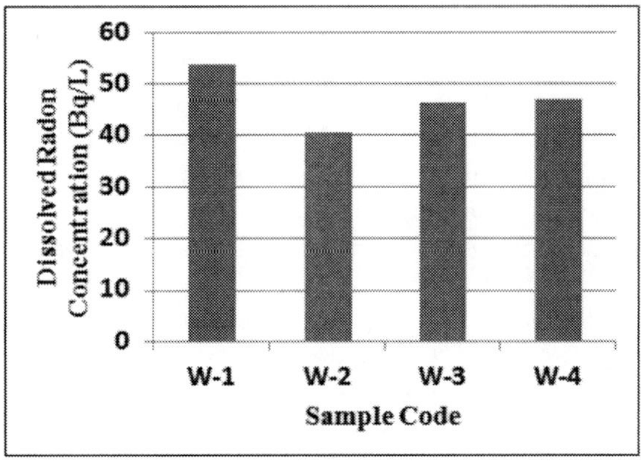

Figure 19. Disolved radon concentration in water samples.

7. Conclusions

In the present work, twenty samples of sediments and twenty samples of water were collected from Al-Husseiniya River in Karbala governorate, and it is clear from the analysis of gamma-ray spectroscopic (HPGe), the radioactivity concentrations of 238U, 232Th, and 40K radionuclides are within normal range. From the measured values, the Radium equivalent activities Ra_{eq} are less than the approved value. The external hazard H_{ext} and the internal hazard H_{in} indices were under the risk limit. The measured gamma dose rate D_{γ} is comparable to the worldwide average were found less than the suggested world average. The average values of D_{out} and D_{in} was below the assigned worldwide values. The values of Radon concentrations appear in the safe limit from the radiation protection point of view were found to be less than the value of 11.1Bq/L given by USEPA (2012). The annual effective dose values were lower than the UNSCEAR and WHO limit for members of the public of 1mSv/y. These values were all below the assigned worldwide values. Hence, the consumption of this water does not cause any health risk to the population around Al-Husseiniya River.

The results obtained showed that the radon concentrations in sediments and water are below the allowed maximum contamination level. The measured radon concentrations in Euphrates River sediments and water will pose none serious health risks and hence the study area is considered safe use of these sediments as building materials and water for irrigation or drinking and other uses for the population. Although these water and sediment materials contain low-level radioactivity but the accumulated dose can be high.

Acknowledgments

The authors wish to thank the Environment Agency in Karbala, to provide administrative facilities in the implementation of this study under the environmental protection program from radiation. Also the authors of the present work would like to thank the college of science and physics department in Al-Mustansiriyah University for their support.

References

[1] Issa S, Uosif M and Elsaman R 2013 Gamma radioactivity measurements in Nile River sediment samples *Turkish J. of Engin. and Envir. Sci.* **37**(1) 109–22 doi: 10.3906/muh-1207-21

[2] Ibrahim M, Shalabiea O and Diab H 2014 Measurement of some radioactive elements in drinking water in Arar city, Saudi Arabia *American J. of Life Sci.* **2**(1) 24-28 doi: 10.11648/j.ajls.20140201.13

[3] Ehsanpour E, Abdi M R, Mostajaboddavati M and Bagheri H 2014 226Ra, 232Th and 40K contents in water samples in part of central deserts in Iran and their potential radiological risk to human population *J. of Envir. Health Sci. and Engin.* **12**(80) 1-7 doi: 10.1186/2052-336X-12-80

[4] Śleziak M, Petryka L and Zych M 2010 Natural radioactivity of soil and sediment samples collected from Postindustrial Area *Polish J. of Envir. Studies* **19**(5) 1095-99

[5] Ibrahim U, Akpa T C and Daniel I H 2013 Assessment of radioactivity concentration in soil of some mining areas in central Nasarawa State *Nigeria Sci. World J.* **8**(2) 7-12

[6] Attiyah M M 2002 Determination of uranium concentration and annual effective dose in drinking water of Baghdad and Nineveh Governorates using Kinetic Phosphorescence Analyzer (KPA) and CR-39 Nuclear Track Detector *M. Sc. Thesis*, Baghdad University, Iraq

[7] Klement A W 1982 *Handbook of Environmental Radiation Boca Raton, Florida* (CRC) Chemical Rubber Company Press 177

[8] Jebur J H and Subber A Rh 2014 A comparative study for indoor radon measurements between CR-39 and Kodak LR115-II detectors *J. of Basrah Researches* (Sciences) **40**(4) 24-34

[9] UNSCEAR 1988 United Nations Scientific Committee on the Effects of Atomic Radiation. Sources, Effects and Risk of Ionizing Radiation. *Report to the General Assembly, with Scientific Annexes* United Nations, New York.

[10] UC 1999 European Commission. Radiological protection principles concerning the natural radioactivity of building materials In *EC radiation Protection 112* Directorate General Environment, Nuclear Safety and Civil Protection.

[11] UNSCEAR 2000 United Nations Scientific Committee on the Effects of Atomic Radiation. Sources and Effects of Ionizing Radiation. *Report to the General Assembly, with Scientific Annexes* Volume **I**: Sources, United Nations, New York.

[12] Tufail M 2012 Radium equivalent activities in the light of UNSCEAR *report Envir. monitoring and assessment* **184**(9) 5663–67 doi: 10.1007/s10661-011-2370-6 PMID: 21915591

[13] Lu X, Li L Y, Wang F, Wang L and Zhang X 2012 Radiological hazards of coal and ash samples collected from Xi'an coal-fired power plants of China *Envir. Earth Sci.* **66**(7) 1925–32

[14] Mamont-Ciesla K, Gwiazdowski B, Biernacka M and Zak A 1982 Radioactivity of building materials in Poland. In: Vohra G, Pillai K C and Sadavisan S (Eds.) *Nat. Radiat. Envir.* Halsted Press, New York 551

[15] Taskin H, Karavus M, Ay P, Topuzoglu A, Hidiroglu S and Karahan G 2009 Radionuclide concentrations in soil and lifetime cancer risk due to gamma radioactivity in Kirklareli, Turkey *J. of Envir. Radioact.* **100**(1) 49–53 doi: 10.1016/j.jenvrad.2008.10.012 PMID: 19038480

[16] ICRP 2012 International Commission on Radiological Protection). Publication 119: Compendium of dose coefficients based on *ICRP Publication 60. Annals of the ICRP 41 (suppl)* **42**(4) 1–130

[17] ICRP 1989 International Commission on Radiological Protection). Age-dependent doses to members of the public from intake of radionuclides *Pt.1. ICRP Publication 56* Pergamon Press, Oxford. Ann. ICRP **20**(2) 1-122

[18] ICRP 1994 International Commission on Radiological Protection). Dose coefficients for intakes of radionuclides by workers *ICRP Publication 68 Ann.* ICRP **24**(4) 1-83

[19] Somogyi G 1990 *Technical Reports Series* IAEA Vienna **1**(310) 229–56

[20] Verma D, Khan M S and Zubair M 2012 Assessment of effective radium content and radon exhalation rates in soil samples *J. Radional. Nucl. Chem.* **294** 267-70 https://doi.org/10.1007/s10967-012-1694-1

[21] Orubu A O 1965 A short geography of bended state offset lithography.

[22] Somogyi G, Hafez A, Hunyadi I and Toth- Szilagyi M 1986 Measurement of exhalation and diffusion parameters of radon in solids by plastic track detectors *Nucl. Track Radiat. Meas.* **12**(1-6) 701-04

[23] Ferreira A O, Pecequilo B R and Aquino R R 2011 Applications of a sealed can technique and CR-39 detectors for measuring radon emanation from undamaged granitic ornamental building materials *Radiol. protection J.* **46** 49-54

[24] UNSCEAR 2012 Effects and Risks of Ionizing Radiation", United Nations Scientific Committee on the Effects of Atomic Radiation *Report to the General Assembly, with Scientific Annexes A and B Series.* Annex A - Attributing health effects to ionizing radiation exposure and inferring risks. Annex B - Uncertainties in risk estimates for radiation-induced cancer. United Nations, New York.

[25] UNSCEAR 1993 United Nations Scientific Committee on the Effects of Atomic Radiation. Sources and effects of ionizing radiation *Report to the General Assembly, with Scientific Annexes*, United Nations, New York.

[26] AL-Ubaidi K H M, AL Nasri S K and AL Saudany Z A 2015 Natural radionuclides and hazards in water and sediment samples of Tigris River in Al- Amara City - Maysan – Iraq *Adv. in Physics Theories and Applic.* **44** 117-22

[27] Kurnaz A, Ku¨ c-u¨ ko¨ merog˘ lua B, Keser R, Okumusoglu N T, Korkmaz F, Karahan G and C-evik U 2007 Determination of radioactivity levels and hazards of soil and sediment samples in Fırtına Valley (Rize, Turkey) *Appl. Radiat. and Isotopes* **65** 1281–89 doi:10.1016/j.apradiso.2007.06.001

[28] Njinga R L, Jonah S A, Oladipo M O A , Ewa I O B and Alfa B 2012 Use of k0-ENAA technique for evaluation of Th, U and K in sediments for archaeometry studies *J. of Sci. and Technol.* **1**(8) 391-99

[29] Vasconcelos D C, Reis P A L, Pereira C, Oliveira A H, Santos T O and Zildete Rocha Z 2013 Modelling natural radioactivity in sand beaches of Guarapari, Espírito Santo State, Brazil *World J. of Nucl. Sci. and Technol.* **3** 65-71 http://dx.doi.org/10.4236/wjnst.2013.32011

[30] Al-Gazali H and Al-Ayashi A 2013 Determination and mapping of activity levels of 40K in Madloom Region in Al-Najaf Al-Ashraf *J. of Kufa-Physics* **5**(1) 49-54

[31] Yussuf N M, Hossain I and Wagiran H 2012 Natural radioactivity in drinking and mineral water in Johor Bahru (Malaysia) *Sci. Res. and Ess.* **7**(9) 1070-75

[32] Kurnaz A, Kucukomeroglu B, Keser R, Okumusoglu N T, Korkmaz F, Karahan G and Cevik U 2007 Determination of radioactivity levels and hazards of soil and sediment samples in Fırtına Valley (Rize, Turkey) *Appl. Radiat. and Isotopes* **65** 1281–89 doi:10.1016/j.apradiso.2007.06.001

[33] Ramasamy V, Sundarrajan M, Suresh G, Paramasivam K and Meenakshisundaram V 2014 Role of light and heavy minerals on natural radioactivity level of high background radiation area, Kerala, India *Appl. Radiat. and Isotopes* **85** 1–10 http://dx.doi.org/10.1016/j.apradiso.2013.11.119

[34] Qureshi A A, Tariq S, Ud Din K, Manzoor S, Calligaris C and Waheed A 2014 Evaluation of excessive lifetime cancer risks due to natural radioactivity in the rivers sediments of Northern Pakistan *J. of Radiat. Res. and Appl. Sci.* **7**(4) 438-47 http://dx.doi.org/10.1016/j.jrras.2014.07.008

[35] Ramasamy V, Suresh G, Meenakshisundaram V and Ponnusamy V 2011 Horizontal and vertical characterization of radionuclides and minerals in river sediments *Appl. Radiat. and Isotopes* **69** 184–195 doi:10.1016/j.apradiso.2010.07.020

[36] Baltasa H, Sirin M, Dalgic G and Cevik U 2018 An overview of the ecological half-life of the 137Cs radioisotope and a determination of radioactivity levels in sediment samples after Chernobyl in the Eastern Black Sea Turkey *J. of Marine Syst.* **177** 21-27. doi:10.1016/j.jmarsys.2017.09.005

[37] Singh J, Singh H, Kumar A, Mahajan S, Bajwa B S and Singh S 2005 Radon measurement in ground water and mineral water of Amritsar district by alpha scintillometry *J. Environ. Geochem.* **1** and **2** 126–29

[38] USEPA 1991 National Primary Drinking Water Regulations for Radionuclides, Washington, DC, *US Governmental printing Office* EPA/570/9-91/700

[39] WHO 1993 Gidelines for drinking water Quality, Recommendations, World Health Organization Geneva, **1**

[40] USEPA 1999 United States Environmental Protection Agency, Office of Water: 40 CFR Parts 141, and 142: National Primary Drinking Water Regulations; radon-222: proposed rule. *US Environmental Protection Agency*, Washington, D.C.
[41] EPA 2000 Environmental Protection Agency. Technical Fact Sheet: Final Rule for (Non-Radon) Radionuclides in Drinking Water *EPA 815-F-00-013* November
[42] UNSCEAR 2011 United Nations Scientific Committee on the Effects of Atomic Radiation, Sources and effects of ionizing radiation. *United Nations, New York*
[43] WHO 2012 World Health Organization, Progress on Drinking Water and Sanitation *World Health Organization* Geneva
[44] Jaber M Q, Subber AR H and Al-Hashimi N H N 2015 Radon concentrations in the marine sediments of Khor-Abdulla northern west of the Arabian Gulf. *Internat. J. of Physics* **3**(6) 239-43 doi: 10.12691/ijp-3-6-3
[45] Binesh A, Mohammadi S, Mowlavi A A and Parvaresh P 2010 Evaluation of the radiation dose from radon ingestion and inhalation in drinking water *Internat. J. of Water Resources and Environ. Engineering* **2**(7) 174-78
[46] Hashim A K 2014 Measurement of radon and radium concentrations in different types of water samples in Al-Hindiyah city of Karbala Governorate, Iraq *J. of Kufa-Physics* **6**(2) 69-77
[47] IAEA 1996 International Atomic Energy Agency. International basic safety standards for protecting against ionizing radiation and for the safety of radiation sources *Safety Series No.115* IAEA, Vienna
[48] ICRP 1993 International Commission on Radiological Protection. Protection against Radon-222 at Home and at Work *ICRP publication 65, Annals of the ICRP* **23**(2)
[49] WHO 2004 World Health Organization *Guidelines for Drinking Water Quality* Third edition, **1** Recommendations, Geneva
[50] EU 1998 European Commission. European drinking water directive 98/83/EC of 3rd November 1998 on the quality of water intended for human consumption. Official J L 330. *Official J. of the European Commission*
[51] ICRP 1991 International Commission on Radiological Protection 1990 Recommendations of the International Commission on Radiological Protection *ICRP Publication 60, Annals of the ICRP* **21**(1-3) Pergamon Press, Oxford

Fabrication and Characteristics Study of Bismuth Oxide/Silicon Heterojunction for Photodetector Applications

[1]**Hanaa Mohammed Ibrahim**, [2]**Sana Juma Ali Al. Tamimi** ، [3]**Ahmed N. Abd**

[1,2]Physics Department, Education Faculty, Mustansiriya University, Baghdad, Iraq
[3]Physics Department, Science Faculty, Mustansiriya University, Baghdad, Iraq

[3]Corresponding Email: ahmed_naji_abd@yahoo.com

Abstract: This work presents the fabrication of bismuth oxide /Silicon Heterojunction for Photodetector applications. Bismuth oxide nanoparticles investigation by simple chemical Tanique. The Bi_2O_3 thin film was deposited onto n-type Silicon and glass substrate by drop casting technique. The structural and optical properties of Bi_2O_3 thin film were studied; also, XRD show that the thin film is polycrystalline. The optical properties showed the having direct optical band gap of 2.5 eV, Dark and illuminated I-V, C-V, spectral responsivity and quantum efficiency of Bi_2O_3/n-Si photodetector ware investigated and discussed.

Keywords: Bismuth oxides, Thin film, solar cell, simple chemical method, Optical properties.

1. Introduction

A bismuth oxide Bi_2O_3 with peculiar properties becomes The main objective is between the bismuth containing substances [1].This composite has large values of refractive index. The polarization of Bi_3 + cation is highly sensitive to light [2]. Also, the energy gap is wide [1.73- 3.98] eV ,it is close to CdS, which is considered a prominent Filter for solar cells application [3-5], besides .These properties have four multi-forms α- Bi_2O_3,β- Bi_2O_3, δ-Bi_2O_3 , γ-Bi_2O_3 [6] and the unequal stages are $Bi_2O_{2.33}$,$Bi_2O_{2.75}$ [7] .These interesting properties have been recommended for a wide range of applications in solid state technology such as optical coatings, photovoltaic and ceramic glass manufacturing [8,9].Different method was used to prepare Bi_2O_3 such as sol-gel, magnetron reaction and sputtering Thermal evaporation [10-12], in this paper, the preparation and characterization of Bi_2O_3 thin film on n-silicon Si in order to study their characterization for the solar cell applications.

2. Experimental work

About 1.6 g of $Bi(NO_3)2$ and 224.98 g/mol (BDH /Chemicals Ltd Pool /England) was dissolved in 45 ml of Polyvinylpyrrolidone (Sigma Aldrich USA) 1%wt. and Re-distilled water was used through this work.

The solution was added in a round flask with flipping. The mixture was yellow. About 20 mL of NaOH (1M) was rapidly added to the mixture, and nanoparticle suspension was formed. The suspension was maintained at 80°C for two hours. After cooling to room temperature, nanoparti-cles were separated by centrifugation, then washed by distilled water to get rid of any pollution:

$Bi(NO3)2 + 2NaOH \rightarrow Bi_2O_3 + Na_2NO_4 + H_2O$........................ (1)

Figure 1 shows that the Bi_2NO_3 molecular nanoparticles, which are prepared by the chemical method, are grounded by the technique of droplet casting on glass and silicon substrates. The solution has been taken by pipette and then drops onto the glass surface and the n-silicone substrate only 3drops, then the particles dried up by using heater at 80 °C, then the thin film is ready. Weight difference method can be used to measure thickness. The difference for the film before and after the deposition on a glass substrate is the mass of Bi_2O_3 the film (m), also, the area of the thin film (A) and density (d) of the film material. The thickness ($t \approx 200$ nm) determined using following relation:

$t = m/Ad$.......... (2)

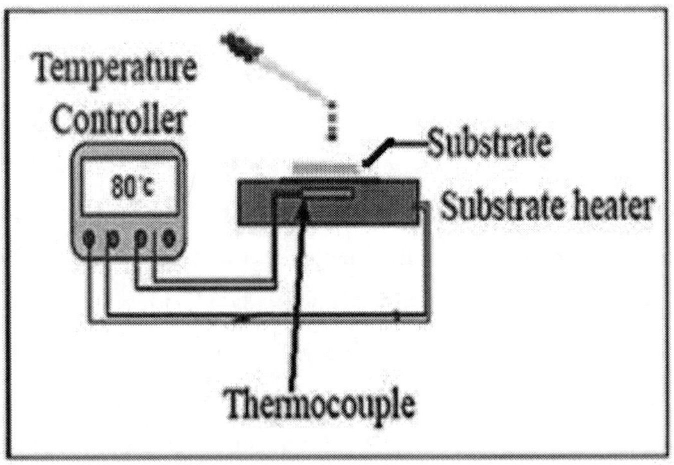

Figure 1. Diagram of the projection method of casting experimental mode

IOP Conf. Series: Journal of Physics: Conf. Series **1032** (2018) 012013 doi:10.1088/1742-6596/1032/1/012013

3. Resultants and desiccation

TEM image shows in Figure 2 confirmed that the Bi_2O_3 colloidal was ball shaped, which different diameter, about 13 to 25 nm, TEM studies were done to find the surface morphology of synthesized Bi_2O_3 .TEM studies shows Bi_2O_3 is in pure form and particles are improve black colored nanoparticles.

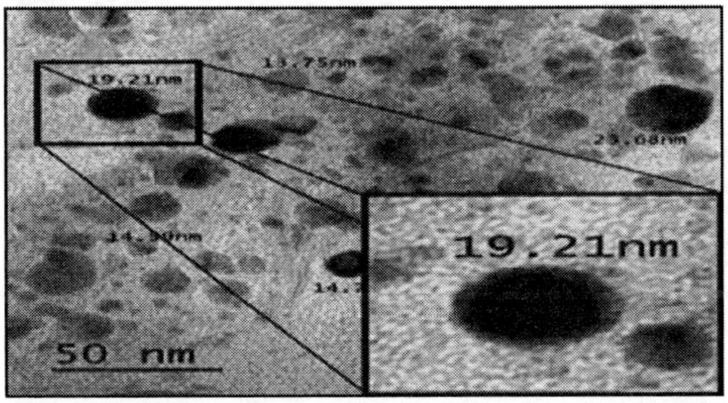

Figure 2. TEM image of synthesized Bi_2O_3 nanoparticles

XRD is an useful experimental technique used to calculate the crystalline structure of the stationary fixed-crystal orientation of the single crystal and the preferred Trends of Polycrystalline Crystals, [13] as shown in Figure 3. The X-ray diffraction model for the high-ready film on a sub-surface of the glass shows a multicolor structure with a peak of bismuth metal at 2 27 27.59, 32.4, 48.2 (111), (200) and (116) aircraft respectively, corresponding d-values were compared with the standard [JCPDS file , plane (Alpha Order): Bi_2O_3 :00-041-1449.

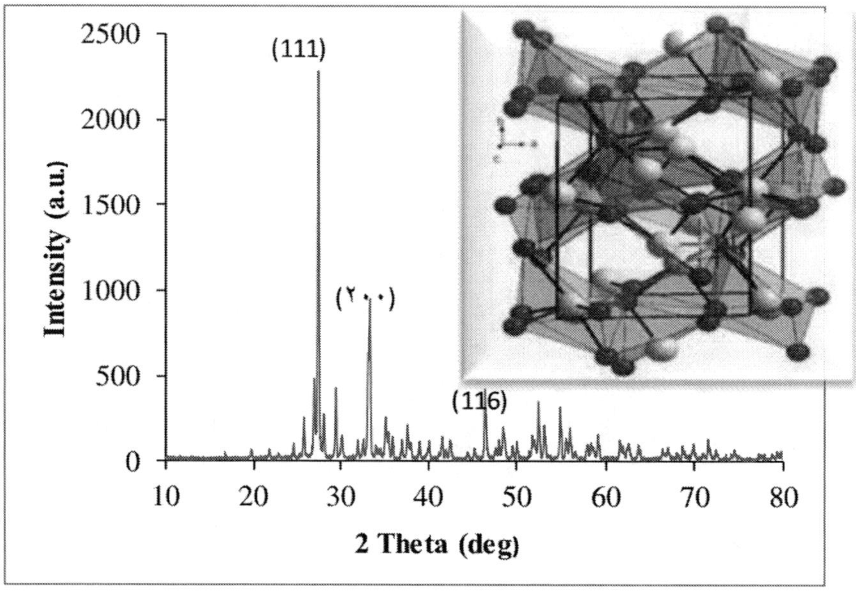

Figure 3. XRD pattern of Bi_2O_3 thin film which prepared by chemical method and deposited by drop casting technique on glass.

Figure 4 shows the magnification of 50 Kx for the SEM image of the Bi_2O_3 thin layer on the glass bottom layer. Also, the surface of the film consists of a regular distribution of anostructured grains with different grain sizes between 60 to 100 nm.

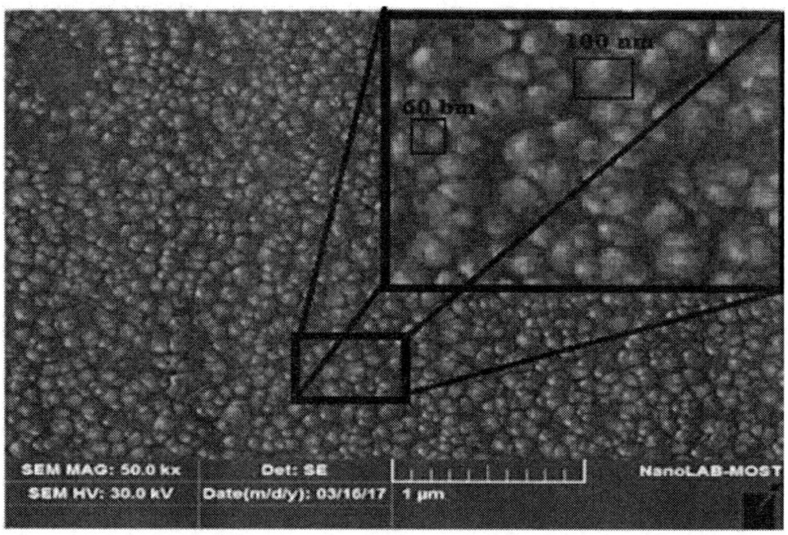

Figure 4. Scanning electron micrographs of the Bi_2O_3 thin film.

IOP Conf. Series: Journal of Physics: Conf. Series **1032** (2018) 012013 doi:10.1088/1742-6596/1032/1/012013

Morphology behavior of the thin film is showed in Figure 5. The film exhibited the lowest grain size of 91 nm, root mean square roughness *(rms)* of 2.14 nm, and the roughness average is about 1.84 nm. This behavior can be explained by the high oxygenation rate in the air, leading to lower roughness and volume in this atmosphere.

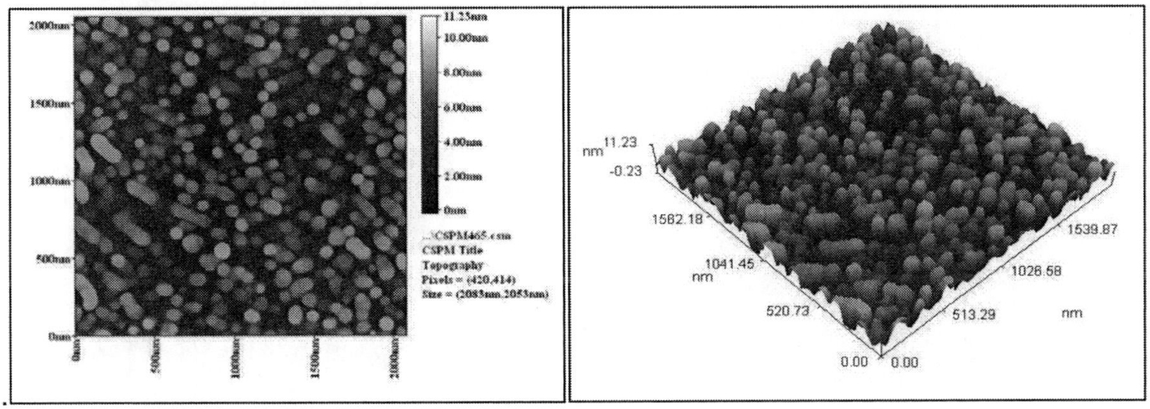

Figure 5. AFM images of Bi_2O_3 thin film

Figure 6 presents their optical absorption spectra in a 300 to 900 nm spectral range, proving their relatively good absorbance in the visible and much higher absorbance at 330 nm.

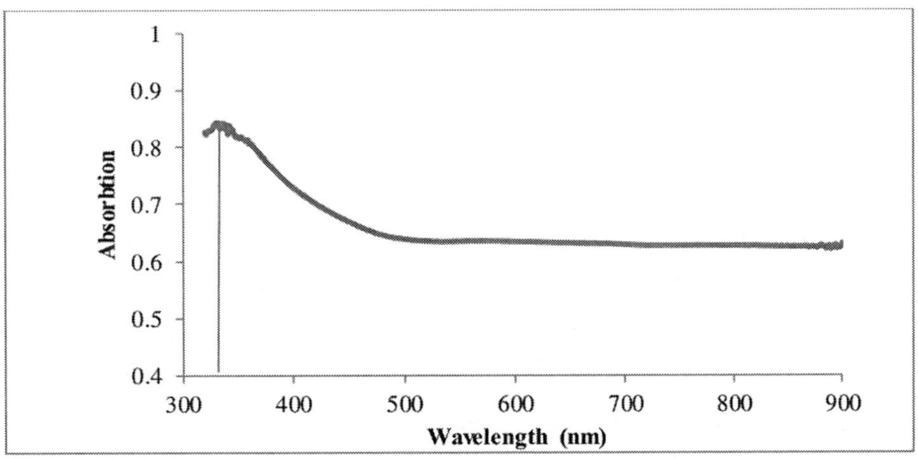

Figure 6. Optical absorption spectrum of Bi_2O_3 thin film

The energy gap can be found through the absorption coefficient by using to it by the following equation [19]:

$$(\alpha h\upsilon)^2 = A(h\upsilon - E_g)^n \quad \text{..........(3)}$$

Where α is the refractive index , hυ is the energy gap of the incident photon,, *A* is a band edge parameter and n is an refractive index that characterizes the optical absorption process which is equal 2 eV or 0.5 eV for indirect and direct allowed conversions, respectively. Eg is calculated by extrapolating the straight line portion $(h\upsilon\alpha)^2 = 0$. The result of the energy gap at the range from 2.5 eV as shown in Fig.7.The small energy gap value may attributed to the bismuth Bi ions and atoms available in the plume thus Fermi level move into the conduction band and thus made the material act as dissolve semiconductor [15].

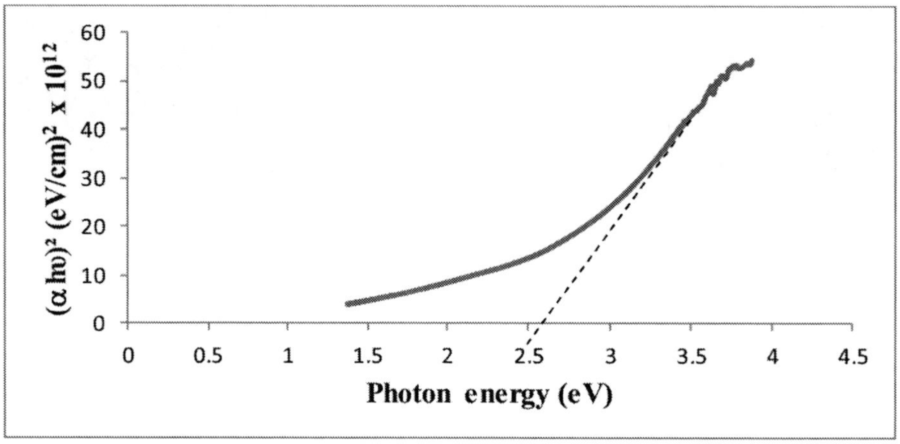

Figure 7. $(\alpha h\nu)^2$ versus photon energy gap of Bi_2O_3 thin film.

Figure 8 shows the dark properties of (I-V) in the front and back direction of Al/Bi_2O_3/n-Si/Al Photodetector. The forward current of photodetector is very small at voltage less than 2V. This current is known as *recombination current* which occurs at low voltages only. It is generated when each electron excited form valence band to get the balance back. The second high-voltage zone represents the propagation or bending region that relies on serological resistance .In this region, the bias voltage will equip the electrons or holes with enough energy to break the barrier between the two sides of the intersection .In this region; the bias voltage will release electrons or holes with enough energy to penetrate the barrier between the two sides of the junction. These results agree with the other workers [16].

IOP Conf. Series: Journal of Physics: Conf. Series **1032** (2018) 012013 doi:10.1088/1742-6596/1032/1/012013

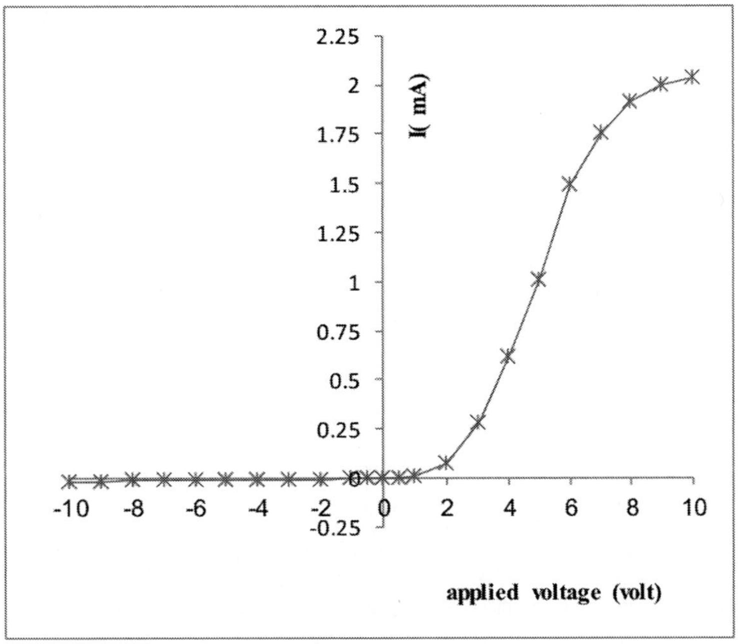

Figure 8. Dark I-V characteristics of Al/ Bi_2O_3/n-Si/Al photodetector

Figure 9 show that the contrast ln (I/I_O) with bias voltage of Bi_2O_3/n-Si prepared. The ideality factors of heterojunctions are estimated and it is found to be around (2.44) for Bi_2O_3/n-Si prepared. When the chassis has a resistive chain and interface state, the ideal factor n becomes higher than the unit; most Schottky diodes process show deviation from the ideal thermionic theory.The fact that such recombination currents are flowing not homogeneously in the structure , But always at local sites. These results agree with the other workers [17].

IOP Conf. Series: Journal of Physics: Conf. Series **1032** (2018) 012013 doi:10.1088/1742-6596/1032/1/012013

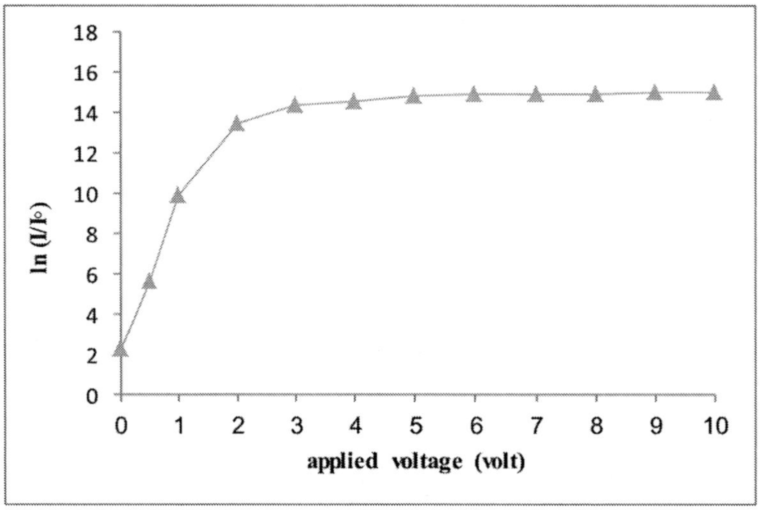

Figure 9. variance of the value of ln (I / I_O) with the deviation voltage of Bi_2O_3 / n-Si.

Optical properties are the best use of photoreceptor mass imaging since these properties determine how light energy is converted to photocurrent. Figure 10 shows that the characteristics of the inverted power of the instrument are measured in the dark and under the lighting of the tungsten ceiling lamp 41W /cm^2. The reverse current value can be observed at a standard voltage of Heterojunction Bi_2O_3/Si under illumination higher than those in the dark. They increase with increasing light intensity. The current that starts at low voltage is a typical thermal emission pattern. In the linear area, the heat emission and the carrier velocity will increase when the sample is lit with a light intensity of varying intensity.

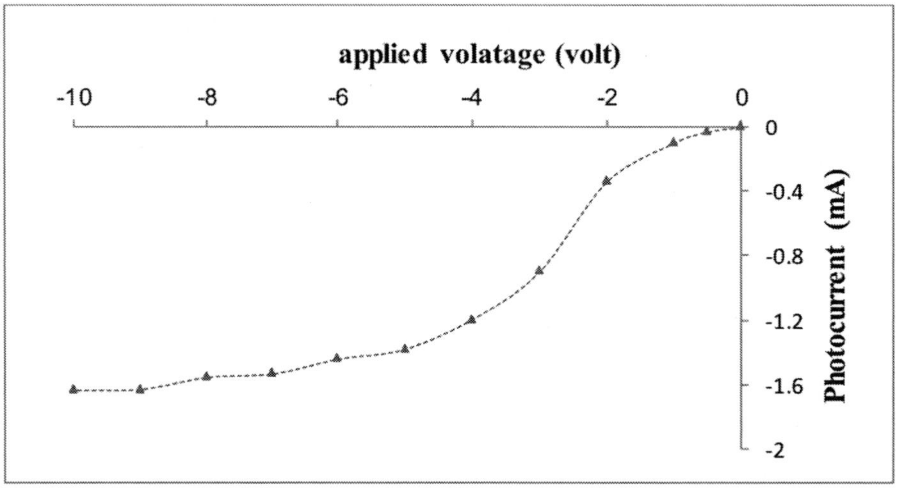

Figure 10. Dark and luminous characterization (I-V) of Bi_2O_3 / n-Si Photodetector

Figure 11 shows that the liner relationship between the C^{-2} and the reverse voltage was obtained to bias the structure. This linear relationship represents the Bi_2O_3/Si image detector. The potential internal voltage values were obtained at 0.4 volts. It represents the energy required of the electron to transfer it from Bi_2O_3 to Si.

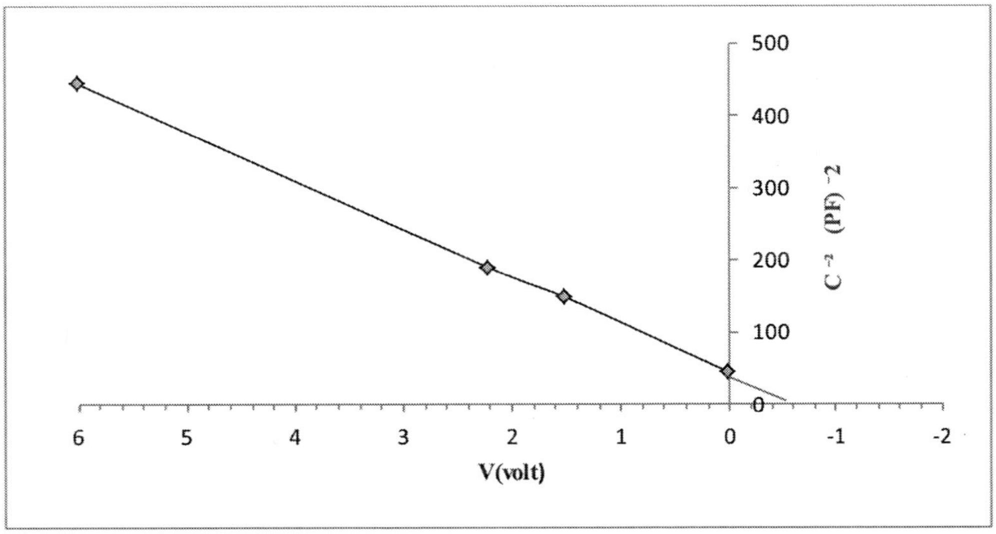

Figure 11. $1/C^2$ versus reverse voltage of Al/ *Bi_2O_3*/Si//Al photodiode.

Figure 12 shows the spectral responsivity and Quantum efficiency of Al/Bi_2O_3/n-Si/Al photodiode is investigated in the wavelength range from 400 nm to the 900nm with 5volt bias، which is calculated by following equation:

$$R_\lambda = \frac{I_{ph}}{P_{in}} \qquad (3)$$

Where Iph is the stream of the optical vessel and the pin is the input power. Spectral responsivity R_λ is an important function to find out the amount of detector signal that will be available for application [16].The structure of the Bi_2O_3 /n-Si sandwich (photodetector) consists of appro-ximately 600 nm peaks representing the absorbance edge of Bi_2O_3 / Si as shown in figure) and the other in 700nm related to Bi_2O_3 / n-Si as shown in Figure 12. The photoresponse magnitude painted in Figure 12 is calculated after subtracting the dark current when aligned opposite. The result of the response means that the portion of high-energy light, such as 600-700 nm, is absorbed by a Bi_2O_3 film (Zone 1) and the portion of light with less energy, such as 800-900 nm (Zone 2) Completely in Si substrate and absorbed. These results due to absorption edges Bi_2O_3 and Si.The R_λ of photodetector Al/ Bi_2O_3/Si/Al is increased from≈0.68

A/W.Quantum efficiency (Q) can be defined as the ratio of generated electrons to the incident photons when the light strikes the surface of photodetector. The Q is the number of photons that occur if the detector is a semiconductor (p-n) junction device, in which (hole-electron) pairs are produced. Over a period of time 10^5 photons are incident on the detector and 10^4 (e-h) pairs are produced, then, the Q is about 10%. The Q of photodetector is basically another way of expressing the effectiveness of the incident optical energy for producing an output of electrical current, the Q (in percent) may be related to the R_λ by the relation:

$$Q = R_\lambda \frac{1.24}{\lambda_{(\mu m)}} X\ 100\% \qquad (4)$$

R (λ): is the responsively (A/W) of the photodetector as a function of wavelength λ (μm) [18].
The quantum efficiency Q found to be 14% at wavelength 740 nm for Bi_2O_3/Si Photodetector.

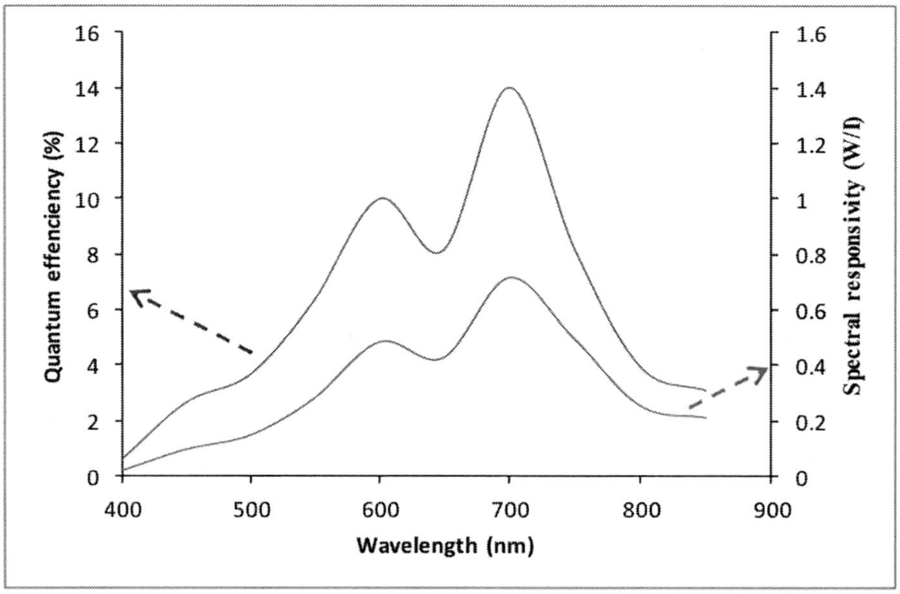

Figure 12. Spectral response and Quantitative efficiency as a function of wavelength for Bi_2O_3 / n-Si photodetectors

4. Conclusions:

The synthesised $Bi_2O_3NP_S$ are in nanosized about 13 to 25 nm prepared by simple chemical method which it .has a good characteristics of Bi_2O_3 $NP_{S,}$ Also, X-ray diffraction detection measurement that Bi_2O_3 NP_S are polycrystalline. Deposition of Bi_2O_3 NP_S on porous silicon gives suspensions photodetector characteristics. The spectral responsivity (R_λ) of Al/ Bi_2O_3/Si/Al photodetector is 0.72 A/W at ≈ 740 nm wavelength due to the absorption edge of Si and 0.6 A/W at ≈ 650nm wavelength due to the absorption edge of Bi_2O_3 NPs.

5. References

[1] Michael Mehring "From molecules to bismuth oxide-based materials: Potential homo-and heterometallic precursors and model compounds" Coordination Chemistry Reviews Vol.251, pages 974–1006, 2007.

[2] L. Leontie, M. Caramana, G. I. Rusu " ON THE PHOTOCONDUCTIVITY OF Bi2O3 IN THIN FILMS"Journal of Optoelectronics and Advanced Materials Vol. 2,No. 4, pages 385-389, December 2000.

[3] H.T. Fan, S.S. Pan, X.M. Teng, C. Ye, G.H. Li, L.D. Zhang " δ-Bi2O3 thin films prepared by reactive sputtering Fabrication and characterization", Thin Solid Films Vol. 513, pages 142–147, 2006.

[4] He Weidong, Qin Wei, Wu Xiaohong, NingHailong "Thin bismuth oxide films prepared through the sol–gel method" Materials Letters, Vol. 61, pages 4100–4102, 2007.

[5] Raid A. Ismail "Fabrication and Characteristics Study of n-Bi2O3/n-Si Heterojunction "Journal Of Semiconductor Technology And Science, Vol.6, No.2,June, (2006).

[6] A. J. Salazar-Pérez, M. A. Camacho-López, Superficies y Vacío"Structural evolution of Bi2O3 prepared by thermal oxidation of bismuth nano-particles,Vol. 18,pages 4-8, septiembre, (2005). [7] S. Yilmaz, O. Turkoglu, M. Ari, I. Belenli"Electrical conductivity of the ionic conductor tetragonal (Bi2O3)1-x(Eu2O3)x" Cerâmica, Vol. 57(2011) 185-192.

[8] H.T. Fan, S.S. Pan, X.M. Teng, C. Ye, G.H. Li, L.D. Zhang " δ-Bi2O3 thin films prepared by reactive sputtering Fabrication and characterization", Thin Solid Films Vol. 513(2006) 142–147.

[9]H T Fan, S S Pan, X M Teng, C Ye and G H Li "Structure and thermal stability of δ-Bi2O3 thin films deposited by reactive sputtering "J. Phys. D: Appl. Phys., Vol. 39(2006) 1939–1943. [10]He Weidong, Qin Wei, Wu Xiaohong, NingHailong "Thin bismuth oxide films prepared through the sol–gel method" Materials Letters, Vol.61 (2007) 4100–4102,2007.

[11] L. LEONTIE " Optical properties of bismuth oxide thin films prepared by reactive agnetron sputtering", JOURNAL OF OPTOELECTRONICS AND ADVANCED ATERIALS, Vol. 8, No. 3, June (2006)1221-1224.

[12] S. Condurache-Bota, N. Tigau, A.P. Rambu, G.G. Rusu, G.I. Rusu "Optical and electrical properties of thermally oxidized bismuth thin films" Applied Surface Science, Vol. 257, pages, (2011)10545–10550.

[13] Simões, A.Z.; Ries, A.; Stojanovic, B.D.; Biasotto, G.; Longo, E.; Varela, J. Electrical properties of lanthanum doped Bi4Ti3O12 thin films annealed in different atmospheres. *Ceram. Int.*, *33*,(2007) 1535–1541.

[14] L. Leontie, M. Caraman, M. Alexe, C. Harnagea " Structural and optical characteristics of bismuth oxide thin films" Surface Science 507–510 (2002) 480– 485.

[15] Sugaki A., Kitakaze A., KitazawaI H., *Synthesized tin and tin-sulfide minerals; Synthetic sulfide inerals (XIII)*, Science Reports of the Tohoku University **16**,3(1985) 199–211

[16] M.J. Sailor , J.L. Heinrich, ***"Luminescent porous silicon: Synthesis, chemistry, and applications"***, Stud. Surf. Sci. Catal. 103, (1997) 209-235.

[17] R. Mahfouz, F. J. Cadete, S. Aires, A. Brenier, B. Jacquier, J. C. Bertolini ,"***Synthesis and physico-chemical characteristics of nanosized particles produced by laser ablation of a nickel target in water***", Applied Surface Science 254, (2008)5181-5190.

[18] L. De Stefano, Buccolieri, G.; De Luca, F.; Plescia, P. Milling ***,"effects upon quantitative determinations of chrysotile asbestos by the reference intensity ratio method",*** *Powder Diffr.15,(* 2000 *)* 26-29.

IOP Conf. Series: Journal of Physics: Conf. Series **1032** (2018) 012014 doi:10.1088/1742-6596/1032/1/012014

Potential of Renewable Energy in Al-Fourat Al-Awsat: Al Najaf City Case Study

Ali H. Al-Hamdani(1), Abeer Faris Dawood(2), Khudhayer N. Abdullah(3), Hashim Mohammed Hasan(4), Mahdi Sarhan(5), Haider Shareef(6), Rawshan Tarik (7)
(1) Laser and optoelectronics Engineering Department, University of Technology/ Baghdad,
Email:ali_alhamdani2003@yahoo.com
(2) Civil Engineering Department, Al-FarabiUniversity College/ Baghdad,
Email: eng.abeerfaris@yahoo.com
(3) Building and Construction Engineering Department, University of Technology/ Baghdad,
Email: khudhayer67@yahoo.com
(4) Building and Construction Engineering Department, University of Technology/ Baghdad,
Email: hashimhero@yahoo.com
(5, 6) Energy and Renewable Energy Technology Center, University of Technology/Baghdad/
(7)Electrical Engineering Department, University of Technology/Baghdad.

Abstract: This paper reviews an analytical study of the hybrid power generation system in the al Najaf city at Iraq Country. The design and analysis of optimum choice to supply specific load (Renewable Energy center) building in the al Najaf city. The steps of this project consist of the measuring and collection of the basic meteorological data of solar radiation for al Najaf city, renewable energy software HOMER used for developing off-grid and on-grid system optimization simulation models. The simulation model has been used to find out the best result optimization based on energy efficient system for the specific load and taking into account costs and environmental emissions. Two reference models used to compare the costs and environmental impact. The optimization results indicate that the appropriate system for al Najaf Renewable energy center is consists of (81 kW PV, 58 kW Converter and 108 kW Grid) which have good components with the city requirements, this system would be cause reduction in CO_2 by (174.38%) percentage.

1. Introduction

Climate change is the most serious environmental risks facing the planet and community. Green house gases, which are responsible for climate change appear from human activity. The most serious of these gases is Carbon dioxide (CO2) which is raising the average temperatures at the surface of the Earth, leading to harsher weather events, droughts, storms, and flood as well as rising sea level and disruption to whole ecosystems[1,2].

The effect on communities across the world could be disastrous if steps to stop this danger did not take soon. To avert dangerous climate changes, rising global CO2 emissions within the next decade must be controlled as well as saving energy and using it more efficiently. For these reasons, switching to forms of energy that do not produce CO2 and there are plenty of them, such as sun, wind, tides, waves, plants and others which are inexhaustible sources of renewable energy. These natural sources could be exploited to create electricity without adding carbon to environmental.

To take a step towards improving the environment in Iraq, an analytical study of a hybrid energy system in al Najaf applied using the Hybrid Optimization Model for Electric Renewable (HOMER) software, which is a software that has been developed by NREL (National Renewable Energy Laboratory). It performs hourly simulations of each attainable combination of parts entered and ranks the systems in step with user-specified criteria, like price of energy or Capital cost.

Several studies have been reported to design optimal alternative energy systems around world Henryson and Svensson [3] proposed strategies of meeting power and energy demands for the Swedish Antarctica Research Program (SWEDARP) underneath the strict natural enactments by a quite way exploitation the free renewable energy assets. The results depend upon analyses and computer simulations (HOMER) on how each primary system component is predicted to work in the harsh climate conditions existed at the station Wasa in Antarctica. It was found that the wind power has the greatest potential as an essential energy source in Antarctica supported by suitable batteries, a reliable and in many meaningfully systems can be obtained.

Mizani and Yazdani [4] used the HOMER software package to put the optimum microgrid configuration and their optimum generation combine. The results showed that the optimum choice of renewable energy sources and energy storage devices in a grid-connected microgrid, the lifetime cost and emissions of the microgrid could significantly reduce in conjunction with an optimal transmutation strategy. Lal, D.K., et al. [5] designed a hybrid power generation system adequate for remote area application by using HOMER software, supplying by a (Micro-hydro, wind turbine, PV array) model, a diesel generator and batteries. The author illustrated the importance of equalizing between environment, economy and energy, by determining the optimum hybrid composition. Additionally, the COE within the planned theme is comparably over the traditional energy sources notwithstanding additional potency and fewer environmental aspect effects.

Al-Hamdani, H. Ali, Dawood. A and et al [6] used weather data in the Homer program to get the optimum renewable energy program in Baghdad city, the result shows that the best chose Photoelectric cells with a total production equal to 8 kW cost of electricity produced 0.903 US$/kW this system were cost 32015 US$.

2. Parameters assumptions Inputs of models:

AC Load:

Figure 1 shows the typical daily load profile of. Hypothetical Renewable center building in the al Najaf city. The energy expendable by the microgrid is 203 kWh/day with a 108 kW peak demand (7 hour /day, 5 days/ week), this load includes (fans, lights, air conditioner and lab devices):

Solar Resource:

Solar radiation information was obtained from, the National Aeronautics and Space, Surface Meteorology and solar power web site [5]. The annual average solar radiation for this city was (5.510 kWh/m^2). Monthly clearness index and daily radiation profile as a monthly average for a year 2013 obtained through HOMER are given in Figure 2:

Photovoltaic modules:

The PV module utilized in the simulations was mounted at a direction straight to the south, and also the slopes of the panels were calculated to be equal latitude of Najaf city 31.9833 (decimal degree).

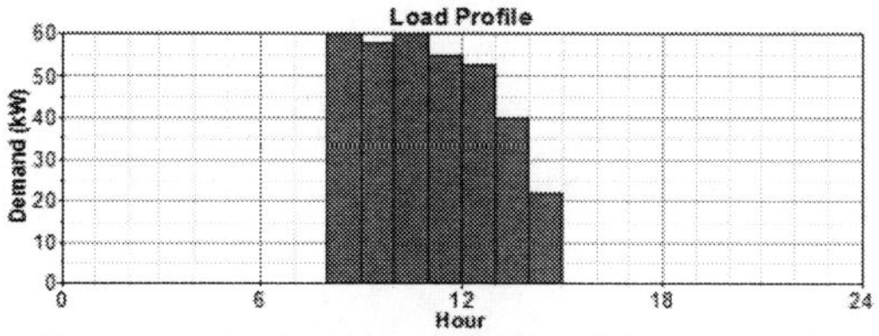

Figures (1.a): Daily load profile of the system for January and February.

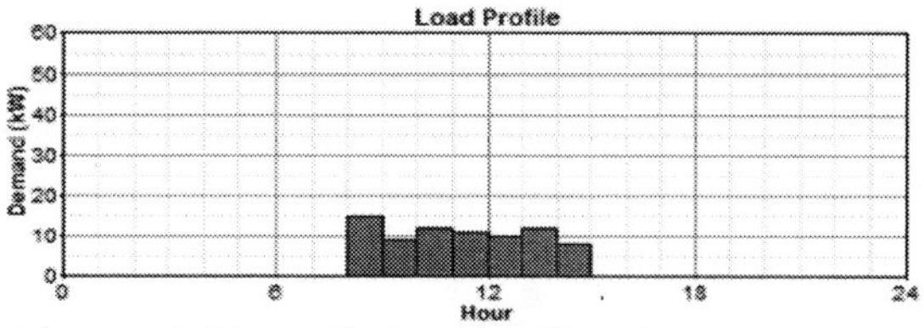

Figures (1.b): Daily load profile of the system for March.

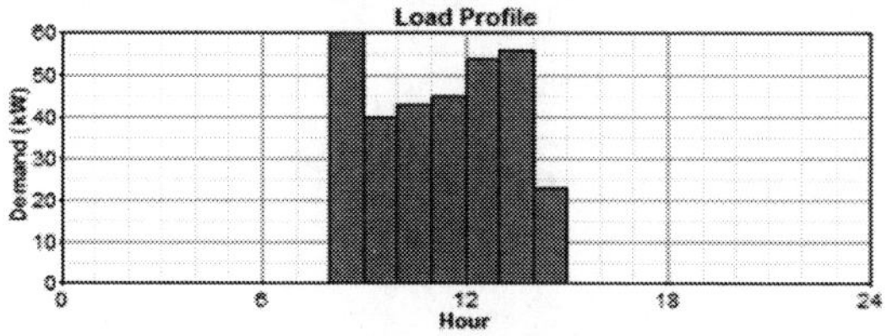

Figures (1.c): Daily load profile of the system from April to October.

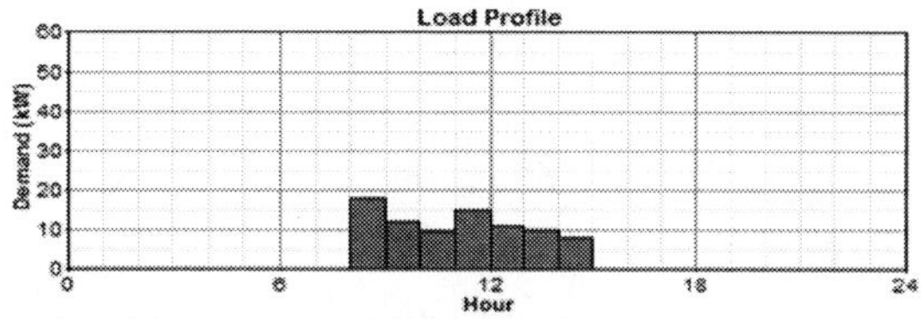

Figures (1.d): Daily load profile of the system for November.

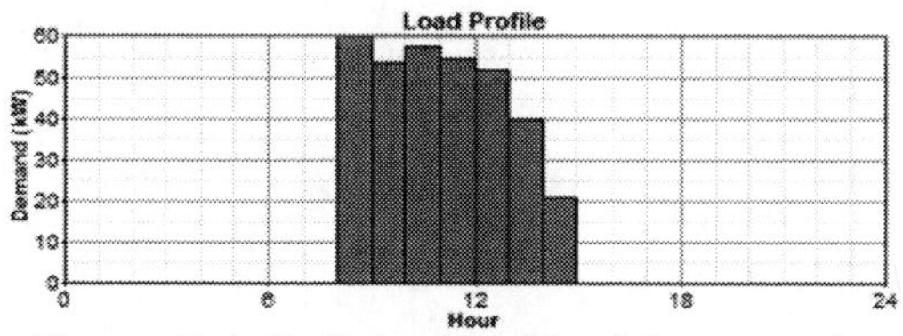

Figures (1.e): Daily load profile of the system for December.

Figure 1. the daily load profile of Hypothetical Renewable centre building in the al Najaf city.

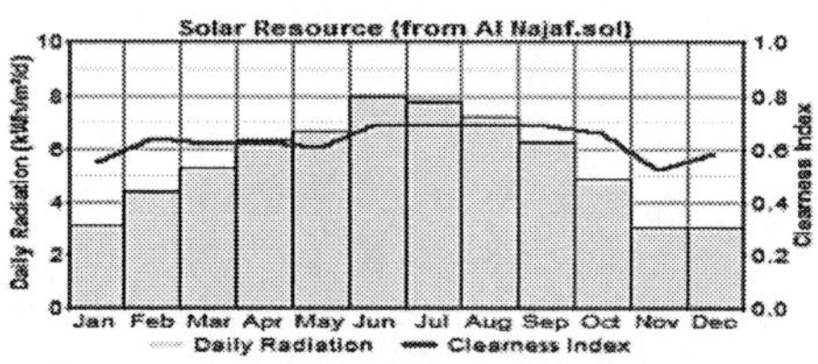

Figures 2. Monthly clearness index and daily radiation of Najaf city.

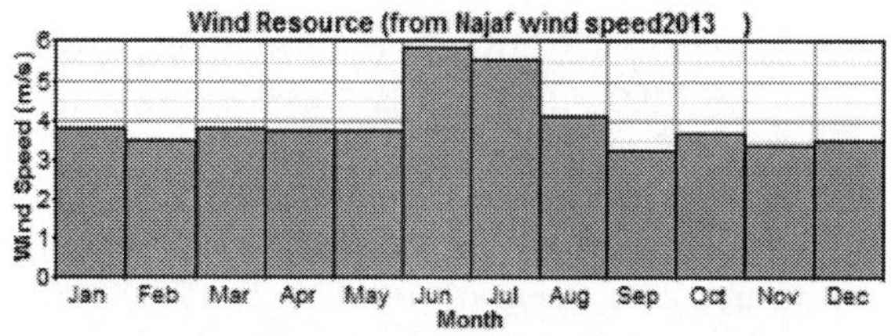

Figure 3. Monthly average wind speed profile of Najaf city.

Wind Resource: Wind data for regions of study are obtained from the NASA global satellite database at 10 m anemometer height. Figure 3 below shows the monthly average wind speed profile over the year of 2013.

Wind Turbine: The chosen type of wind turbine was DC current RCVA-60KW type, which has a hub height of 18 m [8], it has a power curve profile shown in Figure 4. The lifetime of the turbine was assumed to be 15 years.

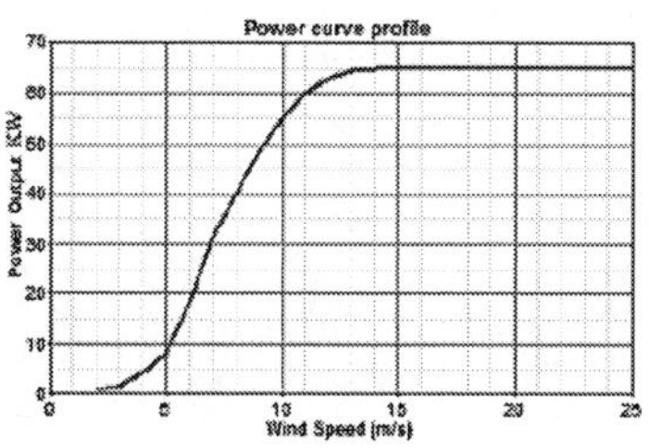

IOP Conf. Series: Journal of Physics: Conf. Series **1032** (2018) 012014 doi:10.1088/1742-6596/1032/1/012014

Figures 4. RCVA-60KW power curve [7].

Converter and battery: The battery chosen is the Vision 6FM200D. It has a nominal voltage of 12 Volts and nominal capacity of 200 Ah (2.4 kWh). While, the converter efficiency is assumed 85 %., and its lifetime is up to 15 years.

Generator: The AC diesel generator was scheduled to be operated all day of the week (except weekend days) from (8 am to 3 pm), which has 15000 operating hour life.

Constraint: The Maximum annual capacity shortage assumed to be 20%

Economic: This paper used the project lifetime of 15 years and interest rate of 6 %, table 1shows the costs parameter.

Table 1. Renewable system components cost

Components	*Cost*		
	Capital	Replacement	*Operating & Maintenance*
Solar[1]	2000 $/kW	2000 $/kW	*10 $/year*
Wind turbine[2]	89200 $/turbine	89200 $/turbine	*10 $/year*
Converter[3]	3000 $/10 kW	3000 $/10 kW	*10 $/year*
Battery[3]	300 $/battery	300 $/battery	*10 $/year*
Diesel generator4	500 $/kW	500$/kW	*0.0136 $/hr.*

Grid: The maximum amount of power that can be drawn from the grid set to be 108 kW with (0.0081799[a], 0.0040899[b]) $/kWh power price and sell back rate, respectively.

Grid and generator emission factor: HOMER multiplies the net grid purchases (in kWh) by the emission factor (in g/kWh) for each pollutant insert. If the system sells additional power to the grid than it buys from the grid over the year, the net grid purchases will be negative sign and so will the grid-related emissions of all pollutants inserted [9,10].

While the emissions factor determined (kg of pollutant emitted per unit of fuel consumed) for each pollutant. After the simulation, it calculates the annual emissions of that pollutant by multiplying the emissions factor by the total annual fuel consumption. Table 2 shows CO_2 emission factors used in system that content Grid or generator.

Table 2. CO2 emission factors

Type of system	CO2 factors
Generator [9]	6.5 (g/l)
Grid [11]	907.685(g/kWh)

1- Iraqi Local market.
2-Qingdao Richuan Precision Machinery Co., Ltd.
3-http://www.maxima.net.au/.
4-Iraq Global Technologies Company, http://www.iraqglobal.com/
[a] 0.0081799 = (10 DIQ /kWh)* (1222.5$/ IQD) [Iraqi ministry of electricity] [9].
[b] 0.0040899=(5 DIQ /kWh)* (1222.5$/ IQD) [assumed].

Case studies:

Six models systems cases study was used in this paper which includes;

1. The first case study consists of PV generators only.(off-grid)
2. The second case study consists of wind turbine only.(off-grid)
3. The third case study of the hybrid system consists of PV and Wind turbine.(off-grid)
4. The fourth case study consists of PV generators only.(on-grid)
5. The fifth case study consists of wind turbine only.(on-grid)
6. The sixth case study consists of wind turbine only.(on-grid)

The process to find the best suitable system for Najaf is by using Homer optimization for each of the above systems individually and evaluate the suitable system (low cost and low CO_2 emission). Then compassion was made with a reference system (diesel and government guide) as shown in figure 5 and 6.

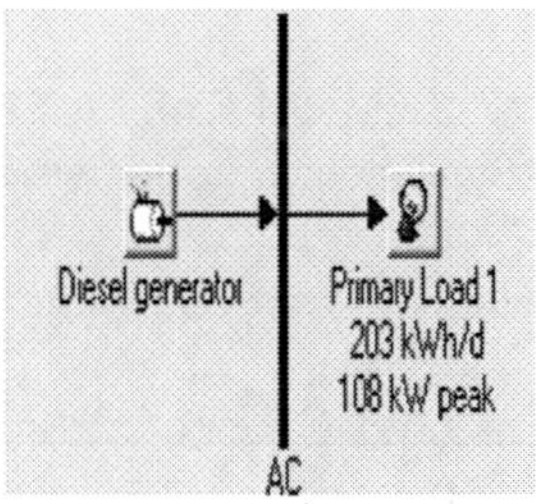

5.a reference model, diesel generator

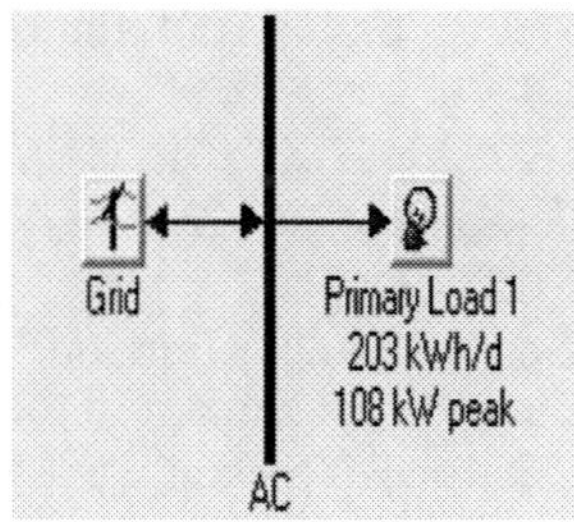

5.b reference model, grid

Figure 5. References model

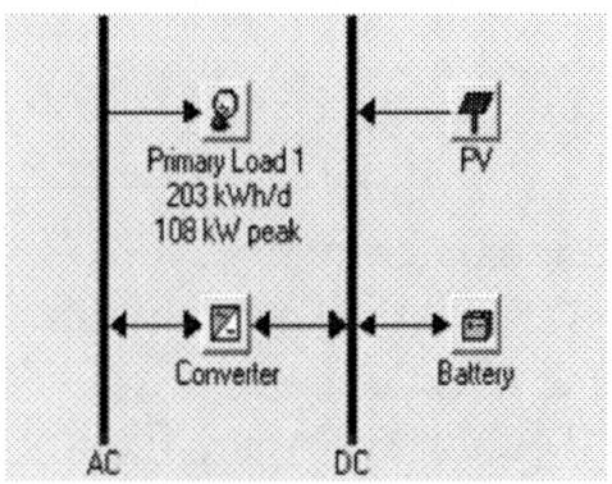

6.a case study consists of PV generators only.(off-grid)

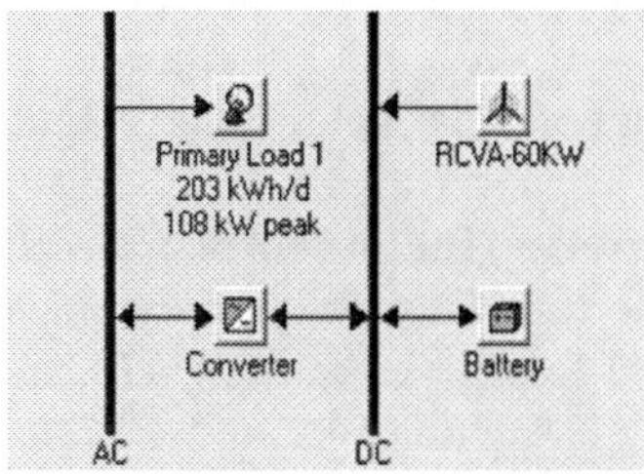

6.b case study consists of wind turbine only.(off-grid)

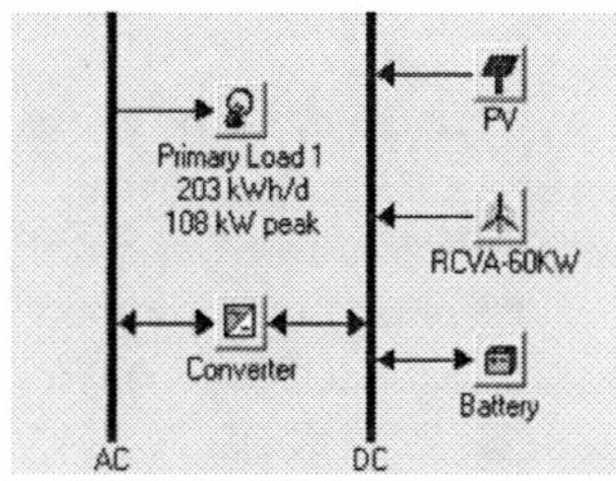

6.c case study of the hybrid system consists of PV and Wind turbine.(off-grid)

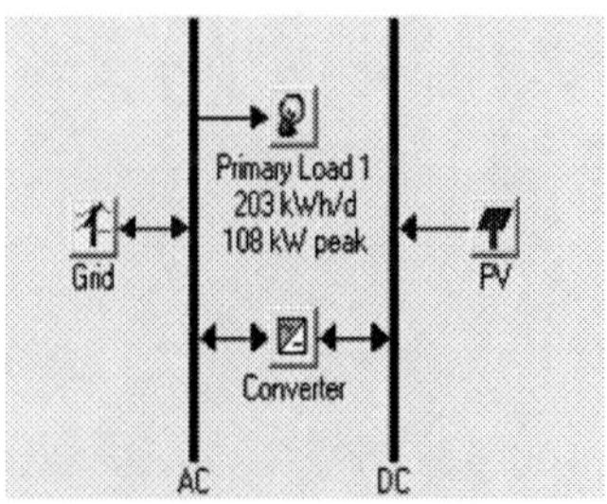

6.d study consists of PV generators only.(on-grid)

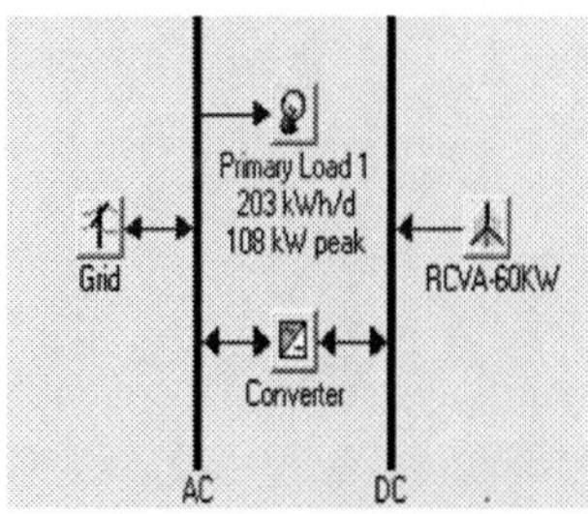

6.e case study consists of wind turbine only.(on-grid)

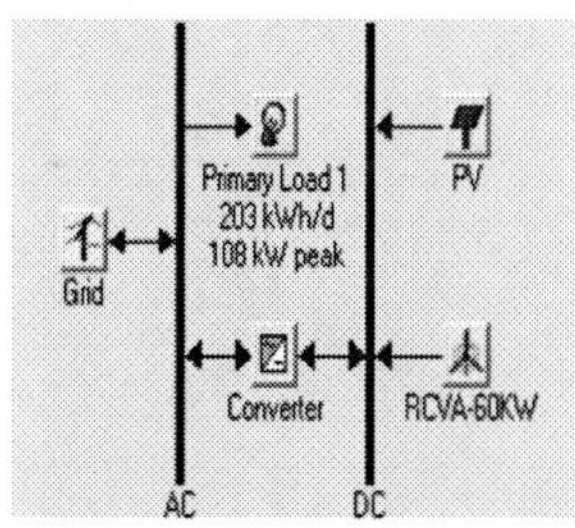

6.f case study consists of wind turbine only.(on-grid

Figure 6. Six design case studies

3. Results and conclusion:

After evaluating and optimization all the case studies mentioned in above. An analysis was done includes the total system prices and the amount of CO2 contribution for each case studies to the environment of the Najaf town. It was found that the system which is suitable for this load in the Najaf is a grid- on system which consists of (81 kW PV, 58 kW Converter and 108 kW Grid). This

The system has good components with the city requirements. The total net present cost of this system is 186729 $, in this city for (15 year). So one concludes that the solar energy (PV) system is more effective than other renewable energies in the Al-Fourat Alawsat or Al Najaf. This system would cause CO_2 reduction by 174.38% as it would produce more renewable energy than consumed.

4. References;

[1] Ali H.A. Al-Waeli, K Sopian, Hussein A Kazem, Miqdam T Chaichan, Photovoltaic/Thermal (PV/T) systems: Status and future prospects, Renewable and Sustainable Energy Reviews, Vol. 77(2017) 109-130.
[2] Chaichan M. T, Kazem H A, Al-Waeli A H A, Al-Mamari A S A, Al-Kabi A. H. K, A "photovoltaic application in car parking lights with recycled batteries: A techno-economic study," Vol. 9, no. 36 (2015) 43-49.
[3] Henryson, M., and Svensson, M., "Renewable Power for the Swedish Antarctic Station Wasa, "Royal Institute of Technology, Stockholm, Sweden, MSc. Thesis, 2004, 127 p.
[4] Mizani, S., and Yazdani, A., "Optimal Design and Operation of a Grid-Connected Microgrid," Proceedings of IEEE Electrical Power & Energy Conference, 2009, pp. 1-6.
[5] Lal, D. K., Dash, B. B., and Akella, A. K., "Optimization of PV/Wind/Micro Hydro/Diesel Hybrid Power System in HOMER for the Study Area," International Journal of Electrical Engineering and Informatics, Vol. 3, No. 3,(2011) 307-324.
[6] Al-Hamdani, H., Dawood .A and et al.,"Optimal sizing of photovoltaic systems using HOMER for Baghdad," International Journal of Computation and Applied Sciences IJOCAAS, Vol. 1, No. 3(2011) 6 .
[7] NASA, "Near Real-time Global Radiation and Meteorology Data Base," https://eosweb.larc.nasa.gov/sse/
[8] Qingdao Richuan Precision Machinery Co., Ltd, Email:benson@richuanchina.com.
[9] The Hybrid Optimization Model for Electric Renewables (HOMER), The Micropower Optimization Model, National Renewable Energy Laboratory US, version 2.42, 2007.
[10] Brander, M., Sood, A., Wylie, C., Haughton, A., and Lovell, J., "Technical Paper: Electricity-Specific Emission Factors for Grid Electricity," Ecometrica, (2011), 22 p.
[11] Iraqi Ministry of Electricity, http://www.moelc.gov.iq/index.php
[12] Ali A Kazem, Miqdam T Chaichan, Hussein A Kazem, "Dust effect on photovoltaic utilization in Iraq, Renewable and Sustainable Energy Reviews," Vol.37 (2014) 734-749.

IOP Conf. Series: Journal of Physics: Conf. Series **1032** (2018) 012015 doi:10.1088/1742-6596/1032/1/012015

Fractional variational iteration method for solving the hyperbolic telegraph equation

H K Jassim, W A Shahab

Department of Mathematics, Faculty of Education for Pure Sciences, University of Thi-Qar, Nasiriyah, Iraq

Hassan Kamil Jassim: hassan.kamil@yahoo.com

Abstract. In this article, an analytical solution procedure is described for solving one dimensional second order hyperbolic telegraph equation using a reliable semi-analytic method so called the local fractional variational iteration method (LFVIM) subject to the appropriate initial conditions. Various illustrate examples are carried out to check the accuracy, efficiency, and convergence of the described method.

1. Introduction

The fractional telegraph equations (FTEs) have been investigated by many authors in recent years. Garg et al. [1] derived a solution of space-time fractional telegraph function in a bounded domain by the method of generalized differential transform and obtained the solution in terms of Mittage-Leffler function. Chen et al. [2] obtained the solution of nonhomogeneous FTE. Ansari [3] derived a formal solution of the FTE by applying a fractional exponential operator. Huang [4] considered the time-fractional telegraph equation for the Cauchy problem and signaling problem, he solved the problem by the combined Fourier-Laplace transforms. Also, Huang derived the solution for the bounded problem in a bounded-space domain by means of Sine-Laplace transforms methods. Das et al. [5] used a homotopy analysis method in approximating an analytical solution for the (FTE). and different particular cases have been derived.

VIM proposed in [6-8] was utilized to deal with the following differential equations: Helmholtz [9], Burger's and coupled Burger's [10], Klein- Gordon [11], KdV [12], the oscillation [13], Schrodinger [14], reaction-diffusion [15], diffusion equation [16], Bernoulli equation [17], and others. The extended VIM, called the fractional VIM, was developed and applied to handle some fractional differential equations within the modified Riemann-Liouville derivative [18] and local fractional derivative operator [19,20].

The present paper describes an analytical scheme, the LFVIM to provide approximate analytical results of the one dimensional telegraph equations. The accuracy and efficiency of the proposed method are demonstrated by several test examples. The biggest benefit of the described method is that it finds the solution of FTE. directly without using any transformation, linearization, discretization or any other restrictive conditions. Further, the method can be easily implemented in multidimensional problems arising in many areas of science and engineering. The paper has the following organization. In Section 2 the concepts of local fractional derivatives and integrals are

IOP Conf. Series: Journal of Physics: Conf. Series **1032** (2018) 012015 doi:10.1088/1742-6596/1032/1/012015

briefly reviewed. In Section 3 the LFVIM is recalled. In Section 4 the analytical solutions for local fractional telegraph equations are presented. Finally, Section 5 outlines the main conclusions.

2. A Brief Review of the Local Fractional Calculus

Definition 1 [19-22]: The LF derivative of $f(x)$ of order α at $x=x_0$ is

$$f^{(\alpha)}(x_0)=\lim_{x\to x_0}\frac{\Delta^{\alpha}(f(x)-f(x_0))}{(x-x_0)^{\alpha}}, \tag{2.1}$$

where $\Delta^{\alpha}(f(x)-f(x_0))\cong\Gamma(\alpha+1)(f(x)-f(x_0))$.

The formulas of LF derivatives are:

$$D_x^{(\alpha)}ag(x)=aD_x^{(\alpha)}g(x), \tag{2.2}$$

$$\frac{d^{\alpha}}{dx^{\alpha}}\left(\frac{x^{n\alpha}}{\Gamma(1+n\alpha)}\right)=\frac{x^{(n-1)^{\alpha}}}{\Gamma(1+(n-1)\alpha)},\quad n\in N \tag{2.3}$$

Definition 2 [19-22]: The LF integral of $f(x)$ in the interval $[a,b]$ is given by

$${}_aI_b^{(\alpha)}f(x)=\frac{1}{\Gamma(1+\alpha)}\int_a^b f(t)(dt)^{\alpha}=\frac{1}{\Gamma(1+\alpha)}\lim_{\Delta t\to 0}\sum_{j=0}^{N-1}f(t_j)(\Delta t_j)^{\alpha}. \tag{2.5}$$

The formulas of LF integrals are as:

$${}_0I_x^{(\alpha)}ag(x)=a\,{}_0I_x^{(\alpha)}g(x), \tag{2.6}$$

$${}_0I_t^{(\alpha)}\left(\frac{x^{n\alpha}}{\Gamma(1+n\alpha)}\right)=\frac{x^{(n+1)^{\alpha}}}{\Gamma(1+(n+1)\alpha)},\quad n\in N \tag{2.7}$$

3. Analysis of the LFVIM

We consider a general nonlinear PDEs:

$$L_{\alpha}u(x,t)+R_{\alpha}u(x,t)+N_{\alpha}u(x,t)=f(x,t),\ 0<\alpha\le 1, \tag{3.1}$$

subject to fractal initial conditions

$$u(x,0)=\varphi_1(x),\quad \frac{\partial^{\alpha}u(x,0)}{\partial t^{\alpha}}=\varphi_2(x), \tag{3.2}$$

and the fractal boundary conditions

$$u(0,t)=\psi_1(t),\quad \frac{\partial^{\alpha}u(0,t)}{\partial x^{\alpha}}=\psi_2(t). \tag{3.3}$$

where L_{α} and R_{α} are linear LFDOs of order 2α and N_{α} denotes nonlinear LFDOs and $f(x,t)$ is the nondifferentiable function.

According to the theory of LFVI algorithm [19,20], we can write the iteration formula for (3.1) is constructed as:

$$u_{n+1}(t)=u_n(t)+{}_0I_t^{(\alpha)}\left(\frac{\lambda(\xi)^{\alpha}}{\Gamma(1+\alpha)}\left[L_{\alpha}u_n(\xi)+R_{\alpha}\tilde{u}_n(\xi)+N_{\alpha}\tilde{u}_n(\xi)-f(\xi)\right]\right). \tag{3.4}$$

Making the LF variation of (3.4), we have

$$\delta^{\alpha} u_{n+1}(t) = \delta^{\alpha} u_n(t) + {}_0I_t^{(\alpha)} \delta^{\alpha}\left(\frac{\lambda(\xi)^{\alpha}}{\Gamma(1+\alpha)} \left[L_{\alpha} u_n(\xi) + R_{\alpha}\tilde{u}_n(\xi) + N_{\alpha}\tilde{u}_n(\xi) - f(\xi)\right] \right). \tag{3.5}$$

The extremum condition of u_{n+1} is given by

$$\delta^{\alpha} u_{n+1} = 0. \tag{3.6}$$

In view of (3.6), we get:

$$1 - \left(\frac{\lambda(\xi)^{\alpha}}{\Gamma(1+\alpha)}\right)^{(\alpha)}\Bigg|_{\xi=t} = 0, \quad \frac{\lambda(\xi)^{\alpha}}{\Gamma(1+\alpha)}\Bigg|_{\xi=t} = 0, \quad \left(\frac{\lambda(\xi)^{\alpha}}{\Gamma(1+\alpha)}\right)^{(2\alpha)}\Bigg|_{\xi=t} = 0. \tag{3.7}$$

So, from (3.7), we get

$$\frac{\lambda(\xi)^{\alpha}}{\Gamma(1+\alpha)} = \frac{(\xi - t)^{\alpha}}{\Gamma(1+\alpha)}.$$

Thus, we have

$$u_{n+1}(t) = u_n(t) + {}_0I_t^{(\alpha)} \left(\frac{(\xi - t)^{\alpha}}{\Gamma(1+\alpha)} \left[L_{\alpha} u_n(\xi) + R_{\alpha} u_n(\xi) + N_{\alpha} u_n(\xi) - f(\xi)\right] \right). \tag{3.8}$$

and

$$u_0(x,t) = \varphi_1(x) + \frac{t^{\alpha}}{\Gamma(1+\alpha)} \varphi_2(x). \tag{3.9}$$

Finally, from (3.8), the solution of (3.1) is:

$$u = \lim_{n\to\infty} u_n. \tag{3.10}$$

4. Applications

Example 1. Let us consider the linear telegraph equation with LFDOs:

$$\frac{\partial^{2\alpha} u}{\partial x^{2\alpha}} = \frac{\partial^{2\alpha} u}{\partial t^{2\alpha}} + \frac{\partial^{\alpha} u}{\partial t^{\alpha}} - u, \quad 0 < \alpha \le 1 \tag{4.1}$$

and

$$u(x,0) = E_{\alpha}(x^{\alpha}), \quad \frac{\partial^{\alpha} u(x,0)}{\partial t^{\alpha}} = -2E_{\alpha}(x^{\alpha}), \tag{4.2}$$

$$u(0,t) = E_{\alpha}(-2t^{\alpha}), \quad \frac{\partial^{\alpha} u(0,t)}{\partial x^{\alpha}} = E_{\alpha}(-2t^{\alpha}). \tag{4.3}$$

The correction functional is given by

$$u_{n+1}(x,t) = u_n(x,t) + {}_0I_t^{(\alpha)} \left\{ \frac{\lambda(\xi)^{\alpha}}{\Gamma(1+\alpha)} \left(\frac{\partial^{2\alpha} u_n(x,\xi)}{\partial \xi^{2\alpha}} - \frac{\partial^{2\alpha} \tilde{u}_n(x,\xi)}{\partial x^{2\alpha}} + \frac{\partial^{\alpha} \tilde{u}_n(x,\xi)}{\partial \xi^{\alpha}} - \tilde{u}_n(x,\zeta) \right) \right\}. \tag{4.4}$$

where $\frac{\lambda(\xi)^{\alpha}}{\Gamma(1+\alpha)} = \frac{(\xi - t)^{\alpha}}{\Gamma(1+\alpha)}$.

Therefore,

$$u_{n+1}(t) = u_n(t) + {}_0I_t^{(\alpha)} \left\{ \frac{(\xi - t)^{\alpha}}{\Gamma(1+\alpha)} \left(\frac{\partial^{2\alpha} u_n(\xi)}{\partial \xi^{2\alpha}} - \frac{\partial^{2\alpha} u_n(\xi)}{\partial x^{2\alpha}} + \frac{\partial^{\alpha} u_n(\xi)}{\partial \xi^{\alpha}} - u_n(\zeta) \right) \right\}. \tag{4.5}$$

Now, we get

IOP Conf. Series: Journal of Physics: Conf. Series **1032** (2018) 012015 doi:10.1088/1742-6596/1032/1/012015

$$u_0 = E_\alpha(x^\alpha)\left(1-\frac{2t^\alpha}{\Gamma(1+\alpha)}\right). \tag{4.6}$$

$$u_1 = u_0 + {}_0I_t^{(\alpha)}\left\{\frac{(\xi-t)^\alpha}{\Gamma(1+\alpha)}\left(\frac{\partial^{2\alpha}u_0(x,\xi)}{\partial\xi^{2\alpha}} - \frac{\partial^{2\alpha}u_0(x,\xi)}{\partial x^{2\alpha}} + \frac{\partial^{\alpha}u_0(x,\xi)}{\partial\xi^{\alpha}} - u_0(x,\zeta)\right)\right\}$$

$$= E_\alpha(x^\alpha)\left(1-\frac{2t^\alpha}{\Gamma(1+\alpha)}+\frac{4t^{2\alpha}}{\Gamma(1+2\alpha)}-\frac{8t^{3\alpha}}{\Gamma(1+3\alpha)}\right)$$

$$= E_\alpha(x^\alpha)\sum_{\varepsilon=0}^{3}\frac{(-2)^\varepsilon t^{\varepsilon\alpha}}{\Gamma(1+\varepsilon\alpha)}. \tag{4.7}$$

$$u_2 = u_1 + {}_0I_t^{(\alpha)}\left\{\frac{(\xi-t)^\alpha}{\Gamma(1+\alpha)}\left(\frac{\partial^{2\alpha}u_1(x,\xi)}{\partial\xi^{2\alpha}} - \frac{\partial^{2\alpha}u_1(x,\xi)}{\partial x^{2\alpha}} + \frac{\partial^{\alpha}u_1(x,\xi)}{\partial\xi^{\alpha}} - u_1(x,\zeta)\right)\right\}$$

$$= E_\alpha(x^\alpha)\left(1-\frac{2t^\alpha}{\Gamma(1+\alpha)}+\frac{4t^{2\alpha}}{\Gamma(1+2\alpha)}-\frac{8t^{3\alpha}}{\Gamma(1+3\alpha)}+\frac{16t^{4\alpha}}{\Gamma(1+4\alpha)}-\frac{32t^{5\alpha}}{\Gamma(1+5\alpha)}\right)$$

$$= E_\alpha(x^\alpha)\sum_{\varepsilon=0}^{5}\frac{(-2)^\varepsilon t^{\varepsilon\alpha}}{\Gamma(1+\varepsilon\alpha)}. \tag{4.8}$$

$$u_3 = u_2(x,t) + {}_0I_t^{(\alpha)}\left\{\frac{(\xi-t)^\alpha}{\Gamma(1+\alpha)}\left(\frac{\partial^{2\alpha}u_2(x,\xi)}{\partial\xi^{2\alpha}} - \frac{\partial^{2\alpha}u_2(x,\xi)}{\partial x^{2\alpha}} + \frac{\partial^{\alpha}u_2(x,\xi)}{\partial\xi^{\alpha}} - u_2(x,\zeta)\right)\right\}$$

$$= E_\alpha(x^\alpha)\left(\begin{array}{l}1-\frac{2t^\alpha}{\Gamma(1+\alpha)}+\frac{4t^{2\alpha}}{\Gamma(1+2\alpha)}-\frac{8t^{3\alpha}}{\Gamma(1+3\alpha)}+\frac{16t^{4\alpha}}{\Gamma(1+4\alpha)}-\\ \frac{32t^{5\alpha}}{\Gamma(1+5\alpha)}+\frac{64t^{6\alpha}}{\Gamma(1+6\alpha}-\frac{128t^{7\alpha}}{\Gamma(1+7\alpha)}\end{array}\right)$$

$$= E_\alpha(x^\alpha)\sum_{\varepsilon=0}^{7}\frac{(-2)^\varepsilon t^{\varepsilon\alpha}}{\Gamma(1+\varepsilon\alpha)}. \tag{4.9}$$

$$\vdots$$

$$u_n = E_\alpha(x^\alpha)\sum_{\varepsilon=0}^{2n+1}\frac{(-2)^\varepsilon t^{\varepsilon\alpha}}{\Gamma(1+\varepsilon\alpha)}. \tag{4.10}$$

$$u(x,t) = E_\alpha(x^\alpha)E_\alpha(-2t^\alpha). \tag{4.11}$$

The plots of the FVIM solutions of (4.1) with initial conditions (4.2) and the boundary conditions (4.3) are shown in the figures 1,2,3 for different values of $\alpha = 1.\frac{1}{2}.\frac{\ln(2)}{\ln(3)}$.

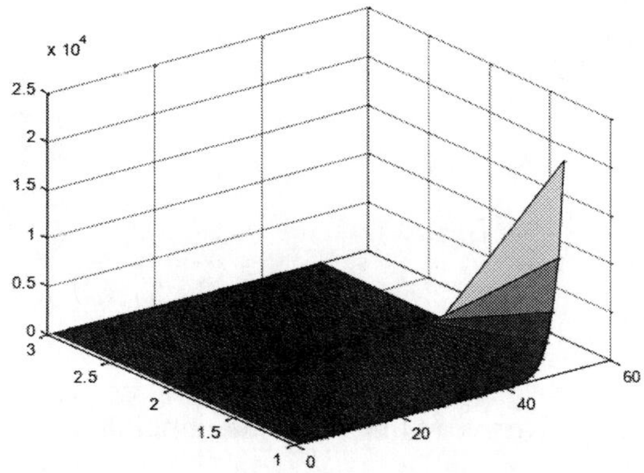

Figure 1: The plot of solution to local fractional telegraph equation with fractal dimension $\alpha = 1$.

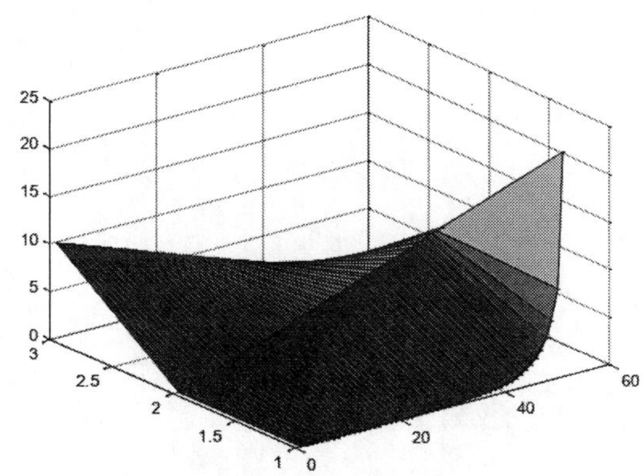

Figure 2: The plot of solution to local fractional telegraph equation with fractal dimension $\alpha = \frac{1}{2}$.

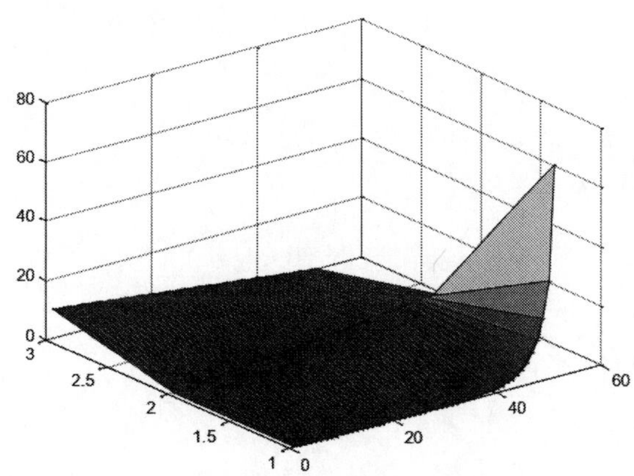

Figure 3: The plot of solution to local fractional telegraph equation with fractal dimension $\alpha = \frac{\ln 2}{\ln 3}$.

Example 2. Consider the nonlinear telegraph equation with LFDOs:

$$\frac{\partial^{2\alpha} u}{\partial x^{2\alpha}} = \frac{\partial^{2\alpha} u}{\partial t^{2\alpha}} + \frac{\partial^{\alpha} u}{\partial t^{\alpha}} + u^2 + E_\alpha(x^\alpha + t^\alpha) - E_\alpha(2x^\alpha)E_\alpha(-2t^\alpha), \quad 0 < \alpha \le 1 \tag{4.12}$$

and

$$u(x,0) = E_\alpha(x^\alpha), \quad \frac{\partial^\alpha u(x,0)}{\partial t^\alpha} = -E_\alpha(x^\alpha), \tag{4.13}$$

$$u(0,t) = E_\alpha(-t^\alpha), \quad \frac{\partial^\alpha u(0,t)}{\partial x^\alpha} = E_\alpha(-t^\alpha). \tag{4.14}$$

Applying Eqs. (3.8) and (4.12), we arrive at the following iteration formula:

$$u_{n+1} = u_n + {}_0I_t^{(\alpha)}\left\{\frac{(\xi - t)^\alpha}{\Gamma(1+\alpha)}\left(\frac{\partial^{2\alpha} u_n}{\partial \xi^{2\alpha}} - \frac{\partial^{2\alpha} u_n}{\partial x^{2\alpha}} + \frac{\partial^{\alpha} u_n}{\partial \xi^{\alpha}} + u_n^2 + E_\alpha(x^\alpha + \xi^\alpha) - E_\alpha(2x^\alpha)E_\alpha(-2\xi^\alpha)\right)\right\} \tag{4.15}$$

By using Eqs. (3.9) and (4.13), we obtain

$$u_0 = E_\alpha(x^\alpha)\left(1 - \frac{t^\alpha}{\Gamma(1+\alpha)}\right). \tag{4.16}$$

Therefore, we obtain:

$$\begin{aligned} u_1 &= u_0 + {}_0I_t^{(\alpha)}\left\{\frac{(\xi - t)^\alpha}{\Gamma(1+\alpha)}\left(\frac{\partial^{2\alpha} u_0}{\partial \xi^{2\alpha}} - \frac{\partial^{2\alpha} u_0}{\partial x^{2\alpha}} + \frac{\partial^{\alpha} u_0}{\partial \xi^{\alpha}} + u_0^2 + E_\alpha(x^\alpha + \xi^\alpha) - E_\alpha(2x^\alpha)E_\alpha(-2\xi^\alpha)\right)\right\} \\ &= E_\alpha(x^\alpha)\left(1 - \frac{t^\alpha}{\Gamma(1+\alpha)} + \frac{t^{2\alpha}}{\Gamma(1+2\alpha)} - \frac{t^{3\alpha}}{\Gamma(1+3\alpha)}\right). \end{aligned} \tag{4.17}$$

$$\begin{aligned} u_2 &= u_1 + {}_0I_t^{(\alpha)}\left\{\frac{(\xi - t)^\alpha}{\Gamma(1+\alpha)}\left(\frac{\partial^{2\alpha} u_1}{\partial \xi^{2\alpha}} - \frac{\partial^{2\alpha} u_1}{\partial x^{2\alpha}} + \frac{\partial^{\alpha} u_1}{\partial \xi^{\alpha}} + u_1^2 + E_\alpha(x^\alpha + \xi^\alpha) - E_\alpha(2x^\alpha)E_\alpha(-2\xi^\alpha)\right)\right\} \\ &= E_\alpha(x^\alpha)\left(1 - \frac{t^\alpha}{\Gamma(1+\alpha)} + \frac{t^{2\alpha}}{\Gamma(1+2\alpha)} - \frac{t^{3\alpha}}{\Gamma(1+3\alpha)} + \frac{t^{4\alpha}}{\Gamma(1+4\alpha)} - \frac{t^{5\alpha}}{\Gamma(1+5\alpha)}\right). \end{aligned} \tag{4.18}$$

$$\vdots$$

$$u_n = E_\alpha(x^\alpha)\sum_{\varepsilon=0}^{2n+1} \frac{(-1)^\varepsilon t^{\varepsilon\alpha}}{\Gamma(1+\varepsilon\alpha)}. \tag{4.19}$$

$$\begin{aligned} u &= \lim_{n\to\infty} u_n(x,t) \\ &= E_\alpha(x^\alpha)\sum_{\varepsilon=0}^{\infty} \frac{(-1)^\varepsilon t^{\varepsilon\alpha}}{\Gamma(1+\varepsilon\alpha)} \\ &= E_\alpha(x^\alpha - t^\alpha). \end{aligned} \tag{4.20}$$

The plots of the FVIM solutions of (4.12) with initial conditions (4.13) and the boundary conditions (4.14) are shown in the figures 1,2,3 for different values of $\alpha = 1.\frac{1}{2}.\frac{\ln(2)}{\ln(3)}$.

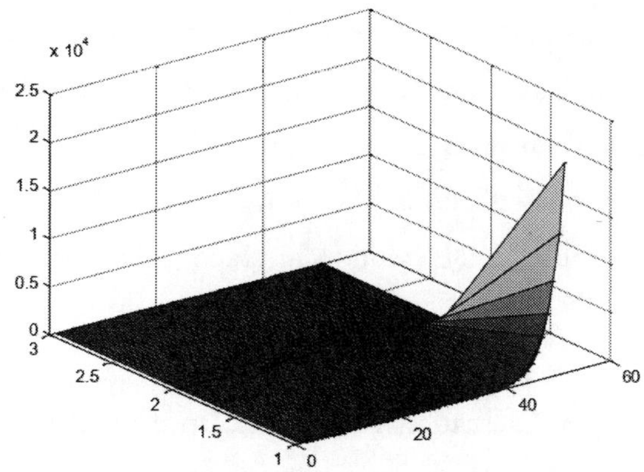

Figure 4: The plot of solution to problem (4.12) with fractal dimension $\alpha = 1$.

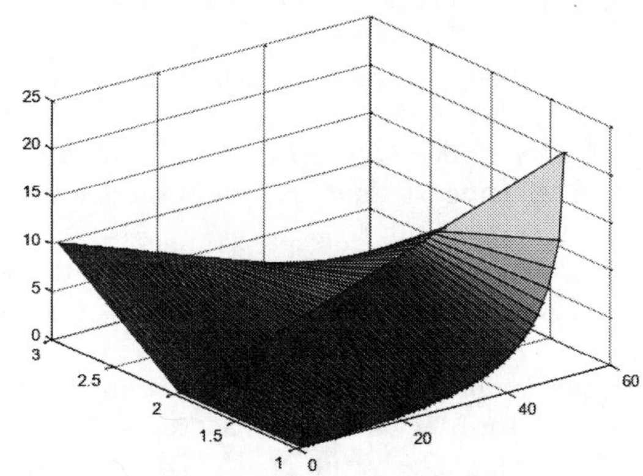

Figure 5: The plot of solution to problem (4.12) with fractal dimension $\alpha = \frac{1}{2}$.

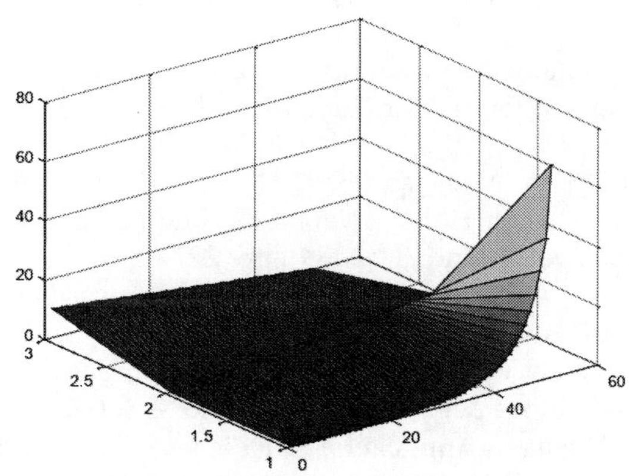

Figure 6: The plot of solution to problem (4.12) with fractal dimension $\alpha = \frac{\ln 2}{\ln 3}$.

5. Conclusion

In this manuscript, utilizing the LFDOs, we investigated the telegraph equations. Based on the LFVIM, the solutions of the local fractional telegraph equations were presented. The iteration functions, which is local fractional continuous, is obtained easily within the fractal Lagrange multipliers, which can be optimally determined by the local fractional variational theory. It is shown that the LFVIM is an efficient and simple tool for handling PDEs with local fractional differential operator.

References

[1] M. Garg, P. Manohar, and S. L. Kalla, Generalized differential Transform method to space-time fractional telegraph equation, International Journal of Differential Equations, vol. 2011, Article ID 548982, (2011) pp. 1-8.

[2] J. Chen, F. Liu, and V. Anh, Analytical solution for the time fractional telegraph equation by the method of separating variables, Journal of Mathematical Analysis and Applications, vol. 338, no. 2, (2008), pp. 1364-1377.

[3] A. Ansari, Fractional exponential operators and time fractional telegraph equation," Boundary Value Problems, vol. 2012, Article 125, (2012), pp. 1-7.

[4] F. Huang, Analytical solution for the time-fractional telegraph equation," Journal of Applied Mathematics, vol. 2009, Article ID 890158, (2009), pp. 1-10.

[5] S. Das, K. Vishal, P. K. Gupta, and A. Yildirim, An approximate analytical solution of time-fractional telegraph equation, Applied Mathematics and Computation, vol. 217, no. 18, (2011), pp. 7405–7411.

[6] J. H. He, Approximate analytical solution for seepage flow with fractional derivatives in porous media, Computer Methods in Applied Mechanics and Engineering, vol. 167, no.1-2, (1998), pp. 57–68.

[7] J.-H. He, Variational iteration method a kind of non-linear analytical technique: some examples, International Journal of Non-Linear Mechanics, vol. 34, no. 4, (1999), pp. 699–708.

[8] J. H. He, G. C. Wu, and F. Austin, The variational iteration method which should be followed, Nonlinear Science Letters A, vol. 1, no. 1, (2010), pp. 1–30.

[9] S. Momani and S. Abuasad, Application of He's variational iteration method to Helmholtz equation, Chaos, Solitons & Fractals, vol. 27, no. 5, (2006), pp. 1119–1123.

[10] M. A. Abdou and A. A. Soliman, Variational iteration method for solving Burger's and coupled Burger's equations, Journal of Computational and Applied Mathematics, vol. 181, no. 2, (2005), pp. 245– 251.

[11] S. Abbas bandy, Numerical solution of non-linear Klein- Gordon equations by variational iteration method, International Journal for Numerical Methods in Engineering, vol.70, no. 7, (2007) pp. 876-881.

[12] S. T. Mohyud-Din, M. A. Noor, and K. I. Noor, Traveling wave solutions of seventh-order generalized KdV equations using He's polynomials, International Journal of Nonlinear Sciences and Numerical Simulation, vol. 10, no. 2, (2009), pp. 227–234.

[13] V. Marinca, N. Heris, and C. Bota, Application of the variational iteration method to some nonlinear one-dimensional oscillations, Meccanica, vol. 43, no. 1, (2008) pp. 75–79.

[14] A. M. Wazwaz, A study on linear and nonlinear Schrodinger equations by the variational iteration method, Chaos, Solitons & Fractals, vol. 37, no. 4, (2008), pp. 1136–1142.

[15] M. Dehghan and F. Shakeri, Application of He's variational iteration method for solving the Cauchy reaction-diffusion problem, Journal of Computational and Applied Mathematics, vol. 214, no. 2, (2008) pp. 435–446.

[16] S. Das, Approximate solution of fractional diffusion equation revisited, International Review of Chemical Engineering, vol. 4, no. 5, (2012), pp. 501–504.

[17] J. Hristov, An exercise with the He's variation iteration method to a fractional Bernoulli equation arising in a transient conduction with a non-linear boundary heat flux, International Review of Chemical Engineering, vol. 4, no. 5, (2012) pp. 489–497.

[18] G.-C. Wu and E. W. M. Lee, Fractional variational iteration method and its application, Physics Letters A General, Atomic and Solid State Physics, vol. 374, no. 25, (2010), pp. 2506–2509.

[19] S. Xu, X. Ling, Y. Zhao, H. K. Jassim, A Novel Schedule for Solving the Two-Dimensional Diffusion in Fractal Heat Transfer, Thermal Science, Vol. 19, Suppl. 1, pp. S99-S103, 2015.

[20] H. Jafari, H. K. Jassim, S. P. Moshokoa, V. M. Ariyan and F. Tchier, Reduced differential transform method for partial differential equations within local fractional derivative operators, Advances in Mechanical Engineering, 8(4), 1-6, 2016.

[21] S. Q. Wang, Y. J. Yang, and H. K. Jassim, Local Fractional Function Decomposition Method for Solving Inhomogeneous Wave Equations with Local Fractional Derivative, Abstract and Applied Analysis, vol. 2014, Article ID 176395, (2014), pp. 1-7.

[22] S. P. Yan, H. Jafari, and H. K. Jassim, Local Fractional Adomian Decomposition and Function Decomposition Methods for Solving Laplace Equation within Local Fractional Operators, Advances in Mathematical Physics, vol. 2014, Article ID 161580, (2014), pp. 1-8.

Water battery cell of carbon - PMMA with Cu - Zn electrodes for green energy

F I Mustafa*, A K Al-Sudani, H H Al-Qazzaz
Solar Energy Research Center, Renewable Energy Directorate
Ministry of Science and Technology, Baghdad, Iraq

E-mail: falah_im@yahoo.com

Abstract. Water battery cell is a green energy, and it is a new innovation in the field of renewable green energy as well as that input is zero energy. It produces electricity without generating any negative carbon effects on the environment. Furthermore, it is a costless innovation, and the principle of work is based on using the tap water as a fuel source for the battery. That is mean, the battery doesn't need power charging process (zero input), and it depends on the electrochemical analysis of water into positive and negative ions. The water battery cell (WBC) consists of 8 cells connected in series to produce a suitable voltage value. Each cell is based on Zn and Cu metal electrodes with carbon- PMMA layer sandwiched between them. The (WBC) was tested in different medias such as normal and hot waterbetween 25°C to 60°Ctemperature water to ensure the strength of the innovation. The (WBC)hastested by recording the current, voltage and power values in each water solution media with load and without load (LED lamp). The maximum electrical power record of the battery is65 mWwhen using tap water without load and 10 mW with load at 25°C, then 43 mW when using hot tap water with load at 60°C. The load represents 4 red LED lampsof 10 mA and 1.6 V each, 4 yellow LEDlamps of 10 mA and 1.8 V each, 4 green LED lamps of 10 mA and 1.8 V each were used as a load separately.

1. Introduction

Water cell like in shape as a galvanic cell whose electrodes are zinc and copper submerged in zinc sulfate and copper sulfate, respectively, and it is known as a Daniel cell. They are known as two half cells,and it is often used due to its simplicity. The anode is the zinc metal which is oxidized (loses electrons) to form zinc ions in solution, and copper ions accept electrons from the copper metal electrode and the ions deposit at the copper cathode as an electrodeposit. This cell forms a simple battery as it will spontaneously generate a flow of electric current from the anode to the cathode through the external connection [1].

Basically, to make a water battery cell that combine any two different kinds of metal can be placed in a conducting solution to make a battery. Familiar water batteries include sheet of metals from copper and zinc strips into water, using lemon solution to make battery. The aluminum-copper-coke battery will produce about three quarters of a volt. Another easy battery to make zinc-air battery. In this battery, the strip of zinc is oxidized by dissolved oxygen ina salty water. Three of the copper-zinc-coke batteries produce about 3 volts, and can replace the 3-volt lithium battery in this small clock/calendar/calculator. This power generates electricity without affecting the environment andreduce the pollution and fight the carbon emissions to the environment.

IOP Conf. Series: Journal of Physics: Conf. Series **1032** (2018) 012016 doi:10.1088/1742-6596/1032/1/012016

2. Theory

Metal atoms are held together by electrical attractions between the nuclei and the electrons around the atoms.When you place a strip of metal in a glass of water, the water molecules interact with the metal atoms on the surface of the strip. Water molecules are polar, meaning the one side is a slightly positive, and the other side is a slightly negative. This is because the two hydrogen atoms are not on opposite sides of the oxygen atom, but are instead about 105° apart. The hydrogen side is positive, and the oxygen side is negative as shown in the figure 1.

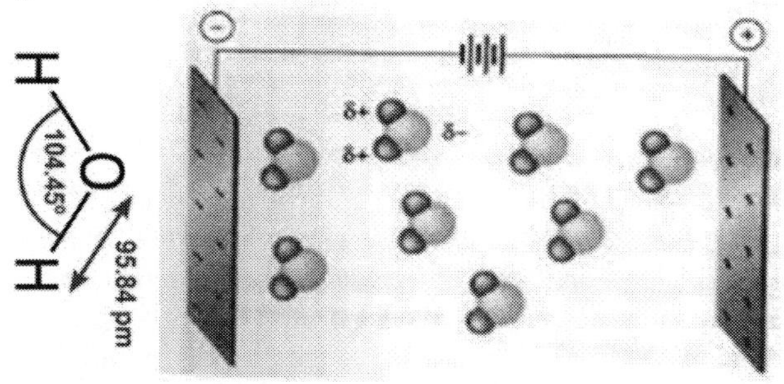

Figure 1. water molecular direction in the cell

At the interface between the water and the metal, some of the metal nuclei are attracted to the negative side of the water molecules. This attraction makes it easier for a metal nucleus to leave one or more of its electrons behind in the metal strip, and migrate away from the strip into the water.
The strip is left with a very small negative electric charge, because it now has one less positive nucleus in it. This tiny charge does not pull very much on the metal ion that has left the strip. In fact, that ion (a metal atom with one or more electrons missing) is quickly surrounded by water molecules, whose negative side is attracted to the positive metal ion. This blanket of water molecules spreads out the positive charge over a larger area, making it even less attracted to the metal strip.This is a very temporary effect, and the metal ion usually gets attracted back to the strip very quickly. But since there are enormous numbers of atoms at the surface of the metal strip, and an enormous number of metal ions are in the water at any given time, the metal strip ends up with quite a few more electrons than metal nuclei. This gives the strip a slight negative charge.

Some metals hold on to their atoms more tightly than others. This means that some metal strips will become more negative when placed in water than others do.If one metal strip has more extra electrons than another one does, those electrons will flow from the first strip to the second, until they both have the same charge. But to flow, the electrons need a conductive path. We give them that path when we connect two strips of different metals with a wire. The electrons then flow through that wire, creating an electric current [2].

The aluminum-air cell belongs to the category of metal-air cells and zinc-air batteries operating in alkaline electrolytes have been developed for industrial and commercial applications. The energy densities of the industrial batteries are 150-300 W h/kg and 175-400 Wh/liter.
The theoretical data for The following table 1 provides the reduction potentials of the indicated reducing agent at 25 °C. [3-4].

Table 1. Metal reduction potentials at 25 °C

Oxidizing agent	Reducing agent	Reduction potential (V)
$Li^+ + e^- =$	Li	−3.04
$Cu^{2+} + e^- =$	Cu^+	+0.16
$Zn^{2+} + 2e^- =$	Zn	−0.76
$Al^{3+} + 3e^- =$	Al	−1.66

2.1. COPPER, ZINC, AND SALT

A different chemistry happens when we have salt instead of acid in the water.Salt breaks up in water to make positive sodium ions and negative chloride ions. These ions reduce the energy needed for water to split into hydroxide ions (OH^-) and hydrogen ions H^+ (the hydrogen ions quickly find another water molecule and create hydronium ions, H_3O^+).At the zinc strip, the zinc ion combines with four hydroxide ions to form one ion of zincate ($Zn(OH)_4^{2-}$), leaving two electrons behind on the zinc strip. The chlorine ions from the salt then combine with the hydronium ions left over when the hydroxide ions were taken away by the zinc, and form hydrochloric acid.

Over on the copper strip, four electrons combine with oxygen dissolved in the water and two molecules of water to form four hydroxide ions. The sodium ions from the salt combine with these hydroxide ions to make sodium hydroxide.The hydrochloric acid and the sodium hydroxide combine back into salt. So the salt is merely in the picture as a way to move charges through the water. It is not used up.We can summarize what happens at the zinc strip (called the anode this way:

$$Zn + 4OH^- \Rightarrow Zn(OH)_4^{2-} + 2e^-$$

$$4Cl^- + 4H_2O \Rightarrow 4HCl + 4OH^-$$

$$Zn(OH)_4^{2-} \Rightarrow ZnO + H_2O + 2OH^-$$

At the copper strip (called the cathode) we have:

$$O_2 + 2H_2O + 4e^- \Rightarrow 4OH^-.$$

$$4Na^+ + 4OH^- \Rightarrow 4NaOH$$

The water battery cell canhave called a zinc-air battery. The oxygen from the air is combining with the zinc [5]. The copper electrode is just there to conduct the electrons, and does not participate in the chemistry. It can be replaced with a carbon rod.You may notice that after a short while, the oxygen in the battery is used up, and the current (and thus the brightness of the LED) begins to drop. Stirring the salt water helps to put more oxygen in the water, and the LED gets bright again [6-11].

3. Experiments and characterization techniques

The (WB) consists of 8 water cells of ($9*5cm^2$) connected in series. as shown in the figure 2, and each cell consists of Zn- Cu electrodes, and a thin layer of carbon- PMMA material sandwiched between the electrodes with thickness about 2mm. The Cu or the positive electrode is a thin plate of pure Cu of ($5*9cm^2$) surface area, and a thickness of 0.2 mm. The Zn or the negative electrode is a thin plate of pure 80% Zn metal and the surface area is about ($5*9cm^2$) with the thickness of 0.3 mm. The carbon- PMMA was prepared by dissolving 2.5g of carbon powder and 2.5g of PMMA polymer in 8 ml of acetone as solvent. The paste was mixed carefully till forming a homogenous thin layer. This layer was rolled and pressed between the two electrodes.

IOP Conf. Series: Journal of Physics: Conf. Series **1032** (2018) 012016 doi:10.1088/1742-6596/1032/1/012016

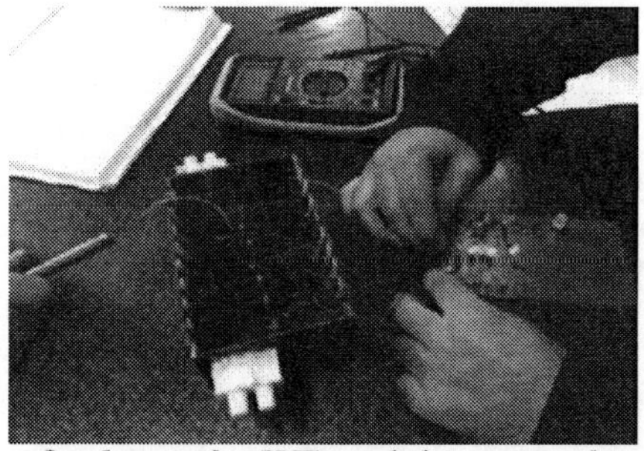

Figure 2. shows the WB and the measuring tools.

After finishing the preparation of the 8 cells, we packed them inside a suitable plastic prepared box and connected them in series by cupper wires. A Fluke multi-meter was used to measure the current (ampere), voltage (volt), and power characteristics by multiply current with voltage (watt). The PH-208 meter was used to test the PH of all solutions. A heat controlled plate heater of Chemat Technology was used to test the performance of the WB under the 60-degree centigrade water and all temperature measurements have tested and recorded by thermometer data logger with sensor type K.

4. Results and discussion

Each cell of the 8 cells consists of Cu plate as a cathode and a Zn plate as an anode. A Thin layer of carbon- PMMA was sandwiched between the electrodes. The carbon- PMMA layer was used to accelerate water analysis into positive and into negative ions to release electrons to generate electricity. The cells were connected in series by using cupper wires to get a suitable voltage values and then the cells were put inside 600ml of tap water to start the electrical testes. The WB was tested inside 600ml of 25°C tap water and it was tested inside 600ml of 60°C tap water to ensure the normality of the electrical performance in a hot environment. We test the WB without using load (open circuit voltage and short circuit current and theoretical power). The WB was also tested under load to record the values of current, voltage, and real power. Thecathode and anode voltage values depends on the oxidation and reduction reactions of the electrodes (REDOXreactions).The difference between the cathodic and the anodic voltages will modify the whole voltage of the cell. The whole electrical current of the cell is highly affected by the ionic content and the dissolved minerals inside the electrolyte (tap water). The tap water consists of the following elements: ((Ca++, Na+, Mg++, K+, Cl-, Al+++, Fe++, SiO2, CO3, SO4) [4].

These ions could react with the metal sheets to release more electrons; as a result, they will increase the value of the total current of the cell. The first case, we tested the WB in 25°C tap water without load and with load. All the voltage, current, and power values were recorded each 5 minutes for 45 min. Figure 3. shows the electrical performance of the WB.

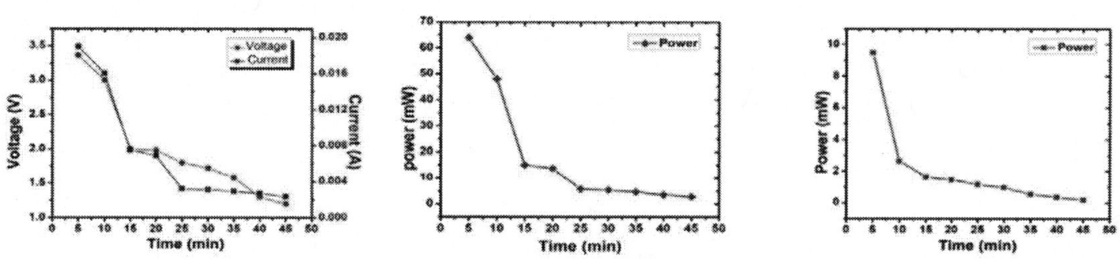

Figure 3. electrical performance of the WB at (25°C) tap water without using load the voltage and current values through 45 mins and the power carve of the water battery.

IOP Conf. Series: Journal of Physics: Conf. Series **1032** (2018) 012016 doi:10.1088/1742-6596/1032/1/012016

In this test, the maximum power of the WB was achieved as (65 mW). We can see the values of the electrical terms start to be decreased with the time due to the chemical analysis of water. The second test was done in 25°C with using load to see the difference in the electrical performance. The WB was tested inside 600ml of 60°C tap waterto check the electrical performance of the WB inside a hot media. The power curve of the water battery working inside 25°C water under 4 red LEDs of 1.6 V and 10 mA each connected in parallel.

The maximum power in figure 4 is about 65 mW in 60°C water without using the electrical load. The increment that observed when using hot water is due to increasing the dynamic movement of the positive and negative ions inside the water.

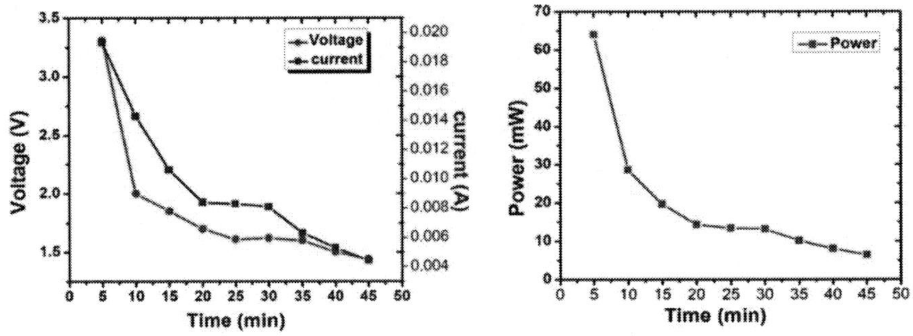

Figure 4. shows the electrical performance of the WB in 60°C without using the electrical load

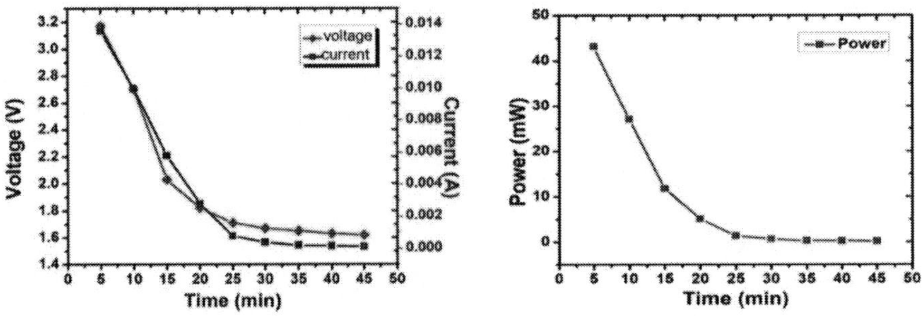

Figure 5. the electrical performance of the WB inside 60°C tap water under electrical load.

In figure 5 the WB works in 60°C water under 4 red LEDs of 1.6 V and 10mA each. The maximum power was recorded here as 43 mW. In order to understand the differences in the electrical performance of the WB inside cooled and hot water with load and without load, figure 6 will make it clear for good understanding.

IOP Conf. Series: Journal of Physics: Conf. Series **1032** (2018) 012016 doi:10.1088/1742-6596/1032/1/012016

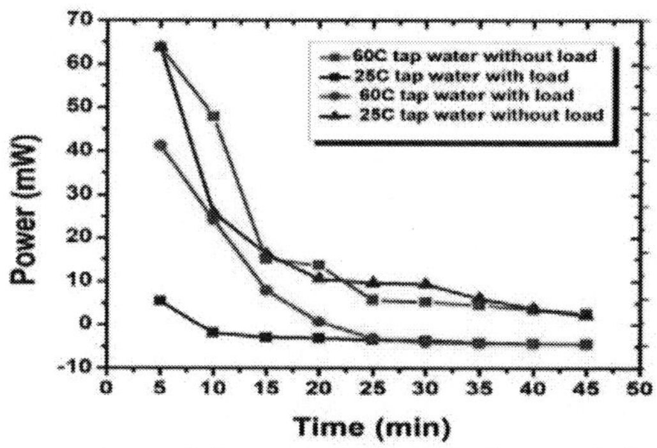

Figure 6. shows the power comparison of the water battery inside cooled and hot water with and without load.

From figure 6 the WB working inside 25°C tap water without load gave the Max power of 65mW, while the WB working inside a 60°C tap water without load gave about 64mW which are the same. The effect of increasing the temperature on the electrical performance was clearly seen on the red line in figure 6. There is a difference in the performance under an electrical load between the red and the black line which clearly reflects the positive effect of increasing temperature. To see the electrical performance against the temperature increasing. Figure 7, shows the effect of temperature on the power of the WB working inside hot water under electrical load. PH factor is very important to be recorded to see what happen to water during the chemical analysis of it. Figure 8, shows the pH value changing with time.

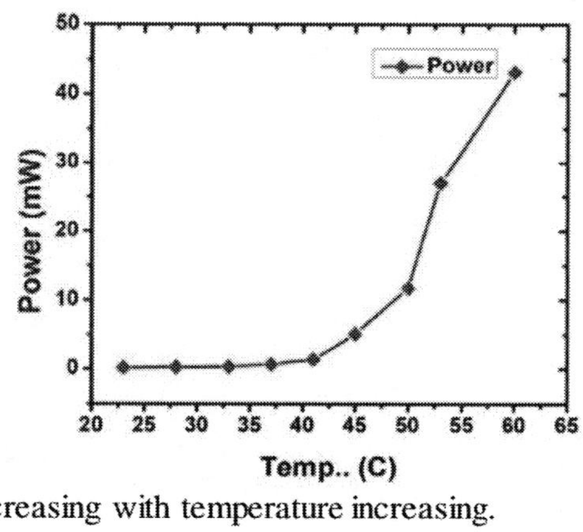

Figure 7. shows the power increasing with temperature increasing.

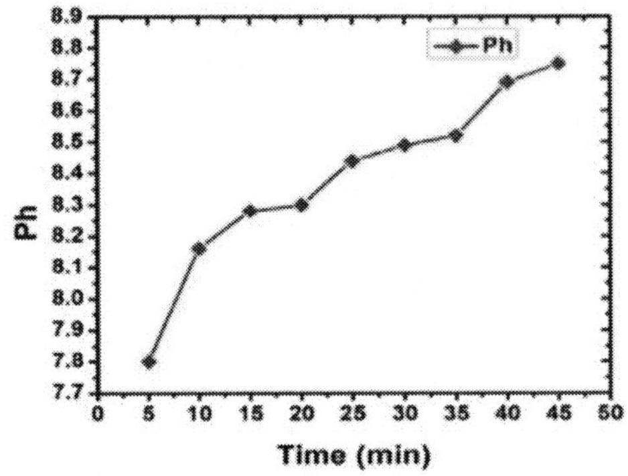

Figure 8. shows the PH variation with the time

From figure 8, we notice the increasing in the PH value with the time which means the water goes toward the based nature due to increasing the (OH^-) ions in the liquid. The positive ions may generate an H_2 gas and escapes from water.

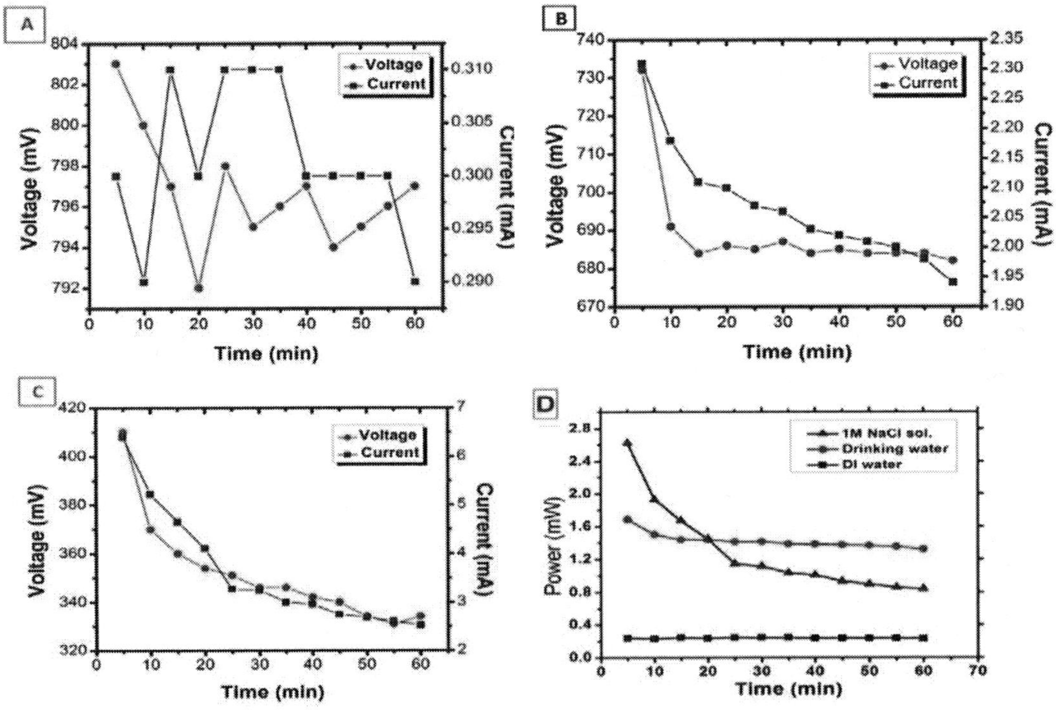

Figure 9. Cu-Zn cell electrical test (voltage, current, and power) at 25°Cat (A) Deionized water. (B) Drinking water. (C) 1M NaCl solution. (D) Power curves of the cell.

5. Conclusion

Water battery (WB) was designed and tested here to open a new way of utilizing the hydropower. WB is a costless device and a very clean energy source. It is also characterized by electrochemically stable and long life time. The WB consist of 8 symmetric water cell connected in series. The WB was tested in different medias with electrical load and without. The max power generated from the device was 65mW when working inside 600ml of 25°C tap water without connecting to an electrical load. While it produced 64mW when working inside 600ml 60°C tap water without using electrical load. The WB produced 43mW power inside 60°C tap water under the electrical load. The electrical load was 4 red LED of 1.6 V and 10 mA each. For the future steps, we will try to optimize the elements of the WB to evaluate the electrical performance of the WB.

6. Acknowledgments

This research was supported by Scientific Projects in Solar Energy Research Center, Renewable Energy Directorate Ministry of Science and Technology of 2018 year in Baghdad Iraq.

References

[1] S James N.; Bodner, George M.; Rickard, Lyman H. 2010 *Chemistry: Structure and Dynamics (Fifth Edition).* John Wiley & Sons. p. 564. ISBN 9780470587119.

[2] Robinson, R Anthony, and R H Stokes 2002 *Electrolyte solutions. Courier Corporation.*

[3] O J Murphy et al 1992 *Electrochemistry in Transition, Plenum Press, New York.*

[4] Burgot, Jean-Louis 2012 *"Redox Reactions and Electrochemical Cells." In Ionic Equilibria in Analytical Chemistry, Springer. New York.* pp. 205-227.

[5] E L Littauer and J F Cooper 1984 *"Metal air batteries," in Handbook of Batteries and Fuel Cells, edited by D. Linden (McGraw Hill, New York*, pp. 30-1–30-21.

[6] Ponrouch, A Goñi, A. & Palacín, M. R 2013 *High capacity hard carbon anodes for sodium ion batteries in additive free electrolyte. Electrochem. Comm.* **27**, 85–88.

[7] M Tamez and J H Yu 2007 *"Aluminum–air battery," J. Chem. Educ.* **84**, 1936A.

[8] D Rathjen and P. Doherty, 2002 *Square Wheels and Other Easy-to-Build Hands-On Science Activities (Exploratorium, San Francisco)*, p.85.

[9] J P Hoare, 1985 *"Oxygen," in Standard Potentials in Aqueous Solution, IUPAC, edited by A.J. Bard, R. Parsons, and J. Jordan (Marcel Dekker, Inc., New York*, pp.54–66.

[10] S M Park 1985 *"Boron, aluminum and scandium," in Standard Potentials in Aqueous Solution, IUPAC, edited by A.J. Bard, R. Parsons, and J. Jordan (Marcel Dekker, Inc., New York*, pp. 566-580.

[11] P N Ross, *"Hydrogen" in Standard Reduction Potentials, IUPAC, edited by A.J. Bard, R. Parsons, and J. Jordan (Marcel Dekker, Inc., New York),* pp. 39-48.

Biocompatibility, Bioactivity and Corrosion Resistance of Stainless Steel316L Nanocoated with TiO_2 and Al_2O_3 by Atomic Layer Deposition Method

Muna Khethier Abbass [1], Sami Abualnoun Ajeel [2] and Haitham Mohammed Wadullah[3]

[1] Dept. of Production Engineering. and Metallurgy , University of Technology, Baghdad, Iraq , Email: dr.munakabbass @gmail.com

[2] Dept. of Production Engineering. and Metallurgy , University of Technology, Baghdad, Iraq , Email: samiabualnon2@yahoo.com
[3] Engineering Technical College, Northern Technical University, Mosul, Iraq, Email: dr.haitham@ntu.edu.iq,

corresponding author
*Email: dr.munakabbass @gmail.com

Abstract. Atomic Layer Deposition (ALD) method has been used to synthesis nanocoatings thin films of Alumina, Titania, and Alumina/Titania multilayer on stainless steel AISI 316L at 250 °C deposition temperatures for the medical applications. SEM and EDX have been used to characterize the morphology of the films and the element analysis of the alloys and the thin films respectively. Open circuit potential, potentiostatic polarization (Tafel extrapolation) and cyclic polarization methods have been used to study the corrosion resistance of the films in Simulation Body Fluid (SBF) at 37 ± 1 °C. Immersion test in SBF for 2weeks at 37± 1 °C has been used to determine the biocompatibility of the films. Numbers of colonies and diffusion zone methods have been used after cultured non-pathogeni E-coli bacteria to demonstrate the bioactivity (toxicity effects) of the thin films. The SEM morphology observations show different particles shape and size of the Alumina (nanotubes shape around 10-20 nm in size) and Titania films (cauliflower particles shape around 20-50nm in size). The results also show that the corrosion resistance can be effectively enhanced by thin films, multilayer proved to be more corrosion protection than single layers, and Alumina has better corrosion resistance than Titania.

Keywords: Alumina, Titania, stainless steel 316L, ALD, thin films, corrosion resistance

1. Introduction

Corrosion is a complex multi-factorial phenomenon that depends on metallurgical, geometric, mechanical, and chemical parameters of the solution environment. Corrosion is the first main considerations for any type of metals to be used in surgical implants. Biomaterial is any natural or synthetic material that interfaces with living tissue and/or biological systems [1, 2]. Biomaterial science is physical and biological study of materials and their interaction in the body. Historically, biomaterials have been used for more than 100 years when the researcher Lane introduced the first metal plate for bone fracture fixation in 1895 [3, 4]. Austenitic Stainless Steel (316 and 316L), Cobalt-Chromium-Molybdenum alloy and Titanium and its alloys are commonly used in metals implants [5, 6, 7] . In 1920s, 18-8 stainless steels had been introduced and immediately attracted the interest of the clinicians due to the grouping of good mechanical properties and far-superior resistance to corrosion, in that time, and were used in a wide range of biomedical applications [8]. The corrosion resistance of stainless steel 316L is due to Chromium and Molybdenum contents, where chromium allows the formation of chro mium oxide (Cr_2O_3) intentionally on the surface of the metal when chromium in the surface layer reacts with oxygen, adherent and coherent oxide thin film (passive layer) envelops the surface and serves as a barrier to corrosion reactions [9, 10]. But, chromium oxide passive film formed on the surface of the implant metal is unstable and prone to localized corrosion in longtime implant applications due to harsh biological effect because the human body is a very complex and aggressive electrolytic environment for biomaterials applications. Also, austenitic stainless steels are not susceptible to intergranular corrosion by the precipitation of chromium carbides at the grain boundaries ($Cr_{23}C_6$), and depleting of the chromium from it, sensitization [11, 12]. Xiao-Yan Zhang et.al [13] has been found that the improvement of 316L SS for orthodontic bow corrosion resistance was depended on the laser power from studied the effect of laser surface remelting on the corrosion resistance of 316L Orthodontic bows and improving its corrosion resistance from artificial saliva.

Nanocoating term is used if the coating thicknesses in nanometer scales (1-100nm), thin films, or the structures are in nanostructure scales. Thin films and substrate are synthesis system with different enhanced properties because thin films materials have vast capacities that do not typically found in conventional coatings [14,15]. Mohammed et al. [16] in 2018 used the electrophoretic deposition (EPD) to produce the nanohydroxyapatite layers on 316L stainless steel substrate. The EPD coatings were prepared by the deposition of hydroxyapatite (HA)-chitosan nanocomposites on different substrate roughness (polish surface, 220 grit SiC grind, and sand blast surfaces). The results showed the deposited coatings on sand blasted substrate has less porosity compared with the polish surface and 220 emery paper SiC grinding substrate respectively. They obtained a good bonding between substrate and coatings, since this bonding have an important role in biomedical applications especially in bone replacement applications.

Currently, nanotechnology enables the production of ultra-thin films and nanocoatings consisting of just one monolayer or a few atomic layers. Atomic Layer Deposition (ALD) Method is a layer-by-layer deposition technique, gas phase chemical process, and a unique method for depositing high quality thin films even with extremely complex shapes. Its special modification of the more common used chemical vapor deposition (CVD) technique [17].

The aim of the present work is synthesis of high quality nanocoatings thin films of Alumina, Titania, and Alumina/Titania multilayer on stainless steel 316L specimens by use ALD method, then characterization and evaluation the structure of the nanocoatings thin films, and study the corrosion behavior, biocompatibility and bioactivity of the thin films deposited.

IOP Conf. Series: Journal of Physics: Conf. Series **1032** (2018) 012017 doi:10.1088/1742-6596/1032/1/012017

2. Experimental Work

Stainless steel type 316L provided by Good-fellow corporation-USA with specimen's area (20×20 mm^2) and thickness 0.9 mm. Table 1 shows the chemical composition of AISI 316L. The samples were ultrasonically cleaned by acetone for 60 minute, washed by methanol and de-ionized water (AMD) to remove all of the organic contaminants before dried by hot air and load into the deposition chamber.
Savannah Atomic Layer Deposition (ALD) reactor manufactured by Cambridge NanoTec Inc was used in Nanotechnology Core Facility (NCF) - University of Missouri-Columbia, USA to deposit Alumina (Al_2O_3), Titania (TiO_2) and Alumina /Titania multilayer.

Table 1 Standard and analytical composition of stainless steel AISI 316L

Element (wt%)	Cr	Ni	Mo	C	Si	Mn	P	Fe
Standard	16-18	10-14	2-3	0.03	0.75	2.0	0.045	Bal.
Analytical	16.5	10	2.0	0.027	0.375	1.32	0.045	Bal.

Titania was deposited by dimethylamido (TDMAT) precursors, chemical compound of Ti $(NMe_2)_4$, Alumina was deposited by Trimethylaluminum (TMA) precursors, chemical compound of Al $(CH_3)_3$. Nitrogen (N_2) was used as a carrier and purging gas. TDMAT, TMA precursors and Nitrogen gas were provided from Sigma–Aldrich (99.9%). Alumina/ Titania multilayer were by deposition of Alumina layer in the range of 25 nm using TMA and then above it deposited Titania layer in the range of 25 nm using TDMAT, H_2O precursors for each one was used, in the same vacuum chamber with continuous process and on different substrates. All the depositions were performed at a temperature of 250°C. The high temperature processes were chosen to obtain defect free, homogenous, and uniform layers in addition to a crystalline structure. The number of deposition cycles of each layer was calculated using a growth rate per cycle for each type which is ~0.39 A° /cycle for Titania and 1.07 A ° /cycle for Alumina, as shown in table 2.

IOP Conf. Series: Journal of Physics: Conf. Series **1032** (2018) 012017 doi:10.1088/1742-6596/1032/1/012017

Table 2 Parameters of deposition thin films.

The parameters	Alumina	Titania
System-growth temperature.	250°C	250°C
Heater 8 and 9 (deposition temperature)	250°C	250°C
Heater 6, 7 and10	150°C	150°C
N_2 gas flow	20 sccm	20 sccm
Pulse TMA (valve 3)	0.015 s, then wait 5s	0.1 pulse , then wait 5s.
Pulse H_2O (Valve 0)	0.015 pulse , then wait 5s	0.015 pulse , then wait 5s
Growth per cycle	~1.07 Angstrom/ cycle (Å / C)	~0.39 Angstrom /cycle (Å/C)
No of cycle	480cycle for 50 nm	1284 cycle for 50 nm

Ellipsometry Spectroscopic type WVASE32-USA instrument with software available for optical data analyses to measure the thickness and the refractive index coefficient (n) for thin films deposited on silicon wafer(100) in Nanotechnology Core Facility - University of Missouri-Columbia, USA. Incident and reflected angles between 65° to 75° were used at wavelengths from 2000 to 10000 nm. Cauchy model was fitted to guess the thickness by Cauchy software optimization.

SEM used is Hitachi S4700-Japan type with an operating voltage of 3.5-10.00 kV at resolution of 400 nm. Quanta 600F is a facility for Energy dispersive X-ray spectroscopy (EDX) with SEM machine used in order to find the element analysis of the thin films by using a working distance of 12 mm , accelerating voltage 10.0 kV and takes off angle 37.3 degree. SEM used in Electron Core Facility (ECF) - College of Veterinary Medicine in the University of Missouri-Columbia, USA .

For Electrochemical corrosion study, Potentiodynamic polarization type WINKING M Lab 200 Potentiostat/Galvanostat from Bank-Elektronik was carried out in Materials Engineering Dept.-University of Technology-Baghdad, Iraq. Simulation body fluid solution (SBF) used, as shown in table 3, at pH adjusted to 7.4 and temperature of 37±1°C. Tafel potential range was ± 250 V of the E_{corr} for uncoated and coated samples. Electrochemical software used at 5 mV/sec scan rate and SCE in potentiostat. The main parameters that can be got from the pote ntiodynamic polarization curves are the corrosion potential (E_{corr}) and the corrosion current density (i_{corr}). The breakdown potential (E_b) and repassivation potential (E_{rep}) were carried out by cyclic polarization at scan rate of 5 mV/s.

Corrosion rate is calculated depended on the current density the equation [18, 19];

$$\text{Corrosion Rate (mpy)} = 0.13 * \text{i corr} * \left[\frac{\text{e.w}}{\rho}\right] \quad (1)$$

where; i corr ; is corrosion current density ($\mu A.cm^{-2}$), e.w ; is equivalent weight (gm), and ρ ; is the density (gm/cm^3). Mpy meaning mils per year (0.001 in) as a penetration range.

Biocompatibility test was in Nanotechnology Core Facility - University of Missouri-Columbia, USA. It was done according to the ASTM F2475 and ASTM F981-04. The samples were immersed in a bio-cell, which was composed of container, pH-meter, thermocouple, hot plate magnetic stirrer, and Ringer solution at 37°C. Duration of immersion was 2-week. In the end of immersion time, the samples were

IOP Conf. Series: Journal of Physics: Conf. Series **1032** (2018) 012017 doi:10.1088/1742-6596/1032/1/012017

raised, dried and investigated by SEM-EDX in Production Engineering and Metallurgy Department-University of Technology- Baghdad, Iraq.

Table 3 Chemical composition of simulation body fluid solution (SBF) .

Formula	**gm. / liter**
Sodium chloride (NaCl)	9.00
Potassium chloride (KCl)	0.43
Calcium chloride 6H_2O ($CaCl_2$)	0.24
Sodium bicarbonate ($NaHCO_3$)	0.20

Bioactivity test was investigated in Nanotechnology Center-University of Technology- Baghdad, Iraq. It was done according to ASTM F895 – 11 (standard test method for agar diffusion cell culture screening for cytotoxicity). One of the most important reasons for culturing bacteria in vitro is to study the toxicity effects of the films on the bacteria, the non-toxic film that is killed fewer cells of non-pathogenic bacteria, bioactive films. In this section, the toxicity effects of alumina, titania and alumina/titania multilayers thin films is evaluated by cultured the E-coli bacterial on the thin films at 37 ℃ for 24 hr.

For a Preparation of Bacterial Microorganism, Non-pathogenic Escherichia coli (E-coli) bacteria used in this test were selected from the collection of the laboratory of Biotechnology branch in Department of Applied Science/University of Technology. The bacteria were sub cultured on nutrient agar (N. agar), England Company, and incubated for 24 h at 37 ℃. The Bacteria were suspended in 50 ml of normal saline to yield a bacterial suspension of around 107 (CFU/ml), Number of Colonies, using plate count. Normal saline solutions used was 0.9% of NaCl, prepared in conical flask by adding 0.9g of NaCl dissolved in 90 ml of distilled water . The pH of the solution was 7 and sterilized by autoclave before used. Number of Colonies [(CFU)/ml] = (Number of colonies for each dilution) / (dilution factor)/ sample volume. Also, estimated the percentage reductions of bacteria using [20]:

$$R = \frac{(Z-A)}{Z} x100\% \qquad (2)$$

Where Z & A is the number of live cell of bacteria per milliliter corresponding to the control samples and corresponding to the thin films samples cultured respectively. Diffusion zone method depended on the measured the area around the samples which has different color almost off white, number of bacteria died.

3. Results and Discussion

Table 4 shows the ellipsometry spectroscopy results of 50 nm of Alumina, Titania, and Alumina/Titania. For Alumina film, the number of cylices is 470 cycle, measured thickness is 47 nm, and refractive index is 1.638. The standard deviation in measured thickness compared to the theoretical thickness is about1.5% for 50 nm of Alumina. For obtaining 50 nm of Titania, the number of cycles is 1284 cycle, measured thickness is 43.04 nm, and refractive index is 2.145. The standard deviation in measured thickness compared to the theoretical thickness is about 3.5% for the 50 nm of Titania respectively. Measured thickness is 49.6 nm for Alumina /Titania multilayer, standard deviation is about 0.2 % from the theoretical thickness, and the refractive index is 1.98. On the other hand the measured thickness for single layer is 26.99 nm for Alumina at 25 nm and 26.45 nm for Titania at 25 nm, a total of 53.33 nm. The presence of multiple layers introduces added complexity in the software modeling used to interpret the ellipsometer results because each layer has different refractive index as denoted in a previous study [21].

IOP Conf. Series: Journal of Physics: Conf. Series **1032** (2018) 012017 doi:10.1088/1742-6596/1032/1/012017

Table 4 Data results by using ellipsometer device for thin film deposited by ALD.

Type	**50 nm Alumina**	**50 nm Titania**	**50 nm Alumina /Titania**
No of cycles (X)	480 (GPC=~1.07Å)	1284 (GPC=~0397Å)	250 cycle of Alumina+ 700 cycle of Titania
Theoretical Thickness	50 nm	50 nm	50 nm
Measured Thickness	47.3nm	43.04 nm	49.6 nm
Standard deviation	1.5%	3.5%	0.2 %
Refractive index (n)	1.638	2.145	1.98

It is concluded from the results of ellipsometry tests that the obtained thin films are characterized by excellent control over the thickness and uniformity. The standard deviation of the theoretical thickness did not exceed 3.5 % as a larger deviation in measured thickness for Titania thin films with 50 nm thicknesses, and the smaller deviation is 0.2 for Alumina/ Titania. Furthermore, the refractive index for Titania is higher around 2.145, and the lower refractive index is 1.618 for Alumina. The refractive index obtained in the acceptable limit as reported by previous authors [22- 25].

Figure (1a, c, & e) shows the SEM results of Alumina, Titania and Alumina /Titania multilayer thin films deposited on stainless steel316L specimens. It can be observed from the SEM images presented that there isn't any defect, micro crack and blistering phenomenon on the thin films for all thin films deposited. The thin films also have excellent homogeneity and large area uniformity. The uniformity of the thin films is related to the type of deposition method used, ALD, at high deposition temperature (250 °C). The Alumina films have nanotubes particles shape with very fine spherical particles size between them around 10-20 nm, while the Titania thin films are similar to cauliflower particles in shape with different particles size ranging between 20-50nm. This means that the particles shape and size of the Alumina films are different from that of Titania films. These results are similar to that reported in the literature [24]. The particles of the Alumina /Titania thin films have the same shape and size as that of the Titania thin film, because it's the upper layer.

From Figure 1 (b, d,& e) shows the EDX results, it can be observed that there are aluminum with oxygen peaks for 50 nm of Alumina, Titanium with Oxygen peaks for 50 nm of Titania, and Aluminum with Titanium with oxygen peaks for Alumina/Titania multilayer thin films.

IOP Conf. Series: Journal of Physics: Conf. Series **1032** (2018) 012017 doi:10.1088/1742-6596/1032/1/012017

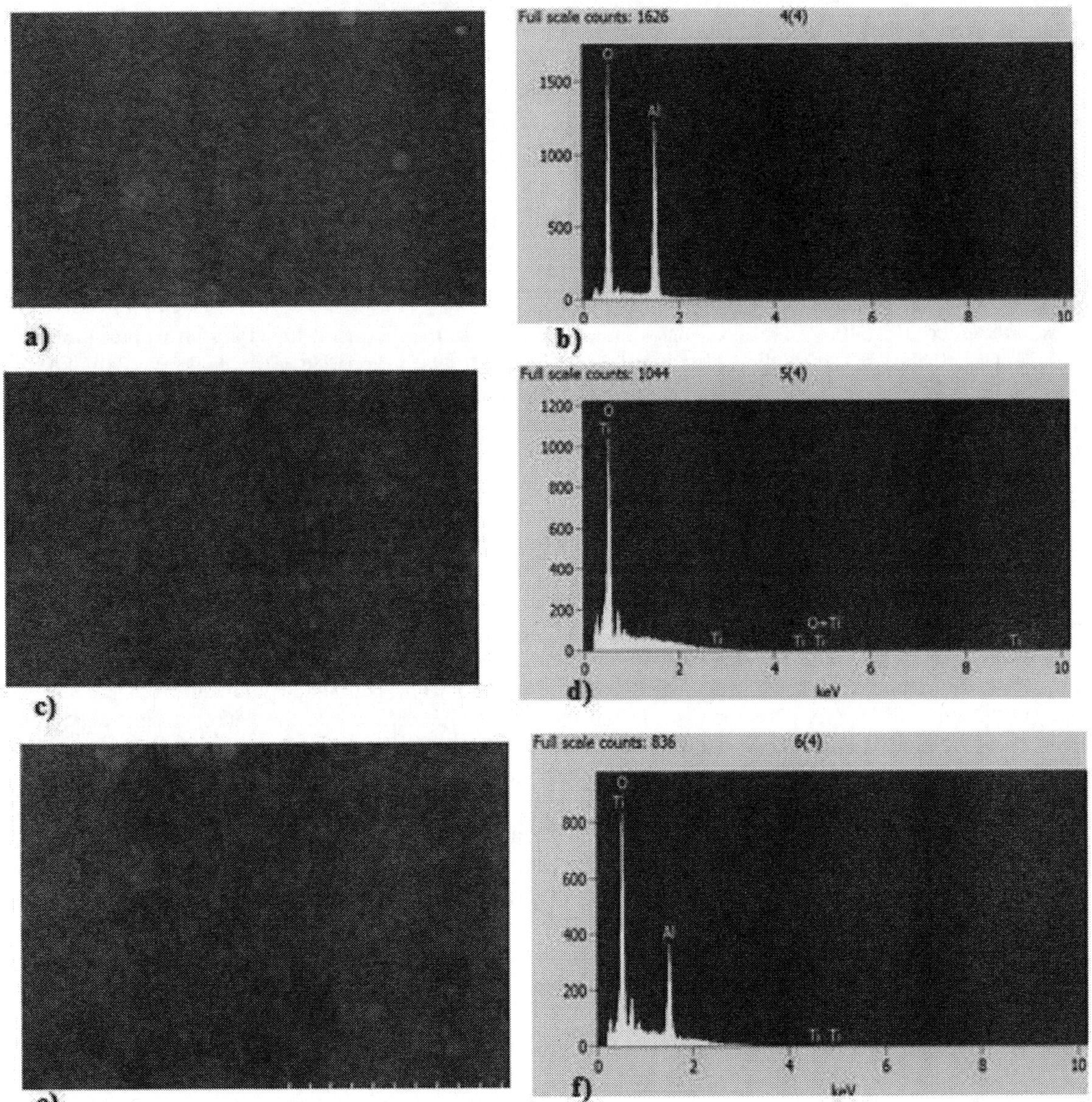

Figure . 1 SEM/EDX results of the thin film deposited on the stainless steel 316L;

a) SEM image 50 nm alumina
b) EDX of 50 nm alumina
c) SEM image of 50 nm titania
d) EDX of 50 nm titania
e) SEM image of alumina/titania multilayer
f) EDX of alumina/titania multilayer

Figure 2 and table 5 show the OCP- time results for 316L stainless steel specimens, the potential is generally changed from initial positive value of 170 mV to the more positive value of 182 mV and then remains stable at this value for uncoated specimens. The improved corrosion behavior of stainless steel 316L can be attributed to Chromium which forms passive Chromium oxide film (Cr_2O_3) as confirmed by [17]. The potentials of the OCP curves also for coated 316L stainless steel specimens by thin films of Alumina, Titania, and Alumina/Titania multilayer generally shift toward more noble direction and then are stable continuously with time. From the results are concluded that the stainless steel 316L specimens have passive behavior and after deposit thin films the corrosion resistance improved.

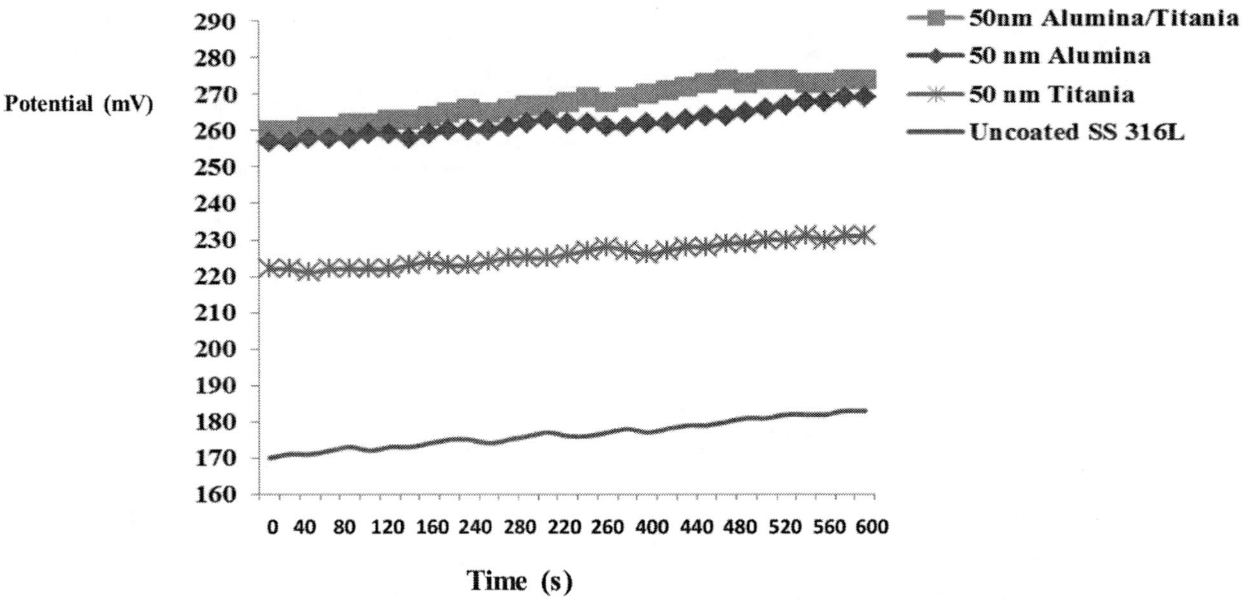

Figure. 2 Open circuit potential results in SBF solution at 37 ℃ ±1 for uncoated **and** coated Stainless steel 316L by alumina, titania and alumina /titania multilayer.

Table 5 OCP potentials of uncoated and coated stainless steel AISI 316L alloy

Type	Uncoated Co-Cr alloy	50nm Titania	50nm Alumina	Alumina / Titania
Potential start (mV)	+182	+231	+270	+275

Figure 3 and table 6 show uncoated and coated stainless steel 316L by Alumina, Titania, Alumina /Titania for investigating the cathodic and anodic polarization curves in SBF solution at 37±1 ℃. The corrosion potentials results of uncoated stainless steel 316L specimens have less positive corrosion potentials values and larger current densities than specimens coated 50 nm of Alumina, Titania and Alumina/ Titania respectively. The rapidly converting of Cr^{6+} to Cr^{3+} of alloys containing chromium decreases the concentration of chromium ions at the surface and then increases the dissolution of metal atoms [22]. It means that the corrosion resistances of coated specimens are better than that for uncoated specimens and the protective coatings by thin films, 50 nm deposited by ALD at 250℃, have effect on the improving the corrosion resistance. Multilayer of Alumina / Titania thin films deposited on stainless steel 316L specimens have a greater decreases in the corrosion current densities and increases in potential values. It means that the Alumina / Titania multilayer provided more protection in corrosion resistance than that of single layer of Alumina and Titania separate at 50 nm thicknesses, and Alumina film was more resistance than Titania. This is due to the chemically abrupt of the Alumina/Titania multilayer introduces as confirmed by reference [25]. The diffusion bonding between Alumina/Titania multilayers created chemically complex barriers being effective in the corrosion resistance [21].

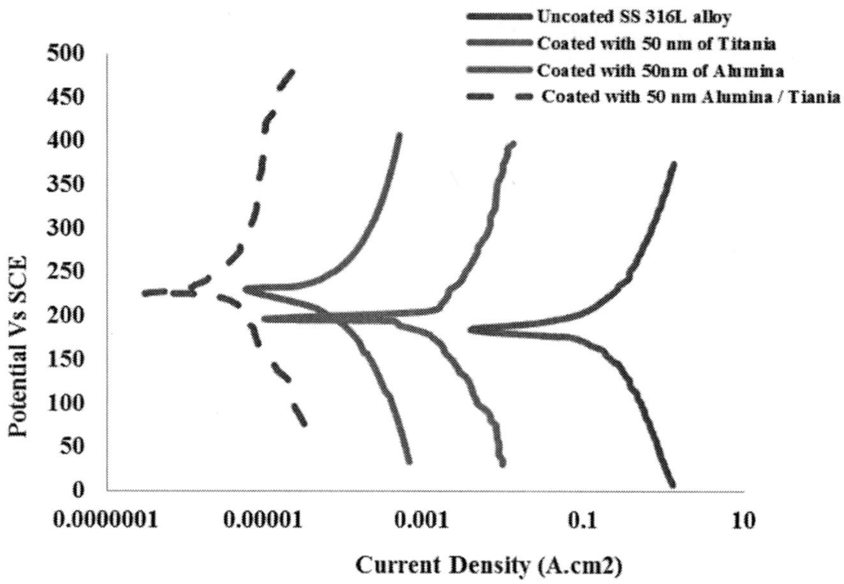

Figure. 3 Polarization results of coated Stainless steel 316L specimens with 50 nm of alumina, titania and alumina/titania in SBF solution at 37±1 ℃.

Table 6 Electrochemical corrosion results of AISI 316L alloy specimens .

Condition	E_{corr} Mv	i_{corr} $nA.cm^{-2}$	-bc mV/Dec	+ba mV/Dec	C.R (mpy)
Uncoated Ss 316L -alloy	180.9	$4.56*10^{0}$	110.3	85.8	1.913
50 nm of Titania	200	$4.8*10^{-1}$	119	134	0.2
50nm of Alumina	200	$4.8*10^{-1}$	188	179	0.17
50nm of Alumina / Titania	246	$3.9*10^{-1}$	88.6	178.6	0.16

Figure 4 and table 7 show cyclic polarization curves for stainless steel 316L specimens without coatings and coated with 50 nm of Alumina, Titania and Alumina/Titania. It shows cathodic and anodic regions and returns to the cathodic region again. It also shows that at the anodic region has rapid increase in potential from critical potential until reaching breaking potential. Then the potential decreases until it intersects the curve at repassivation potential. The cyclic polarization curves results are agrees with the results reported by [26] for stainless steel 316L in simulating human body fluids. Smaller hysteresis loops are found in the results of 50 nm thin films which mean fewer tendencies to the localized corrosion and more resistance to pitting corrosion than that of uncoated samples. No hysteresis loops, higher breakdown and repassivation potential of Multilayer than that of single layer (Alumina or Titania) also indicated from the result which means more resistance to pitting corrosion. Corrosion rate calculated by current density is 1.913 mpy for uncoated sample and decreases after coated to 0.203 mpy for Titania film, 0.174 mpy for Alumina film, and 0.164mpy for Multilayer.

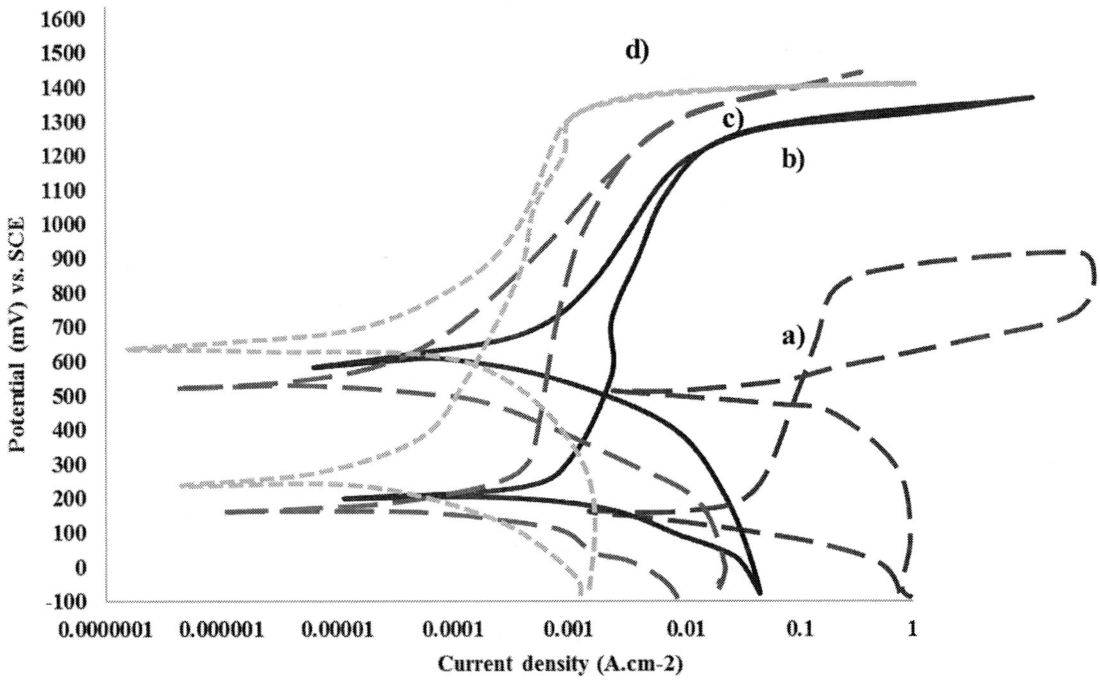

Figure. 4 Cyclic polarization results in SBF solution at 37±1 °C for ;

a) Uncoated stainless steel 316L specimens;
b) Coated with alumina.
c) Coated with titania
d) Coated with alumina/ titania.

IOP Conf. Series: Journal of Physics: Conf. Series **1032** (2018) 012017 doi:10.1088/1742-6596/1032/1/012017

Table 7 Cyclic polarization data for coated and uncoated AISI 316L alloy in SBF solution at 37±1 ℃.

Condition	E_{Break}	$E_{Rep.}$
Uncoated ss316L alloy	800	500
50 nm of Titania	1350	1250
50 nm of Alumina	1350	1300
50 nm of Alumina/ Titania	NO	NO

Figure 5(a & c) shows components of solution have precipitated on the surface of films, where alumina thin film has completed coverage and more adsorption (Figure 5 a) than that of titania thin film (Fig. 5 c), which seems less prevalent. The surface roughness of alumina and titania thin films increases after biocompatibility test. From EDX results (Figure 5 (b & d) it can be seen that the precipitates are found to be strongly enriched in calcium (Ca), approximately 16 % on alumina thin film and around 14.7 % on titania thin film. It means that the value of the calcium /phosphate (Ca/P) ratio is around 1.6 for alumina thin film and 1.47 for titania thin film which results good bonding to the bone during implantation because the ratio of Ca/P in the bone is 1.6 as reported by Qiyi-Zhang [27]. On the other side, alumina thin films have hydrophilic nature compared with the hydrophobic natural of titania thin films, which are because of high concentration of surface-exposed hydroxyl groups in alumina thin film deposited by ALD. The wetting properties of hydrophilic surfaces also have strong influence on the biocompatibility of alumina thin film [28]. Accordingly, titania hydrophobic surface showed a significant decrease of the biocompatibility. The biocompatibility of the alumina thin film deposited by ALD agrees with the results reported by researcher Dudley [29] and Muna et al. [30].

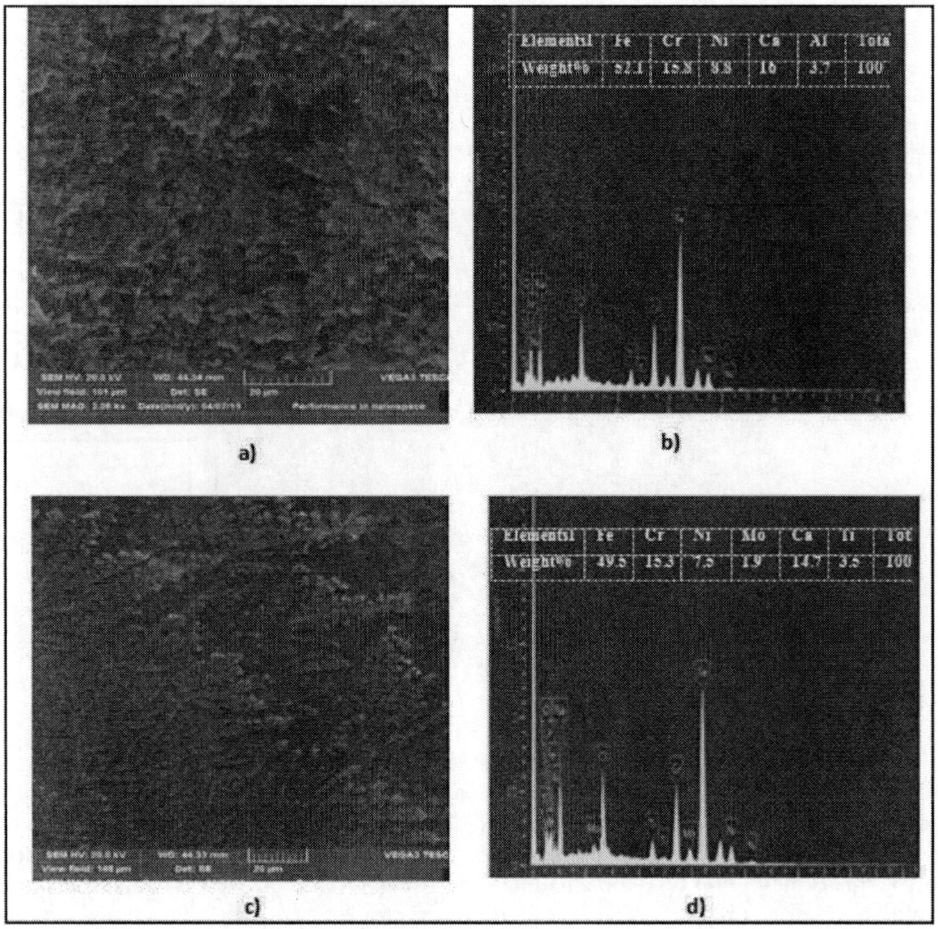

Figure . 5 SEM/EDX results of the biocompatibility test;

a) SEM image of alumina layer
b) EDX of alumina layer
c) SEM image of titania layer
d) EDX of titania layer

The bioactivity of alumina, titania and alumina /titania multilayers deposited on stainless steel 316L alloy with 50 nm thickness is investigated by cultured of Non-pathogenic Escherichia coli (E- coli) bacteria for 24 hr. at 37 ℃ as indications to the toxicity effects of thin films deposited. Thin film has less toxicity effects on non-pathogenic E-coli bacteria cultured means its bioactive film. The number of colonies living on the agar plate, after incubation at 37 °C for 24 h, was counted and studied according to equation (2). The results were expressed as mean colonies forming units per milliliter and antibacterial efficacy was determined on the basics of duplicates test result. The titania thin film has 15 % reduction, alumina thin film has 20 % reduction, and alumina/ titania multilayers have 40% reduction. Another method of antibacterial is the diffusion zone as shown in figure 6. Where, the diffusion zone (number of bacteria died) is around 2mm for titania thin film, 4mm for alumina thin film and more than 6 mm for alumina/titania multilayers thin films respectively.

IOP Conf. Series: Journal of Physics: Conf. Series **1032** (2018) 012017 doi:10.1088/1742-6596/1032/1/012017

The small diffusion zone and less percentage of bacteria reductions mean that the titania thin film has less toxicity effects on E-coli bacteria used, which means more bioactive than that of alumina thin film and alumina/titania multilayers thin films.

Figure. 6 Diffusion zone method of the bacterial cultured

4. Conclusion

The results showed that the good nanocoatings thickness control with excellent homogeneity and high area uniformity can be obtained from using ALD at 250℃ stainless steel 316L specimens and Si-Wafer (100). The surface morphologies of the thin films have different nanoparticles size and shape depend on type of thin films deposited, Alumina or Titania. It was found that the alumina, titania and alumina / titania multilayer thin films deposited with 50 nm nanocoatings thicknesses by ALD at 250℃ have great effect on improving the corrosion resistance of stainless steel 316L compared with the uncoated specimens electrochemically in SBF solution at 37±1℃. Alumina / Titania multi-layer thin films proved to be more corrosion resistant with respect to single film of alumina and titania at 50 nm thickness. While alumina has better corrosion resistance than titania. Biocompatibility of alumina film was more than that of titania film at 50 nm thickness, where more components of solution precipitated on alumina than that of titania according to determine of Ca/p ratio, which is equal to 1.6 for alumina and to 1.47 for titania. Bioactivity of titania film at 50 nm thickness with non-pathogenic E-coli bacteria was more than that of alumina and alumina/titania multilayers thin films, which means less toxicity of titania thin film deposited.

References

[1] Kamachi U. Mudali, T. M. Sridhar and Baldev Raj, 2003 "*Corrosion of bio implants*", Metallurgy and Materials Group, Indira Gandhi Centre for Atomic Research, India, Vol. 28, Parts 3 & 4, pp. 601–637.

[2] Sujata K. Bhatia, 2012 "*Engineering Biomaterials for Regenerative Medicine*", Novel Technologies for Clinical Applications, Springer Science and Business Media, UK.

[3] Miao Yu , 2014 "*Microstructure, Corrosion Behavior, Mechanical Property and Biocompatibility of Compression-molded Zinc-Nano-diamond Composites as a Bio-implant Material* ",PhD Thesis, Material Eng. Dept., Drexel University, USA.

[4] Hendra Hermawan, Dadan R. and Joy R. P. D., 2011 "*Metals for Biomedical Applications*", Biomedical Engineering and Health Science, University of Technology, Malaysia.

[5] Nitech R. Patel, and Piyush P. Gohil, 2012 "*A Review on Biomaterials: Scope, Applications & Human Anatomy Significance*", International Journal of Emerging Technology and Advanced Engineering, Vol. 2, Issue 4.

[6] Kotoka Ruben, 2013 "*Functionalized Thin Film Coatings for Medical Implant Applications*", PhD Thesis, Mechanical Engineering Dept ,North Carolina Technical State University.

[7] Davis J. R., 2000 "*ASM Specialty Handbook : Nickel, Cobalt, and Their Alloys*", 9th edition ASM International, Vol.9.

[8] Ali Parsapour, . Khorasani S N. and Fathi M. H., 2012 "*Effect of Surface Treatment and Metallic Coating on Corrosion Behavior and Biocompatibility of Surgical 316L Stainless Steel Implant*", J. of Mater. Science & Technology, Vol.28, No.2, pp.125-131.

[9] Yoshimitsu Okazaki, and Emiko G., 2005 "*Comparison of Metal Release From Various Metallic Biomaterials in Vitro*", J. of Biomaterials , pp. 11–21.

[10] Dekker Michael Y., Debra J., and Vasif H, 2004 "*Biomaterials in Orthopedics*", Marceld Ekker Inc., USA.

[11] Rolf Sandström , 2002 "Metallic Biomaterials", Savigny Pascale & Giroud Elodie Version 1.

[12] Virtanen S., Milošev L., and Gomez-Barrena E., 2011 "*Special Modes of Corrosion Under Physiological and Simulated Physiological Conditions*", Institute for Surface Science and Corrosion, Department of Materials Science, University of Erlangen-Nuremberg, Germany.

[13] X.-Yan Zhang, Y. Zou, X.-Long Zeng, 2016 "*Effect of Laser Surface Remelting on the Corrosion Resistance of 316L Orthodontic Brackets*", Int. J. Electrochem. Science, Vol.11, pp.2877– 2886.

[14] Injeti G., and Leo B., 2008 "*Electrodeposition of nanostructured coatings and their characterization*", J. of Science and Technology of Advanced materials, Vol.9, pp. 121-129.

[15] Aliofkhazraei Mahmood, 2011 "*Nanocoatings; Size Effect in Nanostructured Films*", Springer, New York.

[16] Mohammed J. Kadhim, Nawal E. Abdulateef , Makarim H. Abdulkareem 2018 "*Evaluation of Surface Roughness of 316L Stainless Steel Substrate on Nanohydroxyapatite by Electrophoretic Deposition*", Al-Nahrain Journal for Engineering Sciences (NJES), Vol.21 , No.1, pp.28-35.

[17] Riikka L. Puurunena, 2005 "*Surface Chemistry of Atomic Layer Deposition ALD: A Case Study for the Trimethylaluminum/Water Process*", Journal of Applied Physics ,Vol.97, pp. 121301-121352.

[18] Hamid Reza, and Majid P., 2011 "*Corrosion Study of Metallic Biomaterials in Simulated Body Fluid*", J. of Metallurgical Engineers of Serbia ,Vol. 17 (1) , pp. 13-22.

[19] Robert Wen and Chun Ch., 2005 "*Electrochemical Corrosion Studies On Co–Cr–Mo Implant Alloy in Biological Solutions*", J. of Materials Chemistry and Physics ,Vol. 93 , pp. 531–538.

[20] Duha S. A.,2014 " *Preparation and Application of Functionalized MWCNTs/ Ag Nanoparticles for Antibacterial Activity*" PhD Thesis in Applied Sciences , University of Technology-Baghdad.

[21] Gerry Triani , Peter J. Evans, and David R.G, 2005 "*Atomic Layer Deposition of* TiO_2 / Al_2O_3 *Films For Optical Applications*", J. of Process of SPIE, Vol. 5870, pp.1-10.

IOP Conf. Series: Journal of Physics: Conf. Series **1032** (2018) 012017 doi:10.1088/1742-6596/1032/1/012017

[22] Alekhin A. P., G. I. Lapushkin, and Markeev A. M., 2010 "*Atomic Layer Deposition of the Titanium Dioxide Thin Film from Tetraethoxytitanium and Water*", J. of General Chemistry, Vol. 80, No. 6, pp. 1091–1096.
[23] George B., 2010 "*The Corrosion of Co-Cr-Mo Alloys For Biomedical Applications*", MSc Thesis, Dept. of Metallurgy and Materials, University of Birmingham, UK.
[24] Xiaofeng Zhang, and Li Li Yu Lei, 2013 " *Structure and Electrochemical Study of Al_2O_3 and TiO_2 Coated Li Alloys Cathode Material Using Al*", J. of Advance Materials, Vol.3, pp.1299-1307.
[25] Mitchell D R G , Triani G., and Attard D J, 2006 "*Atomic Layer Deposition of TiO_2 and Al_2O_3 Thin Films and Nano-Laminates* ", J. of Smart Material Structure, Vol.15, pp.S57–S64.
[26] Kajzer W. , Krauze A., Walke W., and Marciniak J., 2008 "*Corrosion Behavior of AISI 316L Steel in Artificial Body Fluids* ",J. of Achievements in Materials and Manufacturing, Vol. 31, No.2.
[27] Qiyi Zhang , Yang Leng , and E. Renlong Xin , 2007 "*An effective Approach to Activate 316L Stainless Steel For Biomimetic Coating of Calcium Phosphate: Electrochemical Pretreatment* ",J. of Mater Science, Vol.42, pp. 6205–6211.
[28] Dudley S. Finch, and Tammy O., 2007 "*Biocompatibility of Atomic Layer-Deposited Alumina Thin Films*", J. of Biomedical Materials Research Part A, Vol. 6 , No. 3, pp. 243-352.
[29] Nicola Pinna and Mato Knez, 2012 "*Atomic Layer Deposition of Nanostructured Materials*", Wiley, Germany, pp. 301-318.
[30] Muna Khethier Abbass , Sami Abualnoun Ajeel , Haitham M. Wadullah *, 2015 "Atomic Layer Deposition properties of TiO2 and Al2O3 thin films on Stainless Steel 316L and Si-Wafer",* Association of Arab Universities Journal of Engineering Sciences ,University of Baghdad, , Vol.22, No.1. ,pp.27-36.

Acknowledgments

One of the authors (Haitham M.) thanks Chair Professor Dr. Shubhra Gangopadhyay and Dr. Cherian Joseph in College of Electric and Computer Engineering-Nanotechnology Core Facility and Dr. Tomme white in College of Veterinary Medicine -Electron Core Facility (ECF) in University of Missouri- Columbia, USA. Thanks also to the Materials Engineering Dept.-University of Technology-Baghdad, Iraq.

Preparation of high transmittance TiO_2 thin films by sol-gel technique as antireflection coating

Mohammed Hadi. Shinen[1], S A A AlSaati[1] and Fouad Z Razooqi[1]

[1]Department of Science, College of Basic Education, University of Babylon, Iraq
E-mail of the corresponding author: mohammed.shinen@yahoo.com

Abstract. In this present work sol–gel technique with spin coating method is used to deposit Titanium dioxide thin films on glass substrates as anti-reflection coating (ARC) films. Nano crystalline TiO_2 (anatase phase) is obtained after heating at 550°C for 2 h. The optical properties of TiO_2 films were studied by UV-Vis absorption/reflection spectroscopy, the average transmittance of monolayer coated on glass substrates was greater than 95% in the wavelength range of 300–1000 nm.

Keywords. *Sol-gel technique, spin coating method, thin films, antireflection coating.*

1-Introduction

Thin film is a layer of material (molecular or element) which ranging thickness of the nanometer region (fractions of a nanometer to a few micrometers) [1-3]. Films are first made by (Busen and Grove) in 1852 A.D by using chemical reaction deposition. In 1857 A.D, Faraday was able to deposit a thin metal film by using of thermal evaporation vacuum [4]. Thin films have many applications in several fields like protective films, electrically operating coatings, solar cells, batteries, decorative coatings, and many optical applications [5,6]. Deposition techniques are classified into two kinds, depending on the state materials, chemical or physical [7]. Chemical deposition methods are depending on phase of the precursor, the techniques that come from chemical processes are as follow: dip coating, chemical vapor deposition, chemical solution deposition, plasma enhanced CVD, atomic layer deposition, Langmuir-Blodgett and spin coating [8-10]. Physical deposition method uses mechanical, electromechanical or thermodynamic means to make a thin layer of solid material. Chemical reactions are not used here; physical deposition technique tends to require a low pressure evaporation (high-vacuum). The techniques that come from chemical processes are as follow: electron beam evaporator [11], molecular beam epitaxy (MBE) [12], sputtering [13], pulsed laser deposition [14], cathodic arc deposition or Arc-PVD [15] and electrohydrodynamic[16]. In this research, TiO_2 thin films prepared via spin coater which is branch of chemical deposition. One of the most important applications of thin films is anti-reflective which is considered optical applications of high optical transmittance. Anti-reflective coatings (ARC) are applied in a several wide of applications where light passes through a surface with low optical reflection. For examples anti-glare or anti-brightness *coating on corrective lenses, cameras lenses elements, and windows of photo voltaic cells. The first* and simplest form of antireflective coating was discovered by Rayleigh in 1886 AD. [17]. Single-layer interference is the simplest anti-reflective coatings (ARC_s) consist of a single thin film (mono layer) of transparent materials like Titanium dioxide (TiO_2) thin film with refractive index equal to square root of substrate's refractive index in air, such a coatings theoretically give zero reflectance for light with wavelength (in the coating) equal to four times the coating's thickness. Reflectance is also decreased for wavelengths in a broad band around the center. A layer of thickness equal to a quarter of some design wavelength is called a "quarter-wave layer". TiO_2 film is important material with wide applications like optoelectronic devices, photocatalytic systems and anti-reflection coatings [18, 19].

2. Experimental part.

2.1. Preparation of sol-gel solution.

The sol-gel was prepared by using Titanium isopropoxide TIP $Ti\{OCH(CH_3)_2\}_4$,ethanol C_2H_5OH-EtOH and acetic acid CH_3COOH as a catalyst, all materials were obtained from (Sigma Aldrich) with purity=99%. The sol-gel was obtained by dissolving 3.15 ml TIP in 25 ml ethanol and 5 ml acetic acid, the solution was stirred for 1h.

2.2. Preparation and characterizations of samples.

The optical glass substrates (1 mm*25.4 mm*76.2 mm) were ultrasonically cleaned by the mixture of acetone and propanol and then dried at 50 °C. The spinning process by spinner ((Mod- 4000 Electronic Micro System) was carried out at (T=25 °C) and normal atmosphere with the different speeds (1000, 2000, 3000, 4000 and 5000 rpm) on the cleaned substrates, after that treated (annealed) gradually escalating at heat temperature 550°C for (1h) and then let the samples inside the oven cool down as well.

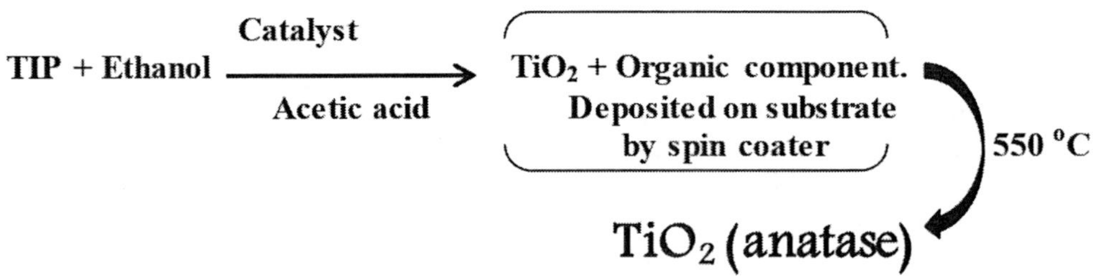

Table (1) shows the samples cods which annealed at 550°C.

Sample code	Spin speed (rpm)	Thickness (nm)
S1	1000	116
S2	2000	92
S3	3000	79
S4	4000	62
S5	5000	46

TiO_2 films were characterized by absorption, transmission and reflection properties as well as by spectrophotometer (UV-VIS).

3. Results and discussions.

3.1. Absorbance Measurements (A).

Fig. (1) refers to absorbance of (TiO_2) thin films of different thickness.. At the visible regions absorption is decreased, it is excellent to be antireflection layer. At high wavelength the incident lights have not enough ability (energy) to make interaction with materials so photons will be transmitted, when (λ) decreases (photon energy increase). The interaction between incident photons and material will occur and then the absorbance will increase [20]. Absorption is direct proportional to the film thickness; this means that optical absorptions are inversely proportional to the rotational deposition speed.

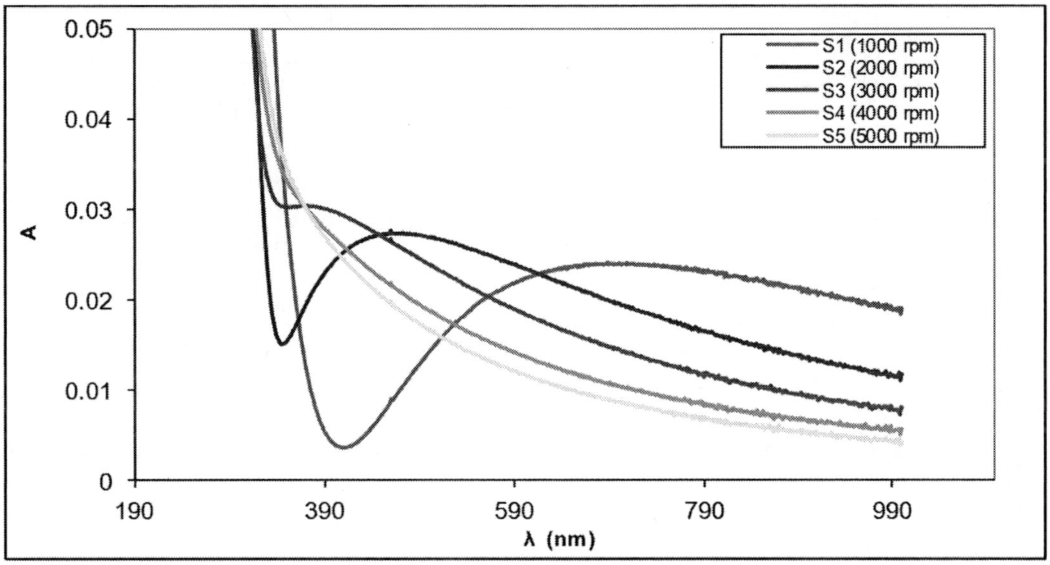

Figure 1. Relation between wavelength and absorptance of TiO_2 thin films.

3.2. Transmittance Measurements (T).

Fig. 2. Illustrates that the optical transmittance of Titanium dioxide thin films. It is observed that maximum transmittance at (S3, S4 and S5) at thicknesses (79, 62 and 46 nm) respectively for wavelength range (420-900 nm) which is suitable for solar cell window. As showed in figure (4) transmittance is inversely proportional with thickness [21]. This behavior can be attributed to the increasing of thicknesses which lead increasing the atoms number per film and thus observe that optical transmittance is decreased because most of the photons have been absorbed by atoms [22].

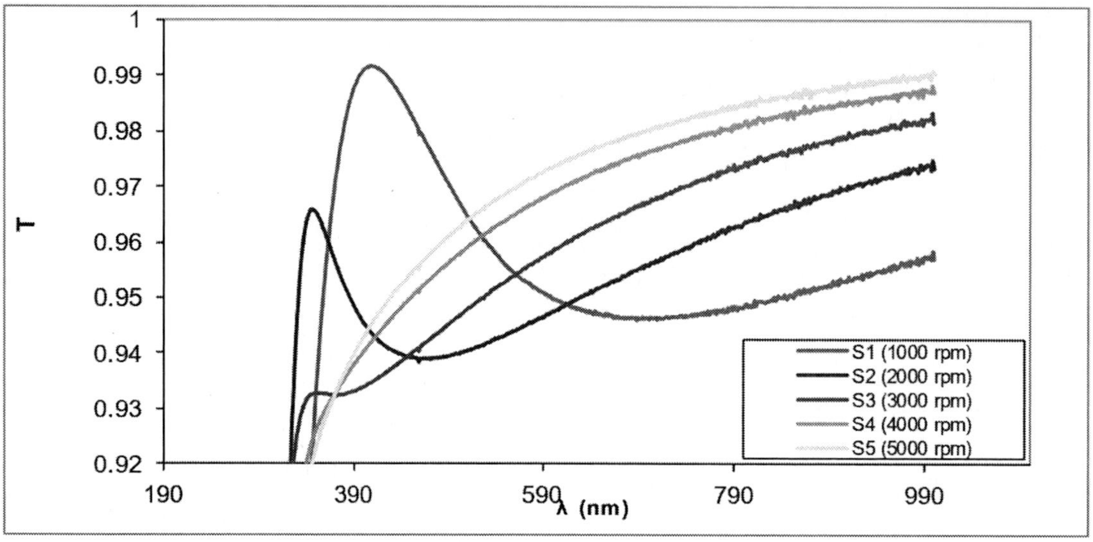

Figure 2. *The relation between transmittance and wavelength of TiO_2 thin films for different rpm.*

3.3. Reflectance Measurements (R).

Reflectance of the surface of a material (thin film) is the fraction of incident lighte that is reflected at an interface [23]. Fig. 3.below shows the spectral reflectance curve as a function of wavelength. Reflectance is calculated from absorptance and transmittance by using eq. (1). We observed that reflectance have high value which decreasing with wavelength increasing for all films.

(R)=1-(A+T)……… (1)

When R is reflectance, A is absorbance and T is transmittance.

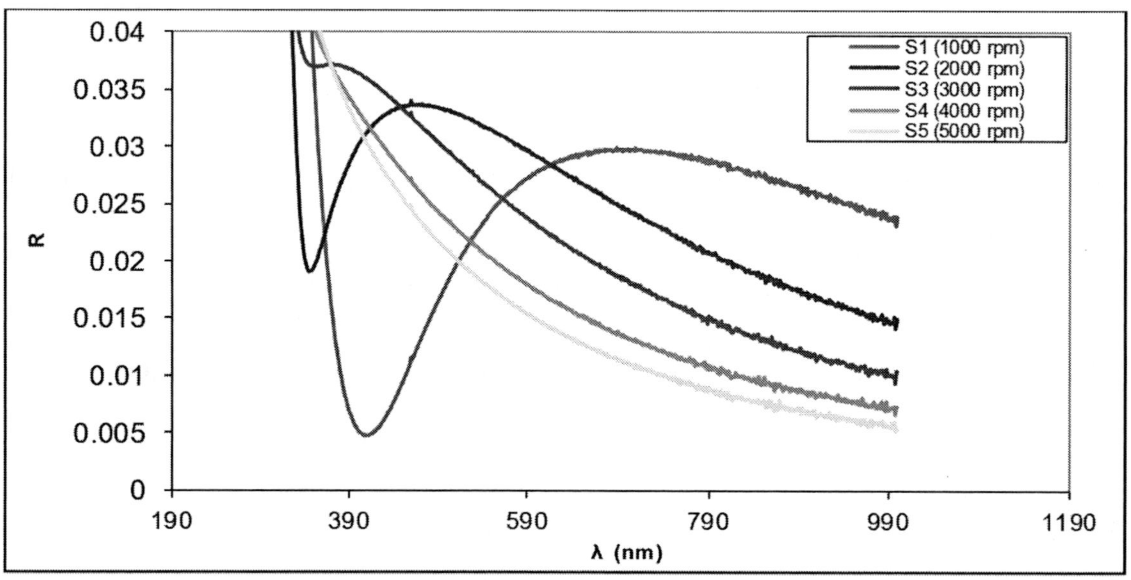

Figure 3. The reflectance at wavelength range (300-990 nm.) of TiO_2 thin films for different rpm.

3.4.Absorption Coefficient Measurements (α).

Absorption coefficient is defined as (the ratio decrement in flux of incident electromagnetic energy relative to the distance unit in the direction of incident wave diffusion. Its depends on incident light energy (hv) and on semiconductor characteristics (n or p) type, [24]. Absorption coefficient can be seen in figure (4) that the absorption coefficient is the smallest at high wavelength, this means that the possibility of electron transition is little because the energy of the incident photon is not sufficient to move the electron from the valence band to the conduction band ($h\upsilon < E_g$), at short wavelength, absorption is bigger; this means that a great possibility for electron transitions consequently, the energy of incident photon is enough to move the electron from the valence band to the conduction band. The energy of the incident photon is greater than the forbidden energy gap, this shows that the absorption coefficient assists in figuring out the nature of electron transition, when the values of the absorption coefficient are high ($\alpha > 10^4$) cm^{-1} at high energies. It is expected that direct transition of electron occur, the electrons and photons maintain the energy and moment. However, when the values of the absorption coefficient are low ($\alpha < 10^4$) cm^{-1} at low energies, it is expected that indirect transition of electron occurs, and the electronic momentum is maintained with the assistance of the phonon [25].

$$\alpha = (2.303 \times A) / t_t \ .... \ (2)$$

When α is absorption coefficient, A is absorbance and t_t is thickness.

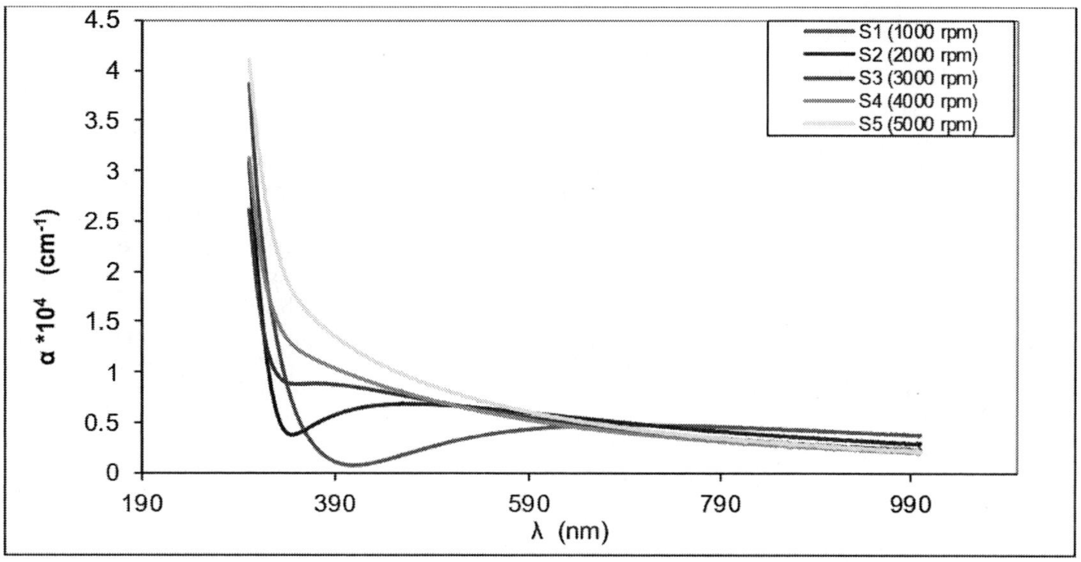

Figure 4. Absorption coefficient as function of wavelength for different thickness of TiO_2 thin films for different rpm.

IOP Conf. Series: Journal of Physics: Conf. Series **1032** (2018) 012018 doi:10.1088/1742-6596/1032/1/012018

3.5. Extinction Coefficient Measurements (K_o) .

Extinction coefficient (k_0) is calculated by using equation (3). The change of the extinction coefficient, as a function of the wavelength, is shown in figures (5). It can be noted that (k_0) has same behavior of absorption coefficient, absorption coefficient has a direct related with (k_0) as in the equation (3) [26].

$$k_0 = \frac{\alpha\lambda}{4\pi} \text{..........(3).}$$

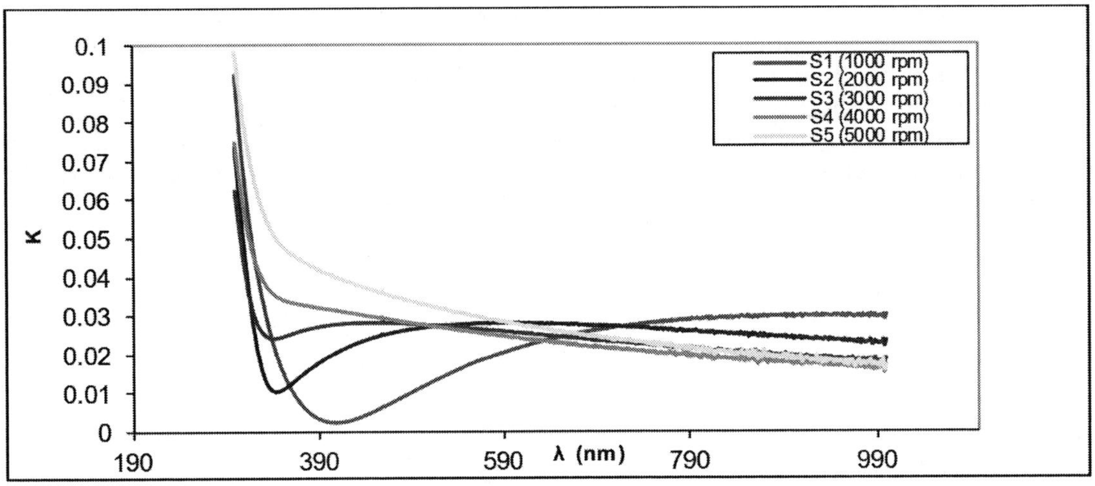

Figure 5. Excitation coefficient as a function of wavelength *for different rpm.*

3.6. Refractive Index (n) .

Refractive index is measured by using relation (4). It has been shown in figure (6) as a function of wavelength of (TiO_2) films. The shape of the curve with wavelength is the same curve reflectance curve because the relation between them is increased with photon energy increasing and then decreases, refractive index increases with thickness increasing. The increase may be attributed to higher packing density and the change in crystalline structure, this increase due to the enhancement of growth crystalline [27].

$$n = \sqrt{\frac{4R - k_0^{\,2}}{(R-1)^2} - \frac{(R+1)}{(R-1)}} \text{.............. (4)}$$

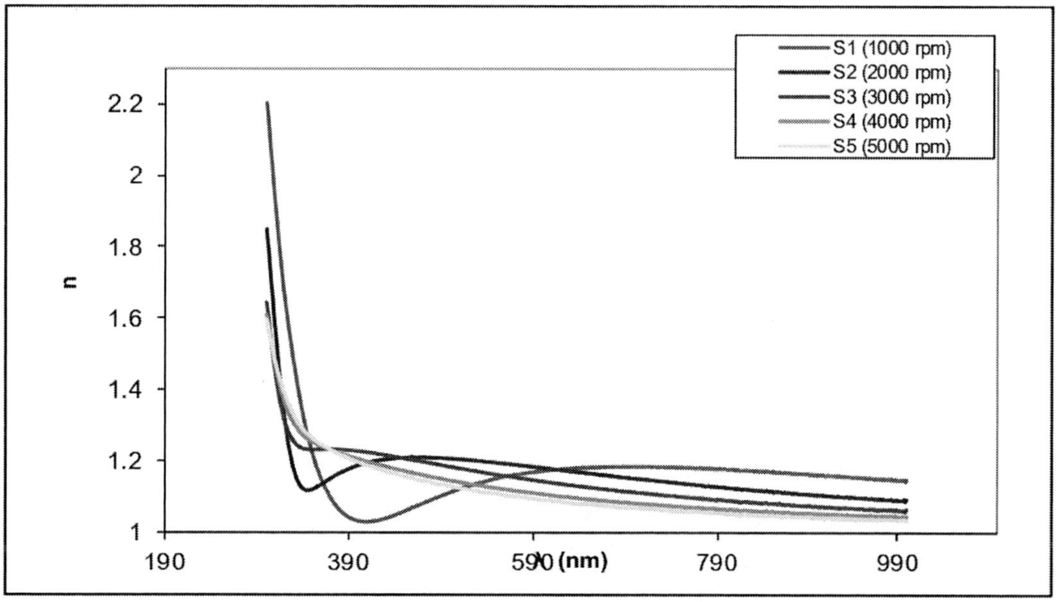

Figure 6. Refractive index as function of wavelength for different thickness of TiO_2 thin films.

3.7. Dielectric Constant Measurements.

Both real (ε_1) and imaginary (ε_2) dielectric constant are measured for prepared films by using relations (5) and (6) respectively. Figures (7) and (8) illustrate variation of (ε_1) and (ε_2) as a function of wavelength. It is observed that their values increase with wavelength to maximum value of (ε_1) and then decrease [60]. It can be seen that ε_1 considerably depends on (n^2) due to low value of (k_0^2) so.

$$\varepsilon_1 = (n^2 - k_0^2) \quad (5)$$

$$\varepsilon_2 = (2nk_0) \quad (6)$$

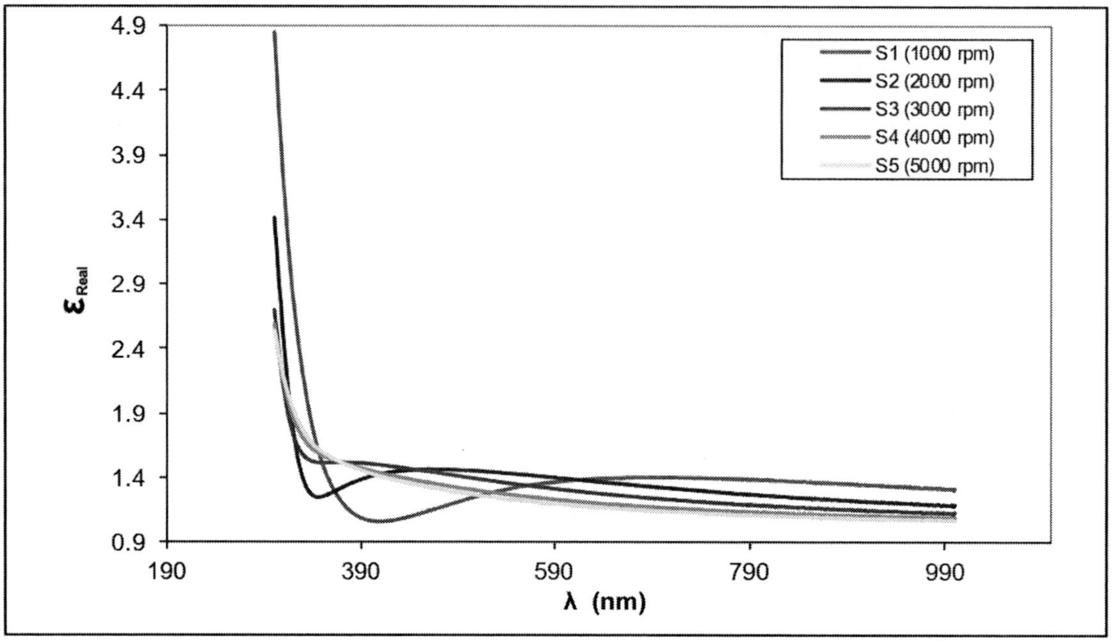

Figure 7. Real dielectric constant (ε_1) of TiO_2 thin films for different rpm.

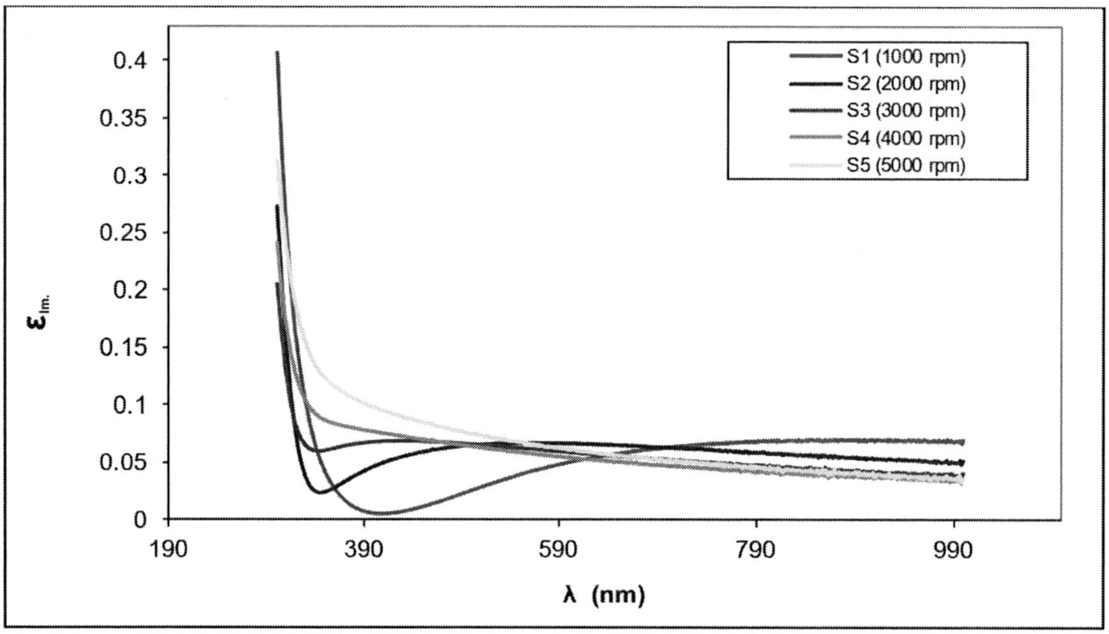

Figure 8. Imaginary dielectric constant (ε_2) of TiO_2 thin films for different rpm.

IOP Conf. Series: Journal of Physics: Conf. Series **1032** (2018) 012018 doi:10.1088/1742-6596/1032/1/012018

3.8. Energy gaps of allowed direct transition (E_g)

The energy gap can be calculated from equation (7), the relation is drawn between $(\alpha h\nu)^2$ and photon energy (hv),as in figure (9) illustrates allowed direct transition electronic.

$$\alpha h\upsilon = B(h\upsilon - E_g^{opt.})^r \ldots\ldots (7)$$

Where $E_g^{opt.}$ the energy gap between direct transitions, B is the constant depended on type of material, r is the exponential constant; its value depended on type of transition. When r =1/2 for the allowed direct transition.

Table 2. Shows direct energy gap for different thickness of TiO_2 thin films

Sample code	*Thickness (nm)*	*Allowed energy gap E_g (eV)*
S1	116	3.66
S2	92	3.7
S3	79	3.8
S4	62	3.6
S5	49	3.56

From previous table and the following figure, it can be observed that (E_g) is decreasing slightly with the increasing of thickness for all films. This is because of the increasing in the carrier concentration, which results in filling the bottom of the conduction band that leads to the decrease in hole between (C.B.) level and (V.B.) level that leads to zoom out the energy gap band [28].

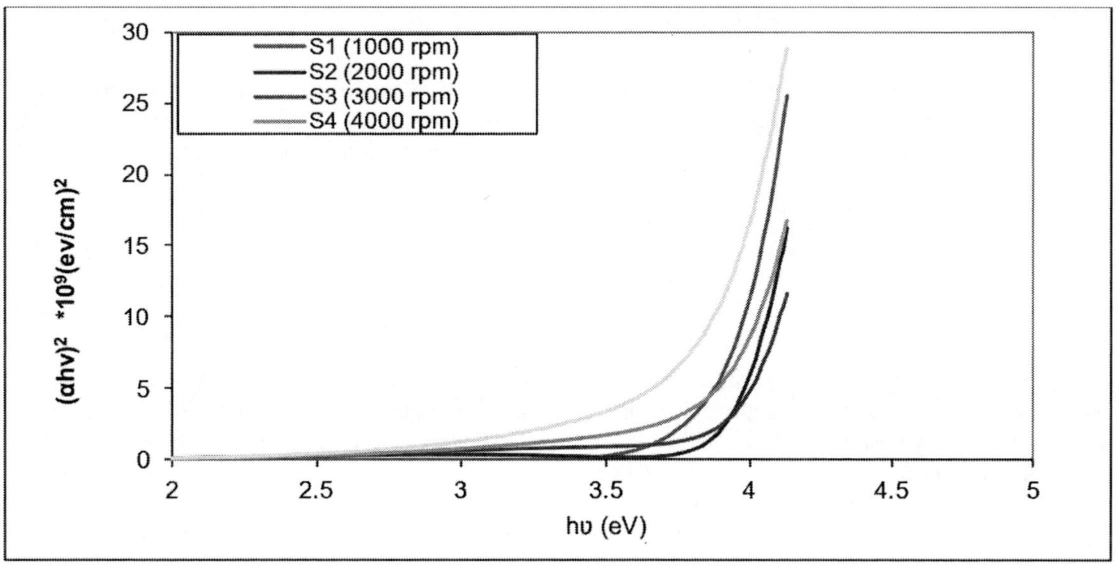

Figure 9. Shows allowed direct electronic transitions of TiO_2 thin films for different rpm.

4. Conclusion

According to a result above we have been concluded an important points, the best transmittance of (TiO_2) film is (95%-~99%) for thickness (49nm.), within wavelength range (400-900nm.), the film is good to be windows of solar cell. But absorptance (A) is high at short wavelength, therefore; the film is good to be photo detector at ultra-violet range and The values of refractive indexes (n) of films are good ranges. This means that the films are good to be antireflection coating. Also we noticed that, Absorption coefficient (α) of (TiO_2) film is about ($\alpha>10^4 cm^{-1}$); thus absorption coefficient has higher increase at wavelength. This converts to a large probability that direct electronic transition will happen and Energy gap for allowed direct transition has been calculated as shown in table (2) . This proves that the transitions happen from the close points in the top of valance band to the close points in the bottom of conduction band.

Referances

[1] A. S. Edelstein and R. C, "Cammarata, in Nanomaterials: Synthesis, Properties and Applications", Institute of Physics Publishing, Bristol, United Kingdom, 1996.

[2] K. Lu, "Nanocrystalline metals crystallized from amorphous solids: nanocrystallization, structure, and properties", Materials Science and Engineering R16, 161 (1996).

[3] S. C. Tjong and H. Chen, "Nanocrystalline materials and coatings", Materials Science and Engineering R45, 1 (2004).

[4] L. Eckortova, *"Physics of Thin Films "*, (plenum press), (1977).

[5] M. Birkholz; K.-E. Ehwald; D. Wolansky; I. Costina; C. Baristyran-Kaynak; M. Fröhlich; H. Beyer; A. Kapp; F. Lisdat (2010). "Corrosion-resistant metal layers from a CMOS process for bioelectronic applications" (PDF). Surf. Coat. Technol. 204 (12–13): 2055–2059.

[6] Flexible Cell Construction. Mpoweruk.com. Retrieved on 2012-01-15.

[7] Functional Polymer Films Eds. R. Advincula and W. Knoll – Wiley, 2011, ISBN 978-3527321902..

[8] Ariga, Katsuhiko; Yamauchi, Yusuke; Mori, Taizo; Hill, Jonathan P. (2013-12-01). "25th Anniversary Article: What Can Be Done with the Langmuir-Blodgett Method? Recent Developments and its Critical Role in Materials Science". Advanced Materials. 25 (45): 6477–6512. ISSN 1521-4095.

[9] Hanaor, D; Triani G.; Sorrell C.C.; (2011). "Morphology and photocatalytic activity of highly oriented mixed phase titanium dioxide thin films". Surface and Coatings Technology:. 205 (12): 3658–3664.

[10] M. Faustini, G. L. Drisko, C. Boissiere, D. "Grosso Liquid deposition approaches to self-assembled periodic nanomasks, Scripta Materiala". 2014.

[11] Harsha, K. S. S, "Principles of Physical Vapor Deposition of Thin Films", Elsevier, Great Britain (2006), p. 400.

IOP Conf. Series: Journal of Physics: Conf. Series **1032** (2018) 012018 doi:10.1088/1742-6596/1032/1/012018

[12] Jaeger, Richard C. "Film Deposition". Introduction to Microelectronic Fabrication (2nd ed.). Upper Saddle River: Prentice Hall. ISBN 0-201-44494-1 (2002).

[13] R. Behrisch (ed.)). Sputtering by Particle bombardment:. Springer, Berlin. ISBN 978-3-540-10521-3. (1981).

[14] Rashidian Vaziri, M R. "Monte Carlo simulation of the subsurface growth mode during pulsed laser deposition". Journal of Applied Physics. 110: 043304.

[15] SVC "51st Annual Technical Conference Proceedings" (2008) Society of Vacuum Coaters, ISSN 0737-5921 (previous proceedings available on CD from SVC Publications).

[16] Wall, Staffan. "The history of electrokinetic phenomena." Current Opinion in Colloid & Interface Science 15.3 (2010): 119-124.

[17] Zhang, Jun-Chao; Xiong, Li-Min; Fang, Ming; He, Hong-Bo (2013). "Wide-angle and broadband graded-refractive-index antireflection coatings". Chinese Physics B. 22 (4): 044201.

[18]A. Verma, S. B. Samanta, A. K. Bakhshi, S.A. Agnihotry, Solar Energy Materials&Solar Cells 88, 47 (2005).

[19] D. Mardare, E. Apostol, J. Optoelectron. Adv. Mater. 8(3), 914 (2006).

[20] B.G.Stereeman, "*Solid State Electronic Devices*", 2nd Ed, Practice Hall, Inc. Engle wood Cliffs, N.J. (1980).

[21] I.Oja, A.Mere, M.Krunks, R.Nisumaa, C.H.Solterbeck and M.Es.Souni, "Structural and electrical characterization of TiO_2 films grown by spray pyrolysis", "*Thin solid films*", (515), (2), (2006), (674-677).

[22] Peter Y.Yu Manual Cardona, "*Fundamentals of Semiconductors*", Springer_verlag Berlin Heidelberg New York, (1996).

[23] "Thermal insulation — Heat transfer by radiation — Physical quantities and definitions". ISO 9288:1989. ISO catalogue. 1989. Retrieved 2015-03-15.

[24]A. N. Donald, "*Semiconductor Physics and Devices*", Irwin, USA, (1992).

[25] M. Crane, translation and Y. Hassan, "Solar Cells", Collage of Education, University of Mousl, (1989).

[26] H. N. Najeeb, G. A. Wahab Ali, A. K. Kodeary and J. Ali, "Doping Effect on Optical Constants of Poly-Vinyl Chloric (PVC)", Academic Research International, Vol.4, pp. 53-62, (2013).

[27] Jinting Jiu, Seiji Isoda, Motonari Adachi and Fumin Wang," Preparation of TiO_2 nanocrystalline with 3–5 nm and application for dye-sensitized solar cell", "*Science Direct*", (189), (2-3), (2007), (314-321).

[28] L.S. Roman "Optical band-edge absorption of oxide compound SnO_2", "*Applied Surface Science*", (252), (2006), (5361–5364).

Effects of Annealing Temperature on Structural and Optical Properties Of CIGS Thin Films for Using in Solar Cell Applications

Mohammed Abdulhur Kadhim Mankoshi[1], Falah I. Mustafa[2], Nora J. Hintaw[1]

[1] Department of Physics, science college, Kerbala, University, Kerbala, Iraq
[2]Solar energy research center, Renewable energy Directorate,Higher education and scientific research ministry
mmmed1963@gmail.com

Abstract. A copper indium0.5gallium0.5selenide2 thin films (CIGS) were deposited at room temperature on glass substrate by thermal evaporation technique. The deposited film were annealed at different annealing temperature (293, 100, 200 and 300)°C. The effect of annealing temperature on structural, optical and electrical properties CIGS films were studied by XRD, UV-Visible absorbance and Hall effect measurements. It is found that the crystallinety enhanced, optical energy gap increases and electrical conductivity decrease with increasing annealing temperature. Three design of solar cells, based on CIGS, were examined to reach the suitable solar cell with higher efficiency .

1. Introduction

CIGS is a I-III-VI_2 compound semiconductor material composed of copper, indium, gallium, and selenium, with a chemical formula of $CuIn_xGa_{(1-x)}Se_2$, where the value of x can vary from 1 to 0.3. It is a tetrahedral bonded semiconductor,with the chalcopyrite crystal structure. The bandgap varies continuously with x from about 1.0 eV (for copper indium selenide) to about 1.7 eV (for copper gallium selenide)[1].

A copper indium gallium selenide solar cell (or CIGS cell) is a thinfilm solar cell. It is fabricated by depositing a thin layer of copper, indium, gallium and selenide on substrates, along with electrodes on the front and back to collect current. Because the material has a high absorption coefficient and strongly absorbs sunlight, a much thinner film is required than of other semiconductor materials. CIGS layers are thin enoughto be flexible, allowing them to be deposited on flexible substrates.

CIGS cells being developedto reach silicon-like efficiencies, while maintaining their low costs, as is thin-film technology[2].

A p-type CIGS absorber layer is deposit in many techniques varied from high-temperature deposition like thermal evaporation [3]sputtering deposition [4]and low-temperature deposition techniques, screen print deposition[5]combined with a thin n-type like cadmium sulfide (CdS) to make hetrojunction for photo-voltaic applications.

CIGS has an exceptionally high absorption coefficient of more than $10^5 cm^{-1}$at higher energy photons[6]. Soda-lime glass [7]or flexible substrates such as polyimide or metal foils[8].

2. Experimental

A copper, indium, gallium, and selenium is made of Germany films were prepared onto different substrates bythermal evaporation technique in a high vacuum system of (10^{-5}) torr using Edward coating unit model (E 306). The deposited films, on glass substrates, were examined by X-ray diffraction (XRD) measurements using (SHIMADZU diffractometer system (XRD-6000)in the range of 2θ from 20° to 60°, UV visible spectroscope, The and atomic force microscopy type AA3000 Angstrom advanced and I-V characteristic MSIV .

3. Results and discussions

The X-Ray diffraction spectra for CIGS powder material, wererecorded as illustrated in Fig (1). Theinter planer distance (d_{hkl}) were calculated and compared with the standard valued in JCPDF as shownin Table (1). XRD pattern show a polycrystalline structure with monoclinic phase with high intensity for (112), (220), (312) and (332) directions and lower intensity for others.The crystalline size wascalculated usingScherrer equation [9]as shown in Table (1).

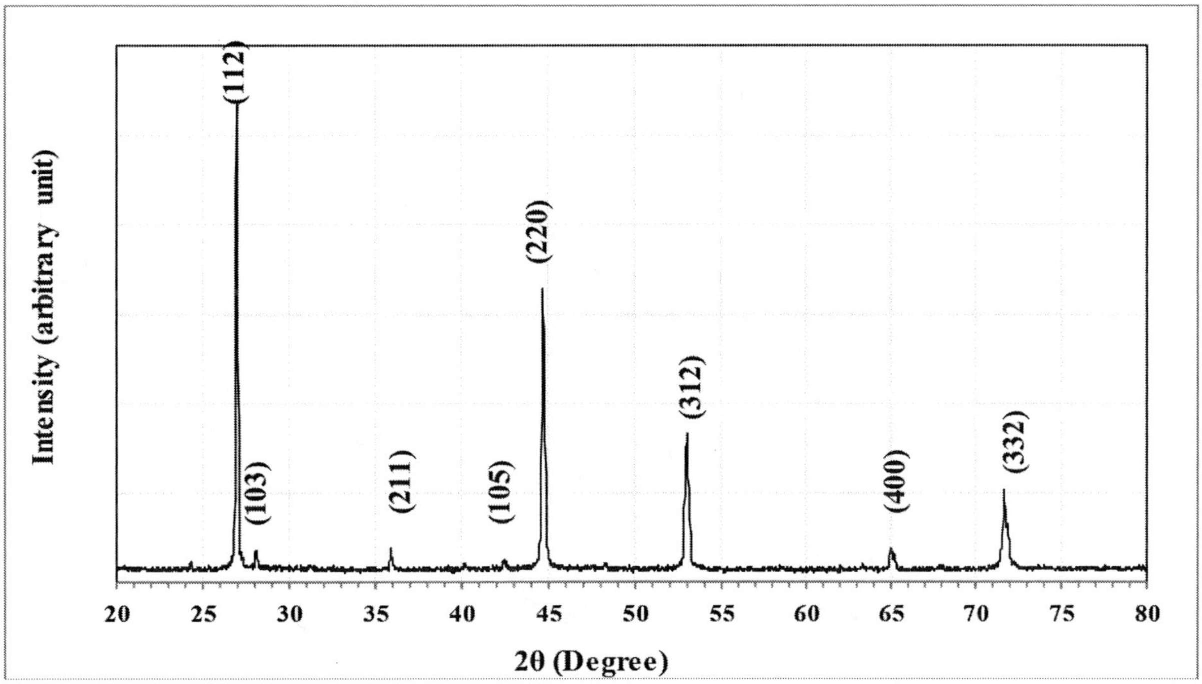

Figure 1: X-ray diffraction of CIGS powder.

Table 1: X-ray diffraction parameters and crystallinesize of CIGS powder.

2θ (Deg.)	FWHM (Deg.)	d_{hkl} Exp.(Å)	G.S (nm)	d_{hkl} Std.(Å)	hkl	card No.
26.9130	0.1522	3.3102	53.7	3.3121	(112)	35-1102
28.0217	0.1739	3.1817	47.1	3.1801	(103)	35-1103
35.8478	0.1957	2.5030	42.7	2.5021	(211)	35-1104
42.4348	0.3044	2.1284	28.0	2.1291	(105)	35-1105
44.6739	0.2173	2.0268	39.5	2.0281	(220)	35-1106
52.9565	0.3261	1.7277	27.2	1.7281	(312)	35-1107
64.9565	0.3044	1.4345	30.9	1.4340	(400)	35-1108
71.6957	0.3695	1.3153	26.5	1.3150	(332)	35-1109

Fig. (2) shows the XRD patterns for (CIGS) thin films prepared by thermal evaporation and annealed at different annealing temperatures. Polycrystalline structure were observed in all samples with two phases tetragonal structure in nature of CIGS and with peaks diffraction at 2θ = 26.98°, and 27.96° with orientation (112) and (103) respectively, and CuSe hexagonal structure with peak located at 2θ =45.89° belong (110) direction for 293K samples. Annealingat 200°C cause to appearance of another peaks corresponding to hexagonal structure at 31.19° for (006) direction. While, at 300°C annealing temperature only CIGS peaks appeared at 2θ = 26.73°, 44.45° with orientation (112) and

(220) with tetragonal structure. It can be concluded that the crystallinity of films is improved at 300°C with pure CIGS structure. The crystallize size (D) of CIGS thin film is estimated using Debye–Scherer's formula

Summary of X-Ray characterization shown in Tables (2) with comparison between the experimental and the standard d_{hkl} values from International Centre for Diffraction Data (JCPDS). It can be concluded that by increasing annealing temperature the crystalline size increases. This means that the crystallinity of the films has improved due to defects reduction.

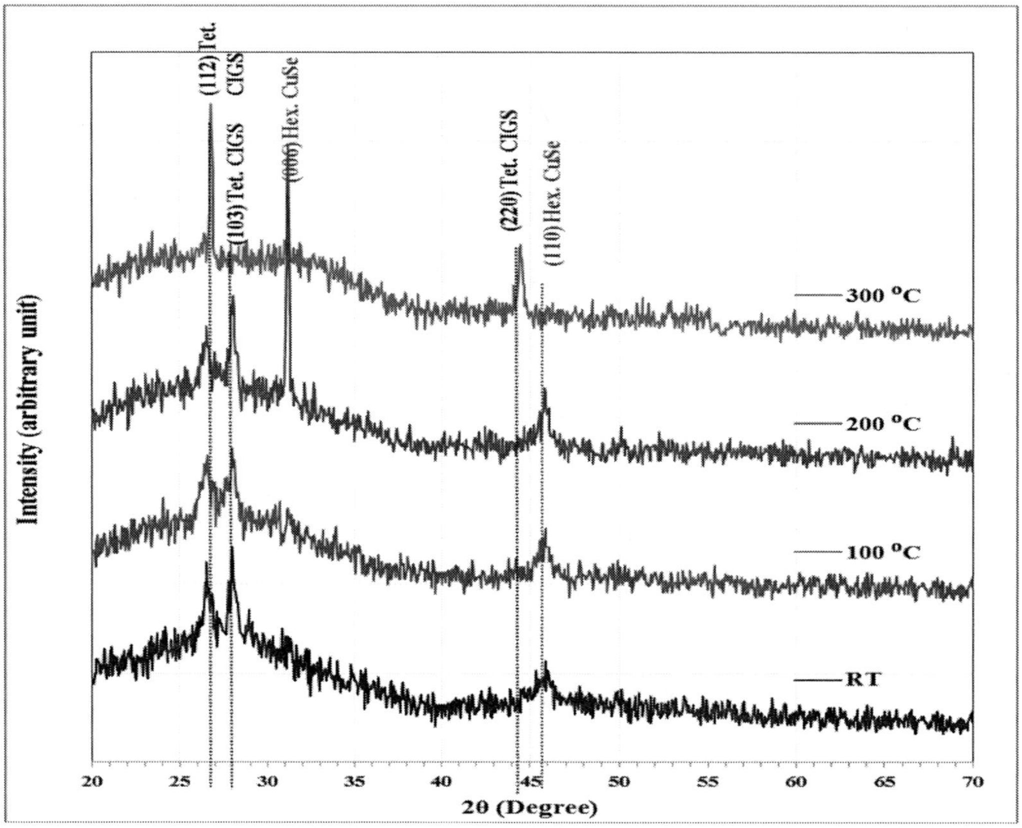

Figure 2. X-ray diffraction of the CIGS thin films with deferent annealing temperature.

Table 2. X-ray diffraction data for CIGS thin films with deferent annealing temperature

2θ (Deg.)	FWHM (Deg.)	d_{hkl} Exp.(Å)	G.S (nm)	d_{hkl} Std.(Å)	hkl	card No.
26.9130	0.1522	3.3102	53.7	3.3121	(112)	35-1102
28.0217	0.1739	3.1817	47.1	3.1801	(103)	35-1103
35.8478	0.1957	2.5030	42.7	2.5021	(211)	35-1104
42.4348	0.3044	2.1284	28.0	2.1291	(105)	35-1105
44.6739	0.2173	2.0268	39.5	2.0281	(220)	35-1106
52.9565	0.3261	1.7277	27.2	1.7281	(312)	35-1107
64.9565	0.3044	1.4345	30.9	1.4340	(400)	35-1108
71.6957	0.3695	1.3153	26.5	1.3150	(332)	35-1109

The 2D AFM images and granularity accumulation distribution chart of CIGS structures synthesized by thermal evaporation technique annealed at different temperatures (Ta=100,200 and 300 °C) are shows in Fig. (3). The CIGS structures have balls or semi-balls -shaped. The grains are homogenous. By using special software the estimated values of root mean square RMS of surface roughness average and average grain size are listed in Table(3). The CIGS structures thin films agglomerate increases as annealing temperatures. The results of average grain size have agreed with those estimated from XRD and it is clearly seen that the surface was very smooth. It was found that the size of grain and the RMS of surface roughness increases as annealing temperatures increase, but with cracks at high temperature.

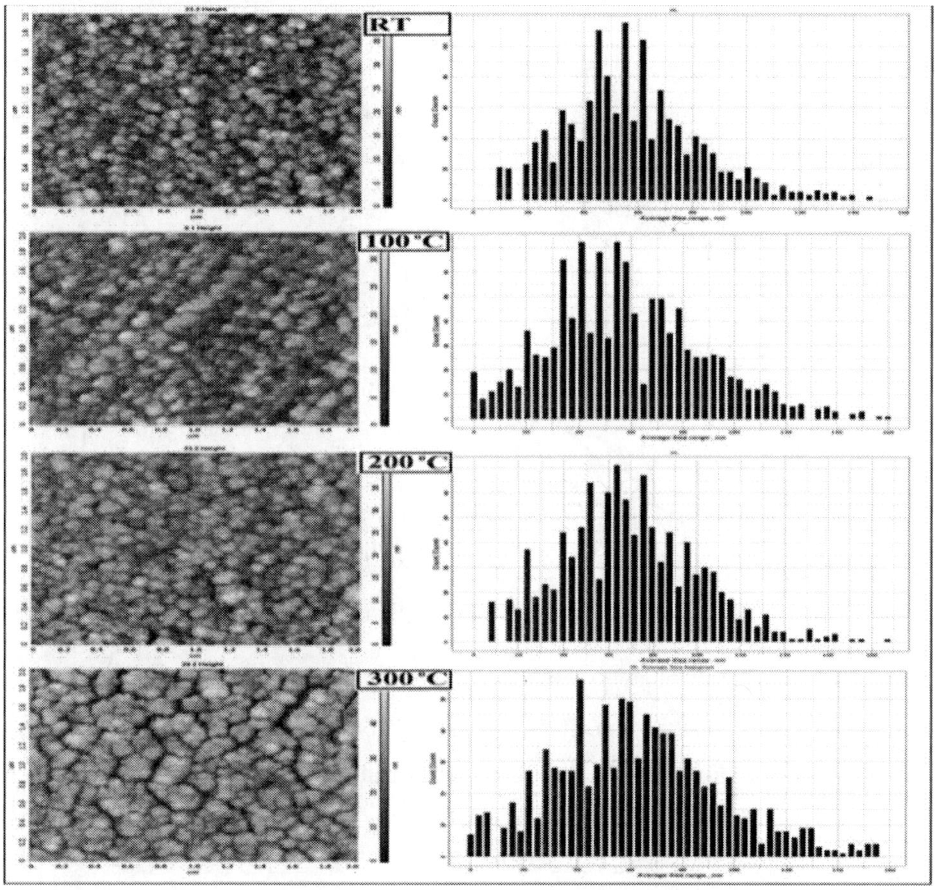

Figure 3. AFM images and granularity accumulation chart for CIGS thin films annealed at different temperature

Table 3: Particle size, average roughness and root mean square for CIGS thin films annealed at different temperature

Ta (°C)	Root mean square(nm)	Average Grain Size (nm)
293	6.154	87.2
100	8.194	110
200	9.157	112
300	18.847	116

Fig. (4) shows the optical transmission for CIGS films, deposited on glass substrate by thermal evaporation deposition and annealed with different annealing temperature, was carried out in the wavelength range 500–1100 nm. The transmittance of all deposited thin films increases with increasing of wavelength (λ) and increases with the increase of annealing temperature in the visible region. There are a Swanapol shape due to high film homogeneity[10]

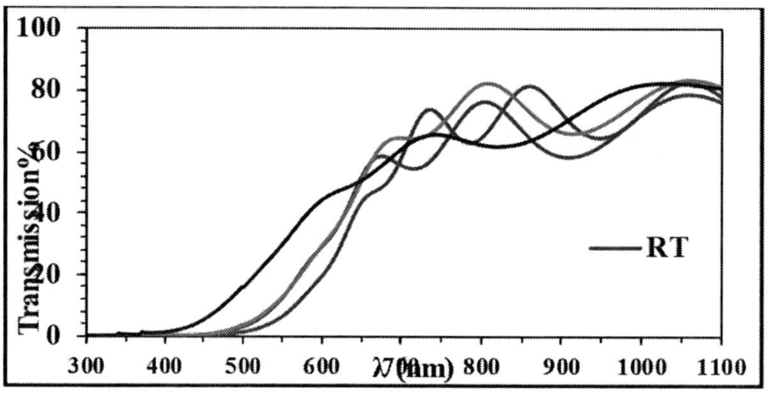

Figure 4. The transmission curves for CIGSthin films annealed at different temperature

The optical energy gap values were determined by Tauc equation.Fig.(5) shows the relation between $(\alpha h\upsilon)^2$with photon energy (hυ).It can be seen from this figure that the energy gap increase from 2.35 eV to 2.6 eV with increasing annealing temperaturefrom RT to 300 °C maybe due to phase transition.

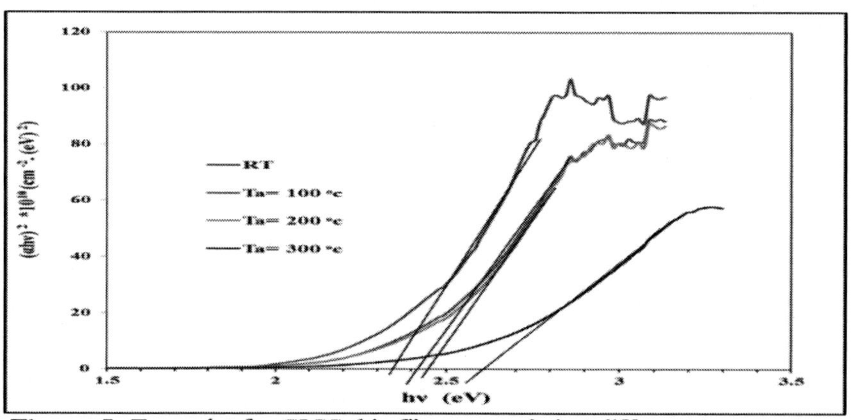

Figure 5. Tauc plot for CIGS thin films annealed at different temperature

Fig. (6) shows the I-V characteristics for solar cells with different configurations Al grid/CdS/CIGS/AL, ITO/CdS/CIGS/AL and Algrid/CdS/CIGS/Si/AL using CIGS films annealed at 300ºC. It can be seen from this figure that the first is the best one with highest Vm, Im and solar cell efficiency, as shown in Table (4).

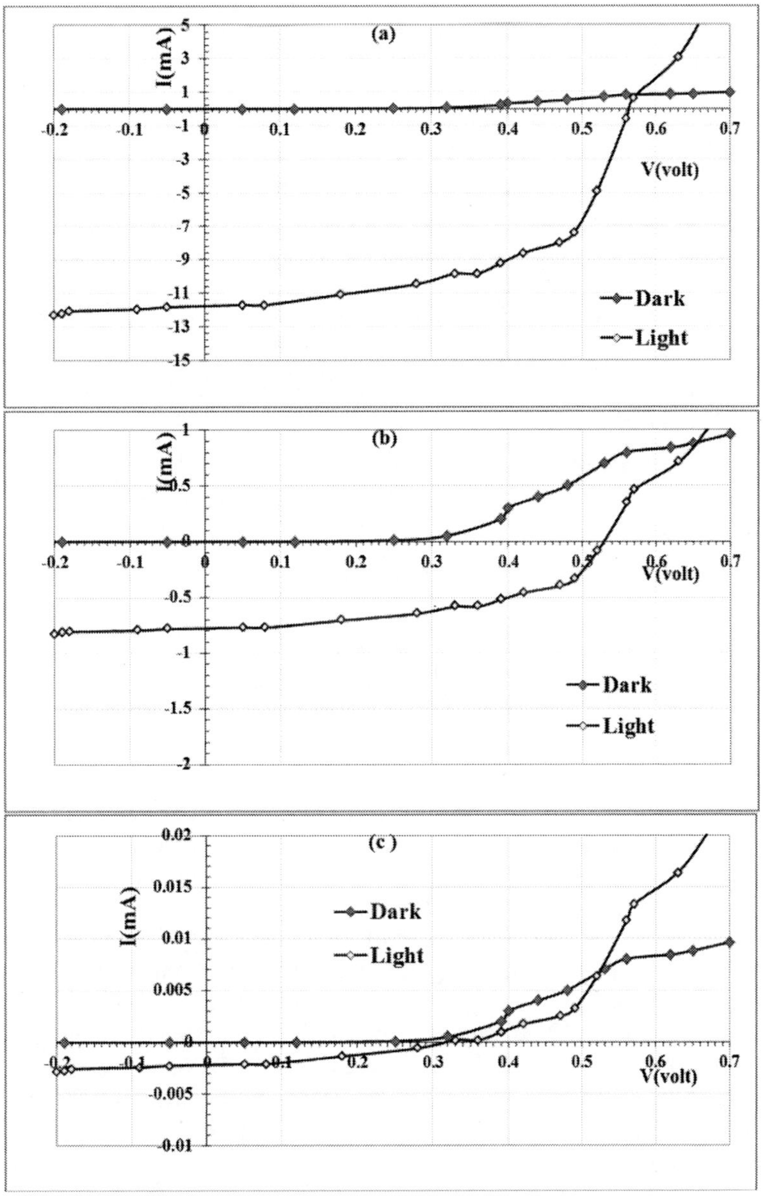

Figure 6. The I-V characteristics for CIGS solar cells with different configurations (a) Al grid/CdS/CIGS/AL (b) ITO/CdS/CIGS/AL and (c) Al grid/CdS/CIGS/Si/AL

IOP Conf. Series: Journal of Physics: Conf. Series **1032** (2018) 012019 doi:10.1088/1742-6596/1032/1/012019

Table 4: Solar cells parameters for CIGSsolar cells with different configurations

Thickness	A(cm^2)	P(mw/cm^2)	V_m(V)	I_m(mA)	Eff.
Aluminum electrode base (Al grid/CdS/CIGS/AL)					
T=1000nm	1	100	0.502	7.78	%3.91
ITO electrode base(ITO/CdS/CIGS/AL)					
T=1000nm	1	100	0.470	0.57	%0.27
AL/Si electrode base(Al grid/CdS/CIGS/Si/AL)					
T=1000nm	1	100	0.148	0.002	%0.02

4. Conclusions

Structural properties and optical for as deposited and annealed CIGS thin films at (100, 200 and 300) °C prepared by thermal evaporation I-V characteristics for solar cell can be summarized as follows:

- The XRD tests of CIGS powder and thin films indicates polycrystalline structure.Two phases tetragonal CIGS and hexagonal CuSe structure were observed at RT samples. Annealing at 300°C cause to convert to pure CIGS structure and enhance films crystallinety. The crystalline size increase with annealing temperature.
- The AFM topographic images, founded that the agglomerate increases as annealing temperatures, but with cracks at high temperature.The surface roughness increased by increasing annealing temperature
- The optical energy gap increase from 2.35 eV to 2.6 eV with increasing annealing temperature.
- I-V characteristics for solar cells with different shapes show that the best solar configuration is Al /CdS/CIGS/AL than the others examined.

References

[1] Tinoco T, Rincón C, Quintero M , and Pérez G (1991), 'Phase Diagram and Optical Energy Gaps for CuInyGa1−ySe2 Alloys'. Physica Status Solidi (a). Vol.124, No. 2, P 427.

[2] Andorka F 2014 'CIGS Solar Cells, Simplified'. Solar Power World. Archived from the original on 16 August 2014. Retrieved 16 August.

[3] Chenene M and Alberts V 2003 ' Structural and compositional properties of Cu(In,Ga)Se2 thin films prepared by the thermal evaporation of compound materials', J.Phys.D Appl.Phys. , p36 51.

[4] Piekoszewski J , Loferski J, Beaulieu R , Beall J, Roessler B and Shewchun J 1980 'RF-sputtered CuInSe2 thin films',Sol. Energ. Mater., Vol. 2, P 363.

[5] Wada T, Matsuo Y, Nomura S , Y. Nakamura, Miyamura A , Chiba Y, Yamada A and Konagai M 2000 Fabrication of Cu(In,Ga)Se2 thin films by a combination of mechanochemical and screen-printing /sintering processes',phys. status solidi (a),Vol. 203, P 2593.

[6] Stanber B Copper Indium Selenides and Related Materials for Photovoltaic Devices'2002. Critical Reviews in Solid State and Materials Science. Vol. 27, No.2,

[7] Hedström J , Ohlsen H , Bodegard M , Kylner A;. Stolt L, Hariskos D , Ruckh M , Schock H 1993 ZnO/CdS/Cu(In,Ga)Se2 thin film solar cells with improved performance'. Proceedings of 23rd IEEE Photovoltaic Specialists Conference: P.364–371.

[8] Dhere G 2007 Toward GW/year of CIGS production within the next decade'. Solar Energy Materials and Solar Cells. Vol. 91, PP.15–16).

IOP Conf. Series: Journal of Physics: Conf. Series **1032** (2018) 012019 doi:10.1088/1742-6596/1032/1/012019

[9] Scherrer P 1918 ,Göttinger Nachrichten Gesell," vol. 2, p. 98,.
[10] Baban C , Rusu G. I. and Prepelita P 2005 On the Optical Properties of Polycrystalline Cdse Thin Films," *J.optoelectron Adv Mater.*, vol. 7, no. 2, p 817–821.

IOP Conf. Series: Journal of Physics: Conf. Series **1032** (2018) 012020 doi:10.1088/1742-6596/1032/1/012020

Characterization of CuInSTe/CdS Heterojunctions

Bushra A.Hasan[1] ,Duua A.Umran[2] and Mohammad Abdulhur Kadhim Mankoshi[2]

[1]University of Baghdad, College of Science, Department of Physics
[2] University of Kerbala, College of Science, Department of Physics
bushra_abhasan@yahoo.com

Abstract. In this work, the role of substrate temperature in the range (293-423)K on the structure and microstructure of CuInSTe thin films was investigated. The results showed that the structure was amorphous for films prepared at ambient temperature Ts=293Kwhile it was polycrystalline for films prepared at high temperature. The grain size and roughness increased with increasing substrate temperature. The frequency and temperature dependencies upon AC conductivity were investigated in the frequency range 100Hz-10MHz and temperature of heat treatment in the range of 303-453K. The AC activation energy was found to increase from 0.1132 to 0.1258 eV with increasing substrate temperature from 293 to 423K and decrease to 0.0962 eV with increasing frequency from 100 to 107 Hz. The exponents show a nonsystematic sequence with the increment of substrate temperature. The obtained results gave indication that there is a relation between the structure and preparation temperature as well as heat treatment. According to the current-voltage characteristic, the CuInSTe/CdS heterojunction prepared at substrate temperature of 423K showed the best electrical characteristics under illumination conditions.
Keywords: quaternary alloy CuInSTe; Thin films; Heterojunction; Activation energy, dielectric properties .

1. Introduction

The quaternary compounds (CuInSeTe, CuInSeS and CuInSTe) are considered as absorbing materials for solar cell applications. There are few data available about their bulk material properties [1-3] as well as thin films [4-7]. The available data refer that there is a strong dependence between their structure and preparation conditions, which may be due to the amorphous nature as well as the dependencies of their properties on the ambient conditions. Solaiman has investigated the electrical properties [8] and optical properties [9] for CuInSTe, CuInSeTe and CuInSeS thin films. Landery et al [10]prepared Chalcopyrite semiconductor compounds, MIMIIE2 (MI = Cu, Ag; MII = Al,Ga, In; E = S,Se, Te using microwave irradiation of appropriate mixtures of the pure elements in a domestic microwave oven and examined by X-ray diffraction (XRD) and energy-dispersive X-ray (EDX) analysis. William et al [11] presented the results of preparing photovoltaic CuInSSe cell using solution processed at law temperature ~350 °C. They indicated that ban gap in the range (1.0-1.14) eV can be tuned by changing the amount of sulphur in the precursor solution. The obtained power conversion of 7.43% at 1-sun intensity .

The present paper is devoted to perform quaternary alloy of CuInSTe and then to prepare thin films at various substrate temperatures (298, 373, and 423) K using thermal evaporation method and point out the dependencies of dielectric and structural properties on deposition condition using relation between the preparation temperature and these properties.

2. Experimental Work

The bulk samples of CuInSTe were prepared by quenching technique. The exact amount of high purity elements cupper , indium , sulphur and tellurium (99.99) supplied from Certified Chemicals , UK accordance with their atomic percentages are weighed using an electronic balance with the least count of (10^{-4} g). The material was then sealed in evacuated (~10^{-5} Torr) quartz ampoule (length ~ 25 cm and internal diameter ~ 8 mm). The ampoules containing material are heated to 1000 °C using controlled oven type Heresies electronic for 5 hours according to the phase diagram. The temperature of the furnace was raised at a rate of 10 °C/min. During heating the ampoules are constantly rocked .This is done to obtain homogeneous glassy alloys. After that the obtained melt was rapidly quenched into water.

The vacuum unit system, which is used to prepare thermally evaporated CuInSTe films at different substrate temperatures (293, 373 and 423K), was Edward Coating unit model 306A. High vacuum inside the enclusure (chamber) obtained using rotary pump to about 10^{-2} mbar and diffusion pump where the pressure decreases to about $4x10^{-5}$ mbar. The deposition rate of thin films is about (10-15nm/min).

The structural properties of the prepared CuInSTe films were studied using Shimadzu XRD-6000 X-ray diffractometer. The microstructure of the deposited films was also examined using AA3000 atomic force microscope and structural parameters such as crystallite size and roughness were obtained.

Aluminum electrodes with thickness of 200nm were deposited on each adjacent surfaces of specimen by thermal evaporation technique under pressure of 10^{-5} Torr using an Edward coating unit. The specimen was fixed in a holder and placed into temperature-controlled oven (type Heresies electronic). High and low holder terminals are connected to dielectric analyzer (type Hewlett-Packard HP4274A & HP4275A), the third holder terminal was connected to the earth. The dielectric parameters like total resistance (RT), total capacitance (CT) and dissipation factor (tanδ) were measured (in parallel mode) under certain frequency range 10^2-10^6 Hz. The device structure fabricated in this work consists of (from the bottom) an Al layer as the back contact (≈ 200 nm), thin film layer of CuInSTe thickness of ~700nm and CdS layer (t~300nm), all these layers are prepared by thermal evaporation. The wires were connected by silver baste. The sunlight incident upon the device from the top .The current-voltage measurements were carried out in dark for the prepared CuInSTe/CdS heterojunctions using a Keithley-616 digital electrometer and DC power supply. The bias voltage was varied from -1 to 1 volt in the forward and reverse bias conditions.

3. Results and Discussion

Figure.1showed the X-ray diffraction patterns for bulk CuInSTe compound. Table .1 showed a comparison between the experimental and the standard peaks from Landry et al [10] and their miller indices since there is no available data for ternary compound to compare our results with them. It shows that the main peaks located at 2Θ=32.91° and 43.993° which are corresponding to the reflection planes (212) and (110) are identically with the $CuInS_{2-x}Te$ phase standard peaks of hexagonal structure, another traces of other miner phases like $Cu_{1.92}S$ and $CuInS_2$ were found. Our results are similar to the finding of Robbins et al[12].

IOP Conf. Series: Journal of Physics: Conf. Series **1032** (2018) 012020 doi:10.1088/1742-6596/1032/1/012020

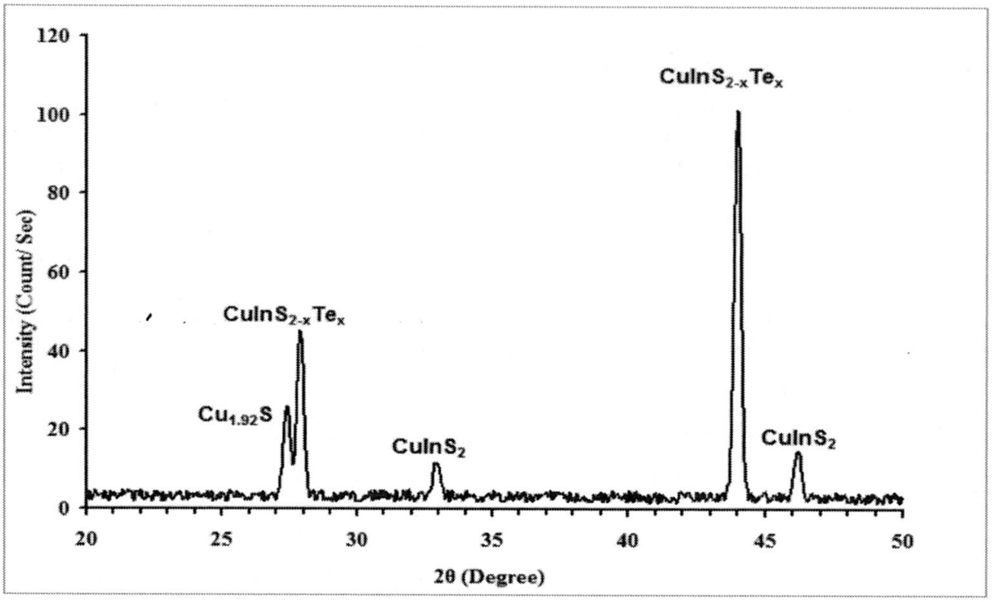

Figure 1. The XRD pattern for CuInSTe powder

Table1. The X-ray diffraction data for CuInSTe powder .

Samples	2Θ (deg)	$d_{exp.}$ (Å)	$(I/I_o)_{exp}$	FWHM (deg)	D (Å)	d x10^{-5} (Å)$^{-2}$
powder	27.381	3.254	23%	0.1956	417.58	0.573
	27.858	3.199	43%	0.1655	494.79	0.408
	32.916	2.718	9%	0.2346	353.26	0.801
	43.993	2.056	100%	0.1620	529.11	0.357
	46.180	1.964	13%	0.1774	487.02	0.422

The XRD patterns of CuInSTe thin films prepared at different substrate temperatures (293, 373 and 423K) are plotted in Figure. 2. The pattern of the thin films deposited at room temperature i.e. 293 K give indication of amorphous structure since it absent from any diffraction peaks, while there were many diffraction peaks belonged to the X-ray patterns of thin films deposited at elevated temperatures i.e. 373and 423 K, this give indication of polycrystalline structure . The main diffraction peak for thin films prepared as 373K belonged to the $CuInS_{2-x}Te$ with hexagonal structure, there was some other traces of another phase like Cu1.96S and In Te, while the main diffraction peaks for thin films prepared as 423 K belonged to the $CuInS_{2-x}Te_x$ with tetragonal structure and there was no any other traces of any phase which identical with the data of Landry et al. and Robbins et al [10 and 12]. It was pointed out that reaction of a stoichiometric mixture of the S, In, Cu and Te elements required for the chalcopyrite, $CuInS_{2-x}Te_x$ does not yield the alloy phase, as determined by XRD, but a mixture of $CuInS_x$ and $CuInTe_x$. This proposed that the alloy, $CuInS_{2-x}Te_x$, represents the limit of sulfide solubility in $CuInTe_x$. In addition, the formation of $CuInS_{2-x}Te_x$ requires the relative concentration of sulfur to be higher than the solubility limit .Thus the obtained structure type at different substrates temperatures can be related with variation of solubility limit of sulfur in the prepared samples which showed

IOP Conf. Series: Journal of Physics: Conf. Series **1032** (2018) 012020 doi:10.1088/1742-6596/1032/1/012020

strongly dependence upon substrate temperatures .Debey –Scherrer's formula [13] is used to estimate the crystal size of the film:

$$D = \frac{0.94\lambda}{FWHM \cos\theta} \quad (1)$$

The estimated parameters are listed in table.2 the inverse relation between the crystal size (D) and dislocation density (δ) is clearly remarked.

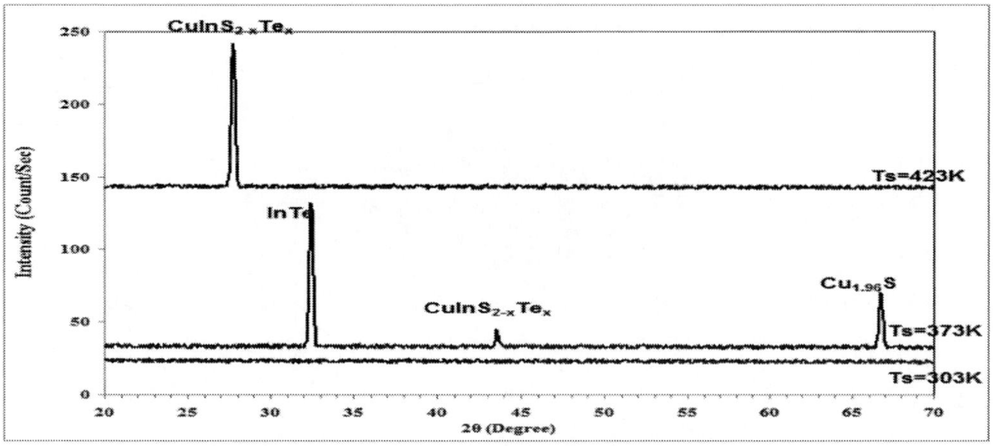

Figure 2. The XRD patterns for CuInSTe thin films thickness of 700nm deposited at different substrate temperatures.

Table 2. The X-ray diffraction data of CuInSTe thin films thickness of 700nm deposited at different substrate temperatures (293, 373 and 423K)

Samples	2Θ (deg)	$d_{exp.}$ (Å)	$(I/I_o)_{exp}$	FWHM (deg)	D (Å)	δ x10^{-5} $(Å)^{-2}$
Films T_S=293K	Amorphous					
Films T_S=373K	32.457	2.756	100%	0.1310	631.89	0.25
	43.542	2.103	9%	0.1700	503.41	0.394
	66.750	1.400	39%	0.2873	331.22	0.911
Films T_S=423K	27.758	3.211	100%	0.24675	33.180	0.910

The grain size (grain diameter) and average roughness of CuInSTe thin films samples deposited on glass substrates with different substrate temperature have been measured using AFM , the obtained images is presented in Figure.3 and the results were listed in table .3.

IOP Conf. Series: Journal of Physics: Conf. Series **1032** (2018) 012020 doi:10.1088/1742-6596/1032/1/012020

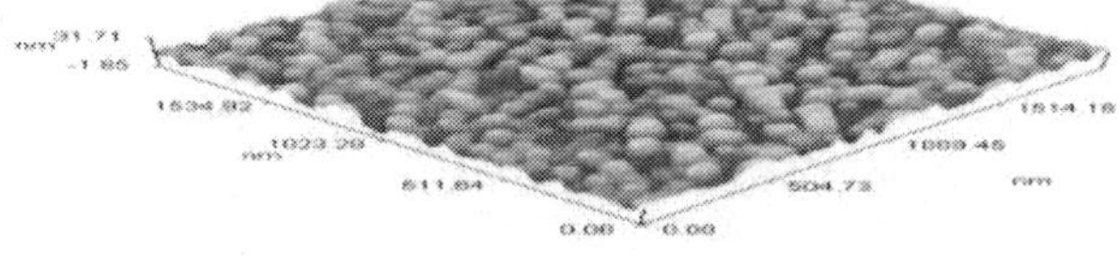

(a) T_S=293

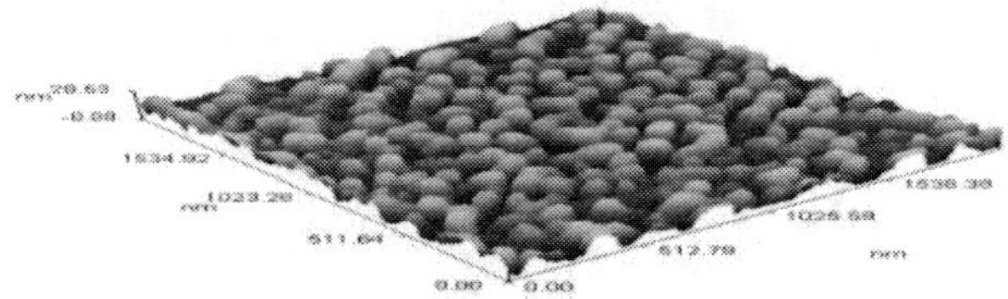

(b) T_S=373

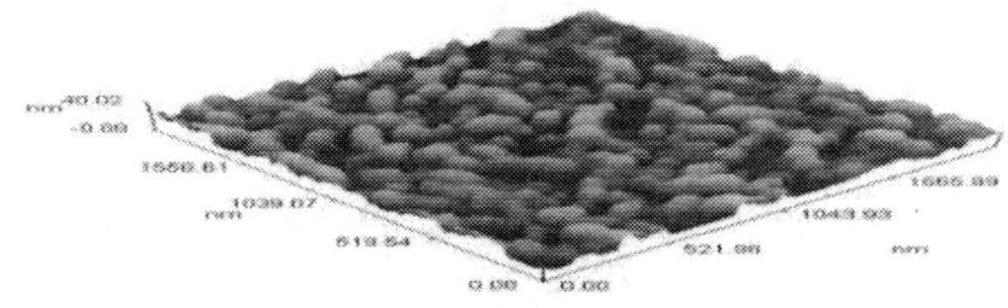

(c) T_S=423

Figure3. AFM images of CuInSTe thin films deposited at different substrate temperatures

Table 3. The average grain size and average roughness of the prepared CuInSTe thin films

Average roughness (nm)	Grain size (nm)	T_S (K)
3.42	66.29	293
4.73	80.53	373
5.25	77.92	423

It is obvious from table .3 that the average grain size (but not in systematic manner) and roughness get to increase with the increasing of Ts, moreover they increases from 66.29 to 77.92 and from 3.42 to 5.25 nm respectively when Ts increases from 293 to 473K due to structure enhancement by the action of heat. As seen from the x-ray diffraction pattern the crystallization takes place at elevated substrate temperatures (i.e., Ts=373 and 423K). The well-pronounced single phase of CuInSTe at high substrate temperature (423K) has produced much more crystallites arranged vertically with a size of 77.92 nm of the film. The total conductivity σ_{tot} at a certain frequency and temperature is defined as:

$$\sigma_{tot}(\omega) = A_t\omega^s + \sigma_{d.c} \quad (2)$$

Where $\sigma_{d.c}$ is the D.C conductivity which depends strongly on temperature and dominate at low frequencies, while $\sigma_{a.c}$ is the A.C. conductivity which is weaker temperature dependence than $\sigma_{d.c}$ and dominate at high frequency and nondependent temperature and (s) is a function of frequency and is determined from the slope of a plot ln $\sigma_{a.c}(\omega)$ versus ln(ω). $\sigma_{tot}(\omega)$ obtained from the equation

$$\sigma_{tot}(\omega) = \frac{t}{R.A} \quad (3),$$

where R is the sample resistance , A is the effective area for capacitance, t is the thickness. Figure (4) shows the angular frequency dependence of total conductivity $\sigma_{tot}(\omega)$ for CuInSTe films deposited at different substrate temperatures(293, 373 and 423)K and thermally treated at (303, 333, 363, 393, 423 , and 453) K. It is clear that there is proceeding increase of $\sigma_{tot}(\omega)$ that increased in the whole frequency range, i.e., the conductivity is frequency dependent or pure A.C and σtot(ω) increases exponentially with frequency in the hole frequency range which give indication of domination of the second term of equation (2)upon the first term (D.C conductivity) ,and the polarization is electronic. The exponent s, which is estimated from the slope of plot ln $\sigma_{tot}(\omega)$ against ln(ω) are calculated and listed in table 4.It is found that s increases with the increase of thermal treatment fro thin films samples deposited at law temperatures i.e. Ts=293 and 373K , indeed s increased from o.58 to 0.79 and from 0.534 to 0.79 for when thermal treatment temperature increases from 303 to 453 K for thin films deposited at Ts=293 and 373K respectively , while s get to vary in contrast manner for thin films samples deposited at elevated substrate temperature i.e. 423K , indeed s decreases from 0.95 to 0.81 when thermal treatment temperature increases from 303 to 453 K .The reduction of s values fro thin films deposited at law Ts reflect the hoping mechanism. Thus, the suitable models for these samples are SP and CBH, respectively, as seen from Table .4.

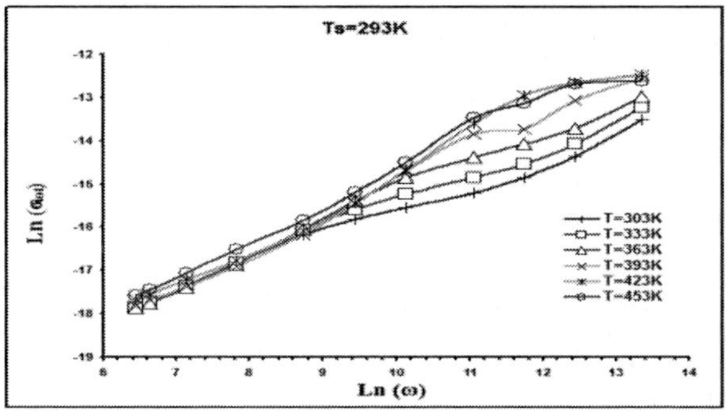

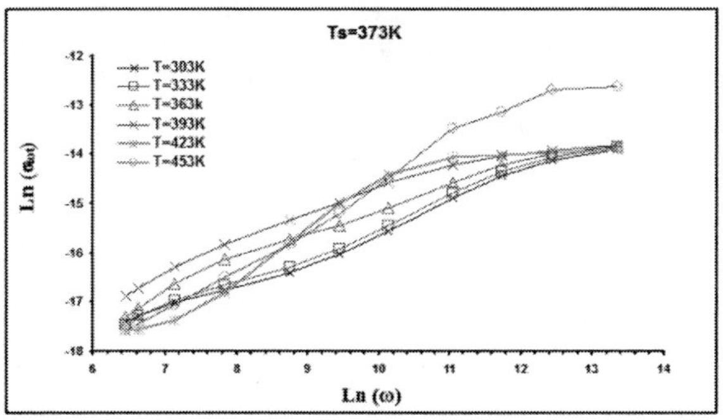

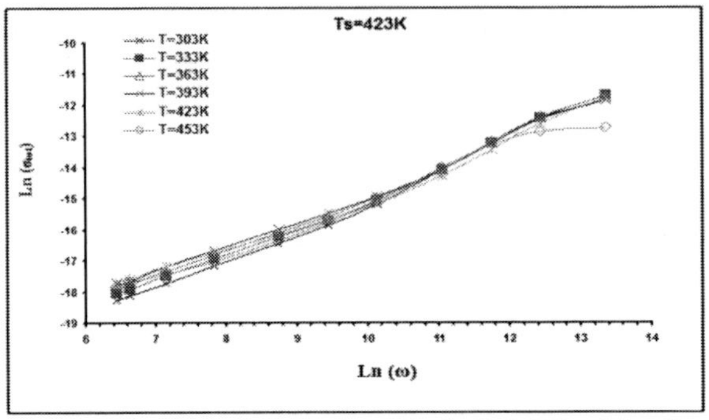

Figure 4. Variation of Ln (σ_{tot}) with Ln(ω) for as deposited CuInSTe thin films and annealed at different temperatures.

The AC activation energy, EAC in the studied for CuInSTe films are estimated from the drawing of Ln σtot (ω) against the reciprocal absolute temperature by fitting different regions with the equation

$$\sigma = \sigma_0 e^{-E_a/k_B T} \qquad (4)$$

as indicated in Figure.5. are calculated.Table.3 illustrates the E_{AC} values at selected frequencies (200Hz, 1kHz, 4 kHz, 10 kHz and 100 kHz). The results show that each sample declared one conductivity mechanism and hence, one activation energy E_{AC}. On the other side, there is a direct relation between E_{AC} and frequency, while there is an inverse relation between activation energy value E_{AC} and T_S. The E_{AC} decreases from 0.1132 to 0.0962eV with the increase of T_S from 293 to 423K, while E_{AC} increases from 0.1132 to 0.1258eV with increasing frequency from 1kHz to 100kHz. The increase in frequency results in an increase of vibrating energy and hence the E_{AC} values will be decreased. This indicated that conductivity is purely AC, while the elimination of the localized states and vacancies explained the increase of E_{AC} with increasing TS. Seto's [14] proposed the presence of large amount of trapping states at the grain boundary that able to capture free charge carries. These charged states at grain boundary create potential barriers, which oppose the passage of carriers from grain to neigh boring ones, where these sites increase with increasing T_S, which resulted in decreasing conductivity or increasing E_{AC} values.

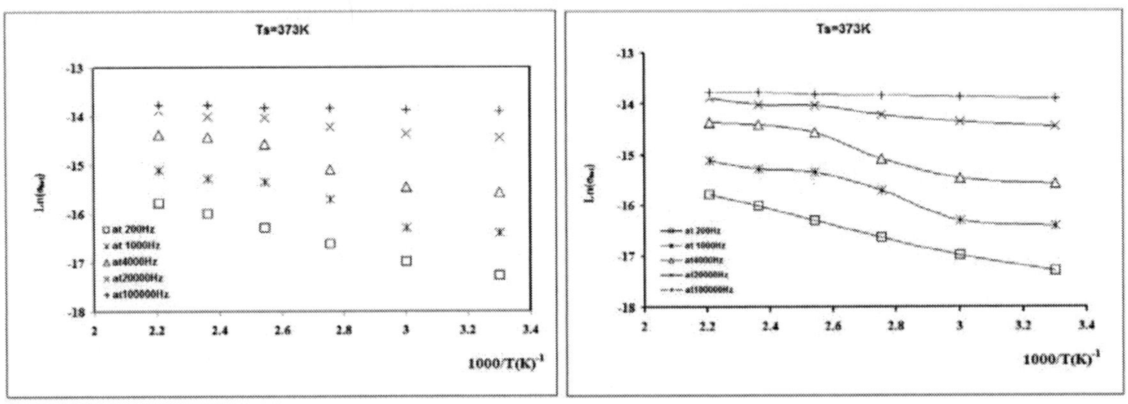

IOP Conf. Series: Journal of Physics: Conf. Series **1032** (2018) 012020 doi:10.1088/1742-6596/1032/1/012020

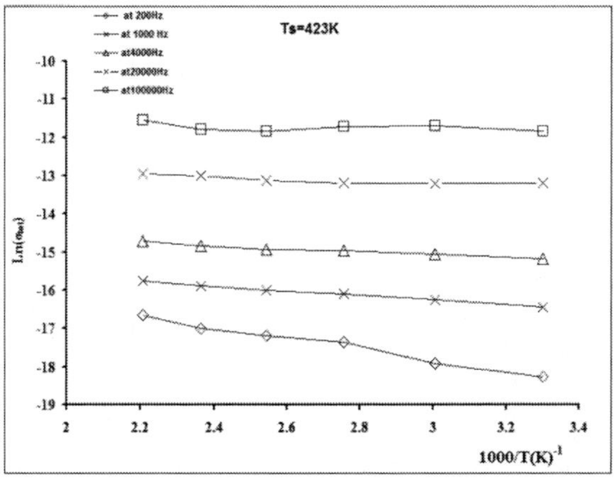

Figure 5. $\ln(\sigma_{tot})$ with 1000/T for CuInSTe thin films deposited at different substrate temperatures.

Table 4. Values of E_{AC} for CuInSTe thin films

E_{AC}(eV)			Frequency(Hz)	Thin Films Samples
T_s(K)				
423	373	293		
0.1258	0.1210	0.1132	200	CuInSTe
0.0517	0.1126	0.1115	1000	
0.0336	0.1089	0.1022	4000	
0.021	0.0449	0.0973	10000	
0.0114	0.0103	0.0962	100000	

The real dielectric constant determined using the following relation : εr= C.t / ε0 .A (5)where C is the capacitance , ε0 is the permittivity of free space,ω is the angular frequency.

The real dielectric constant (ε_r) of CuInSTe films prepared at different Ts(293, 373 , and 423)K and thermally treated at different temperatures(303, 333, 363, 393, 423 , and 453), are measured within the employed frequency range (10^2-10^7 Hz). It is clearly from the εr pattern versus Log(ω), in Figure.6 that ε_r exhibits to increase with the increasing of heat treatment temperature, while it decreases with increasing frequency. This is ascribed to the fact that the electrode blocking layer is dominated, thus, the dielectric behaviour is affected by the electrode polarization, while εr attained minimum values at high frequencies, which indicates that the dielectric signal is not affected by electrode polarization [8]. On the other hand, it is noticeable remarked that the values of ε_r, at frequency of 10^2 Hz, increase with increasing heat treatment and substrate temperatures. Indeed, the ε_r increases from 615 to 1064 and from 615 to 802 when heat treatment temperature increases from 303 to 453K and substrate temperature increases from 293 to 423K, respectively.

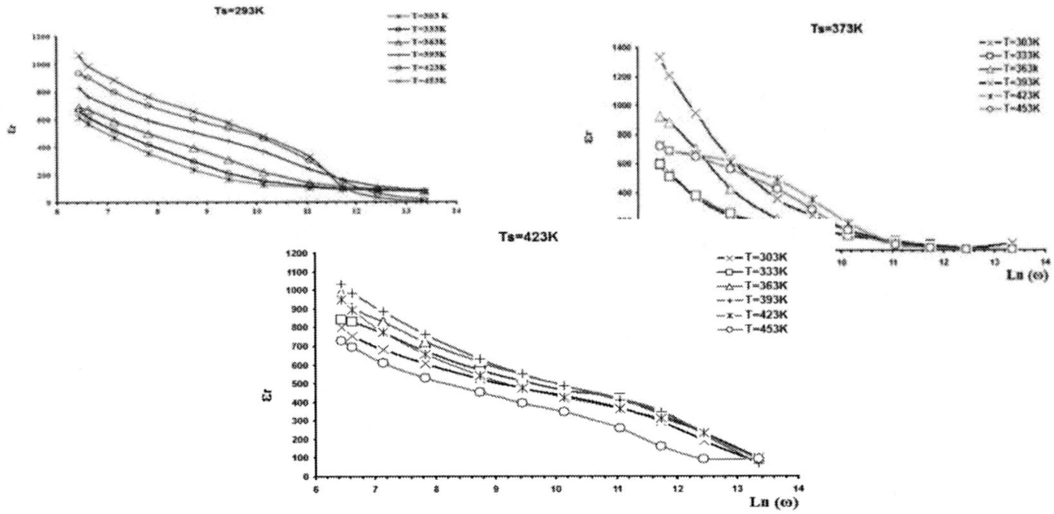

Figure 6. Variation of ε_r with lnω for as deposited CuInSTe thin films and annealed at different temperatures.

The imaginary dielectric constant εi calculated from the relation

$$\varepsilon_i = \frac{\sigma_{AC}}{\varepsilon_0 \omega} \qquad (6)$$

The variation ε_i with Ln(ω) are drawn in Figure.7 for CuInSTe thin films annealed prepared at different Ts(293, 373,and 423)K and thermally treated at different temperatures(303, 333, 363, 393, 423 , and 453). It can be seen that the dielectric loss absorption bands appear for CuInSTe films treated and deposited at high temperatures. While the loss bands disappear from most of the deposited samples. The absorption band observed at characteristic frequency proposed the foundation of relaxation dipoles for samples deposited and thermally treated at high temperatures. It is remarked. The relaxation peak shift to high frequency with the increase of temperature of deposition and treatment, but return to shift to lower frequency at high treatment temperature i.e. 453K. With the increase of temperature of deposition it is considered that there is tendency to crystallization as seen from x-ray diffraction of thin films, figure 2.the increase of crystalline content speed up the segmental motion throughout reducing the free volume. It is illustrated by the loss ban shifting toward high frequency range, resulting in reducing of relaxation time.

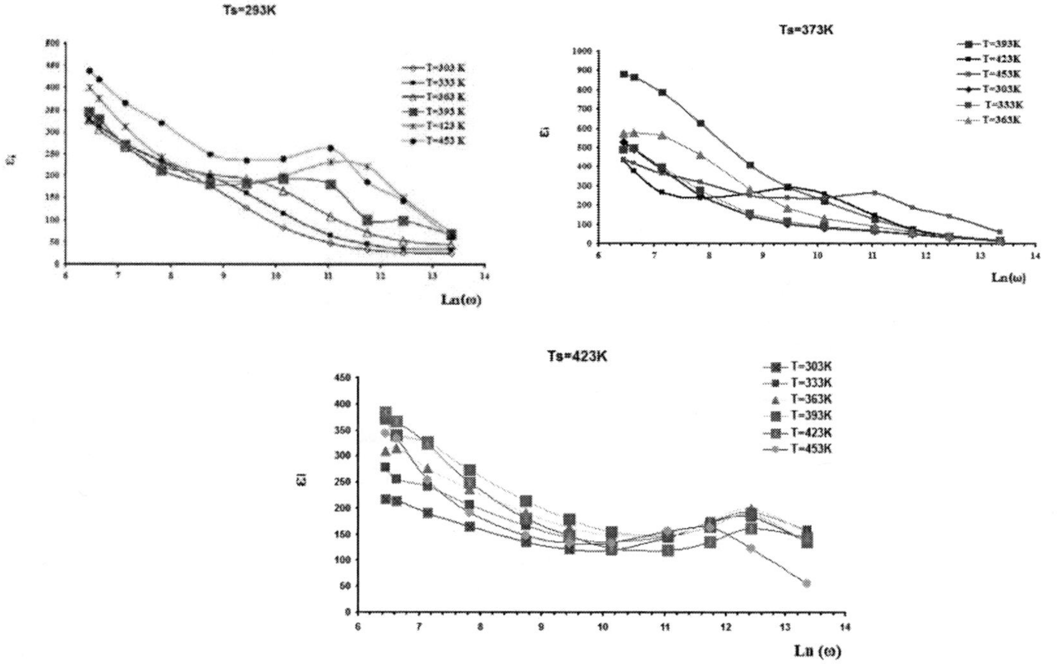

Figure 7. Variation of ε_i with lnω for as deposited CuInSTe thin films and annealed at different temperatures.

It is well known that dielectric data is a contribution of two components: the contribution of conductivity that results an increase of dielectric constant ε_r and dielectric loss εi with the decreasing of frequency and relaxation process which show at ε_i which tend to shift to law frequency side with the increase of temperature. Also the dielectric loss composed from the contribution of two components diffusion of charge carrier and dipolar orientation.

Using the equation $w_D=1/\tau$ the relaxation time were estimated from frequencies corresponding to the loss maxima, where w_{max} is ~ macroscopic relaxation time w_D, the values of τ were listed in Table .4. It is clearly remarked that rising of thermal treatment temperature lead to foundation of more than one absorption loss peak in (ε_i) spectrums.

The appearing of more than one loss peak is related with presence of amorphous and crystalline phase in the thin films samples of CuInSTe,the vanishing of absorption bands in law frequency range of the εi diagram versus frequency give rise to the dominating of one phase .Thus rising of temperature of deposition or treatment caused the overall increase of εr and εi which ascribed to simultaneous contribution of dipolar and free charge. The increase of charge mobility resulted in increase of free charge carriers and the shift of loss peak toward higher frequency side.

Cole-Cole diagrams are shown in Figure.8. The polarizability (α) was determined and listed in Table (2). It can be noticed that the values of α also show a nonsystematic variation with the increasing of TS, while α decreases with the increasing of thermal treatment, moreover α increases from 0.122 to 0.2016, when TS increases from 293 to 373K, while α decreases from 0.3920 to 0.1680, when TS increase from 303 to453K for CuInSTe films that prepared at 373K, see Table (4). The decreasing and increasing of α with the increasing of heat treatment resulted from the rise and weaken of the intermolecular forces respectively [16-18].

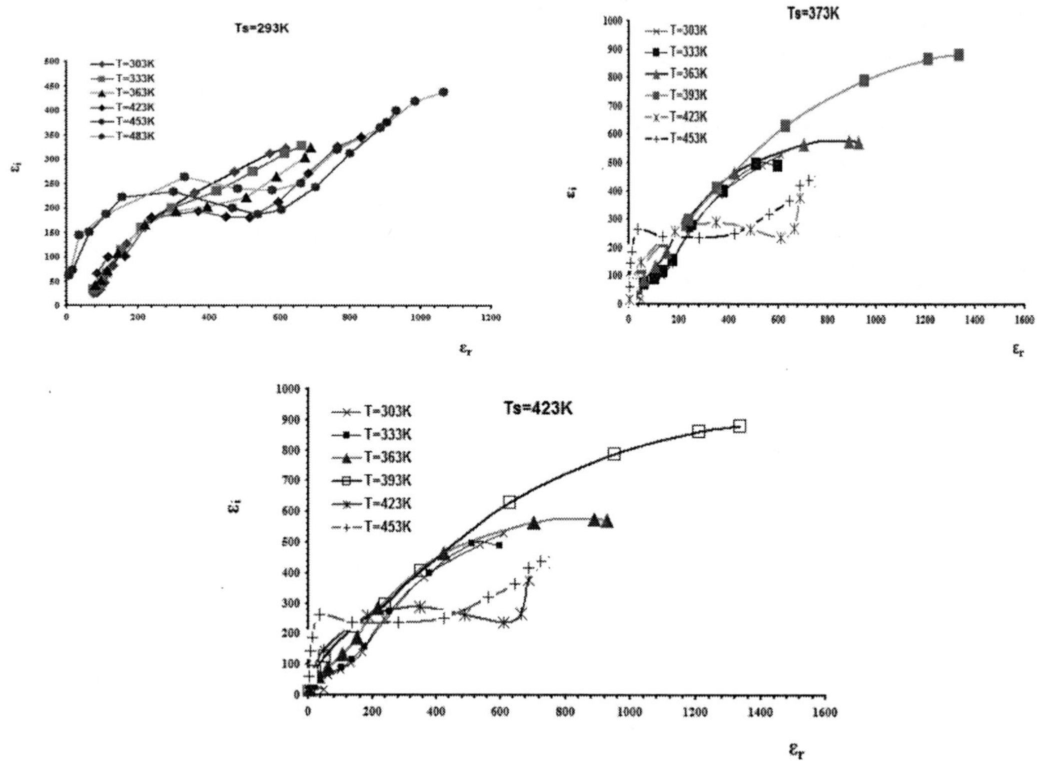

Figure 8. Cole-Cole diagrams for CuInSTe thin films deposited at different substrate

Table 5. Values of s, τ and α for CuInSTe thin films

Ts=423K			Ts=373K			As deposited			Ta(K)
α	$\tau x10^{-4}$(sec)	S	α	$\tau x10^{-4}$(sec)	s	α	$\tau x10^{-4}$(sec)	s	
0.2016 0.2800	--- 0.0398	0.956	0.3920 0.1792	-------	0.534	0.1120	------	0.585	303
0.2240 0.2016	--- 0.0398	0.929	0.1904 0.1008	13.281	0.544	0.1456	------	0.639	333
0.1456 0.2800	13.281 03989	0.903	0.1008 0.1680	13.281	0.508	0.1008	------	0.713	363
---------	--- 03989	0.863	0.2352	------	0.463	0.1232 0.0784	----- 0.398	0.797	393
0.2240 0.1680	--- 0.039	0.845	0.0672 0.0336	----- 0.796	0.621	0.0560 0.1120	----- 0.159	0.834	423
0.2240 0.2016	----- 0.0796	0.815	0.1680 0.1680	----- 0.159	0.791	0.1232 0.1120	------- 0.159	0.791	453

Variation of current under illumination at different power density with biasing voltage of p-CuInSTe/n-CdS thickness of 700nm deposited at various temperatures (293, 373 and 423K) is plotted in Figure.9. The measurements were carried out under illumination with different power intensity. From this figure it is clear that the photocurrent increases with increasing of the bias voltage. At forward biasing voltage the current rises exponentially with forward voltage until the slope becomes more gradual. This can be due to high level injection of carriers such that the applied voltage is no longer totally developed across the depletion region. At the high reverse biasing break down can occur due to impact ionization or Zener tunneling. These mechanisms can be separated by temperature dependence. The I-V parameters V_{oc}, I_{sc}, and η were calculated and listed in table .5. It clear that increasing deposition temperature had significant effect on the illumination current, i.e., there is valuable increment in the illumination current at elevated substrate temperature. This arises from the phase transformation from amorphous to crystalline structure. The illumination current increases with the increase of Ts as a result of increasing grain size. This is in agreement with previous results [19].

It is evident from table .5 that V_{oc} tends to vary in non systematic sequence with Ts, i.e., V_{oc} increases and decreases with increasing of TS. The increase in Voc is resulted from decreasing electron affinity (χ) between the CuInSTe system and CdS layer, while the decreasing of V_{oc} ascribes to increasing electron affinity between the two layers of the junction. The high values of V_{oc} and I_{sc} at TS=423K for CuInSTe/CdS indicate that this junction is very suitable for solar cell fabrication. For photocurrent consideration, which is equal to short circuit current I_{sc}, the smaller band gap is required because more photons are collected. It was found that maximum efficiency can be obtained when the band gap lies in the range 0.8-1.4eV and the ideal factor increases primarily due to the increase of V_{oc} while the photocurrent is increased linearly with the light intensity [20].

4. Conclusions

Depending on the results from this work, one can put the following points:

It can be concluded that the single phase with chalcopyrite structure of CuInSTe becomes more pronounces at elevated temperatures. The quaternary alloy CuInSTe was successfully prepared using quenching from the melt method. The exponent s value of CuInSTe films show progressive increase with the increase of temperature of deposition and thermal treatment giving rise that the conductivity is pure A.C. The increase of deposition temperature result in increasing of polarizability α values and hence reducing of molecular forces while increasing of thermal treatment temperature result in reducing of polarizability α values of CuInSTe films consequently increases the intermolecular force. The continuous variation of α value gives indication that the structure may be more enhanced with further increase of temperature.

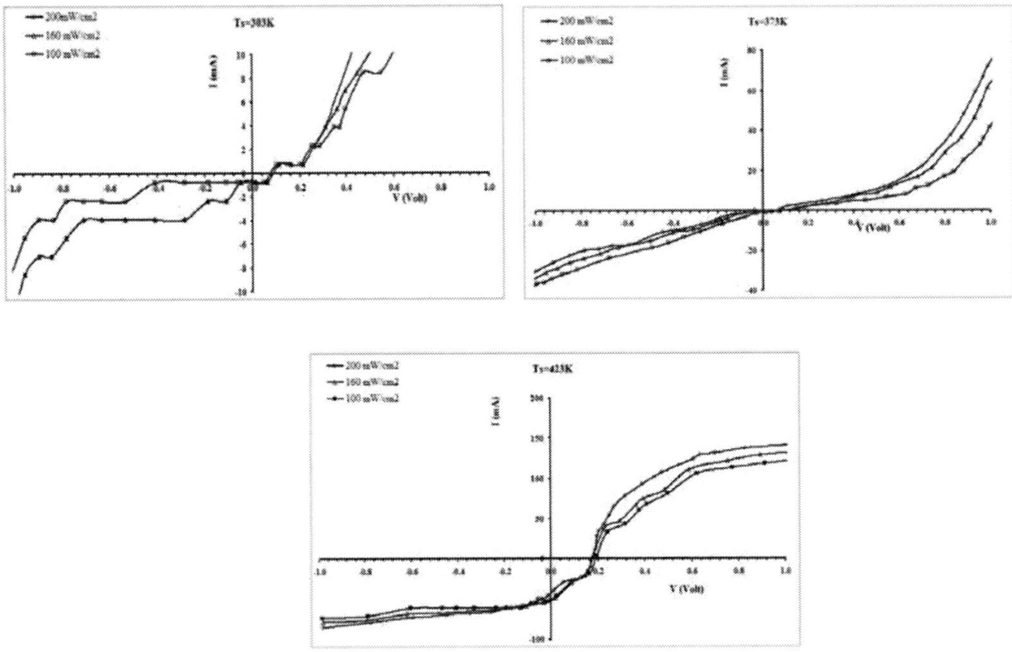

Figure 9. I-V characteristics for p-CuInSTe/n-CdS heterojunction with t~700nm at different substrate temperatures and power densities

Table 6. Illustrated the values of I_{sc} V_{oc} for for p-CuInSTe/n-CdS heterojunction thickness of 700nm at different substrate temperatures (293, 373 and 423 K).

T_s (K)	V_{oc} (V)	I_{sc} (mA)	P (mW/cm^2)	η (%)
293	0.09	0.80	100	0.072
	0.10	0.77	160	0.048
	0.10	0.80	230	0.040
373	0.07	0.09	100	0.004
	0.07	0.90	160	0.032
	0.06	1.00	230	0.06
423	0.18	46.3	100	4.2
	0.19	48.4	160	5.7
	0.20	46.2	230	9.2

Acknowledgment
Author would like to thank Department of Physics, College of Science, University of Baghdad for supporting this work.

References

[1] Hasan B. and Umran D. 2012 Semicond. Sci. Technol. 27 125014.
[2] Hasan B. 2013 Int. J. Thin Film Sci. Tech., 2 29.
[3] Diaz R. Leon M. and Reuda F. 1992 J. Vac. Sci. Technol. A 10 295–300.
[4] Inakoshi K. Ohashi T. Hashimoto Y. and Ito Kentaro 1998 Solar Energy Mater. Solar Cells 50 37
[5] Subbaramiah K. and Raja V. 1992 Thin Solid Films 208 247.
[6] Ohashi T. Ger A. Miyazawa T. Hashimoto Y. and Ito K. 1995 Jap. J. Appl. Phys., 34 4159.
[7] Kuranouchi S. and Nakazawa T. 1998 Solar Energy Mater. Solar Cells, 50 31.
[8] Soliman L. and Hendia T. 1997 Rad. Phys. Chem. 50 175.
[9] Soliman L. Hendlia T. and Zayed H. 2007 Fizika A (Zagreb), 16 39.
[10] Landry C. Lockwood J. and Barron R. 1995 Chem. Mater. 7 699.
[11] William W. Hou B. Sheng-han L. Yang Y. 2009 Thin Solid Films 517 6853.
[12] Robbins M. Phillips J. Lambrecht V. 1973 J. Phys. Chem.Solids 34 1205
[13] Holmes P. 1962 Academic Press, New York.
[14] Seto J. 1975 J. Electrochem. Soc., 122 701.
[15] Fava R. 1973 Polymers Part(C) 16 Academics Press New York 311.
[16] Kumar S. Prajapati B. Tiwari S. and Tiwari V. 2008 Indian J. Pure Appl. Phys. 46 198.
[17] Jonscher A. 1999 J. Phys. D: Appl. Phys. 32 14.
[18] Ahmad Z. 2014 Ch. 1 4.
[19] Bari R. and Patil L. 2011 Sensors and Transducers, 125 213.
[20] Sze S. and Kwok K. 2007 3rd ed., John Wiley & Sons, Inc.

IOP Conf. Series: Journal of Physics: Conf. Series **1032** (2018) 012021 doi:10.1088/1742-6596/1032/1/012021

Effect shapes of nanorods, nanowires and nanobelts TiO_2 on the degradation of MB photo catalyst.

W J Aziz, R S Sabry and A K Jarallah

Al- Mustansiriyah University, College of Science, physics Department
Corresponding Author e-mail: abbaskasoob@yahoo.com

Abstract. Titanium dioxide (TiO_2) was prepared using hydrothermal method. The crystallite structure was characterized by X-ray diffraction (XRD), which showed the TiO_2 have two phases anatase and rutile. Field-emission scanning electron microscopy (FE-SEM) used to investigate the morphology of TiO_2, such the images showed that the product have nanorods hexagonal, nanobelts, and nanowires. The different shapes of nanotitanum dioxide on Photocatalyst were studied on degradation of methylene blue dye under UV irradiation, optimum degradation revealed with nanowires shape compare with nanobelts and nanorod shapes.

1. Introduction

Hydrothermal technique has been realized as an excellent and cost-effective method for the processing of nanostructure materials [1, 2]. Hydrothermal method is one of the chemical synthesis routes with high potential for obtaining nanostructured materials with controlled properties for energy harvesting applications [3]. TiO_2 nanoparticles can be obtained by hydrothermal treatment of peptized precipitates of a titanium precursor with water [4]. It is a versatile material with novel properties suitable for a number of technologically important applications, such as catalysis, white pigment for paints or cosmetics, electrodes in lithium batteries [5]. Several semiconductor nanoparticles (ZnO, SnO_2, WO_3 and TiO_2) have been utilized for the photo catalytic applications [6]. Among the chemical methods, hydrothermal synthesis is often used due to its simplicity, allowing the control of grain size, morphology and degree of crystallinity by easy changes in the experimental procedure [7]. The photo catalytic TiO_2 also revealed antibacterial [8]. Hydrothermal synthesis is an environmentally friendly methodology used to synthesize nanostructured materials [9]. Furthermore, the hydrothermal technique is typically done using an autoclave with Teflon liners under controlled temperature and/or pressure in aqueous solutions [10].

2. Experimental part

TiO_2 has been prepared via hydrothermal method, using titanium isopropoxide as precursor material with (HCL, NaOH and HF). 4 mL of HCL added drop wise to mix component (20 mL of distilled water with 4 mL of titanium isopropoxide) kept in a Teflon-lined hydrothermal with continues stirrer for 15 min. The reaction time and temperature were 3.5 h and 150 °C, after finished the hydrothermal time, the product rinsed with distilled water by centrifuge then dried at

IOP Conf. Series: Journal of Physics: Conf. Series **1032** (2018) 012021 doi:10.1088/1742-6596/1032/1/012021

100 °C. The same steps above have been used with NaOH and HF. Preparing thin films on clean glass substrate was used drop casting method. The activities of photo catalytic evaluated by the degradation of Methylene blue in a 50 mL distilled water, and UV- lamp was selected as light source at wavelength 254 nm. The solution maintained in the dark for 1 h before the irradiation, 5 mL of the sample solution was taken out every 30 min during irradiation, and then calculated the degradation rate of Methylene blue.

3. Results and discussion

Figures (1- A) represented XRD Patterns of pure nanoparticles-TiO_2 synthesized by hydrothermal method using HF solution. The XRD pattern exhibits diffraction peaks (101), (103), (-114), (004), (217), (021), (105), (213), (116), and (301) respectively. The presence of these diffraction peaks indicating that the deposited titanium dioxide pigment present in the form of mixed phases. The crystallite size of (101) peak is very small equal to (10 nm), which determined using Debye–Scherrer equation. It chooses (101) peak in Debye–Scherrer equation because the preferring value in anatase phase of TiO_2. Figure (1- B) XRD patterns of TiO_2 synthesis by hydrothermal method when with NaOH, appear two phases anatas and rutile corresponding to peaks (101), (112), (200),(220), and (215) for anatas phase and peaks (110), (101), (111), (211), and (112) of rutil phase. The intensity of (101) peak decrease after washing, relative to same peak before washing may be caused to replacement of Na^+ with H^+ during the rinsing process. The peaks are fully assigned to TiO_2 and their broad width suggests the presence of small-sized particles. Crystallite size of (101) peak is (20 nm). In Figure (1- C) the X- ray diffraction pattern confirmed that pure TiO_2 has a polycrystalline peaks. The diffraction peaks are (101), (112), (200), (105), (211), (204), (116), (220), and (215) where pointed to anatase phase. No other phases could be detected in the pattern and the crystallite size is (57 nm) of peak (101).

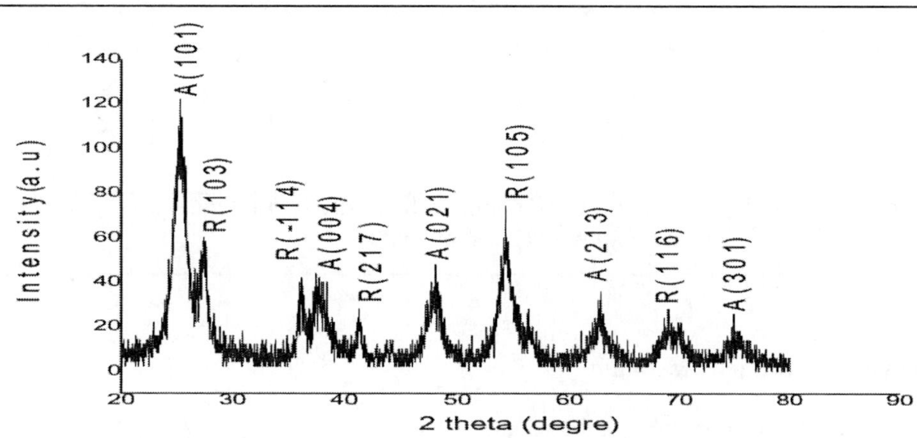

A

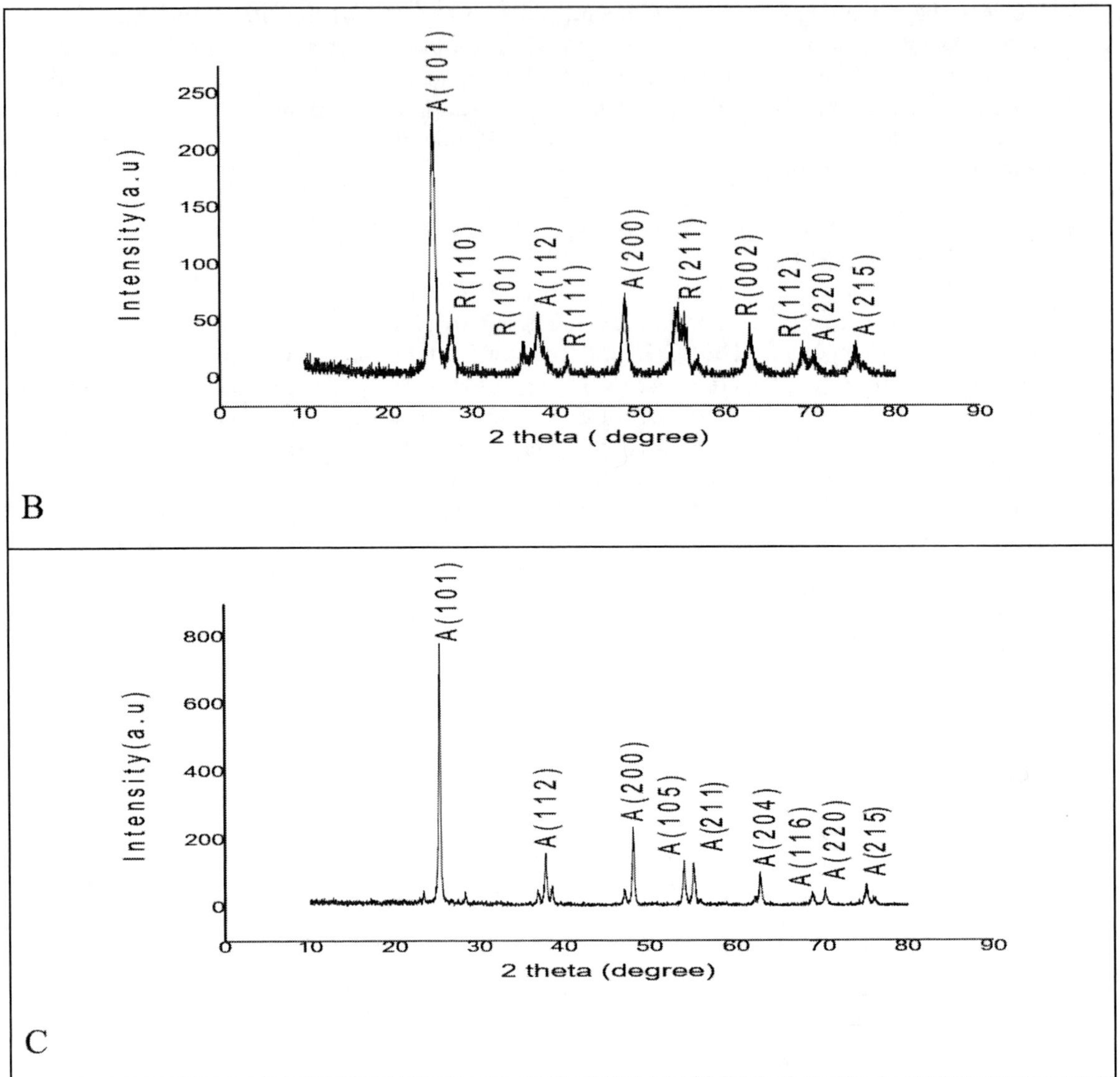

Figure 1. XRD patterns of TiO_2 synthesis by hydrothermal method, A) with HF solution, B) with NaOH solution, and C) with HCL solution.

4. Morphology characterizations

The FESEM photos were obtained to characterize the morphologies of TiO_2. As shown in Figures (2, 3, and 4) images of TiO_2 formed on glass substrate consist of vertically aligned nanorods, nanobelts, and nanowires. Also the figures reveal a high uniform of TiO_2 nanoparticles with average diameter (200 nm) and average height (150–250 nm) of nanorods, (50- 200 nm) in width and (150- 850 nm) in length of nanobelts and finally the length of nanowires is less than 100 nm.

IOP Conf. Series: Journal of Physics: Conf. Series **1032** (2018) 012021 doi:10.1088/1742-6596/1032/1/012021

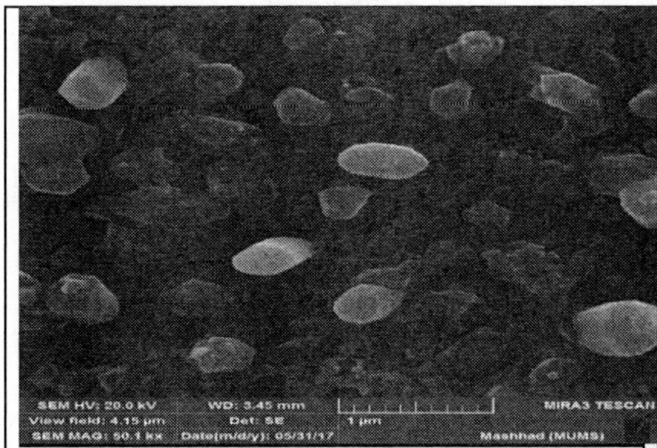

Figure 2˙ FESEM images of nanorods hexagonal TiO_2 synthesis with HCL solution by hydrothermal method.

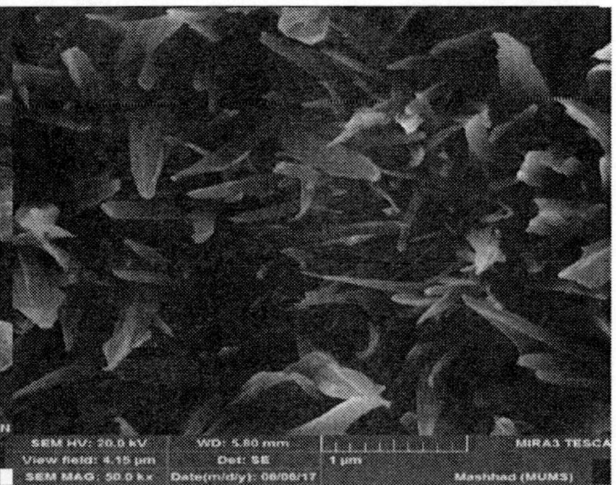

Figure 3. FESEM of TiO_2 nanobelts synthesis with NaOH solution by hydrothermal method.

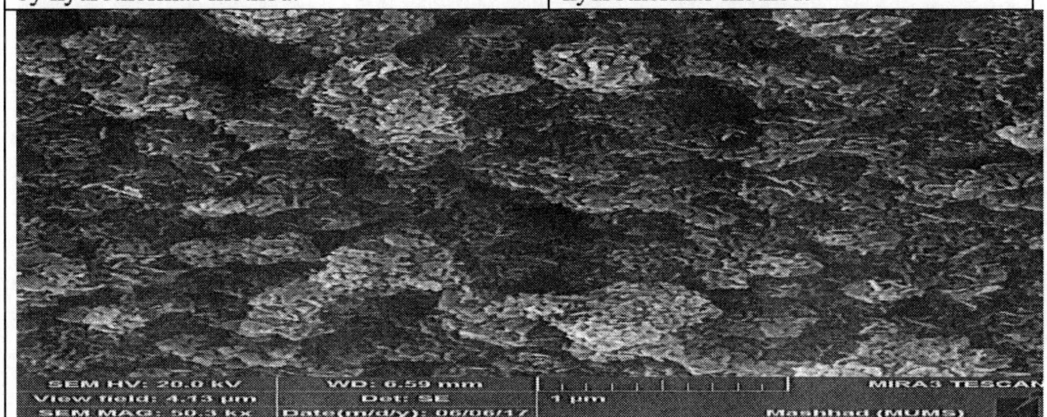

Figure 4. FESEM image of TiO_2 nanwires synthesis with HF solution by hydrothermal method

Figures (5, 6, and 7) show the relation between the absorption and wavelength with effect of irradiation time. The absorption of MB is decreasing with increasing the irradiation time, due to the MB degradation. These Figures also, reveal the effect of shape and size on the MB degradation. Among of these three shapes of TiO_2, nanorod, nanobelt, and nanowire, the highest photocatalytic activity obtained when used nanowires compare with other shapes, because the increasing of degradation and degradation ratio with decreasing the crystallite size of TiO_2.

IOP Conf. Series: Journal of Physics: Conf. Series **1032** (2018) 012021 doi:10.1088/1742-6596/1032/1/012021

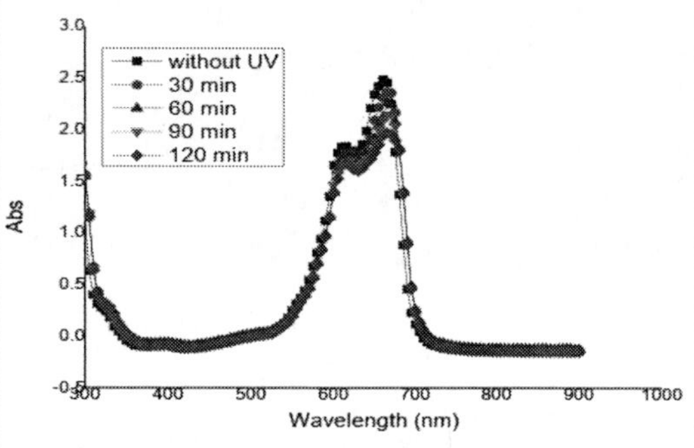

Figure 5. The relation between absorption of MB and wavelength when using pure TiO_2 nanorod preparing by hydrothermal method.

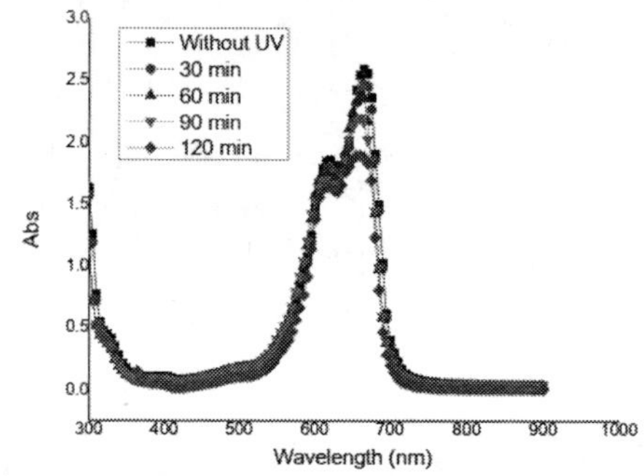

Figure 6. The relation between absorption of MB and wavelength of pure TiO_2 nanobelts preparing by hydrothermal method.

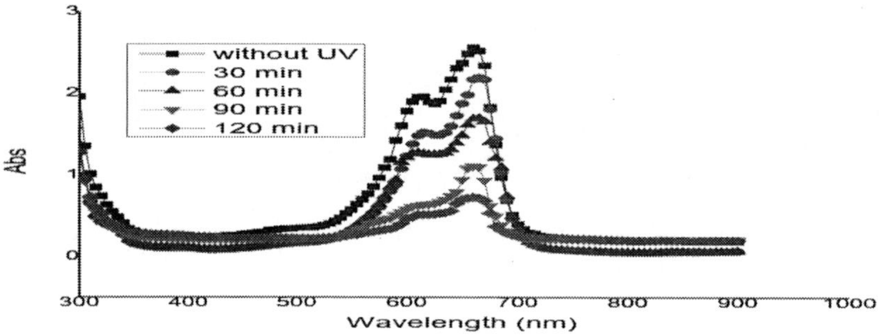

Figure 7. The relation between absorption of MB and wavelength using pure TiO_2 nanwires preparing by hydrothermal method.

Figures (8) and (9) show the effect of the shape and crystallite size respectively on the degradation ratio of MB, which appear the relation between the degradation and crystallite size inversely relation.

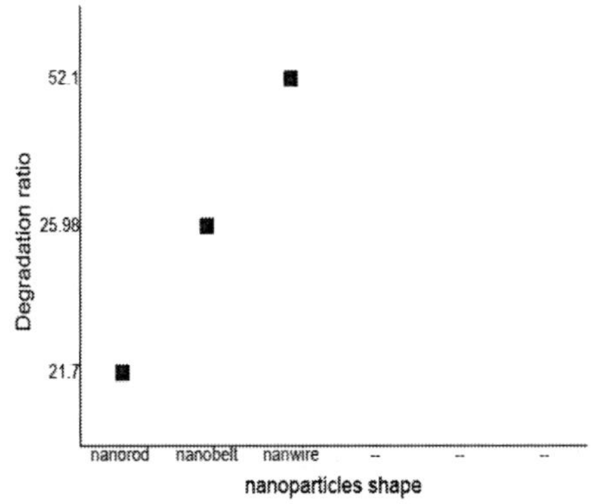

Figure 8. Effect of shape on nanoparticles degradation ratio of MB.

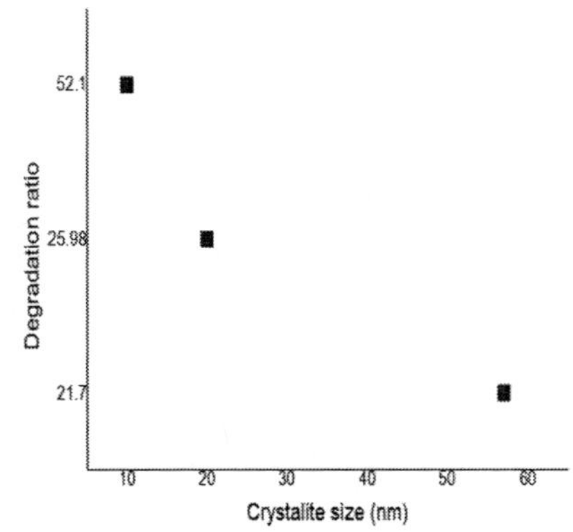

Figure 9. Effect the crystallite size on the degradation ratio of MB.

5. Conclusions

Nanostructured titanium dioxide (TiO_2) was successfully synthesized using the hydrothermal technique. The shapes (nanorod, nanobelt, and nanowire) and size of the product strongly depend on the preparation conditions. The prepared TiO_2 are polycrystalline and exist in two phases (anatase and rutile) when used HF and NaOH solvent, while appear one phase (anatase) when used HCL solvent. The degradation of MB depends on the irradiation time, and depends on (shape and size) of TiO_2 such the optimum degradation of MB was in nanowire shape and the crystallite size of (10 nm), the degradation ratio of MB higher than 52%.

References

[1] K. Byrappa and T.Adschiri, 2007*Progress in Crystal Growth and Characterization of Materials,* **53** 117–166.
[2] M. Yoshimura and K. Byrappa 2008 *Journal ofMaterials Science* **43** 2085–2103.
[3] R. Piticescul, A. Motoc1, A. Tudor1, C. Rustil, R. Piticescul and M. Ramiro-Sanchez 2015 **1** 31-42.
[4] H. Luo, L. Shen, K. Rui, H. Li, X. Zhang J. Am. Chem. 2013 Soc, **572** 37-42.
[5] A.R. Armstrong, G. Armstrong, J. Canales, P.G. Bruce 2005 J. Power Sources **146** 501–506.
[6] K. Santhi, P. Manikandan, C. Rani. and S. Karuppuchamy, x 2015 Applied Nanoscience **5** 373-378 .
[7] **S. Komarneni, Q.H. Li, R. Roy** 1996 J. Mater. Res **11** 1866 .

[8] V. Prince, J. Agrawal, M .Agrawal, L .Nanjannawar, S. Fulari, V .Kagi, S .Shetti 2017 Int J Sci Stud **4** 132-136.

[9] Y. Mao, T. Park, F. Zhang, H. Zhou, S.Wong 2007 *Small* **3** 1122-1139.

[10] A. Nakahira, T. Kubo, C. Numako 2010 *Inorganic Chemistry* **49** 5845-5852 .

IOP Conf. Series: Journal of Physics: Conf. Series **1032** (2018) 012022 doi:10.1088/1742-6596/1032/1/012022

physical properties of CdS / CdTe / CIGS thin films for solar cell application

B A Ahmed[1], I H Shallal[1], F I mustafa AL-Attar[2]*
[1]University of Baghdad, College of Education for Pure Sciences,
[2]Solar Energy Research Center, Renewable Energy Directorate
[1,2]Higher Education and Scientific Research Ministry, Baghdad, Iraq

falah_im@yahoo.com

Abstract. Thin films of CdS, CIGS and CdTe were prepared on the glass substrate by vacuum thermal evaporation technique in the thickness 200,500,500 respectively at room temperature. From the study Structural properties of the thin films by the X - ray diffraction, we found polycrystalline nature for all thin films with a preferred orientation along [111] cubic for CdS and [211] ,[112] hexagonal for CIGS and [111], [202], [311] cubic for CdTe. The Transmittance spectrum was recorded as a function of wavelength range (500-1100) nm. The energy gap measurements record 2.35,1.75.1.5eV respectively. we have found one activation energy for CIGS and two for both of CdS and CdTe that with explored from measure electrical conductivity with temperature. The Hall Effect measurements show that CIGS and CdTe were p-tyes carrier concentration and CdS thin films were n-type carrier. the current -voltage test in dark and light condition show all types of thin films have photoconductivity properties.

1. Introduction

In the last years from 1990 solar cells based on Cd- II-VI semiconductor materials like CdS, CdTe with CIGS are one of the most attractive and promising in thin film photovoltaic's. Solar cells have developed and updated by introducing these semiconductors thin films which provide lower energy devices with high production capacity and consuming minimum material [1]. Attracted considerable attention was focused on Cd-semiconductor compounds due to their applications in photovoltaic devices [2]. All these materials are direct-gap semiconductors that can absorb incoming solar radiation at a thickness much thinner than the required thickness for the silicon wafers in crystalline silicon (Si) solar cells [3].

CdTe with a band gap of 1.45 eV is the material well suited to match the Air Mass (AM) equal 1.5 solar spectrum. Furthermore, its high absorption coefficient causes that only a few microns absorber film is required for solar cell operation and (Cupper Indium Gallium Seleinde) CIS/ CIGS materials are significantly different. The highest value of solar radiation absorption belongs to CIS/CIGS that can absorb almost complete incoming radiation at first 3–4 μm of the material thickness and 95% of the radiation in its first 0.4–0.5 μm [4].

$Cu(In,Ga)Se_2$ solar cells are very interesting because of their high efficiencies, with low cost and many important application possibilities. (CIGS) is one of the most promising semiconductor for use as absorber layer for thin-films solar cells .It is a member of an I-III-VI2 group of chalcopyrite semiconductors , It is used as light absorber material in solar cells. These thin films deposited on different substrates with a variety of growth methods. CIGS deposition yield lower efficiencies, but offer other advantages.

CdS is a II-VI compound , n-type semiconductor with energy gap of 2.45 eV. It is used to fabricate heterojunction with CIGS. the carrier density of CdS is higher than in CIGS, the depletion field is completely in the CIGS layer where it generates electron-hole pairs. So that, it reduces

IOP Conf. Series: Journal of Physics: Conf. Series **1032** (2018) 012022 doi:10.1088/1742-6596/1032/1/012022

recombination of minority carriers at the metallurgical interface It is used in photovoltaics, photonics and optoelectronics [5]. In optoelectronics, CdS is used to manufacture lasers, It works also as a window layer which allows the light to pass through it with relatively low absorption. Cadmium telluride (CdTe) is a stable crystalline compound formed form cadmium and tellurium. It is mainly used as the semiconducting material in cadmium telluride photovoltaics and an infrared optical window . It is usually sandwiched with cadmium sulfide to form a p-n junction solar PV cell. Typically, CdTe PV cells use a n-i-p structure. The important advantage in this type of solar cells is the ability to quickly and easily manufacture these cells at much lower costs than crystalline silicon photovoltaic but typically have efficiencies that are lower than crystalline silicon [6-7].

The aim of this research is to study the properties of thin films to search for the suitability of these films for the manufacture of solar cells based on Cu(In,Ga)Se2 semiconductor material for better utilization of the solar spectrum to yield higher efficiency and performance stability [8-9].

2. Experimental procedure

This research includes the practical steps to depositing (CIGS,CdS and CdTe) thin films. The deposition of $Cu(In_{1-x}Ga_x)Se_2$, CdS, CdTe layers with thicknesses of (500 nm) and deposited Aluminum metal electrode (Al) ohmic contacts with (200 nm) thickness ,were done using thermal evaporation on glass substrates using thermionic model (VE.90, USA) Vacuum system for deposition which reach to 10^{-6} Torr. The boat of molybdenum (Mo) was used to evaporate the materials while using Tungsten(W) boat o evaporate the Al contact electrodes. To examine the crystal structure, orientation, grain size and position of atoms can be calculated by using X-ray diffraction technique. X-ray diffraction (XRD) measurements using (SHIMADZU diffractometer system (XRD-6000)), having the following features: The source of x-ray is Cu with wavelength λ=1.54060A°, Current was 30 mA ,Voltage was 40 KV and the scanning range was between angles (2θ) of (20 – 80) degrees with scanning speed 5 (degree/minute).

The Atomic force microscopy type AA3000 from Angstrom advanced Inc , used to study roughness and surface morphology for the CdS , CIGS and CdTe thin films. The optical properties of the CdS , CIGS and CdTe thin films were studied, the absorption and transmittance spectra for samples are recorded using UV-Visible spectrometer model UV-2610 product by Biotech engineering management company (UK)in the wavelength range (200-1100) nm and V-670 UV-Vis/NIR spectrophotometer . The electrical properties measurements of semiconductor thin films which included DC conductivity and Hall Effect were determined to investigate the activation energies and the concentration of charge carriers, as well as mobility and use these measurements to distinguish whether the semiconductor n-type or p-type using system type HSM-300 made by ECOPIA Co. (Korea).

3. Results and discussion

3-1 Structural properties of thin films:

3-1-1 XRD measurements

All thin films have shown the polycrystalline structure in nature as shown in the figure 1 at RT.

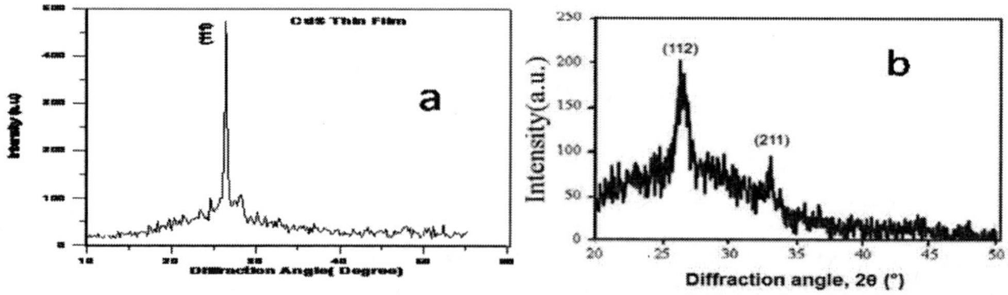

IOP Conf. Series: Journal of Physics: Conf. Series **1032** (2018) 012022 doi:10.1088/1742-6596/1032/1/012022

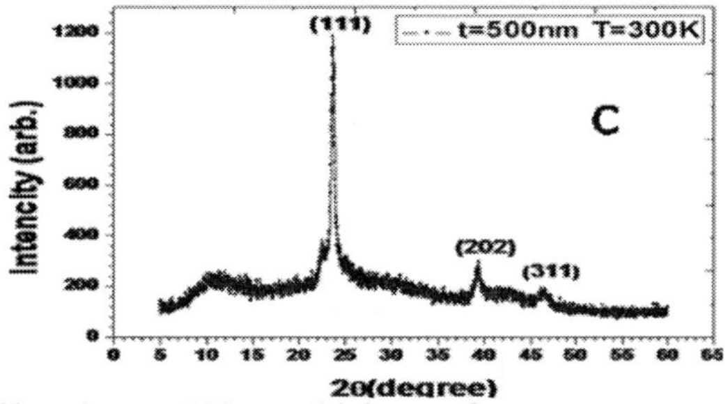

Figure 1. shows the X-ray diffraction at 500 nm thickness for temperatures (300) K, (a) CdS,(b) CIGS, (c) CdTe thin films

3-1-2 Atomic Force Microscopy

We have test the AFM of CIGS,CdS and CdTe thin films respectively and showed in figure 2. that topography of surface polycrystalline with defect according to deposition process.

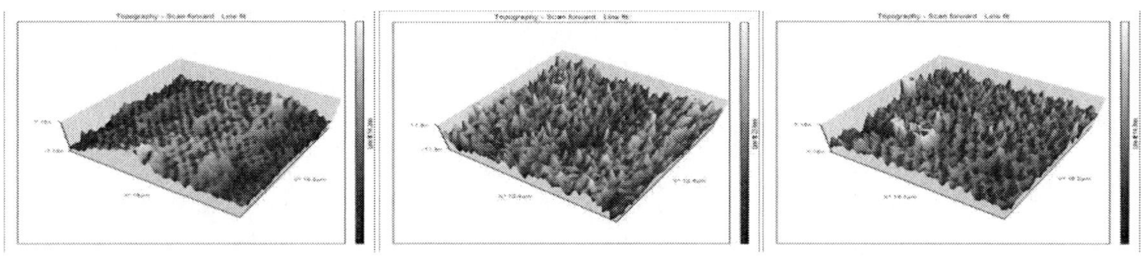

Figure 2. AFM picture of CIGS, CdS and CdTe thin films respectively.

3-2 Optical Properties of thin films

In the figure 3. the absorbance and transmittance curve show all thin films lay in the visible spectrum with different transparency curve

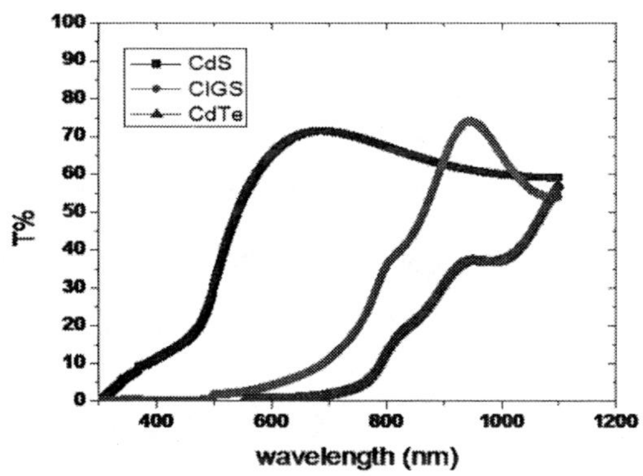

Figure 3. The transmittance curve for CdS, CIGS and CdTe thin films at thickness 200, 500, 500 nm deposited on glass at RT

3-2-1 Energy Gap:

The energy gap measurements record 2.35, 1.75, 1.5eV respectively from transmittance curve for CdS, CIGS and CdTe thin films at thickness 200 500, 500 nm deposited on glass at RT. The best thickness of CdS thin film for give high transmittance and high carrier mobility for solar cell measurements was 200nm (Babkair 2010) . The results of energy gaps of all thin films shows that the

IOP Conf. Series: Journal of Physics: Conf. Series **1032** (2018) 012022 doi:10.1088/1742-6596/1032/1/012022

CdS layer, the first window face light permit the visible spectrum to next layer of CdTe and CIGS to absorbed of photons to generated electron holes inside and separated to electrodes by internal built-in voltage in P_N junction of solar cell as shown in the figure 4.

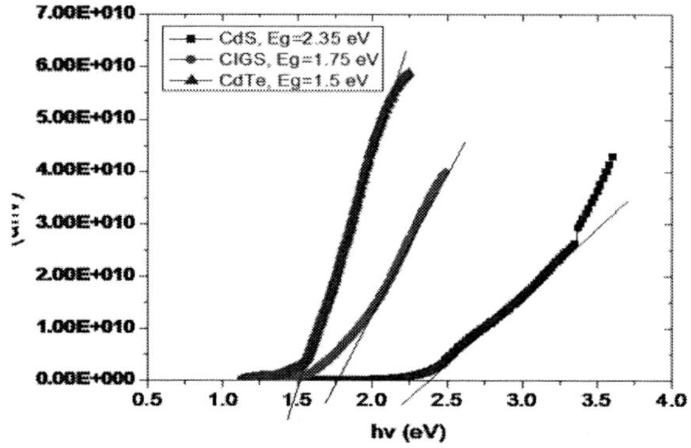

Figure 4. The optical energy gap for CdS, CIGS and CdTe thin films.

3-3 Electrical Properties of Thin Films:

3-3-1 DC conductivity

The DC conductivity have found one activation energy for CIGS and two for both of CdS and CdTe that with explored from measure electrical conductivity with temperature where one activation energy for CIGS and two activation energy for both CdTe and CdS in two range of temperature and conductivity measurements from graph as shown in the figure 5. and the data records in the table 1.

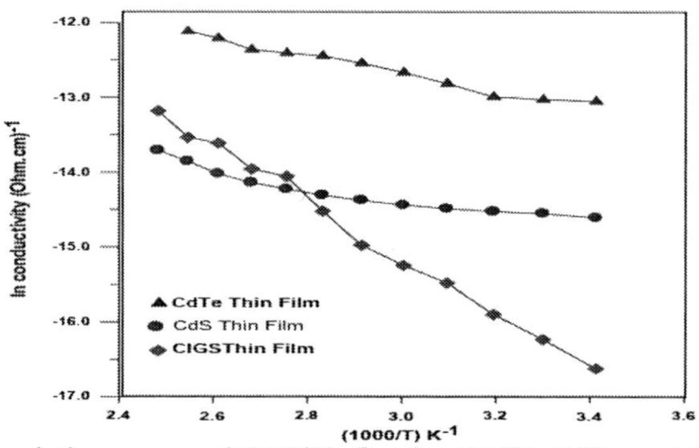

Figure 5. ln conductivity variation versus (1000/T) for the CIGS ,CdS, and CdTe thin

Table 1. Electrical DC conductivity 0f CIGS,CdS,and CdTe thin Films.

Thin Film	$\sigma_{R.T}$ $(\Omega.cm)^{-1}$	Ea_1 (eV)	Temp. Range (K)	Ea_2 (eV)	Temp.Range (K)
CIGS	6.07×10^{-8}	0.8844	403-293	-----	------
CdS	4.52×10^{-7}	0.6923	293-353	0.2439	353-403
CdTe	2.15×10^{-6}	0.2776	293-313	0.0640	313-403

3-3-2 Hall Effect

The hall measurements show that CIGS and CdTe were p-tyes carrier and CdS thin films was n-type carrier. All data about Hall carrier mobility, conductivity and carrier concentration as shown in the table 2.

IOP Conf. Series: Journal of Physics: Conf. Series **1032** (2018) 012022 doi:10.1088/1742-6596/1032/1/012022

Table 2. Hall Effect measurements for CIGS, CdS and CdTe thin films.

Thin Film	R_H (cm^3/c)	μ_H $cm^2/V.sec$	$\sigma_{dc\ at\ R.T}$ $(\Omega.cm)^{-1}$	n_H (cm^{-3})	Type
CIGS	$+2.6*10^2$	$5.2*10^2$	1.79	$3.5*10^{16}$	p-type
CdS	$-2.3*10^1$	$4.6*10^1$	$6.8*10^{-1}$	$4.1*10^{16}$	n-type
CdTe	$+4.5*10^7$	$2.4*10^2$	$3.2*10^{-6}$	$3.4*10^{11}$	p-type

4. Conclusions

All results data as structural, optical and electrical properties showed that energy gap are suitable range in the optical visible range of wavelength that good matching like solar radiation spectrum and have good photoconductivity properties for solar cell application.

References

[1] Su-Huai Weia, S. B. Zhang, and Alex Zunger, 2000 *"First-principles calculation of band offsets, optical bowings, and defects in CdS, CdSe, CdTe, and their alloys", journal of applied physics,* vol. **87,** no. 3.

[2] David Julien Louis Brémaud, 2009 *"Investigation and Development of CIGS Solar Cells on Flexible Substrates and with Alternative Electrical Back Contacts",Doctoral Thesis, ETH Zurich university.*

[3] Mushtak A. Jabbar, 2015 *"Manufacturing and Evaluating of Flexible Nano Solar Cell", Doctoral Thesis, College of Education, University of Al-Mustansiriyah.*

[4] E.R. Shaaban, N. Afify, and A. El-Taher, 2009 *"Effect of film thickness on microstructure parameters and optical constants of CdTe thin films", Journal of Alloys and Compounds* Vol. **482**,pp.(400-404).

[5] Zhou Fang, Xiao Chen Wang, Hong Cai Wu, and Ce Zhou Zhao, 2011 *"Achievements and Challenges of CdS/CdTe Solar Cells", International Journal of Photoenergy,* Volume **2011**, Article ID 297350, p 8.

[6] M. Dharmadasa, P. A. Bingham, O. K. Echendu, H.I. Salim, T. Druffel, R. Dharmadasa, G. U. Sumanasekera, R. R. Dharmasena, M. B. Dergacheva, K. A. Mit, K. A. Urazov, L. Bowen, M. Walls and A. Abbas 2014 *"Fabrication of CdS/CdTe-Based Thin Film Solar Cells Using an Electrochemical Technique" Coatings Journal,* Vol. **4**, PP. 380-415.

[7] M. Dharmadasa, 2014 *" Review of the $CdCl_2$ Treatment Used in CdS/CdTe Thin Film Solar Cell Development and New Evidence towards Improved Understanding, Coatings Journal*,Vol.**4**, PP. 282-307.

[8] S.D. Gunjali, Y. B. Khollam, M. T. Sarode, S. A.Arote, P. N. Shelke, K. C.Mohite, 2014 *"Solar Cell Properties CdS/CdTe Heterojunctions Prepared by using Spray Pyrolysis Technique", International Journal of Chemical and Physical Sciences,* Vol.**3**, PP.102-108.

[9] S.S. Babkair, **2010** *"Charge Transport Mechanisms and Device Parameters of CdS/CdTe Solar Cells Fabricated by Thermal Evaporation", JKAU:Sci.,* Vol.22, No.**1**, pp.21-33.

IOP Conf. Series: Journal of Physics: Conf. Series **1032** (2018) 012023 doi:10.1088/1742-6596/1032/1/012023

Novel Technique for Photovoltaic Solar Cell Efficiency Enhancements by Coating with Chlorophyll

Dhafer Manea Hachim AL-Hasnawi [1], Ali Abdulabbas Abdullah[2], Furkan Kamil[3]
[1,2,3] Engineering Technical College / Najaf, Al-Furat Al-Awsat Technical University, 31001 Najaf, Iraq.

dhafeer_manee@yahoo.com

Abstract. In this paper a novel technique proposed and implemented practically for photovoltaic cell efficiency enhancements. It is known that Iraq is one of the hottest countries in the world. It has a solar insolation average and energy higher than that of most of other countries. Solar photovoltaic cell has a limited efficiency boundary, which is affected by the amount of absorbed solar heat (energy). It is found that; when the solar array temperature excess 70 C^0, the photovoltaic solar cell efficiency will be decreased remarkably. This novel technique which include painting the solar cell surface with chlorophyll only one time and painting it's surface with chlorophyll and polymer mixture another time gives a perfect solution to get a wide band of solar temperature array with a remarkable efficiency enhancement. The experimental results show that coating solar cell surface with chlorophyll only increases its efficiency by 5.5 % which is the maximum increase, while coating solar cell surface with chlorophyll and polymer mixture increases its efficiency by 3.1 % only.which is less than increase in the first case but more than that of solar cell without coating. All cells tested at constant solar irradiation (I=1000 W/m^2), and variation of temperature in range (30 ^oC to 80 ^oC).

1. Introduction

As a great potential renewable energy source, solar energy is becoming one of the most important energies in the future. Recently, there has been an enormous increase in the understanding of the operational principle of photovoltaic devices, which led to a rapid increase in the power conversion efficiencies of such devices. Solar cells vary under temperature changes; the change in temperature will affect the power, output from the cells. Garca M.C. and Balenzategui J.L. (2004) [1] used many different PV modules to simulate nominal operation cell temperature (NOCT)for building integration applications, where the yearly modules temperature and performance for different orientations and tilted angles were presented. Islam M.D. and etal (20090 [2] conducted a study to show that Abu Dhabi has a strong potential for solar energy capture by measuring the solar radiation and surface temperature for one year. Yassine Charabi and Adel Gastli (2013)[3] used fuzzy logic and GIS-based spatial multi-criteria evaluation to conduct an investigation on siting large PV power plant to study the effects of temperature and dust on the performance. Schwingshackl C. and etal (2013)[4] investigated the change on PV panels temperature by the influence of natural wind cooling and showed that including wind data in the analysis leads to better results and prediction. Cuce E. and etal (2013)[5] presented a study addresses statistical and experimental investigation dealing with the effect of temperature within range of 15-60 ◦C and intensity levels of 200-500 W/m^2. The author showed that the current has a proportional relationship to the intensity, whereas the voltage decreases when the temperature increases.

Sanaz Ghazi and Kenneth Ip (2014)[6] conducted an experimental investigation to study the effect of dust density, high humidity, rain and snow on the performance of PV panels in the town of

Brighton in UK, and showed that these parameters have significant effect and could annihilate any system output.

Goverde Hans and etal (2015)[7] proposed numerous models that can be used to predict the relationship between performance of PV system and solar radiation, ambient temperature and wind. Where they showed that the performance is better when the wind speed and solar radiation increase and poor performance when ambient temperature increases.

Chander Subhash and etal (2015) [8] similar study is presented by Cuce E. and etal (2013)[5] with different ranges, namely 25-60 ◦C for temperature and intensity levels of 215-515 W/m^2. Experimental study the effect of cell temperature in the range 25-60 C at constant light intensities 215-515 W/m^2 on the photovoltaic parameters of mono-crystalline silicon solar cell. The results show that the open circuit voltage, maximum power, fill factor and efficiency are decreased with cell temperature.

Yun Da and Yimin Xuan (2015) [9] conducted a study for temperature effect on performance of solar cells that made of silicon thin-film and found that the performance decreases as the ambient temperature increases. Pierrick Haurant and etal (2015) [10] presented a 3D dynamical model to simulate the steady-state dynamic regimes of PVT collector. Wolff Bjrn and etal (2016) [11] Used support vector regression (SVR) to study various methods of PV power output prediction. Khan Firoz and etal (2016) [12] analytically investigated the effect of ambient temperature on set of PV panels under high solar radiation. Saidan Motasem and etal (2016)[13] investigate the dust accumulation on the PV panels performance in Baghdad-Iraq. Mehmood Umer and etal (2017) [14] a recent work which is presented by Mehmood Umer and etal (2017) [14] investigated the effect of dust and dried mud of the PV modules in the area of Kingdom of Saudi Arabia-Dhahran. Hachim Dhafer Manea and etal (2017) [15] they work experimentally to measure the weather ambient temperatures (DBT), solar intensity and wind speed in Najaf city (Iraq 44 °E, 31° N) over a period of one year from April 2015 to March 2016, where in present work obtained the measurement weather information of Najaf city of Hachim Dhafer Manea and etal (2017) [15], to find the maximum solar irradiation and temperature reaches during all year, and this date help to test the PV cell in same conditions in Lab.

In this paper experimental investigation, the effect of coating the PV cell by chlorophyll only and mixing (chlorophyll and polymer) at constant solar irradiation and variable temperature on performance of photovoltaic cell to improve it efficiency.

2. Experimental Setup and Procedure

In this work, the experiment includes two main parts. First, the implementation of the chlorophyll in lab. Second part is investigating the performance of photovoltaic cell by measuring the efficiency of PV cell in the laboratory at a constant solar radiation with and without coating the cell by chlorophyll and mixing (chlorophyll and Nitro Cellulose Polymer) Hubeatir K. Abid and etal (20017) [16] Kamil Furkan and etal (2015) [17].

2.1. Implementation of the chlorophyll

The efficiency of cells layers depends on the homogeneity and purity of the natural dye (Chlorophyll) extract. In order to reach this homogeneity, celery samples are extracted through a very accurate operation. First, the samples are washed with pure water, cut into smaller pieces and finely Immersed in ethanol solution. They are blended then filtered to obtain a homogeneous Chlorophyll extract as shown in **Figure 1**.

2.2. Testing PV cell

IOP Conf. Series: Journal of Physics: Conf. Series **1032** (2018) 012023 doi:10.1088/1742-6596/1032/1/012023

The PV cell used in this study is 39 x 31.2mm polycrystalline silicon solar cell panel, its power: 0.02W, average voltage: 0.5V, and average current:0.4A as shown in **Figure 2**. Where the solar cell tested at constant solar irradiation (I=1000 W/m^2) with and without coating by chlorophyll and polymer at different temperature by using the following devices in labs. of alternative and renewable research unit in Engineering Technical College of Najaf (as shown in **Figure 3**):

1. Solar irradiation sensor: this device used to measure the solar irradiation.
2. Halogen lamp: this lamp used as solar irradiation source to test the cell.
3. Six T-type thermocouples: these thermocouples used to measure the temperature of cell.
4. Data logger: the device used to record the temperature of cell.
5. Load (variable resistance): this device used to change the load on the cell.
6. Digital voltmeter: this device used to measure the voltage of cell.
7. Digital ammeter: this device used to measure the current of cell.

2.3.Fluorescence Spectrum of Chlorophyll Extracted from Green Celery

The fluorescence spectrum was measured from a wavelength of 200 nm to 750 nm as shown in **Figure 4** the performance of chlorophyll could be obtained by testing the emission of the dye and calculate the energy level distribution. The peaks were observed from 220 nm and 700 nm. The Emission of molecules stimulated at different energized wavelengths. The amplitude of the peaks seen in the emission spectrum is proportion directly to the number of photons that are emitted.

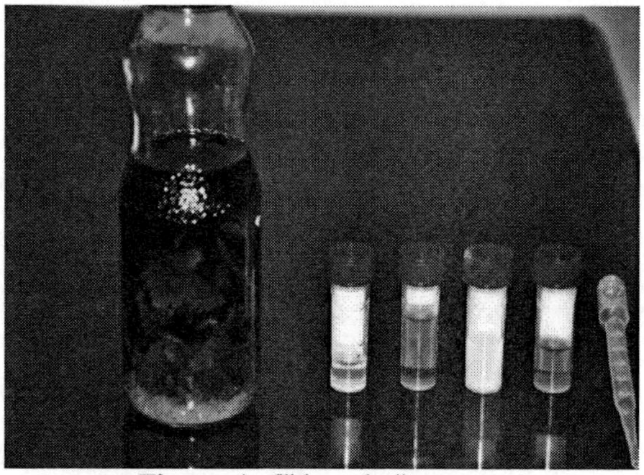

Figure 1. Chlorophyll extract

Figure 2. Polycrystalline Silicon Solar Cell Panel

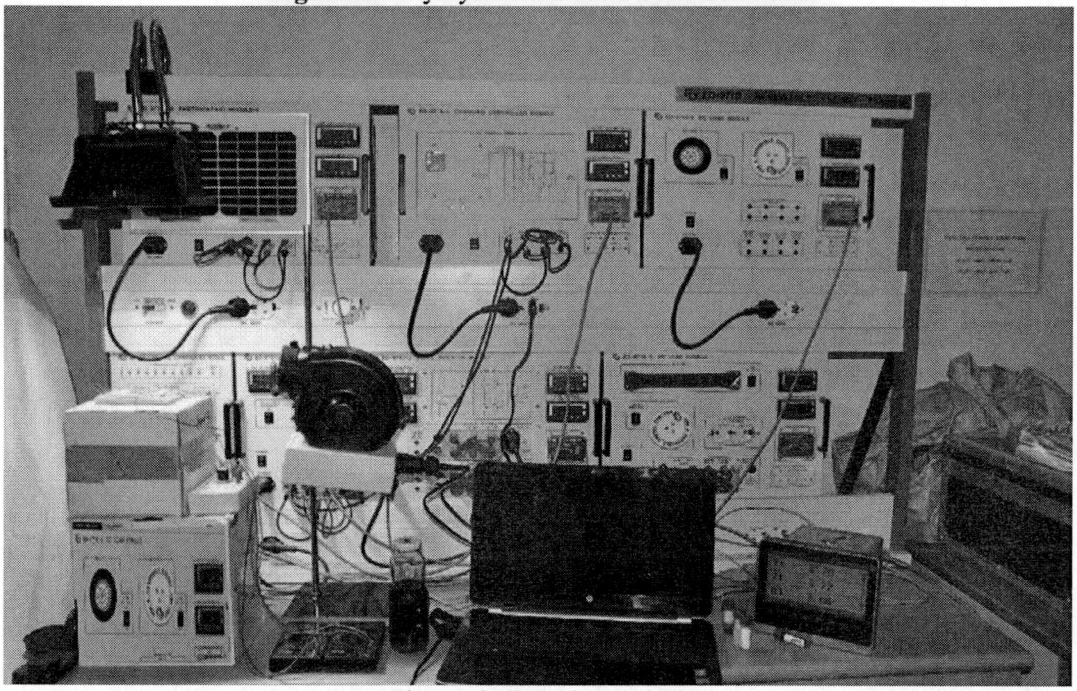

Figure 3. Experimental Rig

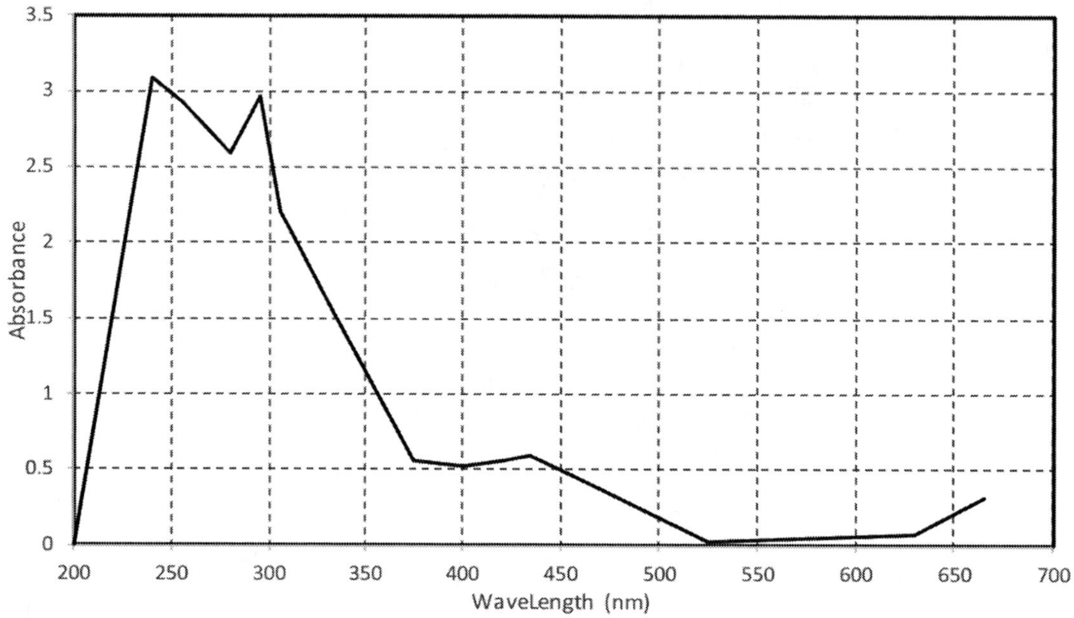

Figure 4. Chlorophyll spectra test results

3. Results and Discussions

The results obtained in this study include the following: Analysis of the data recorded by the photovoltaic system. It also takes data that includes solar radiation, temperatures, voltage, current and power. All this information is drawn and studied its effect on the performance of the photovoltaic system. With and without coating the PV cell by chlorophyll and polymer.

Figure 5 show the relation between the voltage and current output from PV cell at constant solar irradiation and variable load. Where note that the cell coating by chlorophyll only have maximum peak point camper with cell coating with chlorophyll and polymer, and the cell without coating.

Figure 6 and **Figure 7** shows the variation of power with voltage and current at constant solar irradiation and variable load applied on the solar cell. The curves show that the maximum power output is located and this point is called the maximum power point, where this point depending on the load applied on PV cell, so that the curve before and after this point drop, because changing the load applied on PV cell. When tested the three types of cell (coating with chlorophyll only, chlorophyll and polymer, and the cell without coating) at the same conditions. That mean keeping solar radiation and the temperature of solar cell constant. The peak power in power voltage curve for the cell coating by chlorophyll and polymer. While the peak power in power current curve for the cell coating by chlorophyll only. Because the chlorophyll work to decrease the temperature of solar cell and increase the absorbance.

The effect of temperature is the important parameter in present study. So that plot the efficiency variation with temperature at constant solar irradiation and load, by using the line chart as shown in **Figure 8** and using the bar chart for more explanation as shown in **Figure 9**. The cell coating by chlorophyll only have maximum efficiency camper with other cells. Also, the cell coating by chlorophyll and polymer have good efficiency for temperature less than 50 ^{o}C camper with the cell without coating. This happen because the effect of adding chlorophyll and polymer, where the chlorophyll work to improvement the performance of the solar cell at high temperature and increase the absorbance of sunlight, while the polymer work to keep on the chlorophyll day from outside effect, also make the cell more smoothing and that help to preventing adhesion of dust to the solar cell, thus maintaining its cleanliness and increasing its efficiency.

4. Conclusions

The experimental results show that the power output from solar cell increases with coating by chlorophyll only and chlorophyll and polymer mixture, the efficiency of solar cell increases too also at constant solar irradiation, cell temperature and variable load applied on cell. Then when testing the cell at constant solar irradiation and load with different cell temperature the efficiency of cell coating with chlorophyll only is maximum camper with other cell, the maximum increasing of the efficiency is 5.5 %. While the cell coating with chlorophyll and polymer mixture have less efficiency from the cell coating with chlorophyll only and more than the cell without coating, where the maximum increasing its efficiency is 3.1 %. The cell coating by mixing (chlorophyll and polymer) make the cell more smoothing and that help to preventing adhesion of dust to the solar cell, thus maintaining its cleanliness and increasing its efficiency.

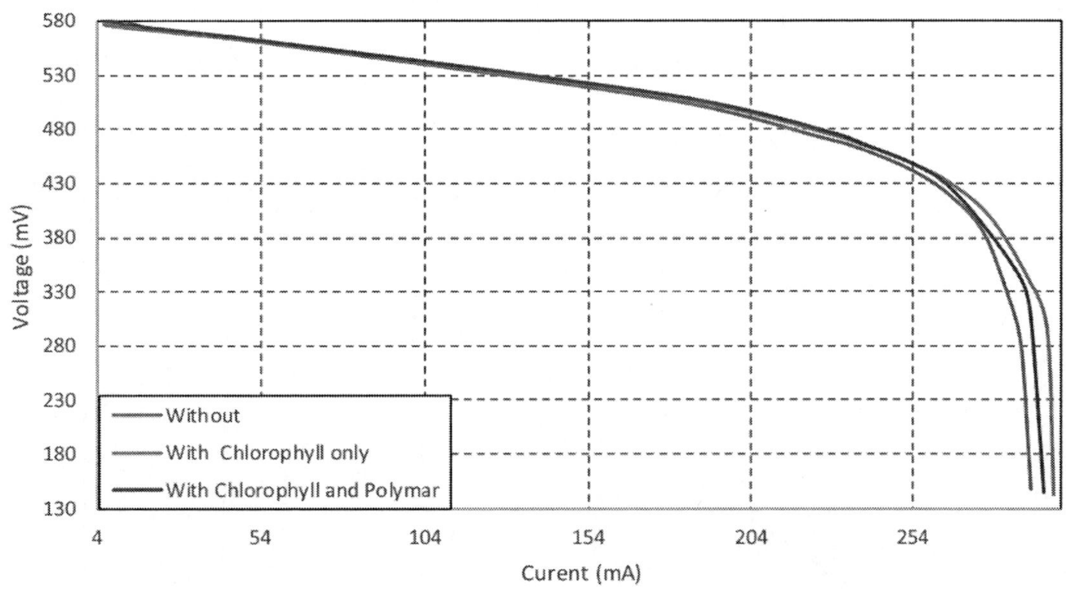

Figure 5. show the relation between the voltage and current output from PV cell at constant solar irradiation and load.

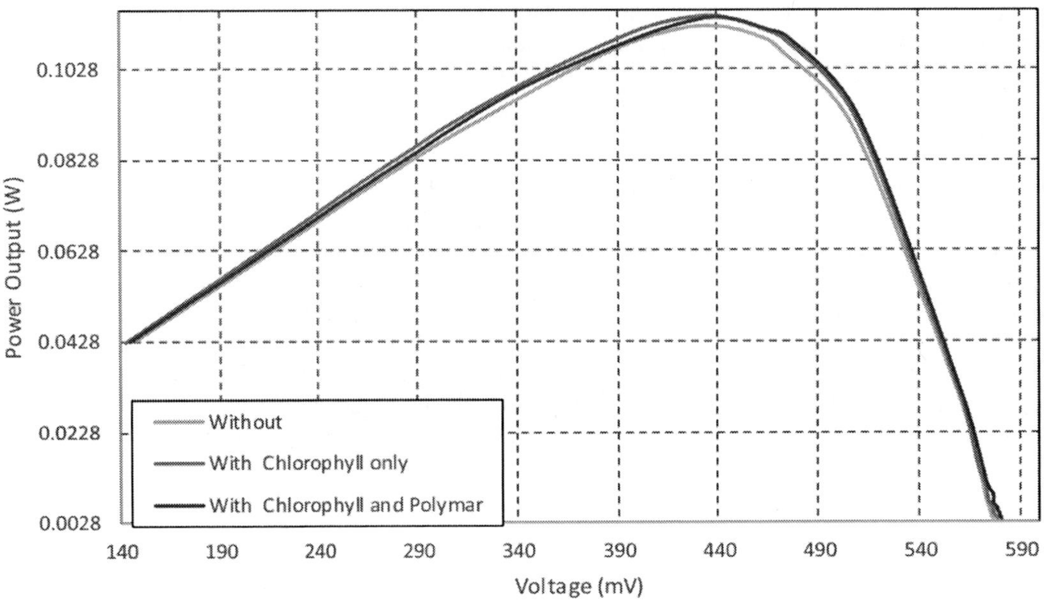

Figure 6. the variation of power with voltage at constant solar irradiation and variable load

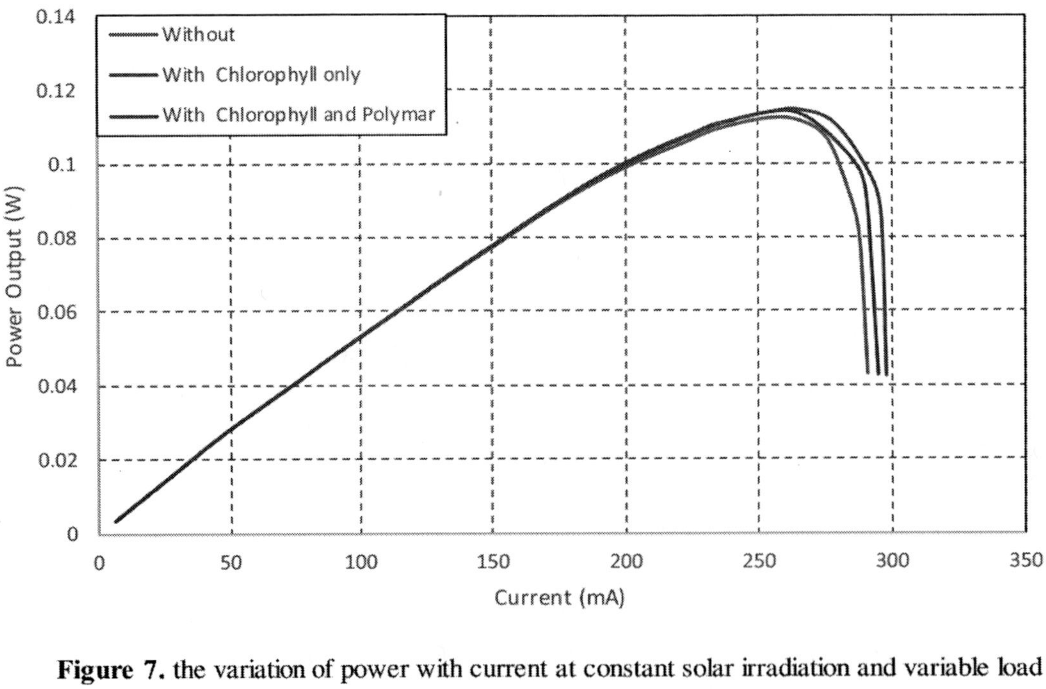

Figure 7. the variation of power with current at constant solar irradiation and variable load

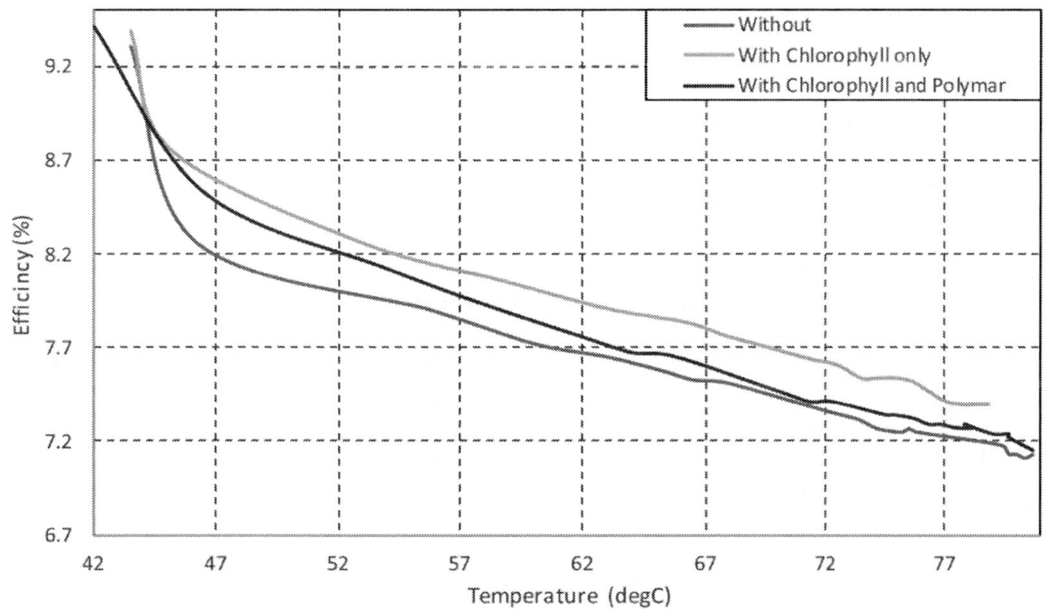

Figure8. The effect of temperature on the efficiency of cell at constant solar irradiation and load

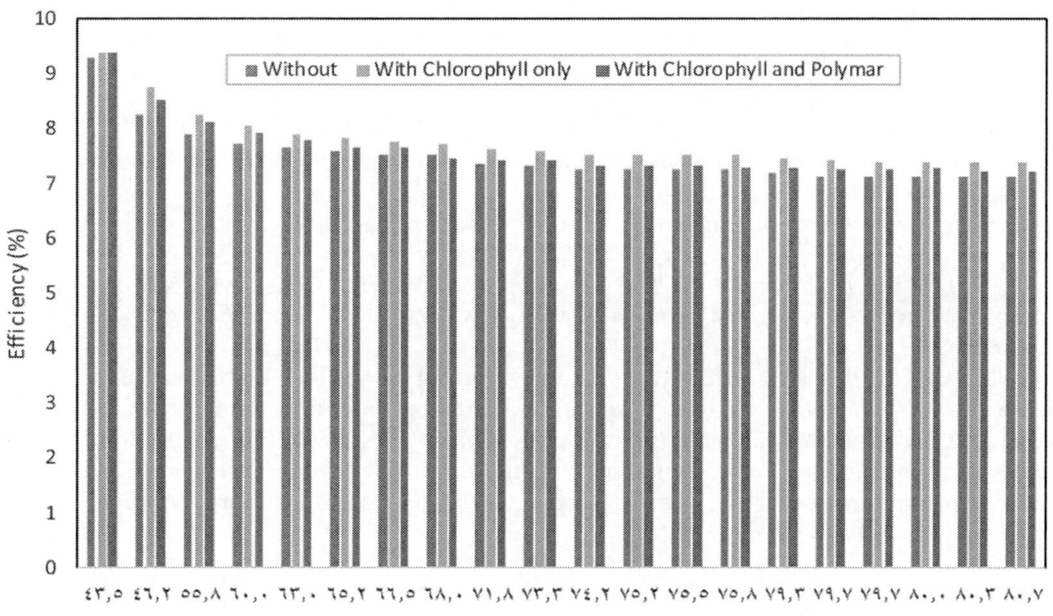

Figure9. The effect of temperature on the efficiency of cell at constant solar irradiation and load

5. Acknowledgements

We thank our colleagues Professor Dr. Dhakil N.Taha and Professor Dr. Mohammed H. Al-Maamori from university of Babylon who provided insight and expertise that greatly assisted the research.

6. References

[1] M.C. Alonso Garca, J.L. Balenzategui "Estimation of photovoltaic module yearly temperature and performance based on Nominal Operation Cell Temperature calculations" *Renewable Energy, Volume 29, Issue 12, 2004, Pages 1997-2010.*
[2] M.D. Islam, I. Kubo, M. Ohadi, A.A. Alili "Measurement of solar energy radiation in Abu Dhabi, UAE" *Applied Energy, Volume 86, Issue 4, 2009, Pages 511-515.*
[3] Yassine Charabi, Adel Gastli "Integration of temperature and dust effects in siting large PV power plant in hot arid area, *Renewable Energy, Volume 57, 2013, Pages 635-644.*
[4] C. Schwingshackl, M. Petitta, J.E. Wagner, G. Belluardo, D. Moser, M. Castelli, M. Zebisch, A. Tetzlaff "Wind Effect on PV Module Temperature: Analysis of Different Techniques for an Accurate Estimation" *Energy Procedia, Volume 40, 2013, Pages 77-86.*
[5] Erdem Cuce, Pinar Mert Cuce, Tulin Bali "An experimental analysis of illumination intensity and temperature dependency of photovoltaic cell parameters" *Applied Energy, Volume 111, 2013, Pages 374-382*
[6] Sanaz Ghazi, Kenneth Ip "The effect of weather conditions on the efficiency of PV panels in the southeast of UK, *Renewable Energy" Volume 69, 2014, Pages 50-59.*
[7] Hans Goverde, Dirk Goossens, Jonathan Govaerts, Vikas Dubey, Francky Catthoor, Kris Baert, Jef Poortmans, Johan Driesen "Spatial and temporal analysis of wind effects on PV module temperature and performance" *Sustainable Energy Technologies and Assessments, Volume 11, 2015, Pages 36-41.*
[8] Subhash Chander, A. Purohit, Anshu Sharma, Arvind, S.P. Nehra, M.S. Dhaka "A study on photovoltaic parameters of mono-crystalline silicon solar cell with cell temperature" *Energy Reports, Volume 1, 2015, Pages 104-109*
[9] Yun Da, Yimin Xuan "Effect of temperature on performance of nanostructured silicon thin-film solar cells" *Solar Energy, Volume 115, 2015, Pages 109-119.*
[10] Haurant Pierrick, Mnzo Christophe, Gaillard Leon, Dupeyrat Patrick "Dynamic numerical model of a high efficiency PVT collector integrated into a domestic hot water system" *Solar Energy, Volume 111, 2015, Pages 68-81.*
[11] Bjrn Wolff, Jan Khnert, Elke Lorenz, Oliver Kramer, Detlev Heinemann " Comparing support vector regression for PV power forecasting to a physical modeling approach using measurement, numerical weather prediction, and cloud motion data" *Solar Energy, Volume 135, 2016, Pages 197-208.*
[12] Firoz Khan, Seong-Ho Baek, Jae Hyun Kim " Wide range temperature dependence of analytical photovoltaic cell parameters for silicon solar cells under high illumination conditions" *Applied Energy, Volume 183, 2016, Pages 715-724.*
[13] Motasem Saidan, Abdul Ghani Albaali, Emil Alasis, John K. Kaldellis " Experimental study on the effect of dust deposition on solar photovoltaic panels in desert environment" *Renewable Energy, Volume 92, 2016, Pages 499-505*

[14] Umer Mehmood, Fahad A. Al-Sulaiman, B.S. Yilbas " Characterization of dust collected from PV modules in the area of Dhahran, Kingdom of Saudi Arabia, and its impact on protective transparent covers for photovoltaic applications" *Solar Energy, Volume 141, 2017, Pages 203-209.*

[15] Dhafer Manea Hachim , Assaad Alsahlani, and Adel A. Eidan "Measurements of Wind and Solar Energies in Najaf, Iraq", *Advances in Natural and Applied Sciences, Vol. 11, No. 8 (June), 2017.*

[16] K. Abid Hubeatir, F. Kamil, A. A. Al-Amiery, A. A. H. Kadhum and A. B. Mohamad, "Polymer solar cells with enhanced power conversion efficiency using nanomaterials and laser techniques", *Advanced Performance Materials Volume 32, No.5, 2017.*

[17] Furkan Kamil, Kadhim Abid Hubeatir, Mohammed Shamel and Ahmed A. Al-Amiery, "Microwave-assisted solvent-free synthesis of new polyimine", *Cogent Chemistry .Vol.1, 2015.*

Energy Harvesting from the Waste Heat of an Electrical Oven via Thermoelectric Generator

Hasan T Hashim

College of Engineering-Kerbala University-Kerbala-IRAQ

* Corresponding author. E-mail address: hasanth2008@yahoo.com

Abstract. This paper presents an attempt to generate the electricity by recovering the waste heat of an electrical oven. An experimental work has been done to investigate the electricity production by using a thermoelectric generator (TEG). The performance characterization of TEG was experimentally investigated in an open circuit mode by connecting the multimeter directly to the TEG and in a closed circuit mode by connecting different load to the TEG. The key results showed that it is possible to generate up to 1 V and 30mW in the open circuit condition, while the TEG could generate a 0.5 V and 25 mW in the closed circuit conditions when the maximum temperature differences across the TEG is about 20 °C. The power output could be increased by using more than one TEG module. This power could be used for charging a cell phone or light emitting diodes LEDs during cooking.

KEYWORDS: Waste heat, heat recovery, thermoelectric generator, electricity generation.

1. Introduction

The major source of electricity is generated through stations based on burring of the fossil fuels, which pollute the environment. Therefore, finding new sources of energy is important to overcome this problem. Renewable energy is considered as a reliable source of energy because of many reasons such as environment-friendly and unlimited source of energy.

Solar energy is one type of renewable energy that converts the sun light into electricity by using photovoltaic.

Thermoelectric generator is made from semiconductors N-type and P-type which is able to convert the waste heat into electricity using the electron as a working fluid by Seebeck effect [1]. Harvesting solar energy by thermoelectric generator was studied by many researchers [2][3][4][5] moreover, the use of biomass cooker stove as a source of heat for thermoelectric was investigated by [6][7]. The recovery of waste heat that dissipated to the environment was studied by using two stroke petrol engine and TEG [8]. An attempt to recharge a battery in a compression ignition engine has been done by experimental and theoretical work [9].

A waste heat recovery system based on a thermoelectric generator was designed by Demir and Dincer [10]. The system was integrated with the exhaust manifold of the automobile to recover the waste heat. Three TEG modules were used in the study. The maximum generated power from the waste heat was 158 W. The energy conversion efficiency was calculated based on the heat transfer rate from the hot side to the thermoelectric generators and the generated power.

There are many ways to harvest the waste heat at home [11][12][13]. However, integrating TEG with home electrical oven has not investigated yet.

The objective of this study is to investigate experimentally the electricity generation by TEG using the waste heat from electric oven. This electricity can be used to charge a battery of a cell phone or to power LEDs inside a house.

IOP Conf. Series: Journal of Physics: Conf. Series **1032** (2018) 012024 doi:10.1088/1742-6596/1032/1/012024

2. Experimental setup

A schematic diagram and actual photograph of the experimental setup is shown in figure 1. Two K-type thermocouples were used to measure the temperature of the hot side (TH) and the cold side (TC) of thermoelectric. The top thermocouple was attached to the surface of the oven by drilling a hole inside a copper plate to measure the hot side temperature. The cold side temperature was measured by using another thermocouple connected to the heat sink.

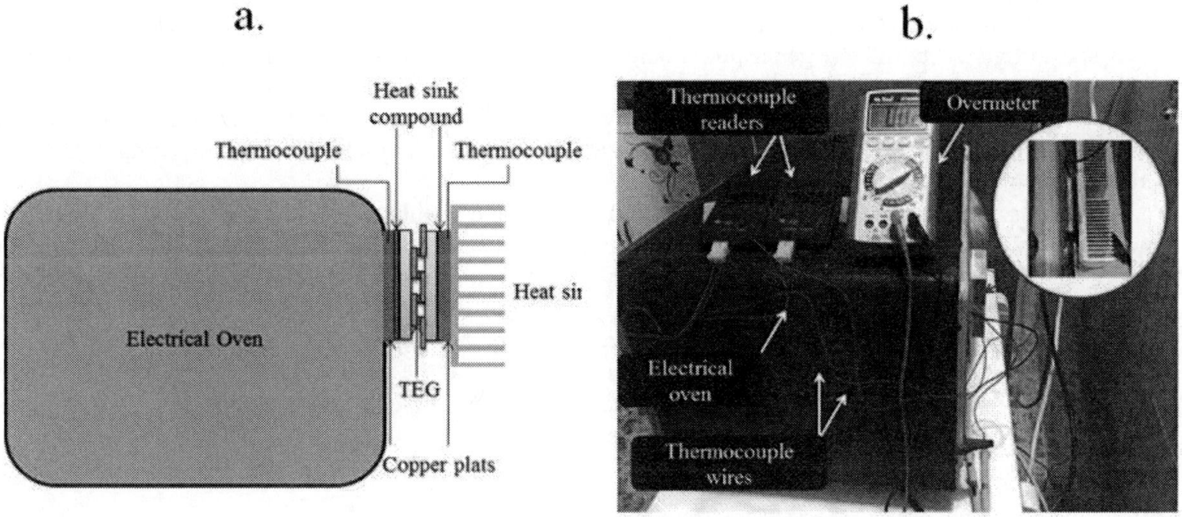

Figure 1. Shows a schematic diagram (a), and actual photograph of the experimental (b).

A thin uniform layer of heat sink compound with a thermal conductivity of 2.9 W/m.K was applied between the oven and the thermoelectric also, between the cold side and the thermoelectric. A thermoelectric device (40 mm x 40 mm geometry and 256 legs) has been used during the experimental setup.

The thermoelectric generator is able to convert the thermal energy into electrical energy directly which is an alternative method to produce potential energy. The efficiency of thermoelectric converters in general, depends on the material and design aspects such as the number of legs and the lengths [2]. Also, the heat transfer conditions have a significant influence on power output generation and the efficiency of TEG. It is worth to mention that the power output in this work was calculated using the following formula which is usually used to calculate the power output in open circuit mode $P = \frac{V^2}{4R}$ [14],

were v is open circuit voltage and R is the internal resistance of the TEG device. The internal resistance of TEG could be obtained by connecting different loads to the TEG, and then the value of the load at the maximum power output represents the internal resistance of the TEG.

The focus of this work is developing tools for the evaluation of thermoelectric power generation from the waste heat of an electrical oven. The power generated from this rig could be used for charging cell phone or other uses.

IOP Conf. Series: Journal of Physics: Conf. Series **1032** (2018) 012024 doi:10.1088/1742-6596/1032/1/012024

3. Results and Discussion

The oven was switched on for twenty minutes using both the upper and lower heater for the maximum temperature. The electrical oven was consumed electrical power (Q_P) approximately 686.4 W, which was calculated by multiplying the voltage by the current. First, the thermoelectric generator characterized using the open circuit method i.e without connecting load to the external circuit and the results are shown in figure 2. It can be shown in figure 2.a, the increase in the hot side temperature (T_H), cold side temperature (T_C) and the difference between them for the TEG with time. It is clear that both T_H and T_C are increasing with the time mainly due to transfer the heat from rear surface of the oven to the hot side of TEG then to the T_C via TEG.

The temperature difference across TEG causes generation of electricity by Seebeck effect as we can see in figure 2.b and the generated voltage is increased by increasing the temperature difference between them. figure 2.c showed the open circuit voltage and the power output as a function of hot side temperature. Both of them were increased with increasing the hot side temperature due to increase in the temperature difference between the hot side (electrical oven) and the cold side temperate (heat sink).

After that, the load was connected to the TEG and the results are shown in figure 3. figure 3.a shows the voltage generated by connecting difference resistance to the circuit. It can be seen that, the TEG was characterised at constant temperature difference when the electrical oven reached to the steady state condition where the T_C and T_H about 87 °C and 101°C respectively. The power output was calculated by multiplying the voltage by the current and the results are demonstrated in figure 3.b and 3.c. This experiment showed that the waste heat from oven can be used to generate useful DC voltage and power, which may be used to charge cell phone or power DC light during the night.

IOP Conf. Series: Journal of Physics: Conf. Series **1032** (2018) 012024 doi:10.1088/1742-6596/1032/1/012024

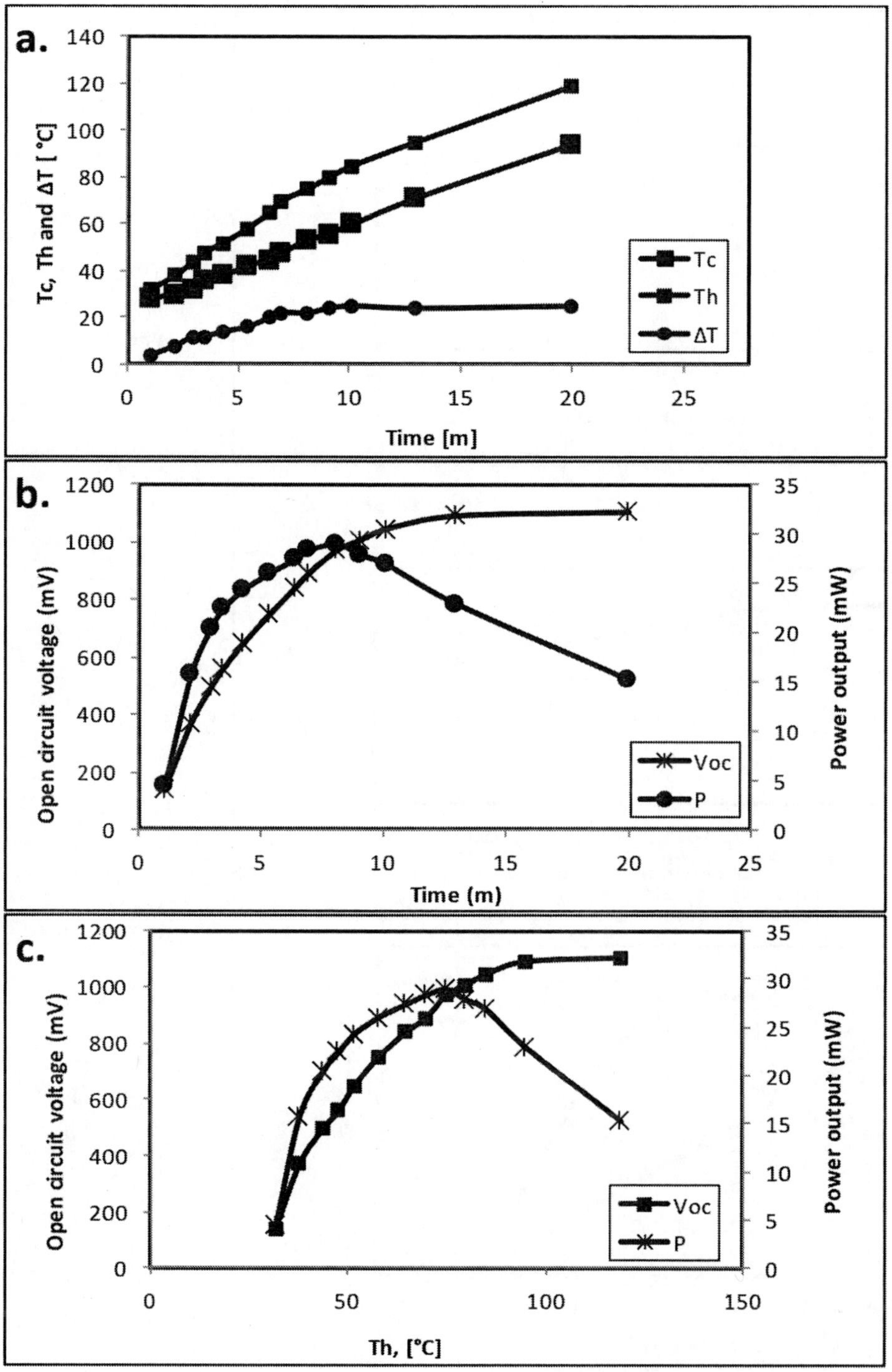

Figure 2. Shows the characterization of thermoelectric generator in open circuit mode, a. temperature Vs time, b. voltage and power Vs time and c. voltage and power Vs hot side temperature.

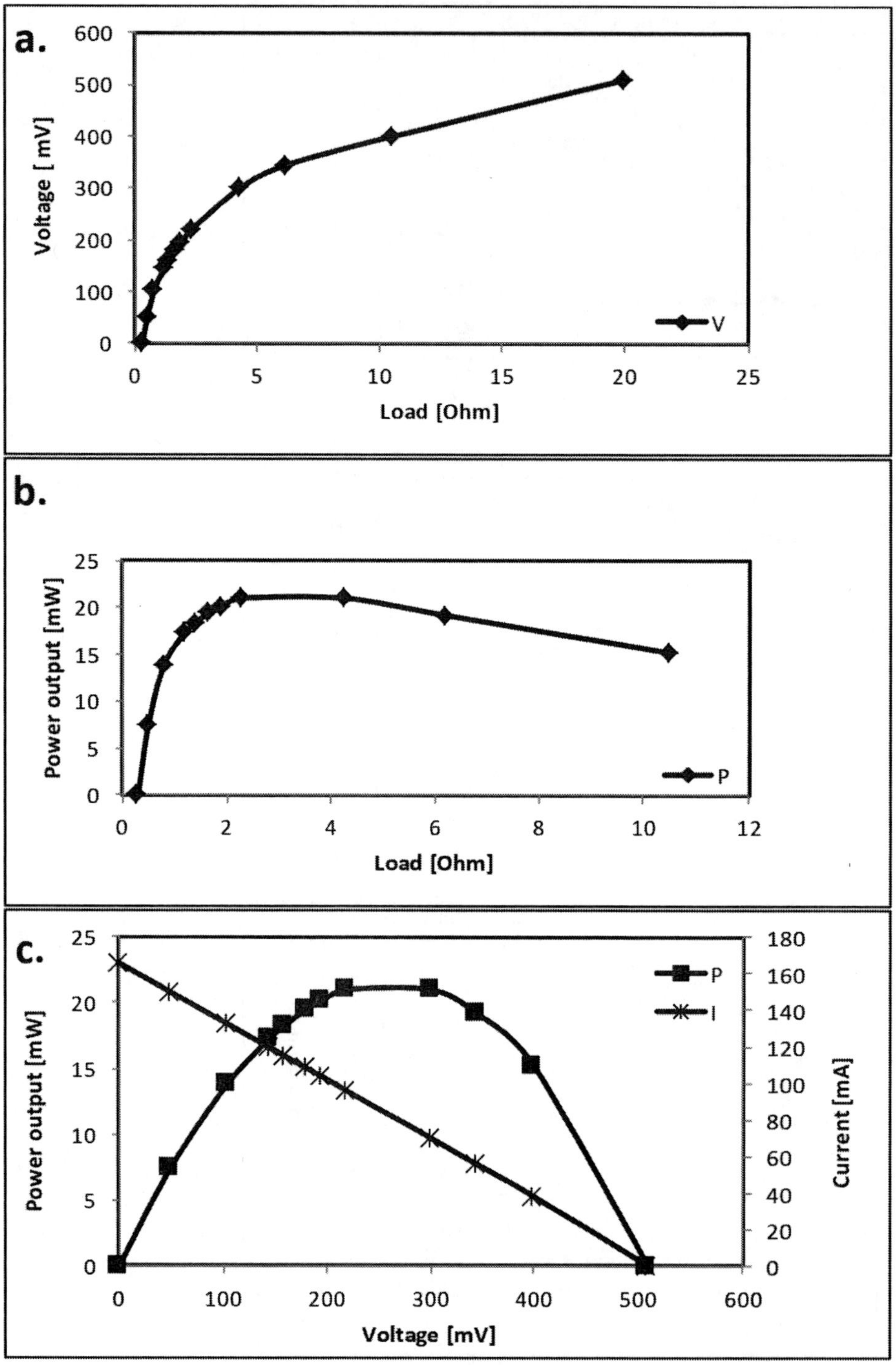

Figure 3. Shows the characterization of thermoelectric generator in closed circuit mode, a. voltage Vs load, b. power output Vs load and c. power output and current Vs voltage.

4. Conclusions

The thermoelectric generator is successfully integrated between the electrical oven and the heat sink which is capable of producing a DC electrical power from an electrical oven. It has been shown that the thermoelectric generator is producing more than 1 volte in open circuit mode and 0.5 volte in closed circuit condition during 20 minutes of the operation of the electrical oven. The generated energy can be store in a battery.

The study shows the availability and possibility of recovering the waste heat from electrical oven. Future work is investigating of; more than one TEG modules, active heat sink and charging portable cell phone charger.

References

[1] G Min and D M Rowe 2007 Conversion efficiency of thermoelectric combustion systems *IEEE Trans Energy Convers* **vol 22** no 2 pp. 528–34

[2] H Hashim, J J Bomphrey, and G Min 2016 Model for geometry optimisation of thermoelectric devices in a hybrid PV/TE system *Renew. Energy* **vol 87** pp 458–63

[3] E A Chávez Urbiola and Y Vorobiev 2013 Investigation of Solar Hybrid Electric/Thermal System with Radiation Concentrator and Thermoelectric Generator *J. Photoenergy* **vol. 2013** no 4 pp 1–7

[4] E A Chávez Urbiola and Y Vorobiev 2013 Investigation of Solar Hybrid Electric/Thermal System with Radiation Concentrator and Thermoelectric Generator Int J Photoenergy no 4 pp 1–7

[5] N Miljkovic and E N Wang 2011 Modeling and optimization of hybrid solar thermoelectric systems with thermosyphons *Sol. Energy* **vol 85** no 11 pp 2843–55

[6] D Champier, J P Bedecarrats, M Rivaletto, and F Strub 2010 Thermoelectric power generation from biomass cook stoves *Energy* **vol 35** no 2 pp 935–42

[7] S M O Shaughnessy, M J Deasy, C E Kinsella, J V Doyle and A J Robinson 2013 Small scale electricity generation from a portable biomass cookstove: Prototype design and preliminary results *Appl. Energy* **vol 102** pp 374–85

[8] P Baskar, S Seralathan, D Dipin, S Thangavel, C F I Norman, and C Arnold 2014 Experimental Analysis of Thermoelectric Waste Heat Recovery System Retrofitted to Two Stroke Petrol Engine *J. Adv Mech Eng* **vol 4** no 1 pp 9–14

[9] G G Kumar 2015 Experimental Investigation on Thermo-Electric Generator Used in Compression Ignition Engines *J. Sci. Res* **vol 4** no 6 pp 196–200

[10] M E Demir and I Dincer 2017 Performance assessment of a thermoelectric generator applied to exhaust waste heat recovery *Appl. Therm. Eng* **no 120** pp 694– 07

[11] M I A Jalil and J Sampe 2013 Experimental Investigation of Thermoelectric Generator Modules With Different Technique of Cooling System *Am. J. Eng. Appl. Sci* **vol 6** no 1 pp 1–7

[12] P M Kumar, I Kathiravan, G Aadhithyan, C Prabu, and N Bharath 2015 Electric Power Generation Using Refrigeration *J. Res Appl Sci Eng Technol* **vol 3** no III pp 207–12

[13] L C Ding, N Meyerheinrich, L Tan, K Rahaoui, R Jain and A Akbarzadeh 2017 Thermoelectric Power Generation from Waste Heat of Natural Gas Water Heater *Energy Procedia* **vol 110** no December 2016 pp 32–37

[14] D M Rowe and G Min 1988 Evaluation of thermoelectric modules for power generation *J. Power Sources* **vol 73** pp 193–8

Evaluate heat loss through windows by using Guarded Hot Box (GHB)

H Humaish

Middle Technical University, Technical Institute, Department of Surveying, Al-Kut, Iraq

E-mail hussein2018humaish@gmail.com

Abstract. Windows are major part of construction design providing light, air and visibility and protect it from dust, it's also provide comfortable environment for occupants. The good design for window according to the heat losses is effect on the energy performance for constructions. Houses losses 15% and gain 50% from the heat through windows. In nowadays design of windows is very important to conserve the energy and to reduce CO2 emission. In this paper we calculate heat loss by using Guarded Hot Box (GHB) for different types of windows. Thermal transmittance has been measured by finding thermal resistance according to the ASTM recommendations by using Guarded Hot Box (GHB). The results of heat loss which have been obtained by using guarded Hot Box (GHB) with different windows were close to the manufacturing results and references.

1. Introduction

Windows supplying the houses with light, heat and ventilation. But in the other hand big amount from heating of houses lost through the windows. Houses losses 10% from the heat through the windows. Windows lose heat during different way, radiation through glazing, conduction through window frame and air leakage around opining lights and frame. Around 70% of the energy lost from a standard window is through the glazing and a small amount of heat is lost through convection. Air leakage is second contributor for heat losses through frame of windows and depends on the state of the windows (old or new) and on the frame material (timber frames perform better than metal). Windows supplying the houses with light, heat and ventilation. But in the other hand big amount from heating of houses lost through the windows.

Heat losses through window are effected by the frame around the window, where the magnitude of the U- value influence with the manufacturing of the frame materials. The frame area will increase for windows with multiple openings multiple small panes of glass. Since the U-value of the frame is more than that of the glass, then window of the smallest frame area have less U-value and less heat losses. Due to the high visibility for occupants and allowing daylight into the building, windows have less thermally efficient compared with walls [1]. Using double and triple glazed windows reduce thermal transmittance more than 70% compare with the single glass. Also, using Argon gas in double glazed windows reduce thermal transmittance by 15% [2].

Hot Box is used to test thermal properties of composition of materials in multilayer, walls cavity walls and windows structures [3]. During the steady state, heat transfer coefficient or a U-value for the sample or thermal resistance can be obtained [4].

There are two types from hot boxes; Calibrated Hot Box (CHB) and Guarded Hot Box (GHB), each types consist of different parts. Guarded Hot Box (GHB) is different of Calibrated Hot Box (CHB)

IOP Conf. Series: Journal of Physics: Conf. Series **1032** (2018) 012025 doi:10.1088/1742-6596/1032/1/012025

with existing the metering box inside guarded box to reduce heat loss and to keep same temperature in the two boxes. Guarded Hot Box (GHB) used in this study to evaluate heat transfer through windows. Guarded Hot Box (GHB) is used to test skylight windows with alternative designs by the National Research Council Canada [5]. University of Ulster used Guarded Hot Box (GHB) according to the BS EN ISO 8990 to test of the insulating properties of evacuated glazing [6]. Guarded Hot Box (GHB) was used to evaluate heat loss in different types of windows single, double and triple glazing.

2. Experimental apparatus

The experimental has been done by using Guarded Hot Box (GHB) which is consist of two parts (hot zone and cold zone), as shown in Figure 1.

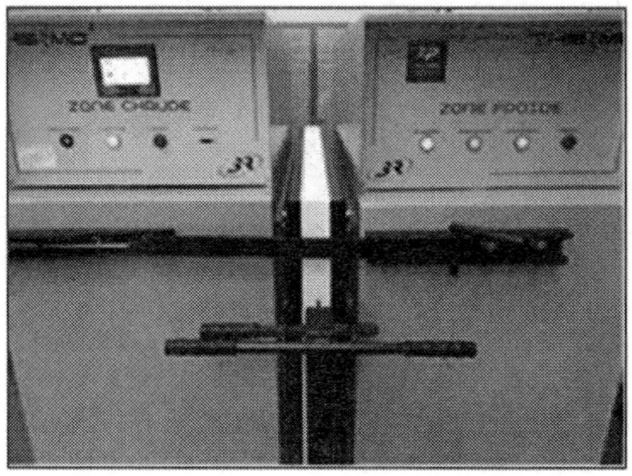

Figure 1. Guarded Hot Box (GHB)

Cold zone has one zone with external dimensions 1200x1200x600 mm and internal dimension is 1000x1000x500 mm, it's includes refrigeration unit and ventilation unit, as shown in Figure 2. Hot zone has two zone guard zone and measured zone with dimension 1200x1200x600 mm and 800x800x300 mm respectively.

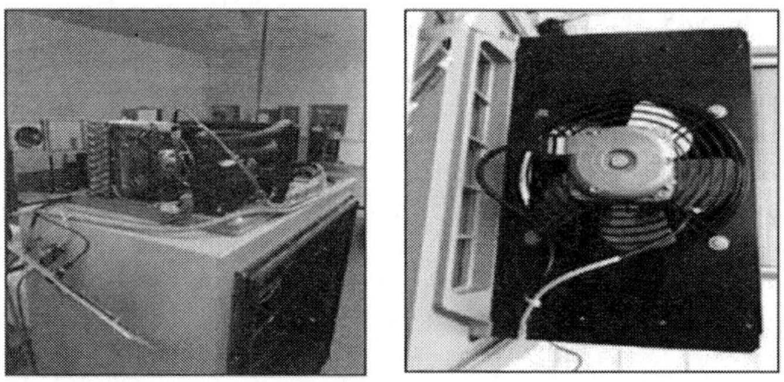

Figure 2. Refrigeration and ventilation units

The walls of two cells are isolated by 100 mm polyurethane. The materials to be studied (wall, door and window) are placed between the two cells and formed with dimension (1200mm width and 1200mm height) and thickness from 10 to 300 mm in order to be installed with the frame of the guarded hot box. The power supply 450 W for cold zone and 200 W for each zone (guard zone and measured zone). The cold zone supplying with 8 digital temperature sensors distributed throughout the volume of the zone while the hot zone supplying with 12 digital temperature sensors (8 for measure zone and 4 for guard zone).

3. Methodology

3.1. Guarded Hot Box (GHB)

Heat transfer through test specimen (Q_{ts})in the measure zone of the hot cell, shown in Figure 3 can be expressed as:

$$Q_{ts} = Q_{in} - Q_p - Q_w \tag{1}$$

WhereQ_{ts} , Q_{in}, Q_p and Q_w represent heat transfer through test specimen [W], total power input [W], heat transfer parallel to test specimen [W] and heat transfer through metering box walls [W] respectively.

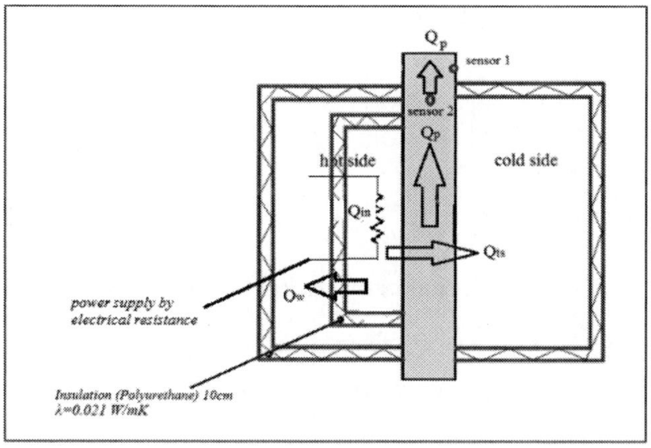

Figure 3. Guarded Hot Box Losses [7]

Influence of the heat loss through the sample outside GHB (Q_p) has been investigated by using two thermocouples type T made of copper and constantan with range of temperature -200 °C to + 400°C). Sensor 1 fixed outside GHB to measures the temperature of the laboratory while sensor 2 fixed inside the wall used. The wall used is composed of three layers, external and internal layer from plaster with thickness 13 mm and glass wool material between them thickness 90 mm with boundary condition as shown in the figure 4.

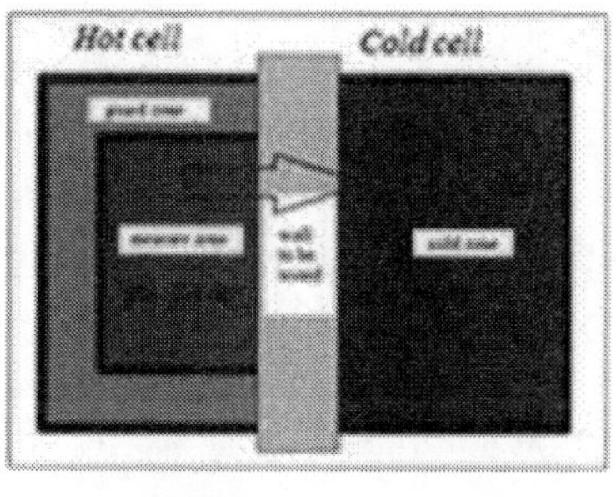

Figure 4. Boundary condition of the two cells during experimental

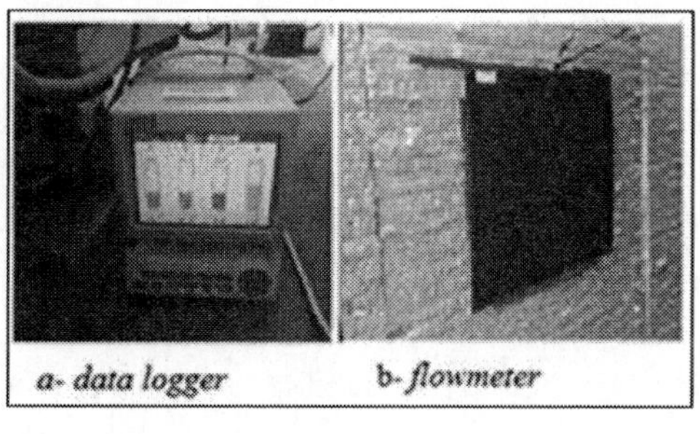

Figure 5. Data logger and flowmeter used in experimental

The sensitivity of flowmeters used in the hot surface of the wall is 264 μV/ (w/m²) with accuracy 0.1 °C. (MV100 by YOKOQAWA) data logger has been used to store the temperature as shown in figure 5.

Figure 6 illustrate the variation of the temperature for two sensors similar, that's mean the temperature inside the specimen is similar of the laboratory temperature. The temperature of the two sensors not affected by the temperature of the hot cell which is regulated at 30°.

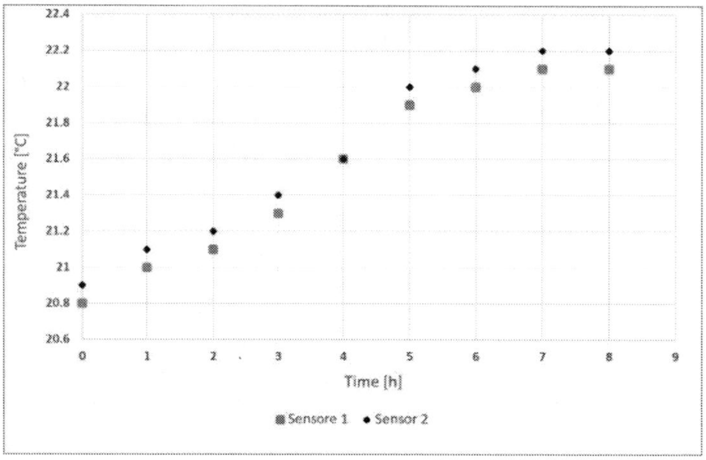

Figure 6. Temperature variation of the sensors

So we can conclude that the heat loss Q_p from measured zone to ambiance of the laboratory can be neglected and the heat transfer through specimen (Qts) from the measured zone can be expressed by:

$$Q_{ts} = Q_{in} - Q_w \tag{2}$$

The value of Q_{ts}, has been obtained by fixing one flowmeter with sensitivity 264 μV/ (w/m²) on the hot surface of the measured zone. The value of Q_w has been calculated by using the expression:

$$Q_w = \frac{A.\Delta T}{R_w} \tag{3}$$

Where A, ΔT and R_w represent surface area of the metering box [m²], temperature differences between measured box and guarded box [°C] and thermal resistance of the wall of the measure box [m² K/w] respectively.

Q_{in} is calculated by Guarded Hot Box according to methodology of supplying heat power from electrical resistance. The value of Q_w has been calculated experimentally from equation (2) [0.02 W] and theoretically equation (3) [0.04 W]. The differences between two values coming from the accuracy of temperature measurement (temperature measured in ambiance not in wall surface and Q_p negligible. During transient state (first hour of the test) the variation of the flow Q_{in} is hug and there is a need to big amount of flow to stabilise temperature in cells. After thermal equilibrium obtained, the flow reached to the stabilised and Q_w can be neglected as shown in figure 7.

When steady state is reached, Q_{in} and Q_{ts} are very high (8W) while Q_w is near (or under) 0.1 W. So, for the hot cell, equation (2) can be expressed as:

$$Q_{ts} = Q_{in} \tag{4}$$

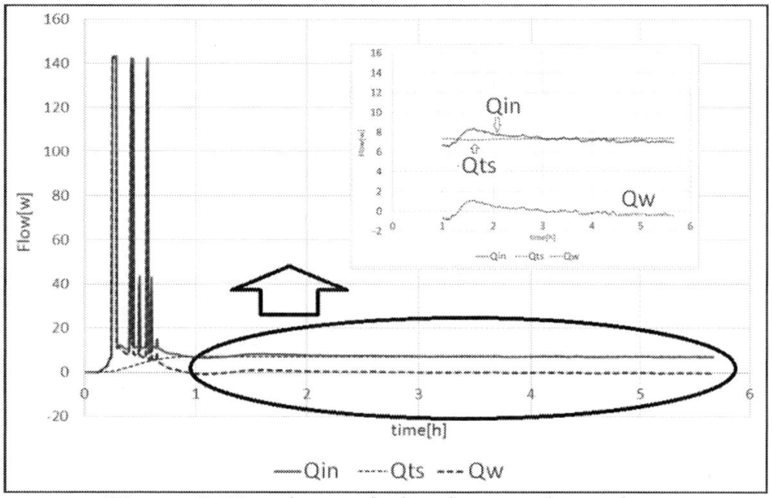

Figure 7. Heat losses during the experimental test

3.2. Thermal transmittance for windows

The experimental has been done to investigate transmittance value for the window, we used two difference window first one PVC single glazed window 4mm, size 650mmx650mm, second one PVC double glazed window 4/16/4, size 650mmx650mm. the window in two case fixing on the polyurethane frame (1200mm x 1200mm and 80mm thickness). Argon gas use between double glazing as shown in figure 8.

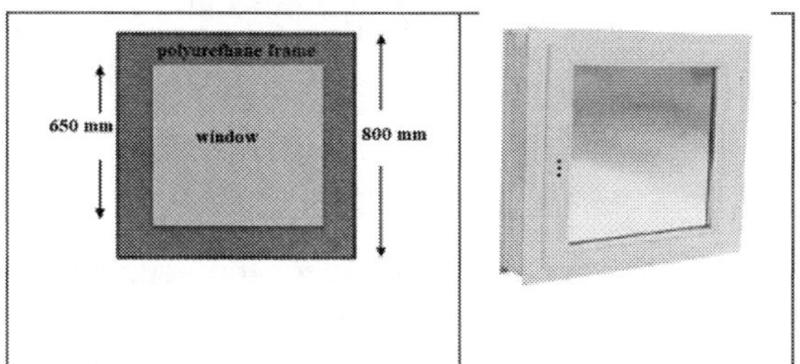

Figure 8. window use in experimental

Guarded Hot Box (GHB) used to measure the thermal resistance of the window then to calculate the value of thermal transmittance Uw. We calculated the total thermal resistance of the window (frame and window) according ASTM recommendation [8].

According to the Fourier law and heat flow on the direction normal to the plane of the window:

$$q_{tot} = \frac{(T_{hot} - T_{cold})}{R_{tot}} A_{tot} \tag{5}$$

Where q_{tot}, T_{hot}, T_{cold} and R_{tot} represent total flow through window and frame [W], hot cell temperature [C°], cold cell temperature [C°] and total thermal resistance for window and frame [m^2 K/w] respectively.

The total flow through the wall is equal to the sum of the flux passing through the frame (polyurethane) portion q_f and the windowq_w.

$$q_{tot} = q_f + q_w = (\frac{A_f}{R_f} + \frac{A_w}{R_w})(T_{hot} - T_{cold}) \quad (6)$$

Hence, by identifying:

$$\frac{A_{tot}}{R_{tot}} = \frac{A_f}{R_f} + \frac{A_w}{R_w} \quad (7)$$

Where A_f, A_w, R_f and R_w represent area of frame [m²], area of window [m²], thermal resistance for frame [m² K/w] and thermal resistance for window [m² K/w] respectively.

By definition:

$$R_f = \frac{d_f}{\lambda_f} \quad (8)$$

$$R_w = \frac{1}{U_W} \quad (9)$$

Where d_f, λ_f and U_W represent thickness of the frame [mm], thermal conductivity for frame [w /m K] and thermal transmittance for window [w /m² K] respectively.

Therefore:

$$U_W = \frac{1}{A_w}\left(\frac{A_{tot}}{R_{tot}} - \frac{\lambda_f . A_f}{d_f}\right) \quad (10)$$

Where U_W, A_{tot} and R_{tot} represent thermal transmittance for window [w /m² K], total area for window and frame and total thermal resistance for window and frame [m² K/w].

The principle of Measurement depends on the heat transfer from the hot cell to the cold cell and observe the temperatures on both sides of the window, we deduce the value of thermal resistance and thermal transmittance.

The window is place between the two cells and select the boundary conditions (25 C° and 10 C°) inside hot and cold cell respectively. We wait until the two cells are reached to the thermal equilibrium then the results can be collected for the variation of temperature, variation of flow and variation of thermal resistance.

As mentioned previously, the most heat losses through windows accrue during the conduction (this study neglected the convection).

The heat losses through the surface (window) is calculated as equation:

$$H = U . A . \Delta T \quad (11)$$

Where H and U represent heat loss [W] and thermal transmittance [W/m².K] respectively.

Using double and triple glazed windows reduce thermal transmittance more than 70% compare with the single glass. Also, using Argon gas in double glazed windows reduce thermal transmittance by 15% [9].

Thermal transmittance for single glass is about 6 [W/m ^{2}K ^{1}] and it's less during use double glass filling with air or Argon [10].

IOP Conf. Series: Journal of Physics: Conf. Series **1032** (2018) 012025 doi:10.1088/1742-6596/1032/1/012025

4. Results and discussion

The experimental procedure measurement stabilization the temperature of the hot side and cold side of the cells at 25 °C and 10C° respectively. The temperature variation reached to the steady state after after half an hour as shown in figure 9.

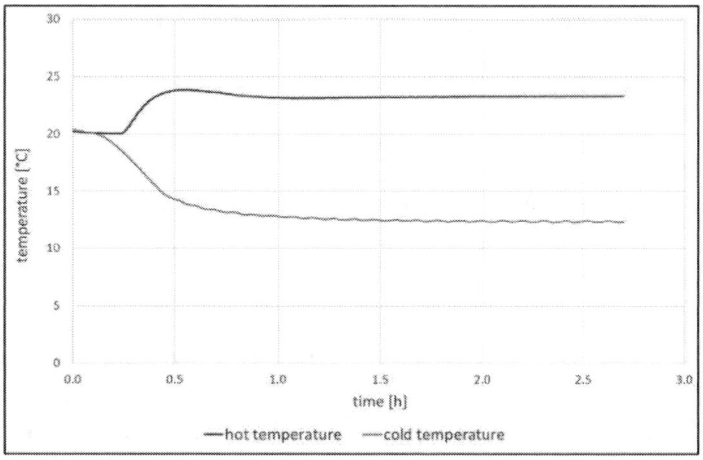

Figure 9. variation of temperature during the test

Heat flow (W) transmitted from hot cell to the cold cell according to the test time shown in figure 10. Heat flow has been reached to the stabilization after one hour. The huge variation in the first half hour due to the regulation between two cells before reached to the steady state.

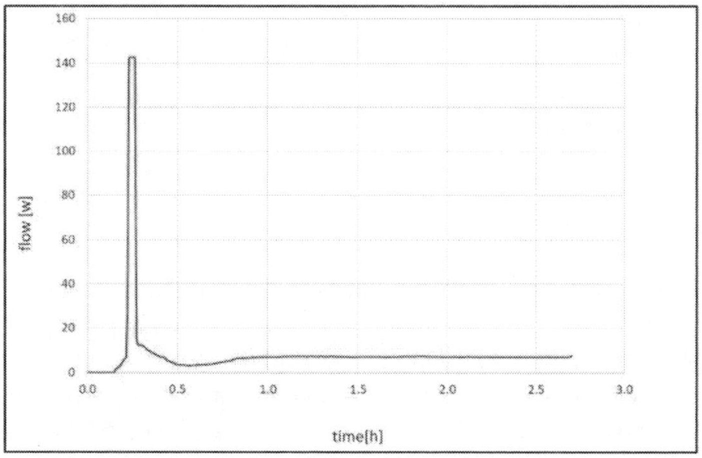

Figure 10. variation of flow (w) during the test period

During experimental procedure, total thermal resistance of the window by using the average value has been investigated. The calculated thermal resistance is 0.92 [m^2K/W] as shown in figure 11.

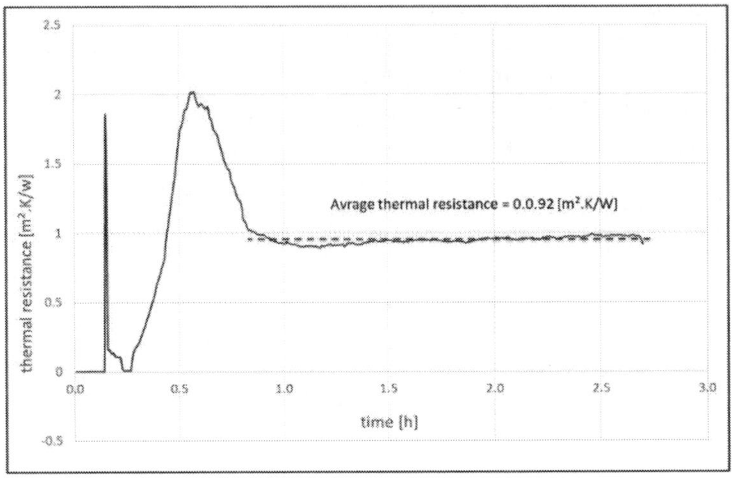

Figure 11. Variation of thermal resistance

Referring to equation (9) and according to the ASTM recommendations, thermal transmittance Uw for single and double glazing window were 5.6 [W/ m^2K] and 1.09 [W/ m^2K] respectively. Heat losses through the surface (window) have been calculated for single and double glazing window according to the equation (11). The heat loss for single glazing window was five times of the double glazing window. Table 1 illustrate the results of thermal transmittance and heat loss for two tests (single and double glazing).

Table 1. Thermal transmittance and heat loss

Glazing types	Uw [W/ m^2K]	Heat loss[W]
Single 4 mm	5.6	53.8
Double with argon (4/16/4)	1.09	10.5

5. Conclusion

Thermal performance of the glazing has been investigated by using Guarded Hot Box (GHB). This device can be considered suitable to create realistic climatic conditions during measurement of thermal properties of windows. Two cells of guarded hot box (GHB) have been calibrated by using multilayer wall (plaster+ glass wool + plaster). Thermal transmittance investigated for single glazing 4 mm and for double glazing 4mm glass, 16 mm gap with argon and 4 mm glass. The experimental results were agreement with the references and showed importance of double glazing for saving heat in buildings.

References

[1] Matthias M Koebel, Manz Heinrich, and Karl Emanuel Mayerhofer, "Service-life limitations in vacuum glazing: A transient pressure balance model," *Solar EnergyMaterials&SolarCells,* vol. **94**, pp. 1015–1024, 2010.

[2] Sabatiuk, P. A, "Review of Gas Filled Window Technology: Summary Report," in *Proceedings of ASHRAE/DOE Conference, Thermal Performance of the Exterior of Buildings 11*, ASHRAE Special Publication No. **38**, 1982, pp. 643-653.

[3] Yuan S., Experimental and Analytical Heat Transfer analysis for a calibrated Hot Box and Fenestration System, University of Massachusetts Amherst. UMI No.3027279, 2001.

[4] Curcija D. PHD, Trends and developments in Window Testing Methods, University of Massachusetts, *Centre for Energy Efficiency and Renewable Energy.*

[5] Elmahdy, A.H.; Haddad, K., Experimental procedure and uncertainty analysis of a guarded hotbox method to determine the thermal transmission coefficient of skylights and sloped glazing, *ASHRAE Transactions,* v. **106**, pt. 2, pp. 601-613, 2000.

[6] Fang Y, An experimental and theoretical investigation into the design, development and performance of evacuated glazing, PhD thesis, the University of Ulster, Northern Ireland, UK, 2002.

[7] Önder, K., R. Yumrutas., & O Arpa. (2009). Theoretical and Experimental Investigation of Total Equivalent Temperature Difference Values for Building Walls and Flat Roofs in Turkey. *Applied Energy*, **86**, 737–747.

[8] Standard Test Method for Steady-Stat Thermal Performance of Building Assemblies by Means of a Guarded Hot Box 1, Designation: C 236 - 89 (Reapproved 1995).

[9] Sabatiuk, P. A, "Review of Gas Filled Window Technology: Summary Report," in Proceedings of ASHRAE/DOE Conference, Thermal Performance of the Exterior of Buildings 11, *ASHRAE Special Publication* No. **38**, 1982, pp. 643-653.

[10] E Cuce and S B Riffat, "State-of-the-artreview on innovative glazing technologies," *Renewable and sustainable energy reviews*, vol. **41**, pp. 695 - 714, 2015.

IOP Conf. Series: Journal of Physics: Conf. Series **1032** (2018) 012026 doi:10.1088/1742-6596/1032/1/012026

Urinary excretion as a function of uranium concentration in bladder cancer patients using kinetic phosphorimetry analyzer

Shafik S Shafik[1], Shaymaa M Qaddoori[1, 2]

[1]Physics department, College of Science, Baghdad University, Iraq

[2]sh_ma_81@yahoo.com

Abstract. Measurement of urinary excretion is most commonly used as a method of assessing internal contamination due to insoluble nuclides. The pulsed-laser Kinetic Phosphorimetry Analyzer (KPA-11 with analysis range 0.01μg/L up to 50mg/L for Uranium) has been used to determine the Uranium concentration (Uc) in urine samples of three groups of persons; Bladder cancer patients, Healthy and Infants (with age less than two month) and their mothers. The range of Uc excreted in all subjects have been found to be0.735 –3.876μgl^{-1} with an overall average of 2.00038 μgl^{-1}, 0.856–1.042μgl^{-1} with an overall average of 0.9464 μgl^{-1} and0.505–0.979μgl^{-1} with an overall average of 0.7742 μgl^{-1},respectively. The obtained results illustrated that there is statistically significant correlation between Uc and residential area. The obtained Mean values of Uc of the different groups were found to be approximately proportional to age up to 50 years. A noticeable drop is observed for subjects greater than 50 years old. A synthetic urine analysis was chosen in this study to proclaim any concern over biohazards, such as acidity of urine, increasing Amorphous urate, Uric acid, and decreasing Mucous.

Key words: Uranium concentration, Bladder cancer, KPA, Synthetic urine.

1. Introduction

Diagnosis of bladder cancer may occur by urine tests, cancer cells, and other signs of disease [1]. A risk factor of bladder cancer is anything that increasing the chance of getting a disease such as smoking, which can be changed by leaving this habit. Others, such as a person's age or family history, can't be changed [2].The risk of bladder cancer increases with age, and about 9 out of 10 people with bladder cancer are older than 55. It is much more common in men than in women. Urinary infections, kidney and bladder stones, bladder catheters left in place a long time, and other causes of chronic bladder irritation have been linked with bladder cancer (especially squamous cell carcinoma of the bladder), but it's not clear if they actually cause bladder cancer [3].

There are many studies indicated that radiation exposure is one of that carcinogenic reasons of bladder cancer or any type of cancer [4-6].The radiation effect may be appear in the exposed person; directly or, take up along time [7]. Evidence of injury from low or moderate doses of radiation may not show up for months or even years. For leukemia, the minimum time period between the radiation exposure and the appearance of disease (latency period) is 2 years. For solid tumors, the latency period is more than 5 years [8].

Till now, the misuse of forbidden weapons in Gulf war I and II caused environmental pollution with uranium in wide areas of Iraq, which lead to different diseases, children malformations, leukemia,

IOP Conf. Series: Journal of Physics: Conf. Series **1032** (2018) 012026 doi:10.1088/1742-6596/1032/1/012026

and cancers [9]. Besides, Mc Diarmid et al. [10] found evidence of neurocognitive impairment in Gulf War veterans who had retained fragments of Depleted Uranium (DU) shrapnel, although these individuals showed little evidence of impaired kidney function. Hindin et al. [11] made a review about the teratogenic effects on animal and human. They concluded that the weight of evidence is consistent with an increased risk of birth defects in the progeny that exposed to DU. Accidental or chronic exposure to DU occur in military use of DU munitions in combat at the 1991 and 2003 Gulf Wars [12].

At the last decade in Iraq, the steady increasing in morbidity cases such as cancers, children with birth defects and abortion of pregnant women have been noted [13-16].Uranium, principally, represents an internal radiation hazard. The risks associated with Uranium in the body are both chemical and radiological [6, 17]. The damage caused by ionizing radiation from radionuclide transformations is a result of the energy absorbed by body tissues [18]. Heavy metals compounds, like carbonate/ bicarbonate compounds, e.g. $[UO_2(CO_3)_2]2$, may cause a number of cytotoxic effects. These compounds are stable at a neutral pH value (pH of blood) and in this form are not very reactive, but the highly reactive uranyl ion UO_2^{+2} is released at low pH values (as urine) [19].

The effects of the radionuclides vary with many parameters, specifically its pathway, before reaching the target. Uranium can travel through the air, water (both groundwater and surface water), soil and through the food chain. The radionuclides may enter the human body by ingestion, inhalation and through the skin, and then played the internal exposure [20]. Moreover, the effect of radionuclides is depending on biokinetic of them inside the body. The behavior of a substance in the body, such as intake, uptake, distribution, excretion, and retention is called biokinetics [21]. These biokinetics are depending on the properties of Uranium compound such as bio-solubility. However, the Gastrointestinal (GI) tract absorption of Uranium at environmental levels is about (1%) [22].The other property of Uranium is biological half-life (T_b), which is the time was taken for the amounts of a particular element in the body to decrease to half its initial value due to elimination by biological processes alone, and hence some of radionuclides can be excrete by urine and/ or feces. This means that a radioactive atom can be expelled before it has had the chance to decay. For a number of radioisotopes of particular medical interest, the rate of excretion has been specified in the form of an effective biological half-life. The overall elimination half-life of Uranium under conditions of normal daily intake has been estimated to be between 180 and 360 days in humans [23].

A synthetic urine analysis has been chosen in this study to proclaim any concern over biohazards. Monitoring for occupational incorporation of Uranium and Thorium are usually carried out by urinary excretion analysis. Uranium is poorly absorbed by the digestive tract and most of the absorbed Uranium is eliminated in the urine. Thus, urine is considered to be the best sample for the detection of excessive intake of Uranium [24].Measurement of urinary excretion is most commonly used as a method of assessing internal contamination due to insoluble nuclides. Analysis of human urine has revealed the presence of various nuclides such as Plutonium, Polonium, Uranium, Carbon-14, Strontium, Barium, Tritium, etc. [25].

The purpose of this research is to get answers of the questions: is the increasing of Uranium concentration (Uc) in urine represents a bio marker indication to get cancer or is a main cause of the cancer reasons?, is there any correlation relation between Uc in urine sample and its bio marker analyses?, and finally, is there any variance between patient and control groups due to their age and home location?.

2. Sample Collection

Urine samples were collected from different group of persons as illustrated in the table 1. The patient group participators were mostly from all districts of Baghdad governorate, from all age groups, and with both genders. The age of control group is asymptotic to that for patient, while the age of the infant is just a few days.

Table 1. Groups of persons from whom urine samples were taken.

Group	Number of samples	Source
Bladder cancer patients	60	Al-Amal National Hospital for Cancer Management, Al-Imamein Al kadhimein city hospital / Al Jawad center for tumor treatment and Medical City / Oncology teaching hospital
Healthy (control)	30	different position
Infants with their mother	10	Al Khadraa Al Ahly hospital and Al Elwea teaching hospital for tocology

2.1.Samples for biological test

The samples those taken for a biological test and from the entire groups that shown in table 1were kept cool until the test was carried out. The sample was taken from the adult person with urine cup contains 60 ml, then was put in cooling box until reach the bio-laboratory, but in the case of the infant, the sample gathering by urine bag and also transfers in cooling box.

2.2.Samples for radiological test

The urine samples for a radiological test from the same participators in table 1don't need stay cooling, and they were accumulated to be 500-1000 ml for almost 24 hour. Complete 24hour urine samples give better precision than spot samples in estimating Uc at these low levels, but presented more logistic difficulties in the collection [26]. In other wards these samples were taken in another day, because of all patient are visitors not residents. Total 24 hour urine samples were collected in polyethylene bottles, and immediately acidized by the addition of 1ml concentrated HCl for each sample to avoid precipitate formation.

3. Sample Analyses

3.1.Biological analyses

It could be done for all groups of participators in table 1 to study the relation between the cancer incidence probability and bio marker analysis of the urine. The biological tests are fixing in tables 2.

Table 2. The biological tests of urine samples.

No.	Gender	pH	pus cell	Epithelial cell	Amorphous urate	Uric acid	Calcium Oxalate	Mucous	others Bacteria

3.2. Radiological analyses

It could be done for all groups of participators in table 1 to study the relation between the cancer incidence probability and Uc measurements. In general, execration of Uc in urine of healthy person was limited by several international agencies, for example World Health Organization (WHO), to about100 ng/L [27]. Therefore, Uc measurements in urine need a detecting system with low minimum detectable activity. Elements in environmental sample are presented in very small amount and their measurements required reliable and sensitive analytical technique. Environmental levels of Uranium in human excreta are highly variable, depending on Uc in air, food and water, and on the health of the individual [28]. Analytical techniques used as part of a radiation protection program should provide

IOP Conf. Series: Journal of Physics: Conf. Series **1032** (2018) 012026 doi:10.1088/1742-6596/1032/1/012026

limits of detection comparable to or below these levels, in order to differentiate between environmental and occupational Uranium exposures.

Concentrations of various elements in different samples were determined using many analytical techniques, whereas, Uranium content as been measured in various foods and drink samples using Gamma-Ray Spectrometry, but this technique needs to use of special food and liquid calibration standards [29]. Neutron Activation Analysis and radiochemical separation can use to estimate Uc in some biological samples [30]. Photometric techniques such as fluorometry and phosphorometry can use to measure Uc with very good resolution and sensitivity. KPA-11 spectrometer, which is one of the updated phosphorometry techniques, has been used to evaluate Uc in urine samples of the present work. KPA-11 provides highly precise and accurate measurements that eliminate the need for internal standards. The advantages of this technique are the extremely low detection limit (10 ng/L), and only required pretreatment of samples. KPA-11 is a bench-top instrument that rapidly performs single-measurement, manufactured by Chemchek Instruments (Richland, USA). This model is equipped with a pulsed nitrogen/dye laser to supply monochromatic ultraviolet light to excite Uranium atoms in the sample solution. It is a fully integrated computerized system for data collection and analysis. Chemchek's KPA Win software controls the KPA-11 along with storing and interpreting the analytical data returned from the KPA [31].

The urine samples were prepared and Uc measured in μg/l in Radiation Protection Center of Iraqi Environmental Ministry.

Uranium octoxide U_3O_8 used to preparing Uranium standard solutions. Firstly, a stock standard solution of 1000 mg/*l* (1000 ppm) was prepared by dissolving 117.9 mg of U_3O_8 in 100 ml of 0.82 M nitric acid (HNO_3) in volumetric flask. A calibration curve for KPA at the low Uc is obtained from 0.05, 0.1, 0.5, 1, 5 and 10 μg/*l* concentrations. Secondly, the desired Uc for the standard sample was prepared by further diluting the stock solution in 0.82M HNO_3 utilizing the dilution formula:

$$1^{st}\text{ volume} \times 1^{st}\text{ concentration} = 2^{nd}\text{ volume} \times 2^{nd}\text{ concentration} \ldots (1)$$

4. Data Processing

Questionnaire data, related to all participators included in table 1, had been done as fixed in table 3. They were taken from participators and checked for validity and consistency with hospital computer files of them. Identities were stored separately from questionnaire data, which were indexed by an anonymous unique participant number. In order to stand about all the effected parameters on the bladder cancer incidence probability, Uc and biological measurements have been gathered with this questionnaire and studied against each other.

The Statistical Analysis System- SAS (2012) program was used to study the effect of factors differences in studied parameters [32]. Chi-square test was used to significant compare between percentage and T-test instead of Least Significant Difference –LSD test (because of the comparison is between two groups) was used to significant compare between Mean of the studied parameters.

Table3. The questionnaire of all participators of study.

No	Gender	age	Home location	Degree of study	The work type	Are you marrying?	Are you having children?	Are you smoking?	Type of drinking water

IOP Conf. Series: Journal of Physics: Conf. Series **1032** (2018) 012026 doi:10.1088/1742-6596/1032/1/012026

5.Results and Discussion

5.1.Results of biological analyses

T-test was used to significant compare between the Mean of each bio marker of urine, related to the patient, healthy (control) and infants with their mothers, who participated in this study. The comparison showed that there are high significant differences between patients and controls in pH, Amorphous urate, Mucous, Uric acid, Epithelial and Pus cell. Also a significant difference appeared in other bacteria, as shown in table 4. The comparison between control participants and infants with their mothers showed no significant differences as illustrated in table 5.

Table 4. The comparison between the bladder cancer patients and control in the studied biological parameters.

Parameters	Mean ± SE		T-Test (P-value)
	Patients	Control	
pH	6.56 ± 0.08	7.17 ± 0.06	0.2457 ** (0.0001)
Pus cell	1.541 ± 0.14	0.766 ± 0.08	0.426 ** (0.0005)
Epithelial	0.233 ± 0.04	0.766 ± 0.07	0.1542 ** (0.0001)
Amorphous rate	2.14 ± 0.16	0.917 ± 0.09	0.4833 ** (0.0001)
Uric acid	0.316 ± 0.07	0.00 ± 0.00	0.217 ** (0.0047)
Calcium oxalate	0.950 ± 0.25	0.716 ± 0.12	0.5026 NS (0.3532)
Mucous	0.00 ± 0.00	0.216 ± 0.04	0.0643 **(0.0001)
Other bacteria	1.10 ± 0.16	0.467 ± 0.11	0.4931 * (0.0120)

** (P<0.01), NS: Non-significant.

Table 5. The comparison between Control and Infants with their mothers in the studied biological parameters.

Parameters	Mean ± SE		T-Test (P-value)
	Control	Infants with their mothers	
pH	7.17 ± 0.06	7.30 ± 0.15	0.3714 NS(0.3742)
Pus cell	0.766 ± 0.08	0.950 ± 0.18	0.3002 NS (0.3239)
Epithelial	0.766 ± 0.07	0.450 ± 0.13	0.3193 NS (0.0501)
Amorphous rate	0.917 ± 0.09	0.950 ± 0.42	0.50907 NS (0.9097)
Uric acid	0.00 ± 0.00	0.00 ± 0.00	0.00 NS (1.000)
Calcium oxalate	0.716 ± 0.12	0.500 ± 0.11	0.3725 NS (0.2509)
Mucous	0.216 ± 0.04	0.200 ± 0.08	0.1874 NS (0.8581)
Other bacteria	0.467 ± 0.11	0.300 ± 0.15	0.4417 NS (0.4496)

** (P<0.01), NS: Non-significant.

However, the bio markers of urine are really considered as indicators on healthiness states, and the infants with their mothers considered as a reference to healthiness states. A comparison among all groups of participants was showed in figure 1. It's clear that the uric salts (Amorphous urate, Calcium oxalate and Uric acid) increase in urine as much as the pH decreasing for all groups, and the infants with their mothers had minimum amount of these salts and highest pH. Furthermore, other bacteria had the same behavior, while Epithelial and Pus cell in the second and third groups had disconcerted behavior due to sample related to women in all most cases.

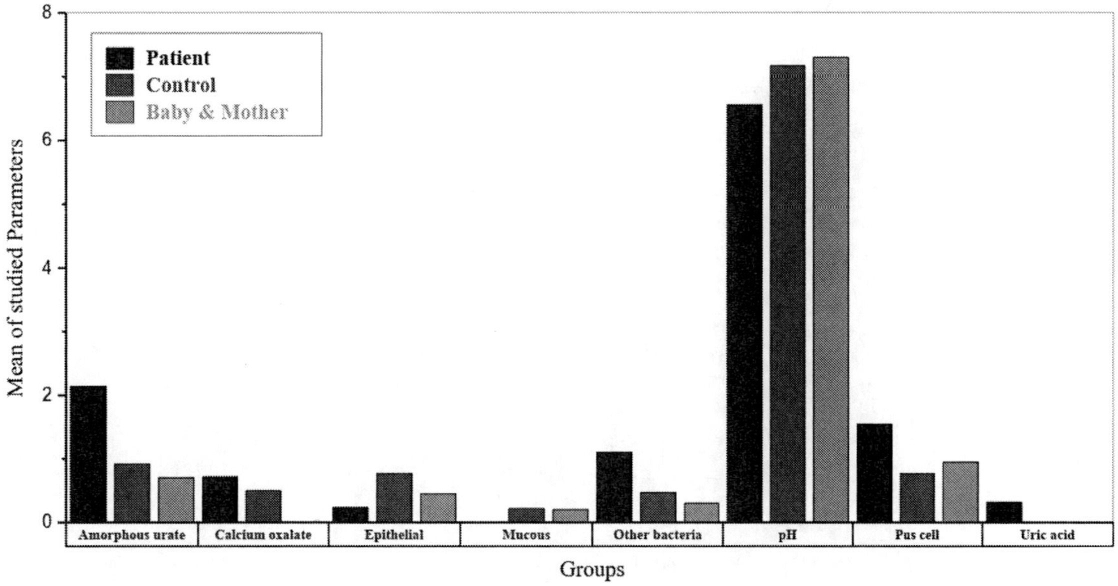

Figure 1. The Mean variation of the biological studied parameters for all groups of participants.

5.2. Results of Radiological Analyses

The Mean of the measured Uc in urine samples of patient and control participants had been compared significantly using T-test. The comparison showed there is high significant difference between participants as in table 6. Additionally, the comparison between control and infants with their mothers showed a high significant difference as in table 7.

Table 6. The Mean of Uc ($\mu g l^{-1}$) in the urine samples of the bladder cancer patients and control participants.

The group	No	Mean ± SE of Uc
Patients	60	2.0003 ± 0.112
Control	30	0.946 ± 0.009
T-Test	---	0.3165 **
P-value	---	(0.0001)
	** (P<0.01)	

Table 7. The Mean of Uc ($\mu g l^{-1}$) in the urine samples of the control and Infants with their mothers.

The group	No	Mean ± SE of Uc
Control	30	0.946 ± 0.009
Infants with their mother	10	0.774 ± 0.06
T-Test	---	0.0732 **
P-value	---	(0.0001)
	** (P<0.01).	

Figure 2 illustrates a comprehensive comparison between the studied groups and the international organizations; World Health Organization (WHO) and International Commission on Radiological Protection (ICRP). Publication 23 of ICRP listed the values 0.04-0.4 μg/L for urinary Uc execration [28]. WHO reviewed data from the early 1990s suggesting that urinary Uc execration in the general population ranged from about 0.04 to 0.57 μg/L [27]. However, comparison shows that Mean of Uc of patient has the highest value against all other groups, while Mean Uc of infants with their mothers is the lowest value, but it is higher than the permissible recommended values of WHO and ICRP.

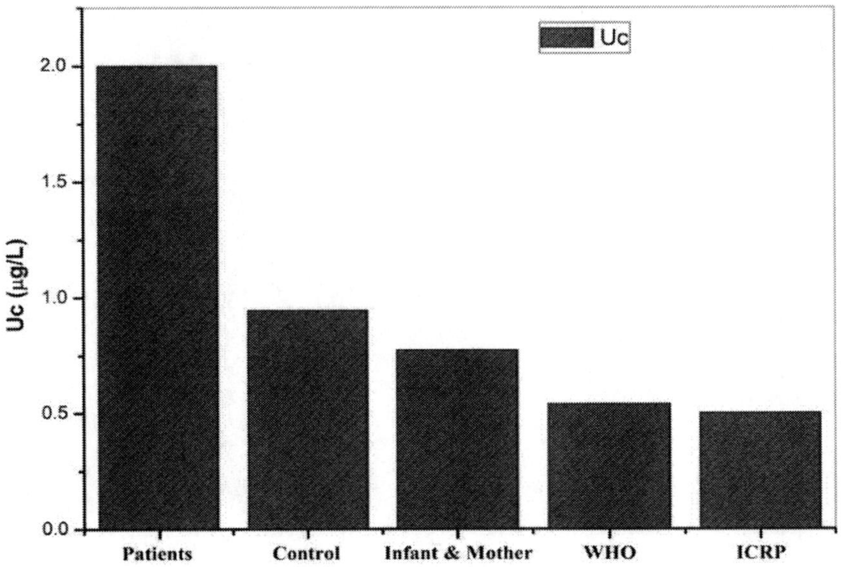

Figure 2. Comparison of Uc of the studied groups and international organizations.

The correlation coefficients between measured Uc and the biological parameters were shown in table 8. The results showed a reverse correlation coefficient with high significant between Uc and pH, Mucous and Epithelial, while a positively proportional correlation coefficient with high significant with Amorphous urate, and a non-significant correlation coefficient with respect to the other parameters.

Table 8. Correlation coefficients between Uc and studied biologicalparameters.

biological Parameters	Correlation coefficient-r	Level of sig.
pH	-0.40	**
Pus cell	0.45	**
Epithelial	-0.30	**
Amorphous rate	0.57	**
Uric acid	0.11	NS
Calcium oxalate	0.09	NS
Mucous	-0.34	**
Other bacteria	0.10	NS

** (P<0.01), NS: Non-significant.

5.3. Results of Questionnaire Data

Chi-square test was used to significant comparison between percentages of distribution of sample which related to patient character according to Questionnaire that fixed in table 3. The results showed that the bladder cancer patients had high percentage in some factors as shown in table 9, such as gender and period of illness of 1 year.

IOP Conf. Series: Journal of Physics: Conf. Series **1032** (2018) 012026 doi:10.1088/1742-6596/1032/1/012026

Table 9. Percentage distribution of bladder cancer patient according to their character.

Factor : character	No.of cases	Percentage (%)	Chi-Square
Male	50	83.33	13.409 **
Primary study	18	30.00	7.271 **
Farmer or Gainer	23	38.33	8.341**
Marred	58	96.67	14.594 **
4-7 (no. of child)	31	51.67	9.025 **
Smoker	38	63.33	9.017 **
Pipe /Filter	29/31	48.33/ 51.677	1.082 NS
1 year / (Period of illness:1-5year)	26	43.33	10.671 **
Family history(No)	42	70.00	10.750 **
	** (P<0.01).		

Bladder cancer patients' distribution with their Mean of Uc according to home location is shown in table 10. The home location with symbols 2, 8 and 9 registered highest numbers of cases with highest Mean of Uc, with taking into consideration that almost the patients living in home location since 10-15 years or from the born. In spite of that City center (symbolized with 1) registered 7cases, but their Mean of Uc is lower although the high population density of this location (Mansour, Ghazaliya, Al-Shurtah, Juafir, Muasalat, America and Hafa Street) against others. However, the archives of the patients in the Al Jwad center for tumor treatment demonstrated that the registered cases, for all types of cancer, got from home location with symbol 2 represent 30-40% of all other home locations. Also Al-Amal National Hospital for Cancer Management registered cancer cases with 30-40% from home location with symbols 8 and 9.

Table 10. Distribution of bladder cancer patients classified according to home locations and their Mean of Uc ($\mu g l^{-1}$)

Home location with symbol	No.of cases	Mean ± SE of Uc	Percentage (%)
City center (1):{Mansour, Ghazaliya, Al-Shurtah, Juafir, Muasalat, Ameria, Hafa Street, Bayaa, Turath, Jihad, Harthiya, Attaifiya, Dora, Karada, Amil District}	7	1.75 ± 0.12	11.67
Shuala, Kadhimiya, Hurriya, Dabbash, Chekok (2)	11	2.27 ± 0.10	18.33
Hossrinia, Taji (3)	4	1.87 ± 0.10	6.67
Abu Ghraib, Sabaa AL-Bour, Khan Dhari (4)	2	1.56 ± 0.08	3.33
Rasheed, Yusufiya, Mahmudiyah (5)	3	2.09 ± 0.11	5.00
Shaab, Hay Ure (6)	3	2.22 ± 0.14	5.00
Mustansiriya, Falastin street (7)	2	1.58 ± 0.11	3.33
Sadr city , Habibiya (8)	9	2.35 ± 0.11	15.00
Baghdad AL Jadeeda , AL Baladiyat (9)	10	2.05 ± 0.08	16.67
Nahrawan (10)	3	2.80 ± 0.16	5.00
AL Zaafraniya, JisrDiyala , Madain (11)	2	2.01 ± 0.09	3.33
AL Fallujah, Ramadi, Baqubah, Amerli, Kut (12)	4	1.69 ± 0.11	6.67
Chi-Square	---		7.271 **
	** (P<0.01).		

To emphasis the results that obtained from table 10, a comprehensive comparison had been made among all participants of studied groups, where control group (included healthy and infant and their mothers), with respect to home location as shown in figure 3-a. The Uc of participants of control group in home location (1) is lower than in other home locations (2, 8 and 9).

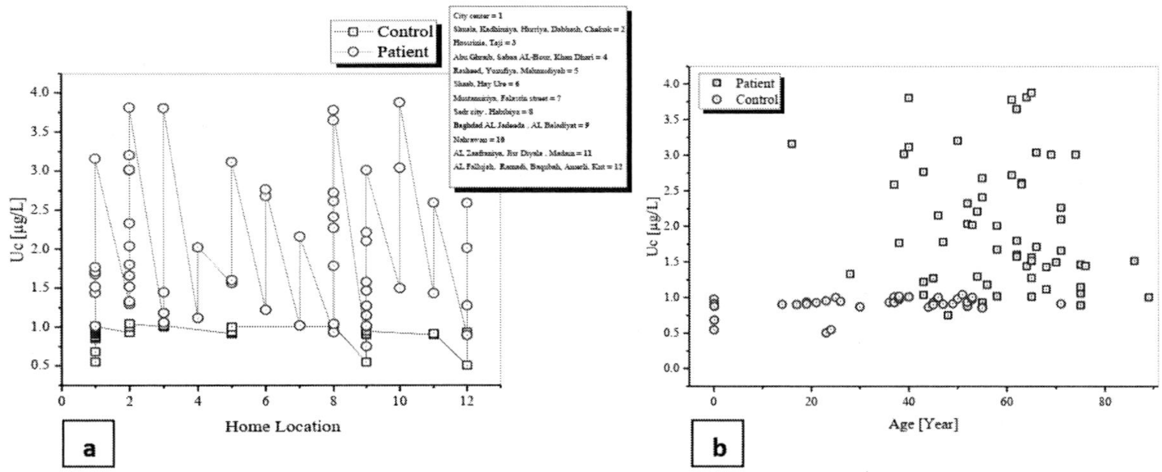

Figure 3. Comparisonof Uc among groups of study according to their: (a) home location, (b) age.

Moreover, bladder cancer patients' distribution with their Mean of Uc according to age groups is shown in table 11. Age group larger than 50 years registered highest number of cases but with lower Mean ± SE of Uc, while the other age groups registered lower number of cases but with highest Mean ± SE of Uc. This result can be explained as fallow: the archives of the patients in the hospital that registered cases with bladder cancer refer to largest number of cases in age 25-50, but they couldn't survive firstly; because of highest Uc contaminated in their body that attributed for many reasons like type of working, food consumed, etc. . Secondly, from table 9, where the period of illness was (1-5) years and as the period increase the probability of death increase, where cases with period of 1 year represent highest percentage in distribution of studied samples. To emphasis the results that deduced from table 11 a comprehensive comparison according to age had been made among all participants of study, where control group (included healthy and infant and their mothers), as shown in figure 3-b.

Table 11. Distribution of bladder cancer patients classified according to age groups and their Mean of Uc ($\mu g l^{-1}$)

Age groups (year)	No.of cases	Mean ± SE of Uc	Percentage (%)
≤ 30	3	2.24 ± 0.05	5.00
31-50	12	2.19 ± 0.07	20.00
>50	45	1.93 ± 0.07	75.00
Chi-Square	---		12.632 **

** (P<0.01).

6. Conclusions

Through the results of this work, one can strongly conclude that the increasing of Uc in bladder cancer patients represents one of the main reasons of the increasing bladder cancer incidence probability in the Iraqi populations. The second important conclusion is the Uc increment in the urine samples of control and infants and their mothers compared with international agencies such as WHO and ICRP, which indicates on the Uc increment in the Iraqi environment and, therefore, the increasing health hazards in populations. However, there are other remarkable conclusions one can be deduced from this work. The decreasing pH of urine caused increasing uric salts especially amorphous urate and this considered as strong indicator to renal failure in particular and generally in urinary system. Additionally, increasing of acidity of urine, which accompanying with activity of Uc, leads to increase the incidence rate of a bladder cancer. Finally, the home location 2, 8 and 9 suffer from highest Uc pollution.

References

[1] Carmack A J, Soloway M S 2006 The diagnosis and staging of bladder cancer from RBCs to TURs. *Urology* **67** (3suppl 1) pp 3-10.https://doi.org/10.1016/j.urology.2006.01.026

[2] Smith A, Balar AV, Milowsky MI, Chen RC, Chapter 83: Bladder Cancer. In: Niederhuber JE, Armitage JO, Dorshow JH, Kastan MB, Tepper JE, eds 2014.. Abeloff's Clinical Oncology. 5th ed. (Philadelphia: Elsevier)

[3] American Cancer Society 2016 *Cancer Facts and Figures 2016* (250Willams street, NW, Atlanta, Ga: American Cancer Society Inc.) GA 30303-1002.

[4] United Nations Scientific Committee on the Effects of Atomic Radiation (UNSCEAR) 2000 *The Sources and Effects of Ionization Radiations* **vol.1** (United Nation, New York: Report to the general assembly with scientific annexes) pp126-127.

[5] International Atomic Energy Agency 2006 *environmental consequences of the Chernobyl accident and their remediation: twenty years of experience* report of the Chernobyl forum expert group 'environment' Radiological assessment reports series (Austria : IAEA STI/PUB/ 1239) ISSN1020- 6566

[6] Durakovi A 1999 *Medical Effects of Internal Contamination with Uranium* (Department of nuclear medicine, Georgetown University School of Medicine, Washington D.C., USA) **40** No.1.

[7] Cameron J R and Skofronick J G 1992 *Medical Physics* (New York, United State of American: John Wiley and sons Inc.)

[8] U.S. Environmental Protection Agency 2005, "Radiation Information: Fact sheets series on Ionizing Radiation **2**

[9] Annual report to Congress Federally Sponsored Research on Gulf War Veterans' Illnesses for 2007. Department of veterans' affairs.

[10] Diarmid Mc M A, Keogh J P , Hooper F J , Mc Phaul K, Squib K, Kane R, Di Pino R, Kabat M, Kaup B, Anderson L, Hoover D, Brown L, Hamilton M, Jacobson-Kraum D, Burrows B and Walsh M 2000 Health effects of depleted uranium on exposed gulf war veterans *Environ. Res.* **82** pp 168–180

[11] Hindin R, Brugge D and Panikkar B 2005, Teratogenicity of depleted uranium aerosols: a review from an epidemiological perspective *Environmental Health*: a global access science source, /http://www.ehjournal.net/content/4/1/17.

[12] Xiongxin Dai 2011 Isotopic uranium analysis in urine samples by alpha spectrometry *J RadioanalNucl Chem*.**289**:595–600

[13] Haider Aabood Neaamah M 2010 Statistical study of depleted uranium effectiveness in increasing cancer disease in Iraq *Journal of Baghdad college of economic sciences university* **no.24** 327-340

[14] Huda Saleh Mahdy Aamash and Bahaa Al-deen Maarouf March 2002 Baghdad, Iraq *Using depleted uranium weaponry against Iraq and legal confronting it at international level* (Baghdad: decisions of scientific conference about effectiveness of using depleted uranium in human and environment in Iraq) part **1** p 32

[15] Awqati Naira A, Ali Mohamed M, Al-Ward Nada J, Majeed Faiza A, Khawla Salman, Mahdi Al-Alak and Naeema Al-Gasseer 2009 Causes and differentials of childhood mortality in Iraq. *Bio Med Central BMC Pediatrics* **9**:40. doi: 10.1186/1471-2431-9-40

[16] Samira Alaani, Muhammed Tafash, Christopher Busby, Malak Hamdan and Eleonore Blaurock-Busch 2011 Uranium and other contaminants in hair from the parents of children with congenital anomalies in Fallujah Iraq. *Conflict and Health* **5** pp 1-15

[17] Diehle P, Fahey D, Bertell R, Robicheau D, Bristow R, Arbuthnot F and Van der kreur H1999 *Depleted uranium: A post-war disaster for environmental and health* (Laka Foundation, , Amsterdam, Netherland: IAEA international nuclear information system) INIS **vol. 30** INIS issue 46

[18] Zarkadas Ch, Karydas A G and Paradellis T 2001 Determination of uranium in human urine by total reflection X-ray fluorescence *Spectrochim. Acta* Part B **56** p 2505-11

[19] Benton E V and Makeover S W S 2001 Proc. of the 20th Int. Conf. Radiation Measurement on Nucl. Tracks in Solids vol 34 pp 1-6

[20] Murray L 1994 *Understanding Radioactive Waste4th Edition* (Columbus, Ohio: Battelle Press)

[21] Fisenne I M, Perry P M and Harley N H 1988 Uranium in humans *Radiation Protection Dosimetry* **24** pp127-131

[22] Wrenn M, Durbin P, Howard B, Lipzstein J, Rundo J, Sill E and Willis D 1985 Metabolism of ingested U and Ra. *Health Physics* **vol. 48**, p 601-603

[23] Berlin M and Rudell B 1986 *Uranium. In: Handbook on the Toxicology of Metals*, (Amsterdam: Elsevier Science Publishers) pp 623-637

[24] AllainP, Berre S, Preamble-Cabic A and Mauras Y 1991 Investigation of the direct determination of uranium in plasma and urine by inductively coupled plasma mass spectrometry *Analytica Chimica Acta* **251** pp 183–185

[25] Chakarvarti S K, NandLal and Nagpaul K K 1980 Content of uranium trace in urine determined from fission track etch technique *international Journal of Applied Radiation & Isotopes* **vol.31**pp 793 – 795

[26] Jones A D, Miller B G, Walker S, Anderson J, Colvin A P, Hutchison P A and Soutar C A2007 A normative value pilot study: Levels of uranium in urine samples from UK civilians *Environmental Research* **104** pp 216–223.

[27] WHO Department of Protection of the Human Environment 2001 *Depleted Uranium: Sources, Exposure and Health Effects* (Geneva, Switzerland: World Health Organization)

[28] Snyder W S and Cook M J 1975 *Report of Task Group on Reference Man* (Oxford: International Commission on Radiological Protection) ICRP Publication 23

[29] Yu K N and Mao S Y 1999 *Health Phys.***77** p 686

[30] Dang H S, Pullat V R and Pullai K C 1992 *J. Radioanal. Nucl. Chem.***162** p 163

[31] Chemchek Instrumentsinc.2006 Operation Manual of Manual and automatic Kinetic phosphorescence analyzer KPA-11 richland, W A 99354 U.S.A.

[32] Statistical Analysis System SAS. 2012 *User's Guide. Statistical.* Version 9.1th ed. SAS. Inst. Inc. Cary. N.C. USA.

IOP Conf. Series: Journal of Physics: Conf. Series **1032** (2018) 012027 doi:10.1088/1742-6596/1032/1/012027

Reflectivity effect of the PS on solar cell efficiency

M.S. Mohammed*, R. A. Shlaga**

* Department of Applied Sciences, the University of Technology, Baghdad, Iraq

** E-mail address: ruqaya12322@gmail.com

Abstract. All articles *must* contain an abstract. The abstract text should be formatted using 10 point Times or Times New Roman and indented 25 mm from the left margin. Leave 10 mm space after the abstract before you begin the main text of your article, starting on the same page as the abstract. The abstract should give readers concise information about the content of the article and indicate the main results obtained and conclusions drawn. The abstract is not part of the text and should be complete in itself; no table numbers, figure numbers, references or displayed mathematical expressions should be included. It should be suitable for direct inclusion in abstracting services and should not normally exceed 200 words in a single paragraph. Since contemporary information-retrieval systems rely heavily on the content of titles and abstracts to identify relevant articles in literature searches, great care should be taken in constructing both.

1. Introduction

The reduction of silicon through the layer thickness decreasing accompanied by low surface reflection is one of the immediate research challenges in photovoltaic applications [1]. First essential goal for photovoltaic power production is to reach grid parity; two ways exist to reach this goal; higher efficiency and lower production cost. Porous silicon (PS) has more possible applications in photovoltaic and hold potential to contribute to desired development [2]. The used of macro porous silicon is one of multi chooses to prepared thin layer of crystalline silicon with low reflectivity [3]. Porous silicon (PS) is silicon with pores inserted into its complex structure in macroscale sized like sponge structures which are used as high potential anti reflection material due to their enhanced absorption properties. It is fabricated when crystalline silicon wafers are etched photo-electrochemically in hydrofluoric acid HF based electrolyte sol this method can be controlled through several parameters such as current density, power density, etching time and Hf concentration etc. [4]. It is a promising material due to the excellent optical, mechanical, thermal properties, chemical stability and the low cost [5] therefore, has a wide range of potential application like photovoltaic devises, chemical sensors, biological sensors etc. [6-10]. Laser assisted etching (LAE) an easy method and no thermal effect, where light or laser illuminated the silicon electrode during the anodization procedure. This illumination leads up to the creation of electron-hole pairs at the top layer due to light absorption, as a result reduction in sizes [11]. PS shows interesting characteristics like low refractive index, good light trapping thereby decreased reflection losses of solar cells; direct band gap, variable reflectivity, randomized morphological structure, and blue photoluminescence make this substance to be interest electric in photo detector applications [12, 13]. The main objective of this work is to study the effect of different illumination of PS formed on solar cell; which plays a significant role in controlling the porous morphology and considered as an important feature for nanostructured anti reflection solar cells.

2. Experimental

PS samples were synthesized via laser assisted etching of (n-type) silicon (100) orientation with a resistivity (10Ω.cm) at constant etching time (8min), power density (20mW/cm2) and current density (10) mA/cm2. The etching method of (n-type) silicon carried out in Teflon cell which does not react with HF showing very offensive nature. The experimental setup is illustrated in figure (1) [14].

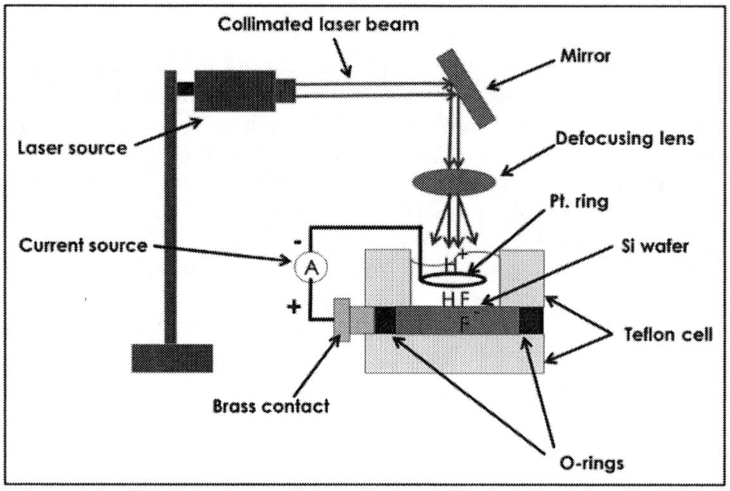

Figure 1. Illustrate experimental setup for laser assisted method.

Silicon acts as the anode to remove the electrons from solution and Platinum acts as cathode to provide electrons to the solution. The samples were immersing in (25%) concentration mixture by (HF) acid to ethanol (C2H5OH) (1:1) rate. Ethanol adding to HF to reduce the surface tension of hydrogen bubbles. Thereby it allows the hydrogen gas formed through the reaction to escape and prevents sticking to the etching surface and improves the homogeneity of porous layer [15]. The setup included a (DC) power supply, AVO meter and illumination source (red and green laser diode with wavelength (645nm and 532nm) respectively. The resulting PS layer was studied via scanning electron microscope (SEM) (Tescan VEGA 3 SB), used to weighting measurement for porosity, the Sartorius (BL210S) digital steelyard with reliability of (10-4gm) instrument. reflectance spectra of prepared samples were recorded TF Prop Spectroscopic reflectometer (SR UV-VIS) series, and He-Cd laser was used as a source of illumination, spectra were collected in the wavelength range 400-700nm.

3. Results and discussion

The change of illumination wavelengths was studied based on the analysis of the SEM images. Used fixed each of the illumination intensity 20mW/cm2, etching time (8min) and current density (10mA/cm2) to prepared porous silicon structures. Figure (2) represents the surface morphological with different illumination wavelengths (green 532nm and red 645nm) laser; the following notes can be summarized based on the SEM images, which randomly distributed pores are observed on the silicon surface. Several pores had star fall, connected and circular shape. Figure (2) illustrates the SEM images (top view) of PS, from this images can be notice that the pore width increases with decreasing of illumination wavelength which attributed to increasing of holes' number on Si surface with short wavelength lead to preferential dissolution between nearest-neighbour pores, thereby promoting the pore-pore overlap. Pores shape in figure (2a) which appearances in non-uniform distribution with thick walls were evidently on the PS surface in multilayer. While figure (2b) illustrates PS layer prepared under illumination wavelength (645nm), star fall and connected pores were appear in semi uniform distributed shaped, this different of PS structures for each wavelengths; due to the different surface absorption of illumination wavelength and to non-uniform of power density distribution of illumination (attributed to that laser beam intensity will decreased from center to its edges (Gaussian beam)), which lead to non-uniform hole-photo generation consequently. The irregular distribution of pores could enhance the photo conversion as a result of the widened PL peaks.

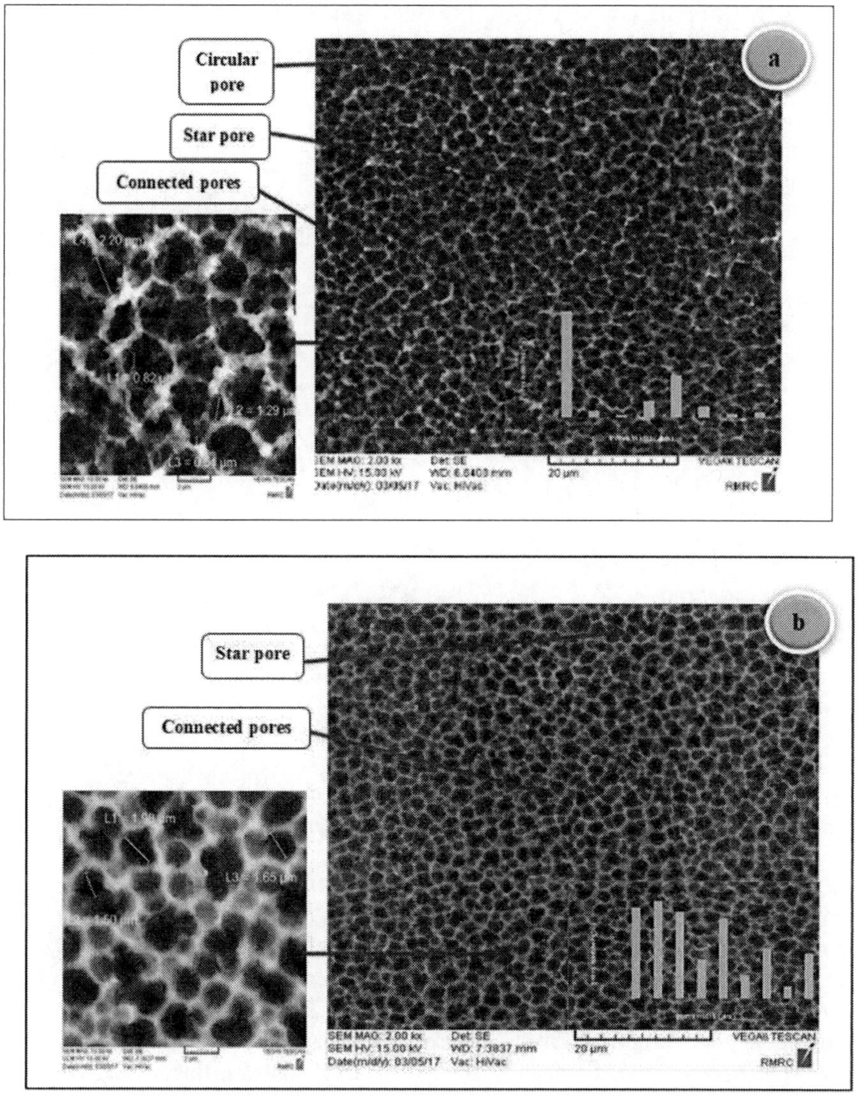

Figure 2. SEM images of PS layer formed by used different illumination of lasers a) green laser (532nm) and b) red laser (645nm).

Pores shape in fig (2a) which appearances in non-uniform distribution with thick walls were evidently on the PS surface in multilayer. While figure (2b) illustrates PS layer prepared under illumination wavelength (645nm), star fall and connected pores were appear in semi uniform distributed shaped, this different of PS structures for each wavelengths; due to the different surface absorption of illumination wavelength and to non-uniform of power density distribution of illumination (attributed to that laser beam intensity will decreased from centred to its edges (Gaussian beam)), which lead to non-uniform hole-photo generation consequently. The irregular distribution of pores could enhance the photo conversion as a result of the widened PL peaks.

The peak position of photoluminescence spectrum for samples prepared for different lasers; intensity is regularly depending on the quantum size influence in nanostructures because porous silicon is included of Si nano-wires and high energies quantum dots. Highest value of photoluminescence as a result of porosity increased, while the blue shifting of peak position is due to the decreasing of nanocrystallite size for silicon, figure (3) displayed that: The PL intensity has been found to appear at (421 and 419 nm) respectively, indicating that there is a blue shift of the PL intensity and can be note it increased. This behaviour illustrates changed total volume of the nanocrystallites on the surface of PS [16]. The visible range of PL emission and a blue shift with the porosity of PS are strong evidence for a quantum confinement in PS.

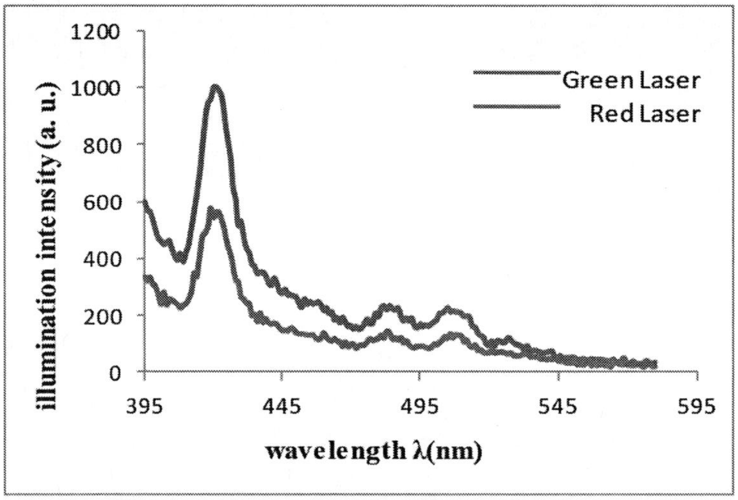

Figure 3. PL spectra of PS samples prepared in different illumination wavelengths (532 and 645) nm at fixed each power density (20 mW/cm^2), current density (10mA/cm^2) and etching time (8min).

The reflectivity was studied for both illumination wavelengths (532nm) and (645nm) it relies on the surface topography which mean porosity of silicon, it decreased with increased porosity figure (4) illustrates that: The variation of optical properties (reflectivity and refractive index) are generally attributed to porous layer nature, lowest effective reflectance was obtained from PS layer prepared under red illumination wavelength (645nm), figure 4(a) illustrate the reflectivity that decreased from (1.4%) to (0.66%), whereas varied from (1.28%) to (0.64%) in (b), this varied depending on porosity and can be calculated easily via using Looyonga equation (1) [17]:

$$n^{2/3}{}_{PS} = (1-p)\, n^{2/3}{}_{Si} + p\, n^{2/3}{}_{air} \qquad (1)$$

Where n_{PS} refractive index of porous silicon, p porosity of porous silicon, n_{Si} refractive index of bulk silicon (3.4) and n_{air} refractive index of air (1.0003), can be determined using reflection spectrum as viewed and using eq. (2) [18], we found the porosity value of green illumination wavelength reach to (88.9%) and for red illumination wavelength about (89.3%).

$$R = (n_{PS} - n_{air}/n_{PS} + n_{air})^2 \qquad (2)$$

Where R: is porous silicon reflectivity at (500nm) So, the refractive index value of bulk silicon can be noted that decreased from (3.4) to (1.217), (1.209) for green and red illumination wavelength respectively this decreasing relies on natural structure of PS layer and controlled by the pores respected to high porosity [13].

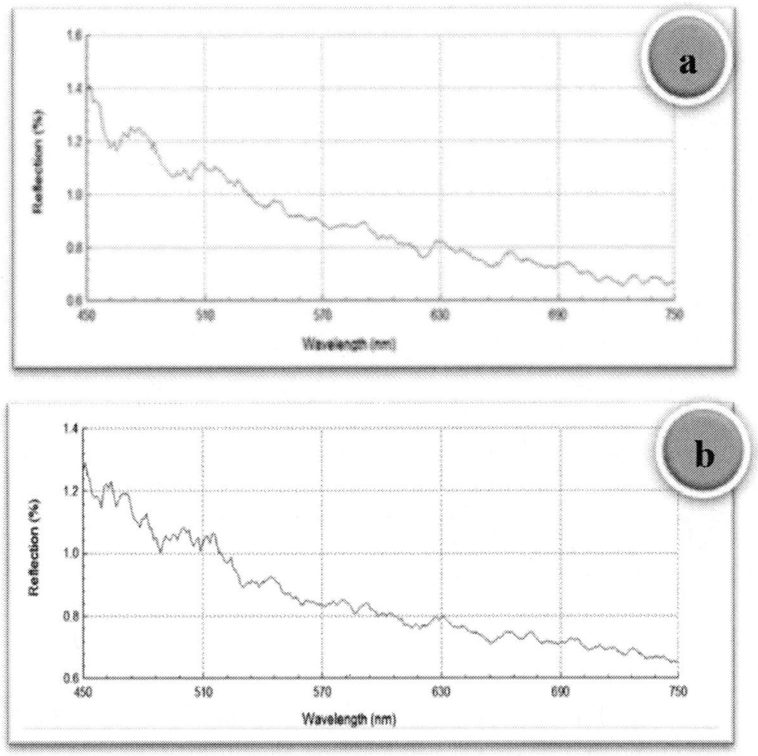

Figure 4 a, b. The reflectivity of PS prepared at different illumination wavelengths a) Green laser (532nm). b) Red laser (645nm).

The peak positions of the reflectance are compared with other reports work and have found good results with enhancement of refractive indices [19, 20]. The effective values of porosity on the reflectivity and refractive index due to fact of effective dielectric constant of porous material system caused decreasing in reflectivity as in antireflection coating for wavelength in range (400-800) nm that is suitable for solar cell applications. The reflectivity effect on solar cell efficiency (η) can be displayed when V_{oc} equal to (595mV), FF about (0.75) and supposed the J_{sc} equal to J_{ph} which can be determined via eq. (3) [21]:

$$J_{ph} = \frac{P\,(1-R)\eta * q}{hv} \qquad (3)$$

Where J_{Ph} Where: (P_{in}) is incident power in (Watt), (R) the reflectivity of PS layer, (η) quantum capability, (v) laser frequency, (h) plank fixed and (q) electric charge., to calculate the efficiency of solar cell by reflectivity effect at (500) nm region for each (532 and 645) nm illumination wavelengths used eq. (4):

$$\eta\,\% = (J_{ph}\, V_{oc}\, FF)/\, P_{in} \qquad (4)$$

· The efficiency values of PS ARC prepared under (532nm) wavelength reach to (14.25%) larger than red illumination wavelength about (14.23%), this mean increasing the light conversion efficiency of solar cell by tripping of the incident radiation, thereby enhancement the photovoltaic properties.

4. Conclusion

When preparing porous silicon by laser assisted etching for different illumination wavelengths results can be concluded by: The surface topographical properties of porous silicon can be easily and effectively controlled by adjacent the laser wavelength, where the morphology of PS prepared at red laser (645nm) which more regularity than the green laser (532nm). The photoluminescence of porous silicon revealed three peaks with blue shift spectroscopy, that is relates with larger pore sizes and hence larger energy gaps. Enhancement optical properties (reflection) of porous silicon layers have been carried out; therefore, reflectivity of PS much lower than Si reaches to less than 1% we found that the reflectivity effect on solar cell efficiency about (14.25%).

5. Acknowledgment:

We wish to thank Professor M. Ahadian for his assistance with the PL measurements, Razi centre for SEM measurements and to the solar energy department in the ministry of science and technology of Iraq for reflectivity measurements. Authors wishing to acknowledge assistance or encouragement from colleagues, special work by technical staff or financial support from organizations should do so in an unnumbered Acknowledgments section immediately following the last numbered section of the paper.

References

[1] K. W. Kolasnski, *in Recent Research Advances in Applied Physics*, edited by S. G. Pandalai 2004, (Transworld Research Network, Kerala, India, Vol. 7, pp. 267-292.

[2] R. Bilyalov, R. Ludemann, W. Wettling, L. Staalmans and c. Levy-Clemet,2000, sol. Energy Mater. Sol. Cell Vol. 60, No.391.

[3] Brendel R. and Ernst M., "*Macroporous Si as an absorber for thin –film solar cell*", 2010, Phys. Status Solid RRL pp. 40-42.

[4] Narayanan, Purnima. *Photoelectrochemical Etching of Isolated, High Aspect Ratio Microstructures in n-Type Silicon (100)*. Msc. Thesis, Department of Chemical Engineering by, University of Mumbai, (2011).

[5] U. M. Nayef and M. W. Muayad, "*Typical of Morphological Properties of Porous Silicon*", International Journal of Basic and Applied Sciences IJBAS-IJENS, 2013, Vol.13, No.02.

[6] J. Torres, H.M Martinez, J.E. Alfonso and L.D. Lopez C, "*Optoelectronic study in porous silicon thin films*", 2008, *Microelectronics Journal*, 39, pp.482-484.

[7] P. Vitanov, M. Kamenova, N. Tyutyundzhiev and V. Gantcheva, "*Application of porous silicon layer in photovoltaic devices*", 1997, Thin solid films, Vol. 297, pp.299-303.

IOP Conf. Series: Journal of Physics: Conf. Series **1032** (2018) 012027 doi:10.1088/1742-6596/1032/1/012027

[8] W. A. Badawy, "*Porous Silicon Modified Photovoltaic Junctions: An Approach to High-Efficiency Solar Cells*", 2007, AIP Conf. Proc. February Vol. 14, No.8, pp.29-35.

[9] Ph. M. Fauchet and S. Member., "*Progress Toward Nanoscale Silicon Light Emitters*",1998, IEEE journal of selected topics in quantum electronics, Vol. 4, No. 6, pp.1020-1028.

[10] P. G. Han, H. Wong, Andy H. P. Chan and M. C. Poon, "*Formation Mechanism of Light-emitting Porous Silicon Prepared by Reactive Ions Etching*", 0-7803-3166-4/96/$5.00© 1996 IEEE.

[11] K. Tsujino, M. Matsumura, Y. Nishimoto, "*Texturization of multicrystalline silicon wafers for solar cells by chemical treatment using metallic catalyst*", 2006, Sol. Eng. Mater. & Solar Cells, Vol. 90, pp.100–110.

[12] L. T. Canham, "*Silicon quantum wire array fabrication by electrochemical and chemical dissolution of wafers*", 1990, Appl. Phys. Lett. Vol.57, No.10, pp.1046.

[13] F. Alfeel, F. Awad, I. Alghoraibi and F. Qamar," *Using AFM to Determine the Porosity in Porous Silicon*", 2012, Journal of Materials Science and Engineering, p.p 579-583.

[14] Mohammed A. and H. Shnieshil, "*Characteristics Study of Porous Silicon Produced by Electrochemical Etching technique*", 2013, International Journal of Application or Innovation in Engineering and Management, Vol.2, Issue 9.

[15] Guozheng Wang, Shencheng Fu, Yanjun Gao, Ye Li, Xin Wang, Qingduo Duanmu "*Optimization of macropore silicon morphology etched by photo-electrochemistry.*" Solid-State and Integrated-Circuit Technology, 2008. ICSICT 2008. 9th International Conference on. IEEE, (2008).

[16] Alwan M. Alwan, Adawiya J. Haider and Allaa A. Jabbar. "*Modifying and fine controlling of sliver nanoparticle nucleation sites and SERS performance by double silicon etching process*", 2017, Plasmonics, pp. 1-12.

[17] Khaldum A. Salman, Z. Hassan and Khalid Omar, "effect of silicon porosity on solar cell efficiency", 2012, International Journal of Electrochemical science, Vol. 7, pp. 376-386.

[18] Andrea Edit Pap, Krisztian Kordas, Jouko Vahakangas, Antti Uisimaki, Seppo Leppavuori, Laurent Pilon, and Sandor Szatmari. "*Optical properties of porous silicon. Part III: comparison of experimental and theoretical results.*",2006, Optical materials, Vol. 28, pp. 506-513.

[19] R.S. Dubey, "Electrochemical fabrication of porous silicon structures for solar cell ", 2013, Nanoscinence and Nanoengineering, Vol. (1), pp. 36-40.

[20] R.Ludemann, M.Damiani, and A. Rohatgi,"novel processing of solar cells with porous silicon texturing", IEEE 2000, conference.

[21] Alwan M. Alwan, Suaad M. Ali, "Gradient-Porosity Porous Silicon (GPSi) as Anti-Reflection Coating in Solar Cell Applications", 2015, J. Eng. and Technology, University of Technology, Baghdad, Iraq, Vol. 33, Part (B), No.1.

IOP Conf. Series: Journal of Physics: Conf. Series **1032** (2018) 012028 doi:10.1088/1742-6596/1032/1/012028

Effect of Thermal Variations on Some Nanostructural and Optical Characterizations of Element Oxides Compound (Y_2O_3/Fe_2O_3) Prepared Using Laser Ablation

[1*]Noor al-huda Yaareb Kraidi , [1]Saif M.Al-Shirefy , [2]Ayad F. Al-Kaim , [1]Rafea Tuama Ahmed and [1]Mohammad H. K. AL-Mamoori

[1]College of Science for Women/Laser Physics Department /University of Babylon/Iraq
[2]College of Science for Women /Chemistry Department/University of Babylon/Iraq
*Corresponding Email: nooryarub1993@yahoo.com

Abstract : In this study have been synthesis nano oxide composites (Y_2O_3/Fe_2O_3) are prepared chemically and affected by Laser Neodymium –YAG ablation by energy (200 mJ), and number of pulses (200 pulses) at two different temperatures (4°C and 65°C) in distilled water . The structural , morphological, and optical properties of synthesized oxides were studied, (XRD) pattern showed that the oxides were highly crystallized in both hot and cold cases with variations in intensity values as they increased in the cold state than in the hot with additional peaks of intermediate compounds [$Y(OH)_3$; PLA], monoclinic and cubic and the two phases (ε-Fe_2O_3 ; β-Fe_2O_3).The (AFM) test indicate the roughness values that increased in the cold state than in the hot state. The results of the (SEM) showed a close approximation in the hot & cold cases in cubic forms, conical sticks (like nanorods , whiskers) as a result of recrystallization. noting that the value of absorbance increased in the cold than in the case of hot. The sample was manufactured in a solar cell ,by obtaining a I-V curve under the dark and its illumination. The results showed a significant change in the performance of the solar cell at the values of (Im=42 mA; Vm=0.21 V), when the Iron oxide overlaps with the yttrium oxide with the best conversion efficiency of 7%.

Keywords: laser ablation, Y_2O_3/Fe_2O_3 nanostructure, optical & structural properties

1. Introduction

Quantitative effects begin to control the behavior of the material in the nanospace, especially in the lower end, affecting the properties of electrical, magnetic and optical materials [1,2].Nanomaterials for semiconductors are one of the richest classes of nanocompounds . Where it has played an important role in advancing research in the field of nanoscience and nanotechnology, leading to a new classes of semiconductor materials. Normally the current flow in conductors are due only to electrons but for the semiconductor material is caused by both the electron as well as any gap (electron-hole) [3].This field is expected to open up new fields for science and technology in the preparation and application of nanostructure for metal oxides, which are one of the classes belonging to semiconductor nanomaterials and nanotechnology. Material oxides have proved to be a very promising future for a variety of practical applications. The good thermal and chemical stability of these inorganic materials enables them to be widely used in the fields of chemistry, physics and materials science [4]. Yttrium oxide is the most common yttrium compound, where the latter is a three-valence metal transition. Three electrons share their equivalence and yttrium oxide is an example. (Y_2O_3) has attracted a lot of attention recently because of its interest in several physical properties, such as its crystalline stability up to 2325 °C (where its melting point is 2450 °C). The

IOP Conf. Series: Journal of Physics: Conf. Series **1032** (2018) 012028 doi:10.1088/1742-6596/1032/1/012028

crystalline structure is cubic but there may also be hexagonal and Monoclonal. The high thermal conductivity is up to (0.131Wcm^{-1} K^{-1}), the large optical gap is almost (5.5 ~ eV), the relatively high electrical insulation constant is in the range (14-18), and the wide range of transmittance (8-0.29) µm. The refractive index is almost high (n=2) where it is perfectly suitable for waveguide applications and is used as a very good protective behavior when coated in interactive [5-6].

Iron oxides are widely used as inexpensive and durable dyes in paint, and concrete. Colors are usually available at the "earthy" end of yellow, orange, red, brown, black scale. Different stages of iron oxide (Fe_2O_3) are alpha (α-Fe_2O_3), beta (β-Fe_2O_3), gamma (γ-Fe_2O_3) and epsilon (ε-Fe_2O_3). It is clear above that iron oxides are metal-related metal oxides, because they possess multiple oxidation states that can be specifically tuned to a particular application [7].

The aim of the research is to prepare nanograins of materials of their oxides by combining the chemical method with laser fragmentation. Also synthesis of nanoparticles with the mixing of semiconductors for use in energy application.

2. Experimental

The monoxides materials (Y_2O_3/Fe_2O_3) were prepared by means of several processes, the first thermal ones to ensure the mixture. The process is by adding (4gm) of Y_2O_3 to (3gm) of NaOH and adding it to (2gm) of $FeCl_3$, Each of these substances dissolves in (100 ml) distilled water with continuous magnetic stirring for 24 hours, Then treat it thermally to produce the mixture as in Fig.(1).

Figure 1. An image the resulting material of the compound

Paragraph (1-1) describes the chemical path of the basic chemicals and mixes them to prepare for the target of the laser bombing and the outcome of this last operation.

$$Y_2O_3 + NaOH + FeCl_3 + H_2O \longrightarrow Y(OH)_3 + Fe(OH)_3$$

Laser

$$OH - OH - Y - O - Fe - OH\text{-}OH \xrightarrow{H_2O} Y_2O_3 / Fe_2O_3$$

Nano Composites (1-1)

The second process is laser ablation in liquids. As it is the main process to synthesize the required oxides. The basic process of laser ablation in the liquid phase is the reaction of the laser beam with the surface of the hard target that causes evaporation of the target and a small amount of liquid surrounding it. Chemical reactions between the removed parts and molecules in the liquid also occur as scattered fragments emitted electronically and transiently, and the reaction products are

IOP Conf. Series: Journal of Physics: Conf. Series **1032** (2018) 012028 doi:10.1088/1742-6596/1032/1/012028

nanoparticles made up of target atoms and surrounding liquid, which is a collide solution in the liquid. As a result of the accumulation of these particles in the surrounding liquid (composed of a suspended solution) and the sustainability of the interaction with the laser beam leads to additional changes in the composition, size and shape of nanoparticles.

The plasma plume is composed of different thermal layers, atomic ionization and plasma plume formed by the laser beam of the material's target. Four groups of liquid and metal species can be identified due to different temperature, pressure and composition. In the process of laser ablation in the liquid, the plasma is confined to the liquid and produces a shock wave that travels through the target. When the target is absorbed for a part of the laser pulse it leads to a constant increment of evaporation of the parts of the solid target. In addition, the shock wave will produce additional pressure in the plasma caused by the laser. The plasma generated by the laser for high temperatures, high pressure and high density is generated at the surface of the solid-liquid surface when the surface of the solid is irradiated in the liquid by the laser pulse. In the case of a liquid solution up to 2ml as shown in Figure(2), then the laser pulse is bombarded at two different temperatures, the first high $65^{\circ}C$ and the second low $4^{\circ}C$ where the three samples of non-dried (solid solution) baker and immersed in (2ml) of distilled water of the same solution with nitrogen gas for 5 minutes to stir bubbles and uniformity of materials.

The morphological characteristics of nano oxides materials synthesized by the (AFM) and (SEM) systems. The structural properties were examined by the XRD in a range of 10-80 degrees. The structural properties are an important feature of the study of the crystalline structure of the oxides of (Fe_2O_3) prepared with the help of Yttrium oxide.

Figure 2. Laser ablation system on liquid contained compound materials

3. Results and Discussion

By Based the chemical formula (1-1), Figure 3 shows the X-ray diffraction pattern of the sample of the composited nano oxides (Y_2O_3/Fe_2O_3) prepared first. In a hydro-thermal method, a medium was recorded to obtain a solid solution in the medium-base liquidity phase to reach a solution containing compound and deposition material formed by the dissociation of compound Y_2O_3 / Fe_2O_3 compound and then to the laser ablation method in distilled water under hot (65 °C) and Cold (4 °C) during laser pulses with the JCPDF universal card to verify the results and identify the material correctly . In the hot case, the iron oxide material when tested for X-rays is more crystalline than the colder case. Where the angles (32.8925°) (13.7274°) (22.1097°) (29.4440°)(31.8073°) at the crystalline planes (-420) (200) (220) (-411) (003)respectively, which belongs to the phase ε- Fe_2O_3 within card PCPFD # 16-0653. Therefore, the form of crystalline structure Orthorhombic system is consistent with the researchers [8] . There is also a small percentage of phase β-Fe_2O_3 at angles (24.8912°) (48.6529°) (19.8832°) for crystalline planes (2013) (017) (104) respectively, for card PCPDF # 40-1139.

In addition to the yttrium oxide, the angles (29.2443°) (31.1180°) (40.3311°) (47.1729°) for crystalline planes (002),(101) , (102) ,(110) respectively corresponding to the world card (20-1412) shown, along with several corners of the card PCPDF # 43-0661, Note that ranges of the crystal size according to Scherrer formula as shown in Table 1 are from (32-80.50) nm in hot case.In the cold case, iron oxide is slightly less than hot at angles (13.2893) (13.6952) (29.5040) (32.9170) at the crystalline plane (200) (200) (-411) (-420) respectively, according to the international card PCPDF # 16-0653. Also corresponds to the card PCPDF # 40-1139 for stage β-Fe_2O_3 at angles (21.4563) (31.0820) (48.6588) degrees at planes (105) (113) (2013) respectively And also a few percentages.
As for Y_2O_3 in the cold state, the peaks are at the angles (29.33042) (40.3361)degrees at the planes (002) (102), respectively. In the cold case, It was noted that $Fe_6Cl_{2-x}(OH)_{12+x}$ is Iron Chloride Hydroxide at the angle (11.4264°) and crystalline plane (003) according to the card (PCPFD # 49-0095) and the iron chloride Fe $Cl_3.7H_2O_2$ According to the card PCPFD #1-0132 . Note that the crystal size according to Scherer formula as in Table 1 ranges from (25.20-57.10) nm.These differences between the peaks lead to the differences in crystalline of the monoxides in the two cases the hot and colds, because of the aggregation characteristics or coalescence property of materials that appear in hot case more than cold case. The (Y_2O_3/Fe_2O_3) compound give a new manner related to the intensity of peaks especially at (200) plan. The crystal size was calculated from Scherer's equation[9].

$$C.S = \frac{0.9\,\lambda}{\beta \cos\theta} \quad (1)$$

Where : C.S is the crystal size , β is the full width at half maximum (FWHM) (in radians) of the peak , θ is the Bragg angle, (λ=1.5406 Å) X-ray wavelength .It was noted that the size have the large crystal in the case hot than in the cold state, because of coalescence or adhesive of nanograins to form greater grains The values of grain sizes of the compound (Y_2O_3/Fe_2O_3) in two thermal cases (hot and cold) , was shown in table (1) The crystalline size of the sample is somewhat large and within the nanoscale of the hot case and smaller than the cold state Due to the effect of the laser,and because here the crystal size of the required materials was calculated only regardless of the other materials formed

IOP Conf. Series: Journal of Physics: Conf. Series **1032** (2018) 012028 doi:10.1088/1742-6596/1032/1/012028

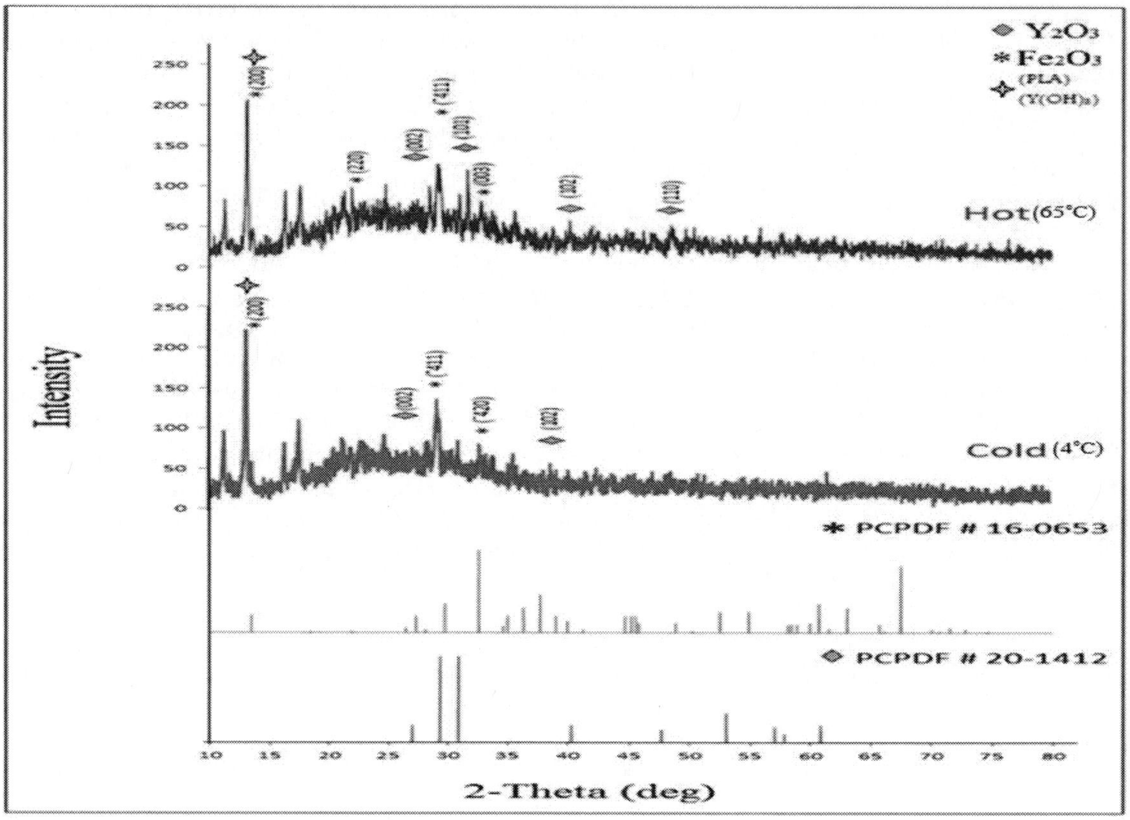

Figure 3. XRD of measured nano-oxide Y_2O_3 / Fe_2O_3 compound prepared composed by laser ablation at two different temperatures.

Table 1. The values of grain sizes of the compound (Y_2O_3/Fe_2O_3) in two thermal cases (hot and cold).

Sample	Thermal Case	2-theta (deg)	FWHM (deg)	C.S (nm)
	HOT	13.3,22.1	0.19 , 0.109	43.6 , 80.5
		31.8 , 32.8	0.12 , 0.2	71.2 , 43.2
		29.24	0.26	32
(Y_2O_3/Fe_2O_3)		40.33	0.12	70.7
	COLD	13.28	0.272	30.9
		29.30	0.346	25.2
		29.5	0.15	57.1

When examining the surface topography of nano-oxide solutions prepared by laser ablation in distilled water on glass using an AFM microscope with high surface-imaging power as shown in Fig. 4 showing AFM images of binary (a) , (b) and the accumulation statistical distribution (c) for hot noting the cumulative statistical distribution of the GCD, the surface roughness (Ra) and the square root rate (Rq). The highest cumulative percentage was 7%. The shape of the grain is spherical and the grain volumes are between 40 and 160nm. The thermal effect on the grain led to adhesions between the grains, which made the values of multiple grain sizes and different diameters. The increase in granular size, square root rate and surface roughness were observed due to the occurrence of covalent bonds of Y_2O_3 and Fe_2O_3 by heat factors (high and low temperature in the two preparation cases) and pressure in the hydrothermal process of the solid solution with the resulting compounds (In hot and cold conditions) through the two compounds ($Y(OH)_3$) and (PLA). Thus, a change in the molecule of the mixture will occur with a combined hybridization of oxidize as a result of the analytical medium. As well as the aggregation status of nanoparticles due to inter-particle intersections. The change in granular volumes due to the breakdown of iron oxide (Fe_2O_3) is due to vapor pressure and chemical reaction products and the association of its components again by laser radiation with saturation in the composition of the superconductor (Y_2O_3/Fe_2O_3) within a crystalline structure as the assembly state increases the granular volume. The granular size of the structure is then increased to a floating majority within the solution.

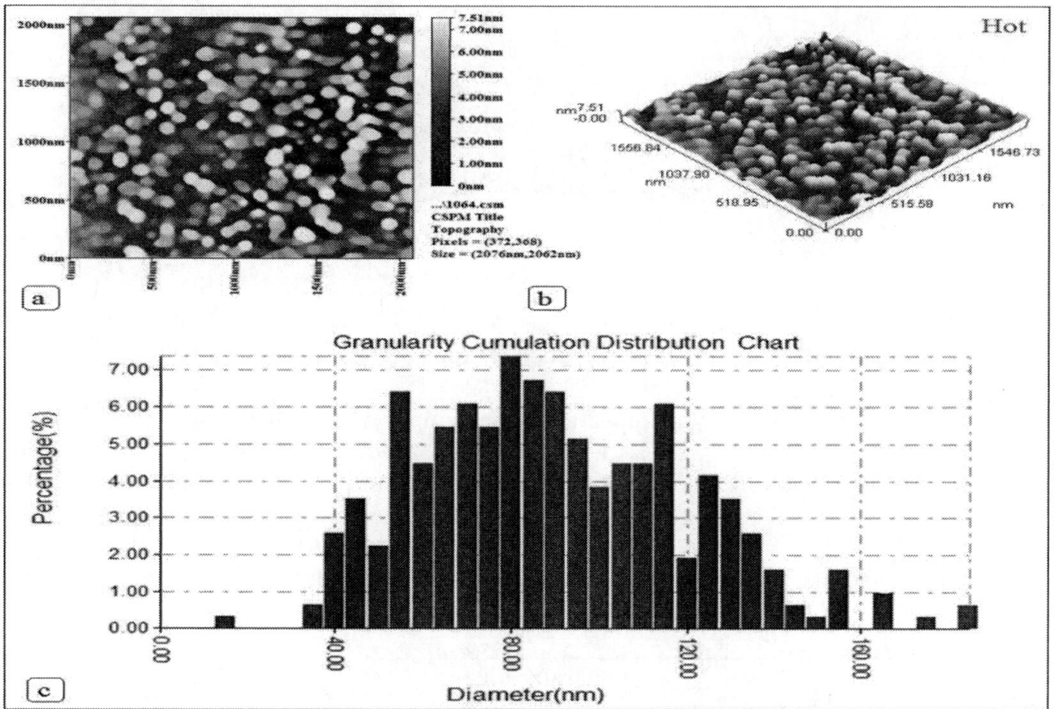

Figure 4. AFM images of nanoparticle oxides for hot sample Y_2O_3/ Fe_2O_3 in (a) two dimensions (b) three dimensions (c) the cumulative statistics

When examining the samples in the cold state of laser ablation and obtaining nano-oxides, it was noted that the crystallization and growth of nanoparticles with specific sizes was 8.5% with 40nm grain diameters where the volumes are close to the proportions as shown in Fig.(5).

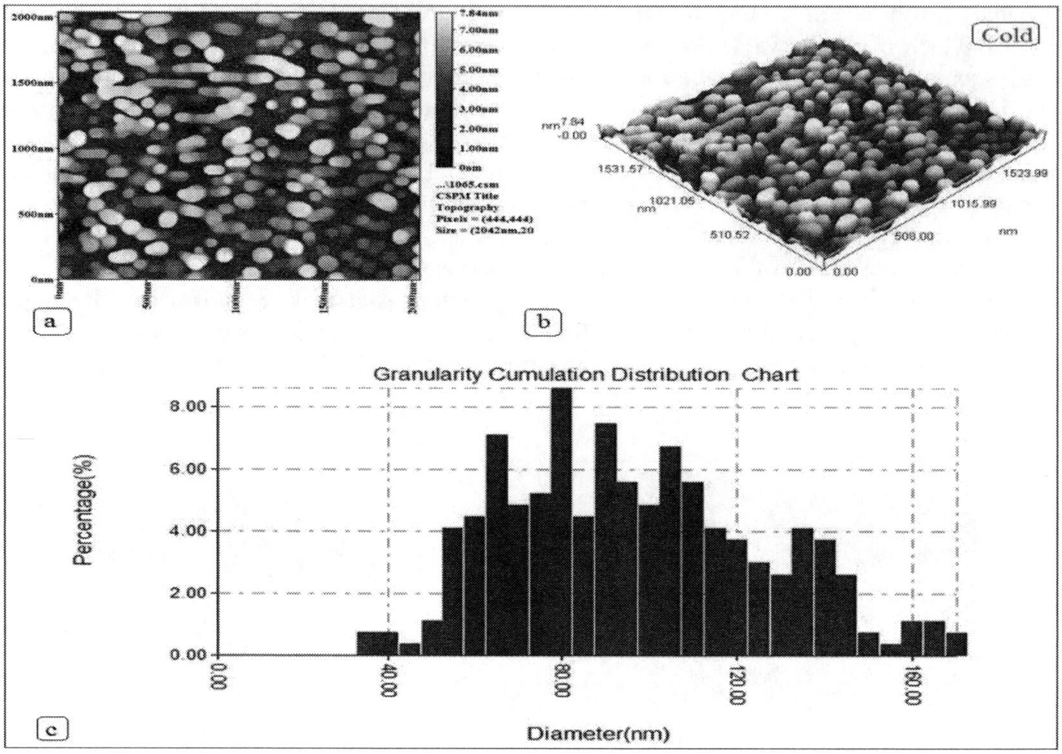

Figure 5. AFM images of nanoparticles of the cold sample (a) in two dimensions (b) three dimensions (c) the accumulation statistical distribution of nanoparticle oxide particles

Table 2 shows average grain size (Any all topography the sample is not only the size of the crystal), percentage, diameter of granule and surface roughness of a sample (Y_2O_3/ Fe_2O_3) As each of the parameters mentioned in the values in the cold state from the warm state because of the impact of low temperature, which led to the difference in granular volumes, thus increasing the roughness.

Table 2. The values of average grain size, percentage, diameter of granule and surface roughness of a sample of Y_2O_3/ Fe_2O_3.

Sample	Thermal Case	Grain size (nm)	G C D	Rq (nm)	Ra (nm)
Y_2O_3/ Fe_2O_3	HOTING	87.52	7.5%	1.86	1.58
	COLDING	93.90	8.6%	2.14	1.84

Fig.6 shows the scanning electron microscopy images of a sample of the nanoparticle oxides of the hot compound Y_2O_3 / Fe_2O_3 at the degrees 4°C and 65°C. The presence of vacancy due to the degradation of the nanotube by the heat, which leads to the possibility of a fusion between the nanoparticles of the compound.

Figure (6) and based on the chemical formula (1), the scanning electron microscopy images of the Y_2O_3 / Fe_2O_3 compound of the hot state at (65)°C show the presence of voids due to the decomposition of the nanotube by heat, The integration of the nanoparticles of the compound and the

IOP Conf. Series: Journal of Physics: Conf. Series **1032** (2018) 012028 doi:10.1088/1742-6596/1032/1/012028

gaps between the nanoparticles and the appearance of the iron oxide by the installation of a clear cube shape.

Figure 6. Scanning Electron microscope image for sample of Y_2O_3/ Fe_2O_3 for hot case .

Fig. 7 shows a change in the crystalline structure by a polycrystalline of the crystalline grains lead to growth, which leads to the formation of crystalline oxides of the crystalline grains around the majority of the grains collected. In addition, the effect of laser ablation affects the cold effect some compact.

In the cold state of 4°C, a change in the crystalline structure shows a convergence between the crystalline granules due to the state of extinction, which leads to the nucleation of crystalline particle oxides, which accumulate around the majority of granules (ie partial shrinkage of the material) The effect of coldness, which helped to form the compounds of precipitation in the solution, which contributed to the development of crystalline volumes and raise their values and make them somewhat monolithic and free of spaces between the granules and thus the emergence of dendritic assembly and a clear nanometer scale and this enhances the results obtained in the examination (AFM), it should be noted that there is a clear crystalline tigers in both hot and cold conditions and in a preferred crystalline direction and is very similar to the whiskers which are formed when moving from the gaseous state to the solid directly, To this advanced growth, which is one of the reasons for crystallization and evidence of some points that represent the beginning of the crystalline development Whiskers , which researchers describe the composition of the cone with advanced growth and this is consistent with what other researchers [10]

Figure 7. SEM for sample of Y_2O_3/ Fe_2O_3 for cold condition.

In order to ascertain the components of the sample obtained by laser excretion in distilled water and in hot and cold cases, and based on the chemical formula (1) and the (SEM) test, the X-ray dispersion analysis measurements in table (3) is high value for each element (Y) at the shell (L) and (O) when the shell is K). The Fe element in both L and K is based on its electronic arrangement (^{26}Fe: $3d^6$ $4s^2$) and the last electron exists in a double state in the secondary orbit (d). In the iron element, there are two peaks with an atomic number (Z =26). The ratios are fairly close in both cold and hot conditions, and very few impurities are present for chlorine (Cl) and sodium (Na). Figure (8) shows the proportions of the elements and their mass and atomic percentages in the hot and cold cases, in addition to the impurities in the sample.

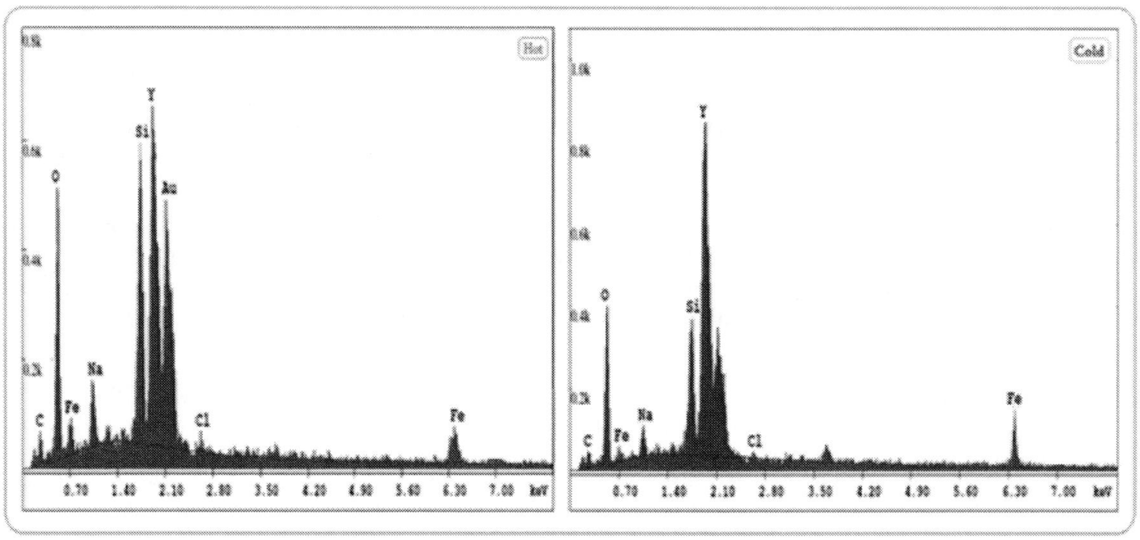

Figure 8. scheme The EDX for the colloid solution (Y_2O_3/Fe_2O_3) .

IOP Conf. Series: Journal of Physics: Conf. Series **1032** (2018) 012028 doi:10.1088/1742-6596/1032/1/012028

Fig. 9 shows the absorption of yttrium oxide after the formation of iron oxide effect on heat and cold cases, noting that the value of absorbance increased in the cold with respect to the case of hot and the reason is that iron contains an abundance of electrons that increase absorption coefficient. In the case of cold operation is contrary to the behavior of the yttrium as the hot operation is higher. This is due to the cold working condition. The nanoparticles are more stable. However, when the temperature is raised, it will lead to disintegration of the material and thus less absorbance. After the wavelength (400 nm) ,the effect of visible radiation as the absorbance is greater than the hot state.

Table 3. The percentage of material elements shown in the EDX test.

Composites	Case	Contents	Elements			
			Y(L)	O(K)	Fe (L)&(K)	Impurities (C, Na, Cl, Ag)
Y_2O_3/Fe_2O_3	Hot	Mass %	38.72	32.11	16.88L 6.19K	6.1
		Atomic %	14	64.52	9.72L 3.56K	8.26
	Cold	Mass %	44.21	27.73	6.84L 2.99K	18.23
		Atomic %	29.29	24.81	13.96L 8.88K	23.06

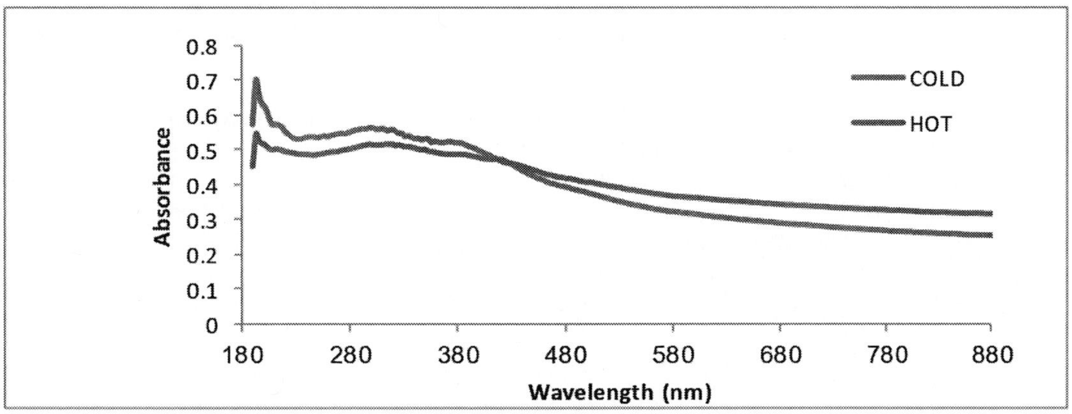

Figure 9. Absorbance as a function of the wavelength of the Y_2O_3/ Fe_2O_3 nanoparticle mixture for hot and cold cases (65 and 4)°C.

It is observed from Fig. 10 that transmittance behaves in the opposite of absorbance behavior because the transmittance is inversely proportional to absorbance. The permeability in the cold state is less than warm due to the crystallization of the oxides due to the low temperature. This may also be due to the formation of levels of Fe and other impurities within the energy gap which lead to increased absorbance as observed in Figure 8 and thus decrease permeability, The spectrum of permeability depends largely on the amount of energy levels, the large role of which is related to the chemical and crystalline structure of the material. An important factor is the presence of surface defects, which

IOP Conf. Series: Journal of Physics: Conf. Series **1032** (2018) 012028 doi:10.1088/1742-6596/1032/1/012028

increase the dispersion of the radiation and thus decrease the permeability of the samples prepared in the cold state.

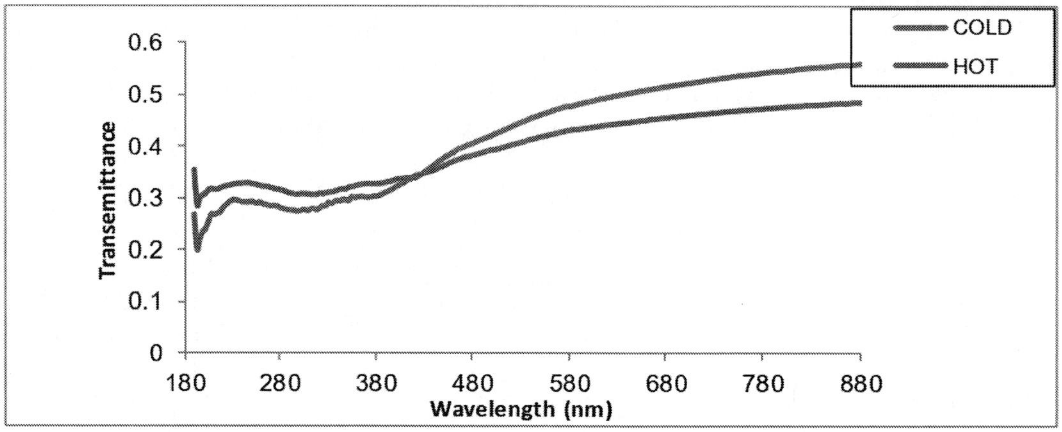

Figure 10. Transmittance as a function of the wavelength of the Y_2O_3/ Fe_2O_3 compound of the hot and cold cases (65 and 4)°C.

In Fig. 11 showed the change in the absorption coefficient within the range (190 - 800) nm is due to the change in the wavelength of the nano-oxide of the sample prepared in experimental conditions. It increases at short wavelengths (less than 300 nm) and decreases in values and stabilizes somewhat with increasing wavelength due to its suitability with absorption (Absorption behavior) and according to equation (2)[11], where the absorption coefficient is calculated from them. Since the absorption coefficient value is $\alpha < 10^4$ cm^{-1}. At the low energies, electronic transitions are indirect.

$$\alpha = 2.303\left(\frac{A}{t}\right) \qquad (2)$$

where: A: Absorbance , t: thickness, α : absorption coefficient. It is observed that the calculated absorption factor values for most samples are between ($1< \alpha <10^4$) cm^{-1} within the exponential absorption region. Therefore, the electron transitions from the valence packet to the conduction beam are within the edge. Where there is weak absorption and the electronic transitions in this region represent the tails within the energy gap. As we can see, the absorption coefficient is slightly at the long wavelengths, the transitions The values of the absorption factor at the edge of absorption are high in the direction of high energies.

IOP Conf. Series: Journal of Physics: Conf. Series **1032** (2018) 012028 doi:10.1088/1742-6596/1032/1/012028

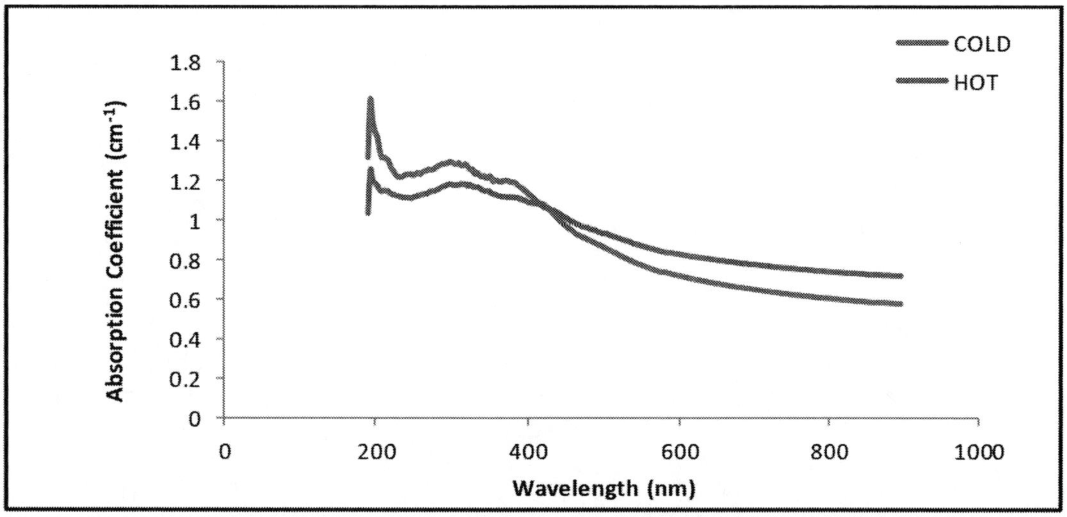

Figure 11. Absorption coefficient as a function of the wavelength of the Y_2O_3/ Fe_2O_3 nanoparticle mixture for hot and cold cases (65 and 4)°C.

The optical energy gap is determined from the absorption edge because it indicates the electronic transitions from the equivalence beam to the coupling. The transitions of these samples are indirect because the absorption coefficient is less than $10^4 cm^{-1}$. The indirect optical power gap is calculated according to equation[12] .

$$\alpha h\upsilon = B^{\backslash}(h\upsilon - E_g^{\backslash} \pm E_p)^m \quad (3)$$

Where: ($B^{\backslash}$:constant, $E_g^{\backslash}$: the optical energy gap for indirect transitions, +E_p: the phonon absorption process, -E_p: phonon emission process)

Variance in the values of the energy gap when the temperature changes and the material changes from one sample to another. In this sample, the decrease in the value of the energy gap, especially the cold state, is indicated by the warm state. This is confirmed by the results of AFM if the granular volume is colder than in the warm state. Decreasing is the entry of the positive ionic element into the crystalline structure to interfere with the atoms of the Y_2O_3, an increase in the number of positive ions. Consequently, internal levels within the energy gap will be formed, causing a decrease in its value, Eg = 2.79 eV for Hot, Eg = 2.5 eV Cold case.

IOP Conf. Series: Journal of Physics: Conf. Series **1032** (2018) 012028 doi:10.1088/1742-6596/1032/1/012028

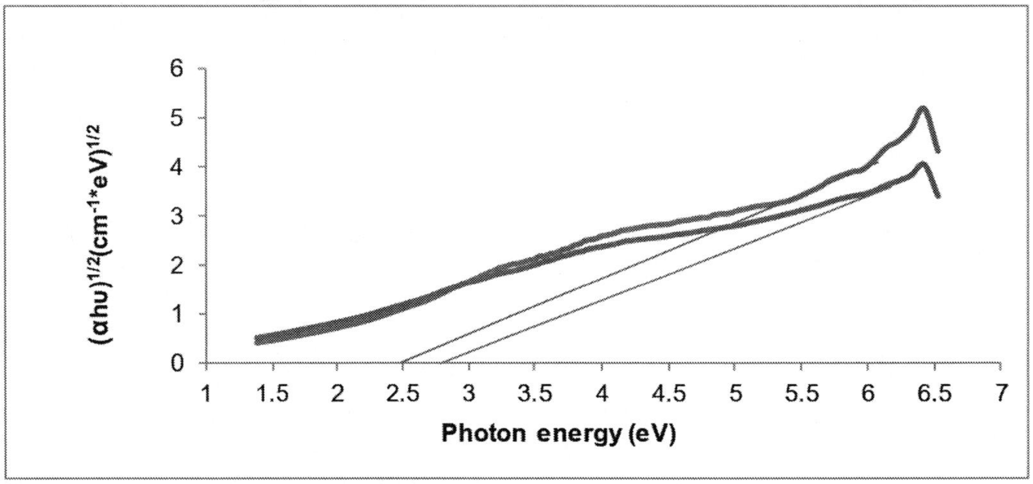

Figure 12. The relationship between the indirect energy gap and the photon energy of colloidal nanoparticles at 4°C (Cold) and 65 °C (Hot)

4. Application solar cell

Table (4) shows the effect of the hall of the sample membrane (Y_2O_3 / Fe_2O_3) in hot and cold conditions for preparation at temperature (65°C, 4°C.) The R_H coefficient depends on temperature and semiconductor type, The conductor of the resulting overlapping oxides in the hot and cold cases is of type **(n)**, so the vast majority of the conductivity is the electrons, while the temperature of the preparation (by increasing the temperature) is lower, because the temperature generates additional energy levels between the valence and the conductance bands. It was obtained from a decrease in the coefficient of Hole in the cold state than in the warm which was caused by a strong thermal shock in the low-temperature solid solution by the continuous laser pulse chain temperature, which gave the chemical bonds of the solution compounds a thermal content within it that is concentrated by the propagation. (μ_H) depends on the concentration of the carriers as the mobility decreases with the increase of the concentration of the carriers because the interstitial distance between the atoms decreases and thus impedes movement even though the traffic increases with increasing temperature. From Hall Impact Measurements, the velocity of the (υd) and resistivity (ρ) and the lifespan of the carriers can be calculated as shown in the same table. The conductivity of the sample veneers (Y_2O_3/Fe_2O_3) in hot and cold conditions to prepare at temperature (4°C, 65°C) increases by increasing the temperature. Therefore, the temperature effect on the resistivity is reversed. The drift velocity increases by increasing the temperature and it is inversely proportional to the concentration of the carriers . The faster the drift, the less the concentration of the charge carriers is affected by the age of the charge carriers, which is increased by increasing the drift speed, making the free carrier paths decrease by increasing the carriers because the interstellar distance decreases with increasing concentration.

Table (4) shows the measurements of the effect of the sample membrane (Y_2O_3 / Fe_2O_3) in hot and cold conditions

Sample Structure	Type	□d (cm/s)	□×□□□□ □□□s□	R_H (cm^{-3}/C)	μ_H (cm^2/V.s)	□ (□.cm)	σ (Ω.cm)$^{-1}$	n_H (cm^{-3})

[Y_2O_3/Fe_2O_3:p-Si] | Hot	**n**	**6880.12**	**616.75**	**11.85**	**$33.83\times10^{+2}$**	**0.07149×10^{-1}**	**137.5**	**$5.267\times10^{+17}$**
[Y_2O_3/Fe_2O_3:p-Si] | Cold	**n**	**6328.31**	**601.82**	**9.04**	**$24.88\times10^{+2}$**	**0.04529×10^{-1}**	**138.9**	**$5.290\times10^{+18}$**

Figure (13) shows the voltage current of the two solar cells of sample [Y_2O_3 / Fe_2O_3: p-Si] in hot and cold conditions in the dark. In the case of anterior bias, the current flows relatively freely while in the opposite direction, the flow of current is very weak due to the flow of the majority and voltages applied to the majority carriers and the width of the region The concentration of the majority and minority carriers is higher than the concentration of the base carriers, which generates the re-union current in the low voltage zone (0-3.0 Volt). This is because the electrons are raised from the valence pack to the conduction beam and then re-combined with the existing holes The parity package and this is what explained a significant increase in the Union re-stream in the region is low-voltage while the case of the tunnel in the high voltage area occur. After that, there is a huge exponential increase in the value of the current with increasing voltage and this is called the propagation current. Which dominate the high voltage (> 0.2 volts). In the reverse bias area, the bias voltage increases the width of the depletion zone, leading to a reduction in the concentration of the majority and minority carriers. In the first low voltages (< 0.1 volts), the current increases slightly with the applied voltage increase and slightly dominates the current. In the second high voltage zone (> 0.1 volts), the current is dominated.

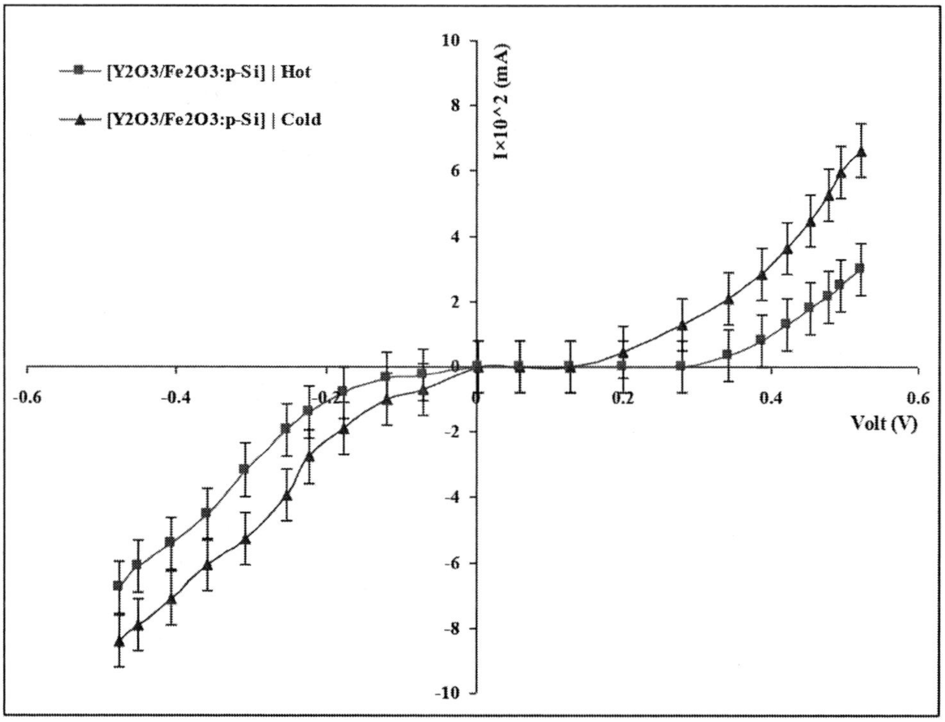

Figure 13. A voltage-current curve of the two solar cells of the sample [Y_2O_3 / Fe_2O_3: p-Si] in the case of darkness

IOP Conf. Series: Journal of Physics: Conf. Series **1032** (2018) 012028 doi:10.1088/1742-6596/1032/1/012028

The voltage-current values of the two solar cells of the sample [Y_2O_3 / Fe_2O_3: p-Si] were examined at the intensity of the lamp (55.5 mW / cm^2) in hot and cold conditions in Figure (14). It shows the relationship of the photovoltaic current at the reverse bias voltage, as the photovoltaic current increases by increasing the voltages, and the optical current (I_{ph}) increases by increasing the intensity of the falling light, thus increasing the number of generated carriers and displaying the depletion area increasing with the increase of reverse bias voltages which increase absorption and generation Electron - hole pairs. At the front bias voltages, the current doubles exponentially with the forward voltage until the slope becomes more gradual. This is due to the high injection levels of the carriers as the voltage is no longer equal throughout the depletion zone. The short circuit current increases with the intensity of the incident light. It has been found that in the case of reverse bias of the voltages, the current is increased by increasing the applied voltages, which in turn increases the width of the attenuation zone. While in the case of the front bias of the voltages increase current by increasing the front voltages. Table (5) shows the voltage of the open circuit (Voc), the short circuit current (Isc), the greatest voltage (Vm), the greatest current (Im), the filling factor coefficient (FF) and efficiency (η) of the sample solar cells [Y_2O_3 / Fe_2O_3: (55.5mW / cm^2), indicating that the solar cell efficiency of the sample [Y_2O_3 / Fe_2O_3: p-Si] prepared in the hot state is (3.9%) which is less than the efficiency of the cell of the sample [Y_2O_3 / Fe_2O_3: p-Si] prepared in the cold state, which is about 7%. "Here, it is worth mentioning the role of the intermediate compounds resulting from the reaction, which enhanced the electronic content of the cells in the case of cold, as explained by the interpretation of the effect of Hall and encircled by the results of X-ray examination of the presence of these compounds, which in turn will generate additional energy levels between the valence and conduction bands, which gives more opportunities to obtain values of the current oscillating accuracy measured depending on the nature of bias and drift speed of carriers Applies even to measurements of current voltage values in the case of darkness as seen in Figs. (13) and (14).

Table (5) current -voltage parameters: open circuit voltmeter, short circuit current, greatest voltmeter, maximum current, fillinf factor and efficiency coefficient for solar cells in the case of 55.5mW / cm^2

Solar Cells	V_{oc} (V)	I_{sc} (mA)	V_m (V)	I_m (mA)	FF	η %
[Y_2O_3/Fe_2O_3:p-Si] | Hot	0.29	42	0.16	34	0.44	3.9
[Y_2O_3/Fe_2O_3:p-Si] | Cold	0.39	44	0.22	45	0.57	7

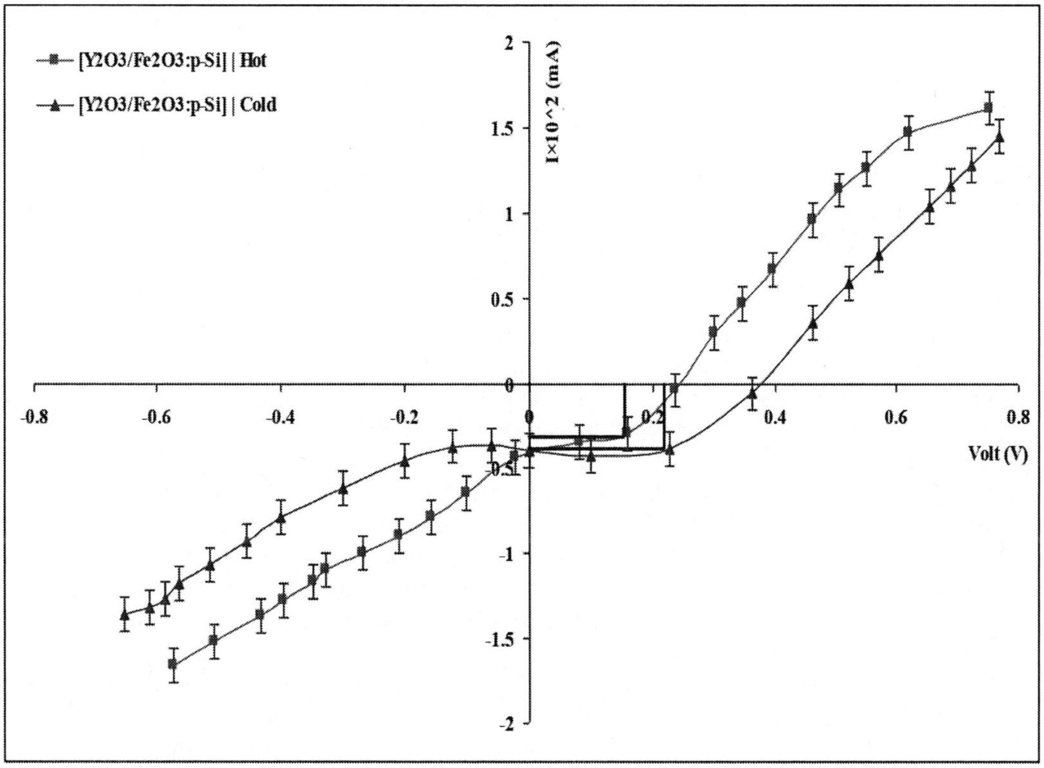

Figure 14. The current - voltage of the two solar cells of sample [Y_2O_3 / Fe_2O_3: p-Si] in the case of a 55.5 mW / cm^2 light bulb

5. Conclusions

It is concluded the following :

1- The thermal effect led to adhesions between the grains, which made the values of grain sizes multiple and different diameters.

2- The presence of vacancy due to the degradation of the nanotube by the heat, which leads to the possibility of a fusion between the nanoparticles of the compound.

3- Absorption increased in the cold than in the case of hot.

4-In the hot and cold cases, the results showed that the Hall coefficient depends on the temperature, the nature of the intermediate compounds formed, and the type of semiconductor of the two samples, which is of type (n). The electrons are the predominant carriers, and the drift velocity increases as the concentration of the charge carriers decreases due to the time life of the charge carriers, which increases with the speed of drift, making the free carrier paths decrease by increasing the carriers.

5-In the hot and cold cases in the dark mode, the behavior of the two cells was found to be non-linear. This indicates that the coupling mechanism is non-ionic and that the current flows relatively freely at the forward bias rather than the flow current at reverse bias, which is very weak due to the applied voltages for injection of the flowing majority carriers as well as the display of the attenuation zone.

6- It is observed the relationship of the photovoltaic current to the reverse bias voltage. The I_h is increased by increasing the voltages The intensity and intensity of the incident light, which increases the number of carriers generated. In the case of front bias voltages, the current doubles exponentially with the forward voltage until the slope becomes more gradual. This is due to the high injection levels of the carriers.

IOP Conf. Series: Journal of Physics: Conf. Series **1032** (2018) 012028 doi:10.1088/1742-6596/1032/1/012028

7-The efficiency of the solar cell in hot state is (3.9%), and the efficiency of the solar cell in the cold state (7%)

6. References

[1] Saha, S. A. T. Y. A. J. I. T., & Bera, S. R., "Growth and characterization of CdTe nanostructures grown by chemical reduction route,.", Int. Journal of Metallurgical of materials Science and Engineering , Vol .3, No. 1, (2013).

[3] José, A. R., & Marcos, F. G, "Synthesis, Properties, and pplications of Oxide Nanomaterials", Willey, New Jersey (2007).

[4] Fierro J. L. G., "Metal Oxides: Chemistry and Applications, CRC Press", Florida Fierro, Jose Luis G., ed. Metal oxides: chemistry and applications. CRC press, (2005) .

[5] Munoz, R., "Co-precipitation of Y_2O_3 powder." M.Sc., Thesis, Division of Ceramics, Department of Materials Science and Engineering, KTH Stockholm, Sweden , INSA Lyon, France, (2011).

[6] Larimi, Z. M., Amirabadizadeh, A., & Zelati, A., "Synthesis of Y_2O_3 nanoparticles by modified transient morphology method." Proceedings of the International Conference on Chemistry and Chemical Process IPCBEE , IACSIT Press, Singapore , pp. 86-90, (2011).

[7] Drbohlavova, J., Hrdy, R., Adam, V., Kizek, R., Schneeweiss, O., & Hubalek, J. , "Preparation and properties of various magnetic nanoparticles", . Sensors, Vol. 9, No. 4, pp. 2352-2362, (2009).

[8] Tanskanen, A., Mustonen, O., & Karppinen, M. , "Simple ALD process for ε-Fe_2O_3 thin films", APL Materials, Vol. 5, No. 5, pp. 56104, (2017).

[9] Ling, C. C., Fung, S., Beling, C. D., & Huimin, W. "Defect study of Zn-doped p-type gallium antimonide using positron lifetime spectroscopy". Physical Review B, Vol. 64, No. 7, pp. 75201, (2001).

[10] Yao, W., Zhuang, W., Ji, X., Chen, J., Lu, X., & Wang, C. , "Solid-state synthesis of $Li4Ti_5O_{12}$ whiskers from TiO_2-B". Materials research bulletin, Vol. 75, pp. 204-210,(2016).

[11]Hollas, J. M., "Modern spectroscopy." John Wiley & Sons, (2004).

[12] Kittel, C., McEuen, P., & McEuen, P., "Introduction to solid state physics" , Vol. 8, pp. 323-324, (1996).

IOP Conf. Series: Journal of Physics: Conf. Series **1032** (2018) 012029 doi:10.1088/1742-6596/1032/1/012029

Reduced Quadruple Transition Probability B(E2) for Some Oxygen Isotopes Using Different Interactions and Effective Charges

[1]**Ahmed H. Ali**, [2]**Akram M. Ali**
[1]College of Medicine, University of Fallujah

[2]College of Sciences, University of Anbar

Author E-mail: dr.ahmedphysics@uofallujah.edu.iq

Abstract. Reduced quadrupole transition rates B(E2; $0_1^+ \rightarrow 2^+$) the effective charges are calculated for even-even oxygen isotopes(A=16,18,20,22)based on p shell and sd shell model spaces. NushellX@ code are using for calculate one body transition matrix (OBTM) for these isotopes. Our calculations theoretical for reduced quadrupole transition rates B(E2) are using the two types of effective interactions to obtain the best interaction compared with the experimental data were adopted at same time different the effective charges.

1. Introduction

Study the reduced transition probabilities of the oxygen isotopes provide diverse examples of progress made in experiment and theory. The B(E2) values for transitions from first excited states to ground states ($0_1^+ \rightarrow 2^+$) in even-even nuclei are important quantities in nuclear physics for determining the collectivity in nuclei [1].

Quadrupole collectivities (reduced electric quadrupole transition rates), or B(E2)↑ values, play an important role in nuclear physics and are in high demand for nuclear model calculations. Originally, these values were compiled by Raman et al. at the Oak Ridge Nuclear Data Project [2, 3].

The general trend of the 2^+ excitation energy Ex (2^+) and the reduced electric quadrupole transition strength B(E2;$2^+ \rightarrow 0_1^+$) for even-even nuclei are expected to be inversely proportional to one another [3].

In shell-model studies, for many of existing conventional interactions, it is difficult to reproduce simultaneously the drip lines of carbon and oxygen isotopes as well as some other properties such as energies of (2_1^+)states and B(E2). From a microscopic study, the inclusion of effects of three-body force is important in describing the drip line of oxygen isotopes [4].

In the present work, shell model calculations are performed with two effective interactions: Millener-Kurath (MK) interaction for region *psd* shell [5] and Brown and Richter (USDB) interaction for region *sd* shell [6].

2. Theory

A microscopic model has been proposed to include effects from outside the model space, which is called core polarization effects [7]. The reduced matrix elements of the electron scattering operator is expressed as a sum of the model space (MS) contribution and the core polarization (CP) contribution, as follows:

$$M(EJ) = \sum_{t_z}\left[e(t_z)\left\langle J_f\left\|\hat{O}_J(\vec{r},t_z)\right\|J_i\right\rangle_{MS} + e\left\langle J_f\left\|\Delta\hat{O}_J(\vec{r},t_z)\right\|J_i\right\rangle_{CP}\right] \quad (1)$$

The core polarization CP electric matrix element can be obtained as follows:

$$\left\langle J_f \| \Delta\hat{O}_J(\vec{r},t_z) \| J_i \right\rangle_{cp} = \sum_{jj'} OBTM(J_i,J_f,J,t_z,j,j')\left\langle j \| \Delta\hat{O}_J(\vec{r},t_z) \| j' \right\rangle \quad (2)$$

Where OBTM is the one body transition matrix times the single particle matrix elements. The single-particle matrix element of the CP term is

$$\left\langle j \| \Delta\hat{O}_J \| j' \right\rangle = \left\langle j \| \hat{O}_J \frac{\hat{Q}}{E_i - H_0} \mathrm{V}_{res} \| j' \right\rangle + \left\langle j \| \mathrm{V}_{res} \frac{\hat{Q}}{E_f - H_0} \hat{O}_J \| j' \right\rangle \quad (3)$$

where the operator is the projection operator onto the space outside the model space. The single particle core-polarization terms given in equation (3) are written as [8]

$$\left\langle j \| \Delta\hat{O}_J \| j' \right\rangle = \sum_{j_1 j_2 \lambda} \frac{(-1)^{j'+j_2+\lambda}}{\varepsilon_{j'} - \varepsilon_j - \varepsilon_{j_1} + \varepsilon_{j_2}} (2\lambda+1) \begin{Bmatrix} j & j' & J \\ j_2 & j_1 & \lambda \end{Bmatrix} \sqrt{(1+\delta_{j_1 j})(1+\delta_{j_2 j'})} \quad (4)$$

$$\times \left\langle j j_1 | \mathrm{V}_{res} | j' j_2 \right\rangle_\lambda \left\langle j_2 \| \hat{O}_J \| j_1 \right\rangle$$

+terms with and exchanged with an overall minus sign,
where the index runs over particle states and over hole states and ε is the single-particle energy. For the residual two-body interaction Vres, the two-body Michigan three range Yukawa (M3Y) interaction of Bertsch et al. [9] is adopted.

The electric matrix element can be represented in terms of only the model space matrix elements by assigning effective charges( ) to the protons and neutrons which are active in the model space,

$$M(EJ) = \sum_{t_z} e^{\mathrm{eff}}(t_z) \left\langle J_f \| \hat{O}_J(\vec{r}, t_z) \| J_i \right\rangle_{\mathrm{MS}} \quad (5)$$

The effective nucleon charge can be obtained as follows

$$e^{\mathrm{eff}}(t_z)\left\langle J_f \| \hat{O}_J(\vec{r}, t_z) \| J_i \right\rangle_{\mathrm{MS}} = e(t_z)\left\langle J_f \| \hat{O}_J(\vec{r}, t_z) \| J_i \right\rangle_{\mathrm{MS}} + e\left\langle J_f \| \Delta\hat{O}_J(\vec{r}, t_z) \| J_i \right\rangle_{\mathrm{CP}} \quad (6)$$

$$e^{\mathrm{eff}}(t_z) = e(t_z) + \frac{\left\langle J_f \| \Delta\hat{O}_J(\vec{r}, t_z) \| J_i \right\rangle_{\mathrm{CP}}}{\left\langle J_f \| \hat{O}_J(\vec{r}, t_z) \| J_i \right\rangle_{\mathrm{MS}}} e = e(t_z) + e\delta e(t_z) \quad (7)$$

where is the nucleon polarization charge.

The reduced electric transition can be obtained as follows [8]:

$$B(EJ) = \frac{|M(EJ)|^2}{2J_i+1} \quad (8)$$

3. Results and Discussion

Reduced transition probabilities in units of e2fm4 are calculated for Oxygen isotopes (Z = 8) with mass number A = 18, 20, 22, 24 and with neutron number N = 10, 12, 14, 16, respectively.

Adopted two interactions such as *psdmk* interaction for *psd* model space [5] and the USDB interaction (universal *sd*-shell interaction B) for sd model space [6] to generate the one body transition matrix element (OBTM) values via the computer code NuShellX [10]. The core polarization effect on the transition strength is based on microscopic theory, which combines shell-model wave functions and configurations with higher energy. The reduced transition probabilities are calculated by using different effective charges such as conventional effective charges (CON) [11], Bohr-Mottelson effective charges (B-M) [12]

$$e^{eff}(t_z) = e(t_z) + e\delta e(t_z), \delta e(t_z) = Z/A - 0.32(N-Z)/A - 2t_z[0.32 - 0.3(N-Z)/A].$$

, also using standard effective charges (ST) ep = 1.36 and en = 0.45 [13].

The radial wave functions for the single-particle matrix elements were calculated with the harmonic oscillator (HO) potential with size parameters calculated for each isotope [14]

$$b = \sqrt{\frac{\hbar}{M_p \omega}}, \quad \text{with} \hbar\omega = 45A^{-1/3} - 25A^{-2/3}$$

Theoretical Calculations of B(E2) are presented using different effective interactions and use different effective charges for each interaction. The results presented for B(E2) in this work and are compared with the available experimental data give in Ref. [15].

3.1. *PSDMK Interaction*

The calculations are performed *psd* -model space with *psdmk* interaction [5] to generate the OBTM elements for the ground state with J=0 and excited state with J=2. The harmonic oscillator size parameter was fixed and calculated for each isotope. Calculations of B(E2) are presented using the conventional effective charges (CON) which are for proton 1.3 e and for the neutron 0.5 e [11], The Bohr-Mottelson effective charges (B-M)[12] and the standard effective charges ST which are 1.36e for the proton and 0.45e for the neutron [13] in a comparison with experimental results [15]. The Bohr-Mottelson effective charges (B-M)[12] are calculated for oxygen isotopes (A=18, 20, 22, 24) as shown in Table1.

The shell-model calculations, ^{4}He is chosen as the core for *psd* model spaces, neutrons and protons are distributed on the *p* and *sd* shell. The size parameter was calculated for ^{18}O isotope (stable) b= 1.75 fm. The $B(E2;2^{+} \rightarrow 0_{1}^{+})$ were calculated for ^{18}O isotope where B(E2)CON=19, B(E2)ST=15.4 and B(E2)B-M=36.8, these values agree with experimental data 43±38 e^2fm^4[15]. The size parameter was calculated for ^{20}O b= 1.751 fm and the $B(E2;2^{+} \rightarrow 0_{1}^{+})$ were calculated for ^{20}O isotope where B(E2)CON=24.1 and B(E2)ST=19.6 and B(E2)B-M=34.3, these values agree with the experimental data 29.8±26 e^2fm^4. The size parameter was calculated for 22O isotope b= 1.795 fm and the $B(E2;2^{+} \rightarrow 0_{1}^{+})$ were calculated for ^{22}O isotope where B(E2)CON=24.68 and B(E2)ST=19.9 and B(E2)B-M=26.1, these values agree with the experimental data 21±8 e2fm4. Also the size parameter was calculated for ^{24}O isotope b= 1.815 fm and the $B(E2;2^{+} \rightarrow 0_{1}^{+})$ were calculated for ^{24}O isotope where B(E2)CON=7.93 and B(E2)ST=6.42 and B(E2)B-M=6.33, the experimental data not available.

The results of B(E2) agree with experimental data for all these isotopes. The B(E2) values calculated with B-M effective charges agree very well with the experimental for all oxygen isotopes considered in this work as shown in figure.1.a. Calculations of the excitation energies are compared with the experimental data and tabulated in Table1 and plotted in figure.1.b (lower panel) which shows a reasonable agreement except for the ^{24}O isotope, where the theoretical value overestimates the experimental data.

3.2. *USDB Interaction*

The calculations are performed with *sd* -model space with *usdb* interaction [6] to generate the OBTM elements for the ground state with J=0 and excited state with J=2 and All isotopes in the present work takes ^{16}O as the core. The harmonic oscillator size parameter was fixed and calculated for each isotope. Calculations of B(E2) are presented using the conventional effective charges (CON) which are for proton 1.3 e and for the neutron 0.5 e [11], The Bohr-Mottelson effective charges (B-M)[12] and the standard effective charges ST which are 1.36e for the proton and 0.45e for the neutron [13] in a comparison with experimental results [15]. The Bohr-Mottelson effective charges (B-M)[12] are calculated for oxygen isotopes (A=18, 20, 22, 24) as shown in Table2. The size parameter was calculated for ^{18}O isotope (stable) b= 1.75 fm and the B $B(E2;2^{+} \rightarrow 0_{1}^{+})$were calculated for ^{18}O isotope B(E2)CON=20.1, B(E2)ST=16.3 and B(E2)B-M=38.96, these values agree with experimental value 43±38 e^2fm^4[15]. The size parameter was calculated for ^{20}O b= 1.751 fm and the $B(E2;2^{+} \rightarrow 0_{1}^{+})$ were calculations for 20O isotope where B(E2)CON=25.4 and B(E2)ST=20.5 and B(E2)B-M=36.1, these values agree with the experimental value 29.8±26 e^2fm^4. The size parameter was calculated for ^{22}O isotope b= 1.795 fm and the $B(E2;2^{+} \rightarrow 0_{1}^{+})$were calculated for 22O isotope B(E2)CON=24.68 and B(E2)ST=19.8 and B(E2)B-M=25.9, these values agree with the experimental value 21±8 e^2fm^4. Also the size parameter was calculated for ^{24}O isotope b= 1.815 fm and the $B(E2;2^{+} \rightarrow 0_{1}^{+})$were calculated for ^{24}O isotope where B(E2)CON=10.28 and B(E2)ST=8.32 and B(E2)B-M=8.20, the experimental value not available.

Figure.2.a show excellent agreement between theoretical results and experimental data for most isotopes. The Calculated B(E2) values show a decrease with increasing neutron number for $N \geq 10$. The calculated B(E2) values, with B-M effective charges are closer to the experimental data than those of the other sets of effective charges . Calculations of the excitation energies are compared with the experimental data and tabulated in Table2 and plotted in figure2.b (lower panel) which shows a reasonable agreement except for the ^{24}O isotope, where the theoretical value overestimates the experimental data.

4. Conclusions

The shell model is less successful in describing the transition rates unless taking the core polarization CP effects into account. The CP is necessary, beneficial and plays a major role for describing nuclear properties.

The calculations in the psd model space were for proton number P= 8 and neutron number N > 2, which covered the orbits $1p_{3/2}$,$1p_{1/2}$ and the orbits $1d_{5/2}$, $2s_{1/2}$ and $1d_{3/2}$. Also calculations in the sd model space were for proton number P≥8 and neutron number N ≥ 8, which covered the orbits $1d_{5/2}$ $2s_{1/2}$ and $1d_{3/2}$.

In this present work, two effective interactions adopted, PSDMK interaction (psdmk int.) in the region psd model space and USDB interaction (usdb int.) in the region sd model space and are employed different effective charges. The B(E2) values calculated with B-M effective charges agree very well with the experimental data than of that the other sets of effective charges for all O-isotopes considered in this work. The results show a systematic decrease in the B(E2) values for $10 \leq N \leq 16$. The small B(E2) values of ^{22}O, ^{24}O indicates that these nuclei are nearly spherical closure in accordance of previous conclusions of the magic number 20. Good agreement is obtained between the theoretical excitation energy with USDB interaction and experimental excitation energy for all O isotopes considered in this work.

References

[1] M.Bircha, B.Pritychenkob, B.Singh. 2016. Nuclear Physics A. 955.145–155
[2] S. Raman, C.H. Malarkey, W.T. Milner, C.W. Nestor Jr., P.H. Stelson. 1987. Data Nucl. Data Tables 36 1.
[3] S. Raman, C. W. Nestor, Jr., and P. Tikkanen. 2001. Atomic Data and Nuclear Data Tables, 78, pp:1-128.
[4] T. Otsuka, T. Suzuki, J. D. Holt, A. Schwenk, and Y. Akaishi.2010. Three-body forces and the limit of oxygen isotopes. Physical review letters. 105, 032501
[5] D. J. Millener and D. Kurath. 1975. The particle-hole interaction and the beta decay of 14B. Nuclear Physics A. 255, 315
[6] B. Alex Brown, W. A. Richter. 2006. New USD Hamiltonians for the sd shell.Physical Review C 74, 034315.
[7] H. Sagawa, B.A. Brown. 1985. Coulomb form factors of collective E4 transitions in s-d shell nuclei. Physics Letters B.150, 247
[8] P.J. Brussaard and P. W. M. Glademans.1977. Shell- model Application in Nuclear Spectroscopy" North- Holland Publishing Company, Amsterdam.
[9] Berstch, G., Borysowicz J., McManus H. Love W. G. 1977. Interactions for inelastic scattering derived from realistic potentials. Nuclear Physics, A 284, pp: 399.
[10] B. A. Brown, W. D. M. Rae. 2014. The Shell-Model Code NuShellX@MSU. Nuclear Data Sheet 120, 115-118.
[11] B. A. Brown. 2001. The nuclear shell model towards the drip lines. Progress in Particle and Nuclear Physics. 47, 517
[12] A. Bohr, B. R.1975. Mottelson.Nuclear Structure. Benjamin, New York. 2
[13] W. A. Richter, S. Mkhize, B. Alex Brown. 2008. sd-shell observables for the USDA and USDB Hamiltonians. Physical Review C, 78(064302).

[14] Brown, Radhi, R. and Wildenthal, B.H. 1983. Electric quadrupole and hexadecupole nuclear excitations from the perspectives of electron scattering and modern shell-model theory. Phys. Rep., 101, 313.
[15] B. Pritychenko, M. Birch, B. Singh, M. Horoi. 2016. Tables of E2 transition probabilities from the first 2+ states in even–even nuclei, Atomic Data and Nuclear Data Tables 107, 139.

Table 1. The reduced electric transition probability B(E2) in units of e2 fm4 and excitation energies for oxygen isotopes (Z = 8). Experimental Ex and B(E2) are taken from Ref. [15]. Calculations B(E2) using psdmk int.[5] and set effective charges, conventional effective charges (CON) ep = 1.3 and en = 0.5 [11], Bohr-Mottelson effective charges (B-M)[12], and standard effective charges (ST) ep = 1.36 and en = 0.45 [13]

A,N $_8$O	$(E_x)_{thor.}$(MeV)	$(E_x)_{exp.}$(MeV)	$B(E2)_{CON}$	$B(E2)_{ST}$	e_p , e_n B-M	$B(E2)_{B\text{-}M}$	$B(E2)_{exp.}$ PSDMK
18,10	1.98	1.982	19	15.4	1.12,0.7	36.8	43±38
20,12	1.77	1.673	24.1	19.6	1.08, 0.6	34.3	29.8±26
22,14	3.25	3.199	24.68	19.9	1.04, 0.51	26.1	21±8
24,16	5.42	4.72	7.93	6.42	1.01, 0.45	6.33	

Table 2.The reduced electric transition probability B(E2) in units of e2 fm4 and excitation energies for oxygen isotopes (Z = 8). Experimental Ex and B(E2) are taken from Ref. [15]. Calculations B(E2) using usdb int.[6] and set effective charges, conventional effective charges (CON) ep = 1.3 and en = 0.5 [11], Bohr-Mottelson effective charges (B-M)[12], and standard effective charges (ST) ep = 1.36 and en = 0.45 [13]

A,N $_8$O	$(E_x)_{thor.}$(MeV)	$(E_x)_{exp.}$(MeV)	$B(E2)_{CON}$	$B(E2)_{ST}$	e_p , e_n B-M	$B(E2)_{B\text{-}M}$	$B(E2)_{exp.}$ USDB
18,10	1.99	1.982	20.1	16.3	1.12, 0.75	38.96	43±38
20,12	1.77	1.673	25.4	20.5	1.08, 0.6	36.1	29.8±26
22,14	3.15	3.199	24.68	19.8	1.04, 0.51	25.9	21±8
24,16	5.04	4.72	10.28	8.32	1.01, 0.45	8.20	

IOP Conf. Series: Journal of Physics: Conf. Series **1032** (2018) 012029 doi:10.1088/1742-6596/1032/1/012029

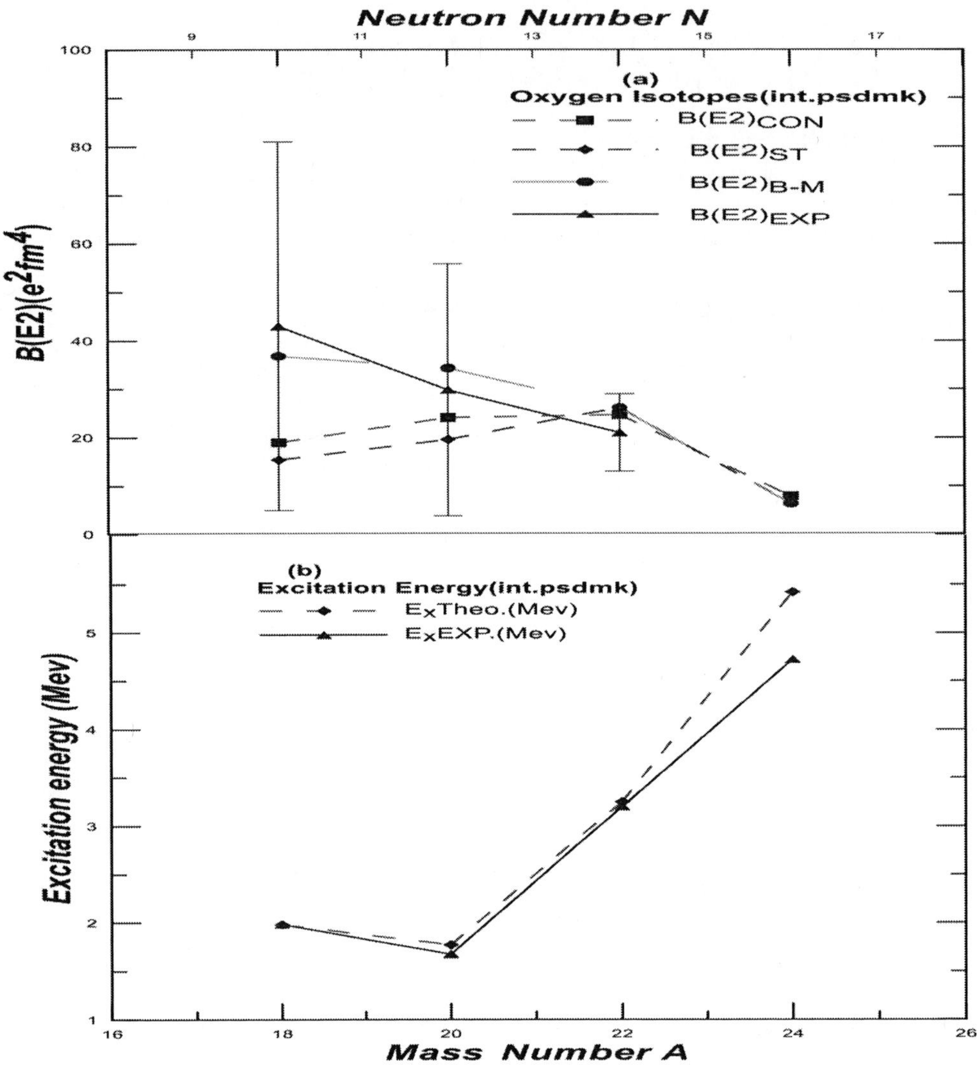

Figure1. Calculated and measured $B(E2;\ 0^+ \rightarrow 2^+)$ and excitation energy of eve- even oxygen isotopes. The experimental values are taken from Ref.[15].

IOP Conf. Series: Journal of Physics: Conf. Series **1032** (2018) 012029 doi:10.1088/1742-6596/1032/1/012029

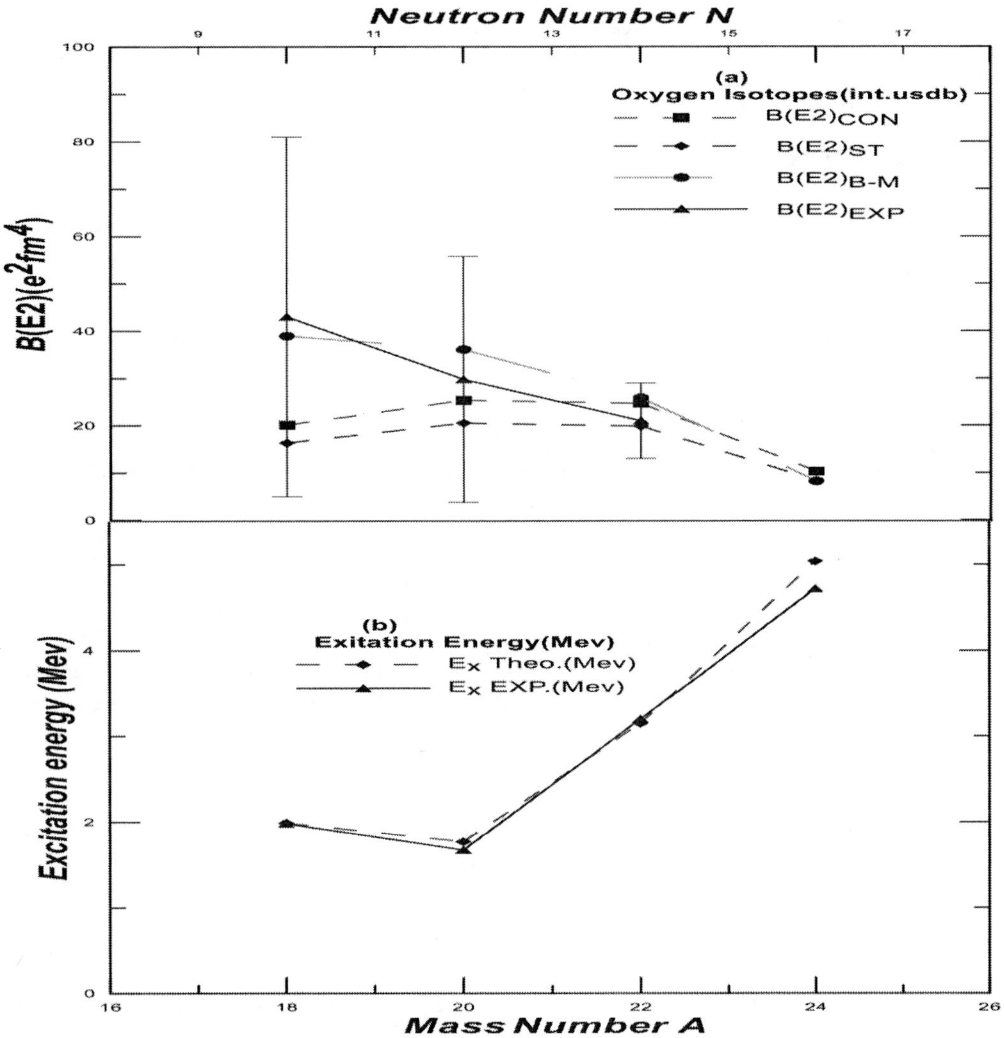

Figure 2. Calculated and measured B(E2; 0+ → 2+) and excitation energy of eve- even oxygen isotopes. The experimental values are taken from Ref.[15].

IOP Conf. Series: Journal of Physics: Conf. Series **1032** (2018) 012031 doi:10.1088/1742-6596/1032/1/012031

Improve the performance of solar modules by reflectors

[1]**Naseer K Kasim1, [2]Ahmed F Atwan, and [3]Fadhil Muhmood Eliewi**

[1]Ministry of electricity, Renewable Energy and Environment Center, Iraq.

[2]Mustansiriyah University, College of Education, Physics Department, Iraq.

[3]Ministry of education-Iraq.

e-mail: irss2004@yahoo.com, ahmed1973@uomustansiriyah.edu.iq, fadhilphy@yahoo.com

Abstract. Solar energy is one of the most significant renewable energy sources, in both its applications thermal and photovoltaic, world needs. Iraq is one of the countries with the abundance of this type of energy, where annual solar activity reaches about 2000kWh / m^2 / year on the horizontal surface. In the present work a solar collectors with V-trough concentrator system were designed and fabricated with geometric concentration ratios two suns to get an extra electric output power. The experiment was carried out in Al-Zwraa Company in Baghdad. The main results indicated that the solar concentrator system caused to increase the short circuit current due to increase the amount of incident solar radiation on the solar modules, and thus increase in output power. As a comparison with reference module, the gain in output power, at solar noon, exceeded 48%. It can be concluded that the use of V-Trough concentrator system coupled with solar module systems can result in a favourable cost-benefit ratio.

Keywords: CPV, V-trough reflector.

1. Introduction:

Solar photovoltaic (PV) technology is a very attractive renewable energy option for clean energy generation but is limited in use due to its prohibitively high cost. However the cost of PV power generation has decreased substantially over the last two decades, and now stabilized.

There are several ways by which the PV material consumption/watt of generated power can be reduced. Some of these include concentrator PV technologies. In the PV concentrator technology, the use of optical concentrators replaces the expensive PV cell area by cheaper reflector material, hence reducing the material consumption. Compared to the non-concentrating PV, the required area of the solar cell is reduced by the factor of concentration ratio, providing significant reduction in the cost of PV system. Although the concept of use of solar PV concentrator with PV modules/cells is simple but it is difficult to implement, especially for high concentration ratio. Based on the above concepts, V-trough concentrator is an attractive option to reduce the price of the PV electrical power using conventional solar PV cells. V-trough's are static concentrators, wherein the light intensity is boosted by placing reflectors to the sides of the PV module. Since, these are the non-imaging type of concentrators hence, diurnal tracking of the sun is not required and the added cost of the V-troughs is low as they are simple to manufacture.

The V-trough with plan Reflectors is one of the types of CPV system. Reflectors are an optical Device with cheap and suitable technology to concentrate the light and increase the amount of sunlight onto the panel [1]. Reflectors are Lower cost, Superior efficiency, less materials availability issues, Ease of recycling, Ease of rapid and high local manufacturing content [2].

PV/T system with and without reflectors has been studied by Kostic et al [3]. These studies show the positive effect of reflectors made of AL sheet and AL foil, considering the additional cost of about 10% for the reflectors, there is an energy gain in the range 20.5–35.7% in the summer period. The present work aims at measuring the effect of V-Trough reflector on the output power of array solar system in Iraqi environment.

2. V-Trough solar concentrator

V-trough geometry is defined by two parameters: the Geometric concentration ratio and the trough angle or vertex angle (*Ψ)* (the angle between reflector and the column on PV). The geometric concentration ratio is the ratio between the area of aperture plane *(A)* and area of the absorber plane a as shown in figure 1, which can be expressed by [4]:

$$C = \frac{A}{a} = \frac{W_{ape}}{W_{abs}} \quad \ldots\ldots\ldots\ldots\ldots\ldots\ldots\ldots\ldots\ldots\ldots (1)$$

Where the W_{ape} is the aperture width and W_{abs} is the absorber width (width of the PV module). The geometric concentration ratio in the system at vertex angle equal 30° is two suns.

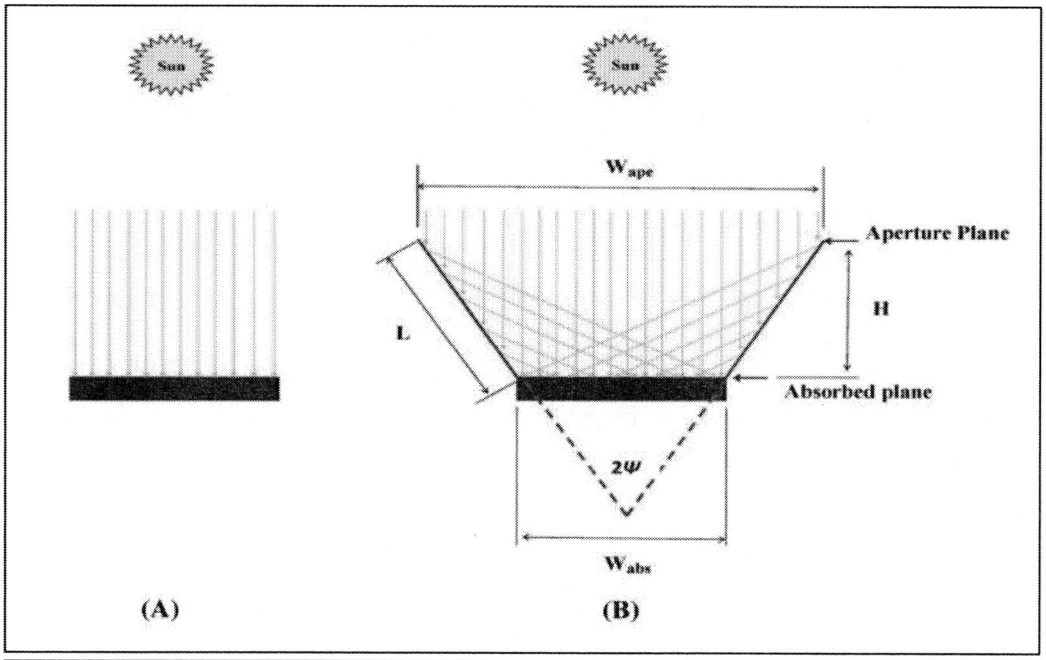

Figure 1. A- PV module, B- V-PV module [5].

An early study by Bannerot and Howell [6] had suggested that static V-trough collectors could achieve an annual average concentration ratio of over 1:2 for locations with a high diffuse solar fraction, and might be suited to applications where the reflectors were off setting the cost of expensive solar absorbers such as photovoltaics. Tabaei and Ameri [7], investigated the aluminum foil and stainless steel reflectors can increase power output around 8.5%-14% of polycrystalline solar panel. The average output power for PV panels with aluminum foil reflector and water film concurrently is represented about 50% improvement in power generation. The effects of V-trough geometric parameters such as slant height of the reflectors and vertex angle on the trough optical performance in terms of effective concentration ratio and irradiance distribution were investigated to determine the optimum design .Results show that the optimum vertex angles of the V-trough concentrator for 1.5,2, 2.5 and 3 are 30,30,22, and 19, respectively [5]. Shneishil [8], study the effect of one and two reflectors on V-trough concentrator on PV output power theoretically and experimentally. The increase on insolation at noon is 61% and 120% respectively, who's showed a good performance of solar cell.

3. Experimental

3.1 System description:

The system consists of array of three poly-crystalline (pc-Si) solar modules and V-Trough reflectors, which consists of six sheets of Aluminum foils called cpv1, cpv2 and cpv3 with total area (3*1.48m*0.67m) as shown in figure 2. There is a single PV module without reflectors as a reference. The characteristics of the PV module are listed in table 1.

Figure 2. CPV system.

Table 1. The characteristics of Sharp solar module 330 W.

Parameter	Value	parameter	Value
P_{max}	330 W	Dimensions	1958mm*1992mm*50 mm
I_{SC}	9.28 A	Temperature coefficient V_{OC}	0.31V/°C
V_{OC}	45.9 V	Temperature coefficient I_{SC}	0.41mA/°C

IOP Conf. Series: Journal of Physics: Conf. Series **1032** (2018) 012031 doi:10.1088/1742-6596/1032/1/012031

I_{pm}	8.85 A	Temperature coefficient P_{max}	-45%/°C
V_{pm}	27.3 V	weight	27 Kg
Efficiency	17%	NOCT	45 °C
Type of cell	Pc-Si	No. of cell	72 in series

3.2 Measurements

The experiment was implemented normally from 9 Am to 13 Pm in August/ 2017. During the experiment duration, ambient temperature (Ta), module temperature (Tm), short circuit current (Is), open circuit voltage (Voc), wind speed and direct with and without reflected solar irradiance (Id &Id,r respectively) were hourly measured and recorded. The recorded values of temperature, Voc, Is, and Id,r with presence of reflectors represent the average value of three modules. Four K-type thermocouples are used to measure the temperatures of PVC modules and reference module. The geometric concentration ratio in the system at vertex angle equal 30° was two suns.

4. Results and Discussion

In the present work each of ambient temperature (Ta) and module temperature with and without reflectors (Tm &Tr respectively) were measured. The results indicated that the temperature of the reference module was increased and reached, approximately, the twice value of the ambient temperature as show in figure 3. This figure also shows that the module with reflectors had the highest mean value of temperature. This is due to interaction of light with PV material and converts a part of incident power into heat. The amount of released heat proportional with amount of incident radiation.

Another important factor that contributes in output power of PV system is the irradiance. The direct incident radiation per unit area of reference solar module (Id) was measured for one hour periodically. The result was predicted. The levels of irradiance were increased with sun elevation as demonstrated in figure 2. This figure, also, shows that the levels of this irradiance rises with using V-Trough reflectors (Id,r).

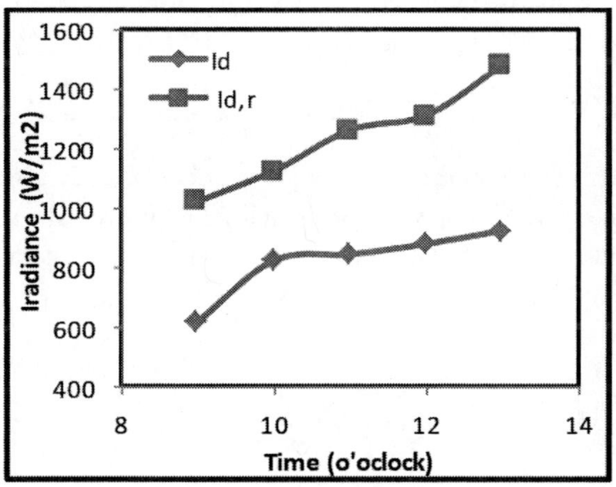

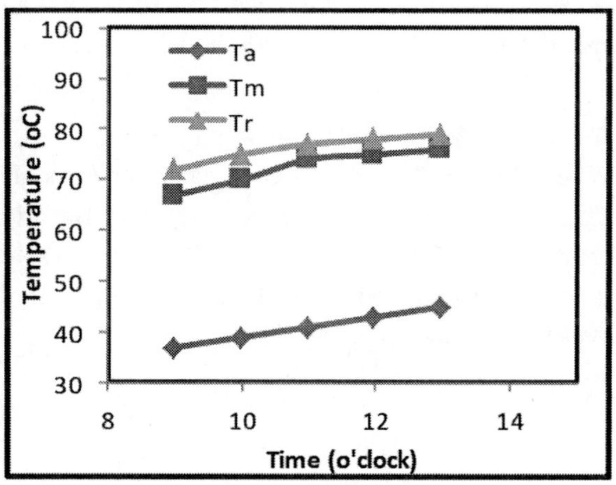

Figure 3. Variation of Ta, Tm, and Tr with day time. **Figure 4.** Variation of irradiance with day time.

From figure 4 it is appear that the reflectors raise the irradiance to, approximately, double value. It is known that and according to the band theory in semiconductor the electrons leave the equivalent band toward valance band when it received photon with energy equal or more than energy gap and the number of free electrons increase with increase of incident photons and thus increase the output current of PV device. This fact is demonstrated in figure 5 by reflectors action. In this figure the current value of module with reflectors is greater than that for reference module (Im). The increment in current levels leads to improve the performance of PV solar module. This result appears the importance of using reflectors with PV solar systems. On the other hand there was dropping in output voltage of PV module with presence of V-Trough reflectors (Vr) as shown in figure 6. This drop is due to rises in module temperature by reflectors.

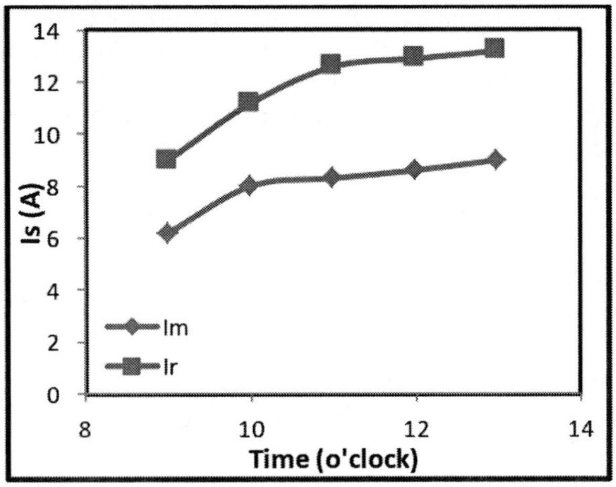

Figure 5. Variation of output current with time.

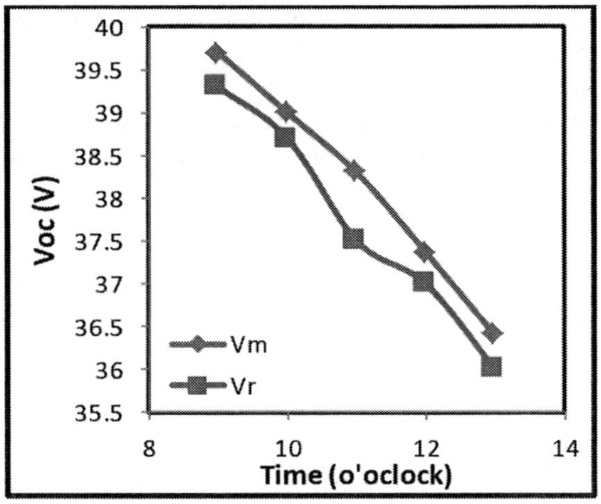

Figure 6. Variation of output voltage with time.

The voltage decrease of a silicon cell is typically 2.3 mV per 1 °C. The temperature variation of the current or the fill factor are less pronounced and usually neglected in the solar cell system design Accordingly, only the voltage variation with temperature is allowed for in practical calculations, and for individual module consisting of N_c cells connected in series is set equal to [9]:

$$V_{oc} = V_{oco} - (2.3\ N_c T_c) \ldots\ldots\ldots\ldots\ldots\ldots\ldots \quad (2)$$

Where V_{oco} is the open circuit voltage under standard test conditions, N_c is the number of cells inside the module and T_c is the cell temperature. Based on the values of Vos and Is the output power of each reference module and that with reflectors was calculated. The results indicated that there was a gain in power due to V-Trough reflectors as demonstrated in figure 7. This gain was 48% at solar noon and, as average, 42.6% along day time.

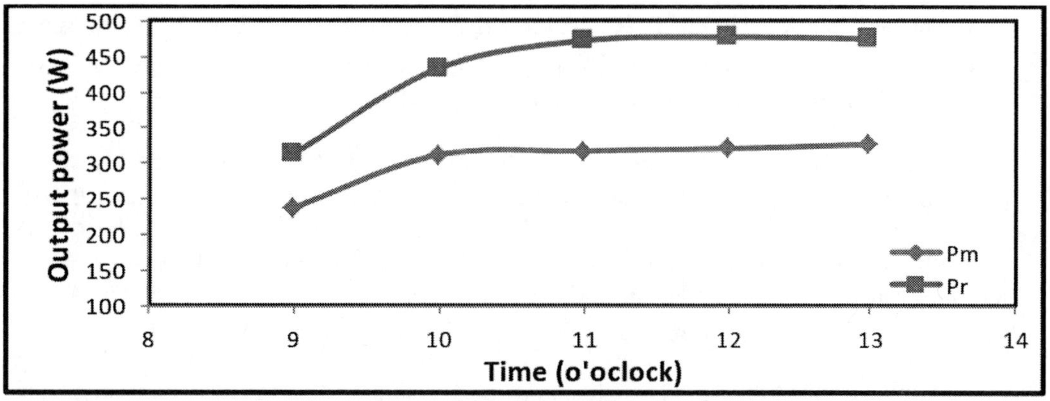

Figure 7. Output power for both modules with (Pr) and without reflectors (Pm).

According to above results, it can be predicted that the area of high-cost solar module can be reduced to, approximately, fifty percent by using low-cost reflectors.

5. Conclusion

From the present work, it can be concluded that:

1. The solar reflectors cause to rise the incident radiation levels and thus increase the PV module short circuit current.

2. The solar reflectors cause to rise the temperature of PV module and thus decrease in its output voltage.

3. The percentage of current increases more than the percentage of droop in voltage and thus there was gain in output power.

4. The use of V-Trough concentrator system coupled with solar module systems can result in a favourable cost-benefit ratio.

5. PVC project is promised in Iraqi environment.

6. References:

[1] Rizk, J., & Nagrial, M. H., (2009)," Impact of reflectors on solar energy systems", International Journal of Electrical, Computer, Energetic, Electronic and Communication Engineering Vol:2, No:5, 2008.

[2] Richard M. Swanso, (2000), "The Promise of Concentrators," Progress in photovoltaics: research and applications Prog. Photovolt. Res. Appl.8, 93±111"

[3] Kostić, L. T., Pavlović, T. M., & Pavlović, Z. T., (2010), "Optimal design of orientation of PV/T collector with reflectors", Applied Energy, 87(10), 3023-3029.

[4] Fraidenraich, N., (1992), "Analytic solutions for the optical properties of V trough concentrators". Appl. Opt. 31 (1), 131–139.

IOP Conf. Series: Journal of Physics: Conf. Series **1032** (2018) 012031 doi:10.1088/1742-6596/1032/1/012031

[5] Al-Shohani, W. A., Al-Dadah, R., Mahmoud, S., & Algareu, A., (2016), "Performance of a V-trough photovoltaic system". In Renewable Energy Research and Applications (ICRERA), 2016 IEEE International Conference , pp. 946-951.

[6] R.B. Bannerot, and J.R., Howell, (1979), "Predicted daily and yearly average radiative performance of optimal trapezoidal groove solar energy collectors", Solar energy, 229-234.

[7] Tabaei, H., and M. Ameri, (2015), "Improving the effectiveness of a photovoltaic water pumping system by using booster reflector and cooling array surface by a film of. "Iranian Journal of Science and Technology. Transactions of Mechanical Engineering 39.M1: 51.

[8] Alaa A. shneishil, (2016), "Desighn and performance Analysisof a v-trough photovoltaic concentrator", Dyala journal for pure science, vol.13, No.1-3,pp.134-116, ISSN.8373222/25189255.

[9] Tomas Markvart,(2000), *"Solar Electricity, second edition"*, University of Southampton, UK, Madrid, Jon Wiley and sons Ltd, ISBN 0-471-98852-9.

IOP Conf. Series: Journal of Physics: Conf. Series **1032** (2018) 012032 doi:10.1088/1742-6596/1032/1/012032

Ambient Decontamination by Non -Thermal Plasma in Atmospheric Pressure Air

Mohammed Ubaid Hussein[1], Rana Talb Mohsen[2]

[1]Department of Medical Physics & Physiology, Medicine College, University of Anbar , Iraq.

[2]Department of Biochemistry, Medicine College, University of Anbar, Iraq.

mmphysics361@gmail.com

Abstract : In this research, Cold Plasma, which is non-thermal, is used as a weak gas and ionized , away from equilibrium ;thermally and dynamically. The needle plasma , a discharge in the non-thermal atmosphere is performed by different low-ionizing gases such as argon. The important properties of this plasma are that it works at approximately room temperature .Decontamination of surfaces contaminated and water Of bacteria (*Salmonella typhoid*, *pseudomonas aeruginosa*) have been investigated in discharge of atmospheric pressure . The generator cold plasma as non equilibrium. Water treated through the needle pole has led to the removal of biological contamination. The experiments provided to separate different plasma factors, oxidation stress measurements in cell membranes, and induced chemical changes in treated water to better understand plasma factors responsible for disabling microbes. Roots and reactive oxygen species appear to be biologically dominant agents.

1.Introduction.

The plasma described as ionized gas , containing free charge carriers (electrons and ions) **[1]**. The temperature of the electron is 1-10 eV, and the electrons are unable to transfer their entire kinetic energy from an electric field applied externally to the particles, so the gas remains non-thermal. Cold plasma is used as a weak gas and ionized , away from equilibrium ;thermally and dynamically **[2]**. The different energy states between electrons, ions and neutral molecules in Non-thermal plasma **[3]**. Neutrals , ions remain relatively non thermal , this provides the possibility of using this plasma in treat as biological tissues , heat-sensitive materials and biological materials **[4]**.

Needle plasma is operate by different noble gas (He, Argon) , the plasma does not cause any thermal damage to the materials that come into contact with it, it possesses important properties at its work at room temperature nearby. This property opens the possibility of using this plasma to process heat sensitive materials, and this type of plasma has the composition of one electrode, also this type developed for deactivation the bacteria **[5]**.

2.Theory.

In non-thermal plasma, free radicals and reactive oxygen species play their primary and important role in this case , to the charged particles. Important mechanisms depend on the formation of non-thermal plasma, treated Microbiology, temperature and the environment (air, surfaces , and water).

In cold air discharges (torch , Dielectric Barrier Discharge, corona), radicals and Reactive oxygen species (ROS) are identified as the dominant Deactivation agents**[6]**.

Escherichia coli bacteria is a member of considered an opportunistic disease and consists of a large family of enterobacteriaceae bacteria. Which live in the intestines of animals especially people ,and may affect stomach upset leading to diarrhea and urinary tract infections.

In epithelial cells of the small intestine, *salmonella typhi* spreads and begins to attack these cells ,is lead in regional lymph nodes and the under - lying connective tissue, begins to multiply. *S. typhi*

IOP Conf. Series: Journal of Physics: Conf. Series **1032** (2018) 012032 doi:10.1088/1742-6596/1032/1/012032

bacteria are transmitted by water and food contaminated with faeces , so this bacteria is a cause of typhoid fever in people. The organism is typically transmitted by fecally contaminated water & food. When enters the bloodstream where it produces an acute germ and thus infects the bone marrow spleen as well as the liver, , and ultimately gall bladder and kidneys. This stage, accompanied by high temperature and diarrhea , as figure 1; shows *Escherichia Coli* and *Salmonella typhi* bacteria **[7]**.

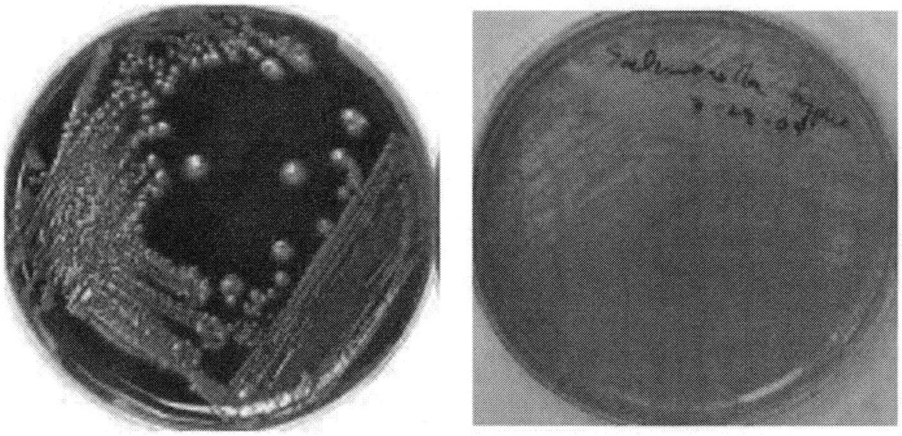

A B

Figure 1. A.) Escherichia Coli on sheep blood agar B.) Salmonella Typhi on salmonella –shigella agar.[7]

3.Setup Experiential.

At 10 kHz frequency, use the high voltage power supply in our study to generate the discharge between two poles with air, one of which is surrounded by insulation . The discharge current and voltage are measured by oscilloscopes with high frequency.

The non-thermal plasma system depends on the conventional plasma discharge which is mainly driven by the AC system . High alternating voltage (H.V) is alternating between two conductors one is surrounded by an insulator to prevent the arc transition.

The discharge was generated by applied voltage **(0 to 5 kV)** (from peak-to-peak) between the grounding base that holds the sample and an insulated high-voltage pole.

Power is connected to stainless steel tube , current power supply and a variable voltage was used for samples. . The barrier prevents flow of current between the electrodes, for this non-thermal plasma is created with active reactive concentrates and little gas heating, the distance varying from (1 to 5cm).

In frequency **(15 kHz)**, that is to prevent transition to an arc and limit the current . Figure 2, shows the installation of experience in our work . The system consists of power supply of high voltage varying from **(0 to 20 kV)** related to a wire to the stainless steel tube. Other part of the system is gradually connected by a piece of mica to stainless steel to prevent the transmission of discharge to the catcher. Discharge occurs between the top surface and bottom surface of tube of the sample. The

IOP Conf. Series: Journal of Physics: Conf. Series **1032** (2018) 012032 doi:10.1088/1742-6596/1032/1/012032

distance at which discharge occurs was controlled from 1 to 5 cm. The diameter of the glass tube was 3 cm. According to the same procedure , all the treatments are at atmospheric pressure and room temperature .

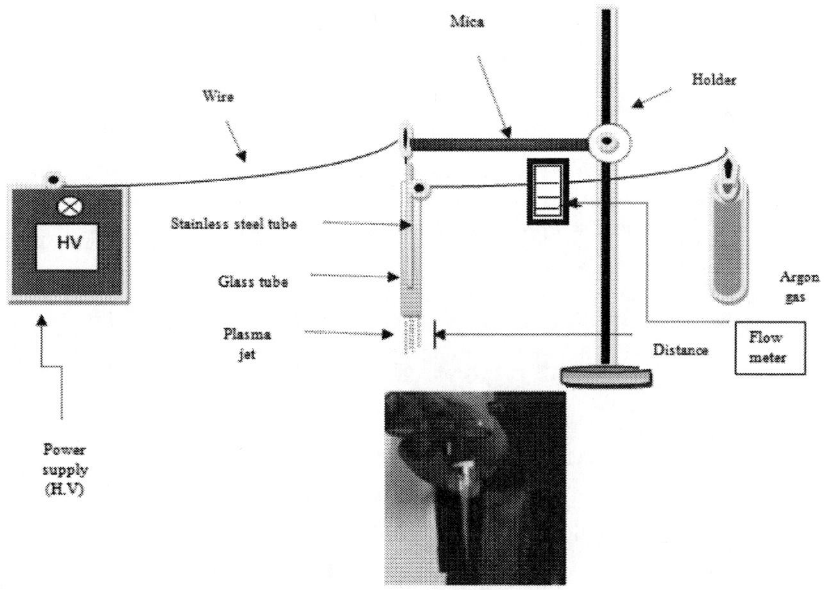

Figure 2. System Non-Thermal Plasma Needle .

4.Method.

These samples were collected from water, sterilization tests were performed by drop from water sample and culturing bacteria before and after plasma needle treatment, while show growth of normal water bacteria (e.g *Salmonella Typhi* bacteria) on the cultures of control sample cultured a large quantity of bacteria obtained from were cultured on artificial media (salmonella –shigella agar, Blood agar) to obtain bacteria colonies, carried out investigation of effects of non thermal plasma on bacterial cultures to determine size of disinfection , determine possible factors responsible. The prepared culture plates were exposure by plasma needle , they are then incubated for (24h).

5.Results & Discussion.

Optimum conditions was selected Voltage (2kV), frequency (15 kHz), distance (3,5 cm), time (10, 20, 30, 40 sec) , and flow rates (1, 2) l/min, by using plasma needle for killing bacteria that inhabit from water. The initial study compared the individual susceptibility of two bacteria belonging to different species *E.Coli bacteria* and *salmonellae typhy bacteria*.

5.1. Effect of flow rate on bacteria killing:

At applied voltage (2kV) and distance (3cm) ,increasing with flow rate of argon gas shows increasing with killing percentage in *salmonella typhy* and *E. Coli* bacteria through increasing plasma treatment time as figures 3 &4.

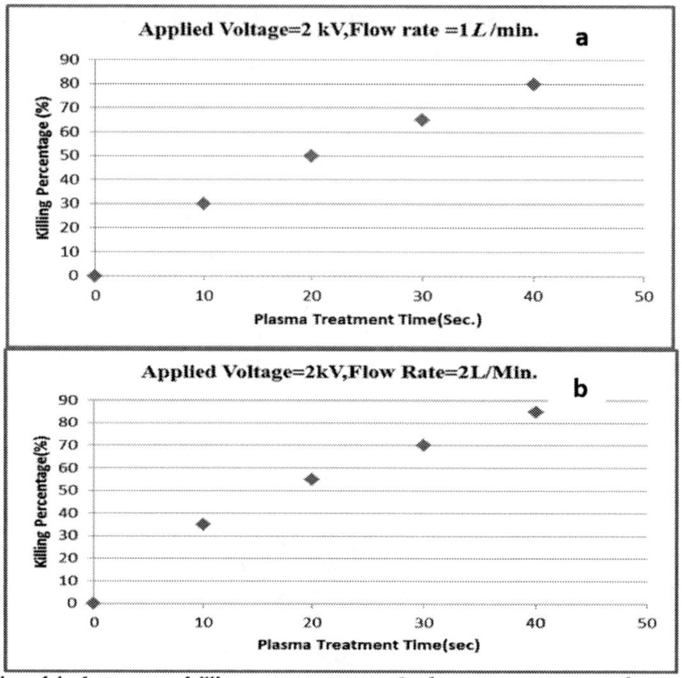

Figure3. The relationship between killing percentage &plasma treatment time of salmonella typhy bacteria at distance 3 cm and Applied voltage =2kv at (a) Flow rate =1L/Min. and (b) Flow rate =2 L/Min.

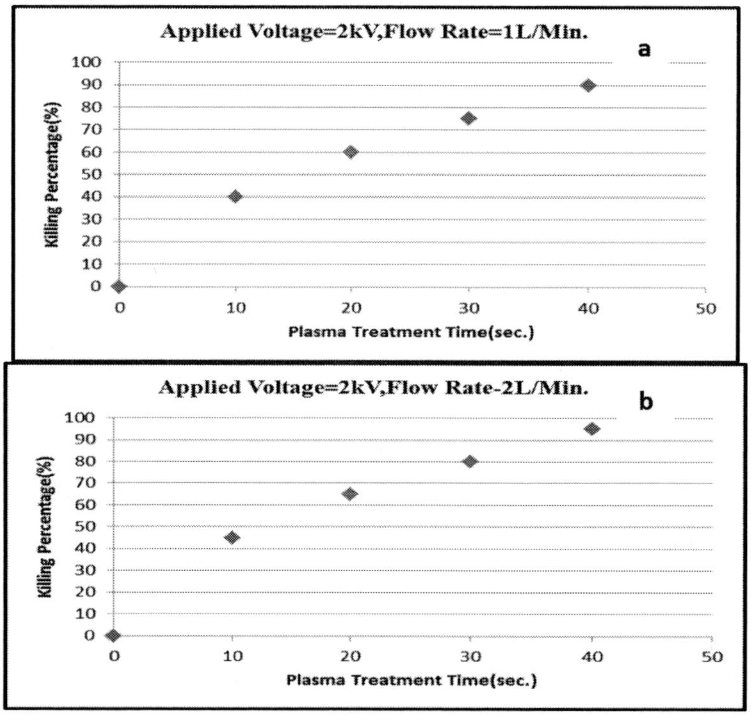

Figure 4. The relationship between killing percentage &plasma treatment time of E.Coli bacteria at distance 3 cm and Applied voltage =2kv at (a) Flow rate =1L/Min. and (b) Flow rate =2 L/Min.

IOP Conf. Series: Journal of Physics: Conf. Series **1032** (2018) 012032 doi:10.1088/1742-6596/1032/1/012032

5.2. Effect of distance on bacteria killing.

At applied voltage (2kV) and distance (5 cm) ,increasing with flow rate of argon gas shows increasing with killing percentage in *salmonella typhy* and *E. Coli* bacteria through increasing plasma treatment time as figures 5 &6.

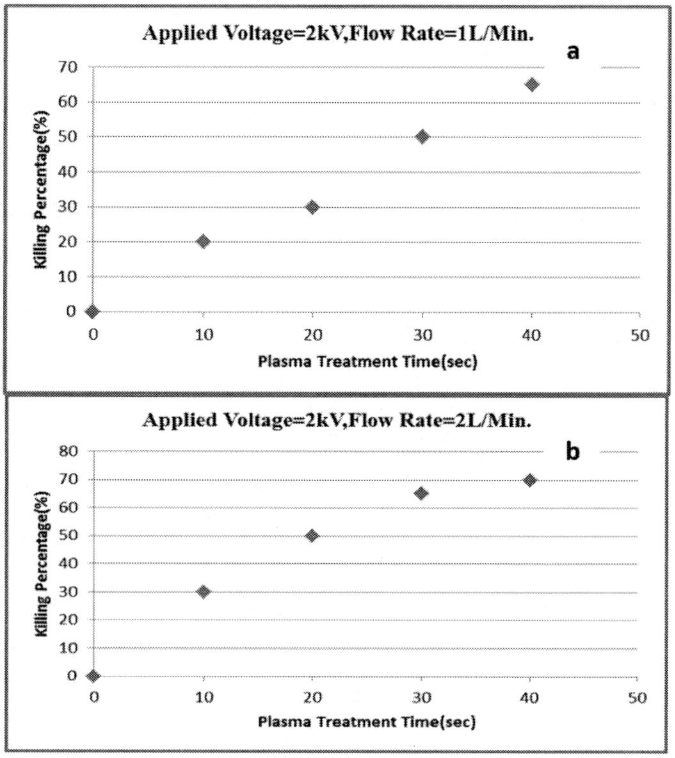

Figure 5. The relationship between killing percentage &plasma treatment time of salmonella typhy bacteria at distance 5 cm and Applied voltage =2kv at (a) Flow rate =1L/Min. and (b) Flow rate =2 L/Min.

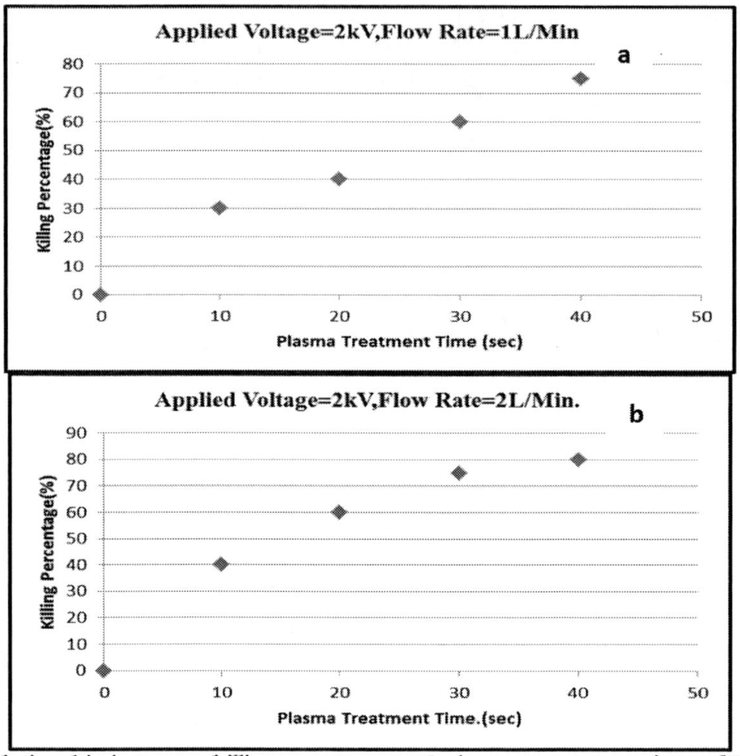

Figure 6. The relationship between killing percentage &plasma treatment time of E.Coli bacteria at distance 5 cm and Applied voltage =2kv at (a) Flow rate =1L/Min. and (b) Flow rate =2 L/Min.

The effects of the increasing gas flow rate and reactive oxygen species penetrate inside the outer membrane of the bacteria which has important function of the deactivation of the bacteria by non thermal plasma . If bacteria are treated with increase gas flow rate, cell structure can be destroyed and the charges distributed on the cell membrane . The release of cytoplasm and the destruction of the cell wall will cause bacterial cell death due to high charged particles .

At operating condition that applied in this work , *Salmonella typhy* bacteria more sensitive to the non thermal plasma treatment than *E.Coli* bacteria;, the partial deactivation time for *E. Coli* bacteria greater than *salmonella typhy* at time (40sec) of same condition. Focuses on determining plasma factors in deactivation the chemical effects induced in water (H_2O_2, NO_2 - and NO_3 -) and their biocidal effects by coupling the electrical discharge , roots and reactive oxygen species (ROS) appear to be biologically dominant agents. Cold plasma produces long living as H_2O_2 , and short lived as OH, ions , electrons and neutral particles. These are likely to be cytotoxic, stimulating low levels of cell membrane rupture and altering signaling pathways between cells. It is therefore possible to create a specific plasma to produce either charged or neutral particles to know the critical mechanism .

IOP Conf. Series: Journal of Physics: Conf. Series **1032** (2018) 012032 doi:10.1088/1742-6596/1032/1/012032

Destruction of the outer membrane of bacterial cells by charged particles which played an important role **[8]**.

6.Conclusion.

The non-equilibrium cold plasma generator, water treated through the electrode needle, led to rapid bio-purification , in non thermal plasma ; the inactivation depends on operating conditions such as treatment time , gas flow as well as distance .

Cold plasma produces long living as NO ,neutral particles , short lived as OH, as well as charged particles(ions and electrons). These are likely to be cytotoxic, stimulating low levels of cell membrane rupture and altering signaling pathways between cells. It is therefore possible to create a specific plasma to produce either charged or neutral particles to know the critical mechanism, reactive oxygen species , radicals were found agents of atmospheric air discharges .

References

[1] Jan, Van. & Dijk, M. 2012 *CAUSA symposium, Plasma technology prospects for biomedical applications*, Eindhoven University of Technology Department of Applied Physics,.

[2] Ibrahim Karim Abbas, Mohammed Ubaid Hussein, Hamid H. Murbat, 2017 *Iraqi Journal of Science,* " The Study of Electrical Description for Non-Thermal Plasma Needle System",, *Vol. 58, No. 3B, pp: 1447-1453..*

[3]. Xiong, Q. and Lu, X.P. 2008 *IEEE Transaction on Plasma Science* ,An Atmospheric pressure Non equilibrium Plasma Jet Device,. **36** (4): 420-428.

[4] Stoffels, E., Sakiyama Y. and Graves D. B. 2013 *IEEE Transactions on Plasma Science*, Cold Atmospheric Plasma: Charged Species and Their Interactions With Cells and Tissues, **36**(4): 1441-1454.

[5] Mohammed Ubaid Hussein & Rana Talb Mohsen 2017 *Al-Kindy College Medical Journal , university of Baghdad,* Urinary Tract Infection Risk Assessment By Non- Thermal Plasma In Iraqis Patients *, : Vol.**13** pp. 86-91, No. 1,.*

[6] K. Oehmingen, M. Hahnel, R. Brandenburg 2010, Plasma Proc. Polym. **7** , 250.

[7] Michael J. Leboffe, Burton E. Pierce 2011" A Photographic Atlas for the Microbiology Laboratory ",4th Edition ,San Diego City College, Morton Publishing Company,.

[8]Mohammed Ubaid Husain 2015 Low-Temperature Plasma Interactions in Medical Applications, University of Baghdad as a partial fulfillment of the requirements for the Degree of Doctor of Philosophy in Medical Physics

IOP Conf. Series: Journal of Physics: Conf. Series **1032** (2018) 012033 doi:10.1088/1742-6596/1032/1/012033

Fabrication of Hydrophobic Nanocomposites Coating Using Electrospinning Technique For Various Substrate

J H Talal [1,1] **, D B Mohammed** [1,2] **and K H Jawad** [2,3]
[1]University of Technology , Department of Applied Science, Iraq
[2]University of Babylon, Department of Poly. Eng. ,Iraq
[1]hassantalal9999@yahoo.com , [2]balkees1957@yahoo.com ,
[3]hanajawad21@yahoo.com

Abstract. Hydrophobic coating were fabricated using electrospinning technique on different substrates (glass, ceramic and metal) , polymer solutions of (Si) were used in different range of composition for each solutions, also nano Al_2O_3 and nano TiO_2 were used to fabricate nanocomposites. Contact angle, surface tension , viscosity , X-ray diffraction and FTIR were calculated for all specimens. (20%Si) owned a larger contact angle about (110.173°) for metal substrate and show a hydrophobicity for ceramic and glass substrate after coating with (20%Si). SEM showed the morphology of the surfaces and show that specimens has good surface morphology after coating with (20%Si/Al_2O_3) and (20%Si/TiO_2). All results were discussed.

Keywords : (Si) Silicone Rubber R.T.V. , (Th) Thinner , (TiO_2) Nano Titanium dioxide , (Al_2O_3) Nano Alumina, (CA) Contact angle , (SEM) Scanning Electron Microscope

1. Introduction

Recently, interested of engineering and nanotechnology increased to the hydrophobic materials and coatings according to their requirements to obtain material with the requested properties which are better than raw materials. This research focused on studies the hydrophobic coating to achieve materials with special surface morphology by calculating wettability of coated surfaces (ceramic, glass and metal) substrate to find out the hydrophobicity and characteristics of coated surfaces, these substrate were picked up because they are very common in use to daily people life and industrial manufacturing which control an special role around this research. Also tested to Scanning electron microscope (to show the morphology texture of coated substrate).

Hydrophobic coating, is the resistance of material surface to the water and according to the contact angle test , hydrophobic must have an angle acceded more than 90o which between (90o - 150o), There are different types of hydrophobic substrate in nature, these examples has hydrophobicity and low adhesions properties. These special characteristics showed a great consideration by science researchers that involving (self-cleaning, anti-corrosion, anti-wetting ..etc.). So this article specialized on fabricate hydrophobic coating to obtain what almost in nature types and achieve these special requested properties[1].

Electrospinning is one of the most practical and better method which become familiar and popular in the last years and interested grown up towards nanotechnology because of the production of nanofibers polymer with diameter (2nm - 5μm). Generally, electrospinning process contains syringe pump, high voltage source and collector, the polymer solution place inside syringe pump and boosted in constant flow and the needle connected to high voltage source in range (3KV - 30KV)[2].

IOP Conf. Series: Journal of Physics: Conf. Series **1032** (2018) 012033 doi:10.1088/1742-6596/1032/1/012033

Parameters of electrospinning process are the scale of the calculated material from electrospinning process that affact on the fibers properties. There are many types of electrospinning parameters (Solution parameters, Instrumental parameters and Ambient or Environmental parameters)[3].

Viscosity is very important in electrospinning technique which determine by the add of polymer concentration to the solutions, factors effect on viscosity is molecular weight and the concentration of the polymer.

Surface tension is also one of the substantial agent effect on electrospinning operation which affact by the composition of polymers solution and this will outcome with changing the surface tension. When there is high amount ratio of free solvent molecules, molecules mixed to each other and form beads[4].

Contact angle, define the wettability of the substrate which are very useful in this article which involve the perusal of contact angle test as primary data, and calculated the degree of wettability when the water drop touch surface that been coated. Contact angle split to many parts and every sections has values of wettability range and resistance of the substrate to the water drop , samples in this research will calculated by this test to define the wettability of the substrate that fabricate by electrospinning process, sections of CA are (figure (1) shows CA range) [5] :

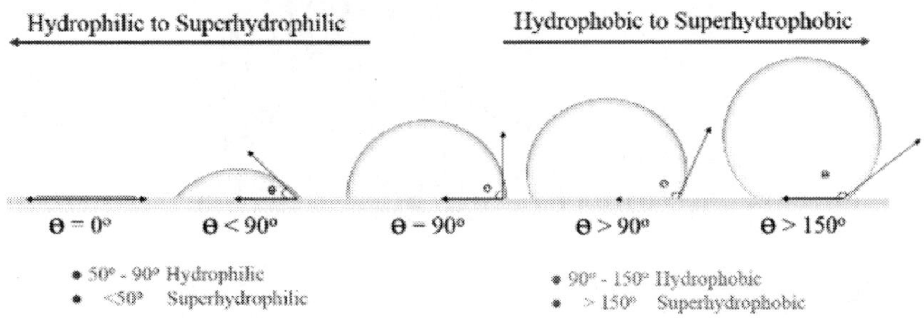

Figure 1. Contact Angle Range.

2. Experimental Part

2.1 Materials Used

1. Silicon Rubber R.T.V.

2. Thinner Solution (Sun Shine Futureline).

3. Nano titanium dioxide (TiO_2) nano particle with diameter (20nm - 40nm) from micxy reagent company.

4. Nano Alumina (Al_2O_3) nano particle with diameter (20nm - 30nm), purity (99.99%).

IOP Conf. Series: Journal of Physics: Conf. Series **1032** (2018) 012033 doi:10.1088/1742-6596/1032/1/012033

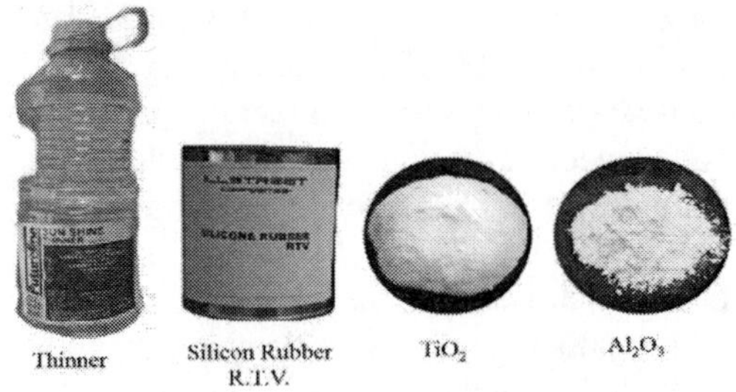

Figure 2. Materials Used.2.2 Preparation of Samples

Substrate used are (ceramic, glass and metal) each of these surface has cleaned with distilled water and alcohol for 5 min then dried in oven at temperature of 35-40° to acheive a smooth substrate without infection and scotch to the surface [6]. Figure (3) shows samples prepared before been coated.

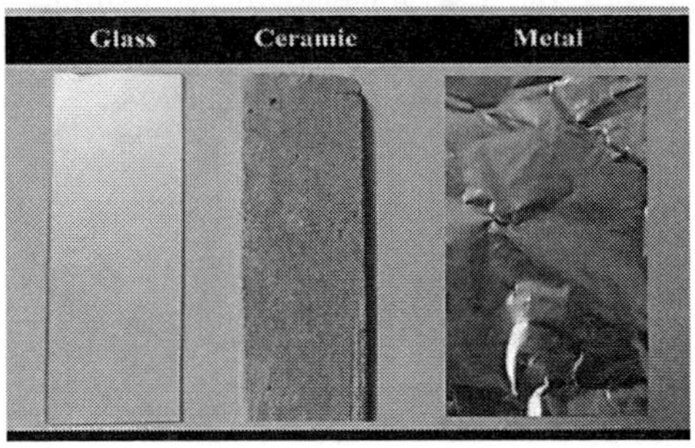

Figure 3. Sample Preparation (Ceramic, Glass and Metal) before coating.

Preparation of polymer solution is done by weighed silicon rubber R.T.V. in various wt% composition and dissolve it with thinner solution , the percentage of (Si/Th) solution prepared in this research are in range of (10%,15% and 20%) of Si , and placed on a magnetic stirrer for 6 hours to make an homogenous polymer solution , and the best contact angle of (Si/Th) solution coated to metal substrate is (20%Si/Th) was depended in this article because it shows the larger CA over other percent which shows about (110.173o) degree and depending on this percent another solution prepared by addition of nano Al2O3 and nano TiO2. Figure (4) shows various polymer solution of (Si/Th.), (20%Si/Th./Al2O3) and (20%Si/Th/TiO2) [7]

Figure 4. Polymer Solutions of (Si/Th), (20%Si/Th/Al_2O_3) and (20%Si/Th/TiO_2).

For metal sample is coated with (Si/Th) solution with different wt% composition of (10%,15% and 20%), and resultant with hydrophobicity properties after been coated and tested to CA test, that provide good enhancement to the surface of the specimen, figure (5) shows metal samples coated with various composition of (Si/Th) solution :

IOP Conf. Series: Journal of Physics: Conf. Series **1032** (2018) 012033 doi:10.1088/1742-6596/1032/1/012033

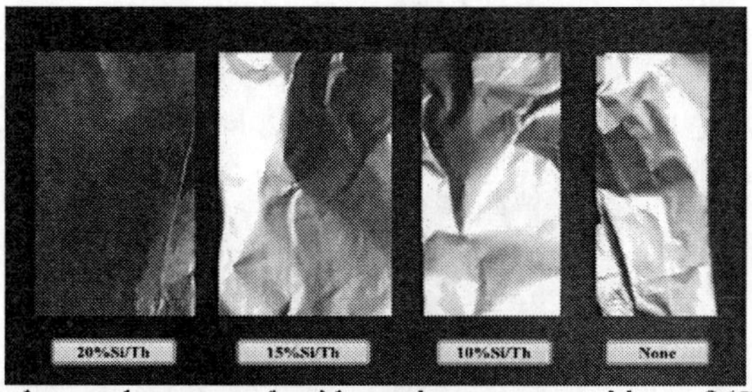

Figure 5. Metal samples coated with various composition of (Si/Th) solutions.

(20%Si/Th) shows larger CA to other coated samples and depending on this percent , a solution of (20%Si/Th/Al_2O_3) and (20%Si/Th/TiO_2) also prepared to fabricate nanocomposite coating by adding a known amount of each nanoparticle used that mentioned , figure (6) shows metal samples coated with (20%Si/Th/Al_2O_3) and (20%Si/Th/TiO_2) solution :

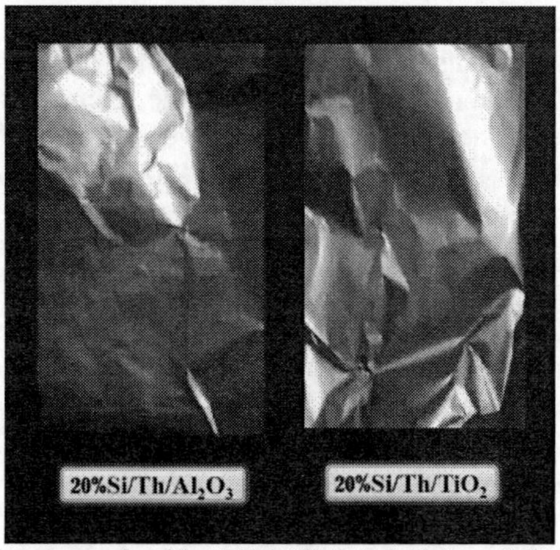

Figure 6. Metal samples coated with (20%Si/Th/Al_2O_3) and (20%Si/Th/TiO_2) solution.

For glass sample is coated with (Si/Th) solution with different wt% composition of (10%,15% and 20%), and resultant with hydrophobicity characteristic after been coated and tested to CA test, which provide good enhancement to the surface of the sample, figure (7) shows glass samples coated with various composition of (Si/Th) solution :

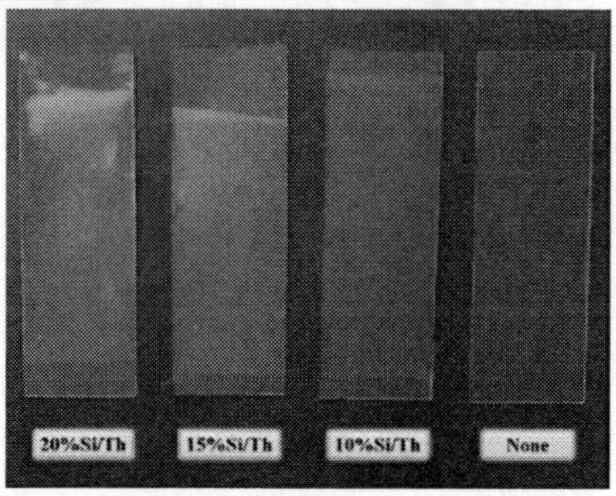

Figure 7. Glass samples coated with various composition of (Si/Th) solution.

IOP Conf. Series: Journal of Physics: Conf. Series **1032** (2018) 012033 doi:10.1088/1742-6596/1032/1/012033

For ceramic sample is coated with (20%Si/Th) solution and resultant with great enhancement to hydrophobicity characteristic from (0°) of CA which mean superhydrophilic to hydrophobic surface of (90°) of CA test after been, figure (8) shows ceramic sample coated with (20%Si/Th) solution:

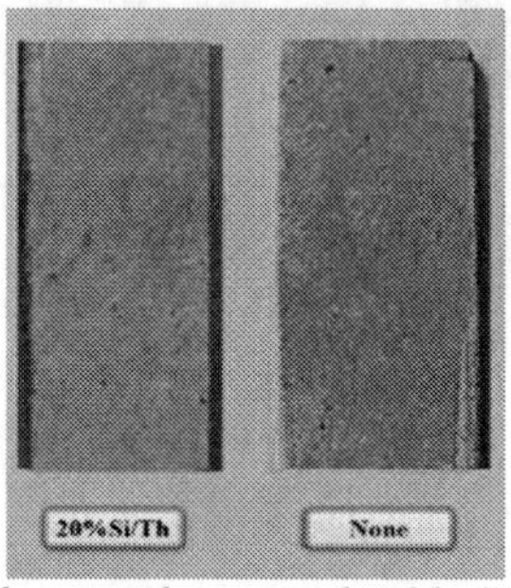

Figure 8. Ceramic sample coated with (20%Si/Th) solution

Installation of electrospinning operation should be set up with advance handle person and care should take. Generally , the operation of electrospinning has syringe pump, high voltage source and collector, initially, this process contains of 3 electrode, the polymer solution that prepared previously explained, placed inside syringe pump, one of the electrode is connected to the needle of the syringe pump, the needle diameter is connected with high voltage source (6.5KV used in this research), the syringe used is 3ml size, which passing in static flow rate. other electrode is connected to the plate that hold samples substrate on it which named collector and the third electrode is connected to earthen to prevent electrical fault shock [8].

When an electrical current is passing to the needle of the syringe pump it generate electro static force that changing the surface tension of (Si/Th) solution and deform the shape of spray on collector plate and when the voltage increased the change in the shape of the spraying increased, when the voltage stabilized it work on evaporate the solute and coating fabricated by this process only stay on the surface of the sample [2-3]. Figure (9) show electrospinning technique method preparation and set up as explained previous section.

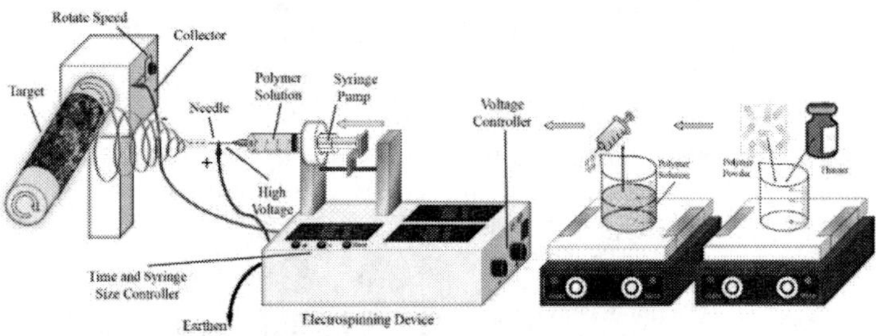

Figure 9. The Electrospinning Technique Method Preparation and Set-up .

3. Results and Discussion

Table (1) shows the result of prepared (Si/Th) solutions after tested to viscosity and surface tension tests. For viscosity show increasing with ascending the percent of Si that mixed with Th solute. By ascending viscosity of the (Si/Th) solution much higher with high molecular weight or ascending the Si concentration in the solution, then a mixture of aggregate and beads will be obtained. Figure (10) shows relation between viscosity and surface tension [9].

Table 1. Viscosity and surface tension of (Si/Th) solutions.

Polymer Solution	Viscosity ($m^2.sec^{-1}$)	Surface Tension ($N.m^{-2}$)
(10%Si/Th)	12.15	10.55
(15%Si/Th)	13.68	12.9
(20%Si/Th)	19.06	17.38

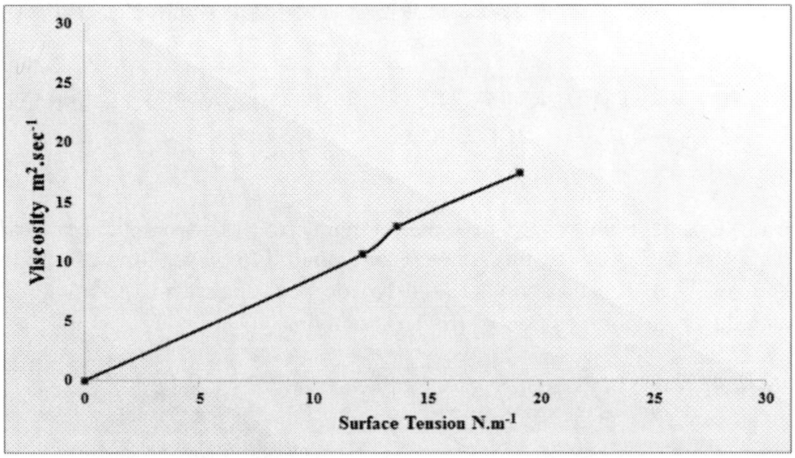

Figure 10. Relation between viscosity and surface tension.

Contact angle has tested for all samples that prepared before and after coating with electrospinning process, For metal substrate CA test shows good enhancement from hydrophilic material to hydrophobic material , (10%,15% and 20%) of (Si/Th) solutions are coated to metal samples and the larger CA was the specimen of (20%Si/Th) which about (110.173o) degree. Figure (11) shows CA result for metal surface coated with different composition of (Si/Th) solutions.

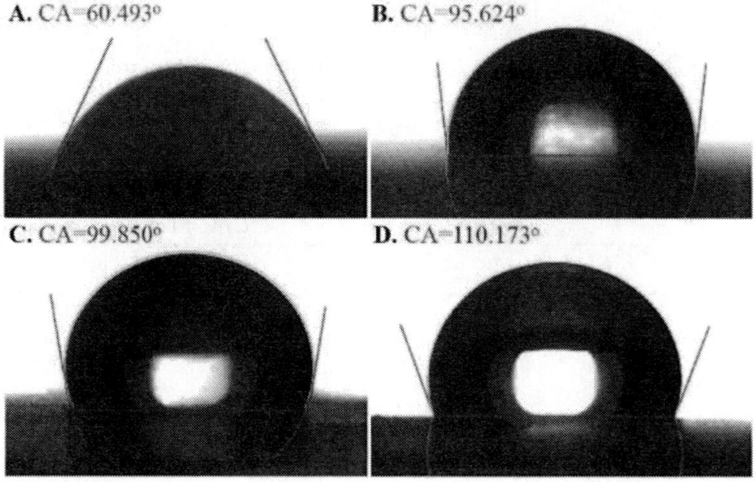

Figure 11. CA. for metal specimens coated with different percent of (Si/Th) solution. A.(None) CA=60.493°, B. (10%Si/Th) CA=95.624°, C.(15%Si/Th) CA=99.850° and D.(20%Si/Th) CA=110.173°

For glass substrate CA test shows good enhancement from hydrophilic material to hydrophobic material , (10%,15% and 20%) of (Si/Th) solutions are coated to glass samples and the larger CA was the specimen of (15%Si/Th) which about (107.491o) degree. Figure (12) shows CA result for glass surface coated with different composition of (Si/Th) solutions.

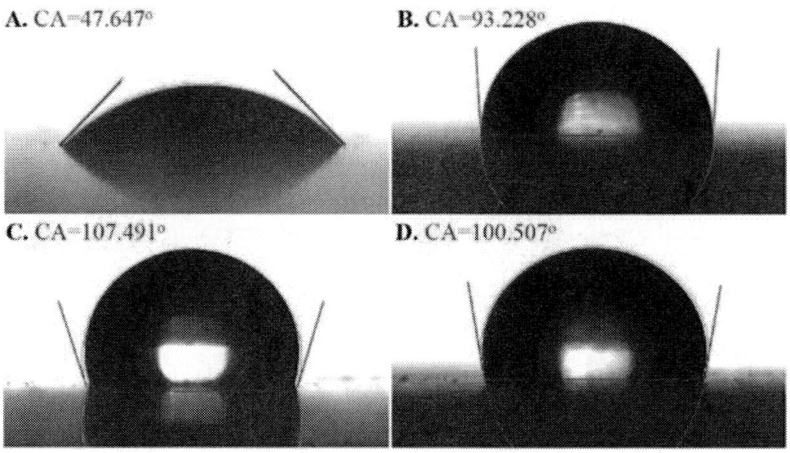

Figure 12. CA. for glass specimens coated with different percent of (Si/Th) solution. A.(None) CA=47.647°, B.(10%Si/Th) CA=93.228°, C.(15%Si/Th) CA=107.491° and D.(20%Si/Th) CA=100.507°

For ceramic substrate CA test shows great enhancement from superhydrophilic material to hydrophobic material , For ceramic (None) substrate shows (0°) CA because of high porosity of the surface which comes out as a superhydrophilic material, (20%Si/Th) solution coated to ceramic sample and shows a hydrophobicity to the surface of the sample. Figure (13) shows CA result for ceramic surface coated with different composition of (Si/Th) solutions [10].

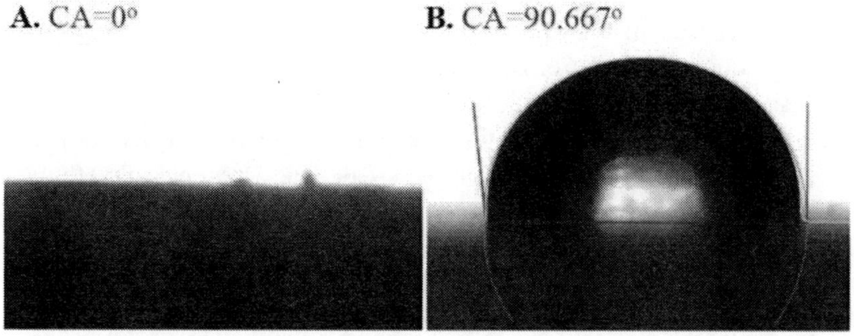

Figure 13. CA. for ceramic specimens coated with (20%Si/Th) solution. A.(None) CA=0°, B.(20%Si/Th) CA=90.667^{o}.

Table (2) shows the balance of CA test for left CA , right CA and average of CA for all specimens coated with (Si/Th) Solutions.

Table (2) : CA test for left CA , right CA and average CA for all specimens coated with (Si/Th) **Solutions.**

Polymer Solution	Contact Angle Left (CA_L)	Contact Angle Right (CA_R)	Contact Angle Average (CA_AV)
(None) - Metal	60.493	60.493	60.493
(10%Si/Th) - Metal	95.624	95.624	95.624
(15%Si/Th) - Metal	99.314	100.386	99.850
(20%Si/Th) - Metal	109.158	111.189	110.173
(None) - Glass	47.647	47.647	47.647
(10%Si/Th) - Glass	95.912	90.544	93.228
(15%Si/Th) - Glass	107.491	107.491	107.491
(20%Si/Th) - Glass	100.507	100.507	100.507
(None) - Ceramic	0	0	0
(20%Si/Th) - Ceramic	92.843	88.490	90.667

As mentioned in previously section metal samples of (None) , (20%Si/Th) , (20%Si/Th/TiO_2) and (20%Si/Th/Al_2O_3), that owns best contact angle as compared to other specimens has also tested to X-ray diffraction to show the structure of fabricated materials or coating. Figure (14) show x-ray diffraction pattern for mentioned specimens and show that specimen without coating has less peak while sample of (20%Si/Th) shows large peaks and also show new three different range of peaks and considered theta steady at 3 position which is about (64.9516^{o}) , (78.0959^{o}) and (44.5536^{o}). Other specimen coated with (20%Si/Th/TiO_2) and (20%Si/Th/Al_2O_3) shows new and different range of peaks, these two sample with nano

IOP Conf. Series: Journal of Physics: Conf. Series **1032** (2018) 012033 doi:10.1088/1742-6596/1032/1/012033

particle coated substrate also owns much larger peaks and higher rate of intensity as compared to (None) and (20%Si/Th) specimens.

Other test was examined for materials used of (Si R.T.V.) , (Nano-TiO_2 Particles) and (Nano-Al_2O_3 Particles) by collecting of remains in coating of (20%Si/Th) , (20%Si/Th/TiO_2) and (20%Si/Th/Al_2O_3) at up of the surface and scratch it to have amount of these powder and test it to FTIR spectrum test, figure (15) shows the FTIR spectra of mentioned materials , for (20%Si/Th) FTIR spectra showed some aggragte of peaks in range of 3100 to 3600 cm^{-1} that have unstable range of multiple peaks confirming that the produced coating is of crystalline form, for (20%Si/Th/Al_2O_3) shows board peak in range of 500-900 cm^{-1} , and for (20%Si/Th/TiO_2) the broad band centered at 500-600 cm^{-1} is assigned to the bending vibration bond in TiO_2 lattice , the broad band centered at 3600-3400 cm^{-1} is assigned to the intermolecular interaction of Si group with metal substrate.
Metal Samples of (20%Si/Th) , (20%Si/Th/TiO_2) and (20%Si/Th/Al_2O_3), that owns best contact angle as compared to other specimens has tested to scanning electron microscope test (SEM) to see the morphology of the surfaces, and show that specimen of (20%Si/Th/Al_2O_3) have higher amount of beads and this due to the moderate viscosity or increasing of Si concentration in the Th solution, then a mixture of aggragate and beads will be obtained, and due to unbalanced applied voltages in range of (6.5KV - 7.8KV) also the use of roller tube collector to obtain alignment coating, as in figure (17).

While in (20%Si/Th) show beads and defect and this forms because (Si/Th) solution has low viscosity also not need a matrix because already adhesion with surface so it appears in this way as in the figure (18), and for (20%Si/Th/TiO_2) shows a great bonding because of the presence of TiO_2 nanoparticles which owns very high surface area that helps to bond togather with less beads. as in the figure (19) [11 - 12].

IOP Conf. Series: Journal of Physics: Conf. Series **1032** (2018) 012033 doi:10.1088/1742-6596/1032/1/012033

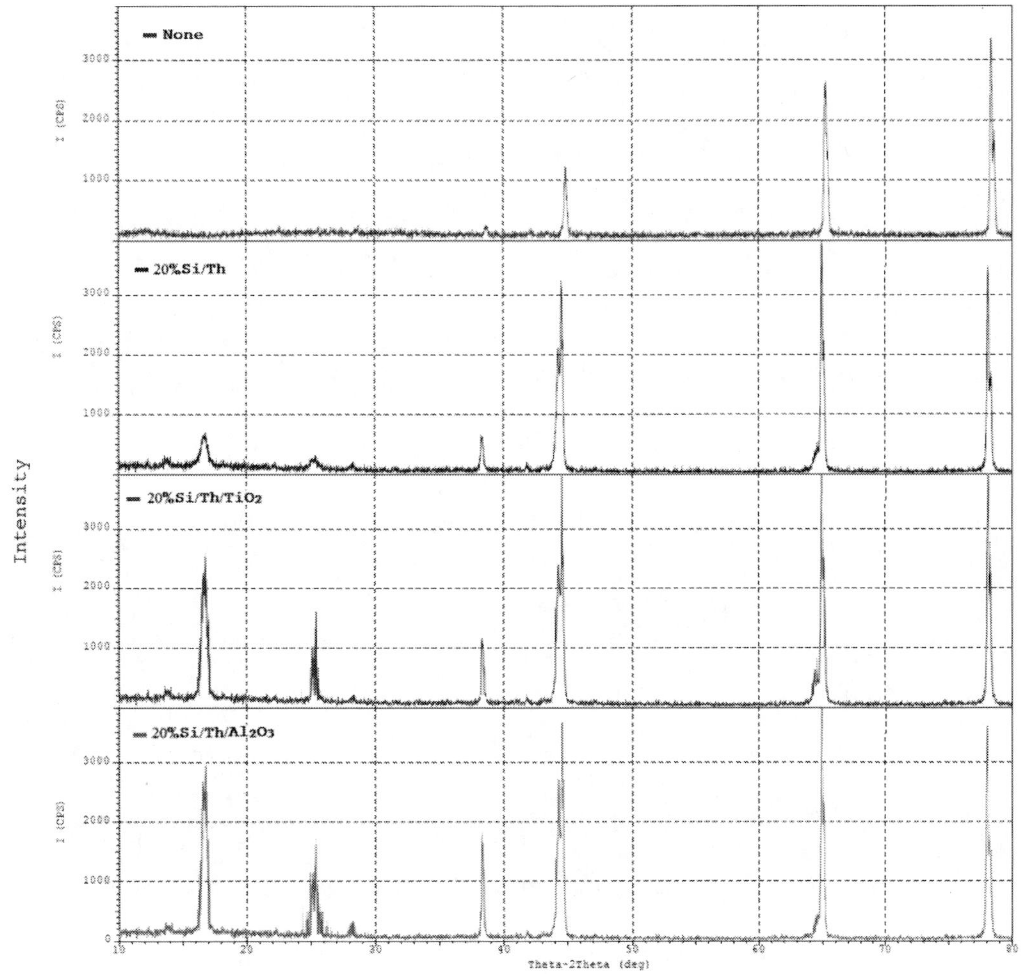

Figure 14. X-Ray Diffraction Pattern for Metal substrate.

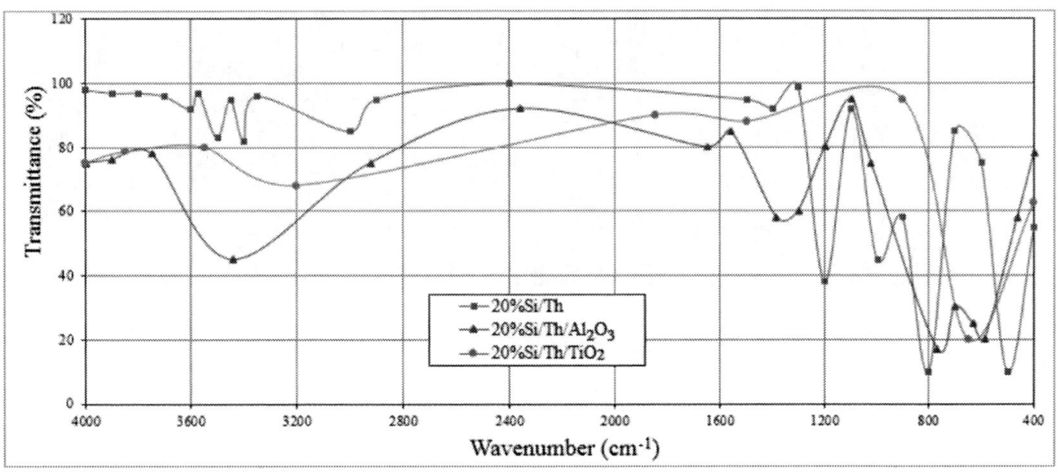

Figure 15. FTIR Spectrum for Collecting Remains of Coating At Metal Substrate.

IOP Conf. Series: Journal of Physics: Conf. Series **1032** (2018) 012033 doi:10.1088/1742-6596/1032/1/012033

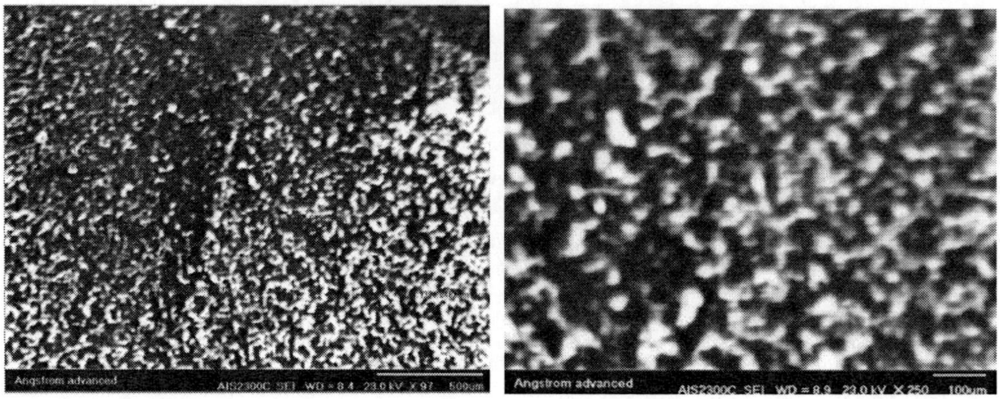

Figure 16. SEM images for (20%Si/Th/Al_2O_3) - Metal substrate.

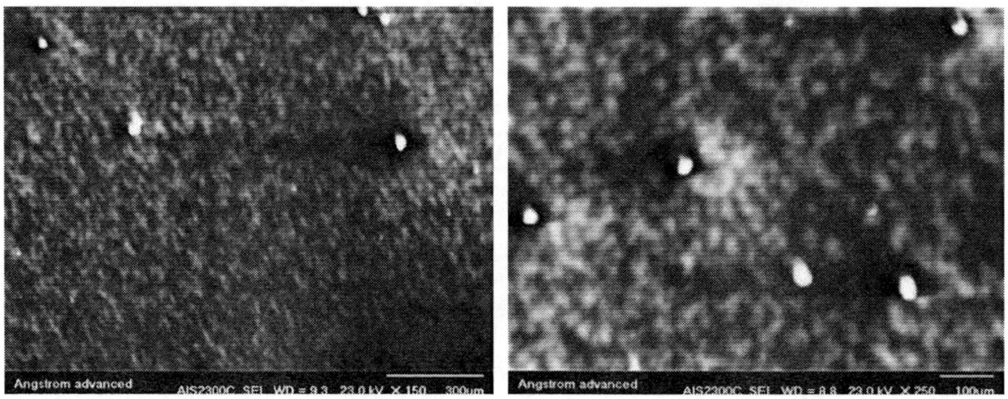

Figure 17. SEM images for (20%Si/Th) - Metal substrate.

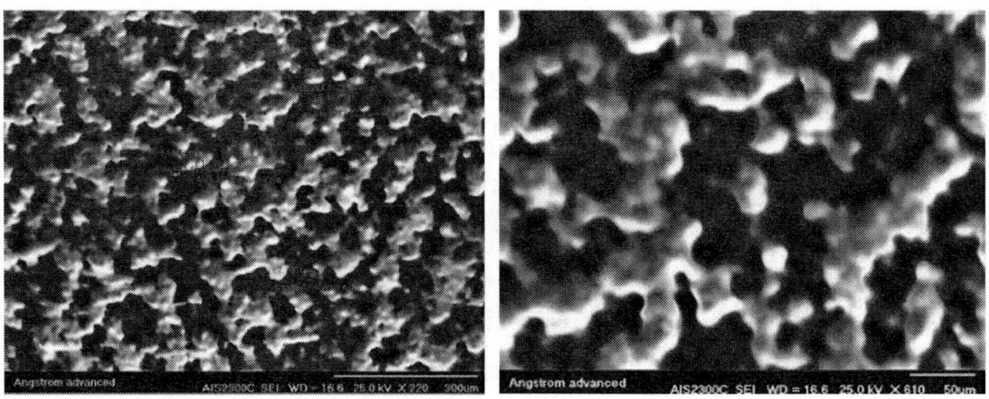

Figure 18. SEM images for (20%Si/Th/TiO_2) - Metal substrate.

4. Conclusions :

- (20%Si/Th) for metal substrate owns the higher contact angle about (110.173^o) as compared to other (Si/Th) coated samples. (15%Si/Th) for glass substrate owns the higher contact angle about (107.491^o) as compared to other (Si/Th) coated samples.
- (20%Si/Th) has higher amount of beads due to the viscosity of (Si/Th) solution.
- (20%Si/Th) has higher viscosity and surface tension.
- Both metal and glass substrate obtains good enhancement from hydrophilic to hydrophobic material.
- Excellent enhancement for (ceramic substrate) after coating by electrospinning process with (20%Si/Th) solution and contact angle changed from superhydrophilic (0°) to a hydrophobic (90.667^o) , and the ceramic surface become repels water instead of absorb water.

5. References:

[1] Zhiping Y., Meimei Wang, Xiaolong Q. and Jijun X. " ***Synthesis and Characterization of Super-Hydrophobic Coating Materials*** " School of Materials Science and Engineering, Hebei University of Science and Technology, 050018 Shijiazhuang, P.R. China *Corresponding author , 2nd International Conference on Green Materials and Environmental Engineering (GMEE 2015).

[2] Thenmozhi, S., Dharmaraj, Kadirvelu N. K., Hak Yong K. " ***Electrospun nanofibers: New generation materials for advanced applications*** " , Review Article , Pages 36-48 , Materials Science and Engineering: B, Volume 217,(March 2017).

[3] Haifeng L., Xili D., Gang Z. , Ping L., Xing W., and Yubo F. ***"Electrospinning of Nanofibers for Tissue Engineering Applications "*** , Hindawi Publishing Corporation, Journal of Nanomaterials, Volume 2013, Article ID 495708, 11 pages, (2013).

[4] Siqi H., Guoxiang L.,, Guangping H., Wanli C., Zongying F.,, Qinglin W., and Qingwen W. " ***Effect of Experimental Parameters on Morphological, Mechanical and Hydrophobic Properties of Electrospun Polystyrene Fibers*** " , Materials, Volume 8, pages 2718-2734; (2015).

[5] NandanaBhardwaj S. and C. Kundu , " ***Electrospinning: A fascinating fiber fabrication technique"*** , Biotechnology Advances , Volume 28, Issue 3, Pages 325-347, May–June (2010).

[6] Yuehua Y. and Randall Lee T. " ***Contact Angle and Wetting Properties"*** , Chapter 1 , springer, surface science techniques, 663 p, ISBN: 978-3-642-34242-4, (2013).

[7] Mohammed S. , Hamza Alaa A., Shatha K. , ***"Fabrication Of TiO2 , V2O5 Thin Film (Super Hydrophobic Surface)By Powder Coating Technique"*** , Journal of Engineering, Number 2 Volume 22 February , (2016).

[8] Azimirad R., and SAFA S., ***"Electrospun polystyrene fibers on TiO2 nanostructured film to enhance the hydrophobicity and corrosion resistance of stainless steel substrates"*** , PRAMANA journal of physics, Indian Academy of Sciences, Vol. 86, No. 3 , March (2016).

[9] GAarg, V., Nath S., " ***Hydrophobic Coating of Polymethylmetha- Crylate (PMMA) on Glass Substrate for Reduced Bacterial Adhesion*** " , Journal of Polymer Materials , Vol. 32 Issue 4, p503-512. 10p. , Oct-Dec (2015).

[10] Seyed A. , Jesse Z. , ***"Superhydrophobic RTV silicone rubber insulator coatings"*** , Applied Surface Science , Volume 258, Issue 7, Pages 2972-2976 , (2012).

[11] Manar A. , Akram R. J., .Emad S. , " ***Preparation and Characterization of an Artificial Tissue Using Polymer Blend by Electrospinning Method"*** Thesis, University of Technology , Iraq, (2015).

[12] Balkees M. D. and Hassan T. J. , " ***Superhydrophobic Nanocomposites Coating Using Electrospinning Technique on Different Materials*** " , University of Technology, Baghdad, Iraq. IJAER , ISSN 0973-4562 Vol.12, No.24 , pp.16032-16038 , (2017).

Acknowledgement

The authors extend their assessment to the University of Technology and University of Babylon / collage of Material engineering polymer Engineering Department for proceed this search.

Pyrolysis Conversion Reaction Mechanism of Sewage Sludge to Biofuel

Itimad K. Al-asfoor, Maysoon J. Zaiter, Ethar M. Jassar, Milad A. Mahmood

Ministry of Science & Technology, Renewable Energies Directorate, Hydrogen & Biofuel Dept.

Abstract: The conversion reactions to produce biofuel include many possible paths and routes leads to chemical changes due to each bath. For example, biofuel produced from solid wastes mainly consists of methane gas (CH4) and carbon dioxide (CO2), in addition to traces of chemicals as impurities.To determine the optimum reactants quantities, and to help engineers in designing the pyrolysis plants we made this research. In this research we will suggest the mechanisms of the internal (sub) reactions that occurs during the conversion of sewage sludge to biofuel via pyrolysis operation according to the results of the analysis of samples taken from different stages of the pyrolysis depending on the products of the research of (Pyrolysis of Sewage Sludge), and its analysis before and after the reaction.

We used and will use many techniques as shown below:

1- Fourier-Transform Infrared (FTIR) analysis in order to know the chromophores to the reactants and products.

2- X-Ray Fluorescence (XRF) analysis for the reactant to determine the metals in the sludge.

3- Carbon-Hydrogen-Nitrogen analysis (C.H.N.) for the samples before and after and within the pyrolysis.

4- Differential Scanning Calorimetry (DSC) analysis to determine the stages of pyrolysis operation and the rate determining step (slow step) that affect the velocity of the reaction.

Key words: Mechanism, slow step, chemical conversion.

3- Carbon-Hydrogen-Nitrogen analysis (C.H.N.) for the samples before and after and within the pyrolysis.

4- Differential Scanning Calorimetry (DSC) analysis to determine the stages of pyrolysis operation and the rate determining step (slow step) that affect the velocity of the reaction.

1. Introduction:

The noticeable amounts of sewage sludge are produced during the industrial and domestic wastes of water treatment by biological and chemical processes and it increases rapidly.

Many improvements have been made on the treatment processes of sewage sludge including; thermal treatment, direct treatment (composted as fertilizer), landfill, and dumping to the sea.

On the other hand, all these methods are conventional and cause many problems throughout the process; e.g. the thermal treatment (by combustion or incineration) causes ash resulted from the operation, a land pollution is caused by toxic metals and organic compounds due to the use of direct process (as a fertilizer) [1].

An alternative method called pyrolysis supplies a solution to most of the problems mentioned above except the poisoning by (cadmium and mercury), also it gives a new source of power and new valuable chemicals by utilizing the stored energy [2].

Analysis of the gas phase must be done; these qualitative and semi quantitative analysis show that the pyrolysis process undergoes formation of new compounds.
In order to understand the intrinsic of the pyrolysis of the sewage sludge, study of the kinetics of the pyrolysis is the important part, kinetic data for fuels (fossil and alternative) and solid wastes were reported [3-11].
Few researches deal with the produced gas quantitatively.
The design of the pyrolysis reactor differs from company to another one, it may be designed as continues, batch, or mixes wastes [12-13].

1.1 Aim of the research:

1-Determination of the optimum quantities of the reaction.
2-Having a comparison leads to the best path according to the available conditions.
3- Help the engineers in their calculations about designing the pyrolysis plants.
4- Treatment and purification of the produced gas by releasing the impurities at the proper step.

2. Materials and Methods:

2.1 Sample preparation:

Our samples were provided from (Al-Rustumiya site for treating the sewage water). First the sample was dried in vacuum oven with temperature about 1000C for 24 hour, and then it was grinded into molecular size within (0.2-2) mm, finally it was sieved out to have different molecular sizes. By using the pyrolysis unit shown in below we did our experiments at different temperatures and molecular size of 0.8 mm.

2.2 Tests and analysis:

In order to determine the chemical properties of the sewage sludge many quantitative and qualitative analysis was applied for the reactant and the products of the pyrolysis: Tables (1-4) show these properties.

2.2.1 FTIR Analysis: Figures (1-5) illustrates the FTIR spectra for the row sewage sludge and for the sample before and after pyrolysis at different temperatures, whereas table (5) summarizes the values of the wavelengths of the whole operation at different temperatures.

2.2.2 Differential Scanning Calorimetry (DSC): In order to complete the results of the pyrolysis of sewage sludge and suggesting the mechanism of the operation, a (DSC) analysis was held to samples of sewage sludge.

To determine the sludge properties, differential scanning calorimetry (DSC); it is a technique measures the difference between the heat needed to increase the temperature (of decomposition and reformation) of the sample relative to a reference, where the sample and the reference exposed almost to the same temperature in a linear relationship as a function of time [14].
The DSC analysis was studied by (Schimadzu DSC-60), at a range of temperature from (0 – 300) °C, and we got the curves (6 and 7).

3. Results and discussion:

The mechanism of pyrolysis of the sewage sludge has been studied via many techniques as mentioned above. By studying these curves, we suggest a mechanism for the sewage sludge pyrolysis and the steps of this reaction, where the reaction occurs in three steps and the second step was the slow step (rate determining step) at a temperature between (141 – 148) °C. By comparing these results with the (FTIR) curves we found that:

- Appearance of a peak at the wavelength range (3600 – 3200) Cm^{-1} refers to (O-H) bonding of water , also we have a peak at (100) °C in the (DSC) curve (fig. 6) shows that the sample undergoes a dehydration (even the water came from the moisture or within the chemical composition of the sample) at the first step of pyrolysis. The evidence was when we dried a sample of sludge and analyze it by (DSC), the peak at (100) ° C disappeared, (fig. 7).

- The peak at (1100 – 1000) Cm^{-1} refers to (C=O) bonding, and the peak at (1520 – 1400) Cm^{-1} refers to (C-H) bonding show formation of CO, CO_2, and some aliphatic compounds from simplest one (CH_4) to bigger. In addition, the (DSC) curve shows a peak at (141 – 148) °C leads to a change in the phase from solid to liquid and gaseous. By noticing the high loss in mass of the reactant at this temperature (at this step) because of the conversion of most of the mass into gases (this result agree with the peak intensity in DSC curve (fig. 8), it shows that the second step is the rate determining step of the pyrolysis reaction, which also shows a highest conversion to (mainly CO, CO_2) according to FTIR peak intensity, then (CH_4), and little quantity of unstable (R-C-NH_2) which broken into (NH_3 + R-CH or R-C-R) [14,15].
- By comparing the second step with the first one we found that it is higher energy than the first step because of the high reactivity compounds resulting from the degradation of oligomers that the sludge consists of, these compounds undergo other reactions to form new stable compounds.
  a. At the range of (220 – 260) °C the curve goes to stability because of these reasons:
  b. The pyrolysis reaction was complete.
  c. Reactions of the unstable transition compounds resulting from the second step, and formation of the final products [16, 17].

According to all the results above, we suggested the mechanism shown in figure (8)

4. Conclusions:

- According to (C.H.N) and (XRF) analysis we found that there are heavy metals in the sample because there is no industrial source near the treatment station.
- FTIR spectra show increasing in the intensity of the peak at the wavelength 1674°A caused by formation of carbonyl group in form of CO_2.
- Appearing of new peaks approved breaking and formations of new bonds in forms of new aliphatic short chain compounds contain (C-H) bonds like CH_4.
- The (DSC) curves shows that the general reaction takes place according to a mechanism with three steps, and the second step was the rate determining step.

5. Figures

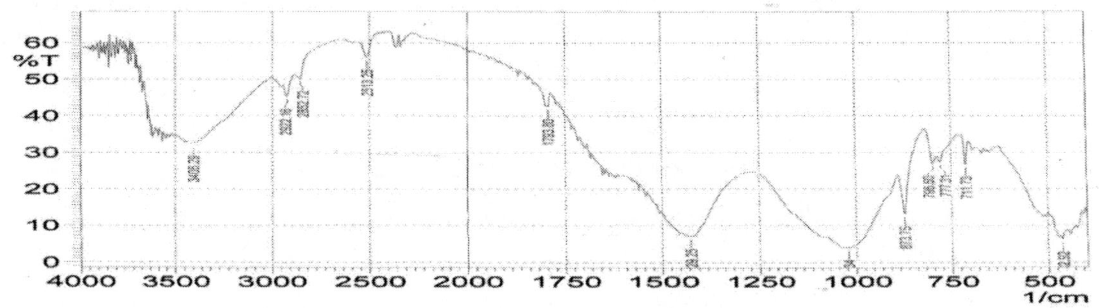

Figure 1. FTIR analysis for the row sewage sludge

IOP Conf. Series: Journal of Physics: Conf. Series **1032** (2018) 012034 doi:10.1088/1742-6596/1032/1/012034

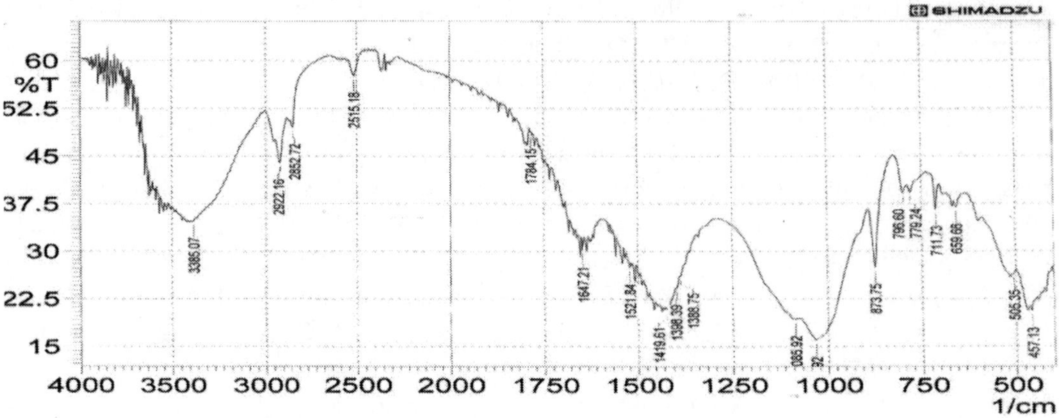

Figure 2. FTIR analysis for dried sewage sludge

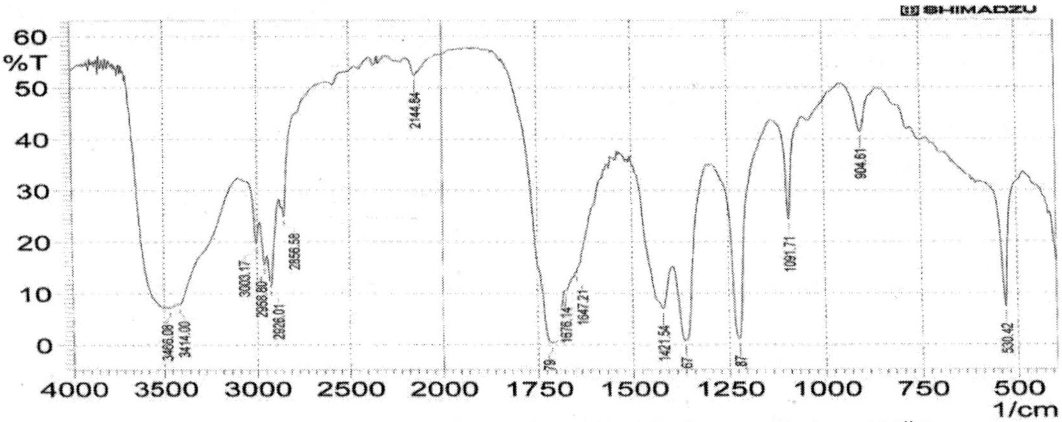

Figure 3. FTIR analysis for the produced bio-oil after Pyrolysis at 290^{0}C

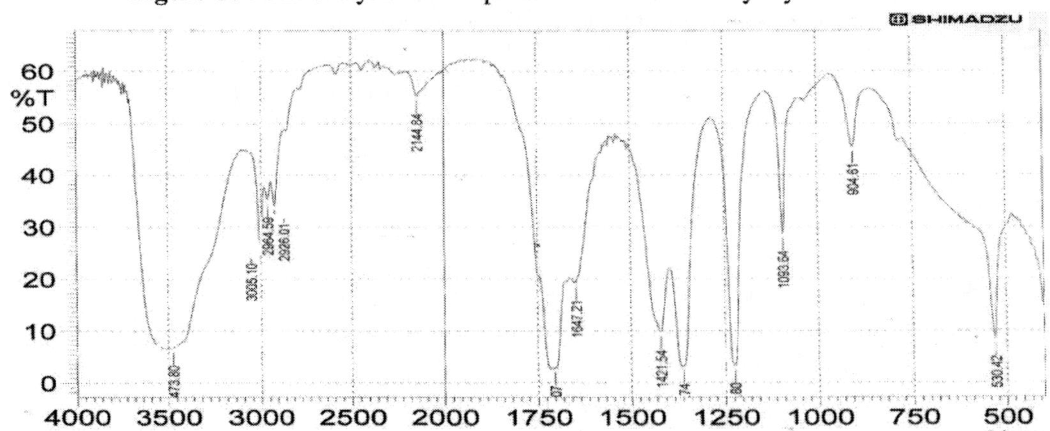

Figure 4. FTIR analysis for the produced bio-oil after Pyrolysis at 275^{0}C

IOP Conf. Series: Journal of Physics: Conf. Series **1032** (2018) 012034 doi:10.1088/1742-6596/1032/1/012034

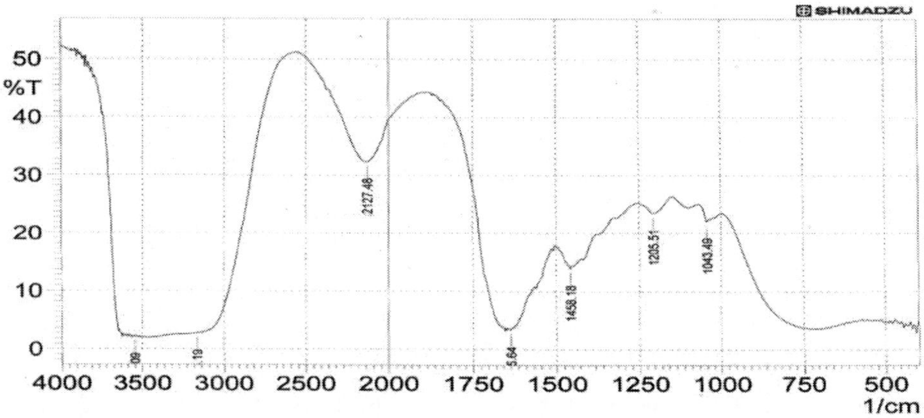

Figure 5. FTIR analysis for the produced bio-oil after Pyrolysis at 250^0

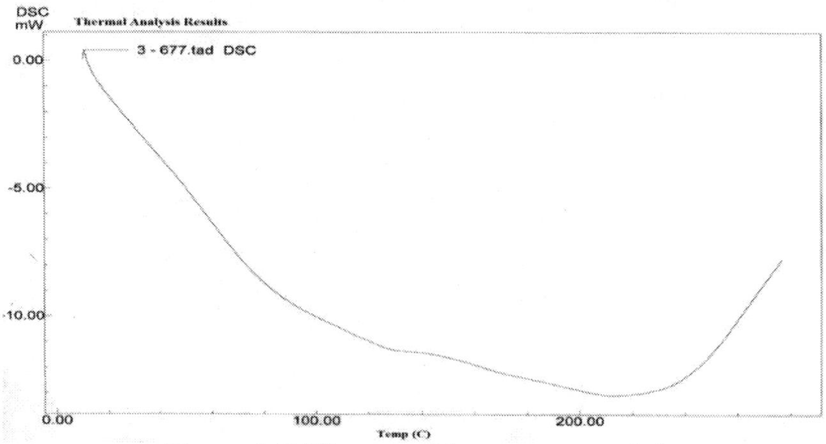

Figure 6. DSC curve of the row sewage sludge

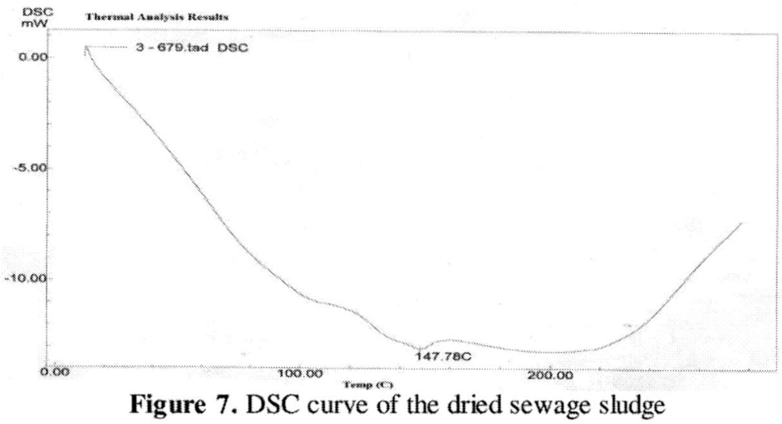

Figure 7. DSC curve of the dried sewage sludge

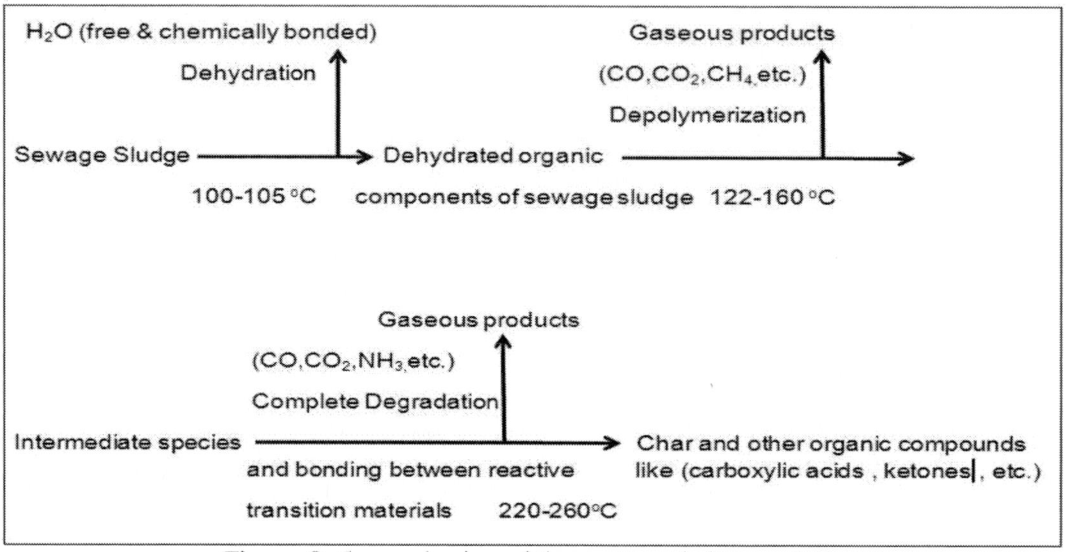

Figure 8. the mechanism of the sewage sludge pyrolysis

6. Tables

Table 1. Basic tests for sewage sludge

analysis	value
moisture	3.7
Volatile compound	57.3
Ash content	34.6
Carbon	4.4

Table 2. Carbon, Hydrogen, Nitrogen (C.H.N) Analysis for sewage sludge

C	H	N	O	S	H/C	H/N
41.13	5.66	4.18	48.02	1.01	1.67	18.89

Table 3. X-Ray Fluorescence for the sample before drying

	Sample No. 1		
Element	Content	Limit	Value
Fe	ND	0	1.395
Kr	ND	0	0.0208
Sn	ND	0	0.2851
Ga	ND	0	0
Ti	ND	0	0.3636

Table 4. X-Ray Fluorescence for the sample after drying

Sample No. 2

Element	Content	Limit	Value
Fe	ND	0	0.4568
Sn	ND	0	0.0599
Al	2.7637	0	2.7637

Table 5. Summarizes the values of the wavelengths of the whole operation at different temperatures

Bonding	Wavelength Cm^{-1}
O-H vibration of carboxylic and alcoholic groups H-bonding N-H group	3600 – 3150
Aliphatic C–H stretching CH_3 , CH_2 , and C-H 2 or 3 bonded	3010 – 2850
S-H thiols	2515 – 2510
C=O stretching	1794 – 1784
C=O, aromatic C=C	1680 - 1645
Aliphatic C–H deformation	1520-1385
C-H, O-H within carbonyl group	1419 – 1398
C–O stretching, C-N	1100 – 1000
=C-H & =CH_2	910 – 450

7. References:

[1] Rio S, Faur-Brasquet L, Coq L, Cloirec LP. Structure characterization and adsorption properties of pyrolyzed sewage sludge. Environ Sci Technol. 2005; 39(11):4249–57.

[2] Campostrini R, Ischia M, Palmisano L. Pyrolysis study of sewage sludge by TG-MS and TG-GC-MS coupled analyses. J Therm. Anal. Calorim.2007; 87:567–74.

[3] Viana MM, Melchert MBM, Morais LC, Buchler PM. Sewage sludge coke estimation using thermal analysis. J Therm. Anal. Calorim.2011; 10:1392–401.

[4] Capana AS, Martins QV, Crespi MS, Ribeiro CA, Barud HS. Thermal behavior of residues (sludge) originated from Araraquara water and sewage treatment station. J Thermal Anal Calorim.2009; 97:601–4.

[5] Ninan KN, Krishnan K, Krishnamurthy VN. Kinetics and mechanism of thermal decomposition of institution generated calcium carbonate. J Therm. Anal. Calorim.1991; 37:1533–43.

[6] Campanella L, Tomassetti M, Tomellini R. Thermal analysis of ancient, fresh and waterlogged woods. J Therm. Anal. Calorim.1991; 37:1923–32.

[7] Jaber JO, Probert SD. Pyrolysis and gasification kinetics of Jordanianoil-shales. Appl. Energy. 1999; 63:269–86.

[8] Ebeling JM, Jenkins HM. Physical and chemical proper- ties of biomass fuels. Trans ASAE. 1985; 28:898–902.

[9] Sorum L, Grønli MG, Hustad JE. Pyrolysis characteristics and kinetics of municipal solid wastes. Fuel. 2001; 80:1217–27.

[10] Conesa JA, Font R, Fullana A, Caballero JA. Pyrolysis of polyethylene in a fluidized bed reactor. Fuel 1998; 77:1469–75.

[11] Conesa JA, Marcilla A, Prats D, Rodriguez M. Kinetic study of the pyrolysis of sewage sludge. Waste Manage. Res. 1997; 15:293–305.

[12] Hussel, R.L., "Proximate Analysis of Coal and Coke ", Dupont Co., Willington.

[13] Olexsey, R.A., "Proc of the 1975 Nat. Conf. On Mun. Sludge Management and Disposal", Anahiem Calif., 1975, pp.139-145.

[14] Antal, M.J. et al.," Design and Operation of Solar Fired Biomass Flash Pyrolysis Reactor", presented at IGT, Symposium "Energy from Biomass and Wastes V", Lake Buena Vista FL., Jan. 1981.

[15] Antal, M.J., Edwards, W.E., Friedman, H.L., and Rogers, F.E., "A Study of Steam Gasification of Organic Wastes", Final progress report to U.S.E.P.A., Princeton Univ. Dept of Mech. & Aero. Engineering, Princeton, NJ.

[16] Conesa, A., Marcilla, A., Moral, R., Moreno-Caselles, J., and Perez-Espinosa, A. (1998). " Evolution of gases in the primary pyrolysis of different sewage sludge." Thermochim. Acta, 313(1), 63-73.

[17] Conesa, A., Marcilla, A., Parts, D., and Rodriguez-Pastor, M. (1997). "Kinetic study of the pyrolysis of sewage sludge." Waste Mgmt. and Res., London, (15)3, 293-305.

8. Acknowledgment:

The researchers want to thank the laboratories of Iraqi Ministry of Science & Technology especially the solar energy research center/ renewable energies directorate, and materials science researches directorate for their help in doing tests and analysis.

The effect of solar eruption promulgation on the propagation of solar energetic particles

Doha Al-Feadh[1], Ali AL-Bekheet[1] and Wathiq Al-Ramdhan[2]

[1] Basra University /college of science /Physics Department/ Iraq
[2] Environmental Pollution Research Unit/ Southern Technical University/Iraq

E-mail: mahmood_doha@yahoo.com

Abstract. Solar eruption events investigated for the period from 2000 to 2014. These events associated with solar energetic particles (SEPs) that composed of halo-type coronal mass ejections (CMEs) and accompanying with intense flares. The incline of the time intensity profile are computing by analysing the eastern and western events through the CMEs source regions. This study shows that the incline angle for west events is smaller than Eastern Events. The western events are seem to be sharper in the acceleration phase than the eastern event. Considerate these phenomena is important to relate the space weather environment near Earth and other destinations in the solar system.

1. Introduction

Expressive the source of the solar proton event is very important where the protons can be accelerated to more than 10 MeV during the propagation of the CMEs and near the front of the shock or solar flare growth. The solar energetic particles (SEPs) acceleration could be due to flaring processes, acceleration in coronal mass ejections (CMEs) and/or interplanetary shocks, or it could be a combination of all [1]. These two sources and their location at the Sun, make the time intensity profile to be different depending on the position of the observation point. In addition, the spiral pattern of the interplanetary magnetic field (IMF) gives rise to the longitude dependent solar proton profiles. It can be interpreted as proton acceleration by an expanding shock in the IMF [2-3].

The heliographic coordinates of the CMEs have been identified as the location of the associated flare, or coronal darkening from various observations [4]. Numerous physical relationships have been studied to show the behaviour of the time intensity profile for the SEPs such as CME angular width, CME location and speed. There is no differences found for the probability effect of the background solar wind speed on the SEP event timescales which studied by Kahler and Vourlidas [5].

Our study is concentrated on the location of the source region of SEPs events and its effect on the behaviour of the time intensity profile. There were about 259 Halo CMEs and intense flares, which originated towards the western and eastern side of the Sun, which compiled in order to infer properties that benefit to study the change and compute the incline's angle in the time intensity profile of the produced SEPs.

2. Data Analyses

The purpose of this study is to compute the incline's angle of the time intensity profile, and how solar energetic particles can propagate, which had to be observed with the CMEs, and were well-connected to the field-lines near L1, where the Energetic and Relativistic Nuclei and Electron (ERNE: [6]) instrument on board the Solar and Heliospheric Observatory (SOHO) is located. For finding the characteristics of CMEs (speed, angular width, and lift-off time) we used The SOHO/Large Angle and Spectrometric Coronagraph (LASCO) CME catalog at (http://cdaw.gsfc.nasa.gov/). The NOAA National Geophysical Data Center (NGDC) are listings at http://www.ngdc.noaa.gov/stp/spaceweather.html was used for flares and its locations. The X-ray data

IOP Conf. Series: Journal of Physics: Conf. Series **1032** (2018) 012035 doi:10.1088/1742-6596/1032/1/012035

were checked from the Geostationary Operational Environmental Satellite (GOES). To study the time intensity profile for associated SEP events, we used the SOHO/ERNE proton data at http://www.srl.utu. /erne data/main english.html.

These CMEs with angular widths of 60° and up, and have speed with more than 850 kms^{-1}. Their associated flares are having M and X classification. On the other hand we exclude the backside and ambiguous events to reduce uncertainties in predicting the eruption source properly. After that, we have examined the time intensity profiles of all the 259 SEPs for protons particles with intensities larger than $10^{-3}cm^{-2}sr^{-1}s^{-1}MeV^{-1}$.

3. Results and Discussion

The solar events which selected are associated with halo and fast CME with angular widths of 60°and up, and intense flares that occurs for areas within ±80° heliographic longitude and ±40° heliographic latitude from all the examined events. These indicated the occurrence of the variation in protons flux from one event to another during the rising phase for low and high-energy channels.

We sort numerical analysis to give a description of the fits in with the issue of the study to reduce the errors as much as possible. The results sorted according to the longitudinal lines with incline's angle and coronal mass ejections velocity, as shown in Figures (1) and (2) respectively.

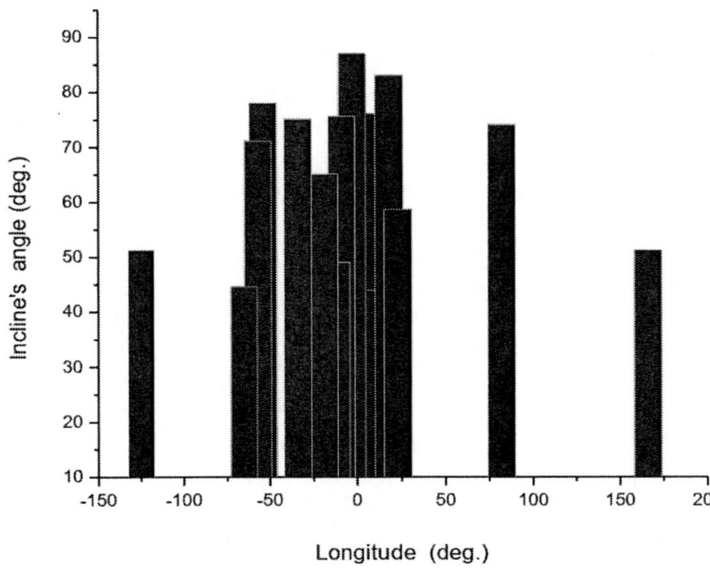

Figure 1. The variation of incline's angle with longitude.

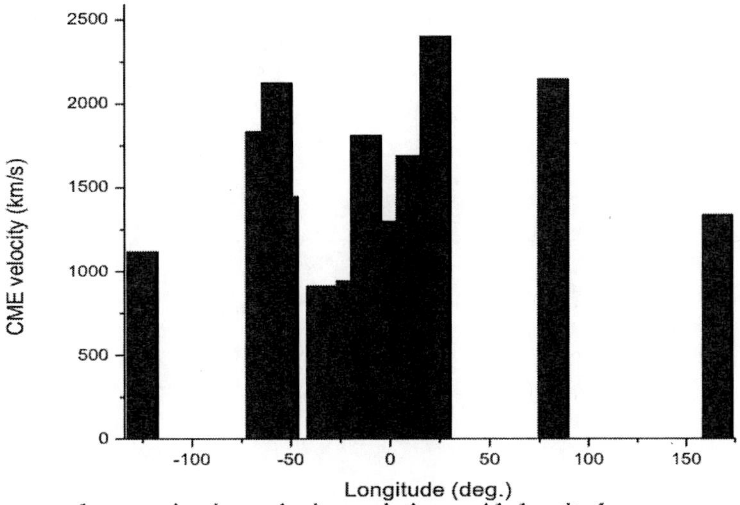

Figure 2. The coronal mass ejection velocity variations with longitude.

Figure (1) shows gradual increase in the intensity with the solar source located in the eastern hemisphere of region with angle (21°to 90 °), fast increase with eastern hemisphere of region within (20° to -20 °), and have swift increase with western hemisphere of region within (-21°to -90 °). While figure (2) shows the SEPs associated with fast CMEs that have velocity from about (900 - 2400) kms^{-1} within western events which were well-connected to the field lines near L1, were the ERNE instrument on board SOHO is located.

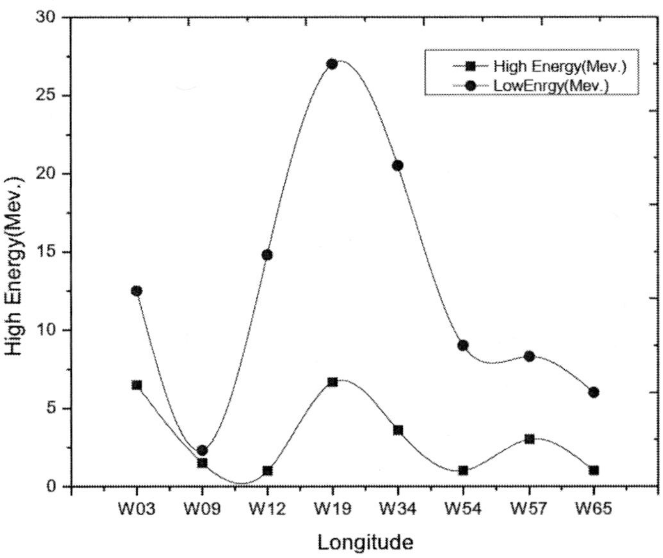

Figure 3. The longitude variation with SEPs intensities time scales (squares) for high energy and (circles) for low energy.

IOP Conf. Series: Journal of Physics: Conf. Series **1032** (2018) 012035 doi:10.1088/1742-6596/1032/1/012035

Figure (3) displays the variations of SEPs intensities with high and low energies which increase within the center of the Sun at hemisphere of region within (-9°to -54 °) western events.

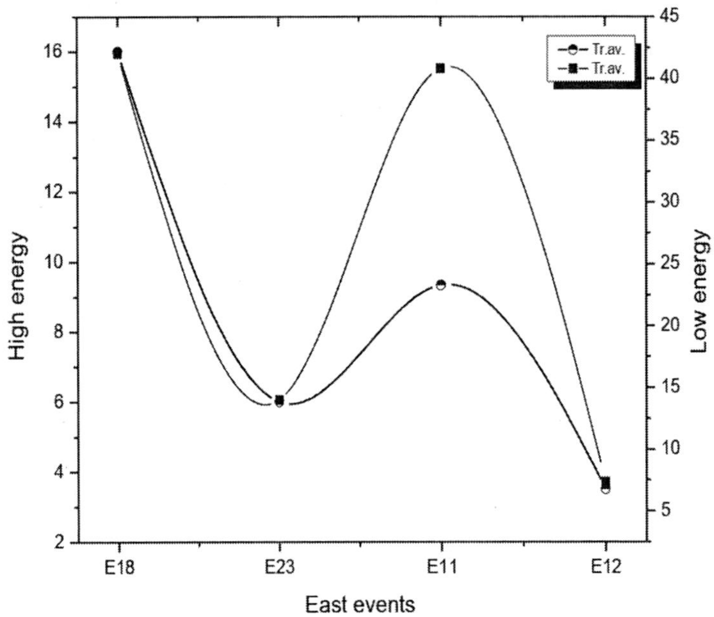

Figure 4. Shows the latitude variation with SEPs intensities time scales (squares) for high energy and (circles) for low energy.

On the other hand the eastern events in figure (4) shows increasing in intensities of SEPs at hemisphere of region within (11°-12°) which indicate the solar cycle's maximum. Figure (5) represents the variation of incline's angle of (50 °) and up at eastern latitude while its increasing of 45°and up for western latitude events. While the figure (6) shows variation of CMEs velocity with latitude which increase from (850 -2000) kms^{-1} with distinction of latitude of (9°to 31°).

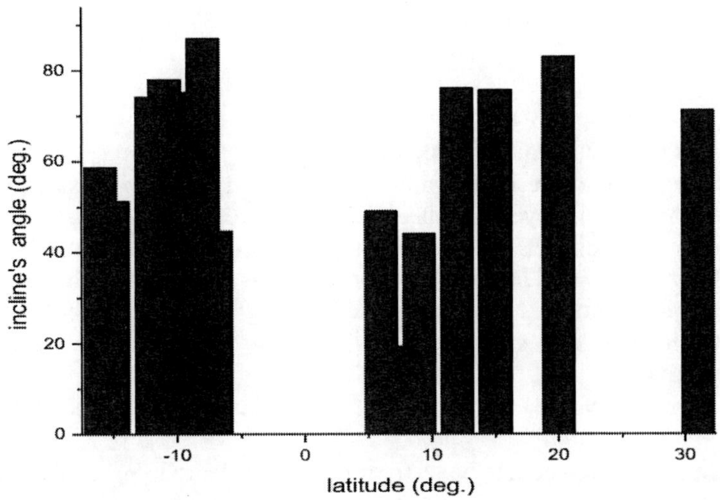

Figure 5. Shows variation of incline's angle with latitude.

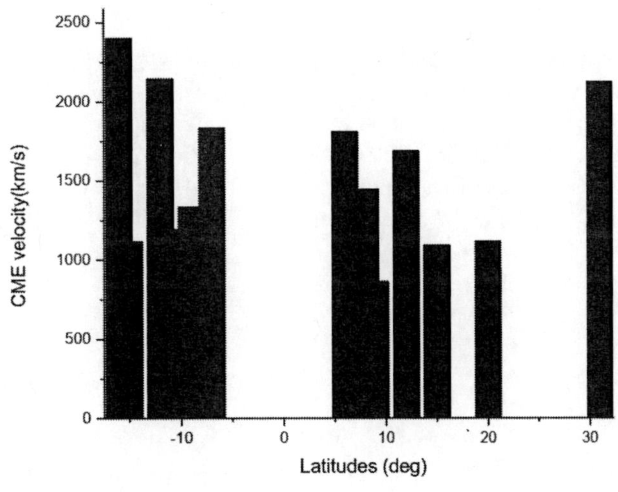

Figure 6. Display coronal mass ejections velocity variations with latitude.

The results show the western events are seem to be sharper in the acceleration phase than the eastern event. The rising phase of the high energies take shorter time than the low energies. This marks that the bulk of acceleration take place near the Sun where the shock wave efficiency is higher due to the perpendicular propagation to the Sun magnetic field lines. Geomagnetic storms are associated with CMEs which is a major disturbance of Earth's magnetosphere and they also can disrupt navigation systems so Considerate these phenomena is important to relate the space weather environment near Earth and other destinations in the solar system.

4. Conclusions

From this study we can conclude that:

1- Solar eruption events were investigated for the period from the maximum of solar cycle 23 to the maximum of solar cycle 24 observing by SOHO, and GOES satellites.
2- These events were chosen with solar energetic particles (SEPs) that accelerated by both (X and M) X-ray flare and high speed coronal mass ejections (CMEs).
3- The incline of the time intensity profile determined by analyzing the eastern and western events through the CMEs source region shows that the incline for west events is smaller than Eastern Events.
4- The solar energetic particle events are fewer regular for solar cycle 24 than those of solar cycle 23.
5- The western events are seem to be sharper in the acceleration phase than the eastern event.

Acknowledgments. We are grateful to all the entities who have donated in forming and updating the various solar event catalogs. The SOHO/LASCO CME Catalog is generated and maintained at the CDAW Data Center by NASA and the Catholic University of America in cooperation with the Naval Research Laboratory. SOHO is a project of international cooperation between ESA and NASA. Grateful are extend to the staff of the Department of Physics and Astronomy, University of Turku, Turku, Finland especially to Proof Dr. Eino Valtonen.

References

[1] Trottet G , Samwel S, Klein K-L, Dudok de Wit T, Miteva R 2015 *Solar Phys.* **290** 819.
[2] Cane H V, Reames D V, Rosenvinge T. T. von J. 1988 *Geophys. Res.* **93** 9555–9567.
[3] Reams D V 1999 *Space Sci.Rev* .**90** 413.
[4] Gopalswamy Akiyama and Yashiro 2007 *JGR* **112** A06112.
[5] Kahler S W , Vourlidas A 2014b *Astrophys. J.* **791** 4.
[6] Torsti J , Valtonen E , Lumme M , Peltonen M P , Eronen T , et al. 1995 *Solar physics* **162** 505- 531.

IOP Conf. Series: Journal of Physics: Conf. Series **1032** (2018) 012036 doi:10.1088/1742-6596/1032/1/012036

Producing High Purity of Metal Oxide Nano Structural Using Simple Chemical Method

M A Aseel[1], F H Itab[1] and F M Ahmed[2]

[1]Physics Department, College of Science, Al-Mustansiryah University, Baghdad, Iraq.
[2]Physics Department, College of Education for Pure Science, University of Baghdad, Baghdad, Iraq.

e-mail: aseelalaziz@uomustansiriyah.edu.iq

Abstract. Calcium oxide nanoparticles and nanorods were prepared by using chemical method at two Molarities. The field emission scanning electron microscopy (FESEM) images suggested the formation of nanoparticles (50-60) nm and nanorods with diameter about 500 nm and length 3μm. The average surface roughness of CaO thin film was imaged by carried out atomic force microscope (AFM) and roughness found to be between (3.52-5.46) nm. XRD patterns show that the CaO was cubic crystalline nature and average grain size was found to be 70 nm. The direct optical band energy gap of all conditions to be ranged between (3.5-3.7) eV and thermal gravimetric analysis (TGA) was studied. Photoluminescence (PL) properties were investigated and founding two emission bands.

1. Introduction
Calcium oxide (CaO) is an important inorganic material with a good physical properties and their power to resist hard process conditions. CaO is of special attention as they are not only constant under hard process conditions but also mostly note as safe material to human beings and animals [1]. CaO is a highly important wide-band gap insulator, least weight, perfect chemical resistance, rise secondary electron emission, rise thermal stability [2] and has the merit of non – toxicity and of being ready from easily available and economical portent and subsequently has large potential as solid bactericidal material [3]. Hence CaO is important of potential applications in different fields, such as catalysts, water treating, gas sensing and superconductors [4]. CaO nanomaterial with high surface area is important attention in adsorption materials. Recently all attempt have been studied CaO with increase surface area in different morphologies such as nanoparticles, nanorods and nanowire [5, 6]. Through different methods, chemicals methods [7], chemical bath depositions [8] and sol-gel [9]. The real dimensions such as diameters of (10-100) nm and lengths of a small micrometers supply a structural devices that can be very small, which have enhance for applications in biological devices, security monitoring and defence technology [10]. For the present work, we have studied CaO nanostructures by chemical methods under two molarities, result their simplicity, cost effective and low reaction temperature.

2. Experimental Part
Calcium oxide nanostructures were prepared by chemical technique at different molarities (0.4 and 0.6) mol using calcium chloride dihydrate ($CaCl_2.2H_2O$) was add to 100 ml of distilled water and use

IOP Conf. Series: Journal of Physics: Conf. Series **1032** (2018) 012036 doi:10.1088/1742-6596/1032/1/012036

sodium hydroxide (NaOH) and urea$CO(NH_2)_2$ were add to 25 ml of distilled water with continuous stirring for 4hr. and then the white precipitate was filtered and washed two times with distilled water and methanol and dried at 100 °C. Finally the white colored material was calcinations at 700 °C for 5 hr. in a vacuum oven to obtain CaO nanostructure. Morphology of CaO nanostructures were analyzed with field emission scanning electron microscope (FESEM) and Atomic Force microscope (AFM). Thermo gravimetric analysis (TGA) of CaO were investigation, crystal structure analysis was carried out by X- ray diffractometer (XRD-Philips) using (cukα) radiation (λ= 1.5406 °A), a complete 2Θ scan was made between 20° to 70°. From XRD data the average grain size of particles was calculated by Scherrer equation. The optical absorption of the colloidal CaO nanostructures were measured using spectrophotometer (CARY, 100 CONC plus, UV-Vis-NIR, Splitbeam Optics, Dual detectors) in the range of (200-900) nm.

3. Results and Discussion

Thermal Gravimetric Analysis spectrum was used to analyze the thermal behaviour of the CaO nanomaterial. Figure 1 shows the relation between the weight losses (%) with the thermal temperature. The starting weight loss up to about 300 °C due to the evaporation of water and organic species. The another major weight loss was show in the temperature range of (300–450) °C concerning to decomposition of $Ca(OH)_2$ after calcinations at 700°C gained white powder of CaO was characterized with XRD [11].

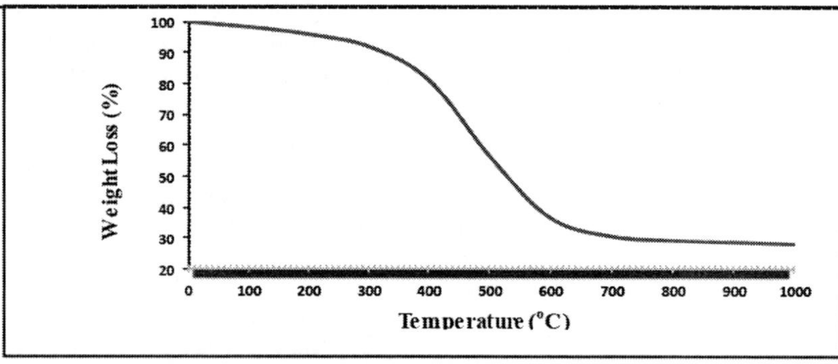

Figure 1. TGA spectra of CaO.

XRD pattern of the CaO nanomaterial at 0.4 M is shown in figure 2. It is seen that peaks appears at 32°, 37.02°, 54° and 64° corresponding reflecting planes are (111), (200), (220) and (311) respectively.

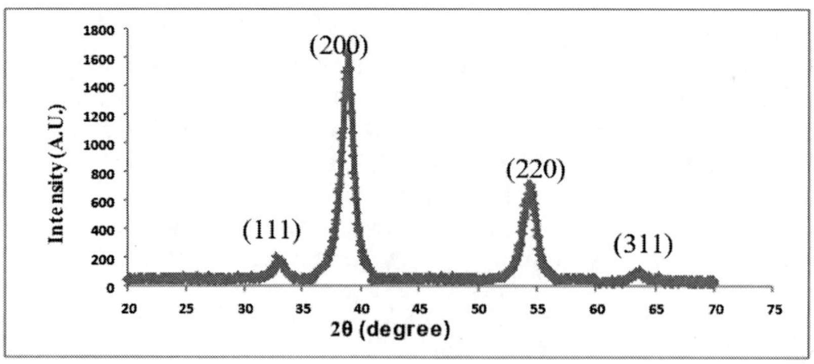

Figure 2. XRD pattern on CaO.

IOP Conf. Series: Journal of Physics: Conf. Series **1032** (2018) 012036 doi:10.1088/1742-6596/1032/1/012036

The diffraction patterns detect a good polycrystalline quality without any foreign material and can be indexed to cubic crystalline structure according to JCPDS file number (37-1497). There is not any other clear phase in the XRD pattern which means high purity of the obtained CaO nanoparticles. The average particle size of CaO nanoparticles can be estimated using Scherrer equation was about 70 nm and same as reported by [12]: Crystallite size $= \frac{\mathbf{0.94\,\lambda}}{\boldsymbol{FWHM\ cos\theta}}$.
Where λ=1.54A is the wavelength of X-ray, FHWM is the breading of the diffracted peak at half maximum and θ is the Bragg angle.
The FESEM of CaO nanomaterial studied by chemical method at different molarities (0.4 and 0.6) is shown in figure 3.

Figure 3. FESEM of CaO. (a) CaO nanoparticales 0.4 mol and (b) CaO nanorods 0.6 mol.

As previously mentioned, increased molarities of nanomaterial that effect on phase formation and morphology of acquired powder. It is clear in sample (a) at 0.4 M was consisted of uniform and smooth CaO nanoparticles with average grain size around (50-60) nm. In sample (b) at 0.6 M was consisted of CaO nanorods are exhibited hexagonal surface, the mean diameter to be about 500 nm and mean length is about 3 µm.
Figure 4 shows the formation of CaO nanostructure at different molarities (0.4 and 0.6). AFM micrograph proves that the grains are uniformly distributed within the scanning area (2000×2000) nm with individual columnar grains extending upwards.

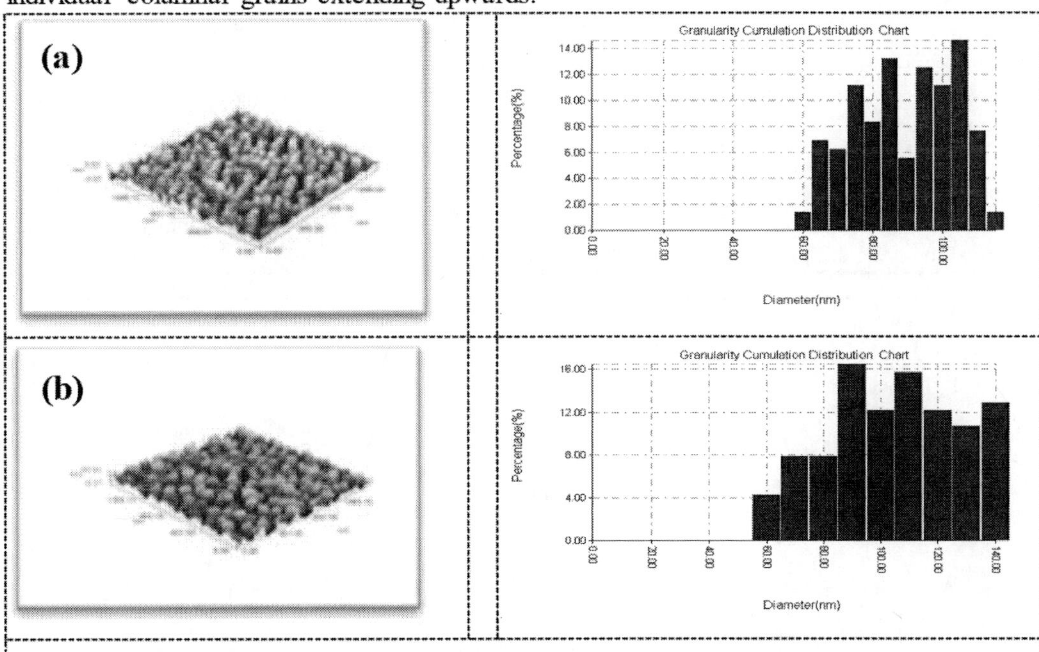

Figure 4. AFM images of CaO with different molarities (0.4 and 0.6) M.

IOP Conf. Series: Journal of Physics: Conf. Series **1032** (2018) 012036 doi:10.1088/1742-6596/1032/1/012036

This surface topography is important for many applications such as for applications in optoelectronic devices, solar cells, gas sensors and catalysts. The CaO nanostructures have spherical shaped with good dispensability, homogeneous grains aligned vertically. The roughness of CaO increasing with the molarities, this was perhaps due to scattering mechanism in which affect in the particle size. Symmetry of the particle size of CaO is important because the nanoparticle size influence to the electrical conductivity of CaO [2]. The estimated values of root mean square of surface roughness average and average grain size are listed in table 1.

Table 1. The size, roughness and root mean square of CaO.

Molarities (mol)	Grain Size (nm)	Roughness Density (nm)	Root Mean Square (nm)
0.4	85.93	3.52	3.52
0.6	100.2	5.46	7.41

The Uv-Vis spectrum of CaO Nanostructures at room temperature with different concentration deposited on quartz substrate is shown in figure 5. The regime absorption peak at 230 nm can be associated to band transition in the band gap region [8]. Absorption of CaO was found to be increased with an increase of molar concentration.

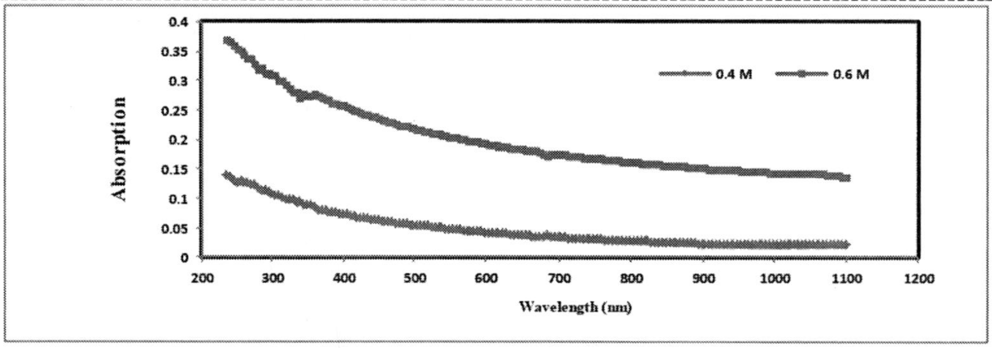

Figure 5. Optical absorption for different molarity of CaO nanostructures vs. with wavelength.

The optical band gap of CaO was determined from $(\alpha h\nu)^2$ vs. $(h\nu)$ plot as shown in figure 6 and values have been obtained to be 3.7 eV and 3.5 eV for the molarities 0.4 and 0.6 mol respectively. The internal figure shows the optical band gap decreases with increase of molar concentration due to the increase of density of localize state in the conduction band [11].

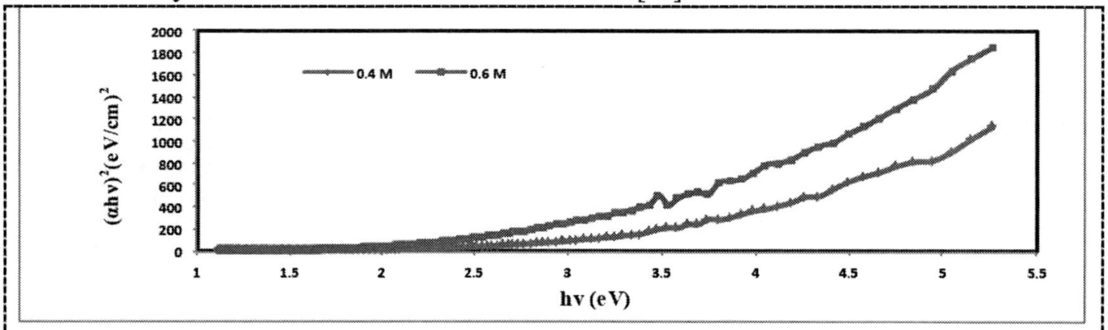

Figure 6. The energy gap (E_g) CaO nanostructures with different molarities (0.4 and 0.6) M.

IOP Conf. Series: Journal of Physics: Conf. Series **1032** (2018) 012036 doi:10.1088/1742-6596/1032/1/012036

Photoluminescence spectrum of these CaO nanostructures is studied. As shown in figure 7, CaO nanoparticles and nanorods correspond to different PL properties, the properties was recorded at 325 nm as an excitation wavelength.

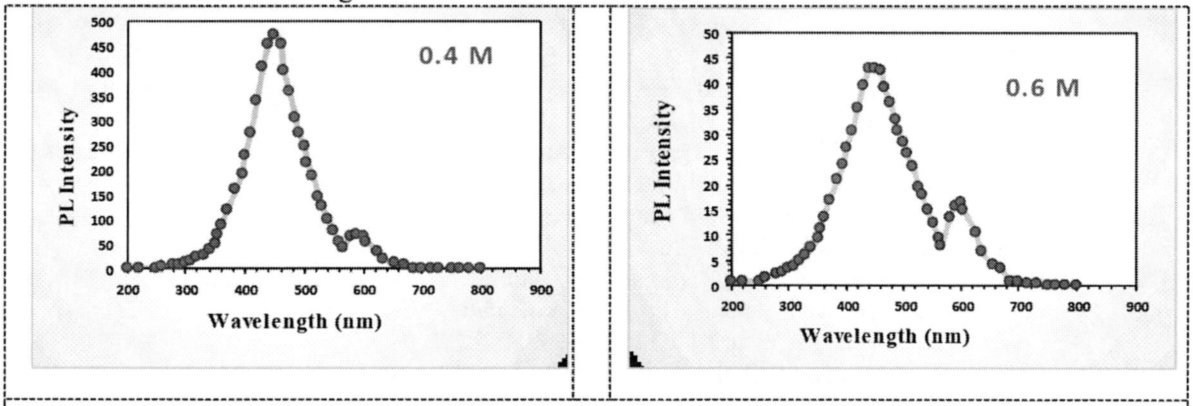

Figure 7. PL intensity with different molarities (0.4 and 0.6) M.

The CaO nanostructures have two strong emission bands, the first peak observed at 440 nm due to exaction emission by the radiative annihilation. The second peak in the green region of the visible spectrum at 580 nm resulted by radiative recombination of an electron in the conduction band [8]. It is found that the intensity of the green visible emission is less than intensity of exciton emission from CaO nanostructures because the green visible have more oxygen vacancies. However, figure show that the intensity of CaO nanorods less than CaO nanoparticles because CaO nanorod have higher crystallization, less oxygen vacancy concentration and the diameter of particles increase than the CaO nanoparticles [13].

4. Conclusions

In this work pure CaO were prepared by chemical method with different molarities (0.4, 0.6) M. FSEM result show the formation two kinds of CaO nanoparticles with size in the range of (50-60) nm and CaO nanorodwith have a diameter of 500 nm and length about 3μm. The PL property show that the two types of bands, the first is an exciton emission in UV region and the second in the green region. The XRD result confirmed the high purity crystallinity of the CaO.

5. References

[1] Jagadesh D, Prashantha K and Shabadi R, " Star shaped surcrose capped CaO nanoparticles from Azadirachtaindica: A novel green synthesis", synthesis and Reactivity in Inorganic and nano-metal chemistry, ISSN: 1553-3174, (2016).

[2] Albuquerque E L and Vasconcelos M S, "Structural Electronics and optical properties of CaO", Journal of Physics: Conference series 100, (2008).

[3] Kazemi H, Zandi K and Momenian H, " Sonochemical synthesis of $Ca(OH)_2$ nanoparticles and its application in preparation of MWCNT- paraloid nanocomposite', Journal of nanostructures, vol. 5, (2015), 25-32.

[4] Mirghiasi Z, Bakhtiari F and Darezereshki E, " preparation and characterization of CaO nanoparticles from $Ca(OH)_2$ by direct thermal decomposition method", Journal of industrial and Engineering chemistry, vol. 20, (2014), 113-117.

[5] Tahmasian A, Morsali A and Woojoo S,"Sonochemical synthesis of a one- dimensional Mg metal-organic framework:Anew precursor for preparation of MgO one-Dimensional nanostructure",Journal of Nanomaterials,3134567,(2013),7.

[6] Bult A R, Ejaz S, Barom J C, "CaO nanoparticles as a potential drug delivery for biomedical

IOP Conf. Series: Journal of Physics: Conf. Series **1032** (2018) 012036 doi:10.1088/1742-6596/1032/1/012036

application", Digest journal of nanomaterials and Biostructures", vol. 10, No. 3, (2015), 799-809.
[7] Amin M A and Morsali A," Ultrasonic-assisted synthesis of $Ca(OH)_2$ and CaO nanostructures", Journal of Experimental nan science, vol. 5, No.2, (2010), 93-105.
[8] Nirmala P N, Suresh, et..," Influence of annealing temperature on optical properties of CaO thin film", International Journal of Recent scientific Research, vol. 4, Issue 4, (2013), 425-427.
[9] Darcanova O, Beganskiene A and Kareiva A, "Sol- gel synthesis of calcium nanomaterial for papperconsrrvation",Chemija, vol. 26, No. 1, (2015), 25-31.
[10] Xing Tang Z, Claveau D, Corcuff R, "Preparatuion of nano- CaO using thermal- decomposition method", Elsevier, vol. 62, (2008), 2096-2098.
[11] Xing Tang Z, "Sonication – Assisted preparation of CaO nanoparticles for Antibacterial agents", Quim Nova, vol. 7, (2013), 933-936.
[12] Sundrarajan M, Suresh J, "A comparative study on Antibacterial properties of MgO nanoparticles prepared under different calcination temperature", Digest Journal of Nanomaterial and Bio structures, 7,3,P.983-989,(2012).
[13] Zhang H, Cao T and Cheng Y, " synethesis of nanostructured MgO powder with photoluminescence by plasma- intensified pyrophdrolysis process of bischofite from brine", DE Gruyter, Green Process synth, vol. 3, (2014), 215-222.

IOP Conf. Series: Journal of Physics: Conf. Series **1032** (2018) 012037 doi:10.1088/1742-6596/1032/1/012037

Effect of thermal annealing on properties of polycrystalline Titanium dioxide (TiO_2) thin film prepared by simple chemical method

[1]Ali Hamid Abdulsada , [1]Khudheir A. Mishjil , [2*]Ahmad N. Abd

[1]Physics Department, Education College, Al-Mustansiriyah University, Baghdad, Iraq
[2]Physics Department,Science College, Al-Mustansiriyah University, Baghdad, Iraq
[*]Corresponding Email: ahmed_naji_abd@yahoo.com

Abstract: Titanium dioxide (TiO_2) thin film prepared in a simple chemical method, on an optically flat a glass substrate. The effect of the annealing temperature on a structural and the optical properties of TiO_2 films is studied systematically by X-ray diffraction (XRD), Atomic Force Microscopy (AFM), UV–VIS spectroscopy . The XRD patterns show that the deposited films and the annealed films are polycrystalline and the average grain size of the crystallite grains ranges from 41 to 50 nm in films. AFM images reveal a regular grain distribution of grains in films and the grains in the nanoscale dimension. The UV–Vis rays results show that increase in transmittance with increase the annealing temperature, also ,the energy gap decreases from 3.6 to 2.6 volts, due to the effect of quantum volume.

Keywords: Titanium dioxide (TiO_2), Optical properties, FTIR ,thermal annealing, AFM

1. Introduction

Titanium dioxide (TiO_2) is one of the important semiconductors that found in nature [1]. TiO_2 exist in three forms of Anatase, Rutile and Prokite [2] TiO_2 owned an energy band gap of (3.2 eV) and refractive index (n=2.52) [3] It has a unique optical properties such as good magnetic properties [4] which has made the material important in many practical applications, including stimuli [5], energy conversion [6] and biomedical applications [7] functionalized hybrid materials [8] and nanocomposites [9] . TiO_2 can be used as solar cell applications (DSSC) [10] electrochemical cells (PEC) [11] chemical sensors [12] and self-cleaning coatings [13]. Thin films of TiO2 are deposited by various methods including electron beam [14,15] magnetron sputtering technique[16], plasma evaporation [17], pulsed laser deposition [18], self-assembly process [19], laser chemical vapor depositing [20,21] and sol-gel method [22]. Ba-Abbad et al. (2012) synthesized the TiO2 nanoparticles catalyst by sol-gel method using the titanium tetra-isopropoxide as starting material. The effects of calcination temperatures on the crystalline structure, surface area and photo catalytic activity of TiO_2 nanoparticles were investigated. The anatase to rutile percentage decreases when temperatures are greater than 500°C and evaluated from XRD intensity of (101) and (110) peak, respectively [23].

Nocun et al. 2011 prepared SiO_2/TiO_2 thin films on glass and quartz substrates by dip coating sol-gel method. They showed that band energy gap (Eg) decreases with increase in TiO_2 content [24]. The aim of this work is to manufacture nanoparticles of titanium dioxide by a simple chemical method and Study structural, chemical and optical properties of titanium dioxide as a thin film deposited on glass substrate.

2. Experimental details

Titanium dioxide is fully prepared by Simple chemical methode of 80 ml of water and adding 1 g of TiO_2 powder to heat the mixture at a temperature below boiling point 80 °C .then tend add adrop (100μl) HF and PVC to speed up the interaction ,The TiO_2 colloidal has been taken from the solution by pipette and then drop on the glass substrate only 5 drops, then waiting for (5-6 hours) until dry under room temperature.

3. Results and Discussion

Figure 1 depicts the x-ray diffraction pattern of TiO2 thin films before and after annealing process. Several major peaks for as-prepared at 2θ = (25.2), (37.78) and (48) corresponding to (101), (004) and (200) respectively. The crystallite size were estimated using Sheerer equation [29]

$$D_{av} = \frac{0.9\,\lambda}{\beta\ cos\theta} \ldots\ldots\ldots \quad (1)$$

Where : D_{av} The crystal size which measured by the nanometer , β the full width at half maxima , θ is the angle of diffraction is measured by (deg) , where λ (1.54056 Å) λ is the wavelength of the incident X-ray .The crystallite size are found to be (41.5nm),(50.3nm) and (45.8nm). And the angle (25.28),(37.78) and (48.02) corresponds to (101),(004) and (200) after annealing . The crystallite size of TiO_2 thin films after annealing at 500 nm were (34.4nm),(46.4) and (41nm) respectively . Their values are listed in Table (1). These data were compared with JEPDS card number 021-1272, which refers to a hexagonal crystal structure.. The process of annealing was not significantly affecting the material.

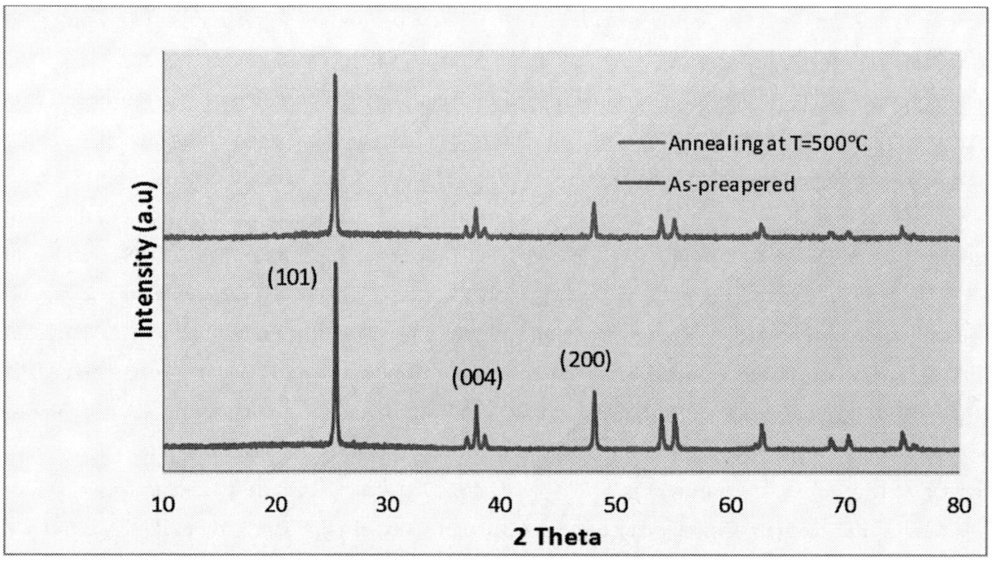

Figure 1. X-ray pattern of TiO_2 film

Both strain value and dislocation density can be calculated using the following equations (2) and (3), their values were listed in Table (1)
[30, 31]

$$\eta = \frac{\beta \cos\theta}{4} \dots\dots..(2)$$

$$\delta = \frac{1}{D_{av}{}^{2}} \dots\dots\dots.(3)$$

Figure (2) shows the 3D AFM image of TiO_2 thin film deposited on glass substrate by drop casing method. The surface of the thin film has a vertically closely packed ball –shaped. homogenous and a good roughness grains of TiO_2 nanostructure within scanning area $(2\times2)\mu m^2$ for as-prepared The average diameter size, the roughness and the root mean square (RMS) was about (83.33nm), (0.267nm), (0.33nm) respectively and annealing at T=500C° the average diameter size, the roughness and the root mean square (RMS) were about (88.94nm), (10.3nm), (12.1nm) respectively

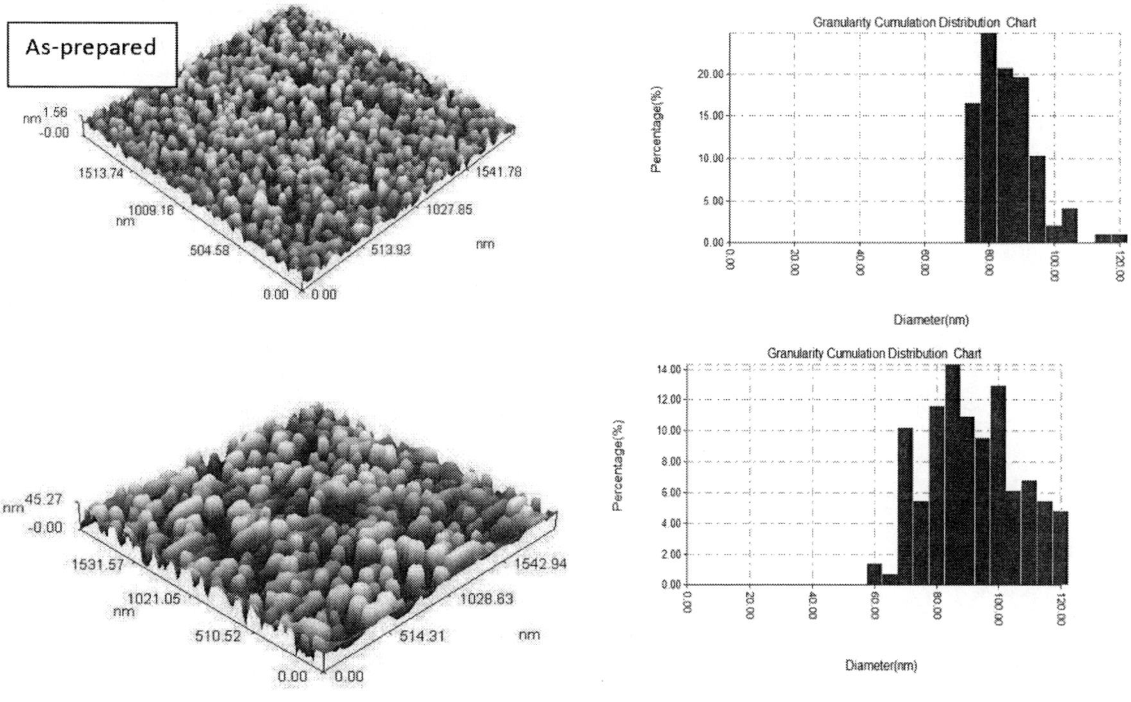

Figure 2. 3D AFM image of synthesized TiO_2 thin films

IOP Conf. Series: Journal of Physics: Conf. Series **1032** (2018) 012037 doi:10.1088/1742-6596/1032/1/012037

Table 1. the values of structural parameters of TiO_2 thin films before and after annealing

Samples	2 Theta (deg)	hkl (Planes)	Beta (deg)	D (nm)	δ x10^{14} (lines.m^{-2})	Strain × 10^{-4}
	25.34	(101)	0.195	41.535	5.7963	8.3422
As-prepared	37.80	(004)	0.166	50.333	3.9472	6.8841
	48.04	(200)	0.174	45.807	4.7657	7.5642
	25.28	(101)	0.235	34.461	8.4202	10.054
Annealing	37.78	(004)	0.180	46.415	4.6416	7.6515
at T=500 °C	48.02	(200)	0.211	41.028	5.9407	8.4454

Figure 3 shows the FTIR spectrum of the compound TiO_2 powder. This indicates strong vibration of Ti=O about 687 cm^{-1}[32] . Absorption peaks around 3640 cm^{-1} to 2500 cm^{-1} due to the presence of O-H stretching vibration (monomer, intermolecular, intermolecular and polymer)[33] . The absorption band range at 1629 cm^{-1} indicates the presence of O-H bending vibration. The Figure shows the (FT-IR) spectrum of as-synthesized TiO_2 thin films. In the case of film, the peak centered around 650 cm^{-1} is due to the Ti=O extended vibrations [34]. The FT-IR spectra confirm the presence of Ti=O vibrations in both cases.

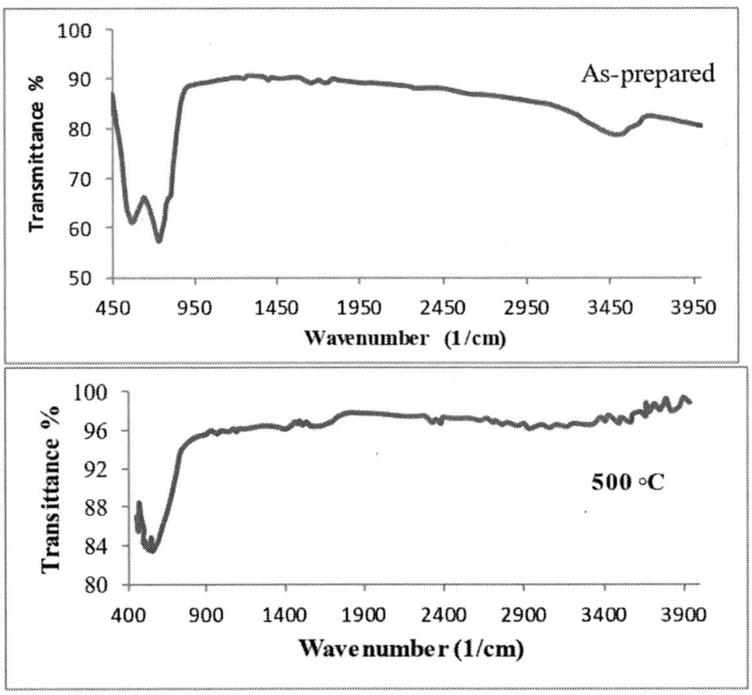

Figure 3. FTIR of TiO_2 powder

Figure (4) explain the transmittance spectrum of TiO_2 nanoparticles, from figure it can be notice that transmittance reaches 0.85 at 720 nm wavelength as shown in fig6-a and 0.24 at 720 nm at annealing at 500 °C as shown in Fig 4-b, then the transmittance increase as the wavelength increased.

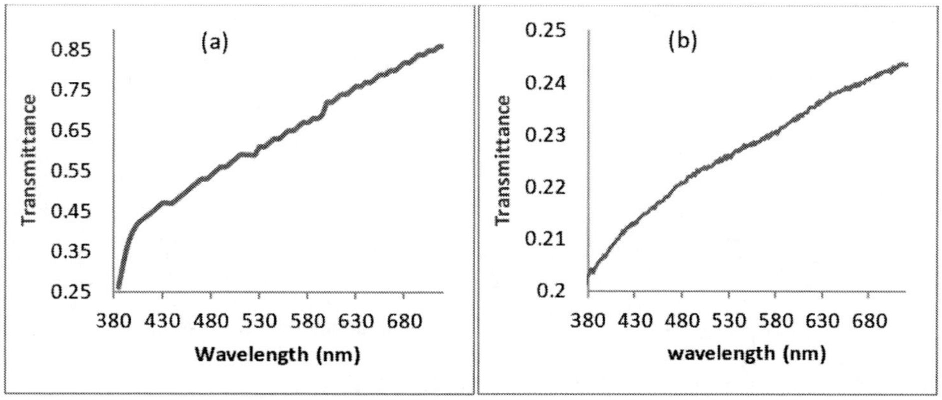

Figure 4. Optical transmittance of Tio_2 thin films

Fig. (5) shows the variation of $(\alpha h\upsilon)^2$ versus $h\upsilon$ for direct band gap which have been determined by the extrapolation of linear portion versus the photon energy axis. It can be seen that the value of the energy gap is about 2.8 eV. The process of annealing affected the material and led to decrease in the value of energy gap and therefore to 3.6 eV ,this may be attributed to quantum confinement as shown in Figure (5-a) and after annealing at 500 °C became 2.8eV as shown in Figure (5-b).

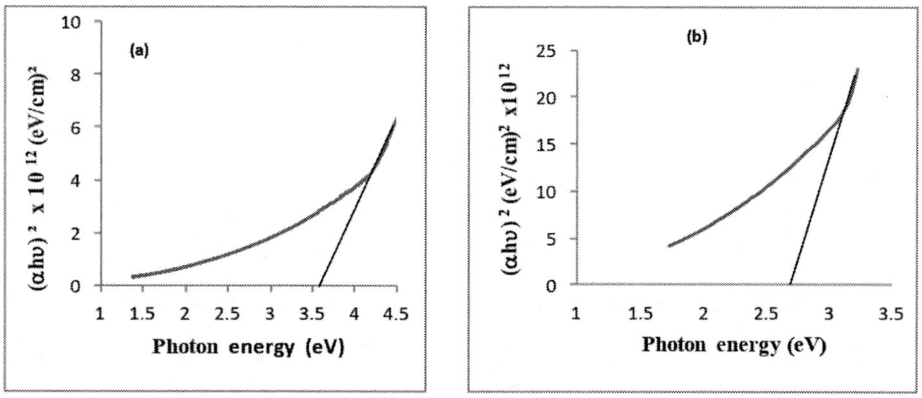

Figure 5. $(ahv)^2$ versus photon energy band gap for TiO2 thin films

The reflectance of TiO_2 thin film increases from 0.35 to 0.47 when the wavelength increases from 380 nm to greater than 720 nm as shown in figure (6-a) and at annealing from 0.170 to 0.185 when the wavelength increases from 380 nm to greater than 720 nm as shown in figure (6-b)

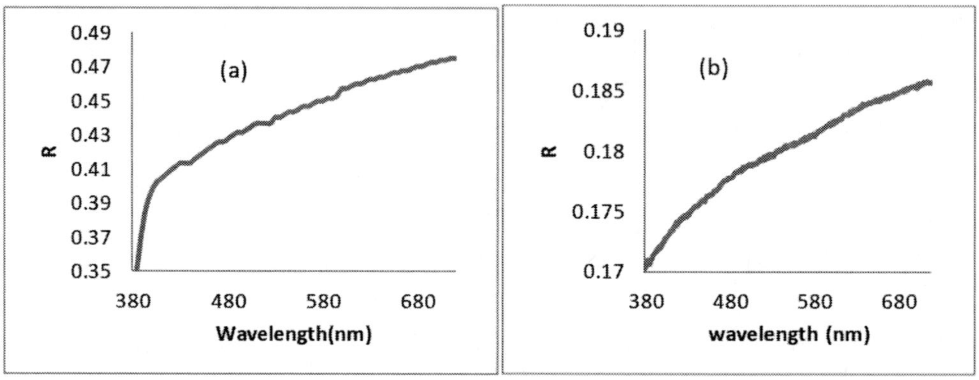

Figure 6. Reflection spectra of TiO_2 thin film a function of wavelength.

The refractive index (n) has been calculated by formula [35]:

$$n = \frac{1+\sqrt{R}}{1-\sqrt{R}} \ldots\ldots\ldots (4)$$

Figure 7 shows the change in the value of the refractive index as a function of the wavelength as in Figure (7-a) , the index of refraction highest value is 1.65 at wavelength of 720 nm .Figure (7-b) Depicts the refractive index after annealing . The maximum value was 2.51 at a wavelength of 720nm.

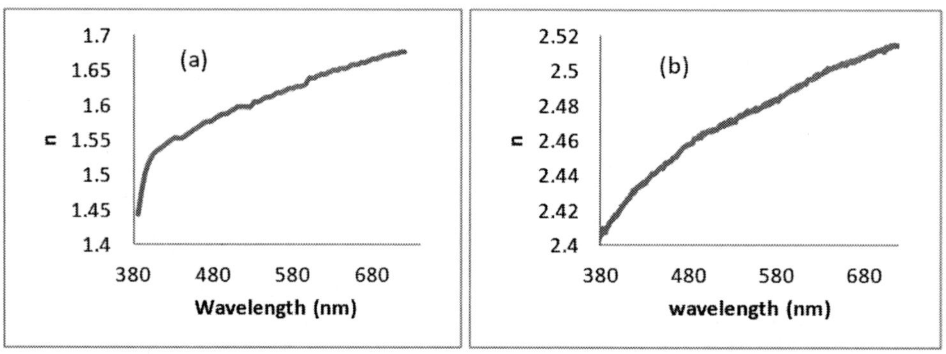

Figure 7. Refractive index spectra of TiO_2 thin film as a function of wavelength

Figure (8-a) shows black dots representing the gaps in this material. It is also clear to us from the pictures that the material is homogeneous and that its atoms are distributed equally On the surface of the base and it is clear to us that the atoms of the material in the form of spherical atoms and this is a confirmation of the tests of AFM where they give the same result. And that the effect of the process of annealing became the surface of the thin film more rough as shown in Figure (8-b) This is an important application of the material to act as a solar cell.

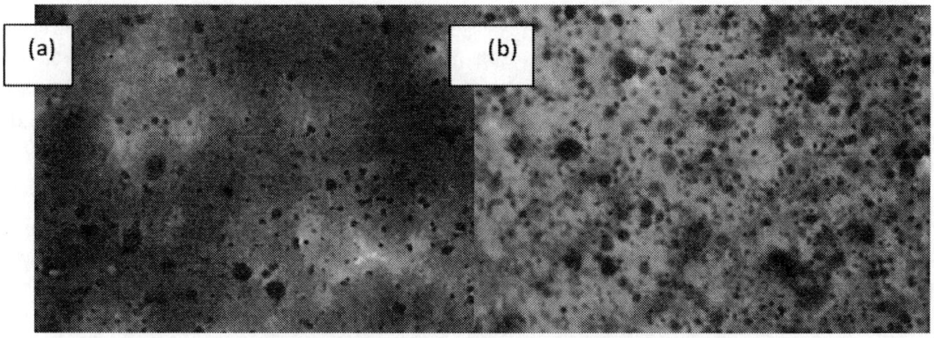

Figure 8. Optical microscopy images of TiO_2 nanoparticles:
(a) As-prepared (b)At annealing at 500 °C/h

Figure (9) shows the alternating current of the solar cell Al/TiO_2/Si/Al in two cases during preparation and at the annealing at 500 °C for one hour. It is observed from the curve that the behavior of the solar cell in the case of darkness takes a non-linear shape and is curved to an exponential function, which means that the mechanism of conductivity is not ohmic. In the forward bias, the current flows very freely and we observe a high value of the current. In reverse bias, the flow of current is very weak due to the flow of charge carriers and the applied voltage to inject the majority carriers and display the area of attrition. The concentration of the majority and minority carriers is higher than the concentration of the basic carriers, which generates the docking current in the region (0-0.2V). This is because the electrons are moved from the valence pack to the conduction beam and then returns to replace the gaps in the valence pack. Docking in the region is low voltage, while the tunnel condition occurs in the high voltage area and then we observe a huge increase in the value of the current with a small increase in voltage. This is called the propagation current which dominates the high voltage area. And In the case of the state of the deflection zone, the bias effort increases the width of the attenuation zone. This is why the concentration of the majority and minority carrier's decreases and the current is increasing slightly by increasing the voltage.

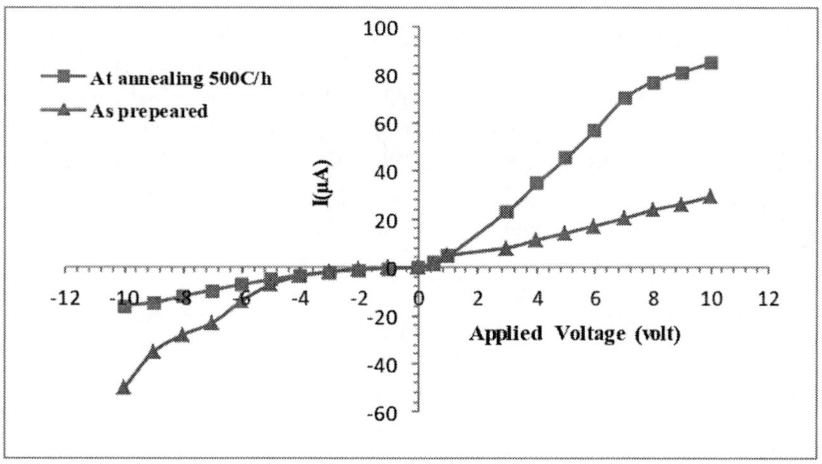

Figure 9. shows the current-voltage curve in the dark for Al/TiO_2/Si/Al

The process of annealing greatly influenced the current-voltage scheme and from the figure we note that the current in the case of bias in the front became larger than after the material was born at 500°C within one hour because the process of dependence affected the material and made its particles to expand and increase the space between the gaps and leading to transit Bulk carriers and their transmission from the parity package to the delivery package.

The diagram(10) shows a current voltage diagram in the case of $Al/TiO_2/Si/Al$ reverse bias in the case of darkness and in the case of lighting with a 15µw/cm^2 lamp and about 30cm away from the sample and studying its characteristics before and after the process of annealing at 500°C/h.It is clear to us that by means of the form, the optical current (I_{ph}) increases by increasing the rated voltage as well as increasing the intensity of the fall, thus increasing the number of charge carriers, as well as displaying the attrition zone increase with the increase of the reverse bias voltage, which leads to increased absorption and generation of electrons.

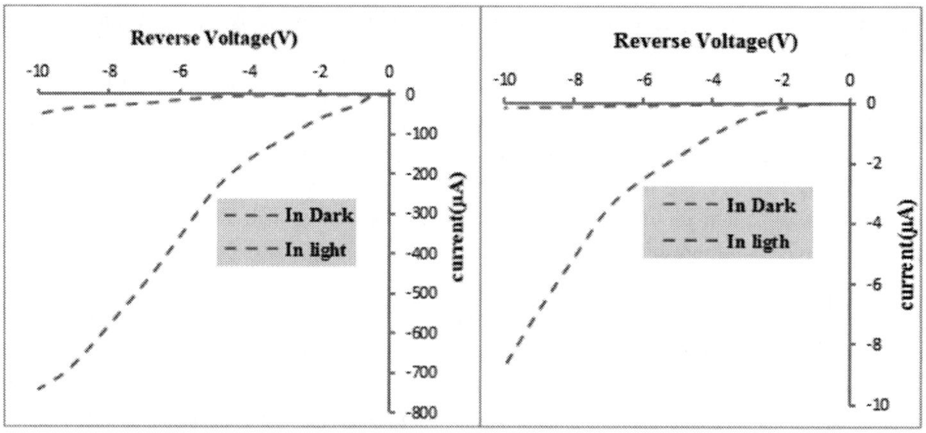

Figure 10. Dark and illuminated (I-V) characteristic of $Al/TiO_2/Si/Al$ Solar Cell (a)at as-prepared and (b)at annealing at 500°C/h

Figure 11 shows the voltage of the open circuit (V_{oc} , I=0 , R=∞), is the voltage output, as the load resistance is much greater than the device resistance. The short circuit current (I_{sc} , V=0 , R=0), is the current output when the load resistance is much smaller than the device impedance, then the Fill Factor (F.F) is the rate of maximum power output (V_m , I_m) respectively, to the product of V_{oc} and I_{sc} . from these parameters, efficiency can be estimated. The conversion efficiency (η) is defined as the ratio of (P_m/P_{in} ×100%) and $F.F=(I_m V_m/I_{sc} V_{oc} \times 100\%)$ Calculated parameters of the solar cell for $Al/TiO_2/Si/Al$ before and after the annealing process, tightness (15µW/cm^2) are listed in Table (2) showing the estimated parameters of solar cell.

Table 2. Parameters of Solar Cell for TiO_2

Sample	V_{oc}(V)	I_{sc}(µA)	V_m(V)	I_m(µA)	F.F%	η%
As-Prepared	2.5	3.5	2.04	2.82	65.7	6.13
At annealing 500°C/h	1.2	26.4	0.94	21.75	64.5	10.5

Table (2) indicates when photons were incident on Al/TiO_2/Si/Al, We note that the value of the efficiency after the effect of the process of annealing, because after the process of annealing will show very thin layers on the surface of the sample and give it the ability to convert photons of light to greater electrical energy.

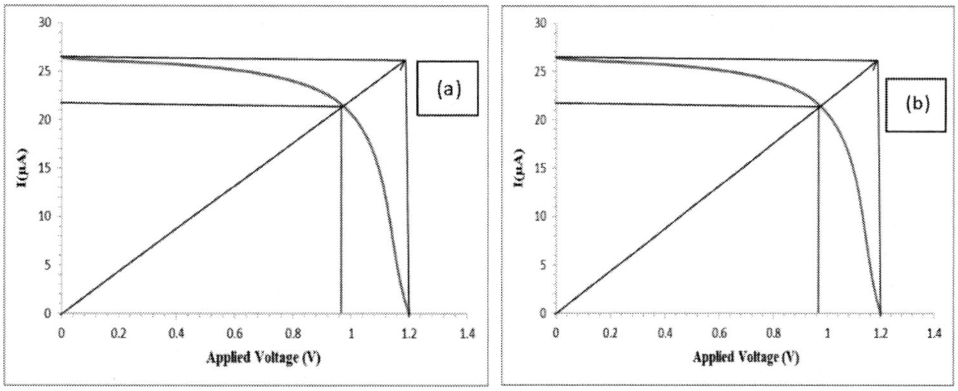

Figure 11. Open-circuit voltage of Al/TiO_2/Si/Al for (a) as prepared (b) at annealing at 500°C/h

4. Conclusions.

The influence of the thermal annealing temperature on the crystalline structure, surface morphology and the optical properties of TiO2 thin films were investigated .The XRD results have shown that the nanocrystalline TiO_2 films with preferred orientation along (101) direction can be observed before annealing and by increasing of annealing temperature to 500 ◦C no significant changes occur in the film structure. The TiO_2 phase and improvement of crystallinity was observed after annealing at temperatures of 500 ◦C. It was found that the annealing temperature significantly influenced the size and shape of the TiO_2 particles. The best crystallization , surface morphology and good absorption were observed at a temperature of 500 ° C .Deposition of TiO_2 NPS on silicon gave suspensions Solar cell characteristics. The conversion efficiency (η) of the solar cell for Al/TiO_2/Si/Al enhances after thermal annealing.

References

1- Y. Leprince-Wang, D. Souche, K. Yu-Zhang, S. Fisson, G. Vuye and J. Rivory, Thin Solid Films, 359 (2000) 171.

2- Karunagara B., Kim K., Mangalaraj D., Yi J. and Velumani S., (2005), Structural, Optical and Raman Scattering Studies on DC Magnetron Sputtered Titanium Dioxide Thin Films, Solar Energy Mater.& Cells, 88, 199-208.

3- P. Zeman and S. Takabayashi, Surface and Coatings Technology, 93 (2002) 153.

4- G. Gupta, P. Shah, X. Zhang, A. Saunders, B. Korgel and K. Johnston, "Enhanced Infusion of Gold Nanocrystals into Mesoporous Silica with Supercritical Carbon Dioxide," Chem Mater, Vol. 17, No. 26, (2005), PP. 6728-6738

5-Do T, "Nano - Particles: Potential Applications in Catalysis and Nanotechnology," Recent Research Developments in Chemistry, Vol. 2, (2004), P. 2.

6- J. Tarascon, S. Grugeon, S. Laruelle, D. Larcher and P. Poizot, "The key Role of Nanoparticles in Reactivity of 3D Metal Oxides Toward Lithium," In Lithium Batteries Science and Technology, Kluwer Academic, Boston, (2004), PP. 220-246

7- C. Martin and P. Kohli, "The Emerging Field of Nanotube Biotechnology," Nature Reviews Drug Discovery, Vol. 2, (2003), PP. 29-37.

8- S. Zhou, L. Wu, M. Xiong, Q. He and G. Chen, "Dispersion and UV-VIS Properties of Nanoparticles in Coatings," Journal of Dispersion Science and Technology, Vol. 25, (2005), PP. 417-433.

9- J. Pena, M. Vallet-Regi and J. San Roman, "TiO2-Polymer Composites for Biomedical Applications," Journal Biomed Mater Res, Vol. 35, (1997), PP. 129-134.

10- G. Mor, K. Shankar, M. Paulose, O. Varghese and C. Grimes, "Use of Highly-Ordered TiO2 Nanotube Arrays in Dye-Sensitized Solar Cells," Nano Letters, Vol. 6, (2006), PP. 215-218.

11- J. Bisquert, D. Cahen, G. Hodes, S. Ruhle and A. Zaban, "Physical Chemical Principles of Photovoltaic Conversion with Nanoparticulate, Mesoporous Dye-Sensitized Solar Cells," Journal Physics Chemistry, Vol. 108, (2003), PP. 8106-8118.

12- S. Yoo, S. Akbar and K. Sandhage, "Nanocarving of Titania (TiO2): A Novel Approach for Fabricating Chemical Sensing Platform," Ceramics International, Vol. 30, (2004), PP. 1121-1126.

13- R. Wang, K. Hashimoto, A. Fujishima, M. Chikuni, E. Kojima, A. Kitamura, M. Shimohigoshi and T. Watanabe, "Light-Induced Amphiphilic Surfaces," Nature, Vol. 388, (1997), PP. 431-432.

14- M. Cevro, G. Carter, J. Phys. D, Appl. Phys. 28 (1995) 1962.

15- F.X. Wang, C.K. Hwangbo, B.Y. Jung, J.H. Lee, B.H. Park, N.Y. Kim,Surf. Coat. Technol. 201 (2007) 5367.

16- S. Bhagwat, R.P. Howson, Surf. Coat. Technol. 111 (1999) 162.

17- H. Huang, X. Yao, J. Cryst. Growth 268 (2004) 564.

18- C. Garapon, C. Champeaux, J. Mugnier, G. Panczer, P. Marchet, A.Catherinot, B. Jacquier, Appl. Surf. Sci. 96-98 (1996) 836.

19- D. Huang, Z.D. Xiao, J.H. Gu, N.P. Huang, C.W. Yuan, Thin Solid Films 305 (1997) 110.

20- A. Watanabe, Y. Imai, Thin Solid Films 348 (1999) 63.

21- B.R. Sankapal, M.C. Lux-Steiner, A. Ennaoui, Appl. Surf. Sci. 239 (2005) 165.

22- Z. Wang, U. Helmersson and Per-Olov Käll, "Optical Properties of Anatase TiO2 Thin Films Prepared by Aqueous Sol–Gel Process at Low Temperature," Science Direct, Vol. 405, (2002), PP. 50-54.

23- Ba-Abbad M.M. , Kadhum A.A.H.,Mohamad A.B. , Mohd S. Takriff and Sopian K., Synthesis and Catalytic Activity of TiO2 Nanoparticles for Photochemical Oxidation of Concentrated Chlorophenols under Direct Solar Radiation, Int. J. Electrochem. Sci., 7, 4871 – 4888, (2012).

24- M.Nocun, S.Kwasny and J.Zontek,"Optical properties of TiO2–SiO2 layers prepared by sol-gel method",Optica Applicata, 1,4(2011)979-987.

25- Brady, John B., and Boardman, Shelby J., 1995, Introducing Mineralogy Students to X-ray Diffraction Through Optical Diffraction Experiments Using Lasers. Jour. Geol. Education, v. 43 #5, 471-476.

26- Binnig, G.; Quate, C. F.; Gerber, C. (1986). "Atomic Force Microscope". Physical Review Letters. 56: 930–933.

27- Griffiths, P.; de Hasseth, J. A. (18 May 2007). Fourier Transform Infrared Spectrometry (2nd ed.). Wiley-Blackwell. ISBN 0-471-19404-2.

28- Skoog, Douglas A.; Holler, F. James; Crouch, Stanley R. (2007). Principles of Instrumental Analysis (6th ed.). Belmont, CA: Thomson Brooks/Cole. pp. 169–173. ISBN 9780495012016.

29- H. A. Macleod, "Thin Film Optical Fillers",3rd ed, Taylor and Francis, (2001).

30- R. Karimzadeh, N. Mansour, "The Effect of Concentration on The Thermo Optical Properties of Colloidal Ag Nanoparticles", Optics and Laser Technology, 42 (2010) 783.

31- N. Semaltianos, S. Logothetidis, W. Perrie, S. Romani, R. Potter, et al., " CdTe Nanoparticles Synthesized by Laser Ablation", Applied Physics let., 95 (2009).

32- D. Arun Kumar, Francis P. Xavier and J. Merline Shyla," Natural dye sensitization of TiO2 Thin Films using Lawsone dye extractedfrom Lawsonia Inermis for Solar Cell Applications", Archives of Applied Science Research, 2012, 4 (5):2122-2132

33- Bedri Erdem,Robert A. Hunsicker,Gary W. Simmons,E. David Sudol,Victoria L. Dimonie, andMohamed S. El-Aasser" XPS and FTIR Surface Characterization of TiO2 Particles Used in Polymer Encapsulation" Lehigh University 2001, 17 (9), pp 2664–2669

34- Flora Boccuzzi, Anna Chiorino, Maela Manzoli, Ping Lu, Tomoki Akita, Satoshi Ichikawa, Masatake Haruta" Au/TiO2 Nanosized Samples: A Catalytic, TEM, and FTIR Study of the Effect of Calcination Temperature on the CO Oxidation" Volume 202, Issue 2, 10 September 2001, Pages 256-267

35- Ahmed. N Abd," Improved photoresponse of porous silicon photodetectors by embedding CdS nanoparticles", World Scientific News (WSN),13(2015)94 -23 .

IOP Conf. Series: Journal of Physics: Conf. Series **1032** (2018) 012038 doi:10.1088/1742-6596/1032/1/012038

Facile method to synthesis of anatase TiO_2 nanorods

Aseel Adnan Ouda[1], Firas K. Mohamad Alosfur[1*], Noor J. Ridha[2], Saleh H. Abud[3], Nibras Mossa Umran[1], Hayder H. Al-aaraji[1], Rajaa A. Madlool[1]

[1]Department of Physics, College of Science, University of Kerbala, Kerbala, Iraq
[2]Department of Chemistry, College of Science, University of Kerbala, Kerbala, Iraq
[3]Department of Physics, Faculty of Science, University of Kufa, Najaf, Iraq

* Corresponding author: frsos2005@yahoo.com, firas.k.m@uokerbala.edu.iq

Abstract. A simple wet chemical method was used to synthesise TiO_2 nanorods using Titanium isopropoxide (TTIP) as a precursor material. The chemical transformation of TTIP by ethylene glycol (EG) was demonstrated as a strategy for regulating and controlling the shape of formation TiO_2 nanorods. The structure of the sample was studied by X-ray diffraction and it revealed that the prepared TiO_2 exhibit a pure anatase phase. While the Fourier transform infrared spectroscopy (FTIR) was showing the vibration patterns in the spectrum of the sample. The morphology of sample was studied by scanning electron microscopy (SEM) and it showed that the synthesised TiO_2 made of nanorods with length about (698 nm) and a diameter (220 nm).

1. Introduction

Titanium dioxide (TiO_2) is a semiconductor material that has a wide energy gap (3.2 eV) [1, 2]. Many studies have been focused in the present time on TiO_2, due to its large surface area and characteristic electrical properties such as physical and chemical stability, high refractive index, low cost, non-toxicity and high photo-catalytic activities…etc. TiO_2 can be found in three crystalline phases: anatase, rutile both tetragonal structure and brookite with orthorhombic structure. Each of these phases have various chemical and physical properties depending on the composition of the atoms, thus leads to a varied performance in its application [3, 4]. Anatase and brookite are metastable, while rutile phase is stable under certain conditions [5]. The anatase and brookite phases are easily turned to rutile phase during a high temperature [6, 7]. Different morphologies of TiO_2 have been synthesized such as nanorods, nanowires, nanotubes, nanoparticles and nanobelts…etc. One dimension (1D) structures of TiO_2 especially nanorods, nanowires and nanotubes have been attracted a lot of attention due to the unique properties and great applications comparison nanoparticles. TiO_2 with (1D) structures have exhibited the faster process of generating electron-hole and a lower recombination rate comparison TiO_2 nanoparticle [8, 9]. Therefore, TiO_2 nanorods have been used in many applications such as self-cleaning coating [10], dye-sensitized solar cells [11], sensors [12], wastewater purification [13], lithium-ion batteries [14], electrochromic devices [15, 16],…etc. On the other hand the most important uses of TiO_2 are in photocatalytic applications, due to the cheap and friendly to the environment and human [17, 18].

There are several methods to prepare TiO_2 nanorods such as sol-gel [19], hydrothermal [20], microwave [21, 22], solvothermal [23] …etc. In this paper, anatase TiO_2 nanorods with high purity

IOP Conf. Series: Journal of Physics: Conf. Series **1032** (2018) 012038 doi:10.1088/1742-6596/1032/1/012038

were synthesized via a simple wet chemical method from Titanium Isopropoxide and Ethylene Glycol at temperature 100 ℃.

2. Experimental

2.1. Raw materials

All the chemicals used in the experiment are used with high purity. Titanium (IV) Isopropoxide (TTIP) Ti [OCH $(CH_3)_2]_4$, Ethylene Glycol (EG), absolute ethanol C_2H_5OH and deionized (DI) water were used in the TiO_2 preparation process.

2.2. Synthesis of TiO_2 powder

TiO_2 powder has synthesized by using the wet chemical method from (TTIP) and (EG) as a precursor materials. A specific amount (60 ml) of EG was placed in a glass vessel and heated up to a temperature 100 °C under magnetic vigorous stirring for 30 min. Next, 2 ml of TTIP was added drop wise into a vessel with vigorous stirring at 700 rpm. The maxing solution was kept under this condition of stirring for 2 h. When the titration process achieved, the resulting solution was changed from transparent color to a milky white, which indicates the formation of the TiO_2 nanoparticle.

After that, the precipitate was washed with deionized (DI) water and absolute ethanol several times, to eliminate any residual organic species remaining in the final products. The precipitate was then dried at temperature 60 °C overnight. Finally, the white fine powder was calcined at (450°C) in the air for 1h to obtain TiO_2 nanoparticle.

Different techniques were used to characterize the synthesized sample such as X-ray diffraction (XRD), Fourier-transform infrared spectroscopy (FTIR) and scanning electron microscope (SEM). The schematic diagram for the whole synthesis procedure and characterization of TiO_2 nanorods is shown in Figure 1.

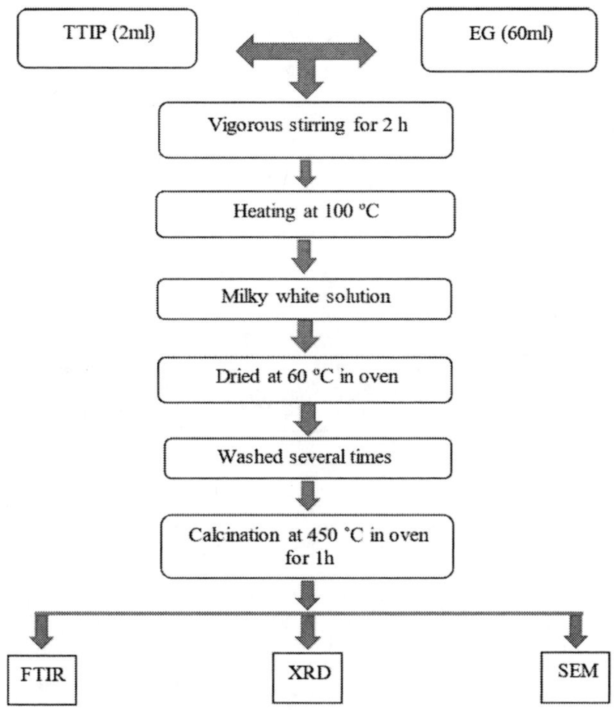

Figure 1. Schematic diagram of the steps of TiO_2 preparation.

2.3. Characterization of the sample

The crystal structure of TiO_2 powder was studied using X-ray diffractometer (Shimadzu- Japan) model 6000 (CuKα1 radiation, λ = 1.54056 Å). While the chemical bounding on the sample was investigated using Fourier transform infrared spectroscopy (Shimadzu-Japan, FTIR- 8400s) with a single beam in the range between (4000-400) cm^{-1}. The morphology of TiO_2 powder was characterized by scanning electron microscopy SEM (Inspect 550, FEI company made in the Netherlands).

3. Result and discussion

The XRD patterns of synthesized TiO_2 at temperature 100 °C were shown in figure 2. The XRD diffraction of the prepared sample detects a sharp domain (101) peak at about 2θ=25.3° ascribed to anatase phase of TiO_2. Comparing the XRD pattern with the standard (JCPDS file No. 21-1272), it can be observed that there are no peaks of brookite or rutile phase in the pattern, indicating the high purity of the synthesized sample. The crystal size was calculated by applying the Debye-Scherrer equation (1) on diffraction peaks of anatase phases [24].

$$D = K\lambda / (\beta\cos\theta) \qquad (1)$$

where D is the mean crystallite size of the sample. K is the shape factor which varies depending on the actual shape of the crystallite, but typically has a value of 0.9. λ is the X-ray wavelength, in this case, it was CuKα radiation (λ = 1.54056 A°). β is the full width of the peak at half maximum (FWHM) in radians and θ is defined as Bragg angle.

The average crystallite sizes of the sample of the anatase TiO_2 prepared at 100 °C and calcined at 450 °C for 1h was appeared to be 10.76 nm.

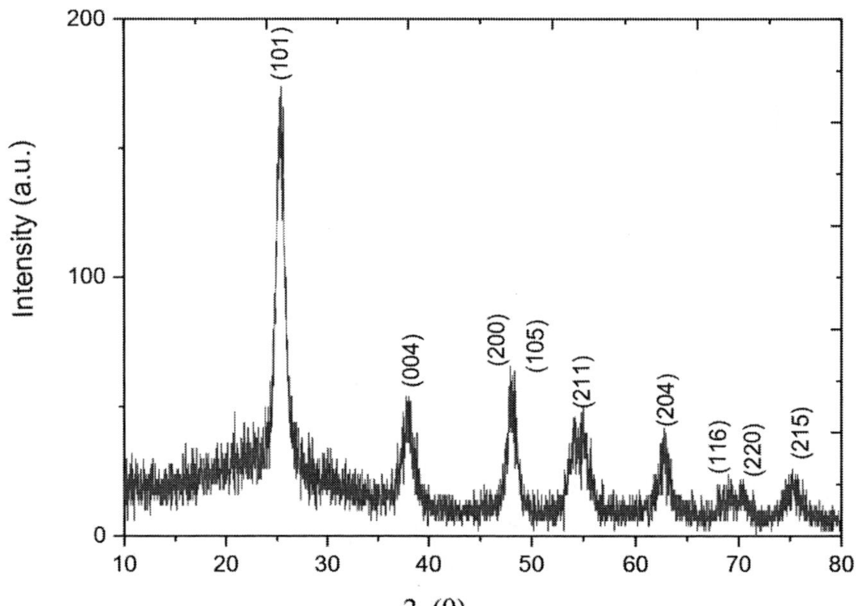

Figure 2. XRD pattern of the anatase TiO_2 nanorods sample.

Figure 3 was shown the FTIR spectrum of the prepared anatase TiO_2. The broad and strong absorption band of IR spectrum between (3000-3500) cm^{-1}, was attributed to the stretching vibration of O-H groups which is due to absorption of water molecules during the post-installation period because of the higher surface area of interaction with the organic additives and also small particle size [25]. The weak bending vibration at (1651.12 cm^{-1}) and (2802.46 cm^{-1}) indicated the existence of water and C-H aldehydic respectively, which may be derived from the absorption of the products when exposed in the atmosphere. Another peak at (1384.94 cm^{-1}) attributed to carboxylate stretch with

IOP Conf. Series: Journal of Physics: Conf. Series **1032** (2018) 012038 doi:10.1088/1742-6596/1032/1/012038

symmetric vibration. While the absorption band below (950 cm^{-1}) attributed to the characteristic vibration of TiO_2. The peaks at (736.83 cm^{-1}) and (538.16 cm^{-1}) were observed the growth of Ti-O-Ti linkage in spectrum [26].

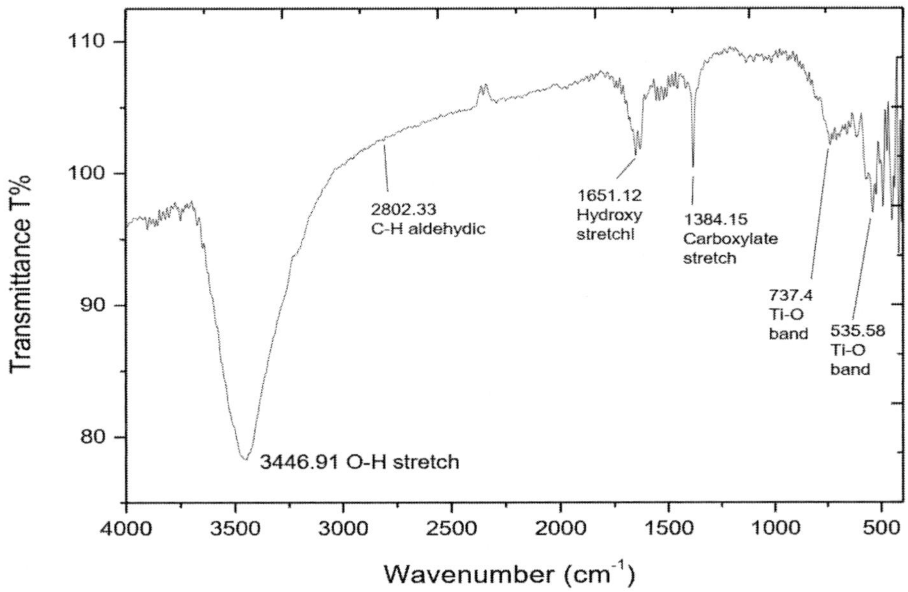

Figure 3. FTIR spectrum of TiO_2 nanorods.

The morphology of synthesized TiO_2 nanoparticles was characteristic using SEM as shown in Figure 4. According to SEM image, it was observed that the synthesised TiO_2 have uniform rod-like shapes with the average length and diameter about 698 nm and 220 nm respectively.

Figure 4. Shows SEM image of TiO_2 nanorods prepared at 100 ℃.

4. Conclusions

In this study, a simple wet chemical method was used for preparation TiO_2 nanorods using Titanium (IV) Isopropoxide as a precursor and Ethylene Glycol as a solvent. The X-ray diffractions results detected that TiO_2 nanoparticles are highly crystalline with anatase phase only. FTIR results have clarified that Ti-O bonds formed between 400- 900 cm^{-1} in the spectrum. SEM image was showed that the morphology of prepared TiO_2 in nanorods shape with an average length (698 nm) and diameter of (220 nm).

References

[1] Tang H, Berger H, Schmid P, Levy F and Burri G 1993 Photoluminescence in TiO_2 anatase single crystals Solid State Communications 87 847-50
[2] Jumali H, Hafizuddin M, Firas K, Radiman S and Umar A A 2012 Dressing of MWCNTs with TiO_2 nanoparticles using modified microwave method. In: Advanced Materials Research: Trans Tech Publ) pp 228-31
[3] Tahir M N, Theato P, Oberle P, Melnyk G, Faiss S, Kolb U, Janshoff A, Stepputat M and Tremel W 2006 Facile synthesis and characterization of functionalized, monocrystalline rutile TiO_2 nanorods Langmuir 22 5209-12
[4] Jumali M H H, Alosfur F K M, Radiman S, Ridha N J, Yarmo M A and Umar A A 2014 Rapid synthesis of TiO_2/MWCNTs nanocatalyst with enhanced photocatalytic activity using modified microwave technique Materials Science in Semiconductor Processing 25 207-10
[5] Rath C, Mohanty P, Pandey A and Mishra N 2009 Oxygen vacancy induced structural phase transformation in TiO_2 nanoparticles Journal of Physics D: Applied Physics 42 205101
[6] Chen Q, Liu H, Xin Y and Cheng X 2013 TiO_2 nanobelts–Effect of calcination temperature on optical, photoelectrochemical and photocatalytic properties Electrochimica acta 111 284-91
[7] Alosfur F K M, Ridha N J, Jumali M H H and Radiman S 2018 One-step formation of TiO_2 hollow spheres via a facile microwave-assisted process for photocatalytic activity Nanotechnology 29 145707
[8] Xia X, Luo Y, Wang Z, Liang Y, Fan J, Jia Z and Chen Z 2007 Ultrasonic synthesis and photocatalytic activity investigation of TiO_2 nanoarrays Materials Letters 61 2571-4
[9] Kim I-D, Rothschild A, Lee B H, Kim D Y, Jo S M and Tuller H L 2006 Ultrasensitive chemiresistors based on electrospun TiO_2 nanofibers Nano Letters 6 2009-13
[10] Kaida T, Kobayashi K, Adachi M and Suzuki F 2004 Optical characteristics of titanium oxide interference film and the film laminated with oxides and their applications for cosmetics Journal of cosmetic science 55 219-20
[11] Bach U 2000 Solid-state dye-sensitized mesoporous TiO_2 solar cells
[12] Trouiller B, Reliene R, Westbrook A, Solaimani P and Schiestl R H 2009 Titanium dioxide nanoparticles induce DNA damage and genetic instability in vivo in mice Cancer research 69 8784-9
[13] Tang C and Chen V 2002 Nanofiltration of textile wastewater for water reuse Desalination 143 11-20
[14] Barbe C J, Arendse F, Comte P, Jirousek M, Lenzmann F, Shklover V and Grätzel M 1997 Nanocrystalline titanium oxide electrodes for photovoltaic applications Journal of the American Ceramic Society 80 3157-71
[15] Hagfeldt A, Vlachopoulos N and Grätzel M 1994 Fast electrochromic switching with nanocrystalline oxide semiconductor films Journal of the Electrochemical Society 141 L82-L4
[16] Radhi M M, Alosfur F K M and Ridha N J 2018 Voltammetric Characterization of Grafted Polymer Modified with ZnO Nanoparticles on Glassy Carbon Electrode Russian Journal of Electrochemistry 54 27-32

[17] Perron H, Vandenborre J, Domain C, Drot R, Roques J, Simoni E, Ehrhardt J-J and Catalette H 2007 Combined investigation of water sorption on TiO_2 rutile (1 1 0) single crystal face: XPS vs. periodic DFT Surface Science 601 518-27

[18] Alosfur F K M, Jumali M H, Radiman S, Ridha N J and Umar A A 2013 Visible light photocatalytic activity of TiO_2/MWCNTs nanocomposite prepared using modified microwave technique. In: Sensing Technology (ICST), 2013 Seventh International Conference on: IEEE) pp 777-81

[19] Del Ángel-Sanchez K, Vázquez-Cuchillo O, Aguilar-Elguezabal A, Cruz-López A and Herrera-Gómez A 2013 Photocatalytic degradation of 2, 4-dichlorophenoxyacetic acid under visible light: Effect of synthesis route Materials Chemistry and Physics 139 423-30

[20] Zhang Q and Gao L 2003 Preparation of oxide nanocrystals with tunable morphologies by the moderate hydrothermal method: insights from rutile TiO_2 Langmuir 19 967-71

[21] Jia X, He W, Zhang X, Zhao H, Li Z and Feng Y 2007 Microwave-assisted synthesis of anatase TiO_2 nanorods with mesopores Nanotechnology 18 075602

[22] Ridha N J, Jumali M H H, Umar A A and Mohamad F K 2013 Ethanol sensor based on ZnO nanostructures prepared via microwave oven. In: Sensing Technology (ICST), 2013 Seventh International Conference on: IEEE) pp 121-6

[23] Gross S, Vittadini A and Dengo N 2017 Functionalisation of Colloidal Transition Metal Sulphides Nanocrystals: A Fascinating and Challenging Playground for the Chemist Crystals 7 110

[24] Ahmed F, Kumar S, Arshi N, Anwar M, Koo B H and Lee C G 2011 Rapid and cost effective synthesis of ZnO nanorods using microwave irradiation technique Functional Materials Letters 4 1-5

[25] Sun X and Li Y 2003 Synthesis and characterization of ion-exchangeable titanate nanotubes Chemistry-A European Journal 9 2229-38

[26] Yu J, Su Y, Cheng B and Zhou M 2006 Effects of pH on the microstructures and photocatalytic activity of mesoporous nanocrystalline titania powders prepared via hydrothermal method Journal of Molecular Catalysis A: Chemical 258 104-12

Ablation of ZnO in liquid by Nanosecond Laser

Hiba Basim Abbas Kadhim[1], Noor J. Ridha[2*], Firas K. Mohamad Alosfur[1], Nibras M Umran[1], Rajaa Madlol[1], Khawla J. Tahir[1] and Rafea T. Ahmed[3]

[1]Department of physics, college of science, University of Kerbala
[2]Department of Chemistry, college of science, University of Kerbala
[3]College of Science for Women, Laser Physics Department, University of Babylon

*Corresponding author: nooraboalhab@yahoo.com, noor.jawad@uokerbala.edu.iq

Abstract. Zinc oxide (ZnO) was synthesized by laser ablation in liquid. The preparation steps included immersing ZnO bulk pellet in ethanol and then ablating by Nd-Yag laser. The used laser has a wavelength ranging from 1064 nm to 532 nm with 140 mJ pulse intensity. Structural, morphological and optical properties of the prepared material were investigated by X-ray diffraction, scanning electron microscopy and UV-Vis spectroscopy, respectively. The XRD results revealed the formation of hexagonal ZnO structure with (101) as dominant peak. The SEM studies demonstrated the formation of spherical ZnO nanoparticles which agglomerated to form nanocluster. Typical exciton absorption was observed at 376 nm in the absorption spectrum of UV-Vis spectroscopy at room temperature and the energy gap was equal to (3.48 eV).

1. Introduction

Recently, nanotechnology has involved many potential investigations due to their unique chemical, physical, electrical and mechanical properties [1]. Thus nanomaterials revealed improvements in many fields, such as medicine [2], chemical sensors [3], gas sensing [4], solar cells [5], water treatment [6, 7], light emitting diodes [8] and other applications. In recent years, domestic wastewater, industrial waste, dyes and other sources have been major sources of water pollution [9]. The pollution can be treated using complex and traditional methods including thrombosis, adsorption on activated carbon, carbon nanotubes [10], chemical deposition and separation. Recently, metal oxide nanomaterials such as WO_2, Fe_2O_3, ZnS, TiO_2, ZnO have been used as photocatalysts in water treatment due to high efficiency, low cost, high stability and non-toxicity [11, 14].

There are many techniques to synthesis nanomaterials such as sol-gel [15], hydrothermal [16, 17], physical and chemical vapor deposition [18] microwave method [19, 20] and other techniques. These techniques can be classified according to their working principle into two different approaches, top-down and bottom-up [21]. The bottom-up approach is to gather materials atom by atom or molecule by molecule. The top-down approach is to cut material into small pieces to obtain nanoparticles. For example, laser ablation is a typical top-down technique for preparation nanoparticles [22]. Laser ablation (LA) technology is a simple and effective technique for producing nanomaterials. This technique also works when the laser beam is focusing on the target that immersed in liquid which is called laser ablation in liquid (LAL) to produce nanoparticles [23]. The importance of this technique increased due to the absence of secondary products as well as it considered as simple, economy and environmentally friend method [24].

Metals and semiconductors nanoparticles have strong size-dependent properties [25]. Thus, at the moment, preparation these materials in liquid medium are carefully considered for their prospective

IOP Conf. Series: Journal of Physics: Conf. Series **1032** (2018) 012039 doi:10.1088/1742-6596/1032/1/012039

development in many fields. ZnO is a highly used substance because it has a large energy gap (3.37 eV) and high exciton binding energy (60 meV). Other properties of ZnO nanoparticles such as non-toxic, high sensitivity to various gases, easy to manufacture and inexpensive made it widely used materials [26]. Thus, in this work the control of particle size is the key factor in the synthesis of ZnO colloids, which were prepared by nanosecond laser ablation. The purpose of preparation ZnO nanoparticles is to use as a photocatalyst material for water treatment.

2. Experimental

ZnO bulk powder of purity (99%) was purchased from Fluka chemicals. Using hydraulic system, 3 grams of ZnO powder was pressed into pellet form with dimensions of (2.5 cm diameter and 0.2 cm thickness). In order to remove the inner stresses and to make the material more durable and homogeneous, the pellets were annealed at temperature of (400 °C).The used laser source is Nd-YAG with 1064-532 nm wave length, power of (160-1000 mJ) and frequency of (1-6 Hz). A (30 cm) lens was used to focus the laser beam on the sample. The prepared pellet was located in the bottom of a 100 mL baker contained 4 mL ethanol. The ZnO pellet was removed from the reactor after (200) laser pulses. The color of the prepared liquid was changed to gray due to the ZnO nanofluid formation. The preparation setup is shown in Figure (1).

The structural properties of as prepared sample were examined using X-ray diffractometer (XRD). The used instrument was (6000 Shimadzu - Japan) X-ray diffractometer of Cu-Kα radiation with wavelength of (λ= 1.5406 Å). The system voltage, current and scan speed is 40.0 kV, 30.0 mA, and 8.0000 (deg/min) respectively. The morphology of the sample was monitored by scanning electron microscopy (SEM) model (Inspect S50). The optical properties were measured using UV/Vis spectroscopy model (CECIL 7200 - Korea) with wavelength range from (190-1100 nm).

Calculation the energy gap experimentally using UV/Vis spectrum can be done according to the equation below [27]:

$$E_g = \frac{hC}{\lambda_c} = \frac{1240}{\lambda_c} \qquad (1)$$

where Eg represented the energy gap of the sample, h is the planck's constant = $6.626070040\times10^{-34}$, C is the light speed in vacuum = 3×10^{8} m/s and λ_c = cutoff wavelength of the photon which can be determined from the U/Vis spectrum.

The energy gap of nanosize semiconductor can be also calculated using Kubelka-Munk function and Tauc plots .

$$(\alpha h\nu)^{1/n}=k(h\nu- Eg) \qquad (2)$$

where h is Planck's constant, ν the frequency of vibration, α represents the absorption coefficient, Eg is the band gap and k is the proportional constant. The value of the exponent n denotes the nature of the sample transition and it is n = ½ for direct allowed transition.

The crystallite size of the studied material could be estimated from the full width at half maximum (FWHM) of the most intense diffraction line by Debye Scherrer's formula as follows.

$$D_S = \frac{k\lambda}{\beta\cos\theta} \qquad (3)$$

where, D_S is crystallite size, k=0.94 is the shape factor and β is (FWHM) in radians. The X- ray diffraction data can also be used to determine the dimension of the unit cell [28].

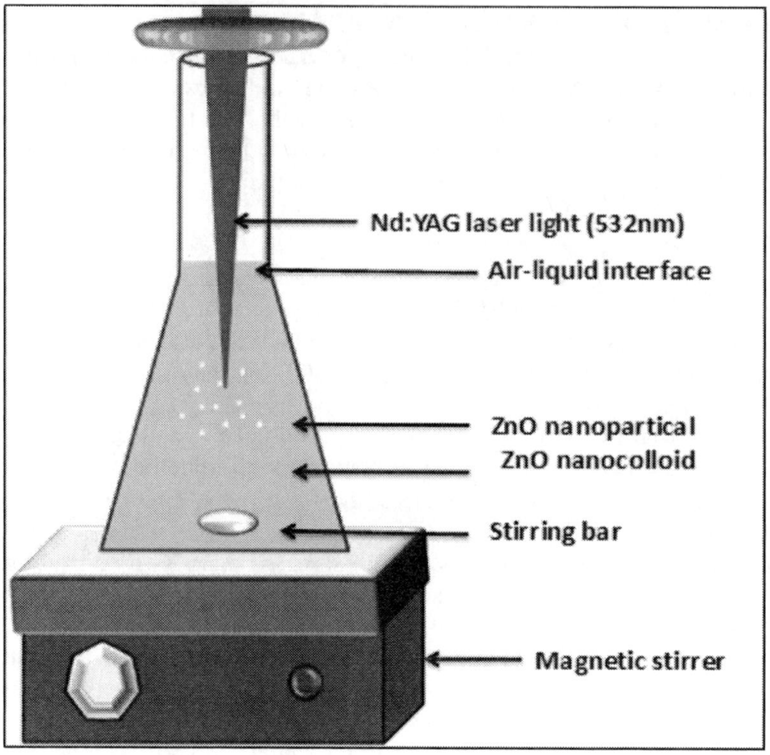

Figure 1. Schematic diagram of the experimental setup of PLAL.

3. Result and Discussion

3.1. Structure characterization by XRD

Primarily, to identify the prepared material, the as prepared sample was examined by XRD as shown in Figure (2). The qualitative investigation was performed by comparing the diffraction peaks of as prepared sample with the ICDD PDF database. This analysis indicated that the sample can be indexed to the wurtzite phase of ZnO (hexagonal, *P63mc*) with lattice constants of $a = 0.325$ nm and $c = 0.520$ nm, which match well with the standard XRD data file of ZnO (JCPDS 00-036-1451).

In addition, the strong relative intensity of the (101) line reveals a preferred orientation. Subsequently, the search was extended to another two peaks that not matching with the ZnO. The result revealed that the unidentified peaks were matched by a mineral Zn (Card-00-004-0831). This result indicates that during the ablation process ZnO was redox into Zn. The reduction of oxygen concentration is due to the reaction of atomic oxygen with ethanol which become much faster in the presences of ultrafast laser pulse [29]. The appearance of Zn peaks in the XRD patterns is attributed to the fact that the post annealing process was not used in this study. Further study will be conducted on effects of annealing on the structure in our next manuscript. Table (1) shows angles, Miller coefficients, the inter planer spacing (d) and full width at half maximum (FWHM). The experimental values of (d) are matching well with that of standards with error of 1% as listed in Table (1). It is obvious that d decreased with increasing the diffraction angle which followed the Bragg law ($n\lambda=2d\sin\theta$). From the figure it is clear that the prevailing peak is (101) at angle equals to 36° with FWHM = 0.1730. This indicates that the crystal size of the sample is very small which was calculated by equation (3). The average crystal size is equal to 50 nm. These results were compared with previous studies of ZnO prepared using laser ablation where Zn were also found [30].

IOP Conf. Series: Journal of Physics: Conf. Series **1032** (2018) 012039 doi:10.1088/1742-6596/1032/1/012039

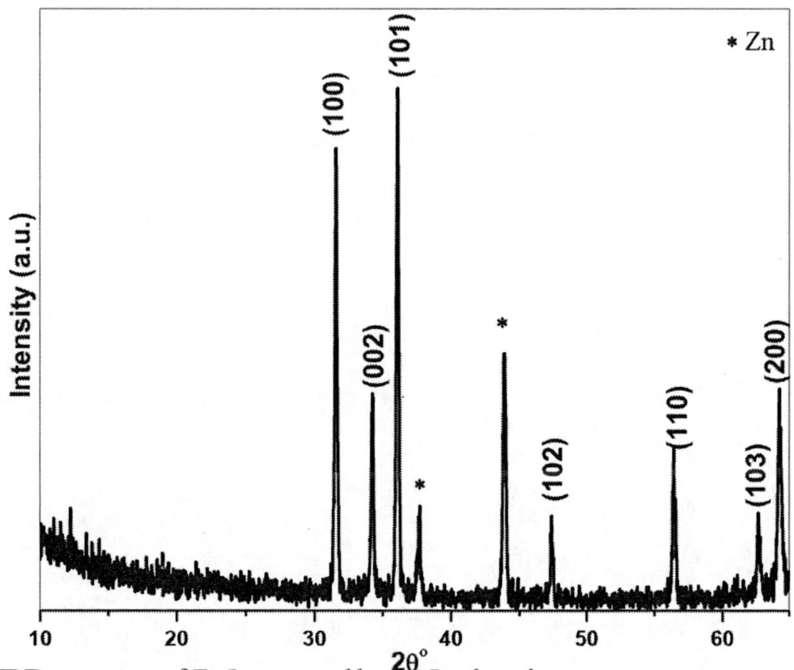

Figure 2. The XRD spectrum of ZnO prepared by LAL ethanol

Table 1. Experimental and standards value of d- spacing of as-prepared sample using LAL method.

2Θ (deg)	hkl	I/I_1	FWHM (deg)	d (Å) Experimental	d (Å) standards	Rate error	Crystal size (nm)
31.5996	100	85	0.1698	2.8291	2.814	0.0151	50.82
34.2711	002	40	0.1691	2.6144	2.603	0.0114	51.39
36.0914	101	100	0.1730	2.4866	2.475	0.0116	50.48
47.3990	102	17	0.1490	1.9164	1.9111	0.0053	60.86
56.4449	110	31	0.1853	1.6289	1.6247	0.0042	50.86
62.8624	103	18	0.1600	1.4 805	1.4771	0.0034	60.82

3.2 SEM studies

SEM is one of the instruments that used to monitor the surface morphology as shown in Figure (3). The sample was monitored under different high magnification namely (1000 and 10000X).The micrographs demonstrated the formation of spherical ZnO nanoparticles which agglomerated to form nanocluster. The dimensions of nanoparticles range from 219.7 nm to 330 nm.

Figure 3. Morphology of ZnO prepared by LAL in ethanol.

3.3. Optical properties by UV-Vis spectroscopy

In this study, UV-Vis spectroscopy was used to determine the optical properties of the sample. The spectrum in Figure (4) revealed that the highest absorption of as prepared sample was at 199 nm due to the natural absorption of ZnO. The absorption reduced as the wavelength increased. Typical exciton absorption at 376 nm was observed in the absorption spectrum at room temperature, which is considered blue shifted compared to the bulk ZnO. The blue shift in UV spectrum is caused by the reduction in size of the nanoparticle. The appearance of this strong peak in the absorption edge is due to the quantum size effect. The energy gap was experimentally calculated from equation (1), which is equal to 3.3 eV. These results were compared with previous ones of ZnO which were equivalent to 3.3 eV for average value of crystallite size of (40-60 nm) [31].

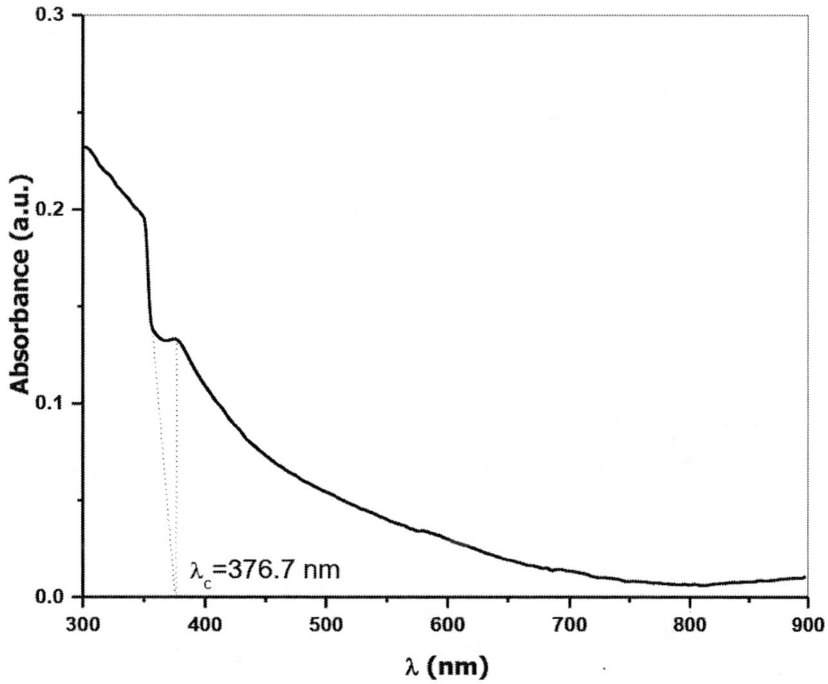

Figure 4. The absorption spectrum of the as prepared sample in ethanol by LAL.

Figure (5) shows the energy gap of ZnO which is equal to (3.48 eV). These results are consistent with the results of the previous studies (3.49eV) [32].

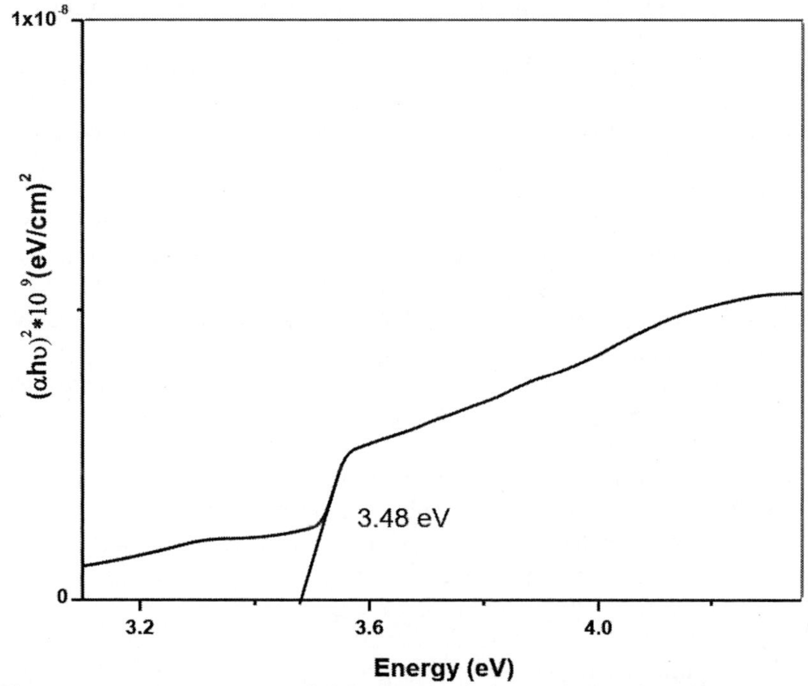

Figure 5. The energy gap spectrum of the as prepared sample in ethanol by LAL.

4. Conclusion

ZnO nanoparticles were prepared by laser ablation in liquid (ethanol). The used laser energy was 140 mJ with 300 pulses. The XRD spectrum revealed that ZnO was successfully prepared using LAL method. The SEM images showed the growth of spherical nanoparticles. The appearance of ZnO absorption peaks in the ultraviolet portion of light spectrum (376.7 nm) with energy gap to equal (3.48eV) which make it good candidate for photocatalyst applications

References

[1] Buzea C, Pacheco I I and Robbie K 2007 Nanomaterials and nanoparticles: Sources and toxicity Biointerphases 2 MR17-MR71
[2] Nikalje A P 2015 Nanotechnology and its applications in medicine Med chem 5 185-9
[3] Radhi M M, Alosfur F K M and Ridha N J 2018 Voltammetric Characterization of Grafted Polymer Modified with ZnO Nanoparticles on Glassy Carbon Electrode1 Russian journal of electrochemistry 54
[4] Zeng W, Wang H and Li Z 2016 Nanomaterials for sensing applications Journal of nanotechnology 2016 1-2
[5] Sethi V, Pandey M and Shukla M P 2011 Use of nanotechnology in solar PV cell International journal of chemical engineering and applications 2 77
[6] Alosfur F, Jumali H, Radiman S, Ridha N J, Yarmo M A and Umar A A 2013 Visible light-responsive TiO_2 coated MWCNTs as a hybrid nanocatalysts Int J Electrochem Sci 8 2977-82
[7] Tayeb A M and Hussein D S 2015 Synthesis of TiO_2 nanoparticles and their photocatalytic activity for methylene blue American Journal of Nanomaterials 3 57-63
[8] Tan Z-K, Moghaddam R S, Lai M L, Docampo P, Higler R, Deschler F, Price M, Sadhanala A, Pazos L M and Credgington D 2014 Bright light-emitting diodes based on organometal halide perovskite Nature Nanotechnology 9 687-92
[9] Alosfur F K M, Jumali M H, Radiman S, Ridha N J and Umar A A 2013 Visible light photocatalytic activity of TiO_2/MWCNTs nanocomposite prepared using modified microwave technique. In: Sensing Technology (ICST), 2013 Seventh International Conference on: IEEE pp 777-81
[10] Jumali H, Hafizuddin M, Firas K, Radiman S and Umar A A 2012 Dressing of MWCNTs with TiO_2 nanoparticles using modified microwave method. In: Advanced Materials Research: Trans Tech Publ) pp 228-31
[11] Jumali M H H, Alosfur F K M, Radiman S, Ridha N J, Yarmo M A and Umar A A 2014 Rapid synthesis of TiO_2/MWCNTs nanocatalyst with enhanced photocatalytic activity using modified microwave technique Materials Science in Semiconductor Processing 25 207-10
[12] Zhou Q, Wen J Z, Zhao P and Anderson W A 2017 Synthesis of vertically-aligned zinc oxide nanowires and their application as a photocatalyst Nanomaterials 7 9
[13] Alosfur F K M, Ridha N J, Jumali M H H and Radiman S 2018 One-step formation of TiO_2 hollow spheres via a facile microwave-assisted process for photocatalytic activity Nanotechnology 29 145707
[14] Hassena H 2016 Photocatalytic degradation of methylene blue by using Al_2O_3/Fe_2O_3 nano composite under visible light Modern Chemistry & Applications
[15] Daraio C and Jin S 2012 Nanotechnology for Biology and Medicine: Springer) pp 27-44
[16] Ridha N J, Jumali M H H, Umar A A and Alosfur F 2013 Defects-controlled ZnO nanorods with high aspect ratio for ethanol detection Int. J. Electrochem. Sci 8 4583-94
[17] Hayashi H and Hakuta Y 2010 Hydrothermal synthesis of metal oxide nanoparticles in supercritical water Materials 3 3794-817
[18] Rajput N 2015 Methods of preparation of nanoparticles-A review International Journal of Advances In Engineering & Technology 7 1806

[19] Ridha N J, Umar A A, Alosfur F, Jumali M H H and Salleh M M 2013 Microwave assisted hydrothermal method for porous zinc oxide nanostructured-films Journal of nanoscience and nanotechnology 13 2667-74
[20] Alosfur F K M, Jumali M H H, Radiman S, Ridha N J, Yarmo M A and Umar A A 2013 Modified microwave method for the synthesis of visible light-responsive TiO_2/MWCNTs nanocatalysts Nanoscale research letters 8 346
[21] Whitesides G M, Kriebel J K and Mayers B T 2005 Self-assembly and nanostructured materials Nanoscale Assembly 217-39
[22] Zheng Z 2009 Synthesis and modifications of metal oxide nanostructures and their applications. Queensland University of Technology)
[23] Honda M, Goto T, Owashi T, Rozhin A G, Yamaguchi S, Ito T and Kulinich S A 2016 ZnO nanorods prepared via ablation of Zn with millisecond laser in liquid media Physical Chemistry Chemical Physics 18 23628-37
[24] Bastianini F 2015 Synthesis of ZnO-based nanostructured materials via pulsed laser ablation in liquid
[25] Said A, Sajti L, Giorgio S and Marine W 2007 Synthesis of nanohybrid materials by femtosecond laser ablation in liquid medium. In: Journal of Physics: Conference Series: IOP Publishing) p 259
[26] Hussain M 2014 Synthesis, Characterization and Applications of Metal Oxide Nanostructures. Linköping University Electronic Press)
[27] Fahad D B 2014 Structural, Electrical and Optical Properties of $CuLayFe_2$-y Ferrite System. University of Baghdad)
[28] Anitha S and Jayakumari I 2015 Synthesis and Analysis of Nanocrystalline Fe 2 Mn 2 Ni 0.5 Zn 1.5 O 9 at Different Treating Temperatures Journal of Nanoscience and Technology 26-31
[29] Uddi M, Jiang N, Mintusov E, Adamovich I V and Lempert W R 2009 Atomic oxygen measurements in air and air/fuel nanosecond pulse discharges by two photon laser induced fluorescence Proceedings of the Combustion Institute 32 929-36
[30] Ghosh A and Choudhary R 2010 Optical emission and absorption spectra of Zn–ZnO core-shell nanostructures Journal of Experimental Nanoscience 5 134-42
[31] Ghazai A J, Aziz W J and Yassen N T 2017 Effect of Laser Pulses and Energy on the Structural Properties of ZnO Thin Film Prepared Using PLA Technique
[32] Yildirım A K and Altıokka B 2015 Effect of potential on structural, morphological and optical properties of ZnO thin films obtained by electrodeposition

IOP Conf. Series: Journal of Physics: Conf. Series **1032** (2018) 012040 doi:10.1088/1742-6596/1032/1/012040

The Role of PU Ratio on the Dielectric Properties of EP /PU Blends

Ahmad A. Hasan

University of Baghdad, College of Science, Department of Physics
ahmadabhasan27@yahoo.com

Abstract. Electrical A.C conductivity Epoxy/polyurethane EP/PU blends with different blend ratio were synthesized and characterized. A.C conductivity measurements of EP/PU blends were studied in the frequency and temperature ranges ($4x10^4$Hz) and (293-373K) respectively. Effect of PU loading levels and temperature on dielectric properties was investigated. The EP/PU blends were diagnosed using FTIR. The results showed that AC conductivity σA.C to be proportionate to ωs .The behavior of the exponent s explained in terms of the correlated barrier hopping and small polaron models. The value of dielectric constant 29723 and dielectric loss 705 of neat epoxy were exhibit to decrease with frequency and increase with temperature. The dependence of activation energy and conduction behavior on the blend ratio also studied. The results showed that the estimated activation energy varied with PU content. The polarizability (α) decreases with increase of PU content and temperature which refer to weaken the intermolecular forces.The addition of PU to EP was responsible about the increase of charge carriers.

Keywords: A.C conductivity, Dielectric properties, EP/PU Polymer blends, Universal power law, Dielectric loss.

1. Introduction

There is progressive increase of tendency to combine physical properties of molecular materials science, such as magnetism and electrical conductivity in a synergistic way in order to produce materials of high technical interest [1,2]. In recent years, the DC electrical properties of different organic and inorganic materials were extensively investigated, but relatively little work done on their AC electrical properties [3-6]. Information on the conduction process involved in the material can be provided from D.C measurements [7]. By varying the type of the electrode metal, the applied electric field strength, and the operating temperature it is possible to estimate whether the process is electrode or bulk limited. Measurements of AC conductivity, however, provide information which can be used to determine which of the three main conduction models applicable to organic and inorganic materials, namely the hopping model, the variable hopping model, and band theory [5], is operative in the system under study. The present study, we have therefore studied the AC conductivity and dielectric properties of epoxy /polyurethane (EP/PU) blends in order to determine the effect of blend ratio on mechanism responsible for the conductivity of this system.

2. Samples Preparation and Procedure

Epoxy resin (EP),type (Thortex) with hardener Isophorone diamine(IPD),supplied by (Thortex Division of E.Wood Ltd.) ,and thermoset Polyurethane resin(PU), type (Swiss Chem.) with hardener

IOP Conf. Series: Journal of Physics: Conf. Series **1032** (2018) 012040 doi:10.1088/1742-6596/1032/1/012040

Isocyanate (HYG), supplied by (Swiss Chem. Company Ltd.) were used to prepare the samples of this work. An proper amount of special hardener was added to the epoxy resin with weight ratio of resin to hardener (2:1) by using a sensitive electronic balance of sensitivity (0.01 gm).The epoxy/polyurethane blends were prepared with (weight%/weight%)ratios for both polymers. Blends were synthesized by simultaneuos mixing of both constituents with suitable mixing ratios of resin to hardner for both polymers (for epoxy 2:1 and for polyurethane 7:1).The content was mixed by a fan type stirrer before casting it as a sheets (of dimensions 25x20cm^2) by using glass mould .The sheets were stored at room temperature for 72 hours, and then for post-curing, the sheets were left for 4 hours in an oven at temperature 353°K .The blend samples were prepared with the following ratios:100%EP,95%EP+5%PU, 90%EP+10%PU,85%EP+15%PU, 80%EP+20%PU, 75%EP+25%PUand 70%EP+30%PU.

Fourier transform infrared spectroscopy (FTIR) was measured in the wavenumber range (4000-400) cm^{-1} on a Perkin- Elmer 1710 spectrophotometer using KBr pellets at room temperature.

A.C measurements were performed under certain conditions as follows:

1- The frequency range concerned in this work was 10^2- $4x10^4$ Hz

2- Temperature range between (293- 373)K, the temperature was changed by constant rate of 2K/min

3-A constant voltage of (1V) was applied in all frequency range and temperature those are indicated in this work.

Aluminum electrodes with thickness 2000°A were deposited on each adjacent surfaces of samples using thermal evaporation method under pressure of 10^{-5} mbar using coating unit type Edward. The specimen was fixed in specimen holder and placed into temperature controlled oven type (Heresies electronic). High and low holder terminals are connected to dielectric analyzer type Hewlett Packard model (HP4274A & HP4275A), the third holder terminal was connected to the earth. Three dielectric parameters were measured directly from above setup total resistance (R), total capacitance (C) and dissipation factor tanδ with an accuracy of 0.1%. A. C. conductivity has been evaluated from dielectric data in accordance with the relation, σ_{AC} is the A.C conductivity given by the relation:

$$\sigma_{A.C} = \frac{t}{R.A} \qquad (1)$$

Where: t is thickness of film, R is resistance of film, A is effective area for capacitance.

The dielectric constants (ε_1,ε_2) can be calculated using the following relations:

$$\varepsilon_1 = C.t / \varepsilon_0 .A \qquad (2)$$

where: C is capacitance, ε_0 is the permittivity of free space =8.854x10^{-14} (F/cm).

$$\varepsilon_2 = \sigma_{AC} / \omega\varepsilon_o \qquad (3),$$

where ω is angular frequency.

3. Results and discussion

3.1 FTIR Spectroscopy

The FTIR spectra of neat epoxy and epoxy / polyurethane blends were shown in Fig 1.The spectrum of neat epoxy A, 3487cm^{-1} (-OH) is one of the characteristic absorption peaks of epoxy resin, while the peak at 1296cm^{-1} is the absorption band of the quaternary carbon atom in the epoxy. On the other hand peaks at 829cm^{-1} and 916cm^{-1}represent epoxy group.As PU introduced to the epoxy matrix,the typical absorption peaks of at 3286cm^{-1} (N-H), 2875cm^{-1} (CH_2 and CH_3),1716cm^{-1} (C=O), 1539cm^{-1} (N-H), and 1107cm^{-1} (C-O-C), of urethane polymer will come to appear in the blends curves . Additionally, the other typical absorption bands at 1668cm^{-1} (C=C),1454cm^{-1} (=CH_2) and 833cm^{-1} (=CH) referred that C=C bonds incorporated in the polyurethane chains. It evidence that the intensity of absorption peak at 2300cm^{-1} (-NCO) will be decreases , and it confirms the good degree of reaction of the -NCO group. Compared with pure epoxy spectra , the absence of the absorption peaks at 3487cm^{-1}, 829cm^{-1} and 916cm^{-1} implied that the -OH group and the epoxy groups of epoxy resin had already been reacted with -NCO group.The existence of the peak of the quaternary carbon atom at 1300cm^{-1} confirms that epoxy group had been grafted in polyurethane chain. As it is shown in Figure 6, the intensity of the characteristic band of C=O stretching vibration at 1736 cm^{-1} increases with the increase of PU content .This explained the enhancement of conductivity as PU content will be raised in the blends samples. It was pointed out that the viscosity of PU increased with the increasing of -COOH content. Since as -COOH content increased, the content of carboxylate anions in the polyurethane chains raised, and intermolecular interaction increased accordingly. The movement of polyurethane chains will be hindered. Hence, the viscosity of PU increased. The previous researchers showed that there is direct relation between the viscosity and epoxy content .It can be observed the intermolecular interaction force of polyurethane chains decreases obviously with reducing epoxy content, since little rigid benzene rings which were introduced into polyurethane chains, which caused the reducing steric hindrance effects. As, a result of viscosity will be reduced and hence, the movement of polyurethane molecule chains was accelerated. On the other hand the previous results showed there is relationship between hardness and epoxy content, such as with increasing of PU content, the opaque of epoxy turned to be transparent. As a result, crosslinking structure may be worsted with the increase of PU content [8].

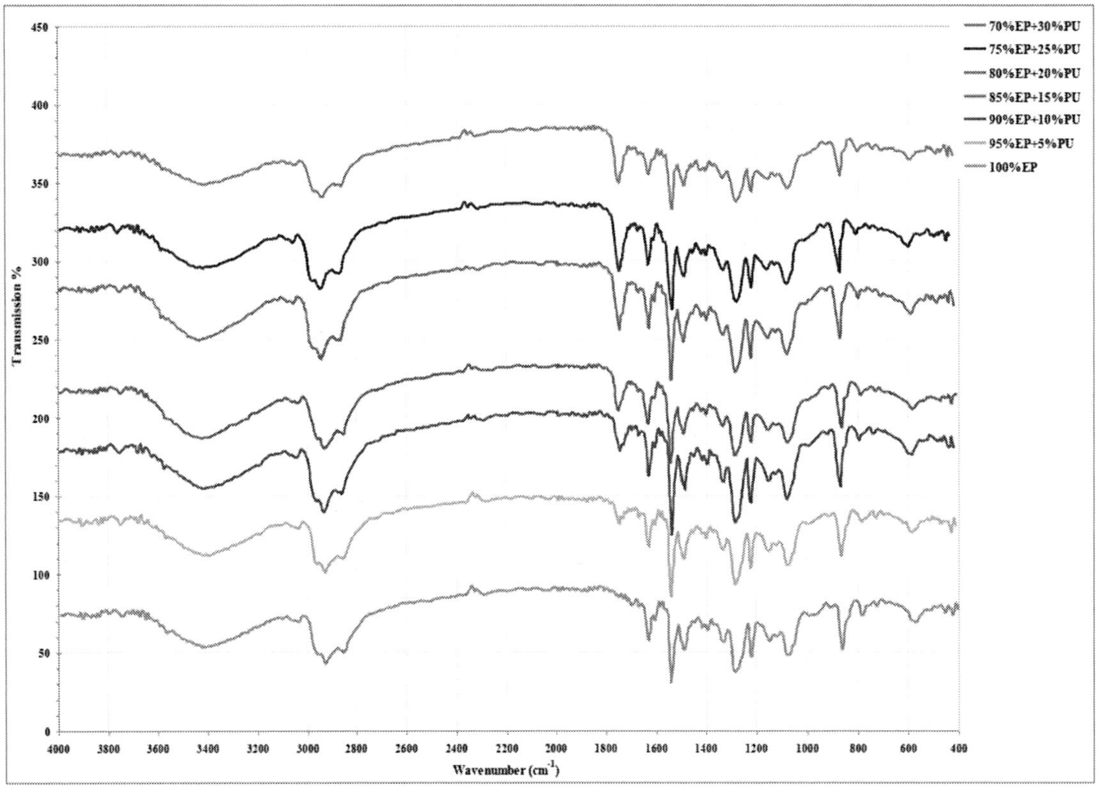

Figure.1 FTIR transmittance spectra for EP/ PU Blends with different blend ratios.

3.2 A.C Conductivity

Dielectric properties of EP/PU blends with different blends ratio were investigated on samples with parallel plate capacitor configuration. Fig. 2 shows the variation of $\sigma_{A.C}$ with angular frequency for EP/PU blends with different blends ratio and temperatures . The AC conductivity patterns for neat

IOP Conf. Series: Journal of Physics: Conf. Series **1032** (2018) 012040 doi:10.1088/1742-6596/1032/1/012040

epoxy show frequency dependence in the hole frequency range. This behavior obeys the universal power law , $\sigma(\omega) = \sigma_{d.c} + A\omega^{s}$, where $\sigma_{d.c}$ is the dc conductivity (frequency independent plateau in the low frequency region). This behavior get to vanish with the introduction of PU to the EP matrix and the AC conductivity become frequency independent i.e. the frequency independent plateau of A.C conductivity become the dominated behavior in the hole frequency range. The A.C electrical conductivity ($\sigma_{A.C}$) increases with the increase of frequency and temperatures for neat epoxy. Indeed $\sigma_{A.C}$ at F=100Hz for EP and 95/5 EP/PU samples increases from 7.5 x 10^{-11} to 3.17 x 10^{-9} and to 9.21 x 10^{-10} $(\text{ohm.cm})^{-1}$ with the increase of frequency from 100 to 4x10^{4} Hz and temperature from 293 to 373K respectively. A.C electrical conductivity ($\sigma_{A.C}$) showed significant increase with temperatures for 90/10, 85/15 EP/PU blends samples where $\sigma_{A.C}$ increases three and two ordered of magnitude while residual blends samples showed temperature independent relation and remain nearly constant with the increase of temperatures. Moreover ($\sigma_{A.C}$) of the mentioned samples increases from 1.47 x 10^{-8} $(\text{ohm.cm})^{-1}$ to 5.05x 10^{-5} $(\text{ohm.cm})^{-1}$ and from 2.58 x $10^{-8}$$(\text{ohm.cm})^{-1}$ to1.38 x 10^{-6} $(\text{ohm.cm})^{-1}$. The exponent (s) was calculated from the equation $s = d(\ln \sigma)/d(\ln \omega)$ and illustrate in Tables.1 .It obvious that s decreases with the increase of blend ratio , indeed s decreases from 0.85 to 0.13 when PU increase from 0 to 30% .While s exhibits to change in nonsystematic manner with the increase of temperature i.e. increases in the first stage and then decreases with further increases of temperature , indeed s for 85/15 EP/PU increases from 0.13 to 0.36 when T increase from 293 to 313 K and then return to decrease to 0.047. The most suitable models to explain our results are small polaron S.P and correlated barrier hopping(CBH)models.

The activation energy for conduction (E_{ac}) in the temperature range (293-373K) obtained from the plot of $\ln\sigma_{A.C}$ with reciprocal temperature at definite frequency fitting different regions with the equation $\sigma = \sigma_0 e^{-E_a/k_B T}$. The activation energy is found to reduce with increase of PU content in the blend samples in the first stage and then get to increase with further increase of PU content in the blends samples, the calculated activation energy are illustrated in table(2) .Indeed E_{ac} decreases from 0.201 to 0.175eVwhen PU content increases from 0 to 10% . This can be possible due to the increase of charge carriers and enhances the electrical conductivity.

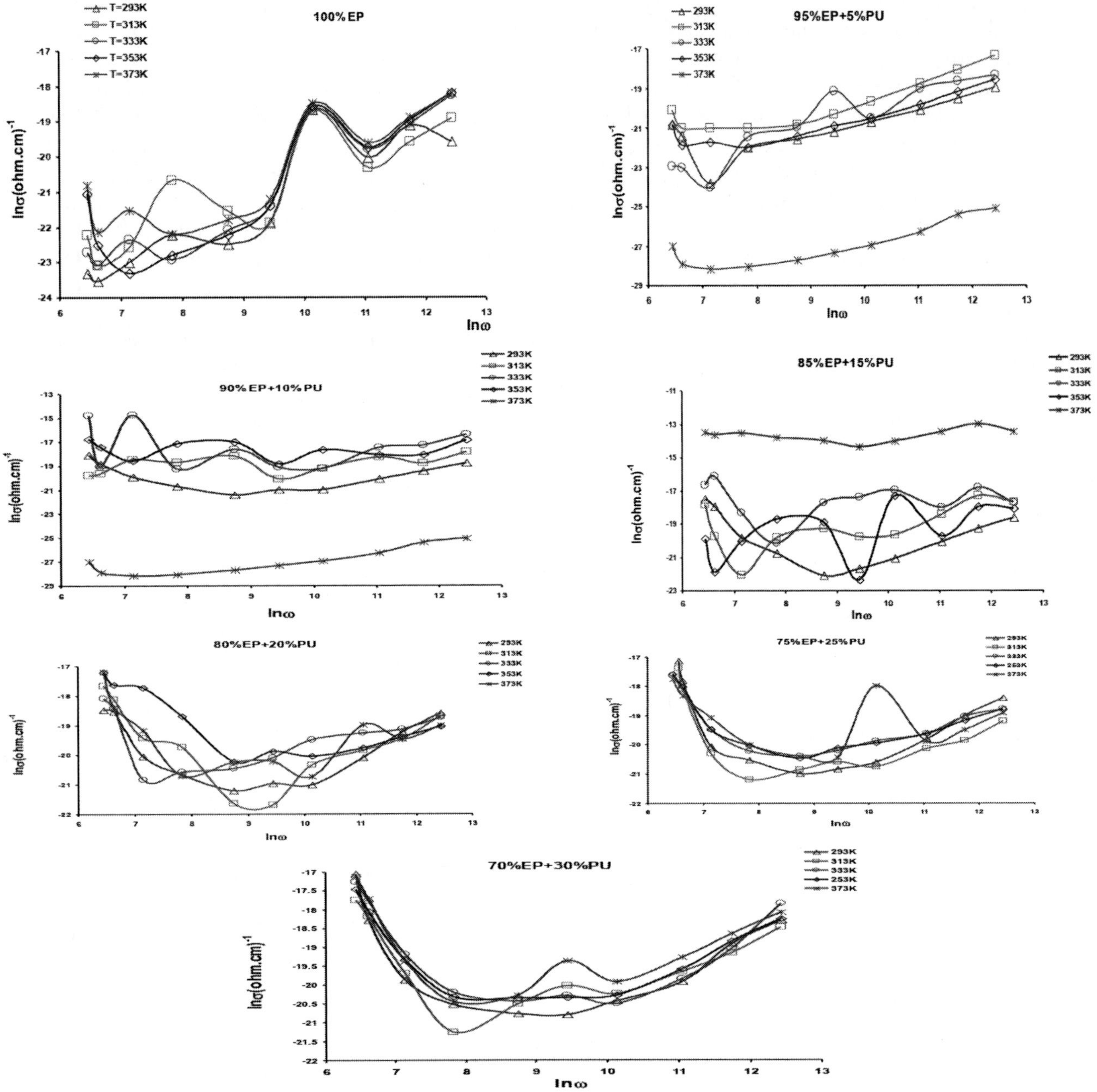

Figure 2. Relation between Ln σ_{AC} and Ln (ω) for EP/PU Blends with different blend ratios and different temperatures.

IOP Conf. Series: Journal of Physics: Conf. Series **1032** (2018) 012040 doi:10.1088/1742-6596/1032/1/012040

Table1. The values of exponent s and α for EP/ PU blends

The Blend ratio %	Temperature (K)	s	α
100EP	293	0.85	-
	313	0.58	-
	333	0.81	-
	353	0.73	-
	373	0.59	-
95EP+5PU	293	0.48	-
	313	0.54	-
	333	0.87	-
	353	0.48	-
	373	0.43	-
90EP+10PU	293	0.065	-
	313	0.16	-
	333	0.10	-
	353	0.04	-
	373	0.43	-
85EP+15PU	293	0.13	0.1555
	313	0.36	0.1555
	333	0.0028	0.1333
	353	0.038	0.1333
	373	0.047	0.1111
80EP+20PU	293	0.025	0.1111
	313	0.15	0.1111
	333	0.041	0.2222
	353	0.35	0.0666
	373	0.15	0.0222
75EP+25PU	293	0.10	0.0222
	313	0.17	0.1111
	333	0.10	0.1333
	353	0.13	0.0666
	373	0.12	0.1111
70EP+30PU	293	0.13	0.3333
	313	0.36	0.3333
	333	0.0028	0.0222
	353	0.38	0.2222
	373	0.047	0.2222

IOP Conf. Series: Journal of Physics: Conf. Series **1032** (2018) 012040 doi:10.1088/1742-6596/1032/1/012040

Table2. The values E_{ac} EP/ PU blends.

The Blend ratio	E_{ac}(eV)	Temp.Range K
100EP	0.201816	293-373
95%EP+5%PU	0.194942	293-373
90%EP+10%PU	0.175588	293-373
85%EP+15%PU	0.193536	293-373
80%EP+20%PU	0.17946	293-373
75%EP+25%PU	0.180064	293-373
70%EP+30%PU	0.18737	293-373

3.3The Dielectric Constants

The angular frequency dependence of (ε_1) for EP/PU blends at different PU content in the temperature range (293-373K) is plotted in Fig.3. The frequency dependence of permittivity which obvious in the low frequency region and followed by a nearly frequency independent behavior above 1 kHz is clearly observed in the spectra of neat epoxy but is found to vanish in the spectra of 95/5, 90/10 and 85/15 EP/PU blends samples and then return to appear for residual blends samples . The reduction of (ε_1) with increase in frequency ascribed to the electrical relaxation processes, but at the same time the material electrode polarization cannot be ignored. The material electrode interface polarization superimposed with other relaxation processes at low frequencies. The data showed that (ε_1) exhibit to increase with the increase of PU content in the first and then decreases with continues addition of PU, indeed (ε_1) increases from539 to15759 when PU increases from 0 to 10%. The loading of PU may lead in more localization of charge carriers along with mobile ions causing higher conductivity. This may be the reason for higher (ε_1) and strong low frequency dispersion samples with high PU content which show temperature independent behavior. On the other side the dielectric constant exhibit to change in different manner with temperature, i.e. decreases or increases with temperature , indeed ε_1 for 90/10 EP/PU increases from 15759 to 991628 when temperature increases from 293 to 333 K and then decreases to 519540 when temperature increases 373K . Fig.4 show the frequency dependence of (ε_2) for EP/PU blends with different blend ratios and temperatures in the range (293-373K) respectively. The spectra of some samples characterizes by peak appearing at a definite frequency proposed the presence of relaxing in all the samples. The strength and frequency of relaxation depend on relaxation type. The relaxation peaks get to vanish at high temperature and especially for samples rich with PU. The addition of PU is believed that there is an increase in the free charge carriers in the materials. The small mobile elements PU speed up the segmental motion. Better transport properties on composites can be achived by the relatively fast segmental motion, but with further increase of PU content results in the reduction of the mobility of free charge which in turns reducing $\sigma_{A.C}$ and ε_2 .

More addition of PU to EP results in overall increase of (ε_1) and (ε_2) due to free charge contributions. So the increase of the molecular mobility is reflected both by increasing of free charge mobility. The enhancement of conductivity take place with increase of loading with PU up to 10% , but then the opposite case take place .

IOP Conf. Series: Journal of Physics: Conf. Series **1032** (2018) 012040 doi:10.1088/1742-6596/1032/1/012040

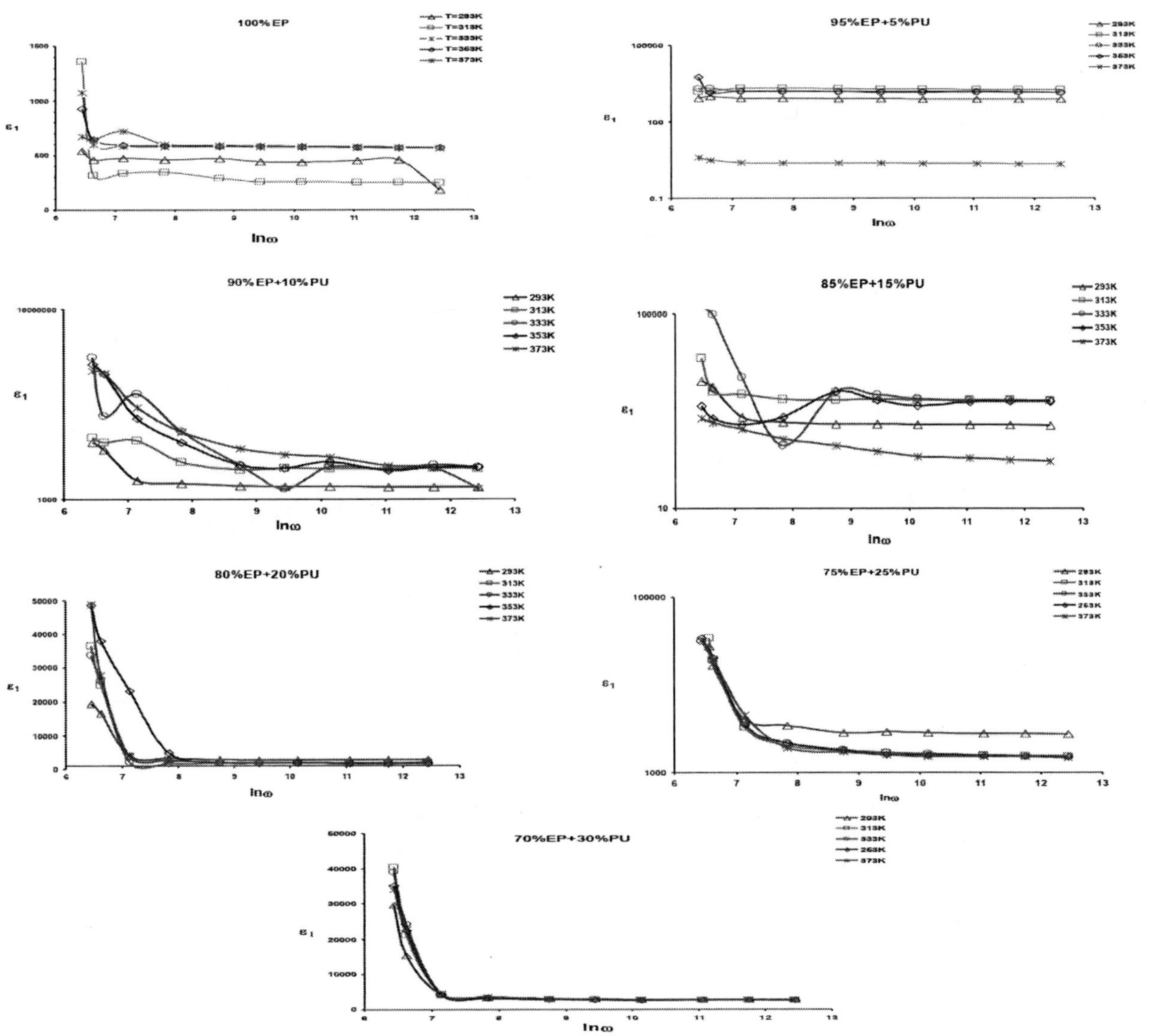

Figure 3. Variation of ε_1 with Ln(ω) for EP/PU Blends with different blend ratios and different temperatures.

IOP Conf. Series: Journal of Physics: Conf. Series **1032** (2018) 012040 doi:10.1088/1742-6596/1032/1/012040

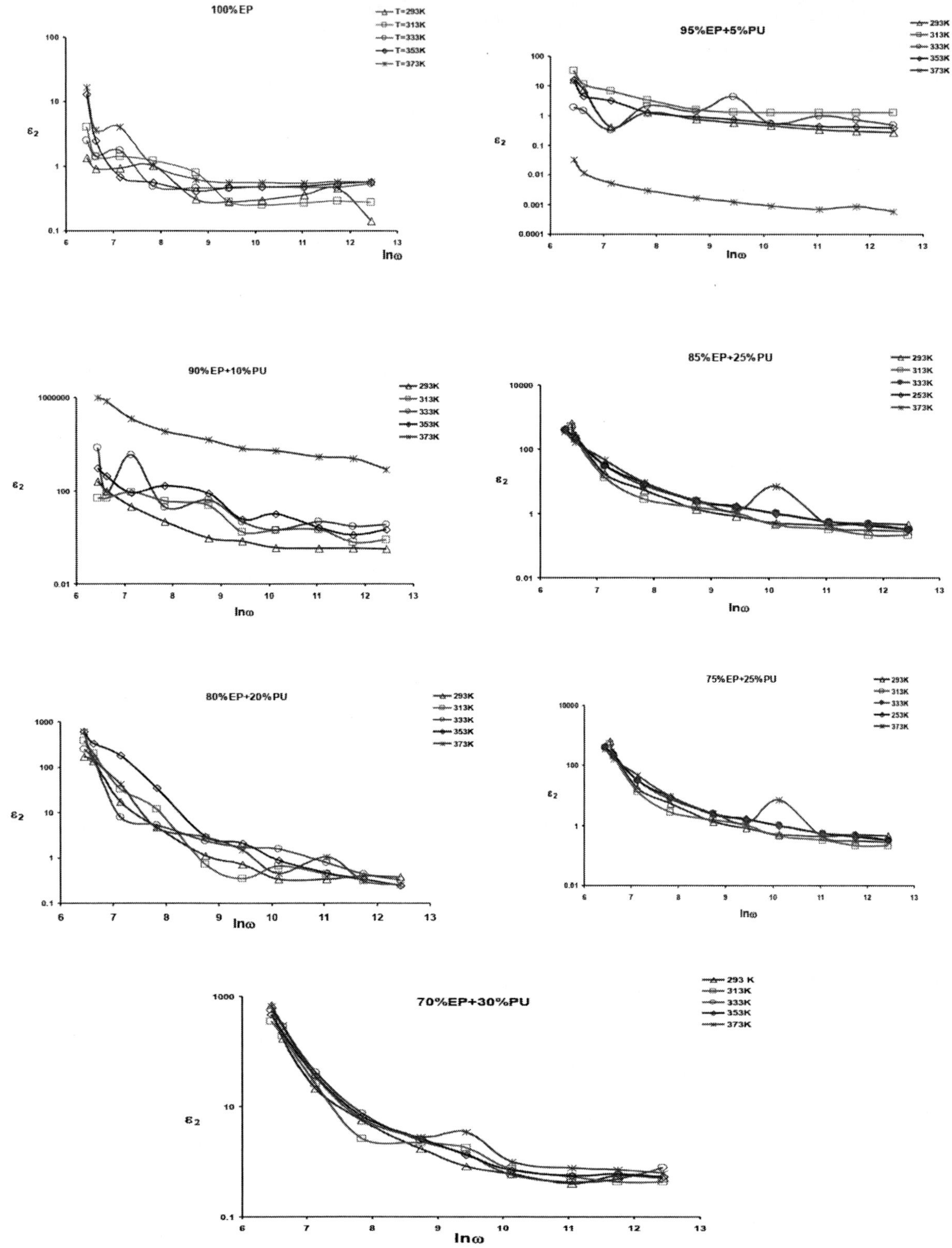

Figure 4. Variation of ε_2 with Ln(ω) for EP/PU Blends with different blend ratios and different temperatures.

3.4 Cole-Cole diagrams

The existence of multi-relaxation time in EP/ PU blends is obtaining by plotting Cole-Cole diagrams as shown in Fig.5 for different blends ratio . It is clearly obvious from arc of circles having their centers lying below the absicca axis for all samples. This proves the existence of distribution of (τ) in the samples. polarizability (α) had been determined by determining the angles (απ/2)the vales of the polarizability and are listed in Table.1.The values of (α) shows non regular variation with the increase of PU content(i.e. increases and decreases with loading of PU), indeed α decreases with the addition of PU and then increases, moreover (α) decreases from 0.155 to 0.022 when PU content increases from 15 to 25% and then increase to 0.333 when PU ratio increases to 30% , the reduction of (α) with the increase of PU content results from strength of the forces of the intermolecular while the increase of(α) is related with the weaken of intermolecular forces[11,12] ,on the other side (α) tends to increase with the increase of temperature which results from the weaken the forces.

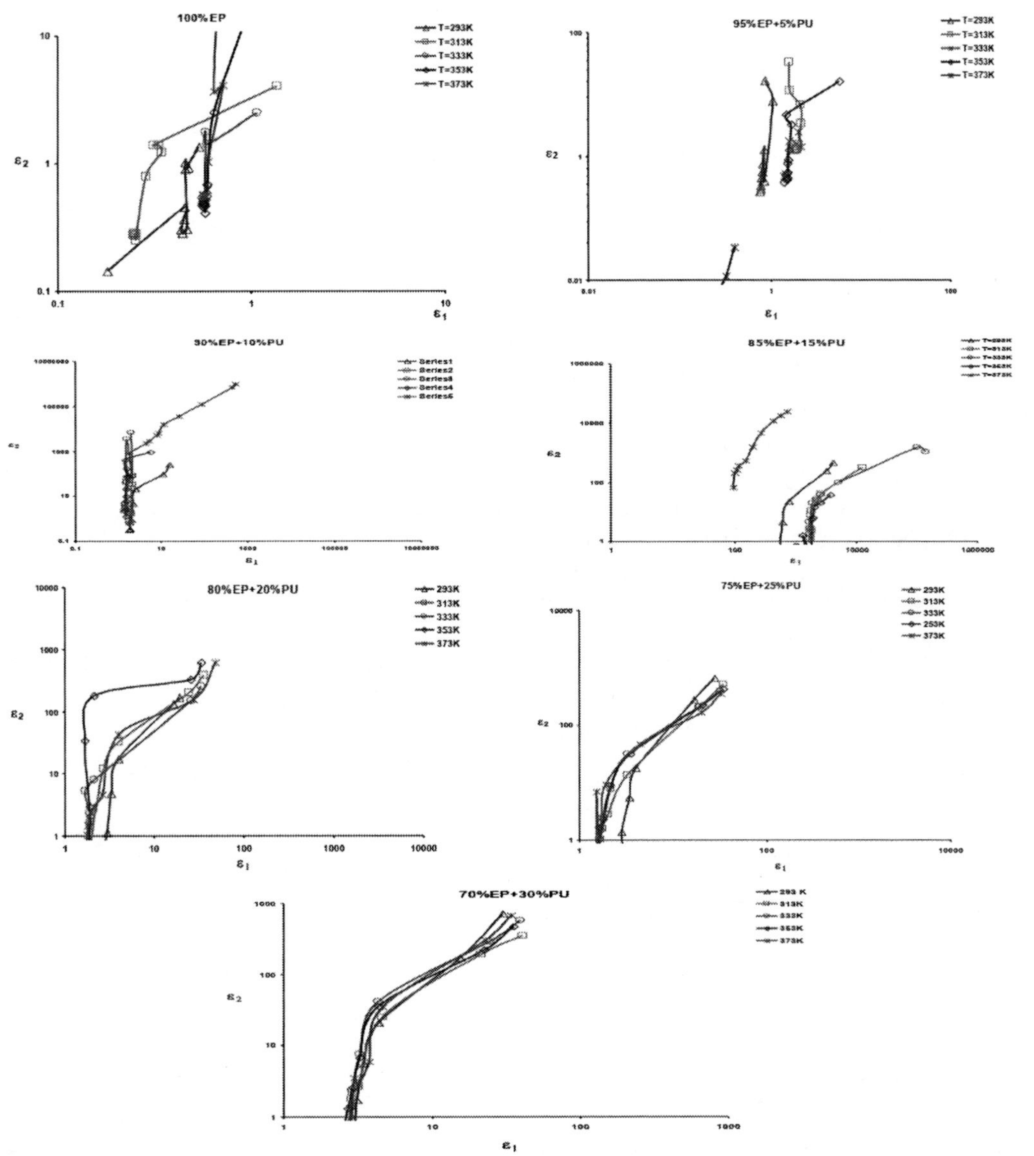

Figure 5. Cole – Cole diagrams for EP/ PU Blends with different blend ratios and different temperatures.

IOP Conf. Series: Journal of Physics: Conf. Series **1032** (2018) 012040 doi:10.1088/1742-6596/1032/1/012040

4. Conclusions

A.C conductivity of EP/PU blends with different blend ratios was studied and the following results can be deduced :-

The A.C conductivity for EP/PU blends changed with blends ratio and accordingly with frequency, the dielectric properties of the blends can be modifed by the loading level of PU. The values of (ε_1) and (ε_2) changed with the EP/PU blend ratios. The exponent s increases with the increasing of loading level of PU but decreases with temperature. Loading EP with PU enhanced the conductivity (EP/PU) blends. The enhancement of stretching band of the epoxy matrix lead to enhanced dielectric properties of the blends.

Acknowledgements

The Author thankful University of Baghdad, College of Science, Department of Physics for supporting this work.

References

[1] Ouahab L. 1997 Chem. Mater 9 1909.
[2] Gatteschi D. 1994 Mater. 6 635.
[3] Ahmad A. 1991 Phys. Status Solidi A 126 411.
[4] Roberts G and Schmidlin F 1969 Phys. Rev. 180 785.
[5] Azim-Araghi M.E. and Campell D. 1998 Thin Solid Films 320 320.
[6] Gould R. 1986 Phys J and Appl. Phys D 19 1785.
[7] Yakuphanoglu F Ph.D Thesis 2002 University Firat and Elazig Turkey.
[8]. Li W, Cao Y and Zhou X 2013 Applied Mechanics and Materials 249 842.
[9] Li W, Cao Y and Zhou X. *2013 Applied Mechanics and Materials 249 842.*
[10] Patell R. Shah M. Hirani A. June 2014 International Journal of Innovative Research in Science,Engineering and Technology 3, 6.
[11] Jing Y. Meng X. Yiu W. Liu J. Ding J. Lee C. Lee S. 2004 J. Phys. Chem. B. Jing, X. Meng, W. Yiu, J. Liu, J. Ding, C. Lee, S. Lee, J. Phys. Chem. B 108 2784 .
[12] Bower D. 2002 Introduction to Polymer Physics Chs. 4 6 9.

IOP Conf. Series: Journal of Physics: Conf. Series **1032** (2018) 012041 doi:10.1088/1742-6596/1032/1/012041

The Co-fermentation of Organic Substrates: A Review Performance of Biogas Production under Different Salt Content

[1]A A Alhraishawi , [2]W K Alani

[1]Department of Civil Engineering, College of Engineering, Misan University, Iraq
[2]Renewable Energy Technology Center, University of Technology, Iraq
[1]Corresponding Author: alihussin2294@yahoo,.com
[2]Corresponding Author: eng.waleed.khalid@gmail.com

Abstract. Sustainable development is the projected demand of all nations at present. Only 20 percent of world's primary energy requirement is met by renewable sources like solar and wind energy, hydropower, biomass, municipal and agri-wastes. Biomass, can be utilized to produce biogas due to its high biodegradability, calorific value and nutritive value to microbes, which will reduce our dependency on fossil fuels so that These organic wastes require being managed in a sustainable way to avoid depletion of natural resources, minimize risk to human health, reduce environmental burdens and maintain an overall balance in the ecosystem. The annal of the National Office of Statistics (NOS) indicated that municipal solid waste overtook more and more million tons in 2016, almost 3.5 times the 1992 rates. About 40–70% of civilian solid waste CSW is composed of civilian biomass waste (CBW), such as kitchen waste (KW), fruit/vegetable remains (FVR) and waste water sewage sludge (WWSS). There was a need to find several ways to deal with such wastes. One of these methods is the anaerobic digestion process, as a waste management system, has risen due to it has the usefulness of lowering volume and getting back energy. Several physical and chemical factors influence the creation of methane, and inhibition of bio-methane efficiency by certain materials (salts, oils, ammonia and toxic substrates) may take place when inhibitors levels appear in high concentrations. In addition, high salt in the organic substrates can inhibit the activity of fermentation microbes by impacting the osmotic pressure of the cell wall of bacteria in the anaerobic system. The purpose of this study was to discuss the effect the types of substrates and salt content on biogas generation in anaerobic digestion.

Keywords: Anaerobic digestion, Inhibition of methane, High saline content, Biogas

1. Introduction

International comfort and community expansion driven by pride and energy, which were especially as a basic requirement of humanity [1]. More than 77% of the world's energy depletion stems from the combustion of traditional sources [2]. The rapid increase in population and the combustion of fuels (oil) have extremely shared to the rise in earth warming caused greenhouse gas releasing [3]. Sustainable (renewable) ability sources could be a applicable alternative to fossil energies. The energy of bio-waste has been widely enhanced without impacting the ecosystem and human performance [4] , organic waste has become the main origin, of smelling , poisonous gas, and groundwater impurity , which discomposed major threat to the ecosystem if process improperly [5]. One such well renowned and exceedingly utilized manner for bio-conversion of trash into bio-fuel is anaerobic digestion (AD), which is consider as the modest technique because it's very confined ecosystem effect [6, 7,8] and high energy recapture potential [6,7,9]. Although comprehensive research has been carried on bio-gas of wastes by AD system, moreover

IOP Conf. Series: Journal of Physics: Conf. Series **1032** (2018) 012041 doi:10.1088/1742-6596/1032/1/012041

research is presently in go ahead in order to reinforce the clean energy production, get high bio-degradation averages and decline the amount of eventual organic waste to be disposed [7,8,9,10]. Scientifically biogas is mixing from CH4 and CO2 in the ratio of 3:1, with confirmed contaminant leftover of SO2, NO2 etc. So, bio-energy of organic waste avails both the targets: i. organic waste dropping ii waste recovery, iii and biofuel production as substitutional and renewable source. The leftovers of waste after AD procedure performances as superior muck that can be facilely applied in agricultural soils as a replace for fertilizer [11]. AD system is conditional on specific microbial assembly for devolution of bio-waste out of main steps namely hydrolysis, acetogenesis and methanogenesis as shown in Fig 1. Investigators have indicated that methanogenesis is the rate-limiting stage for easily degradable materials whereas hydrolysis is the average shackling step for complex organic wastes due to formation of toxic by-products like complex heterocyclic synthesises and un-likeable volatile fatty acids(VFAs). Biogas system is greatly dependent on environmental parameters like pH, T, C / N ratio, C / P , dimension of particle, inhibitors and type of substrate being applied to get back energy from it [12]. The anaerobic handling of organic waste is not as common, as the aerobic process, fundamentally due to the long-term period desired to realize stabilization [13]. The Interaction is also sensible to raise levels of free ammonia, oils (fatty acids, grease) conducting from anaerobic devolution of the nitrogen /protein ratios components [14] . A considerable number of agents that inhibited the AD performance have been considerable reported, such as ammonia, heavy minerals, oil composition and high salt content.

Figure 1: Mechanism of anaerobic digestion[15]

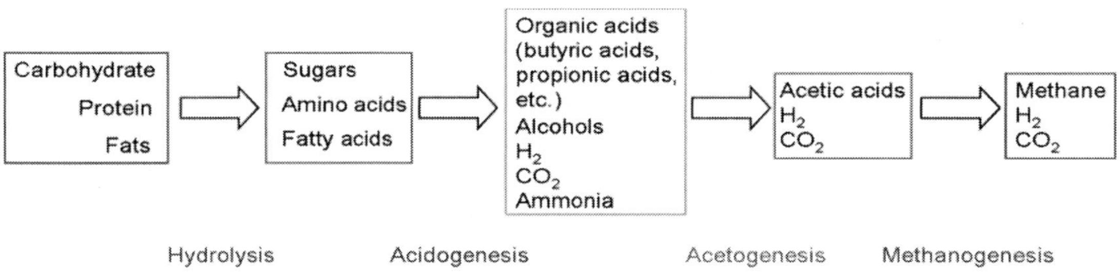

2. Anaerobic digestion of bio waste :

2.1. Anaerobic digestion for mono substrates

With the preamble of both commercial and anaerobic digestion station projections through the early 1990s, AD of bio-waste has received worldwide [16]. Practicability, by which nearly any kitchen waste (KW) can be biologically transformed into another portion, in the deprivation of oxygen. The diverse microbial communities degrade organic waste, which outcomes in the generation of biogas and other energy-rich bio-compounds as limit products [17]. An AD is usable for a wide range of single substance including civilian, agricultural and manufacturing wastes, and plant scraps. Moreover, this operation has

IOP Conf. Series: Journal of Physics: Conf. Series **1032** (2018) 012041 doi:10.1088/1742-6596/1032/1/012041

some features over aerobic system due to a low energy demand for operation and a low biomass output [18], and it is considered a fertile technique renewable energy [19]. The AD (with single substrates) has some side effects such as long retention times and low abstraction performance of organic compositions. A considerable raise, in ammonia concentration and salt inhibition, occurs due to fermentation of wastes alone. This problem turns into especially serious when reactors are fed with single wastes rich in protein and salt content without dilution, also didn't enhance carbon to nitrogen ratios [9].

2.2. Anaerobic co-digestion co-substrates.

Bio-waste composed of tiny-nutrients like carbohydrates, proteins, oils, organic wastes are convenient for bio-methane. Oils are the main benefactor of biogas with long-term retention time due to retarded bio-decomposition, whilst proteins and carbohydrates show quicker transference rates. Co-fermentation also was known as anaerobic co-digestion (ACoD) that raises the capacity of mixed organic materials and speed up the degradation of macro-molecules in substrates by bio-energizing by mix-up substances with one another in various ratios and proportion keeping the C/N (carbon: nitrogen) ratio within the accepted range of 25 – 30. Particularly mix-up organic wastes have indicated in the acceptable range of 20 – 30 [20, 21]. Higher yield of biogas by co-digestion is related to the synergistic influence of the bacteria (microorganisms). Hence, co-digestion is preferred than AD of waste with mono substrates because of its various benefits [22, 23] like reduction of toxic synthesis present in any one of the organic waste has high correlating toxicity, gets better carbon to nitrogen (C/N) ratio of a waste, it also leads to the reduction of the salt content of organic waste and sewage because salt is one of the inhibitors, sludge adjustment, PH enhancement and moisture enhancement content compared to AD with mono substrate [24]. It has been spotted that co-digestion of multi-materials confirm the feed to the bioreactor, thereby improving the C/N ratio and decreasing the concentration of nitrogen, salt [22]. The use of a co-digestion system with a low nitrogen and oil import, waste increases the generation of gas because integral features of both kinds of bio-waste, thus reducing problems related with the piling up of moderate, volatile synthesis , high salt, and ammonia concentrations [25]. Various studies have indicated that blends of agrarian, domestic and industrial wastes can be digested totally and efficiently with one another. in conclusion, the anaerobic system (AD) of individual substrates presents some problems related to substrate features as municipal sewage Sludge (MSS), described by low organic capacities, the municipal solid waste (MSW) could have contained inappropriate substances as well as high levels of heavy metals and salts , therefore The anaerobic co-digestion (multi substrates) is the at same time AD of multi substrates is a convenient option to overcome the inhibitions of mono-digestion[26,27] as shown in Table 1 .

Table 1: Proportional biogas production and methane from the co-digestion of bio waste.

Substrate	Co-substrate	Biogas production rate (l/d) , methane production rate (l/d)	Comments	References
kine excreta	mill waste	1.10, 179	The co-fermentation system generated more than 300% higher biogas than that of excreta single	[28]
Cattle muck	Agricultural waste	2.70, 620	the increment in biogas creation from the co-digestion was spotted	[30]
vegetable waste	Bucher's wastewater	2.53, 611	The addition of Bucher wastewater to the feedstock rose biogas yield up to more than 50%	[31]
domestic solid waste	Fly ash	6.50, 222	Utilizing of fly ash significantly promoted biogas creation yields of the domestic solid waste	[32]
civilian biowastes	Fat, oil, grease (FOG)	13.6, 350	Co-fermentation outputted in a rise of 72% in biogas creation and 46% methane generation yield in comparison with municipal solid waste	[27]
Animal manure	Fish bio-waste	16.4, 620	Maximum biogas generation rate was got by a blend of wastes	[33]
Residual potato waste	Sugar beet	1.63, 680	Co-fermentation enhanced methane creation up to 60% compared to the digestion of single waste	[34]
Slaughterhouse waste	Sewage sludge	8.60, 500	Biogas creation of the co-fermentation systems more than digestion of Slaughterhouse waste alone	[22]

3. Effect the salt content on co-digestion performance:

3.1. Effecting of salt content on biogas volume

The biogas creation is affected various concentration of salt through the digestion period. A massive of factors that causing anaerobic system failure has been formally recorded, such as ammonia, thick minerals, oils and saline content salinity fundamentally consisted of several cations, which could limit the performance of the AD [35]. In past study found ramifications of rising NaCl concentrations . Where a decrease in the biogas volume rate and an increase in the length of the fermentation period, pointed according to microbial activity. A previous study has shown that when adding different concentrations of salt to common precipitation, gas production will be significantly affected by these concentrations Where salt and gas concentrations were respectively (0 ,16,30,60)gm NaCl , (45,21,5 ,2) ml/d from gas rate, Through the chart as shown in Fig 2 , we can observe the rate of the gas reached the high level with a concentration of NaCl zero and became small or almost non-zero concentration with 60 / g NaCl [36]. This height saline could lead cell osmotic fatigue imponderables, causing in dissolution of cytoplasm and/or loss of effectiveness of cells and inhibition biogas rate yield [37]. A comparatively high NaCl concentration of 3,100–5,100 mg/L would lead to average inhibition, but at 9,000 mg/L, it would lead to a terrible inhibitory in the gas output [34]. The effect of organic matter in seawater has a negative impact on the performance of gas production because the organic matter in seawater has a high salt content of more than 10 g/NaCl [38]. The volume of the gas was measured by Equation (1) [39] .

$$V_{biogas} = \Delta_P \cdot V_{head} \cdot \frac{c}{R*T} \quad (1)$$

- Wheresoever V biogas means for diurnal biogas volume (L)
- ΔP means absolute compressing various
- V head indicates to the volume of the top vacuum (L).

Figure 2: Biogas generation rate through criterion periods with various NaCl concentrations [36].

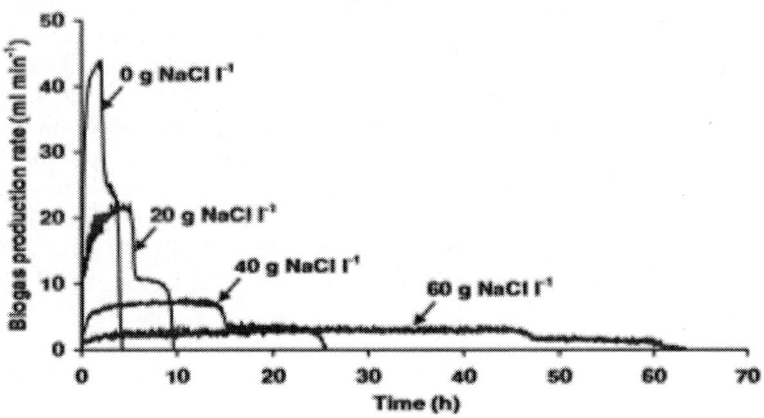

IOP Conf. Series: Journal of Physics: Conf. Series **1032** (2018) 012041 doi:10.1088/1742-6596/1032/1/012041

Figure 3: Methane productions at various concentrations of NaCl. Blank (0.0 g/L), L (5 g/L), M (8 g/L) and H (≥ 20 g/L) at mesophilic temperature of 37 °C [40].

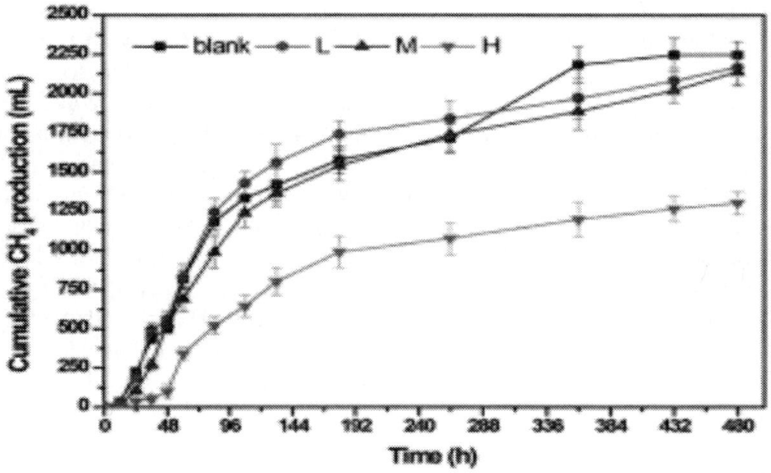

3.2. *Effecting of salt content on methane generation*

Studies have shown that joint digestion enhances the production of biogas, but this enhancement is affected by several factors, including the saline effect inhibited by the proportion of methane produced after fermentation. One of the past research pointed the effect of salt on co-digestion from food waste and sewage sludge with the concentrations of salt 0.8~4.5 g/L NaCl would raise the methane generation while 7 and 12 g/L NaCl recorded in a lessening of methane gas generation by 31% and 45%, respectively[38]. The formalization of methane gas from acetate would be discouragement by 12, 55 and 95 % respectively at salt concentrations of 6.5, 14.2 and 22 g/L [40]. Also, research indicated on how the salinity would influence the biomethane creation has been depiction rising interest. It was pointed that a saline concentration extend from 3 to 7 g/L would reasonably inhibit the biomethane activity, whilst, exceeding 7 g/L would highly dampen methanogenesis creation [41]. Decreasing 55 % of bio-methane performance with uninterrupted insinuation to salt in co-digestion reactors was spotted when sodium concentrations were more than 25 gL−1, but in another side, these inhibition levels were in the zone 7–18g from the salt ion therefore anaerobic co-fermentation is not contemplated proper, for food waste with salt at high condensations [42]. Bacteria of methanogenesis initiate to be influenced at a NaCl concentration of 6 g/L however bacteria of acidogenesis was hardly impacted only at NaCl level exceeding more than 20g/L as shown in Fig 3. The effect of salt on digestion is not limited to methane production but also to the efficiency of determinants such as pH. In same this study in the same study, the researchers pointed the increment of the addition of NaCl influenced on the pH value. The value in all reactors diminished with an rising salt content, with the ultimate, pH of 7.44, 7.40, 7.38 and 7.31 for the blank, L, M and H reactors respectively as shown in Fig 4, referencing that increasing NaCl could cause a higher VFAs and a lower pH [43].

Figure 4 : pH variation at various concentrations of saline[44]

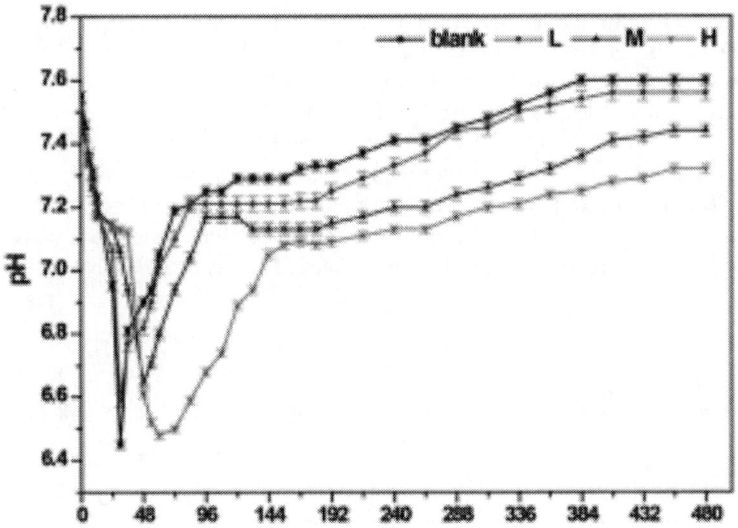

High concentration of salt will activate carbon dioxide and inhibit methane[41].Some previous researchers pointed that decreasing salinity was advantageous for the growth bacteria of methane with value (0-0.8.5 g/L NaCl) and decrease CO_2, while 9~14 g/L NaCl would cause a stumbling block for methane creation and values more than 25 g/L and improving CO_2 generation as in Fig 5.

Figure 5: Comparing between carbon dioxide and methane under different salt concentrations [42]

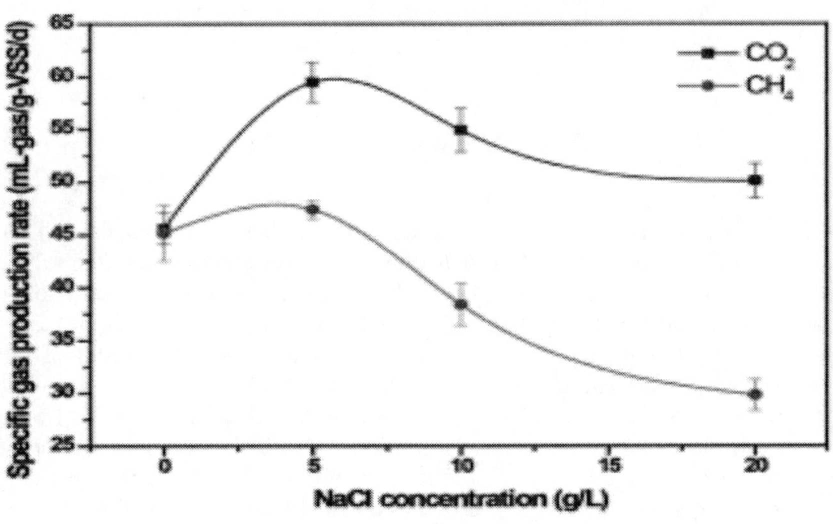

IOP Conf. Series: Journal of Physics: Conf. Series **1032** (2018) 012041 doi:10.1088/1742-6596/1032/1/012041

Five anaerobic tanks (1-5) were investigated the different levels of salt (0-15 gm/NaCl) on methane creation from the co-fermentation of multi-organic matter The outcome of bio-methane creation in the lack of NaCl (0 gm) caused mostly increase from methane generation through time period through 23 days, However, the proportion of methane has decreased significantly with the increase in salt doses between 10-15 / g / salt as in Fig 6. The existence of Na+ and cl- at the high the level would cause a negative influence on the production of bio-methane from multi-organic anaerobic co-digestion. Bio-methane creation of organic substrates reduced with the level of NaCl rose. For example, salt levels 0, 11 and 36 g/ l, methane created were 84.0, 43.7 and 13.4 milliliters of methane g/l, respectively. Methane was zero at all when the salt level was above 65 gL-1 several previous studies have confirmed that sodium levels average from 6.6 to 7.9gL-1 inhibited 50 percent methane effectiveness. High methane inhibitor level of sodium (13.0 gl-1) was indicated in the anaerobic treatment of salty municipal sludge in water remediation. The discrepancy can be referred to multi-factors, including inverse effects, synergies, and acclimation [45,46].

Figure 6: The proportion of methane has decreased significantly with the increase in salt doses between 10-15 / g / salt [47].

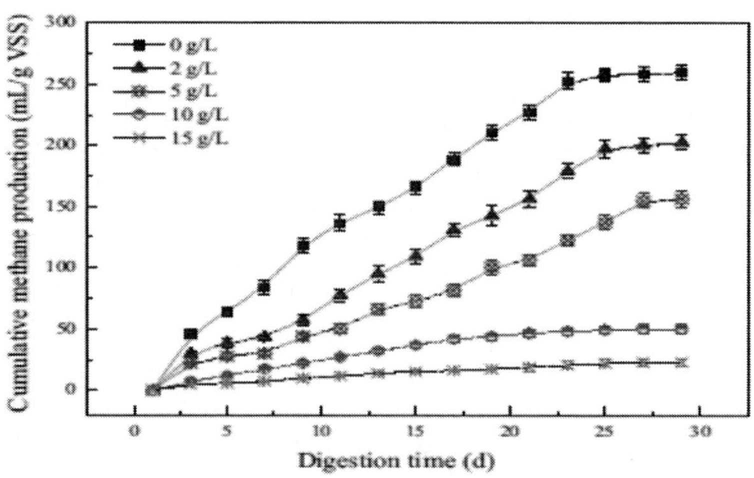

The effect of salts on the performance of total volatile fatty acid (TV) to total alkaline (TA) was studied. In a previous study, 9 tanks containing a mixture of biomaterials were used where the salt concentrations were from 0-8 g/l tank 1- tank 4, while concentration 10-17 g/l tank 5- tank 9. The creation and allocation of VFAs/methanol at each salt concentration. The effluent from the methane fermentation tank fundamentally contained acetate, ethanol, butyrate, and propionate. With the excess of the sodium salt concentration, there was a slow increase of the total V.F.A/methanol (TV) from tank 1-tank4, and a heavy accumulation from tank 5- tank9. The critical, TV was found in tank 9 (which was 3541.8 mg/L), and the TV piling up matched well with the methane diminution. The TV concentration of >8 g/L could demoralize the gas generation CH4. TA plays a serious part in buffering the reduce pH of the anaerobic co-fermentation because of rapid hydrolysis and acidification of the food waste (FW). The rate of TV/TA could be utilized to estimate the system steadiness, with the excess of the ratio; the system steadiness will be inhibited. So that, in the current study, the TV/TA the rate was also examined and is summed up in Table 2. It can be seen that the TV/TA rates of tank 1–tank4 were <0.1111 which agreed well to the highly efficient and steady reactors while the TV/TA rates of tank 5–tank 9 were all above 0.111, which was in matching with the unsettled methane creation procedures. On the fundamental of the outcomes, it can be suggested that when the TV/TA rate is above 0.1111, attentiveness should be driven to avoiding unsteady through the AD of KW.(kitchen wastes) [48,49].

Table 2 : Features of effluents during fermentation with salt content NaCl [50, 51, 52].

Tanks	Salt conc. g/l	pH	Total Alkalinity (TA,	Total VFA/methanol (TV)	TV/TA ratio
Tank 1	0	7.56	6,661 ± 657	155 7.5	0.022
Tank 2	4.5	7.43	6,779 ± 665	206.1 ± 15.3	0.029
Tank 3	5.8	7.42	7,121 ± 732	245.0 ± 17.2	0.033
Tank4	8	7.41	6,341 ± 715	349.1 ± 23.7	0.049
Tank5	10.2	7.39	6,677 ± 684	761.2 ± 52.0	0.119
Tank6	12.1	7.33	5,889 ± 597	1357.4 ± 83.5	0.212
Tank7	14.0	7.31	6,001 ± 598	1876.2 ± 126.0	0.311
Tank8	15.4	7.25	5,677 ± 576	2788.0 ± 174.8	0.477
Tank9	17	7.08	6,342 ± 705	3541.8 ± 203.0	0.543

Previous studies have indicated that the cumulative methane rate and methane inhibition rate can be measured according to the equations below [53]. Biomethane outputs (MOs) out of the complete AD process, was applied according to the Equation 2.

$$B = B_0 \cdot exp\left[-exp\left[\frac{\mu_m . e}{B_0}\right](\lambda - 1) + 1\right] \qquad (2)$$

- where B means the methane yield (MY) (mL/g-VSadded)
- Bo is the methane yield potency (MYP) (mL/g-VSadded)
- µm the superior methane output rate (mL/(g- VSaddedd))
- λ means to the lateness step period (in days)
- e is a constant that equals 2.72 and t means the incubation period (in days)

The inhibition point was estimated by Equation 3 [54].

$$Inhibition\ degree\,\% = (A - B)/A \cdot 100\% \qquad (3)$$

- Where A is the heaped-up methane output without adding inhibitor into any tank

- B is the heaped-up methane output after adding a certain condensation of inhibitor in any tank.

4. Conclusion

Organic substrates can be considered as a resource due to its utilization to enhance sustainable and eco-friendly renewable outputs like biogas using anaerobic co-fermentation. Mechanism investigation showed that the digested of co-substrates enhanced the digestion parameters such as (PH, TS, VS, C/N) and enhanced the consumption of carbohydrate thereby elevating the rate of "hydrolysis, acidification, and methanogenesis" as compared with the mono food waste anaerobic digestion. But, some of the organic wastes have a massive amount of salt, which would inhibit fermentation system high salinity extremely leads inhibition and also stopped in the anaerobic system. The conclusion of this research is that most researchers have found inhibitors of methane generation from the anaerobic co-substrate of food waste under sodium chloride. More than 65 g/l sodium chloride, no bio-methane was created at all. Also in 11gl-1 sodium chloride, only half of the gas was created compared to salt minus organic waste. Low levels of NaCl increased the hydrolysis, and acidification processes but inhibited methanogenesis while both methanogenesis and acidification processes were seriously inhibited by high salt concentrations. Therefore high salinity materials in forms (sulfurite, sulfide, calcium chloride, sodium chloride, etc.) contained in anaerobic reactors should be less than 65g/l for control of methane inhibitors and poor reactor performance.

Acknowledgment

This review was carried out in the framework of the efforts at the University of Missan and all thanks to the Associate Fellows.

References

[1]. Aragaw T. and Andargie M. et al. (2013). Co-digestion of cattle manure with organic kitchen waste to increase biogas production using rumen fluid as inoculums, *Int J Phys Sci, 8, 443-450.*

[2].Goldemberg J. and Johansson TB (2004). *World energy assessment overview 2004 update. NewYork: UNDP. P. 88.*

[3]. De Schamphelaire L. and Verstraete W. (2009). Revival of the biological total biogas energy conversion system, *Biotechnology and bioengineering, 103(2), 296-304.*

[4].Rittmann B.E. (2008). Opportunities for renewable bioenergy using microorganisms, *Biotechnology and bioengineering, 100(2), 203-212.*

[5].Jianwei Zhao a,b, Yiwen Liu c, Dongbo Wang a,b , Potential impact of salinity on methane production from food waste anaerobic digestion journal of *Waste Management 2017*.

[6].Edelmann W. and Baier U. et al. (2005). Environmental aspects of the anaerobic digestion of the OFMSW and agricultural wastes, *Water Sci Technol, 52, 553-559.*

[7].CarrÃ¨re H. and Dumas C. et al. (2010). Pretreatment methods to improve sludge anaerobic degradability: A , *J Hazard Mater. 2010 Nov 15;183(1-3):1-15*

[8].Sonesson U. and BjÃrklund A. et al. (2000). Environmental and economic analysis of management systems for biodegradable waste, *Resources, conservation and recycling, 28(1), 29-53.*

[9].Esposito G, Frunzo L, Liotta F, Panico A and Pirozzi F (2012). BMP tests to measure the biogas production from the digestion and co-digestion of Complex organic substrates. *Open J Environ Eng, 5, 1–8.*

[10].Hartmann H. and MÃller H.B. et al. (2004). Efficiency of the anaerobic treatment of the organic fraction of municipal solid waste: collection and pretreatment. *Waste Management and Research, 22(1), 35-41.*

[11].Nielsen H.B., Mladenovska Z., Westermann P. and Ahring B.K. (2004). Comparison of two stage thermophilic (680 C / 550 C) anaerobic digestion with one stage thermophilic (550 C) digestion cattle manure. *Biotechnol. and Bioeng. 86(3), 291-300.*

[12].Mata-Alvarez J, Mace S and Llabres P (2000), Anaerobic digestion of organic solid wastes. An overview of research achievements and perspectives. *Bioresour Technol, 74, 3–16.*

[13].Fernandez, J., Perez, M., Romero, L.I., 2010. Kinetics of mesophilic anaerobic digestion of the organic fraction of municipal solid waste: influence of initial total solid concentration. *Bioresour. Technol. 101, 6322–6328.*

[14]. Fountoulakis, M.S., Drakopoulou, S., Terzakis, S., Georgaki, E., Manios, T., 2008. Potential for methane production from typical Mediterranean agro-industrial by-products. *Biomass Bioenergy 32, 155–161.*

[15].Insam, H., Franke-Whittle, I., Goberna, M., 2010. Microbes in Aerobic and Anaerobic Waste Treatment. In: H. Insam, I. Franke-Whittle, M. Goberna (Editors), Microbes at Work, From *Waste to Resources. Springer-Verlag, Berlin, 1–34.*

[16]. Karagiannidis, A., Perkoulidis, G., 2009. A multi-criteria ranking of different technologies for the anaerobic digestion for energy recovery of the organic fraction of municipal solid wastes. Bioresour. Technol. 100, 2355–2360.

[17].Lastella, G., Testa,C., Cornacchia, G., Notornicola, M., Voltasio, F., Sharma, V.K., 2002. Anaerbic digestion of semi-solid organic waste: biogas production and its purification. *Energy Conversion and Management 43, 63–75.*

[18].Wang, Q., Kuninobu, M., Kakimoto, K., Ogawa, H.I., Kato, Y., 1999. Upgrading of anaerobic digestion of waste activated sludge by ultrasonic pretreatment. *Bioresour. Technol. 68, 309–313.*

[19].De Baere, L., 2006. Will anaerobic digestion of solid waste survive in the future. Water Sci. Technol. 53, 187–194. Dela-Rubia, M.A., Perez, M., Romero, L.I., Sales, D., 2002. Anaerobic mesophilic and thermophilic municipal sludge digestion. Chem. *Biochem. Eng. Qual. 16, 119–124.*

[20].Sosnowski P. and Wieczorek A. et al. (2003). Anaerobic co-digestion of sewage sludge and organic fraction of municipal solid wastes. *Advances in Environmental Research, 7(3), 609-616.*

[21].Hawkes D.L. (1980). Factors affecting net energy production from mesophilic anaerobic digestion. Anaerobic digestion:[proceedings of the first International Symposium on Anaerobic Digestion, held at University College, Cardiff, Wales, September 1979]/edited by DA Stafford, BI Wheatley and DE Hughes.

[22].Cuetos M.A.J. and Gomez X. et al. (2008). Anaerobic digestion of solid slaughterhouse waste (SHW) at laboratory scale: influence of co-digestion with the organic fraction of municipal solid waste (OFMSW). *Biochemical Engineering Journal, 40(1), 99-106*

[23].Cristancho D.E. and Arellano A.V. (2006). Study of the operational conditions for anaerobic digestion of urban solid wastes. *Waste management, 26(5), 546-556.*

[24].Patil V.S. and Deshmukh H.V. (2015). A review on codigestion of vegetable waste with organic wastes for energy generation. International Research Journal of Biological Sciences, 4(6), 83-86.

[25].Castillo, E.F.M., Cristancho, D.E., Arellano, V.A., 2006. Study of the operational conditions for anaerobic digestion of urban solid wastes. *Waste Manage. 26, 546–556.*

[26]. Martín-González, L., Colturato, L., Font, X. and Vicent, T. Anaerobic co-digestion of the organic fraction of municipal solid waste with FOG waste from a sewage treatment plant: Recovering a wasted methane potential and enhancing the biogas yield. *Waste Management, 30(10), pp. 1854-1859, 2010. DOI:10.1016/j.wasman.2010.03.029*.

[27].Mata-Alvarez, J., Dosta, J., Romero-Güiza, M., Fonoll, X., Peces, M. and Astals, S., A critical review on anaerobic co-digestion achievements between 2010 and 2013. *Renewable and Sustainable Energy Reviews, 36, pp. 412-427, 2014. DOI:10.1016/j.rser.2014.04.039*

[28].Goberna, M., Schoen, M.A., Sperl, D., Wett, W., Insam, H., 2010. Mesophilic and thermophilic co-fermentation of cattle excreta and olive mill wastes in pilot anaerobic digesters. *Biomass Bioenergy 34, 340–346.*

[29].Yenigün, O. & Demirel, B. Ammonia inhibition in anaerobic digestion: a review. *Process Biochemistry 48 (5–6), 901–911.*

[30]. Cavinato, C., Fatone, F., Bolzonella, D., Pavan, P., 2010. Thermophilic anaerobic codigestion of cattle manure with agro-wastes and energy crops: comparison of pilot and full scale experiences. *Bioresour. Technol. 101, 545–550.*

[31].Bouallagui, H., Lahdheb, H., Romdan, E., Rachdi, B., Hamdi, M., 2009a. Improvement of fruit and vegetable waste anaerobic digestion performance and stability with co-substrates addition. *J. Environ. Manage. 90, 1844–1849.*

[32].Lo ,H.M., Kurniawan, T.A., Sillanpaa, M.E.T., Pai, T.Y., Chiang, C.F., Chao, K.P., Liu, M.H., Chuang, S.H., Banks, C.J., Wang, S.C., Lin, K.C., Lin, C.Y., Liu, W.F., Cheng, P.H., Chen, C.K., Chiu, H.Y., Wu, H.Y., 2010. Modeling biogas production from organic fraction of MSW co-digested with MSWI ashes in anaerobic bioreactors. *Bioresour. Technol. 101, 6329–633.*

[33].Alvarez, J.A., Otero, L., Lema, J.M., 2010. A methodology for optimizing feed composition for anaerobic co-digestion of agro-industrial wastes. *Bioresour.Technol. 101, 1153–1158.*

[34].Parawira, W., Murto, M., Zvauya, R., Mattiasson, B., 2004. Anaerobic batchdigestion of solid potato waste alone and in combination with sugar beet leaves. Renew. *Energy 29, 1811–1823.*

[35].Feijoo G, Soto M, Mendez R, Lema JM (1995) Sodium inhibition in the anaerobic digestion process: antagonism and adaptation phenomena. *Enzyme Microb Technol 17:180–188.*

[36].OLefebvre, O. et al. Impact of increasing NaCl concentrations on the performance and community composition of two anaerobic reactors. Appl. Microbiol. *Biotechnol. 75, 61–69, doi:10.1007/s00253-006-0799-2 (2007).*

[37].Mendez, R., Lema, J. M. & Soto, M. 1995 Treatment of seafoodprocessing wastewaters in mesophilic and thermophilic anaerobic filters. *Water Environment Research 67 (1), 33–45.*

[38].Lee, D. H., Behera, S. K., Kim, J. W. & Park, H. S. Methane production potential of leachate generated from Korean food waste recycling facilities: a lab-scale study. *Waste Manage. 29, 876–882, doi:10.1016/j.wasman.2008.06.033 (2009).*

[39]. El-Mashad, H. M. & Zhang, R. 2010 Biogas production from codigestion of dairy manure and food waste. *Bioresource Technology 101 (11), 4021–4028.*

[40].Rinzema, A., Van Lier, J. & Lettinga, G. Sodium inhibition of acetoclastic methanogens in granular sludge from a UASB reactor. *Enzyme Microbiol. Technol. 10, 24–32, doi:10.1016/0141-0229(88)90094-4 (1988).*

[41].Soto M, M´endez R and Lema JM, Sodium inhibition and sulphate reduction in the anaerobic treatment of mussel processing wastewaters. *J Chem Technol Biotechnol 58:1–7 (1993).*

[42]. Shaojie Wang, Xiaocong Hou & Haijia Su , Exploration of the relationship between biogas production and microbial community under high salinity conditions *Scientific Reportsvolume 7, Article number: 1149(2017)*

[43]. Omil, F., Mendez, R. & Lema, J. M. Anaerobic treatment of seafood processing waste waters in an industrial anaerobic pilot plant. *Water S.A. 22, 173–181 (1996).*

[44]. Ren, L., Wu, Y., Ren, N., Zhang, K. & Xing, D. Microbial community structure in an integrated A/O reactor treating diluted livestock wastewater during start-up period. *J. Environ. Sci. 22, 656–662, doi:10.1016/S1001-0742(09)60159-5 (2010).*

[45].D. Zhao, J., Zeng, G., Chen, Y., Bond, P.L., Li, X., 2015b. How does poly (hydroxyalkanoate) affect methane production from the anaerobic digestion of waste-activated sludge?.Environ. *Sci. Technol. 49, 12253–12262.*

[46].Chen WH, Han SK and Sung S, Sodium inhibition of thermophilic methanogens. *J Environ En129:506–512 (2003).*

[47].Jianwei Zhao a,b, Yiwen Liu c, Dongbo Wang a,b, Fei Chen a,b, Xiaoming Li a,b, Guangming Zeng a,b, Qi Yang a,b,: Potential impact of salinity on methane production from food waste anaerobic digestion, *journal homepage of Waste Management 2017*

[48].Namdong, Chuingu, Yongin, Kyonggido Osmoprotectants enhance methane production from the anaerobic digestion of food wastes containing a high content of salt Giyang Oh, Lei Zhang and Deokjin Jahng∗ Myongji University, Environmental Engineering and Biotechnology, San 38-2, Namdong, *Chem Technol Biotechnol 83:1204–1210 (2008).*

[49]. Siegert, I. & Banks, C. The effect of volatile fatty acid additions on the anaerobic digestion of cellulose and glucose in batch reactors. *Process Biochemistry 40 (11), 3412–3418.*

[50].Callaghan, F. J., Wase, D. A. J., Thayanithy, K. & Forster, C. F. Continuous co-digestion of cattle slurry with fruit and vegetable wastes and chicken manure. *Biomass & Bioenergy 22 (1), 71–77.*

[51].Mosey, F. E. & Fernandes, X. A. 1988. Patterns of hydrogen in biogas from the anaerobic digestion of milk-sugars. *Water Science and Technology 21 (4–5), 187–196.*

[52].Liu, C. F., Yuan, X. Z., Zeng, G. M., Li, W. W. & Li, J.2008 Prediction of methane yield at optimum pH for anaerobic digestion of organic fraction of municipal solid waste. *Bioresource Technology 99 (4), 882–888.*

[53].Kafle, G. K., Kim, S. H. & Sung, K. I. 2013 Ensiling of fish industry waste for biogas production: a lab scale evaluation of biochemical methane potential (BMP) and kinetics. *Bioresource Technology 127, 326–336.*
[54].Fang, C., Boe, K. & Angelidaki, I. 2011 Anaerobic co-digestion of desugared molasses with cow manure; focusing on sodium and potassium inhibition. *Bioresource Technology 102 (2), 1005–1011.*

IOP Conf. Series: Journal of Physics: Conf. Series **1032** (2018) 012042 doi:10.1088/1742-6596/1032/1/012042

Borosilicate and Polyurethane as Materials for Lenses to Correct Human Presbyopia

Ali H. Alhamdani*, Hayder H. Al-aaraji, Maha. R. Abdul-Hussein and Rajaa A. Madlool

*Department of Laser and Optoelectronics Engineering/University of Technology, Iraq, Baghdad.

Department of Physics/College of Science/University of Kerbala Iraq, kerbala

*Correspondence: ali_alhamdani2003@yahoo.com

Abstract. Start your abstract here The novel of this paper is, using gradient refractive index and spherical surfaces to design a bifocal polyurethane lens with single power instead of the previous designs, which consist of two lens with different powers to correct the vision of near and medium distances at the same time. Presbyopia occurs naturally as eyes age and the structure of the lens responsible for focusing images changes. Presbyopia treatments include corrective lenses like eyeglasses known as progressive lenses. The aims of this work were to evaluate the performance of the human eye after the treatment by using different gradient refractive glass materials. Zemax was used to design an optical system for human eye based on Liu & Brennan model. Two types of optical materials, namely, borosilicate and polyurethane, were selected to design a multifocal lens. The effects of these materials on the performance of the human eye were compared to select the best material. Three different criteria were used; minimum root mean square error, the modulation transfer function and the diffraction image analysis. Modulation transfer function and root mean square Results show that the polyurethane lens is more suitable than the borosilicate lens for correcting presbyopia. Polyurethane material lens gave a better eye performance in three distances, namely, farther, intermediate, and near than the borosilicate lens. Polyurethane lens could be used as trifocal while borosilicate work only as Bifocal lens. The Results suggested that by using only one polyurethane lens power with gradient refractive index and a variable conic for lens curvature in Zemax corrected the vision of distance, near and medium distances at one time.

1. Introduction

Refractive errors are vision problems that occur when the shape of the eye prevents light from focusing, particularly on the retina. For example, the eyeball is either too long or too short), or the shape of the cornea is wrong. The aging of the lens can also cause refractive errors. The most widely recognized refractive errors can be remedied with eyeglasses, contact lenses, or refractive surgery [1]. Presbyopia, which is the aging of the eye and the loss of near vision as a person ages, is another type of hyperopia. This condition leads to the gradual deterioration of the capability of the eye to focus clearly on nearby objects; it is part of the natural aging process of the eye [2]. The hardening of the eye lens causes the eye to focus light behind rather than on the retina when looking at nearby objects;

IOP Conf. Series: Journal of Physics: Conf. Series **1032** (2018) 012042 doi:10.1088/1742-6596/1032/1/012042

the lens cannot change its shape sufficiently to enable the eye to focus on nearby objects because the accommodation strength of the eye decreases with aging [3].

Glasses are the simplest, safest, least expensive, and most common way to correct vision problems caused by refractive errors [4]. Most cases of presbyopia can be corrected by using simple reading glasses with convex lenses. Individuals suffering from myopia and hyperopia regularly need multifocal lenses [5]. Such lenses have more than one focal point and can be found in lined bifocals, lined trifocals, and progressive lenses. These lenses have either two (bifocals) or three (trifocals) focal points. They are designed for distance vision and reading (bifocals) or for distance vision, intermediate vision, and reading (trifocals).

In the past, eyeglasses were made exclusively of glass. At present, most eyeglasses are made of plastic. Plastic offers the best combination of strength, scratch resistance, and cost. Glass lenses are breakable and are approximately twice as heavy as plastic lenses. Although glass lenses are the most scratch resistant and the heaviest lens type, they are the least impact resistant [6]. The selection of material that will be used for the correction of refractive defects is important in visual design because the material plays a role in the visual aspects, such as efficiency and capacity, and in other areas, such as weight and density. In this regard, manufacturers of lenses have to consider the appropriate properties of plastic and glass materials in the design of lenses. In this study, two types of materials, namely, borosilicate and polyurethane plastic, were selected to model the correction of presbyopia.

2. Material properties

Borosilicate is a typical optical glass material and is commonly used to produce high-quality optical components. This type of material is characterized by the presence of substantial amounts of silica and boric oxide (>8%) [7]. Borosilicate exhibits excellent chemical durability, high permeability in the visible light spectrum, and low thermal expansion. It is also relatively easy to procure. Borosilicate glass is less subject to thermal stress as illustrated in Table 1 [8].

Table 1. Physical properties of Borosilicate [8]

The Properties	The Symbol	The value
Refractive index	n_d	1.5168
Abbe Number	V_d	65
Density	ρ	2.51 g/cm^3
Young Factor	E	82.10^3 N/mm^2
Thermal expansion coefficient	T_e	8.3 .10^{-6} k^{-1}
Glass transition temperature	T_{glas}	557 °C

Polyurethane is a versatile plastic material that exists in various forms. It is unique because it provides the flexibility of rubber and the durability of metal. Polyurethane exhibits excellent resistance to oils, solvents, fats, grease, and benzene; high corrosion resistance; and low temperature resistance [9]. It is lightweight, flexible, nonconductive, and has low fabrication cost (i.e., low tooling cost). Polyurethane was first produced by Dr. Otto Bayer in 1937 as illustrated in Table 2 [10].

IOP Conf. Series: Journal of Physics: Conf. Series **1032** (2018) 012042 doi:10.1088/1742-6596/1032/1/012042

Table 2. Physical properties of polyurethane [10]

The Properties	The Symbol	The value
Refractive index	n_d	1.6585
Abbe Number	V_d	36
Density	ρ	1.4 g/cm^3

3. Optical System Design

To build the human eye model, Zemax optical design software is first placed in sequential mode, which is preferred when designing imaging systems or any system that requires optimization and detailed image analysis. Wavelengths are set within the range of visible light, and the dimensions are in millimeter. The design of the human eye model in Zemax requires eight surfaces, which starts from the surface of the object (Surface 0) and ends on the surface of the retina (image surface) in the following section;

Surface 0: The surface of the object, which is labeled "OBJ" in Zemax Lens Data Editor.
Surface 1: The first surface is a dummy plane, which represents the input beam.
Surface 2: The surface of the outer cornea.
Surface 3: This surface represents the area between the cornea and the glass liquid.
Surface 4: This surface is labeled "STO" in Zemax Lens Data Editor; it is the pupil plane of the eye model.
Surface 5: The front part of the crystalline lens of the model; a surface type (gradient 3) provided by the Zemax program is used.
Surface 6: The back part of the crystalline lens of the model.
Surface 7: The interface between the lens and the vitreous body of the eye.
Surface 8: The retina of the model, which is labeled "IMA" in Zemax Lens Data Editor.

The aforementioned model is the Liou and Brennan (1997) eye model, in which the Lens Data Editor should appear similar to Figure 1. The 3D layout of the system, in which Surface 1 was set as the input beam surface instead of Surface 0, is shown in Figure 2.

IOP Conf. Series: Journal of Physics: Conf. Series **1032** (2018) 012042 doi:10.1088/1742-6596/1032/1/012042

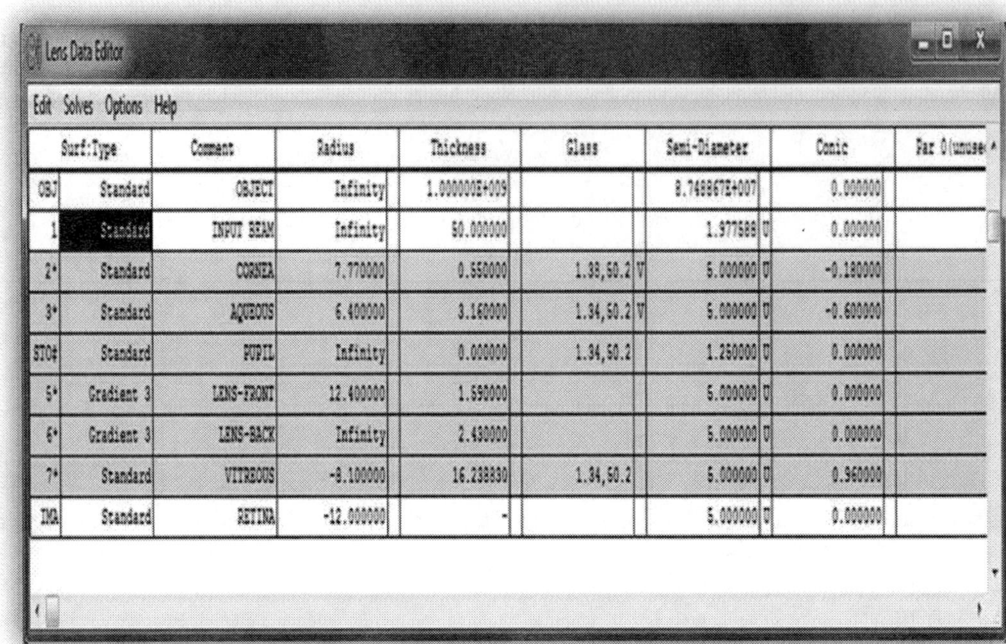

	Surf:Type	Comment	Radius	Thickness	Glass	Semi-Diameter	Conic	Par 0(unuse
OBJ	Standard	OBJECT	Infinity	1.000000E+009		8.748867E+007	0.000000	
1	Standard	INPUT BEAM	Infinity	50.000000		1.977588 U	0.000000	
2*	Standard	CORNEA	7.770000	0.550000	1.38,50.2 V	5.000000 U	-0.180000	
3*	Standard	AQUEOUS	6.400000	3.160000	1.34,50.2 V	5.000000 U	-0.600000	
STO*	Standard	PUPIL	Infinity	0.000000	1.34,50.2	1.250000 U	0.000000	
5*	Gradient 3	LENS-FRONT	12.400000	1.590000		5.000000 U	0.000000	
6*	Gradient 3	LENS-BACK	Infinity	2.430000		5.000000 U	0.000000	
7*	Standard	VITREOUS	-8.100000	16.238830	1.34,50.2	5.000000 U	0.960000	
IMA	Standard	RETINA	-12.000000	-		5.000000 U	0.000000	

Figure 1. Optical data of Liou& Brennan eye model in Zemax optical design software.

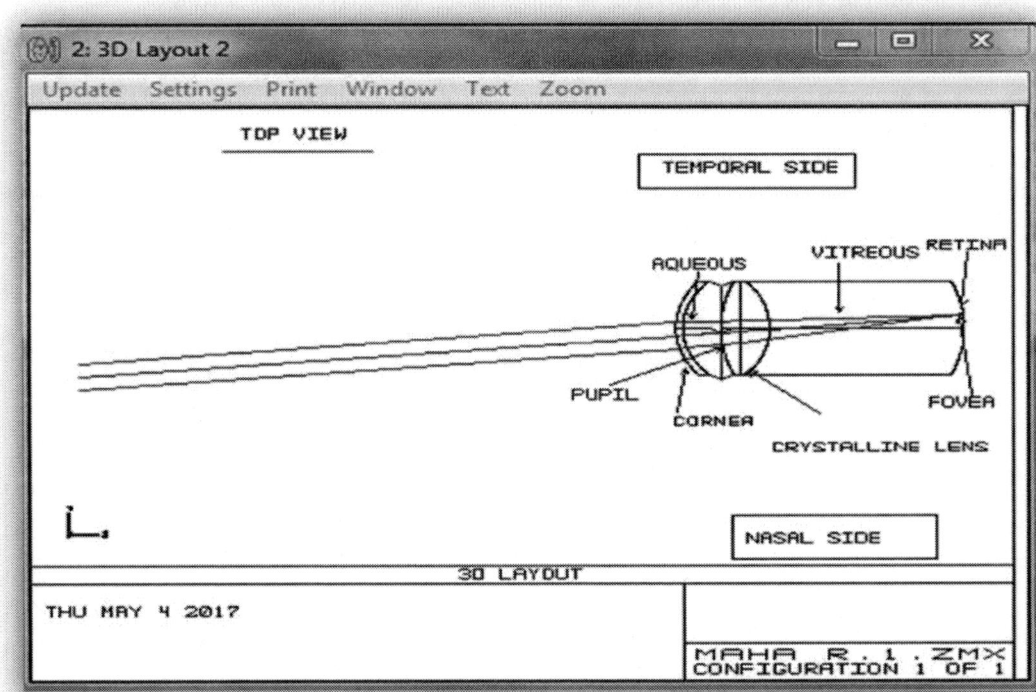

Figure 2. 3D layout of Liou& Brennan human eye

After designing the eye model in Zemax, the model will be used to design an eyeglass lens for presbyopia patients by using the multi-configuration function in the Zemax optical design program.

IOP Conf. Series: Journal of Physics: Conf. Series **1032** (2018) 012042 doi:10.1088/1742-6596/1032/1/012042

The eyeglass lens is designed by placing the lens on the front of the eye model and then optimizing the lens to provide perfect vision for close (mm), midrange (mm), and far (mm) objects. To achieve this, three surfaces were inserted between the input beam and the cornea. These surfaces will represent the front and rear surfaces of the eyeglass lens, and a coordinate break to tilt the eye below the three new surfaces.

Surface 2: The front surface of the eyeglass lens. On this surface, the type of material polyurethane or borosilicate used in the design of the lens will be determined by inserting the refractive coefficient and the Abbe number of each material into the glass cell located in the lens data editor window (see figures 3-6).

Surface 3: The rear surface of the eyeglass lens.

Surface 4: The coordinate break placed at the centre of the eyeball.

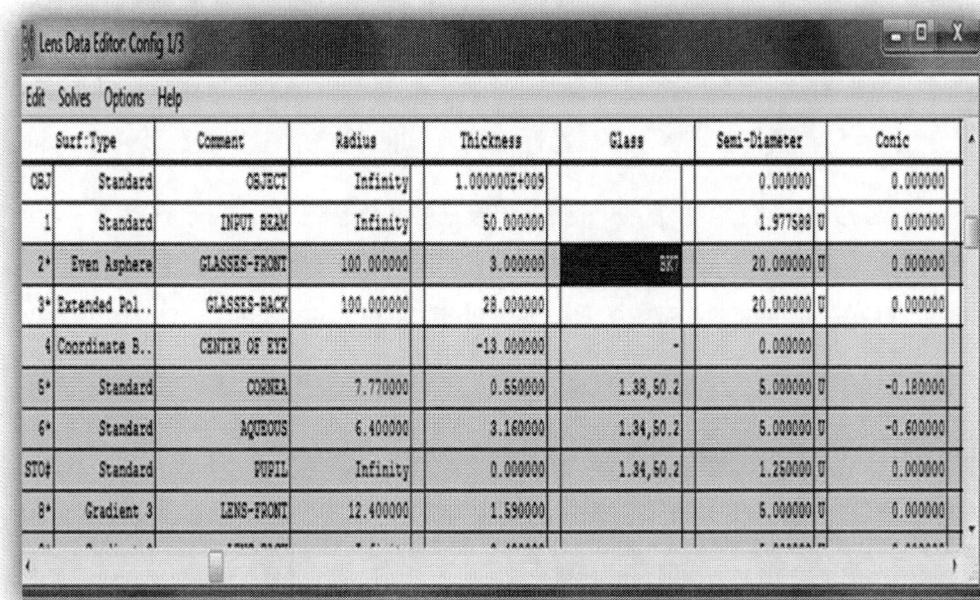

	Surf:Type	Comment	Radius	Thickness	Glass	Semi-Diameter	Conic
OBJ	Standard	OBJECT	Infinity	1.000000E+009		0.000000	0.000000
1	Standard	INPUT BEAM	Infinity	50.000000		1.977588 U	0.000000
2*	Even Asphere	GLASSES-FRONT	100.000000	3.000000	BK7	20.000000 U	0.000000
3*	Extended Pol..	GLASSES-BACK	100.000000	28.000000		20.000000 U	0.000000
4	Coordinate B..	CENTER OF EYE		-13.000000	-	0.000000	
5*	Standard	CORNEA	7.770000	0.550000	1.38,50.2	5.000000 U	-0.180000
6*	Standard	AQUEOUS	6.400000	3.160000	1.34,50.2	5.000000 U	-0.600000
STO#	Standard	PUPIL	Infinity	0.000000	1.34,50.2	1.250000 U	0.000000
8*	Gradient 3	LENS-FRONT	12.400000	1.590000		5.000000 U	0.000000

Figure 3. Optical data of the modified eye model with borosilicate lens in Zemax optical design.

IOP Conf. Series: Journal of Physics: Conf. Series **1032** (2018) 012042 doi:10.1088/1742-6596/1032/1/012042

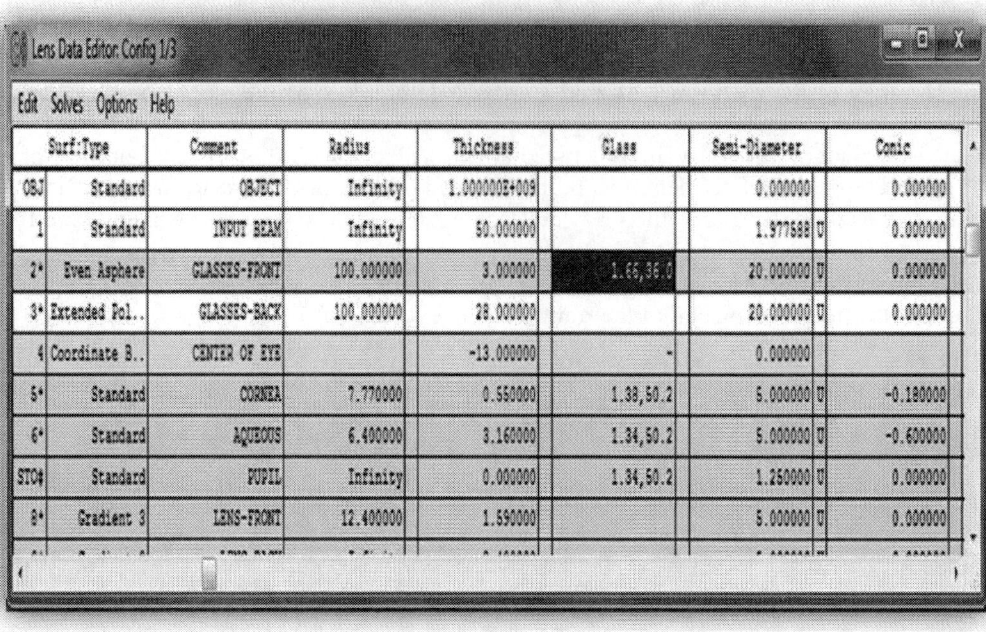

	Surf:Type	Comment	Radius	Thickness	Glass	Semi-Diameter	Conic
OBJ	Standard	OBJECT	Infinity	1.000000E+009		0.000000	0.000000
1	Standard	INPUT BEAM	Infinity	50.000000		1.977588 U	0.000000
2*	Even Asphere	GLASSES-FRONT	100.000000	3.000000	1.66,36.0	20.000000 U	0.000000
3*	Extended Pol..	GLASSES-BACK	100.000000	28.000000		20.000000 U	0.000000
4	Coordinate B..	CENTER OF EYE		-13.000000	-	0.000000	
5*	Standard	CORNEA	7.770000	0.550000	1.38,50.2	5.000000 U	-0.180000
6*	Standard	AQUEOUS	6.400000	3.160000	1.34,50.2	5.000000 U	-0.600000
STO*	Standard	PUPIL	Infinity	0.000000	1.34,50.2	1.250000 U	0.000000
8*	Gradient 3	LENS-FRONT	12.400000	1.590000		5.000000 U	0.000000

Figure 4. Optical data of the modified eye model with polyurethane lens in Zemax optical design software.

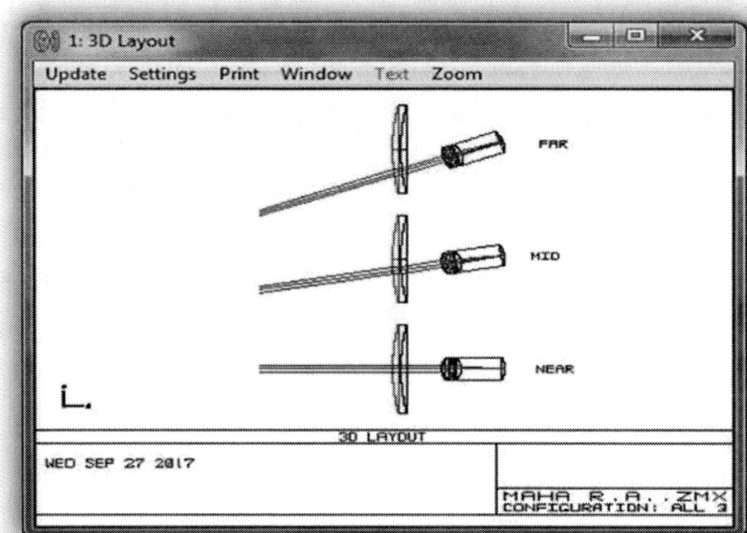

Figure 5. Shows the 3D side-view layout of the system from three configurations (near, intermediate, and far distances) with borosilicate Lens.

IOP Conf. Series: Journal of Physics: Conf. Series **1032** (2018) 012042 doi:10.1088/1742-6596/1032/1/012042

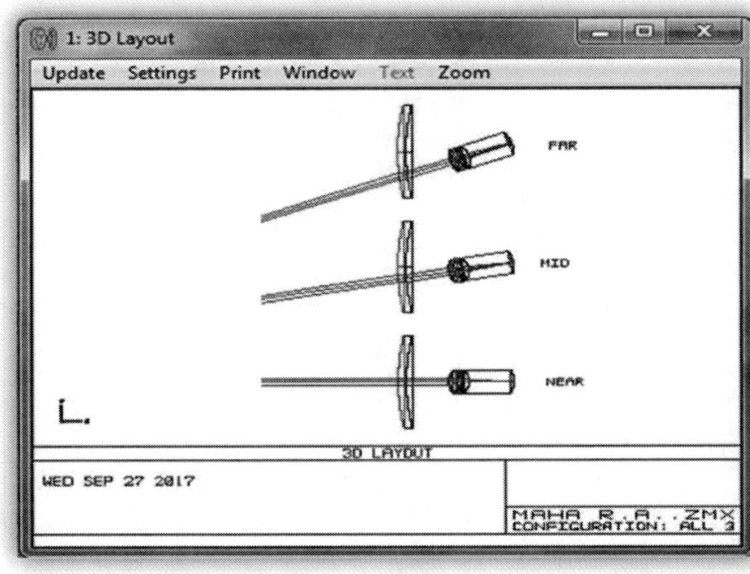

Figure 6. 3D side-view layout of the system from three configurations (near, intermediate, and far distances) with polyurethane Lens.

4. Result and Analysis

Figure 7 shows the Liou model root mean square spot diagrams at the far distance (1.00E-9 mm), intermediate (1000 mm), and near (500 mm) in the visual ranges (400-650 nm). Table.3,colum.2 shows that as for far object, the root mean square is 8.202 μm and it increased for med and near distance (16.377 and 28.787 μm) respectively. Figures 8,9 indicate the modified model by inserting borosilicate and polyurethane in Liou respectively. Figures 8, 9 and Table.3, Colum 3 and show that the insertion of polyurethane or borosilicate lens in the modified model reduce the root mean square for the far, intermediate, and near distances. The reduction in root mean square means a reduction in wave front aberrations. The results indicate that the use of polyurethane improve the optical system more than borosilicate usage.

The modified transfer function is an important method for describing the contrast in an image within a certain spatial frequency presented in the scene being viewed. Modified transfer function plots for the three configurations are provided in Figures 10 and 11. The Sagittal modified transfer function before and after inserting borosilicate and polyurethane lens for the three distances are shown in figure 10. Sagittal modified transfer function after inserting the polyurethane lens is higher than that without lens at far and near distances. Furthermore, the borosilicate lens defocus error appears at near distance. Figure 11 shows the Tangential modified transfer function at far, intermediate, and near distances using the modified transfer function technique before and after the insertion of the polyurethane and borosilicate lenses. The modified transfer function value at far distance before inserting the lenses is less than the modified transfer function value after inserting the polyurethane lens, whereas at an intermediate distance, the modified transfer function value with the borosilicate lens is higher than the modified transfer function value with and without the polyurethane lens, thereby indicating that the polyurethane lens achieves higher contrast at far and near distances and can be used as bifocal lens.

IOP Conf. Series: Journal of Physics: Conf. Series **1032** (2018) 012042 doi:10.1088/1742-6596/1032/1/012042

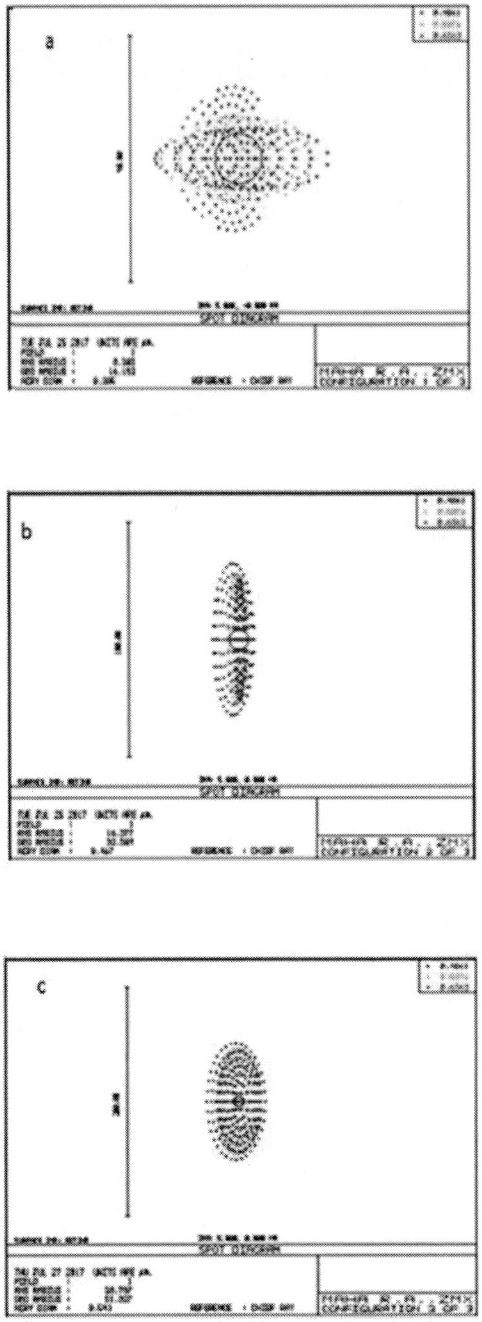

Figure 7. Root mean square for Liou& Brennan model at far distance (a), intermediate (b) and near distance (c).

IOP Conf. Series: Journal of Physics: Conf. Series **1032** (2018) 012042 doi:10.1088/1742-6596/1032/1/012042

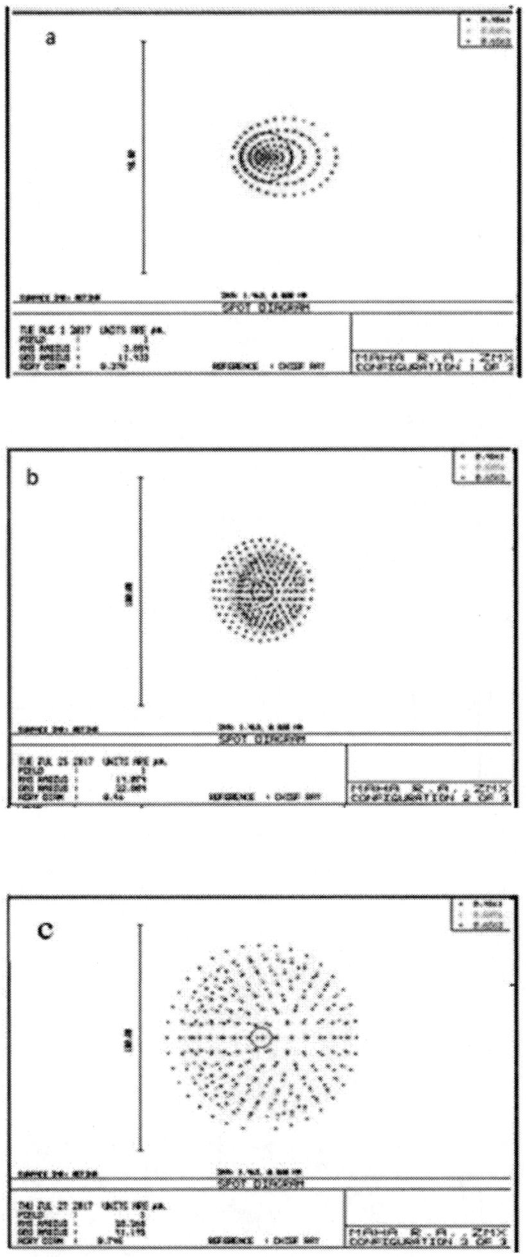

Figure 8. Root mean square for modified model with borosilicate lens at far distance (a), intermediate (b) and near distance (c).

IOP Conf. Series: Journal of Physics: Conf. Series **1032** (2018) 012042 doi:10.1088/1742-6596/1032/1/012042

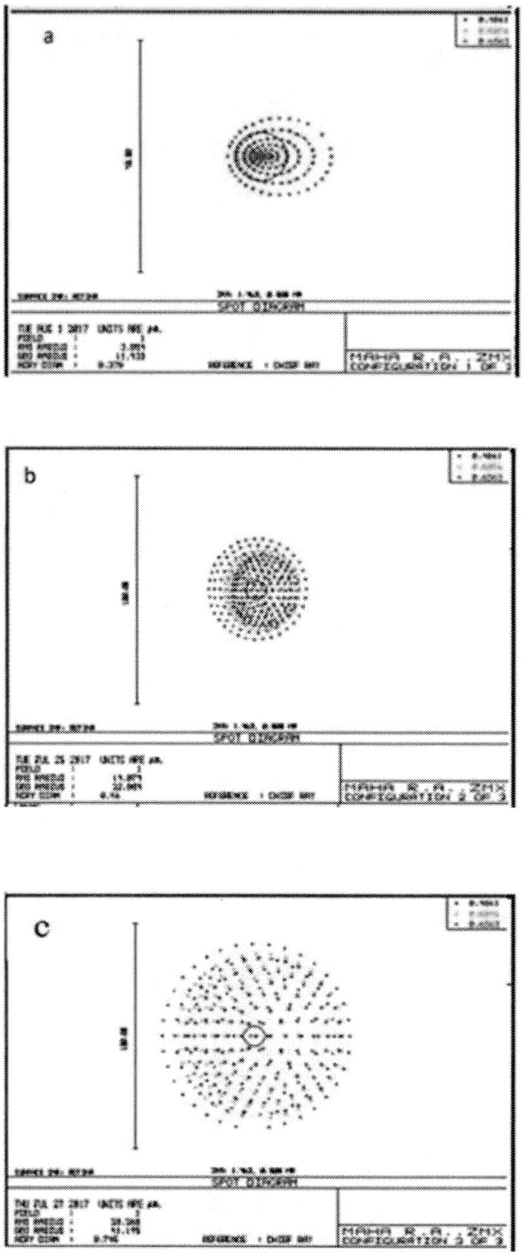

Figure 9. Root mean square for modified model with polyurethane lens at far distance (a), intermediate (b) and near distance (c).

IOP Conf. Series: Journal of Physics: Conf. Series **1032** (2018) 012042 doi:10.1088/1742-6596/1032/1/012042

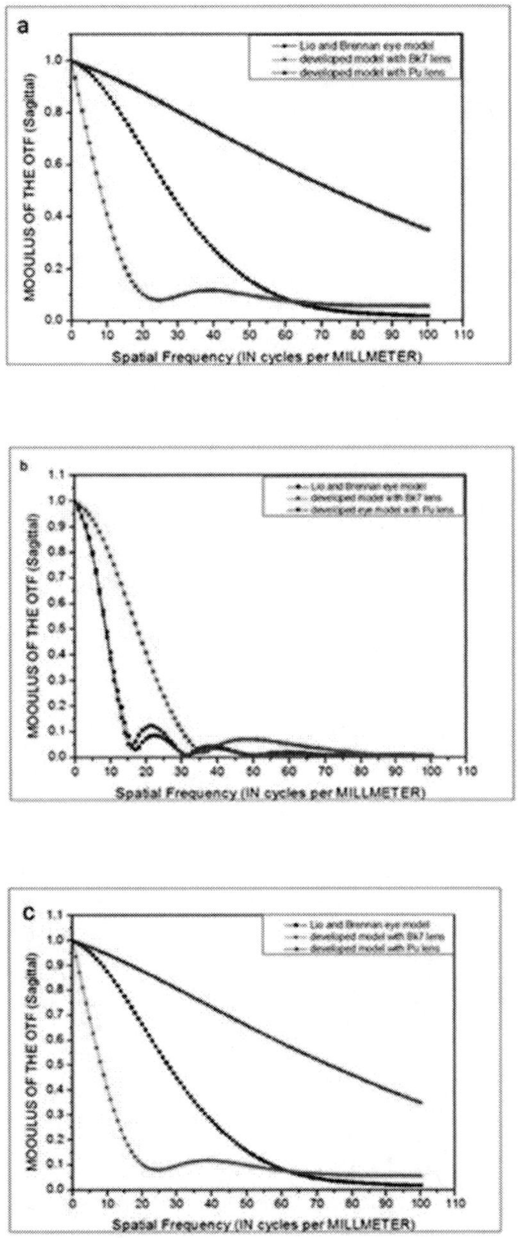

Figure 10. Sagittal modified transfer function for Liou& Brennan model, modified model with borosilicate lens and modified model with polyurethane lens at far distance (b) and near distance (c).

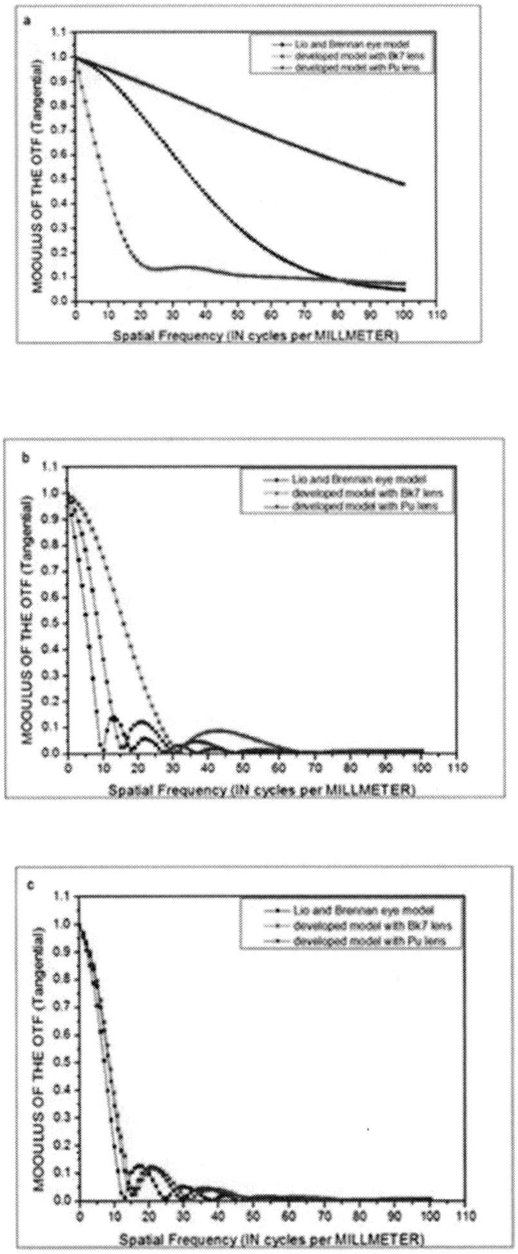

Figure 11. Sagittal modified transfer function for Liou& Brennan model, modified model with borosilicate lens and modified model with polyurethane lens at far distance (b) and near distance (c).

The image analyses of the three configurations with and without borosilicate lens are shown in Figures (12-15). A quick assessment of image quality is provided, such that the eye vision of objects is simulated and an idea of the shape and content of an image is obtained in terms of clarity, colors, and contrast without providing digital data on the characteristics of an image. The image of the letter "F" at far and intermediate distances are clearer after placing the lens than before placing the lens. the

IOP Conf. Series: Journal of Physics: Conf. Series **1032** (2018) 012042 doi:10.1088/1742-6596/1032/1/012042

image at far distance is clearer than the images at other distances, and the image with the polyurethane lens is better than the image with the borosilicate lens.

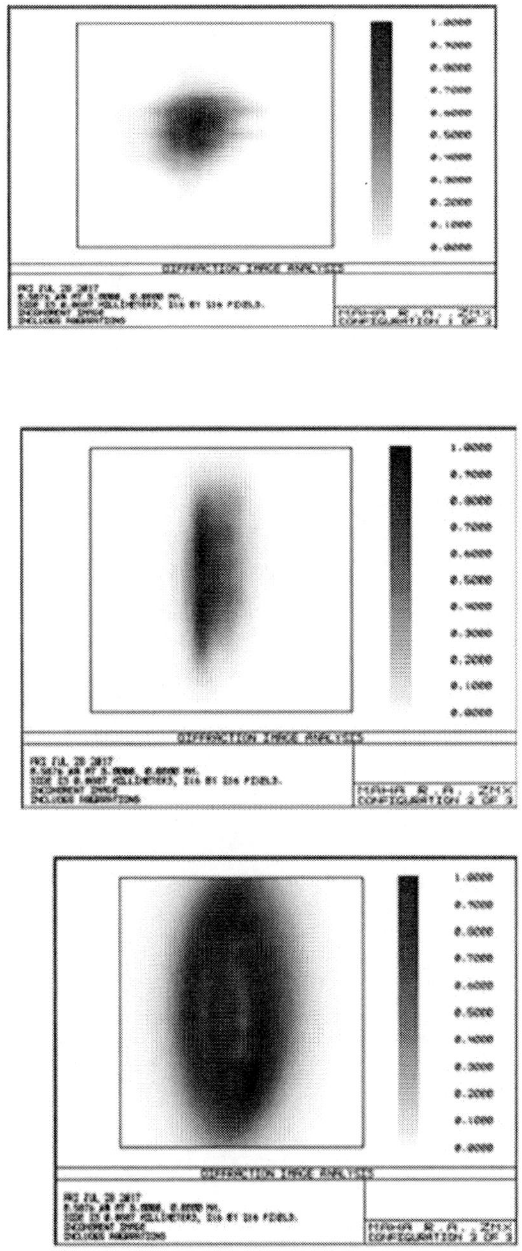

Figure 12. Tangential modified transfer function for Liou, model, modified model with borosilicate lens and modified model with polyurethane lens at far distance (a) intermediate distance (b) and near distance (c).

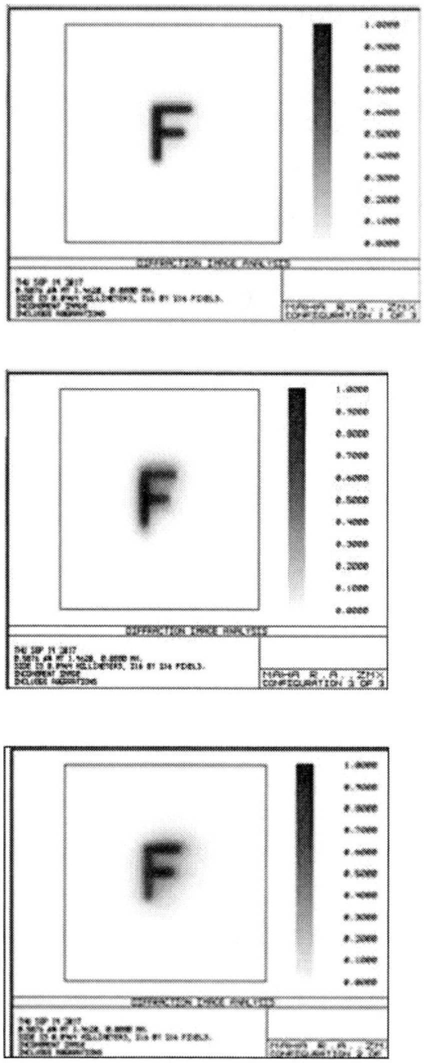

Figure 13. The diffraction image for Liou& Brennan model at far distance (a), intermediate (b) and near distance (c).

References

[1] Fedtke C, Ehrmann K, Holden BA. A Review of Peripheral Refraction Techniques. Optometry and vision science : official publication of the American Academy of Optometry 2009;86:429-46.

[2] Curtin BJ. Physiologic Vs Pathologic Myopia: Genetics Vs Environment. Ophthalmology 1979;86:681-91.

[3] Charman WN, Radhakrishnan H. Peripheral Refraction and the Development of Refractive Error: A Review. Ophthalmic & physiological optics : the journal of the British College of Ophthalmic Opticians (Optometrists) 2010;30:321-38.

[4] Verkicharla PK, Mathur A, Mallen EA, et al. Eye Shape and Retinal Shape, and Their Relation to Peripheral Refraction. Ophthalmic & physiological optics : the journal of the British College of Ophthalmic Opticians (Optometrists) 2012;32:184-99.

[5] Elliott DB. Clinical Procedures in Primary Eye Care, 4th ed. Philadelphia: Saunders/Elsevier Health Sciences; 2013.

[6] Curtin BJ. Physiologic Vs Pathologic Myopia: Genetics Vs Environment. Ophthalmology 1979;86:681-91.

[7] Miller AD, Kris MJ, Griffiths AC. Effect of Small Focal Errors on Vision. Optometry and vision science : official publication of the American Academy of Optometry 1997;74:521-6.

[8] Liu WD, Zhang LC, Mylvaganam K. Relaxation Oscillation of Borosilicate Glasses in Supercooled Liquid Region. Scientific reports 2017;7:15872.

[9] Bahl S, Peuget S, Pidchenko I, Pruessmann T. Pu Coexists in Three Oxidation States in a Borosilicate Glass: Implications for Pu Solubility. 2017;56:13982-90.

[10] Abed AS, Ziadankm ,Qbdulah AQ. Some properties of polyurethane. J. of polymers 2014; 17:18-28.

IOP Conf. Series: Journal of Physics: Conf. Series **1032** (2018) 012043 doi:10.1088/1742-6596/1032/1/012043

Influence of heat and mass transfer on inclined (MHD) peristaltic of pseudoplastic nanofluid through the porous medium with couple stress in an inclined asymmetric channel

Mohammed R. Salman[1], **Ahmed M. Abdulhadi**[2]
University of Baghdad-College of Science-Department of Math-Baghdad-Iraq
Email address:mawb1967@gmail.com (Mohammed R. Salman),
ahm6161@yahoo.com (Ahmed M. Abdulhadi)

Abstract. The paper is intended for studying the Influence of heat and mass transfer on inclined magnetohydrodynamics peristaltic of pseudoplastic nanofluid through the porous medium with couple stress in an inclined tapered asymmetric channel.The governing equations of two-dimensional are the motion and energy equations are resolved and simplified under acting the low "Reynolds number" long wavelength by using perturbation technique series solutions.And have been obtained the effects of different parameters which are "Hartmann number", "Darcy number", "Brinkman number", "Reynolds number", "Froude number", "Prandle number" and "Eckert number" and as well the "couple stress", inclination of magnetic field, mean flow rate and incline angle of the channel on the "pressure gradient", velocity, "pressure rise", "Nusselt number", "temperature", "heat transfer coefficient" and survey the "trapping" phenomena inspection" through different graphs. And in order to illustrate the validity of the theoretical study, numerical results have been computed by using MATHEMATICA software.
Keywords: Peristaltic transport,Reynolds number, Hartmann number, Froude number, Porous Medium.

1. Introduction

The study a non-Newtonian fluid flows has a great interest among researchers Including medical, biological sciences, physiology, blood. Also on the industrial field where has encouraged the complex nature of fluids of which, oils, chemicals, petroleum and other fluids, has encouraged extensive studies and research into the properties of these fluids. where many researchers presented basic equations for various non-Newtonian liquids [1-5].
The couple stress fluid theory is one of the most important topics in the study of Newtonian fluids and represents the excellent study presented written by Stokes "Theories of fluids with micro-structure an introduction" [6,7]. And also there are many important studies of the couple stress fluid that explain the behavior of complex fluids such as blood,thrombosis due to blood clotting , liquid crystals, colloidal liquids and synovial liquids [8–10].
The peristaltic flows, due to extensive applications in industry, physiology, and agriculture in modern irrigation methods, The movement peristaltic flows of the Newtonian and non-Newtonian fluids, which results from contraction and expansion along the tube (channel),for examples, the movement of food from the stomach to the small intestine and large intestine as well as the transfer of the ovum in the fallopian tube and the flow of urine flowing from the kidney to the bladder, the passage of the yellow matter from the gallbladder to the twelve, etc. They are the most common examples in the body. Through these and other applications, many several researchers have studied the of peristaltic flow in different cases[11-17].
The magnetohydrodynamics is currently involved in the detection and treatment of many diseases, the most important of which is the treatment of cancer tumors and control of bleeding caused by surgical

IOP Conf. Series: Journal of Physics: Conf. Series **1032** (2018) 012043 doi:10.1088/1742-6596/1032/1/012043

currencies, as well as MRI to diagnose the disease,magnetic field effect can also be used as a blood pump during operations of cardiovascular surgery (stenosis and arteriosclerosis).From these applications, many researchers studied transport problems in asymmetric channels and studied the effect of the magnetic field on fluid flow behavior, Nadeem and Akram [18,19] have studied the peristaltic flow of a couple stress fluid under the influence of an induced magnetic field in an asymmetric channel and effect of inclined magnetic field on a Williamson fluid model in an inclined asymmetric channel.
The concept of mass transfer is one of the most important topics that have wide applications in biological processes in the body of living things, including humans,such as transferring nutrients in the blood to nearby tissues,and kidney dialysis processes to purify blood from toxins and high urea As well as in many industrial processes such as separation of the membrane, separation of petroleum components, distillation systems, water purification and the process of separating or adding chemical impurities.For example, Saleem and Haider [20] studied theoretical analysis of heat and mass transfer of the non-Newtonian fluid (Maxwell's fluid) in peristaltic channels with creeping flow.Hayat et al. [21] examined and studied the Simultaneous influences of convective conditions and nanoparticles on peristaltic motion. M.R.Salman and A.M.Abdulhadi [22] have analyzed theoretically Effects of MHD on Peristalsis Transport and Heat Transfer with Variables Viscosity in Porous Medium. Also many researchers worked in this field [23,24]. K.Ramesh [25] he studied the effect of heat and mass transfer on peristaltic flow of a couple stress fluid in an inclined asymmetric channel with there exist porous medium under the influence of an inclined magnetic field
In this paper, we will study, discuss and interpret Influence of heat and mass transfer on inclined (MHD) peristaltic of pseudoplastic nanofluid through the porous medium with couple stress in an inclined asymmetric channel,this is done by simplifying and solving the governing equations under the influence of the very small Reynolds number and a large wavelength by use numerical results have been computed by using MATHEMATICA software.and then we analyze these results through shapes and graphs for pressure gradient and temperature and pressure rise and stream function.

2. Problem description

The flow of an incompressible magnetohydrodynamic peristaltic couple stress fluid in an inclined asymmetric channel in a two-dimensional tapered width $(2d)$ through a porous medium. The channel is inclined at an angle (α) to the X-axis. The lower and upper walls $Y=\tilde{H}_1$ and $Y=\tilde{H}_2$ are at the distance (d) from the central line of the channel. Wave speed of peristaltic waves is denoted by (c). We choose a Cartesian coordinate system in such a way that X-axis is taken along the axial direction and Y-axis is normal to the flow direction. to it (see Fig.(1)). The fluid is electrically conducting in uniform magnetic (T_0) field (B_0) and is inclined at an angle (Θ) to the Y- axis, where a uniform magnetic field $B=(0,B_0,0)$. The induced magnetic field is ignored here by making a very small magnetic Reynolds number. Heat and mass transfer is examined through convective conditions. The flow is generated by sinusoidal waves propagating along the compliant walls of the channel. The upper and lower walls are maintained at the temperatures (T_0) and (T_1). The geometry of the wall surfaces is given as follows:

$$\tilde{Y}=\tilde{H}_1(\tilde{X},t')=d+m'\tilde{X}+b_1 Sin\left[\frac{2\pi}{\lambda}(\tilde{X}-ct')\right] \tag{1}$$

$$\tilde{Y}=\tilde{H}_2(\tilde{X},t')=-d-m'\tilde{X}-b_2 Sin\left[\frac{2\pi}{\lambda}(\tilde{X}-ct')+\phi\right] \tag{2}$$

Where b_1 , b_2 are the amplitudes of upper and lower walls, (c) is the "velocity of the wave", (λ) is the wavelength, is the non-uniform parameter, the phase variance (ϕ) varies in the range $(0\le\phi\le\pi)$

, where $(\phi = 0)$ consider to symmetric channel with waves out of phase i.e. both walls move towards outward or inward the same time and further b_1, b_2, d and (ϕ) satisfy the following condition at the inlet of divergent channel

$$b_1^2 + b_2^2 + 2b_1 b_2 \, Cos\,\phi \le (2d)^2 \tag{3}$$

Here we assume the fluid to be electrically conducting in the presence of a uniform inclined magnetic field $B = (0, B_0, 0)$.To compute the "Lorentz force" we will put on a magnetic field just in $\tilde{Y} - direction$ and then we study the influence of it on the fluid flux.
Expression of an extra stress tensor in pseudoplastic fluid is [26] :

$$\tilde{S} + \lambda_1 \frac{D\tilde{S}}{Dt'} + \frac{1}{2}(\lambda_1 - \mu_1)(A_1\tilde{S} + \tilde{S}A_1) = \mu A_1$$

In which λ_1 and μ_1 are the relaxation times. Also $A_1 = \left[\nabla\tilde{V} + (\nabla\tilde{V})^T\right]$, A_1 is the first Rivlin-Ericksen tensor with the velocity gradient, and

$d\tilde{S}/dt' = \partial\tilde{S}/\partial t' + \tilde{V}.\nabla\tilde{S}$ and $\frac{D\tilde{S}}{Dt'} = d\tilde{S}/dt' - (\nabla\tilde{V})\tilde{S} - \tilde{S}(\nabla\tilde{V})^T$ then

$$\frac{D\tilde{S}}{Dt'} = \partial\tilde{S}/\partial t' + \tilde{V}.\nabla\tilde{S} - (\nabla\tilde{V})\tilde{S} - \tilde{S}(\nabla\tilde{V})^T$$

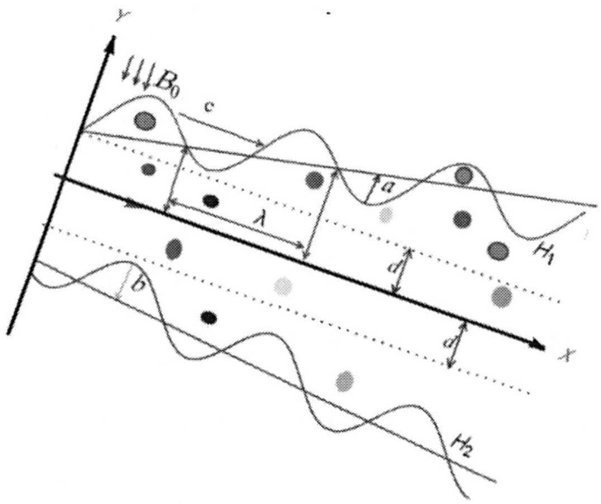

Figure 1. Channel diagram asymmetric and tapered.

3. The governing equations

In the laboratory frame $(\tilde{x}, \tilde{y})$, the "energy equation" is written as follows

$$\frac{\partial\tilde{U}}{\partial\tilde{X}} + \frac{\partial\tilde{V}}{\partial\tilde{Y}} = 0 \tag{4}$$

$$\rho\left[\frac{\partial\tilde{U}}{\partial t'} + \tilde{U}\frac{\partial\tilde{U}}{\partial\tilde{X}} + \tilde{V}\frac{\partial\tilde{U}}{\partial\tilde{Y}}\right] = -\frac{\partial\tilde{P}}{\partial\tilde{X}} + \frac{\partial}{\partial\tilde{X}}(\tilde{S}_{\tilde{X}\tilde{X}}) + \frac{\partial}{\partial\tilde{Y}}(\tilde{S}_{\tilde{X}\tilde{Y}})$$
$$-\sigma' B_0^2 Cos\Theta\left[\tilde{U}Cos\Theta - \tilde{V}Sin\Theta\right] - \frac{\mu}{K_0}\tilde{U} + \rho g Sin\alpha \tag{5}$$

$$\rho\left[\frac{\partial\tilde{V}}{\partial t'}+\tilde{U}\frac{\partial\tilde{V}}{\partial\tilde{X}}+\tilde{V}\frac{\partial\tilde{V}}{\partial\tilde{Y}}\right]=-\frac{\partial\tilde{P}}{\partial\tilde{Y}}+\frac{\partial}{\partial\tilde{Y}}\left(\tilde{S}_{\tilde{Y}\tilde{Y}}\right)+\frac{\partial}{\partial\tilde{X}}\left(\tilde{S}_{\tilde{Y}\tilde{X}}\right)$$
$$+\sigma' B_0^2 Sin\Theta\left[\tilde{U}Cos\Theta-\tilde{V}Sin\Theta\right]-\frac{\mu}{K_0}\tilde{U}-\rho g Sin\alpha \tag{6}$$

$$\rho c_p\left[\frac{\partial\tilde{T}}{\partial t'}+\tilde{U}\frac{\partial\tilde{T}}{\partial\tilde{X}}+\tilde{V}\frac{\partial\tilde{T}}{\partial\tilde{Y}}\right]=\kappa\left[\frac{\partial^2\tilde{T}}{\partial\tilde{X}^2}+\frac{\partial^2\tilde{T}}{\partial\tilde{Y}^2}\right]+\mu\left[2\left\{\left(\frac{\partial\tilde{U}}{\partial\tilde{X}}\right)^2+\left(\frac{\partial\tilde{V}}{\partial\tilde{X}}\right)^2\right\}+\left(\frac{\partial\tilde{U}}{\partial\tilde{Y}}+\frac{\partial\tilde{V}}{\partial\tilde{X}}\right)^2\right]$$
$$+\eta\left[\left(\frac{\partial^2\tilde{U}}{\partial\tilde{X}^2}+\frac{\partial^2\tilde{U}}{\partial\tilde{Y}^2}\right)^2+\left(\frac{\partial^2\tilde{V}}{\partial\tilde{X}^2}+\frac{\partial^2\tilde{V}}{\partial\tilde{Y}^2}\right)^2\right]+Q_0 \tag{7}$$

The extra stress components $\tilde{S}_{\tilde{X}\tilde{X}}$, $\tilde{S}_{\tilde{X}\tilde{Y}}$ and $\tilde{S}_{\tilde{Y}\tilde{Y}}$ atisfy the following relations:

$$\tilde{S}_{\tilde{X}\tilde{X}}+\lambda_1\left[\left(\frac{\partial}{\partial t'}+\tilde{U}\frac{\partial}{\partial\tilde{X}}+\tilde{V}\frac{\partial}{\partial\tilde{Y}}\right)\tilde{S}_{\tilde{X}\tilde{X}}-2\tilde{S}_{\tilde{X}\tilde{X}}\frac{\partial\tilde{U}}{\partial\tilde{X}}-2\tilde{S}_{\tilde{X}\tilde{Y}}\frac{\partial\tilde{U}}{\partial\tilde{Y}}\right]$$
$$+\frac{1}{2}(\lambda_1-\mu_1)\left[4\tilde{S}_{\tilde{X}\tilde{X}}\frac{\partial\tilde{U}}{\partial\tilde{X}}+2\tilde{S}_{\tilde{X}\tilde{Y}}\left(\frac{\partial\tilde{U}}{\partial\tilde{Y}}+\frac{\partial\tilde{V}}{\partial\tilde{X}}\right)\right]=2\mu\frac{\partial\tilde{U}}{\partial\tilde{X}} \tag{8}$$

$$\tilde{S}_{\tilde{X}\tilde{X}}+\lambda_1\left[\left(\frac{\partial}{\partial t'}+\tilde{U}\frac{\partial}{\partial\tilde{X}}+\tilde{V}\frac{\partial}{\partial\tilde{Y}}\right)\tilde{S}_{\tilde{X}\tilde{X}}-2\tilde{S}_{\tilde{X}\tilde{X}}\frac{\partial\tilde{U}}{\partial\tilde{X}}-2\tilde{S}_{\tilde{X}\tilde{Y}}\frac{\partial\tilde{U}}{\partial\tilde{Y}}\right]$$
$$+\frac{1}{2}(\lambda_1-\mu_1)\left[4\tilde{S}_{\tilde{X}\tilde{X}}\frac{\partial\tilde{U}}{\partial\tilde{X}}+2\tilde{S}_{\tilde{X}\tilde{Y}}\left(\frac{\partial\tilde{U}}{\partial\tilde{Y}}+\frac{\partial\tilde{V}}{\partial\tilde{X}}\right)\right]=2\mu\frac{\partial\tilde{U}}{\partial\tilde{X}} \tag{9}$$

$$\tilde{S}_{\tilde{Y}\tilde{Y}}+\lambda_1\left[\left(\frac{\partial}{\partial t'}+\tilde{U}\frac{\partial}{\partial\tilde{X}}+\tilde{V}\frac{\partial}{\partial\tilde{Y}}\right)\tilde{S}_{\tilde{Y}\tilde{Y}}-2\tilde{S}_{\tilde{Y}\tilde{X}}\frac{\partial\tilde{V}}{\partial\tilde{X}}-2\tilde{S}_{\tilde{Y}\tilde{Y}}\frac{\partial\tilde{V}}{\partial\tilde{Y}}\right]$$
$$+\frac{1}{2}(\lambda_1-\mu_1)\left[2\tilde{S}_{\tilde{X}\tilde{Y}}\left(\frac{\partial\tilde{U}}{\partial\tilde{Y}}+\frac{\partial\tilde{V}}{\partial\tilde{X}}\right)+4\tilde{S}_{\tilde{Y}\tilde{Y}}\frac{\partial\tilde{V}}{\partial\tilde{Y}}\right]=2\mu\frac{\partial\tilde{V}}{\partial\tilde{Y}} \tag{10}$$

In which (ρ) is the fluid density, ($\tilde{U}$) and ($\tilde{V}$) are the respective velocity components in ($\tilde{X}$) and ($\tilde{Y}$) directions, ($\tilde{P}$) is the pressure, (σ') is the electrical conductivity of the fluid, (g) is the acceleration due to the gravity, (μ) is the "viscosity", (η) is the "couple stress" viscosity parameter, (K_0) is the permeability parameter, (B_0) is the magnetic field, (κ) the "thermal conductivity" of fluid, (α) is the inclination angle of the channel, (Θ) is the inclination of the magnetic field, ($\tilde{T}$) is the temperature, (c_p) is the specific heat at a constant pressure, and (Q_0) is the heat generation parameter. And $\tilde{S}_{\tilde{X}\tilde{X}}$,$\tilde{S}_{\tilde{X}\tilde{Y}}$ and $\tilde{S}_{\tilde{Y}\tilde{Y}}$ are the components of the additional stress tensor. Respectively. And in the laboratory frame ($\tilde{x}$, $\tilde{y}$), the flow is steady.

Now, for ($\tilde{U}$) and ($\tilde{V}$) be the respective velocity components along $\tilde{X}$ and $\tilde{Y}$ directions in the fixed frame, respectively. For the unsteady two-dimensional flow, the velocity components may be written $\tilde{V}=\left(\tilde{U}\left(\tilde{X},\tilde{Y}\right),\tilde{V}\left(\tilde{X},\tilde{Y}\right),0\right)$ and the temperature function may be written $T=T\left(\tilde{X},\tilde{Y}\right)$.

The suitable boundary conditions comprising wall no-slip and convective boundary conditions are approaching as follows:

$$U=0\,,\ \tilde{T}=T_0 \quad at\ \tilde{Y}=\tilde{H}_1 \tag{11}$$

$$U=0\,,\ \tilde{T}=T_1 \quad at\ \tilde{Y}=\tilde{H}_2 \tag{12}$$

If we introduce the wave frame having coordinates $(\tilde{x},\tilde{y})$ which march in the horizontal direction with the same wave speed (c). Then in the laboratory frame $(\tilde{X},\tilde{Y})$, the unsteady flow can be considered as steady. The coordinates and velocities in the two frames by following relations

$$\tilde{x}=\tilde{X}-ct',\tilde{y}=\tilde{Y},\tilde{u}=\tilde{U}-c,\tilde{v}=\tilde{V},\tilde{p}=\tilde{P} \tag{13}$$

Where $\tilde{u},\tilde{v}$ are the velocity components in the wave frame $(\tilde{x},\tilde{y})$.

4. Dimensionless parameter

For a dimensional analysis process we set up the following non-dimensional quantities:

$$x=\frac{\tilde{x}}{\lambda},y=\frac{\tilde{y}}{d},t=\frac{ct'}{\lambda},u=\frac{\tilde{u}}{c},v=\frac{\tilde{v}}{c},\delta=\frac{d}{\lambda},h_1=\frac{\tilde{H}_1}{d},h_2=\frac{\tilde{H}_2}{d},\theta=\frac{T-T_\circ}{T_1-T_\circ},p=\frac{d^2\tilde{p}}{\lambda\mu c},$$

$$S_{ij}=\frac{d}{c\mu}\tilde{S}_{\tilde{ij}},K=\frac{K_0}{d^2},a=\frac{b_1}{d},b=\frac{b_2}{d},Re=\frac{\rho cd}{\mu},m=\frac{m'\lambda}{d},P_r=\frac{\mu c_P}{\kappa},M=\sqrt{\frac{\sigma'}{\mu}}dB_0, \tag{14}$$

$$\gamma=\sqrt{\frac{\mu}{\eta}}d\ ,\ Fr=\frac{c^2}{gd}\ ,\ Ec=\frac{c^2}{c_p(T_1-T_0)},Br=EcPr,\beta=\frac{Q_0d^2}{\kappa(T_1-T_0)},u=\frac{\partial\psi}{\partial y},v=-\delta\frac{\partial\psi}{\partial x}$$

Where (δ) a "wave number", (Re) a "Renold number", (γ) represent "couple stress" parameter, (K) a "Darcy number", (M) a "Hartman number", (Fr) a"Froude number", (θ) is a "temperature distribution", (Ec) a"Eckert number" and (Pr) is a"Prandle number".

By using the transformations in Eqs.(13) and (14) into a governing Eqs.(4)-(7) we have the following form.

$$\delta\frac{\partial u}{\partial x}+\frac{\partial v}{\partial y}=0 \tag{15}$$

$$\begin{aligned}Re\,\delta\left[\left(\frac{\partial\psi}{\partial y}+1\right)\frac{\partial^2\psi}{\partial x\,\partial y}-\frac{1}{\delta}\frac{\partial\psi}{\partial x}\frac{\partial^2\psi}{\partial y^2}\right]&=-\frac{\partial p}{\partial x}+\delta\frac{\partial}{\partial x}(S_{xx})+\frac{\partial}{\partial y}(S_{xy})\\&-M^2Cos\Theta\left[\left(\frac{\partial\psi}{\partial y}+1\right)Cos\Theta+\delta\frac{\partial\psi}{\partial x}Sin\Theta\right]-\frac{1}{K}\left(\frac{\partial\psi}{\partial y}+1\right)+\frac{Re}{Fr}Sin\,\alpha\end{aligned} \tag{16}$$

$$\begin{aligned}Re\,\delta^2\left[\frac{\partial v}{\partial t}+(u+1)\frac{\partial v}{\partial x}+\frac{1}{\delta}v\frac{\partial v}{\partial y}\right]&=-\frac{\partial p}{\partial y}+\delta\frac{\partial}{\partial y}(S_{yy})+\delta\frac{\partial}{\partial x}(S_{xy})\\&-\frac{\sigma'B_0^2d_1^2\delta}{\mu}Sin\Theta\left[(u+1)Cos\Theta+v\,Sin\Theta\right]-\frac{\delta}{K}v+\frac{\rho gd_1^2}{\mu}\delta Cos\,\alpha\end{aligned} \tag{17}$$

$$Re\,Pr\left[\delta(u+1)\frac{\partial\theta}{\partial x}+v\frac{\partial\theta}{\partial y}\right]=\left[\delta^2\frac{\partial^2\theta}{\partial x^2}+\frac{\partial^2\theta}{\partial y^2}\right]$$
$$+Ec\,Pr\left[2\left\{\delta^2\left(\frac{\partial u}{\partial x}\right)^2+\left(\frac{\partial v}{\partial y}\right)^2\right\}+\left(\frac{\partial u}{\partial y}+\delta\frac{\partial v}{\partial x}\right)^2\right] \quad (18)$$
$$+\frac{Ec\,Pr}{\gamma^2}\left[\left(\delta^2\frac{\partial^2 u}{\partial x^2}+\frac{\partial^2 u}{\partial y^2}\right)^2+\left(\delta^2\frac{\partial^2 v}{\partial x^2}+\frac{\partial^2 v}{\partial y^2}\right)^2\right]+\beta$$

Similarly Eqs.(11) and (12) become

$$\frac{\partial\psi}{\partial y}=0\ ,\ \theta=0 \quad at \quad y=h_1=-1-m(x+t)-aSin[2\pi x+\phi] \quad (19)$$

$$\frac{\partial\psi}{\partial y}=0\ ,\ \theta=1 \quad at \quad y=h_2=1+m(x+t)+bSin[2\pi x] \quad (20)$$

Eqs.(16)-(18), under long wavelength (δ □ 1) and low "Reynolds number" approximation is given a following equations:

$$\frac{\partial p}{\partial x}=\frac{\partial}{\partial y}(S_{xy})-M^2(Cos\Theta)^2\left(\frac{\partial\psi}{\partial y}+1\right)-\frac{1}{K}\left(\frac{\partial\psi}{\partial y}+1\right)+\frac{Re}{Fr}Sin\alpha \quad (21)$$

$$\frac{\partial p}{\partial y}=0 \quad (22)$$

$$\frac{\partial^2\theta}{\partial y^2}+Br\left[\left(\frac{\partial^2\psi}{\partial y^2}\right)^2+\frac{1}{\gamma^2}\left(\frac{\partial^3\psi}{\partial y^3}\right)^2\right]+\beta=0 \quad (23)$$

After removing the dimensions (by Eq.(14)), the boundary conditions are as follows:

$$\psi=-\frac{F}{2}\ ,\ \frac{\partial\psi}{\partial y}=0\ ,\theta=0 \quad at\ y=h_1 \quad (24)$$

$$\psi=\frac{F}{2}\ ,\frac{\partial\psi}{\partial y}=0\ ,\theta=1 \quad at\ y=h_2 \quad (25)$$

Where (F) represents the dimensionless mean flows . Here

$$F(x,t)=Q+aSin[2\pi x+\phi]+bSin[2\pi x] \quad (26)$$

In which (Q) is the dimensionless mean flows in the wave frame, Here

$$F=\int_{h_1(x)}^{h_2(x)}\frac{\partial\psi}{\partial y}dy=\psi(h_2)-\psi(h_1)\ ,Q\equiv\frac{\tilde{q}}{cd} \quad (27)$$

$$\tilde{q}=\int_{\tilde{H}_1(\tilde{X},t)}^{\tilde{H}_2(\tilde{X},t)}\tilde{u}(\tilde{x},\tilde{y})d\tilde{y} \quad (28)$$

Here it is referred that the conditions of (ψ) which is satisfy Eq.(27) and the conditions on $\partial\psi/\partial y$ means that the flow is no-slip.

Now we have $\left\{ S_{xy} = \dfrac{\dfrac{\partial^2\psi}{\partial y^2}}{1-\xi\left(\dfrac{\partial^2\psi}{\partial y^2}\right)^2} \right\}$, where $\xi = \left(\mu_1^2 - \lambda_1^2\right)$ is the pseudoplastic fluid parameter[27].

By recompense this equation in Eq.(21)

$$\frac{\partial p}{\partial x} = \frac{\partial}{\partial y}\left(\frac{\dfrac{\partial^2\psi}{\partial y^2}}{1-\xi\left(\dfrac{\partial^2\psi}{\partial y^2}\right)^2}\right) - M^2(Cos\Theta)^2\left(\frac{\partial\psi}{\partial y}+1\right) - \frac{1}{K}\left(\frac{\partial\psi}{\partial y}+1\right) + \frac{Re}{Fr}Sin\alpha$$

$$= \frac{\partial}{\partial y}\left[\frac{\partial^2\psi}{\partial y^2}\left(1+\xi\left(\frac{\partial^2\psi}{\partial y^2}\right)^2+\dots\right)\right] - M^2(Cos\Theta)^2\left(\frac{\partial\psi}{\partial y}+1\right) - \frac{1}{K}\left(\frac{\partial\psi}{\partial y}+1\right) + \frac{Re}{Fr}Sin\alpha$$

$$\Rightarrow \quad \frac{\partial p}{\partial x} = \left[\frac{\partial^3\psi}{\partial y^3} + 3\xi\left(\frac{\partial^2\psi}{\partial y^2}\right)^2\frac{\partial^3\psi}{\partial y^3}\right] - M^2(Cos\Theta)^2\left(\frac{\partial\psi}{\partial y}+1\right) - \frac{1}{K}\left(\frac{\partial\psi}{\partial y}+1\right) + \frac{Re}{Fr}Sin\alpha \qquad (29)$$

By Eq.(22) and Eq.(29) we get :

$$\frac{\partial^4\psi}{\partial y^4} + 3\xi\left(\frac{\partial^2\psi}{\partial y^2}\right)^2\frac{\partial^4\psi}{\partial y^4} + 6\xi\left(\frac{\partial^3\psi}{\partial y^3}\right)^2\frac{\partial^2\psi}{\partial y^2} - \left[M^2(Cos\Theta)^2 + \frac{1}{K}\right]\frac{\partial^2\psi}{\partial y^2} = 0 \qquad (30)$$

5. Coefficient of heat transfer

The coefficient of heat transfer at the upper wall is offered by the following law

$$Z(x) = \frac{\partial h_1}{\partial x}\left(\frac{\partial\theta}{\partial y}\right)_{y=h_1} \qquad (31)$$

And the coefficient of heat transfer at the lower wall is given by

$$Z(x) = \frac{\partial h_2}{\partial x}\left(\frac{\partial\theta}{\partial y}\right)_{y=h_2} \qquad (32)$$

6. Average pressure rise

The non-dimensional average rise in pressure (Δp) per wavelength is given by

$$\Delta p = \int_0^1 \frac{\partial p}{\partial x}\,dx \qquad (33)$$

7. Solution technique

In order to solve the equations (23), (29) and (30), we expand the flow quantities in a power series of (ξ) as follows

$$\begin{aligned} \psi &= \psi_0 + \xi\psi_1 + \xi^2\psi_2 + \dots \;, \\ P &= P_0 + \xi P_1 + \xi^2 P_2 + \dots \;, \\ \theta &= \theta_0 + \xi\theta_1 + \xi^2\theta_2 + \dots \;, \\ Z &= Z_0 + \xi Z_1 + \xi^2 Z_2 + \dots \;. \end{aligned} \qquad (34)$$

Substituting Eq.(34) into Eqs.(23),(29) and (30) we get:

7.1. Zero order

The corresponding systems at zero order are

$$\frac{\partial^2\theta_0}{\partial y^2}+Br\left[\left(\frac{\partial^2\psi_0}{\partial y^2}\right)^2+\frac{1}{\gamma^2}\left(\frac{\partial^3\psi_0}{\partial y^3}\right)^2\right]+\beta=0 \tag{35}$$

$$\frac{dP_0}{dx}=\frac{\partial^3\psi_0}{\partial y^3}-N_1^2\left(\frac{\partial\psi_0}{\partial y}+1\right)+\frac{Re}{Fr}Sin\alpha \tag{36}$$

$$\frac{\partial^4\psi_0}{\partial y^4}-N_1^2\frac{\partial^2\psi_0}{\partial y^2}=0 \tag{37}$$

Where $N_1^2=M^2(Cos\Theta)^2+\frac{1}{K}$ with corresponding boundary conditions

$$\psi_0=-\frac{F_0}{2},\frac{\partial\psi_0}{\partial y}=0\,,\theta_0=0\ \ at\ \ y=h_1 \qquad and \quad \psi_0=\frac{F_0}{2},\frac{\partial\psi_0}{\partial y}=0\,,\theta_0=1\ \ at\ \ y=h_2\,.$$

7.2. First order

The corresponding systems at first order are

$$\frac{\partial^2\theta_1}{\partial y^2}+2Br\left[\frac{\partial^2\psi_0}{\partial y^2}\frac{\partial^2\psi_1}{\partial y^2}+\frac{1}{\gamma^2}\frac{\partial^3\psi_0}{\partial y^3}\frac{\partial^3\psi_1}{\partial y^3}\right]=0 \tag{38}$$

$$\frac{dP_1}{dx}=\frac{\partial^3\psi_1}{\partial y^3}+3\left(\frac{\partial^2\psi_0}{\partial y^2}\right)^2\frac{\partial^3\psi_0}{\partial y^3}-N_2^2\frac{\partial\psi_1}{\partial y} \tag{39}$$

$$\frac{\partial^4\psi_1}{\partial y^4}+3\left(\frac{\partial^2\psi_0}{\partial y^2}\right)^2\frac{\partial^4\psi_0}{\partial y^4}+6\frac{\partial^3\psi_0}{\partial y^3}\frac{\partial^2\psi_0}{\partial y^2}-N_1^2\frac{\partial^2\psi_1}{\partial y^2}=0 \tag{40}$$

Where $N_2^2=M^2+\frac{1}{K}$ with corresponding boundary condition

$$\psi_1=-\frac{F_1}{2},\frac{\partial\psi_1}{\partial y}=0\,,\theta_1=0\ \ at\ \ y=h_1 \text{ and } \ \psi_1=\frac{F_1}{2},\frac{\partial\psi_1}{\partial y}=0\,,\theta_1=0\ \ at\ \ y=h_2$$

8. Perturbation Solutions

By using the series solution (regular perturbation technique), it is found that the solution of the zeroth and first Solutions system is given:

8.1. Zeroth- Order Solution

The solutions of Eqs.(35),(36),(37) are given by

$$\psi_0=C_3+C_4y+\frac{C_1e^{-yN_1}+C_2e^{yN_1}}{N_1^2} \tag{41}$$

$$u_0=C_4+\frac{-C_1e^{-yN_1}N_1+C_2e^{yN_1}N_1}{N_1^2} \tag{42}$$

$$\frac{\partial P_0}{\partial x}=\frac{Re\,Sin\,\alpha}{Fr}+\frac{-C_1e^{-yN_1}N_1^3+C_2e^{yN_1}N_1^3}{N_1^2}-N_1^2\left(1+C_4+\frac{-C_1e^{-yN_1}N_1+C_2e^{yN_1}N_1}{N_1^2}\right) \tag{43}$$

$$\theta_0=d_1+d_2y+\frac{\left(\begin{array}{c}-\dfrac{BrC_1^2e^{-2yN_1}\left(\gamma^2+N_1^2\right)}{2N_1}-\dfrac{BrC_2^2e^{2yN_1}\left(\gamma^2+N_1^2\right)}{2N_1}\\ \hline -y^2N_1\left(\gamma^2\left(2BrC_1C_2+\beta\right)-2BrC_1C_2N_1^2\right)\end{array}\right)}{2\gamma^2N_1} \tag{44}$$

$$Z_{0h_1}=\frac{\partial h_1}{\partial x}\left(\frac{\partial\theta_0}{\partial y}\right)_{y=h_1}$$

$$=\left(-m-2a\pi Cos\left[2\pi x+\phi\right]\right)(d_2+\frac{1}{2\gamma^2N_1}(BrC_1^2e^{-2yN_1}\left(\gamma^2+N_1^2\right) \tag{45}$$

$$-Br\ C_2^2e^{2yN_1}\left(\gamma^2+N_1^2\right)-2yN_1(\gamma^2\left(2BrC_1C_2+\beta\right)-2BrC_1C_2N_1^2)))$$

Moreover, the non-dimensionless average rise in (Δp_0) per wavelength at this order is

$$\Delta p_0=\int_0^1\left(\frac{\partial p_0}{\partial x}\right)_{y=0}dx \tag{46}$$

8.2. First-Order Solution

Substituting the zeroth-order solution into Eqs.(38), (39),(40) and then we solve the resulting system with the existence of "boundary conditions", we get

$$\psi_1=C_7+yC_8+\frac{1}{8N_1}\left(\begin{array}{l}-2e^{-yN_1}\left(-6C_1^2C_2y-\dfrac{15C_1^2C_2+4C_5}{N_1}\right)\\ +2e^{yN_1}\left(-6C_1C_2^2y+\dfrac{15C_1C_2^2+4C_6}{N_1}\right)-\dfrac{C_1^3e^{-3yN_1}}{N_1}-\dfrac{C_2^3e^{3yN_1}}{N_1}\end{array}\right) \tag{47}$$

$$u_1=\frac{\partial\psi_1}{\partial y}=C_8+\frac{1}{8N_1}\left(\begin{array}{l}3C_1^3e^{-3yN_1}+12C_1^2C_2e^{-yN_1}-12C_1C_2^2e^{yN_1}-3C_1^3e^{3yN_1}\\ +2e^{-yN_1}\left(-6C_1^2C_2y-\dfrac{15C_1^2C_2+4C_5}{N_1}\right)N_1\\ +2e^{yN_1}\left(-6C_1C_2^2y+\dfrac{15C_1C_2^2+4C_6}{N_1}\right)N_1\end{array}\right) \tag{48}$$

$$\frac{dp_1}{dx}=\frac{3\left(C_1e^{-yN_1}N_1^2+C_2e^{yN_1}N_1^2\right)^2\left(-C_1e^{-yN_1}N_1^3+C_2e^{yN_1}N_1^3\right)}{N_1^6}$$
$$+\frac{1}{8N_1}(27C_1^3e^{-3yN_1}N_1^2+36C_1^2C_2e^{-yN_1}N_1^2-36C_1C_2^2e^{yN_1}N_1^2-27C_2^3e^{3yN_1}N_1^2$$
$$+2e^{-yN_1}\left(-6C_1^2C_2y-\frac{15C_1^2C_2+4C_5}{N_1}\right)N_1^3+2e^{yN_1}\left(-6C_1C_2^2y+\frac{15C_1C_2^2+4C_6}{N_1}\right)N_1^3) \quad (49)$$
$$-(C_8+\frac{1}{8N_1}(3C_1^3e^{-3yN_1}+12C_1^2C_2e^{-yN_1}-12C_1C_2^2e^{yN_1}-3C_2^3e^{3yN_1}$$
$$+2e^{-yN_1}\left(-6C_1^2C_2y-\frac{15C_1^2C_2+4C_5}{N_1}\right)N_1+2e^{yN_1}\left(-6C_1C_2^2y+\frac{15C_1C_2^2+4C_6}{N_1}\right)N_1))N_2^2$$

$$\theta_1=d_3+yd_4+\frac{1}{16\gamma^2N_1}Br(\frac{9C_1^4e^{-4yN_1}\left(\gamma^2+3N_1^2\right)}{4N_1}+\frac{9C_2^4e^{4yN_1}\left(\gamma^2+3N_1^2\right)}{4N_1}$$
$$-8y^2N_1((3C_1^2C_2^2+2C_2C_5+2C_1C_6)\gamma^2+\left(3C_1^2C_2^2-2C_2C_5-2C_1C_6\right)N_1^2)$$
$$+2C_1e^{-2yN_1}(-6C_1^2C_2y\left(\gamma^2+N_1^2\right)-\frac{9C_1^2C_2\gamma^2+8C_5\gamma^2+33C_1^2C_2N_1^2+8C_5N_1^2}{2N_1}) \quad (50)$$
$$+2e^{2yN_1}(6C_1C_2^3y\left(\gamma^2+N_1^2\right)-\frac{C_2\left(9C_1C_2^2\gamma^2+8C_6\gamma^2+33C_1C_2^2N_1^2+8C_6N_1^2\right)}{2N_1}))$$

The heat transfer coefficient at this order is given by

$$\boldsymbol{Z}_{1h_1}=\frac{\partial h_1}{\partial x}\left(\frac{\partial\theta_1}{\partial y}\right)_{y=h_1}=\left[-m-2a\pi Cos\left(2\pi x+\phi\right)\right](d4+\frac{1}{16\gamma^2N_1}Br(-12C_1^3C_2e^{-2yN_1}$$
$$\left(\gamma^2+N_1^2\right)+12C_1C_2^3e^{2yN_1}\left(\gamma^2+N_1^2\right)-9C_1^4e^{-4yN_1}\left(\gamma^2+3N_1^2\right)+9C_2^4e^{4yN_1}\left(\gamma^2+3N_1^2\right)$$
$$-16yN_1\left(\left(3C_1^2C_2^2+2C_2C_5+2C_1C_6\right)\gamma^2+\left(3C_1^2C_2^2-2C_2C_5-2C_1C_6\right)N_1^2\right) \quad (51)$$
$$-4C_1e^{-2yN_1}N_1\left(-6C_1^2C_2y\left(\gamma^2+N_1^2\right)-\frac{9C_1^2C_2\gamma^2+8C_5\gamma^2+33C_1^2C_2N_1^2+8C_5N_1^2}{2N_1}\right)$$
$$+4e^{2yN_1}N_1\left(6C_1C_2^3y\left(\gamma^2+N_1^2\frac{C_2\left(9C_1C_2^2\gamma^2+8C_6\gamma^2+33C_1C_2^2N_1^2+8C_6N_1^2\right)}{2N_1}\right)\right)$$

The non-dimensional the pressure rise (Δp_1) a per wavelength is given as follows:

$$\Delta p_1=\int_0^1\left(\frac{\partial p_1}{\partial x}\right)_{y=0}dx \quad (52)$$

9. Nusselt number

The non-dimensionless Nusselt number is obtained by the relation $Nu=-\frac{\partial\theta}{\partial y}$, this leads

$$Nu = -d2 - \frac{1}{2\gamma^2 N_1}(BrC_1^{\ 2}e^{-2yN_1}\left(\gamma^2 + N_1^{\ 2}\right) - BrC_2^{\ 2}e^{2yN_1}\left(\gamma^2 + N_1^{\ 2}\right)$$
$$-2yN_1(\gamma^2\left(2BrC_1C_2 + \beta\right) - 2BrC_1C_2N_1^{\ 2})) - \xi(d4 + \frac{1}{16\gamma^2 N_1}Br(-12C_1^{\ 3}C_2e^{-2yN_1}$$
$$\left(\gamma^2 + N_1^{\ 2}\right) + 12C_1C_2^{\ 3}e^{2yN_1}\left(\gamma^2 + N_1^{\ 2}\right) - 9C_1^{\ 4}e^{-4yN_1}\left(\gamma^2 + 3N_1^{\ 2}\right) + 9C_2^{\ 4}e^{4yN_1}$$
$$\left(\gamma^2 + 3N_1^{\ 2}\right) - 16yN_1(\left(3C_1^{\ 2}C_2^{\ 2} + 2C_2C_5 + 2C_1C_6\right)\gamma^2 + \left(3C_1^{\ 2}C_2^{\ 2} - 2C_2C_5 - 2C_1C_6\right)N_1^{\ 2})$$
$$-4C_1e^{-2yN_1}N_1\left(-6C_1^{\ 2}C_2y\left(\gamma^2 + N_1^{\ 2}\right) - \frac{9C_1^{\ 2}C_2\gamma^2 + 8C_5\gamma^2 + 33C_1^{\ 2}C_2N_1^{\ 2} + 8C_5N_1^{\ 2}}{2N_1}\right)$$
$$+4e^{2yN_1}N_1\left(6C_1C_2^{\ 3}y\left(\gamma^2 + N_1^{\ 2}\right) - \frac{C_2\left(9C_1C_2^{\ 2}\gamma^2 + 8C_6\gamma^2 + 33C_1C_2^{\ 2}N_1^{\ 2} + 8C_6N_1^{\ 2}\right)}{2N_1}\right)))$$

10. Numerical results and discussion

In this section, we will be discussing the behavior and graphical description of various parameters like "Hartmann number" (M), "Darcy number" (K) , "couple stress" parameter (γ), "heat generation parameter" (β), "inclination of magnetic field" (Θ), "Brinkman number" (Br), "Reynolds number" (Re), "Froude number" (Fr), "Prandle number" (Pr), "Eckert number" (Ec), "mean flow rate"(Q) and "incline angle" of the channel (α) and their effect on pressure gradient, temperature, "Nusselt number", heat transfer coefficient, pressure rise and the behavior of variously wave forms, and discussed in the form of streamlines. And has been using Mathematica program to find the results numerically and graphically.

10.1. The Pressure gradient distribution

In this section, the Figs.(2),(3),(6) and (7) Turns out that the pressure gradient (dy/dx) increases with the increase in the (Re), (Q) ,(K) and (ϕ) .While the behavior is reflected where the gradient pressure (dy/dx) is decreasing with an increase in (Fr), (M) As shown in the Figs.(4) and (5) .

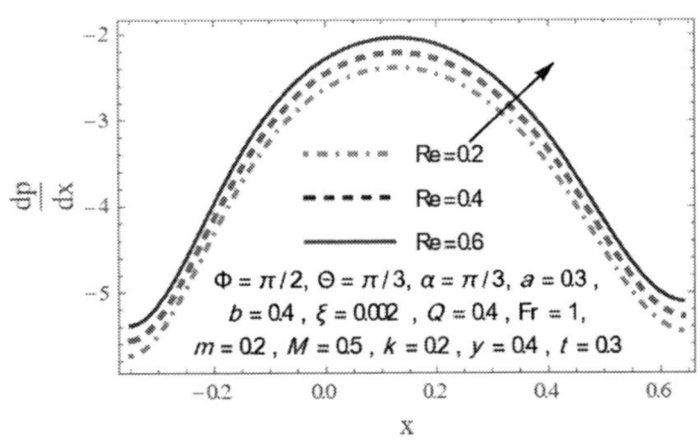

Figure 2. Different of pressure gradient with variation of Re

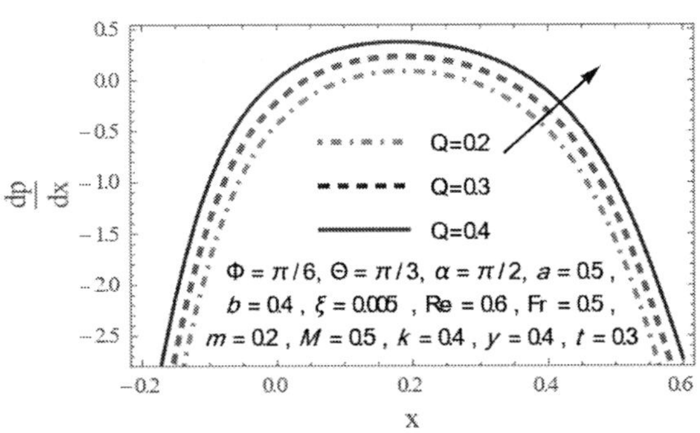

Figure 3. Different of pressure gradient with variation of Q

IOP Conf. Series: Journal of Physics: Conf. Series **1032** (2018) 012043 doi:10.1088/1742-6596/1032/1/012043

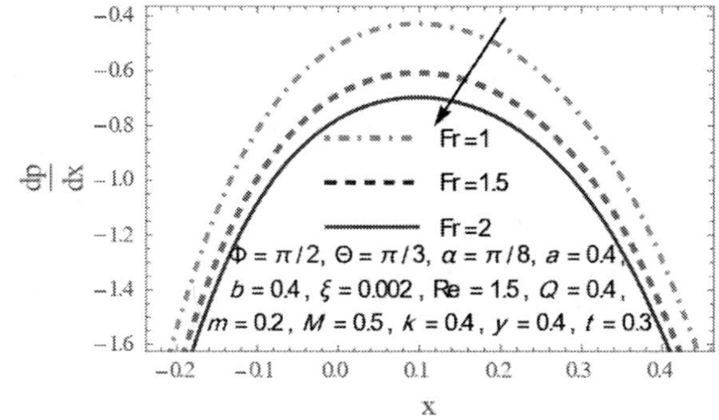

Figure 4. Different of pressure gradient with variation of Fr

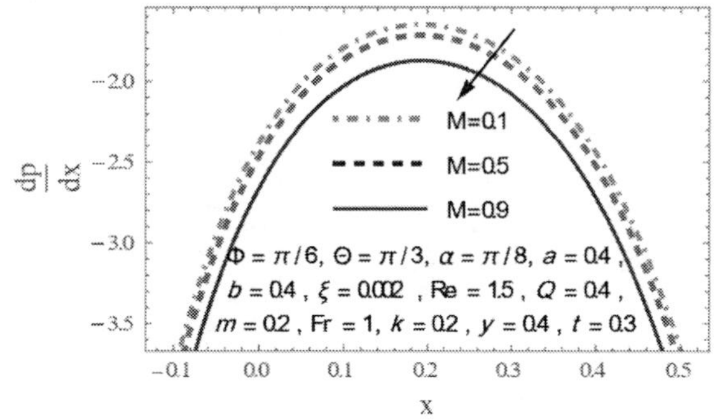

Figure 5. Different of pressure gradient with variation of M

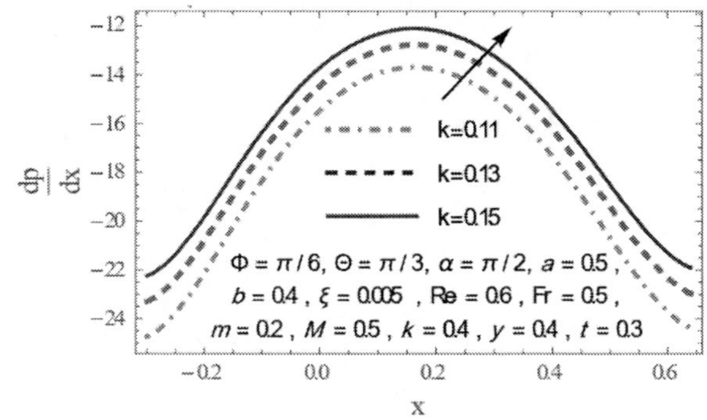

Figure 6. Different of pressure gradient with variation of K

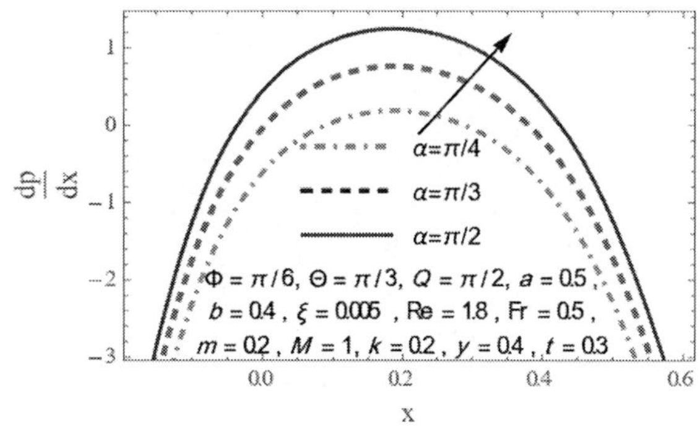

Figure 7. Different of pressure gradient with variation of α

10.2. Pumping Characteristics

The Figs.(8)-(13) are observed the behavior of average pressure rise (Δp) against mean flow rate (Q) for different flow parameters. We partition the region into five parts as follows: **(I)** Peristaltic pumping area (Δp > 0, Q >0). In this area, the positive value of (Q) is wholly due to the peristalsis, and overcoming the pressure rise.**(II)** Augmented pumping region when (Δp < 0,Q > 0). In this region, a negative pressure difference assists the flow due to the peristalsis of the walls.**(III)** (Δp < 0, Q < 0) is the region where there is no flow . **(IV)** The region (Δp > 0,Q < 0) is a backward pumping region. **(V)** If Δp = 0 then this region is corresponds to the free pumping region.This flow occurs due to peristalsis in the walls. It is seen in Fig.(8), the increase of (K) leads to that (Δp) increasing in a co-pumping region (Δp < 0,Q > 0), but the case is reverse in the peristaltic pumping region(Δp > 0,Q > 0), which (Δp) is a decreased with an increase (K) . Fig. (9) shows that is inverse to what happened in Fig. (8) completely,where (Δp) decreases in a co-pumping region (Δp< 0,Q > 0) and (Δp) is an enhance in the peristaltic pumping region (Δp > 0,Q > 0) with increase of (M). The Figs.(10), (12) and (13) demonstrates the effect of (Re, α, ξ) on (Δp). It is noted that (Δp) is increasing in all the regions mentioned above respectively.And we note the opposite in Fig. (11) shows an influence of (Fr) on average "pressure rise" (Δp). We notice from the graph the pumping rate (Δp) is a decreasing with increasing of (Fr) in all the regions mentioned.

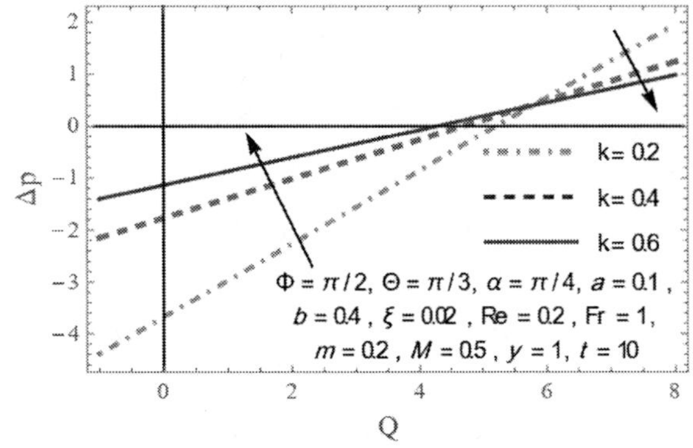

Figure 8. Pressure rise Δp versus Q with variation of K

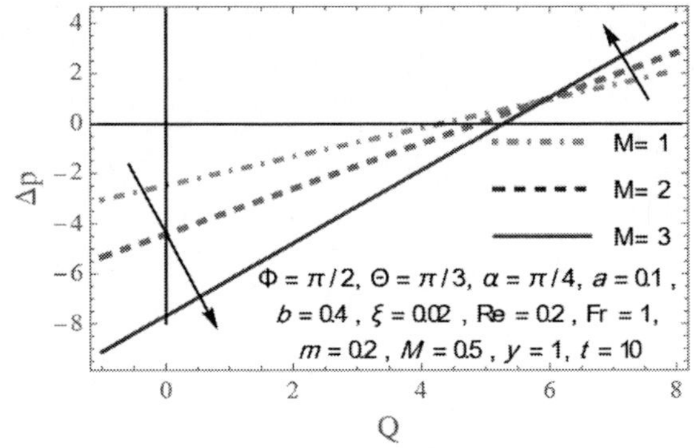

Figure 9. Pressure rise Δp versus Q with variation of M

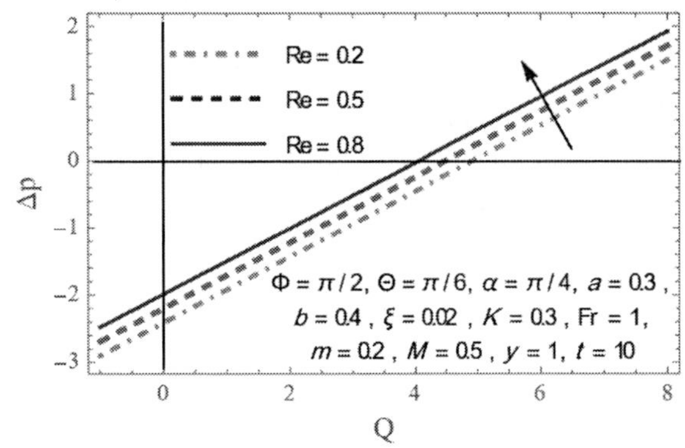

Figure 10. Pressure rise Δp versus Q with variation of Re

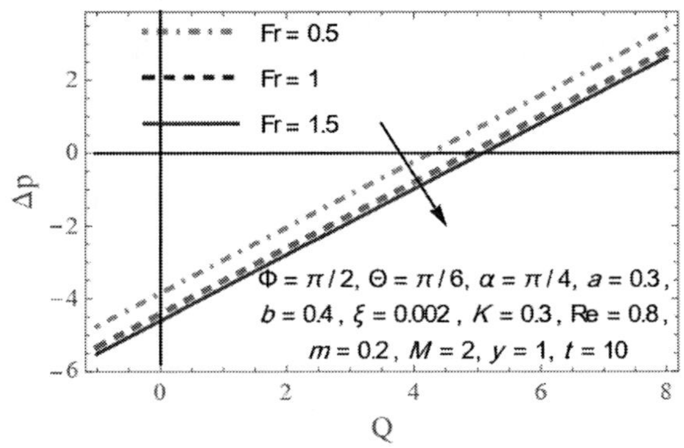

Figure 11. Pressure rise Δp versus Q with variation of Fr

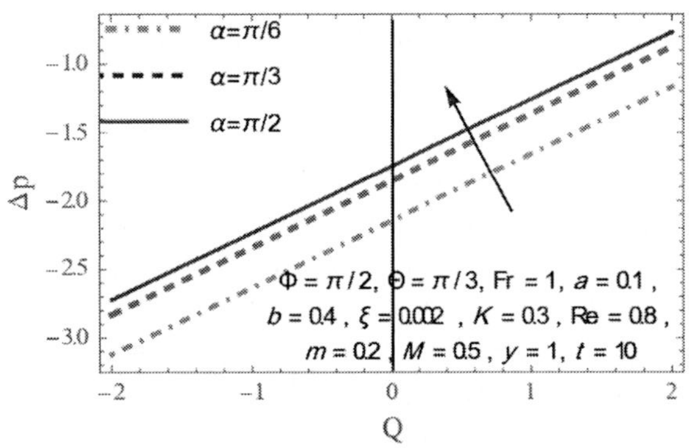

Figure 12. Pressure rise Δp versus Q with variation of α

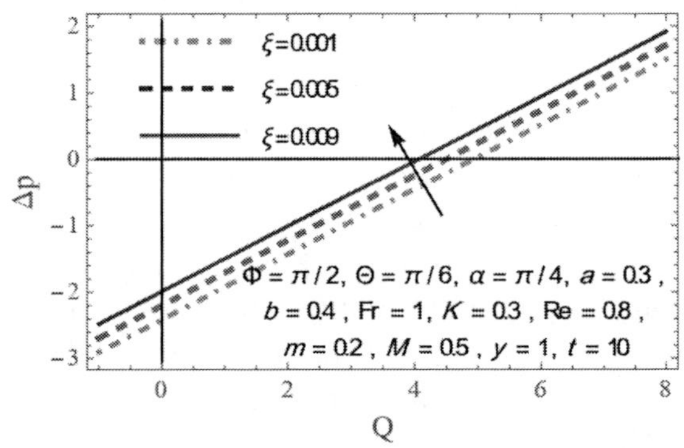

Figure 13. Pressure rise Δp versus Q with variation of ξ

10.3. Velocity Profile

The graphs are sketched in Figs.(14)-(19) at the fixed values of $x = 0.3$ and $t = 5$ .The aim of this part is to discuss the influence of the non-uniform parameter (m), Hartmann number (M), the inclination angle of the magnetic field (Θ), phase angle (ϕ), the pseudoplastic fluid parameter (ξ) and mean flow rate (Q) on the velocity (u).It is note that from figures, the velocity profile traces a parabola form trajectory with maximum value occurring in mid of the channel.From the Fig.(14), the velocity increases with increase in (m). Fig.(15) that, the velocity decreasing near the mid of the channel with an increase in (M). When observing Fig.(16) that, the behavior of (Θ) on the velocity profile is completely reverse compared to the case in (M). Furthermore it, we notice the increase in velocity when the center of the channel with the increase in (Θ). It is clear from Fig. (17) that, the velocity decreases with enhance in (ϕ). The velocity drop in the middle of the channel and increasing near the walls with an increase of parameter (ξ) (see Fig. (18)). This confirms that the velocity decreases in the middle of the channel with a porous medium in it, so the more porous in a porous medium will be less resistance to the flow of fluid,this makes a velocity is the greater extent.We observed from Fig.(19) the velocity profile for the parameter (Q) which depicts that the velocity(u) enhance with an increase in (Q).

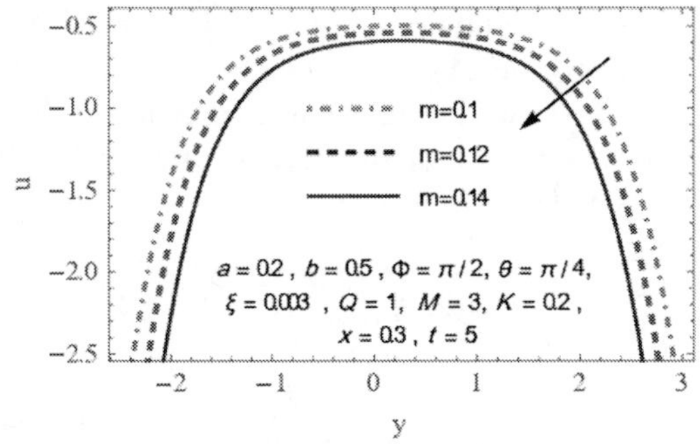

Figure 14. Velocity u versus y- direction for different values m

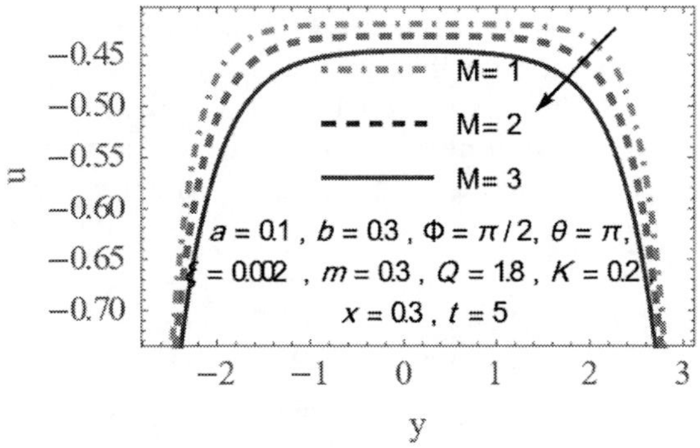

Figure 15. Velocity u versus y- direction for different values M

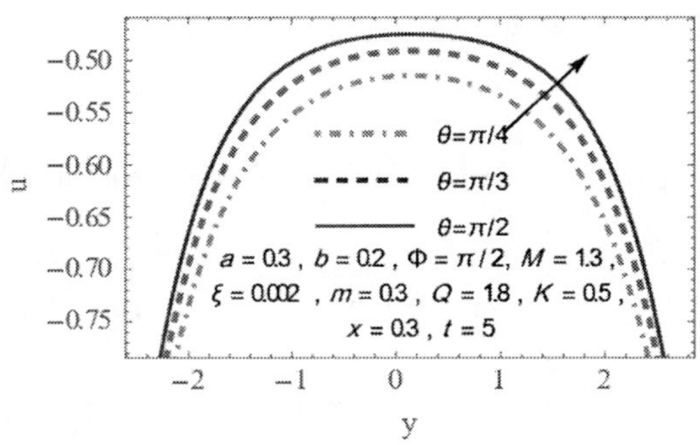

Figure 16. Velocity u versus y- direction for different values θ

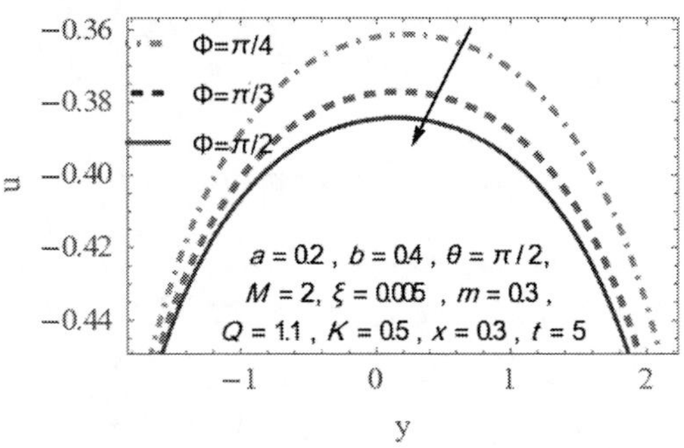

Figure 17. Velocity u versus y- direction for different values ϕ

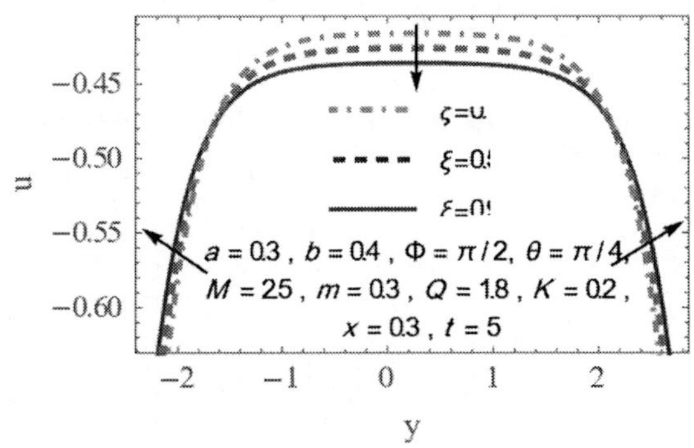

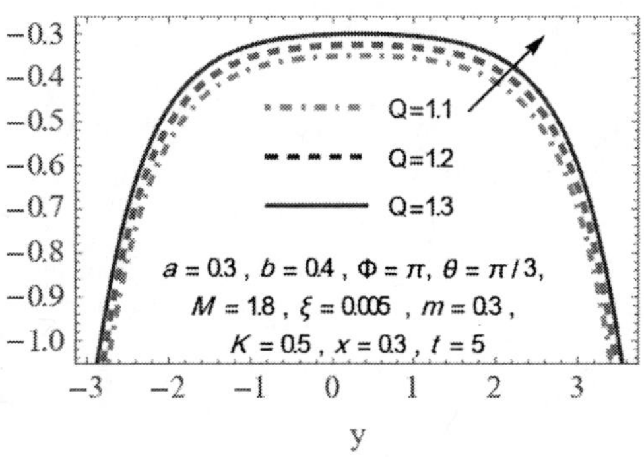

Figure 18. Velocity u versus y- direction for different values ξ

Figure 19. Velocity u versus y- direction for different values Q

10.4. Temperature distribution

This subsection devotes the effect of Hartmann number (M), inclination of the magnetic field (Θ), Darcy number (K), heat generation parameter (β), Brinkman number (Br) and couple stress fluid parameter (γ) on the temperature of the fluid under the influence of peristalsis is appears in Figs.(20)-(25) for the fixed values of $x = 0.3$ and $t = 5$, which are depicts in the middel part of the channel.It is noticed from Fig.(20), the temperature increases with an increase of (M). Also, the temperature of the fluid is increased with an increase of (β) (see Fig.(21)).We see for Fig.(22), the temperature decreasing with increase in (K) in the center part of the channel and, while opposite behavior is observed near channel boundaries.The temperature can be higher when there is a porous medium.This applies to the rest of the parameters(Θ) and (γ), where it is noticed from Figs.(23) and (24), the temperature decreases with increases of the inclination of the magnetic field (Θ) and couple stress parameter (γ). It is observed from Fig.(25) that, the temperature increases with increasing of (Br) in core (center) channel.This case came because an increase in values of (Br) rises the resistance offered by shear inflow which in turn increases the heat generation due to the viscous dissipation effects and hence the temperature of the fluid increases, but the opposite near the boundary of the channel.

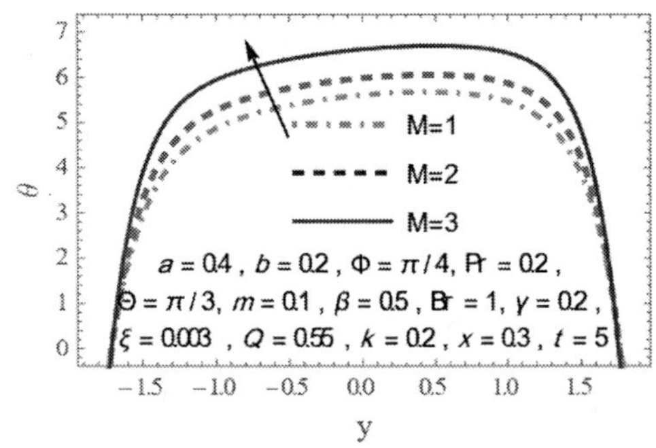

Figure 20. Temperature distribution θ for different values M

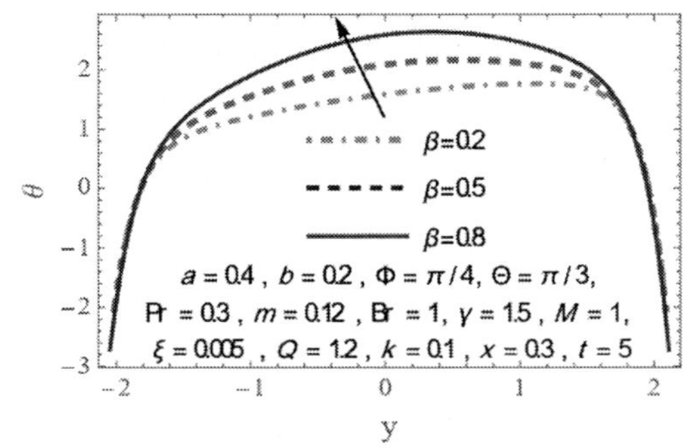

Figure 21. Temperature distribution θ for different values β

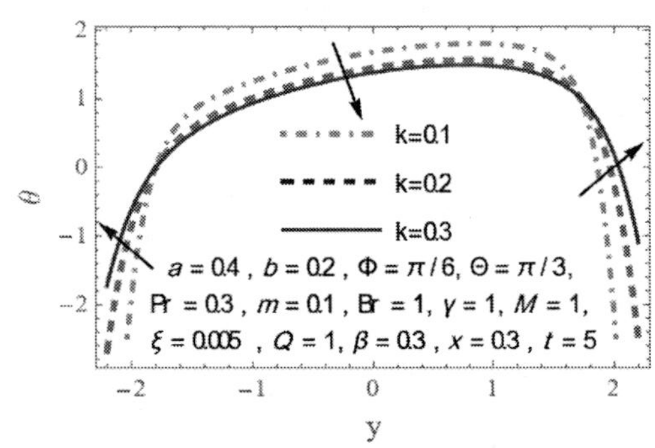

Figure 22. Temperature distribution θ for different values K

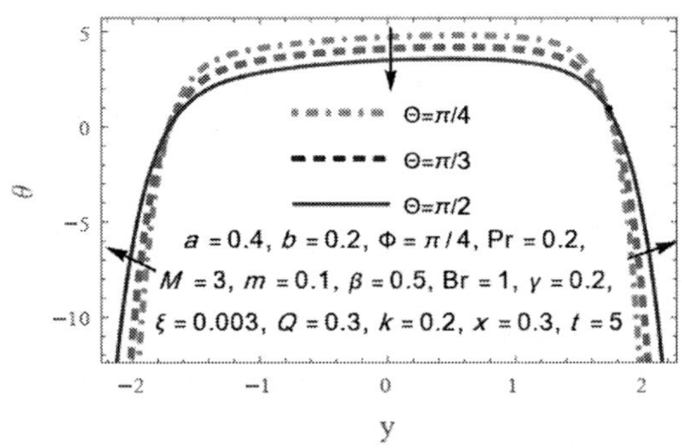

Figure 23. Temperature distribution θ for different values Θ

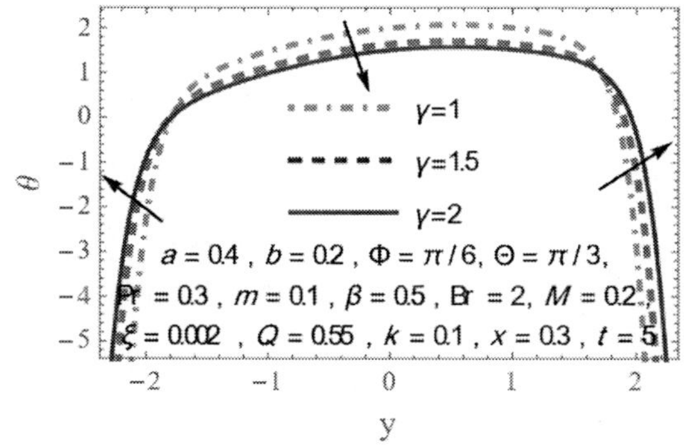

Figure 24. Temperature distribution θ for different values γ

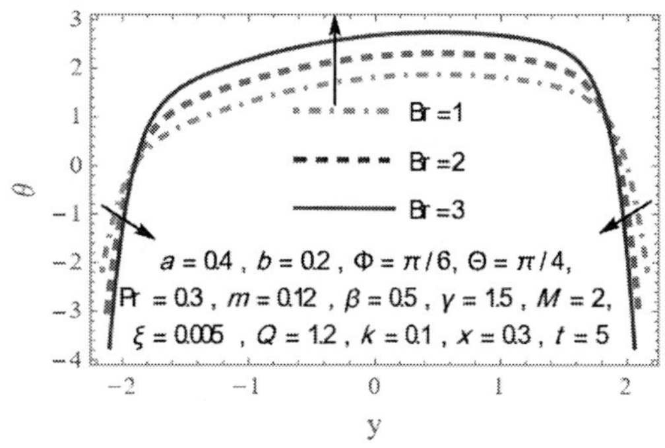

Figure 25. Temperature distribution θ for different values Br

10.5. Heat transfer coefficient

This section devotes the effect of "Hartmann number" (M), heat generation parameter (β), "Brinkman number" (Br) and "couple stress" fluid parameter (γ) on the "heat transfer coefficient" Z_{h1} at the wall, The visualization of heat transfer coefficient at the upper wall under the influence of the peristalsis in a channel. It is shown from the figures (26),(27),(28) that, heat transfer coefficient Z_{h1} is wobbling. This is due to spread of sinusoidal waves along the channel walls, the heat transfer coefficient Z_{h1} increases with the increase of (Br), (M), (β) respectively. It is noticed from Fig.(29), the absolute value of heat transfer coefficient Z_{h1} decreases with an increase of (γ).

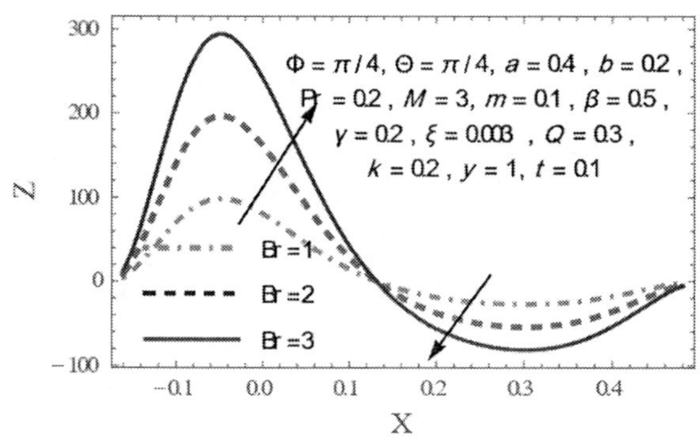

Figure 26. Heat transfer coefficient Z_{h1} for different values of M

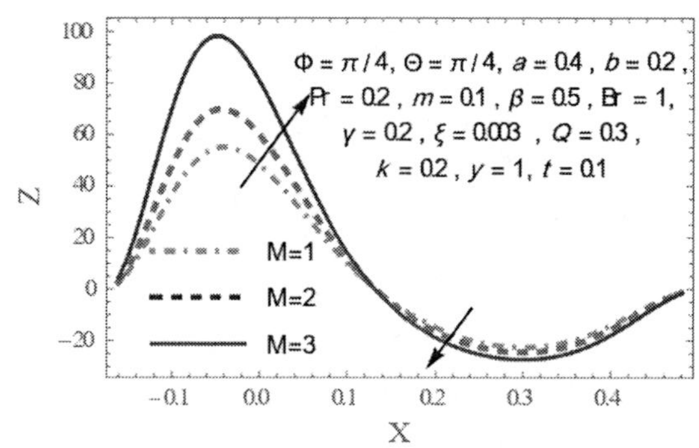

Figure27. Heat transfer coefficient Z_{h1} for different values of M

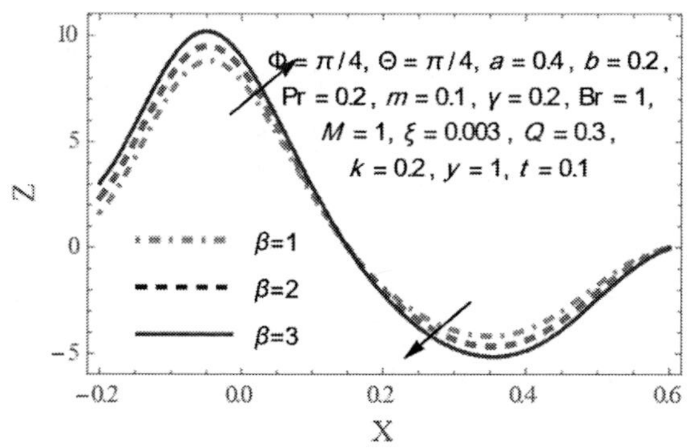

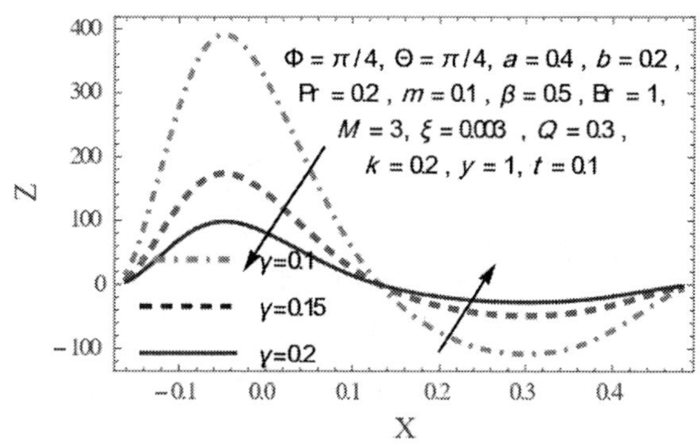

IOP Conf. Series: Journal of Physics: Conf. Series **1032** (2018) 012043 doi:10.1088/1742-6596/1032/1/012043

Figure 28. Heat transfer coefficient Z_{h1} for different values of β

Figure 29. Heat transfer coefficient Z_{h1} for different values of γ

10.6. Nusselt number

In this paragraph the Figs.(30)-(35) denotes the effect of Hartmann number (M), the inclination of the magnetic field (Θ), Darcy number (K), mean flow rate (Q), Brinkman number (Br) and couple stress fluid parameter (γ) on the "Nusselt number". It is noticed of a Figs.(30)-(33), the "Nusselt number" grow with enhance of (Br), (K), (Θ) and (Q) respectively. While in Figs.(34) and (35) the case is inverse, with an enhancing of (γ) and (M) the "Nusselt number" (Nu) is decreasing.

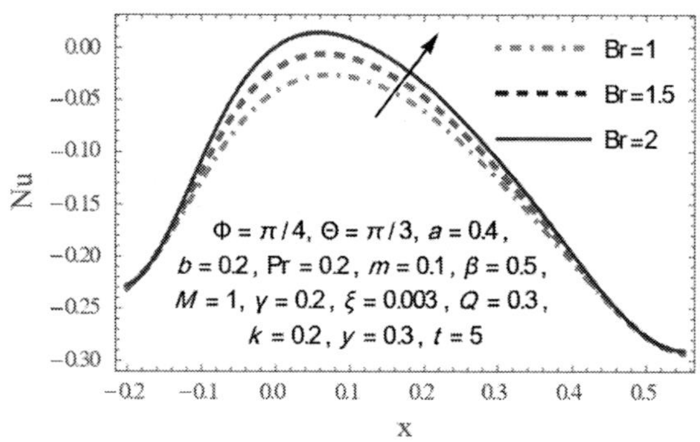

Figure 30. Nusselt number Nu for different values of Br

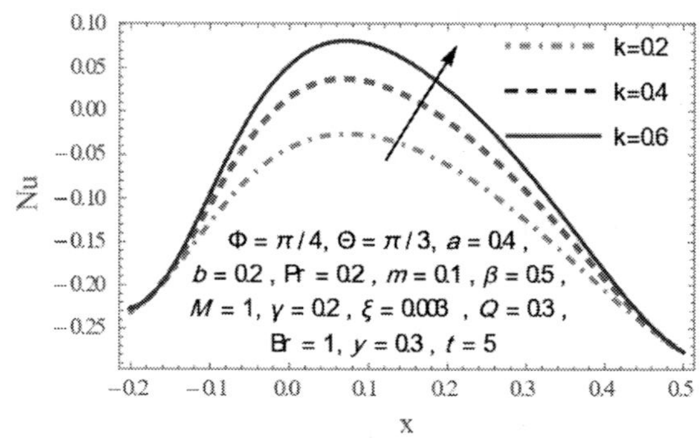

Figure 31. Nusselt number Nu for different values of K

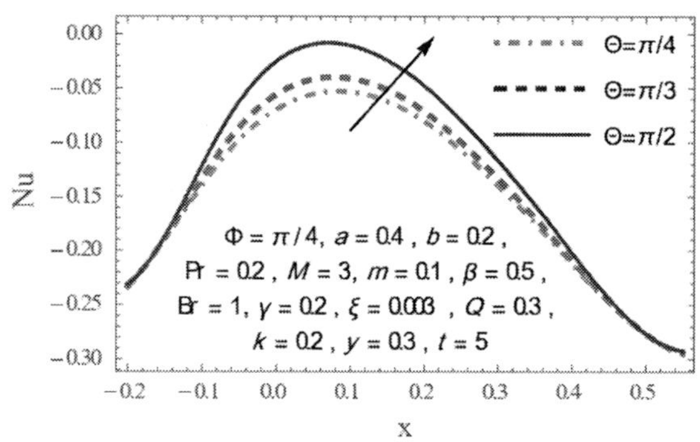

Figure 32. Nusselt number Nu for different values of Θ

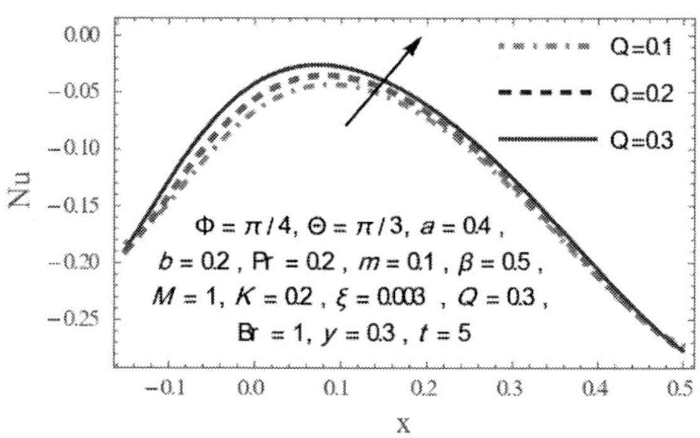

Figure 33. Nusselt number Nu for different values of Q

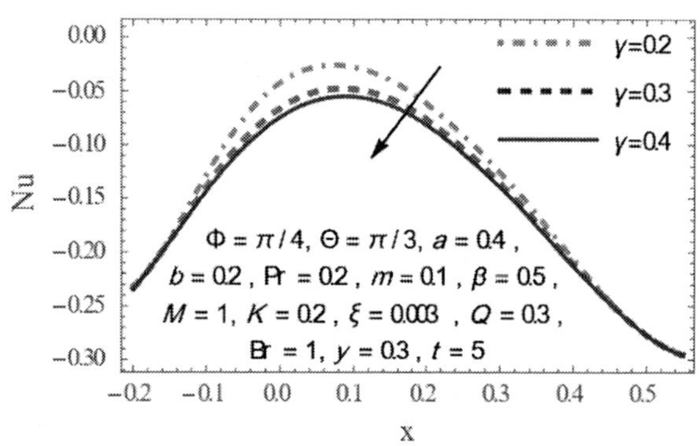

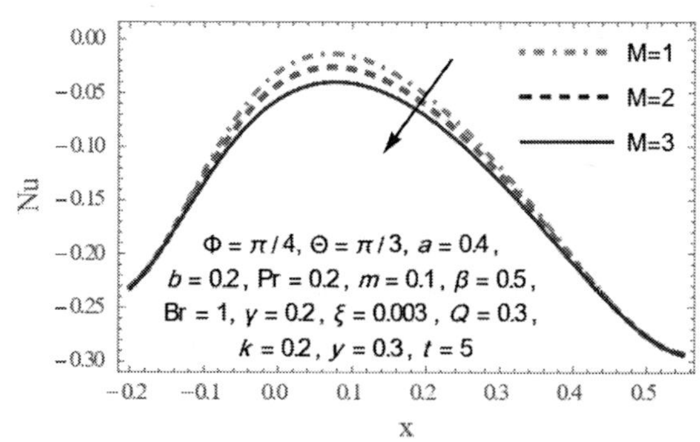

Figure 34. Nusselt number Nu for different values of γ

Figure 35. Nusselt number Nu for different values of M

11. Trapping phenomena

The trapping is a phenomenon interesting in peristaltic motion, where they are considered from a physical phenomenon which may be responsible for thrombus formation in blood and the movement of food bolus in the gastrointestinal tract.In this paper, we study trapping phenomenon for various values parameters, and a walls is peristaltic and the flow is no-slip. this trapped bolus is reinforcing along with the peristaltic wave. we observed that a bolus also in the fixed frame which may likely to be due the impression of non-zero time-average of the flow over one period of the wave. The trapping for different values of (m, Q , ξ, M and K) are shown in Figs.(36)-(40) at fixed value of $t = 0.5$. In this figures, we note that the significant variation and the strong impact of occurs bolus near on walls with high deformations. The Fig.(36) shows the effect non-uniform parameter (m) on the trapping, when is examined we found the size of the trapped bolus is decreasing with an enhance in (m). And the effect of the "mean flow rate" (Q) on the trapping is observed of streamlines (see Fig.(37)). We found that the trapping bolus increase in size with enhancing of values of (Q). The Fig.(38) we observe impact pseudoplastic fluid parameter (ξ) on the trapping (As shown in the drawing streamlines) with increasing value of (ξ) the size of the trapped is increasing. As for the variance values of the "Hartmann number" (M) in the streamlines,.The increase in the values of (M), leads to an increase in the size of the trapping bolus (see Fig.(39)). It is noticed from Fig.(40) that, the trapping bolus decreases with an increase of the Darcy number (K) in asymmetric channels.

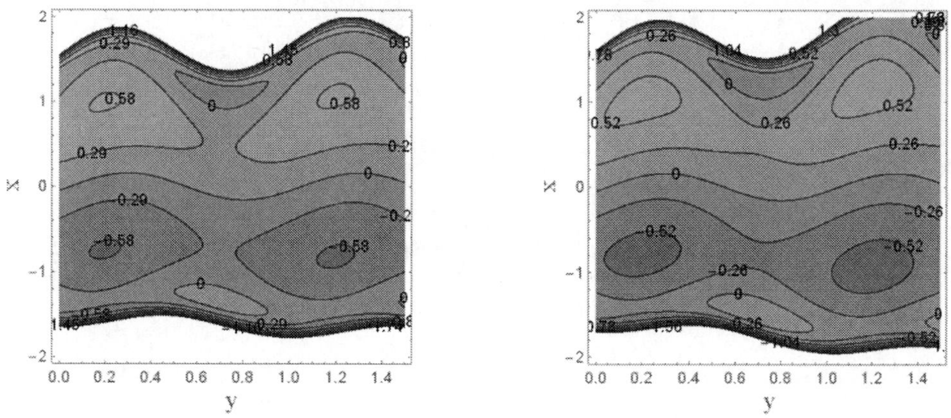

Figure 36. Streamlines for $b = 0.3$, $\phi = \pi / 2, \theta = \pi / 3, a = 0.1$, $\xi = 0.05, Q = 0.5, M = 1, K = 0.2$, $t = 0.5$ and for different m: (a) m=0.1 (b) m=0.2.

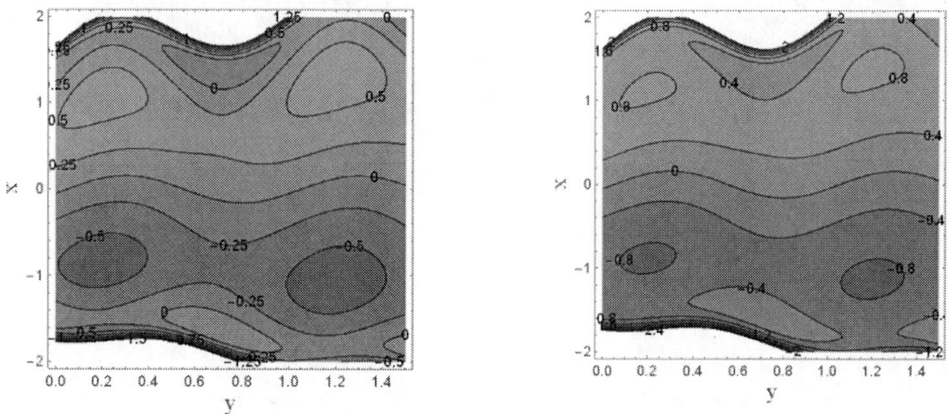

IOP Conf. Series: Journal of Physics: Conf. Series **1032** (2018) 012043 doi:10.1088/1742-6596/1032/1/012043

Figure 37. Streamlines for $b = 0.3, \phi = \pi/2, \theta = \pi/3, a = 0.1, \xi = 0.05, m = 0.3, M = 1, K = 0.2, t = 0.5$ and for different Q : (a) $Q = 0.5$ (b) $Q = 1$

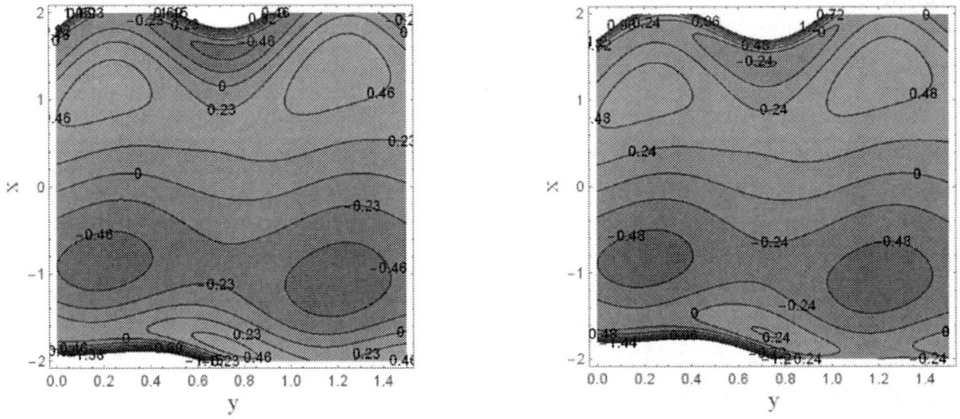

Figure 38. Streamlines for $b = 0.3, \phi = \pi/2, \theta = \pi/3, a = 0.1, m = 0.3, Q = 0.5, M = 1, K = 0.2, t = 0.5$ and for differen ξ **(a)** ξ =0.02 **(b)** ξ =0.04 .

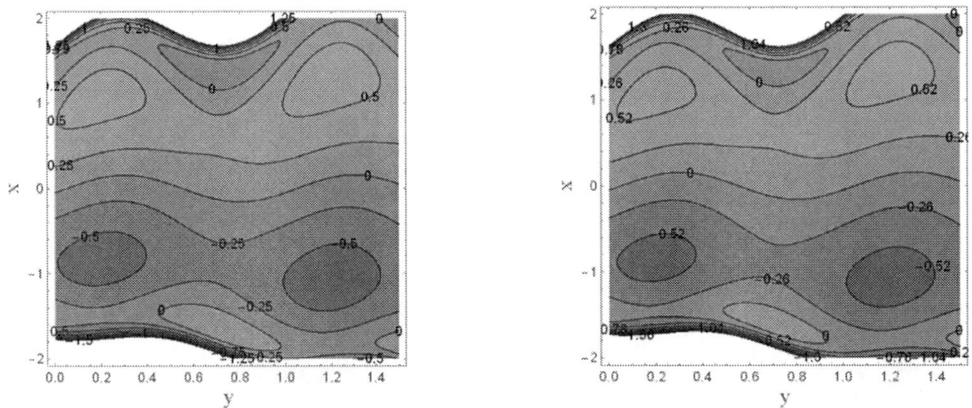

Figure 39. Streamlines for $b = 0.3, \phi = \pi/2, \theta = \pi/3, a = 0.1, \xi = 0.05, m = 0.3, Q = 0.5, K = 0.2, t = 0.5$ and for different M: **(a)** M=0.5, **(b)** M=2 .

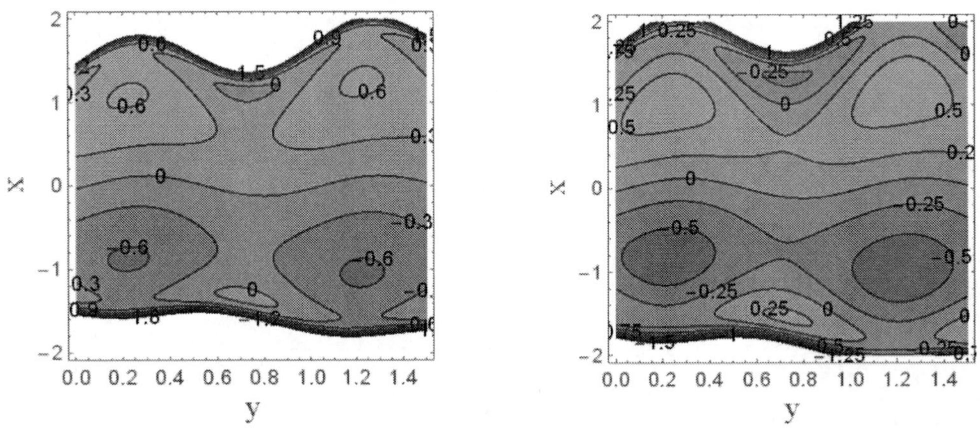

Figure 40. Streamlines for $b = 0.3, \phi = \pi/4, \theta = \pi/3, a = 0.1, \xi = 0.05, m = 0.2, Q = 0.5, M = 1, t = 0.5$ and for different K: **(a)** K=0.1 (b) K=0.3.

12. Conclusions

In this paper, we studied a mathematical model of the peristaltic transport of a pseudoplastic nanofluid in the tapered asymmetric channel with the effect of magnetic field and porous medium.And for heat effect with peristaltic transport through slanted tubes for the non-uniform walls.The effects of Hartmann number (M), Darcy number (K) , couple stress parameter (γ),heat generation parameter (β), inclination of magnetic field (Θ), Brinkman number (Br), Reynolds number (Re), Froude number (Fr), Prandle number (Pr), Eckert number (Ec), mean flow rate (Q) and incline angle of the channel (α) . We discussed the main findings as follows:

- The pressure gradient increases with increase in (Re), (Q) , (K) and (ϕ) while it decreases for increasing of (Fr), (M)
- It has been found that the average rise in pressure (Δp) increasing with the increase of (K), (M), (Re) and (ξ) while it decreases by increasing (Fr) and (b).
- The velocity profile increases with an increase (m), (Θ) and (Q) while it decreases with increasing (M), and (ϕ) .
- The axial velocity decreases with an increase in (ξ) at the core part of the channel. However opposite behavior on near channel boundaries.
- The temperature distribution is increased with an increase (β) and (M).
- The temperature decreases with an increase in (K), (Θ) and (γ) at the core part of the channel. However opposite behavior on near channel boundaries.
- Heat transfer coefficient is a decreasing function with an increase in (Br),(M) and (β) while it enhances by increasing (γ) .
- The Nusselt number increases with increasing of (Br), (K), (Θ) and(Q) .While (Nu) is decreasing with the increase of (γ) and (M).

If there is no-slip flow then the trapping bolus decreases with increasing of parameters (m) and (K) while it has a reverse behavior with increasing of(Q) , (ξ) and (M).

References

[1] S. Rashidi, M. Dehghan, R. Ellahi, M. Riaz, M.T. Jamal-Abad, Study of stream wise transverse magnetic fluid flow with heat transfer around an obstacle embedded in a porous medium, *J. Magn. Magn. Mater.* 378 (2015) 128–137.

[2] A. Zeeshan, R. Ellahi, M. Hassan, Magnetohydrodynam-ic flow of water/ethylene glycol based nano fluids with natural convection through a porous medium, *Eur. Phys. J.Plus* 129 (12) (2014) 1–10.

[3] M. Sheikholeslami, R. Ellahi, H.R. Ashorynejad, G. Domairry, T. Hayat, Eff-ects of heat transfer in flow of nano fluids over a permeable stretching wall in a porous medium,*J. Comput. Theor. Nanosci.* 11 (2) (2014) 486 –496.

[4] R. Ellahi, X. Wang, M. Hameed, Effects of heat transfer and nonlinear slip on the steady flow of Couette fluid by means of Chebyshev spectral method, *Z.Naturforsch.*,A 69(1–2) (2014) 1–8.

[5] R. Ellahi, E. Shivanian, S. Abbasbandy, T. Hayat, Anal-ysis of some magnetohydrodynamic flows of third order fluid saturating porous space, *J. Porous Med.* 18 (2) (2015) 89 –98.

[6] V.K. Stokes, Couple stresses in fluids, *Phys.Fluids* 9 (9) (1966) 1709–1715.

[7] V.K. Stokes, Theories of Fluids with Microstructure, *Springer, New York,* 1984.

[8] T. Hayat, M. Mustafa, Z. Iqbal, A. Alsaedi, Stagnation point flow of couple stress fluid with melting heat transfer, *Appl. Math. Mech.* 34 (2) (2013) 167–176.

[9] N.S.Akbar, S.Nadeem, Intestinal flow of a couple stress nano fluid in arteries, *IEEE Trans. Nanobioscience* 12 (4) (2013) 332–339.

[10] D. Srinivasacharya, N. Srinivasacharyulu, O. Odelu, Flow and heat transfer of couple stress fluid in a porous channel with expanding and contracting walls, Int. Commun.Heat Mass Transfer 36 (2009)180–185.

[11] S. Nadeem, A. Riaz, R. Ellahi, Peristaltic flow of viscous fluid in a rectangular duct with compliant walls, *Comput. Math. Model.* 25 (3) (2014) 404 – 415.270 K. Ramesh / Journal .

[12] A. Riaz, R. Ellahi, S. Nadeem, Peristaltic transport of a Carreau fluid in a compliant rectangular duct, *Alex. Eng. J.* 53 (2) (2014) 475–484.

[13] M. Mishra, A.R. Rao, Peristaltic transport of a Newt-onian fluid in an asy-mmetric channel, *J. Appl. Math. Phys.* 54 (2003) 532–550.

[14] N. Ali, T. Hayat, Peristaltic flow of a micropolar fluid in an asymmetric channel,*Comput. Math. Appl.* 55 (2008) 589–608.

[15] P. Naga Rani, G. Sarojamma, Peristaltic transport of a Casson fluid in an asymmetric channel, *Australas. Phys. Eng. Sci. Med.* 27 (2) (2004) 49–59.

[16] T. Hayat, A. Afsar, N. Ali, Peristaltic transport of a Johnson-Segalman fluid in an asymmetric channel, *Math. Comput. Model.* 47(2008) 380–400.

[17] T.Hayat, M. Javed, Exact solution to peristaltic transport of power-law fluid in asymmetric channel with compliant walls, *Appl. Math.Mech.*31(10) (2010) 1231–1240.

[18] S. Nadeem, S. Akram, Peristaltic flow of a couple stress fluid under the effect of induced magnetic field in an asymmetric channel, *Arch. Appl. Mech.* 81 (2011) 97–109.

[19] S. Nadeem, S. Akram, In fluence of inclined magnetic field on peristaltic flow of a Williamson fluid model in an inclined symmetric or asymmetric channel, *Math.Comput. Model.* 52 (2010) 107–119.

[20] M. Saleem, A. Haider, Heat and mass transfer on the peristaltic transport of non-Newtonian fluid with creeping flow, *Int. J. Heat Mass Transf.* 68 (2014) 514 –526.

[21] T. Hayat, H. Yasmin, B. Ahmad, B. Chen, Simultaneous effects of convective conditions and nanoparticles on peristaltic motion, *J. Mol. Liq.* 193 (2014) 74–82.

[22] M.R.Salman, A.M.Abdulhadi, Effects of MHD on Peristalsis Transport and Heat Transfer with Variables Viscosity in Porous Medium, *Int.J. of Sci. and Res.(IJSR)* Volume 7 Issue 2, February (2018) pages 612-623.

[23] S. Nadeem, N.S. Akbar, Effects of heat and chemical reactions on peristaltic flow of Newtonian fluid in a diverging tube with inclined MHD, Asia Pac. J. Chem. Eng. 6 (2011) 659 – 668.

[24] S. Nadeem,S.Akram,In fluence of inclined magnetic field on peristaltic flow of a Jeffrey fluid with heat and mass transfer in an inclined symmetric or asymmetric channel, *Asia Pac. J. Chem. Eng.* 7 (2012) 33 – 44.

[25] K.Ramesh, Influence of heat and mass transfer on peristaltic flow of a couple stress fluid through porous medium in the presence of inclined magnetic field in an inclined asymmetric channel, *J. Mol. Liq.* 219 (2016) 256-271

[26] T. Hayat, S. Noreen, M.S. Alhothuali, S. Asghar, A. Alhomaidan, Peristaltic flow under the effects of an induced magnetic field and heat and mass transfer, *Int. J. Heat Mass Transf.* 55 (2012) 443–452.

[27] N.S. Akbar, T. Hayat, S. Nadeem, S. Obaidat, Peris-taltic flow of a Willia-mson fluid in an inclined asymmetric channel with partial slip and heat transfer, *Int. J. Heat Mass Transf.* 55 (2012) 1855 – 1862.

IOP Conf. Series: Journal of Physics: Conf. Series **1032** (2018) 012044 doi:10.1088/1742-6596/1032/1/012044

Concentrations of some heavy elements in fly ash samples from thermal electric power stations in the middle and south of Iraq

Israa kamil Ahmed

Information &communication Department, College of information Engineering, Al-Nahrain University, Iraq
esraa_nuc_med@yahoo.com

Abstract. The objective of this present work is determining some heavy elements concentrations like (Cd, Co, Mg, Mn, Fe, Cu, Ni, Pb & Zn) in thermal electric power stations in the middle and south of Iraq. 18 Fly Ash samples were collected from thermal electric power stations in the middle and south of Iraq ,the concentration values of (Cd, Zn , and Fe ,Mg, Cu) in some samples were over the international limits in the Fly Ash samples .In general the increasing of the heavy metals in some fly ash samples due to the releasing of these metals is the combustion of the fuel and fuel type which provides thermal electric power station different from units to other , and geology of the region collected samples from it and the geochemical characteristics of the crude oil.

1. Introduction

A heavy element is any relatively dense metal or metalloid that is noted for its potential toxicity, especially in environmental contexts. The term has particular application to cadmium, mercury, lead, zinc and arsenic, all of which appear in the organization's list of 10 chemicals of major public concern [1].

Toxic heavy elements are found naturally in the earth. They become concentrated as a result of human caused activities and can enter plant, animal, and human tissues via inhalation, and manual handling. Then, lead-acid batteries; they can bind to and interfere with the functioning of vital cellular components. The toxic effects of arsenic, mercury, and lead were known to the ancients, but methodical studies of the toxicity of some heavy metals appear to date from only 1868. In humans, heavy element poisoning is generally treated by the administration of chelating agents. Some elements regarded as heavy metals are essential, in small quantities, for human health [1].

Toxic heavy elements are found naturally in the earth, and become concentrated as a result of human caused activities. Common sources are from mining and industrial wastes; vehicle emissions; fertilizers; paints; treated woods; and aging water supply infrastructure. Arsenic, cadmium and lead may be present in children's toys at levels that exceed regulatory standards. Lead can be used in toys as a stabilizer, color enhancer, or anti-corrosive agent. Cadmium is sometimes employed as a stabilizer, or to increase the mass and luster of toy jewelry. Arsenic is thought to be used in connection with coloring dyes [2].

IOP Conf. Series: Journal of Physics: Conf. Series **1032** (2018) 012044 doi:10.1088/1742-6596/1032/1/012044

Lead is the most prevalent toxic heavy element contaminant. As a component of tetraethyl lead, $(CH_3CH_2)_4Pb$, it was used extensively in gasoline during the 1930s–1970s. Lead levels in the aquatic environments of industrialized societies have been estimated to be two to three times those of pre-industrial levels. Although the use of leaded gasoline was largely phased out in North America by 1996, soils next to roads built before this time retains high lead concentrations [3]. Heavy elements enter plant, animal and human tissues via air inhalation and manual handling. Motor vehicle emissions are a major source of airborne contaminants including arsenic, cadmium, cobalt, nickel, lead, antimony, vanadium, zinc, platinum, palladium and rhodium. Stations are exposed to heavy elements through the uptake of water; animals eat these plants; ingestion of plant- and animal-based foods are the largest sources of heavy elements in humans. Absorption through skin contact, for example from contact with fly ash from electrics power station is another potential source of heavy element contamination. Toxic heavy metals can bio accumulates in organisms as they are hard to metabolize [3].

Although there is no clear definition of what a heavy element is, density is in most cases taken to be the defining factor. Heavy elements are thus commonly defined as those having a specific density of more than 5 g/cm^3. The main threats to human health from heavy elements are associated with exposure to lead, cadmium, mercury and arsenic (arsenic is a metalloid, but is usually classified as a heavy metal).

Inhalation of heavy elements fumes or particles can be life threatening, and although acute pulmonary effects and deaths are uncommon, sporadic cases still occur. Exposure may cause kidney damage. Several reports have since shown that kidney damage and/or bone effects are likely to occur at lower kidney levels. Long-term high heavy elements exposure may cause skeletal damage, The exposure was caused by heavy elements -contaminated water used for irrigation of local rice fields experiments have suggested that may be a risk factor for cardiovascular disease, heavy elements exposure may give rise to lung damage or may cause contact eczema, These intriguing contradictory findings need to be followed up by more studies of other similarly exposed populations [4].

The symptoms of lead poisoning are headache, irritability, abdominal pain and various symptoms related to the nervous system. Lead encephalopathy is characterized by sleeplessness and restlessness. Children may be affected by behavioral disturbances, learning and concentration difficulties. In severe cases of lead encephalopathy, the affected person may suffer from acute psychosis, confusion and reduced consciousness. People who have been exposed to lead for a long time may suffer from memory deterioration, prolonged reaction time and reduced ability to understand [5]. Atomic absorption spectrometry (AAS) is an analytical technique made in Australia that measures the concentrations of heavy elements us we show in fig (2). Atomic absorption is so sensitive that it can measure down to parts per billion of a gram (ppp or µg/ L) in a sample [6]. The technique makes use of the wavelengths of light specifically absorbed by an element. They correspond to the energies needed to promote electrons from one energy level to another, higher, energy level. Atomic absorption spectrometry has many uses in different areas of chemistry [7].

Atoms of different elements absorb characteristic wavelengths of light. Analyzing a sample to see if it contains a particular element means using light from that element. For example with lead, a lamp containing lead emits light from excited lead atoms that produce the right mix of wavelengths to be absorbed by any lead atoms from the sample. In AAS, the sample is atomized – i.e. converted into ground state free atoms in the vapor state – and a beam of electromagnetic radiation emitted from excited lead atoms is passed through the vaporized sample. Some of the radiation is absorbed by the lead atoms in the sample [7, 8].

The greater the number of atoms there is in the vapor, the more radiation is absorbed. The amount of light absorbed is proportional to the number of lead atoms. A calibration curve is constructed by running several samples of known lead concentration under the same conditions as the unknown. The amount the standard absorbs is compared with the calibration curve and this enables the calculation of the lead concentration in the unknown sample. Consequently an atomic absorption spectrometer needs the following three components: a light source; a sample cell to produce gaseous atoms; and a means of measuring the specific light absorbed [9].

2. Material and Methods

Fly ash samples from thermal electric power station after block out the chimney of it to measure the concentrations of studied heavy elements using (ICP-MS) equipment, after collecting samples stage, all the samples were taken to the laboratories of Baghdad university / chemistry lab by The chemical process of digestion to prepare the fly ash samples for measuring the studied Heavy elements, preparing as follow:

fly ash sample are taken from different thermal electric power station in the middle and south of Iraq, Crumbled and put it in the oven till the dryness and then Crush them to small volumes by mill and after harmonizing these samples, (10) gm from each sample was taken for analysis, A (7) ml of concentrated Hydrochloric acid (HCI) and a (8) ml of concentrated Nitric acid (HNO3) were added to the sample after putting it in a Beaker (100) ml ,then the mixture was heated to the end drought and avoid boiling, Returned to the previous steps .Then the acid was added for the second times and the samples were placed on Hot Plate and bring it up to the boiling point only and Put this mixture on the magnetic stirrer for (24) hours then Filtered the mixture by Filtering paper (No. 42) and put it in volumetric flask (100) ml and Finally were wished by Filtered water and adding the washing water to the filter and complete the Volume to (100) ml us we show in figure (1) then sending the prepared samples were sent to the analysis by (Optical Emission Spectrometer, PerkinElmer, Optima 2100DV) equipment's (AAS) us we show in figure (2) [9,10].

Figure 1. prepared the fly ash samples for AAS

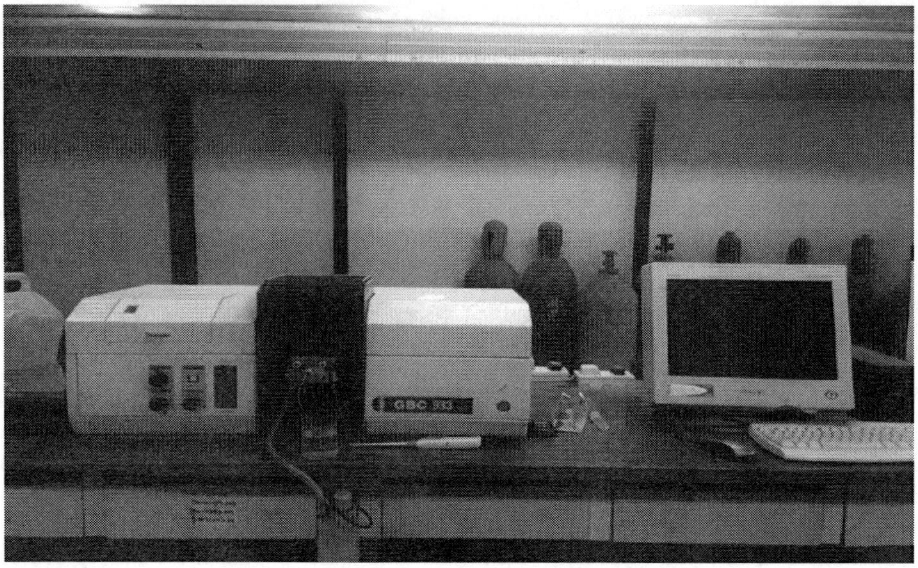

Figure 2. Atomic Absorption Spectrometer

3. Results and Discussion

After collecting fly ash samples from thermal power stations in the middle and south of Iraq all samples were taken to the laboratory to preparing these samples for analyzing by (AAS) equipment's (Buck Scientific Model 210 VGP, Atomic Absorption Spectrophotometer).

S No	(Ni) µg/ml	(Zn) µg/ml	(Pb) µg/ml	(Mg) µg/ml	(Cu) µg/ml	(Fe) µg/ml	(Cd) µg/ml	(Mn) µg/ml	(Co) µg/ml
S1	3.27	12.9	1.893	High	1.211	487.6	0.927	0.095	0.804
S2	3.929	0.303	0.422	0.031	0.418	6.56	0.432	3.829	0.07
S3	4.675	2.39	0.232	0.27	0.246	1.201	0.242	0.9	0.074
S4	6.499	0.531	0.639	High	0.975	964.5	0.937	0.135	0.224
S5	2.227	31.925	6.197	0.058	2.092	3.242	2.043	3.339	0.155
S6	0.473	0.037	0.112	High	0.039	0.105	0.025	0.031	-0.013
S7	3.562	0.932	0.577	0.02	0.316	1.114	0.052	2.455	0.194
S8	2.539	0.58	0.458	0.423	1.339	7.416	0.052	0.104	0.128
S9	4.902	0.132	0.363	0.049	3.849	6.575	0.106	0.222	0.15
S10	5.579	0.005	0.814	0.256	1.489	4.352	0.073	0.09	0.255
S11	4.818	0.118	0.512	0.054	3.956	9.407	0.051	0.12	0.146
S12	3.644	0.134	0.842	0.273	1.698	1.283	0.066	1.455	0.193
S13	0.911	1.195	0.432	0.035	4.8	1100.5	0.051	0.099	0.102
S14	3.163	1.515	1.085	0.305	1.764	344.7	0.047	3.053	0.076
S15	3.97	0.104	1.017	0.15	0.597	0.745	0.042	1.137	0.143

IOP Conf. Series: Journal of Physics: Conf. Series **1032** (2018) 012044 doi:10.1088/1742-6596/1032/1/012044

S16	3.045	2.00	0.253	0.058	0.105	0.647	0.026	0.772	0.101
S17	1.087	0.009	0.363	0.029	2.375	4.475	0.033	2.845	0.095
S18	2.524	0.06	2.241	High	104.7	6.052	0.069	26.8	0.146
ave	3.378	4.04	1.025	0.143	7.33	163.9	0.293	2.637	0.169

Table 1. Heavy elements concentration

The concentration of Cadmium in the samples is ranging (2.043_0.025µg/ml) with a mean (0.293 µg/ml), the greatest value in S5 which is in fact represents S3 in Al-Dura thermal power station

It is within the range (0.5 µg/ml) [11, 12] in fly ash and the smaller concentration represented S5 in Al-Rasheed thermal power station. Table 1 shows us the concentration of Cadmium and fig. 3 shows us the greatest and smaller concentration.

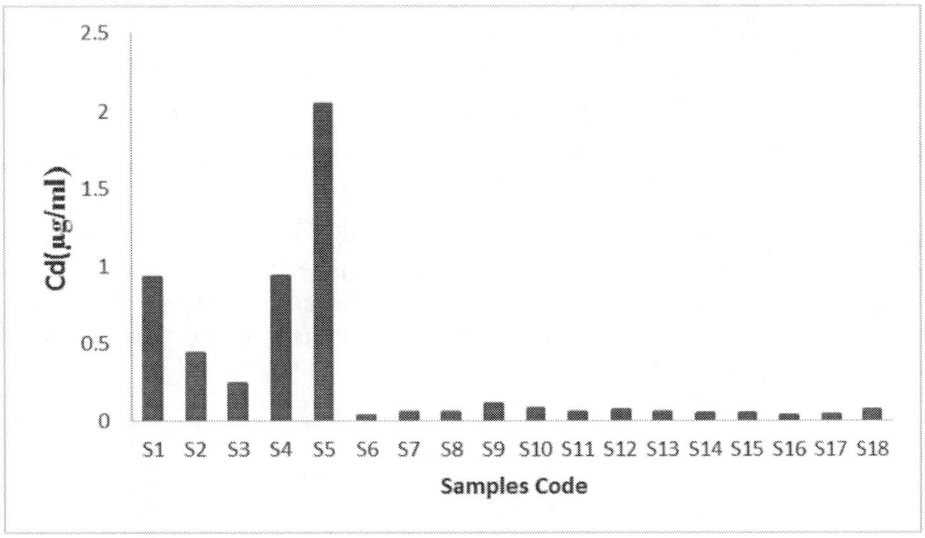

Figure 3. Cd concentration

The concentration of Cobalt in the samples is ranging (0.804_-0.013µg/ml) with a mean (0.169 µg/ml), the greatest value in S1which is in fact represent S2 in Al-Dura thermal power station it is less than the range (8 µg/ml) [11, 12] in fly ash and the smaller concentration in S6 represented S5 in Al-Rasheed thermal power station. Table 1 shows us the concentration of Cobalt and fig 4 shows us the greatest and smaller concentration.

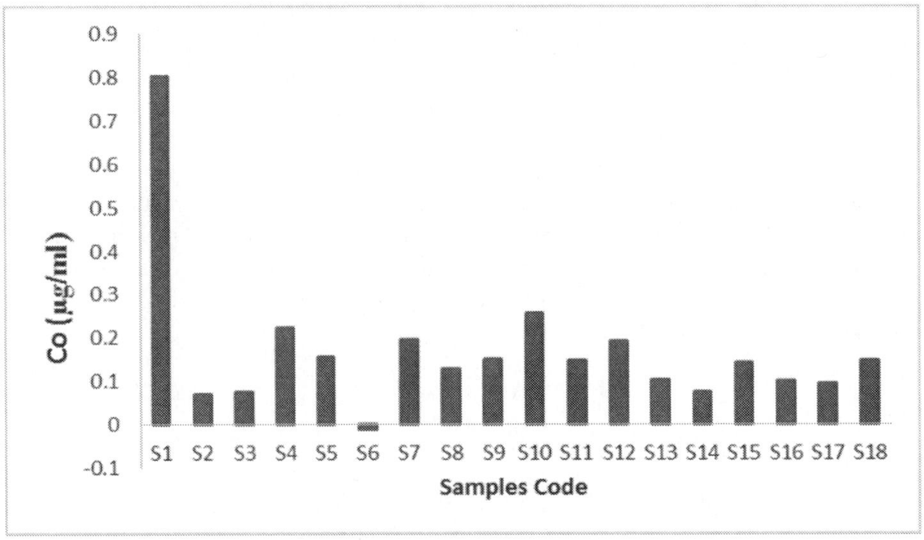

Figure 4. Co concentration

The concentration of Manganese in the samples is ranging (28.6_0.031μg/ml) with a mean (2.637 μg/ml), the greatest value is S18which is in fact representing S10 in Musaeb thermal power station. It is less than the limit (3.85 μg/ml)[11,12] in fly ash and the smaller concentration is S6 represented S5 in Al-Rasheed thermal power station Table1 show us the concentration of Manganese and fig 5 show us the greatest and smaller concentration.

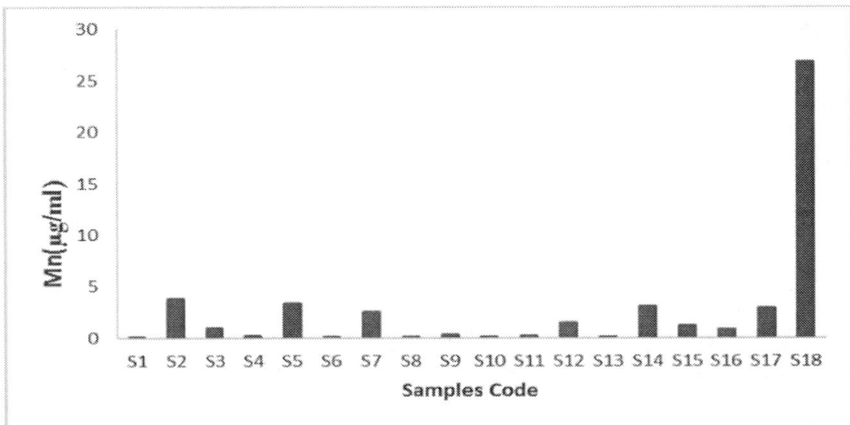

Figure 5. Mn concentration

The concentration of Nickel in the samples is ranging (5.579_0.473μg/ml) with a mean (3.378 μg/ml), the greatest value is S10which is in fact representing S6 in Naseraya thermal power station. It is less than the limit (7.15 μg/ml)[11,12] in fly ash and the smaller concentration is S6 represented S5 in Al-Rasheed thermal power station Table1 show us the concentration of Manganese and fig 6 show us the greatest and smaller concentration.

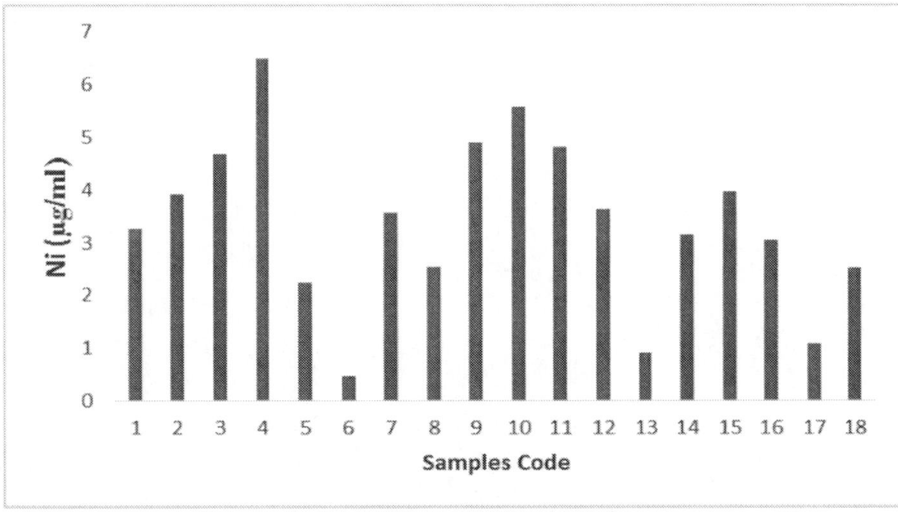

Figure 6. Ni concentration

The concentration of Zinc in the samples is ranging (31.92_0.005µg/ml) with a mean (4.04 µg/ml) the greatest value is S5which is in fact representing S3 in Al-Dura thermal power station. It is more than the limit (1.65 µg/ml)[11,12] in fly ash and the smaller concentration is S10 represented S6 in Naseraya thermal power station Table1 show us the concentration of Manganese and fig 7 show us the greatest and smaller concentration.

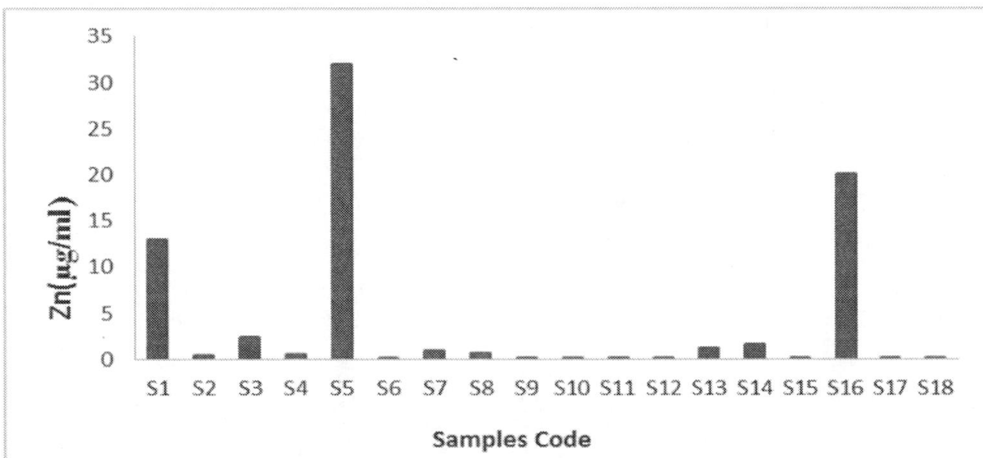

Figure 7. Zn concentration

The concentration of Lead in the samples is ranging (2.241_0.112µg/ml) with a mean (1.025 µg/ml), the greatest value is S5which is in fact represents S3 in Al-Dura thermal power station. It is less than the limit (9.35 µg/ml)[11,12] in fly ash and the smaller concentration is S6 represented S5 in Al-Rasheed thermal power station Table1 show us the concentration of Manganese and fig 8 show us the greatest and smaller concentration.

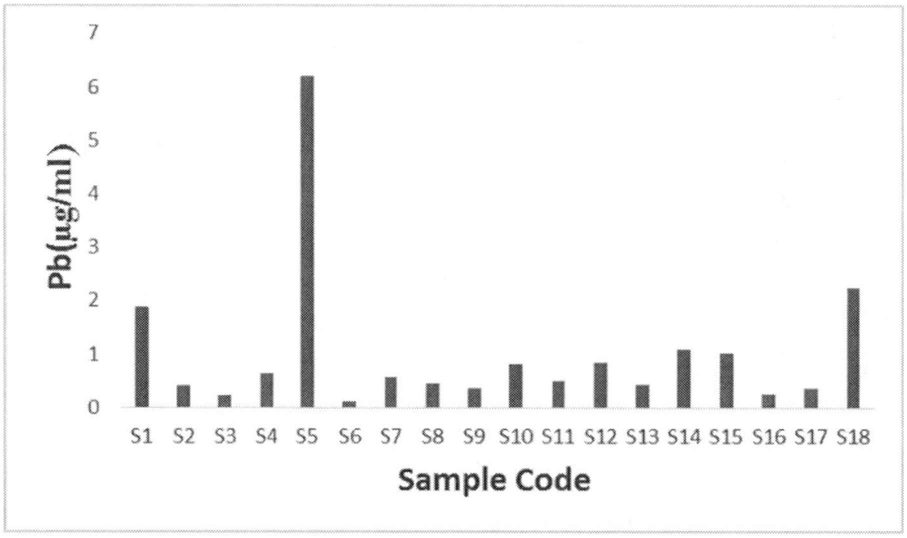

Figure 8. Pb concentration

The concentration of Magnesium in the samples is ranging (High_0.029µg/ml) with a mean (0.143 µg/ml)),the greatest value is S1, in fact represents S2 in Al-Dura thermal power station, S6 in fact represents S5 in Al-Rasheed thermal power station,S18which is in fact represent S10 in Musaeb thermal power station. It is more than the limit (0.440 µg/ml)[11,12] in fly ash and the smaller concentration is S17 represented S3 in Hartha thermal power station Table1 show us the concentration of Manganese and fig 9 show us the greatest and smaller concentration.

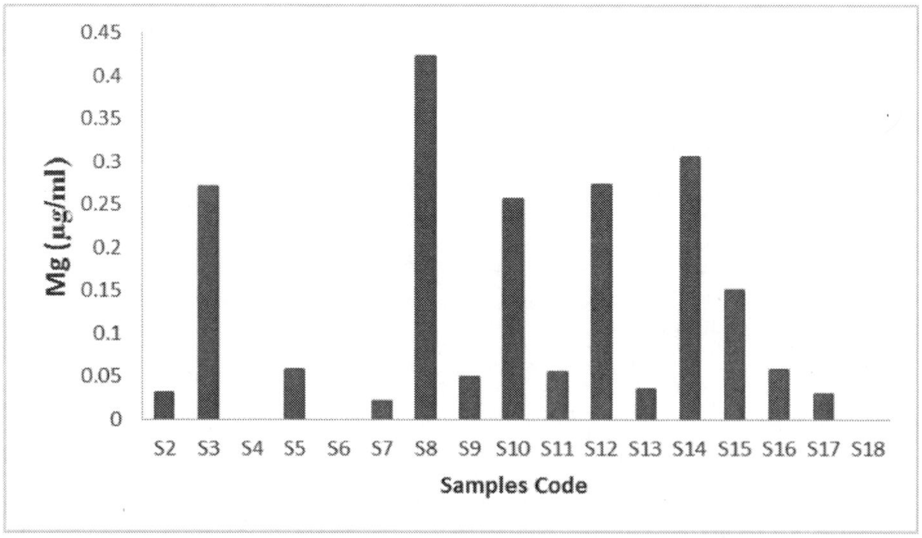

Figure 9. Mg concentration

The concentration of Copper in the samples is ranging (104.7_0.039µg/ml) with a mean (7.33 µg/ml), the greatest value is S18which is in fact representing S10 in Musaeb thermal power station. It is more than the limit (1.98 µg/ml)[11,12] in fly ash and the smaller concentration is S6 represented S5 in Al-Rasheed thermal power station Table1 show us the concentration of Manganese and fig 10 show us the greatest and smaller concentration.

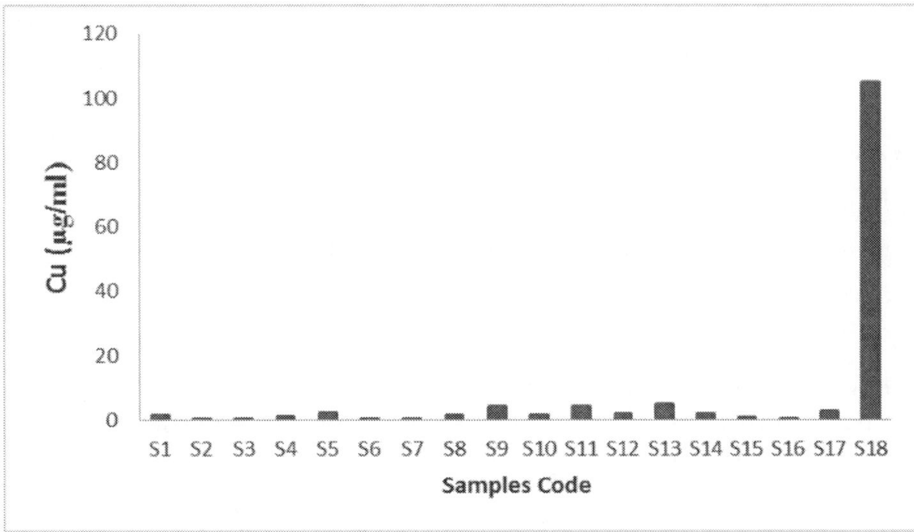

Figure 10. Cu concentration

The concentration of Iron in the samples is ranging (1100.5_0.105µg/ml) with a mean (163.9 µg/ml), the greatest value is S13which is in fact represents S2 in Hartha thermal power station. It is more than the limit (9.90 µg/ml)[11,12] in fly ash and the smaller concentration is S6 represented S5 in Al-Rasheed thermal power station, Table 1 show us the concentration of Manganese and fig 11 show us the greatest and smaller concentration

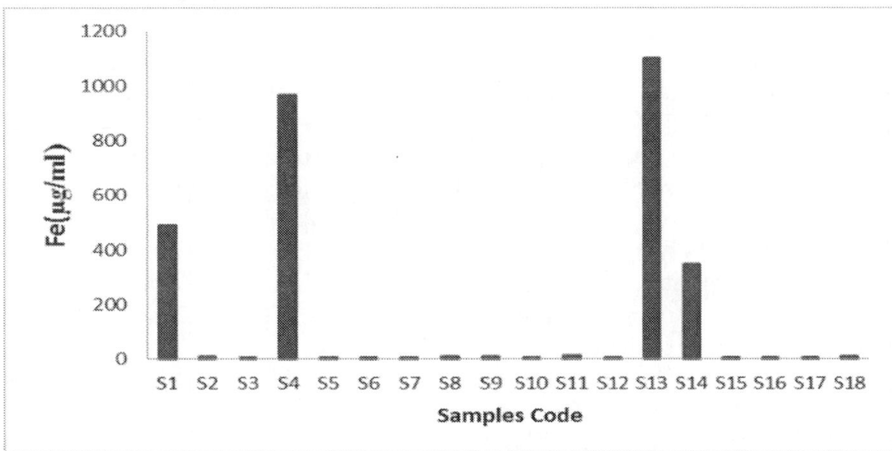

Figure 11. Fe concentration

4. Conclusion

The results which obtained from the present work show that fly ash samples from thermal electric power stations in the middle and south of Iraq were found to be significantly contaminated with metals like Cd, Zn, and Fe, Mg, Cu at levels above the background concentration in the international limits, which may give rise to various health hazards, while the concentrations other heavy elements were under the background concentration in the international limits. The high concentration of some heavy elements can be due to the type of fuel used in the thermal power station. The increase in the concentration of some heavy metals can be due to the type of fuel used in the thermal power plant. The increase of some heavy elements can lead to serious diseases, especially lung cancer, so it is necessary to get rid of fly ash in other ways or replace the fuel with another.

Acknowledgment

I thank University of Baghdad, College of Science, and Department of chemistry for supporting this work. I thank University of Kerbala, College of Science, and Department of physics for supporting this work exactly the Head of Physics Department, Dr. Fadhil I. Sharad.

References

1. Krishna, A. K. and Govil, P. K 2007 "crude oil contamination due to heavy metals from an industrial thermal power plant of Surat, Gujarat, Western India, Springer, Environ. Monit. Assess 124, pp. 263–275.
2. The central organization for statistics and information technology, 2005, internal report about heavy elements in crude oil.
3. Aziz, A. M.1995, Effect of some heavy metals in the solid waste and waste water on the growth of lettuce and soil contamination, MSc. Thesis, University of Baghdad, College of Agriculture, P. 70.
4. Al-Janabi, A.Y, Al-Saadi, N.A, Zainal,Y. M, Al-Bassam, K. S., and Al-Delaimy, M. R,2002 Work procedures of the S.E of Geological survey and mining. State establishment of Geological survey and mining (geosurv) part 21, No, B, pp. 59-65.
5. ICRP, 1993. "Age-Dependent Doses to Members of the Public from Intake of Radionuclides." Part 1. ICRP Publication 56. Annals of the ICRP: Pergamon Press, Oxford.
6. United Nations scientific committee on the effects of atomic radiation, "Sources and effects of ionizing radiation". UNSCEAR 2006 report Vol.1 to the general assembly, with scientific annexes, United Nations sales publication, United Nations, New York.
7. Roiter, Y; Minko, S, 2005,"AFM single molecule experiments at the solid-liquid interface: in situ conformation of adsorbed flexible polyelectrolyte chains, Journal of the American Chemical Society **127**, 2005, (45).
8. A. Báez, R. Belmont, R.M. García, H. Padilla, M.C.Torres,2007 'Chemical composition of rainwater collected at a southwest site of Mexico City, Mexico". Rev. Atmospheric Research , 86: 61-75
9. WHO (World health organization) 2007, "International heavy elements project survey on heavy elements guidelines, programmers and activities". WHO, Geneva.
10. ICRP, 2012 "compendium of dose coefficients based on ICRP publication 60." Part 1. ICRP Publication 119. Annals of the ICRP: volume 41, Oxford, 2012.
11. WHO, 2004 "Guideline values for chemicals that are of health significance in fly ash from thermal power plants in Japan.
12. Lindsay, W. L, 2008, Chemical equilibrium of crude oil. John Wiley and Sons, p.449.

IOP Conf. Series: Journal of Physics: Conf. Series **1032** (2018) 012045 doi:10.1088/1742-6596/1032/1/012045

Effect of substitution (Al and P) atoms in ZnO nanosheet on structural and electronic properties

Nibras M Umran[1]*, Haneen A Rashed[1], Firas K. Mohamad Alosfura[1] and Noor J. Ridha[2]

[1]Department of Physics, College of Science, University of Kerbala, 56001, Karbala, Iraq
[2]Department of Chemistry, College of Science, University of Kerbala, 56001, Karbala, Iraq

E-mail: nibrasmossa@uokerbala.edu.iq

Abstract. We explore the possibility of formation of two different Al_n and P_n atoms (n=1-6) are substitute doped of ZnO nanosheet has been investigated based on ab initio density functional theory. We study the effect of Al and P doping on the structural and electronic properties of a ZnO nanosheet. This study explains the effect of several doping on structural and electronic properties. The results indicate that Al and p substituted on the O site. The binding energy of doped system is negative, implying a stable integration of Al and P. We also found that the least energy gap lies in the Al at two atoms either in P at three atoms. These findings demonstrate that ZnO nanosheet doped of Al and P likely to be applied for various applications such as photocatalytic reactions, optoelectronic, solar cells, etc.

1. Introduction

Zinc Oxide (ZnO) is a semiconductor; it has a direct wide band gap 3.37 eV and a large exciton binding energy of 60 meV for the bulk hexagonal wurtzite crystal structure. The wurtzite crystal of ZnO has a normal dipole moment and spontaneous polarization resulting from the positively charged (0001)-Zn and negatively charged $(000\bar{1})$-O polar surfaces. ZnO has continued to attract great attention due to enters into a variety of practical applications such as optoeletronics, piezo-electric devices, solar cells, gas sensors, and biomedical science [1-5]. ZnO has multiple nanoscale structures such as nanosheet, nanorods, nanowire, and nanotubes have been explored in both experimental and theoretical areas. Two-dimensional (2D) materials have attracted great attention in recent times because of their unique properties, such as ultrathin thickness, large lateral size, the electron confinement in two-dimension interlayer interactions, and high optical transparency[6]. In particular, experimental observed the planar sheets of ZnO monolayer has been prepared by deposition on Ag (0001) and predicted theoretically a two-dimensional layer phase of ZnO thin films are less stable than two dimension ZnO sheets[7, 8]. Very recently, nanosheet structures of ZnO have been obtained which prepared on Pd(111)substrate[9]. Previous studies have shown that ZnO nanosheet has distinct properties when compared with thinfilm ZnO due to the effects of quantification. By doping the ZnO nanosheet gives an effective approach to adjust the physical properties

IOP Conf. Series: Journal of Physics: Conf. Series **1032** (2018) 012045 doi:10.1088/1742-6596/1032/1/012045

of the material[6, 10]. ZnO is n-type, in experimental study of P doped ZnO nanorods observed ZnO nanorods transition to p-type[11].

Theoretical studies have shown that substitutional doping is an important and effective strategy for improving chemical activity. This makes two-dimensional materials have a large amount in the fields of sensing and catalysis[12, 13]. The main objective of this study is to understand the effect mechanism of Al and P dopant the stability and electronic properties of the ZnO nanosheet by preforming the first principles calculation based on density functional theory (DFT).

2. Computational approach

All calculations for this work have been implemented by using the Spanish Initiative for Electronic Simulation with Thousands of Atoms (SIESTA)[14] code, in which the density functional theory (DFT)[15]. The electronic calculation was performed using generalized gradient approximation (GGA) with the exchange- correlation potential in the form of Perdew –Burke-Ernzerhof (PBE)[16, 17]. The pseudo-potential standards are built by using Trouiller- Martins schemes that are described as interaction of valence electrons with atomic central. Pseudopotential are used valance electron configurations for Zn 3d10 4s2 and O 2s2 2p4. We note that the electronic arrangement for Al 3s2 3p1 and 3s2 3p3 for P. The initial structure of the ZnO sheet cleaved from a (0001) layer of bulk ZnO. Using the conjugate gradient algorithm, structural improvements were made to ZnO nanosheet so that the remaining forces in the optimization were smaller than 0.001 eV/ Å. The Brillouin zone is used by Monkhorst Pack scheme[18]. A plane wave basis set for cut off energy was used as 200eV. We constructed the ZnO nanosheet supercell from a bulk w-ZnO structure with lattice parameters a = 3.29 Å and c = 5.3 Å these values are both experimental and theoretical results[19, 20]. The supercell of ZnO nanosheet consisting of 84 atoms has been used that take hexagonal structures with a vacuum layer 19Å along the perpendicular direction to avoid interlayer interactions.

3. Result and discussion

The structure and electronic properties of doped ZnO nanosheet have been investigated by DFT calculations. First, we have performed a full geometry optimization then the impurities were introduced by substituting Al, and P atoms into the O atoms to create the doping atom (AlO or PO). the bond length of Zn-O has been calculated as 1.9 Å, which agree well with the experimental and theoretical value of 1.92 Å[21, 22]. The Zn-O bond length in bulk ZnO is 2.01 Å is longer than bond length in ZnO nanosheet. In fact, the contraction of bond length ZnO nanosheet due to the atoms has the lowest number of neighbors compared to bulk. As a result, atoms in ZnO nanosheet are closely related, led to a smaller Zn-O bond length compare with bulk. We have done doping of Al atoms in ZnO nanosheet, and in figure 1, shown the ground state structures of Aln-ZnO (n= 1-6). The bond length between the Al and nearest Zn is 2.31 Å, it is clear that the Al-Zn is increased compared of the Zn-O bond of the ZnO nanosheet.

In Al_2-ZnO having Al-Al bond length is 2.28 Å in ZnONS. In the same way, we doped the P atoms the relaxed structures of Pn-ZnO (n= 1-6) are presented in figure 2.

The bond length of P-Zn in relaxed structure is 2.18 Å this value agreed well with the theoretical value in case (6×6) is 2.18 Å[23]. In both cases it is found that the complexes doping ZnO are stable up to n=1-6. To demonstrate the relative stability of Al and P atoms doping ZnO nanosheet, the binding energy can be expressed as

$$E_b(ZnO{:}M) = E_{tot}(ZnOM_n) - nE_{Zn} - nE_O - nE_M/N \qquad (1)$$

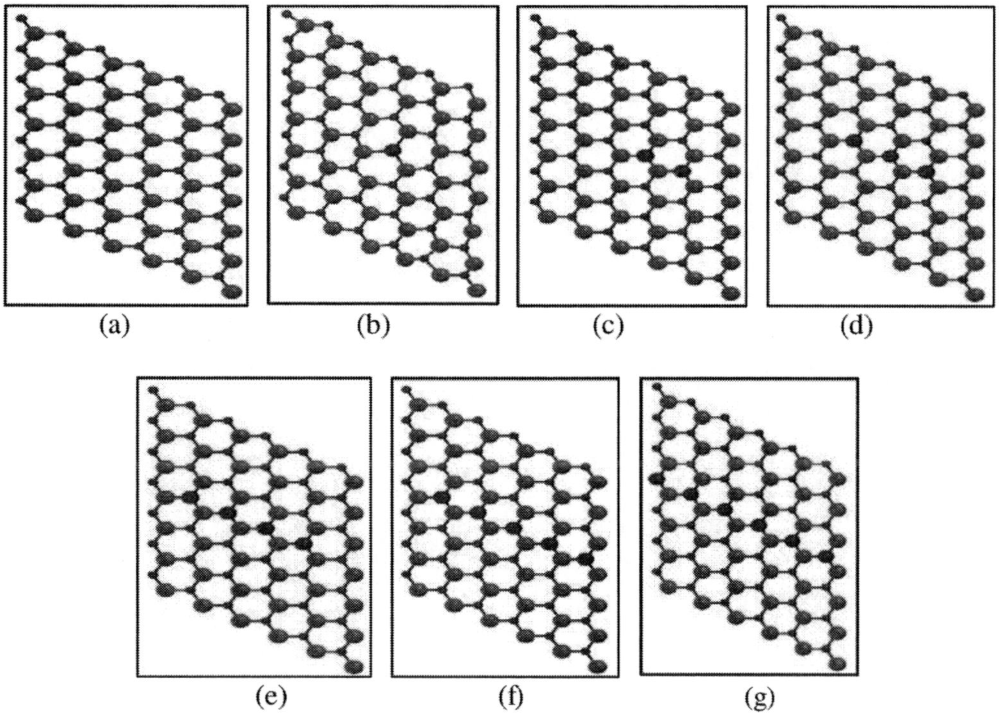

Figure 1. ZnO nanosheet, the red and purple spheres refer to O and Zn, atoms respectively. The blue sphere refers to the dopants Al atoms. (a) pristine and (b-g) optimized structure of Al doped in the ZnO nanosheet respectively.

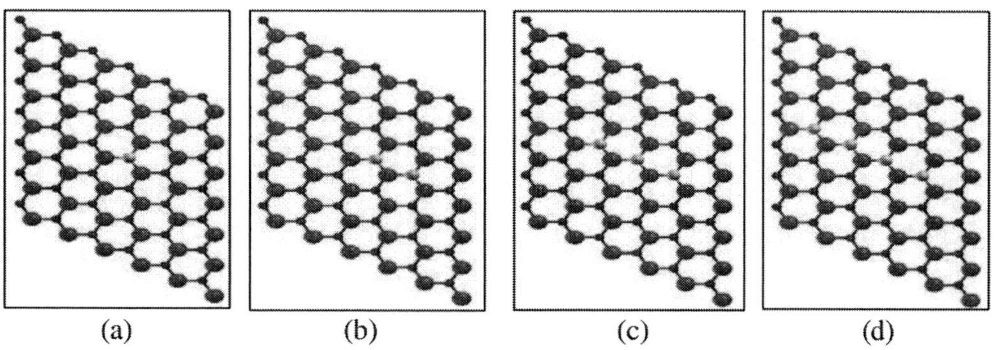

IOP Conf. Series: Journal of Physics: Conf. Series **1032** (2018) 012045 doi:10.1088/1742-6596/1032/1/012045

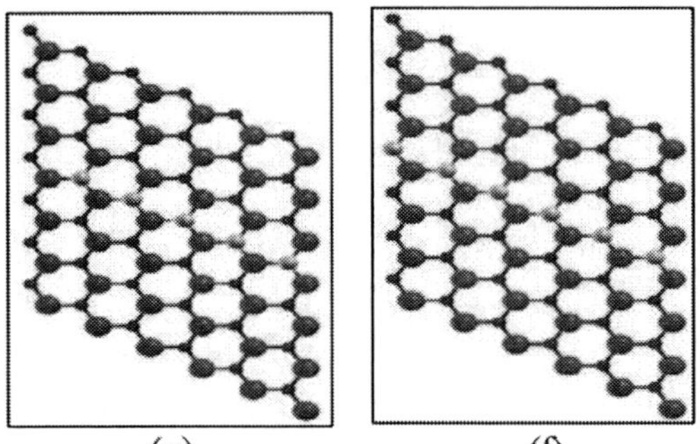

Figure 2. (a-f) Optimize structure of P doped ZnO nanosheet, the gold spheres refer to P atoms.

Where$E_{tot}(ZnOM_n)$, E_{Zn}, E_O, and E_M are the total energies one atom of Zn, O, and dopant n representing the number of atoms of each Zn, O, and dopant respectively, and N are the total number of atoms. The negative value of binding energy corresponds the system is a stable, and observed the system is metastable when add the dopant. This makes the system in a state of relaxation in relation to all degrees of freedom. In figure 3, it was observed that binding energy will increase when the doping one atom, then decreases gradually with increasing doping dose of Al and P atoms in nanosheet. This is due to the association of both Al and P atoms with Zn atoms will weaken Zn-O bonds.

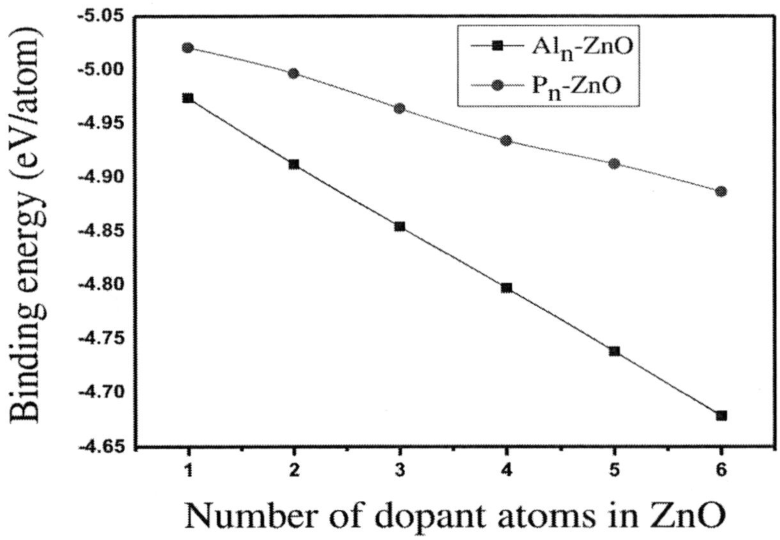

Figure 3. The variation binding energy for Aln-ZnO and Pn-ZnO (n= 1-6).

The ionization process for Vertical Ionization Potential (VIP) and Vertical Electron Affinity (VEA) is an instantaneous process occurs when removing or adding an electron in addition does not relax of molecular geometry. The Vertical Ionization Potential of ZnO nanosheet doping Al and P atoms is shown in fig. 4. In the case of Al we observed that there is a slight stability with the increase of doping and then get a sharp decrease in Al_5-ZnO and then returns in increase. The decrease in the ionization potential due to the fact that the Al contains one electron in the secondary casing p, which facilitates the loss of this electron. In the case of p the potential of ionization was found to vary between 3.5eV and 4.1eV and take a form oscillatory. the ionization potential for p higher of ionization potential for Al, this is due to the secondary casing for P is saturated with more than that in the Al. Figure 5 shows our calculated electron affinity of doped ZnO nanosheet for different Al and P atoms. We observed that the electron affinity for Al dopant takes a form of oscillator and shows the similar trend for P dopants. The lowest value of electron affinity at Al_3-ZnO and P_2-ZnO while highest value Al_3-ZnO and P_6-ZnO, in general the behavior is oscillatory. The high electron affinity for Al and P dopants is found to very 3.0eV and 3.28eV; this indicates that ZnO nanosheet is likely to be a good electron acceptor. So it can be used in applications such as photovoltaics as well as charge mobility and re-regulation energy.

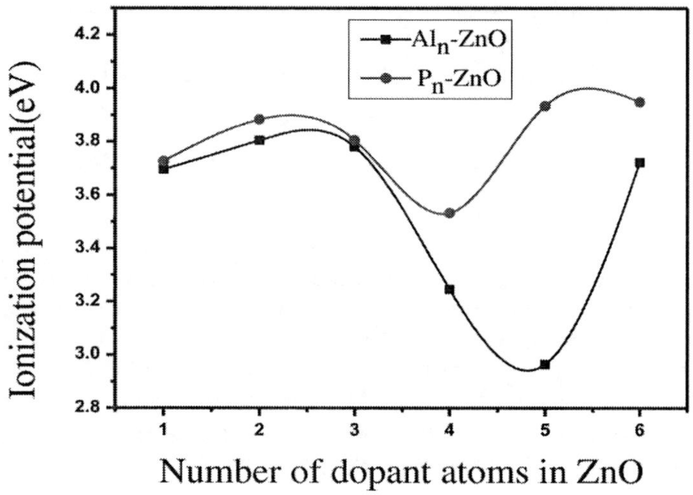

Figure 4. The variation ionization potential for Al_n-ZnO and P_n-ZnO (n= 1-6).

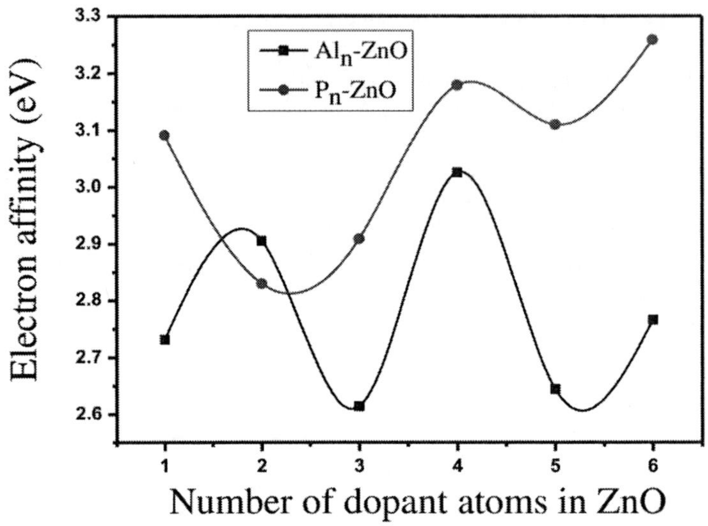

Figure 5. The variation electron affinity for Al_n-ZnO and P_n-ZnO (n= 1-6).

The energy band gap of nanosheet can be determined by the difference between highest occupied molecular orbital and lowest unoccupied molecular orbital (HOMO-LUMO gap), and energy gap is also considered one of chemical stability indicators for nanosheet. A plot of the variation of energy with number of atoms is shown in Fig. 6. The energy gap has been calculated at the GGA-PBE level that shows the lowest energy gap has 0.03eV and 0.012 eV for Al and P doped. It is less than the energy gap in pure ZnO nansheet. This is due to the fact that the electrons reside in the conduction band of a semiconductor and get full saturation. While we get the highest energy gap 0.10eV at Al atom and 0.11eV for P atoms this leads to higher stability kinetic.

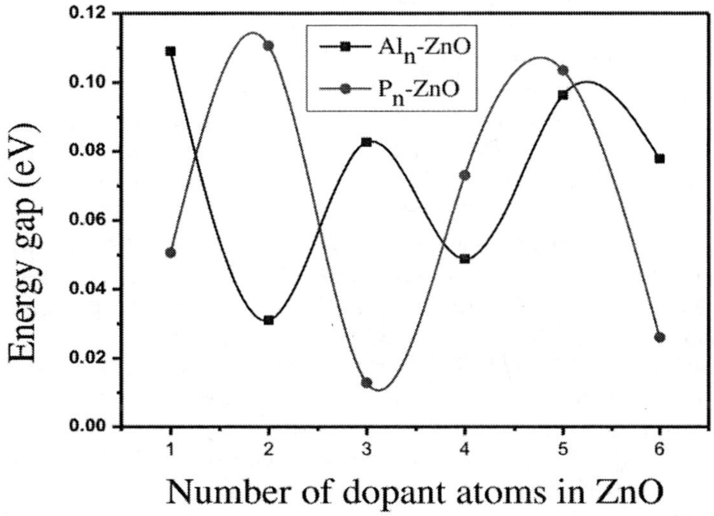

Figure 6. The HOMO-LUMO gap versus the number of doped Al and P atoms.

4. Conclusions

In summary, we investigated the electronic structure, and structural stability of ZnO nanosheet doped by Al, and P using the ab initio calculation in generalized gradient approximation (GGA). Our results show that the lowest energy gap in Al_2-ZnO, and P_3-ZnO are smaller than theof ZnO nanosheet. This refers to the kinetic reactive of these complexes. Also we observed that the P doped ZnO nanosheet is slightly more reactive compared to Al because it has the highest value of a vertical electron affinity and ionization potential.

5. Acknowledgement

Authors are thankful to the SIESTA group for providing their computational code and greatly acknowledge the Department of Physics, Science of college, Kerbala University, Karbala, Iraq. Authors are grateful to organizers conference in Department of Physics, Science of college, Kerbala University, Karbala, Iraq.

Reference

[1] Shinde K, Pawar R, Sinha B, Kim H, Oh S and Chung K 2014 Study of effect of planetary ball milling on ZnO nanopowder synthesized by co-precipitation *Journal of Alloys and Compounds* **617** 404-7
[2] Hayashi H, Ishizaka A, Haemori M and Koinuma H 2003 Bright blue phosphors in ZnO–WO_3 binary system discovered through combinatorial methodology *Applied Physics Letters* **82** 1365-7
[3] Wan Q, Li Q, Chen Y, Wang T-H, He X, Li J and Lin C 2004 Fabrication and ethanol sensing characteristics of ZnO nanowire gas sensors *Applied Physics Letters* **84** 3654-6
[4] Janotti A and Van de Walle C G 2009 Fundamentals of zinc oxide as a semiconductor *Reports on progress in physics* **72** 126501
[5] Al-Sunaidi A and Goumri-Said S 2011 Investigating the adsorption of H_2O on ZnO nanoclusters by first principle calculations *Chemical Physics Letters* **507** 111-6

[6] Guo H, Zhao Y, Lu N, Kan E, Zeng X C, Wu X and Yang J 2012 Tunable magnetism in a nonmetal-substituted ZnO monolayer: a first-principles study *The Journal of Physical Chemistry C* **116** 11336-42
[7] Claeyssens F, Freeman C L, Allan N L, Sun Y, Ashfold M N and Harding J H 2005 Growth of ZnO thin films—experiment and theory *Journal of Materials Chemistry* **15** 139-48
[8] Tusche C, Meyerheim H and Kirschner J 2007 Observation of depolarized ZnO (0001) monolayers: formation of unreconstructed planar sheets *Physical review letters* **99** 026102
[9] Weirum G, Barcaro G, Fortunelli A, Weber F, Schennach R, Surnev S and Netzer F 2010 Growth and surface structure of zinc oxide layers on a Pd (111) surface *The Journal of Physical Chemistry C* **114** 15432-9
[10] Feng X-Y, Wang Z, Zhang C-W and Wang P-J 2013 Electronic structure and energy band of IIIA doped group ZnO nanosheets *Journal of Nanomaterials* **2013** 61
[11] Fan D, Zhang R and Wang X 2011 Effect of phosphorus incorporation on morphology and optical properties of ZnO nanorods *Materials Research Bulletin* **46** 596-600
[12] Lin S, Ye X, Johnson R S and Guo H 2013 First-principles investigations of metal (Cu, Ag, Au, Pt, Rh, Pd, Fe, Co, and Ir) doped hexagonal boron nitride nanosheets: stability and catalysis of CO oxidation *The Journal of Physical Chemistry C* **117** 17319-26
[13] Song E, Wen Z and Jiang Q 2011 CO catalytic oxidation on copper-embedded graphene *The Journal of Physical Chemistry C* **115** 3678-83
[14] Soler J M, Artacho E, Gale J D, García A, Junquera J, Ordejón P and Sánchez-Portal D 2002 The SIESTA method for ab initio order-N materials simulation *Journal of Physics: Condensed Matter* **14** 2745
[15] Giannozzi P, Baroni S, Bonini N, Calandra M, Car R, Cavazzoni C, Ceresoli D, Chiarotti G L, Cococcioni M and Dabo I 2009 QUANTUM ESPRESSO: a modular and open-source software project for quantum simulations of materials *Journal of Physics: Condensed Matter* **21** 395502
[16] Perdew J P, Burke K and Ernzerhof M 1996 Generalized gradient approximation made simple *Physical review letters* **77** 3865
[17] Junquera J, Paz Ó, Sánchez-Portal D and Artacho E 2001 Numerical atomic orbitals for linear-scaling calculations *Physical Review B* **64** 235111
[18] Monkhorst H J and Pack J D 1976 Special points for Brillouin-zone integrations *Physical Review B* **13** 5188
[19] Benali Kanoun M, Goumri-Said S, Manchon A and Schwingenschlögl U 2012 Ferromagnetism carried by highly delocalized hybrid states in Sc-doped ZnO thin films *Applied Physics Letters* **100** 222406
[20] Catti M, Noel Y and Dovesi R 2003 Full piezoelectric tensors of wurtzite and zinc blende ZnO and ZnS by first-principles calculations *Journal of Physics and Chemistry of Solids* **64** 2183-90
[21] Ye X, Song H, Zhong W, Xu M, Qi X, Jin C, Yang Z, Au C and Du Y 2008 The effect of nitrogen incorporation on the magnetic properties of carbon-doped ZnO *Journal of Physics D: Applied Physics* **41** 155005
[22] Tu Z 2010 First-principles study on physical properties of a single ZnO monolayer with graphene-like structure *Journal of Computational and Theoretical Nanoscience* **7** 1182-6
[23] Kökten H and Erkoç Ş 2015 X-Doped (X= C, N, F, P) ZnO Sheet: Density Functional Theory Calculations *Journal of Computational and Theoretical Nanoscience* **12** 395-8

IOP Conf. Series: Journal of Physics: Conf. Series **1032** (2018) 012046 doi:10.1088/1742-6596/1032/1/012046

Low-Lying States Properties of the Even-Even ^{78}Se and ^{80}Kr Isotopes

Wisam N. Hussain and Fadhil I. Sharrad*

Department of Physics, College of Science, University of Kerbala, 56001 Karbala, Iraq

*fadhil.altaie@gmail.com

Abstract. The interacting boson model has been applied to calculate the low-lying bands in Se and Kr nuclei with neutron number (N=44) and proton numbers Z=34 and 36. Reasonable agreement with available energies and B(E2) transition rates. The potential energy surfaces (PESs) to the IBM Hamiltonian have been obtained using the intrinsic coherent state.

Keywords: Interacting Boson Model; Energy levels; B(E2), PES.
PACS : 21.60.Ev; 27.70.+q; 23.20.-g; 23.20; Lv 21.10.-k;42.40. Ht; 42.30. Kq

1. Introduction

Arima and Iachello (1979) [1] have developed the interacting boson model (IBM), which is based on the well-known shell model and on geometrical collective model of the atomic nucleus.
The simplest form of the interacting boson model (IBM) describes even-even nuclei in term of an inner core plus the valence particles outside the nearest closed shells considered as boson. The number of boson is equal to half of the total number of valence protons and neutrons. The ingredients of the IBM are monopole s and quadrupole d bosons, which correspond respectively to collective nucleon-pairs with angular momenta $J^{\pi} = 0^{+}$ and 2^{+} [1, 2]. Different reductions of the unitary group U(6) give three dynamical symmetry limits, known as harmonic oscillator, deformed rotor and asymmetric deformed rotor, which are labeled by U(5), SU(3), and O(6), respectively [3–5]. Another phenomenological study indicated that nuclei might have an intermediate structure of the U(5)–SU(3), U(5)–O(6) and SU(3)–O(6) limits [6-8]. IBM-1 is a successful model for describing structures of the low excited states in the medium and heavy even-even nuclei [9, 10].
The neutron-proton interaction is known to play a dominant role in quadrupole correlations in nuclei. As a consequence, the excitation energies of collective quadrupole excitations in nuclei near a closed shell are strongly dependent on the number of nucleons outside the closed shell. The nuclei ^{78}Se and ^{80}Kr, have atomic number Z = 34 and 36, and same neutron number N = 44, these nuclei with Z > 28, N < 50 protons and neutrons are allowed to occupy $g_{9/2}$, $p_{1/2}$, $p_{3/2}$ and $f_{5/2}$ orbitals. In the past few years, low-lying energy states and electromagnetic transition properties in several isotopes like Se and Kr have been measured for the first time or re-measured with higher precision in the A $\approx$ 80 mass region [11- 18]. Recently, in the same region of O(6) symmetry, some nuclei studied like the even–even Pt isotopes for A=190 to 196 [19], some even $^{120-126}$Xe isotopes [20], even-even Pd isotopes for A = 102 to 106[21] and the low-lying positive parity yrast bands in $^{190-198}$Hg nuclei[22].

IOP Conf. Series: Journal of Physics: Conf. Series **1032** (2018) 012046 doi:10.1088/1742-6596/1032/1/012046

The aim of the present work by application of IBM-1 to predict the low-lying levels, reduced transition probabilities and PES to understand the type of dynamical symmetry which exist in Se and Kr nuclei for neutron N = 44.

2. Method of Calculations

The IBM-1 Hamiltonian can be expressed as [1, 23, 24]:

$$\begin{aligned} H &= \varepsilon_s(s^\dagger.\tilde{s}) + \varepsilon_d(d^\dagger.\tilde{d}) \\ &+ \sum_{L=0,2,4} \frac{1}{2}(2L+1)^{\frac{1}{2}} C_L\left[\left[d^\dagger \times d^\dagger\right]^{(L)} \times \left[\tilde{d} \times \tilde{d}\right]^{(L)}\right]^{(0)} \\ &+ \frac{1}{\sqrt{2}}v_2\left[\left[d^\dagger \times d^\dagger\right]^{(2)} \times \left[\tilde{d} \times \tilde{s}\right]^{(2)} + \left[d^\dagger \times s^\dagger\right]^{(2)} \times \left[\tilde{d} \times \tilde{d}\right]^{(2)}\right]^{(0)} \\ &+ \frac{1}{2}v_0\left[\left[d^\dagger \times d^\dagger\right]^{(0)} \times [\tilde{s} \times \tilde{s}]^{(0)} + \left[s^\dagger \times s^\dagger\right]^{(0)} \times \left[\tilde{d} \times \tilde{d}\right]^{(0)}\right]^{(0)} \\ &+ \frac{1}{2}u_0\left[\left[s^\dagger \times s^\dagger\right]^{(0)} \times [\tilde{s} \times \tilde{s}]^{(0)}\right]^{(0)} + u_2\left[\left[d^\dagger \times s^\dagger\right]^{(2)} \times \left[\tilde{d} \times \tilde{s}\right]^{(2)}\right]^{(0)} \end{aligned} \quad (1)$$

This Hamiltonian is specified by nine parameters, two appearing in the one body terms (ε_s and ε_d), and seven in the two-body terms [c_L (L = 0, 2,4), v_L (L = 0, 2), u_L (L = 0, 2)], where ε_s and ε_d are the single-boson energies. However, the total number of boson N_b (pairs) is conserved, $N_b = n_s + n_d$ [24].

Then the IBM-1 Hamiltonian in equation (1) can be written in general form as [24, 25]:

$$\hat{H}=\varepsilon\hat{n}_d+a_0\hat{P}.\hat{P}+a_1\hat{L}.\hat{L}+a_2\hat{Q}.\hat{Q}+a_3\hat{T}_3.\hat{T}_3+a_4\hat{T}_4.\hat{T}_4 \quad (2)$$

where ε is the boson energy, and the operators are:

$$\left.\begin{aligned} \hat{n}_d &= (d^\dagger.\tilde{d}) \\ \hat{p} &= 1/2\left[(\tilde{d}.\tilde{d}) - (\tilde{s}.\tilde{s})\right] \\ \hat{L} &= \sqrt{10}\left[d^\dagger \times \tilde{d}\right]^1 \\ \hat{Q} &= \left[d^\dagger \times \tilde{s} + s^\dagger \times \tilde{d}\right]^{(2)} + \chi\left[d^\dagger \times \tilde{d}\right]^{(2)} \\ \hat{T}_r &= \left[d^\dagger \times \tilde{d}\right]^{(r)} \end{aligned}\right| \quad (3)$$

Here, $\hat{n}_d$ is the total number of d_{boson} operator, $\hat{p}$ is the pairing operator, $\hat{L}$ is the angular momentum operator, $\hat{Q}$ is the quadrupole operator (χ is the quadrupole structure parameter and take the values 0 and $\pm\frac{\sqrt{7}}{2}$ [4, 26, 27]) and $\hat{T}_r$ is the octupole (r=3) and hexadecapole (r=4) operator. The phenomenological parameters a_0, a_1, a_2, a_3 and a_4 represent the strengths of the pairing, angular momentum, quadrupole, octupole and hexadecapole interaction between the bosons, respectively. In that case, one says that the Hamiltonian (H) has a dynamical symmetry. These symmetries are called U(5) vibrational, SU(3) rotational and O(6) γ-unstable[24].

The eigenvalues for these three limits are given by [24]:

$$\left.\begin{aligned} E &= \varepsilon nd + \beta nd\,(nd + 4) + 2\gamma\, \upsilon(\upsilon + 3) + 2\delta L\,(L + 1) \ldots\ldots\ldots\ldots \mathrm{U}(5) \\ E &= \frac{a_2}{2}\left(\lambda^2 + \mu^2 + \lambda\mu + 3(\lambda + \mu)\right) + \left(a_1 - \frac{3a_2}{8}\right)L(L+1) \ldots\ldots\ldots\ldots SU(3) \\ E &= a_0/4\,(N - \sigma)(N + \sigma + 4) + a_3/2\ \tau(\tau + 3) + (a_1 - a_3/10)\,L(L + 1) \ldots O(6) \end{aligned}\right| \quad (4)$$

According to the Hamiltonian [1, 24], we can be discussed the calculated results separately, by plotting the potential energy surface $((N, \beta, \gamma))$, which gives a final shape to the nucleus

The technique described by Dieperink et al. [28] allows one also to give an algebraic description of the nature of the transition between one phase and another. Simpler expressions, which display the essential dependence on β and γ, have been given [4, 24, 29]:

$$E(N_b, \beta, \gamma) = \varepsilon\, N_b[\beta^2/(1+\beta^2)], \ldots \mathrm{U}(5) \quad (5)$$

$$E(N_b, \beta, \gamma) = a_2 N_b(N_b - 1)[(1 + 3/4\beta^4 - \sqrt{2}\beta^3 \cos 3\gamma)/(1 + \beta^2)^2], \ldots \mathrm{SU}(3) \quad (6)$$

IOP Conf. Series: Journal of Physics: Conf. Series **1032** (2018) 012046 doi:10.1088/1742-6596/1032/1/012046

$$E(N_b, \beta, \gamma) = a_0 N_b(N_b - 1)[(1 - \beta^2)/(1 + \beta^2)]^2, \ldots O(6), \quad (7)$$

where β and γ are the intrinsic deformation parameters which determine the geometrical shape of the nucleus. These expression give (for large N_b) $\beta_{min} = 0$, $\sqrt{2}$, and 1 for U(5), SU(3), and O(6), respectively.

3. Results and discussion

Se and Kr isotopes have neutron number N = 44 and have an even atomic number Z =34 and 36 (Z values near mid shell and N value near closed shell would suggest structure the O(6) symmetry and O(6) – U(5) transition). The degree (and type) of collectivity can be expressed in terms of the energy ratio $R_{4/2} = E_{4_1^+}/E_{2_1^+}$, which used as a starting point and is a good indicator of the shape deformation of the nucleus and its value is 10/3 for the well-deformed nuclei SU(3), 2.5 for O(6) or γ-unstable nuclei and 2 for vibrational U(5), 2.2 for the analytically solvable symmetry E(5) on the U(5) - O(6) path and 2.9 for the approximate X(5) symmetry on the U(5)-SU(3) path[24, 30-34].
The experimental values of $R = E4_1^+/E2_1^+$ of low–lying energy levels of ^{78}Se and ^{80}Kr nuclei are shown in Table 1. From this Table, $R_{4/2}$ attains the O(6) value of ~ 2.5 in ^{78}Se at Z= 34 and in Z= 36 ^{80}Kr lie close to the E(5) symmetry .

Table 1. The ratio $R_{4/2} = E_{4_1^+}/E_{2_1^+}$ for ^{78}Se and ^{80}Kr nuclei [35-37].

Nucleus	^{78}Se	^{80}Kr
$R4/2$	2.44	2.32

The calculations have been performed using IBM with PHINT code [38] and, hence, no distinction made between neutron and proton bosons which calculated from the sum of the proton bosons and the neutron bosons of the close shells (28 and 50). The number of bosons and the parameters of the IBM-1 Hamiltonian (2) which give the best fitting between theoretical and experimental energy levels [35-37] of the above isotopes are shown in Table 2.

Table 2. Adopted values for the parameters used for IBM-1 calculations. All parameters are given in MeV, excepted N_b for Se and Kr isotopes.

Isotopes	N_b	ε	PAIR	ELL	OCT
^{78}Se	6	---	0.107	0.080	0.053
^{80}Kr	7	0.1	0.083	0.079	0.054

(ELL = $2a_1$ and QQ = $2a_2$, CHQ= $\sqrt{5}\chi$)[24].

Figure 1 show that the calculated and experimental values of ground (GSB), β- and γ- bands are plotted for even-even Se and Kr isotopes. In the Figure 1, the calculated energy levels are in good agreement with the experimental [35-37] ones for Se and Kr isotopes. Levels with "( )" in g, γ and β states correspond to cases for which the spin and/or parity of the corresponding states are not well established experimentally.

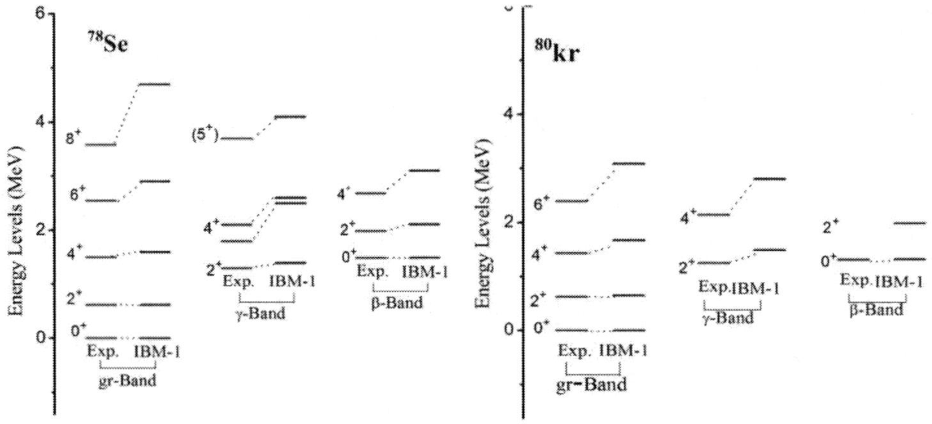

Figure 1. (Color online) comparisons between the calculated IBM-1 and the experimental data [35-37] for ^{78}Se and ^{80}Kr with N=44.

Now, we discuss other information on the structure of nuclei is the transition strength between the excited states and can be expressed in terms of the reduced E2 matrix element which must be a Hermitian tensor of rank two when N_b must be conserved. The $B(E2)$ strength for the $E2$ transitions is given by [1, 24, 39]

$$T^{E2} = \alpha_2 [d^\dagger s + s^\dagger d]^{(2)} + \beta_2 [d^\dagger d]^{(2)} = e_B \hat{Q} \qquad (8),$$

where $(s^\dagger, d^\dagger)$ and (s, d) are creation and annihilation operators for s and d bosons, respectively, while α_2 and β_2 are two parameters, and $\left(\beta_2 = \chi\alpha_2, \alpha_2 = e_B(\text{effective charge of boson})\right)$.
The reduced transition probability for the U(5) and O(6) limits are given by[24, 34]:

$$\text{U(5)} \quad B(E2; L \to L-2) = e_B{}^2(n_d+1)(N-n_d) \qquad (9)$$

$$\text{O(6)} \quad B(E2; L \to L-2) = e_B{}^2(N-\tau)(N+\tau+4)\frac{\tau+1}{2\tau+5} \qquad (10)$$

where L is the angular momentum. From the given experimental value B(E2) of transition $(2_1^+ \to 0_1^+)$, one can calculate the value of the parameter $e_B = \alpha_2$for each isotope. This value is used to calculate the reduced transition probabilities $B(E2; L \to L-2)$. Table 3 shows the values of the e_B parameter, which was obtained in the present calculations. Table 4 shows the B(E2) in the low lying states for Se and Kr nuclei with the neutron number N=44 and compared with the experimental data [35-37]. The reduced transition probabilities B(E2) are increase as proton number increases and most of the calculated results in IBM-1 reasonably consistent with the available experimental data, except for few cases that deviate from the experimental data [35-37] in Se and Kr isotopes.

Table 3. Parameter (in eb) used to reproduce B(E2) values for ^{78}Se and ^{80}Kr isotopes.

A	N_b	e_B
^{78}Se	6	0.074
^{80}Kr	7	0.070

Table 4. The IBM-1 and Experimental [35-37] values of B(E2) for ^{78}Se and ^{80}Kr isotopes (in $e^2 b^2$).

$L_i \to L_f$	^{78}Se		^{80}Kr	
	EXP.	IBM	EXP.	IBM
$2_1^+ \to 0_1^+$	0.066	0.0670	0.0760	0.0751
$2_2^+ \to 2_1^+$	0.0439	0.0868	0.051	0.1003
$2_4^+ \to 2_1^+$	0.0089	0.0000	--	0.0005
$2_3^+ \to 0_2^+$	0.0198	0.0203	--	0.0645
$3_1^+ \to 2_1^+$	0.0015	0.0023	0.0010	0.0090
$3_1^+ \to 2_2^+$	0.0495	0.0631	0.0690	0.0755
$3_1^+ \to 4_1^+$	--	0.0006	0.1020	0.0235
$4_1^+ \to 2_1^+$	0.0975	0.0868	0.1420	0.1003
$4_2^+ \to 4_1^+$	--	0.0765	0.1020	0.0854
$5_1^+ \to 4_1^+$	--	0.0025	0.0020	0.0061

In Figure 4, the contour plot of the potential energy surfaces, E(N$_b$,β,γ), show that ^{78}Se is a deformed and has γ-unstable-like characters ($\gamma \approx \frac{\pi}{6}$), while ^{80}Kr, the shape phase transition from γ-unstable O(6) to vibrational U(5) symmetry.

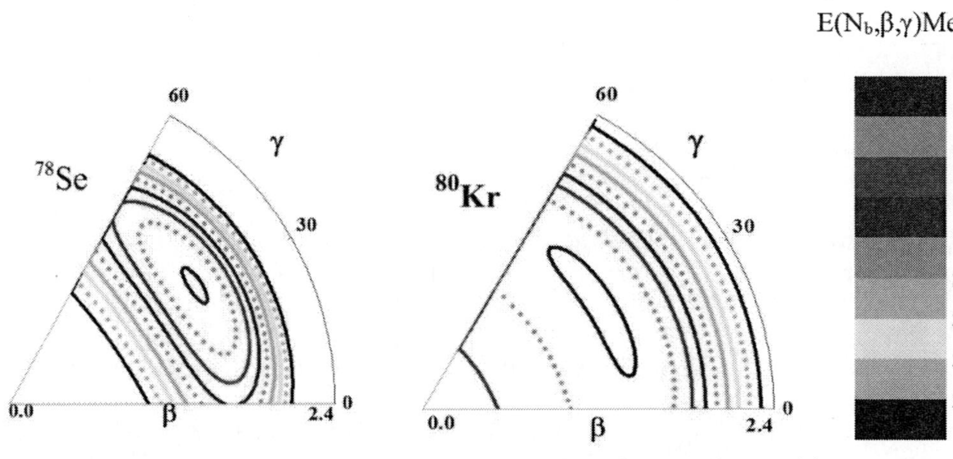

Figure 2. (Color online) the potential energy surfaces for Se and Kr nuclei.

4. Conclusions

The low-lying bands are calculated using IBM-1 for ^{78}Se and ^{80}Kr nuclei with neutron number N=44. The result shows good agreement with published experimental data. The reduced transition probabilities B(E2) values have been calculated using Interacting Boson Model (IBM). A good agreement is obtained for all the observed able studied. The contour plot of PES show, that ^{the} ^{78}Se is a deformed and has γ-unstable-like characters, while ^{80}Kr lie close to the E(5) symmetry.

Acknowledgments
We thank University of Kerbala, College of Science, Department of Physics for supporting this work.

References

[1] Iachello F and Arima A 1987 Cambridge University Press, Cambridge.
[2] Khudher H H, Hasan A K and Sharrad F I 2017 *AIP Conference Proceedings* **1888** 020030.
[3] Arima A, Iachello F 1976 *Ann. Phys. (N.Y.)* **99** 253.
[4] Arima A, Iachello F 1978 *Ann. Phys. (N.Y.)* **111** 201.
[5] Arima A, Iachello F 1979 *Ann. Phys. (N.Y.)* **123** 468.
[6] Iachello F 2000 *Phys. Rev. Lett.* **85** 3580.
[7] Sharrad F I, Abdullah HY, AL-Dahan N, Mohammed-Ali A A, Okhunov A A and Abu-Kassim H 2012 *Rom. Journ. Phys.* **57** 1346.
[8] Casten R F and McCutchan E A 2007 *J. Phys. G* **34** R285.
[9] Sharrad F I, Hossain I, Ahmed I M, Abdullah HY, Ahmad S T and Ahmed A S 2015 *Braz J Phys* **45** 340.
[10]] Sharrad F I, Abdullah HY, AL-Dahan N, Umran N M, Okhunov A A and Abu-Kassim H 2013*Chinese Physics, C* **37** 034101.
[11] Hossain I, Ahmed I M, Sharrad F I, Abdullah HY, Salman A D and AL-Dahan 2015 *Chiang Mai J. Sci.* **42** 996.
[12] Shelley A, Hossain I, Sharrad F I, Abdullah HY and Saeed M A 2015 *PROBLEMS OF ATOMIC SCIENCE AND TECHNOLOGY* **64** 38.
[13] Hossain I, Kassim H H, Sharrad F I and Ahmed A S 2016 *ScienceAsia* **42** 22.
[14] Al-Jubbori M A, Kassim H H, Sharrad F I and Hossain I 2016 *Nucl. Phys. A* **955** 101.
[15] Hossain I, Sharrad F I, Saeed M A, Abdullah HY and Mansour S. A 2016 *Maejo Int. J. Sci. Technol.* **10** 95.
[16] Khudher H H, Hasan A K and Sharrad F I 2017 *Chinese Journal of Physics* **55** 1754.
[17] Bo Bai H, Wei Li X, Jun Lü L, Fei Dong H, Wang Y and Fu Zhang J 2016 *Chinese Physics C* **40** 074103.
[18] Kassim H H and Sharrad F I 2015 *Nucl. Phys. A* **933** 1.
[19] Kassim H H and Sharrad F I 2014 *International Journal of Modern Physics E* **23** 1450070.
[20] Khudher H H, Hasan A K and Sharrad F I 2017 *Ukrainian Journal of Physics* **62** 152.
[21] Waheed M O and Sharrad F I 2017 *Ukrainian Journal of Physics* **62** 757.
[22] Hussain K A, Mohsin M K and Sharrad F I 2017 *Ukrainian Journal of Physics* **62** 653.
[23] Waheed M O and Sharrad F I 2017 *NUCLEAR PHYSICS AND ATOMIC ENERGY* **18** 313
[24]Casten R F and Warner D D 1988 *Rev. Mod. Phys.* **60** 389.
[25] Al-Jubbori M A, Kassim H H, Sharrad F I Attarzadeh A and Hossain I 2018 *Nuclear Physics A* **970** 438.
[26] Okhunov A A, Sharrad F I, Al-Sammarea A A and Khandaker M U 2015 *Chinese Physics C* **39** 084101.
[27] Kassim H H, Mohammed-Ali A A, Al-Jubbori M A, Sharrad F I, Ahmed A S and Hossain I 2018 *J. Natn. Sci. Foundation Sri Lanka* **46** 3.
[28] Dieperink A E L, Scholten O and Iachello F 1980 *Phys. Rev. Lett.* **44** 1747.
[29] Ahmed I M, Flaiyh G N, Kassim H H, Abdullah H Y, Hossain I and Sharrad F I 2017 *Eur. Phys. J. Plus* **132** 84.
[30] Iachello F 2000 *Phys. Rev. Lett.* **85** 3580.
[31] Iachello F 2001 *Phys. Rev. Lett.* **87** 052502.
[32] Casten R F and Zamfir NV, 2001 *Phys. Rev. Lett.* **87** 30.
[33] Casten R F, 2005 *Roma. Reports in Phys*. **57** 515.
[34 Al-Jubbori M A, Radhi F Sh., Ibrahim A A, Abdullah Albakrid S A, Kassim H H and Sharrad F I 2018 *Nuclear Physics A* **971** 35.
[35] http://www.nndc.bnl.gov/chart/getENSDFDatasets.jsp.

[36] Farhan A R and Singh B 2009 *Nuclear Data Sheets* **110** 1917.
[37] Singh B 2005 *Nuclear Data Sheets* **105** 223.
[38] Scholten O 1980 Computer code PHINT, KVI; Groningen, Holland.
[39] Al-Jubbori M A, Al-Mtiuty K A, Saeed K I and Sharrad F I 2017 *Chinese Physics C* **41** 084103.

Assessing land cover/use changes in Karbala city (Iraq) using GIS techniques and remote sensing data

Enas Ali Mohammed[1], Zahra Yarub Hani[1] and Ghufran Qais Kadhim[1]

[1]Department of Computer Science, College of Science, University of Kerbala, Iraq.

the_programmer86@yahoo.com

Abstract. In this research, land cover changes between 2000 to 2017 were studied in Karbala city (Iraq) using geographic information system techniques and remote sensing data. Satellite imagery data -Landsat7 (2000) and Landsat 8 (2017) - were processed using the supervised maximum likelihood algorithm, and the post-classification comparison method in QGIS 2.14. Four land cover categories, namely, built-up, vegetation, water, and soil have been specified to produce land cover maps for each acquisition date. The overall accuracies were 99.99% (2000), and 99.98% (2017) with Kappa statistics of 0.99, and 0.99. The evaluation of land cover changes shows that built-up, vegetation and soil have been increased by 1.9% (93.5 km2), 0.4% (20.8 km2) and 4.4% (221.3 km2) while water land has a shrink by 6.7% (335.6 km2) respectively. These changes conclude a serious threat to water land which it effects on population growth, human activities and agricultural activities in the study area.
Keywords: Remote sensing, Karbala, Maximum Likelihood, Post-classification ,Land use, Land cover, GIS.

1. Introduction

Effective analysis and monitoring of land cover/use changes are necessary for environmental sustainability. Several factors have triggered rapid unplanned and uncontrolled changes to earth's land cover over the last two decades such as population growth, urban expansion, cropland loss, climate variability, and etc. These changes have resulted in deforestation, degradation, biodiversity loss, mismanagement of agricultural, natural disasters, and increase global warming[1],[2]. In the recent years, there has been an increasing interest in studying the land cover status at regional and global scales, as well as the assessment of land cover changes is fundamental for a sustainable management of natural resources [3,2,4,5,6,7].

Land cover/use change detection is necessary for better understanding variations in the nature of a phenomenon by studying it during a known period of time. Land cover/use change is a common process, essentially induced by anthropogenic activities and natural phenomena, which in turn make changes that would affect natural ecosystem [8]. The study of land cover change detection is a major subject in remote sensing applications because of its adverse effect on the ecology of the area and vegetation. So, good planning and management of land cover change play a significant role in climate change mitigation, moderation of extreme events and land degradation [9]. Understanding landscape patterns, changes, and interactions between natural phenomenon and human activities are necessary for good land management and decision improvement. Currently, earth resource satellites data are very useful and applicable for land cover/use change detection studies [6].

IOP Conf. Series: Journal of Physics: Conf. Series **1032** (2018) 012047 doi:10.1088/1742-6596/1032/1/012047

Change analysis of features of earth's surface is necessary for better understanding of relationships and interactions between natural phenomena and human activities. This understanding is necessary for improved resource management and improved decision making. Change detection involves applying multi-temporal remote sensing information to analyze the historical effects of an occurrence quantitatively and thus helps in determining the changes associated with land cover and land use properties with reference to the multi-temporal datasets. The monitoring of land cover/use change dynamics can be applied by comparing a land cover/use map at present with the conditions in prior years as multi temporal [10,11,3].

In recent years, Geographic information system (GIS) and remote sensing technologies have progressed significantly because these technologies provide information to study and monitor the dynamics of natural resources for sustainable development and environmental management. Now the land cover/use change can be assessed or modeled using these technologies by analyzing multi temporal remotely sensed data to identify, detect, monitor, and map differences in LU/LC patterns over time [6]. The availability of historic spatiotemporal data and high-resolution satellite images from many satellites such as MODIS, CBERS, TM and ETM Landsat are necessary for decision makers and planners to get accurate information about land covers' changes [1].

The primary purpose of this study was to locate the range of changes occurred in Karbala-Iraq over 17 years' time period. However, the specific purposes included (1) to delineate various main land cover/use categories in Karbala city from 2000 to 2017 (2) to test the possibility of combining GIS with remote sensing in analyzing the spatial distribution of various land cover/use changes (3) to determine and quantify the variation in land cover/use categories through spatial comparison of the land cover/use maps generated.

2. Study Area

Karbala city is placed in the middle of Iraq about 110Km south-west Baghdad. It is located between latitude 32N to latitude 33N and longitude 43E to longitude 44E, its area about 5.034 km². The Karbala city includes three provinces which are the center of Karbala, Hindiyah, and Ean-tamer. In the last two decades, Karbala has witnessed a great incensement in urbanization as a result of the annual growth in people and migration to it. The selected area could be recognized in figure (1), which illustrates the Iraq map with a selected window represents the study area. The lands of Karbala are urbanization and agriculture. Agricultural areas include the cultivation of date palms, barley, wheat, fruit trees and summer crops. Urban areas include industrial plants, educational facilities, and residential units.

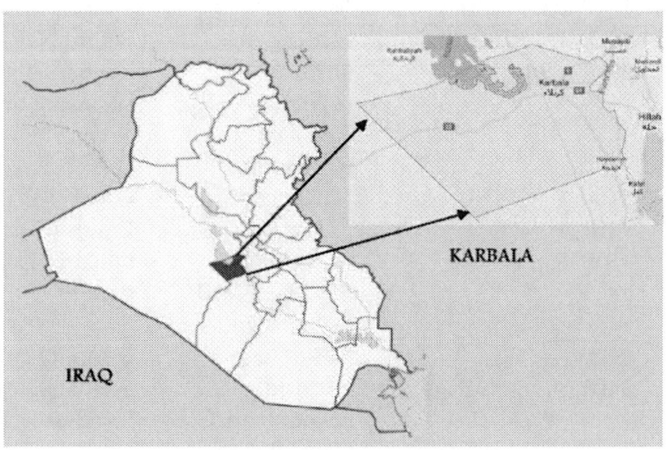

Figure .1 Iraq map with a selected window represents Karbala city.

3. Material and Methods

IOP Conf. Series: Journal of Physics: Conf. Series **1032** (2018) 012047 doi:10.1088/1742-6596/1032/1/012047

3.1. Data Preparation

Multi-temporal Landsat images were obtained from USGS Earth Explorer site[12]. The satellite imagery data for year 2000 was selected from Landsat 7 and for year 2017 was selected from Landsat 8. The satellite data was obtained in the same time period to avoid differences occurred in land cover change after implementing the mosaic process on the classification results. The field surveys were implemented from April to August 2017. Through the field surveys, 300 reference points within the study area were enrolled using global positioning system technology. While old maps the reference points determined from the satellite image taken in 2000.

3.2. Data Pre-Processing and Land Cover Classification

QGIS 2.14 software was used in this study, which is a free and open source geographic information system. The satellite data was imported in QGIS to calculate surface reflectance using DOS1 atmospheric correction tool. The resulted data was clipped to specify the study area.
In this study, the supervised maximum likelihood algorithm was applied to implement the land cover/use classification process. This algorithm is popular in processing remotely sensing image data, which its work based on the probabilities of pixels [5,1,6].

Four land cover/use types are specified in the study vis., the delineated types are (i) built-up land, (ii) vegetation land, (iii) water land, (iv) soil land. For each identified land cover/use type, training areas/regions of interest (ROIs) were selected by delimiting polygons over homogeneous areas of the image. The spectral signatures of the land cover/use are calculated according to the pixels values corresponding to these ROIs. Then the spectral signature list was used by the maximum likelihood algorithm to classify the entire satellite image.

Maximum likelihood is one of the most popular classifications and has been implemented in several land use/ cover change studies[11,3]. This algorithm supposes that the statistics for each class in each band are normally distributed and calculates the probability that a given pixel belongs to a specific class. Each pixel is allocated to the class that has the highest probability. To implement this algorithm, an adequate number of pixels is required for each training area to calculate the covariance matrix.

During work, some indices were calculated to improve the classification results, these indices are illustrated in table (1).

Table 1. **Spectral indices' formula used in this study.**

Eqs.		Formula	Remarks	
1	Normalized difference vegetation index	$NDVI=(NIR_B-R_B)/(NIR_B+R_B)$	NIR_B: near infrared reflectance, R_B: Red light Reflectance	where NIR_B and R_B in Landsat7 are band4 and band3 while in Landsat8 are band5 and band4
2	Normalized difference built-up index	$NDBI=(SWIR_B-NIR_B)/(SWIR_B+NIR_B)$	NIR_B: near infrared reflectance, $SWIR_B$: SWIR reflectance	where $SWIR_B$ and NIR_B in Landsat7 are band5 and band4 while in Landsat8 are band6 and band5
3	Normalized difference water index	$NDWI=(G_B-NIR_B)/(G_B+NIR_B)$	NIR_B: near infrared reflectance, G_B: Green light reflectance	where G_B and NIR_B in Landsat7 are band2 and band4 while in Landsat8 are band3 and band5

3.3. Accuracy Assessment
After implementing the land cover/use classification, it is necessary to evaluate the quality of the classified maps and determine the level to which they corresponds to real world conditions. The accuracy was evaluated using error matrices.

In this study, the accuracy assessment was based on a stratified random sampling method of a minimum of 20 reference points per category from ground truth data and visual interpretation. For the image of 2000, the ground truth data were generated from the satellite images that are suitable for detecting vegetation, water lands, and etc. For the Image of 2017, the ground truth data were acquired from field surveys as well as satellite images. The accuracy of each product was evaluated using several statistics such as user accuracy, producer accuracy, Kappa hat, and overall accuracy, as used in several land cover and land use studies.

3.4. Land Cover/Use Change Assessment
For assessing land cover/use changes, post classification comparison was implemented in QGIS. Post classification comparison is one of the most common methods for calculating land cover changes. This method works on comparing the land cover maps from different classifications and calculating the number of pixels that alter from one land category to another [13]. The result is a land cover map with a change matrix included a quantitative data of the overall changes in each category between 2000 and 2017.

4. Results and Discussion
Land cover maps resulted from the classification process appear good agreement with the real world as showed by overall accuracies of 99.99% for 2000 and 99.98% for 2017. The corresponding Kappa statistics are 0.99 for both 2000, and 2017.

Results obtained from the classification process and change detection are graphically explained in figures (2–4) and data are recorded in tables (2-3). Figure (2) describes land / land cover status, figure (3) describes land cover/use change in different land cover categories and figure (4) explains the magnitude of change in different land categories.

The figure (2-a) depicts the spatial distributional pattern of LU/LC of Karbala for the year 2000, while figure (2-b) for the year 2017. These data reveal that in 2000, about 9.36% (466.98km2) the area of Karbala was under vegetation, 75.49% (3764.35km2) under Soil, 3.87% (193.21km2) under built-up land and 11.26% (561.5km2) under water. During 2017, the study area in the land cover categories was found about 9.78% (487.75km2) under vegetation, 79.93% (3985.66 km2) under Soil, 5.75% (286.74km2) under built-up land and 4.53% (225.89km2) under water (table 1).

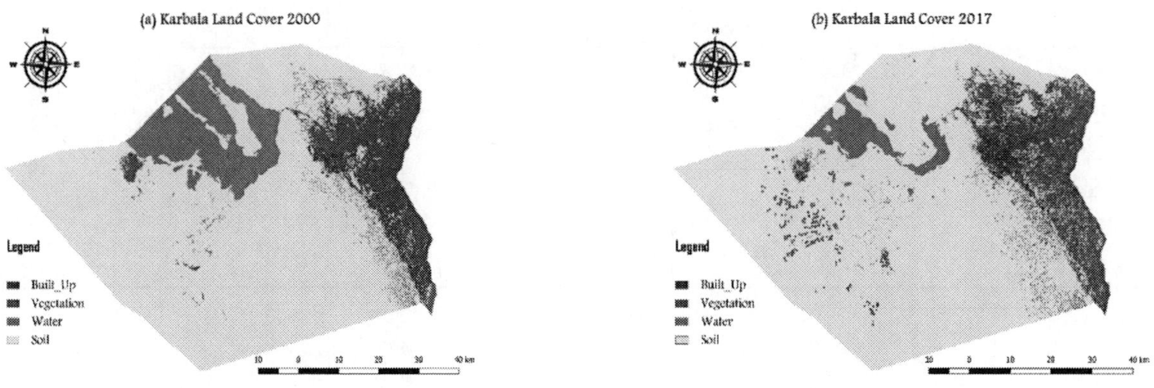

Figure 2. Land cover/use status of Karbala; (a) in 2000, (b) in 2017.

Data recorded in table (1) and figures (3-4) show both negative and positive shifts that have happened in the land cover/use pattern of Karbala.

IOP Conf. Series: Journal of Physics: Conf. Series **1032** (2018) 012047 doi:10.1088/1742-6596/1032/1/012047

In the last two decades, the vegetation in the study area has slightly increased from 466.98 km2 in 2000 to 487.76 km2 in 2017 which accounts for 0.4% of the total study area. The built-up area has increased from 193.21km2 in 2000 to 286.74km2 in 2017 which accounts for 1.9%. The soil land has been increased from 3764.35km2 in 2000 to 3985.66 km2 in 2017 which accounts for 4.4%. The water land of the study area has decreased from 561.5km2 in 2000 to 225.9 km2 in 2010 which accounts for 6.7%.

Table 2. The area and the amount of shift in the identified land cover/use categories in Karbala over the last period (2000 – 2017).

LU/LC categories	2000		2017		Change 2000–2017	
	Km^2	%	Km^2	%	Km^2	%
Built-up	193.217	3.875	286.74	5.75	93.523	1.9
Vegetation	466.985	9.366	487.76	9.78	20.775	0.4
Water	561.504	11.261	225.9	4.53	-335.604	-6.7
Soil	3764.354	75.498	3985.66	79.94	221.306	4.4
Total	4986.06	100	4986.06	100	0	0

Table (3) includes land cover/use change matrix that shows land crawling for different land classes during the last two periods, which reveals that:

i. About 22.77% the area of built-up has been turned into vegetation, 47.25% into soil and 1.1% in water;

ii. About 10.65% the area of vegetation covered has been turned into built-up, 2.14% area under water area and 22.8% area under soil land;

iii. About 1.4% of the area of soil has been turned into built-up, 2.18% area under vegetation and 8.43% area under water land;

iv. About 1.03% area of water covered has been turned into built-up, 0.36% in vegetation and 4.72% in soil.

Table 3. Land cover/use change matrix showing land crawling (in %) of Karbala.

LU/LC categories	Year 2000			
Year 2017	Built-up	Vegetation	Water	Soil
Built-up	28.88	10.65	1.03	1.4
Vegetation	22.77	64.41	0.36	2.18
Water	1.1	2.14	93.89	8.43
Soil	47.25	22.8	4.72	87.99
Class Total	100	100	100	100

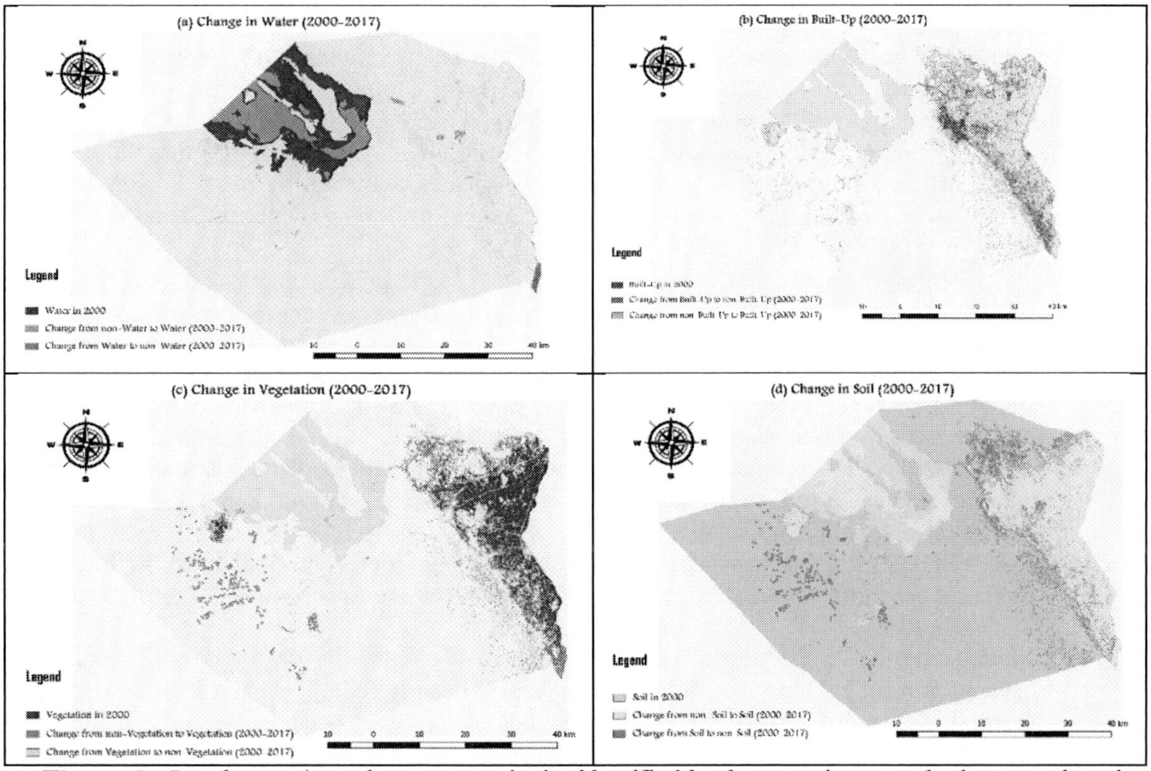

Figure 3. Land cover/use change maps in the identified land categories over the last two decades in Karbala development (2000–2017); (a) water (b) built-up (c) vegetation (d) soil (based on Landsat imagery data).

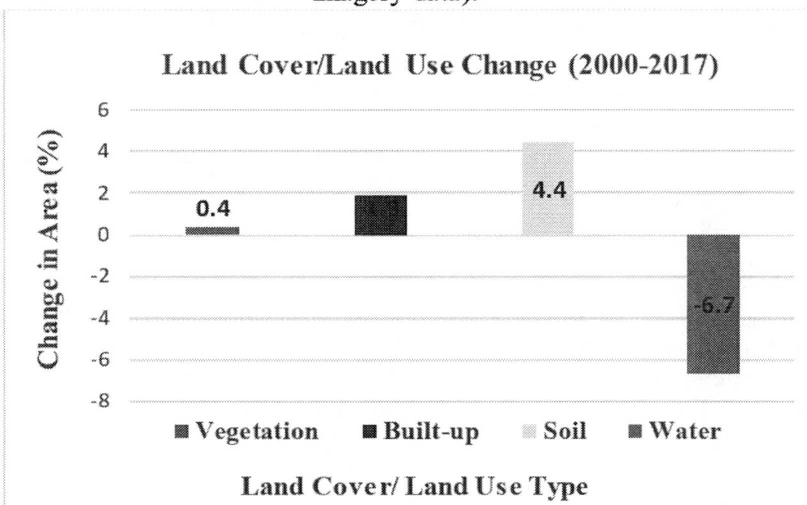

Figure 4. Land cover/use change in percent over the last two periods (2000–2017) in Karbala.

5. Conclusion

The study conducted in one of the important holiest cities in Iraq advocates that multi temporal satellite imagery and remote sensing together with GIS techniques play a vital role to monitor urban expansion effectively over time by producing accurate Land cover/use maps and change statistics. The study shows that the main land use in the study area is built-up. The area of built-up has increased by

1.9 % (93.52 km2) due to expansion during 2000 to 2017. The second main category of land is vegetation which has also increased. During the study period (i.e., 2000–2017), vegetation land has been slightly increased by 0.4% (20.78km2). The third main category of land is water which was decreased by 6.7% (335.6 km2) due to conversion in vegetation, soil land and built-up land. The area under the fourth category of land, i.e., soil land has increased by 4.4% (221.3km2) due the decline in the water land in the last two decades. Based on the results achieved by using GIS and Remote Sensing techniques to obtain the specific research aims for dealing with the dynamics of land cover/use change in Karbala, it's concluded that the land cover/use practices in the study area have altered significantly in 17 years.

6. References

[1] A. Butt, R. Shabbir, S. S. Ahmad, and N. Aziz, "Land use change mapping and analysis using Remote Sensing and GIS: A case study of Simly watershed, Islamabad, Pakistan," Egypt. J. Remote Sens. Sp. Sci., vol. 18, no. 2, pp. 251–259, 2015.

[2] Y. T. Mustafa, R. T. Ali, and R. Najimaldeen, "Monitoring and evaluating land cover change in the Duhok city, Kurdistan region-Iraq, by using remote sensing and GIS," Int. J. Eng. Invent., vol. 1, no. 12, pp. 28–33, 2012.

[3] M. M. El-Hattab, "Applying post classification change detection technique to monitor an Egyptian coastal zone (Abu Qir Bay)," Egypt. J. Remote Sens. Sp. Sci., vol. 19, no. 1, pp. 23–36, 2016.

[4] R. Madugundu, K. A. Al-Gaadi, V. C. Patil, and E. Tola, "Detection of Land Use and Land Cover Changes in Dirab Region of Saudi Arabia Using Remotely Sensed Imageries," Am. J. Environ. Sci., vol. 10, no. 1, pp. 8–18, 2014.

[5] P. K. Mallupattu, J. Reddy, and S. Reddy, "Analysis of Land Use / Land Cover Changes Using Remote Sensing Data and GIS at an Urban Area , Tirupati , India," Sci. world J., vol. 2013, no. Figure 1, pp. 1–7, 2013.

[6] J. S. Rawat and M. Kumar, "Monitoring land use/cover change using remote sensing and GIS techniques: A case study of Hawalbagh block, district Almora, Uttarakhand, India," Egypt. J. Remote Sens. Sp. Sci., vol. 18, no. 1, pp. 77–84, 2015.

[7] J. O. Owoeye and O. A. Ibitoye, "Analysis of Akure Urban Land Use Change Detection from Remote Imagery Perspective," Urban Stud. Res., vol. 2016, pp. 1–9, 2016.

[8] A. Singh, "Review Articlel: Digital change detection techniques using remotely-sensed data," Int. J. Remote Sens., vol. 10, no. 6, pp. 989–1003, 1989.

[9] F. J. Gallego, "Remote sensing and land cover area estimation," Int. J. Remote Sens., vol. 25, no. 15, pp. 3019–3047, Aug. 2004.

[10] S. Sinha, L. K. Sharma, and M. S. Nathawat, "Improved Land-use/Land-cover classification of semi-arid deciduous forest landscape using thermal remote sensing," Egypt. J. Remote Sens. Sp. Sci., vol. 18, no. 2, pp. 217–233, 2015.

[11] S. K. McFeeters, "Using the normalized difference water index (ndwi) within a geographic information system to detect swimming pools for mosquito abatement: A practical approach," Remote Sens., vol. 5, no. 7, pp. 3544–3561, 2013.

[12] U.S. Geological Survey, "USGS Earth Explorer." [Online]. Available: https://earthexplorer.usgs.gov/.

[13] Q. Weng, "A remote sensing – GIS evaluation of urban expansion and its impact on surface temperature in the Zhujiang Delta , China.," Int. J. Remote Sens., vol. 22, no. 10, pp. 1999–2014, 2001.

A comparison of resource allocation process in grid and cloud technologies

Huda Hallawi [1], Jörn Mehnen [2] and Hongmei He [3]

[1] Computer Science Department, University of Kerbala, Iraq
[2] Digital Manufacturing Department, University of Strathclyde, Glasgow, UK
[3] Manufacturing Department, Cranfield University, Bedford, UK

huda.f@uokerbala.edu.iq

Abstract. Grid Computing and Cloud Computing are two different technologies that have emerged to validate the long-held dream of computing as utilities which led to an important revolution in IT industry. These technologies came with several challenges in terms of middleware, programming model, resources management and business models. These challenges are seriously considered by Distributed System research. Resources allocation is a key challenge in both technologies as it causes the possible resource wastage and service degradation. This paper is addressing a comprehensive study of the resources allocation processes in both technologies. It provides the researchers with an in-depth understanding of all resources allocation related aspects and associative challenges, including: load balancing, performance, energy consumption, scheduling algorithms, resources consolidation and migration. The comparison also contributes an informal definition of the Cloud resource allocation process. Resources in the Cloud are being shared by all users in a time and space sharing manner, in contrast to dedicated resources that governed by a queuing system in Grid resource management. Cloud Resource allocation suffers from extra challenges abbreviated by achieving good load balancing and making right consolidation decision.

Keywords. Cloud Resources Allocation; Grid Resources Allocation; Resources Consolidation; Load Balancing; Energy Consumption; Scheduling Strategies, Performances.

1. Introduction

Nowadays, Cloud and Grid computing are widely used in solving scientific problems; those technologies are based on providing computing services on demand just like conventional power and water National Grids; both technologies were developed with the goal of creating a scalable and powerful virtual computer out of a large collection of homogenous or heterogeneous systems that share various combinations of resources [1, 2, 3].

Resources allocation, also known as resource management is one of the major fields in both technologies, since it controls the way that resources and services are made available to use by entities like users, applications, or services, and to make sure of the efficient utilization of computing resources and to optimize the performance of the submitted tasks [4,5]. For instance, a grid manger in one geographical area may have limited access over the system components or there may be a

IOP Conf. Series: Journal of Physics: Conf. Series **1032** (2018) 012048 doi:10.1088/1742-6596/1032/1/012048

discrepancy of resource availability in the highly distributed cyber physical systems. This discrepancy may lead to severe increase in the energy consumption during the execution of grid applications on one side, and to the realization of tasks submitted by gird users on the other side [6,7].

In Cloud Computing, scheduling and resource allocation is a widely considered problem due to its crucial role in the whole of Cloud paradigm. Generally, it controls a number of conflicting objectives; computing resources must be well-managed to prevent overloading and waste of bandwidth, processing unit, memory, etc. This waste relates directly to significant financial loss for large Cloud service providers with regards to energy, operational cost as well as dissatisfaction of the Cloud service user [6].

This paper addresses the comparison between the Resources Allocation processes in Grid and Cloud technologies while characterizing the differences between both technologies in detail. The paper is structured as follows: Grid and Cloud models are defined in Section II. The processes of resources allocation and scheduling for both Grid Computing and Cloud Computing are reviewed in Section III. The common associated problems in both models are discussed in Section IV, and finally the conclusion comes in Section V.

2. Background

2.1. Grid technology

Based on the idea of electricity gird the Grid Computing (GC) introduced to the world with the aim of integrating the grid with existing computing technologies like web and virtual reality computing to serve some complex scientific problems [8,9]. Grid computing is one of the many computer scientific disciplines which can be defined as a large-scale and multidisciplinary infrastructure based on computational and data intensive platform for solving large-scale scientific problems. GC is known with its outstanding characteristics such as reliability, security, dependability and coordination [2,8,10].

Generally, a Grid system is seen as a multi-layer architecture with a hierarchical management system that consists of two or three levels depending on the privilege access to system data, system services, and resources.

Buyya et al. in [11] defines the grid components in four layers as below:

- Grid fabric layer, that is composed of the grid resources, services, and local resource administration systems.
- Grid core middleware, which is in charge of services related to security, management authority, scheduling, and remote tasks submission.
- Grid user layer that comprises a set of end-users, service providers, and system administrators.
- Grid applications layer.

These layers present fundamental Grid architecture; both of middleware layer and user layer have effective impacts in the multi-criterion scheduling and resource management. The most critical characteristics of GC can be summarized as in [1,8,10,12]: the heterogeneity and the multiple administration controls. The computing resources and internetworking connections are almost heterogeneous, whereas the conventional Grid consists of different institutions, each of which has its own policy to control its resources, and hence, it provides a large scale distributed platform.

2.2. Cloud Technology

Cloud Computing is known as the cutting edge of distributed system that initially comprises of Cluster Computing and GC. Cloud computing has emerged as a computing model aimed at providing resources as a service according to pay-per-use paradigm [4]. It has been defined by (NIST) as "a model for enabling convenient, on demand network access to a shared pool of configurable computing resources (e.g. network servers, storage, applications and services) that can be rapidly provisioned or released with minimal management effort or services provider interaction. This Cloud model promotes availability." [13]. The sole principle of Cloud computing is that data is not stored in locally but in the

IOP Conf. Series: Journal of Physics: Conf. Series **1032** (2018) 012048 doi:10.1088/1742-6596/1032/1/012048

data centers via the Internet. The key attributes which distinguish Cloud Computing from the other traditional computing can be described as in [6], [14], [15]:

- Underlying infrastructure and software is abstracted and offered as a service.
- Build on scalable and flexible infrastructure.
- Offers an on-demand service provisioning and quality of service (QoS) guarantees.
- Pay for use of computing resources without up-front commitment by Cloud users.
- Shared and multitenant.
- Accessible over the Internet by any device.

Due to all these attributes there has been a rapid growth in the demands for Cloud services. This brings more challenges for Cloud providers to provide resources to the Cloud subscribers, hence, emerging new paradigm "Federated Cloud", where many providers are collaborated to fulfil clients' requests [13], [16]. This collaboration can be seen as a large scale Cloud environment.

3. Resources Allocation in Grid And Cloud Technology

Grid and Cloud resources allocation and scheduling deals with a large number of processes (tasks) racing to obtain resources (CPU, memory, storage, network) in order to complete their tasks. This requires an optimal mapping between the competing processes and the underlying computing resources. Moreover, scheduling and resource allocation in Grid Computing is considered as an NP-complete optimization problem [17], [18], The complexity of this problem depends on the number of the objectives to be minimized such as task inter-relations, makespan, resources utilization and energy consumption ,while Cloud resources allocation problem is approved as NP-hard in a strong sense [19,20].

3.1. Resources Allocation in Grid Technology

In Grid Resource Allocation, a local resource manager (LRM) manages the computing resources for a Grid site which contains a pool of resources. When users submit their jobs, these jobs will be considered as batch jobs. Each job needs to be executed by leasing some of the free resources for some time specified by the user. The hosted resources could not be shared with other jobs for that time figure 1 presents a basic batch scheduler in Grid. Resource Allocation can be classified into three phases [15,17,21]:-

Resources discovery represents the first phase, where the Grid scheduler should continuously monitor the status of the participated resources to update its resource availability database to know which resources are free to serve the upcoming requests. This phase is a challenge in its own rights and needs a specific algorithm to cover it. Many algorithms have been proposed to deal with this phase, for example "Scalable Grid Resource Discovery through Distributed Search" has been introduced by Butt et al. [22] and "semantic based scalable decentralized grid resources discovery framework" was proposed by Hassan et al. [23].

The second phase is resources selection, where Grids are usually composed of heterogeneous resources which are completely varied in their number of cores, computation speed and memory size. Therefore, the resource selection phase is focusing on selecting the best resource among the available ones to serve the incoming request. Each request specifies the amount of computing power it needs and for how long.

The third phase is focusing on resources usage; after allocating a suitable resource for the incoming task, the task will then be sent to the selected resource over a wide area network. The resource will be reserved to serve this task for the specified amount of time and it could not be used to host another request during this allocated period.

IOP Conf. Series: Journal of Physics: Conf. Series **1032** (2018) 012048 doi:10.1088/1742-6596/1032/1/012048

Resources selection is the critical phase in the whole process. For example, if the incoming requested task needs 3 cores CPU for 3 hours and it was hosted in 4 CPU cores node, one un-needed CPU core would run for 3 hours which affects the total Resource Utilization [24]. This leads to lose much of computation power and energy. , where and represent the number of used processors and total processors on resource respectively.

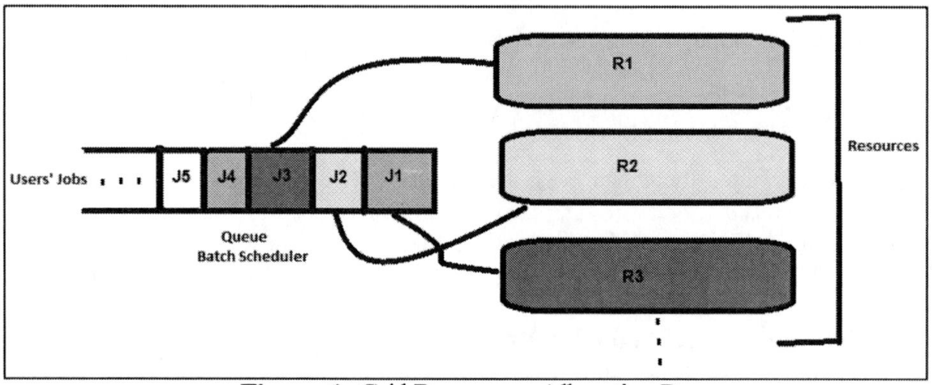

Figure 1. Grid Resources Allocation Process

3.2. Resources Allocation in Cloud Technology

Resource allocation and provisioning is the process of mapping the available resources to different applications of the Cloud over the Internet. It must be managed precisely in a way that prevents overloading and resources wastage (wastage of bandwidth, processing unit, memory, etc.). The service provider is responsible for managing its resources, while the Cloud's users are responsible for specifying their application requirements and the Service Level Agreement (SLA) [13],[24],[25]. Cloud resource allocation can be identified by four phases. Figure 2 illustrates abstracted resources allocation process:

Phase 1: Virtual Machine (VM) Creation Each VM is created according to user's request, which is usually specified in SLA document. Given that Cloud could accept hundreds of users' requests simultaneously, hundreds of VMs can be created simultaneously too.

Phase 2: VM Deploying This is the process of mapping a VM to a suitable available physical machine. The scheduler should have detailed information about the physical machines: which machines are overloaded and which ones are able to host new VMs. Each VM will be accommodated by one host (Physical Machine (PM)). This process is similar to resource selection in Grid resource allocation with extra difficulties. Many VMs can be deployed to one PM. Therefore, the deploying decision should be handled in a way that maintains good resources utilization and avoid performance degradation.

Phase 3: Task Encapsulation this is the process of encapsulating tasks to be processed into the VM. In reality, the VM may be allocated for a specific user or it can be used for multiple users depending on Cloud service level. User task should be encapsulated in its VM. This commonly happens in the Cloud models of Infrastructures-as-a-Service (IaaS) and Platform-as-a-Service (PaaS). Pre-created VMs can be used to serve user incoming tasks, which are usually taking place in the Software-as-a-Service (SaaS) model. Generally, Cloud applications are executed on VMs. Each application has specified requirements of processing power, and the created VM must provide the processing power to the application.

Phase 4: VM Usage In this phase VM will be running to process the encapsulated tasks.

Public, private and hybrid Clouds all have the ability to provide VMs with unlimited scalability. The cost of purchasing, operating and maintaining the physical resources of these Cloud datacenters is enormous [26]. Thus Cloud providers intend to optimize the usage of these resources, through planned allocation of VMs in hosts. Cloud service providers are also keen to maintain a positive client experience. The difficulty of resources allocation problem comes in twofold: maintaining the performance requirement of VMs on one side while reducing the operational cost on the other side [26], [27]. An optimal resource allocation should avoid the following criteria [19,20,27]:

- Resources contention is the situation when two or more VMs try to use the same resources at the same time.
- Scarcity of resources arises when the available resources are limited.
- Resources fragmentation is the situation where the resources are isolated.
- Over provisioning arises when the application acquires surplus resources more than the required one.
- Under-provisioning of resources happens when the application is allocated with fewer numbers of resources than the demands.

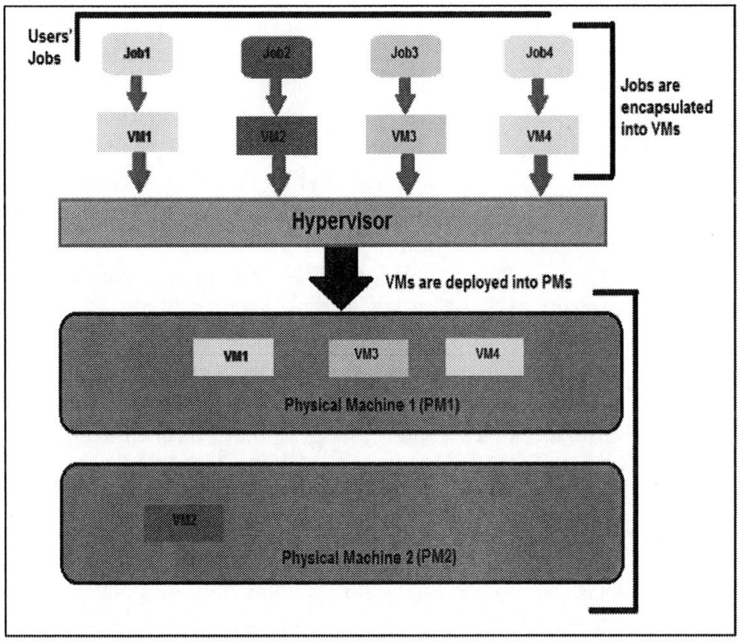

Figure 2. Cloud Resources Allocation Process

4. Comparison of Resources Allocation Problems Identified on Grid and Cloud Computing

In both Cloud and Grid models, there are a number of factors connected to resource allocation problem such as the level of virtualization, the applications hosted, scheduling policies and some of special associative problems. Table 1 summaries all these factors. This section provides a comprehensive classification of all associated problems and surveys the existing the techniques to deal with these issues.

4.1. Full virtualization

Virtualization plays a crucial role in management resources of Cloud platform; given that Cloud computing is based on full virtualization technology where a single PM is able to host several VMs which are completely isolated. Each VM has its own virtual processor, memory, and its own operating system which is called guest operating system [28]. The virtualization process is performed through a Hypervisor. Renowned implementations are e.g. Xen, Kernel based Virtual Machine (KVM), Open Nebula and VMWare [26,29]. A Hypervisor is dedicated software running on top of the physical hardware enabling to create virtual environment to operate virtual machines. The reason behind using fully virtualized hosts in Cloud is to get efficient hardware utilization, lower energy consumption, and improve fault tolerance [4,25,26,27,30]. However, VMs are provisioned in two ways: direct

IOP Conf. Series: Journal of Physics: Conf. Series **1032** (2018) 012048 doi:10.1088/1742-6596/1032/1/012048

provisioning in which the computing power (VM) provides directly to the customers; this usually is the provisioning method in an IaaS, and indirect provisioning which is carried by wrapping the provisioned applications in a VM this case is widely used in Software as-a-Service (SaaS) and PaaS. Popular Clouds use different resource allocation strategies: greedy first fit and rotating scheduling are used e.g. in Eucalyptus [4].

On the other hand, full virtualization concept has not been used in Gird systems yet. Each individual organization in Grid maintains full control of their resources. However, dedicated resources allocation is a likely strategy for the job hosting as illustrated in figure1, scalable provisioning or unplanned increase for the job's requirements are not allowed in Grid [1,2].

4.2. Scheduling Strategies

Batch scheduler is the well-known scheduler in GC specifically in computational Grid in which the LRM such as Portable Batch System (PBS) is a job resource manager in a batch cluster environment [31], [32]. Condor [33] is governing resources by using a queueing system. Bouyer et al. (2013) in [34] developed a novel queuing algorithm for local Grid scheduling called market-oriented job scheduling and applied it to the grid scheduling problem. This algorithm is different from the previous queuing algorithms such as First Come First Serve (FCFC), Round Robin (RR), and Back Filling. The proposed algorithm uses the coming tasks' requirements and restrictions like minimum completion time (MCT), reliable completion time), and minimum execution cost to make the queue.

According to Section 2.2, mapping services with complex computing requirements to physical machines with specified capacities is not a trivial issue. It is referred to as a classic vector (or multi-capacity) bin packing problem (VBPP) [30,35]. Each dimension of the problem is corresponding to a type of resources such as CPU, RAM memory, disk space and bandwidth. Several approaches have been used to solve this problem like approximation algorithms (FF, FFD, PP) [35], [36] or Evolutionary Algorithms for Examples Genetic Algorithms (GA) and Ant Colony Optimisation [37]. Haizea is used as VM based lease management architecture scheduler in Open Nebula [29], and PBS (Portable Batch System) and SGN (Sun Grid Engine) are the scheduler of Nimbus [4,16].

4.3. Application Models

Recently Grid Systems have been developed far beyond their original intention. Modern GC can be applied to serve various complex applications, for example collaborative engineering, data processing and exploration, and e-Science applications. Due to the expensive scheduling decisions, data staging in and out, and potentially long queue times, many Grids do not natively support interactive applications. However, there are efforts in the Grid community to enable lower latencies to resources via multi-level scheduling, to allow applications with many short-running tasks to execute efficiently on Grids [12], [11,19].

Clouds are able to host a range of applications, including interactive applications and patch processing applications, such as HPC jobs, scientific workflow, or periodical jobs [13]. Highly available applications are the applications within high availability rate about 99.9999% (availability is the ratio between the hosted requests and the total number of required requests). They also comprise several components such as (web server, communication server, and data store). The resources allocation within the highly available applications needs extra efforts to prevent resources failure [16].

4.4. *Resources Consolidation and- VM migration*

Dynamic Server Consolidation (DSC) is a special problem which derives from virtualized hosted platforms like Cloud platform. It aims at consolidating maximum number of VMs onto minimum number of physical machines (PMs), thus improving resources utilization [20,38]. DSC is accompanied by over-provisioning and under-provisioning problems due to the Cloud characteristic of dynamic resources provisioning (elastic workloads). In case of workload changes, the consolidation decision will be modified through live migration of VMs. If a server (PM) is underutilized, all VMs running in this server should be migrated to another server then the underutilized server can be switched to low level mode. If workloads have been increased by Cloud users thus overloading the hosted server, then the migration decision should be done to avoid overprovisioning problem which may lead to performance degradation.

Consolidation decision can be taken in two ways: long term consolidation decision and short term consolidation decision. The short term decision was used in [20] and [39]. This decision is taken without considering the historical information of nodes' load which may lead to a wrong decision in terms of high migration cost. The authors of [24] employed the Markov Decision Process (MDP) to generate long term precise migration decision which aims to improve the profit by avoiding the wrong decisions which may have an adverse effect on the total profit. By comparing the performance of the proposed MDP algorithm and the existing Dynamic Management Algorithm (DMA) the conclusion is driven that the migration decisions of MDP policy are more valuable than the DMA decisions, thus producing better profit.

An integral thermal and compute controlled heuristic approach was introduced in [40] to improve the energy consumption of IaaS Cloud. The proposed approach integrates several of IaaS cloud characteristics such as server heterogeneity, workload dynamicity, VM migration, server processor SLEEP state, server processor state transition power and latency time overheads to configure a new metric called StaticPPMMax that is used for the VM allocation algorithm. The results show that the StaticPPMMax allocation metric and allocation heuristic approach improves the datacenter energy consumption savings by 12% in comparison to an independent thermal and compute controlled scheme. Considering the server processor SLEEP state and transition power and time latencies increase the datacenter energy savings (around 0.5% to 1%).

Grids use a different allocation policy (see Section III.A) that does not enable simultaneous time and space resources sharing. Accordingly, no resource consolidation problem has been introduced in Grid resource allocation yet.

4.5. Load Balancing

Efficient load balancing is one of the major challenges in Cloud resources allocation process. Basically, it is the process of assigning the load to different nodes in such a way which maintains no overloaded or under-loaded nodes. Load balancing aims to improve system performance substantially and maintain stability of the system. Many algorithms were proposed to tackle this problem for example Biased Random Sampling [41], Honeybee Foraging Algorithm, and ACCLB [42].

Some of these algorithms deal with static load whereas others consider dynamic load. In static environments, scheduling decision is made before submitting the jobs; this approach assumes that all the information about resources is known in advance and that resources properties do not vary over time. The main disadvantage of this is the danger of nodes failure. Round Robin and static greedy are used in Eucalyptus to solve static load balancing problem [9].

In dynamic environment, the decisions are taken while the job is being executed. This approach takes into account the dynamism of the cloud and it is generally preferred as resources properties are flexible and may change suddenly. The advantage of the dynamic load balancing over the static one is that the system does not need to be aware of the run-time behavior of the job before execution. LBMM (Load balancing Min Min), ACCLB (Ant Colony and Complex Network Load Balancing), and OLB (Opportunistic Load Balancing) [43] are famous dynamic load balancing algorithms. Based on OBL and LBMM the Three Level load balancing algorithm has been developed in [44] with the aim of getting better executing efficiency and preserve load balancing.

Grid resource allocation does not suffer from load balancing as it is based on reserving physical node for a specified amount of time. Space sharing is not allowed in Grid resource allocation which prohibits load balancing problems.

4.6. Energy Consumption

Energy-aware management of large-scale distributed systems has attracted significant attention by researchers in the last decade. Datacenters are considered as one of main environmental danger causes.

IOP Conf. Series: Journal of Physics: Conf. Series **1032** (2018) 012048 doi:10.1088/1742-6596/1032/1/012048

There was 63% increase in power consumed by datacenters for only one year between 2011 and 2012, and it was around 38 Giga Watt in 2012 [45]. The low-power hardware techniques that have been used to reduce the energy consumption are not enough to handle this problem. They must be accompanied by intelligent decision making, resource management and task scheduling mechanisms [16],[26].

Developing an efficient scheduler in both distributed systems (Cloud and Grid) is a challenging issue if energy optimization is to be considered as a scheduling objective. Shu et al. in [46] improved immune clonal selection algorithm to design Cloud and Grid schedulers with a goal of simultaneous optimization of the energy utilization. ACO was used by Suphalakshmi in [45] to reduce energy consumption based on VMs allocation. The proposed approach considers the expected behavior of each user requests even dynamic user requests to make the resources allocation decision instead of using the actual user requests. The resources allocation decision aims to minimizing the number of used servers therefore; reducing energy consumption associated with that decision.

Energy consumption, in a nutshell, is an active researching field in Cloud and Gird resource management. Ramani and Bohara in [47] defined a temperature threshold of the hosts in Cloud data center. It reduces the consumption of the maximum resources and controls the processor temperature at the same time. The proposed approach leads to minimize the overall energy consumption.

4.7. *Performance*

The extensive use of virtualization in Cloud computing has exaggerated the performance problem inside the Cloud model which adds extra challenges for resource allocation and scheduling [48]. Virtualization has had significant performance losses for some applications, which is considered as one of the primary disadvantages of using virtualization. Cloud resources manager should ensure covering the application SLA requirements which is usually done by efficient scheduling. Although performance problem is affecting tangibly in Cloud computing model, it has not been considered widely by Cloud scheduling researches. Maintaining a good Quality of Service level is widely considered in Cloud Resources Allocation research [48], [49].

Grid computing seems to supply better performance than Cloud computing, Grid performance problem has been extensively studied by a great number of Grid resources management studies. However, there is a lack of studies that compare Grid and Cloud performance.

Table 1. Comparison of Grid and Cloud Resources Allocation

Resources Allocation Aspects and Issues	Grid Technology		Cloud Technology	
Full Virtualisation	Not found		Xen KVM Open Nebula VMWare	
Scheduling strategy	Batch Scheduling	Market-Oriented RR FCFS	Bin Packing	 FF FFD GA ACO
Application Models	E-Science applications HPC Jobs Scientific workflow		Interactive applications HPC Jobs Highly available applications	
Load Balancing	Not found		Static Round Robin static greedy	Dynamic LBMM ACCLB OLB

Resource Consolidation	Not found			Long term decision MDP StaticPPMMax	Short term decision DMA
Energy Consumption	Immune Clonal Selection			Immune Clonal Selection ACO Temperature Threshold	
Performance	Grid resources management			Good Quality of Service level SLA	
Multi-Site Administration	Global Hierarchical multi- level	Inter-site level Resource allocation decisions and Scheduling	Intra-site level Resource allocation decisions and Scheduling	Federated Clouds	

4.8. Multi- Site Administration

The global grid management system is defined as cooperation between the centralized and decentralized resources and service management systems; therefore, the scheduling and resource allocation decisions are delivered either at global, inter-site or intra-site levels [16], [19].

The Grid is known to have a multi-site administration problem since it comprises various collaborated virtual organization each one has its own policy to schedule its resources by using a specific scheduler. Huge efforts have been made to solve this problem; a hierarchical multi- level scheduling is one of multi-administration solution in which the global Grid management system can be defined as a compromise between the centralized and decentralized resources and service management systems, where the scheduling and resource allocation decisions are provided at global, inter-site and intra-site levels [50].

Large scale Clouds (it is also referred to as Federated Clouds) are also suffering from multi-administration problem as they consist of different Cloud services provider with complete resource management policy. This problem resulted in adding a new challenge to large scale Cloud resource allocation and scheduling. Research results in this field are quite similar to Grid resources allocation; however, they disregard Cloud resources management problems and focused on the unification of collaborated providers.

5. Conclusions

This paper presents a thorough comparison of two distributed system models - grid and cloud - in terms of the resource allocation problem, as the two computing models often get confused due to their similar conceptual properties. Resource Allocation plays a crucial role in both models as it directly affects their performance in respects to resources utilisation, energy consumption, and/or load balancing. The main findings of this paper are:

Firstly, the resource allocation problem is completely different in both models in terms of associated challenges, scheduling algorithms, and deploying strategies. In Cloud Computing, the applications are hosted indirectly, i.e. first the applications are allocated to VMs and then the VMs get deployed into physical resources; this strategy enables simultaneous time and space sharing of the underlying resources. This feature allows latency sensitive applications to operate efficiently on Cloud. Whereas, Grid Computing resources are directly and independently deployed to the coming applications, but the deployment decision is also restricted by time and space requirements.

Secondly, the comparison shows that Cloud resources are rented under full virtualisation concept which leads to improve the overall resources utilisation but it adds extra challenges in making the right allocation decision. It also indicates that both resources consolidation and load balancing are major challenges related to single Cloud resources allocation and not found in Grid resources allocation. On the other hand, Grid resources allocation suffers from the multi-site administration problem; wrong resources may be allocated with high energy consumption or low utilisation.

Thirdly, there is notable lack of performance analysis and fault tolerance research in Cloud resources allocation; ensuring that a good level of QoS is delivering to the end users is likely to be one of the major challenges in Cloud Computing since the Cloud is applicable to scalable user demands and dynamic providing scheme.

Fourthly, Clouds come in two types namely Single Cloud and Federated Cloud. Research of resource allocation and scheduling in single Cloud environment is different from that in Federated Cloud. In the Federated Cloud the major resources allocation challenge is multi-site administration.

6. References

[1] Foster, Y. Zhao, I. Raicu, and S. Lu, "Cloud Computing and Grid Computing 360-Degree Compared," Grid Computing Environments Workshop, IEEE, Austin, TX, pp:1–10, 2008. DOI: 10.1109/GCE.2008.4738445.

[2] N. Sadashiv, "Cluster, Grid and Cloud Computing: A Detailed Comparison," 6th International Conference on Computer Science & Education (ICCSE), pp:477 – 482, Singapore, , 2011.DOI: 10.1109/ICCSE.2011.6028683.

[3] S. Di, D. Kondo, and W. Cirne, "Characterization and comparison of cloud versus grid workloads," Proc. - 2012 IEEE International Conference in Cluster Computing, pp. 230–238, 2012.

[4] K. Radha, B. Rao, S. Babu, K. Rao, V. Reddy, and P.Saikiran, "Allocation of Resources and Scheduling in Cloud Computing with Cloud Migration", *International Journal of Applied Engineering Research* ISSN 0973-4562, vol 9 (19) 2014. [Online]. http://news.bbc.co.uk/1/hi/world/americas/4808342.stm

[5] A. Andreazza, A. Annovi, and D. Barberis, "Computing infrastructure for ATLAS data analysis in the Italian Grid cloud," J. Phys. Conf. Ser., vol. 331, no. 5, p. 052001, 2011.

[6] Z. A. Mann, "A Taxonomy for the Virtual Machine Allocation Problem*", *International Journal of Mathematical Models and Methods in Applied Sciences,* 2015, volume 9, pp: 269-276.

[7] M. Hu and B. Veeravalli, "Requirement-aware scheduling of bag-of-tasks applications on grids with dynamic resilience," IEEE Trans. Comput., vol. 62, no. 10, pp. 2108–2114, 2013.

[8] C. Weng and X. Lu, "Heuristic scheduling for bag-of-tasks applications in combination with QoS in the computational grid," Futur. Gener. Comput. Syst., vol. 21, no. 2, pp. 271–280, 2005.

[9] I. Brandic and S. Dustdar, "Grid vs Cloud — A Technology Comparison," it - Information Technolgy., vol. 53, no. 4, pp. 173–179, 2011.

[10] J. Kolodzieja,, S.Khanb , and L. Wangc , "Security, Energy, and Performance-aware Resources Alocation Mechanisms for Computational Grids", Future Generation and Computer Systems, pp:77-92 , 2014.

[11] R. Buyya and M. Murshed, "GridSim: A Toolkit for the Modeling and Simulation of Distributed Resource Management and Scheduling for Grid Computing Concurrency and Computation: Practice and Experience. Vol.14, pp:1175–1220, 2002. DOI: 10.1002/cpe.710.

[12] M. Frincu, "Adaptive Scheduling for Distributed Systems", Ph.D. thesis, West University of Timisoara, Faculty of Mathematics and Computer Science, May 2011.

[13] E. Carlini, M. Coppola , P. Dazzi ,L. Ricci , and G. Righetti, "Cloud Federations in Contrail", chapter in a book "Euro- Par 2011: Parallel Processing Workshops", part 1, Sipringer, LNCS 7155, pp. 159–168, 2012.

[14] W. Tian, X. Liu, C. Jin, and Y. Zhong, "LIF: A Dynamic Scheduling Algorithm for Cloud Data Centers Considering Multi-dimensional Resources*", *Journal of Information & Computational Science*, vol.10, pp:3925–3937, 2013.

[15] E. Amiri, H. Keshavarz, N. Ohshima, and S. Komaki, "Resource Allocation in Grid: A Review," Procedia - Social Behavior Sciences, vol. 129, pp. 436–440, 2014.

[16] B. Malet and P. Pietzuch, "Resource allocation across multiple cloud data centres", 8th International Workshop on Middleware for Grids, Clouds and e-Science, Copyright 2010 ACM 978-1-4503-0453-5, 2010. DIO:10.1145/1890799.1890804.

[17] A. Yousif, A. Abdullah, M. S. Abd Latiff, and M. B. Bashir, "A Taxonomy of Grid Resource Selection Mechanisms", *International Journal of Grid and Distributed Computing,* vol. 4, no.3, pp. 107-118, September 2011.

[18] K. Etminani and M. Naghibzadeh, "A Min-Min Max-Min selective algorithm for grid task scheduling," 2007 3rd IEEEIFIP Int. Conf. Cent. Asia Internet, pp. 1–7, 2007.

[19] M. Zolt and E. February, "Allocation of Virtual Machines in Cloud Data Centers – A Survey of Problem Models and Optimization Algorithms", ACM Computing Surveys (CSUR), vol.48, no.1, 2015,
DOI:10.1145/2797211

[20] F. Hermenier, X. Lorca, J.-M. Menaud, G. Muller, and J. Lawall, "Entropy: a Consolidation Manager for Clusters," Proc. 2009 ACM SIGPLAN/SIGOPS International Conference on Virtual Execution Environments, no. 41, pp: 41-50, 2009.

[21] A. A. Haruna., N. B. Zakaria and N.Haron, "Grid Resource Allocation: A Review ". *Research Journal of Information Technology*, vol. 4(2), pp: 38-55, 2012.

[22] Butt, F., Bokhari, S.S., Abhari, A., and Ferworn, "A Scalable Grid Resource Discovery through Distributed Search", *International Journal of Distributed and Parallel Systems*, pp: 1-19, 2011.

[23] M. Hassan and A. Abdullah, "A New Grid Resource Discovery Framework," *the International Arab Journal of Information Technology*, vol. 8, no. 1, pp. 99-107, 2011.

[24] M. E. Frincu, "Scheduling highly available applications on cloud environments," Future Generation and Computer Systems, vol. 32, no. 1, pp. 138–153, 2014.

[25] SI. Version, D. Sivapriyanka, and S. Santhanalakshmi, "Allocation of Resources Dynamically In Cloud Systems", *International Journal of Engineering Research and Applications*, vol. 4, no. 5, pp. 113–118, 2014.

[26] A. Hameed, A. Khoshkbarforoushha, R. Ranjan, and P. P. Jayaraman, "A Survey and Taxonomy on Energy Efficient Resource Allocation Techniques for Cloud Computing Systems," "Computing", Springer, pp.1-24, 2014.

[27] K. Radha, B. Thirumala Rao, S. M. Babu, K. Thirupathi Rao, V. Krishna Reddy, and P. Saikiran, "Allocation of Resources and Scheduling in Cloud Computing with Cloud Migration," *International Journal of Applied Engineering Research.,* vol. 9, no. 19, pp. 5827–5837, 2014.

[28] A. S. Tanenbaum, Modern Operating Systems, 3rd edition, Person Prentice Hall, US, 2009.

[29] S. S. Manvi and G. Krishna, "Resource Management for Infrastructure as a Service (IaaS) in cloud computing : A survey," *Journal of Network and Computer Applications*., vol. 41, pp. 1–17, 2013.

[30] A. Alahmadi, A. Alnowiser, M. M. Zhu, D. Che, and P. Ghodous, "Enhanced First-Fit

Decreasing Algorithm for Energy-aware Job Scheduling in Cloud," Proc. - 2014 Int. Conf. Comput. Sci. Comput. Intell. CSCI 2014, vol. 2, pp. 69–74, 2014.

[31] R.L. Henderson. "Job Scheduling under the Portable Batch System" In D. Feitelson and L. Rudolph, editors, Job Scheduling Strategies for Parallel Processing (Pro- ceedings of the 1st International JSSPP Workshop; LNCS no. 949), pp:178– 186. Springer-Verlag, 1995.

[32] Research computing and cyber infrastructure, http://rcc.its.psu.edu/user_guides/system_utilities/pbs/, accessed August 07, 2011.

[33] M. Litzkow, M. Livny, M. Mutka, "Condor – A Hunter of Idle Workstations", in: 8th International Conference of Distributed Computing Systems, pp:104–111, June 1988.

[34] A. Bouyer, B. Arasteh, F. Nasrollahi, "Maximizing Job Submission Rate in Market-Oriented Grids", AWERProcedia,Information Technology & Computer Science, vol. 3, pp:1877-1885, 2013. http://www.world-education-center.org/index.php/P-ITCS.

[35] W. Tian, X. Liu, C. Jin, and Y. Zhong, "LIF: A Dynamic Scheduling Algorithm for Cloud Data Centers Considering Multi-dimensional Resources*", *Journal of Information & Computational Science*, vol.10, no.12, pp:3925–3937, 2013.

[36] A. Alahmadi, A.Alnowiser, M. Zhu, D. Che and P. Ghodous, "Enhanced First-fit Decreasing Algorithm for Energy Aware Job Scheduling in Cloud", International Conference on Computational Science and Computational Intelligence, vol.2, pp: 69-74, 2014.

[37] K. Chandrasekaran, and U. Divakarla, "Load Balancing Virtual Machine Resources in Cloud Using Genetic Algorithm", ICCN 2013, pp.156-168, Elsevier2013.

[38] H. Ferdaus and M. Murshed, "Virtual Machine Consolidation in Cloud Data Centers Using ACO Metaheuristic", Springer, Euro-Par 2014: Parallel Processing: 20th International Conference, pp. 306–317, 2014.

[39] G. Jung , K. Joshi, M. Hiltunen , R. Schlichting, and C. Pu, "A Cost-Sensitive Adaptation Engine for Server Consolidation of Multitier Applications", chapter in "Middleware 2009", , pages 163–183, International Federation for Information Processing, 2009.

[40] M. Kumara, and S. Raghunathan, "Heterogeneity and Thermal Aware Adaptive Heuristics for Energy Efficient Consolidation of Virtual Machines in Infrastructure Clouds*", Journal of Computer and System Sciences*, vol. 82, pp:191–212, 2016.

[41] M. Randles, D. Lamb, A. Taleb-Bendiab, " A Comparative Study into Distributed Load Balancing Algorithms for Cloud Computing", IEEE International Conference on Advanced Information Networking and Applications Workshops, IEE Computer Society, pp: 551-556, April 2010.

[42] Al-Jaroodi, and N. Mohamed, "DDFTP: Dual-Direction FTP," IEEE Computer Society Washington, DC, USA, pp: 504-513, 2011.

[43] X. Ren, R. Lin, and H. Zou, "A Dynamic Load Balancing Strategy for Cloud Computing Platform based on Exponential Smoothing Forecast", Cloud Computing and Intelligence Systems IEEE, pp: 220 - 224, 2011.

[44] S. Wang, K. Yan, W. Liao, and S. Wang, "Towards a Load Balancing in a Three-level Cloud Computing Network" Computer Science and Information Technology (ICCSIT), 3rd IEEE International Conference IEEE, vol. 1, pp:108 - 113, 2010.

[45] A. Suphalakshmi and S. M, "An Intelligent, Energy Conserving Load Balancing Algorithm for the Cloud Environment Using Ant's Stigmergic Behavior", *International Journal of Communications and Engineering,* vol.4, no. 3, March2012.

[46] W. Shu, W. Wang, and Y. Wang, "A Novel Energy-Efficient Resource Allocation Algorithm based on Immune Clonal Optimization for Green Cloud Computing", *Journal on Wireless Communications and Networking,* vol. 2014, no. 1, p. 64, 2014.

IOP Conf. Series: Journal of Physics: Conf. Series **1032** (2018) 012048 doi:10.1088/1742-6596/1032/1/012048

[47] M. Ramani, and M. Bohara, "Energy Aware Load Balancing In Cloud Computing Using Virtual Machines", *Journal of Engineering Computers & Applied Sciences*, vol. 4, no.1, pp:1-5 , 2015.

[48] A. Tchernykh, L. Lozano, U. Schwiegelshohn, P. Bouvry, J. E. Pecero, S. Nesmachnow, and A. Y. Drozdov, "Online Bi-Objective Scheduling for IaaS Clouds Ensuring Quality of Service," *Journal of Grid Computing*., pp. 911–918, 2015.

[49] J. W. Lin, C. H. Chen, and C. Y. Lin, "Integrating QoS Awareness with Virtualization in Cloud Computing Systems for Delays Applications", Future Generation Computer Systems, vol. 37, pp. 478–487, 2014.

[50] N. Bessis, S. Sotiriadis, V. Cristea, and F. Pop, "Modelling Requirements for Enabling Meta-Scheduling in Inter-Clouds and Iinter-enterprises", Third International Conference on Intelligent Networking and Collaborative Systems, pp. 149–156, 2011.

Acknowledgment

This work has been funded by the Ministry of Higher Education and Scientific Research of Iraq. The authors gratefully acknowledge this support.

IOP Conf. Series: Journal of Physics: Conf. Series **1032** (2018) 012049 doi:10.1088/1742-6596/1032/1/012049

Cryptographic system based on Unicode

Noor D. AL-Shakarchy[1], Huda F. AL-Shahad[1] and Dhamyaa A. AL-Nasrawi[1]
[1]Computer Science Department, College of Science, University of Kerbala, Iraq

dh.alnasrawy@uokerbala.edu.iq

Abstract. During recent years, information security plays major roles with the wide use and development in communications networks. Stronger encryption and decryption represent the important demand in most applications. This strength can be achieved by data encryption and decryption algorithms which are very hard to crack and accomplish. The CIA security aims are Confidentiality, Integrity and Authenticity. In this paper, a novel cryptographic system was proposed based on principles of Unicode and crossover; This new combination add high level of security to proposed system. The English plaintext was converted to symbols in Unicode, key generated based on Arabic alphabet(also converted to Unicode), while the encryption and decryption method implemented depend on crossover principle by encryption and decryption keys K1,K2. Two parents (English plaintext, and encryption key) selected for mating using Average Crossover method (AC) which product one child only (Encrypted text) and vice versa in decryption method using decryption key. The security was achieved by confusion and diffusion concepts; variations and randomness in the encrypted text make the features of the original language are unclear; so the prediction of the plain text is difficult.
Keywords: Crossover, Cryptographic, Ciphertext, Key generation, Randomness, Unicode.

1. Introduction

Cryptography is the science and study of secret writing where by plaintext (or clear text) is transformed based on key in to ciphertext (cryptogram) by using an encipherment (encryption) process. The reverse process of encipherment of transforming ciphertext based on key into plaintext can be known as decipherment (decryption). The wide use of the internet which can be considered insecure network increase the demand to development of cryptography methods which can be able to store, transmit and deal with sensitive and important information[1]. Shannon presented the characteristics of good ciphers by:

- The amount of appropriate encryption and decryption work should be determined by the amount of secrecy needed.
- The complexity of the set of keys and enciphering algorithm should be free.
- The cipher system should be as simple implementation as possible.
- The error propagation in ciphering message should as less as possible so that if an error appear in early message this must not cause corruption of further information in the message.
- The ciphertext size should be no longer than the original message.

To design any encryption and decryption algorithm the Shannon's characteristics of good ciphers must be taken in account. In the other hand, improve the security level of encryption and decryption algorithm is major purpose to any design[2].

In this paper, new cryptograph system based of Unicode and crossover concepts was proposed. Many features were considered to make the system achieves confidentiality, integrity and authenticity such as randomness and variations. The contributions of this paper are shorten as below:

1. Create (char, int) mapping table for English alphabet, that mean give number for each character in English alphabet. In this step, int of b does not necessarily come after int of a, this called (Alphabet interleaving).

2. Convert English plaintext to Unicode plaintext, two English characters to one Unicode character, which can be hide the natural features of the original language, this step adds more robust against the statistical cryptanalysis.

3. Key management implemented based on Arabic alphabets, also converted to Unicode based on mapping table for Arabic alphabet. The generated key K1 used in encryption method, while the decryption key K2 which constructed from initial key K1, used for decryption method.

4. Used crossover principle in genetic algorithm, the resulted child represent the Unicode encrypted text.

The organization of this paper is: Principles of Unicode presented in section2, Section 3 introduces the related works. Section 4 show the methodology and GUI. Statistical cryptanalysis and results shown in section 5. Finally, conclusions in section 6.

2. Principles of Unicode

A character is the smallest composition of a text. Characters are traditionally represented by single byte values (allows for 256 characters) in most languages. When a system is used for a new language, the encoding has to be adapted to use that language 's characters[3,4].
Unicode is one Universal Code for every character no matter what the program, platform, or language. Unicode started out using 16-bit characters instead of 8-bit characters. That means 2^16 = 65,536 different values available, making it possible to represent many different characters from many different alphabets; an initial goal was to have[5].

"characters" is the basic elements of Unicode which called code points. Code points are specified by number, usually written in hexadecimal with the prefix "U+", such as U+006E "n" Latin small letter or U+0020 "space" . Each code point also has a short name, and quite a few other properties, specified in the Unicode Character Database. Codespace mean the set of all possible code points. The Unicode codespace consists of 1,114,112 code points which partition into 17 planes, from 0 to 16[6].

The proposed system concentrated in the first plane, Plane 0 "Basic Multilingual Plane", or BMP. All the characters needed for modern text in any script contains basically in BMP, including Latin, Cyrillic, Greek, Arabic, Japanese, Chinese, Korean, Devanagari (Indian), Hebrew, and many more. See roadmap of Basic Multilingual Plane in figure 1.

3. Related Works

Maram Balajee(Maram Balajee,2011) introduces a new method for cryptography by using UNICODE and colors. This work based on Private-key cryptography. The range of colors will be decided by the sender which will be assigned to 1,00,000 UNICODE characters. The color selected in such a way from three ranges of colors to generate the Unicode character range. The sample binding between character/symbol/digit, UNICODE and Color determined by create a dynamic mapping table. In this table the Unicode for each character of alphabet is determined then the corresponding color is determined. This table created before the encryption process.
The encryption process based of this dynamic mapping table. A color-chart of each page of text can be created. The drawback of this method can be detected by: The Unicode based in this method determined equaling of each character, This method used the predefined mapping to give the corresponding color for each UNICODE character. according to the predefined mapping[7].

IOP Conf. Series: Journal of Physics: Conf. Series **1032** (2018) 012049 doi:10.1088/1742-6596/1032/1/012049

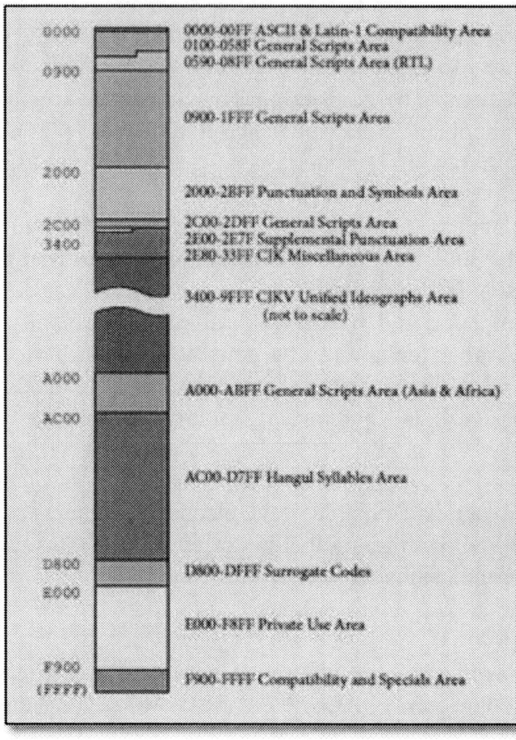

Figure 1. Basic Multilingual Plane Roadmap

A.Joseph and V.Sundaram (A.Joseph and V.Sundaram, 2012) presented an algorithm of "Data Encryption through Fibonacci Sequence and Unicode Characters". The plaintext (original) message are converted into the cipher text by searching each character in the message and interchanging it with character based on the Fibonacci number generated. Further, the cipher text is converted into the Unicode symbols[8].

In proposed system, create mapping table for English alphabet and for Arabic alphabet, then convert English plaintext to Unicode plaintext, key generation based on Arabic alphabets which converted to Unicode. The principle of crossover used to produce the Unicode encrypted text.

4. Methodology of Proposed Cryptography System

In this section, new cryptographic system proposed based on converting plaintext to Unicode, then use crossover in encryption and decryption process. the methodology of new cryptographic system is illustrated in detail; many steps must be implemented to encrypt and decrypt English plain text. Show figure 2.

The details of steps (that has been implemented in C#.net language). demonstrated in the following subsections.

4.1. Input English Plain Text

This is a first step in proposed system, the allowed plain text are English alphabet, English numbers, and some punctuations symbols. This step contains alphabet interleaving process, which create (char, int) mapping table for English alphabet. Table 1 show the mapping table for English alphabet in proposed system.

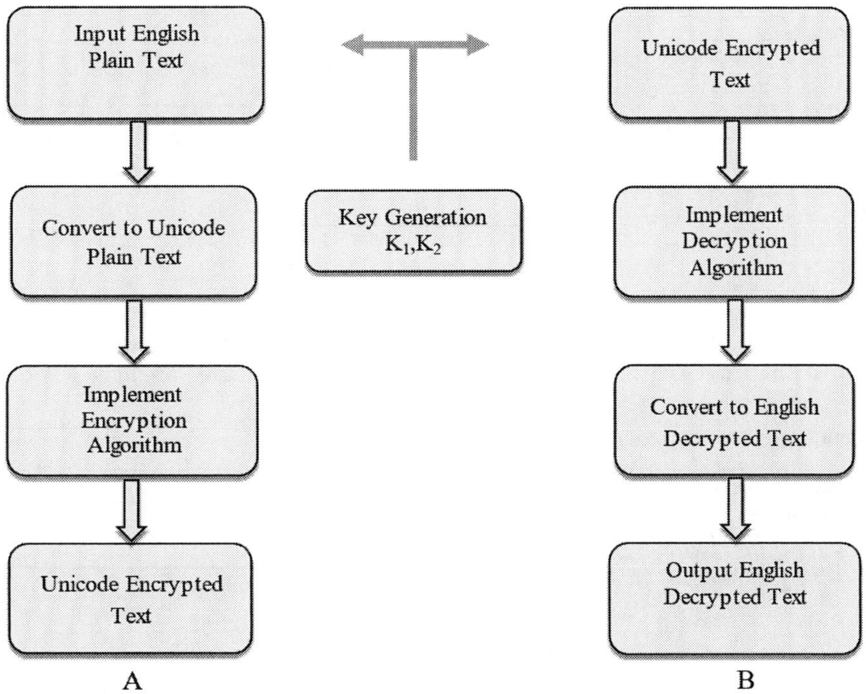

Figure 2. Steps of Proposed Cryptographic System, A: Encryption Process, B: Decryption Process

Table 1. Mapping Table of English Alphabet

char	*int*	*char*	*int*
a	38	v	49
b	82	w	76
c	60	x	35
d	44	y	32
e	13	z	56
f	64	?	66
g	68	.	54
h	81	,	83
i	59	$	78
j	61	space	30
k	79	:	72
l	42	0	33
m	80	1	73
n	47	2	52
o	18	3	23
p	20	4	21
q	36	5	70
r	26	6	75
s	69	7	55
t	15	8	12
u	63	9	67

Dictionary data structure in c# used to implement this step, see the following pesedo code that explains mapping process:

IOP Conf. Series: Journal of Physics: Conf. Series **1032** (2018) 012049 doi:10.1088/1742-6596/1032/1/012049

Mapping process
Inputs: English plain text
Output: get int for each char in English plain text
foreach char in English plain text: if char in dictionary: get int (value of char) else: do nothing.

4.2. Convert to Unicode Plain Text

The principle of Unicode was used to introduces more complexity since Unicode is committed not just to supporting texts in any single language, but also to letting multiple languages coexist within one text, which is the objective of this step.

The converting process explained in pseudo code as follow:

Convert to Unicode plain text process
Inputs: English plain text
Output: Unicode plain text
foreach *two char* in English plain text: - Get *int* (value of char) - Combine two *int* of two *char* together. - Get Unicode char of two *int* - Display the result

The details of this step explained in example below:

Let the English plain text: ***no pain no gain.***

1. By mapping table of English alphabet , get the *int* of each *char* are:

n	o	space	p	a	i	n	space	n	o	space	g	a	i	n	.
47	18	30	20	38	59	47	30	47	18	30	68	38	59	47	54

2. Combine two *int* together:

n	o	space	p	a	i	n	space	n	o	space	g	a	i	n	.
4718		3020		3859		4730		4718		3068		3859		4754	

3. Get Unicode and result

4718	3020	3859	4730	4718	3068	3859	4754
䜘	〠	㡙	䜰	䜘	と	㡙	䝔

In traditional converting to Unicode "n" = U+006E Latin small letter, while in proposed converting "n" combined with next letter to outcome Unicode as shown in above table "no" = U+4718, as well as "n" combined with another letter result new Unicode, example "n." = U+4754

4.3. Key generation

The process of generating key that used to encryption and decryption called Key generation, at first both keys are equal. In proposed system, this process implemented depending on Arabic alphabet by (abjad) order not on (hija'ai) order. The name (abjad) is based on the old Arabic alphabet's first four letters (أ, ب, ج, د). All Arabic alphabet collected in one string, followed by letters of (surat AL-Fath, aya 29 from Holy Quran) which contains all Arabic letters.

As described in the previous steps, Arabic alphabet has mapping table also, see table 2, then converted to Unicode in same as before. See the part of key generation process in the following example.

Table 2. Mapping Table of Arabic Alphabet

char	*int*	*char*	*int*
ا	16	ف	50
ب	19	ق	51
ت	24	ك	53
ث	25	ل	57
ج	27	م	58
ح	28	ن	62
خ	29	ه	65
د	31	و	71
ذ	34	ي	74
ر	37	ى	10
ز	39	ء	11
س	40	آ	14
ش	41	ا	17
ص	43	إ	77
ض	45	ؤ	85
ط	86	ة	22
ظ	87	ئ	84
ع	46	space	30
غ	48		

The details of key generation process that generate encryption key K1 explained in example below:
Let the part of Arabic alphabet in (abjad) order: أبجدهوزحطيكلمن

1. By mapping table of Arabic alphabet, get the int of each char are:

ا	ب	ج	د	ه	و	ز	ح	ط	ي	ك	ل	م	ن
16	19	27	31	65	71	39	28	86	74	53	57	58	62

2. Combine two int result these codes:

ا	ب	ج	د	ه	و	ز	ح	ط	ي	ك	ل	م	ن
1619		2731		6571		3928		8674		5357		5862	

3. Get Unicode and resulted key:

1619	2731	6571	3928	8674	5357	5862
ᘙ	✱	敱	㤨	虴	南	塢

As it's known, Unicode range from 0621 to 063A and from 0641 to 064A for the Arabic letters , while diacritics and images, range from 0600 to 06FF[3,9].
So, the letter "ج" as example = U+062C in traditional Unicode, , while in proposed converting "ج" combined with next letter such that "جد" = U+2731.

4.4. Implement Encryption Algorithm

This step focuses on using principle of crossover in genetic algorithm. Crossover is a genetic operator used to modify the chromosomes from one generation to the next. It works by taking more than one parent and producing a child from them.

In proposed system, crossover operator is the encryption algorithm, two parents represent Unicode plain text and Unicode encryption key, the produced child represent Unicode encrypted text. Average crossover method was used that reproduction one child only by average of the two parents, each gene in a child is produced by averaging genes from both parents[10]. The equation of encoding and construction decryption key are done together, see equation of encoding in details where, Parent1 ,Parent2 ,Child represent Unicode plain text, Unicode encryption key (K1) and Unicode encrypted text respectively:

$$\text{Child} = (\text{Parent1}+\text{Parent2})/2 \qquad \ldots(1)$$

$$\text{If (Parent1+Parent2) is odd} \begin{cases} \lceil \text{Child} \rceil & , K2 = K2+1 \\ \lfloor \text{Child} \rfloor & , \text{no change on } K2 \end{cases} \qquad \ldots(2)$$

We note that after the encryption process the decryption key K2 was constructed, the encryption process explained in pseudo code as follow:

Encryption process
Inputs: Unicode plain text, Unicode encryption key
Output: Unicode encrypted text
foreach *int* in Unicode plain text, *int* in Unicode encryption key K_1: - Compute average crossover from equations(1,2) - Replace new *int* (gene) to produced child. Get Unicode char of produced child Display the result

The details of implementation encryption process shown in below:

$Parent_1$ Unicode plain text	4718	3020	3859	4730	4718	3068	3859
	䜘	〠	㡙	䜰	䜘	と	㡙

Parent2 Unicode key K1	1619	2731	6571	3928	8674	5357	5862
	ᘙ	✱	敱	㤨	虴	南	塢

Produced Child Unicode encrypted text	3169	2876	5215	4329	6696	4213	4861
	ㅩ	⡶	刕	䌩	暖	䈓	䡡

4.5. *Implement Decryption Algorithm*

In the other side, the process is opposite, the decoding process implemented on Unicode encrypted text using Unicode decryption key K2 by the following equation:

$$\text{Parent1} = (\text{Child} * 2) - \text{Parent2} \qquad \ldots(3)$$

Where, Parent1 ,Parent2 ,Child represent Unicode decrypted text, Unicode decryption key K2 and Unicode encrypted text respectively.

The decryption process explained in pseudo code as follow:

Decryption process
Inputs: Unicode encrypted text, Unicode decryption key
Output: Unicode decrypted text
foreach *int* in Unicode encrypted text, *int* in Unicode decryption key: - Retrieve Unicode plain text from equation(3) - Replace new *int* to Unicode decrypted text Get Unicode char of decrypted text Display the result

The details of decryption process explained below:

Unicode encrypted text	3169	2876	5215	4329	6696	4213	4861
	ᆳ	∷	劦	纖	暖	蔦	鞜

Unicode decryption key	1620	2732	6571	3928	8674	5358	5863
	ര	✻	鼓	供	蚝	単	塗

Unicode decrypted text	4718	3020	3859	4730	4718	3068	3859
	讖	[illegible]	幌	徲	讖	と	幌

4.6. Convert to English Decrypted Text

This step is opposite of step 2, the Unicode number was split to produce two int according to mapping table of English alphabet.

The converting process shown in the following pseudo code:

Convert to English plain text process
Inputs: Unicode decrypted text
Output: English decrypted text
foreach *value* in Unicode decrypted text: - Split into two *int* - Get *char* of each *int* - Display the result

The details of this step explained as example below:

讖	[illegible]	幌	徲	讖	と	幌	獾
4718	3020	3859	4730	4718	3068	3859	4754

4718		3020		3859		4730		4718		3068		3859		4754	
↓	↓	↓	↓	↓	↓	↓	↓	↓	↓	↓	↓	↓	↓	↓	↓
n	o	space	p	a	i	n	space	n	o	space	g	a	i	n	.

47	18	30	20	38	59	47	30	47	18	30	68	38	59	47	54
n	o	space	p	a	i	n	space	n	o	space	g	a	i	n	.

Figures 3,4 and 5 show the GUI of proposed system with three examples.

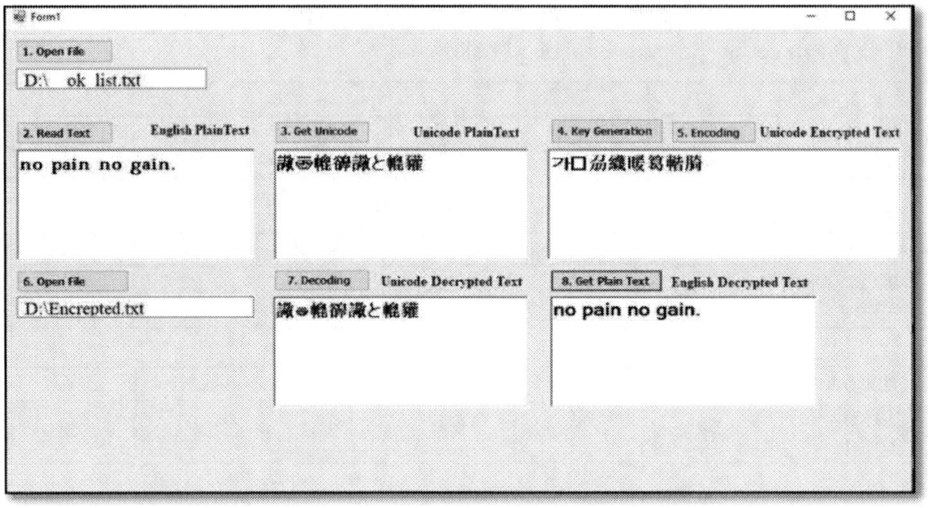

Figure 3. GUI of Proposed Cryptographic System

Now in another example:

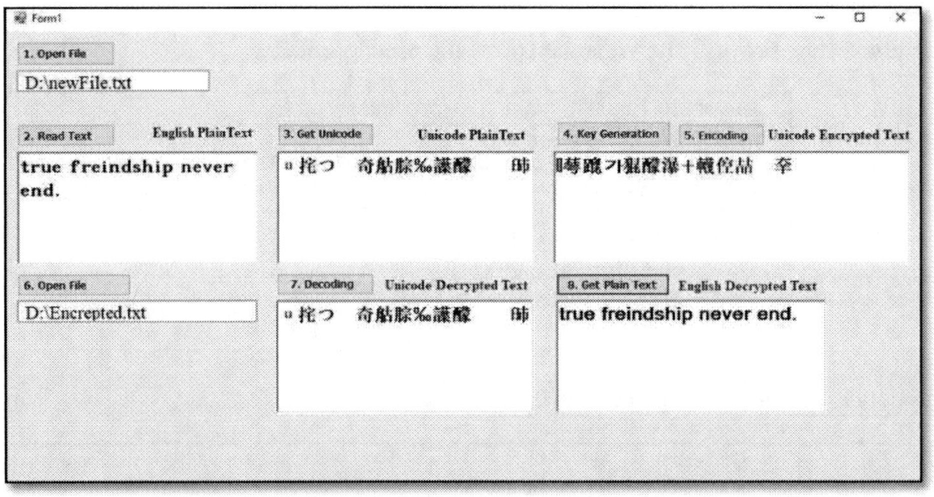

Figure 4. GUI of Proposed Cryptographic System with Second Example

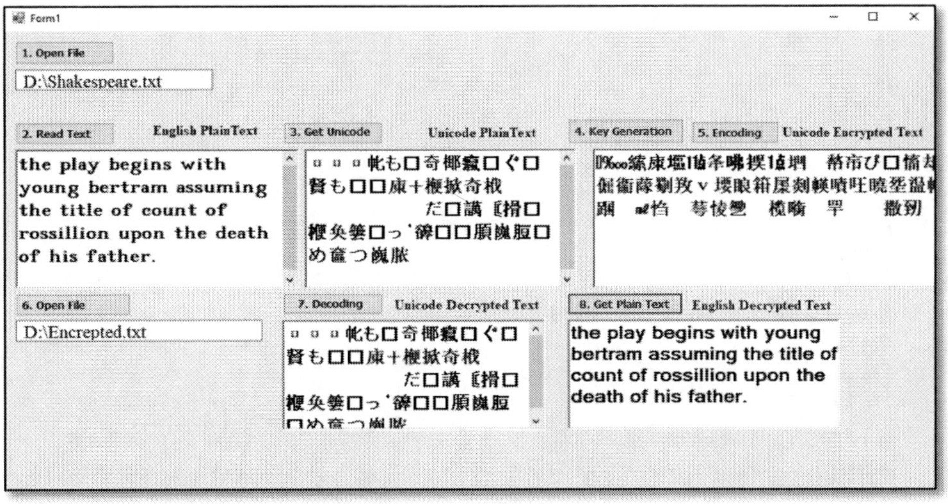

Figure 5. GUI of Proposed Cryptographic System with Third Example

5. Statistical Cryptanalysis and Results

All natural language have statistical characteristics, which means that each character in the alphabet has it's own frequency in any text. Since these frequencies are so consistent, then an approximate probability can be attached to each letter. The most significant letter in a monoalphabetic cipher is equivalent for letter "e". However, English characters can be grouped in to five sets according to their frequencies, see table 3[11]:

Table 3 Frequencies of English Characters

Group	*Letter*	*Frequency*
1	e	9.614
2	t, a, o, i, n, s, h, r	6.855 – 4.532
3	d, l	3.219 – 3.047
4	c, u, m, w, f, g, y, p, b	2.106 – 1.129
5	v, k, j, x, q, z	0.741 – 0.056

The frequency of letters appearance satisfies the probability law: $\sum_{n=z}^{a} \mathbf{P}\ n = 1$. Cryptanalysis can be described as the art and science of breaking ciphertext. Therefore any cipher system can be measured by evaluating the security of cryptosystem. This measure concerns the computational effort required to break a cryptosystem. We might define a cryptosystem to be computationally secure if the best algorithm for breaking it requires at least N operations, where N is some specified, very large number so that it requires a large amount of computer time.

Now, in this section, we will evaluate the security level of the proposed system according to it's ability to pass the statistical analysis. Three experimental results were implemented with plaintext length 759, 239, 248 respectively, see figures 6,7, and 8.

IOP Conf. Series: Journal of Physics: Conf. Series **1032** (2018) 012049 doi:10.1088/1742-6596/1032/1/012049

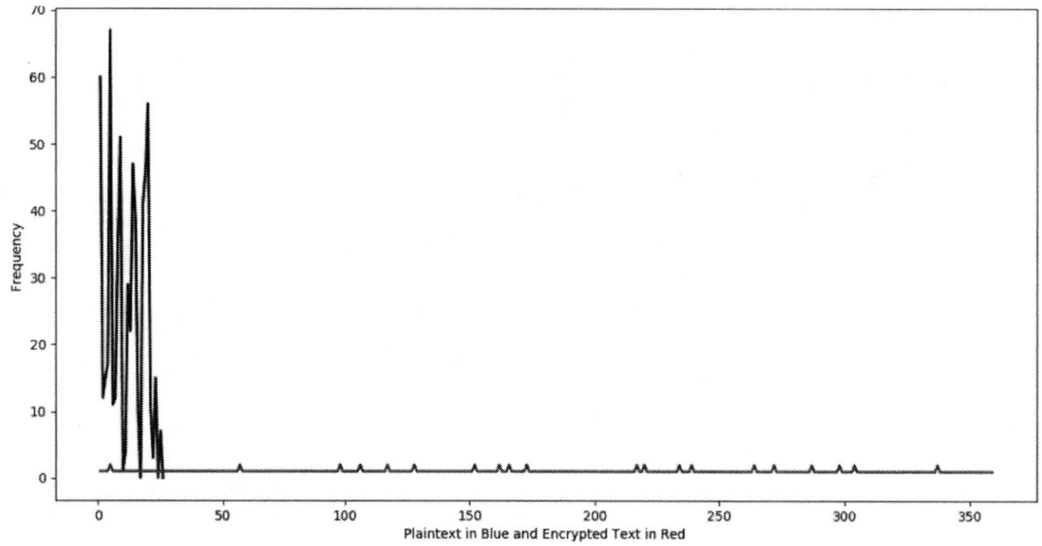

Figure 6. Histogram of Plaintext with Length 759 and Encrypted Text

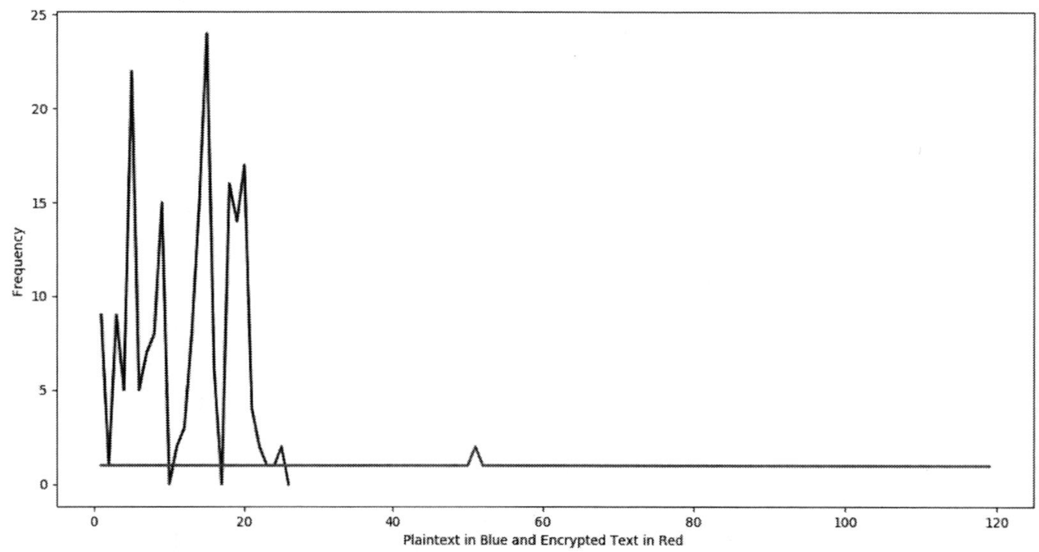

Figure 7. Histogram of Plaintext with Length 239 and Encrypted Text

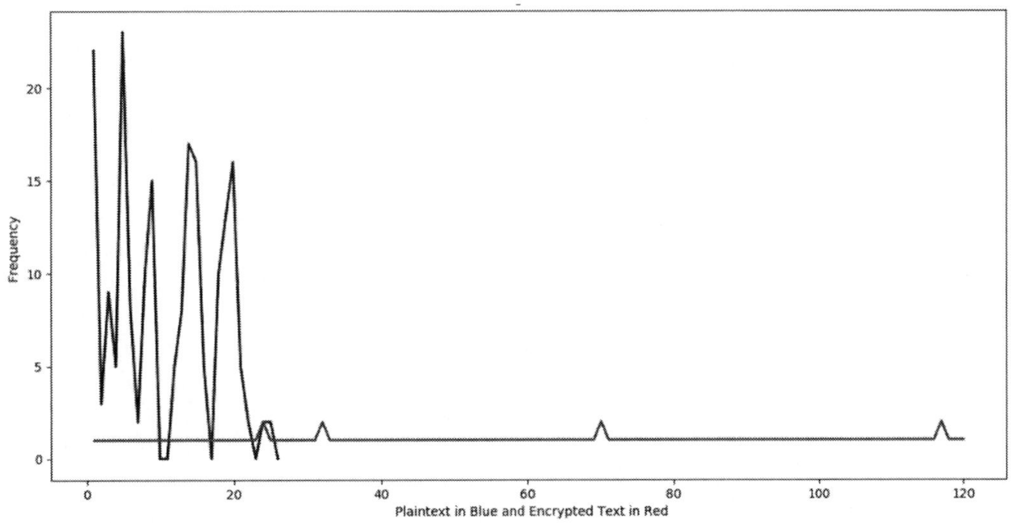

Figure 8. Histogram of Plaintext with Length 248 and Encrypted Text

As shown in figures, the alphabet of encrypted text was increased as well as no high frequency for each character. That mean the effort required to predict the plaintext, key or both from encrypted text is much more.

6. Conclusions

A novel method of cryptographic system was proposed based on combination of principles of Unicode and crossover. Sufficient amount of secrecy can be introduced by many steps, firstly in alphabet interleaving process which create mapping table for English alphabet used in plaintext and mapping table for Arabic alphabet used in key generation. Secondly in converting to Unicode process; this implemented on English plaintext and Key. These steps make the proposed system more robust against the statistical analysis. The statistical characteristics of plaintext message (natural language) scattered such that the attacker can't predict them. Thirdly in crossover process by mixing two parents and produce new child only.

The converting to Unicode process in proposed system provides:

- Generality and flexibility; it can be exceeds the limitation of alphabet; the encrypted text has more alphabet than plaintext.
- Randomness and variation; by convert two plaintext character to one Unicode character with very small frequency.
- Compression; presented in encrypted text. Plaintext with length N encrypted to encrypted text with length N/ 2. That makes it faster in transmission.

In addition, the proposed method is not needed especial hardware capability.

7. References

[1] Bruce Schneier , " Applied Cryptography Protocols, Algorithms, and Source Code in C ", Second Edition, John Wiley&Sons,Inc., 784 p., 1996.

[2] Henk C.A. van Tilborg, "Cryptology: A Professional Reference and Interactive Tutorial ", First Edition, Kluwer Academic Publishers Norwell, MA, USA, ,503 p., 1999.

[3] The Unicode Consortium, "The Unicode Standard", Available from: http://www.unicode.org.

[4] Yannis H. "Fonts & Encodings". Sebastopol, CA, USA: O'Reilly Media, Inc. 1040 p., 2007.

[5] Dhamyaa A. AL-Nasrawi et al., " Unicode Text Editor for Ancient Egyptian Hieroglyphs Writing System", *Egyptian Computer Science Journal* Vol. 38 No. 2 ,pp 48-55, May 2014.

[6] Jukka K. "Unicode Explained", USA: O'Reilly Media, Inc. 678 p., 2006.

[7] Maram Balajee, "UNICODE and Colors Integration tool for Encryption and Decryption", *International Journal on Computer Science and Engineering* (IJCSE), Vol. 3 No. 3 , pp 1197-1202, Mar 2011.

[8] A.Joseph and V.Sundaram, "Secured Communication through Fibonacci Numbers and Unicode Symbols", *International Journal of Scientific & Engineering Research*, Vol. 3, Issue 4, pp. 1-5, April-2012.

[9] Dhamyaa A. AL-Nasrawi, Ahmed F. Almukhtar, Wafaa S. AL-Baldawi, " From Arabic Alphabets to Two Dimension Shapes in Kufic Calligraphy Style Using Grid Board Catalog", Communications in Applied Sciences, Vol. 3, No. 2, pp 42-59, 2015.

[10] A.J. Umbarkar , P.D. Sheth, " Crossover Operators in Genetic Algorithms: A Review", *ICTACT JOURNAL ON SOFT COMPUTING*, Vol. 06, ISSUE 01, pp 1083 - 1092 , October 2015.

[11] William Stallings, "CRYPTOGRAPHY AND NETWORK SECURITY, PRINCIPLES AND PRACTICE", FIFTH EDITION, Pearson Education, Inc., publishing as Prentice Hall, 2011

Designing and conducting an empirical experiment comparing two web sites

Wafaa S. AL-Baldawi

Computer Science Department, College of Science, University of Kerbala, Karbala 56001, Iraq

wafaashamh@gmail.com

Abstract. This paper seeks to present a usability evaluation of the two websites (www.bmw.co.uk, www.ford.co.uk) by performing three tasks related with ability to access, navigate, and find the information to complete the tasks without using the search engines with answering some paper-pencil questions after each task. A sample of 16 users is used to collect the actual data. There are three types of questions have been asked including open ended, closed-ended and partially closed items. While the user is doing the task, many factors have been measured to determine which website is more usable than other. Designing the empirical experiment is to identify the major strengths and weaknesses of the site and to incorporate the results and user feedback.
Keywords: Website usability, satisfaction, effectiveness, efficiency.

1. Introduction

This paper reported to compare two websites (www.bmw.co.uk, www.ford.co.uk) in term of usability by performing the three tasks related with ability to access, navigate, and find the information to complete the tasks without using the search engines with answering some paper-pencil questions after completed each task by participating the actual users.

The usability can be defined as the effectiveness, efficiency and user satisfaction by determining if a sample set of users can complete the specific set of tasks in a lab environment [1].

When the user is doing the task, the number of errors, time taken to complete a specific task, verbal protocol, visual protocol, visual scanning patterns, and attitude had been measured to know which website is easier to use than another. Also, if the users satisfy with a specific website separately and if the task can be completed or uncompleted in a formal laboratory experiment. The goals of the evaluation had been defined by identifying the objectives of an evaluation to make sure the methods which should be used and the factors that should be measured are considered [2].

Three evaluation measurements are used: the efficiency, the effectiveness, and the users' satisfaction. These aspects illustrates in the following:

1. The efficiency:

In evaluating the usability of a website, it is important to calculate the time required to perform a specified task as well as the number of errors encountered to retrieve a particular information and their typologies [3][4][5]. Badre [5] mentioned that errors are the hardest decision in the usability measures and that includes identifying the sort of errors happening and there frequency of occurrence. To solve this issue, the experimenter can compute the misdirected or incorrect activities that will not let the user to complete the requested task.

IOP Conf. Series: Journal of Physics: Conf. Series **1032** (2018) 012050 doi:10.1088/1742-6596/1032/1/012050

2. The effectiveness of the web site:

Navigation can negatively affect the usability of any website and this can be related to the web-user's expectations, and the positioning and the availability of the controls in the specified page [5].

Krug states that many experts in the usability and designing websites waste a lot of time calculating and relating the numbers of clicks required to get the desired information without irritating the user. Even set a maximum number of clicks that should not be exceeded to get a piece of information when designing the website [6]. The concentrate should be on how easily the navigation design (such as content structure and adequacy) is used to allow the user to obtain the information demanded [7].

3. The Users' Satisfaction:

Collecting, and analyzing self-reported metrics such as user satisfaction, attitude, and feelings about the explored website is necessary to complete the overall evaluation of any website since high performance measurements, in many cases, do not mean positive satisfaction (vice verses) [8]. Good user satisfaction results are what a website requires. These will gain the users' attention and frequent visiting [9]. These subjective/qualitative metrics may be found by asking users carrying out survey techniques after finishing the experiment [7].

The organization of this paper is as the following: section 2 introduces research model and hypotheses that includes users, experimental design, variables, apparatus and materials, description of user interaction, procedure, discussion and conclusions, and limitation and recommendations. In section 3 references are listed. Finally, appendices are the section 4 which includes: tasks' sheet, questionnaire that includes two parts: questionnaire for users' background and users' subjective opinions questions.

2. Research Model and Hypotheses

To test the usability, designers should collect the qualitative and quantitative data from real users. The final basis of this work is depending on what in the literature review. Effectiveness, efficiency, and satisfaction are proposed the three first-order factors usability construct. Thus, the following hypotheses are posited:

H0a: There is no difference between Ford website and BMW website in navigation.

H0b: There is no significant difference in the effectiveness level of user interaction in the BMW website and the Ford websites.

H0c: The ford website is the same as the BMW website in terms of user satisfaction.

2.1. Users / Respondents

Users must meet the following criteria in order to participate:

Users are selected from different nationalities who are experienced driver with a valid driving license as the website to be tested is aiming to attract driver.

1. Not sober, not drunk and not drug trials.
2. The age more than 18.
3. Drivers and/or interested with cars.
4. Do not work in car companies or as website designers because they may be working as the designers of the websites for competitors companies.

The number of users is difficult to determine and the sample of population which are available but the truly representative sample should use rather than from the population at large [10]. In this study, the 16 users are observed for testing the websites. They are undergraduate, postgraduate students and graduated workers. Users are from different nationalities and their ages between 20 and 40. They are 5 females and 11 males.

2.2. Experimental Design

IOP Conf. Series: Journal of Physics: Conf. Series **1032** (2018) 012050 doi:10.1088/1742-6596/1032/1/012050

The majority of website evaluation tests are carried out using comparative research. In such research a new website is compared with its old version of it or a website is compared with another. In this later case, the two websites are compared to find which one is more usable than another. The two websites are www.BMW.co.uk and www.ford.co.uk. There are Within subjects design and Between subjects design can be used.

The Within subjects design is preferred because the number of users is fewer than the number of users in Between subjects design. Despite the more testing is needed in within subject for each user. There is a benefit from using the within subjects such as briefing, practicing, scheduling, and so on are easier than in between users design.

Also, the variant in measurements is more acceptable due to the differences between the websites than to behavioral differences among users while using the Between-subjects design. However, in Within subjects, the users learn something while they are testing in the first website. Therefore, using a Latin square which is a design in each test is assigned to each time period the same number of times and to each subject the same number of times [11]. Latin square is used by dividing the user in two groups and presenting the two websites in different order [12]. While the users are performing the tasks, the BMW website is presented first to Group1 then Ford website. In contrast, the Ford website is presented first to Group2 then the BMW website.

The users are treated by the experimenter in the same way and in the same environment but not all the users are tested in the same day. The time of the tests which are taken depended on the free time of each user separately. The experimenter controlled the situation of the session. However, the Internet speed and the network traffic would significantly affect the results.

The time for each usability study session is 50 minutes in which the evaluation test is carried out by each user and the quantities and qualities performances are measured.

2.3. *Variables*

In this study, two independent variables are considered throughout two set of tasks:

•In the first task, the user has to find a certain page and the extract details of a particular dealer.

•In the second task, the user has to check the specific feature of the exact car model in each website.

The feedback is collected by the help of questionnaire. The dependent variables came in use to measure the efficiency and the effectiveness of the website. A number of errors that occurred during the entire surfing of website are calculated (such as accessing wrong pages, etc.) at the end of the whole experiment by the experimenter with the help of software (For details, please refer to 'Apparatus and Materials' Section).

The method is used to calculate the efficiency, effectiveness and user satisfaction. These are achieved by carrying out a questionnaire, by which users are asked to share their perception by scoring the question.

The user's satisfaction mode can be judged on the basis of their questionnaire scores, and the experimenter's observation. The confounding variables may become a hurdle in the middle of the experiment. The experiment is conducted in an entirely separate room to avoid all the means of distraction to the user.

2.4. *Apparatus and Materials*

1. The equipment uses in the experiment is a laptop running Windows 7,3GB RAM, Intel core i7 processor with an internet connection (Wi-Fi internet connection) to surf both the websites named bmw.co.uk and ford.co.uk. The websites are surfed in the In Google Chrome browser.
2. A built-in laptop camera.
3. Task based Testing summary sheet is needed for measuring the tasks achievement (see Appendix).
4. Microsoft Excel 2007
5. Digital stopwatch is used for time measurement.
6. To reduce the environment effects, the user is asked to set the mobile to the silent mode, and a water bottle and beverages/refreshments are arranged for the user in case of requirement.

IOP Conf. Series: Journal of Physics: Conf. Series **1032** (2018) 012050 doi:10.1088/1742-6596/1032/1/012050

2.5. Description of User Interaction

At the beginning of the test, the experimenter welcomes user with introduction about the objectives and the aim of the session and give to the user one instruction paper to know how to complete all the activities (doing the tasks and answering the questionnaire). When the session is started the experimenter gives the user one conform form to fill. While the user fills the form and the biographic questions to know her/his experiences. The experimenter prepares the tasks cards, the questionnaire papers and pens to be ready. During the session the experimenter writes some notes, time which is taken to finish each completed task and which task is uncompleted and the entire errors. The user is asked to visit the both websites one after another to accomplish the two tasks. When the two tasks finishes, a questionnaire is provided to the user. The main focus of preparing questions is to evaluate the efficiency of website, effective of website and the user's satisfaction. Above all, a guideline sheet is also introduced to the user, describing the two tasks.

2.6. Procedure

One experiment conducted by 16 users by visiting the websites and filling the questionnaire in the end of each task. The total time spend of each experiment is set.

On seeking the permission of the user verbally and through consent form as well, the experiment started. The laptop with the network connection is presented to the user. Then, the experimenter starts the stopwatch. Throughout the task, the experimenter times each task and observes the user's activities. On completion of the first task, the user is asked to write down the result on a sheet of paper, already prepared by the experimenter.

This criterion is followed after each sub task. Once each task is accomplished, the user is given a questionnaire, contains a set of questions organized in category wise (please refer to Appendix for details of the questionnaire).

2.7. Experimental Results (Statistical Analysis)

• Statistical results of the effeciency of the two websites:

BMW	Ford	Diff	t-Test: Paired Two Sample for Means	α	0.05	
02:02.0	02:03.5	#######				
06:00.0	03:49.6	02:10.4		*BMW*	*Ford*	
03:30.7	01:50.0	01:40.7	Mean	0.001754	0.001567	
03:20.1	04:30.0	#######	Variance	1.17E-06	1.3E-06	
05:00.2	06:03.9	#######	Observations	16	16	
03:00.3	03:54.1	#######	Pearson Correlation	0.747011		
02:08.0	00:23.0	01:45.0	Hypothesized Mean Difference	0		
00:54.0	01:00.0	#######	df	15		
01:59.0	01:00.0	00:59.0	t Stat	0.945		
00:30.0	01:47.0	#######	P(T<=t) one-tail	0.180		Accept Null Hypothesis because p > 0.05 (Means are the same)
01:04.0	01:05.0	#######	T Critical one-tail	1.753		
01:36.0	00:25.0	01:11.0	P(T<=t) two-tail	0.360		Accept Null Hypothesis because p > 0.05 (Means are the same)
01:00.0	00:31.0	00:29.0	T Critical Two-tail	2.131		
04:15.5	02:52.4	01:23.2	The two values of p are greater than 0.05 so the Null Hypothesis is accepted			
02:05.0	02:50.0	#######				
02:00.0	02:01.9	#######				
			Time Task1 - Result			

Figure 1. Statistical Result of the Time Spent for Task1

IOP Conf. Series: Journal of Physics: Conf. Series **1032** (2018) 012050 doi:10.1088/1742-6596/1032/1/012050

BMW	Ford	Diff	t-Test: Paired Two Sample for Means	α	0.05	
06:45.2	05:06.0	01:39.2				
02:37.0	06:41.9	######		*BMW*	*Ford*	
08:00.0	03:31.7	04:28.3	Mean	0.003486	0.003387	
10:00.0	08:00.1	01:59.9	Variance	4.46E-06	1.72E-06	
07:00.5	05:01.0	01:59.5	Observations	16	16	
10:00.9	05:01.7	04:59.2	Pearson Correlation	0.414169		
03:48.0	07:55.0	######	Hypothesized Mean Difference	0		
03:12.0	02:42.0	00:30.0	df	15		
01:29.0	02:30.0	######	t Stat	0.202		
03:37.0	05:46.0	######	P(T<=t) one-tail	0.421		Accept Null Hypothesis because p > 0.05 (Means are the same)
02:15.0	06:18.0	######	T Critical one-tail	1.753		
03:22.0	03:09.0	00:13.0	P(T<=t) two-tail	0.842		Accept Null Hypothesis because p > 0.05 (Means are the same)
01:25.0	02:15.0	######	T Critical Two-tail	2.131		
09:17.4	06:19.3	02:58.0	The two values of p are greater than 0.05 so the Null Hypothesis is accepted			
02:30.0	02:50.0	######				
05:00.6	04:55.0	00:05.6				
			Time Task 2 - Result			

Figure 2. Statistical Result of the Time Spent for Task2

BMW	Ford	Diff	t-Test: Paired Two Sample for Means	α	0.05	
06:51.3	04:10.0	02:41.3				
05:31.0	05:24.0	00:07.0		*BMW*	*Ford*	
06:45.0	03:10.6	03:34.4	Mean	0.002397	0.002309	
02:03.0	04:00.3	######	Variance	2.18E-06	1.56E-06	
05:09.0	04:00.0	01:09.0	Observations	16	16	
06:37.0	08:29.1	######	Pearson Correlation	0.734298		
02:00.0	02:00.0	00:00.0	Hypothesized Mean Difference	0		
01:27.0	02:14.0	######	df	15		
03:00.0	02:10.0	00:50.0	t Stat	0.347		
02:02.0	03:12.0	######	P(T<=t) one-tail	0.367		Accept Null Hypothesis because p > 0.05 (Means are the same)
00:01.0	00:53.0	######	T Critical one-tail	1.753		
02:19.0	02:03.0	00:16.0	P(T<=t) two-tail	0.734		Accept Null Hypothesis because p > 0.05 (Means are the same)
02:00.0	01:45.0	00:15.0	T Critical Two-tail	2.131		
04:20.4	04:11.0	00:09.5	The two values of p are greater than 0.05 so the Null Hypothesis is accepted			
03:00.0	03:00.0	00:00.0				
02:08.0	02:30.0	######				
			Time Task 3 - Result			

Figure 3. Statistical Result of the Time Spent for Task3

BMW	Ford	Diff	t-Test: Paired Two Sample for Means	α	0.05	
0	1	-1				
2	0	2		*BMW*	*Ford*	
0	0	0	Mean	0.625	0.3125	
0	2	-2	Variance	0.783333	0.3625	
1	1	0	Observations	16	16	
1	1	0	Pearson Correlation	-0.14075		
1	0	1	Hypothesized Mean Difference	0		
1	0	1	df	15		
3	0	3	t Stat	1.098		
0	0	0	P(T<=t) one-tail	0.145		Accept Null Hypothesis because p > 0.05 (Means are the same)
0	0	0	T Critical one-tail	1.753		
0	0	0	P(T<=t) two-tail	0.289		Accept Null Hypothesis because p > 0.05 (Means are the same)
0	0	0	T Critical Two-tail	2.131		
1	0	1	The two values of p are greater than 0.05 so the Null Hypothesis is accepted			
0	0	0				
0	0	0				
			Number of Errors Task1 Result			

Figure 4. Statistical Result of the Number of Errors in Task1

IOP Conf. Series: Journal of Physics: Conf. Series **1032** (2018) 012050 doi:10.1088/1742-6596/1032/1/012050

BMW	Ford	Diff	t-Test: Paired Two Sample for Means	α	0.05	
1	2	-1				
2	2	0		*BMW*	*Ford*	
3	0	3	Mean	4.5	3.8125	
4	3	1	Variance	6.533333	14.69583	
7	0	7	Observations	16	16	
3	1	2	Pearson Correlation	0.09185		
8	10	-2	Hypothesized Mean Difference	0		
2	11	-9	df	15		
3	1	2	t Stat	0.624		
7	3	4	P(T<=t) one-tail	0.271		Accept Null Hypothesis because p > 0.05 (Means are the same)
1	7	-6	T Critical one-tail	1.753		
9	1	8	P(T<=t) two-tail	0.542		Accept Null Hypothesis because p > 0.05 (Means are the same)
7	10	-3	T Critical Two-tail	2.131		
5	1	4	The two values of p are greater than 0.05 so the Null Hypothesis is accepted			
6	7	-1				
4	2	2				

Number of Error Task 2 Result

Figure 5. Statistical Result of the Number of Errors in Task2

BMW	Ford	Diff	t-Test: Paired Two Sample for Means	α	0.05	
0	0	0				
1	0	1		*BMW*	*Ford*	
0	1	-1	Mean	0.0625	0.0625	
0	0	0	Variance	0.0625	0.0625	
0	0	0	Observations	16	16	
0	0	0	Pearson Correlation	-0.06667		
0	0	0	Hypothesized Mean Difference	0		
0	0	0	df	15		
0	0	0	t Stat	0.000		
0	0	0	P(T<=t) one-tail	0.500		Accept Null Hypothesis because p > 0.05 (Means are the same)
0	0	0	T Critical one-tail	1.753		
0	0	0	P(T<=t) two-tail	1.000		Accept Null Hypothesis because p > 0.05 (Means are the same)
0	0	0	T Critical Two-tail	2.131		
0	0	0				
0	0	0	The two values of p are greater than 0.05 so the Null Hypothesis is accepted			
0	0	0				

Number of Error Task 3 Result

Figure 6. Statistical Result of the Number of Errors in Task3

- Statistical results of the effective of the two websites:

BMW	Ford	Diff	t-Test: Paired Two Sample for Means	α	0.05	
3	2	1				
5	4	1		*BMW*	*Ford*	
3	3	0	Mean	3.5	2.375	
4	6	-2	Variance	1.333333	2.516667	
3	5	-2	Observations	16	16	
5	4	1	Pearson Correlation	0.29115		
5	2	3	Hypothesized Mean Difference	0		
3	1	2	df	15		
6	1	5	t Stat	2.697		
2	1	1	P(T<=t) one-tail	0.008		Reject Null Hypothesis because p < 0.05 (Means are Different)
3	1	2	T Critical one-tail	1.753		
2	1	1	P(T<=t) two-tail	0.017		Reject Null Hypothesis because p < 0.05 (Means are Different)
3	1	2	T Critical Two-tail	2.131		
3	2	1	The two values of p are less than 0.05 so the Null Hypothesis is rejected			
3	2	1				
3	2	1				

Number of Clicks Task1 - Result

Figure 7. Statistical Results of Number of Clicks in Task1

BMW	Ford	Diff	t-Test: Paired Two Sample for Means	α	0.05	
6	7	-1				
12	5	7		*BMW*	*Ford*	
10	5	5	Mean	11.125	8.1875	
36	9	27	Variance	52.11667	15.89583	
11	10	1	Observations	16	16	
5	4	1	Pearson Correlation	0.253915		
16	16	0	Hypothesized Mean Difference	0		
10	18	-8	df	15		
8	6	2	t Stat	1.608		
12	8	4	P(T<=t) one-tail	0.064		Accept Null Hypothesis because p > 0.05 (Means are the same)
10	7	3	T Critical one-tail	1.753		
11	6	5	P(T<=t) two-tail	0.129		Accept Null Hypothesis because p > 0.05 (Means are the same)
7	10	-3	T Critical Two-tail	2.131		
7	5	2	The two values of p are greater than 0.05 so the Null Hypothesis is accepted			
11	10	1				
6	5	1				

Number of Clicks Task2 Result

Figure 8. Statistical Results of Number of Clicks in Task2

3	2	1	Mean	2.4375	2.25	
2	1	1	Variance	0.529167	1.133333	
2	1	1	Observations	16	16	
2	2	0	Pearson Correlation	0.193694		
2	1	1	Hypothesized Mean Difference	0		
2	4	-2	df	15		
2	3	-1	t Stat	0.643		
3	4	-1	P(T<=t) one-tail	0.265		Accept Null Hypothesis because p > 0.05 (Means are the same)
4	2	2	T Critical one-tail	1.753		
3	2	1	P(T<=t) two-tail	0.530		Accept Null Hypothesis because p > 0.05 (Means are the same)
2	3	-1	T Critical Two-tail	2.131		
1	2	-1	The two values of p are greater than 0.05 so the Null Hypothesis is accepted			
3	4	-1				
2	1	1				

Number of Clicks Task 3 - Result

Figure 9. Statistical Results of Number of Clicks in Task3

Effective's Questions Summary (Mean)

	Ford	BMW
Easy to complete task1	4.75	4.0625
Difficult to complete task2	2.875	3.625
Mean	3.8125	3.84375

Task1 &Task3 were completed
Task2 was uncompleted as shown in the table below:

Completed Task2

BMW/Ford	BMW	Ford
U1	1	0
U2	0	0
U3	0	1
U4	1	0
U5	0	1
U6	0	1
U7	1	1
U8	1	1
U9	1	1
U10	1	1
U11	1	1
U12	1	1
U13	1	1
U14	0	1
U15	1	1
U16	0	1

Figure 10. Effective of the two Websites Summary and Completed Task2

BMW	Ford	Diff	t-Test: Paired Two Sample for Means	α	0.05	
1	0	1				
0	0	0		BMW	Ford	
0	1	-1	Mean	0.625	0.8125	
1	0	1	Variance	0.25	0.1625	
0	1	-1	Observations	16	16	
0	1	-1	Pearson Correlation	-0.04134		
1	1	0	Hypothesized Mean Difference	0		
1	1	0	df	15		
1	1	0	t Stat	-1.145		
1	1	0	P(T<=t) one-tail	0.135		Accept Null Hypothesis because p > 0.05 (Means are the same)
1	1	0	T Critical one-tail	1.753		
1	1	0	P(T<=t) two-tail	0.270		Accept Null Hypothesis because p > 0.05 (Means are the same)
1	1	0	T Critical Two-tail	2.131		
0	1	-1				
1	1	0	The two values of p are greater than 0.05 so the Null Hypothesis is accepted			
0	1	-1				
				Completed Task2 - Result		

Figure 11. Statistical result of completed task2

- Statistical results of the satisfcation of the two websites:

Ford	BMW	Diff	t-Test: Paired Two Sample for Means	α	0.05	
3.5	3	0.5				
3.8125	3.25	0.5625		Ford	BMW	
3.5625	3.6875	-0.125	Mean	3.175	2.9625	
1.1875	2.125	-0.9375	Variance	1.254688	0.339063	
3.8125	2.75	1.0625	Observations	5	5	
			Pearson Correlation	0.768685		
			Hypothesized Mean Difference	0		
			df	4		
			t Stat	0.618		
			P(T<=t) one-tail	0.285		Accept Null Hypothesis because p > 0.05 (Means are the same)
			T Critical one-tail	2.132		
			P(T<=t) two-tail	0.570		Accept Null Hypothesis because p > 0.05 (Means are the same)
			T Critical Two-tail	2.776		
			The two values of p are greater than the significant level value so the Null Hypothesis is accepted			
			Satisfication Result			

Figure 12. Statistical Result of Satisfication Summary

2.8. Discussion and Conclusions

On the basis of statistical data of number of clicks in task1, ford is revealing a better effectiveness comparison with BMW. However, the two tested websites are same in terms of efficiency, effectiveness and user satisfaction. Null hypothesis (0a, 0b, 0c) on the basis of three different conducted t-tests is accepted except in the statistical data mentioned above. After evaluating the entire 16 users' responses, in one of the question, Ford got more scores with the statistical significance in comparison to BMW on the ground of "easily identifying of links".

Three type of errors occurred in the typing of the postcode during the first task in the both web sites. On comparing the responses of users, experimenter finds few conclusions regarding the two tested websites:-

Firstly, in an open question, most of the users find frustrated in completing the second task, in the case of BMW. But they are positive in finding the result while surfing the Ford website. Secondly, the BMW website is eye catching and interesting as it possess variety of colors and attractive slideshow. On the other hand, Ford does not find more appealing on the aforesaid grounds. Thirdly, while surfing Ford website, most of the users complained about the small font size and their grey color theme. On the contrast, BMW colors and theme are recognized positively by most of the users. Lastly, the experimenter found few errors in different pages of both the websites. In the case of BMW:-

•On clicking the BMW logo, it's not redirecting to the homepage as it always happens in normal scenario.For example, in the "find the dealer" page, the Search indicator (▶) is not working at all.

In the case of Ford:-

•At the bottom of the page, few links like "Focus" in the first column are not redirecting to desired page.

2.9. Limitation and Recommendations

The limitation in this experiment can be:

- The limited number of evaluated criteria.
- The learning effect on the efficiency criteria because of the use of "within users" approach.
- Struggling in a part of the selected tasks in one website can affect the users' attitude toward the next website.
- The recommendations suggested by the users in the multiple choice questions and in open question as well can be described as below. The experimenter also observes and faces the same problem during the surfing.
- The BMW logo should be navigated to the home page.
- In BMW, the linking of pages to reach a proper model can be improved as user gets confuse while choosing the model in a series.
- In BMW, the menu items should be at the top of the page instead at the bottom to make it more usable.
- In Ford, grey theme could be changed to a bright color as it may turn to more appealing. Color blind people may face difficulty while surfing this website.
- In Ford, the menu style seems to be more traditional according to the 21st century.

3. Appendices

3.1. Tasks' Sheets

Task1: finding a dealer

A: Initial stage: open http://www.bmw.co.uk
Script: please do not use search engine.
Goal state: find the nearest dealer for the post code M5 4WT
Additional criteria: task completion<=5 minutes, total click<=3
B: Initial stage: open http://www.ford.co.uk
Script: please do not use search engine.
Goal state: find the nearest dealer for the post code M5 4WT
Additional criteria: task completion<=5 minutes, total click<=2

Task2: finding a front lights

A: Initial stage: Browser located at http://www.bmw.co.uk
Script: please do not use search engine.
Second stage: find the 116i sport model.
Goal state: find out if this model has the Front Fog Light as a standard
Additional criteria: task completion<=7minutes, total click<=4
B: Initial stage: Browser located at http://www.ford.co.uk
Script: please do not use search engine.
Second stage: find the New Focus Titanium X model.
Goal state: find out if this model has the Front Fog Light as a standard

IOP Conf. Series: Journal of Physics: Conf. Series **1032** (2018) 012050 doi:10.1088/1742-6596/1032/1/012050

Additional criteria: task completion<=6 minutes, total click<=3

Task3: Website Appearance
A: Initial stage: Browser located at http://www.bmw.co.uk
Script: please look at the website in general and in details by providing your likes and dislike about it, please begin describing by talking.
Second stage: check the website's screen layout, text, colors, and pictures.
Goal state: find out if the website is more liked than disliked and readable text or not, attractive layout or not and so on.
Additional criteria: task completion<=6 minutes, total click<=5
B: Initial stage: Browser located at http://www.ford.co.uk
Script: please look at the website in general and in details by providing your likes and dislike about it, please begin describing by talking.
Second stage: check the website's screen layout, text, colors, and pictures.
Goal state: find out if the website is more liked than disliked and readable text or not, attractive layout or not and so on.
Additional criteria: task completion<=6 minutes, total click<=5

3.2. Questionnaire
3.2.1. Demographic Questionnaires
About you:
□□Eyesight: normal/corrected
□□Occupation: Job description:
□□Age: under 20 20-29 30-39 40 or over
□□Gender: female/ male
□□Education level: high school collage bachelor's degree graduate work post-graduate

Your experience:
□□How long have you used computers?
Under 1 year 1-3 years more than 3 years
□□Which operating system has you used regularly?
DOS Windows Mac Unix Other
□□Which browser has you used regularly?
Mozilla Firefox Google Chrome Internet Explorer Other
□□Do you currently drive a car?
Yes No
□□When you see a new model car and like it, do you search features of that car to get more information?
Yes No
□□How you search the features of the cars?
In the internet dealers store from
□□How much are exterior features are important for you?
1 2 3 4 5
3.2.2. Users' Subjective Opinions

Task 1 questions: (www.ford.co.uk)
□□Have you ever search any dealer for any car before?
Yes, many times Yes, just once No, never
□□How many times did you browse until you find the dealer?
1-3 4-6 more than 6

□□Do you think if you go out and ask someone about the position of nearest dealer is better than using Ford website? Why?
□□Is it easy for you to find dealer locator from the main page?
Yes, completely Yes, Partially No
□□It is better to mention about the distance after the nearest dealer search.
1=strongly disagree, 2=disagree, 3=agree somewhat, 4=agree or 5=strongly agree
□□It is more useful to see the link of the nearest dealer’s official website after the search.
1=strongly disagree, 2=disagree, 3=agree somewhat, 4=agree or 5=strongly agree

Task 1 questions: (www.bmw.co.uk)
□□How many times did you browse until you find the dealer?
1-3 4-6 more than 6
□□Is it easy for you to find dealer from the main page?
Yes, completely Yes, partially No
□□It is better to mention about the distance after the nearest dealer search.
1=strongly disagree, 2=disagree, 3=agree somewhat, 4=agree or 5=strongly agree
□□It is more useful to see the link of the nearest dealers’ official website after the search.
1=strongly disagree, 2=disagree, 3=agree somewhat, 4=agree or 5=strongly agree
□□Which website is best to find the dealer? Why?
BMW Ford
□□Do you think if you go out and ask someone about the position of nearest dealer is better than using BMW website? Why?

Task 2 questions: (www.ford.co.uk)
□□Have you ever search any front fog lights for any model before?
Yes No
□□Did you find the information you are looking for in the Ford website?
Yes No
□□Did you know any features of New Focus Titanium X models’ before?
Yes No
□□It is easy to find the New Focus Titanium X model features information.
1=strongly disagree, 2=disagree, 3=agree somewhat, 4=agree or 5=strongly agree
□□It is easy to find the New Focus Titanium X model has front fog lights.
1=strongly disagree, 2=disagree, 3=agree somewhat, 4=agree or 5=strongly agree
□□It is difficult to complete this task for Ford website?
1=strongly disagree, 2=disagree, 3=agree somewhat, 4=agree or 5=strongly agree
If it is difficult, what could you advice to the developer to navigate it easier?
□□How many times did you browse until you find feature of front fog light for New Focus Titanium X model?
4.6 7-10 more than 10

Task 2 questions: (www.bmw.co.uk)
□□Did you find the information you are looking for in the BMW website?
Yes No
□□Did you know any features of BMW 116i Sport model before?
Yes No
□□It is easy to find the BMW 116i Sport model features information.
1=strongly disagree, 2=disagree, 3=agree somewhat, 4=agree or 5=strongly agree
□□It is easy to find the BMW 116i Sport model has front fog lights.
1=strongly disagree, 2=disagree, 3=agree somewhat, 4=agree or 5=strongly agree
□□Did you know the front fog light belongs of the safety and technology group of features?

IOP Conf. Series: Journal of Physics: Conf. Series **1032** (2018) 012050 doi:10.1088/1742-6596/1032/1/012050

Yes No
□□It is difficult to complete this task for BMW website?
1=strongly disagree, 2=disagree, 3=agree somewhat, 4=agree or 5=strongly agree
If it is difficult, what could you advice to the developer to navigate it easier?
□□How many times did you browse until you find feature of front fog light for BMW 116i Sport model?
4.6 7-10 more than 10
□□Which website is easier than another to find the front fog light?
BMW Ford

Task 3 questions:
□□The screen layout works well for you?
1=strongly disagree, 2=disagree, 3=agree somewhat, 4=agree or 5=strongly agree
□□The writing is clear and easy to understand?
1=strongly disagree, 2=disagree, 3=agree somewhat, 4=agree or 5=strongly agree
□□The text and lettering is legible and readable?
1=strongly disagree, 2=disagree, 3=agree somewhat, 4=agree or 5=strongly agree
□□The colors are appropriate for the content and messages being presented? 1=strongly disagree, 2=disagree, 3=agree somewhat, 4=agree or 5=strongly agree
□□Graphics and pictures are not distracting?
1=strongly disagree, 2=disagree, 3=agree somewhat, 4=agree or 5=strongly agree
□□The theme for the site is a good representation for the content?
1=strongly disagree, 2=disagree, 3=agree somewhat, 4=agree or 5=strongly agree
□□What do you particularly like about www.ford.co.uk?
□□And what do you dislike? What could be improved? What is missing? (for Ford website)
□□What do you particularly like about www.bmw.co.uk?
□□What do you dislike? What can be improved? What is missing? (for BMW website)
□□What are the problems faced in each website?
BMW Ford
□□Which website from these two do you like? Why?
BMW Ford
□□Which website from these two do you unlike? Why? What do you think to improve it?
BMW Ford
□□What do you think the purpose of the ford website? BMW website? Is the design matching to the purpose perfectly?
□□Did you find any confuse in the websites? Which one?
BMW Ford

IOP Conf. Series: Journal of Physics: Conf. Series **1032** (2018) 012050 doi:10.1088/1742-6596/1032/1/012050

3.3. Task-based Testing Summary

TASK	COMPLETED TASK?	COMMENTS
Finding the nearest dealer : - BMW		
Ford		
Finding out if the model has front fog lights : -116i sport (BMW)		
-New Focus Titanium X (Ford)		
General appearance (Readability, Colour, Font, Images etc…): -BMW		
-Ford		

IOP Conf. Series: Journal of Physics: Conf. Series **1032** (2018) 012050 doi:10.1088/1742-6596/1032/1/012050

4. References

[1] P. Sherman and W. Quesenbery, "Engineering the user experience: UX and the Usability Professionals' Association'. Interactions 12, 3 (May. 2005), 38-40.

[2] T. T. Hewett, The role of iterative evaluation in designing systems for usability, Proceedings of the Second Conference of the British Computer Society, human computer interaction specialist group on People and computers: designing for usability, p.196-214, December 1986, York, United Kingdom.

[3] A. Auinger, A. Aistleithner, H. Kindermann and A. Holzinger, Conformity with User Expectations on the Web: Are There Cultural Differences for Design Principles? In: Marcus, A. (ed.) Design, User Experience, and Usability. Theory, Methods, Tools and Practice. Lecture Notes in Computer Science LNCS 6769, vol. 6769, pp. 3-12. Springer, Heidelberg, Berlin, New York ,2011.

[4] M. Leavitt and B. Shneiderman, Research-based web design & usability guidelines. Background and Methodology. US Department of Health and Human Services. Washington, 2006.

[5] A. Badre, Shaping web usability: interaction design in context. Indianapolis, IN: Pearson, 2002.

[6] S. Krug, Don't make me think! ,2nd ed. A common sense approach to web usability. Indianapolis: New Riders, 2006.

[7] M. Matera, F. Rizzo and G. Carughi, Web usability: Principles and evaluation methods. In E.Mendes & N. Mosley (Eds.), Web engineering (pp. 143–180). Berlin, Germany: Springer, 2006.

[8] T. Tullis and B. Albert, Measuring the user experience: collecting, analyzing, and presenting usability metrics. Burlington, MA: Elsevier, 2008.

[9] P. Alpar, Satisfaction with a Web Site: Its Measurement, Factors and Correlates. In A.-W. Scheer & M. Nüttgens (Hrsg.), Electronic Business Engineering /4. International Tagung Wirts chafts informatik. Heidelberg: Physica Verlag, 1999.

[10] C. Courage and K. Baxter, Understanding Your Users: A Practical Guide to User Requirements. Morgan Kaufmann Publishers, San Francisco, CA, 2005.

[11] A. Dean and D. Voss. Design and analysis of experiments. Springer Verlag, New York, NY, USA,1999.

[12] D. Martin, Doing psychology experiments ,4th ed.]S.I.] Brooks/Cole Pub. Co, Pacific Grove, Calif, 1996.

IOP Conf. Series: Journal of Physics: Conf. Series **1032** (2018) 012051 doi:10.1088/1742-6596/1032/1/012051

Impact of an on-line health safety exam on the students of college of science laboratories

Ammar A. Altameemi[1] and Ahmed H. Alrubai[1]

[1]University of Kerbala, College of Science, Computer Science Department
ammar77@outlook.com,

Abstract. In spite of education institutes subject their staff and students to a compulsory health safety course; however, many events have proven that college students do not follow the correct health safety instructions in case of emergencies. In this paper, an on-line exam of health safety was designed and implemented. PhP programming language was used to design the web pages of the exam while the data base was manipulated by MySQL language. The exam consists of about 20 specialized questions for each of all College of Science departments. The user should be able to access the exam through the website of the University or the College. Also, the user should register by entering his name, email, and password. After the exam practicing, the user will be notified about his score and if he/she put the correct or wrong answer for every question individually. The user can retry the exam again until he has got the full mark, which is the only mark that allows the examinee to pass the exam and then will be authorized to access the college laboratories and other workplaces. The student can know his mark through his email directly. The admin will also know all the examined students and their marks. Consequently, the college departments could be provided with a list of practiced students. The aim of this research focuses on increasing the education level of the health safety of the students and staff inside the laboratories of universities. So it ensures the entire workers at our laboratories will be conscious in case of any emergencies. Empirical results have been received by subjecting about 20 students in the college of science by measuring four knowledge parameters. Results have discussed the difference between the conventional manner of learning the health safety course through a brochure only and the proposed manner which is based on e-learning and on-line practicing.
Key words: Web programming, Web design, Web applications, e-Learning, OHS, HIS.

1. Introduction

In the College of Science, the students sometime deal with dangerous materials and emergency cases. Obligating students to read and follow the health safety instructions (HSI) and signposts may not give the required level of conscious, especially like these cases require a swift reaction to overcome the problems. So an e-learning procedure has been suggested here for HSI to be followed through just simple, repeatable, and educating web pages that could help to increase the level of awareness of the health safety inside university laboratories and workplaces.

In most large and intermediate enterprises, factories and our educational institutes as well, there is a responsible person caring about health safety in a specific workplace. This person is in charge to assess and recover the risks, training the representatives of entire subsections of the workplace, and monitoring how the health safety going on inside the workplace. This professional person has to be trained and educated to deal with multiple situations that associated with his enterprise environment. For example, how the nurse is educated enough to recover and minimize the risks that face the old people such as falling, lost their way, or eat from unallowable staff [1].

Nowadays, occupational health and safety (OHS) major becomes one of seriously important subjects in the most of organizations. So the organizations are seeking to be with one of standard OHS for getting a good sign mark and competitive reputation. OHSAS 18001 is considered as one standard benchmarking of OHS certifications, which can be awarded by some registered authorities for companies after passing many OHS tests, demonstration and verification of the obligations of the all safety aspects [2].

Some research teams have found there is a significant difference between large and small institutes in the area of OHS education. The blame could be pointing toward budges, the priority of the managers' attitudes, human and technical resources quality, the effect of the workers guild, and the type of OHS management system [3].

Nowadays, health safety curriculums are adopted as one of compulsory or optional materials that must be given to students especially for children in the elementary schools in many developed and developing countries [4]. However, the geographical distance between the educational and industrial institutes has prevented the adequate communication between headmasters and teacher in vocational schools, and supervisors of OHS in workplaces. The traditional training also prevents the headmaster to follow up the teachers and students in the OHS course. In addition, it has been demonstrated a misunderstanding of each responsible person role. This lead to decline level of pupils and students knowledge about OHS in these schools and colleges [5].

2. Related Work

Most of around the world, universities and institutes practice some examinations to assess the health safety awareness of the workers inside their laboratories. Many companies and enterprises even do not admit their employees without health safety certification, whereas there are only specific third-party authorities are eligible to issue health safety cards for employees in the western countries [6].

The global financial crisis (GFC) has affected the aspect of the workplace health promotion (WHP). The GFC actually has affected the small and medium enterprises (SME) more than the larger enterprises. Involving and participating a wider workforce by including OHS professionals, managers, owners, and all other workers is a worthy proposal to be investigated to increase the level of OHS priority in SME workforce. The tool that used to achieve wider subscription was Healthy Together e-learning training course. While the problem of WHP in SME is related to the building capacity, the e-Learning method has demonstrated positive results and could overcome the issues of the building resources limitation [7].

Some surveys have been reviewed on various sized enterprises and concluded the OHS knowledge management system should be conducted through an e-learning platform. They have discovered the large and small businesses should be turned through different procedure to acquaint and retain tacit knowledge. The research has focused on small to medium-sized enterprises (SME). The suggested procedure to follow with these enterprises is to narrate the methods and instructions in simple and easy communication between workers and e-learners [8].

Another investigation based on questionnaire that has included physicians, nurses, and the professional experience people in occupational medicine for both genders who are working in variant fields of Medicare institutes shows that the occupational health that provided to employees are not enough [9].

E-Learning based OHS training and education would be an effective manner to raise the knowledge, behavior, and interaction of workers. It always reflects positive results to the workers regardless of their age, gender, level of education, language, number of employees. Instructional design factors such as motivation to learn, learning content, interaction with learners and instructors, previous experience related to e-learning are all factors that influent the degree of learning satisfaction OHS education based on e-learning procedure [10].

Vinodkumar and Bhasi have conducted a comparison of OHSAS 18001, ISO 9001, and acts managements without any health safety certification to measure the risk accidents between them. The researchers have found the management systems with OHSAS 18001 certification is saver than others.

IOP Conf. Series: Journal of Physics: Conf. Series **1032** (2018) 012051 doi:10.1088/1742-6596/1032/1/012051

In contrary, they have demonstrated the management systems without health safety certification could not pass the requirements of the work safety. Also they have recommended to enforcing the employees to be obligated with safety rules could give better results [11].

A comparison has been published of three OHS standards, which are OHSAS 18001, the International Labour Organization (ILO) guidelines, and Oregon state OHSA guidelines, based on many aspects such as policy, planning and implementation, evaluation, and action of improvement. This work has concluded the integration of these three health and safety benchmarks in assessments [12].

In conclusion of this survey, the facilities and benefit of learning using computer OHS software and programming management system has not demonstrated or investigated clearly toward the conventional training course in colleges and universities. This research aims to show the advantages amount of using information technology in the field of OHS. It also demonstrates the awareness acquisition of students and personnel consciousness comparing with the traditional courses of OHS.

3. Implementation of the On-Line Exam

Although there are available open-source e-Learning platforms such as Moodle, ILIAS, and Sakai specified to work efficiently with OHS environment [13]. However, this research is based on the implementation of a tailor-made web application to be more specific and suitable for colleges of science students or to the university students in general.

Instead of using the traditional method, like test on paper, we have developed this online website to test the students experience in the field of occupational health safety (OHS) system to be qualified of entry into the college laboratories. The test was designed to cover all the prospective materials, scenarios, sudden events and situations that could happen in the college of science laboratories and workplaces.

The implementation has cared two key requirements, which are the database and the security.

It can be considered that there are many database requirements as follow:

Database system must be used to store the users' information and the examination questions. MySQL has been used for this purpose in this research. In addition, there should be an internet browser to run the on-line Health Safety system. Finally, there is a need to install WAMP server to build the system database.

With regarding to the security aspects, the health safety system must provide a safe way to allow the registered users to be authorized. So it provides a security mechanism to prevent the intruders from accessing system. It also provides a way to allow the users to do what they want if that is compatible with their permissions.

3.1. *The Administrator-User Interaction*

The system's designer has created the structure of the user's profile which can be considered as a container of the user's information that is submitted by the user in registration process. There are three types of users: Database Administrator, System Administrator and normal user. The last two types of user can access the system via web browser, which requires the internet connection on their computers. Their requests will be sent to web server. The OHS software system displays questions from the database when a person begins the OHS test. Furthermore, the user has been enabled to review its information and result if a pre-exam had been practiced. Figure 1 shows how the system interacts with all the types of users in general.

The administrator can edit, delete, view and add any extra questions or users when it logins as seen in Figure 2.

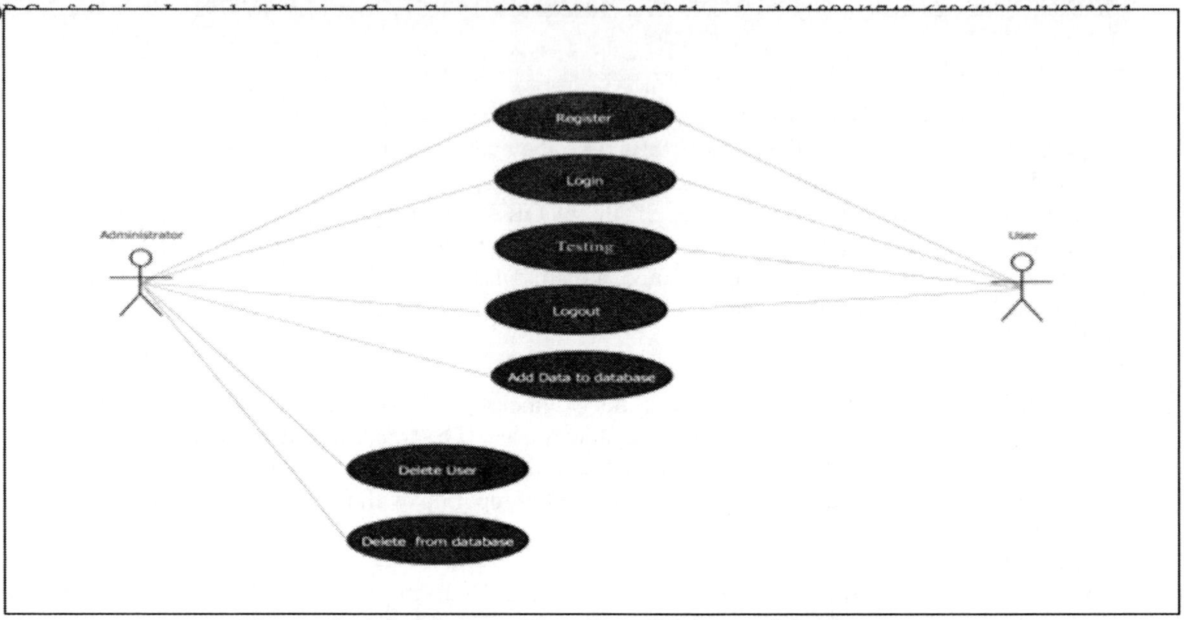

Figure 1. Administrator-User Case Diagram of the System

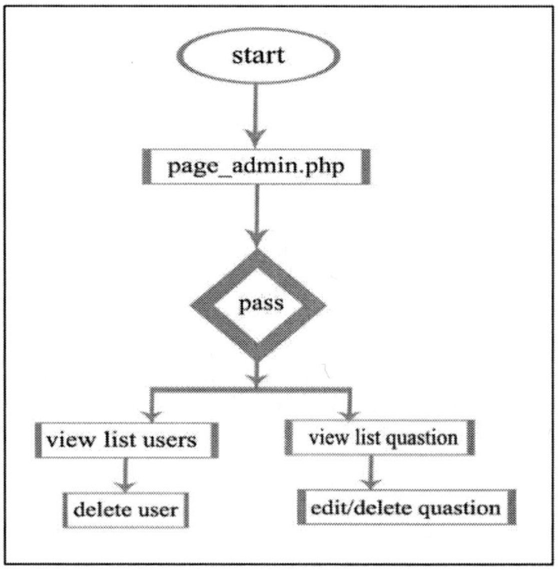

Figure 2. The sequence of administrator activities in the system

3.2. Database Design
The database of the proposed system has been manipulated using MySQL language. Following tables illustrate how the database sections are arranged.

3.2.1. User Table
To store the data and meta-data of the system, the database has to be designed. Therefore, there are three tables have been added to the table that was created during implementation, as illustrated in Figure 3.

3.2.2. Question Table

IOP Conf. Series: Journal of Physics: Conf. Series **1032** (2018) 012051 doi:10.1088/1742-6596/1032/1/012051

This table has made four tables all with the same structure and to facilitate the work when one of the sections is completed information. The database was easily designed by putting all the scattered tables in one table only. See figure 4.

Column
id
username
ymail
deprt
password
degress
admin

Figure 3. User Table

Column
id
t
q
a1
a2
a3
a4
t_a
q_status

Figure 4. Question Table

3.2.3. Image Table
Since the question is divided into four tables. So to facilitate the transition to the next question, the table was divided to four images. All the tables are with same form of structural. See figure 5.

3.2.4. Answer Table
Answer table is associated to the user name, so every examinee person must have his nominated answer table to evaluate his answers and record his degree. However, when the test is finished, this table will be deleted. See figure 6.

Column
id
name
size
type
content
fk

Figure 5. Image Table

Column
id
t
quse
chack_answer

Figure 6. Answer Table

IOP Conf. Series: Journal of Physics: Conf. Series **1032** (2018) 012051 doi:10.1088/1742-6596/1032/1/012051

3.2.5. Counter table

This table counts the person trails to avoid interfering with another person who tested at the same time and also to make the site more flexibility in moving from a question to another. This table will be deleted after the test is finished.

Column
id
count

Figure 7. Counter Table

4. OHS software GUI and windows

The OHS software exam can be logged in using a unique username, which will specify the user as an examinee or an administrator, see figure 8. However, the user can create his/her account if he/she has not registered yet, see figure 9.

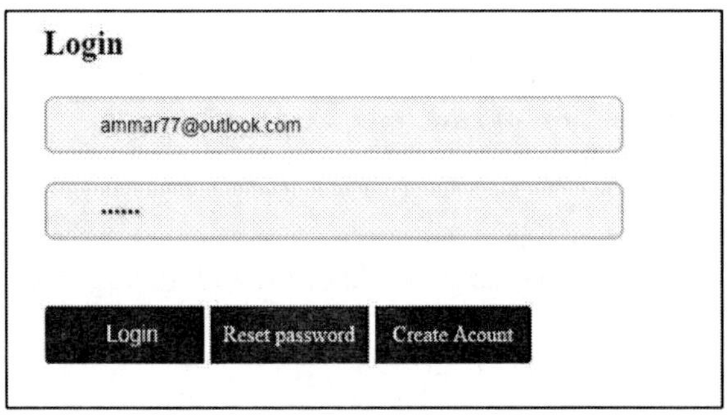

Figure 8. Log in page

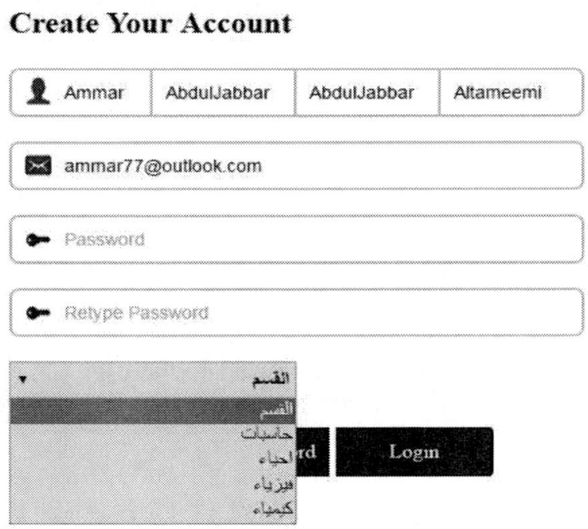

Figure 9. Creating the account page

The user who has entered as administrator can edit, add, and delete the questions. Also, he has the right to explore the name of examinees and their scores, see figure 10 and 11. However, he could not edit or modify the scores or names. As shown in figure 10, the administrator should enter to examinees section, while the examinee has to enter into the questions section after specifying his/her department.

There are four different sorts of questions. Every sort contains 20 questions categorized according to the departments of the college of science. The examinee must practice all the questions and select the right answer. Otherwise, he will not pass the exam and get the required certification of OHS. Anyway, the user can pause the exam and return back later. Figure 11 shows a question example for all departments' students.

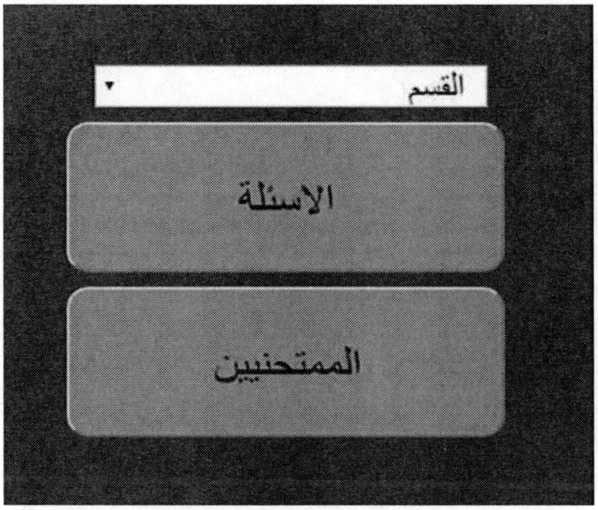

Figure 10. Beginning Page

Figure 11. A question Form

After the examinee passing the exam successfully, his/her name and email will be added to the qualified students list, as shown in figure 12.

Finally, the successful students will be notified about their result through his/her registered email and get a certification manuscript. As we mentioned, all the results can be explored by the administrator, or could be provided to the person in charge in the department.

5. Results and Discussion

Results have been obtained through making a test of two categories of students, each group contains 10 students. The first students' category has read the health safety brochure without practicing any exam. While the second category has read the brochure then practiced the proposed on-line exam. A comparison has been made between the two students' categories to show the degree of knowledge, which is compiled as choosing the accurate answers and how long the information is kept in mind as well. The speed of the correct reaction was the second comparison parameter, while the thirds parameter was the willingness to spread the information and participating to give help in emergency situations. Figure 13 illustrates a diagram of these results comparison between the two categories.

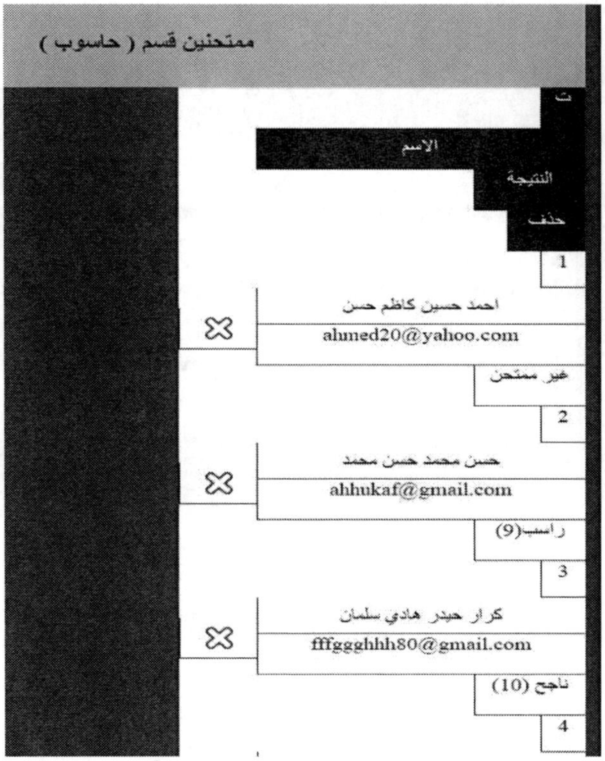

Figure 12. List of Examinees

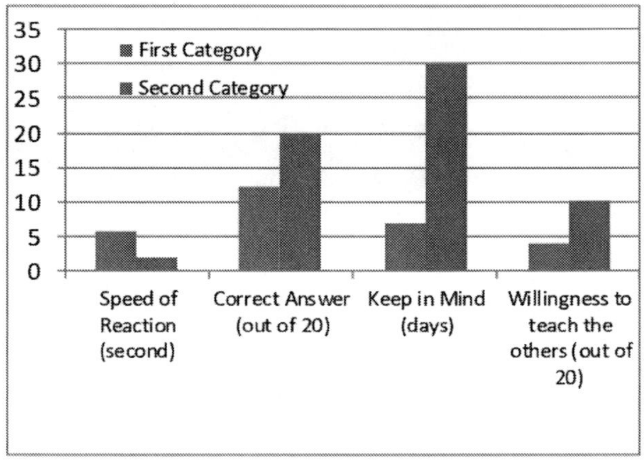

Figure 13. Diagram of the Whole Results

6. Conclusion

The person in charge who is working in our university laboratories must be responsible to take a rapid action in case of accidents. So he/she must be trained well to be able to identify the problem, control the situation, and then treat the injured persons. However, all the students inside the laboratory should have been educated enough about the health safety and way to be followed.

However, we are ambitious to make all the Iraqi institutions are going to be seeking in the same ways. So we recommend these following prospective works:

1. The project can be extended to include the colleges of Medicine, Engineering, or other colleges of the University.
2. Legislation should be put to enforce all the people working inside some dangerous workplaces of the Iraqi institutions to follow them. So we can improve the safety and keep people souls.
3. The aim of this project is educational, so we recommend being like this websites in the other educational and industrial companies.
4. The online exams can help the students to understand the curriculum topics more easily. The students in primary and high schools of Iraq must be educated very well in aspect of health safety. So this types of exam my help the mission of this field of education.

7. References

[1] Kent Family Mediation Service. (2015). Medical Law in the UK, Health and Safety Legal Requirements. Available: http://www.unsystemceb.org/health-safety-legal.htm

[2] SGS Group Management, "OHSAS 18001 Occupational Health and Safety Management Systems," 2015.

[3] P. Arocena and I. Núñez, "An empirical analysis of the effectiveness of occupational health and

IOP Conf. Series: Journal of Physics: Conf. Series **1032** (2018) 012051 doi:10.1088/1742-6596/1032/1/012051

safety management systems in SMEs," International Small Business Journal, vol. 28, pp. 398-419, 2010/08/01 2010.

[4] D. B. Buller, A. M. Taylor, M. K. Buller, P. J. Powers, J. A. Maloy, and H. Barbara, "Evaluation of the Sunny Days, Healthy Ways Sun Safety Curriculum for Children in Kindergarten through Fifth Grade," Pediatric Dermatology, vol. 23, pp. 321-329, July 2006.

[5] I.-M. Andersson, K. Gunnarsson, and G. Rosèn, "Role of Headmasters, Teachers, and Supervisors in Knowledge Transfer about Occupational Health and Safety to Pupils in Vocational Education," Safety and Health at Work, vol. 6, pp. 317-323, 2015/12/01/ 2015.

[6] The official site for the Construction Skills Certification Scheme. (2015). About CSCS. Available: http://www.cscs.uk.com/

[7] M. Hodgins, B. Battel-Kirk, and A. G. Asgeirsdottir, "Building capacity in workplace health promotion: the case of the Healthy Together e-learning project," Global Health Promotion, vol. 17, pp. 60-68, 2010/03/01 2010.

[8] A. Floyde, G. Lawson, S. Shalloe, R. Eastgate, and M. D'Cruz, "The design and implementation of knowledge management systems and e-learning for improved occupational health and safety in small to medium sized enterprises," Safety Science, vol. 60, pp. 69-76, 2013/12/01 2013.

[9] M. C. Rodríguez-Jareño, E. Molinero, J. de Montserrat, A. Vallès, and M. Aymerich, "How much do workers' health examinations add to health and safety at the workplace? Occupational preventive usefulness of routine health examinations," Gaceta Sanitaria, vol. 29, pp. 266-273, 2015/07/01/ 2015.

[10] Y. J. Lee and D. Lee, "Factors Influencing Learning Satisfaction of Migrant Workers in Korea with E-learning-Based Occupational Safety and Health Education," Safety and Health at Work, vol. 6, pp. 211-217, 2015/09/01/ 2015.

[11] M. N. Vinodkumar and M. Bhasi, "A study on the impact of management system certification on safety management," Safety Science, vol. 49, pp. 498–507, March 2011.

[12] A. J. Dababneh, "Effective Occupational Safety and Health Management System: Integration of OHSAS 18001, ILO-OSH 2001, and OR-OSHA.," 2011.

[13] N. Apostolopoulos, H. Hoffman, and J. Dietz. (2009). ELISA e-learning platform evaluation report: e-learning for improving access to information society for SMEs in the SEE Area. ELISA project. Available: http://www.elisa-project.net/

IOP Conf. Series: Journal of Physics: Conf. Series **1032** (2018) 012052 doi:10.1088/1742-6596/1032/1/012052

Arabic text classification using master-slaves technique

Zinah Abdulridha Abutiheen[1], Ahmed H. Aliwy[2] and Kadhim B. S. Aljanabi[2]

[1]Department of Computer Science, Faculty of Science, University of Kerbala, Kerbala, Iraq
[2] Department of Computer Science, Faculty of CS and Mathematics, University of Kufa, Kufa, Iraq

z.aboaltaheen@uokerbala.edu.iq

Abstract. Text classification (TC) is an essential field in both text mining (TM) and natural language processing (NLP). Humans have a tendency to organize and categorize everything as they want to make things easier to understand. Therefore, text classification is an important step to achieve this goal. Arabic text classification (ATC) is a difficult process because the Arabic language has complications and limitations resulting from the nature of its morphology. In this paper, a proposed approach called the Master-Slaves technique (MST) was used to improve Arabic text classification. It consists of two main phases: in the first phase, a new Arabic corpus of 16757 text files was collected. These text files were classified into five categories manually. In the second phase, four different classifiers were implemented on the collected corpus. These classifiers are Naïve Bayes (NB), K-Nearest Neighbour (KNN), Multinomial Logistic Regression (MLR) and Maximum Weight (MW). Naïve Bayes classifier was implemented as Master and the others as Slaves. The results of these slave classifiers were used to change the probability of the Naïve Bayes classifier (Master). The four classifiers used were implemented individually and the simple voting technique was implemented among them too on the collected corpus to check the effectiveness and efficiency of the proposed technique. All the tests were applied after the pre-processing of Arabic text documents (tokenization, stemming, and stop-word removal) and each document was represented as vector of weights. For the reliability of the results, 10-fold cross-validation was used in this paper. The results showed that the Master-slaves technique gives a good improvement in accuracy of text document classification with accepted algorithm complexity compared to other techniques.
Key words: Arabic corpus, text classification, text pre-processing, Arabic text classification, combined classifier.

1. Introduction

Text Classification (TC) is a very useful task in many applications of natural language processing (NLP) and text mining (TM) like information retrieval (IR). It is used as a basis for employment in text processing, web mining, text filtering (as email spam), text organization (as documents archive), and so on. TC (also known as text categorization, topic spotting, document categorization or document classification) is a process of assigning and labeling documents to a set of predefined categories based on their contents [1]. The categories may be literature, arts, economy, sport, etc.

The nature of Arabic language morphology is very complicated. Thus, classification is very difficult and requires time because the language needs preprocessing[2]. The preprocessing is done twice, the first time for data set and the second time for preparing it to input to classifiers. Many classifiers used for Arabic documents classification such as Decision Trees [3, 4], K-Nearest Neighbor[5], Naïve Bayes[6], Maximum Entropy [7, 8]and others which give good accuracy.

2. Related Work

Almost all the research carried out in this field showed that the results obtained from the combination of classifiers are much better than the results obtained by the same classifiers individually for text categorization.

In [9] proposed the combination of three classifiers: K-Nearest Neighbor (KNN), Relevance feedback and Bayesian independence classifiers on the corpus that consists of 11,599 documents divided into a training set of 10,902 documents and a testing set of 187 documents in the medical domain for automatic assignment of ICD9 codes. The performance of the classifiers was measured on the basis of document ranks. The result showed improving accuracy.

In [10] combined many monolingual classifiers (the sum of SVMs output values, F1 weighted sum of SVMs output values, F1 weighted sum of SVMs decisions) on multilingual text classification. A set of 2714 documents, two Anglo-Saxon languages (English and German) and two Roman (Italian and Portuguese) were used. Each document was represented by using the bag-of-words approach, a vector space model (VSM) and TF-IDF weighting function. The result of combined classifiers showed better accuracy than the single classifier.

In [11] combined classification approach: first, lexicon based method is used which classifies some documents, then the classified documents are used as training set for maximum entropy model which subsequently classified some other documents. After that, k-nearest model was used to classify the rest of the documents. The documents collected in Arabic from three different domains: "education", "politics" and "sports" forum, and a total of 1143 posts contain 8793 Arabic statements with average of 7.7 statements in each post were used. The accuracy was moved (almost) from 50% when using only lexicon based method to 60% when using lexicon based method and maximum entropy together, to 80% when using the three combined methods.

In [12] combined probabilistic classifiers (Naive Bayes classifier and Maximum Entropy). They used two merging operators, max and harmonic mean. The proposed method was evaluated using the "ModApte" split of the Reuters-21578 data set. The evaluation results showed a measurable improvement in the final evaluation accuracy.

In [13] combined Naive Bayes and modified Maximum Entropy classifiers. Reuters-21578 and the DB-World data set were used. During this work, the proposed combination classifier with max combining operator gives the best accuracy.

In [14] used a hybrid classifier to analyze text review. Naïve Bayes and MaxEnt classifiers have been combined using combining operations like mean, max, and average etc. These classifiers were applied to the Blitzer's unprocessed data set and showed a better accuracy.

In [15] built four models by combining different approaches to improve Arabic text documents classification. The first combined model is built using fixed combination rules. The best classification accuracy, 95.3%, is achieved by using majority voting rule with seven classifiers. The second combination approach is stacking. In their experiments, they used different numbers of base classifiers and two different meta classifiers: Naïve Bayes and linear regression. Stacking achieved very high classification accuracy, 99.2% and 99.4%, using Naïve Bayes and linear regression as meta classifiers, respectively. The third model they used AdaBoost to boost a C4.5 classifier with different number of iterations. Boosting improves the classification accuracy of the C4.5 classifier; 95.3%, using 5 iterations, while the accuracy is 99.5% using 10 iterations. The fourth model uses bagging with decision tree. The accuracy is 93.7% achieved when using 5 iterations, and 99.4% when using 10 iterations. They used three datasets, to test the combined models, BBC Arabic of 4763 text files, CNN Arabic of 5070 text files , and OSAC datasets of 22429 text files.

This work, compared to the related works, focuses on: (i) introducing the technique called the Master-Slaves technique for solving the problems of classifying Arabic text documents (ii)

IOP Conf. Series: Journal of Physics: Conf. Series **1032** (2018) 012052 doi:10.1088/1742-6596/1032/1/012052

constructing a new Arabic corpus taken from Iraqi media, which can be used by researchers in the future, (iii) testing the corpus with four classifiers: NB, KNN, LR and MW, (iv) comparing the results of six classifiers.

3. Pre-processing

Preprocessing represents the first step in TC in which the required documents are prepared as input to the classification process by applying machine learning techniques with high efficiency and accuracy. Documents should, firstly, be preprocessed which is an important stage in TC.

The aim of pre-processing is to reduce dimensions and to distinguish between the most important documents by selecting the significant or relative words and ignore the irrelevant words. This will turn the document into more suitable format.

Algorithm: Pre-processing
Input: D: collection of documents
Output: V: collection of vectors
Step 1: Let V= List of vectors (size=collection size)
Step 2: for each document in D do:

- Tokenization
- Stop-Word Removal
- Stemming
- Calculate the weights of each word
- Add the vector of weights to V.

End

After cleaning and organizing the corpus files, they must be preprocessed to get a new version of the documents suitable for the next phase. In other words, the corpus must be suitable to input for machine learning. The general idea of preprocessing is explained in algorithm below.

Where in Tokenization is the act of breaking up a sequence of strings into pieces such as words, keywords, phrases, symbols and other elements called tokens. In Stop-Words Removal is simply removing the words that occur commonly across all the documents in the corpus and unimportant words that appear in a text. In Stemming is the process for reducing inflected words to their word stem (base form) which can be removed any affixes (prefixes that added to the beginning of the word, infixes that added to the middle of the word, and suffixes that added to the ending of the word) from the words to reduce these words to their stems or roots under the assumption that words sharing the same stem. In Weight is calculated from term frequency (TF) and inverse document frequency (IDF) as follows:

$$\text{Weight}(t) = tf*idf \qquad \ldots(1)$$

Where:
tf: Term Frequency tf(t,d) of term t in document d is defined as the number of times that t occurs in d.
idf: Inverse Document Frequency estimates the rarity of a term in the whole document collection. (If a term occurs in all the documents of the collection, its IDF is zero.)

$$\text{Idfi}=\log\left(\frac{\text{Number of documents}}{\textit{Number of documents that contain word } i}\right) \qquad \ldots(2)$$

Each word with idf equal to zero is deleted when computing weights, thus reduces the size of the text file.

4. Classifiers

IOP Conf. Series: Journal of Physics: Conf. Series **1032** (2018) 012052 doi:10.1088/1742-6596/1032/1/012052

Many classification techniques can be used to classify texts such as Naïve Bayes, K-Nearest Neighbor and many others. Like other tasks of NLP, TC needs preprocessing (tokenization, stop-word removal, normalization and stemming). Supervised machine learning techniques used in this research include labeling texts for training and predefining texts for testing to evaluate the performance of the classifiers. In this work, four classifiers were used, which are Naïve Bayes, K-Nearest Neighbor, Logistic Regression, and Maximum Weights.

4.1. Naïve Bayes

Naïve Bayes (NB) classifier is a simple probabilistic classifier based on Bayes' theorem with strong independence assumptions between the features. For this reason, it is called naive. NB is the most commonly used generative classifier[16]. In general, most researchers employ the NB method by applying Bayes rule:

$$C_{MAP} = \underset{c_j \in C}{\operatorname{argmax}_c} \hat{P}(c_j) \prod_i \hat{P}(x_i | c_j) \quad (3)$$

Where:

$$\hat{P}(c_j) = \frac{\text{doccount}(C = c_j)}{N_{doc}} \quad (4)$$

$$\hat{P}(x_i | c_j) = \frac{\text{count}(x_i, c_j)}{\sum_{x \in V} \text{count}(x, c_j)} \quad (5)$$

Where

p(cj): probability of class cj among set of classes C.,

p(×i|cj): probability that the term xi occurs in class cj which may be zero in training data, so the smoothing technique (as Laplace) is chosen to estimate it[17].

There are two classes of models commonly used for NB classification based on the distribution of the words in the document. The major difference between these classes is an assumption in terms of taking (or not taking) word frequencies into account, in the Multinomial Model (MNB), the document is represented by a bag of words and frequencies of terms are computed in it[18]. The probability of occurrence of each feature i in a category c can computed as p(×i|y)[19]. Another class is Multivariate Bernoulli Model (BNB), in which the presence or absence of words in a text document can be used as features to represent a document. Thus, the word features in the text are assumed to be binary[18].

The Naïve Bayes is used as a master with modifying probabilities of the classes or the features for a given class in the Master-Slaves technique.

4.2. K-Nearest Neighbor

The k-nearest neighbor algorithm (k-NN) is a statistical learning algorithm based on the distance or similarity function for pairs of observations, such as the Euclidean distance or Cosine similarity measures between "k" training data and the testing data. The degree of similarity between documents and k training data depend on a value of k [20]. The best choice of k depends upon the data. Usually, the greater value of k reduces the effect of noise on the classification, but it makes the boundaries between classes less distinct. If k is greater than one, then the object is classified by a majority voting of its neighbors according to the value of k, if k = 1, then the object is assigned to the class of its nearest neighbour [21].

4.3. Multinomial Logistic Regression

Multinomial Logistic Regression (MLR) is also known as Log-linear Models (linear if log is taken), and a conditional exponential classifier or logistic regression classifier. It includes polytomous LR,

IOP Conf. Series: Journal of Physics: Conf. Series **1032** (2018) 012052 doi:10.1088/1742-6596/1032/1/012052

multiclass LR, softmax regression, multinomial logit, maximum entropy (MaxEnt) classifier and conditional maximum entropy model [5, 18].

Multinomial logistic regression modelling is a general and an intuitive way for estimating a probability from the data and it has been applied successfully in various natural language processing tasks [22].

Conditional distributed p(c|d) is computed as follows:

$$p(c|d) = \frac{1}{Z}\exp\left(\sum_{i} w_i f_i(c,d)\right) \quad (6)$$

Where Z(d) is the normalization function that is computed as:

$$z(d) = \sum_c \exp(\sum_i \alpha_i f_i\,(d,c)) \quad (7)$$

α is a weight of term; fi is a feature .

4.4. *Maximum Weights*

Maximum Weight (MW) is a new classifier suggested by Zena Abd Al-Retha Abo-Altaheen. It is a very simple method for text classification, which works by selecting the highest weight of the term among the categories and only these values are used to predict the best class for any input example.

Formally:

Let classes C = {c1,c2,…,cn}

Suppose that:

Wimax: is the maximum weight of the term i among all these classes.

Cij: is a binary value which indicates the class j and contains the maximum weight of the term i

$$c_{ij} = \begin{cases} 1, & \text{if } w_{imax} \in c_j \\ 0, & \text{otherwise} \end{cases} \quad (8)$$

Then the best class can be estimated by:

$$c_{best} = \text{argmax}_{c_j \in C} \sum_{i=1}^{m} c_{ij} * w_{imax} \quad (9)$$

Where:

m=set of words in the test document.

Obviously, this method depends on features; each feature in the testing data is searched in all categories. Maximum weight in the training data in one class is taken, and then the maximum frequency of the category is computed as a result.

This method can be implemented easily, where each feature is associated with one category only. Each class has a subset of all features set in the dataset, so the feature reduction does not matter in this method.

4.5. *Voting Technique*

This classifier works on building multiple models (typically of different types), and simple statistics (like calculating the mean) are used to combine predictions. This work depends on NB, MW, MLR, and KNN in its results.

4.6. *Master-Slaves Technique*

Master-Slaves technique (MST) is suggested in [23],[24] on parts of speech. In this paper, this classifier was updated and implemented on Arabic text classification. MST consists of one classifier as a master and several other classifiers as slaves. NB is selected as the master and MLR, KNN and MW as the slaves. The master classifier modifies its probability according to the results of slaves by multiplying each probability by a factor according to Eq. 8. This factor reflects the weight of those slaves.

The master classifier multiplies the probability of all features with category except the results of KNN or MLR or MW. In other words, it takes the result of KNN classifier, multiplies the probability of each feature by a factor except this result and repeats this multiplication in accumulative for other slaves classifiers. Thus, it gives a result that depends on three classifiers and modifies the probability of document with each class.

$$fi=1- Ai+C \qquad (10)$$

Where fi is the factor for slave i, Ai is the accuracy of slave i, and c is a constant value less than 0.5 used because of the eq. (10) may give a very small number; may be less than 0.2.

The Master-Slaves technique has been explained in Figure 1 below:

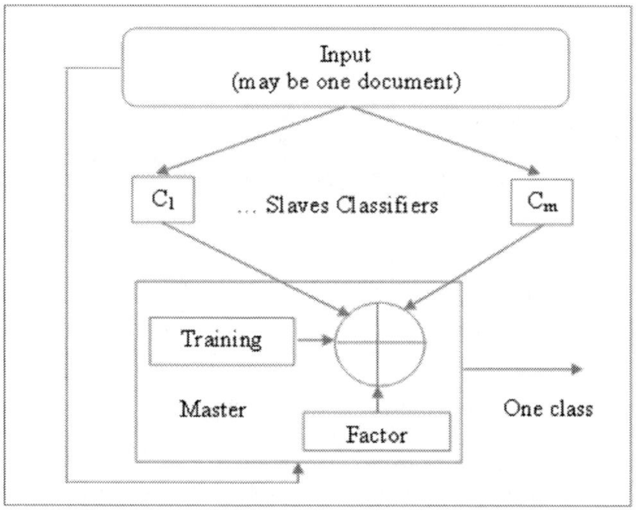

Figure 1. Master-Slaves Technique

5. Data Set

The corpus is built from "Al-Sabah" newspaper. 16757 Arabic documents are used for the first time in supervised machine learning. This corpus is organized manually into five different categories as in figure 2. Dataset is divided into two parts: 90% as training data and 10% as testing data.

6. Implementation and Results

Initially, two text classification algorithms (KNN, and NB) are applied on the new Arabic corpus by two scenarios:

- The first scenario is conducted by merging all documents under the same category into one file, and then the classifier is applied to evaluate the relationship between test document and the entire category.
- The second scenario is achieved by taking each document separately and the similarity is calculated with each file. Then the average of all documents similarity is calculated to classify the test document and see to which class it belongs.

The results calculated by the average are 86.495% and 84.30% for first and second scenarios respectively. The higher results are obtained from the first scenario so it will be relied on in the next tests.

All six classifiers are implemented in a 10-fold cross-validation. The results are explained in table 1:

a. Each classifier was applied separately and tested alone.
b. The four classifiers are combined in the voting technique.
c. These classifiers are used in master-slaves technique where NB represents the master and the other three classifiers as slaves with constant factor f=0.1. This test called is Master_Slaves1.
d. A second test called Master-Slaves2 was conducted using the variable factor fi, according to Eq. 10.

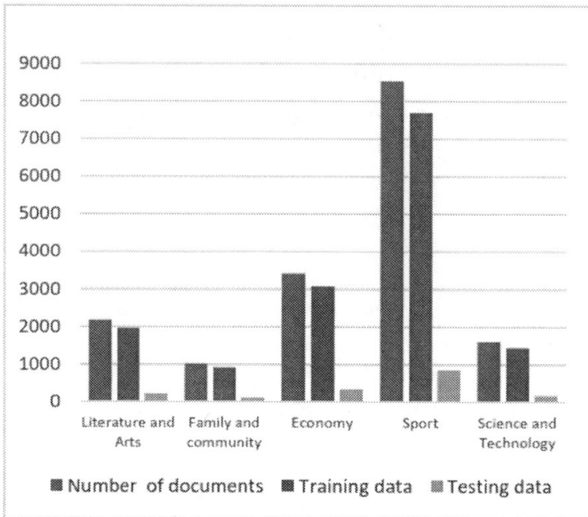

Figure 2. Number of Text Documents

e. The experiments showed that MST can be used for document classification. It is also in control of the result of classification even if found very weak as a result of a classifier. That is, there is no dropping at all in the accuracy results from pure classifier.
f. MST always makes the classification with a new probability and the training process is new. This probability is always different and therefore the training is not fixed, and this procedure does not exist in any other way by the mixed classifier.

Table 1. Results of Implementing Six Different Classifiers.

#Fold	# training files	# testing files	Accuracy % NB	KNN	LR	MW	Voting	Master Slaves1	Master slaves2
1	15080	1677	85.45	78.903	76.937	81.585	82.41	86.524	86.524
2	15080	1677	89.326	81.63	77.34	83.13	83.96	89.45	89.326
3	15080	1677	87.12	79.73	76.63	82.11	82.41	87.359	87.24
4	15080	1677	78.295	72.81	69.53	74.84	75.194	79.43	79.43
5	15080	1677	89.92	83.07	79.96	85.39	85.927	90.46	90.28
6	15082	1675	90.042	81.515	76.923	83.363	84.079	90.34	90.28
7	15082	1675	91.711	84.258	77.877	85.868	86.524	91.95	91.77
8	15082	1675	89.267	80.739	78.354	83.304	83.781	89.922	89.624
9	15082	1675	88.67	82.409	80.143	84.496	85.092	89.863	89.863
10	15082	1675	88.193	80.382	77.818	82.469	82.887	88.611	88.611
Average			87.80	80.54	77.15	82.66	83.23	88.39	88.29

7. Conclusion

The method that used in Arabic text classification (Master-Slaves Technique) for first time can be used for document classification as showed in the results. It also improves the performance of documents classification. As we can see, the accuracy of Master-Slave technique is much better than the other classification techniques (single and combined classifiers).

We can see that there is no dropping at all in the accuracy results of the Master classifier when compared to pure classifier results. It still has the control for selecting the right class.

Finally, Master Slaves always makes the classification with a new probability and the training process is new. This probability is always different and therefore the training is not fixed, and this procedure does not exist in any other way in the mixed classifier.

8. References

[1] A. Abu-Errub, "Arabic Text Classification Algorithm using TFIDF and Chi Square Measurements," *International Journal of Computer Applications*, vol. 93, 2014.

[2] A. Al-Badarenah, E. Al-Shawakfa, K. Al-Rababah, S. Shatnawi, and B. Bani-Ismail, "Classifying Arabic text using KNN classifier," 2016.

[3] S. Bahassine, M. Kissi, and A. Madani, "New stemming for Arabic text classification using feature selection and decision trees," in Proceedings of the 5th International Conference on Arabic Language Processing, 2014, pp. 26-27.

[4] H. F. Witschel, "Using decision trees and text mining techniques for extending taxonomies," in Learning and Extending Lexical Ontologies by using Machine Learning Methods, Workshop at ICML-05, 2005, pp. 61-68.

[5] R. Alhutaish and N. Omar, "Arabic text classification using k-nearest neighbour algorithm," Int. Arab J. Inf. Technol.(IAJIT), vol. 12, pp. 190-195, 2015.

[6] N. Sharma, "CLASSIFICATION USING NAÏVE BAYES-A SURVEY."

[7] A. M. El-Halees, "Arabic text classification using maximum entropy," *IUG Journal of Natural Studies*, vol. 15, 2015.

[8] S. Zhu, X. Ji, W. Xu, and Y. Gong, "Multi-labelled classification using maximum entropy method," in Proceedings of the 28th annual international ACM SIGIR conference on Research and development in information retrieval, 2005, pp. 274-281.

[9] L. S. Larkey and W. B. Croft, "Combining classifiers in text categorization," in Proceedings of the 19th annual international ACM SIGIR conference on Research and development in information retrieval, 1996, pp. 289-297.

[10] T. Gonalves and P. Quaresma, "Multilingual text classification through combination of monolingual classifiers," in Proceedings of IV Workshop on Legal Ontologies and Artificial Intelligence Techniques, 2010, p. 29.

[11] A. El-Halees, "Arabic opinion mining using combined classification approach," 2011.

[12] K. Fragos, P. Belsis, and C. Skourlas, "Combining Probabilistic Classifiers for Text Classification," Procedia-Social and Behavioral Sciences, vol. 147, pp. 307-312, 2014.

[13] H. Kashyap and B. Buksh, "Combining Naïve Bayes and Modified Maximum Entropy Classifiers for Text Classification," 2016.

[14] M. Shah, S. Shinde, R. S. Sawant, and P. Wagh, "Analysis of Text Review using Hybrid Classifier," *International Journal of Engineering Science,* vol. 10914, 2017.

[15] I. S. Abuhaiba and H. M. Dawoud, "Combining Different Approaches to Improve Arabic Text Documents Classification," *International Journal of Intelligent Systems and Applications*, vol. 9, p. 39, 2017.

[16] a. K. K. Nisha Mariam Daniel, "Survey on Text Classification Methods," *International Journal of Advanced Research in Computer Science and Software Engineering,* vol. 6, pp. 585--588, 2016.

[17] D. Shen, Z. Chen, Q. Yang, H.-J. Zeng, B. Zhang, Y. Lu, et al., "Web-page classification through summarization," in Proceedings of the 27th annual international ACM SIGIR conference on Research and development in information retrieval, 2004, pp. 242-249.

[18] C. C. Aggarwal and C. Zhai, "A survey of text classification algorithms," Mining text data, pp. 163-222, 2012.

[19] M. Usman, Z. Shafique, S. Ayub, and K. Malik, "Urdu Text Classification using Majority Voting," *INTERNATIONAL JOURNAL OF ADVANCED COMPUTER SCIENCE AND APPLICATIONS,* vol. 7, pp. 265-273, 2016.

[20] A. Khan, B. Baharudin, L. H. Lee, and K. Khan, "A review of machine learning algorithms for text-documents classification," *Journal of advances in information technology*, vol. 1, pp. 4-20, 2010.

[21] M. K. Saad, "The impact of text preprocessing and term weighting on arabic text classification," Gaza: Computer Engineering, the Islamic University, 2010.

[22] A. Jain and J. Mandowara, "Text Classification by Combining Text Classifiers to Improve the

Efficiency of Classification," *International Journal of Computer Application* (2250-1797), vol. 6, 2016.

[23] A. H. Aliwy, "Arabic morphosyntactic raw text part of speech tagging system," 2013.

[24] A. H. Aliwy, "Combining POS taggers in master-slaves technique for highly inflected languages as Arabic," in Cognitive Computing and Information Processing (CCIP), 2015 International Conference on, 2015, pp. 1-5.

Mapping the Spatial Distribution of Crimes in Iraq

Enas Ali Mohammed[1]

[1]Department of Computer Science, College of Science, University of Kerbala, Iraq.

the_programmer86@yahoo.com

Abstract. In the last years, Iraq has witnessed challenges in security have led to rice crime's rates. So there is need to map crimes concentrations in place and time from trusted sources and provide crime incidence reports for crime reduction efforts in the field of law enforcement. This study concerned with designing and implementing a web-based GIS system for crime mapping and decision support in Iraq. This system is being developed in Kerbala University with sample databases.The proposed system provides the general public with information of recent crime activity near to their places within a radius of any location (such as place of residence, school, or business). Display the distance from any crime to any address of interest. Filter by crime type as well as date recorded in order to focus on specific patterns. Plot only occurrences for the agency you need to view. Plot all crime incidents as heat map for a specified date.
Keywords: Iraq, Crime Map, Crime Locations, Google Maps API, GIS Cloud.

1. Introduction

In recent years, several factors grouped with the challenges of insecurity have led to rising incidences of crime in both developed and developing countries, these factors are the increased rate of poverty, loss of life and assets, fear of victimization, unemployment, drop in income, displacements, evictions, emotional depression. Despite, all countries apply crime's prevention measures to control crime's rate. Crime's rate becomes the main challenge for the economic and social development of countries all over the world [1]. Manual crime's recording systems haven't returned reliable and accurate data round the clock can help in decision support and crime trend prediction. Therefore, they failed to meet the demands of the current crime situation[2].

Geographic Information System (GIS) can help a great deal through crime analysis and mapping processes to handle the spatial analysis of crimes problems[1]. A GIS is a set of computer-based tools for analyzing and mapping spatial data. GIS uses geography and computer-generated maps as an interface for integrating and reaching large amounts of location-based data[3].

Crime mapping process is an important part of crimes management, monitoring, and control. This process can assist police to protect civilians more effectively. Simple maps that outline positions where crime's concentrations have happened can be manipulated to support direct patrols to exist in areas they are needed[1]. Therefore, an identification of where and why crimes happen can develop efforts to support crime mapping. Crime mapping process offers crime analysts graphic descriptions of crime-related concerns to identify crime hot spots (high-crime-density areas), along with other trends and patterns. Hotspot analysis supports police recognize high-crime places, varieties of crime being committed, and the best action to counter[2].

The growth of internet technologies, especially web-based GIS technologies, is presenting new possibilities for use of crime mapping to raise crime prevention. Web-based GIS becomes technology distributed for conventional GIS applications because interactive maps and fascinating cartography are available now for any standard web browser without any plugins and spatial installation [4]. A web-based GIS application makes the users access to GIS and its essential functionalities like receiving directions, navigation, finding locations, panning and zooming maps via web browsers[5]. Current applications of web-based crime mapping concentrate on raising society policing rather than analytical purposes such as pattern analysis and estimate.

A few countries are using such web-based GIS applications to map crime data for people security purposes. The users of those applications are appreciating these applications. These applications are playing an essential role in the security of the community. Hence, a web-based GIS application can be developed to show locations of crimes. The application that will be developed can be manipulated by the general public and law enforcement agencies. Recently, a number of different crime mapping and analyzing techniques are implemented in different regions in the world [6,7,8,9].

2. Problem Statement

The general people want the attention of their areas. Everybody cautions for the lives of kids and their loved ones so they should be aware of the current situations in the area. This study aims towards public to help them in their lives by notifying them the current situation of their areas. Law enforcement offices may use this type of applications to protect the society aware of the present statutes and to assist them to live jointly by communicating and sharing information with each other.

3. Research Methodology

Law enforcement agencies will have the ability to access and inquire data directly in order to form plans, tactical analysis (e.g. crime forecasting), as well as for making better decisions. The general public will use this application to map, visualize, analyze crime incident patterns, and get reports of the crimes in their surrounding areas. Also, it allows you to get detailed information on crimes in your area, such as burglary, assaults, theft or vandalism.

So, the proposed system consists of three main functionalities (data storage, data process, and web service) as shown in figure (1). So, in order to obtain system functionalities, a number of techniques were used as well as a number of interfaces were designed.

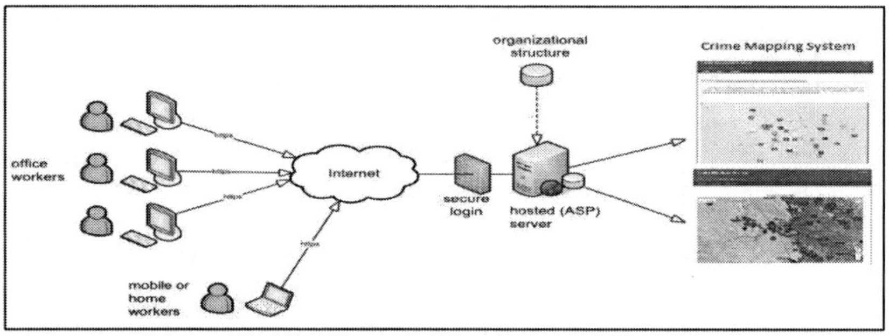

Figure 1. The Proposed System Architecture.

3.1. Tools and Techniques

3.1.1. Microsoft SQL Server Language

The perfect crime's mapping system includes a compatible database platform (e.g., MySQL, SQL server) for the real-time analysis. SQL server was chosen to create system's database. Because of it is considered an effective tool for data management and compatible with most web applications[10].

3.1.2. Microsoft Visual Studio 2015

Microsoft Visual Studio 2015/ Visual C#/ ASP.net web application was used to implement the proposed system. Usually, Microsoft Visual Studio is an integrated development environment enables

IOP Conf. Series: Journal of Physics: Conf. Series **1032** (2018) 012053 doi:10.1088/1742-6596/1032/1/012053

programmer/developers to write various types of codes and add new various packages (e.g., Subgurim Google Maps API)[11].

3.1.3. JavaScript Language

JavaScript which is the most popular language for programming codes called Jquery instructions, it was used as part of web applications whose implementations client-side scripts to interact with the user. Writing Jquery codes in ASP.NET web pages allow the creation of rich browser-based functionality with Google Maps[12].

3.1.4. Google Maps Application Programming Interface (API)

Google Maps API supplies developer interfaces to access Google satellite images and street maps as well as their geospatial services such as routing and geocoding[13] [14] [15]. Google Maps API's are commonly used code libraries to enable the programmers and developers to create web applications that read and write data from the Google services. Google Maps API provides four types of maps (Roadmap, Satellite, Hybrid, and Terrain). Each type consists of several layers (e.g., data layer, heat map layer, and cloud layer) where they are objects on the map composed of one or more independent items operate as one unit[13]. A heat-map layer is a thematic map for graphical representation of data. It uses colors to indicate the level of activity, how a measurement varies geographically or to depict the intensity of data at geographical points. It uses gradients or a color scale to depict data per area on the map. Usually, brighter colors to indicate low activity and darker colors were used to indicate high activity [5].

3.1.5. GIS Cloud's Mobile Data Collection

GIS Cloud's Mobile Data Collection is a simple application for web and mobile devices for real-time field location, attribute and multimedia data collection. It provides a low-cost and feature-rich services[16].

This application allows you to gather real-time, media-enriched location information in the field. It uses the cloud to send data points and attributes directly to a custom layer on a personalized map. Access, analyze, share, and publish your data in the GIS Cloud Map Editor (available in both free and paid editions) through any web browser in real time. Features include custom form building for data collection through the mobile app and the ability to assign media (image and audio) attributes[16].
Law enforcement agencies may use this kind of applications in data gathering process. In this study, a new project will be designed and created called Crime Data Collection in Iraq. This project will facilitate policemen's work in gathering crime data using mobile.

3.2. Workflow

The proposed system consists of a number of steps, some of these are listed below. The flowchart of crime mapping process illustrated in figure (2).

Step1: Data gathering process for crimes incidents play a crucial role in enhancing the work of law enforcement and crime analyses to track the movements and changes of crimes events and to increase the validity of crime mapping systems in all regions by identifying more accurate data and best prediction models.

Therefore, the proposed system can receive crime data from many police stations and georeferenced quickly. Admins of police stations can log-in by entering their information (username and password). Then, they can upload the data, edit, and update and delete crimes information from the central database.

Also, they can use the mobile application to upload crimes data. When policemen have to report a crime then they will open application to attach or take photo and send that photo with crime information to GIS Cloud server using mobile data collection as shown in figure (3). The information of the policemen will be verified at the time of signup and then the data will be automatically transferred to system data base.

IOP Conf. Series: Journal of Physics: Conf. Series **1032** (2018) 012053 doi:10.1088/1742-6596/1032/1/012053

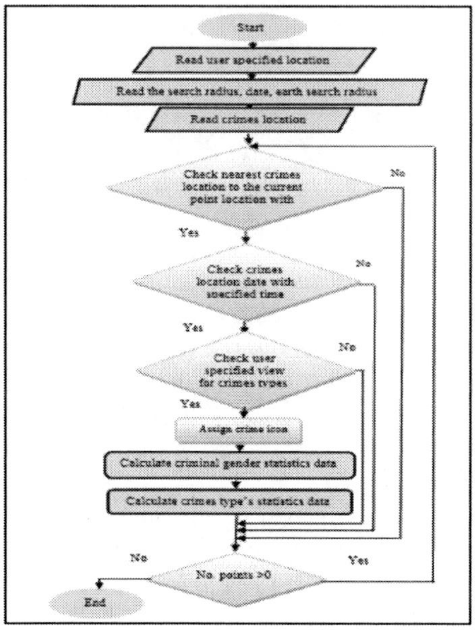

Figure 2. Flowchart of crime mapping process

System database contains data about both crimes and criminals with the following main attributes:

- Crime-Id: individual crimes are labeled by unique crime Ids.
- Crime-Type: indicates crime type.
- Crime-Date, time: indicate when a crime happened.
- Criminal-Id: individual criminals are designated by unique Ids.
- Criminal-Age: age of individual criminal.
- Criminal-Gender: belongs to which gender.
- Criminal-Job: job of individual criminal.
- Crime-Address, location of individual criminal (longitude, latitude).
- Marital status of individual criminal.
- Crime-Description.

Step2: users can specify a particular location or an agency to request information about recent crime activity as shown in figure (4).

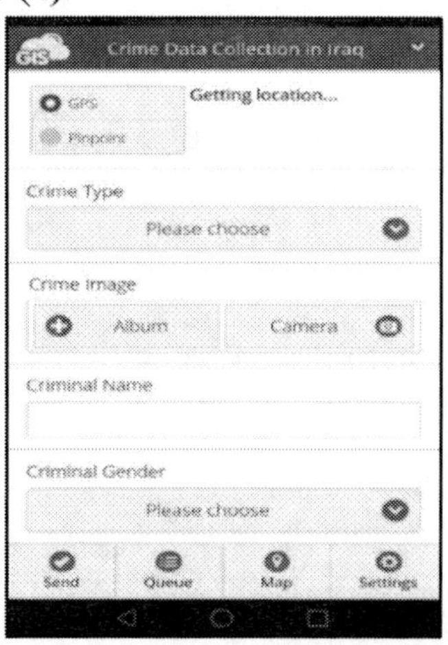

Figure 3. The main window of crime data collection in Iraq project

Step3: citizens can use this application to view crimes data by various search parameters, to include crime type, a date range (yesterday, last week, last month, or a specified date), or distance from a specific address. The data will be illustrated on a map and the various crimes will be depicted by various crime type icons, each containing general details of the incident as shown in figure (5).

Step4: Google heat map layer was used to generate crime heat map that shows the density of crime in any region and any specified date in Iraq as shown in figure (6).

Figure 4. System main page

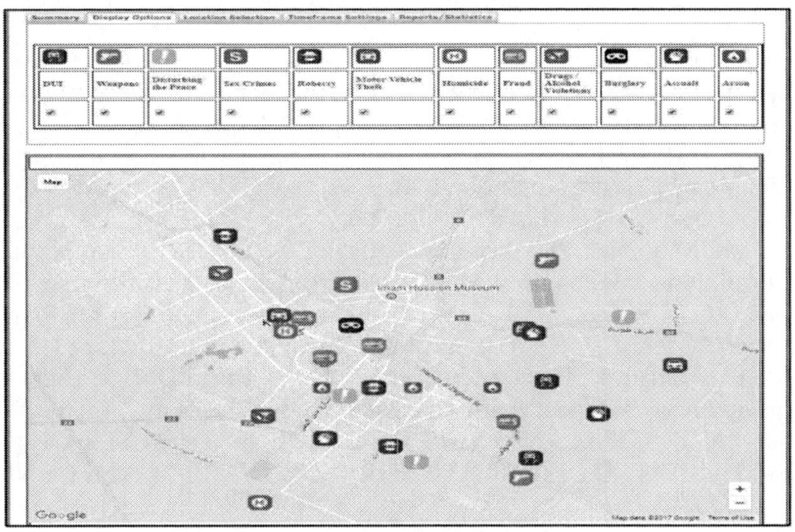

Figure 5. The search-view page

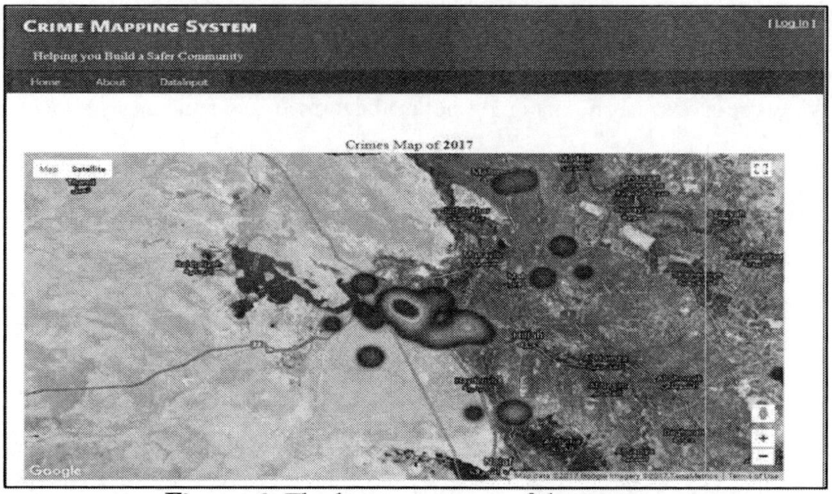

Figure 6. The heat map page of the system.

4. Conclusions

Under this system, police stations can upload crimes events to the central system database. These data will help law enforcement protects citizens more effectively in the areas they serve and take preventive measures. Common people can view crimes events daily to avoid some places by the understanding that their chances of being a victim are greater in some of these places than in others. Also, they can identify their choices of places, streets, schools, and neighborhoods. Crime analysts can use the proposed system to make crime forecasts, mapping criminal networks and identifying possible suspects, and finding relationships and possible explanations to make better decisions.

5. References

[1] A. Olajuyigbe, K. Omole, T. Bayode, and A. Adenigba, "Crime Mapping and Analysis in the Core Area of Akure, Nigeria," vol. 5, no. 4, pp. 21–24, 2016.

[2] K. D. Harries, Mapping crime: principle and practice. Diane Pub Co (March 1999), 2009.

[3] C. P. Johnson, "Crime Mapping and Analysis Using GIS," Geomatics 2000 Conf. Geomatics Electron. Gov., no. January, pp. 1–5, 2000.

[4] Z. Khan and M. Adnan, "Usability evaluation of web-based GIS Applications," Proc. 11th Int. Conf. Inf. Integr. Web-based Appl. Serv. - iiWAS '09, no. January, p. 557, 2009.

[5] F. Cheong, C. Cheong, and F. Jie, "Re-Purposing Google Maps Visualisation for Teaching Logistics Systems," J. Inf. Technol. Educ. Innov. Pract., vol. 11, pp. 159–177, 2012.

[6] K. Patel, P. Thakkar, L. Patel, and C. Parekh, "GIS based Decision Support System for Crime Mapping , Analysis and identify Hotspot in Ahmedabad City," vol. 4, pp. 32–35, 2014.

[7] C. Bunn, "A SPATIAL ANALYSIS OF CRIME IN THE CITY OF SAN LUIS OBISPO USING FREE AND OPEN SOURCE GIS SOFTWARE," CALIFORNIA POLYTECHNIC STATE, 2014.

[8] S. M. Raza, "A Proposed Solution for Crime Reporting and Crime Updates on Maps in Android Mobile Application," vol. 124, no. 1, pp. 28–31, 2015.

[9] M. A. Oyinloye, I. O. Olamiju, and V. K. Otokiti, "Spatial Distribution of Crime in Akure , Nigeria : The GIS Perspectives," vol. 2, no. 2, pp. 21–38, 2017.

[10] Microsoft Company, "SQL Server." [Online]. Available: https://www.microsoft.com/en-us/sql-server/.

[11] Microsoft Company, "ASP.NET." [Online]. Available: https://www.asp.net/.

[12] Code School team, "JavaScript." [Online]. Available: https://www.javascript.com/.

[13] Google Developers, "Google Maps APIs." [Online]. Available: https://developers.google.com/maps/.

IOP Conf. Series: Journal of Physics: Conf. Series **1032** (2018) 012053 doi:10.1088/1742-6596/1032/1/012053

[14] I. N. Piarsa, A. A. K. Oka, and G. W. M. Gunadi, "Web-based GIS by using Spatial Decision Support System (SDSS) Concept for Searching Commercial Marketplace-Using Google MAP API," Int. J. Comput. Appl., vol. 50, no. 7, pp. 1–5, 2012.
[15] a. K. Akanbi and O. Y. Agunbiade, "Integration of a city GIS data with Google MAP and Google API for a web based 3D Geospatial Application. , 2(11), pp. 200-203.," Int. J. Sci. Res., vol. 2, no. 11, pp. 200–203, 2013.
[16] GIS Cloud team, "GIS Cloud/Mobile Data Collection." [Online]. Available: https://www.giscloud.com/.

Performance evaluation of CCM and TSCP routing protocols within/without data fusing in WSNs

Haydar Abdulameer Marhoon[1], M. Mahmuddin[2], Shahrudin Awang Nor[2], Mohsin Hasan Hussein[1] and Ahmed Sileh Gifal[1]

[1]College of Sciences, Computer Department, University of Kerbala, Iraq
[2]InterNetWorks Research Lab, School of Computing, College of Arts and Sciences Universiti Utara Malaysia, UUM Sintok, Kedah, Malaysia

haydar@uokerbala.edu.iq

Abstract. Wireless Sensor Network (WSN) is a large number of small sensor devices that can connect each other wirelessly. WSNs applications are rapidly growing in last decades, furthermore, in WSN research, energy is one of the important issues that must consider when designing a new protocol. Due to the fact, almost all of nodes' energy deplete in the communication part, and the data fusing directly impact the performance of routing protocol. This paper studies the impact of data fusing for chain based routing protocols. In this study, ns-3 simulator used to evaluate Chain-Cluster Mixed (CCM) and Two Stage Chain Protocol (TSCP) routing protocols with deterministic nodes deployment. The experiments show that TSCP overcomes CCM in network lifetime when data fusing applied while CCM overcomes TSCP in the same metric with non-fusing of data for First Node Die (FND), 10%, 25%, 50% and Last node (LND). Furthermore, CCM is still playing a good behavior in delay for both approaches. The main conclusion for this paper is non-fusing of data must be applied when design new routing protocol to study all the packets traffic behaviors in the path from source to destination.
Keywords: Chain based, Routing Protocols, CCM, TSCP, Data fusion, WSN

1. Introduction

WSNs have become a challenging area for researchers in different perspectives. Its applications are growing in the last few decades in different areas [1] such as habitat monitoring, industrial company, health and medical, military issues, disasters prediction and management, security, agriculture and others [2]–[4]. Sensors network makes connections between computational, physical and human environment, and data collected from environment by sensors and delivered to the base station using node networking and this process in every round. In general, WSN consists of large number of small devices called Sensor Node (SN). All sensor nodes have the ability of sensing data, processing and communicating wirelessly with each other, and these sensor nodes have limitations in memory, power resources, bandwidth, and computational capability [1]. Second one is a super node which has unlimited resources call Base Station (BS) that works as a sink.

Basic sensor node architecture consists of four units [5]: the first one is sensing unit that is responsible for sensing the outside environment according to its capability for example temperatures, humidity, light and so on. The second unit is for processing, memory, and all computing and processing operations, it is also different according to nodes types and it has almost limited ability. Third is communication unit that makes the necessary connections and network. Furthermore, this unit

IOP Conf. Series: Journal of Physics: Conf. Series **1032** (2018) 012054 doi:10.1088/1742-6596/1032/1/012054

has the largest power consumption among all node units. Finally, power (battery) unit is working as energy supplier for all units in sensor node. Figure 1 shows the basic architecture for sensor node in WSNs.

Many factors can directly affect the performance of WSN. These factors are including the ways sensors nodes will be deployment in the sensing area (randomly or deterministic deployment), the routing protocols that will be used to create a suitable directions to the base station and is the sensors will use data fusing or not that's depend on when the data collecting and the type of data.

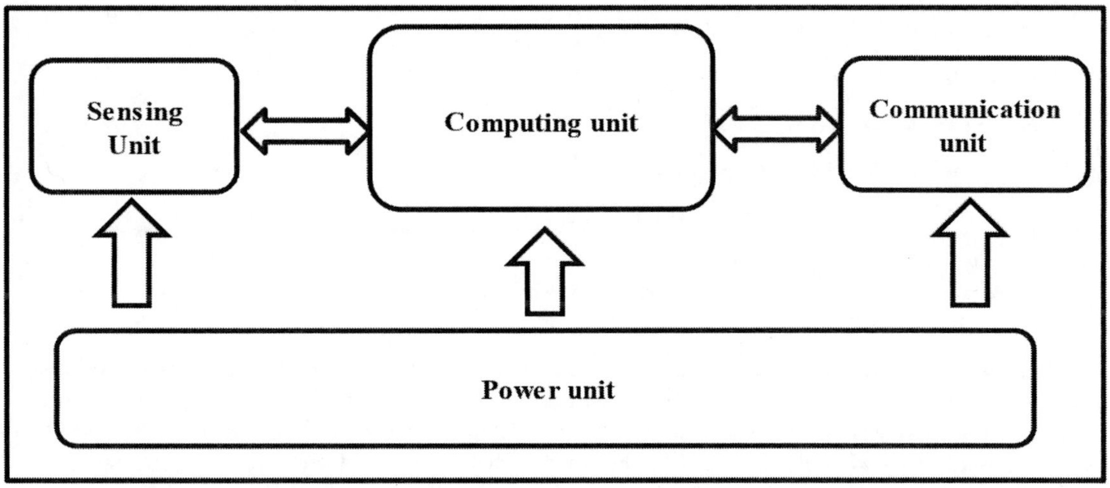

Figure 1. Basic Architecture for Sensor Node in WSNs (Adopted From [6])

This paper studies the differences between data fusing and non-fusing in chain based routing protocol and it will use CCM [7] and TSCP [8] as examples to investigate their behavior in both approaches.

2. Sensors Deployment in WSNS

The sensors deployment method can affect the performance of whole WSNs. Choosing the good sensors deployment can reduce the node redundant, minimize the network overall cost, prolong the network lifetime and reduce the complexity of data fusing and routing [5], [9]–[11]. So, the main issue in sensors deployment is to use effective way in order to increase the coverage area, to provide the efficient nodes connection and energy saving.

Damuut and Gu [12] classified node deployment into two main types first is deterministic and second is non-deterministic nodes placement. A node placement scheme depends on the following three things:

1. Application area: deterministic is more suitable for healthcare, scientific measurements, domestic appliance and it is common in surveillance applications for example in agricultural area [12]. However, non-deterministic is preferred in military, forest fire detection and disasters application.
2. Type of sensor: In some cases nodes deployment depends on nodes characteristics such as weight, size or materiel.
3. Cost: nodes cost, maintenance cost, installation cost is important parameters in choosing deterministic or non-deterministic deployment.

Figure 2 shows the number of nodes that deployed in Deterministic and Random ways in the sensing area:

IOP Conf. Series: Journal of Physics: Conf. Series **1032** (2018) 012054 doi:10.1088/1742-6596/1032/1/012054

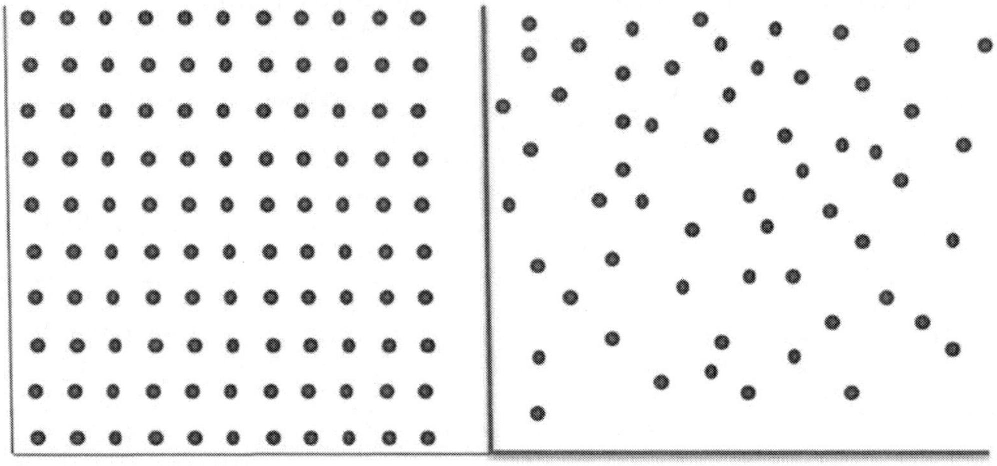

Figure 2. Deterministic and Random Nodes Deployment

Deterministic strategy is the best placement of nodes in the sensing area and sensor location which will not change during network lifetime [9]. Moreover, Deterministic Sensor Placement scheme (DSP) [13] is a common scheme in nodes deployment to meet some specific performance objectives after good planning for nodes position [14]. The main advantages of DSP are its utilization for sensors devices, controllable network topology, more efficient routing protocols and coverage performance. However, time for consuming installation is still the main drawback in deterministic nodes deployment.

Furthermore, random deployment (or non-DSP) strategy is used in some area that have time-sensitive application due to its quick deployment and self-organization which has very important drawback in network lifetime in terms of the number of nodes death in every round [12].

3. WSNS Applications

The sensors technologies are widely used in many applications related with particular life of the human nowadays. Figure 3 illustrate some of the most important applications of WSNs in different area and domain [15][16][17]. The specific characteristics of application are required specific type of sensors, routing protocol, and deployment strategies (deterministically or randomly). The environments and energy consumption is important parameters to choosing the proper types of sensor for applications, due to the energy is serious problem for the network designer [16].

4. Data Fusing in WSN

Sensors node may generate redundant data, so it applies a data aggregation to prevent the duplication for the same data and decreasing the number of packet transmissions. The aggregated (fusing) data is a combination of packets that are collected from different nodes and put together to decrease the number of packet and its size [18]. Many functions use for fusing data such as maxima, minima, duplicate suppression and average. Data fusing techniques are very efficient to increase the lifetime in WSN, especially when multi-hop routing applied in that network. However, non-fusing must be considered when designing a new routing protocol to study the real behavior of packet transmitting from source to destination.

In this paper, the way of data fusing applied based on data fusion technique. Data fusion is the processes of combination n packets with size k and the result will be one packet of size k instead of one packet of size nk [19].

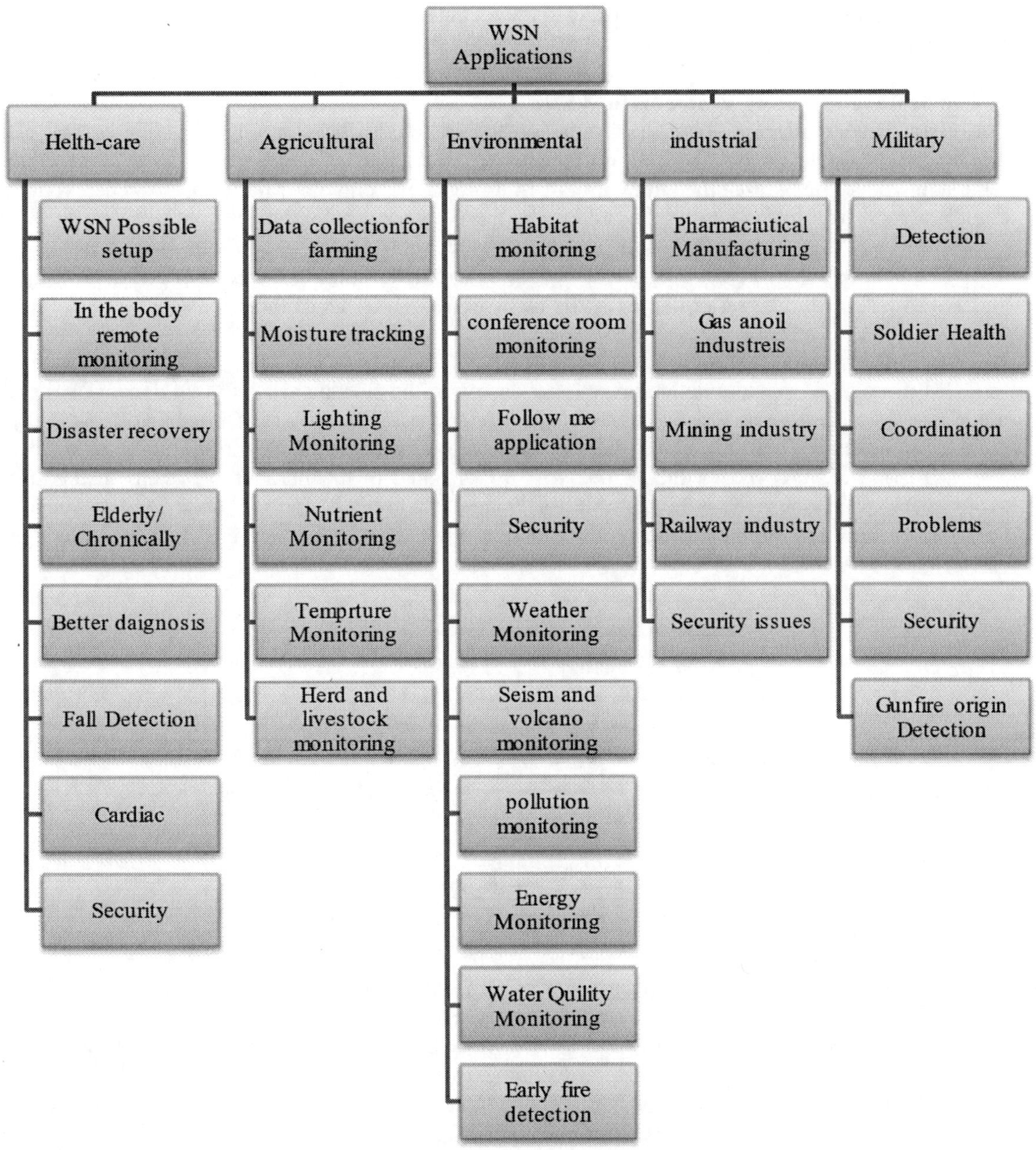

Figure 3. WSNS Applications in Different Area

The main goal of data fusion is to decrease the number of the packets that transmitted in the network. If the same data transmitted without data fusing, the size and the number of the packet will be increase gradually and energy consumption will dramatically increase. Depend on the data fusing energy consumption that proposed in [20], the cost of energy for data aggregation is 5nJ/bit/message. Whereas, the equation to calculate the amount of energy for each packet is below:

$$E_{fusion}(k) = E_{fn} * k \text{ ....(1)}$$

Where, E_{fusion} means the energy consumption for data fusion for k-bit per packet, Efn is the energy consumption for fusing 1-bit message. Moreover, for experimental comprehensive study this

IOP Conf. Series: Journal of Physics: Conf. Series **1032** (2018) 012054 doi:10.1088/1742-6596/1032/1/012054

paper will exams the impact of data fusing for the routing protocols and focusing on TSCP and CCM chain based routing protocols.

5. Chain Based Routing Protocols in WSNS

The chain base routing protocol approach considered as the best among all other energy efficient routing protocols in WSNs [8][21]. Furthermore, deterministic nodes deployment can reduce the redundant node, minimize the network overall cost, prolong the network lifetime, reduce the complexity of data fusing and routing and make the network topology more controllable [9][12][5].

Therefore, many protocols used in chain-based routing approach with deterministic sensors deployment in WSN to achieve efficient energy consumption during the network lifetime [22]. CCM and TSCP will be discussing in details in the next sections in terms of their phases.

5.1. Chain Construction

In CCM protocol all sensors nodes are evenly deploy in the sensing area, so two dimensional assigned name can be taken to every nodes as its id like S(i,j) where i refers to the number of row and j is the number of column. Then, the chain will be constructed among all nodes in the same row (for S(i,1), S(i,2), S(i,3), ...), this means the number of row is equal to the number of chains in this protocol. From every chain, one node is responsible for being a chain head, and the chain head makes a cluster (one hop cluster) and the main head sends its data to the base station. Figure 5 shows the chain and the cluster built by CCM.

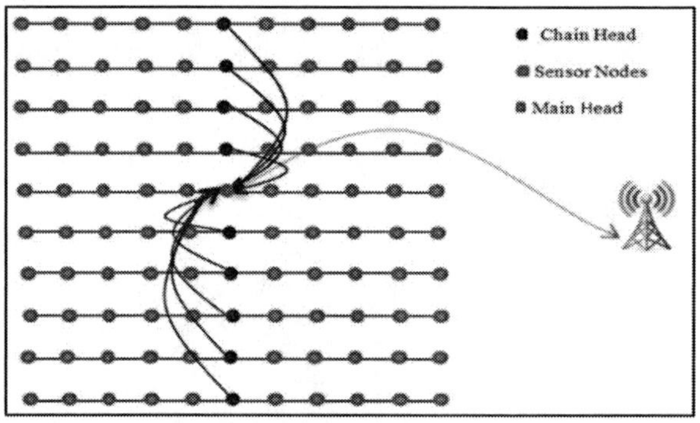

Figure 4. Chain and Cluster Formation in CCM

CCM can reduce the power consumption in intra connection by playing chain concept every node will tune its power radio to hear two neighbors only. However, it conserves more energy when cluster approach is applied in inter connection and when nodes heads are far away from each other. On the other hand, TSCP as in Figure 6 takes advantage from CCM when it built in tow stage chain. The chain is horizontal and as if CCM has intra connection where every node in the same row will connect to two neighbors only (that means the number of the horizontal chain is equal to the number of rows in the network).

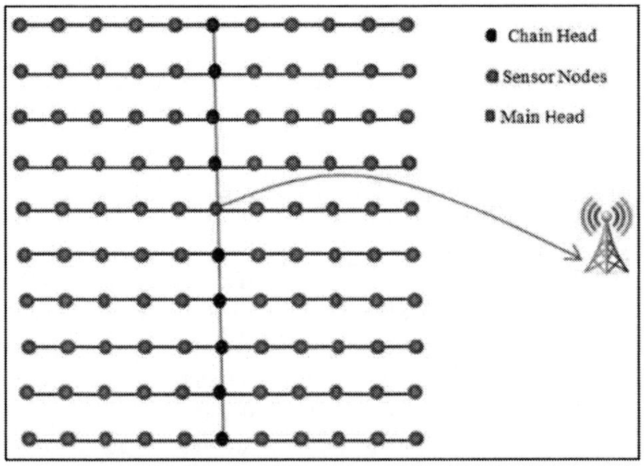

Figure 5. Chains Constructed by TSCP Protocol

TSCP successfully reduce energy dissipation and made energy balances sensor nodes during sequential moving of inter chain (vertical chain for chain heads). However, author does not clearly mentioned about dies nodes, and how the network will deal with these issues. TSCP protocol applies a new method in the network when nodes share almost of their energy in vertical chain. Chain construction will depend on nodes that have maximum energy to build the main chain and this method really can affect the network performance when vertical nodes are far away from each other.

5.2. Chain Head and Main Head Selection

Typically, there are two ways to select CH in WSNs depending on the parameters that are used: deterministic way which depends on fix parameters and adaptive way that depends on variable parameters like remaining energy (weight-based method) [23]. CCM has assigned chain head in each row (horizontal chains) sequentially for every round (node 1 will be chain head in first round and node 2 will be chain head in the second round and so on). Moreover, CCM will choose the main head based on residual energy in chains heads, for which nodes has the highest energy. It will then be the main head and it will be responsible for delivering all network data to the.

Therefore, choosing sequential method to select chain head can reduce overhead on the network and minimize energy dissipation in computation processing. However, ignoring the nodes remaining energy, and this will make some nodes that have few energies to become the chain heads and these nodes will drain their energy quickly where, if these nodes die, the network will lose chains data in this round. As such, chain head is responsible for sending all chain data to the main head. Moreover, CCM uses residual energy only when selecting the main head has critical cases. Specially, when the main head is far away from the base station, while, some chains heads have little bit less energy but with very good position according to base station. In these cases, this node (main node) will spend its energy to deliver all network data to the sink and maybe die earlier than others. In TSCP, chains heads will be selected by sequential way like CCM, for round 1, first node in the chain will be assigned as the chain head in every row (horizontal chains), so a vertical chain heads will construct vertical chain and the chain head that has maximum residual energy will be the main head in this round.

As mention above, the sequential method has critical drawbacks when ignoring the remaining energy to select chain head, but TSCP will put another way to select CH when the network nodes drains their energy by choosing the chain head depending on the maximum energy for all nodes in the same row. However, this way will add more drawbacks to TSCP because vertical chain will affected by long chain if chains heads are far from each other and this will make nodes spend their energy more quickly than the sequential way. Moreover, for main head in TSCP, it is not enough to consider energy only for selection. Distance from base station is very important factor for the main head selection because distance d2 will increase by long distance and d2 is the main factor in energy consumption.

Additionally, single main node in all of these protocols caused a bottleneck problem, since, one node only plays as gateway for network. This research takes bottleneck problem from power

IOP Conf. Series: Journal of Physics: Conf. Series **1032** (2018) 012054 doi:10.1088/1742-6596/1032/1/012054

consumption perspective not from congestion side because all network's data must be delivered to the base station by this node (main node) as result main node will drain its energy very quickly.

5.3. Next hop Selection and Data Aggregation

Intra-connections in CCM, and TSCP are the same, connection starts from the first node in the row, this node will select the next hop by distance only and this connection will be repeated sequentially (for example node S(i,1) will be connect to its neighbor S(i,2) and so on). This type of choosing does not have flexibility for any change in the network so, if any node dies for example node S(i,2) they will make S(i,1) connection to S(i,3) directly though S(i+1,1) is nearer than other. Greedy algorithm uses distance only to select next hop connection and this considered inefficient method because some nodes are not suitable to be in the chain due to their low energy.

CCM uses chain head to send message sent by to every end node in the chain in order to inform them to start sending data to their neighbors. This neighbor will fuse receiving data with its data then forward it to the next hop. Simple way used in TSCP to send nodes data. Every node senses data and fuses it with received data then transmits it to the next hop. TSCP way is simple but it ignores data collisions without any arrangement for data sending.

6. Radio Model for Energy Consumption

This research apply the First Order Radio Model as energy model, which is the same radio model discussed in [24][20][25]. In this model, the energy that is needed for running transmitting or receiving circuit is E_{elec} = 50 nJ/bit and the energy that is required by the transmitting amplifier is E_{amp}= 100 pJ/bit/m2. So, Equation 2 is used to transmit k-bit from any node to other with d distance between them and Equation 3 is to receive k-bit in any node.

For Transmitting k-bit

$$E_{TX}(k,d) = E_{TX-elec(k)} + E_{TX-amp}(k,d)$$

$$\boldsymbol{E_{TX}\ (k,d) = E_{elec} * k + E_{amp} * k * d^2} \text{ .... } \mathbf{(2)}$$

For Receive k-bit

$$E_{Rx}(k) = E_{Rx-elec}(k)$$

$$\boldsymbol{E_{Rx}(k) = E_{elec} * k} \text{ ... ...} \mathbf{(3)}$$

where E_{TX} is energy for transmitting, E_{elec} is energy to run transmitting circuit for 1 bit, E_{amp} is energy required for amplifier for 1 bit for m2, k is number of bit and E_{Rx} is energy required to receive k bits. Figure 7 shows the basic elements of the first order radio model [26].

7. Experimental Evaluation

Network simulation 3.22 version (ns-3.22) [27] is used to make a comparison between CCM and TSCP in tow case, first when data fusing approach is applied to aggregate the data from all sensor nodes, second when data transferring without data fusing for all nodes. ns-3 is select for two reasons, first because it is realistic, open source network simulation, and second because it became very rising in WSN research area.

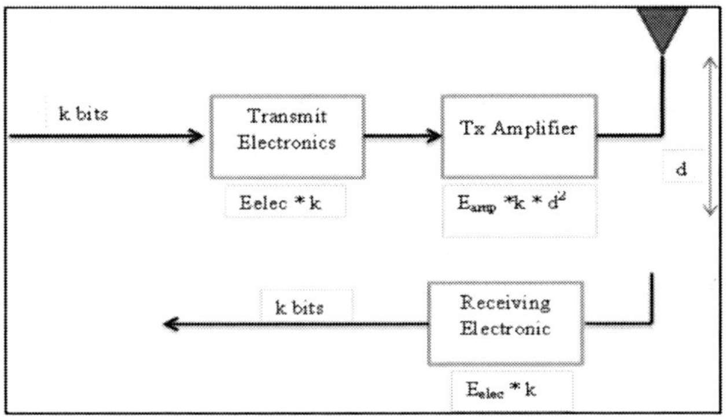

Figure 6. First Order Radio Model

7.1. Performance Evaluation with Data Fusing

In this approach, every node will receive one packet from its neighbour then fuse this data with its own data and the output will be only one packet with same size. Therefore, CCM and TSCP chain based routing protocols compared by using ns-3.22 simulator with the following simulation settings:

Table 1. Simulation Settings

Parameters	Setting
Simulator	ns3.22
Number of nodes	90
Initial energy	0.25 J
Energy model	First order radio model
Base Station	Static and single BS
Packet length	1024
Dist. between Nodes	10 meter
Sensing area	100 m X 100
Routing Protocol	CCM, CCBRP
Energy cons. For data fusing	5nJ/bit
Node deployment	Deterministic with Static Mobility Model

To study the impact of data fusion in WSN there are two important metrics need to calculate, first is the network lifetime and second is the End-to-End delay.

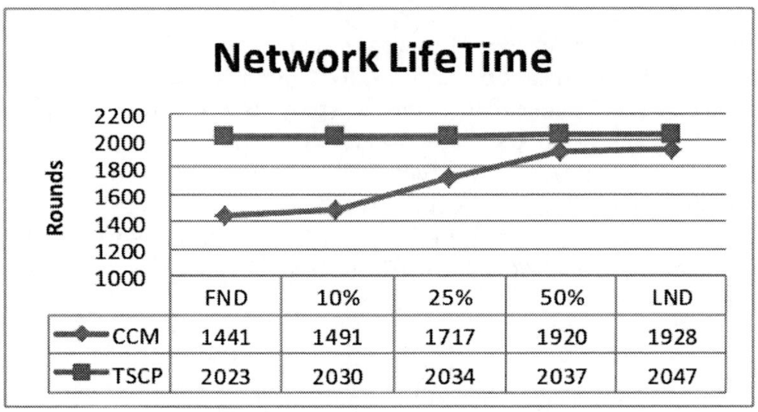

Figure 7. Network Lifetime for CCM and TSCP with Data Fusing

IOP Conf. Series: Journal of Physics: Conf. Series **1032** (2018) 012054 doi:10.1088/1742-6596/1032/1/012054

The Figure 8 illustrates the difference of network life time between CCM and TSCP, where y axis refer to number of round and x axis for percentage of nodes die. This is easy to show that TSCP outperform CCM in the first node die, 10%, 25%, 50% and last node die. This is very important to calculate the robustness of the routing protocol; furthermore, how it is keep the sensors node live as long as possible.

As mentioned above, CHs in TSCP are connected to each other as a chain, so it is keep the energy that spend by the long distance (distance consider very important factor). While all CHs in CCM connected to the main head directly, it will spend more energy and reduce the network lifetime for all nodes.

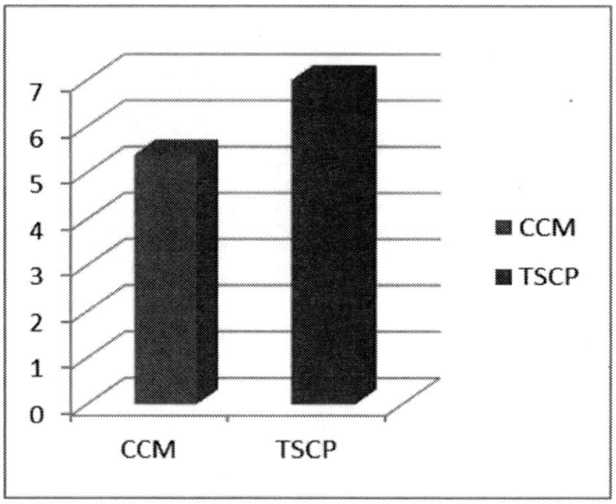

Figure 8. Average End-to-End Delay for CCM and TSCP with Data Fusing.

End-to-End delay is an important metric in WSN, which measure the speed of data delivery from the source to the destination. From Figure 9, TSCP has more delay than CCM this is coming from the internal behavior of second mechanism of TSCP. Data redundant will occur when every CH transmitting its data to the nearest CH until reach the base station. While CCM is outperforming TSCP in delay metric coming from cluster base behavior of CHs connection.

7.2. Performance Evaluation without Data Fusing

Data fusion is an active technique for reducing the data that delivering or transferring among network nodes but it is also has a lot of disappoint that is need to discus and study like delay for data fusing, network behavior, energy consideration, fusing algorithm and weight method that used if non data fusing is applied. So in this section paper will exam same routing protocols CCM and TSCP (chain based routing protocols) but without data fusing to study different with the previous results. Weight method will be used in both protocols that's mean every node responsible to transmit what receive with its own data. Figure 10 shows the network lifetime for CCM and TSCP routing protocols with same simulation setting except the initial energy that will be 2j for more comprehensive studying. For non-data fusing approach CCM outperform TSCP in the all metrics in the network lifetime section. This is coming from reducing the amount of data redundancy in CHs connection mechanism (cluster based).

IOP Conf. Series: Journal of Physics: Conf. Series **1032** (2018) 012054 doi:10.1088/1742-6596/1032/1/012054

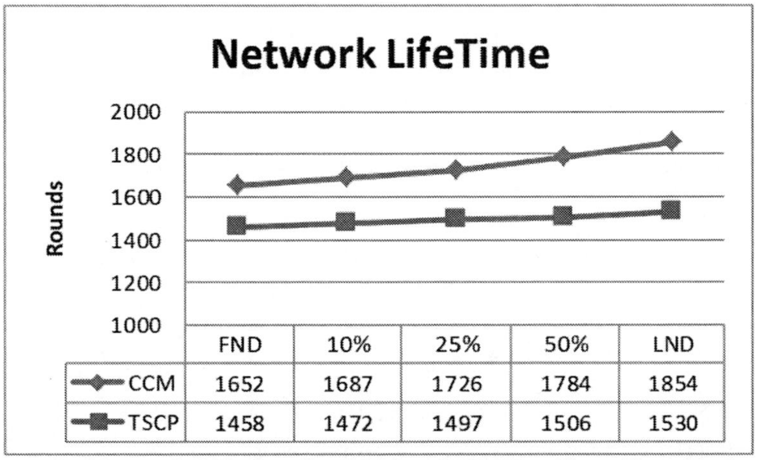

Figure 9. Network Lifetime for CCM and TSCP without Data Fusing.

Furthermore, End-to-End delay is important metric needs to exam with data fusion approach to calculate the behavior of both protocols from delay perspective, because of delay consider the main problem in the all chain based routing protocols in WSN.

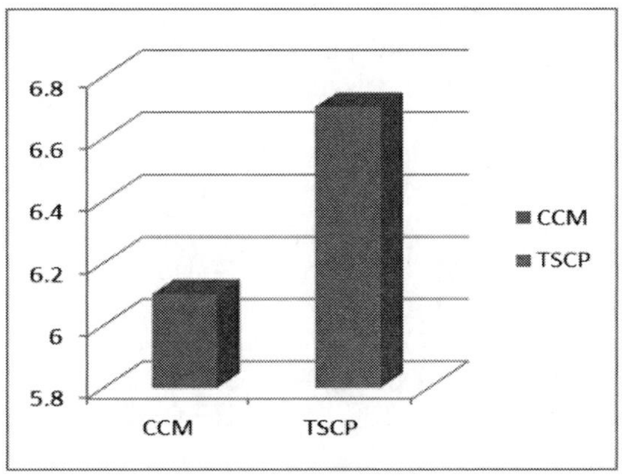

Figure 10. Average End-to-End Delay for CCM and TSCP without Data Fusing.

Figure 11 shows that CCM still outperforms TSCP in delay metric, because of reducing the length of CHs by using single hop connection between them and must be mention here that the delay causing by data fusing very difficult to calculate in WSN due to it is consider as processing delay.

8. Discussion and Conclusion

Energy consumption is main consideration in all WSN research special in routing protocols designing because of the almost of node's energy deplete in data transmitting between nodes. In this paper CCM overcome TSCP in delay metric in both approach. This is coming from behavior of the second phase in CCM when CHs connect each other using cluster form to reduce the delay cause by long link. This result lead the researcher to mixing chain based to reduce power consumption with cluster based to reduce the delay. While TSCP overcome CCM in the network lifetime when data fusing approach applied because of few number of packets traveling with smallest distance between CHs while, CCM play good performance with non-fusing approach because of the intermediate CHs are not responsible to deliver the previous data and this will reduce the power consumption for CHs and prolong the network lifetime. From all these point, there is not big difference by data fusing or non-fusing when

IOP Conf. Series: Journal of Physics: Conf. Series **1032** (2018) 012054 doi:10.1088/1742-6596/1032/1/012054

the researcher (protocol designer) interested with delay metric and anything related. While it must consider if designing related with extend the network lifetime.

For future work, designing new routing protocol must take the advantage of both chain and cluster based and makes fair balancing to reduce the power consumption, extend network lifetime and reduce delay. In addition, non-fusing for data recommended applying for make real packets traffic.

9. References

[1] X. Liu, "A survey on clustering routing protocols in wireless sensor networks.," Sensors (Basel)., vol. 12, no. 8, pp. 11113–53, Jan. 2012.

[2] M. Eslaminejad and S. A. Razak, "Fundamental lifetime mechanisms in routing protocols for wireless sensor networks: a survey and open issues.," Sensors (Basel)., vol. 12, no. 10, pp. 13508–44, Jan. 2012.

[3] Y.-K. Chiang, N.-C. Wang, and C.-H. Hsieh, "A cycle-based data aggregation scheme for grid-based wireless sensor networks.," Sensors (Basel)., vol. 14, no. 5, pp. 8447–64, Jan. 2014.

[4] H. A. Marhoon, M. Massudi, and A. N. Shahrudin, "CHAIN-BASED ROUTING PROTOCOLS IN WIRELESS SENSOR NETWORKS: A SURVEY," ARPN J. Eng. Appl. Sci., vol. 10, no. 3, 2015.

[5] W. Y. Poe and J. B. Schmitt, "Node Deployment in Large Wireless Sensor Networks : Coverage , Energy Consumption , and Worst-Case Delay," in AINTEC'09 Conference Proceding, Bangkok, Thailand, 2009.

[6] S. Rana and A. Aggarwal, "Study and Comparison of Chain Based Protocol to Enhance Energy Utilization in Wireless Sensor Network," Int. J. Comput. Sci. Mob. Comput., vol. 3, no. 8, pp. 11–17, 2014.

[7] F. Tang, I. You, S. Guo, M. Guo, and Y. Ma, "A chain-cluster based routing algorithm for wireless sensor networks," J. Intell. Manuf., vol. 23, no. 4, pp. 1305–1313, May 2010.

[8] H. Kareem, S. J. Hashim, S. Suberamaniam, and A. Sali, "Energy Efficient Two Stage Chain Routing Protocol (TSCP) For Wireless Sensor Networks," J. Theor. Appl. Inf. Technol., vol. 59, no. 2, pp. 442–450, 2014.

[9] H. Zhang and C. Liu, "A Review on Node Deployment of Wireless Sensor Network," IJCSI Int. J. Comput. Sci., vol. 9, no. 6, pp. 378–383, 2012.

[10] G. Huang, D. Chen, and X. Liu, "A Node Deployment Strategy for Blindness Avoiding in Wireless Sensor Networks," IEEE Commun. Lett., vol. 19, no. 6, pp. 1005–1008, 2015.

[11] X. Liu, "A Deployment Strategy for Multiple Types of Requirements in Wireless Sensor Networks," IEEE Trans. Cybern., vol. 45, no. 10, pp. 2364–2376, 2015.

[12] L. P. Damuut, "A Survey of Deterministic Vs . Non-Deterministic Node Placement Schemes in WSNs," in The Sixth International Conference on Sensor Technologies and Applications, 2012, no. c, pp. 154–158.

[13] S. Meguerdichian, F. Koushanfar, G. Qu, and M. Potkonjak, "Exposure in wireless Ad-Hoc sensor networks," Proc. 7th Annu. Int. Conf. Mob. Comput. Netw. - MobiCom '01, pp. 139–150, 2001.

[14] J. Li, L. L. H. Andrew, C. H. Foh, M. Zukerman, and H.-H. Chen, "Connectivity, coverage and placement in wireless sensor networks.," Sensors (Basel)., vol. 9, no. 10, pp. 7664–93, Jan. 2009.

[15] M. Kuorilehto, M. Kohvakka, J. Suhonen, H. Panu, H. Marko, and T. D. H, Ultra-Low Energy Wireless Sensor Networks in Practice Theory, Realization and Deployment. U.S.A: John Wiley &Sons, Ltd, 2007.

[16] S. Khan, A.-S. K. Pathan, and N. A. Alrajeh, Wireless Sensor Networks, Current Status and Future Trends. Taylor & Francis Press, 2013.

[17] B. Krishnamachari, Networking Wireless Sensors: Sleep-oriented MAC – Efficient Routing Data-centric Concepts – Congestion Control. New York: cambridge university press, 2005.

[18] K. Akkaya, K. Akkaya, M. Younis, and M. Younis, "A survey on routing protocols for wireless

sensor networks," Ad Hoc Networks, vol. 3, pp. 325–349, 2005.
[19] H. Ö. Tan and I. Körpeoğlu, "Power efficient data gathering and aggregation in wireless sensor networks," ACM SIGMOD Rec., vol. 32, no. 4, p. 66, Dec. 2003.
[20] C. S. Lindsey, Stephanie Raghavendra, "PEGASIS : Power-Efficient GAthering in Sensor Information Systems '," in Aerospace conference proceedings IEEE, 2002, pp. 3–1125.
[21] K. S. Ahn, D. G. Kim, B. S. Sim, H. Y. Youn, and O. Song, "Balanced Chain-Based Routing Protocol (BCBRP) for Energy Efficient Wireless Sensor Networks," 2011 IEEE Ninth Int. Symp. Parallel Distrib. Process. with Appl. Work., pp. 227–231, May 2011.
[22] H. A. Marhoon, M. Mahmuddin, and S. A. Nor, "DCBRP: A Deterministic Chain-Based Routing Protocol for Wireless Sensor Networks," Springerplus, vol. 5, no. 2035, pp. 1–21, 2016.
[23] C. Wei, J. Yang, and Y. Gao, "Cluster-based routing protocols in wireless sensor networks: A survey," Proc. 2011 Int. Conf. Comput. Sci. Netw. Technol., pp. 1659–1663, Dec. 2011.
[24] W. R. Heinzelman, A. Chandrakasan, and H. Balakrishnan, "Energy-Efficient Communication Protocol for Wireless Microsensor Networks," in Proceedings of the 33rd Hawaii International Conference on System Sciences, 2000, vol. 0, no. c, pp. 1–10.
[25] S. SUMITHRA and A. A. VICTOIRE, "An Energy Efficient Linear Chain-Based Clustering Routing Protocol For Wireless Sensor Networks," Theor. Appl. Inf. Technol., vol. 63, no. 3, 2014.
[26] W. B. Heinzelman, "Application-Speci c Protocol Architectures for Wireless Networks," Cornell, 2000.
[27] "Network Simulator 3," 2015. [Online]. Available: https://www.nsnam.org/. [Accessed: 28-Feb-2015].

Smart system for dust detecting and removing from solar cells

Hussein A. Mohammed[1], Baha'a A. M. Al- Hilli [2] and Intisar Shadeed Al-Mejibli[3]

[1]University of Information Technology and communication, College of management information.
[2]Univ. of Al Mustansiriyah, College of Education, Physics Dept.
[3]Informatics Institute for Postgraduate Studies

dr.hussein.a.mohammed@uoitc.edu.iq

Abstract. In the last decades, solar energy gained the most demand as it represents a green and sustainable source of energy. One of the most important obstacles for solar energy production in Iraq is the dirtiness of the panel surface as it causes a shadow that reduces its performance. From other hand, Iraq location has a large frequency of sand dust storms. This research proposed a method to eliminate the impact of dust and dirtiness on the performance of the solar panel in power production. The proposed method monitors the power generation and cleans the photovoltaic surface when required. Arduino Uno microcontroller was used to control the washing task with windscreen wiper tool which energized when the output power dropped to 50% from its rate value. In addition, the proposed method presented a comparison on the solar panel performance with and without the proposed method. The results show the ability of the washing mechanism in maintaining the solar panel performance efficiently after the exposure to dust accumulation.
Keywords: Solar Cell, Dust Detection, Self-Cleaning, Arduino.

1. Introduction

Several factors may affect the generation of solar power generation. Shadows, snow, high temperatures, dust, dirt, bird's dropping pollen and sea-salt are regarded as the main factors that minimize or impede the generation of power in the Photovoltaic PV panels [1] [2] [3]. Generally, the world supports using renewable energy such as solar and wind energy, because the generated electricity is without giving rise to emissions of any carbon dioxide. Thus the demand on clean energy has grown resulting in increasing the production of PV panel's. This is represented the result of increase awareness about the environment damage that caused by using sources of fossil fuel such as coal, petroleum and natural gas over the years.

Using PV panels to produce power efficiently requires removing the dirt from them regularly, particularly, dust that is caused by pollen, sea salt and dirt particles [1].

There are many different sources that cause dust accumulation on the PV panels' surface, and may have significant impact on production of electricity. The efficiency of power generation from the solar panel is minimized in a dusty environment and this attributed to the fact of: dust reduces the amount of sun light received by the PV panel [4].

In Baghdad at 2010, the authors of [4] found that the transmittance is minimized by approximately 50% on average, because of dust deposition on PV panels.

IOP Conf. Series: Journal of Physics: Conf. Series **1032** (2018) 012055 doi:10.1088/1742-6596/1032/1/012055

Thus, there is a need for monitoring and cleaning panel's surfaces. The frequency of removing the dust from PV panels depends on the environment of the solar location.

This research focuses on the detection and removing dust by simple mechanism. The proposed mechanism based on using Arduino microcontroller with software to drive simple mechanics.

The rest of this paper is organized as following: methods of dust removal from PV is described in section 2. Section 3 presents the solar panel performance. Proposed system is described in section 4. Section 5 explains the system implementation. Results and discussion are explained in section 6. Conclusion is described in section 7.

2. Methods of Dust Removal from PV

Many methods were suggested and developed for automatically self-cleaning of solar panels, such as wiper and sprayer [6], using motorized brushes, self-cleaning with piezo-ceramic actuation [7], electrostatic cleaning [8] and robotic cleaning [9].

Generally the PV cleaning methods can be classified as follows:

- Rainfalls Cleaning Method: despite it is does not cost anything, it seasonally volatile. This will impact on the reliability of Rainfalls cleaning method passively particularly in Iraq when rainfall is not enough either in quantity or in intensity to wash off the soil [10].
- Manual Cleaning Method: This method depends on laborer to clean the PV surface so it is same as the method of cleaning buildings windows. The laborer needs to use brushes to remove the soil off the surface. In addition, this brushes need to be designed with special bristles to prohibit scratching the surface [10].
- Mobile Cleaners Method: machinery is used in this method of cleaning. In addition, this method may require to use a storage for water supply or sprinkler system [11].

From other hand, the authors in [12] review the recommended cleaning cycle to reduce effect of dust, during dry seasons it's recommended the weekly cleaning while during intensive dust accumulation the daily washing is recommended.

An effective cleaning method should be performed when required with minimum requirements such as water or brushes.

3. Solar Panel Performance

The performance of the PV panel can calculated from the current- voltage (I-V) curve. The PV panel parameters for instance open circuit voltage (V_{oc}) and short circuit current (I_{sc}) are obtained from (I-V) curve. The power curve is obtained from the values of current and voltage. The maximum power point represented by the point of the maximum delivered power (P_{mpp}). The maximum power voltage and current (V_m, I_m) represents by the current and voltage at the maximum power point.

From the aforementioned parameters, the device fill factor (FF) and conversion efficiency (η) can be obtained as represented in equation 1 and 2.

$$F.F. = \frac{V_m I_m}{V_{oc} I_{sc}} \quad \dots\dots\dots\dots\dots\dots\dots\dots\dots (1)$$

$$\eta = \left[\frac{V_{oc}\ I_{sc}\ FF}{A\ \ G}\right] \times 100\% \quad \dots\dots\dots\dots\dots (2)$$

Where (A) is area of the panel and (G) is the irradiance falling on the surface of the panel. The performance of the studied panel is measured by the amount of energy it produces over period of time in (kWh) unit. Finally the panel performance can be represented by Yield (Ys) factor (kWh/kW$_p$) which is the energy generated normalized by the installed capacity in (W_p) as shown in equation 3:

$$Y_s = \frac{E_{total}}{P_m} \quad \dots\dots\dots\dots\dots\dots (3)$$

4. Proposed System

The proposed system consist of three main parts as shown in figure (1). The first part is sensing, processing the dust density and sending SMS warning massage to the operators if required. The second part is the current and voltage sensors with Arduino Uno microcontroller to measure the PV panel output in real time, and then processing the accumulated dust effects on the output power. The third part is windscreen wiper system that operate from microcontroller signal via relay which energized when output power value approach to 50% from its rate values.

In this research the effects of dust that accumulated on the solar panel is investigated by referring the results which are obtained from sensors and microcontroller in dusty atmosphere. Also, an auto cleaning windscreen wiper system works as an auto cleaner, which is equipped on the flat solar panel. The windscreen wiper system consists of washing water sprayer, electrical D.C. power which is needed to drive the windscreen wiper system supplied from the solar plate itself by using D.C. to D.C. step down converter.

The light dependent resistance LDR sensor (that connect to microcontroller as shown in figure (1)) is used to distinguish between the day and night. Thus, the washing system can work in case of accumulation of dust depending on panel output electrical power.

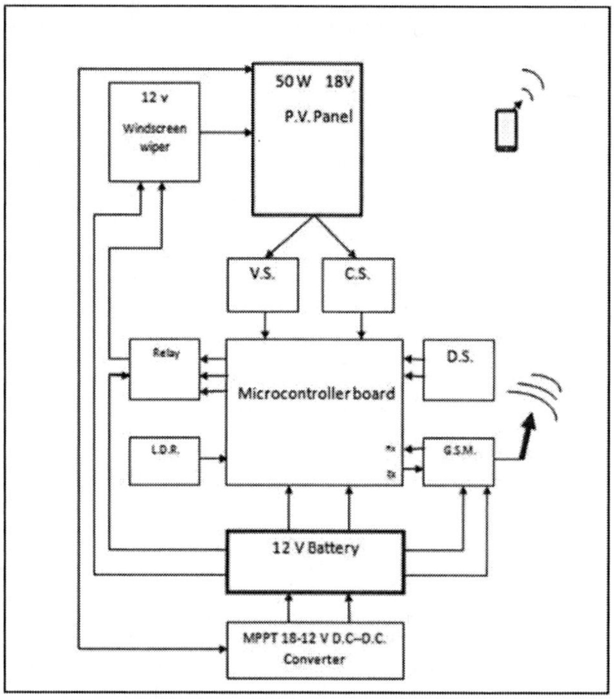

Figure 1. The Proposed System

5. System Implementation

The implementation of the proposed system consist of the following devices:

5.1. Solar PV panel

Solar PV panel from FORTUNER CORP where its specifications are described in table (1).

Table 1. P.V. Solar PV Panel Specification [13]

Module	**50Wp**	**Pmp(W)**	**52.0457**
Manufacturer	FORTUNER	Ambient temp	25
Model	3589.74	Imp(A)	2.805

area(cm2)			4
Irradiance(mW/cm2)	100	Isc(A)	3.105 4
Vmp(V)	18.552	Module eff.(%)	14.51
Voc(V)	21.8465	Est.shunt resistance(ohm)	70.61 89
F.F.	76.72	Est.series resistance(ohm)	0.296 1

5.2. Arduino Uno Microcontroller

Arduino Uno is a microcontroller board based on the ATmega328Pas is used to measure dust density and calculate plate output power. The specifications of employed Arduino microcontroller are described in table 2.

The microcontroller is connected to personal computer (PC) via universal serial bus (USB) in order to upload the software. Figure 2 shows the arduino Uno microcontroller.

Table 2. Arduino Uno Specifications [14]

Microcontroller	**ATmega328**
Operating Voltage	5V
Input Voltage (recommended)	7-12V
Input Voltage (limits)	6-20V
Digital I/O Pins	14 (of which 6 provide PWM output)
Analog Input Pins	6
DC Current per I/O Pin	40 mA
DC Current for 3.3V Pin	50 mA
Flash Memory	32 KB of which 0.5 KB used by bootloader
SRAM	2 KB
EEPROM	1 KB
Clock Speed	16 MHz

Figure 2. Arduino Uno Microcontroller

5.3. Dust Sensor

IOP Conf. Series: Journal of Physics: Conf. Series **1032** (2018) 012055 doi:10.1088/1742-6596/1032/1/012055

The used dust sensor is of module DSM501A and it is specified as a low cost with compact size for a particle density sensor. The principle of operation of this sensor is shown in figure (3).

The quantitative particle density can be measured with the dust particle counter where particles of bigger than (1 μ m) could be detected and the heater induces air inflow located in to the sensor. The sensor connected to the microcontroller as show in figure (4).

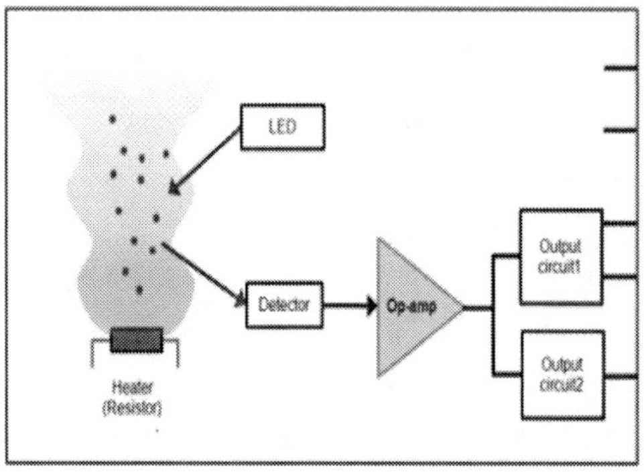

Figure 3. Diagram of dust sensor block

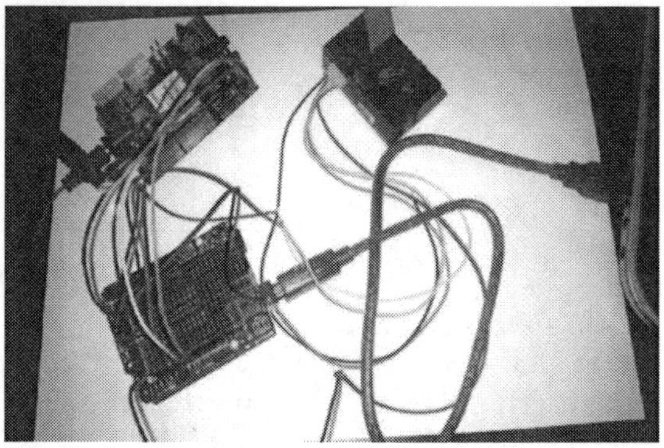

Figure 4. The connections of dust sensor, G.S.M and microcontroller

5.4. *Global System for Mobile Communication (GSM)*

The SIM900 GSM/GPRS modem is used to connect with the GSM network. This modem is responsible on sending warning SMS message to the maintenance officer at the power station when a dust storm occurs. This modem connected to microcontroller as shown in figure (4).

5.5. *Current Sensor*

A30 amp D.C current sensor type ACS712 is used to measure the (Isc.) which is the load current of the solar cell in order to calculate the output solar power. This sensor is connected to the microcontroller as show in figure (4).

5.6. *Voltage Sensor*

The Arduino Uno microcontroller consist of five D.C. inputs and the output voltage of the solar cell that equal to 18 Volts, therefore the D.C voltage senor is used. This sensor consists of voltage divider in order to measured voltage between (0-25) Volts. The sensor is connected to microcontroller as shown in figure (5).

Figure 5. The Connections of Current Sensor, Voltage Sensor, Relay and Microcontroller

5.7. *Relay*

A 5 volt relay is connected to microcontroller at a (digital output). When the accumulated dust on solar panel is increased over the maximum limit the relay switching ON. Then the windscreen wiper system works for (10 sec) period of time, in order to clean the solar cell and return the output power to the rated value.

5.8. Photovoltaic Panel P.V.
The results are obtained in Baghdad city (latitude of 33°13'N) during January month. The panel is mounted and tilt to 30°angel according to (reference) to raise the effectiveness of panel. figure (6. a) the windscreen wiper (for vehicle) consist of 12v D.C motor that control wiper movement along the panel ,12v D.C. water pumping motor and two nozzle used to sprayedwater during the survey of the panel surface. Figure (6. b) dirty Photovoltaic Panel P.V.

(a) Clean

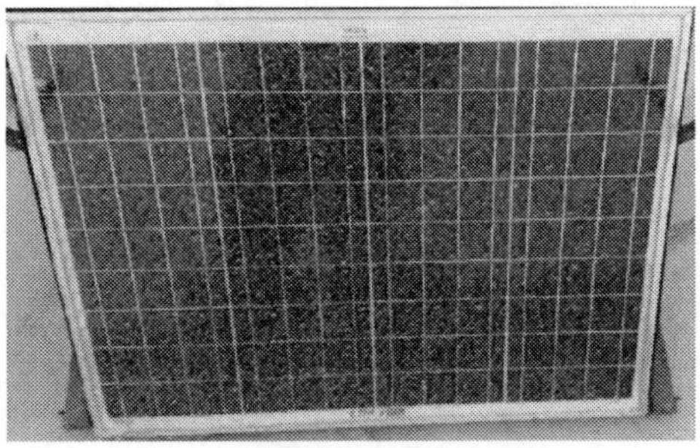
(b) Dirty

Figure 6. The Photovoltaic Panel P.V.

5.9. Maximum Power Point Tracker (MPTT)
MPPT is specific types of charge and controller that will utilize the PV panel to its maximum potential. It is derive 5A step-down (18-12) constant voltage and constant current to charge the battery. Figure 7 shows the procedure of the proposed system based on day light while figure 8 shows the procedure of the proposed system based on dust in atmosphere.

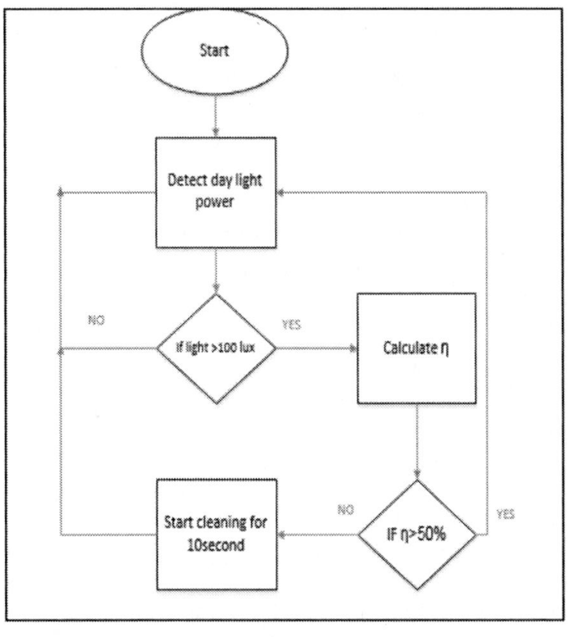

Figure 7. The Proposed System Procedure based on Daylight

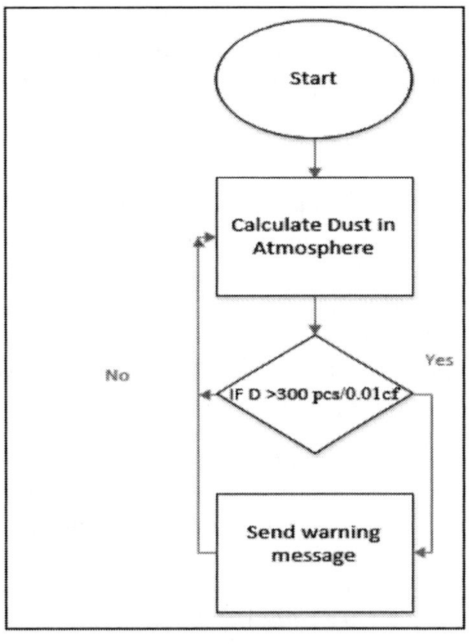

Figure 8. The Proposed System Procedure based on Dust

6. Results and Discussion

IOP Conf. Series: Journal of Physics: Conf. Series **1032** (2018) 012055 doi:10.1088/1742-6596/1032/1/012055

6.1. Short Circuit Current

The short circuit current produce from panel (I_{sc}) and sensing by current sensor, record by Arduino microcontroller software (control panel) as show in figure (9).

```
Current Isc :1.71A
Current Isc :2.36A
Current Isc :2.44A
Current Isc :2.44A
Current Isc :2.44A
Current Isc :2.44A
Current Isc :2.44A
Current Isc :2.44A
Current Isc :2.44A
Current Isc :2.45A
Current Isc :2.45A
Current Isc :2.45A
Current Isc :2.46A
Current Isc :2.45A
Current Isc :2.44A
Current Isc :2.45A
Current Isc :2.45A
Current Isc :2.44A
Current Isc :2.45A
```

Figure 9. The current sensor output

6.2. The Open Circuit Voltage of the Panel

The data obtained for solar output power which accomplished at 2nd September 2017 for two cleaned and dust accumulated panels at constant tilt angle (35°) and clear sky. The temperature were found to be vary from 23 to 40 °C and the solar light intensity to be 800 to 1500 lux conditions. The amount of accumulated dust on the solar panel are obvious as the difference between the short circuit current of clean and dusty solar panels is shown in figure (4). The accumulated dust effects the solar panel performance because it caused the optical attenuation, i.e. the electromagnetic radiation is attenuated by the interaction with the dust solid particles in the external environment. The attenuation is the sum of different kinds of interactions, absorption, elastic scattering and inelastic scattering [9].

A comparison on the I_{sc} generation through a day is presented in figure (10). When the new mechanism is applied (clean PV), the I_{sc} generation is maintained higher level than without using it. The difference between the two cases is approximately 0.8 I_{sc} (Amp) along the day hours. The proposal improve the performance of the solar panel more than 50%.

7. Conclusion

The dust shadowing effects on the solar panel passively. It is considerable due to the fact that the output power is a complex parameter that is influenced by different environmental and weather conditions. The fixed and reliable solar panel power generation always requires clean PV panel. This research designed and implemented a mechanism that enables solar panel from cleaning itself when its performance getting less 50% its average generation rate. The proposal satisfies the opportunity of an efficient automatic self-cleaning with low frequency maintenance in order to improve solar panel performance.

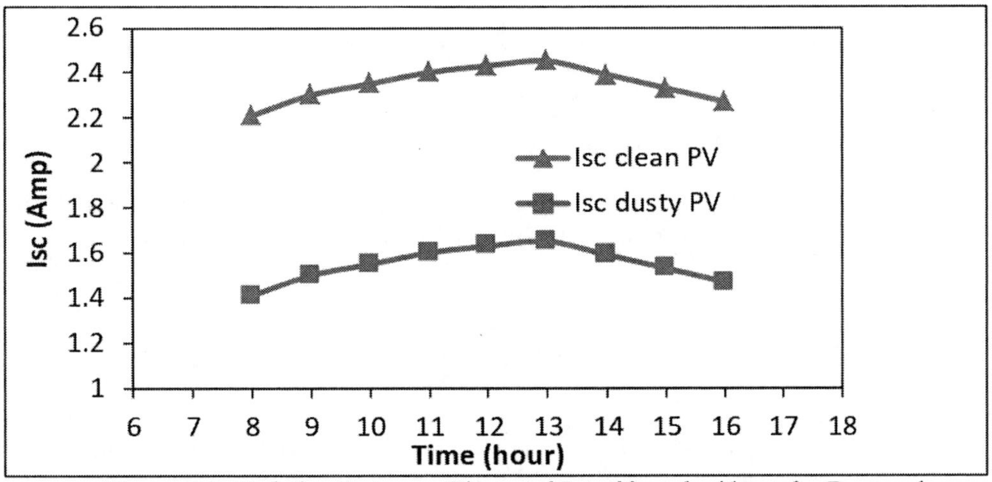

Figure 10. Relation Between Time and I_{sc} with and without the Proposal

8. References

[1] N. Albaqawi, "Development of an automatic robotic cleaning system for photovoltaic plants", Electrical Engineering Development Project, Waikato Institute of Technology, (2014).

[2] Sh. A. Sulaiman, H. H. Hussain, N. S. H. Nik Leh, and M. S. I. Razali, "Effects of Dust on the Performance of PV Panels", International Journal of Mechanical and Mechatronics Engineering, Vol:5, No:10, (2011) 2028-2033.

[3] A. Ibrahim, "Effect of Shadow and Dust on the Performance of Silicon Solar Cell", J. Basic. Appl. Sci. Res., 1(3)222-230, (2011), 222-230.

[4] N.K.Kassim, "New Technique for Treatment of the dust accumulation from PV solar panels surface", Iraqi Journal of Physics, Vol. 8, No.12, (2010), 54 -59.

[5] S. M. R. Saidur, M. Kamalisarvestani, "Effect of dust, humidity and air velocity on efficiency of photovoltaic cells", Renewable and Sustainable Energy Reviews, Vol. 16, Issue 5, June(2012), 2920-2925.

[6] A. Sivan, L. Priya, S. Mathew, "Automatic Self Cleaning Solar Panel", Int. Research J. of Eng. And Tech. (IRJET), Vol.4, Issue 5, May (2015).

[7] R. B. Williams, R. Tanimoto, A. Simonyan, and S. Fuerstenau, "Vibration Characterization of Self-Cleaning Solar Panels with Piezoceramic Actuation", 48th AIAA/ASME/ASCE/AHS/ASC Structures, Structural Dynamics, and Materials Conference, 15th 23-26 April (2007), Honolulu, Hawaii.

[8] P.E. Clark, S.A. Curtis, F.A. Minetto, J. Keller, "Finding A Dust Migitation Strategy that Works on the Lunar Surface", Lunar and Planetary Science XXXVIII (2007).

[9] C.Elisa, Ph.D thesis, "Scattering and Absorption of X-ray by Interstellar Dust", Munchenuniversity, Physics department, Italien, (2004).

[10] Mohammad Reza Maghami, Hashim Hizam, Chandima Gomes, Mohd Amran Radzi, Mohammad Ismael Reza dad, Shahrooz Hajighorbani, "Power loss due to soiling on solar panel: Areview", Renewable and Sustain able Energy Reviews 59 (2016) 1307–1316.

[11] D. G. f. r. Sonnenenergie, "Planning and installing photovoltaic systems: a guide for installers", architects and engineers: Earth scan; 2008.

[12] Monto Mani *, Rohit Pillai, "Impact of dust on solar photovoltaic (PV) performance: Research status, challenges and recommendations", Renewable and Sustainable Energy Reviews 14 (2010) 3124–3131.

[13] Fortuner catalogue, https://www.fortuners.net/uploads/fortuner_product%20_booklet_2017.pdf.

[14] RobotShop, "Arduino Mega 2560 Datasheet", https://www.robotshop.com/media/files/pdf/arduinomega2560datasheet.pdf.

IOP Conf. Series: Journal of Physics: Conf. Series **1032** (2018) 012056 doi:10.1088/1742-6596/1032/1/012056

Photocatalytic Hydrogen Production on Nanocomposite of Carbon Nanotubes and TiO_2

Firas H. Abdulrazzak[a] and Falah H. Hussein[b]

[a]Chemistry Department, College of Education for Pure Sciences, Diyala University, Diyala, Iraq. [b]College of Pharmacy, Babylon University, Hilla, Iraq.

Corresponding author's E-mail: firas_habeb2000@yahoo.com

Abstract. Sonochemical/hydration–dehydration techniques were used to synthesize binary composites CNT/TiO_2 consisting single wall carbon nanotubes (SWNTs) and multiwall carbon nanotubes (MWNTs) with TiO_2. Different percentages of CNT/TiO_2 were prepared (0.25, 0.5, and 1%). The morphological and physiochemical properties were investigated by powder X-ray diffraction (XRD), Raman spectroscopy and UV–vis diffuse reflectance spectroscopy. It was found the TiO_2 nanoparticles were more homogenized with SWNTs compared with MWNTs. 7.5% methanol solution was used as a sacrificial agent to produce hydrogen gas with 65 mg of synthesized binary composites at light intensity 40 mW cm^{-2}. The results show that SWNTs increased the activity of TiO_2 more than for MWNTs. The hydrogen production shows higher activity for TiO_2 with lower ratio of SWNTs, while the activity was raised for TiO_2 with increasing ratio of MWNTs.

1. Introduction.

Energy sources are one of the most important studies that interest researchers not to produce energy but to looking for clean and renewable energy for all the world. Recently, fossil fuels represent the main source, which used in all fields of life for a long time ago. May be the combustion of fossil fuels which produce CO and CO2 with the risk of depleted completely cause bearing the world to search for alternative sources and environmentally friendly [1]. Hydrogen represents the ideal for the world due to easily produce a large amount of energy which equals to 122 kj/g without any harmful product [2]. Many research reported that hydrogen as a source of energy for huge applications [3]. As we reported in our previous work hydrogen can produce by photocatalytic water splitting and photocatalytic reforming of biomass by semiconductor [4].

Photocatalysis reaction, used to produce hydrogen when the chemical reaction was induced by the catalyst in existing a convenient source of irradiation. Semiconductors are the common materials behave as a catalyst in the photocatalytic reaction such TiO_2, ZnO, and CdS [5]. The efficiency of semiconductors depend on three parameters, absorption of photons, prevent recombination and consuming of the excited electrons [6]. The energy of absorption photons responsible to translate the electrons from valence band (VB) to conduction band (CB), and forming positive charge h^+ while CB produces negative charge e^-. The prevention of e^- to recombine with h^+ and consuming the excited electrons at the CB by acceptance species together, that produce the abilities of semiconductor towards hydrogen production.

Titanium dioxide (TiO_2) represent the ideal semiconductor for many applications due to inexpensive, higher activity and friendly to the environment [7]. The limitation for use TiO_2 related to reacting with less than 4% of solar light which represent by UV-light [8]. Many attempts were done to enhance the activity of redox reaction. All the strategies depend on addition different species such as metal/nonmetal doping, photosensitization with dyes and coupling with semiconductors [9]. After 1991 the attentions were brought toward using carbon nanotubes CNTs in hydrogen production research when characterized by rarer and amazing physiochemical properties [10]. Carbon nanotubes CNTs realize as graphene or graphite sheets wrapped in forming cylindrical structures such single–

IOP Conf. Series: Journal of Physics: Conf. Series **1032** (2018) 012056 doi:10.1088/1742-6596/1032/1/012056

walled carbon nanotubes SWNTs or multi–walled MWNTs [11]. The TiO_2/CNT composite can be prepared by sol-gel [12], chemical vapor deposition [13] and electro-spinning methods [14] which characterized by specific properties for the synthesized product. Herein, titanium dioxide was doped with SWNTs or MWNTs by a sonochemical/hydration–dehydration process. The synthesized binary composites were characterized by X-ray diffraction (XRD), Raman spectroscopy and UV–vis diffuse reflectance spectroscopy. The activity of synthesized different ratios and types of CNTs in composites were tested in hydrogen production from aqueous alcohol solution.

2. Experimental.

Materials.

Two types of carbon nanotubes were used, SWNTs and MWNTs which purchased from Aldrich and fabricated by chemical vapor deposition. The SWNTs consist of more than 90% carbon and 77%wt SWNTs, with a diameter 0.7–1.1 nm, while the MWNTs are 95% carbon nanotubes with a mode diameter 4.5 nm. The TiO_2 sample was supplied as Degussa, Germany (TiO_2-P25) (consisting of 20% Rutile and 80% Anatase, with a BET area 51 m^2/g). Nitric acid (65 wt% HNO_3) and sulfuric acid (37 wt% H_2SO_4) were obtained from Fluke and Sigma- Aldrich, respectively. Methanol with purities (99.99%) as the sacrificial agent was purchased from Fluke.

3. Preparation of Binary composite.

The binary composite CNT/TiO_2 were prepared as reported in our previous work [7]. Briefly; an ultrasonic water bath was used for 7 hours when the two types of SWNTs and MWNTs were dispersion in (1/3) a mixture of two concentric nitric and sulfuric acid. The TiO_2 was suspended in 100 mL of required weight of CNTs aqueous solution using an ultrasonic water bath. The two suspensions were washed and dried by a vacuum evaporator (Rota vapor re121 BUSHI 461 water Bath) at 45 °C then thermal treatment overnight in an oven at 90 °C.

4. Characterization of SWNT/TiO2 and MWNT/ TiO_2.

Energy sources are one of the most important studies that interest researchers not to produce energy but to looking for clean and renewable energy for all the world. Recently, fossil fuels represent the main source, which used in all fields of life for a long time ago. May be the combustion of fossil fuels which produce CO and CO_2 with the risk of depleted completely cause bearing the world to search for alternative sources and environmentally friendly [1]. Hydrogen represents the ideal for the world due to easily produce a large amount of energy which equals to 122 kj/g without any harmful product [2]. Many research reported that hydrogen as a source of energy for huge applications [3]. As we reported in our previous work hydrogen can produce by photocatalytic water splitting and photocatalytic reforming of biomass by semiconductor [4].

Figure 1 shows the XRD patterns for SWNTs with two characteristic peaks at 23.7°, 42.9° and many noises which related to the structure of SWNTs. In the same figure reflection planes of MWNTs were (002) and (100) which related to 26.3° and 43.7° respectively with more regular and less noise as compared with SWNTs.

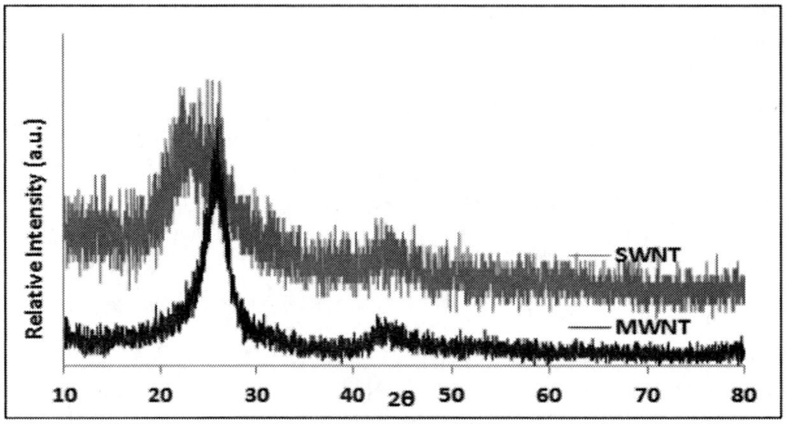

Figure 1. X-ray patterns diffraction for two types of CNTs.

Figure 1 shows the XRD patterns for SWNTs with two characteristic peaks at 23.7∘, 42.9∘ and many noises which related to the structure of SWNTs. In the same figure reflection planes of MWNTs were (002) and (100) which related to 26.3∘ and 43.7∘ respectively with more regular and less noise as compared with SWNTs.

The XRD patterns of pristine TiO_2 and modified with different ratios of CNTs are shown in Fig.2 and 3. The anatase crystallites were determined by the (200) reflection at (2θ= 48.1°), which there is no interference from CNTs. Table 1, shows the particle sizes of composites catalyst were estimated using Scherrer's equation which decreases gradually with the increasing the ratios of CNTs, and showed less intense and wider for binary composite CNTs/TiO_2 [15].

Table 1. The crystal size for the synthesized TiO_2/CNTs calculated from Scherrer's equation.

Ratios of CNTs	SWNT Particle size (nm)	MWNT Particle size (nm)
0.0%	23.09	23.0
0.25%	19.97	22.14
0.5%	15.41	20.84
1.0%	14.19	19.07

The change for peaks in intensity and area were occurred with SWNTs more than MWNTs while the shift was more with MWNTs, which may related to influence number of layers in MWNTs. The low ratios of CNTs causing limited change in peaks which covered by the much more intense peaks for the large ratio of TiO_2.

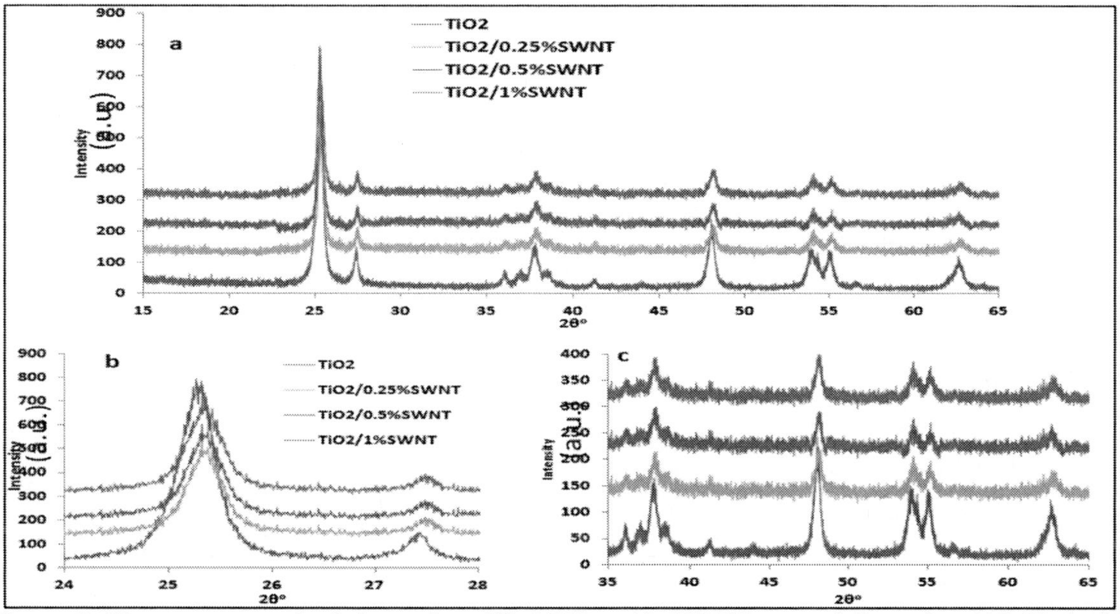

Figure 2. X-ray patterns of pristine and modified TiO_2 with SWNTs a. total figure from 15-65°, b. 24-

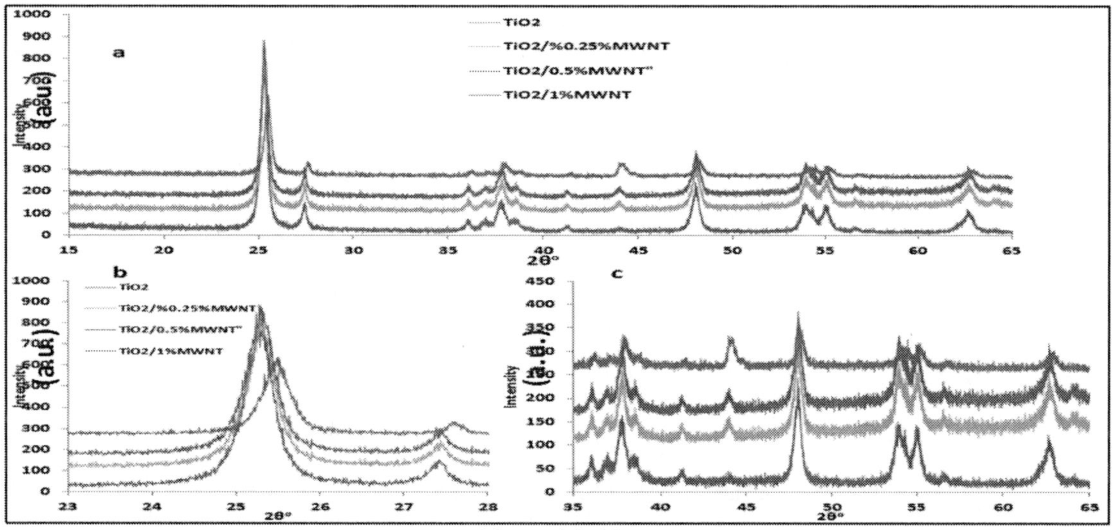

28°, c. 35-65°.

Figure 3. X-ray patterns of pristine and modified TiO_2 with MWNTs a. total figure from 15-65°, b. 24-28°, c. 35-65°.

The Raman spectra in Fig.4, explines the Raman spectrum for SWNTs when characterized with five peaks such RBM, G, D, G', and RBM+G while with MWNTs shows three peaks. The pristine TiO_2 recognize two phases' Anatase and Rutile which represent the continents of this type. The region at 150 cm^{-1} (E_g), 395.1 cm^{-1} (B_{1g}), 512.5 cm^{-1} (A_{1g} + B_{1g}), and 636.7 cm^{-1} (E_g) which refer to orientations (E_g), (B_{1g}), (A_{1g} + B_{1g}), and (E_g) respectively represent the modes of Anatase. The rutile phase appears in two peaks at 143, 235 cm^{-1}, for the B_{1g}, and 445 cm^{-1}, and 612 cm^{-1} for Eg and A_{1g}, respectively [16]. The last behavior indicates that as expected the nano-composite preserved the structures of both the MWNTs and anatase TiO_2. Figure 5 with four parts a, b, c, and d refer to Raman spectroscopy of SWNT/TiO_2 which shows red shift for all anatase and rutile peaks with increase the width more than the effect of MWNT in composites. Fig. 6, for MWNT/TiO_2 composites, distinguish with bands of anatase and MWNTs when did not change, except that the anatase [16] are slightly broadened compared to the pure TiO_2. This peak broadening is consistent with the decreasing in the average of crystallite size. It is worth noting that the peaks assigned to CNTs within the composites exhibit broadening and a slight blue-shift relative to the acid-treated CNTs, indicating the interfacial interaction between the TiO_2 and acid-treated CNTs [15].

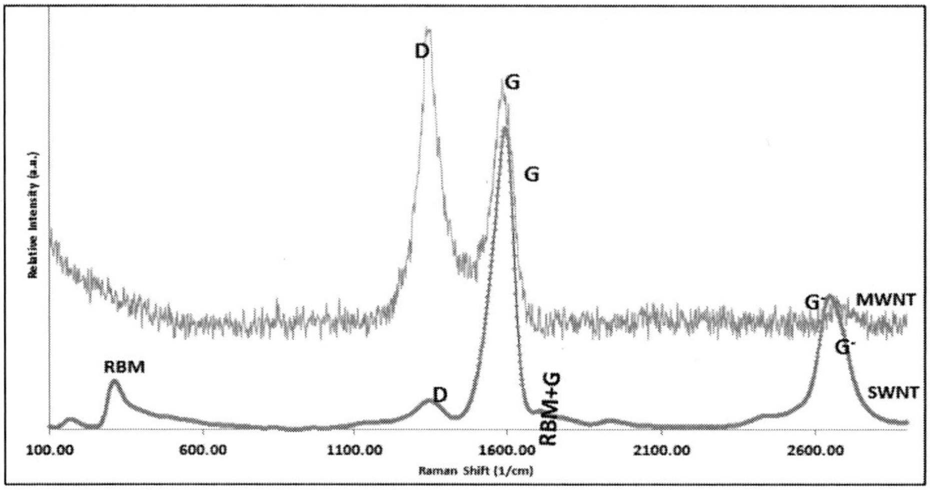

Figure 4. Raman spectra of SWNTs and MWNTs from 100 to 2900 cm^{-1}

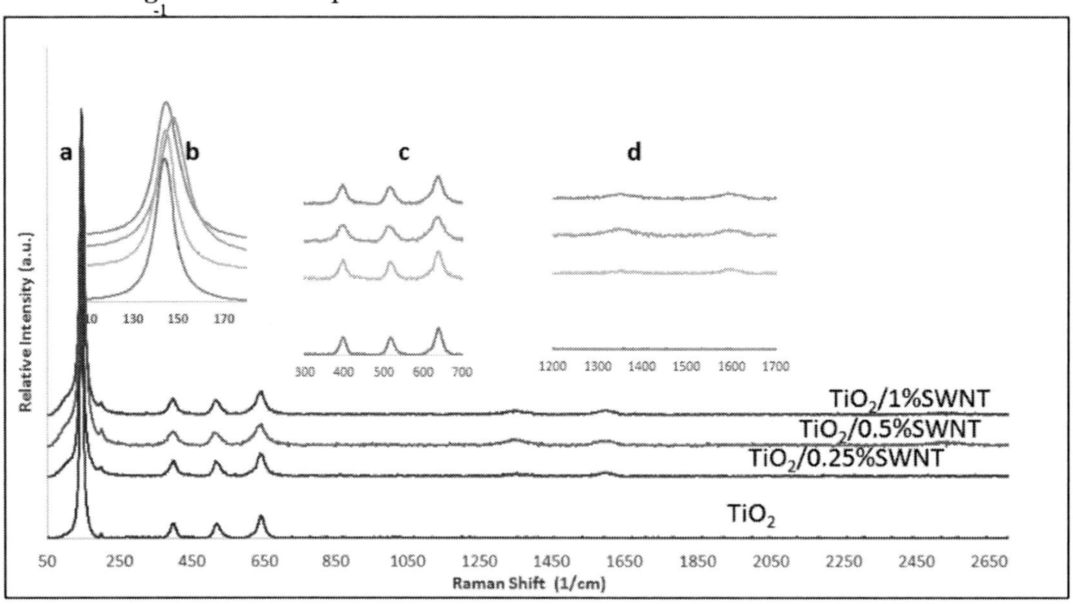

Figure 5. Raman spectroscopy for pristine and modified TiO_2 with SWNTs, a.total skim 50-2650 cm^{-1}, b. for (110-170) cm^{-1}, c. (300-700) cm^{-1}, d. (1200-1700) cm^{-1}.

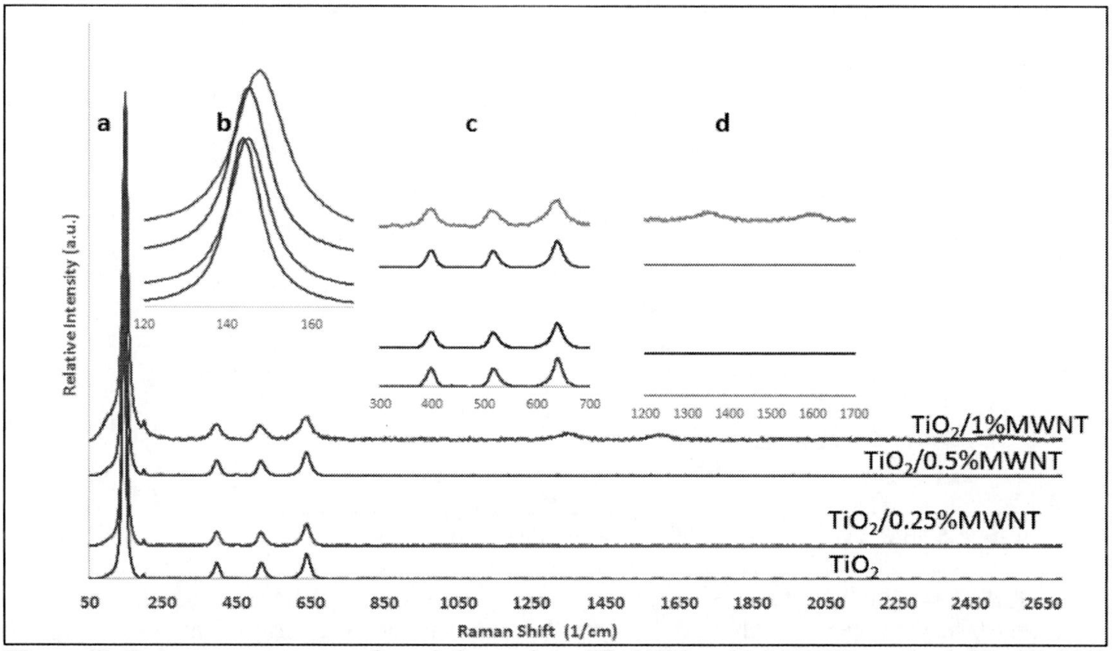

Figure 6. Raman spectroscopy for pristine and modified TiO_2 with MWNTs, a. total skim 50-2650 cm^{-1}, b. for (110-170) cm^{-1}, c. (300-700) cm^{-1}, d. (1200-1700) cm^{-1}

The band gap measurements were done for pristine TiO_2 and with different ratios of CNTs (0.25%, 0.5%, and 1% of SWNTs or MWNTs) which plotted in figure.7 and 8.

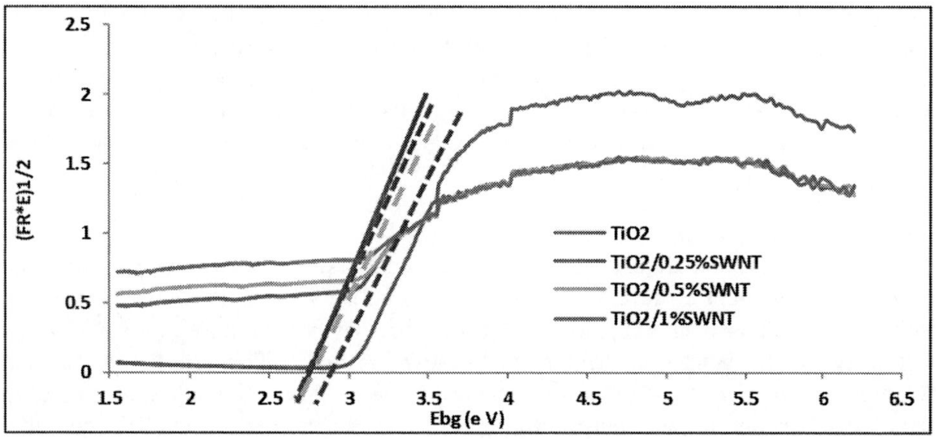

Figure 7. Band gap E_{bg} for pristine and modified with 0.25, 0.5, and 1% SWNTs.

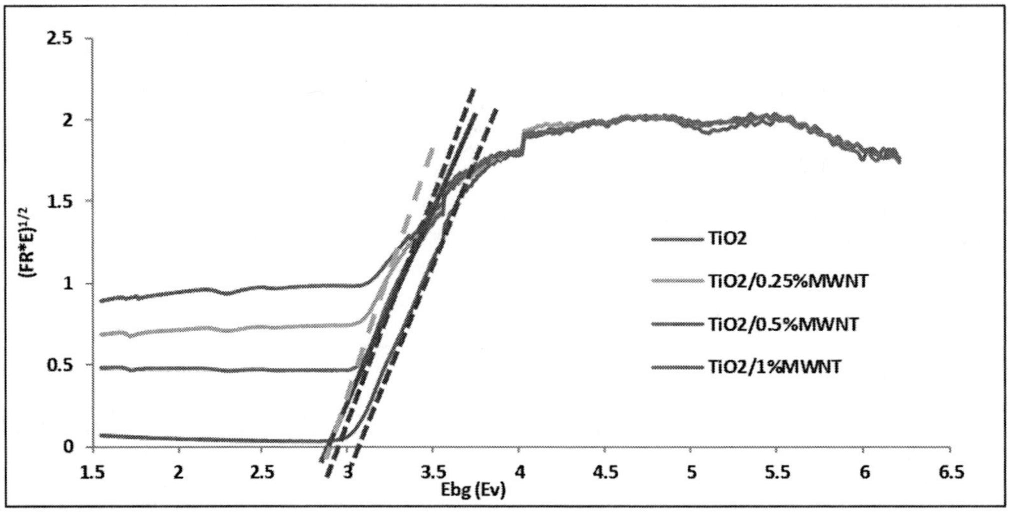

Figure 8. Band gap E_{bg} for pristine pristine and modified with 0.25, 0.5, and 1% MWNTs.

The Kubelka–Munk function [17] were used to estimate the band-gap energies Eg for the synthesized binary composites. Theories of optical absorption for direct transitions which depend on the using equations:

$$hv = A\ (hv - Eg)^{1/2}$$

where A is refer to the absorption coefficient which related to the nature of the material, (hv) refer to photon energy. The extrapolating $(FR \times hv)^{1/2}$ vs. hv curves to FR = 0 within linear portions showed the value of Eg as shown in Figure 7 and 8. The TiO_2 absorbance occurred below 400 nm, whereas for SWNTs and MWNT broad peaks of absorption can be observed between 450-1000 nm. The absorbance of composites TiO_2/CNT appeared red shifted in absorbance towards the visible light which increases with increase ratios of CNTs. The redshift can be related to an intrinsic property of CNTs which more shifted for composite with SWNT due to higher conductivity as compared with MWNTs.

5. Hydrogen Production

Figure 9, shows the reactor that used for hydrogen production. Briefly a xenon lamp (Osram XBO) with a 1000-WattUV-B light at the 240-1000nm wavelength was use as source of energy which fixed horizontally in the direction of the reactor. The light source was switched on 15 min before irradiation to reach stability for the lamp with light intensity 40 mW cm–2. The reactor vessel consisting of a double-jacket Pyrex-glass reactor (110 mL volume) equipped with a quartz disc for light penetration. Prior to irradiation, Argon gas was purged through the suspension for 30 min. the reactor was cooled to room temperature with a Land Nds. Uni Hancooler system to keep the temperature of reaction at 298.15K. During irradiation, the headspace gas (40 mL) of the reactor was intermittently sampled (0.5 µL) and analyzed for H2 using a gas chromatograph (Shimadzu GC – 8A) equipped with a thermal conductivity detector and a carboxen 1000 packed column. The photocatalytic reactions of hydrogen production from 7.5% of an aqueous methanol solution (0.131mol methanol) were carried out using 65 mg of different binary composites of TiO2 with SWNTs or MWNTs. It is obvious that pristine TiO2 without CNTs did not give any activity for producing hydrogen under dark or illuminated conditions.

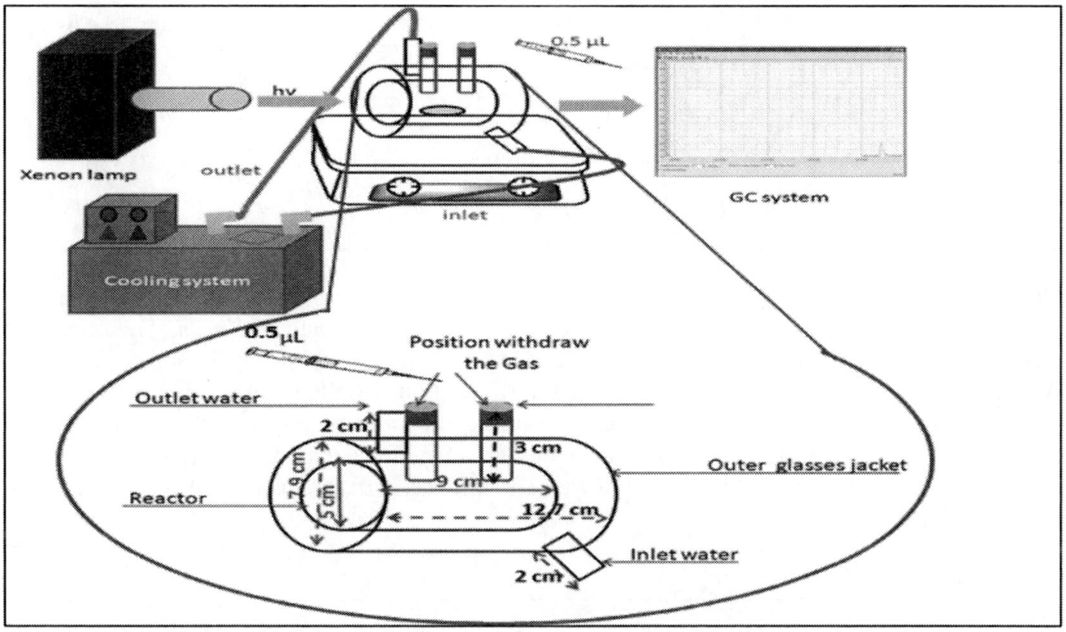

Figure 9. Schematic diagram for the system of hydrogen production.

The work was include studied the effect types and ratio of CNTs. The results were plotted in figure 10 for different binary composites of TiO_2 with SWNTs and MWNTs.

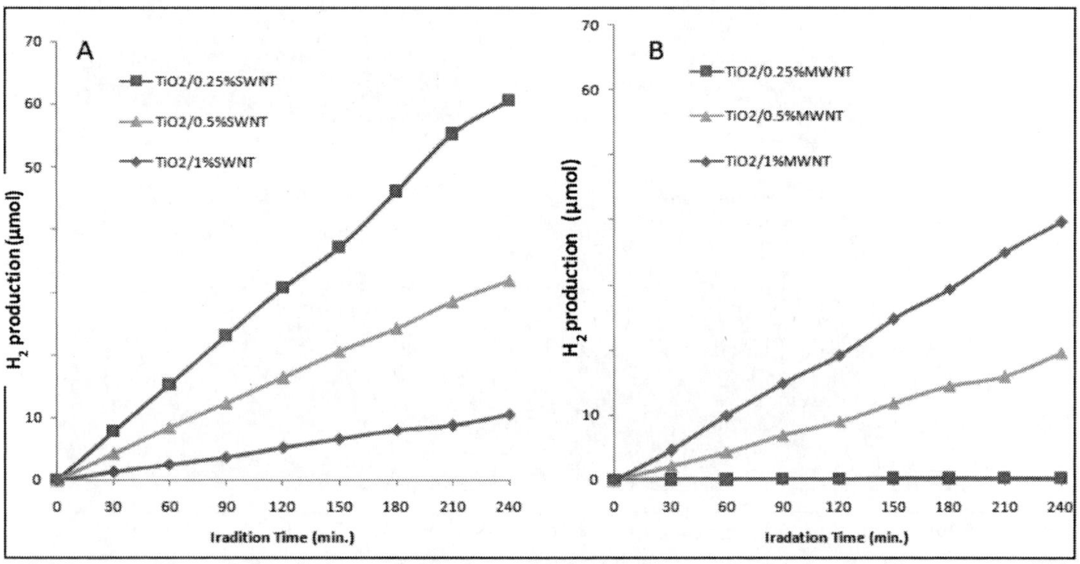

Figure 10. The values of H_2 evolution for 240 minutes from 7.5vol% methanol aqueous solution suspended with 65 mg of a- TiO_2/ SWNTs, b- TiO_2/ MWNTs.

From figure10., and table 2 presence of CNTs directly affect the activities of TiO_2, when producing hydrogen gas in all ratios and types of CNTs. According to the results, CNTs were succeeded to behave as ideal captures for excited electrons causing reduce the recombination of the electrons from conduction band to valence band [18].

Table 2. The rate of H2 production for different ratio of (a)- MWNT, (b)-SWNT with TiO2.

a-%MWNT	k(μmole/min)	b-%SWNT	k(μmole/min)
0.25	0.0007	0.25	0.6381
0.5	0.0363	0.5	0.3366
1.0	0.2297	1.0	0.1081

The influence of CNTs types in the value of emitted hydrogen gas can be related mainly to two causes: the first represent by distribute of SWNTs with TiO_2 more than MWNTs. The second related to lower densities and diameters of SWNTs than MWNTs (1:3) make SWNTs with TiO_2 particles more regular with homogenizing and more darkness [4] than MWNTs as shown in figure 11. The activities of TiO_2/SWNT were reduced with increasing the ratio of SWNTs due to shielding effect which prevents UV-light to reach for the surfaces of TiO_2. This behavior was reduce with MWNTs, however; the increases in H_2 gas were shown with an increase in the ratios of MWNTs, causing reduce in shielding effect.

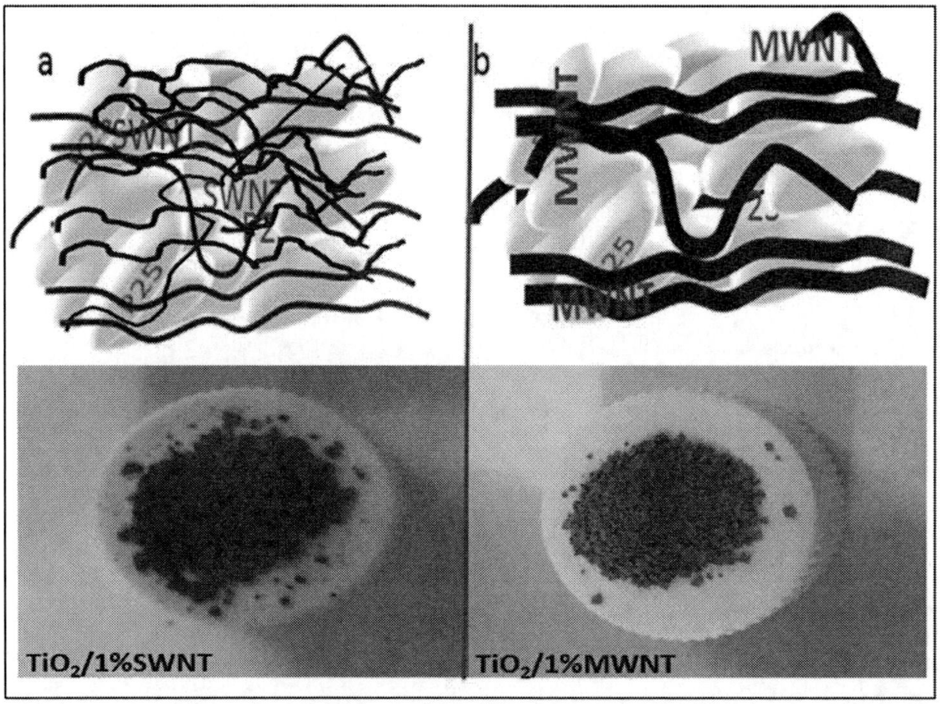

Figure 11. Schematics structures of (a) TiO_2/ 1%SWNTs, (b) TiO_2/ 1%MWNTs.

6. Recovery of Titanium Dioxide.

TiO_2 Degussa P25 physically consists of nano powder which is imposible to remove from solution easily and as mention before there is no activity for hydrogen production. Figure 12, refer to the results which produced from this work when referring to the quantities of hydrogen production efficiency (HPE) with the ability to reuse the catalyst. Under the same experimental conditions, the modified TiO_2 catalyst with CNTs appears to be much easier to separate than pristine TiO_2. The reason behind that can be related to large agglomerate particle precipitation was observed and to complete separation was achieved approximately less than 30 min.

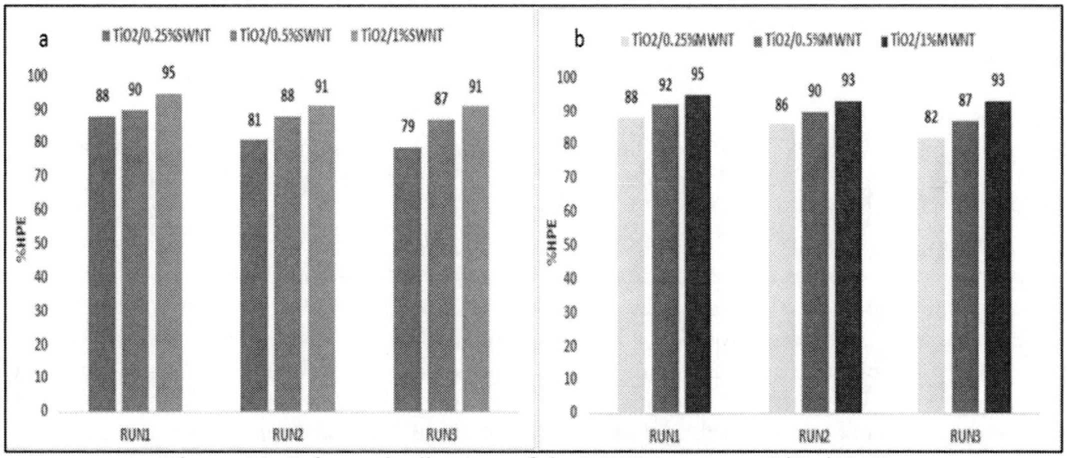

Figure 12. Schematic diagram of the HPE% to reuse TiO_2/CNTs.

7. Mechanism of Hydrogen Production.

The activity of binary composite were plotted in figure 13 which shows maximum rate of hydrogen gas evolved with TiO_2/0.25%SWNT and TiO_2/1% MWNT. The mechanism of activity for TiO_2 with SWNTs or MWNTs depend on transferring the electrons from TiO_2 to CNTs, which agree with reported in many literature [18- 19]. The active site in the binary matrix stimulates under UV-light, to transfer the exited electrons from the TiO_2 surface to the graphitic network of CNTs. Increase photo-catalytic activity produce from synergistic effect between TiO_2 and CNTs and on finding the optimum weight ratios of TiO_2/CNTs to create the best arrangement for the ideal behavior of composites [20].

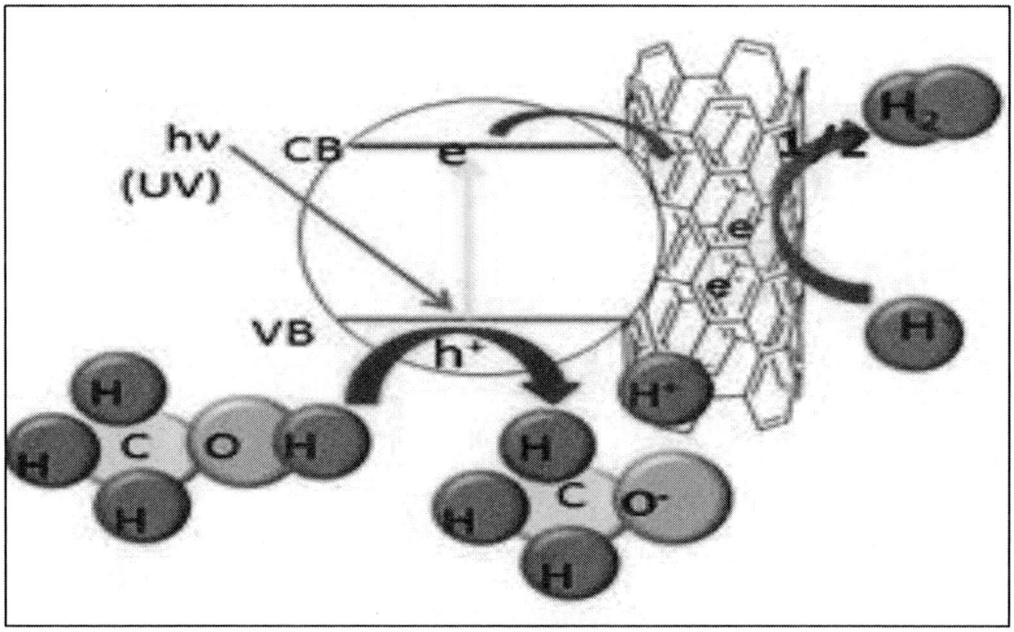

Figure 13. Schematic diagram of the proposed mechanism for the activity of TiO_2/ CNTs.

It should be referred to the work function of $TiO_2 \approx 4.3$ eV [21] and 3.12 eV for the band gap [8-9], while the work function of CNTs is 4.8 eV and band gap is 0.5 eV [22,18]. The composites, a N-/p-type hetero-junction can form an ideal bridge for withdrawing and transferring the electrons from TiO_2

IOP Conf. Series: Journal of Physics: Conf. Series **1032** (2018) 012056 doi:10.1088/1742-6596/1032/1/012056

to a network of carbon nanotubes. The behavior refers to increase the electric charge on the surface of semiconductor oxide with catalysts composite as a result of large surfaces area of higher conductivity CNTs as carrier electrons [23].

8. Conclusions.

Binary composite TiO_2/ CNTs were successfully prepared, by impregnated SWNT and MWNT with TiO_2. The high conductivity and low density of SWNTs enhance the activity of TiO_2 more than MWNTs. Increase the ratios of SWNTs to 1% with TiO_2 reduce the activity of semiconductor due to shielding effect which prevents UV-light reach to TiO_2. The Raman and XRD analysis for TiO_2/ SWNTs show clear change in intensity and area of the peaks than TiO_2/ MWNTs which shows less average crystallite size. The Raman spectroscopy for Anatase and CNT did not change significantly only the Anatase Raman bands for TiO_2/ MWNTs composites are slightly broadened as compared to the pure TiO_2, which in agreement with XRD analysis that shows the decreasing in the average crystallite size. The ideal method of preparation and ratios with types of CNTs can produce the best binary catalyst can behave excellent activities towards hydrogen production reaction.

References

[1] Liao C. H., Huang and C.W. Jeffrey C. S. 2012. Hydrogen Production from Semiconductor based Photocatalysis via Water Splitting, *Catalysts* **2** 490.

[2] Prakash D. V. and Jose A. L. 2017. Review of Hydrogen Production by Catalytic Aqueous-Phase Reforming. *Chemistry Select.* **2(22)** 6563.

[3] Sharif S.A., Barbir F., and Vizroglu T.N. 2003. Principles of hydrogen energy production, storage, utilization. *Journal of scientific and industrial research* **62** 46.

[4] Firas H. A., Falah H. H., Ayad F. A., Irina I,Alexei V. E. and Detlef W. B. 2016. Sonochemical/hydration–dehydration synthesis of Pt–TiO_2 NPs/decorated carbon nanotubes with enhanced photocatalytic hydrogen production activity. *Photochem. Photobiol. Sci.***15** 1347.

[5] Adel A. I. and Detlef W. B. 2014. Photochemical splitting of water for hydrogen production by photocatalysis: A review. *Solar Energy Materials & Solar Cells* **128** 85.

[6] Chi-Hung L., Huang C.W. and Jeffrey C. S. 2012. Hydrogen Production from Semiconductor-based Photocatalysis via Water Splitting. *Catalysts* **2** 490.

[7] Firas H. A.2016. Enhance photocatalytic Activity of TiO2 by Carbon Nanotubes. *International Journal of Chem Tech Research.***9** **(3)** 431.

[8] Hoffmann M. R., Martin, S. T., Choi, W. and Bahnemann, D. W., 1995. Environmental Applications of Semiconductor Photocatalysis. *Chem. Rev.,* **95** 69.

[9] Kumar S.G. and Devi L.G. 2011 Review on modified TiO_2 photocatalysis under UV/visible light: selected results and related mechanisms on interfacial charge carrier transfer dynamics. *J Phys Chem A.* **115(46)** 13211.

[10] Iijima S, 1991. Helical Microtubules of Graphitic Carbon. *Nature* **354** 56.

[11] Kauffman D. R. and Star A., 2008. Carbon nanotube gas and vapor sensors. *Angew. Chem., Int. Ed.* **47** 6550.

[12] Feng W., Feng Y., Wu Z., Fujii A., Ozaki M. and Yoshino K, 2005. Optical and electrical characterizations of nanocomposite film of titania adsorbed onto oxidized multiwalled carbon nanotubes. *J. Phys.:Condens Matter.* **17** 4361.

[13] Fan W, Gao L, Sun J, 2006. Anatase –TiO_2 coated multi-walled carbon nanotubes with the vapor phase. *J. Am.Ceram.Soc.* **89** 731.

[14] Frontera P., Trocino S., Donato A., Antonucci P. L., Lo F. M., Squadrito G., and Ner G. 2014. Oxygen-sensing properties of electrospun CNTs/PVAc/TiO_2 composites. *Electron. Mater. Lett.* **10** 305.

[15] Xie Y., Heo S. H., Yoo S. H., Ali G., and S. O. Cho. 2010. Synthesis and Photocatalytic Activity of Anatase TiO_2 Nanoparticles-coated Carbon Nanotubes. *Nanoscale Res Lett.* **5** 603.

[16] Firas H. A. and Falah H. H.2015. Effects of Nanoparticle Size on Catalytic and Photocatalytic Activity of Carbon Nanotubes-Titanium Dioxide Composites. *Environmental Analytical Chemistry* **12** 2.

[17] Kokhanovsky A , 2007. Physical interpretation and accuracy of the Kubelka–Munk theory. *J. Phys. D: Appl. Phys.* **40** 2210.

[18] Chai B. , Peng T. , Zhang X. , Mao J. , Li K. , and Zhang X. 2013. Synthesis of C60-decorated SWCNTs (C60-d-CNTs) and its TiO_2-based nanocomposite with enhanced photocatalytic activity for hydrogen production. *Dalton Trans.* **42** 3402.

[19] Xu Y. N., Zhuang Y. N., and Fu X. 2010. New insight for enhanced photocatalytic activity of TiO_2 by doping carbon nanotubes: A case study on degradation of Benzene and Methyl Orange. *J. Phys. Chem. C.* **114** 2669.

[20] Khan G., Kim Y. K., Choi S. K., Han D. S., Wahab A. A., and Park H. W.2013. Evaluating the Catalytic Effects of Carbon Materials on the Photocatalytic Reduction and Oxidation Reactions of TiO_2. *Bull. Korean Chem. Soc.* **34 (4)** 1137.

[21] Bickley R., Carreno G. T., Lees J. S., Palmisano L., and Tilley R. 1991. A structural investigation of titanium dioxide photocatalysts. *J. Solid State Chem,* **92 (1)** 178.

[22] Santangelo S., Faggio G., Messina G., Fazio E., Neri F., and Neri G. 2013. On the hydrogen sensing mechanism of Pt/TiO_2/CNTs based devices. *Sensors and Actuators B*, **178** 473.

[23] Kiselev N., Hutchison J., Ryabenko A., Rakova E., Chizhov P., Zhigalina O., Artemov V., and Grigoriev V. 2005. Two structural types of carbon bi-filaments. *Carbon*, **43** 1897.

IOP Conf. Series: Journal of Physics: Conf. Series **1032** (2018) 012057 doi:10.1088/1742-6596/1032/1/012057

Synthesis, Characterization and Antimicrobial Study of nickel(II),palladium(II),platinum(II),rhodium(III), cadmium(II) and zirconium(IV) complexes with (E)-1-(benzo[d]thiazol-2-yl)-4-(hydroxy(2-hydroxyphenyl)methylene)-3-methyl-1H-pyrazol-5(4H)-one

Taghreed M.Musa, Mahmoud Najim Al-jibouri and Redhah I. Hussain Al-bayati
Chemistry department, college of science, university of Mustansiriyah, Baghdad,Iraq.

taghreedmohya@uomustansiriyah.edu.iq
taghreedmohya@yahoo.com
Corresponing authers E-mail: mahmoud_inor71@uomustansiriyah.edu.iq

Abstract. New transition metal complexes of (E)-1-(benzo[d]thiazol-2-yl)-4-(hydroxy(2-hydroxyphenyl)methylene)-3-methyl-1H-pyrazol-5(4H)-one was prepared and their structures and formulas were confirmed by elemental analyses, EIMS ,FT-IR and NMR spectra .The ligand (LH) was prepared through the ring closure of 2-hydrazino benzothiazole and 3-acetyl-4-hydroxy cumarine in dry methanol. The complexes of nickel(II), palladium (II), platinum(IV), cadmium(II), rhodium(III) and zirconium(IV) were isolated in solid state up on the direct reaction of their hydrated chlorides or nickel(II) nitrate with the free ligand solution. The solid structures of metal complexes were supported on the basis of elemental analyses, IR, UV-Visible spectra, as well as the molar conductivity and magnetic susceptibility. The thermal analyses of TG-DSC were done for the LH ligand and their rhodium (III) and palladium(II) complexes in argon atmosphere in order to conclude their structures and thermal stability and the thermo grams observed for the studied compounds have revealed the proposed structures. The data found from preliminary studies revealed that the nickel(II), palladium(II) and platinum(II) were square planner geometry in[$MLCl(H_2O)$] formula, M= Ni, Pd, Pt(II) , whereas the rhodium(III) ,zinc(II) and cadmium(II) were octahedral and tetrahedral geometry respectively in the formulas, $[RhL_2(H_2O)_2]Cl$ and $[MLCl(H_2O)]$, M=Zn and Cd(II) ions. However the zirconium(IV) complex was square pyramid with five-coordination number,$[ZrOL_2]$.The H NMR spectra of the cadmium(II) and platinum(II) complexes exhibited remarkable changes in the peaks of –NH- pyrazolinone ring and much less shifts in the other functional groups of phenolic and aromatic moiety indicating that the the active sites of coordination with the metal ions are the nitrogen atom of benzothiazole ring and the hydroxyl group linked at C3 at C3 of pyrazole ring. The biological activity of the solutions of the prepared compounds in DMSO as control was estimated against some positive and negative gram bacteria and the inhibition zone measured by diffusion method showed that the metal complexes have greater activity in their spectrum of antibiotics in compared with the employed standard drug in 20 ppm concentrations.

1. Introduction.

The pyrazoline based ligands are considered as main type of biologically important nitrogen containing heterocycles, which have received considerable attention in coordination chemistry [1-3]. The interesting subjects of transition elements involved the 2-pyrazole and benzothiazole derivatives have been exhibited many antimicrobial properties especially their chelates with copper (II), palladium (II) and platinum (II) [4]. The isomerization of ruthenium (II) complexes with 5-phrenyl-3-(2-pyridyl)-1H-pyrazole have also interested in supra molecules due to their sensitization with light energy employed in synthesis [5]. The rare earth metal complexes of 2-pyrazoline ligands endowed with anti-

IOP Conf. Series: Journal of Physics: Conf. Series **1032** (2018) 012057 doi:10.1088/1742-6596/1032/1/012057

HIV activity [6,7]. As well as the luminescence properties of aluminum (III) and copper (II) chelates involving pyrazoline-thiazole based ligand have studied and showed considerable applications in synthesis of fluorescent compounds [8,9]. The presence of pyrazoline moiety besides other heterocyclic rings like benzothiazole and which have wide applications in the field of colorimetric determination of heavy metals and light-diodes emitting industry.

2. Experimental section.

2.1. Instrumentation.

The identification of the synthesized ligand with it's metal complexes were carried out with the help of micro-elemental analyses using Carlo-Erba 1106 Elemental analyzer. The electronic spectra were recorded using Shimadzu spectrometer in the range (200-1000) nm in DMF solvent. The 1H and 13C NMR spectra were carried at Albait University, Jordan on Bruker 300 MHz spectrometer in DMSO-d6 solvent. The Fourier transform infrared spectra of the prepared complexes were recorded in KBr and CsI discs using Shimadzu model FT-IR 8400 spectrometer at the Laboratories of Chemistry Department, College of Science, university of Mustansiriyah, Iraq. The molar conductance measurements were made on an Hanna conductivity with a cell having a cell constant of 1.1 cm-1 in DMSO solutions. The proposal structures and formulas of the metal complexes were also confirmed by measuring the metal content (M%) with flame atomic absorption spectroscopy on Shimadzu A.A-670 spectrometer at instrumental analyses laboratory, Chemistry Department, university of Mustansiriyah, Baghdad, Iraq.

2.2. Materials.

The hydrated metal chlorides $RhCl_3·6H_2O$,$PdCl_2·6H_2O$,K_2PtCl_4,$CdCl_2$, $ZrOCl_2.8H_2O$ and Ni(NO3)2·6H2O were purchased from Sigma-Aldrich company and used without purification. The 3-acetyl-4-hydroxy cumarine and 2-mercapto-benzothiazole were supported by Sigma-Aldrich Company (99% purity). The 2-hydrazino benzothiazole was prepared according to the method reported in literature [13]. The molar ratio of metal to ligand was carried out according to the modified method in literature [14] as mentioned.

A series of solutions containing constant volume (5 ml) of the metal ion solution, prepared in the optimum concentration (10-3 M), and variable volumes of the ligand solution prepared in the same concentration (10^{-3} M) were mixed together. The absorption spectra were measured and M:L ratios were obtained by plotting the absorbance versus the mole ratio of the reactants, two straight lines of different slope are obtained the intersection occurs at a mole ratio that corresponds to the combining ratio in the complex.

2.3. Methods.

2.3.1. Synthesis of (E)-3-(4-(hydroxyl (phenyl) methylene)-3-methyl-5-oxo-4,5-dihydro-1H-pyrazole-1yl)quinoxalin-2(1H)-one. The free ligand (LH) was prepared according to the procedure established in literature [15] as shown below.

2-Hydrabenzothiazole (1.65 g, 10 mmole) in dry methanol (20 mL) was added to a solution of 3-acetyl-4-hydroxycumarine (2.053 g, 10 mmole) in methanol (15 mL) and the mixture was refluxed for 3 h. The solid crude product formed was filtered off, dried in air and recrystallized from hot methanol as a pale yellow solid.

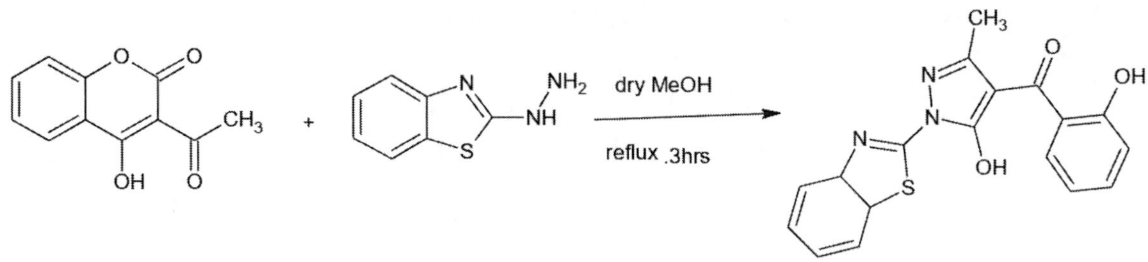

Scheme 1.

Yield: 0.377 g, 65 %.
M.p.: 198-200 °C.
Rf: 0.75 (ethanol:hexane, v:v, 2:1).
FT-IR (KBr, ν, cm−1): 3300-3670(Ar-OH,-C=C-OH), 1640(C=O, Pyrazoline),1630(-C=N-Pyrazole),1390(C-N,N-N),650-670(C-S)
Proton NMR (300 MHz, DMSO-d6, δ, ppm): 2.66 (s, 3H, CH3), 6.2(t, 1H, HC=C-), 6.50-7.09 (m, 3H, J = 3 Hz, Ar-H), 7.70-8.09 (m, 4H, Ar-H), 12.00 (s, 1H, HN-C-OH pyrazole).
MS (m/z (%)): 352 (100) [M1+], 351 (90) [M+]

2.4. Synthesis of the metal complexes.
A methanolic solution (50 mL) $Ni(NO_3)_2\cdot6H_2O$ (0.01 mole, 0.455 g), $CdCl_2$ (0.01 mole,0.183 g) and K_2PtCl_4 (0.01 mole,0.415 g) was added with stirring to an ethanol solution of the ligand (0.01 mole,0.253 g) followed the adjusting the pH of the mixture to 8.0 by addition (1 ml) of 5% KOH solution then refluxed the mixture on water bath for 2 hours except the platinum(II) and palladium(II) which required stirring on water bath (7O 0C) for 30 minutes and removing the side product of potassium chloride led to afford orange to brown crystals.. The obtained coloured solids were separated by filtration under suction, washed with methanol and dried in air overnight then recrystallization from mixture of acetonitrile: absolute ethanol to get purified coloured solids. The complexes of rhodium(III) and zirconium(IV) were prepared in molar ratio 2:1 (LH:M) according to the above mentioned method except ZrOL complex which was prepared by precursor the starting ZrO_2+ by mixing sodium acetate solution of 10%(w/v) in deionized water with (1.22 gm,0.023 mole) of $ZrOCl_2.8H_2O$. The filter containing $ZrO(OAC)_2$ then was dealt with the hot methanolic solution of LH (0.02 mole, 0.316 gm). The mixture was stirring on water bath for 6 hours then kept overnight at room temperature and the separated crude filtered, washed several times with methanol, distilled water and dried in oven to afford white off powder of [ZrOL2] complex, Table(1).

2.5. Biological Activity.
Four types of bacteria strains were used, Gram Negative bácteria, Escherichia Coli(E-Coli), proteus Mirabillus(pro.) , and Gram Positive bacteria, Staphylococcus aureas (Staph), Bacillus(Ba.). These bacteria were diagnosed and cultured on Eosin Methylene Blue (EMB) and Muller Hinton Agar medium for use in this experiment, and in measuring the deactivating capacity of the prepared compounds. The deactivating capacity agent of the isolated bacteria of these compounds was tested by using the holes method. In this method the holes were saturated with 100μm of (20 ppm) of the given compound and left for about 15 min in order to spread on the medium then incubated at 25oC for 24 hour .The deactivation diameters were measured by special ruler designed for this purpose.

3. Results and discussion.
The prepared metal complexes in the recent work were non-hygroscopic (stable at the room temperature) and in the form of amorphous solids. The solubility of the metal complexes were tested in common organic solvents and it is obtained that all complexes were sparingly soluble in methanol,

ethanol and acetonitrile whereas their solubility was high in DMSO and DMF. The results obtained from (CHNS) analyses are in well-agreement with their calculated values then confirm the expected structures and molecular formulas. As well as the molar conductance data of complexes solutions in DMF showed no electrolytic properties except the rhodium(III) complexes which was conductive in the (70 S.cm-1.mole-1), Table 1. The two different tautomeric forms (LA and LB) [10] can be envisaged for ligand of these, A and B are stabilized by intramolecular hydrogen bonding, Scheme(2).

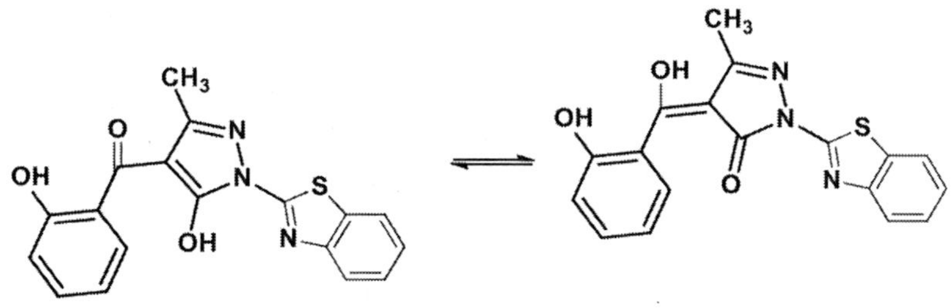

Scheme 2. Tautomeric forms of LH ligand

However, the Job method was applied to reach the optimum conditions for preparation of metal complexes in solution. Here, the figure(1) displays the formation of Rh(III) complex in 2:1 molar ratio LH/Rh III in ethanol solution with adjusting the PH of medium to 8.50 which assists in deprotonation the ligand LH and gives anionic L1- form that is more capable of binding through enol form HC=C-OH moiety and –C=N- of pyrazole ring.

3.1. IR spectral study.

The important IR spectral bands of ligand and corresponding complexes along with assignments are presented in Table 2. The absence of a band in the region 1735 cm-1 related to lactone starting material[15] indicates the attack of HN- moiety on the carbonyl of cumarine in 2-C position [13,15]. The strong absorptions at 1640, 1630, 1480 and 3500-3670 cm^{-1} could be assigned to lactone -C=O, -N=C-, C-N and –phenolic -OH [5,13]. However, the appearance of strong absorption in the free ligand at 970-1020 $cm^{-1}$and 1390-1370 cm^{-1} reveals the –C-N and N-N modes of pyrazoline ring. As well as the medium intensity band at 650 cm-1 reveals the C-S bond of thiazole ring [11,15]. In the complexes, it is obviously that remarkable shifts to lower frequency region (1648-1677 cm^{-1}) and (1600-1550) cm-1 due to the formation of M-O and M-N bonds since the drift of lone pairs on oxygen and nitrogen atoms moved to vacant orbitals of the metal ions ,Figures(1, 2 and 3). In the spectra of all metal complexes, a weak bands were appeared in the far-infrared regions 312-388 cm-1, this may be attributed to (M-Cl)bond and support the participation of chloride ions in inner sphere of metal complexes structures [18]. Furthermore, the nickel(II) complex displayed medium band around 1490 and 970 cm-1 which are characterized to the asymmetric and symmetric stretching of –N=O bond of nitrate –NO3- present in the inner sphere of Ni(II) complex suggesting the linkages of Ni-N=O via nitrogen atom of the nitrate anion. The bands due to ν(NH) are broadened in all the complexes due to the overlapping of ν(OH) bands of enol form of the ligand to H-N-of pyrazoline ring [11,17]. The new bands in the far-IR spectra for all metal complexes in the 500-560 cm-1 and 422-460 cm^{-1} regions are assigned to ν(M–N) and ν(M–O), respectively [18]. The complexes exhibited a broad band in the region (3400-3600) cm^{-1} that are assigned to coordinated water molecules [13,15]. The zirconium(IV) complex showed strong absorption around 890 cm-1 assigning to Zr=O bond [16].

3.2. Mass spectra study.

The absorptions of ionization impact mass spectra of the LH recorded a main peak at M^{+} = 352due to the molecular ion related to free ligand LH that is remarkably gives strong evidence for ring closure of hydrazine-intermediate formed via condensation of acetyl moiety of cumarine with amino one of hydrazine benzothiazole, Scheme 2. The other peaks at 285 and 167 were assigned to cleavage of benzothiazole moity and phenoxy groups. The other peaks in the MS spectrum may have ascribed to the breaking of weak points of methyl, hydroxyl and phenyl groups in the chemical structure of the ligand.

3.3. Electronic spectra and magnetic moments.

The UV-Vis spectrum of the ligand (LH) in methanol within the range (200-1000) nm displayed mainly three peaks. These transitions at 220,370,470 could be ascribed to the moderate energy $\pi\rightarrow\pi^*$transition of the aromatic rings. Furthermore, the sharp (λmax) peaks at (370,470 nm) were related to the $\pi\rightarrow\pi^*$ transition of intermolecular charge-transfer taken place through the azo group (-C=N-, -C=C and |-C=O), [10,15]. The pale coloured solution of zirconium (IV) and cadmium complexes showed high intensity absorptions around (240-390) and (322-470) nm, these absorptions could be assigned to LMCT of L→ Cd and Zr=O respectively[18]. The brown solution of nickel (II) complex in DMF showed two d-d transitions around 498 and 509nm that are assigned to the $^1A_1g\rightarrow {}^1B_1g$ and $^1A_1g\rightarrow {}^1B_2g$ respectively, which are typical transitions for square planner Ni(II) complexes[18]. The pale red solutions of palladium(II) and platinum(II) complexes in DMF solvent recorded low intensity peaks in regions (470-400) and (480-520) nm assigning to $^1A_1g\rightarrow {}^1B_1g$ and $^1A_1g\rightarrow {}^1B_2g$ respectively [18,19] thus ascribing the low spin square planner environments around Pd(II) and Pt(II) ions[20]. However, the absorptions of Rh(III) complex in DMF at 326, 562 and 889 and 630 confirms the octahedral rhodium(III) complexes [21].

3.4. ^1H NMR study.

The ligand LH displays two sharp singlets at 12.00 ppm attributed to the resonance of de shielded protons of pyrazole ring N-C-OH in 5-position of pyrazole ring. As well as the proton of -OH group attached to phenyl group that mainly belonged to cumarine derivative was appeared at 7.41 ppm as singlet peak which support the ring closure occurred by attaching of lone pair of -NH- on the carbon atom C-O of lacton moiety [11,13].However the absorption of aliphatic $-CH_3$ was observed in the region 2.10-2.50 ppm as singlet peak which is strong proof for the keto-enol forms of the free ligand in solution.

3.5. Effect of pH and Concentration on Pd(II)-L Solutions.

The UV-Vis spectra of Pd(II)-ligand mixed solution with a concentration range of (10^{-3}-10^{-5}) M were studied. The concentration of 10^{-4} M was obeyed the Lambert-Beer's law and showed a clear intense color. A straight line, with correlation factor (r = 0.98), was obtained when the absorbance plotted against the molar concentration. The data obtained were tabulated in Table (4).The optimum concentration was chosen for complex solutions giving rise to a constant (λmax) at different pH values. Table (4) includes the absorbance and (λmax) of the complex solution at different pH and concentrations. It is obvious that the concentration of ($0.75\ x10^{-4}$M) for Pd(II) solution are subjected to be the optimum concentration. The influence of the pH was studied at pH range (3.5-8.5). The absorbance-pH curve was carried out at certain (λmax) and several concentrations. The curve exhibit two broad plateaus at pH range (4-5) and (7.5-8.5) represent the presence of two species in the solution at these two pH ranges. The descending part of the curves may represent the dissociation of the two complexes formed at more basic medium. However, the results suggest pH 4.5 and 8 as an optimum pH for the two formed species in solution. At the same time the descending part of the curve may represent the dissociation of the formed complex at more basic medium.

3.6. Thermal analyses (TG-DSC).

The thermo gram curve of LH at temperature range (100-595.5) ^{0}C showed one decomposition step. The TGA peak observed at 595.5^{0}C indicated the loss ($C_{11}H_{10}N_3O_2S$) fragment (det.=11.38mg.71.17%, calc.=11.24mg) Figure(4). The endothermic a peaks may be indicate combustion of the organic ligand in an argon atmosphere. The TG curves of $[RhL_2(H_2O)_2Cl]$ complex displayed thermal decomposition with loss weight percent (9.85mg,46.9%, calc.=10 mg) in the temperature range (100-324.37)°C indicating the cleavage of chloride ions present in the outside the coordination sphere and two coordinated molecules of water in rhodium(III) complex. As well as the degradation of the organic molecule started before the temperature 324°C. The second step of TG curve exhibited (4.55mg, 21.68.9%, calc4.45mg), Figure(5).The thermogram of [PdLCl(H_2O)] is shown in Figure(6). The TGA peak at 225.49°C assigned to the loss of ($C_9H_9N_3OSCl$) (det.= 9.8mg, 46.53%, calc.= 9.9mg). The second step recorded at 326.58 °C(10.43mg, 49.7 %, calc10.27mg), The differences between the observed and calculated may be related to sublimation process. The DSC curve indicated endothermic decomposition processes [15, 18] at temperatures 248.7°C.

3.7. Biological Activity.

The screening of cytoxicity of the prepared compounds against some negative and positive bacteria were carried out via diffusion method [23] and the results are presented in Table (6).The results of the present study show that the complexes have a relatively strong deactivating capacity against the G+staph. and G-pro.Specimen of bacteria, while different deactivating capacity was recorded for G+Bacillus and G-E.Coli specimen of bacteria. However, introduction of metal in complexes increase the deactivating capacity of the ligand [23,24]. For a comparison between the biological activity of Ciproflaxine and the ligand and its complexes which were prepared by us, it was noted that the effectiveness ligand and its complexes was the biggest ones of the standard drug in (20 μg/ml) and it is observed that cadmium(II) and platinum(II) complexes showed the maximum inhibition for growth the micro-organisms understudy[24].

Table 1. The physical properties and elemental analysis of the prepared metal complexes.

Compound	Color	M.p. (°C) [a]	C% Calculated (Found)	H% Calculated (Found)	N% Calculated (Found)	S% Calculated (Found)	M% [b]
LH	Pale yellow	200-202	64.22 (63.60)	5.11 (4. 88)	14.77 (16.90)	10.10 (10.1)	-
[NiL(NO_3)(H_2O)]	Brown	298	44.02 (43.88)	3.28 (3.03)	11.41 (11.59)	6.53 (6.26)	11.95 (11.19)
[PdLCl(H_2O)]	Brown	277	42.20 (41.88)	3.15 (2.01)	8.20 (9.03)	6.26 (6.11)	20.77 (19.51)
[PtLCl(H_2O)]	Orange	301	35.98 (46.81)	2.68 (2.90)	6.99 (8.11)	5.34 (4.90)	32.46 (31.80)
[$RhL_2(H_2O)_2$]Cl	Green	299	49.18 (45.72)	3.67 (3.19)	9.56 (9.90)	7.29 (7.11)	11.70 (10.86)
[$ZrOL_2$]	White off	290-292	53.25 (51.88)	3.48 (2.47)	10.35 (10.80)	7.90 (6.99)	11.23 (1.83)
[CdLCl(H_2O)]	yellow	292dec	41.71 (40.56)	3.11 (2.90)	8.11 (8.50)	6.19 (6.02)	21.69 (20.11)

[a] Dec: Decomposed.
[b] Content of metal was done by flame atomic absorption spectroscopy.

IOP Conf. Series: Journal of Physics: Conf. Series **1032** (2018) 012057 doi:10.1088/1742-6596/1032/1/012057

Table 2. FT-IR absorptions of the LH and its metal complexes.

Compound	νNH,OH	νC=O	νC=N	νC-H	νC-N	νM–N	νM–O	νM-Cl
LH	3383(br)	1666 (s)	1624 (s)	2947, 2862	1377(s)	-	-	-
[NiL(NO3)(H2O)]	3400(br)	1655	1590 (s)	3050, 2964	1370(s)	510	460	1300,950, N=O
[PdLCl(H2O)]	3300	1640	1577 (s)	3010, 2987	1377(m)	530	440	322(w)
[PtLCl(H2O)]	3500(br)	1648	1580 (s)	3010, 2967	1389(m)	522	450	374(w)
[RhL2(H2O)2]Cl	3460(br)	1654	1616 (s)	3010, 2977	13929w)	550	459	322(w)
[ZrOL2]	3519(br) 3400(br)	1659 1650	1600 (s) 1578(s)	3010, 3000 3000.2963	1375(m) 1372(w0	500 502	438 432	890 (Zr=O)

s = strong, m = medium, br = broad, w = weak.
s = strong, m = medium, br = broad, w = weak.

Table 3. The electronic spectra and molar conductance of the prepared complexes.

Compound	UV-visible, λ (nm)	μ (BM)	Λm (S.mol-1.cm2) *	Geometry
LH	220,370,470(CT)	-	-	-
[NiL(NO3)(H2O)]	300, 498,509	0.00	11	Square planner
[PdLCl(H2O)]	280,470,400	0.00	17	Square planner
[PtLCl(H2O)]	350,480,520	0.00	10	Square planner
[RhL2(H2O)2]Cl	326,562,889	0.00	70	Octahedral
[ZrOL2]	240,390(CT)	0.00	10	Square pyramid
[CdLCl(H2O)]	322, 470(CT)	0.00	12	Tetrahedral

* Λm = molar conductance was measured in DMF solutions.

IOP Conf. Series: Journal of Physics: Conf. Series **1032** (2018) 012057 doi:10.1088/1742-6596/1032/1/012057

Table 4. Absorbance at λ_{max} for molar concentration of Pt (II)-Ligand complex solutions at pH = 8.5.

Molar Conc. × 10^{-4}	Absorbance
0.25	0.232
0.50	0.443
0.75	0.695
1.00	0.940
1.25	1.150

Table 5. TGA / DSC data of LH ligand and some metal complexes.

Compound	Decomposition step	TG Temperature	Removes species	Wt. loss (Calc.)mg	Wt. Found mg
[LH]	1st	71-595.5	($C_{11}H_{10}N_3O_2S$)	11.24	11.38
$C_{18}H_{15}N_3O_3S$	(Residue)	Over 595	(C_7H_5O)	4.76	4.62
[$RhL_2(H_2O)_2Cl$]	1st	100-324.37	($C_{18}H_6N_4O_2S_2Cl$)	10	9.85
$C_{36}H_{32}N_6O_8S_2RhCl$	2nd	324.3-595.2	($C_{12}H_8N_2O_2$)	4.55	4.55
	Residue	600	($C_6H_8N_2O_4Rh$)	6.55	6.60
[$PdLCl(H_2O)$]	1st	115-225	($C_9H_9N_3OSCl$)	9.9	9.8
$C_{18}H_{16}N_3O_4SPdCl$	2nd	225-326	$C_9H_7O_3Pd$	10.27	10.43
	Residue	595	Decomposition of complexes		

Table 6. Diameters (cm) of deactivation of bacteria of the LH and it's complexes.

No.	Compound	G+(Staph.)	G+(Ba.)	G-(E-Coli)	G-(Pro.)
1	**DMSO**	0.0	0.0	0.0	0.0
2	**LH**	0.5	0.55	0.72	0.76
3	**$NiLNO_3(H_2O)$**	0.78	0.74	1.00	2.6
4	**$PdLCl(H_2O)$**	0.92	1.2	1.21	2.9
5	**$PtLCl(H_2O)$**	0.99	1.90	1.23	1.7
6	**[$ZrOL_2$]**	0.92	1.01	0.73	0.5
7	**$CdLCl(H_2O)$**	0.99	1.80	1.09	1.0
8	**Ciproflaxine**	0.8	2.5	0.66	0.90

IOP Conf. Series: Journal of Physics: Conf. Series **1032** (2018) 012057 doi:10.1088/1742-6596/1032/1/012057

4. Conclusions.

According to the results obtained from elemental analyses, proton NMR, EIMS,IR and UV-Visible spectra, the complexes of nickel(II),palladium(II) and platinium(II) with the new LH pyrazoline based-ligand were square planner with diamagnetic properties. However, the rhodium(III) complex was formed as octahedral geometry in 2:1 molar ratio L/M in basic medium, whereas, the structure of cadmium(II) complex was tetrahedral in 1:1 mole ratio. Furthermore, the zirconyl complex was square pyramid with [ZrOL2] formula due to the molar ratio 2:1 of ligand LH to metal Zr(IV) in methanolic solution. The vibration frequencies of the metal complexes were compared with the IR spectra of free LH ligand and it is found that LH behaved as monobasic bi dentate Lewis base via nitrogen of thiazole ring besides acidic –OH in 3-C position of pyrazolinone moiety. As well as the ligand and rhodium(III) and palladium(II) complexes were confirmed their thermal stability with the help of TG-DSC analyses. The scheme 3 shows the proposed geometry of the prepared complexes. The solutions of LH and it' complexes in DMSO were screened their biological activity against some bacteria and it is found that the complexes of cadmium(II) and platinum(II) have the greatest activity due to the lipophilicity of their structure which make them more penetrate to cell wall of these bacteria.

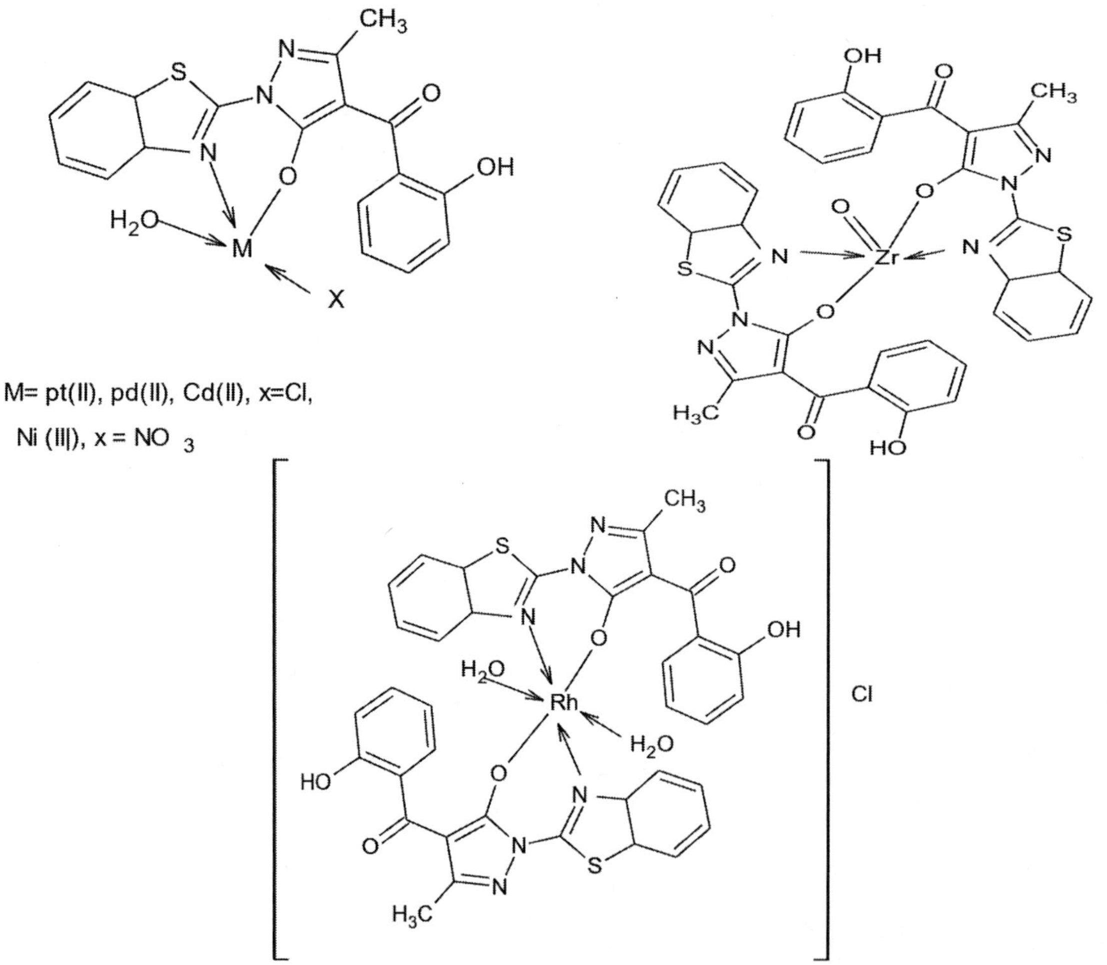

Scheme 3. Square planner on Ni(II), Pd(II), Pt (II) , square Zr(IV), octahedral Rh(III) and tetral hydra of Cd(II) complexes.

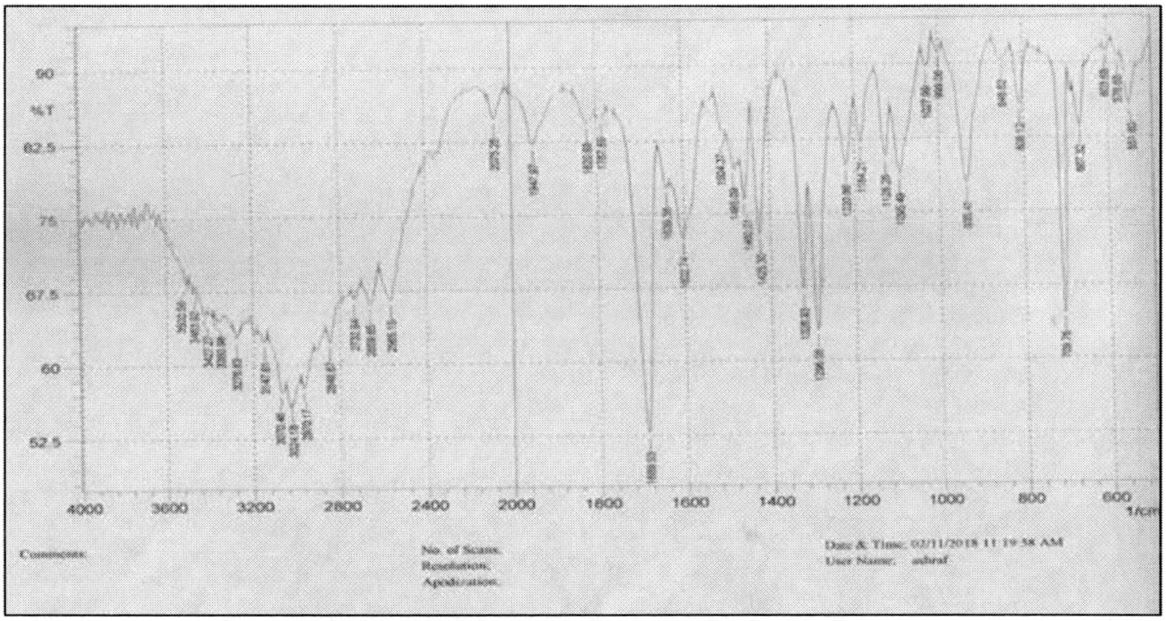

Figure 1. FTIR spectra of LH complex in CsI disc.

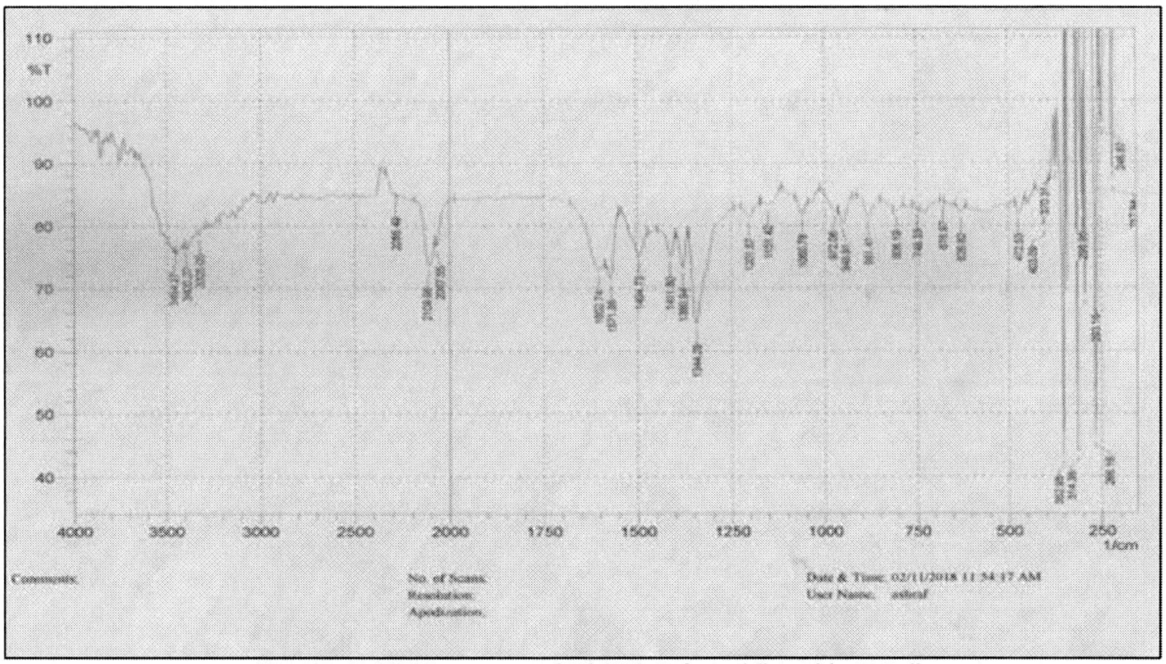

Figure 2. FTIR spectra of Rh(III)-L based ligand in CsI disc.

IOP Conf. Series: Journal of Physics: Conf. Series **1032** (2018) 012057 doi:10.1088/1742-6596/1032/1/012057

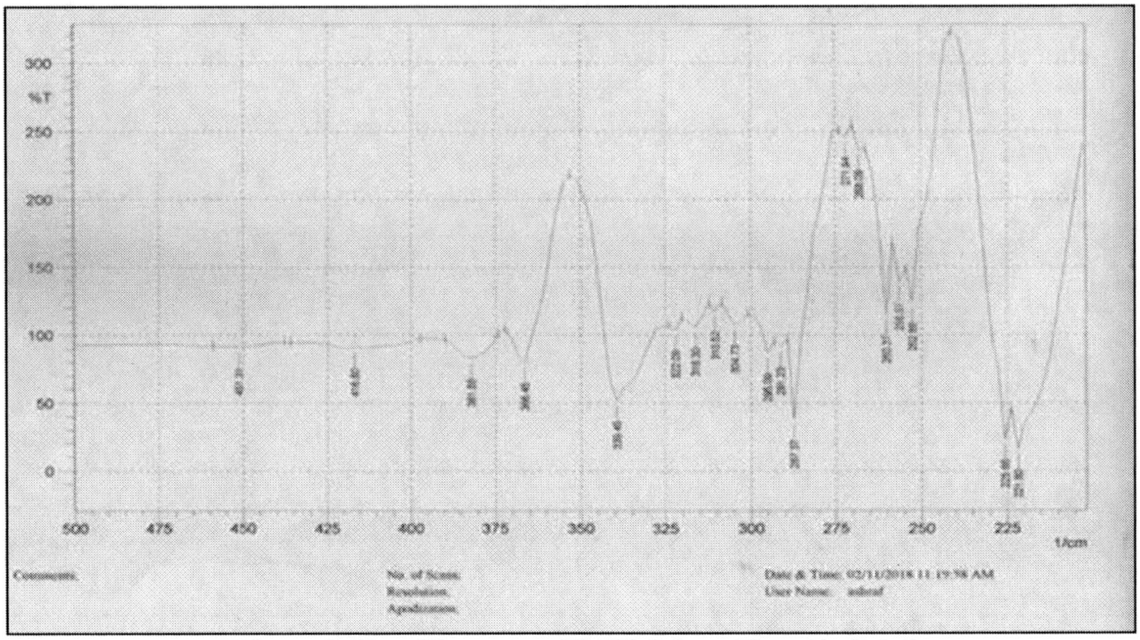

Figure 3. Near IR spectra of Pt(II)-L based ligand in CsI disc.

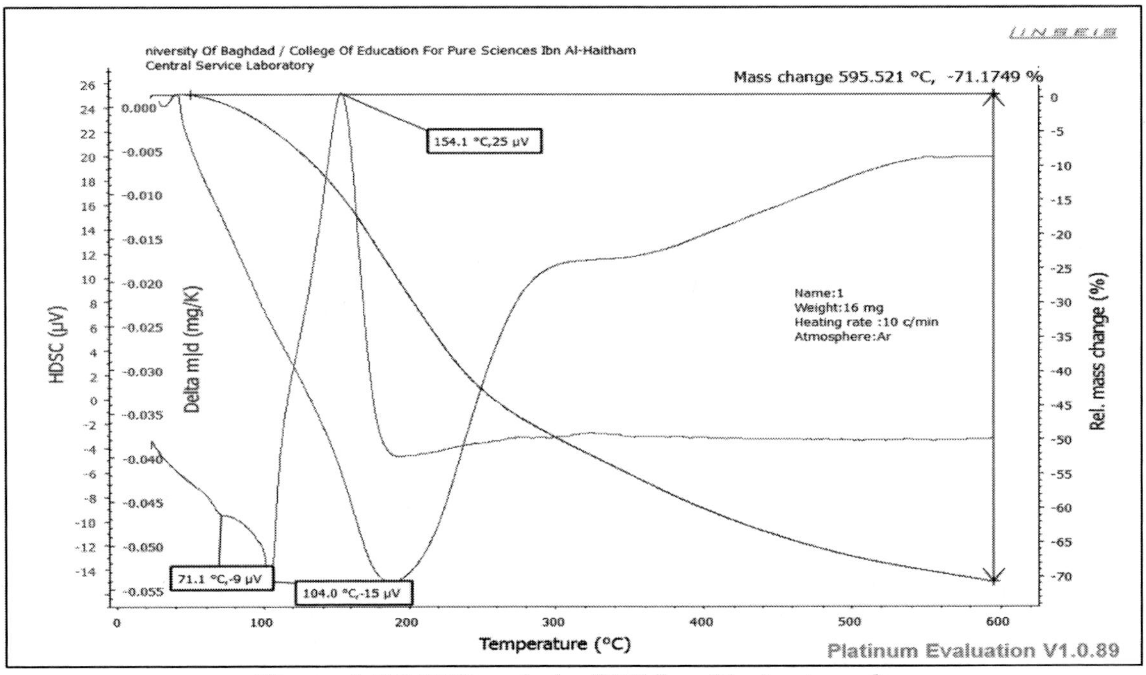

Figure 4. TG-DSC analysis of LH ligand in Ar atmosphere.

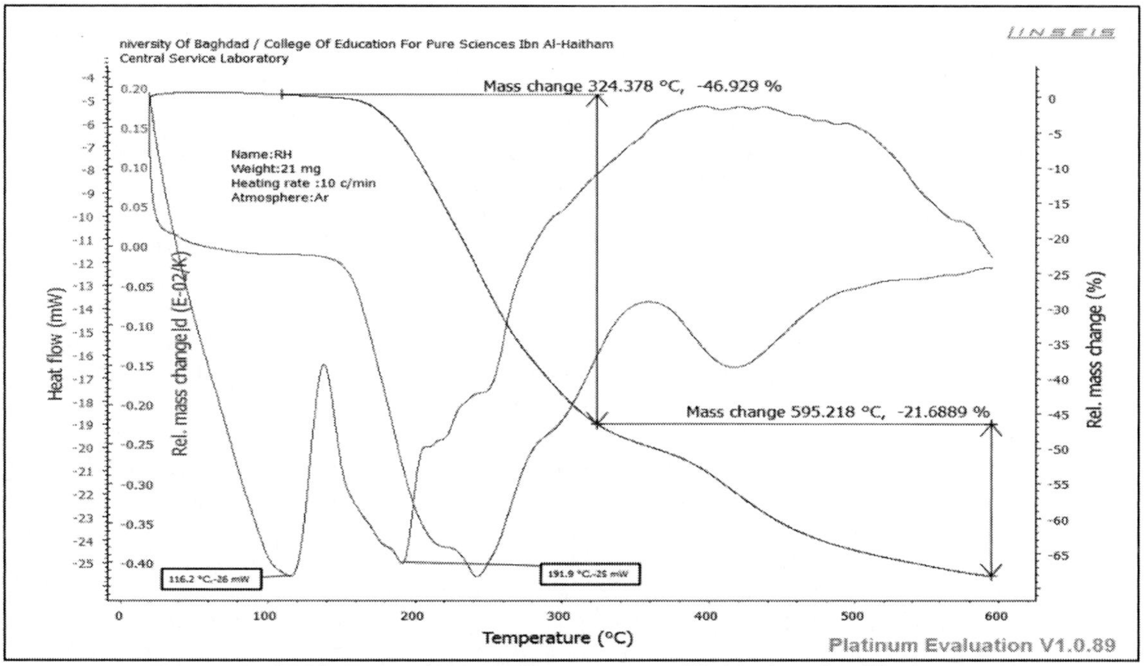

Figure 5. (TG/DTG and DSC)Thermo gram of $[RhL_2(H_2O)_2Cl]$in argon gas.

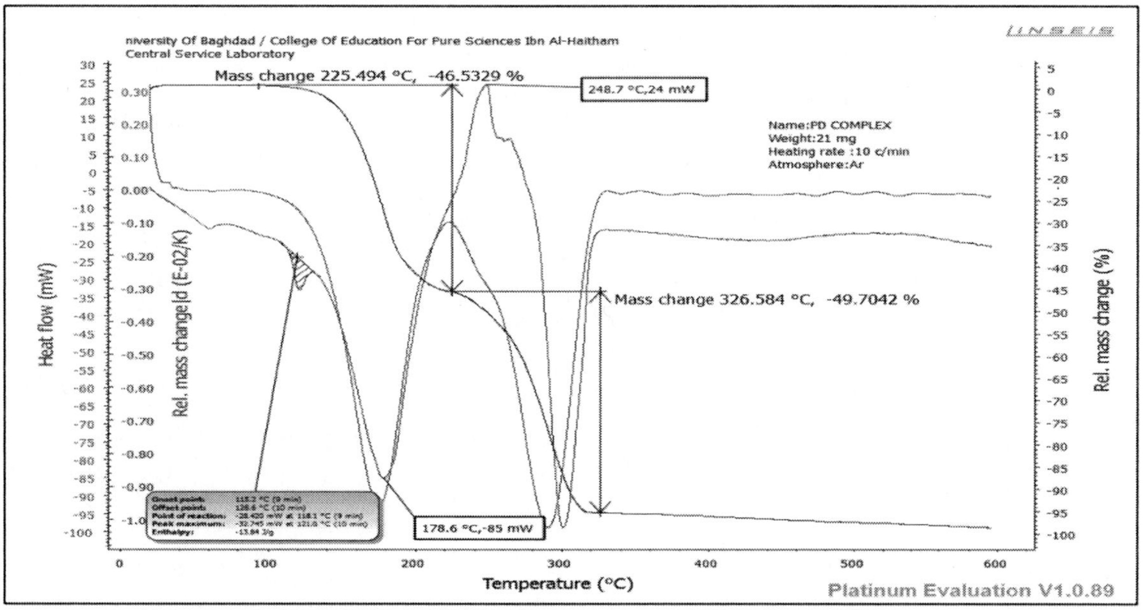

Figure 6. (TG/DTG and DSC)Thermo gram of $[PdLCl(H_2O)]$ in argon gas.

Acknowledgement

The authors thank Department of Chemistry, College of Science, university of Mustansiriyah for providing spectra and analytical facility. Recording of magnetic susceptibility and elemental analyses are gratefully acknowledged. Furthermore, the authors also thanked Albait University (Jordan) for facilitating the NMR analyses.

IOP Conf. Series: Journal of Physics: Conf. Series **1032** (2018) 012057 doi:10.1088/1742-6596/1032/1/012057

References

[1] L.D. Fader, R. Bethell, P. Bonneau, M. B€os, Y. Bousquet, M.G. Cordingley,R. Coulombe, P. Deroy, A.-M. Faucher, A. Gagnon, N.Goudreau, C. Grand- Maître, I. Guse, O. Hucke, S.H. Kawai, J.-E. Lacoste, S. Landry, C.T. Lemke, E. Malenfant, S. Mason, S. Morin, J. O'Meara, B. S imoneau, S. Titolo and C. Yoakim 2011 *Bioorg. Med. Chem. Lett.* **21** 398-404.

[2] M. Abid, A.R. Bhat, F. Athar and A. Azam 2009 *Eur. J. Med. Chem.* **44** 417–425.

[3] F. Meyer, A. Jacobi, B. Nuber, P. Rutsch and L. Zsolnai 1998 *Inorg. Chem.* **37** 1213-1218.

[4] Elzbieta B.;MaleckaM.;BernhardK.;Vladimir B.;Grzegorz A.;UrszulaK and Marek R. 2007 *Eur.J.Inorg.Chem.* 3728-3735.

[5] Ghaida K. Neama, 2017 Synthesis, Characterization and Biological Study of New Melas Complexes Derived from Pyrazole Ligands, M.Sc. thesis, University of Mustansiriyah.

[6] Bekhit AA and Abdel-Aziem T. Design 2004 *Bioorg Med Chem.* **12** 1935-45.

[7] Zhou Y., WangF., KimY. and Kim S.J. 2009 *Org.Letters.* **11** 4442-4445.

[8] Xu Z., Zang L., GuoR., XiangT., WuC. and Zheng Z. 2011 *Sens Actuators.* **156** 546-552.

[9] Bekhit A.A.; Hymete A.; Asfaw H. and Bekhit A. 2012 *Arch Pharm.* **345** 147-54.

[10] Manolov I.,ReleviS., Genova P.,Savov A., Froloshka L., Dandarova D. and Argirova R. 2006 *Bioinorg. Chem. Applications.* 1-3.

[11] Ronconi, I. and Sadler, P. 2007 *J.Coord.Chem. Rev.* **251** 1633-1642.

[12] LI DF, Jiang GJ. and Li JS. 1990 *Chem.J.Chin.Univ.* **11** 205-207.

[13] E.Y. Begantsova, L.N. Bochkarev, V.A. Ilichev and M.N. Bochkarev 2012 *Russ. Chem. Bull.* **61** 1671.

[14] H. J. Mohammed, A. Y. Muhi and H. AL Meisselman 2011 *E.J. Chem.* **8(S1)** S425-S433.

[15] Mohammad A., SadiqueAJ. and Mohammad ZH. 2011 *Med.Chem.Res.* **40** 58.

[16] Geary, W.J. 1971 *CoordChem. Rev.* **7** 81–122.

[17] Nakamato K. 1998 *"Infrared and Raman spectra of Inorganic and Coordination Copounds"* 5th.Edn.John &Wiley Sons (New York) 183-187.

[18] Anurag Joshi, R.B. Shashi Verma, R.R. Gaur and Sharma 2005 *Bioinorg. Chem.* **3** 201.

[19] Lever A.B.P. 1968 *"Inorganic electronic spectroscopy"* Elsevier Publishing Company (New York).

[20] Cotton, F. A.; Wilkinson, G.; Murillo, C. A. and Bochmann, M. 1999 "*Advanced Inorganic Chemistry*" 6th edition, John Wiley & Sons.

[21] Gouda,M.A.Berghot,M.AShoieb, K.AM. 2010 *Eur.J,Inorg.Chem.* **45** 1843-1848.

[22] Hammond C. R. 2004, *The Elements, in Handbook of Chemistry and Physics* CRC press. ISBN, **7** 8493, 0485.

[23] P.V. Anantha Lakshmi, P. Saritha Reddy and V. Jayatyaga Raju 2009 *Spectrochimica Acta part.* **A74** 52-57.

[24] Naveen, V.Kulkarni and Vidyan,K.R. 2011 *J.Coord.Chem.* **64(4)** 725–741.

IOP Conf. Series: Journal of Physics: Conf. Series **1032** (2018) 012058 doi:10.1088/1742-6596/1032/1/012058

Microwave Synthesis of 2,3-Disubstituted-5-methyl-1,3-imidazolidines-4-one Bearing Benzothiazole Moiety and Elementarily Assessment of Their Antibacterial Action

Zeid Hassan Abood [*], **Hussein Ali Qabel and Hayder Raheem Ali**

Departmen of Chemistry, Faculty of Science, University of Kerbala, Kerbala-Iraq

Corresponding author's E-mail: zeid.ab2013@yahoo.com

Abstract. Diazotization of 2-aminobenzothiazole **1** through using sulfuric acid and sodium nitrite resulted in the formation of diazonium salt which reacted with alkaline solution of salicylaldehyde to produce azo-aldehyde derivative of benzothiazole **2**. The resulting aldehyde **2** was condensed with the primary aromatic amines involving (4-nitroaniline, 3-nitroaniline, 4-hydroxyaniline, 4-methoxyaniline, 2-methoxyaniline, 4-bromoaniline, 4-chloroaniline and 2,4-dichloroaniline) using microwave irradiation method in absolute ethanol to produce eight imines of benzothiazole **3a-h**, respectively. The imine compounds **3a-h** have been treated with L-alanine using microwave irradiation in tetrahydrofuran afforded eight new imidazolidines **4a-h** containing benzothiazole moiety, respectively. Preliminary antibacterial activity of the target compounds was investigated in vitro using two kinds of bacteria, *Escherichia coli (Gram-negative)* and *Staphylococcus aurous* (Gram-positive). The results showed that the newly prepared imidazolidines (compounds **4c**, **4d**, **4f**, **4g**, and **4h**) exhibited greater activities than gentamycin against Gram-positive bacteria. On the other hand, compounds **4c** and **4f** were also showed better activities against Gram-negative bacteria when compared with that of the control drug.

1. Introduction

Green chemistry includes the invention, design and application of products and processes to reduce the use and generation of hazardous substances[1,2]. The microwave irradiation technique afforded clean, simple, fast, efficient, and economic for the preparation of a wide number of organic compounds, have provided the momentum for many chemists to switch from conventional method to microwave assisted chemistry[3]. In the latter years microwave assisted organic reaction has emanated as new device in organic synthesis[4].

Imidazolidines (saturated imidazoles), also known as tetrahydroimidazoles are biologically active nitrogen contaning heterocyclic moiety which have been reported to shown wide array of significant bioactivities[5]. Imidazolidines-dione (hydantoins) and their derivatives are well known for their medicinal and many important non-medicinal applications[6]. The discovery of imidazolidine-dione and successfully developing novel hydantoins-based drugs for the treatment of epilepsy[7]. Imidazolidine-dione derivatives are reported to show the extensive range of biological activities such as anti-inflammatory[8], analgesic agents[9], antibacterial[10], antifungal drugs[11], antimicrobial[12], antidepressant drugs are widely used[13], commonly used in the treatment of epilepsy[14] and antiarrhythmic[15].

Benzothiazole comprises of thiazole ring fused with benzene and is also dubbed 1-thia-3-azaindene. It has been found to possess interesting biological activities[16], antimalarial[17], antihelmintic[18], antibacterial and antifungal abilities[19]. Benzothiazole derivatives have been known for different biological properties including antimalarial, antitubercular, antihelmintic, anticonvulsant,

IOP Conf. Series: Journal of Physics: Conf. Series **1032** (2018) 012058 doi:10.1088/1742-6596/1032/1/012058

analgesic, anti-inflammatory, antidiabetic and antitumor activities[20], Further industrial applications as antioxidants[21].

Consequently, in this project, we reported here the microwave synthesis of imidazolidines bearing the biologically vigorous benzothiazole moiety that possibly have some biological activities.

2. Experimental

2.1. General.

All chemicals were utilized as provided from sigma Aldrich, Fluka and Merck. Microwave reactions have been performed on Domestic microwave oven in crucible. Analytical TLC has been done with silica gel 60 F_{254} plates. Melting points have been measured on an Electro thermal Stuart SMP 30 capillary melting point apparatus and are uncorrected. IR spectra have been recorded on SHIMADZU FTIR–8400S Infrared Spectrophotometer as potassium bromide discs. ^{1}H NMR spectra have been obtained using Avance III 500, NMR spectrometer, Bruker, Germany at 500 MHz in DMSO–d_6 as solvent and TMS as an internal reference at a Faculty of Science, University of Tarbiat Modares, Iran. Elemental Analyses were performed with Perkin Elmer 300A Elemental Analyzer at a Faculty of Science, University of Tarbiat Modares, Iran.

2.2. Chemical methods.

2.2.1. (E)-5-(benzo[d]thiazol-2-yldiazenyl)-2-hydroxy benzaldehyde (2). has been prepared by method described in the literature[22] as dark brown solid, mp 141-143 °C, yield 60 %; IR (cm^{-1}): 3429$_{br}$ (νO-H), 3068 (νC-H, benzene), 2856 and 2715 (νC-H, aldehyde), 1645 (νC=O, aldehyde), 1610 (νC=N, benzothiazole), 1502 and 1475 (νC=C, benzene), 1375 (νN=N), 759 (δo.o.p.C-H, benzene).

2.2.2. General procedure to prepare of imine compounds 3a-h[23]. All reactions were carried out on Domestic microwave oven in crucible. Reactions contained the aldehyde derivative 2 (0.283 g, 1 mmol), equimolar amount (1 mmol) of aniline derivatives (4-nitroaniline, 3-nitroaniline, 4-hydroxyaniline, 4-methoxyaniline, 2-methoxyaniline, 4-bromoaniline, 4-chloroaniline and 2,4-dichloroaniline respectively) and absolute ethanol (1 mL). The crucible was introduced to the center of a Domestic microwave oven and then heated (300-350W) for 25-30 minutes. TLC (n-hexane: EtOAc) indicated that the reactions have been completed. The products were washed with diethyl ether and recrystallized from absolute ethanol.

4-((E)-benzo[d]thiazol-2-yldiazenyl)-2-((E)-) ((4-nitrophenyl)imino)methyl)phenol (3a): IR (cm^{-1}): 3371 (νO-H), 3064 (νC-H, benzene), 1608 (νC=N, imine), 1589 (νC=N, benzothiazole), 1514 (νas.NO_2), 1448 (νC=C, benzene), 1404 (νN=N), 1336 (νs.NO_2), 758 (δo.o.p.C-H, benzene).

4-((E)-benzo[d]thiazol-2-yldiazenyl)-2-((E)-((3-nitrophenyl)imino)methyl)phenol (3b): IR (cm^{-1}): 3246 (νO-H), 3088 (νC-H, benzene), 1604 (νC=N, imine and νC=N, benzothiazole, vib. coupling), 1573 and 1502 and 1444 (νC=C, benzene), 1525 (νas.NO_2), 1396 (νN=N), 1354 (νs.NO_2), 758 (δo.o.p.C-H, benzene).

4-((E)-benzo[d]thiazol-2-yldiazenyl)-2-((E)-((4-hydroxyphenyl)imino)methyl)phenol (3c): IR (cm^{-1}): 3369 and 3298 (νO-H), 3032 (νC-H, benzene), 1610 (νC=N, imine and νC=N, benzothiazole, vib. coupling), 1514 and 1460 (νC=C, benzene), 1379 (νN=N), 827 (δo.o.p.C-H, benzene).

4-((E)-benzo[d]thiazol-2-yldiazenyl)-2-((E)-((4-methoxyphenyl)imino)methyl)phenol (3d):

IR (cm^{-1}): 3371 and 3298 (νO-H), 3037 (νC-H, benzene), 2949 (νas.CH_3), 1622 (νC=N, imine), 1595 (νC=N, benzothiazole), 1512 and 1456 (νC=C, benzene), 1392 (νN=N), 821 (δo.o.p.C-H, benzene).

4-((E)-benzo[d]thiazol-2-yldiazenyl)-2-((E)-((2-methoxyphenyl)imino)methyl)phenol (3e):
IR (cm^{-1}): 3462 and 3265 (νO-H), 3064 (νC-H, benzene), 2968 (νas.CH_3), 2870 (νs.CH_3), 1633 (νC=N, imine), 1608 (νC=N, benzothiazole), 1533, 1500 and 1444 (νC=C, benzene), 1363 (νN=N), 831 (δo.o.p.C-H, benzene).

4-((E)-benzo[d]thiazol-2-yldiazenyl)-2-((E)-((4-bromophenyl)imino)methyl)phenol (3f)
IR (cm^{-1}): 3369 and 3261 (νO-H), 3066 (νC-H, benzene), 1633 (νC=N, imine), 1608 (νC=N, benzothiazole), 1506 and 1489 (νC=C, benzene), 1363 (νN=N), 821 (δo.o.p.C-H, benzene).

4-((E)-benzo[d]thiazol-2-yldiazenyl)-2-((E)-((4-chlorophenyl)imino)methyl)phenol (3g):
IR (cm^{-1}): 3369 and 3263 (νO-H), 3064 (νC-H, benzene), 1633 (νC=N, imine), 1606 (νC=N, benzothiazole), 1537, 1496 and 1442 (νC=C, benzene), 1367 (νN=N), 821 (δo.o.p.C-H, benzene).

4-((E)-benzo[d]thiazol-2-yldiazenyl)-2-((E)-((2,4-dichlorophenyl)imino)methyl)phenol(3h):
IR (cm^{-1}): IR (cm^{-1}): 3470 and 3265 (νO-H), 3061 (νC-H, benzene), 1633 (νC=N, imine), 1608 (νC=N, benzothiazole), 1570, 1533, 1500 and 1444 (νC=C, benzene), 1363 (νN=N), 831 (δo.o.p.C-H, benzene).

2.2.3. General procedure to prepare of imidazolidines 4a-h. A mixture of equimolar amounts of imine derivatives **3a-h** (1 mmol) and L-alanine (0.089 g, 1 mmol) in tetrahydrofuran (1 mL) was heated (550-610W) in microwave oven for 20-25 min. TLC (n- hexane: EtOAc) indicated that the reactions have been completed. The products were washed with diethyl ether and recrystallized from absolute ethanol.

(E)-2-(5-(benzo[d]thiazol-2-yldiazenyl)-2-hydroxyphenyl)-5-methyl-3-(4-nitrophenyl) imidazolidin-4-one (4a): IR (cm^{-1}): 3234_{br} (νO-H and νN-H, imidazolidine, vib. coupling), 3057 (νC-H, benzene), 1664 (νC=O and δN-H, imidazolidine, vib. coupling), 1597 (νC=N, benzothiazole), 1510 (νas.NO_2), 1452 (νC=C, benzene), 1421 (νN=N), 1286 (νs.NO_2), 754 (δo.o.p.C-H, benzene); ^{1}H NMR: δ(ppm) = 1.24 (d, J = 6.7 Hz, 3H, CH_3), 3.84 (q, J = 6.7 Hz, 1H, **CH**–CH_3, imidazolidine), 6.52 (d, J = 7.8 Hz, 1H, N–CH–N, imidazolidine), 6.67–7.61 (m, 11H, Ar–H), 8.07 (s, 1H, N–H, imidazolidine), 9.51 (s, 1H, O–H). The signals at 2.49 ppm and 3.42 ppm assigned to DMSO and absorbed H_2O in DMSO, successively; Anal. Calcd. for $C_{23}H_{18}N_6O_4S$: C, 58.22; H, 3.82; N, 17.71; S, 6.76; Found C, 57.86; H, 3.49; N, 17.42; S, 6.67 .

(E)-2-(5-(benzo[d]thiazol-2-yldiazenyl)-2-hydroxyphenyl)-5-methyl-3-(3-nitrophenyl) imidazolidine-4-one (4b): IR (cm^{-1}): 3203_{br} (νO-H and νN-H, imidazolidine, vib. coupling), 3064 (νC-H, benzene), 1668 (νC=O, imidazolidine), 1653 (δN-H, imidazolidine), 1604 (νC=N, benzothiazole), 1558 and 1454 (νC=C, benzene), 1510 (νas.NO_2), 1423 (νN=N), 1288 (νs.NO_2), 825 (δo.o.p.C-H, benzene); ^{1}H NMR: δ(ppm) = 1.24 (d, J = 6.7 Hz, 3H, CH_3), 3.78 (q, J = 7.0 Hz, 1H, **CH**–CH_3, imidazolidine), 6.52 (d, J = 8.1 Hz, 1H, N–CH–N, imidazolidine), 6.66–7.61 (m, 11H, Ar–H), 8.06 (s, 1H, N–H, imidazolidine), 9.49 (s, 1H, O–H). The signals at 2.49 ppm and 3.42 ppm assigned to DMSO and H_2O in DMSO, successively; Anal. Calcd. for $C_{23}H_{18}N_6O_4S$: C, 58.22; H, 3.82; N, 17.71; S, 6.76; Found C, 58.57; H, 3.52; N, 17.35; S, 6.62 .

(E)-2-(5-(benzo[d]thiazol-2-yldiazenyl)-2-hydroxyphenyl)-3-(4-hydroxyphenyl)-5-methylimidazolidin-4-one (4c): IR (cm^{-1}): 3209_{br} (νO-H and νN-H, imidazolidine, vib. coupling), 3093 (νC-H, benzene), 2987 (νas.C-H_3), 1668 (νC=O and δN-H, imidazolidine, vib. coupling), 1599 (νC=N, benzothiazole), 1514 and 1456 (νC=C, benzene), 1413 (νN=N), 833 (δo.o.p.C-H, benzene); ^{1}H NMR: δ(ppm) = 1.24 (d, *J* = 6.9 Hz, 3H, CH_3), 3.84 (q, *J* = 7.0 Hz, 1H, **CH**–CH_3, imidazolidine), 6.51 (d, *J* = 8.1 Hz, 1H, N–CH–N, imidazolidine), 6.66–7.66 (m, 11H, Ar–H), 8.07 (s, 1H, N–H, imidazolidine), 10.12 (s, 2H, 2×O–H). The signals at 2.49 ppm and 3.42 ppm due to DMSO and H_2O in DMSO, successively; Anal. Calcd. for $C_{23}H_{19}N_5O_3S$: C, 62.01; H, 4.30; N, 15.72; S, 7.20; Found C, 61.61; H, 4.16; N, 15.59; S, 6.91.

(E)-2-(5-(benzo[d]thiazol-2-yldiazenyl)-2-hydroxyphenyl)-3-(4-methoxyphenyl)-5-methylimidazolidin-4-one (4d): IR (cm^{-1}): 3203 (νO-H), 3171 (νN-H, imidazolidine), 3070 (νC-H, benzene), 2943 (νas.CH_3) 1674 (νC=O, imidazolidine), 1658 (δN-H, imidazolidine), 1606 (νC=N, benzothiazole), 1516 and 1454 (νC=C, benzene), 1427 (νN=N), 827 (δo.o.p.C-H, benzene); ^{1}H NMR: δ(ppm) = 1.23 (d, *J* = 6.7 Hz, 3H, CH_3), 3.61 (s, 1H, O–CH_3), 3.85 (q, *J* = 6.6 Hz, 1H, **CH**–CH_3, imidazolidine), 6.52 (d, *J* = 8.0 Hz, 1H, N–CH–N, imidazolidine), 6.74–7.60 (m, 11H, Ar–H), 8.06 (s, 1H, N–H, imidazolidine), 9.51 (s, 1H, O–H). The signals at 2.49 ppm and 3.34 ppm attributed to DMSO and H_2O in DMSO, successively; Anal. Calcd. for $C_{24}H_{21}N_5O_3S$: C, 62.73; H, 4.61; N, 15.24; S, 6.98; Found C, 62.44; H, 4.32; N, 15.21; S, 6.75 .

(E)-2-(5-(benzo[d]thiazol-2-yldiazenyl)-2-hydroxyphenyl)-3-(2-methoxyphenyl)-5-methylimidazolidin-4-one (4e): IR (cm^{-1}): 3196$_{br}$ (νO-H and νN-H, imidazolidine, vib. coupling), 3059 (νC-H, benzene), 1674 (νC=O, imidazolidine), 1660 (δN-H, imidazolidine), 1612 (νC=N, benzothiazole), 1548, 1512 and 1452 (νC=C, benzene), 1425 (νN=N), 821 (δo.o.p.C-H, benzene); ^{1}H NMR: δ(ppm) =1.24 (d, *J* = 6.7 Hz, 3H, CH_3), 3.84 (q, *J* = 6.8 Hz, 1H, **CH**–CH_3, imidazolidine), 4.16 (s, 1H, O–CH_3), 6.53 (d, *J* = 8.1 Hz, 1H, N–CH–N, imidazolidine), 6.70–7.73 (m, 11H, Ar–H), 8.06 (s, 1H, N–H, imidazolidine), 9.64 (s, 1H, O–H). The signals at 2.49 ppm and 3.42 ppm assigned to DMSO and H_2O in DMSO, successively; Anal. Calcd. for $C_{24}H_{21}N_5O_3S$: C, 62.73; H, 4.61; N, 15.24; S, 6.98; Found C, 62.36; H, 4.35; N, 15.15; S, 6.81 .

(E)-2-(5-(benzo[d]thiazol-2-yldiazenyl)-2-hydroxyphenyl)-3-(4-bromophenyl)-5-methylimidazolidin-4-one (4f): IR (cm^{-1}): 3188$_{br}$ (νO-H and νN-H, imidazolidine, vib. coupling), 3061 (νC-H, benzene), 2943 (νCH_3), 1668 (νC=O and δN-H, imidazolidine, vib. coupling), 1604 (νC=N, benzothiazole), 1504 and 1452 (νC=C, benzene), 1425 (νN=N), 823 (δo.o.p.C-H, benzene); ^{1}H NMR: δ(ppm) = 1.24 (d, *J* = 6.8 Hz, 3H, CH_3), 3.84 (q, *J* = 7.0 Hz, 1H, CH–CH_3, imidazolidine), 6.64 (d, *J* = 8.0 Hz, 1H, N–CH–N, imidazolidine), 7.00–7.61 (m, 11H, Ar–H), 8.06 (s, 1H, N–H, imidazolidine), 9.50 (s, 1H, O–H). The signals at 2.49 ppm and 3.42 ppm due to DMSO and H_2O in DMSO, successively; Anal. Calcd. for $C_{23}H_{18}N_5O_2SBr$: C, 54.34; H, 3.57; N, 13.78; S, 6.31; Found C, 53.96; H, 3.33; N, 13.43; S, 6.09 .

(E)-2-(5-(benzo[d]thiazol-2-yldiazenyl)-2-hydroxyphenyl)-3-(4-chlorophenyl)-5-methylimidazolidin-4-one (4g): IR (cm^{-1}): 3419$_{br}$ (νO-H and νN-H, imidazolidine, vib. coupling), 3063 (νC-H, benzene), 1680 (νC=O, imidazolidine) 1651 (δN-H, imidazolidine), 1618 (νC=N, benzothiazole), 1558, 1539, 1512 and 1454 (νC=C, benzene), 1421 (νN=N), 817 (δo.o.p.C-H, benzene); ^{1}H NMR: δ(ppm) = 1.24 (d, *J* = 6.7 Hz, 3H, CH_3), 3.84 (q, *J* = 6.8 Hz, 1H, **CH**–CH_3, imidazolidine), 6.59 (d, *J* = 7.5 Hz, 1H, N–CH–N, imidazolidine), 7.02–7.61 (m, 11H, Ar–H), 8.06 (s, 1H, N–H, imidazolidine), 9.53 (s, 1H, O–H). The signals at 2.49 ppm and 3.37 ppm attributed to DMSO and H_2O in DMSO, successively; Anal. Calcd. for $C_{23}H_{18}N_5O_2SCl$: C, 59.54; H, 3.91; N, 15.10; S, 6.91; Found C, 58.19; H, 3.90; N, 15.01; S, 6.51 .

(E)-2-(5-(benzo[d]thiazol-2-yldiazenyl)-2-hydroxyphenyl)-3-(2,4-dichlorophenyl)-5-methylimidazolidin-4-one (4h): IR (cm^{-1}): 3443$_{br}$ (νO-H), 3213 (νN-H, imidazolidine), 3064 (νC-H, benzene), 1681 (νC=O, imidazolidine), 1651 (δN-H, imidazolidine, vib. coupling), 1610 (νC=N, benzothiazole), 1556, 1539, 1510 and 1456 (νC=C, benzene), 1423 (νN=N), 823 (δo.o.p.C-H, benzene); ^{1}H NMR:

δ(ppm) = 1.23 (d, J = 6.8 Hz, 3H, CH_3), 3.84 (q, J = 7.0 Hz, 1H, CH–CH_3, imidazolidine), 6.59 (d, J = 7.1 Hz, 1H, N–CH–N, imidazolidine), 6.81–7.73 (m, 10H, Ar–H), 8.06 (s, 1H, N–H, imidazolidine), 9.65 (s, 1H, O–H). The signals at 2.49 ppm and 3.33 ppm due to DMSO and H_2O in DMSO, successively; Anal. Calcd. for $C_{23}H_{17}N_5O_2SCl_2$: C, 55.43; H, 3.44; N, 14.05; S, 6.43; Found C, 54.06; H, 3.29; N, 13.94; S, 6.41 .

2.3. Elementarily antibacterial test.

The antibacterial activities of the latterly synthesized imidazolidines 4a-h have been scanted by the agar diffusion method[24] using representative Gram (+) and Gram (–) bacteria on agar media. The test bacteria to assess the potential antibacterial action of the newly synthesized imidazolidines were Staphylococcus aurous (Gram-positive) and Escherichia coli (Gram-negative). The imidazolidines have been liquefied in dimethylsulfoxide to prepare the test solutions of 10 mg/mL concentration. Gentamycin have been used as a standard and the activities have been submitted as zones of inhibition for each compound (Table-2).

3. Results and Discussion

3.1. Chemistry.

The target compounds have been synthesized in order to Scheme-I. Eeight intermediates were prepared as the platforms for the designated compounds through two steps. In the first step, initiator aldehyde **2** was prepared by reacting the diazonium salt of 2-aminobenzothiazole **1** with salicylaldehyde dissolved in sodium hydroxide solution using the method described previously[22]. In the second step, the aldehyde derivative **2** was reacted with the primary aromatic amines including (4-nitroaniline, 3-nitroaniline, 4-hydroxyaniline, 4-methoxynitroaniline, 2-methoxyaniline, 4-bromoaniline, 4-chloroaniline and 2,4-dichloroaniline) using microwave irradiation in absolute ethanol to produce eight Schiff bases of benzothiazole **3a-h** respectively[23] . In the following step, the compounds **3a-h** have been allowed to react with L-alanine using microwave irradiation to yield imidazolidine derivatives of benzothiazole **4a-h** in good yields (Table-1).

The chemical constructs of the target compounds synthesized have been deduced from IR, ^{1}H NMR spectral means and elemental analysis which showed good congruity with the suggested structures.

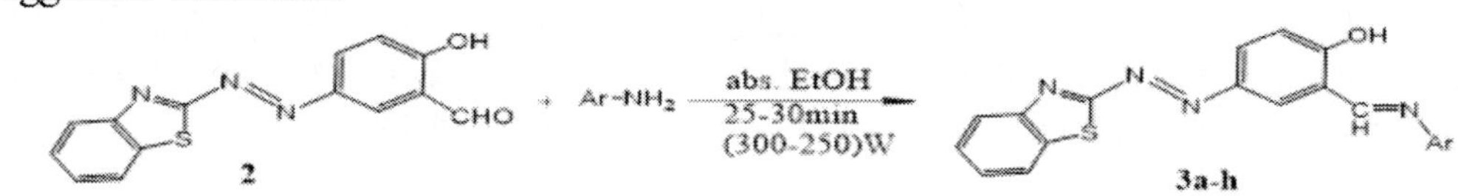

Ar = p-$NO_2C_6H_4$-, m-$NO_2C_6H_4$-, p-OHC_6H_4-, p-$OCH_3C_6H_4$-, o-$OCH_3C_6H_4$-, p-BrC_6H_4-, p-ClC_6H_4-, 2,4-di-ClC_6H_4-
4a 4b 4c 4d 4e 4f 4g 4h

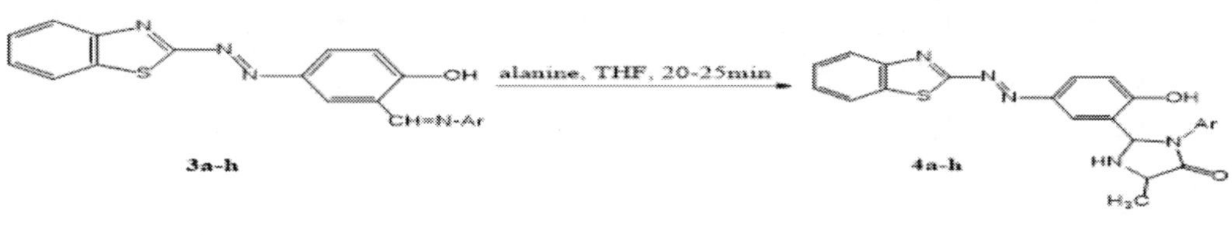

Ar = p-$NO_2C_6H_4$-, m-$NO_2C_6H_4$-, p-OHC_6H_4-, p-$OCH_3C_6H_4$-, o-$OCH_3C_6H_4$-, p-BrC_6H_4-, p-ClC_6H_4-, 2,4-di-ClC_6H_4-
4a 4b 4c 4d 4e 4f 4g 4h

Scheme 1. Synthesis of 2,3-disubstituted-5-methyl-1,3-imidazolidines-4-one

IOP Conf. Series: Journal of Physics: Conf. Series **1032** (2018) 012058 doi:10.1088/1742-6596/1032/1/012058

The IR and ^{1}H NMR spectra of the demanded compounds **4a-h** have been prescribed in minutiae in the Experimental part. The IR spectrum of azo-benzothiazole derivative **2** showed the loosing of a doublet band at 3375 and 3306 cm^{-1} for (NH_2)str in 2-aminobenzothiazole **1** and arising of the following significant bands: the medium band at 1375 cm^{-1} belong to azo group (N=N)str, the band at 3429 cm^{-1} attributed to (O-H)str, the band at 1645 cm^{-1} assigned to aldehydic (C=O)str., the benzothiazolic (C=N)str arose as strong band at 1610 cm^{-1}. IR spectra of benzothiazolic-imines **3a-h** showed disappearing the band at 1645 cm^{-1} attributed to aldehydic (C=O)str., also absence the doublet band for (NH_2)str in the starting amines at the commonalty range 3400-3250 cm^{-1} and appearing absorption band at the scope 1604-1633 cm^{-1} due to the iminic (C=N)str. The benzothiazolic (C=N)str appeared at the range 1589-1610 cm^{-1}. The IR spectra of the benzothiazolic-imidazolidines **4a-h** showed the appearance of strong band at the range 1664-1710 cm^{-1} attributed to the (C=O)str of the imidazolidine ring. The spectra also showed the appearance of absorption band at the scope. 1651-1668 cm^{-1} belong to the (N-H)bend of the imidazolidine ring. The benzothiazolic (C=N)str appeared at the range 1597-1618 cm^{-1}.

The structures of the imidazolidine compounds **4a-h** have been confirmed by their ^{1}H NMR spectra that appeared doublet at δ 1.24, 1.24, 1.24, 1.23, 1.24, 1.24, 1.24 and 1.23 ppm, respectively belong to the methyl (CH_3) protons, the peak for the methine proton $(\mathbf{C}\mathbf{H}\text{-}CH_3)$ of imidazolidine ring as a quartet at 3.84, 3.78, 3.84, 3.85, 3.84, 3.84, 3.84 and 3.84 ppm, respectively. The (N-CH-N) proton of imidazolidine appeared as a doublet at 6.52, 6.52, 6.51, 6.52, 6.53, 6.64, 6.59 and 6.59 ppm, respectively. The (Ar-H) protons appeared at δ 6.66 –7.73 ppm, the (N-H) proton of imidazolidine ring as a singlet at 8.07, 8.06, 8.07, 8.06, 8.06, 8.06, 8.06 and 8.06 ppm, respectively. The (O-H) proton appeared as a singlet at 9.51, 9.49, 10.12, 9.51, 9.64, 9.50, 9.53 and 9.65 ppm, respectively. The methoxy protons $(O\text{-}CH_3)$ in compounds **4d** and **4e** arose as a singlet at δ 3.61 and 4.16 ppm, consecutively.

Moreover, the (CHNS) elemental analysis findings were subsume ± 0.4% of the calculated values and in good congruity with the suggested chemical structures for compounds **4a-h** given in the experimental part.

3.2. Antibacterial activities.

The antibacterial actions of the synthesized imidazolidines **4a-h** have been evaluated via the agar diffusion method[24] using two strains of Gram (+) and Gram (–) bacteria on agar media, as indicated in Table-2. Dimethylsulfoxide has been used as solvent for the assayed compounds.

IOP Conf. Series: Journal of Physics: Conf. Series **1032** (2018) 012058 doi:10.1088/1742-6596/1032/1/012058

Table 1. SOME PHYSICAL PROPERTIES OF THE PREPARED COMPOUNDS

Com.	Physical state	Rf (developer)	Mp (oC)	Yeild (%)	Time (min.)	Mw (W)
2	Dark brown solid	0.69 (n-hexane/ EtOAc, 1:2)	143-145	60	-	-
3a	Brown solid	0.75 (n-hexane/ EtOAc, 1:2)	223-225	78	30	350
3b	Dark brown solid	0.77 (n-hexane/ EtOAc, 1:2)	186-188	75	28	320
3c	Dark brown solid	0.79 (n-hexane/ EtOAc, 1:2)	178-180	79	27	300
3d	Brown solid	0.89 (n-hexane/ EOAc, 1:2)	197-199	73	26	300
3e	Dark brown solid	0.79 (n-hexane/ EtOAc, 1:2)	201-203	77	30	310
3f	Dark brown solid	0.86 (n-hexane/ EtOAc, 1:2)	175-177	69	30	320
3g	Brown solid	0.75 (n-hexane/ EtOAc, 1:2)	208-210	76	28	320
3h	Brown solid	0.79 (n-hexane/ EtOAc, 1: 2)	165-167	77	30	310
4a	Yellow solid	0.89 (n-hexane/ EtOAc, 1: 2)	249-251	74	24	610
4b	Dark orange solid	0.81 (n-hexane/ EtOAc, 1: 2)	243-245	73	23	600
4c	Orange solid Dark	0.81 (n-hexane/ EtOAc, 1: 2)	299-301	70	22	600
4d	Orange solid	0.79 (n-hexane/ EtOAc, 1: 2)	234-236	74	21	560
4e	Orange solid	0.78 (n-hexane/ EtOAc, 1: 2)	204-206	77	20	550
4f	Dark orange solid	0.83 (n-hexane/ EtOAc, 1: 2)	281-283	76	24	570
4g	Yellow solid	0.79 (n-hexane/EtOAc , 1:2)	319(dec.)	78	22	600
4h	Dark brown solid	0.76 (n-hexane/ EtOAc, 1: 2)	288-290	79	22	600

Table 2. THE ANTIBACTERIAL ACTIVITY OF COMPOUNDS **4a-h** AND GENTAMYCIN AS CONTROL DRUG

Product	Staphylococcus aureus (Gram-positive)	Escherichia coli (Gram-negative)
4a	0	0
4b	11	0
4c	20	17
4d	19	0
4e	12	0
4f	20	16
4g	18	0
4h	22	0
DMSO	0	0
Gentamycin	15	15

4. Conclusions

The microwave irradiation is considered efficient and economic technique including short reaction time and high yield in comparison with the conventional method. The rates of reactions of imines with alanine for formation of imidazolidines are approximately equal. All synthesized imidazolidines could be converted to the corresponding phenol salts which are completely soluble in water. The synthesized imidazolidines appeared higher biological action against Gram-positive bacteria than that of Gram-negative bacteria. Imidazolidine compounds 4c, 4d, 4f, 4g and 4h were found to be greater activity than gentamycin against Gram-positive bacteria, while compounds, 4c and 4f showed better activity to the control drug against Gram-negative bacteria.

Acknowledgements

Great thanks for the Faculty of Science, University of Tarbiat Modares, Iran for their considerable assistance in ^{1}H NMR and Elemental Analysis tests of the target compounds.

References

[1] Rajendran, A. and Kumar, G.V. 2015 *Indian Journal of Advances in Chemical Science* **3** 133–140.
[2] Jiménez, G.I.; Bravo, J.A. and Vila, J.L. 2016 *Bolivian Journal of Chemistry* **33** 179–182.
[3] Ravichandran, S.and Karthikeyan, E. 2011,, *ChemTech Res*, **3** 466–470.
[4] Tambakhe, S.P.; Berad, B.N. and Raghuwanshi, P.B. 2015 *International Journal of Chemical and Physical Sciences* **4** 18–22.
[5] Husain, A.; Ahmad, A.; Khan, S.A.; Asif, M.; Bhutani, R. and Al-Abbasi, F.A. 2016 *Saudi Pharmaceutical Journal* **24** 104–114.
[6] Bhadreshkumar, R.S. and Vikas, A.D.S. 2017 *International Journal of Chem.Tech. Research* **10** 441–445.
[7] Khodaee, Z.; Mahmoodi, N.O. and Yahyazadeh, A. 2017 *Org. Chem. Res.* **3** 112–118.
[8] Al-haideri, M.R. and Rasool, S.R. 2017 *International Journal of Chem.Tech. Research* **10** 742–748.
[9] Santacruz-Juárez, E.; Höpfl, H.; Huelgas, G.; Romero, O.; Castro, A.; Juaristi, E.; de Parrodi, C.A.; Morales-Meza, and Sanchez, M. 2017 *Arkivoc* **v** 89–99.
[10] Pıhtılı, G. and Tuncer, H. 2017 *Hacettepe Journal of Biology and Chemistry* **2** 163–174.
[11] Ammar, Y.A.; Abbas, S.Y.; El-Sharief, M.A.M.S.; Salem, M.A.E.-R. and Mohamed, A.R. 2017 *European Journal of Chemistry* **8** 76–81.
[12] Marinov, M.N.; Naydenova, E.D.; Prodanova, R.Y.; Tsoneva, S.H. and Stoyanov, N.M. 2017 *Bulgarian Chemical Communications* **49** 40–45.
[13] Ramadhan, U.H. and Al-salihi, N.J. 2017 *BEMS Reports* **3** 1-6.
[14] Arrowsmith, M.; Hill, M.S. and Kociok-köhn, G. 2015 *Chemistry - A European Journal* **21** 10548–10557.
[15] Beyzaei, H.; Aryan, R. and Keshtegar, Z. 2014 *Advances in Chemistry* **2014** 1–4.
[16] Mene, D. and Kale, M. 2016 *Current Organic Synthesis* **13** 41–57.
[17] Gill, R.K.; Rawal, R.K. and Bariwal, J. 2015 *Archiv der Pharmazie* **348** 155–178.
[18] Cano, N.H.; Ballari, M.S.; López, A.G. and Santiago, A.N. 2015 *Journal of Agricultural and Food Chemistry* **63** 3681–3686.
[19] Maddila, S.; Gorle, S.; Seshadri, N.; Lavanya, P. and Jonnalagadda, S.B. 2013 *Arabian Journal of Chemistry* **9** 681–687.
[20] Peng, X.; Xie, G.; Wang, Z.; Lin, H.; Zhou, T.; Xiang, P.; Jiang, Y.; Yang, S.; Wei, Y.; Yu, L. and Zhao, Y. 2014 *Cell Death and Disease* **5** 1–10.
[21] Azizi, N.; Amiri, A.K.; Baghi, R.; Bolourtchian, M. and Hashemi, M.M. 2009 *Monatshefte fur*

Chemie **140** 1471–1473.
[22] Acton, Q.A. 2011 "*Azo Compounds: Advances in Research and Application*" Scholarly Paper Edition (Atlanta).
[23] Abood, Z.H.; Rasheed, O.H. and Mohammed, M.J. 2016 *International Journal of Science and Research* **5** 1741–1749.
[24] Egorove, N.S. 1985 "*Antibiotics Scientific Approach*" Mir. Publishers (Moscow).

IOP Conf. Series: Journal of Physics: Conf. Series **1032** (2018) 012059 doi:10.1088/1742-6596/1032/1/012059

Synthesis and study the behavior of liquid crystalline phases of new heterocyclic compound from folic acid

Eman Abdul Wahab Alkuwaity and Adraa Abdul Kadhim Wasaf

Department of Chemistry, Faculty of Education for girls,University of Kufa,Iraq.

Corresponding author's E-mail: iman.alkuwaiti@uokufa.edu.iq

Abstract. This work includes syntheses of new heterocyclic compounds of folic acid, The p-hydroxy benzaldehyde reaction with (pentanol, hexanol and heptanol) to produce (4-pentloxy ,4- hexaloxy and4- hepteloxy) benzaldehyde,and compounds (p-hydroxy,p- chloro,p-nitro ,4-N,N-dimethyl) benzaldehyde and Vanillin reaction of folic acid with to produce Schiff bases. The Schiff base of benzaldehyde and Folic acid reaction with (4-pentloxy ,4- hexaloxy and4- hepteloxy) to produceEster. The compounds have been identified using (FT-IR and ^{1}H NMR) .The misomorphic phases were identified by polarized light microscope (POM) and measured transition temperatures for compound by differmtion scanning calorimeter (DSC) then study of liquid crystal where it was found that compounds of (1-8) gave liquid crystalline properties, where the compounds (1,2,3,5,6,7and8) gave the nematic phase ,but the compound (4) gave the smectic phase in addition to the nematic phase,while compounds (9,10and11) where oily compounds and showed not liquid crystalline properties.

1. Introduction.

Folic acid is one of the most important biochemical heterocyclic compound, which is a cyclic compound having six-memberedring and containing heteroatom's of nitrogen (N) and consist of long chains of heterocyclic[1], it has wide range properties in pharmaceutical chemistry ;that's used to prepare other chemicals like Schiff bases because it contains primary amine group[2].

Liquid crystal are organic compounds pass in a medium state between regular solid state and random liquid state ,found two categories by how they are configured, lyotropic liquid crystals (Amphiphilic)[3] ,this type consists of the solubility of organic compounds in volumes of polarity solvents such as water and alcohol , thermo tropic liquid crystals which depend on the temperature , founds many compound pass middle phase by temperature increases[4]. classified according to partial organization to three phase , noematic phase was simplest state because its molecules take the same direction but move around the sides (shape bars)[5] , cholesteric phase called (Twisted noematic phase)[6] ,sematic phase ;which looks like soap because of viscosities ,this phase found many types S_A , S_B ,........S_K , alphabetical sequence back to the precedence of discovery[7,8].

2. Experimental.

All chemical were supplied from(Fluka& Sigma)company,the measurement were carried out by :FT.IR &H^1NMR spectra , recorded on shimadzu 8300 and (Bruker DMSO-500 NMR) spectrophotometer system 400 MHz using DMSO-d_6 solvent in India , the liquid crystalline determined polarized light microscope (POM) &differrntion scanning calorimeter (DSC),melting points were determined by digital-electro thermal,LTD,UK.

2.1. Preparation of Alkali bromide.

Alkyl bromide was prepared from HBr acid (71 mL) and added (16.5mL)of H_2SO_4 acid then added (0.5mol) of the (pentanol, hexanol and heptanol) then added (2.5mL)of H_2SO_4 ,it is heated for an (5-10) hour.The mixture was separation and dried over anhydrous calcium chloride[9].

2.2. Preparation of 4- alkoxybanzaldehayd(6,7and8).

IOP Conf. Series: Journal of Physics: Conf. Series **1032** (2018) 012059 doi:10.1088/1742-6596/1032/1/012059

4- hydroxyl banzaldehayd (0.025 mol) was dissolved in ethanol) 10 mL) then potassium hydroxide (0.033mol) was added slowly, after that the proper alkyl bromide (0.025mol) the reaction mixture was then refluxed for (1hr) the product was extracted by petroleum ether and then evaporated the pure product was liquid yellow color[10].

2.3. Synthesis of Schiff basis compounds (1,2,3,4and5) from folic acid[11,12].

Folic acid compound (0.005mole) was dissolved in(50mL) of absolute ethanol and (50mL) of hot glacial acetic acid, then added (0.005mole) of (4-dimethyl benzaldehyde,4- hydroxy benzaldehyde,4-chloro benzaldehyde,4- nitirobenzaldehyde&vanillin) which dissolved in absolute ethanol the reaction mixture was refluxed with stirring (16hrs) at 120 ^{0}C the precipitate was filtered and re crystallized .

2.4. Synthesis of di ester derivatives from Schiff basis compound (2)of Folic acid(9,10and 11)[13].

Schiff basis compound (2)of Folic acid (0.005mole) was dissolved in (0.01) mole of (pentanol, hexanol& heptanol) then added (0.5mL) of sulphuric acid .The reaction mixture was refluxed with stirring for (12hrs) at 110C^0 and put to product ester calcium chloride anhydrous the yielded was filtered to give pure esters compound (9,10,11).

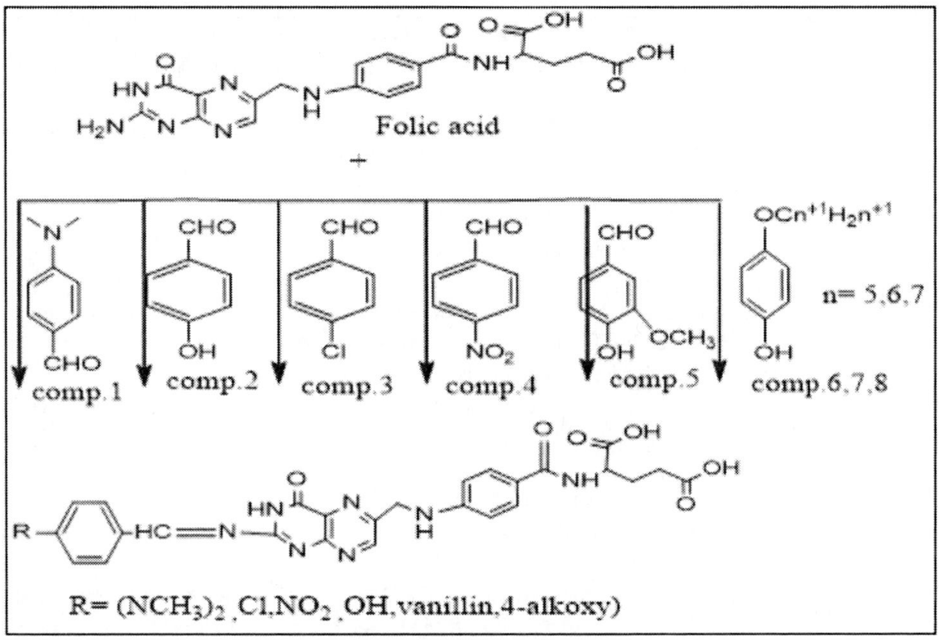

And syntheses of new ester from Schiff bases of folic acid show in figure

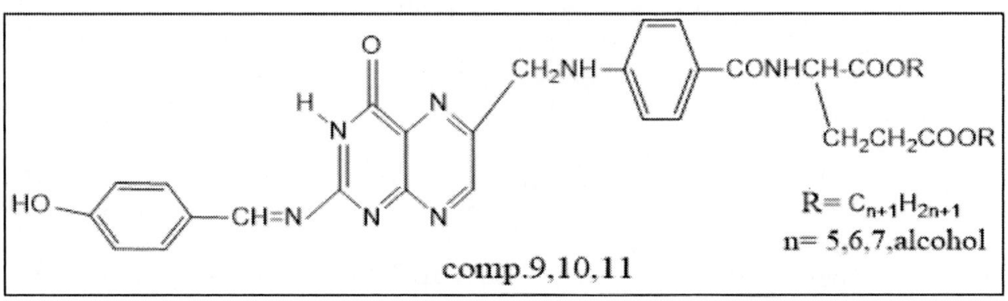

IOP Conf. Series: Journal of Physics: Conf. Series **1032** (2018) 012059 doi:10.1088/1742-6596/1032/1/012059

3. Results and Discussions.

The structure of Folic acid drug consisted from important of fictional group (two carboxylic acid group ,one amino grope and two amid group)therefore is used as starting material in synthesis of Schiff basis derivatives, in this work 4-hydroxy benzaldehyde react with different compounds of alcohol to yield new alkyl derivative from Folic acid like compounds (6,7,8) and compound (2) react with different alcohol to product new ester derivatives of compound (9,10,11),All synthesis compounds (1-11) have been identification by spectroscopic methods like (FTIR ,^{1}HNMR).

(FT.IR) spectrum: showed an absorption bands at (1568-1647)cm^{-1}due to imines group in compound (1-11) and spectra of these compounds showed disappearance of two absorption bands ,of the asym and sym stretching vibration of (NH_2) group of folic acid, the ester compounds absorption band at (1732-1693)cm^{-1} due to(-CO-O-R),(2988-2822)cm^{-1} to(-CH_2-) of ester[14,15] ,The spectra of Alkyl compounds showed absorption bands at (1715-1689)cm^{-1} due to carbonyl group,(1225-1157)cm^{-1} absorption band due to (C-OR) group ,and other data in Table(1)[16].

Table 1.FT.IR-data (cm^{-1}) of compounds (1-11).

Comp.No	FT.IR (only important groups)
1	(CO)carbonyl of Folic acid(1697),(OH)hydroxyl of carboxylic acid:(3425),(NH) group (3365),(CH3) group (2937-2872),C=N of imine (1637),C=C aromatic(1587)
2	(CO)carbonyl of Folic acid(1693),(OH)group:(3408),(CH) aromatic group (3080),C=N of imine (1588),C=C aromatic(1514)
3	(CO)carbonyl of Folic acid(1703),(OH)hydroxyl of carboxylic acid:(3363),(NH) group (3283),(Cl) group (721),C=N of imine (1637),C=C aromatic(1527),CH aromatic (3042)
4	(CO)carbonyl of Folic acid(1680),(OH)hydroxyl of carboxylic acid:(3425),(NH) group (3358),(C-NO2)aromatic group (1442),C=N of imine (1581),C=C aromatic(1492),CH aromatic (3039)
5	(CO)carbonyl of Folic acid(1743),(OH)hydroxyl group:(3433),(NH) group (3389),(CH3) group (2933-2899),C=N of imine (1647),C=C aromatic(1587)
6	(CO)carbonyl of Folic acid(1697),(OH)hydroxyl of carboxylic acid:(3418),(NH) group (3325),(CH2) group (2922-2865),C=N of imine (1533),C=C aromatic(1517), C-O group (1168)
7	(CO)carbonyl of Folic acid(1689),(OH)hydroxyl of carboxylic acid:(3433),(NH) group (3260),(CH2) group (2931-2873),C=N of imine (1589),C=C aromatic(1523),C-O group (1157)
8	(CO)carbonyl of Folic acid(1705),(OH)hydroxyl of carboxylic acid:(3441),(NH) group (3319),(CH2) group (2923-2822),C=N of imine (1599),C=C aromatic(1556),C-O group (1225)
9	(CO)carbonyl of ester(1712),(OH)hydroxyl of phenol:(3431),(NH) group (3253),(CH2) group (2988-2811),C=N of imine (1528),C=C aromatic(1511),C-O ester 1167
10	(CO)carbonyl of ester (1728),(OH)hydroxyl of phenol :(3378),(NH) group (3288),(CH2) group (2978-2870),C=N of imine (1605),C=C aromatic(1577),C-O ester (1166)
11	(CO)carbonyl of ester (1693),(OH)hydroxyl of phenol :(3408),C-O ester (1145),(CH2) group (2926-2881),C=N of imine (1568),C=C aromatic(1514)

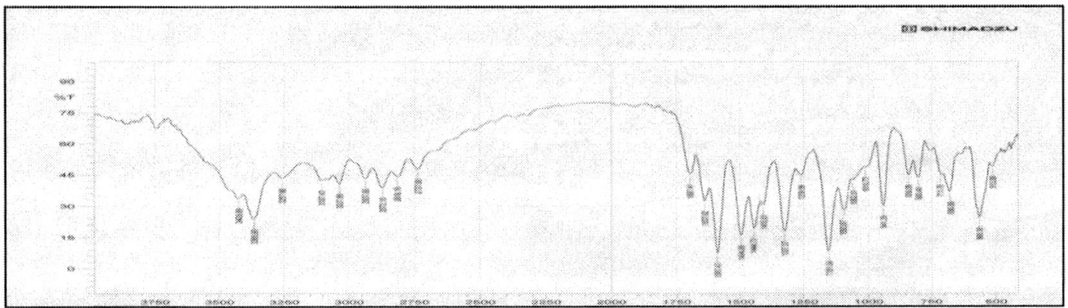

Comp.1

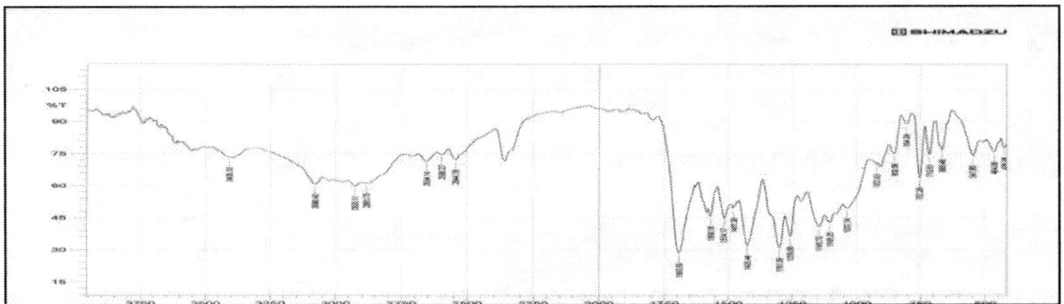

Comp.2

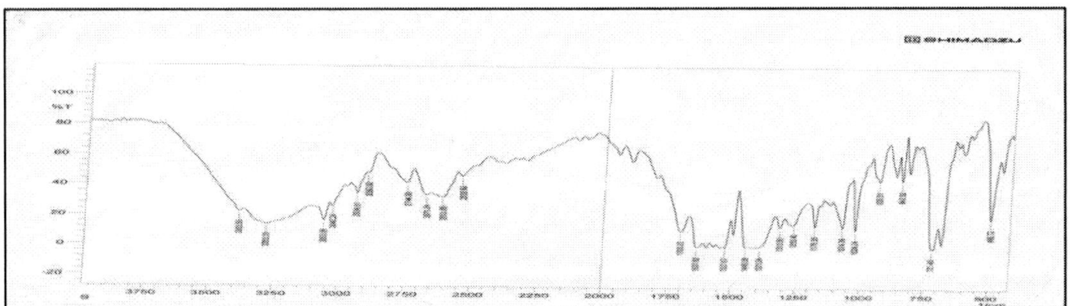

Comp.3

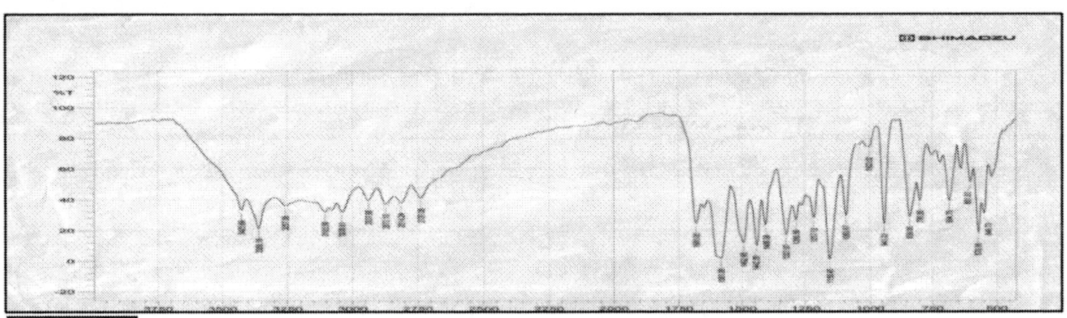

Comp.4

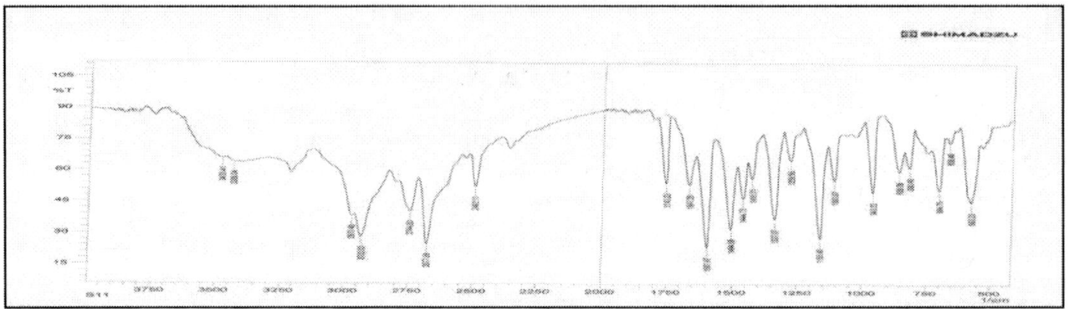

Comp.5

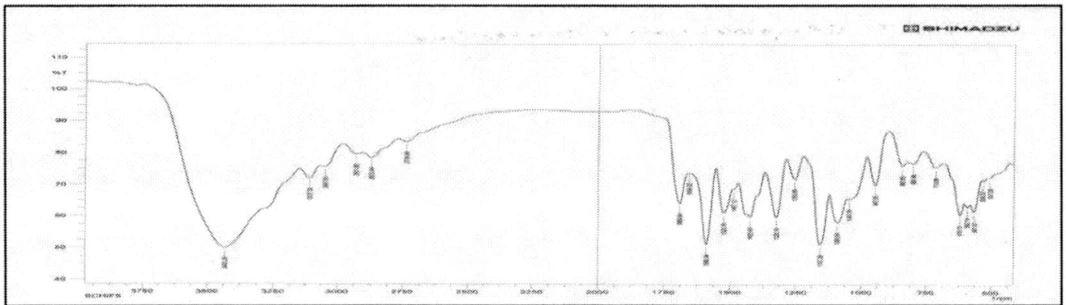

Comp.6

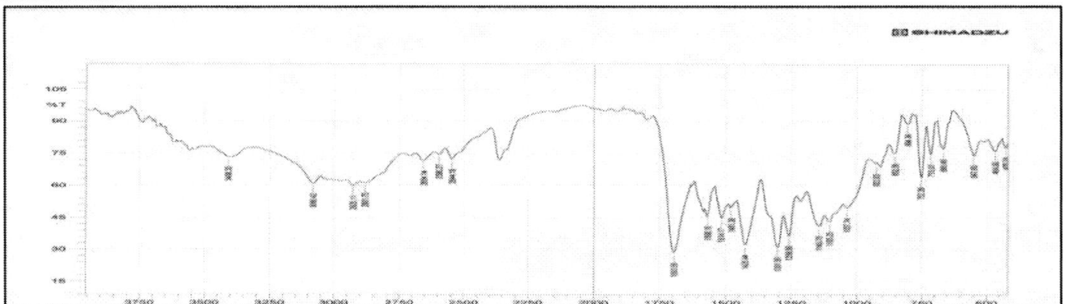

Comp.7

H^1NMR- Spectrum: showed signal at δ(9.77-8.206) due to (CH=N) proton of imine group in compound (1,2,5,8,10)[17] ,signal at (14.93-11.24) due to (COOH) carboxylic acid proton group of folic acid in this compound & showed these signal proton disappeared in compound (10) which appear signal at (0.99-3.76) due to (CH2) of ester[18,19] and other single shown in Table (2).

IOP Conf. Series: Journal of Physics: Conf. Series **1032** (2018) 012059 doi:10.1088/1742-6596/1032/1/012059

Table 2. H[1]nmr-data (δ ppm, DMSO) of some compounds.

Comp.No	1HNMR (only important groups)
1	(14.39-11.24) proton of (OH) hydroxyl of carboxylic acid,(8.47)proton of (C=O-NH), (CH2, (CH3)2) proton group at (1.01-3.42),(CH=N) proton of imine group (9.67),Ar-H proton of phenyl ring (5.91-7.61).
2	(13.50-13.47) proton of (OH) hydroxyl of carboxylic acid,(8.50)proton of (C=O-NH), (CH2,) proton group at (1.13-3.48),(CH=N) proton of imine group (8.206),Ar-H proton of phenyl ring (8.00-7.03),(6.59) proton of phenol group.
5	(11.67-10.31) proton of (OH) hydroxyl of carboxylic acid,(8.45)proton of (C=O-NH), (CH3,CH2,) proton group at (2.05-3.82),(CH=N) proton of imine group (8.49),Ar-H proton of phenyl ring (7.00-7.91).
8	(11.25-9.67) proton of (OH) hydroxyl of carboxylic acid,(7.611)proton of (C=O-NH), (,CH2,) proton group at (2.25-3.42),(CH=N) proton of imine group (8.47),Ar-H proton of phenyl ring (5.9-7.58).
10	(8.47)proton of (C=O-NH), (CH2,) proton group at (0.99-3.76),(CH=N) proton of imine group (9.77),Ar-H proton of phenyl ring (6.5-7.76),(5.99) proton of phenol group.

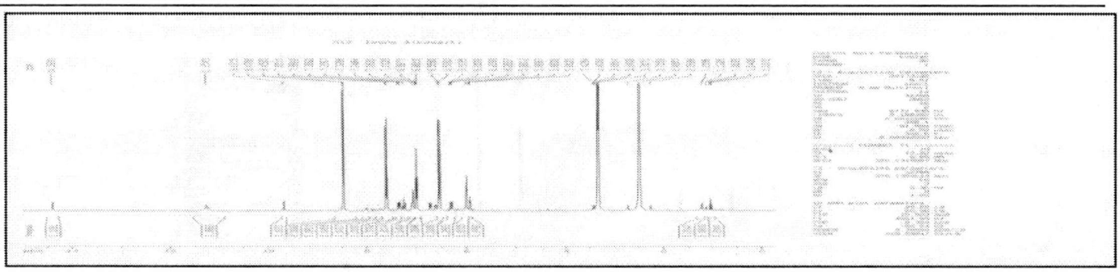

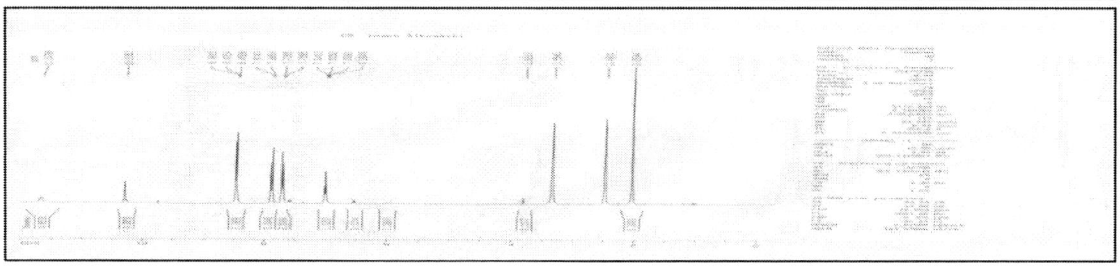

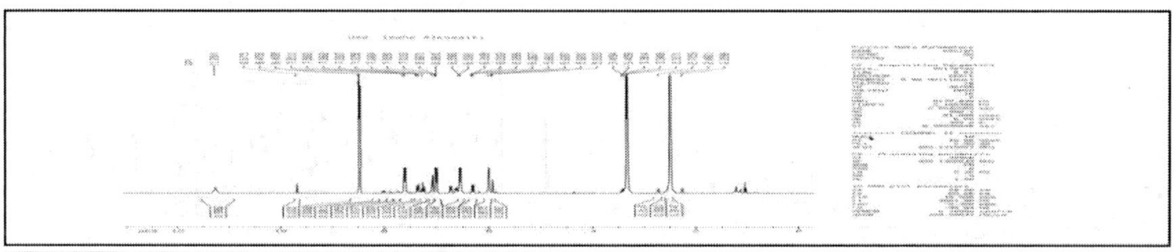

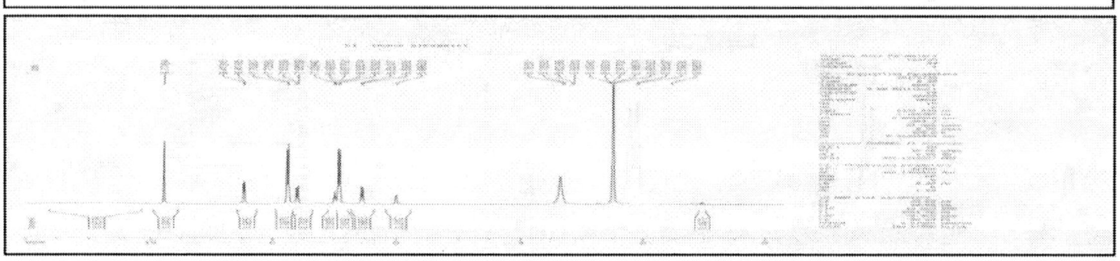

IOP Conf. Series: Journal of Physics: Conf. Series **1032** (2018) 012059 doi:10.1088/1742-6596/1032/1/012059

liquid crystals:

The liquid crystals compounds syntheses have law melting point ,these thermal degree can be using in the industry ,that the nematic phase is more sensitive to electric field ,this feature is useful for work advanced displays .

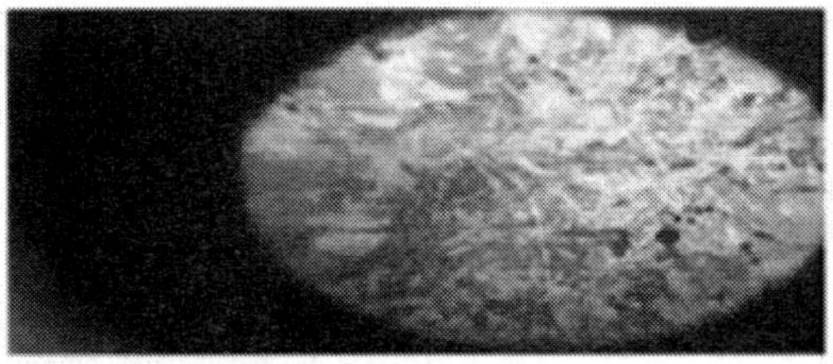

COMP. 1 the appearance of the N

Hemotrophic by heating 78.44 ^{0}C

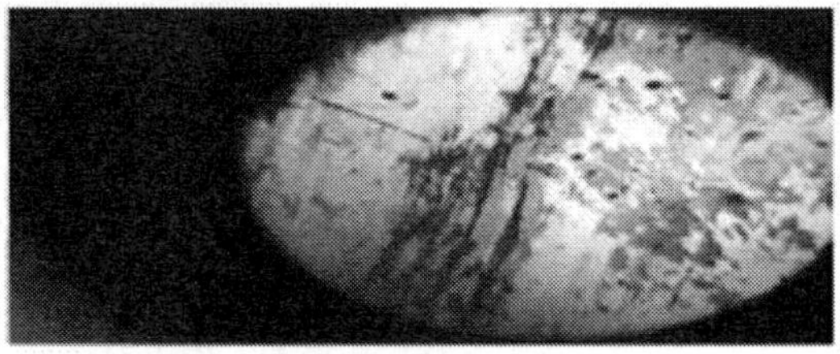

COMP.2 start melting at 65.12 ^{0} C

from Cr to N

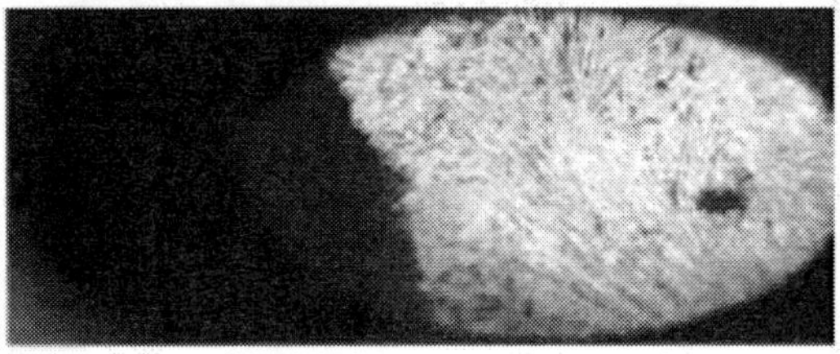

Transformation(Cr to N) COMP.3

99.9 ^{0}C at

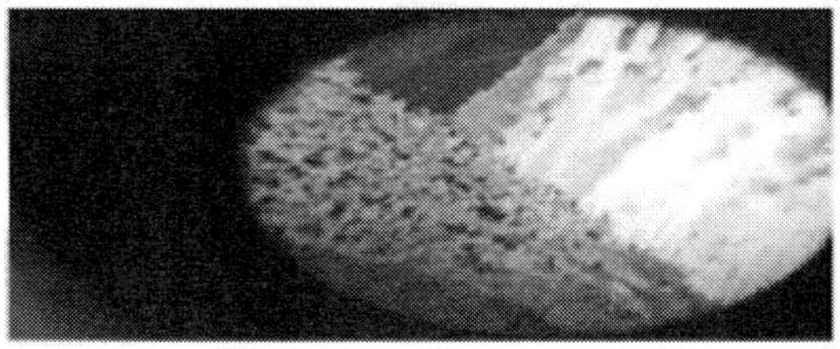

TransformationCOMP. 4

S_c to S_A at 96.94^0 C

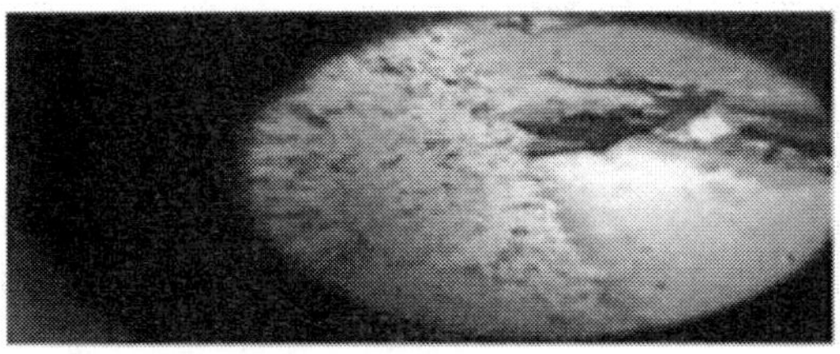

COMP.5 the appearance N

102.86 ^{0}C

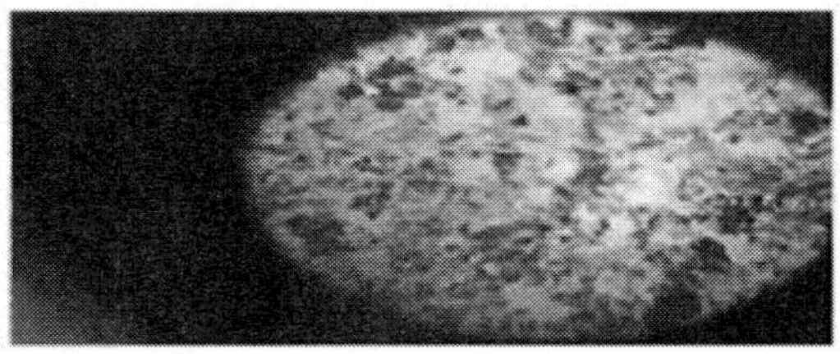

COMP.6Transformation (Cr to N)

phase at 81.4 ^{0}C

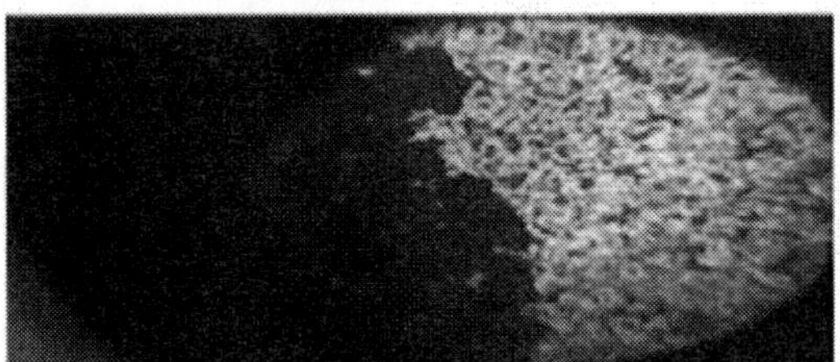

)

COMP. 7 Transformation to Is)N

at 93.24 ^{0}C

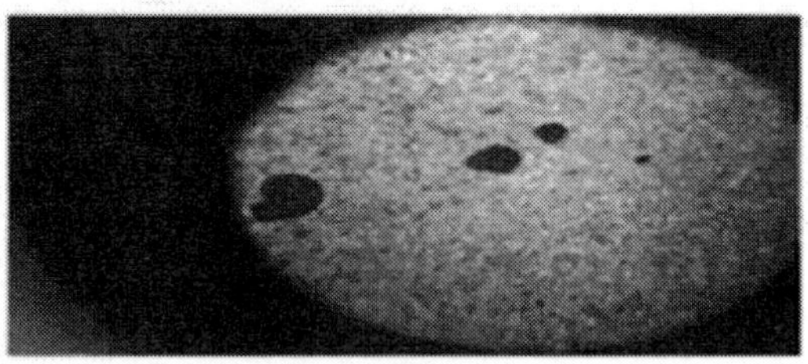

COMP. 8the appearance N

At 87.3 ^{0}C

IOP Conf. Series: Journal of Physics: Conf. Series **1032** (2018) 012059 doi:10.1088/1742-6596/1032/1/012059

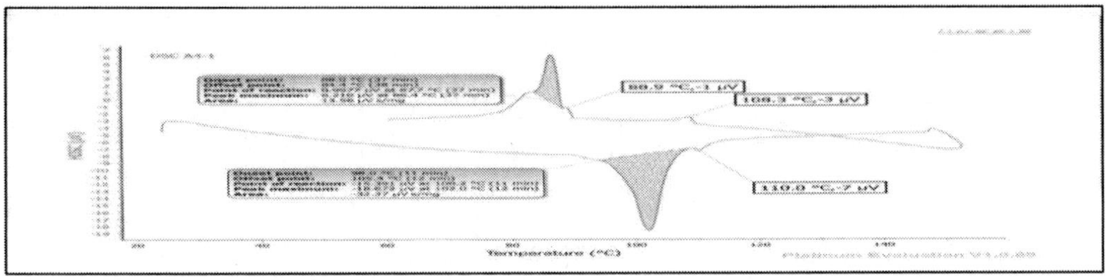

Figure 1. Data of differrntion scanning calorimeter (DSC) of comp. 1

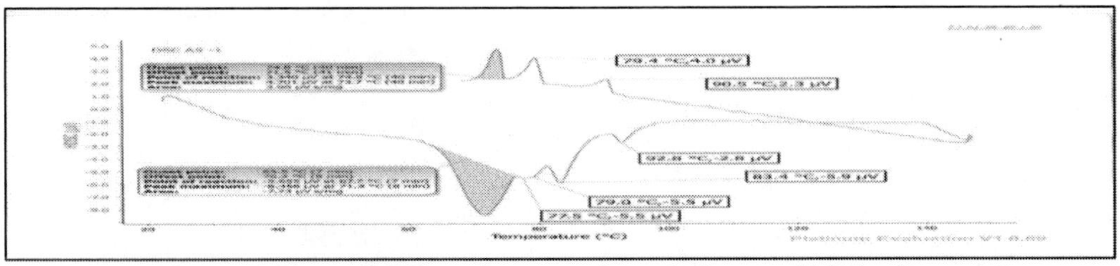

Figure 2. Data of differrntion scanning calorimeter (DSC) of comp. 2

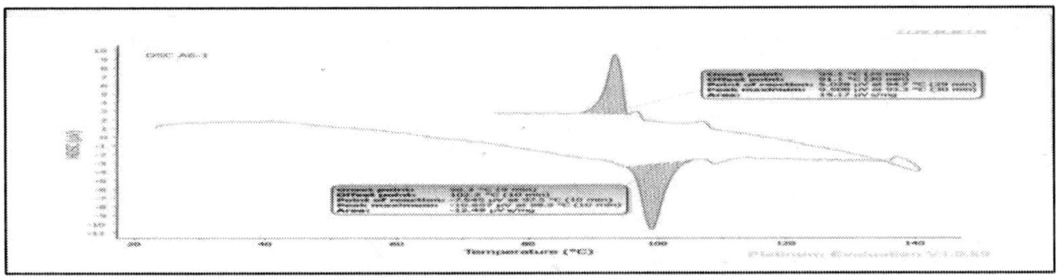

Figure 3. Data of differrntion scanning calorimeter (DSC) of comp. 3

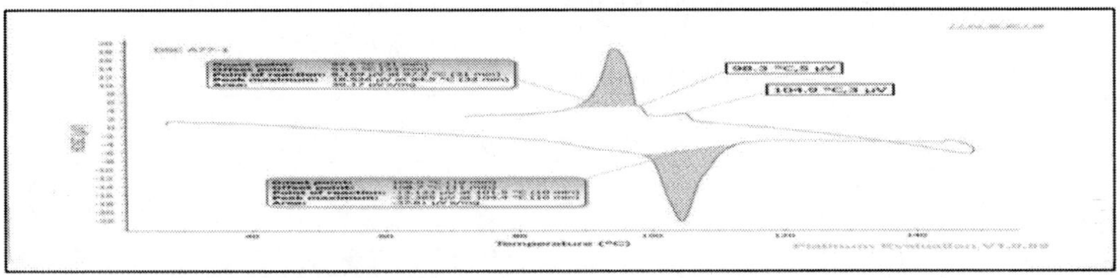

Figure 4. Data ofdifferrntion scanning calorimeter (DSC) of comp. 4

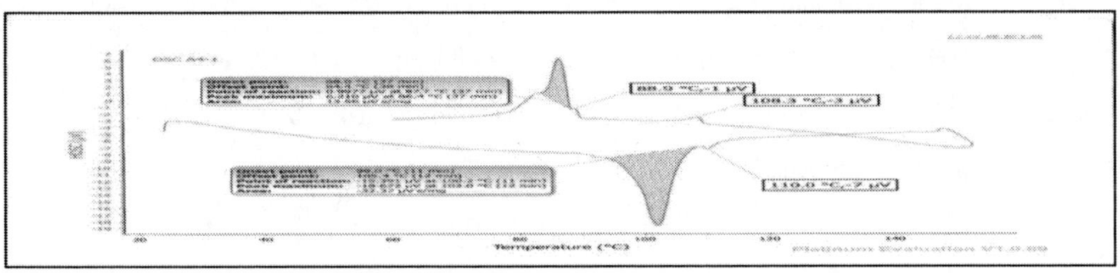

Figure 5. Data ofdifferrntion scanning calorimeter (DSC) of comp. 5

IOP Conf. Series: Journal of Physics: Conf. Series **1032** (2018) 012059 doi:10.1088/1742-6596/1032/1/012059

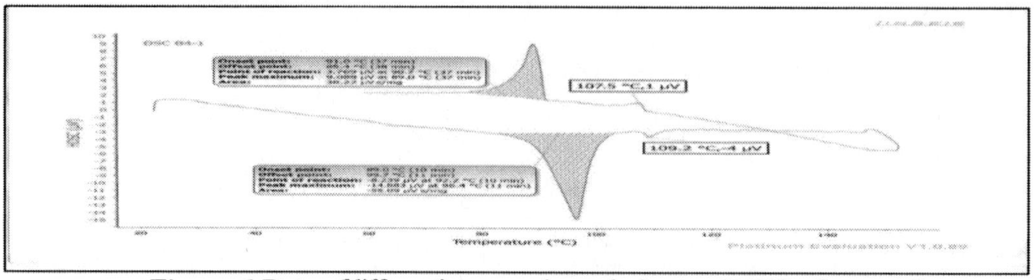

Figure 6. Data ofdifferrntion scanning calorimeter (DSC) of comp. 6

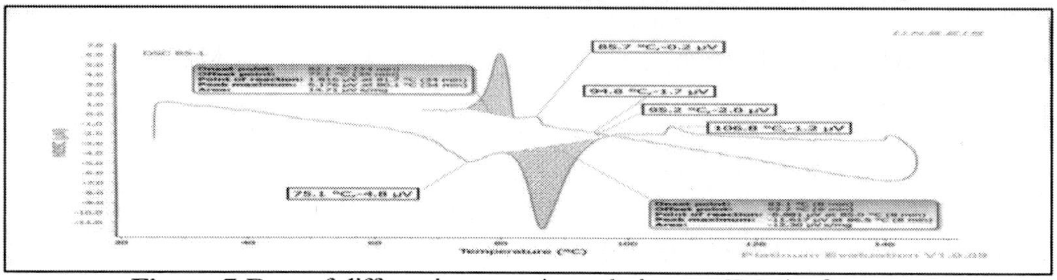

Figure 7. Data of differrntion scanning calorimeter (DSC) of comp. 7

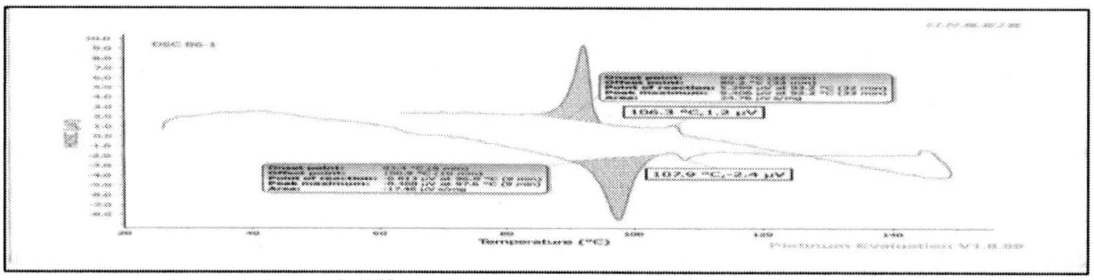

Figure 8. Data of differrntion scanning calorimeter (DSC) of comp. 8

Table 3. Data of thermodynamic compounds (1-8).

No comp.	M.P	Color	Yield	Transition					
				Cr	–Cr		N	–Cr	
				T c$_{o}$	ΔH	ΔS	T	ΔH	ΔS
1	111	Orange	81%	98	32.37-	87.25-	110	7-	18.27-
2	85	yellow	78%	65.5	7.73-	22.83-	83.4	5.9-	16.55-
3	109	yellow	65%	96.3	12.49-	33.82-	110	4.8-	12.53-
4	102	Yellow	68%	97.6	39.17	105.69	104	3	6.23
5	65	White	70%	62.1	36.72	109.57	99	2	5.37
6	91	Yellow	87%	89	55.09-	152.18-	109.2	4 -	10.46-
7	85	Yellow	89%	83.1	12.30-	34.54-	106.8	1.2-	3.15-
8	95	Yellow	88%	93.4	17.46-	47.65-	107.9	2.4-	6.3-

(ΔH=KJ/mol ; ΔS =J /mol .K°)

The calculation of the tow compounds(4,5) were applied in the case of cooling. addition the compound (4) show smectic phase in cooling at temperature 98.3C° , ΔH =5 , ΔS=13.46 .

IOP Conf. Series: Journal of Physics: Conf. Series **1032** (2018) 012059 doi:10.1088/1742-6596/1032/1/012059

References

[1] Richard A.Harvey and Denise Ferrier 2011 *"Biochemistry ,Lippincotts IIlustrated Reviews"* ,Wiliams &Wilkins **5th .ed.**
[2] Guyton and Hall 2011 *"Textbook of Medical Physiology"* **13 th ed**.
[3] G.H.Brownand J.Jwolken 1979 *"Liquid crystels and Biological stercuter"* Academic press (London) **Ch.2**.
[4] D. Yang and S.Wu 2006 *" Fundamentalsof Liquid Crystal Devices"* Wiley (England)**Ch.1**2.
[5] P.J.Collings 2002 *"Liquid Crystals Nature Delicate Phase of Matter"* Princeton University press**2nd edCh.5**75.
[6] S.E .San , O. Koysal and F.N.Ecevit 2003 *Opt.Commun.* **212**405.
[7] N.K.Oleg and G. G.Yuri 2007*electronic-Liquid Crystal Communications January* 22.
[8] G.W .Gray and J.W. Good 1984*"Smectic Liquid Crystals Texture and Structures"* (London Hill).
[9] G.W. Gray,M. Hird,D.L. Acq and K.I.Toyne 1989 *Mol. Cryst. Liq. Cryst.* **172**165.
[10] P. Keller and L. Liebert 1978 *Solid Stute.Phys.Supply.* Academic press (New York) **14**19.
[11] Sridevi Chigurupati 2015 Journal of Medical and Bioengineering**4**5.
[12] Hanaa Hameed Haddad 2016 *American Journal of Analytical Chemistry*7446-451.
[13] Fatemeh Rajabi ,Mohammad Abdollahi and Rafael Luque 2016 *Journal of Materials* **9**557.
[14] R.M.Silverstein,F.X.Webster and D.J.Kiemle 2005 *"Spectrometric Identification of Organic compounds"* **7th ed**. John Wiley and sons Inc. (New York).
[15] D.N.Pavia,G.M.Lampman and G.S.Kriz 2001 *"Introduction to spectroscopy"* **3th ed**Thomson Learning, Inc, (United states of America).
[16] G.Socrates 1980 *"Infrared Characteristic group Frequncies"* John Wiley & Sons ,Ltd (Toronto).
[17] L.D.Fied,S.Sternhell and J.R.Kalman 2008 *"Organic structures from spectra"* **4th ed.**John Wiley & sons (LTD).
[18] J.b.Lambert and E.P.Mazzola 2002 *"Nuclear Magnetic Resonance Spectroscopy"* person education Icc. (Neujersey).
[19] Eman Abdul Wahab Alkuwaity and Ezzat Hussian Zimmam 2016 *J.Der Pharma Chemica* **8(6)**29-39.

IOP Conf. Series: Journal of Physics: Conf. Series **1032** (2018) 012060 doi:10.1088/1742-6596/1032/1/012060

Corrosion conduct of Austenitic stainless steel 316L subjected to surface treatment

(1)Kharia Salman Hassan, (2)Ahmed Ibrahim Razooqi and (3)Munaf hashim Ridha

(1)Department of Mechanic Technician / Institute of Technology -Baghdad, and
(2&3)Middle Technical University- Engineering Technical College/Baghdad

Corresponding author's E-mail: Almaden20002000 @yahoo.com

Abstract. The influence of low temperature liquid nitriding as a surface heat treatment analogy with laser peening treatment at the various of throbs on pitting corrosion of the "AISI 316L Austenitic stainless steel" is investigated in this paper. According to typical ASTM (G71-31) a number of corrosion examination samples are equipped with the measurements of (15 * 15 * 3) mm which distributed into the many groups. Three sets were exposed to liquid nit riding process at temperatures of (500, 400, 300) C° for one hour. The specimens (without coating) were exposed to a number of the throbs (1,2,3) by laser peening. Microstructure variations, compression residual stress, hardness, were inspected in this work. The corrosion and its variables (potential cell, current density) were also evaluated using the potential stat examination and applying the Tafel method using saltwater solution (3.5% NaCl). Tafle equation was used to compute the corrosion degree. The results revealed that the liquid nitriding participated to raise the corrosion rate at (500) C°, compared to the original metal because of chromium nitride and also leaser peening participated to the increase in the corrosion rate due to plastic deformation which led to the heterogeneity in the microstructure but liquid nitriding at temperature (400) °C gave the best result where it was closer to the parent metal's, also laser peening at one throb showed the lower corrosion rate.

1. Introduction.

The Austenitic stainless steels are a secondhand in manufacturing of diverse parts due to their admirable corrosion resistance and microstructure, though these materials have meager wear counteraction and relatively low hardness. An effort was earlier exerted to elevate " the surface hardness and wear" counteraction of "austenitic stainless steels" using " chemical surface treatment" like nit riding , nitro-carburizing which made to increase in corrosion resistance due to the nitrogen attachment or the carbon with the chromium making nitrides or carbides, and decreasing its amount in the solid solution. Corrosion behavior can be improved by nitriding method at low temperature which is performed at temperatures less than 450°C. These low treatment temperatures release a nitride layer from chromium nitride sedimentation CrN and create a nitride film which contains iron nitrides and a nitrogen supersaturated solid solution, named "extended Austenite phase" (αN) by analogy with the previously expanded austenite [1,2]. stainless steel is subjected to Pitting corrosion due to Acid chlorides which respond with chromium element to form the very soluble chromium chloride ($CrCl_3$) therefore chromium is extracted from the this film neglecting lone the effective metal . As the chromium is disbanded, the electrically motivated the chlorides to made hole into the stainless steel, originated a round smooth fence pit. The remaining solution in the rap is ferric chloride ($FeCl_3$), that is in charge of the corrosion to stainless steel [3].

Three impact factors on the pitting corrosion: chloride content, pH, and temperature of the media. In common the probability of pitting corrosion increases when temperature and chloride content are higher and the pH is lower. Leaser shot peening (LSP)is a type of "a mechanical cold working"

IOP Conf. Series: Journal of Physics: Conf. Series **1032** (2018) 012060 doi:10.1088/1742-6596/1032/1/012060

technique that throbs shocks all the face by great energy and produced residual compressive stresses nearby, 4–5 times deeper and greater in intensity with homogeneity through the surface that in consequence encourages the refining of the microstructure. The laser spot extent and shape, are significant factors in the peening procedure. Circular shaped laser spots are widely used .more than greater compressive residual stresses are awaited for smaller patches because of the reversed proportionality of spot scope and intensity of energy. Manifold laser jolts and overlapping of spots, meant to protect great zones, have exposed important effect on the remaining stress shape. This is because of the actuality that the greater quantity of jolts induced the larger plastic deformation until a soakage dot .In addition, overlapped areas show a relatively constant distribution of compressive residual stresses after LSP, higher laser throbs showed deeper residual stresses than in the case of on its own laser shock, deeper stresses can be attributed to the plastic deformation that created more dislocation movements. The remaining stresses at the face of the plate are lesser in the dual laser shocks, whereas it was greater for on its own laser shock. This can be defined as the stress fullness point is on its own take the laser shock so the stresses would be relaxed due to rise in shocks numbers. Moreover, the sucker film may have been ruined through multiple jolts [4]. Many researchers took on the subject as:-

E. Haruman [1] studied the surface film resulting from liquid nitride of AISI- 316L austenitic stainless steel at low temperatures (400-500 C°) by using X-ray diffraction and the identification of phases as well as the study of electrochemical corrosion behavior at scan potential from (-1000 + 1000 mv) relative to the open circuit voltage. The obtain result show that the liquid nitrate generated a film containing the extended austenite at a t emperature of less than 500C° which contributed to improving the corrosion rate.

Askar Triwiyanto [5] studied the importance of chemical, thermal factors on the surface hardness of Austenitic stainless steel type 316L. Thermal ,chemical treatments of carbonate, nitrate and carbo-nitride at low temperatures of less than 450 ° C to get an oxidized film of extended Austenite phase without sedimentation of chromium carbide or nitride during carbonization or Nitrate and the thickness of the layer is irregular as it reached 83 µ in by nitriding for a time of 8 hours while by carbonization and carbonate give more depth for the same period and confirmed that the depth of the film can be greater than before when increasing the time for all thermal dealings and the resulted surface hardness amounted to 1600HV for the method of nitride and 1150 HV and 500HV for carbonizing which participated to improving wear resistance and corrosion compared to metal without treatment.

Abdullahi K. Gujba [6] take the surface hardening by the laser and its effect on the surface finishing and the depth of the hardened layer and the amount of stresses produced and focused on the impact of laser to improve the properties of the metal, including the structure of microscopic and corrosion resistance, and the age of fatigue and compared the results with ultrasound, The selection of shock conditions were mainly velocity, wavelength, frequency.

G. S. Frankel [4] make an offers overview of the criticality elements which effected on the pitting corrosion of metals. The phenomenology of pitting corrosion is debated, containing the special effects of alloy composition, environment, potential, and temperature. abstraction is then assumed of studies that have focused on many steps of the pitting procedure, containing the fracturing of the inactive layer, metastable pitting, and pit growth.

Yuji SANO*, Koichi AKITA [7]. studied the improvement of fatigue characteristics using rotating-bending for several metals, austenitic stainless steel 316L, titanium alloy(Ti-6Al-4V) and cast aluminum alloy using LPWC which informs compressive resisdual stress deprived of any external arrangements. Materials are peened in waterish setting with laser pulses of about 100mJ.Fatigue lives increase even though the rise in surface eliminated SCC susceptibility of sensitized austenitic stainless steels, nickel-based alloys and their weld metals. LPWC has been utilized to prevent SCC.

The influence of nitride surface treatment at different temperature and laser peening at different pulse on pitting corrosion of AISI 316L- stainless steel are manipulated.

2. Experimental work.

2.1. Metal select.
The AISI 316L Austenite stainless steel have a hug kinds of applications such as the chemical industry of petroleum, food industry, shipping and house equipment. The analysis of the elements which are exposed in table.1 were made using ARL spectrometer tool.

Table 1. The elemental composition analyses of 316 L

Element %wt	Fe	C	Si	Mn	Cr	Mo.	Ni	Al	Co	Cu	N	Ti	V	W
Actual value	68.8	0.068	0.429	1.63	16.9	2.57	8.49	0.005	0.363	0.379	0.006	0.012	0.179	0.082
Normal value	-	0.3	0.75	2	18.5	2-3	10-14	-	-		0.1	-	-	-

2.2. Samples Preparation.
Numerous samples were prepared from AISI 316L stain less steel sheet metal for pitting corrosion examination depending on the ASTM (G 71-31) standard with the dimensions of (15*15*3 mm).

2.3. Samples distribution.
Pitting corrosion specimens were divided to seven groups which were listed in table 2.

Table 2. Groups of samples

Specimens symbol	stat
A	as received
B	liquid nit riding at 500C°
C	liquid nit riding at 400C°
D	liquid nit riding at 300C°
E	laser peening one pulse
F	Tow pulse laser peening
G	Three pulse laser peening

2.4. Laser peening.
The specimens having symbols (E, F, G) in table (2) are subjected to laser peening at wave length (λ =1064nm), power=500m/ joule, plus time 10sec. number of (1,2,3) pulse respectively .

2.5. Salt Bath Nitriding.
Liquid nitriding was performed on groups (B,C,D) of specimens as shown in Table(2). The nitriding solution was prepared from (61% sodium cyanide , 24% sodium chloride and 15% potassium carbonates) salts by weight, which were put in a stainless steel vortex then in a furnace at 500°C to melt together before inserting the specimens, which were suspended by thin steel string then were heated in another furnace at 150 °C to become free from moisture to avoid agglomeration due to moisture in case of placing it inside to melt , then the vortex was taken out from furnace, and the specimens put in it and returned to the furnace for one hour and quench in water.

2.6. Residual stress measurements.

X-ray diffraction sin2ψ method was used to determine the remaining stresses. the process is used to evaluated the strains at particular diffraction angles and lattice planes, according to "Bragg's law". Hardness by Rockwell method using steel ball was also used for hardness test. The obtained results are shown in table 3.

Table 3.

Symbol	Comparesion Residule stress Map	Surface Roughness (μm)	Rockwell (B) Hardness in Kg/mm^2
A	-18	0.03	38
B	-265.889	0.029	78
C	-241.917	0.018	75
D	-273.636	0.017	68
E	-287	0.006	84
F	-342	0.009	84.5
G	-372.732	0.001	87

2.7. Microstructure Examination.

All specimens were prepared in a series of steps comprised of, grinding by using SiC paper having the diverse grits of (240,320,600,800 &1000), Polishing which was done using cloth made of polishing with alumina oxide (Al_2O_3) of size (0.3μm) after that they were washed with water and alcohol and dried), then Etching being performed using solution consists of (49 $CuSO_4$ml + 20ml HCl + 20ml distilled water). After that the samples were washed with water and alcohol and dried by air. The specimens were tested using Nikon ME-600 optical microscope provided with a NIKON camera, DXM-1200F as shown in figure(1A,B) .

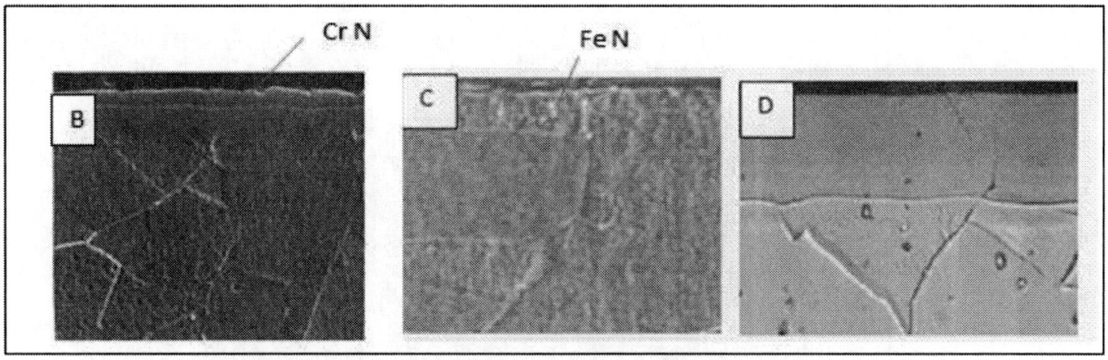

Figure 1. ESM micrograph of nitride specimens (B, C and D).

IOP Conf. Series: Journal of Physics: Conf. Series **1032** (2018) 012060 doi:10.1088/1742-6596/1032/1/012060

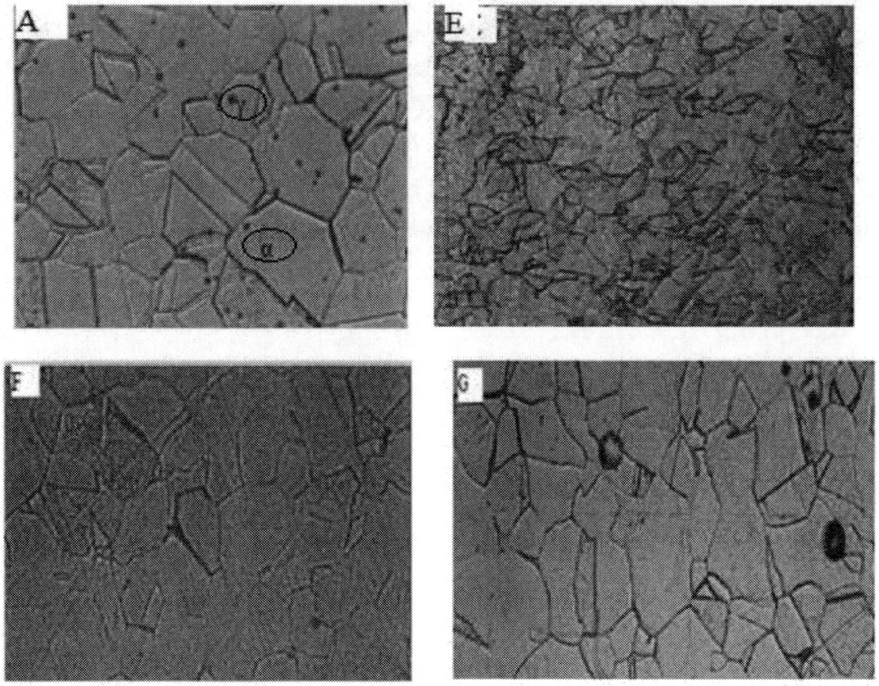

Figure 2. Microstructure for 316L stainless steel specimens40X.

2.8. Corrosion test.

Specimens for corrosion examination were prepared according to standard ASTM G71-31 immersed in the (sea water) 3.5% NaCl solution .The current cell perusals were getting through scan of the potential from (-250 to + 250) mV relative to (OCP). Scan rate describes the rapidity of the potential sweep in mV/sec and it is fasting (10 mv). The examinations were made by a WENKING Mlab with potentiostat and SCI-Mlab corrosion measuring system from Bank Electronics-Intelligent control GmbH, Germany 2007, which shown in figure 2. Tafle equation was used to compute the corrosion degree. The results revealed in table (4) and figure 4.

Figure 3. Corrosion test tool.

IOP Conf. Series: Journal of Physics: Conf. Series **1032** (2018) 012060 doi:10.1088/1742-6596/1032/1/012060

Table 4. Corrosion examination consequence

Specimens symbol	I corr.µA	Potential mv	Corrosion rat m.p.y
A	1.57	-304	0.69
B	47.39	-559.7	20.85
C	1.65	-97.5	0.726
D	17.13	-59.7	7.53
E	2.06	-359.2	0.91
F	19.82	-546	8.72
G	25.99	-428	11.435

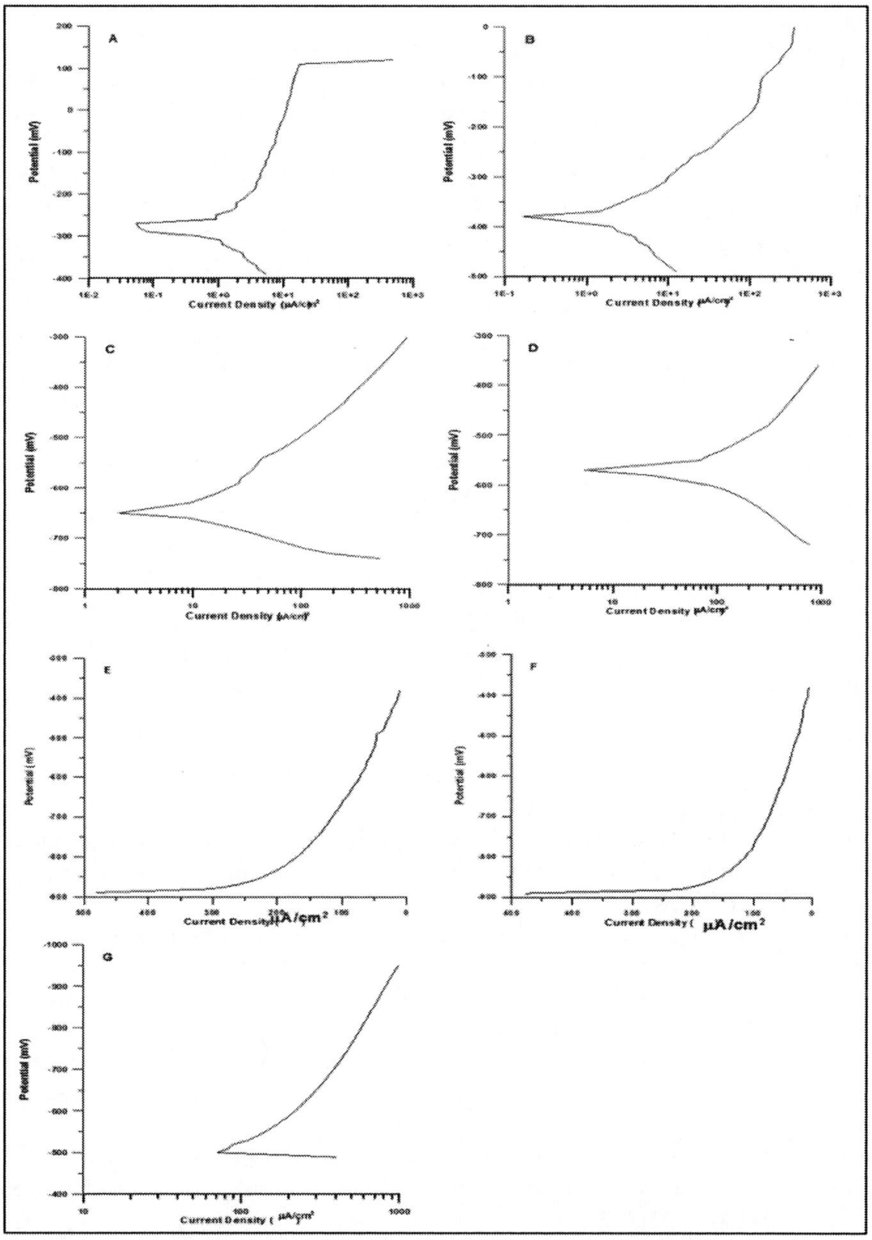

Figure 4. Polarization Curve for all specimens.

3. Discussion.

Figure1. demonstrates the influence of liquid nitriding at various temperatures (300,400,500 C°) on the microstructure in comparison with parent metal (1.2 A) composed of austenite and ferrite phases that obtainable high corrosion resistance, table 3. Nitriding surface treatments at great temperature 500° C and with increasing existence of nitrogen and from the hardening medium exuberant sedimentation of chromium nitrides happens as a film on the surface. This layer is white colored and has high hardness due to the presence chromium nitride (CrN) as shown in figure1., sample (B). But the microstructure under this layer is similar to that (1.2) of sample (A), chromium nitride (CrN) causes appreciable deterioration of the corrosion counteraction of Austenitic stainless steel and this is clear in sample (B) table 3. this feature can be enhanced by the nitriding method at low temperature (300,400) °C to abolish the formation of chromium nitrides and alternatively, the strengthening influence will be changed by great satiety of interstitial species in austenite matrix which leads to the hardening of the surface area several tens micro meter thick. This sedimentation -free nitride film not only exhibits relative high hardness but as well possesses good corrosion resistance due to the availability of retaining chromium in solid solution for corrosion protect. Extended austenite without nitrides is gained as soon as great quantities of atomic nitrogen are melted in stainless steel at temperature under 450 °C. The nitrogen atoms are supposed to reside in the octahedral interstices of the F.C.C. lattice. Because of the efficiency of the inciter compressive remaining stresses using LSP in refining the microstructure, the corrosion behavior of materials will be developed. Studies have shown that this behavior depends on the LSP process variables like the numbers of pulse and its effect on corrosion behavior which gives less corrosive damage at one pulse that presented in sample (E) in table 3. This is due to the fact that the greater numbers of shocks the larger induced plastic deformation until a saturation point is reached this is clear in samples (F, G) in the same table as a result of the compressive RS. The common consensus confirms that the used of leaser shot peening (LSP) is an significant technique for optimization the corrosion behavior of materials by decreasing the current of corrosion at the anodic with reduced the number of pulses. This can be attributed to the refinement of the grains and the reduction of the effect of the compressive residual stress (RS) film.

4. Conclusion.

From the investigated results in this paper showed that surface layer formed using nitriding surface treatment at (400 C°) was contributed in decreasing the pitting corrosion rate nearly from the base of the stainless steel AISI 316L and laser peening at one pulse gave the same effect on corrosion rate compared with base metal.

References

[1] E. Haruman and Y.Sun 2006 "Low Temperature Fluidized Bed Nitriding of Austenitic Stainless Steel" *Solid State Phenomena* **118** 125-130.

[2] Lakh winder singh1 2010 "Effect of shot peening on hardening and surface roughness of nitrogen austenitic stainless steel" *International Journal of Engineering Science and Technology* **2(5)** 818-826.

[3] V. Toshkov 2007 "On low temperature ion nitriding of austenitic stainless steel AISI 316" *Journal of Achievements in Materials and Manufacturing Engineering* **25** 1.

[4] G. S. Frankel 1998 " Pitting Corrosion of Metals A Review of the Critical Factors " *Journal of the Electrochemical Society* **145(6)** 2186-2198.

[5] Askar Triwiyanto1, Patthi Husain1 2016 "Low Temperature Thermochemical Treatments of Austenitic Stainless Steel Without Impairing Its Corrosion Resistance www.intechopen.compp 318-337.

[6] Abdullahi K. Gujba and Mamoun Medraj 2014 "Laser Peening Process and Its Impact on Materials Properties in Comparison with Shot Peening and Ultrasonic Impact Peening" *Materials* **7(10)** 7925-7974.

[7] Yuji SANO and Koichi AKITA 2006 "Laser Peening without Coating as a Surface

Enhancement Technology" *Journal of Laser Micro/Nanoengineering* **1(3)** 161-166.

[8] P.PEYRE, R.FABBRO, L.BERTH and C.DUBOUCHET 1996 "Laser Shock processing of materials,physical processes involved and examples of applications " *Journal of laser applications* **8** 135-141.

[9] Wenquan Zhang 2016 "Residual Stress Distribution and Microstructure at a Laser Spot of AISI 304 Stainless Steel Subjected to Different Laser Shock Peening Impacts" *Article Metals* **6(6)** 1-10.

[10] RTOMIR Donik and Aleksandra Kocijan 2014 "Comparison of the corrosion behavior of austenitic stainless steel in seawater and in a 3.5 % NaCl solution" Materials and technology **48(6)** 937–942.

[11] F. Cajner 2015 "Influence of nit riding on corrosion resistance of martensitic X17CrNi16-2 stainless steel" *Mat.-wiss. u. Werkstofftech.* **46** 1.

IOP Conf. Series: Journal of Physics: Conf. Series **1032** (2018) 012061 doi:10.1088/1742-6596/1032/1/012061

Using Low Cost Membrane in Dual-Chamber Microbial Fuel Cells (MFCs) for Petroleum Refinery Wastewater Treatment

Shrok Allami, Basher Hasan, Mutawer Redah, Hayder Hamody, Zainab D Abd Ali

Department of Hydrogen and Biofuel, Ministry of Science and Technology, Iraq.

Corresponding author's E-mail: shrokabdullah@yahoo.com

Abstract. There are enormous methods to treat petroleum refinery wastewaters (PRW) that contain water-soluble hydrocarbons which cannot be separated by physical methods. Using microbial fuel cell (MFC) is a new PRW treatment method. Using potassium permanganate as cathodic electron acceptors in the cathode apartment of MFC with cheap proton exchange membrane instead of expensive Nafion was used in this work to treat PRW from Al Dura refinery. Potassium permanganate was used as cathodic electron acceptor to enhance MFC power production. So the effects of potassium permanganate amount on the MFC performance and PRW treatment results were investigated. Maximum power production at the room temperature of 1.0032 W/m2 using 0.125 g/L of permanganate concentration, the maximum COD removal efficiency was 71.24 % during 48 hours.

1. Introduction.

In the past 30 years, research had expanded to find renewable energy sources rather than fossil fuels. Developments in the various fuel cell systems that benefit from bioenergy and renewable energy have been consistent with general trends to find these alternatives to these cells are microbial fuel cells.

Microbial fuel cells are electrophysiological cells of the Calvinistic type. The cells that have the potential to generate electricity from the chemical and biological reactions in which they occur. It usually consists of two chambers of anode and a cathode connecting them between a brine bridge or an ion exchange membrane (e.g. Nafion).

Figure 1 shows a typical microbial fuel cell consisting of two chambers of anode and cathode separated by an ion exchange membrane that allows protons to pass while oxygen and other compounds are obscured [1].

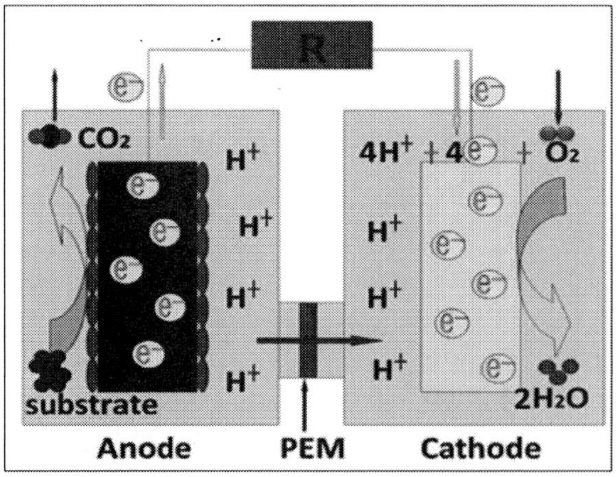

Figure 1. Schematic diagram showing a preferred design for microbial fuel cell, PEM (e.g: Nafion) [1]

IOP Conf. Series: Journal of Physics: Conf. Series **1032** (2018) 012061 doi:10.1088/1742-6596/1032/1/012061

The microbes adhering to the anode electrode dissolve the organic matter present in the anodized solution producing electrons, protons and carbon dioxide. The electrons and protons produced by the microbes move to the opposite side of the cell (cathode) through the outer circle and the ion exchange membrane respectively. In the cathode chamber, the protons and electrons react with the existing oxygen to form the water and an electric current as described in the following reactions [2].

Anodic reaction:

$$CH_3COOH + 2H_2O \longrightarrow 2CO_2 + 8H^+ + 8e^- \quad (1)$$

Cathodic reaction:

$$8H^+ + 8e^- + 2O_2 \longrightarrow 4H_2O \quad (2)$$

Overall reaction:

$$CH_3COOH + 2O_2 \xrightarrow{microbe} 2H_2O + 2CO_2 \quad (3)$$

PRW has two aqueous and oily phases. Two PRW treatment stages were traditionally used. Physical methods to separate oily phase. The main output of the physical process is aqueous phase that contains suspended oil and also solid particles. Then, during the second treatment stage, the suspended particles are separated by coagulation. The output is a homogenous aqueous phase containing water dissolved hydrocarbons including phenol, benzene, xylenes, and other pollutants, which failed to separate by physical methods. Biological approaches are traditional methods to treat this wastewater [3]. Traditional biological process utilizes microorganisms to oxidize organic matter to simple products (CO_2, H_2O, and CH_4) [4]. In the recent years, biological PRW treatment process based on direct oxidation of dissolved hydrocarbons have been developed with aid of photocatalytic and photo-Fenton methods [5-6].

All of the mentioned methods, however, require catalyst, large amount of oxidant, and high energy consumption. PRW treatment by microbial fuel cell as an indirect oxidation method can be considered as a cost effective method not only because of low energy demands but also due to its power output to supply energy for treatment unit [7]. Degradation of hydrocarbons soluble in PRW using dual-chamber MFC has been investigated by Luoa [8].

In the present work, a dual-chamber microbial fuel cell with low cost membrane was used to treat petroleum refinery wastewater from Al Dura, Iraq. In the MFC, As a result of degradation process, electrons travel through external resistance and protons pass through membrane. The corresponding protons migrate to the cathodic chamber. At the cathode, an electron acceptor is reduced by the electrons via the circuit and the protons via the membrane [9]. The effects of potassium permanganate amount on treatment of PRW and MFC performance were investigated.

2. Experimental work: MATERIALS AND METHODS

A.Materials

- Two PVC pipe chosen as a cathode and anode chambers for MFC with the volumes of 300 ml and 7 cm diameter. Each pipe contain 5 and 3 mm holes for solution addition and wires inlets respectively. The holes in anode side were sealed with small cup to prevent air admittance.
- Cathode: 6 cm diameter anodizing aluminum.
- Anode: the MFC used in this experiment utilized non-catalyzed graphite sheet 2*4 cm.
- Membrane: 38.5 cm^2 hydrophobic LowDensity Polyethylene LDEP as a low cost material utilized in this MFC between the two chambers.
- Solution containing organic matter at anaerobic anode side received from wastewater treatment lab, 7 pH and 500 mg/L COD (chemical oxygen demand).
- Potassium permanganate aqueous solution used in cathode side as an electron acceptor, with 0.0625, 0.125, and 0.25 g/l concentrations.
- In conventional MFC platinum coated carbon cloth and carbon cloth and activated with biofilm were used as cathodes and anodes respectively. In most MFCs expensive cost Nafion®, Agar and Gore-Tex membranes were used as membrane materials.

B. Fuel cell configuration

The dual-chamber fuel cell used in the present paper constructed by joining the cathode and anode chambers between two square Perspex glasses by 4 screws. The distance between anode and cathode electrodes was constant in all experiments (3 cm). Copper wires were used for contact with electrodes after sealing the contact area with 'epoxy' material. A 38.5 cm^2 proton exchange membrane was embedded between two chambers. Voltage recorder using data logger, Hantek365A embedded parallel to external resistance, 10 Ω. All the experiments carried out at room temperature.

C- Electric parameters calculations Current and power densities are the most important parameters that should be measured. Ohm law: used to calculate current intensity [10].

$$I = V/R \qquad (4)$$

Where V and R denote, respectively, potential difference and external resistance.

In addition, power obtained using the following equation:

$$P = IV \qquad (5)$$

Power density (W/m^2) and current density (A/m^2) are calculated by dividing power and current intensity to the anodic surface area (m^2).

D- Chemical oxygen demand measurements

COD is the main characteristic of the wastewater, evaluated as an index of MFC performance. In the present research, APHA standard method was used to measure COD concentrations.

3. Results and Discussions

A-Operational MFC power generation

Immediately after cell filling with wastewater and potassium permanganate aqueous solution voltage reached 685, 657, and 100 for cells with cathode side contain 0.25, 0.125, 0.0625 g/L potassium permanganate aqueous media. The proper selection of aluminum and graphite electrodes according to their difference potential in their solutions and oxidizing of the organic matter by microorganisms assumed to be the chemical and the biological factors might have been the reason of the high initial voltage results. Thereafter, the voltage rapidly increased due to biological activity. The instability that can be attributed to the instability of the biofilm on the anode surface during the working period that include growth and degradation [11]. Cell voltages stabilized at 823, 836, and 782 mV after 48 hours, voltage variation as shown in Fig 2. Increasing potassium permanganate concentration in cathode side increase electrolyte conductivity and decrease ohmic and activation losses [10], that increase power density 0.3292, 1.0032, and 0.7038 W/m^2 with potassium permanganate concentration as shown in Fig.3. Decreasing power density at o.25 g/l potassium permanganate concentration related to the formation of $Al–KMnO_4$ as the main product of the aluminum surface and potassium permanganate, cathode solution, reaction that limits electron transfer to the solution and complete the cell reaction to generate electricity.

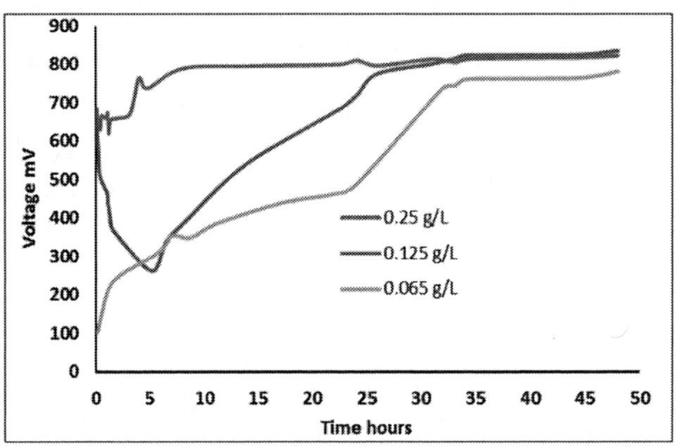

Figure 2. Voltages variation with time for cells with different potassium permanganate in the cathode chamber

Voltage inclined when using the cell after few months without new solutions injection into the chambers, as shown in Figure 4 for the cell with cathode contain 0.25% g/L potassium permanganate aqueous media, this attributed to the formation of brown layer over aluminum surface consist from Al–$KMnO_4$ complex [12] which classified as a protective film. As well as a fully degradation of organic matter in the PRW, as COD measurement confirmed, reducing electron generation and increasing the system resistance.

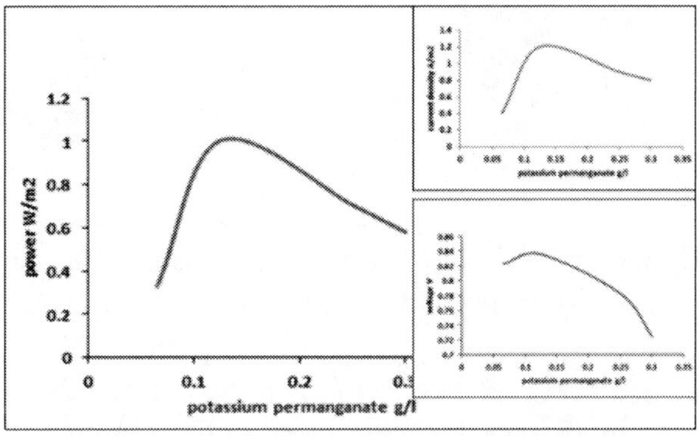

Figure 3. Power densities for cells with different potassium permanganate in cathode side.

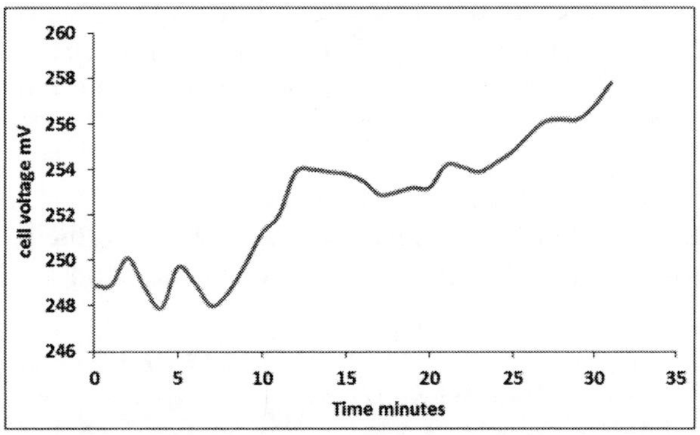

Figure 4. voltage variation with time for cell contain 025 g/l potassium permanganate after few months.

Under typical atmospheric conditions, a native oxide or passive film naturally forms on aluminum. The native oxide layer is nonuniform, thin and noncoherent. The native oxide film imparts a certain level of corrosion protection. Anodizing aluminum is capitalizing on the natural phenomenon of passive film formation on aluminum in a production environment. Anodizing yield a uniform and continuous protective oxide film. It is highly ordered columnar nanoscale structure with central pore [13]. Pores reactivity enables adsorption of the electrolyte ions, as the immersion period increases, the protective film is ruptured and a complex Al–KMnO4 formed that closed the pores and a thick, nonconductive, inhibitor layer formed. Formation of the nonconductive layer increase the double layer and the whole system resistance that have great effect on the cell potential.

B- Chemical oxygen demand COD removal

COD is amount of required oxygen to convert total organic matter present in wastewater into carbon dioxide and water. Treatment process reduces COD. Figures 4 show, COD removal of PRW versus time using potassium permanganate as cathodic electron acceptors, with 100 Ω of external resistance. As can be seen in Figures 5, COD removal efficiencies were 69.254% and 71.24 % and 69.75% during 48 hours in the presence of 0.0625, 0.125, and 0.25 g/l concentrations potassium permanganate respectively.

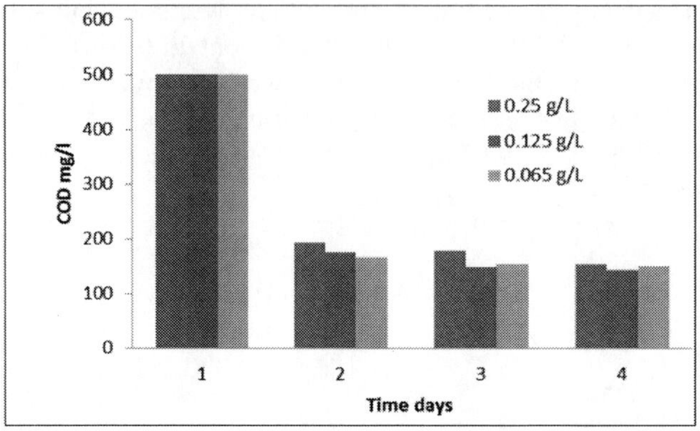

Figure 5. COD removal vs. time using different potassium permanganate in cathode side

During PRW treatment into MFC, COD was found to be reduced extensively in the first day. To explain COD rapid reduction in the first day return back to the variation of voltage with time, it was instable during the first 24 hours that referred to the biofilm instability. Biofilm degradation and rebuilding consume the organic matter to produce electron during adhering of the microorganisms on the electrode. When the biofilm became stable after that time it resist electron transferring to the anode surface through its thickness and inhabit COD reduction. Increasing potassium permanganate concentrations in cathode chamber, as electron acceptor, enhance the electron flow from the anode to the cathode through the external circuit and the protons through the membrane that increasing organic matter degradation reaction and reducing COD.

C- **MFC efficiency and performance calculations**

We depend in our design and electrodes materials selection into their electrochemical properties into the chamber solutions. For theoretical calculations the anode polarization potential found using eq.6 [14]

$$E_{anode} = E^0_{anode} - \frac{RT}{8F}\ln\left(\frac{[CH_3COO^-]}{[HCO_3^-]^2[H^+]^9}\right) \quad (6)$$

Where $E^{\circ}_{anode} = 0.187V$

Calculation produces a potential of -0.296 V. Permanganate redox potential is a function of its concentration in cathode chamber and pH. By using Nernst equation and theoretical potential (E°) of 1.51 V, polarization potential of cathode can be described as follows [15]:

$$E_{cathode} = E^{\circ}_{cathode} + 0.0052 \log [MnO_4] - 0.0422 \quad pH \quad (7)$$

$$MnO_4^+ + 8H_2O + 5e^- \longrightarrow Mn^+ + 12H_2O \quad (8)$$

$$E^{\circ}_{cathode} = 1.662\ V$$

Calculation produces a potential of 1.197, 1.199, and 1.2 V for 0.0625, 0.125, and 0.25 g/l of potassium permanganate aqueous solution used in cathode side.

$$E_{emf} = E_{cathode} - E_{anode} \quad (9)$$

Theoretical E_{emf} were calculated using eq. 9 were found to be 1.493, *1.49*, and 1.496 V for 0.0625, 0.125, and 0.25 g/l of potassium permanganate aqueous solution used in cathode side. In practice, increasing the permanganate concentration increases the electrolyte conductivity and decreases ohmic and activation losses that increase power of MFC. Also, manganese dioxide is the main product of redox reaction and its reaction with cathode electrode, as discussed above, limits electron transfer and MFC performance.

In practice, due to irreversible losses, the actual output voltage is less than the one resulted from the ideal equation. The significant losses that affect MFC operation are activation loss ηact, ohmic loss ηohmic, and mass transfer loss ηconc that take place in both the anode and cathode. The operational voltage output as a function of current density can be calculated as follows [16], taking open circuit voltage VOC into account:

$$V = OCV - \eta act \_\eta ohmic \_\eta conc$$

MFC efficiency measured by equation 10

$$\eta MFC = V_{measured} \cdot (n \cdot F)/E_{emf} \cdot (n \cdot F) = V_{measured} / E_{emf} \qquad (10)$$

Efficiency for cells with different $KMnO_4$ concentrations, 53.44, 54.28 and 50.77%. IT was found that the optimum power generation efficiency with cathode side contain 0.125 g/l $KMnO_4$.

Columbic efficiency is defined as the ratio between the charge transported to anode and maximum charge, which might be obtained from total substrate degradation. This parameter is evaluated as follows [17]:

$$CE = \frac{8\int_0^{dt} Idt}{F\, Van\Delta(COD)} \qquad (11)$$

Where I is the electrical current intensity, F the Faraday's constant, V_{an} the anaerobic chamber volume, and Δ (COD) the overall removal of substrate during t_b.

Using Eq. (11), the values of Columbic Efficiency (CE) with 100 Ω external resistance were 7.56%, 22.05%, and 16.89% during 48 hours with cathode side contain 0.25, 0.125, 0.0625 g/L potassium permanganate aqueous media respectively. Thus, it is obvious that using permanganate as electron acceptor has advantages such as higher values of removal COD and CE. As pointed earlier, the reason is that the rate of oxidation of wastewater by microorganisms is higher due to effective e^- discharge and higher current generation using permanganate as electron acceptor [18].

4. Conclusions.

Petroleum wastewater from Al Dura Refinery (COD = 500.125 mg/L, pH = 7) treated in a dual-chamber microbial fuel cell MFC using potassium permanganate as cathodic electron acceptors. To increase the values of columbic efficiency and wastewater treatment, it is essential to use MFC with optimum potassium permanganate amount. Selecting the electrodes materials with high electrochemical potential difference play a big role in the power generated form MFC. Long time exposure of Aluminum electrode into permanganate aqueous media will produce resistance layer on its surface which increase whole system resistance and reduce the generated power.

Acknowledgments

Main author would like to thank her home institute to their support and Directory of Environment and Water Research in Ministry of science and technology for their support in COD investigations achievement.

References

[1] Gil GC1, Chang IS, Kim BH, Kim M, Jang JK, Park HS, Kim HJ 2003 "Operational parameters affecting the performance of a mediator-less microbial fuel cell." *Biosens Bioelectron.* **18(4)** 327-34.

[2] Thauer RK, Jungermann K, Decker K. 1977 "Energy conservation in chemotrophic anaerobic bacteria" *Bacteriol Rev.* **41(1)** 100-80.

[3] Ishak, S., Malakahmad, A. and Isa, M.H. 2008 "Refinery waste water biological treatment: A short review" *J. Sci. Indus. Res.* **71** 251-256.

[4] Ma, F., Guo, J.B., Zhao, L.J., Chang, C.C. and Cui, D. 2009 "Application of bioaugmentation to improve the activated sludge system into the contact oxidation system treating petrochemical wastewater" *Biores. Technol.* **100(2)** 597-602.

[5] Shahrezaei, F., Mansouri, Y., Zinatizadeh, A.A.L. and Akhbari, A. 2012 "Process modeling and kinetic evaluation of petroleum re_nery wastewater treatment in a photocatalytic reactor using TiO_2 nanoparticles" *Powder Technol.* **221** 203-212.

[6] Priambodo, R., Shih, Y.J., Huang, Y.J. and Huang, Y.H. 2011 "Treatment of real wastewater

using semi batch (photo)-electro-Fenton method" *Sustain. Env. Res.* **21** 389-393.

[7] Logan, B.E. 2008 "*Microbial Fuel Cells*" John Wiley & Sons Inc., Hoboken, (NJ).

[8] Luoa, H., Liua, G., Zhanga, R. and Jinb, S. 2009 "Phenol degradation in microbial fuel cells" *Chem. Eng. J.* **147** 259-264.

[9] Mohan, S.V., Veer Raghavulu, S., Peri, D. and Sarma, P.N. 2009 "Integrated function of microbial fuel cell (MFC) as bio-electrochemical treatment system associated with bioelectricity generation under higher substrate load" *Biosen. Bioelec.* **24** 2021-2027.

[10] M. Bagheri-Esfe, and H. Bagheri-Esfeb 2016 "Petroleum refinery wastewater treatment in two-chamber Microbial Fuel Cells (MFCs)" *Scientia Iranica C* **23(6)** 2791-2799.

[11] Zhang, L., Zhu, X., Li, J., Liao, Q. and Ye, D. 2011 "Biofilm formation and electricity generation of a microbial fuel cell started up under different external resistances" *Journal of Power Sources* **196(15)** 6029-6035.

[12] S. Agnesia Kanimozhi and S. Rajendran 2010 "Aluminum inhibition by potassium permanganate–Zn2+ system" *The Arabian Journal for Science and Engineering* **35** 1A.

[13] Jude M. Runge, Aaron J. Pomis 2006 "Understanding Aluminum Anodic Oxide Film Formation: Improving Properties Through Microstructural Modification" *Presented at the XII Ebrats Brasilian Surface Treatment Meeting and II Latin – American Interfinish Sao Paulo.*

[14] Brucee. Logan et al 2006 "Microbial Fuel Cells: Methodology and Technology" *environmental science and technology* **40** 17.

[15] You, S., Zhao, Q., Zhang, J., Jiang, J. and Zhao, S. 2006 "A microbial fuel cell using permanganate as the cathodic electron acceptor" *J. Power Sources* **162(2)** 1409-1415.

[16] Wen, Q., Wu, Y., Cao, D., Zhao, L. and Sun, Q. 2009 "Electricity generation and modeling of microbial fuel cell from continuous beer brewery wastewater" *Bioresource Technology* **100(18)** 4171-4175.

[17] Min, B., Kim, J., Oh, S., Regan, J.M. and Logan, B.E. 2005 "Electricity generation from swine wastewater using microbial fuel cells" *Water Research* **39(20)** 4961-4968.

[18] Mohan, S.V., Mohanakrishna, G., Velvizhi, G., Babu, V.L. and Sarma, P.N. 2010 "Bio-catalyzed electrochemical treatment of real _eld dairy wastewater with simultaneous power generation" *Biochem. Eng. J.* **51(1)** 32-39.

IOP Conf. Series: Journal of Physics: Conf. Series **1032** (2018) 012062 doi:10.1088/1742-6596/1032/1/012062

Ammonia as Hydrogen Storage Media, Sustainable Method to Hydrogen Evolution

Zainab D.Abd Ali, Shrok Allami, Basheer H. Jwad

Ministry of Science and Technology, Baghdad, Iraq.

Corresponding author's E-mail: zdakheel78@gmail.com

Abstract. The aim of this study is to highlight hydrogen energy as an alternative energy for fossil fuel regarding its producing possibility from cheap, easy and available sources. Since hydrogen gas molecules are smaller than all other gases, they can be leakage through many materials despite considerations taken, which is so dangerous. Therefore, looking for hydrogen-carrier chemicals such as oils, ammonia, and other materials have been attracting the attention of the scientist around the world. Here we review a study about hydrogen energy and ammonia as a source and storage media, focusing on the evolution of hydrogen to be used into advanced power generation device and vehicles applications.

1. Hydrogen

Hydrogen gas, or as it is sometimes called "hydrogen", is one of the elements of the first cycle of the periodic table, It is the smallest element, It is consist of proton and electron., its chemical symbol H, atomic number one, neck above the elements of the first group as well. In the standard conditions of pressure and heat, hydrogen is in the gaseous state, a colorless and odorless gas, a fast-flammable, nontoxic, and also a binary, monoclonal. Hydrogen gas is found in nature in the form of "H2" molecules, which is one of the lightest and abundant elements in the universe. Hydrogen is three-quarters of the size of the universe, which enters most covalent compounds.

It is characterized by its hydrogen boiling point "-252 ° C" and its melting point "-259 ° C". It has a water solubility of 1.6 mg/liter. It has the largest dispersion capacity among all gases. Has the highest conductivity of the temperature, and the largest flow capacity. The degree to his wife is low. In the Palmer series, hydrogen shows four spectral lines in the visible spectrum, located at 656, 468, 434, and 410 nm. In the Hydrogen Spectrometer series, there are three strings in the infrared field: the Pashin series, the Prakt series, the Pond series, and the UV series, the Leman series. Hydrogen gas is an insulating gas for electricity.

Hydrogen is the most committing energy carrier's futurity. It is a high effectiveness, pollution is low in fuel that can be utilized for transportation, power, heating, and generation in places where it is hard to use the electro.

2. Contributing to solving energy problems by using hydrogen derived from renewable energy sources

The energy currently being used as a fuel source is energy-efficient such as oil and fossil fuels. Studies have confirmed that this type of fuel has been introduced by 2053, where oil reserves have been found to last for 50 years, opening the way for a sustainable and environmentally friendly energy source.to solve this problem has been concerned Towards renewable energies.

Sources of Renewable Energy Hydrogen was diverted as a source of renewable energy, clean and sustainable next-generation energy source—is receiving increased attention as an activated secondary energy source for advancing the use of renewable energy.

IOP Conf. Series: Journal of Physics: Conf. Series **1032** (2018) 012062 doi:10.1088/1742-6596/1032/1/012062

3. Source of Hydrogen

Hydrogen is the ninth rank among the elements on the surface of the earth in terms of abundance if the elements are arranged according to the weight ratio, but if the order according to the number of atoms, it ranks third after the oxygen and silicon. Most of this hydrogen is in the form of water, but much of it is found in oil and living matter of various kinds. In the rocks, the hydrogen is relatively rare except what is found in the form of water crystallized with salts or absorbed by the clay. This element is abundant in stars and giant gaseous planets but is very narrow in the Earth's crust (1 ppm in size). The most common source of this element is water, which consists of hydrogen and oxygen atoms (H_2O). Other sources include most organic forms (all known forms of life), including coal, natural gas, and other fossil fuels. Methane (CH_4) is an important source of hydrogen.

Hydrogen can be prepared in several ways, such as passing steam on hot carbon, hydrolyzing hydrocarbons, strong base reactions in its water solutions with aluminum, electrolysis of water and acid exchange reactions with metals.

Hydrogen is produced largely by re-forming steam for natural gas at high temperatures (700-110 ° C), where steam reacts with methane to produce carbon monoxide and hydrogen. Sunlight uses an energy source.in photo electrolysis cell where it is split water to H_2 and O_2. Scientists sought to simulate plants in their ability to analyze water to its first two radicals with the presence of sunlight and chlorophyll, where oxygen and hydrogen are obtained. The latter is used by plants to interact with carbon dioxide to form carbohydrates, and plants have been imitated in the laboratory to produce hydrogen gas.

Although these methods are important for the production of hydrogen, they are expensive and economically inefficient. This has reinforced the idea of using biotechnology to produce hydrogen using some bacteria. This gas is made up of some algae that can produce hydrogen gas naturally from water and with sunlight.[3]

4. Hydrogen Today and Future

Modern civilization is closely associated with oil for the operation of its powerful machines and the development of its industries and agriculture, which are essentially dependent on oil. But the accumulation of emissions, which are metaphorically the waste of burning oil derivatives, affects the environment in a negative way and threatens the climatic changes, particularly in terms of warming the planet Earth, due to the greenhouse effect resulting especially from the accumulation of carbon dioxide.

So, on the threshold of the 21st century, humanity finds itself facing political, economic and environmental crises, security risks and devastating wars no less dangerous than catastrophic geological changes.

More oil consumption will lead to faster threats to the environment and the life itself caused by pollution and, on the other, to the accelerated depletion of this precious resource, which leads to the intensification of conflicts and the loss of these stocks of useful material in the industrial and agricultural fields, Any other material necessary for the survival of humankind at this time of its qualitative and quantitative development.

Many environmentalists and energy scientists believe that only a few decades before the disaster, mankind has to make the necessary and significant transition from the l economy of fossil fuel to the hydrogen economy. This requires the exclusion of fossil fuels from all its derivatives and replacement of sustainable alternative energy sources with the use of hydrogen as an energy vector.

Hydrogen Today, we can operate airplanes, cars, trains, ships, factories, heating homes, offices, hospitals, schools and others. Hydrogen, in its gas state, can transport electricity as long distances and through pipelines, efficiently and at the lowest possible cost. Hydrogen can be powered by energy fuel technology or other energy-generating machinery to provide consumers with electricity, clean drinking water and hydrogen as a chemical element with different uses and applications other than electricity. Hydrogen fuel uses can be classified mainly into the following four main fields:

1. Fuel for transport media (cars, aircraft) operating on hydrogen fuel cell technology and its wider applications for future use in power plants.

2. It can be used as a battery with a range of devices ranging from the small ones used in mobile personal computers to the hydrogen transport vessels that transport them from renewable power stations to remote power stations to solve the problems and costs of long networks and energy losses through them.

3. Generated fuel for thermal energy by burning it directly in boilers in power stations, in addition to being used as a propellant in rockets.

4. Nuclear fuel is a factor in nuclear reactors, especially the ITER technology, which works on the principle of power generation on the surface of the sun.[1]

5. Methods and problems of hydrogen Storage

Hydrogen storage safely and efficiently are the major technological difficulties that currently prevented barring the widespread adoption of hydrogen as an energy carrier and transmission to a so-called hydrogen economy.

1. The most general ways of hydrogen storage is gas *compression* of at high pressure > 200 bars. at ambient temperature, gas in- and out-flow are simple. However, the density of storage is low than other ways.

2. Liquid H_2, 25% to 45% of the energy consume to store it is liquefied with hydrogen. The storage density of the hydrogen is very high since hydrogen boils at -253 C° the necessary to maintain it is low temperature a large thermal insulation. Energy density of hydrogen can be improved. *3. Hydrogen storage as Solid state*

a- chemical storage: material absorb hydrogen under high pressures to form hydride, the process is reversible. The high weight is the main problem in this method of the absorbent material [1].

b- Physical storage: micro materials - the most recent topics of hydrogen storage. At the applied pressure nanotubes، zeolites (microporous aluminosilicate) Activated carbon and metal-organic frameworks (MOFs) can be bonded weakly with hydrogen. [3]

6. Ammonia as an alternative to hydrogen sources

Extract hydrogen from ammonia and use it as clean fuel Ammonia has been utilized for a long time to produce fertilizers and animal feed. It is also used as domestic detergents. Scientists wanted to use it as a kind of fuel where we get a clean source of energy, Scientists have made great efforts to do so, but their efforts have been hampered by the strength bonds between N2 and H2 atoms in the ammonia particles ... but they can be found a solution where:

At Princeton University three researchers discovered a method to weaken these bonds and not only that, but also released the hydrogen atoms, which are then binding together to form hydrogen gas.

Their research papers were published in the journal Matte Pezdek, in which one of them, Paul Churking, described the possibility of using this method as a new source of hydrogen. Another, Jessica Hoover, presented a model of this work in the same article published in the journal with the University of Virginia. Will also outline the impact of these results and will most likely be able to store, exploit and use hydrogen energy.

The new method also used ammonia with phosphine and ammonia. Hoover noted that the bond between nitrogen and hydrogen adheres homogenously. When it breaks down, it produces a one of the hydrogen atom that is free and combines with another like hydrogen gas, which can then be marge and eventually burned to supply multiple types of engines and this is likely to leave an important impact on energy systems. This method is more efficient than other methods such as extracting hydrogen or adding an oxidizing element, and we may be capable to access more efficient methods afterward if we work on ammonia preparation methods.

despite the large numbers of complex compounds of transitional metals resulting from ammonia or water merging so that it gives a couple of electrons and a covalent bond, only a few know the extent to which consistency affects the bonding strength between nitrogen and hydrogen as well as between

IOP Conf. Series: Journal of Physics: Conf. Series **1032** (2018) 012062 doi:10.1088/1742-6596/1032/1/012062

oxygen and hydrogen. "Molybdenum ammonium supported by phosphine and terpyridine", which reduces the heat needed to break the nitrogen-hydrogen bond from 99.5 gaseous to a measurable value of 45.8 kcal per mole, Hydrogen heating on an insignificant addition to hydrogenation styrene; the compounds of similar Molybdenum They promote the bilateral liberation of hydrogen from water and harmonious hydrazine. The electrochemical and theoretical study has demonstrated the contribution to the influence of the oxidative stress reduction of minerals and ammonia acid.

The cost of ammonia energy is more than the cost of hydrogen unless hydrogen has minimal energy stored per volume, ammonia is the inexpensive energy source listed in Table 1. Note that these values are established on the HHV of the fuel and do not have the cost of conversation on to the useful forms the estimated of energy production from Ammonia a cost is US$1.2/kWh while with methanol is US$3.8/kWh and US$25.4/kWh for H_2 [5]. Thus, ammonia offers a very applicable and cost-efficient fuel for fuel cells. [6]

Table 1. Energy Storage Capabilities of different Fuels [3].

Fuel Storage System	P(bar)	Energy Density (G/m^3)	Specific Volumetric coast (USS/m^3)	Specific Energy coast (USS/GJ)
Ammonia gas/prusserized tank	10	13.6	181	133
Hydrogen/metalhydride	14	3.6	125	35.2
Gasoline (C_8H_{18})/liquid tank	1	34.4	1000	29.1
LPG (C_3H_8)/ prusserized tank	14	19.0	542	28.5
CNG (CH_4)/integrated storage system	250	10.4	400	38.3
Metanol/liquid tank	1	11.4	693	60.9

6.1. Ammonia definition

Ammonia is a nitrogen and hydrogen compound with the formula NH3. It is lighter than the air in weight and has a sharp taste, lighter than the air to reduce its mass volume in normal conditions 0.5863 g / Cm 3. The definition of the mullet at -8.77 ° C, and boil at the grade (-33.4°C).

Constant value resistance this displacement is 1.8 × 10-5. The solution has weak alkaline properties. Because of this weak base, salts solutions derived from strong acids such as NH4Cl are weakly acidic.

Ammonia degeneration in water decreases with high temperature; ammonia can be completely released from its solutions by heating. Ammonia is very soluble in water (0.907 g ammonia in one gram of water) and forms a solution known as ammonia hydroxide (NH4OH) or ammonia. It contains NH3, NH4 +, OH- and ammonia particles that are not very effective when interacting with many chemicals. Most of the ammonia molecules in water remain intact.

Let the ammonia hydroxide a lot of acids and form the corresponding ammonium salts. If hydrochloric acid (HCl) is applied to ammonia hydroxide (NH4OH), ammonium chloride solution (NH4Cl) is used. When ammonia hydroxide is combined with some metals. When adding ammonia hydroxide to the blue solution (Cu) (NH4) 4SO4).

Ammonia gas becomes liquid at -33.35 ° C. The ammonia is boiled at the same temperature, and the freezer turns into a solid solid substance at -77.7 ° C. In its transition from liquid to gas again, ammonia absorbs a great deal of heat from the outside, so that one gram of ammonia has 327 calories. Actually, fall into the cooling devices.

Pressure and Temperature: This substance cannot be obtained in liquid condition at atmospheric pressure and at a temperature higher than 28 degrees Fahrenheit.

IOP Conf. Series: Journal of Physics: Conf. Series **1032** (2018) 012062 doi:10.1088/1742-6596/1032/1/012062

Ignition: This gas ignites if mixed with air venting: This vent can be detected by the smell of smell and reduce the venting place by igniting the fingers of sulfur near and repeat near the place of doubt and the presence of vent and showing white smoke in the case of a vent Lubrication: Ammonia is mixed thoroughly with lubricating oils.

6.2. Ammonia definition

6.2.1. Ammonia Electrolysis. Ammonia Electrolysis technology is electrochemical cell called Ammonia Electrolytic Cell (AEC).The AEC works as described: watery ammonia (NH3/H2O) in the existence of (KOH) is fed into the anode compartment of the AEC where NH3 is oxidized in the existence of OH- in accordance with:

$2NH_3 + 6OH^- \text{------->} N_2 + 6H_2O + 6e^-$

At the cathode, a solution of KOH provides and water is decreasing in alkaline medium in accordance with:

2H2O + 2e- -------> H2 + 6OH-

All reaction is given by

2NH3 -------> N2 + 3H2

The production voltages of hydrogen at 25 in alkaline media through of the ammonium electrolysis is 0.058 Vs theoretically and consuming energy is 1.55 W-h per gram of H2 production. Potassium hydroxide is the electrolyte of the system and water works as the solvent, that is, neither KOH nor water are consumed through the operation of the cell [9] show figure 1, 2.

6.2.2. Ammonia decomposition. Ammonia decomposition (cracking) is simply the reverse of the synthesis reaction

NH3 (g) $\rightarrow$ 1/2 N2 (g) + 3/2 H_2 (g) $\Delta H = +46$ kJ/mol

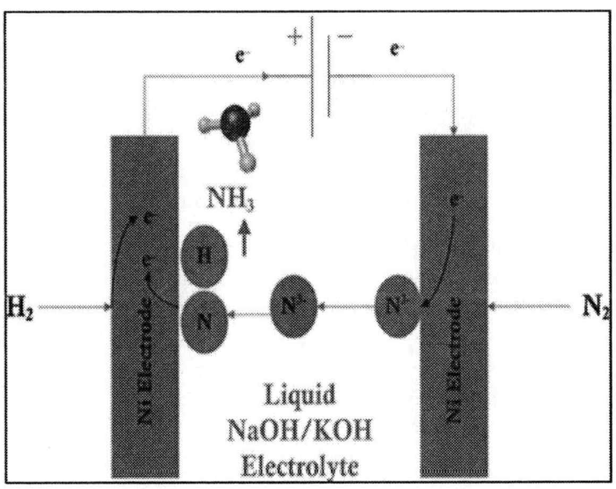

Figure 1. show Electrical cell for Ammonia production.

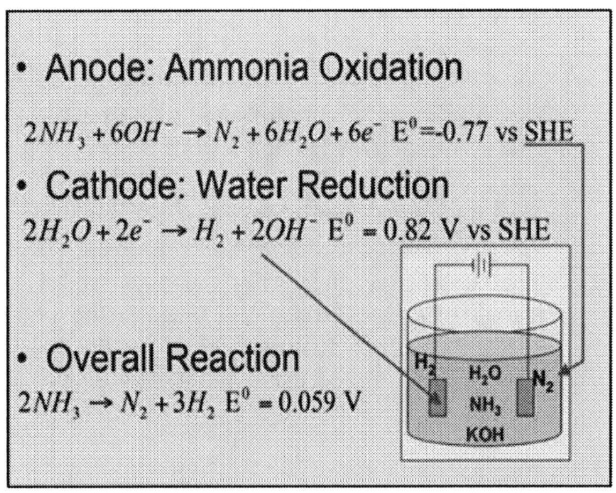

Figure 2. show ammonia Electrolysis.

Note that the reaction is endothermic. The temperature need for efficient cracking depends on the catalyst. There is a wide kind of materials that have been found to be effective, but some (e.g., supported Ni catalysts) need temperatures above 1000°C. Others have high transformation; efficient of temperatures is extended from 650 to 700°C. A probable design for an ammonia cracking reactor is demonstrated in Figure. 4 illustrative system of ammonia decomposition.

The ammonia Liquid would be pumped from a storage tank through a heat exchanger to take waste heat from the hot gases out of the cracking reactor. The heated gases would then go through a furnace or catalytic combustor to heat them to the temperatures is necessary for the reaction. The flow exiting the reaction would go to a portions system which would be optimized to produce a very pure flow of hydrogen while still leaving sufficient hydrogen with the nitrogen and un-reacted ammonia to equip heat for the endothermic cracking reaction. [10] Figure 3. show system of ammonia decomposition.

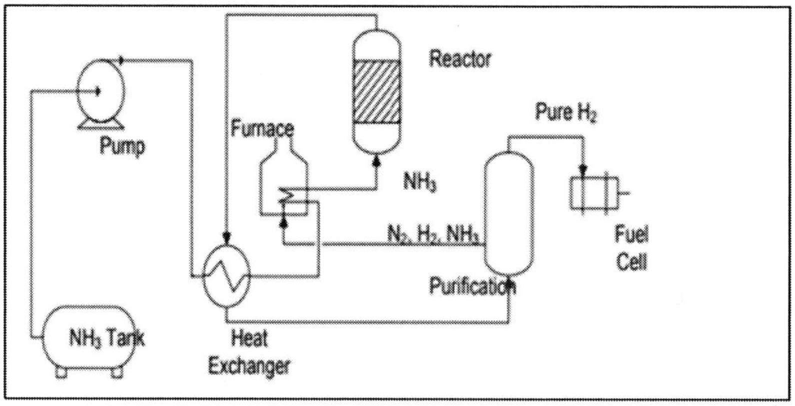

Figure 3. Ammonia decomposition systems.

6.2.3. Produce hydrogen from Ammonia by the solar cell.

A group Australian National University which working in the field of solar thermal energy has been the attention of separation ammonia system by using solar energy, their experimental titled solar driven ammonia based closed loop thermos-chemical energy storage system. The system uses a whole receiver including 20 reactor tubes filled with iron-based catalyst material exposed to collect the radiation from a 20m ^{2} dish solar concentrator that dissociates ammonia under the effect of concentrated solar energy so that the products can be stored and recycled through a conventional ammonia synthesis converter to achieve 24 h power production. They also demonstrated the reliability of the operation over a range of conditions including cloud transients has been demonstrated [11].show figure 4.

7. Ammonia Applications

7.1. Applications of ammonia in electric power production units (fuel cell)

The ammonia fuel cells operate under the same principles as conventional alkaline cells by transferring hydroxide ions through the cell solution with an electrochemical membrane. The reaction at the cathode is given by [12]:

$$O_2 + 2H_2O + 4e^- \rightarrow 4OH^- + O_2 + 2H_2O + 4e^- \rightarrow 4OH^- \quad (1)$$

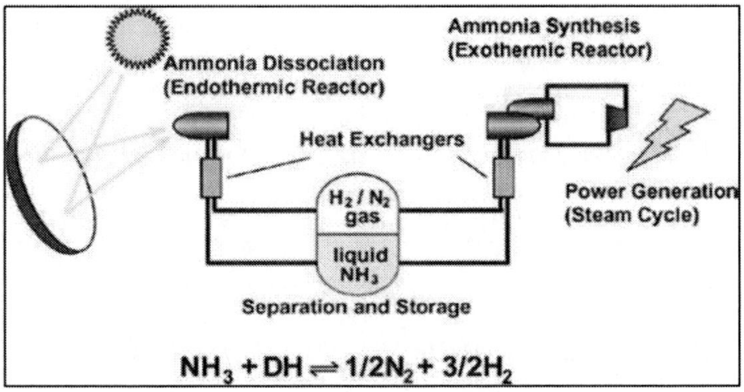

Figure 4. Closed loop thermochemical storage of solar energy using ammonia.

At the anode:

$$2NH_3 + 6OH^- \rightarrow N_2 + 6H_2O + 6e^- + 2NH_3 + 6OH^- \rightarrow N_2 + 6H_2O + 6e^- \quad (2)$$

This gives an overall reaction of:

$$4NH_3 + 3O_2 \rightarrow 2N_2 + 6H_2O + 4NH_3 + 3O_2 \rightarrow 2N_2 + 6H_2O \quad (3)$$

Figure 6 shows direct ammonia fuel cell show working mechanism of ammonia alkaline exchange membrane fuel cells (AEFC) Figure 5 direct ammonia fuel cell show working mechanism of ammonia (AEFC) [13] Ammonia is involved in many applications of hydrogen fuel cells as a source of direct and indirect energy.

Ammonia is involved in many applications of hydrogen fuel cells as a source of direct and indirect energy. Figure 6 shows ammonia applications in the unit of electric power.

7.2. Applications of ammonia in internal combustion engines.

Liquid ammonia is suitable for vehicle fuel in the era of sustainable H_2 society. Liquid ammonia has a twice higher energy density in storage and half expanded energy in delivery compared with liquid hydrogen.

Ammonia is capable to spark ignition internal combustion engines.

Stable combustion is carried out by using outside by thermal cracker which can control hydrogen fraction.

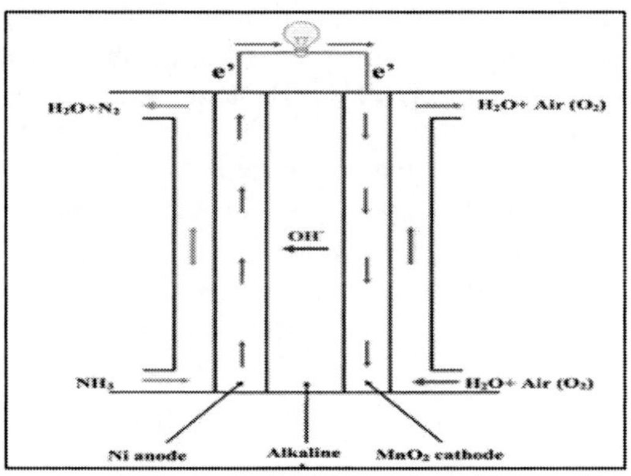

Figure 5. shows ammonia applications in the unit of electric power.

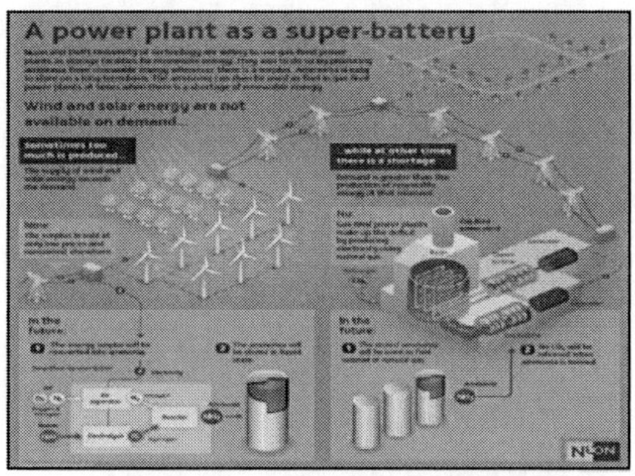

Figure 6. shows ammonia applications in the unit of electric power [14].

IOP Conf. Series: Journal of Physics: Conf. Series **1032** (2018) 012062 doi:10.1088/1742-6596/1032/1/012062

Pure ammonia is available at high load and overcomes knocking even under highly charged condition. Ammonia fed in SI engine seems to be a good alternate for heavy-duty diesel engines the inclusion of an example of research shows the use of ammonia in internal combustion engines this test was performed on the Lombardini engine. It is a four-stroke engine consisting of two cylinders sparking the ignition engine with water cooling and injection system. Details of the engine specifications are present in Table 2[15].

The tests were measured in the brand hydrokinetic dynamometer. This device gives the power, torque and engine speed. The experiments were performed at 30/70 NH_3/H_2, 10/90 NH_3/H_2, 100% hydrogen, 100% methane and gasoline fuel by changing the ignition timing.

Table 2. Details of the engine specifications.

Cylinder number	2
Cylinder volume	505cc
Electronic Engine Unite	Bosch Motronic
Cylinder Bore	72mm
Stroke	62mm
Engine Power	20.4HP 5000d/d
Engine Torque	34NM 2150d/d
Compression Ratio	10.7:1

Fuel is supplied to the inlet-outlet by injection. The tests were performed at 1400 ± 10 rpm, and CO, THC, NO emissions were measured with the MGA 1500 gas analyzer. All test measurements were analyzed at a constant load of 7 ± 0.35 nm and 1400 ± 20 rpm. Figure 7 shows the plot diagram of the engine and gasoline fuel by changing the ignition timing. Fuel is supplied to the inlet-outlet by injection. The tests were performed at 1400 ± 10 rpm, and CO, THC, NO emissions were measured with the MGA 1500 gas analyzer. All test measurements were analyzed at a constant load of 7 ± 0.35 nm and 1400 ± 20 rpm. Figure 7 the pilot diagram of the engine

Figure 8 depicts the values of thermal braking efficiency (BTE) versus spark timing for H_2, CH_4, H_2/NH_3, and gasoline.

The highest values (BTE) are obtained about 15% H_2/NH_3 (70/30). The minimum (BTE) values are obtained about 10.84% in hydrogen fuel. When the timing of a spark increases, BTE first increases and then the BTE values are reduced. When BTE is compared to H_2, H_2/NH_3, and BHT, hydrogen is less benzene because of a volumetric decrease in hydrogen.

If NH_3 is added to hydrogen, the volumetric efficiency is higher than hydrogen, BTE) than H_2/NH_3 mixtures. However, NH_3 was compared with gasoline, which showed that efficiency and energy increased due to the possibility of high compression ratio Torque and Power.

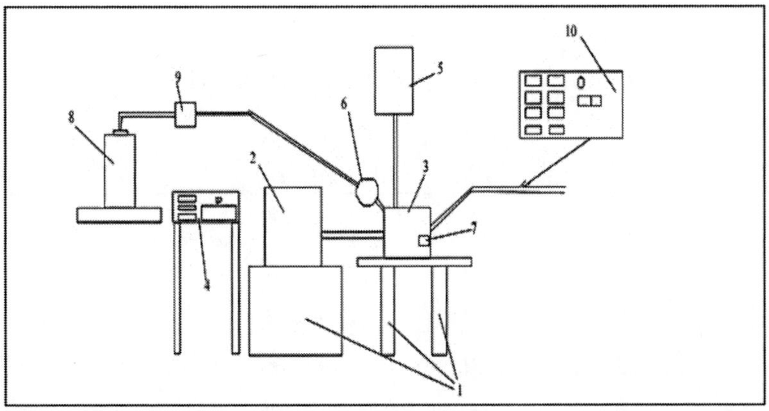

Figure 7. The pilot diagram of 1- Engine Test Chassis 2- Hydrokinetic Dynamometer 3-Engine 4- Control Unit 5- Main Fuel Tank 6-Regulator 7-Fuel Selection Switch 8- Gas Fuel Tank 9- Mass Flow Meter 10- Exhaust Gas Analyzer.

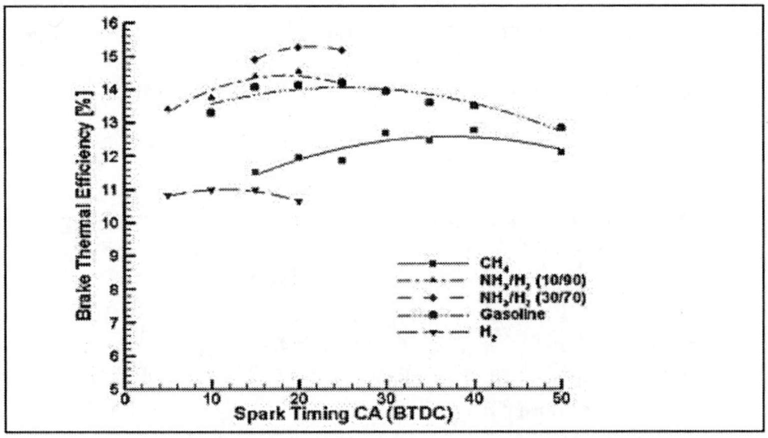

Figure 8. The values of thermal braking efficiency (BTE) versus spark timing.

Table 3 shows average torque values [N × m] and power values [kW]. Torque and power values are 7 ± 0.35 Nm, 1.06 ± 0.4 kW. Almost all measurements were made in the same conditions (approximately ± 1 temperature and atmospheric pressure) for 70/30 (H_2/NH_3, (90/10 H_2/NH_3), hydrogen (100%) and gasoline fuel in the internal combustion engine.

Emission Parameters of CO, THC, NO and sun MGA 1500 gas analyzers are measured using H_2 and the H_2/NH3 mixture in the engine.

Table 3. average torque and power values for NH_3, H_2, CH_4, and gasoline.

	CH_4	H_2/NH_3 (70/30)	H_2/NH_3 (90/10)	Gasoline	H_2
Torque (Nm)	6.83	7.03	7.02	7.34	6.65
Power (kW)	1.02	1.06	1.05	1.10	1.04

Carbon, carbon dioxide and HC emissions do not occur. Emission values are not included in Table 4 for each fuel [16].

Table 4. No emission values versus ignition timing for CH_4, NH_3/H_2, gasoline, and H_2.

NO. CA (BTDC)	CH_4	H_2/NH_3 (70/30)	H_2/NH_3 (90/10)	Gasoline	H_2
5	-	-	+5000	-	2779
10	-	-	+5000	657	3576
15	347	+5000	+5000	926	3918
20	425	+5000	+5000	971	4980
25	719	+5000	+5000	1103	-
30	851	-	-	1285	-

8. Summary.

In this study, ammonia was treated as a source of energy because it is possible to obtain hydrogen from it in different ways such as ammonia cracking and electrolysis of ammonia, in addition to the use of solar energy to produce ammonia and the cost is less expensive and easier to produce. We also referred to the applications of ammonia as an energy source for the production of electrical energy. Internal combustion engines were concluded.

References

[1] National energy education development hydrogen at glance" http://www.need.org
[2] Toshiba Group CSR Report 2015,"Highlight of FY2014 Activities 1", https://www.toshiba.co.jp/csr/en/highlight/2015/pdf/enr_pdf.pdf
[3] A Sherif and et al 2003 " Principle of hydrogen Energy Production, storage and Utilization‖, *Journal of Scientific & Industrial Research* **62** 46-63.
[4] K. L. Lim et al. 2010 " Solid-state Materials and Methods for Hydrogen Storage: A Critical Review" *Chem. Eng.* Technol **33** **(2)** 213–226.
[5] Zamfirescu C and Dincer I. **2008**. *J. Power Sources* **185** 459–465.
[6] Kaye IW and Bloomfield DP. 1998 *Conference of Power Source* (Cherry Hill) 408–409.
[7] Archived from the original on 27 November 2007 "Ammonium hydroxide physical properties" (PDF).
[8] F. Vitse, M. Cooper, and G.G. Botte 2005 "On the use of ammonia electrolysis for hydrogen production" *J. Power Sources* **142** 18.
[9] Ammonia electrolyzer cell AEC.
[10] George Thomas "*A Study of Issues Related to the Use Ammonia for On-Board Vehicular Hydrogen Storage U.S. Department of Energy*"
[11] *Keith Love grove ―Concentrating Solar Fuels for export‖* http://www.itpowergroup.com
[12] Potential Roles of Ammonia in a Hydrogen Economy‖
[13] Electrochem 2010 *Solid-State Lett.* **13(8)** B83-B86.
[14] The power plant as super battery https://visual.ly/community/Infographics/technology/powerplant-superbattery
[15] EMTIAZ ALI BROHI 2014 *"Ammonia as fuel for internal combustion engines "* Department of Applied Mechanics Division of Combustion CHALMERS UNIVERSITY OF TECHNOLOGY Gothenburg, Master thesis Sweden.
[16] ãÕÏÑ Úä ÇáÇÍÊÑÇÞ ÇáÏÇÎáí ááãÍÑß *"Ammonia as a hydrogen storage and the feasibility for the internal combustion engine".*

IOP Conf. Series: Journal of Physics: Conf. Series **1032** (2018) 012063 doi:10.1088/1742-6596/1032/1/012063

Preparation Poly 4-Bromoaniline Thin Films by Electro Deposition Technique for Hydrogen gas sensor

Ashwaq .A. Jabor, Dr. Ahlam.M.Farhan, Dr.Abdul- Hussain K. Elttayef* and Afrah .A. Jabor
Ministry of High Education- Baghdad University, Ministry of Science and Technology *, Iraq.

Corresponding author's E-mail: ashwak.alobaidy@yahoo.com

Abstract. Thin films of Poly 4-bromoaniline organic Semiconductor polymer were prepared using electro deposition technique through oxidation polymerization by adding (2ml) HCl concentrated and (50) ml of Ammonium per sulfate concentrated (0.05M) to (50)ml of 4-bromoaniline (0.04M) at room temperature, The polymer was deposited on Silicon n-type substrate , Structural and optical properties of The thin films were characterized by technique: X-ray diffraction (XRD), (FE-SEM), Fourier transform infrared (FT-IR) spectroscopy, Field emission scanning electron microscopy, uv-visb spectroscopy and sensing of thin film to H2 gas.

1. Introduction

Very large quantities of hydrogen are safely used in industry today, but it's hard to contain and difficult to detect leaks Hydrogen is a very large industrial commodity I with a growing market; its anticipated use as fuel could quadruple the market to more than 2itrillion cubic meters per year[1,2], and The introduction of hydrogen as a consumer fuel has caused heightened concern over its safety with a corresponding increased interest in hydrogen sensors and leak detection[3,4].

Among all conducting polymers have attracted much interest worldwide. Because of chemical stability, simple polymerization, high conductivity, polyaniline derivatives have been used in various application, like, optoelectronics, bio-sensors, gas sensors, microelectronics etc. Several methods of preparation of PANI derivatives film such as, spin coating, drop coating, electrochemical deposition , thermal evaporation, emulsion polymerization. [5-8], this sensor can be easily prepared and fabricated at low cost along with high stability.

In conventional gas sensors, metal oxides are used to fabricate gas sensors (Sivalingam et al., 2012) to sense ammonia for better sensitivity. But metal oxides can be operated only at high temperatures and also it consumes more power. The electrically conducting polymer polyaniline derivatives have high sensitivity like a metal oxides, and it can be operated at room temperature itself (Pant et al., 2007; Abraham et al., 2004). In addition, this sensor can be easily prepared and fabricated at low cost along with high stability [9,10].

Polymeric materials, used in the production of a great percentage of the goods used commonly in our daily lives, have gained considerable importance. Today, The conducting polymers and their derivatives are consider materials which use as sensor since,.1980 electronic conducting polymer Were developed. The interest of these materials has been recognized by the awarding of the Nobel prize in Chemistry in 2000 Generally polyaniline derivatives have been synthesized by chemical oxidative polymerization of aniline compound [11, 12], or by electrochemical method [13].

Most of the studies related to synthesis of polyaniline derivatives use chemical or electrochemical. oxidation or reduction that it may switch over the material from fully reduced to fully oxidized form The researches deal with synthesis and characterization of polyaniline derivatives thin films by galvonostatic. mode of electrochemical deposition method at room temperature. Different preparative parameters, such.as deposition potential, concentration of monomer, deposition time and current density were optimized to get uniform and well adherent polyaniline derivatives thin films Polyaniline is a type of conducting polymer which received the most attention due to the discovery of its high

IOP Conf. Series: Journal of Physics: Conf. Series **1032** (2018) 012063 doi:10.1088/1742-6596/1032/1/012063

electrical conductivity [4,5] reversible acid-base chemistry in aqueous solution, thermal and environmental stabilities and easiness of synthesis.[6] Among all the methods which can be used for the synthesis of polyaniline derivatives such as oxidative synthesis Polyaniline derivatives i have been studied extensively as a special member of the conducting polymer family because of its stability in the presence of air and humidity. Polyaniline derivatives have shown many promising applications in industries related to high technologies. PANI and its

derivatives can be used in active electrodes,

1- indicators recharge
2- able batteries.
3- electrochromic devices.
4- microelectronics.

However, its applications are strongly limited by its poor processibility. Electro polymerization of aniline derivatives have been widely investigated for the improvement of processibility and other properties of conductive PANI. Substituted PANI are used to increase the processibility of the polymer. which can be prepared by modifying the polymer chain in the following ways ii [7,8]:

(i) the post-treatment of parent PANI

(ii) the chemical or electro-chemical polymerization of aniline derivatives

(iii) the copolymerization of aniline I with ring or Ni-substituted derivatives. The effect of electron-donating groups (alkoxy, alkyl, etc.) on the solubility and conductivity. There are also a few studies about the polymerization of aniline containing electron with drawing groups as substituents [9-12].

HI group on the PANI I chain affects the properties of the parent PANI without substantially changing the conductivity and is of specific interest for several reasons such as solubility, environmental stability, and processibility [12-15].

2. Experimental

Materials and Equipment

The Silicon n-type substrate was used as substrate for the electrode matrix., aniline, methanol, and other chemicals were of analytical grades from Merck. All the aqueous solutions were prepared (2mL) HCl concentrate and(50 mL) Ammonium per sulfate concentration of (0.05M) to(50 mL) (0.04M) of 4-bromoaniline at room temperature. graphite as anode electrodes, D.C power supply, Digital multimeter

Electrodeposition of P-4-BrANI Thin Film

Electro polymerization of 4-bromoaniline at the silicon electrodes was achieved by potenti dynamic method. The electrode potential was swept between 7-9 V using an aqueous solution by (2ml) HCl concentrated adding to solution Consists of (50mL) of potassium per sulfate concentration of (0.05M) and (50mL) (0.04M) of 4-bromoaniline at room temperature. The thickness of th deposited films (P4-BrANI film thickness (600 nm)

Electropolymerization of P4-BrANI offers the possibility of controlling the thickness and homogeneity of the conducting polymer film on the electrode surface. the oxidative polymerization of 4-Bromoaniline subsequent growth of the polymer films at substrate. Figure 1. shows electropolymerization system.

IOP Conf. Series: Journal of Physics: Conf. Series **1032** (2018) 012063 doi:10.1088/1742-6596/1032/1/012063

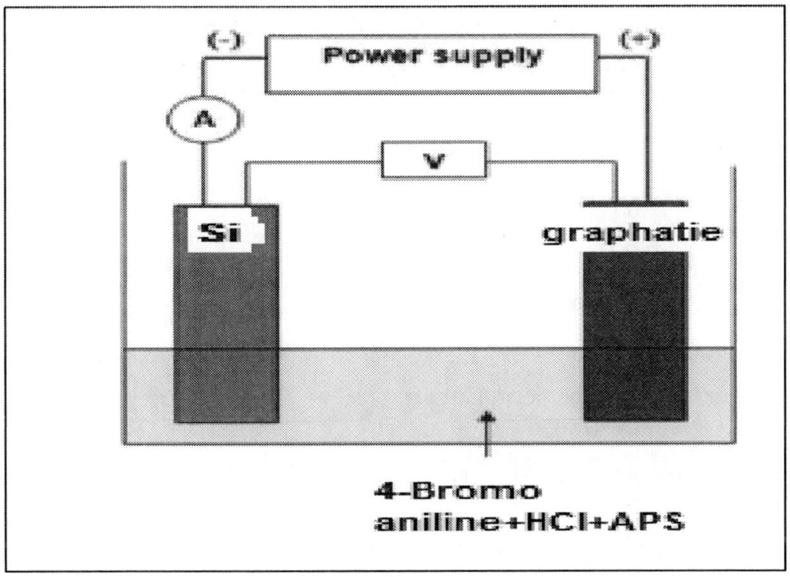

Figure 1. Electro polymerization system.

The polymer thin film was brown gray color, chemical composition is shown in figure (2):

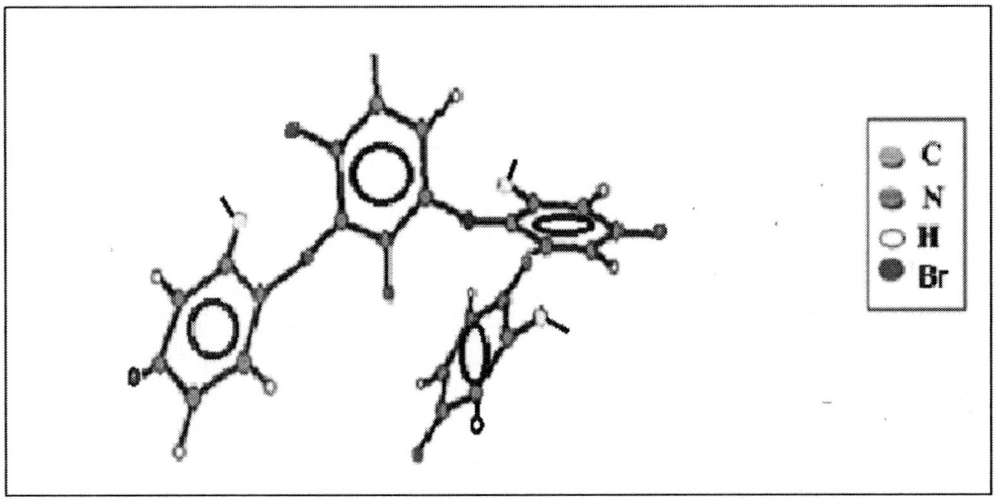

Figure 2. Shows chemical Structure of Poly 4-bromoaniline.

Diagnosis the Prepared Polymer in FT-IR Spectrum

The Prepared Polymer was diagnosed by using iPerkin Elmer FT-IR Spectrophotometer model 1720X.i

FT-IR Spectrum of Poly 4-bromoaniline

FT-IR Spectral measurements of Poly 4-bromoaniline for FT-IR spectrum appeared that there are absorptions bands as shown in the table (1), figure (3) illustrate the absorption of FT-IR spectrum

IOP Conf. Series: Journal of Physics: Conf. Series **1032** (2018) 012063 doi:10.1088/1742-6596/1032/1/012063

Table 1. The absorption bands of the prepared polymer.

Vibration groups	Absorption bands(Cm- 1)
Stretching vibration band (C-H) aromatic	3051
Stretching vibration (N-H)	2970
Stretching vibration band (C=C) to the benzene ring	1469 , 1566
Stretching vibration band (C– N)	1342
bending band (C-H) to the benzene ring	1288
Stretching bend N-ph	1072
Stretching band (C-H)to the benzene ring triplet compensation	671
Stretching vibration C-Br	505

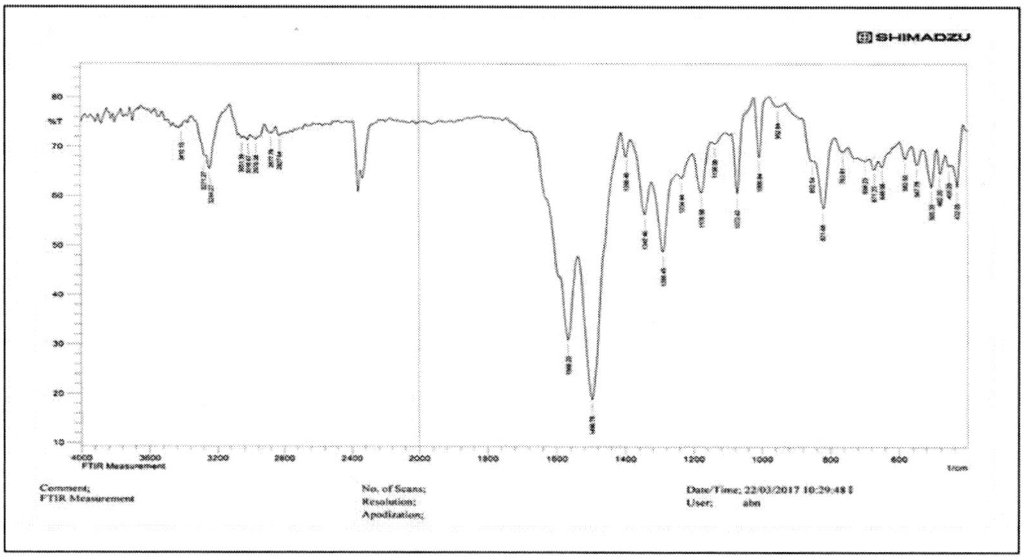

Figure 3. FT- IR spectrum for Poly 4-bromoaniline thin film.

Preparations of Poly 4-bromoaniline sensor

Sensor was prepared by deposit aluminum electrode on Silicon substrate in vacuum evaporation method, and then was pouring molten polymer between the electrodes as shown in following figure (4):

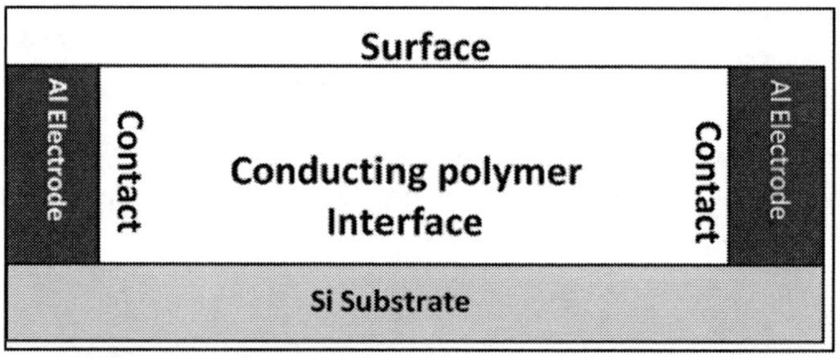

Figure 4. shows polymeric sensor design.

IOP Conf. Series: Journal of Physics: Conf. Series **1032** (2018) 012063 doi:10.1088/1742-6596/1032/1/012063

3. Results and testes

Structural properties measurements

X-ray tests results

The structure of polymer thin film Of thickness (600) ±5nm was analyzed using X-ray diffraction (XRD) system (Shimadzu X-ray diffraction,with CuKα radiation of λ=1.54A^0 .

All films are polycrystalline with tetragonal structure and the peak position of the films (110,101,200,211) were found to be in good agreement with the established standers (JCPDS).

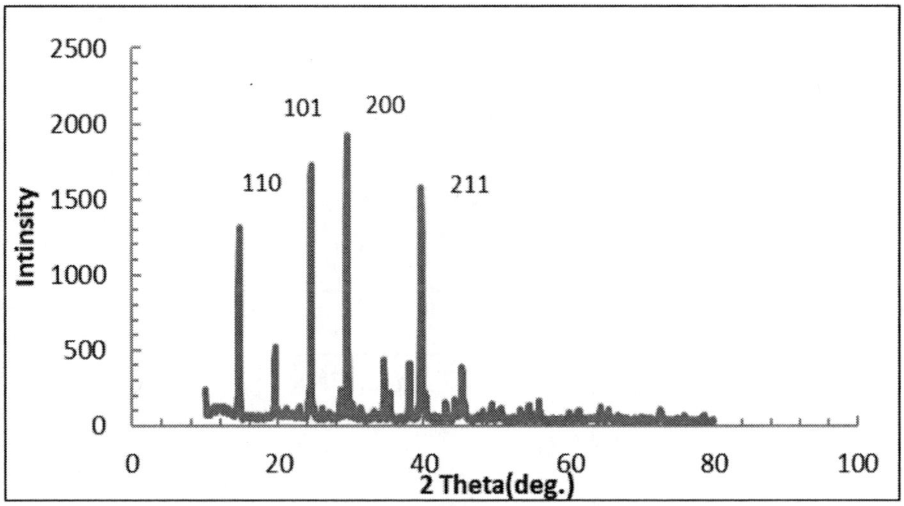

Figure 5. X-ray diffractions of Poly 4-bromoaniline thin.

Optical properties measurements

Transmission and Absorption spectrum

Optical properties is one of the most important studies in semiconductors, providing information on the value of the energy gap, the quality of the electronic transitions occurring in the material and the values of the other optical constants. The optical properties of the semiconductor membranes are generally based on the method and conditions of preparation such as the base temperature, any change in one of them will cause the absorption edge to shift to higher or lower energies.

Transmission and Absorption for prepared thin films measurement were made within Spectral range wavelengths (300-800 nm) by using (UV-VIS Spectrophotometer/1800)which provides from (BIOTECH) England company .

Absorption Coefficient Calculation

The absorption coefficient α was calculated using the follow equation

$$\alpha = 2.303 \frac{A}{t} \quad \text{......}(1)$$

Where α absorption coefficient, A absorbance , t thickness of thin film

Figure (6) shows the change the absorption coefficient Absorbance of thin films polymers with varying wavelengths in general the decrease with increase wavelengths. the absorption edge can by determine by drawing a long net to the curve and the intersection point with X-axis (λ) represent the cut off wavelength and it is value was 365 nm.

The band gap of polymer thin film is calculated by following equation [2]:

$$= A' (h\nu - E_g)^r \text{........} (2) \quad \alpha$$

Where α is the absorption coefficient.

h is the Planck's constant
ν is the frequency of fallen light
E_g is the optical energy gap
r is factor controlling the direct and indirect transition of electron from the valence band to conduction band
A constant
Figure (7) shows the relationship between $(\alpha^*E)^2$and energy gap was 2.27 eV.

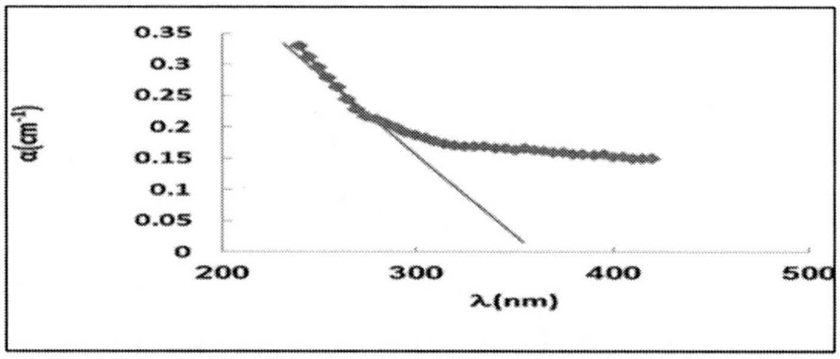

Figure 6. Absorption coefficient as a function of wavelength for polymer thin film.

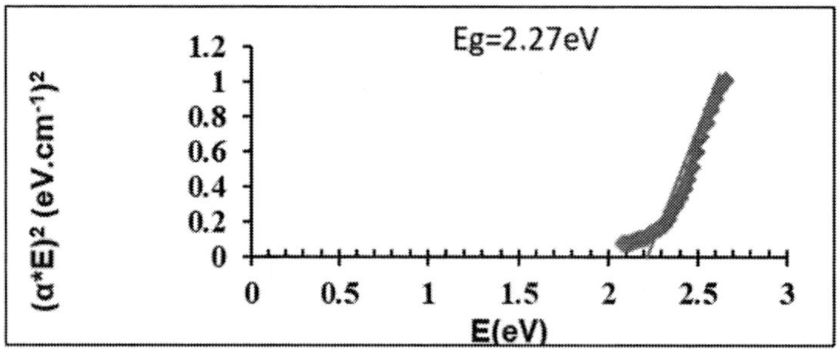

Figure 7. energy gap as a function of wavelengths for polymer thin film.

Sensitivity of thin film Measurements

The polymer sensor was fabricated by evaporated aluminum foil as ohmic contact electrodes on the polymer thin film using metal mask in order to test and measure the sensor Sensitivity ammonia (H_2) gas The following figure (8) was used to find Sensitivity of thin film for various gases, which consist of the following parts- :

1- Vacuum pump (rotary) get pressure value (10^{-2}mbar).
2- Connecting tubes
3- Discharge measurement sensor (negative pressure) PRM Manufacture by Edwards.
4- Pressure reader.
5- chamber manufactured locally with dimensions (20 x 20) cm that the sample was put in it and several holes including :
   A- Hole to pump and discharge gases.
   B- Hole use as a glass window.
   C- Fead throw to enter and take the signal.
6- D.C Power supply.
7- Digital multimeter.
8- Iron stand and holder to formation gas Container.

Figure 8. Gas sensor System.

After connecting deposited aluminum electrodes on the film by connected wires, sample is placed on the base inside the chamber, then Gas is pumped, the resistance will be change with the time (per 5 seconds) as well as photoconductivity (I-V) characteristics was measured in case of the gas pump And at the exit of gas to know effect of gas on these thin film.

Sensitivity measurement

Sensitivity can be calculated (S) from equation below [9]:

$$S = (R_g - R_a) / R_a \times 100\% \ldots\ldots\ldots\ldots\ldots(5)$$

R_g: the electrical resistance of thin film in the case of the gas pump.
R_a: the electrical resistance of the thin film sensor in the air.
Figure (9) shows the Sensitivity of H_2 gas.

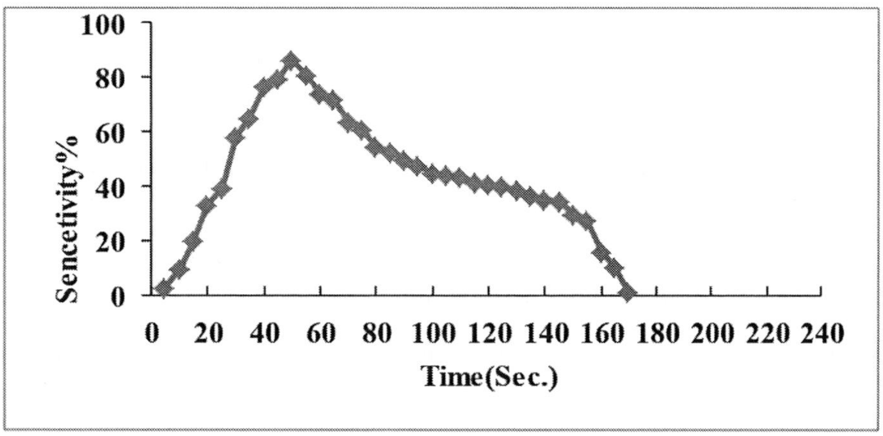

Figure 9. Sensitivity of H_2 gas.

FESM test

The nano fiber structure of polymer appear in Field Emission Scanning Electron Microscopy (FESEM) show in figure (10):

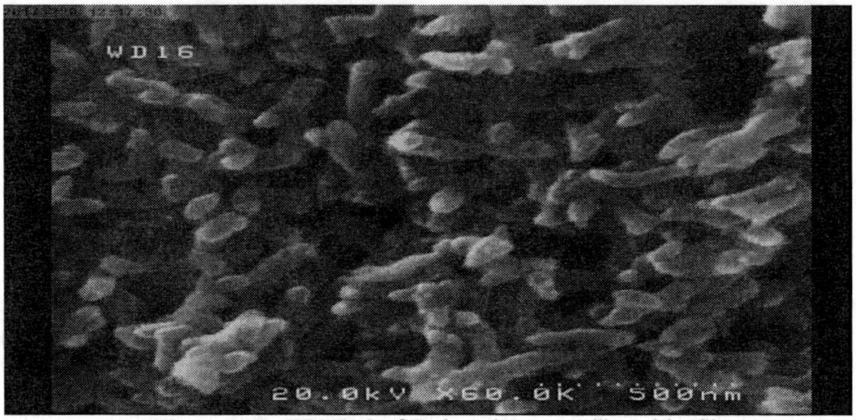

Figure 10. FESEM of Poly 4-bromoaniline thin film.

4. Discussion.

The adsorption of gases on the surface of solids is one of the most important types of adsorption. Depending on the type of forces that bind gas molecules to the surface .Both types of semiconductors (n-type) and (p-type) are used as gas sensors, but type (n) is preferred on type (p) because it gives a resistance change from the highest value to the lowest value in the case of gas, while type (p) Change of resistance is the opposite (from the lowest value to the highest value) Poly 4-bromo aniline appears to be a good candidate to realize gas sensors because of its sensitivity at room temperature, its ease of synthesis by various methods and its low cost.. Poly 4-bromo aniline polymers has been Classified as a good electrical conductivity due to the electronic substituting between amin groups and the benzene ring and the existence of the electrons mobility and thus become as semiconductor when passes a H_2 gas thin film of polymer ,that polymer adsorb through Reversible bond , This bond is increasing from the electrical resistance due to the electron substituting process ,these mechanism is the principle of gases sensor to gas sensing.

References

[1] Jun LiXuefeng HuJichang Wang 2013 "*Electrochemical Recognition of Chiral Molecules with Poly(4-bromoaniline) Modified Gold Electrode*" **25(8)** 1805-2029.

[2] Umesh S. WawareEmail authorA. M. S. HamoudaEmail authorA. Shahul HameedG. J Summers 2017 "Synthesis and characterization of electrically conductive poly(aniline-co-3-bromoaniline)" *Journal of Materials Science: Materials in Electronics* **28(14)** 10693–10699.

[3] Kleijn, S. E. F.; Lai, S. C. S.; Koper, M. T. M.; Unwin, P. R. 2014 "Electrochemistry of Nanoparticles Angew" *Chem. Int. Ed.* **53** 3558-3586.

[4] X.P. Chen, C.K.Y. Wong, C.A. Yuan and G.Q. Zhang 2012 "Impact of the functional group on the working range of polyaniline as carbon dioxide sensors" *Sensors and Actuators B: Chemical* **175** 15– 21.

[5] Z. Liu and P.C. Searson 2006 "Single nanoporous gold nanowire sensors" *The Journal of Physical Chemistry B* **110** 4318-22.

[6] T.Y. Wei, P.H. Yeh, S.Y. Lu and Z.L. Wang 2009 "Gigantic enhancement in sensitivity using Schottky contacted nanowire nanosensor" *Journal of the American Chemical Society* **131(126)** 17690-5.

[7] N. Schroeder, 2016 "Major findings of the new report on markets, strategies, technologies:

sensors markets" *Intechno consulcting*, (Basel)Switzerland, **15** 2012.

[8] X.P. Chen, C.K.Y. Wong, C.A. Yuan, L. Shen and G.Q. Zhang, 2013 "Molecular model for the charge carrier density dependence of conductivity of polyaniline as chemical sensing materials" *Sensors and Actuators B: Chemical* **177** 856- 61.

[9] P.N. Bartlett and S.K. Ling-Chung, 1989 "Conducting polymer gas sensors Part III: Results for four different polymers and five different vapours" *Sensors and Actuators* **20** 287-92.

[10] P.N. Bartlett and S.K. Ling-Chung, 1989 "Conducting polymer gas sensors part II: response of polypyrrole to methanol vapour" *Sensors and Actuators* **19** 141-50.

[11] H. Bai and G. Shi, 2007 "Gas sensors based on conducting polymers" *Sensors* **7** 267-307.

[12] P.N. Bartlett, P. Archer and S.K. Ling-Chung, 1989 "Conducting polymer gas sensors part I: fabrication and characterization" *Sensors and Actuators* **19** 125-40.

[13] H. Liu, J. Kameoka, D.A. Czaplewski and H. Craighead, 2014 "Polymeric nanowire chemical sensor, Nano Letters" **4** 671-5.

[14] X.P. Chen, C.A. Yuan, C.K.Y. Wong, S.W. Koh and G.Q. Zhang, 2011 "Validation of forcefields in predicting the physical and thermophysical properties of emeraldine base polyaniline" *Molecular Simulation* **37** 990-6.

[15] G. Prasad, T. Radhakrishnan, D.S. Kumar and M.G. Krishna, 2005 "Ammonia sensing characteristics of thin film based on polyelectrolyte templated polyaniline" *Sensors and Actuators B: Chemical* **106** 626-31.

IOP Conf. Series: Journal of Physics: Conf. Series **1032** (2018) 012064 doi:10.1088/1742-6596/1032/1/012064

Spectrophotometric Determination of Mebeverine hydrochloride in pharmaceutical preparation via Ion Association Reaction

Abdulbari Mahdi, Zainab Abas
Department of pharmaceutical chemistry, Pharmacy collage, Kerbala University, Iraq.

Corresponding author's E-mail: foxyabd@gmail.com

Abstract. We developed colorimetric method to determine Mebeverine hydrochloride. Briefly, the principle of our method is the formation of ion associated complex between the drug and phenol red indicator. The spectrum of the complex was measured with maximum absorption at 396 nm – Beers law is followed over the range over 2 – 25 ug/ml with molar absorptivity of (7.57×10^4 L/ Mol. Cm), a sandal sensitivity of (0.00615 µg. Cm $^{-2}$), and the relative standard deviation RSD % (3.3974%), mean recovery was found to be in the range (96-99 %), and correlation coefficient of (R^2=0.9941). However, the method accurately and successfully was performed to determine Mebeverine hydrochloride in bulk and pharmaceutical preparations.

1. Introduction.

Mebeverine hydrochloride(MBV) is a white crystalline powder, molecular weight 466 g/Mol, molecular formula $C_{25}H_{35}NO_5$.HCl Fig (1). MBV is freely soluble in water and 96% ethanol, but it is insoluble in di ethyl ether (1). The IUPAC name of Mebeverine hydro Chloride is 3/4- Dimethoxy benzoic acid (ethyl 2, 4 methoxy 4-phenyl -1-methyl ethyl) amino – butyl ester. MBV widely uses as a relaxant and antispasmodic medication for the gastrointestinal tract especially colonic spasm and irritable bowel syndrome (2, 3).

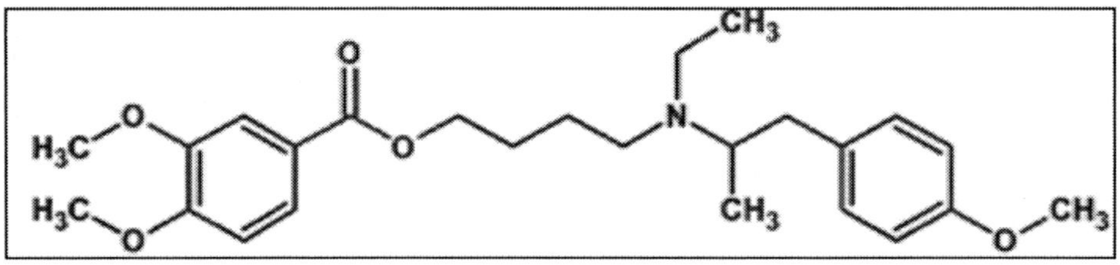

Figure 1. Represents the chemical structure of Mebeverine hydrochloride.

Actually there are several analytical methods including spectrophotometric (4-10), Construction Spectroscopy (11), HPLC (12-19) ion selective Electrodes (20), Carbon Paste Electrode (21) to determine the pharmaceutical dosage of MBV. Therefore, we designed our experiment to develop colorimetric method, which characterized as easier, faster and more accurate to measure the dose of MBVHCL in bulk and pharmaceutical preparation.

2. Experimental parts.

2.1. Apparatus.

We used Shimadzo model 1800, which has a double beam UV visible spectrophotometer, and two matched quartz cells that allow 1 cm light path.

2.2. Material and reagent.

We made sure that all the chemical we used were analytical grade and did not need further purification. Mebeverine HCl was provided from SDI –Iraq. Distilled H2O was used throughout.

- **Stock solution of** Mebeverine HCl **MBV** was prepared by transfer (100 mg) of the drug in 100 mL volumetric flask, dissolved in small amount of DW and make up to the volume with same Solvent.
- **Buffer Solution of 2.3 pH:** was prepared by dissolving (1.287 gm Na OH) in dH2O and finish the volume to 250 ml with dH2O, and (3.4206 gm KH_2PO_4) was dissolving in (D. W) and make up to 250 ml in volumetric flask with same solvent. The pH was adjusted with small amount of 1 M HCl.

Phenol red to get (3.203x10 -3 M), we dissolved 0.056 gram of phenol red in 25 ml ethanol and finish the volume with dH2O to 250 ml in volumetric flask.

2.3. Recommended steps.

Transfer a suitable concentration 2µg/ml equal to 0.2 ml of a standard solution of MBV into a separating funnel (size equal to 125 ml) containing 8 ml of buffer solution (PH=2.3) and 5 ml of 3.203×10^{-3} M of phenol red. The above solution, shaking well for 2 minutes, then 10 ml of Chloroform was added to shake the contents for 2 min to allow the two aqueous phases to separate. We removed the organic layer and kept in 10ml volumetric flask. Then we diluted the mixture to the mark with chloroform. A separated Chloroform layer was standing for 5 minutes and measured at 396nm against a reagent blank.

2.4. Procedure for pharmaceutical preparations.

Ten pills were weighted and grinded to get the medication in powder shape. We dissolved 100 mg of the powder form in 100 ml dH_2O in 100 ml volumetric flask to get 100 µg /ml concentration. The solution was stirred for 10 min then filtered through the filter paper #5 to place the solution new volumetric flask (the size equal 100 ml), then we finished the volume with dH_2O to 100 ml.

3. Results and Discussion.

In order to get the sensitive and the stabile product number of factors that influence of formation complex were carefully studied. The spectrum graph of the colored reaction compound was measured against the blank reagent (absorbance at λ_{max} 396 nm) see figure 2.

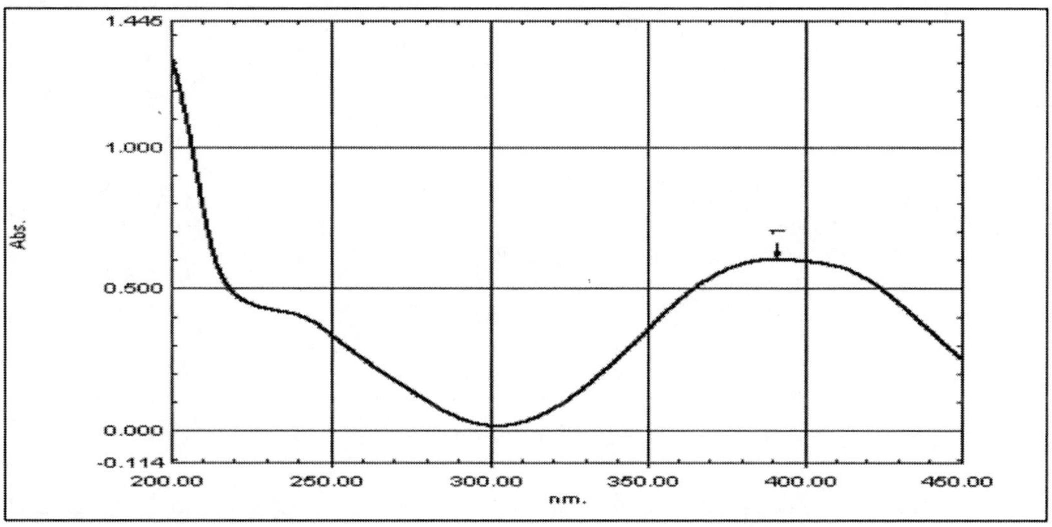

Figure 2. Absorption spectra of the colored compound (2.14×10^{-5} *M*) Of MBV with phenol red (3.2×10^{-3}M).

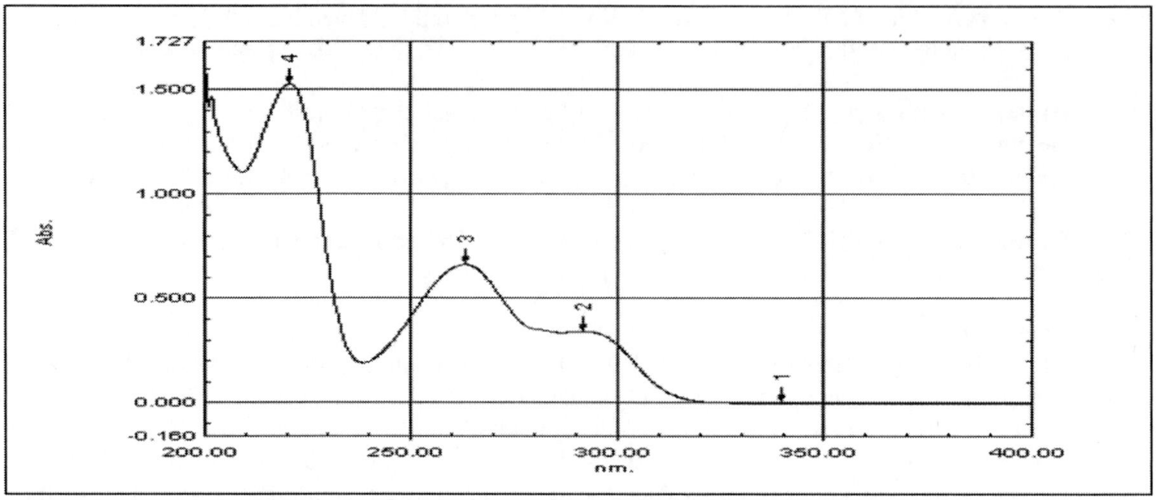

Figure 3. The spectrum of 8 μg/ml of Mebeverine HCl.

3.1. Optimization of the Experimental condition.
We optimized the reaction conditions by measuring various parameters that effect on the absorption intensity of the reaction complex.

3.2. Effect of phenol red concentration.
Various concentration of the phenol red solution was added to fixed amount of MBV and was found that 5 mL of 3.203x10 -3 M of phenol red was the final concentration which could give us full intensity with a minimum blank value of the observance. And this consider to be optimum for measuring the concentration range between (0.2 - 25 μg /ml) of the drug as in figure 4.

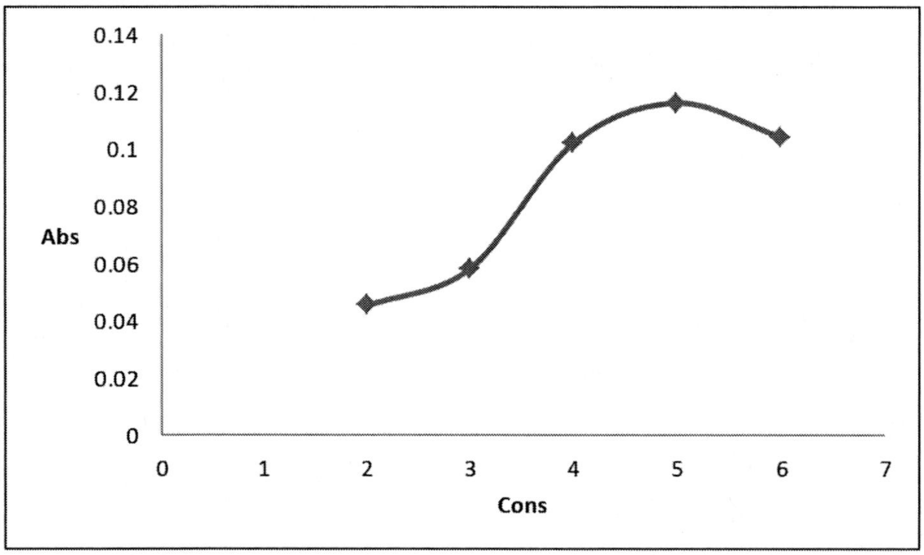

Figure 4. Effect of different volume of $3.203x10^{-3}$ M phenol red on absorption of the complex between MBV and phenol red.

3.3. Study temperature effect.

The temperature effect on the observance of the formed compound using water bath was studied and found that 25 C^ogave the highest absorption at λ max 396 nm, therefore it is recommended that the colored compound should be carried out at 25 C^oas in figure 5.

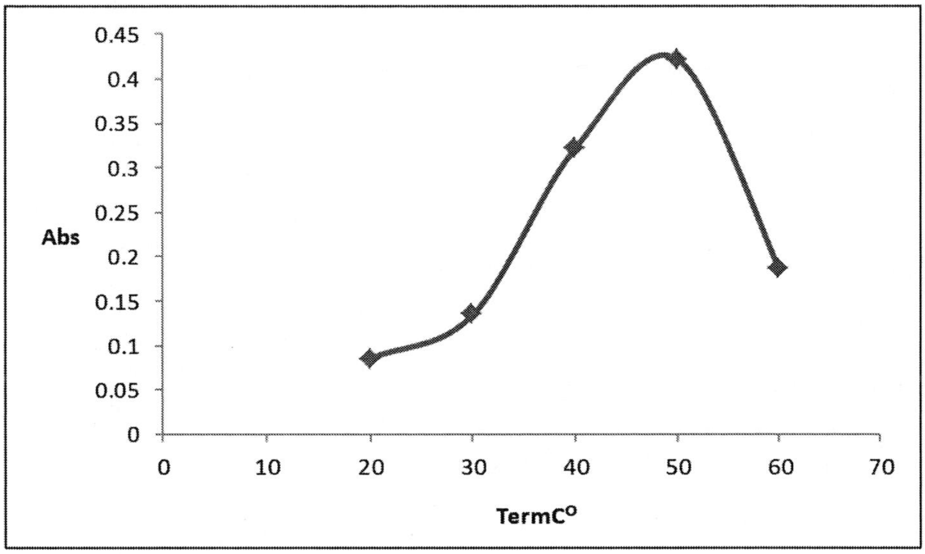

Figure 5. Effect of Temperature absorption spectra of MBV.

3.4. Study time effect on the absorption intensity.

Interesting the color intensity reached a maximum absorption after Mebeverine has been reacted with phenol red at 20min. Therefore, 20 min development time was chosen for further experiment. The colored compound was studied of test 2hour. The results obtained are shown in figure 6.

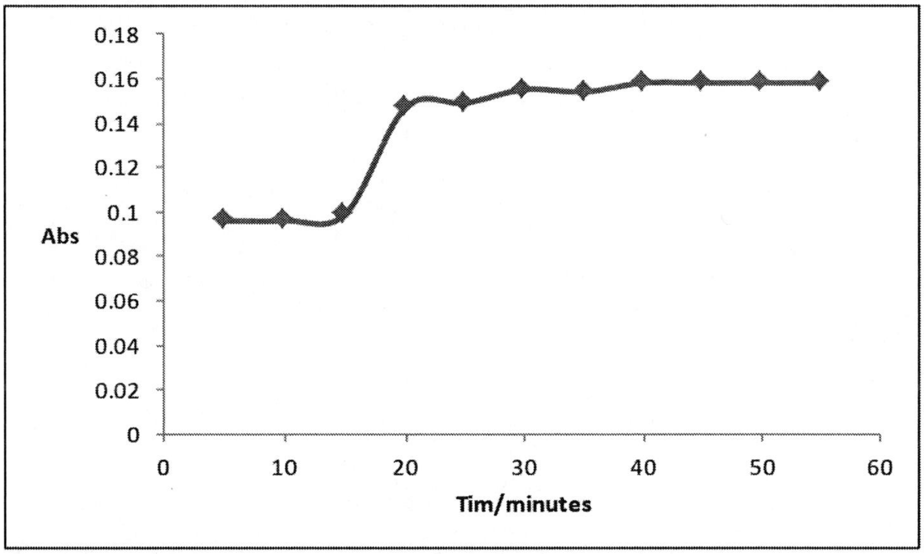

Figure 6. Represents the effect of time on the absorption intensity of MBV. HCL

IOP Conf. Series: Journal of Physics: Conf. Series **1032** (2018) 012064 doi:10.1088/1742-6596/1032/1/012064

3.5. Calibration Graph Study time effect on the absorption intensity.
A linear calibration graph was obtained when employing the condition describe above for determination of MBV which shows that Beer's law obeyed to the concentration range (0.2-25) ug/ml with a correlation coefficient (0.9955). The molar absorptivity of the formed compound was found to be 1.87 x10^4 L/Mol.Cm and a standard sensitivity of 0.0615 μg.cm^{-2}. The result that we obtained is shown in figure 7.

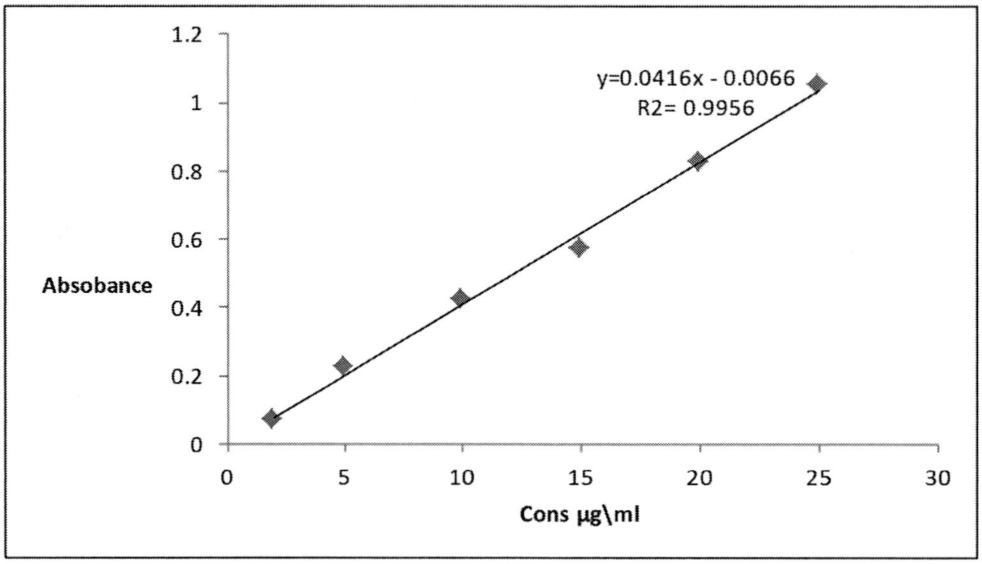

Figure 7. represents calibration curve to measure MBV.HCL according to Beer's law.

3.6. The stoichiometry of the MBV.HCL and Phenol red reaction product.
We followed the mole ratio and Jobs method to test the nature of MBV.HCL and phenol red reaction products. Interesting about 2:1 reaction product was formed between the drug and the reagent (see figure 8 A and B). Schematic figure 1 may explain the mechanism of dye formation in MBV.HCL and phenol red reaction.

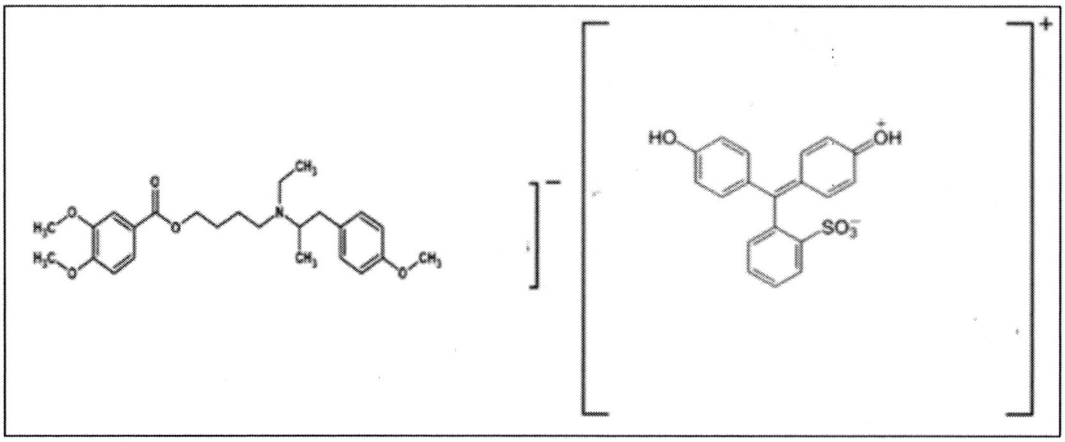

Scheme 1. Represents the mechanism of MBV- Ph. Red ion-pair formation.

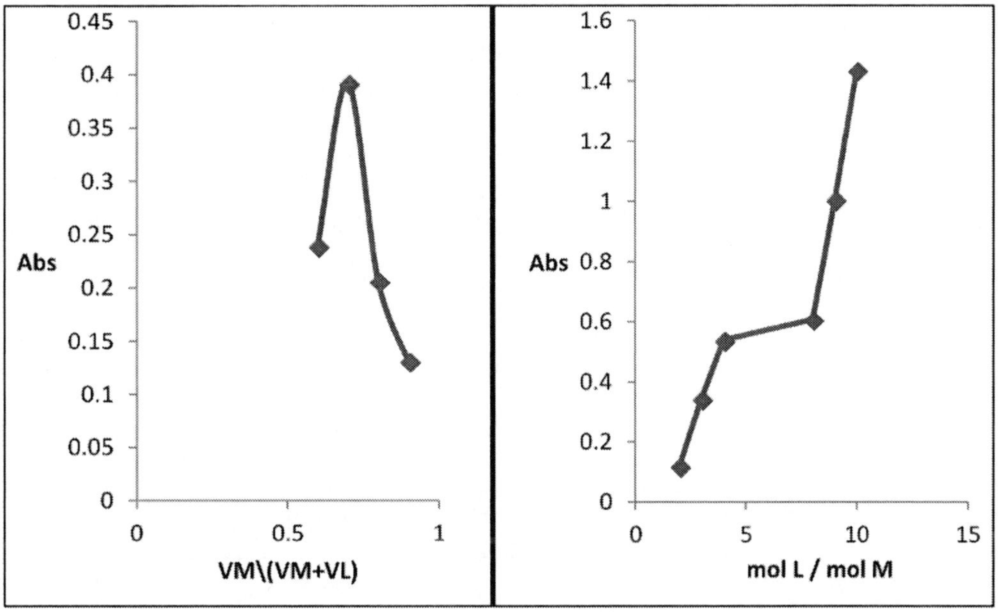

Figure 8. A mole fraction of drug and B represented the mole ratio.

3.7. Method of evaluating the stability of the ion pair complex.
We tasted the stability of the complex that formed between MBV and phenol red. Although the constant observance readings were obtained at short time 20 min and 25 C^0. The formed complex was constant for at least two hours without observance intensity changing or in λ_{max}. The conditional constant was calculated according to literature (22), and found to $1.315x10^7$, and this supported the stability of the forming complex.

3.8. Steps that followed to choose the Organic solvent.
A different of organic solvents as chloroform, carbon tetrachloride, dichloromethane, dichloromethane and ether were tested to choose the suitable solvent to give a high absorbance. Chloroform was the optimal choice in comparing other solvent types for its highest absorbance feature. It was found that one extracted is adequate to obtain high retrieval of the complex and short time to touch the equilibrium between the two phases. Furthermore, the two minutes shaking time provided the stable observation.

3.9. Evaluation of the proposed method.
Evaluation the precision and accuracy of the proposed method.
The result shows that proposed method was accurate according the values of the recovery with the range of (96.74-99.39) when the three levels of concentration were measured also the low values of R.S.D% indicated that the method was high precise (0.479-1.986) %.

IOP Conf. Series: Journal of Physics: Conf. Series **1032** (2018) 012064 doi:10.1088/1742-6596/1032/1/012064

3.10. Interference studies.

The results showed that no interferences were found in the presence of 40 µg/ml for each excipient (Glucose, Lactose, Starch, Sucrose, Magnesium Stearate, Talk, Aacia) with these concentration different concentrations of the drug up to (2,10,20) µg/ml in the determination of MBV.

Table 1. Represents the concentration of some materials that interfered with the pharmaceutical preparation of the drug.

Concentration of MBV. µg/ml	Mean Recovery (%)
2	97.5903
10	96.1445
20	99.3975

N	Concentration In µg/mL	Found	R.S.D%	Recovery (%)
1	8	7.7000	0.479	96.75
2	16	15.6263	0.608	97.66
3	20	19.5180	1.989	97.59

3.11. Detection and quantification limits.

LOD refers to the limit of detection while LOQ refers to the limit of quantification. LOD= 3SD/K and LOQ= 10SD/K, where SD means the standard deviation of replicate determination values under the same conditions in the absent of the drug and K is the slop of the calibration graph (22). According to these forms, the LOD equal to 0.213ug/ml and LOQ equal to 0.213 ug/ml.

3.12. Analytical Applications.

The results of our proposed modified method suggested that the proposed method si easier, accurate and précised as compare to the other measuring methods. The low value of (R.S. D %), which means relative standard division indicated good reproducibility and precision. The mean of percent recoveries obtained were in the range of (97.18-99.55) indicating high accuracy of the suggested method.

Table 2. The obtained results from the application of the proposed method.

Company	Claimed	Found	R.S.D%	Recovery (%)
Duspatalin	135mg	131.2	1.3212	97.18
EVACOL	135mg	133.5	1.4900	98.88
MEVA	135mg	134.4	1.3800	99.55

4. Conclusion.

The proposed method shows a good accuracy, sensitive for estimation of MBV in bulk and pharmaceutical preparation. The proposed has some advantage like fast determination of the drug in

IOP Conf. Series: Journal of Physics: Conf. Series **1032** (2018) 012064 doi:10.1088/1742-6596/1032/1/012064

its pure and in pharmaceutical preparation also the uses of 25 c made the proposed method simple and not require heating. The wide ranges of the linearity of the method gave a good application for the pharmaceutical preparation. The absent of the effect of the interference materials in the reaction of the MBV.HCL and phenol red reaction, which suggests that the method is beneficial for routine analysis tests and quality control assay of the drug in pure materials and in tablet forms.

References

[1] Arena MS, Sultana N and Siddiqui FA. 2005 "A New RPHPLC Method for Analysis of Mebeverine Hydrochloride in Raw Materials and Tablets" *Pak J Pharm Sci* **18(2)** 11-4.

[2] Walash MI, Sharaf El-Din MMK, El-Enany NM, Eid MI, Shalan SM. 2012 "Simultaneous Determination of Sulpiride And Mebeverine By HPLC Method Using Fluorescence Detection: Application to Real Human Plasma" *Chem Cent J* **6(13)**1-12.

[3] Souri E, Aghdami AN. and Adib N. 2014 "A Stability Indicating HPLC Method for Determination of Mebeverine In the Presence of Its Degradation Products and Kinetic Study of Its Degradation in Oxidative Condition" *Res Pharm Sci* **9(3)** 199-206.

[4] M. E. Hassan, A. A Gazy and M. M. Bedair 1995 "Colorimetric Determination of Mebeverine Hydrochloride in Tablets by Charge Transfer Complexation", *Drug Dev. Ind. Pharm.* **21(5)** 633.

[5] T. Rajash N., Lakkshmi A. and Ashok Kumar L. 2014 "Simultaneous Quantitative Estimation of Mebeverine Hydrochloride and Chlordiazepoxide In Capsules Using Spectrophotometry" *Nter. J. Of Pharm.And Pharm. Scie* **6** 96.

[6] K. Rajitha, N. Lakshmi Prasanna, G. Vasundhara, R. Naveen, and A. Ashok Kumar 2014 "U.V Spectrophotometric Method Development and Validation for The Simultaneous Quantitative Estimation of Mebeverine Hydrochloride and Chlordiazepoxide In Capsules" *Lnter. J. Of Pharm.And Pharm. Scie* **6** 345.

[7] A.A. Othman, R.I. Bagary, E.F. Elkady and M.M. EL-Kerdawy 2006 "Development and Validation of Spectrophotometric Methods for Simultaneous Determination of Mebeverine Hydrochloride and Chlordiazepoxide In Bulk and In Dosage Form" *Pharm. Anal. Acta.* **7** 8.

[8] T.H. Sonia, A.M. Bahia, A.H. Hashim, A. and EL. Asmaa "Determination of Certain Antispasmodic Drugs as Single Ingredient, Mebeverine Hydrochloride, And in Two Component Mixtures" *Mebeverine Hydrochloride–Sulpiride And Isopropamide Iodide–Trifluoperazine Hydrochloride* **35(1)** 43-61.

[9] N. Safila, W. Nimra, N. Safeena and R. Hina 2015 "Method Development of Mebeverine Hydrochloride by Using U. V Spectrophotometric Method" *Lnter. J. Of Applied Sci.* 001-005.

[10] K. Sreedhar, C.S.P. Sastry, M. Narayana Reddy and D.G. Sankar 1997 "Spectrophotometric Determination of Mebeverine Hydrochloride" *Mikrochim Acta* **126** 131-135.

[11] S.E. Marwa, S.E. Manal and S.H. Wafaa, "Conductometric And Spectroscopic Determination of Mebeverine Hydrochloride and The Solubility Products of Its Ion Recognition Species" *Int. J. Electrochem.Sci.* **7** 9781-9794.

[12] E. Souri, A. Negahban and N. Adib 2014 "A Stability Indicating HPLC Method for Determination of Mebeverine Hydrochloridein In the Presence of Its Degradation Products and Kinetic Study of Its Degradation in Oxidative Condition" *Pharm. Sci.* **9(3)** 199-206.

[13] W. Mohamed, M.S.H. Mohie, M. Nahed and E. Manal 2012 "Simultaneous Determination of Sulpiride And Mebeverine By HPLC Method Using Fluorescence Detection: Application to Real Human Plasma" *Chem.Cent.J.* **6** 13.

[14] E. Souril, A. Negahban Aghdami and N. Adib 2014 "A Stability Indicating HPLC Method for Determination of Mebeverine In the Presence of Its Degradation Products and Kinetic Study of Its Degradation in Oxidative Condition" *Pharm. Scie. J.* **9(3)** 199-206.

[15] P. Ravikumar, D.R.V. Krishna, E. Sasikiran and Y.R. Amachandra 2014 "Development And Validation Of A Ultra Performance Liquid Chromatographic Method For Assay Of Mebeverine Hydrochloride" *Lnter. J. Of Pharm.And Pharm. Scie*. **6(2)**.

[16] R.M. Youssef 2014 "Development of Gradient HPLC–DAD Method for Assay of Ternary Mixture Containing Amebicide And Analgesic Drugs" *Acta Chromatographic* **26** 67–80.

[17] S. Ramadevi, A. Srikanth and K. Ashok 2015 "RP-HPLC Method Development and Validation for Simultaneous Quantitative Estimation of Mebeverine Hydrochloride and Chlordiazepoxide In Capsules" *Lnter. J. Of Pharm.And Pharm. Scie.* **(7)20** 975-1491.

[18] K. C. Usmangani, M. P. Nishma, A. S. Dimal, A. M Falgun and K. B. Kashyap 2015 "Liquid Chromatographic Method for Simultaneous Quantification of Alprazolam and Mebeverine In Pharmaceutical Dosage Form" *Pharm. Anal. Acta*. **6** 6.

[19] K.R. Senthil, S.N. Meyyanathan and B. Gowramma 2015 "A New Validated Chiral Rp-Hplc Method for The Determination of Stability of Mebeverine Enantiomers in The Presence of Its Degradation Products" *Lnter. J. Of Pharm.And Pharm* **7(5)**.

[20] K. Sujana, M.Z.V. Hamuthal, V.S.N. Murthy and N. Shravani 2015 "A Novel Validated Analytical Method Development for The Binary Mixture of Mebeverine And Chlordiazepoxide In Pharmaceutical Formulation and Its Application to Stress Studies" *Pharm. Anal. Acta,* **6** 1.

[21] A. S. Muthana 2014 "Construction of Membrane Selective Electrodes for Determination of Mebeverine Hydrochloride Drug" *J. For Pure Sci* **(10)**3.

[22] Meier P.C and Zund R.E, 2002 "*statistical methods in analytical chemistry 2nd Ed.*", A Wiley intersciece publication. (USA) 48 and 69.

IOP Conf. Series: Journal of Physics: Conf. Series **1032** (2018) 012065 doi:10.1088/1742-6596/1032/1/012065

Comparative study of nano iron catalysts in the presence of the crude polymer matrix and their cross link polymer in the Fischer-Trophsch synthesis

Abdulqadier Hussien Al khazraji[1], O.S. Dementyeva[3], Z. Pastukhova[2,] M.V. Kulikova[3], Vitaly. R. Flid[2]

[1]Department of Chemistry, College of Education for Pure Science, University of Diyala, Iraq. [2]Moscow Technological University (Institute of Fine Chemical Technologies), Moscow, 119571 Russia. [3]A.V. Topchiev Institute of Petrochemical Synthesis, RAS (TIPS RAS), Moscow, 119991 Russia.

Corresponding author's e-mail: ahnhkm@yahoo.com

Abstract. A new metal-polymer catalyst in a stable colloidal system, were prepared. Iron-paraffin catalysts, nano sized particles, with different polymers have been matrixes were used to examine the effects of polystyrene and their derivatives on chemical/structural properties through Fischer–Tropsch synthesis (FTS) and the yield as well. Nano size iron-containing particles distributed are suspended in the hydrocarbon medium of the slurry-reactor were found to have great impact and high percent conversion of CO to more than 70%. Polystyrene derivatives found to interact with Fe species by the formation of Fe_3O_4 oxide and amorphous hydrolyzed δ-FeOOH which were formed at the given conditions. All catalysts with a polymer found to have a bimodal distribution particle. In structural aspects, polystyrene derivatives increases the dispersion of Fe species and inhibits the coagulation of active iron particles, and on other hand, the chemical and structural effects would cause the increase of the FTS efficiency towered the selectivity of heavy hydrocarbons and olefins during the FTS. The state of iron and polymer-paraffin matrix were examined utilizing XRD and AFM techniques which showed the forms of ground state of iron as Fe_3O_4 oxide and amorphous hydrolyzed δ-FeOOH were exists. When the polystyrene derivatives are introduced, the proportion of amorphous paraffin found to be less than that in a crystalline state. The maximum conversion of CO (71%) and the yield of liquid hydrocarbons ($63 g.m^{-3}$) based on a cross-link polymer such as a polystyrene divinylbenzene (PS-DVB) used. A very interesting fact that the risen temperature found to have a small decrease in selectivity and an increase in conversion efficiency, this was observed in all reactions.

1. Introduction.

Fischer-Tropsch synthesis (FTS) was discovered nearly 95 years ago. It have a great industrial importance due to its wide applications and a great variety of products obtained, such as paraffin's, olefins and alcohols. It converts syngas ($CO+H_2$) to hydrocarbons in the range C_1 to C_{100} [1]. The benefits of FTS were well documented. One of its application include the generation of liquid fuels free of sulfur and aromatic compounds which normally found in the crude oil distillates and fuels. Iron-based catalysts provide three advantages: cheaper, less selective towards methane and promotes the water gas that compared to cobalt based catalyst [2].

There are many methods used in reforming includes, slurry and fixed reactors, slurry reactors were applied in a wide range of chemical processes, at both laboratory and industrial scales [3]. The main uses of FTS that for the generation and reforming sulfur free hydrocarbons as well as aromatic contains products than those found in crude oil distillates.

During the last two decades, nanoparticles, received a great attention, due to their properties and the wide applications. Indeed, extensive and intensive work has been done on the synthesis of polymer/nanoparticles composite materials as a catalyst, due to their direct importance reforming and cracking hydrocarbons [4]. It is therefore very important to find synthetic methodology to produce nanoparticles that have significant impact on changing of physical and chemical properties of such materials, so they should be well investigated [5-9].

The effect of using polymers as stabilizers in the synthesis of cobalt and iron based nanoparticles was studied by Claver [10]. The difficulty of syntheses the metal's nanometer-sized particle is to keep it stable and free standing nanoparticles. Such materials are highly reactive and have tendency to combined together, leading to lose their unique properties. Therefore, to keep their main properties without any change it was necessary to embed them in a matrix polymer [11-14].

The aim of the present work is prepare FTS's catalysts over iron nano-sized particles inserted in the paraffin and polymers to be loaded in dispersion medium to investigate the catalytic properties: activity, selectivity, and conversion of CO efficiency in FTS.

2. Experimental.

To produce iron Nano-sized catalyst Fe–paraffin/polystyrene (Fe-P/PS) and compare it with its derivatives: Fe–paraffin/polystyrene cross-linked hosting polymer as with divinylbenzene (Fe-P/PS-DVB) and Fe–paraffin/butadiene-styrene rubbers (Fe-P/SBR) systems. The Fe-paraffin/polymer catalysts were prepared via a thermal decomposition method[14].Which in brief, add solutions containing Fe $(NO_3)_3.9H_2O$ to a paraffin (mark - 2) and 6 wt % of one of the following polymers (PS), (PS-DVB) and (SBR) matrixes at 280°C under stirring and argon gas as an inert atmosphere stream.

The experiments were conducted the catalyst suspended in a slurry reactor at a pressure of 2 MP. All catalysts were activated by CO gas at 300˚C for 24h, that for all the experiments. A 100 ml of catalyst was loaded into the reactor under syngas load of 1–2 L/(g cat h) and molar ratio of CO: H_2 = 1 : 1. Temperature ranges of 220–320°C were maintained. The temperature was increased stepwise (by 20°C each 12 h period).The gas was analyzed on line by gas chromatographs (GC) (Models Kristallyuks- 4000 M) with a thermal conductivity detector array and helium as a carrier gas.

The liquid products in the cold traps were collected over 72 h and analyzed using liquid chromatographs (LC) (Models Kristallyuks- 4000 M).

The particle size was determined by Malvern Zetasizer Nano ZS instrument for each synthesized catalyst. In succinct, a 0.01 g dissolved from suspended catalyst in 10 ml of hexane with the addition of 5 wt % of a surfactant (sodium dioctyl sulfosuccinate).

In order to obtain the catalysts, the following reagents were used:

a. Paraffin grade P-2 (GOST 23683-89) - a mixture of high-molecular hydrocarbons C_{18}-C_{35}, mainly aliphatic structure, obtained by distillation of oil.

b. Iron (III) nitrate (Fe $(NO_3)_3{\cdot}9H_2O$) nanohydrate, "extra pure" brand, manufactured by Scharlau Chemie S.A.

c. polymers : PS $=[CH_2CH(C_6H_5)]_n$, SBR=[(C_6H_5) CH-CH_2-CH_2-CH=CH-$CH_2)_n$ and PS- DVB= $[CH_2CH(C_6H_5)]_x[CH_2CH[C_6H_4(CHCH_2)]]y$.

3. Results and discussion.

3.1. Particle size distribution.

The particle sizes were examined using dynamic scattering technique in order to evaluate the changes in the particle size distribution of the dispersed phase of the synthesized suspensions. Nano catalyst includes SBR polymers and PS-DVB which brought into the required metal nanoparticles size, that have clear effect in the range of 170 to 660 nm in diameters as compared to the Nano catalyst structure, which is falls on the range from 2 to 670 nm diameter for the (Fe-P/PS) system [14] (Table 1).

In all period of production a systems were gave bimodal distribution[14].

PS introducing has lead to the formation of smaller particles (2 nm), it is going to affects their actions as compared with the catalysis based on SBR and PS-DVB polymers that has active Nano catalyst. It could be implicit that bimodality is connected with the stabilizing Fe possibility with each of polymer and paraffin. By another way, we demonstrate that Nano catalyst suspensions treatment with the CO:

H_2 combination in the temperature range 220-320 ° C for 5 days was the risen of the particles to the size mentioned in Table 1.

Table 1. The effect of polystyrene derivatives on the particles size.

Nano catalyst type	Particles size, nm			
	Before FTS		After FTS	
	nm	% content	nm	% content
Fe-P/PS	2	9	220	2
	677	91	842	98
Fe-P/PS-DVB	171	10	1283	84
	664	81	1643	16
Fe–P/SBR	312	22	540	14
	644	78	953	86

3.2. FTS performance.

The products, liquid and Gas have been sampled for an analysis. The catalysts activity had evaluated in terms of the next parameters:

The catalysts activity is calculated according to the following representation:

$$\text{Specific activity} = \frac{\dfrac{(\text{gram of CO})_{in} - (\text{gram of CO})_{out}}{\text{M.Wt of CO}}}{(\text{gram of Cat.* second})}$$

And the CO conversion (%) can be calculated according to:

$$CO_{\text{conversion (\%)}} = \frac{(\text{gram of CO})_{in} - (\text{gram of CO})_{out}}{(\text{gram of CO})_{in}} \times 100$$

Also the selectivity of liquid hydrocarbons (%) can be calculated according to the following equation:

$$\text{Selectivity of } C_{+5}\ (\%) = \frac{(\text{gram of Carbon})_{in} - (\sum \text{gram of Carbon})_{out}}{(\text{gram of Carbon})_{in} - (\text{gram of Carbon in C5+})_{out}} \times 100$$

The Product yield also calculated according to:

$$\text{Product yield} = \frac{\text{Amount of the product in grams formed}}{\text{gram of Cat. * h}}$$

The entire synthesized catalysts display high catalytic activity in hydrocarbons liquid preparation from CO and H_2 throughout the progression. Whereas the catalyst introduction depend on SBR and

IOP Conf. Series: Journal of Physics: Conf. Series **1032** (2018) 012065 doi:10.1088/1742-6596/1032/1/012065

PS-DVB which led to an increase of the amount of CO gas conversion as compared with the catalysis structure PS polymer in the Fe-Paraffin-Polymer system.

The maximum CO gas conversion shown in (Fig 1) has been achieved was 71%, with the system Fe- P/PS-DVB and the hydrocarbons liquid formation was (63 g/m^3) as in (Fig 2). Whereas utilizing the system of Fe-P/SBR the conversion of CO gas had reached 65%, and the liquid hydrocarbons formation reached (63 g/m^3). Catalyst based on PS had been led to 15% conversion of CO, and the liquid hydrocarbons formation was (26 g/m^3) (Fig 2). Significant rise in conversion of CO in systems based on the SBR and PS-DVB can be attributed cross link structure of the system, which had the capability to react with the metal (iron) much better than that inserted in the PS polymer. That also believed the iron nanoparticals have tendency of agglomeration in PS system compared with the others.

We believe also system based on cross-link acts as a stabilizer which could help to prevents sedimentation and leads to the formation of a completely new active catalyst system and that these catalysts could give some selectivity towards the control on distribution of the nanoparticals over the cross-link systems network structure of polymers and its concentration can be conceded evenly distributed as mentioned previously.

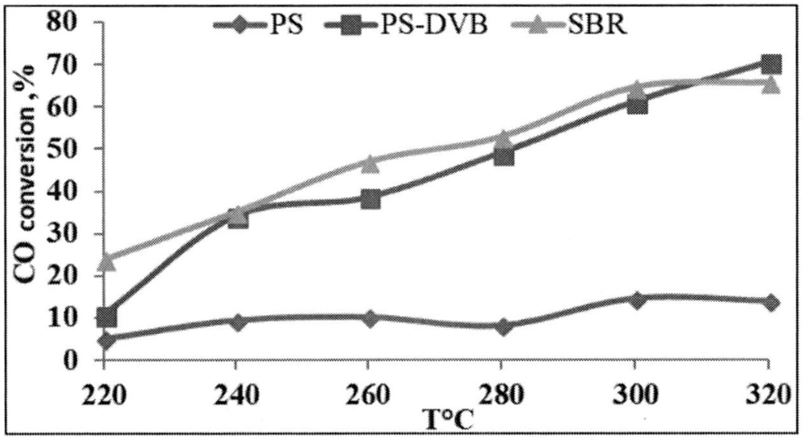

Figure 1. Temperature dependence of CO conversion of the Fe–paraffin–polymer Nano catalyst system: PS, PS-DVB and SBR.

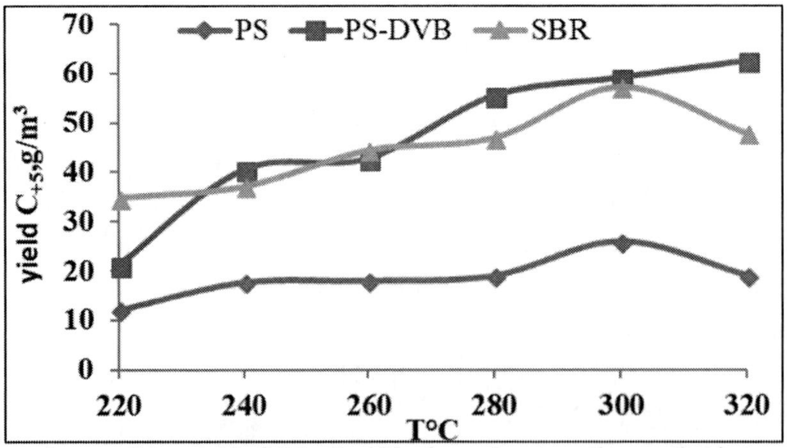

Figure 2. Dependence of the yield of C5+ liquid hydrocarbons on the synthesis temperature in the presence of the Fe–paraffin–polymer Nano catalyst system: PS, PS-DVB and SBR.

IOP Conf. Series: Journal of Physics: Conf. Series **1032** (2018) 012065 doi:10.1088/1742-6596/1032/1/012065

This network structure may plays the center rules of adsorption of the reactive gases ($CO:H_2$).The polystyrene derivatives influence on the essential parameters of the Fischer-Tropsch catalysts based on Fe-paraffin-polymer are represented in Table 2. As well known at such conditions of optimum metal content in classical catalysts for the three-phase Fischer-Tropsch synthesis shall not more than 20% because would suffers from agglomeration. That what we have done to solve such problem by using polymer systems as a stabilizer.

Table 2 and figure (3,4) , shows that the synthesized samples with nano catalysts based on PS it's important to notice that the yield of the main byproduct formed in the presence of iron Nano catalysts gaseous hydrocarbons (C_1, C_2-C_4 and CO_2) yield found to depend on the type of the polymer used . The nanocatalysts depend on polystyren used have a low yield gaseous hydrocarbons: C_1 and (C_2-C_4) from 4 to 33 gm^{-3} , CO_2 from 21 to 208 g m^{-3} receptively.

Table 2. The influence of polystyrene derivatives on the products of the Fischer-Tropsch synthesis (T=220-320°C, P=20 atm , $CO:H_2$ = 1:1).

Nano catalysts	*X_{CO}, %	Yield of Hydrocarbons ,g m^{-3}				Liquid hydrocarbon selectivity,%
		C_1	C_2-C_4	C_{5+}	CO_2	
Fe- P/PS	15	4	4	26	21	88
Fe- P/PS-DVB	71	33	33	63	208	74
Fe- P/ SBR	65	30	31	57	190	54

***X_{CO}, %** = Conversion of CO, %.

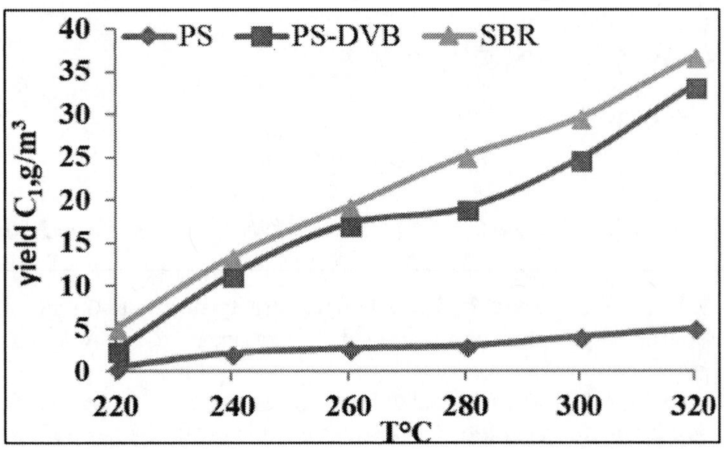

Figure 3. Dependence of the yield of CH4 gas hydrocarbons on the synthesis temperature in the presence of the Fe–paraffin–polymer Nano catalyst system: PS, PS-DVB and SBR.

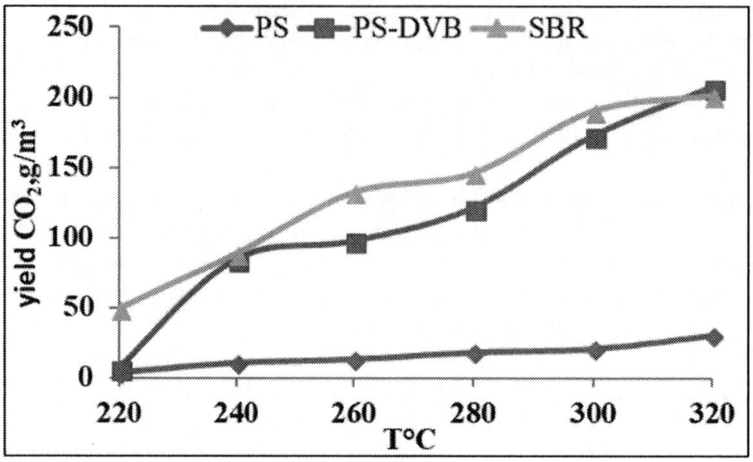

Figure 4. Dependence of the yield of CO2 gas on the synthesis temperature in the presence of the Fe–paraffin–polymer Nano catalyst system: PS, PS-DVB and SBR.

The Fe–paraffin suspensions include the various polymers displayed different actions and different selectivity for the aimed product it could be implicit that the promoting effect had attributed directly to the structure and features of the polymer elements [14] figure 5.

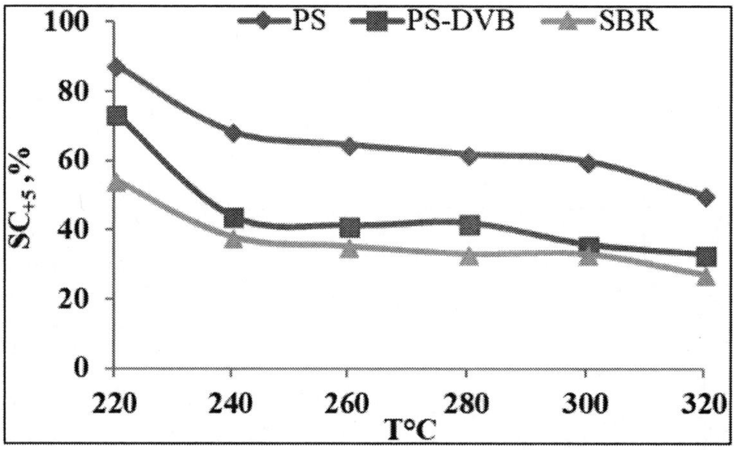

Figure 5. Dependence of the C5+ liquid hydrocarbon selectivity on the temperature in the presence of the Fe–paraffin–polymer Nano catalysts based on: PS, PS-DVB and SBR.

In table 3, we displayed the hydrocarbons liquid analysis which it obtained in the temperature range 220-320 °C. Which is clearly showing the PS derivatives (n-paraffin, iso-paraffin and olefin) have an influence on the composition of liquid hydrocarbon production and on the incomplete composition (gasoline C_5-C_{10}, kerosene C_{11}-C_{18} and wax C_{19+}).

Polystyrene derivatives compared into the Nanocatalyst suspension composition had lead to reduce the unsaturated compounds fraction that in contrast with the polystyrene sample and the paraffin fraction had also changed. Whereas using SBR copolymer had led to an increase in the paraffin yields (from 60 to 75%), whilst using PS polymers had led enhance the olefins yield from 7 to 18%. Additives of copolymer (PS-DVB and SBR) also had an influence on the liquid products fractional composition table 3. the quantity of the C_{19+} heavy fraction in the composition of C_{5+} hydrocarbons increased from 31 to 57%.

IOP Conf. Series: Journal of Physics: Conf. Series **1032** (2018) 012065 doi:10.1088/1742-6596/1032/1/012065

Table 3. Listing the selectivity effect of group composition associated with catalyst type on hydrocarbon yields.

Iron Nano catalyst	Group composition, %			Yield Fractional composition, %		
	n-paraffins	isoparaffins	olefins	C5-C10	C11-C18	C19+
Fe- P/PS	60	21	18	13	56	31
Fe- P/PS-DVB	52	31	17	20	46	34
Fe- P/SBR	75	18	7	6	37	57

3.3. Atomic Force Microscope (AFM).

The surfaces morphology of the synthesized materials reveled to see the texture of the surface Figures 6, 7 and 8. The sample was dissolved in hexane to get rid of the wax layer which coats the polymer. Clearly reviled the wavy textured surfaces in all figures, compared with that washed with hexane of less featured about plane surface. Thus in the case of a polystyrene with catalyst Figure 6, which shows the nanoparticles had been still fixed after treatment with hexane, that to explain its activity might be connected to polystyrene to alter the synthesized catalysts surface. Though most of the surface shows variation in partials size as compare with each other, while on the PS-DVB the one could found the surface turn to be more homogenous compared with surface before washing, Fig. 7. So, any changes, in the catalytic surface would lead to variance in activities and selectivity as well, that because change the active sites either shape or area diameters. In Figure 8, iron had fixed and have more featured sets and regular distribution on both x and y-axis on the catalyst surface compared the that washed and the later shows two folds of waves and three meniscuses along the y-axis of the surface of course all such variations can cause grate differences on the yield of the catalyst system used.

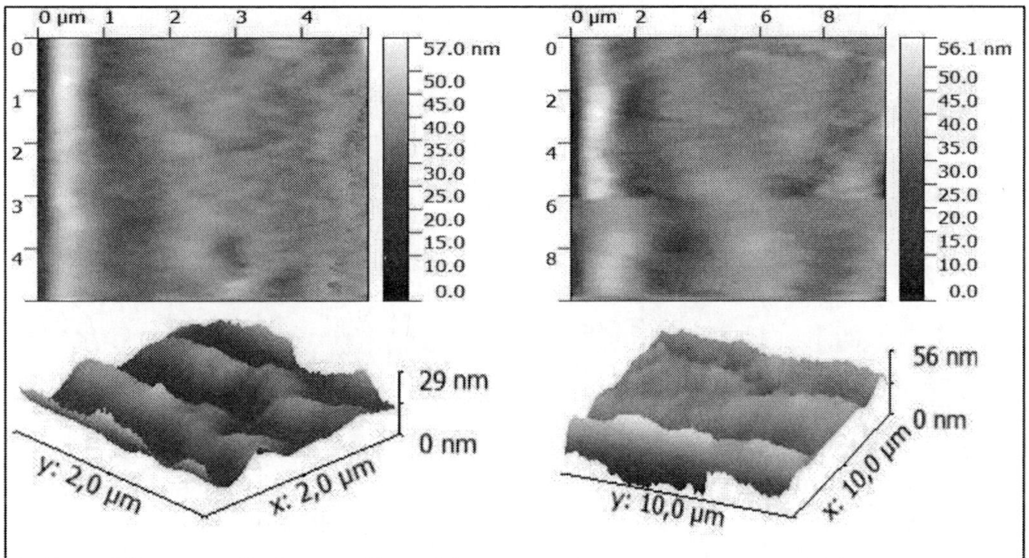

Surface of the system before washing with hexane

after washing system with hexane , note: less featured surface

Figure 6. AFM morphology, for iron-containing solid catalyst Fe-paraffin-PS before and after washing down with hexane.

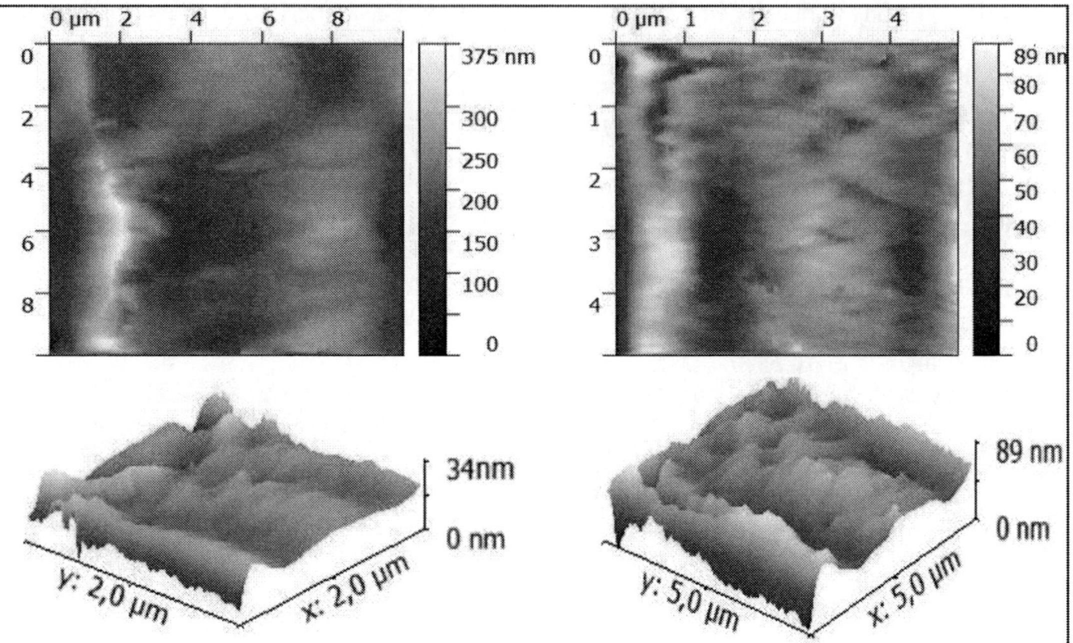

Surface of the system before washing with hexane after washing system hexane

Figure 7. AFM morphology, of iron-containing solid catalyst Fe-paraffin-PS-DVB.

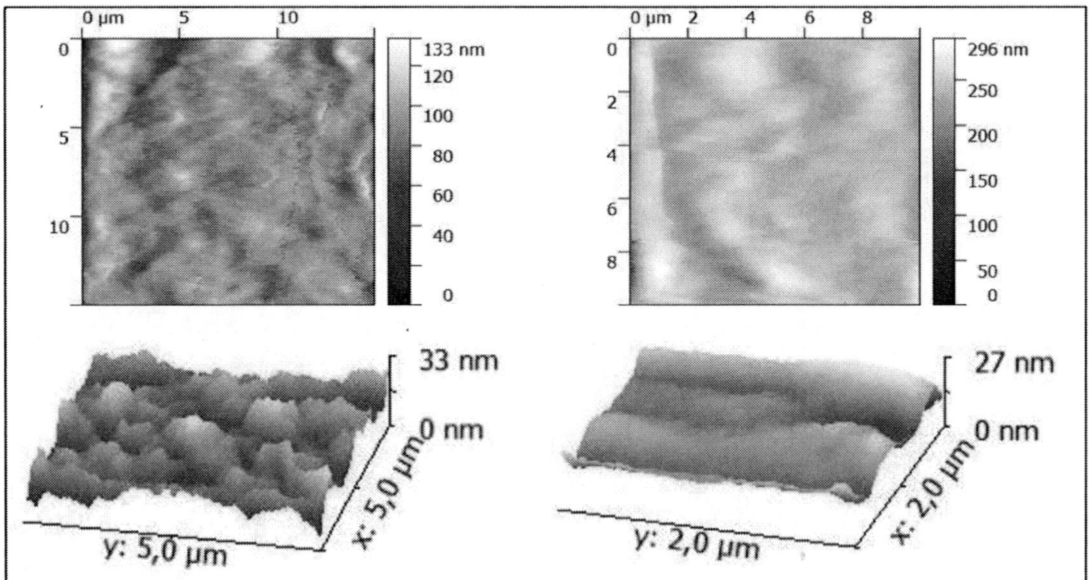

Surface of the system before washing with hexane after washing system hexane

Figure 8. AFM morphology, of iron-containing solid catalyst Fe-paraffin-SBR. The surface morphology AFM.

3.4. XRD Pattern

The iron state and polymer-paraffin matrix had studied using XRD technique (Figure.9). The basic states of the catalysts used throughout the experiment were in amorphous state. At reduced temperature, catalysts reviled a crystalline phase. The iron ground state as Fe_3O_4 oxide found to have crystal properties and hydrolyzed δ-FeOOH to be in amorphous form. In factor this system iron-paraffin-polymer the iron oxide nanoparticles stabilized in PS polymer and paraffin. In the case, we investigated the nanoparticles iron oxide Fe_3O_4 stability gained from paraffin and polymer which it in charge for selectivity and activity.

In the composites of XRD patterns of the polystyrene (Fig. 9a), there are no peaks to refer to any crystallinity of the sample. This is mean there is only δ-FeOOH which might be due to the low contact with the metal-containing with polystyrene in the composite and may be caused by a change of matrix. The amorphous paraffin presence was evident, that iron nanoparticles catalyst- poorly stabilized with polymer and paraffin. Also, the particles size proves this attribution. Another results had gained for a catalyst with styrene butadiene rubber (SBR) and a polystyrene cross-link with divinely benzene (PS-DVB) (Fig. 9b,9c) respectively that we could observe the iron oxide presence, which may be connected to the high magnetite of (Fe_3O_4) presence [15]. These peaks will be more intensive in cases of the composite studied includes such magnetite of quantity. For reactions based on SBR, weak peaks at $2\theta = 30°$ and sharp peaks are obviously appeared at $2\theta = 36°$; were observed, those might be connected to the attendance of magnetite (Fe_3O_4) quantity [16-18]. For sample based on PS-DVB, weak peaks at $2\theta = 36°$ and 42° were observed, which can be related to the attendance of magnetite (Fe_3O_4) too.

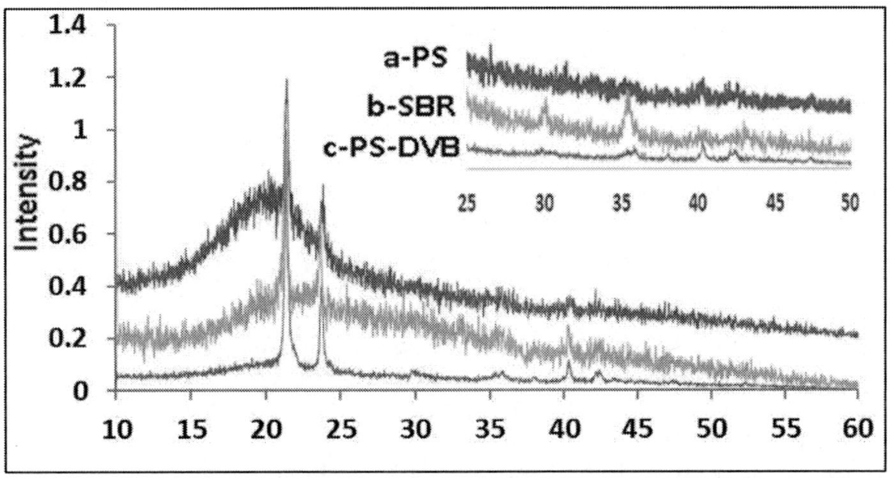

Figure 9. XRD patterns of Nano iron -paraffin-polymer catalysts : a- pristine PS, b- pristine SBR, and c-PS-DVB.

4. Conclusions.

The goal behind this work was to compare and synthesis of new Nano size iron-polymer catalysts for the three-phase Fischer-Tropsch process. Suspensions system (Iron-Paraffin-Polystyrene or Polystyrene derivatives) were applied in Fischer-Tropsch process as a Nano-catalysts; and throughout the present time polymers suggested for the synthesis of Nano-catalysts were used in FTS to be stable at high temperatures. Which reverberate on the steadiness of the Nano-catalyst as long as it not cause the Nano catalyst poisoning. The influence of polystyrene derivatives on the Nano catalyst structure has also proven by XRD and AFM measurements. We explained also the raise in the formation of targeted product and CO conversion. we notice, that Nano catalysts are more valuable in the process of Fischer Tropsch with the cross linked polymers concerned in the composition, compared with the raw polymer .The selectivity of various systems on the production of liquid hydrocarbons also proven.

IOP Conf. Series: Journal of Physics: Conf. Series **1032** (2018) 012065 doi:10.1088/1742-6596/1032/1/012065

In another way, reveled the cross linked polymers nature have an important influence on the iron nanoparticles stabilization properties and changes iron nanoparticles size in particularly.

ACKNOWLEDGMENTS

This work was performed using the equipment of A.V.Topchiev Institute of Petrochemical Synthesis, Russian Academy of Sciences, and it was supported by The President of Russian the Federation.

References

[1] Carlo Pirola, Antonieta Di Fronzo, Federico Galli, Claudia L. Bianchi, Alberto Comazzi, Flavio Manenti, 2014 *Chemical engineering transactions* **37** 595-600.

[2] Kenji Asami , Aiki Iwasa,. Naoko Igarashi, Shoko Takemiya,Katsutoshi Yamamoto, Kaoru Fujimoto, 2013 *Catalysis Today* **215** 80– 85.

[3] Fabiano A. N. Fernandes, 2006 *Ind. Eng. Chem. Res.* **45** 1047-1057.

[4] Nikolay M. Ushakov, Gleb Yu. Yurkov , Leonid V. Gorobinskii Oleg V.Popkovb, Igor D.Kosobudskii, 2008 *Acta.Materialia* **56** 2336-2343 .

[5] Gleb Yu. Yurkov , Alexander V. Kozinkin,Yury A. Koksharov, Alexander S. Fionov,Nikolay A. Taratanov, Valery G. Vlasenko, Irina V. Pirog, Oleg N. Shishilov, Oleg V. Popkov, 2012 *Composites: Part B* **43** 3192–3197.

[6] S P Gubin, Yu A Koksharov, G B Khomutov, G Yu Yurkov, 2005 *Russian Chemical Reviews* **74** 489 – 520.

[7] Marco Bernabo , Andrea Pucci , Hasina Harimino Ramanitra ,Giacomo Ruggeri, 2010 *Materials* **3** 1461-1477.

[8] Zhen Liu, Lin Zhang, Selcuk Poyraz, Xinyu Zhang, 2013 *Current Organic Chemistry* **17** 2256-2267.

[9] Aitor Gual, Cyril Godard, Sergio Castillon , Daniel Curulla-Ferre, Carmen Claver, 2012 *Catalysis Today* **183** 154– 171.

[10] S.P.Gubin , G.Yu.Yurkov, 2005 *Int. J. Materials and Product Technology* **23**.

[11] Qingliang He , Tingting Yuan , Jiahua Zhu, Zhiping Luo, Neel Haldolaarachchige, Luyi Sun,Airat Khasanov , Yutong Li , David P. Young , Suying Wei, Zhanhu Guo, 2012 *Polymer* **53** 3642-3652.

[12] S.P. Gubin, G.Yu. Yurkov, M.S. Korobov, Yu.A. Koksharov, A.V. Kozinkin, I.V. Pirog, S.V. Zubkov, V.V. Kitaev,D.A. Sarichev, V.M. Bouznik, A.K. Tsvetnikov, 2005 *Acta Materialia* **53** 1407–1413.

[13] Г. Н. Бондаренко, М. В. Куликова, А. Х. Аль Хазраджи,О. С. Дементьева,М.И.Иванцов,М.В.Чудакова, 2016 *Наногетерогенный Катализ* **1** 122–128.

[14] M. V. Kulikova, A. H. Al Khazradji,O. S. Dement'eva, M.I.Ivantsov, V. R. Flid, S. N. Khadzhiev, 2015 *Petroleum Chemistry* **55** 537–541.

[15] Hosik Park,Perla Ayala ,Mar A.Deshusses, Ashok Mulchandani, Heechul Choi ,Nosang V.Myung , 2008 *Chem. Eng. J* **139** 208-212.

[16] R.Y.Hong, T.T.Pan, Y.P.Han, H.Z.Li,J.Ding,S ijinHan,J. 2007 *Magnetism Magnetic Mater* **310** 37-47.

[17] R.Y.Hong, T.T.Pan, H.Z.Li, 2006 J. *Magnetism Magnetic Mater* **303** 60-68.

[18] A.N. Volkov , O.V. Popkov, YU.M. Bolbukh, K.V. Katok,G.YU. Yurkov, V.A. Tertykh, 2009 *Journal of Nanostructured Polymers and Nanocomposites* **5/3** 70-78.

IOP Conf. Series: Journal of Physics: Conf. Series **1032** (2018) 012066 doi:10.1088/1742-6596/1032/1/012066

Synthesis, characterization of mixed- ligand complexes containing 2,2-Bipyridine and 3-aminopropyltriethoxysilane

Hayder Hamied Mihsen*, Nesser Kadham Shareef

Department of Chemistry, College of Science, University of Kerbala, Kerbala, Iraq.

Corresponding author's E-mail: hayderalhmedawy@gmail.com

Abstract. A mixed-ligand metal complexes with divalent ion Co, Ni and Cu containing 3-aminopropyltriethoxysilane(3-APTES) as a primary ligand and 2,2-bipyridine as a secondary ligand were prepared. The new complexes were characterized by, Molar conductance measurement, magnetic susceptibility, Uv-visible spectra (Uv-Vis) ,infrared spectroscopy (FT-IR), Nuclear magnetic resonance(^{1}H-NMR), Thermogravimetric analysis (TGA-DTA) and atomic absorption spectroscopy(ASS). FT-IR and ^{1}H-NMR spectra indicate that both ligands 3-APTES and 2,2-bipyridine are coordinated to the metal ions via nitrogen amino group in 3-APTES and nitrogen atoms for bpy. Molar conductance value for the complexes indicate that the complex are non-electrolyte. The TGA-DTA analysis shows that the degradation of complexes occurs between 200-593 °C. The complexes assume octahedral geometry in which the Co(II),Ni(II),Cu(II) have two molecules of 3-APTES and one molecule from 2,2-bipyridine in the coordination sphere.

1. Introduction.

2,2-bipyridine is a compound where two pyridine rings are connected together, the bond between the pyridine rings in the α-position of the nitrogen[1]. 2,2-bipyridine is a chelating component it has been used as bridging ligand so it found different application in coordination chemistry. It form 5-membered chelate ring which is stable upon coordination of metal.

2,2-bipyridine have widely used as a chelating donor site during such bridging ligand because its robust redox stability and relative ease of functionalization[2].Mixed ligand which contain 2,2-bipyridine and other bidentate ligand are antineoplastic agents ,these compounds exhibit cytotoxicity, antitumor effect and genotoxicity also have been bactericidal, bacteriostatic toward many gram-positive bacteria but they are ineffective against gram-negative organism[3-4].

Generally the organosilane is active at both end ,R is amino group(NH_2), mercapto (SH), or isocyanato (NCO).This functionality reacted with other functional groups in industrial or biomolecule for example peptides or oligonucleotides[5-7].While the other end consist of alkoxy silane. This functionality by hydrolysis converted to active group silanols. Which react with themselves to generated oligomeric. This silanol reacted with active surface that contain hydroxyl group. By regard ethoxy silane is less reactive than methoxy silane due to non-produce toxic ethanol while the other compound formed toxic methanol[8].

The reaction of the 3-Aminopropyltriethoxysilane (3-APTES) with transition metal ions must be performed in non-aqueous phase such as dry ethanol or dry benzene or hexane to avoid the hydrolysis of 3-APTES to ethanol and trisilanol [9-10]. 3-APTES can be used in the general reaction for synthesis functionalized ligands such as reaction of salicylaldehyde with 3-aminopropyltriethyoxysilane. The salicylaldimine ligands were reacted with either Cu(II) or Ni(II) salts to form both the model and functionalized Cu(II) and Ni(II) complexes [11]. The objective of the current study is to prepared the new mixed-ligand metal ion complexes; Co(II), Ni(II), Cu(II) containing 3-APTES as a primary ligand and 2,2-bipyridine as a secondary ligand .

2. Experimental part.

IOP Conf. Series: Journal of Physics: Conf. Series **1032** (2018) 012066 doi:10.1088/1742-6596/1032/1/012066

2.1. Materials and Sample characterization.

All chemical material used were AR grade or of a high purity were used directly without further purification,3-aminoopropyltriethoxysilane (Sigma–Aldrich,,98%), 2,2-bipyridyl (BDH, 99%), Cobalt chloride hexahydrate(BDH, 99%), Nickel chloride hexahydrate (Fluka, 99%), Copper chloride dihydrate(BDH, 99%), ethyl acetate(Sigma–Aldrich,,99.8%), ethanol absolute (Fluka,99%), methanol (99%).

The molar conductance measurements of the complex were measured in methanol at room temperature using[Digital conductivityMeter-WT-700-inolab]. The electronic spectra of solution of the complex and ligand in methanol solution(10^{-3}M)concentration were recorded in Uv-visible spectrophotometer-1800, shimadzu. FTIR spectra of the ligand and complexes were recorded using KBr pellets in shimadzu spectrophotometer. The ^{1}H-NMR spectra were recorded in DMSO-d^{6} solution using TMS as the internal standard in 500 MHz NMR spectrophotometer[Bruker Germany].Thermal analysis of sample were performed at heating rang(25-600°C) by using STAPT-1000Linseis-Germany.Magnetic susceptibility measurements were measured using Auto magnetic susceptibility balance Sherwood scientific. Metal content of the complex were determined by atomic absorption technique using Japan A.A-67G Shimadzu.

2.2. Preparation of Complexes.

To methanolic solution of $CoCl_2.6H_2O$ or $NiCl_2.6H_2O$ or $CuCl_2.2H_2O$ (1 mmole), a solution of primary ligand 3-APTES (2 mmole) in 10 mL methanol was added drop wise with constant stirring. 1 mmole of bpy dissolved in the same solvent in (10 mL) was added, the mixture was reflexed for 4-6 hours. The formed precipitate was filtered and washed with cold absolute methanol several times and dried under oven. The reaction was monitoring by TLC using a mixture of ethanol : ethyl acetate (1:1) as an eluent. The data of metal analysis were obtained using flame atomic absorption technique. The calculated values are in a good agreement with the experimental values. Table(1) shows weights, yield% , metal analysis and some physical properties of prepared complexes.

Table 1. Yield%, metal analysis and some of physical properties of prepared complexes.

Compound	Weight(gm) of metal salt in gm	Weight(gm) of 3-APTES	Weight(gm) of Bpy in gm	Yield%	Color	Melting point oC (de)	Metal analysis% Cal.(Exp.)
[Co(3-APTES)2(bpy)Cl2]	0.5	0.9	0.3	70	Light-blue	320	8.08(8.70)
[Ni(3-APTES)2(bpy)Cl2]	0.5	0.9	0.3	55	Bluish-gray	375	8.05(7.73)
[Cu(3-APTES)2(bpy)Cl2]	0.5	1.3	0.46	62	Blue-green	330	8.6(7.60)

3. Results and discussion.

3.1. FT-IR spectroscopy.

Coordination of the cobalt(II),nickel(II),and copper(II) ions with functional groups of the mixed ligands 3-APTES and bpy are established in Table (2) and Figures (1-3).From the data spectra of the complexes. It can be seen that the bands of υ_{N-H} in 3-APTES were shifted and changed in the shape due to coordination of amine group at the nitrogen atom with metal ions[12]. The absorption band of the (C=N)group in bpy appeared at(1454)cm^{-1}[13].This band was shifted and changed in complexes to lower and higher frequencies because of coordinated with metal ions via the nitrogen atom[14].New bands appeared in the spectra of all metal complexes in the low frequency region between (300-500)cm^{-1} ,those bands characteristic to M-N stretching vibrations[15].

Table 2. Characteristic FT-IR bands of the ligand and metal complexes.

Compound	Frequency(cm-1)			
	C=Nυ	υNH2	υC=C	υM-N
bpy	1454		1579	
3-APTES		3369,,3298		
[Co(3-APTES)2(bpy)Cl2]	1514	3227,3417	1604	449
[Ni (3-APTES)2(bpy)Cl2]	1508	3273,3396	1600	459
[Cu(3-APTES)2(bpy)Cl2]	1444	3132,3227	1602	445

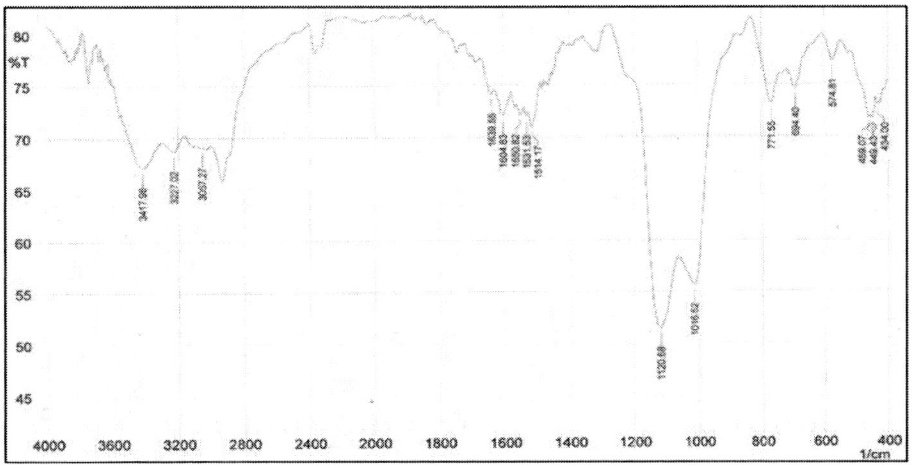

Figure 1. FT-IR spectrum of [Co(bpy)(APTES)$_2$Cl$_2$].

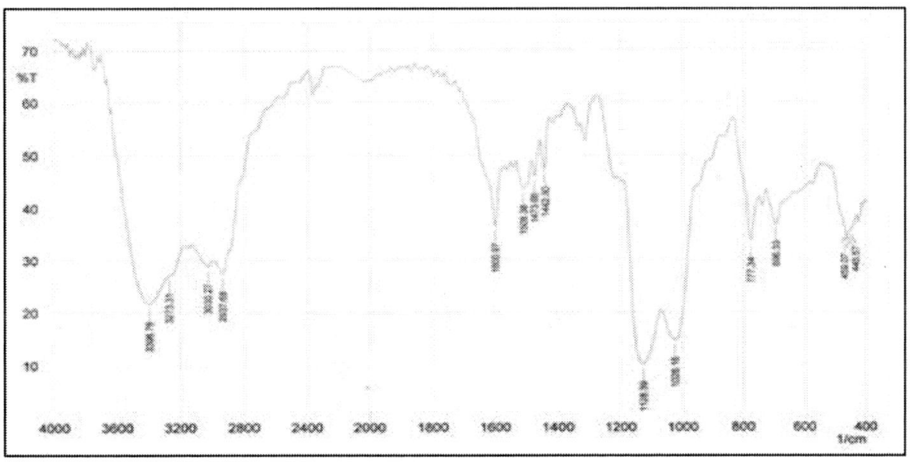

Figure 2. FT-IR spectrum of[Ni(bpy)(APTES)$_2$Cl$_2$].

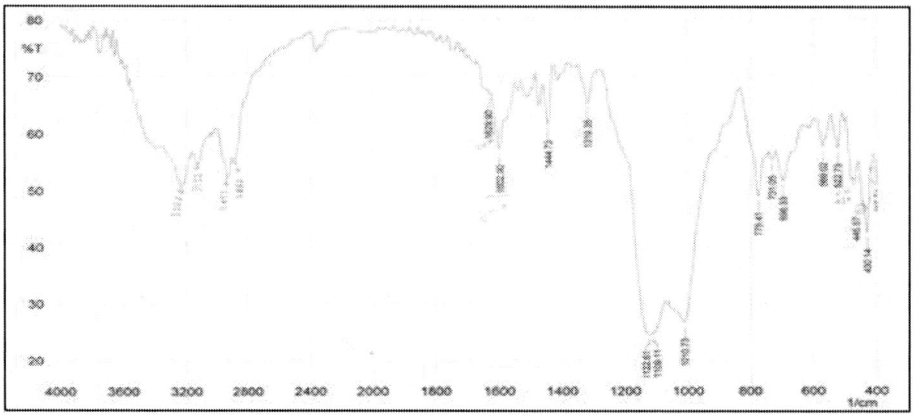

Figure 3. FT-IR spectrum of $[Cu(bpy)_2(APTES)_2Cl_2]$.

3.2. Electronic spectra.

Electronic spectra of the prepared complexes are recorded in warm methanol. The assignments for the electronic spectra are given in Table(3) . The electronic spectrum of Co(II) complex Fig.4 displayed one broad absorption band between (450-550)nm (22222-18181)cm^{-1} due to combination two transition υ_3 $^4T_1g\rightarrow{}^4T_2g_{(P)}$ and υ_2 $^4T_1g\rightarrow{}^4A_2g_{(f)}$ respectively for octahedral geometry[16-18] . The spectrum of nickel(II) complex Fig.5 don't show any d-d transitions ,while ligand metal charge transfer LMGT ($Л_{(bipy)}\rightarrow d_{Ni}$) is appear as absorption band at higher intensity about 300nm(33333cm^{-1}) since obscure d-d transition , the shape may be is octahedral[17,19].The electronic spectrum of Cu(II) complex (not shown) give single broad absorption of high intensity(ϵ =200$L^{-1}.M^{-1}.cm^{-1}$) is observed at 620nm (16129cm^{-1}) which belong to $^2Eg\rightarrow{}^2T_2g$ thus the shape may be distortion octahedral[20,21]. Table(3) also molar conductance of all the complexes. The values were showed that prepared complexes are non-electrolytic. Fig.6 shows general structure for prepared complexes.

Table 3. Electronic spectral data for prepared complexes.

Complex	Band absorption (cm^{-1})	Transition	Conductivity μs/cm
[Co(3-APTES)2(bpy)Cl2]	22222-18181	4T1g(f)→4T1g(P) 4T1g→4A2g(F)	67
[Ni(3-APTES)2(bpy)Cl2]	33333	Π(bipy)→dNi	32
[Cu(3-APTES)2(bpy)Cl2]	16129	2Eg→2T2g	48

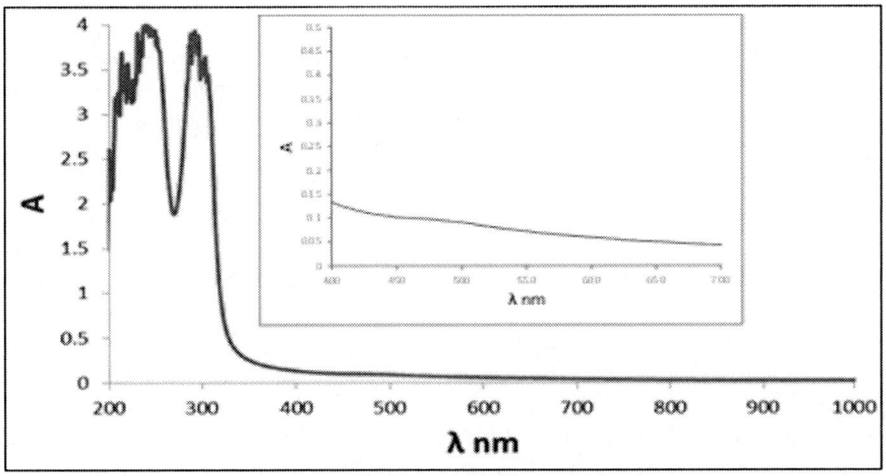

Figure 4. Uv-Vis spectrum for [Co(3-APTES)$_2$(bpy)Cl$_2$].

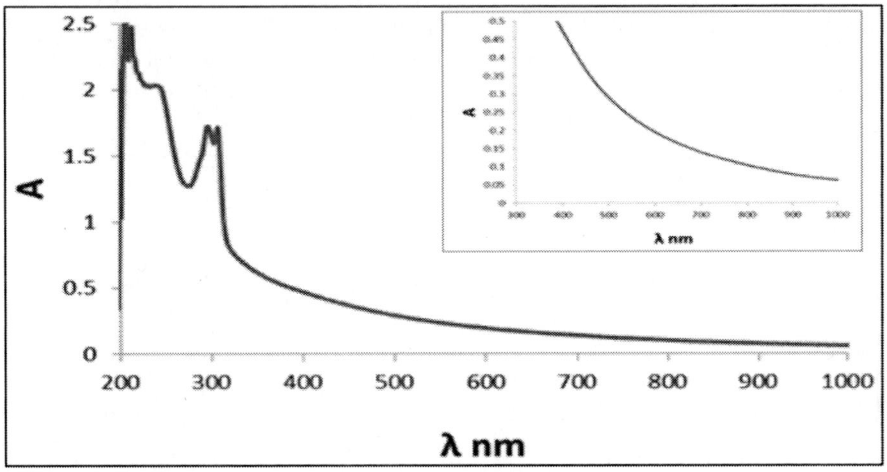

Figure 5. Uv-Vis spectrum for [Ni(3-APTES)$_2$(bpy)Cl$_2$].

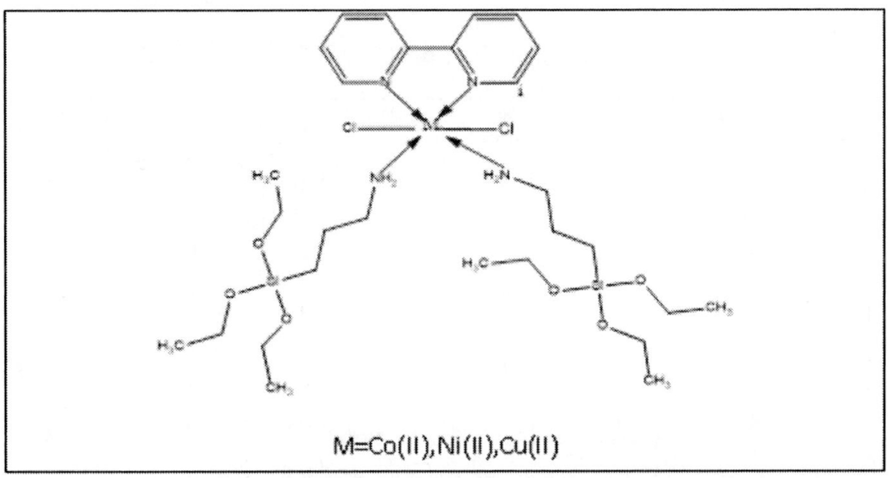

Figure 6. Proposed structure of the complexes.

3.3. ^{1}H-NMR spectroscopy.

For the Co(II) complex in DMSO-d^6 , the signals related to the 3-APTES protons appeared at 0.915-0.990 ppm (4H,CH_2) belong to (Si–**CH_2**–CH_2–CH_2), 1.017-1.095 ppm (18H,CH_3) belong to –Si–OCH_2–**CH_3**,1.118-1.272 ppm (4H,CH_2) belong to (Si–CH_2–**CH_2**–CH_2), 2.448-2.521 ppm (4H,NH_2) belong to (Si–CH_2–CH_2–CH_2-**NH_2)**, 3.303-3.337 ppm (4H,CH_2) belong to Si–CH_2–CH_2–**CH_2**), 3.401-3.500 ppm (12H,CH_2) belong to (–Si–O**CH_2**–CH_3**)** [23,24], 7.22 -8.812 ppm belong to protons of bpy [3,25].

The ^{1}H-NMR spectrum of the Ni(II)complex, in DMSO-d^6, exhibited the signals related to 3-APTES protons located at 0.259-0.401 ppm (4H,CH_2) belong to (Si–**CH_2**–CH_2–CH_2), 1.017-1.230 ppm (18H,CH_3) belong to –Si–OCH_2–**CH_3**, 1.633-1.752 ppm (4H,CH_2) belong to (Si–CH_2–**CH_2**–CH_2), 2.467-2.631 ppm (4H,CH_2) belong to (Si–CH_2–CH_2–CH_2-**NH_2)**, 3.314-3.364 ppm (4H,CH_2) belong to Si–CH_2–CH_2–**CH_2**), 3.409-3.512 ppm (8H,CH_2) belong to(–Si–O**CH_2**–CH_3**)** [26,27], 7.245-12.003 ppm belong to protons of bpy [3,25].

The ^{1}H-NMR spectrum of the Cu(II) complex in DMSO-d^6 Figures. (7-8), show the signals related to 3-APTES protons appeared at 0.975-1.083 ppm (4H,CH_2) belong to (Si–**CH_2**–CH_2–CH_2),, 1.123-1.694 ppm (18H,CH_3) belong to –Si–OCH_2–**CH_3**, 2.054-2.137 ppm (4H,CH_2) belong to (Si–CH_2–**CH_2**–CH_2), 2.487-2.585 ppm (4H,NH_2) belong to (Si–CH_2–CH_2–CH_2-**NH_2)**, 3.144-3.298 ppm (4H,CH_2) belong to Si–CH_2–CH_2–**CH_2**), 4.052-4.119 ppm (8H,CH_2) belong to(–Si–O**CH_2**–CH_3**)** [28,26], 7.101-11.623 ppm belong to protons of bpy[3,25].The peaks observed for all complexes illustrated the signals of amine group and other protons for 3-APTES and bpy are shifted and some it were changed intensity, this good evidence that complexes were formed[22].

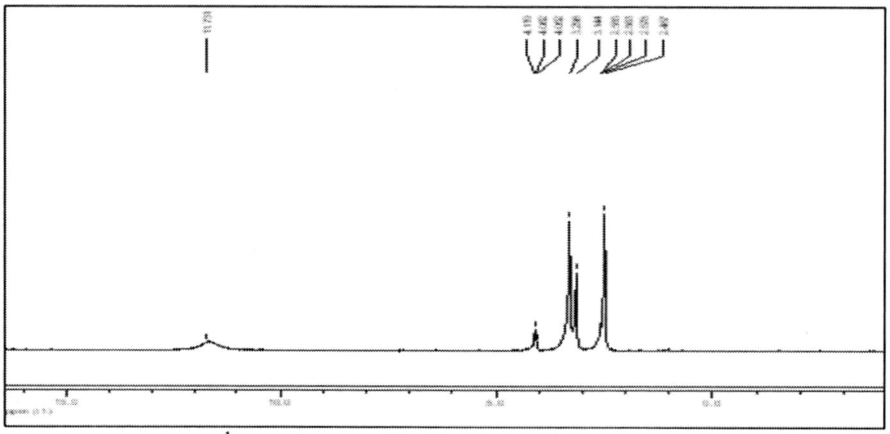

Figure 7. ^{1}H-NMR spectrum of [Cu(3-APTES)2$(bpy)_2Cl_2$].

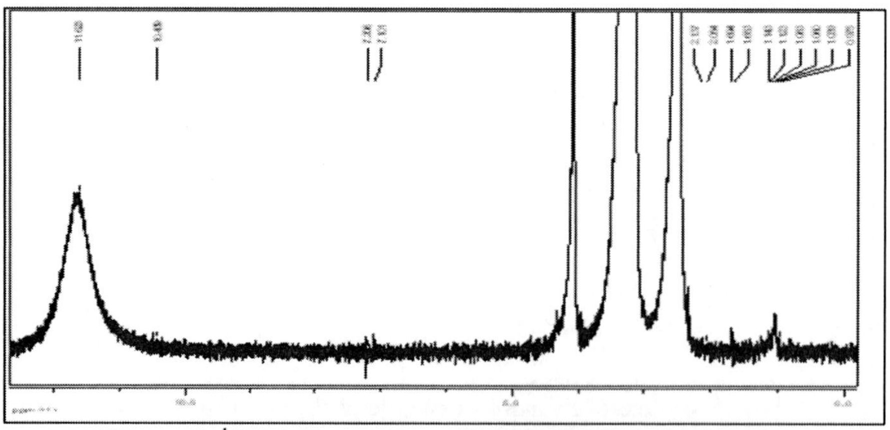

Figure 8. ^{1}H-NMR spectrum of [Cu$(3\text{-APTES})_2(bpy)_2Cl_2$].

3.4. Thermogravimetric analysis(TGA/DTA).

Thermal analysis for many metal complexes was studied by TGA/DTA techniques through the rang (20-600 °C) at heating rate 10 °C/min under nitrogen atmosphere. The results of TGA/DTA are summarized in Table (4) and represented thermo grams is given in Figures (9,10 and 11). TGA curves refer to thermal stability of these complexes and decomposition steps ,in the first step the weight loss corresponding to the removal of molecules with low molecular weight such as water and ammonia which is demonstrated which agreement with IR analysis. While The second and third steps involved the thermal decomposition of the remaining parts of complexes[29,30].On the other hand DTA curves record the changes that causing by temperature such as melting, sublimation, vaporization or changes that attributed weight losses are represented an endothermic character[31,31] .While the processes crystallization or change in the crystal structure and oxidation are represented exothermic character[33].Generally many DTA curves for prepared complexes suggestion have a high thermal stability because the presence of 3-APTES as ligand in these complexes increase both the initial and final decomposition temperatures due to inherent high heat resistances of 3-APTES[34,35].

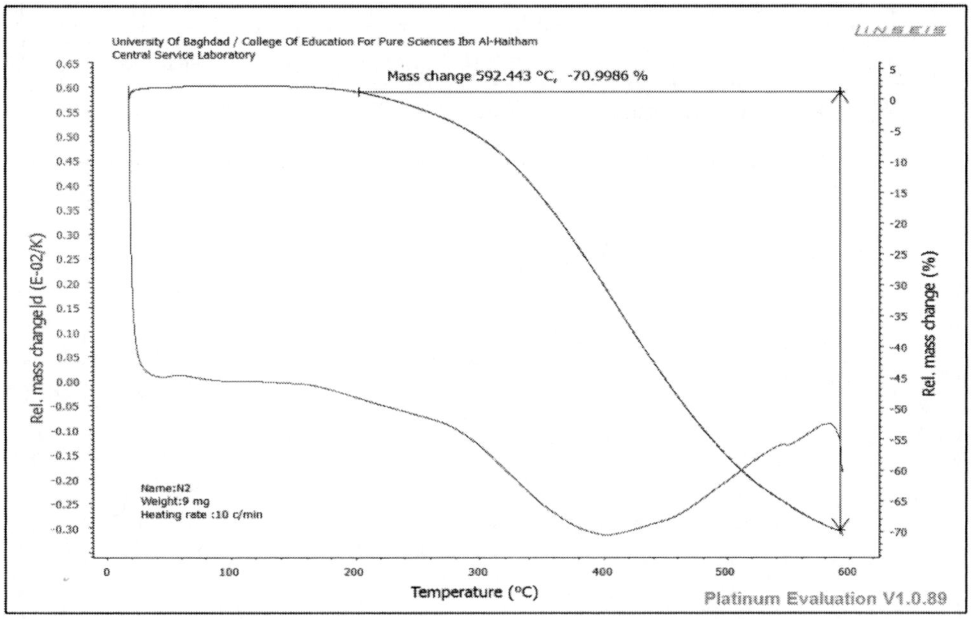

Figure 9. Thermogravimetric analysis(TGA/DTA)of [$Co(3\text{-}APTES)_2(bpy)Cl_2$].

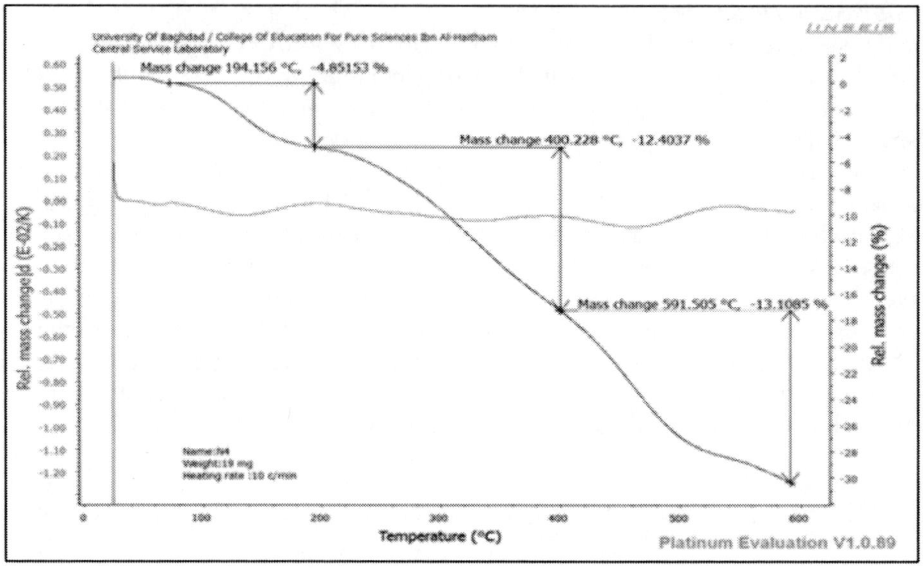

Figure 10. Thermogravimetric analysis (TGA/DTA)of [Ni(3-APTES)$_2$(bpy)Cl$_2$].

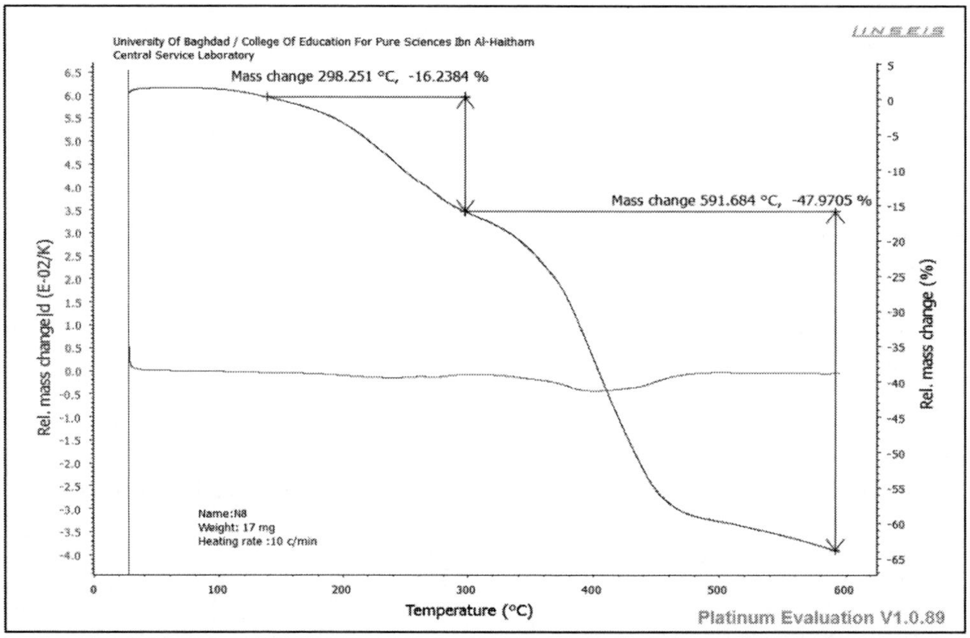

Figure 11. Thermogravimetric analysis (TGA/DTA) of [Cu(3-APTES)$_2$(bpy)Cl$_2$].

IOP Conf. Series: Journal of Physics: Conf. Series **1032** (2018) 012066 doi:10.1088/1742-6596/1032/1/012066

Table 4. Thermo analytical result of prepared metal complexes.

Compound	Molecular weight	Weight loss% found	Calc.	Thermal rang	Compound decomposition	Compound product
[Co(3-APTES)$_2$(bpy)Cl$_2$]	728	70.9	64.3	20-592	$2NH_3$,2Cl,byp $2CH_3CH_2OH$ CH_3CH=CH-CH_3 $CH_3CH_2CH_2CH_3$	$2SiO_2$,$2CH_2CH_2$ CH= Co
[Ni(3-APTES)$_2$(bpy)Cl$_2$]	728	4.85	4.6	20-194	$2NH_3$	bpy,2Cl,$2CH_2C$ H=CH_2- $Si_2(OCH_2CH_3)_6$, Ni
		12.4	12.6	194-400	$2CH_3CH_2OH$	byp,2Cl,$2CH_2$- CH=CH_2 $2SiO_2$,CH_3CH_2- ,$2CH_2CH_2$- Ni
		13.1	13.7	400-591	2Cl,CH_3CH_2-	byp,$2CH_2CH$=C H_2,$2SiO_2$, CH_3CH_2,2- CH_2CH_2-,Ni
[Cu(3-APTES)$_2$(bpy)Cl$_2$]	733	16.2	17.1	20-298	$2NH_3$,$2CH_3CH_2OH$	$2SiO_2$,byp,$2CH_2$ CH_2CH= 2Cl,$2CH_3CH_2$- ,$2CH_2$=CH_2 Cu
		47.9	46.4	298.2-591	2Cl,byp $2CH_3CH$=CH_2CH_3	$2SiO_2$,$2CH_2CH$ =CH_2 Cu

3.5. Magnetic susceptibility measurements of complexes.
The magnetic moments Table (6). of complexes for cobalt(II),nickel(II) and copper(II) ions were 5.14, 3.9 and 1.92 B.M, respectively. This values are shown that the complexes have high spin momentum [36,37].

Table 5. magnetic properties of prepared complexes at 298 K.

Complex	Xg×10-6	XM×10-6	D×10-6	XA×10-6	μeff B.M
[Co(3-APTES)2(bpy)Cl2]	14.8	10700	364.35	11060	5.14
[Ni(3-APTES)2(bpy)Cl2]	8.3	6046.7	364.2	6410.9	3.9
[Cu(3-APTES)2(bpy)Cl2]	1.6	1170	366	1536	1.92

4. Conclusion.

3-APTES and 2,2-bipyridine was reacted with metal(II)(M=Co,Ni,Cu) to form mixed ligand complexes. 3-APTES act as monodentate ligand and 2,2-bipyridine act as bidentate and the coordinated with metal ions through the nitrogen atoms .

Atomic absorption spectroscopy (ASS) technique was used to quantitative determination of metal in the complexes . The results showed that there is a great affinity between practical and theoretical percentages. Molar conductivity measurement show that the complexes was nonelectrolyte. Infrared spectroscopy, UV-Visible spectroscopy, ^{1}H-NMR spectroscopy were used to characterization the structure of the complexes that were produced. The TGA/DTA shows the complexes could be stable and it may decompose above 334°C. The evidence from the all spectroscopy showed the successful formation of these complexes. The proposed structures for prepared complexes with ions of Co(II), Ni(II) and Cu(II) are octahedral geometry.

Acknowledgments

We would like to thank Kerbala University, Republic of Iraq for financial support.

References

[1] Peter J., 1990 "Aromatic nitrogen heterocycles as bridging ligands; a survey" *Coordination Chemistry Reviews* **106** 227-265.

[2] Selvaganapathy, M. and Raman, N., 2016 "Pharmacological Activity of a Few Transition Metal Complexes: A Short Review" *Journal of Chemical Biology & Therapeutics*.

[3] Maurya R. C. , Patel, P. a & Rajpu,S.,2003 "Synthesis and characterization of Mixed-Ligand Complexes of Cu(II), Ni(II), Co(II), Zn(II), Sm(III), and U(VI)O2, with a Schiff Base Derived from the Sulfa Drug Sulfamerazine and 2,2'-Bipyridine" *Synthesis And Reactivity In Inorganic And Metal-Organic Chemistry* **33(5)** 801–816.

[4] Chandraleka, S., Ramya, K., Chandramohan, G., Dhanasekaran, D., Priyadharshini, A. and Panneerselvam, A., 2014 "Antimicrobial mechanism of copper (II) 1, 10-phenanthroline and 2, 2′-bipyridyl complex on bacterial and fungal pathogens" *Journal of Saudi Chemical Society*" **18(6)** 953-962.

[5] Schickle, K., Zurlinden, K., Bergmann, C., Lindner, M., Kirsten, A., Laub, M., Telle, R., Jennissen, H. and Fischer, H., 2011 "Synthesis of novel tricalcium phosphate-bioactive glass

composite and functionalization with rhBMP-2." *Journal of Materials Science: Materials in Medicine* **22(4)** 763-771.

[6] Rovira-Bru, M., Giralt, F. and Cohen, Y., 2001 "Protein adsorption onto zirconia modified with terminally grafted polyvinylpyrrolidone" *Journal of colloid and interface science* **235(1)** 70-79.

[7] Huckel, M., Wirth, H.J. and Hearn, M.T., 1996 "Porous zirconia: a new support material for enzyme immobilization" *Journal of biochemical and biophysical methods* **31(3-4)** 165-179.

[8] Witucki, G.L., 1993 "A silane primer: chemistry and applications of alkoxy silanes" *Journal of coatings technology* **65** 57-57.

[9] Simon, A., Cohen-Bouhacina, T., Porté, M.C., Aimé, J.P. and Baquey, C., 2002 "Study of two grafting methods for obtaining a 3-aminopropyltriethoxysilane monolayer on silica surface." *Journal of colloid and interface science* **251(2)** 278-283.

[10] Malakooti, R., Bardajee, G.R., Hadizadeh, S., Atashin, H. and Khanjari, H., 2014 "An iron Schiff base complex loaded mesoporous silica nanoreactor as a catalyst for the synthesis of pyrazine-based heterocycles" *Transition Metal Chemistry* **39(1)** 47-54.

[11] Murphy, E.F., Ferri, D., Baiker, A., Van Doorslaer, S. and Schweiger, A., 2003 "Novel routes to Cu (salicylaldimine) covalently bound to silica: combined pulse EPR and in situ attenuated total reflection-IR studies of the immobilization" *Inorganic chemistry* **42(8)** 2559-2571.

[12] Fayad, N.K., Al-Noor, T.H., Mahmood, A.A. and Malih, I.K., 2013 "Synthesis, Characterization, and Antibacterial Studies of Mn (II), Fe (II), Co (II), Ni (II), Cu (II) and Cd (II) Mixed-Ligand Complexes Containing Amino Acid (L-Valine) And (1, 10-phenanthroline)" *Synthesis* **3(5)**.

[13] Soliman, A.A. and Mohamed, G.G., 2004 "Study of the ternary complexes of copper with salicylidene-2-aminothiophenol and some amino acids in the solid state" *Thermochimica Acta,* **421(1)** 151-159.

[14] Anupama, B. and Kumari, C.G., 2013 "Cobalt (II) complexes of ONO donor Schiff bases and N, N donor ligands: synthesis, characterization, antimicrobial and DNA binding study" *International Journal of Research in Chemistry and Environment* **3(2)** 172-180.

[15] AL-Hashime, S.M., Sarhan, B.M. and Alazawi, S.A. "Synthesis and studies of som mixed-ligand metal complexes containing benzotriazol with some other ligands" *transition* **18** 19.

[16] Figgis, B.N. and Lewis, J., 1960 "*The Magneto Chemistry of Coordination polymer in Modern Coordination Chemistry*". Interscience, (New York).

[17] Shukla PR.(2012) "*Advance Coordination Chemistry*" Himalya publishing House, (Newdelhi) 165-204.

[18] Shirodkar, S.G., Mane, P.S. and Chondhekar, T.K., 2001 "Synthesis and fungitoxic studies of Mn (II), Co (II), Ni (II) and Cu (II) with some heterocyclic schiff base ligands".

[19] Sutton, D., 1968 "*Electronic spectra of transition metal complexes: an introductory text*" (McGraw-Hill).

[20] Figgis, B.N., 1966. "*Introduction to ligand fields*". Interscience Publishers.

[21] Lakshmi, P.A., Reddy, P.S. and Raju, V.J., 2009 "Synthesis, characterization and antimicrobial activity of 3d transition metal complexes of a biambidentate ligand containing quinoxaline moiety" *Spectrochimica Acta Part A: Molecular and Biomolecular Spectroscopy* **74(1)** 52-57.

[22] Shen, W.Z., Trötscher-Kaus, G. and Lippert, B., 2009 "^{1}H NMR spectroscopic identification of binding modes of 2, 2′-bipyridine ligands in complexes of square-planar d 8 metal ions" *Dalton Transactions* **39** 8203-8214.

[23] Tataurova, Y., Sealy, M.J., Larsen, R.G. and Larsen, S.C., 2012 "Surface-selective solution NMR studies of functionalized zeolite nanoparticles" *The journal of physical chemistry letters* **3(3)** 425-429.

[24] İspir, E., Kurtoğlu, M., Purtaş, F. and Serin, S., 2005 "Synthesis and Antimicrobial Activity of

New Schiff Bases Having the–SiOR Group (R= CH 3or CH2CH3), and their Transition Metal Complexes" *Transition metal chemistry* **30(8)** 1042-1047.

[25] Chandraleka, S. and Chandramohan, G., 2014 "Synthesis, characterization and thermal analysis of the copper (II) complexes with 2, 2-bipyridyl and 1, 10-phenanthroline" *African Journal of Pure and Applied Chemistry* **8(10)** 162-175.

[26] Harmaraj, P., Kodimunthiri, D., Sheela, C.D. and Shanmuga Priya, C.S., 2009 "Synthesis, spectral characterization, and antimicrobial activity of copper (II), cobalt (II), and nickel (II) complexes of 3-formylchromoniminopropylsilatrane" *Journal of Coordination Chemistry* **62(13)** 2220-2228.

[27] İspir, E., Kurtoğlu, M. and Toroğlu, S., 2006 "The d10 metal chelates derived from schiff base ligands having silane: synthesis, characterization, and antimicrobial studies of cadmium (II) and zinc (II) complexes" *Synthesis and Reactivity in Inorganic and Metal-Organic Chemistry* **36(8)** 627-631.

[28] Joubert, C., 2013 "*Heterogenization of Schiff base complexes on mesoporous silica and their application as catalysts in the oxidative transformation of alcohols* "(Doctoral dissertation, Stellenbosch: Stellenbosch University).

[29] Lippincott, E.R., Van Valkenburg, A., Weir, C.E. and Bunting, E.N., 1958 "Infrared studies on polymorphs of silicon dioxide and germanium dioxide" *Journal of Research of the National Bureau of Standards* **61(1)** 61-70.

[30] Thana J.,A, Hayder H., M.,and Kasim M., H.,2017 "Catalytic esterification via silica immobilized p-phenylenediamine and dithiooxamide solid catalysts" *Arabian Journal of Chemistry* **10** S1492–S1500

[31] Sarmento, V.H.V., Schiavetto, M.G., Hammer, P., Benedetti, A.V., Fugivara, C.S., Suegama, P.H., Pulcinelli, S.H. and Santilli, C.V., 2010 "Corrosion protection of stainless steel by polysiloxane hybrid coatings prepared using the sol–gel process" *Surface and Coatings Technology* **204(16)** 2689-2701.

[32] Chang, T.C., Wang, Y.T., Hong, Y.S. and Chiu, Y.S., 2000 "Organic–inorganic hybrid materials. V. Dynamics and degradation of poly (methyl methacrylate) silica hybrids" *Journal of Polymer Science Part A: Polymer Chemistry* **38(11)** 1972-1980.

[33] Laurén, T., 2007 "*Methods and instruments for characterizing deposit buildup on heat exchangers in combustion plants*" Licentiate thesis. Åbo Akademi, Faculty of Chemical Engineering, Process Chemistry Center.

[34] Roland, C.M., Kallitsis, J.K. and Gravalos, K.G., 1993 "Plateau modulus of epoxidized polybutadiene" *Macromolecules* **26(24)** 6474-6476.

[35] Vadivel, M., Kumar, M.S.C., Selvam, V. and Alagar, M., 2012 "*Studies on Thermal and Morphological Behavior of 3-Aminopropyltriethoxysilane Grafted Epoxidized Ethylene-Propylene-Diene Terpolymer".*

[36] Abdulghani, A.J. and Hussain, R.K., 2015 "Synthesis and Characterization of Schiff Base Metal Complexes Derived from Cefotaxime with 1H-indole-2, 3-dione (Isatin) and 4-N, N-dimethyl-aminobenzaldehyde" *Open Journal of Inorganic Chemistry* **5(04)** 83.

[37] Sharma, A. and Shah, M., 2013 "Synthesis and characterization of some transition metal complexes derived from bidentate Schiff base ligand" *Journal of Applied Chemistry* **3** 62-66.

Determination of Minor elements of Human Scalp Hair from Karbala, Iraq

Joda A. Baker[1]* and Ward N.I [2]

[1]Faculty of Sciences, University of Kerbala, Karbala, Iraq
[2]Department of Chemistry, University of Surrey, Guildford, United Kingdom
Corresponding author's E-mail: phdchemistry2008@googlemail.com

Abstract. In recent years elemental analysis of human biological samples has become more advanced in terms of human health. This study has investigated the chemical analysis of Iraqi human hair samples in relation to differing forms of smoking (active and non-smokers) collected from Karbala, Iraq. Hair samples were obtained from 236 individuals (32 females, 204 males) aged between 3 to 70 years. Methods were developed and validated for the analysis of minor elements (namely Na, Ca, Mg and Fe) in human scalp hair samples. Different washing procedures and digestion methods were studied. The effect of gender and smoking habit on elemental levels was examined and the results compared with published values in the literature. Significant correlations were found between the gender and the levels of these elements in scalp hair samples at a probability level P of 0.05. In addition, significant differences were found between the Ca levels in smokers and non smokers at P of 0.05.

1. Introduction.

The connection between inorganic elements and human health has long been recognized, particularly when in the 17th century it was discovered that iron (Fe) was essential for normal human health [1]. Human and other living organisms are exposed to essential, non-essential and toxic elements that are introduced into the environment from natural sources as well as a result of anthropogenic (or man-made) activities. Therefore, the levels of these elements in environmental samples (soil, sediment, water, etc) may be a potential risk factor in assessing the quality of human health. Eventually, these elements are transported from aquatic media, air, foods etc. to the human body and can then be stored in different tissues and fluids [2], as shown in Figure 1.

In the last century, human scalp hair has been used an alternative biological material for blood and urine as a biomarker in the assessment of exposure to various pollutants in an occupational and/or environmental setting [4,5]. This is due to that a blood test provides different results from a hair analysis as the former only reflects the composition at the time of sampling, there are difficulties in blood collection and the stress factor for the patient associated with taking the sample [6-8]. In addition, human scalp hair is a long-term growth material which may provide useful data in determining health status of an individual for long periods and several trace elements being accumulated in hair as well [6,8].

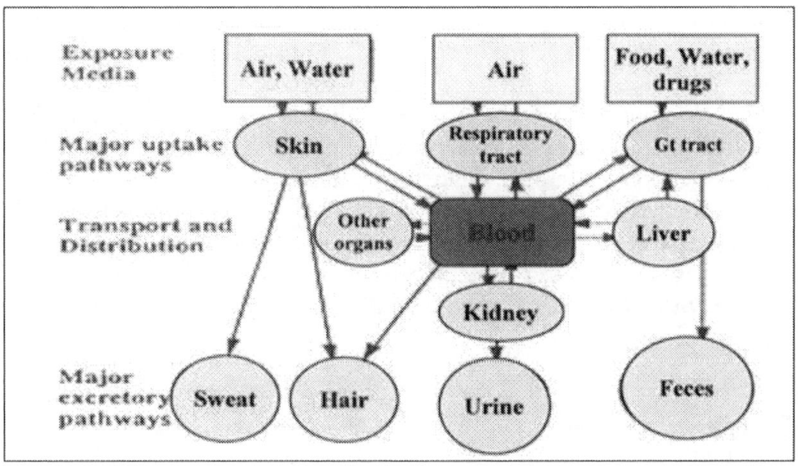

IOP Conf. Series: Journal of Physics: Conf. Series **1032** (2018) 012067 doi:10.1088/1742-6596/1032/1/012067

Figure 1. Exposure and metabolic pathway for elements [3].

The role of minor elements in the human body is not yet fully understood. Previous study has reported that the minor elements maintain electrolyte balance processes [9,10]. Metallic and trace element analysis of biological samples has improved especially over the last forty-seven years providing more scientific evidence of the important role of these elements in the human body [11,12]. Therefore, any excess or deficiency of essential elements (Na, Mg, Ca and Fe) can relate to possible serious problems in terms of the physiology of the body [13]. On the other hand, the concentration of these elements can be used to investigate dietary exposure to chemical contaminants, especially toxic metals, including the effects of smoking and drug usage of an individual or populations [14]. Previous studies have reported that human scalp hair samples can be used to evaluate the effect of age, sex and healthy on element levels inside human body [15]. Furthermore, hair analysis may evidence of raised chemical levels in domestic and occupational exposure cases [4,16].

In Iraq, only blood, saliva and urine analysis are used to determine human minor element levels. However, no study has been published on the use of human hair material in Karbala, Iraq and therefore, this study will be valuable in establishing a database of 'control' levels for comparison with other body tissues and for comparative analysis for future environmental pollution or health disorders and disease studies. The aims of this study are: to determine the levels of Na, Mg, Ca and Fe in the human scalp hair for individuals living in Karbala, Iraq and to investigate the levels of these elements in relation to smoking activity (non and active), and gender of an individual under investigation.

2. Methods and Materials.

2.1. Subjects and Methods.

Karbala is a city in Iraq located about 60 miles south west of Baghdad with approximately one million inhabitants. This town was subjected to military weapons during the two wars in 1991 and 2003[17]. Therefore, air, soil and water have been used to report levels of chemical contamination, especially trace and minor elements. In recent years, many individuals in Karbala have suffered from different health disorders such as cancer, sclerosis, diabetes, asthma, heart diseases, leukemia and various unknown diseases [18]. As part of this study, males (204) and females (32) were studied. The study population consisted of active (100), passive (32) and non-smokers (104), varying in age from three to seventy years. The subjects completed a brief questionnaire to provide information about their age, body mass, health state, profession, diet program, life-style, smoking activity, drinking water and alcohol consumption.

2.2. Sample Collection and Treatment.

Human scalp hair samples were collected from the same region of the scalp for each individual, namely, from the back of head, less than 1 cm from the scalp. At least 0.5 g of scalp hair were collected and cut into pieces of about 1 cm in length using special scissors and plastic tweezers. The samples were then stored in polyethylene bags in a dry place at room temperature [19, 8, 20]. Previous studies have reported that an ideal washing procedure should remove only external contaminants and leave the endogenous elements in the sample for analysis [21, 3]. Therefore, the sequential washing procedure (method A: acetone-water-water-water-acetone) was adopted in this study [20,5, 22]. A wet digestion method using nitric acid in a Kjeldhal™ tube was used to digest the hair sample. In this method, concentrated nitric acid (1 ml) was added to the hair sample, heated at 160° C ± 10° C in an electric digestor within fume until the solution became clear (about 30 min). Samples were then dissolved with deionised water in polypropylene volumetric flasks using constant dilution factor method [23,24].

2.3. *Instrumentation.*

The concentrations of Na, Mg, Ca and Fe were determined by flame atomic absorption spectrometry (FAAS) Perkin Elmer model 400.

3. Results and Discussion.

3.1. *Elemental levels of human scalp hair.*

The results were reported as mean, standard deviation (SD), and of Na, Mg, Ca and Fe in human scalp hair. Sodium, Mg, Ca levels were found to be 41 ± 42, 69 ± 84, 929 ± 563 mg/kg dry wt, respectively, as shown in Table 1. Similar results have been reported by other researches for hair samples [26,27,28,19,29]. In some cases, the value of Fe is higher than literature value [15,10].

Table 1. Elemental levels in human washed scalp hair from Karbala, Iraq (n = 236) mg/kg dry wt .

Variables	Elemental concentrations (mg/kg)			
	Na	Mg	Ca	Fe
Mean	41	69	929	79
SD	42	84	563	104
Range	1 - 323	8 - 874	113 - 2849	26 - 414

: n is the number of samples

The confidence limits (95 %) of the mean value are given by the following formula:

$$\textbf{95\% confidence limits} = \mathbf{x \pm zs / \sqrt{n}}$$

Where: x is mean value, z is the degree of confidence required, s is standard deviation and n is the number of samples. The value of 95% confidence limits of sodium are (35.6 to 46.5 mg/kg), including the mean value of Na (41 mg/kg). This result has confirmed that there is not any systematic error in this analysis [30]. Similar results were also reported for Ca, Mg and Fe, as shown in Table 2. The Grubb's outlier test was used in order to check any normally distribution data for outliers. It was found that the lowest value of sodium (1 mg/kg Na) was not an outlier, but the highest value (323 mg/kg Na) was an outlier (low G_{calc} 0.95; high G_{calc} 6.7). The critical value (G_{crit}) ($P = 0.05$) is 3.62, therefore, the maximum measurement is rejected at the 5% significance level. The Grubb's results for Na, Mg, Ca and Fe are shown in Table 2.

Table 2. Grubb's test and 95% confidence limit for Na, Mg, Ca and Fe in washed scalp hair samples from Karbala, Iraq.

Element	95% confident limit (mg/kg)	Grubb's test			Test result
		Gcalc (min)	Gcalc (max)	Gcrit	
Na	36 to 47	0.95	6.7	3.45	Maximum value is rejected
Mg	58 to 80	0.73	9.58	3.5	Maximum value is rejected
Ca	855 to 999	1.44	3.4	3.5	Significant result
Fe	66 to 92	0.73	3.2	3.4	Significant result

Gcalc= Grubb's calculate, Gcrit=Grubb's critical.

3.2. Influence of Gender

The effect of gender on the concentration of Na, Mg, Ca and Fe was investigated in this study. In fact, approximately, 14 % of the hair samples were taken from females and 86% from males. Table 3 reports the elemental levels as a function of gender. In female's hair, the levels of Na (146 ± 150 mg/kg), Mg (157 ± 89 mg/kg) and Ca (1708 ± 606 mg/kg) were higher when comparing with males hair. In contrast, male's hair samples have higher levels of Fe (44 ± 44 mg/kg). Surprisingly, the content of the majority of elements in hair of females was higher than in the hair of males from Karbala, Iraq. A possible explanation is that the majority of women living in the study area have long term exposure to exogenous contamination; namely permanent solutions, dyes and bleaches [15]. Similar findings have been previously reported for Na, Ca, Mg and Fe [24,28]. Previous study has found that the compositions of human scalp hair are varied in individual having a natural color hair comparison with those that were artificially colored. This study was found that colored hair contained more Ca (270%), Mg (260%), than naturally colored hair [15]. In the case of Fe, mean values were lower in females than in males. This is happening may be due to the menstruation which is a woman's monthly bleeding [31]. In order to assess whether the effect of age makes any significant contribution on the elemental levels of hair samples, the results were examined by an F-test and a two tailed t-test, as shown in Table 3. The findings show that there is a significant effect of gender on the levels of Na ($t_{(234)} = 3.87$, $t_{crit} = 1.97$, $P < 0.05$), Mg ($t_{(234)} = 7.09$, $t_{crit} = 1.97$, $P < 0.05$), Ca ($t_{(234)} = 10.08$, $t_{crit} = 1.97$, $P < 0.05$), and Fe ($t_{(234)} = 3.29$, $t_{crit} = 1.97$, $P < 0.05$), (where the number in brackets is the number of degrees of freedom and the critical value (t_{crit}) is determined at the probability level of $P = 0.05$), as shown in Table 3. As a result, on the basis of our results, female hair is not an appropriate biomarker for most elements such as those evaluated in the present study. In that case, human toe nails could be considered as a useful material for this purpose since most women usually cover their feet when they leave the home.

Table 3: Elemental levels in washed human scalp hair for males and females.

Element	Mean ± SD (mg/Kg)		F-test		Two – tailed t-test			
	Male n_1=204	Female n_2=32	F_{cale}	Sig.	t_{cale}	*df*	Sig.	t_{crit}
Na	69 ± 91	146 ± 151	2.97	<0.05	3.87	234^+	<0.05	1.97
Mg	55 ± 74	157 ± 89	1.50	<0.05	7.09	234^+	<0.05	1.97
Ca	805 ± 447	1708 ± 606	1.84	<0.05	10.08	234^+	<0.05	1.97
Fe	44 ± 44	19 ± 14	9.24	<0.05	3.29	234^+	<0.05	1.97

SD is standard deviation; n_1, n_2, are the number of sample, *df*= degrees of freedom at n_1-1 and n_2-1 for F-test, + degrees of freedom for t-test (n_1+n_2-2), F_{cale} and t_{cale} are the calculated values for F-test and t-test, respectively, t_{crit} is critical value at *P*= 0.05.

IOP Conf. Series: Journal of Physics: Conf. Series **1032** (2018) 012067 doi:10.1088/1742-6596/1032/1/012067

3.3. *Influence of Smoking.*

The determination of the elemental levels in biological and environmental samples is very important due to they play an important role in physiological processes of human and other living organisms [32]. However, hair analysis recently becomes a powerful diagnostic tool to monitor the environmental exposure, human health and nutrition studies of individuals. Therefore, it is necessary to establish standard ranges of hair composition for variable groups of population with respect of age, gender and smoking habits. In general tobacco leaves have widely been used to manufacture smoking. Typically, tobacco plants are absorb many of an essential, non-essential and toxic elements from the soils, fertilizers, pesticide treatments, storage, processing, packing and other processes [32]. Previous study has been identified the concentrations of 21 elements in 16 cigarette brands that commonly used in Karbala, Iraq [33]. The obtained results were namely: 84 – 383 mk/kg Na dry wt, 7852 – 8768 mg/kg Mg dry wt, 12392 – 19600 mg/kg Ca dry wt, 143 – 301 mg/kg Fe dry wt, In this study the effect of smoking habits on the elemental composition of hair was investigated by using two subject groups, smokers and non-smokers. Therefore, 50 % of samples were taken from male smokers and 50 % from male non-smokers who were of the same age range. The findings in Table 4 show that a highly significant difference was found for calcium between smokers (981±488 mg/kg) and non-smokers (769±375). In the case of Mg, mean values were slightly higher in smokers than non-smokers, although the difference was not significant (at $P = 0.05$). Conversely, there are no significant differences for the other elements between smoker and non-smoker groups. A higher level of Ca and Mg and lower levels of Na and Fe in smokers was also found in other study [15,24]. One study has reported that the levels of Ca (16000 – 27800 mg/kg dry weight) and Mg (8950 – 16750 mg/kg dry weight) in commercial cigarette tobacco are higher than other elements [10]. The results were evaluated by using F-test and two tailed t-test, as shown in Table 4.

Table 4. Elemental level in washed human scalp hair for smokers and non-smokers.

Element	Mean ± SD (mg/Kg)		F-test		Two – tailed t-test			
	Smokers n_1=100	Non-Smokers n_2=104	F_{cale}	Sig.	t_{cale}	*df*	Sig.	t_{crit}
Na	93 ± 90	114 ± 114	1.63	<0.05	0.99	202^+	<0.05	1.97
Mg	71 ± 116	50 ± 21	29.79	<0.05	1.27	202^+	<0.05	1.97
Ca	981 ± 488	769 ± 375	1.66	<0.05	2.47	202^+	<0.05	1.97
Fe	38 ± 26	47 ± 60	5.36	<0.05	1.08	202^+	<0.05	1.97

SD is standard deviation; n_1, n_2, are the number of sample, *df*= degrees of freedom at n_1-1 and n_2-1 for F-test, + degrees of freedom for t-test (n_1+n_2-2), F_{cale} and t_{cale} are the calculated values for F-test and t-test, respectively, t_{crit} is critical value at P= 0.05.

4. Conclusions.

The optimum washing procedure was carried out by using acetone-water-water-water-acetone which produced the highest level of element removal. Wet digestion using Kjeldhal™ tubes was found to be the preferred digestion method for most elements under investigation resulting in acceptable percentage recoveries.

The highest mean values in the two gender groups are found for Ca followed by Mg, Na and iron for females,. The order of increasing trace element levels in the tear drops for females is: Fe < Na < Mg < Ca, whilst for males is Fe < Mg < Na < Ca. The findings show that there is a significant effect of gender on the levels of Ca, Mg, Na and Fe. In the case of smoking activity effect, no significant effect ($P = 0.05$) was found for Mg, Na and Fe. In contrast, there is a significant effect for Ca at $P = 0.05$.

Acknowledgment
The authors wish to thank the Iraqi government for its financial support. The authors also wish to acknowledge Al-Hussein hospital, Karbala, Iraq, University of Surrey, UK and the University of Kerbala, Iraq.

References

[1] Iyengar, G.V. 1989 "*Elemental Analysis of Biological System, Volume 1: Biomedical, Environmental, Compositional and Methodological Aspects of Trace Elements*" CRC Press, Boca Raton (Florida) 11-12.
[2] Arain, M.B., Kazi, T.G., Jamali, M.K., Jalbani, N., Afridi, H.I. & Sahah, A. 2008 "Total dissolved and bioavailable elements in water and sediment samples and their accumulation in Oreochromis mossambicu of polluted Manchar Lake" *Chemosphere* **70(10)** 1845-1856.
[3] Apostoli, P. 2002 "Elements in environmental and occupational medicine" J*ournal of Chromatogram* **B 778** 63-97.
[4] Amaral, A.F.S., Arruda, M., Cabral, S. & Rodrigues, A.S. 2008 "Essential and non-essential trace metals in scalp hair of men chronically exposed to volcanogenic metals in the Azores" *Portugal, Environment International* **34** 1104-1108.
[5] Gault, A.G., Rowland, H.A.L., Charnock, J.M., Wogelius, R.A., Gomez-Morilla, I., Vong, S., Leng, M., Samreth, S., Sampson, M.L. & Polya, D.A. 2008 "Arsenic in hair and nails of individuals exposed to arsenic-rich groundwater in Kandal province, Cambodia" *Science of the Total Environmen*t **393** 168-176.
[6] Wang, T., Fu, J., Wang, Y., Liao, Ch., Tao, Y. & Jiang, G. 2009 "Use of scalp hair as indicator of human exposure to heavy metals in an electronic waste recycling area" *Environmental Pollution* **157(8-9)** 2445-2451.
[7] Batista, B.L., Rodrigues, J.L., Nunes, J.A., Tormen, L., Curtius, A.J. & Barbosa, F. 2008 "Simultaneous determination of Cd, Cu, Mn, Ni, Pb, and Zn in nail samples by inductively coupled plasma mass spectrometry (ICP-MS) after tetramethyammonium hydroxide solubilization at room temperature" *Talanta* **76(3)** 575-579.
[8] Sukumar, A. & Subramanian, R. 2007 "Relative element levels in the paired samples of scalp hair and fingernails of patients from New Delhi" *The Science of the Total Environment* **372** 474-479.
[9] Thomson, C.D. 2004 "Assessment of requirements for selenium and adequacy of selenium status: a review" *European Journal of Clinical Nutrition* **58** 391-402.
[10] Ward, N.I. 2000 "*Trace Elements, In: Fifield, F.W. & Haines, P.J. (Editors), Environmental analytical chemistry*". 2nd Edition, Blackwell Science Ltd, (Oxford), UK, 360-392.
[11] Manso, M., Carvalho, M.L. & Nunes, M.L. 2007 "Characterization of essential and toxic elements in cephalopod tissues by EDXRF and AAS" *X-ray Spectrometry* **36(6)** 413-418.
[12] Villanueva, R. & Bustamante, P. 2006 "Composition in essential and non-essential elements of early stages of cephalopods and dietary effects on the elemental profiles of Octopus vulgaris paralarvae" *Aquaculture* **261(1)** 225-240.
[13] Goldhaber, S.B. 2003 "Trace element risk assessment: essentiality vs. toxicity" *Regulatory Toxicology and Pharmacology* **38(2)** 232-242.
[14] Esteban M, Castano A 2009 "Non-invasive matrices in human biomonitoring: A review" *Environ Int 35* 438-449.
[15] Chojnacka K, Gorecka H, Gorecki H 2006 "The effect of age, sex, smoking habit and hair color on the composition of hair" *Environ Toxicol Pharmacol* **22** 52–57.
[16] Ashraf W, Jafaar M, Anwer K, Ehsan U 1995 "Age and sex based comparative distribution of selected metals in the scalp hair of an urban population from two cities in Pakistan" *Environmental Pollution* **87** 61-64.
[17] Senofonte O, Violante N, Caroli S 2000 "Assessment of reference values for elements in human

hair of urban schoolboys" *J. Trace Elements Med. Biol* **14** 6 -13.

[18] United Nations Environment Programme 2003 "*Desk study on the environment in Iraq*" (Geneva) UNEP.

[19] Senofonte O, Violante N, D'Ilio S, Caimi S, Peri A, Caroli S 2001 "Hair analysis and the early detection of imbalances in trace elements for members of expeditions in Antarctica" *Microchem J* **69** 231-238.

[20] Rodrigues JL, Batista BL, Nunes JA, Passos CJS, Jr FB 2008 "Evaluation of the use of human hair for biomonitoring the deficiency of essential and exposure to toxic elements" *Sci Total Environ* **405** 370-376.

[21] Hawkins, D.P. &. Ragnarsdóttir, K.V. 2009 "The Cu, Mn and Zn concentration of sheep wool: influence of washing procedures, age and colour of matrix" *Science of the Total Environment* **407(13)** 4140-4148.

[22] IAEA (International Atomic Energy Agency) 1978 "Co-ordinated research programme on trace element pollutants" IAEA/RL/50, (Vienna) Austria.

[23] Kazi, T.G., Afridi, H.I., Kazi, N., Jamali, M.K., Arain, M.B., Jalbani, N. & Kandhro, G.A. 2008 "Copper, chromium, manganese, iron, nickel, and zinc levels in biological samples of diabetes mellitus patients" *Biological Trace Element Research* **122(1)** 1-18.

[24] Forte, G., Alimón, A., Violante, N., Di Gregorio, M., Senofonte, O., Petrucci, F., Sancesario, G. & Bocca, B. 2005 "Calcium, copper, iron, magnesium, Silicon and zinc content of hair in Parkinson's disease" *Journal of Trace Elements in Medicine & Biology* **19** 195-201.

[25] Bass, D.A., Hickok, D., Quig, D. & Urek, K. 2001 "Trace Element Analysis in Hair: Factors Determining Accuracy, Precision, and Reliability" *Thorne Research, Inc*, **6(5)**: 472-482.

[26] Skalnaya, M.G. & Demidov, V.A. 2007 "Hair trace element contents in women with obesity and type 2 diabetes" *Journal of Trace Elements in Medicine and Biology* **21** 59-61.

[27] Chojnacka, K., Gorecka, H., Chojnacki, A. & Gorecki, H. 2005 "Inter-element interactions in human hair" *Environmental Toxicology and Pharmacology* **20(2)** 368-374.

[28] Rodushkin, I. & Axelsson, M.D. 2003 "Application of double focusing sector field ICP-MS for Multielemental characterization of human hair and nails. Part III. Direct analysis by laser ablation" *The Science of the Total Environment* **305(1-3)** 23-39.

[29] Miekeley, N., Dias Carneiro, M.T.W. & de Silvera, C.L.P. 1998 "How reliable are human hair reference intervals for trace elements" *The Science of the Total Environment* **218** 9-17.

[30] Miller, J.N. & Miller, J.C. 2010 "*Statistics and Chemometrics for Analytical Chemistry*" 4th Edition, Pearson Education Limited, (England) 1-272.

[31] Jian, J., Yang, Q., Dai, D., Eckard, J., Axelrod, D., Smith, J. & Huang, X. 2011 "Effects of iron deficiency and overload on angiogenesis and oxidative stress–a potential dual role for iron in breast cancer" *Free Radical Biology and Medicine* **50(7)** 841-847.

[32] Martinez, T., Lartigue, J., Zarazua, G., Avila-Perez, P., Navarrete, M. & Tejeda, S. 2008 "Application of the Total Reflection X-ray Fluorescence technique to trace elements determination in tobacco" *Spectrochimica Acta Part B* **63(12)** 1469–1472.

[33] Joda, B. 2012 "*Trace Element Levels of Human Fluids and Tissues for Iraqi Individuals*" Ph.D. Thesis, Department of Chemistry, University of Surrey, Guildford, (Surrey) England.

IOP Conf. Series: Journal of Physics: Conf. Series **1032** (2018) 012068 doi:10.1088/1742-6596/1032/1/012068

The QTAIM Approach to Chemical Bonding in Triruthenium Carbonyl Cluster: [Ru$_3$ (μ -H)(μ_3-κ^2-Haminox-*N,N*)(CO)$_9$]

Shatha Raheem Helal[a], Muhsen Abood Muhsen Al-Ibadi[b], Ahlam Hussein Hasan[b], Abrar Taha[a].

[a]Department of physical, [b]Department of Chemistry, College of Science, University of Kufa, Najaf P.O. Box (21), Iraq.

Corresponding author's E-mail: hassantalal9999@yahoo.com

Abstract. The bonding in the hydride triruthenium carbonyl cluster **[Ru$_3$ (μ -H)(μ_3-κ^2-Haminox-*N,N*)(CO)$_9$]** is investigated by using the Quantum Theory of Atoms-in-Molecules (QTAIM). Calculated metal–metal and metal–ligand bond critical points (bcp) properties, electron density $\rho(b)$, Laplacian $\nabla^2\rho(b)$, local energy density $H(b)$, local kinetic (Lagrangian) energy density $G(b)$, local potential energy density $V(b)$ ellipticity $\varepsilon(b)$, and bond delocalization indices $\delta(A, B)$ are consistent with the relevant transition metal clusters in the literature. The topological data recognizes bond paths (bp) and bond critical points for metal-metal interactions in the core of the hydride triruthenium cluster [Ru(1)-Ru(3) and Ru(2)-Ru(3)]. However, no direct bond path is found for the interaction between the hydride bridged Ru atoms, while a non-negligible delocalization index δ(Ru(1)...Ru(2)) is obtained for this non-bonding interaction. An interaction of 5-center 6-electron type is proposed for the core. The topological parameters of Ru-N oxazoline ring bond suggest a pure σ-bond.

1. Introduction.

The interest in synthetic chemistry is focused on the preparation of novel chiral metal catalysts for asymmetric processes [1]. For example of chiral triruthenium clusters reported so far correspond to species containing mono- or bidentate chiral phosphane ligands. These latter complexes have been used as catalyst precursors in asymmetric isomerization, hydroformylation or hydrogenation of unsaturated substrates. In most catalytic reactions, the catalyst precursors have been prepared in situ and their nature remains unknown. No other cluster complexes containing chiral N-donor ligands have been reported to date. During the last decade, reported by Cabeza and his co-workers has thoroughly studied the synthesis and reactivity of carbonyl clusters derived from 2-aminopyridines[2]. In this paper we study one of these catalyst by QTAIM theory The important development in topological system and its ability to identify a bond between any two atoms in terms of bond paths (bps) and bond critical points (bcps)[3].

On the basis of QTAIM, the relation between bonding modes and topological properties of the electron density and its Laplacian cannot be straightforwardly extended to organometallic compounds, because of the nature of the bonding interactions between transition-metal atoms are not completely understood yet and much argue is still occurring on its actual presence, role and mechanism and the bonds in the transition metal display a different and much narrower spectrum of topological indexes. [4][5][6][7][2].

At now, a many papers have been published to study the topology of the electron density in cluster compounds that have a three or multitransition metal–metal interactions between their atoms[8][9][10][11][12][13][14][15][16][17][18][19][20]. So, more QTAIM studies applied on this class of complexes in order to discribe the relationship between metal–metal bonds and the topology of their associated electron density[21].

In this study the QTAIM topological analysis of the electron density employed in the N-Heterocyclic, mono hydride tri ruthenium clusters **[Ru$_3$ (μ -H)(μ_3-κ^2-Haminox-*N,N*)(CO)$_9$** (Fig. 1), which was previously reported by Cabeza and his co-workers[22]. The amido fragment spans an edge

IOP Conf. Series: Journal of Physics: Conf. Series **1032** (2018) 012068 doi:10.1088/1742-6596/1032/1/012068

above the metal triangle, while the N atom of the oxazoline ring is attached to the remaining metal atom. In addition, the bridging NH and H ligand spans on the Ru(1)-Ru(2), this allowing interesting comparisons between the topological properties of related but different atom-atom bonding interactions within the same molecule.

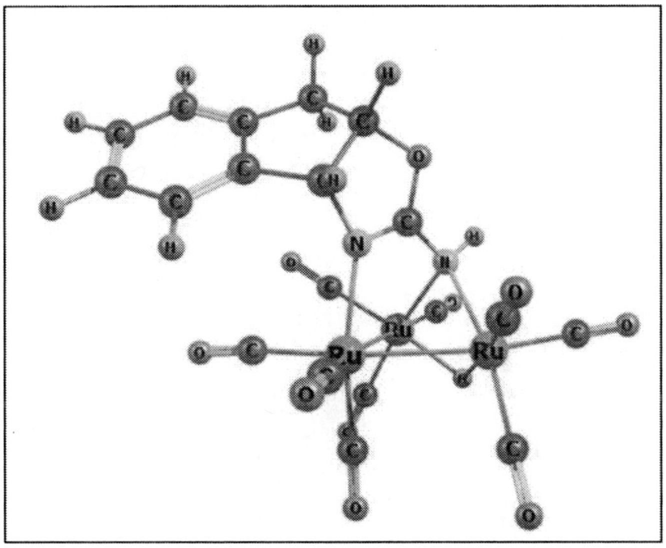

Figure 1: geometry optimization structure of [Ru_3 (μ -H)(μ_3-κ^2-Hamphox-*N,N*)$(CO)_9$].

2. Computational Methods

The X-ray diffraction structures of this compound was used as starting points for geometry optimization, which were performed with the GAUSSIAN03[23] and GAUSSIAN09[24] packages at the DFT level of theory, using PBE1PBE functional. 6-31G (d, p) basis set were employed for H, C, N and O atoms. The LANL2DZ effective core potential (ECP) basis set was used for Ru atoms. These calculations included both local and integral properties and were carried out with the AIM2000 programs[25]. The accuracy of the integrated properties was finally set at least at 1.0 x 10^{-4} (from the Laplacian of the integrated electron density), whereas the accuracy of the local properties was 1.0 x 10^{-10} (from the gradient of the electron density at the bond critical points). The AIM results described in this work corresponding to calculations performed with the PBE1PBE/WTBS/6-31G (d, P)[26][27].

3. Results and Discussion.

3.1. *Topological Properties of the Electron Density*

The main feature of atoms in molecules (AIM) theory of Bader[28], with the AIM2000 program package, is that the topology of electron density ρ(r) contains information about the bonding situation. Specifically, at the critical points (the points in space at which the gradient of the density vanishes [$\nabla\rho$ (r) = 0]), carries important information about the multiplicity and the covalence of the bond. The critical point characterized by two negative curvatures (maximum) and one positive curvature (minimum), labelled as (3,-1), is referred to as a bond critical point (bcp) and is considered to be indicative of the presence of chemical bond, in contrast, a ring critical point (rcp) has one negative curvature and two positive curvatures, labelled as (3,+1). Some structures necessarily contain other critical point of the density: the (3,+3) is produced inside cages (cage critical point, ccp). A chemical bond between two atoms is characterized by a line of maximum electron density (the bond path, bp) which links the two atomic nuclei and intersects an interatomic surface at the bond critical point. The

IOP Conf. Series: Journal of Physics: Conf. Series **1032** (2018) 012068 doi:10.1088/1742-6596/1032/1/012068

chemical structure according to these descriptions is simply the topological graph produced by the maxima of the electron density and the bond paths. The image shown in (Fig. 2) was obtained by applying the QTAIM approach to this compound, it's easily observe bond and ring critical points, together with the bond paths that connect the bonded atoms through their corresponding bond critical points. The bcps and bps for the Ru-N, Ru-C, Ru-H, Ru-Ru, N-H, C-O, C-N, C-C, and C-H bonds were clearly found. Importantly, the bcp and bp related to Ru(1) and Ru(3) or between Ru(3) and Ru(2) bonds were found science no bcps or bp's were observed between Ru(1) and Ru(2). Six rcps were clearly observed, corresponding to the C(4)-C(5)-C(6)-C(7)-C(8)-C(9), C(2)-C(3)-C(4)-C(9)-C(10), O(1)-C(1)-N(2)-C(3)-C(2), Ru(3)-N(2)-C(1)-N(1)-Ru(2), Ru(3)-Ru(1)-[N(1) or H(1)]-Ru(2), and Ru(1)-N(1)-Ru(2)-H(1) rings.

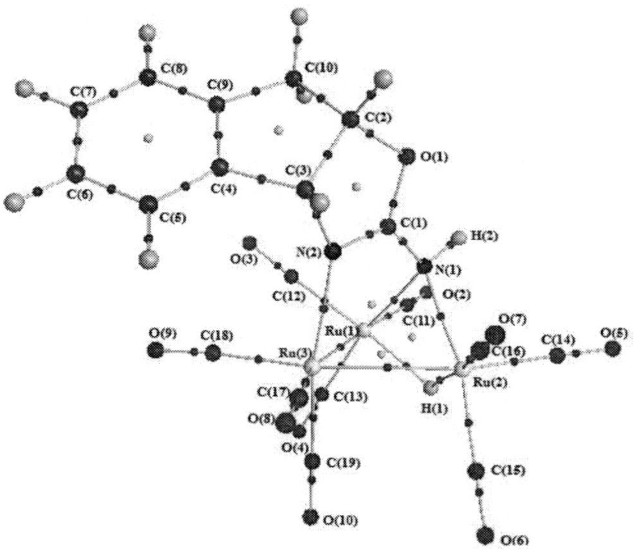

Figure 2: Molecular graph of the triruthenium cluster, showing the bond paths (gray lines) with the bond critical points (small red circles between two atoms) and ring (yellow circles) critical points.

A gradient trajectory map of the total electron density in the Ru(1)-Ru(2)-Ru(3) plane which described in (Fig. 3), showing the bcp and bp between Ru(1)-Ru(3) and Ru(3)-Ru(2) together with the atomic basins for these and other atoms located in the same plane. Three CO ligands, located nearly in the same plane, are also visible in the plot. The bp's and bcps found between H(1) and Ru(1) and Ru(2) metal atoms from the other side, the bps and bcps found between N(1) and Ru(1) and Ru(2) metal atoms also located in these plane can be observed.

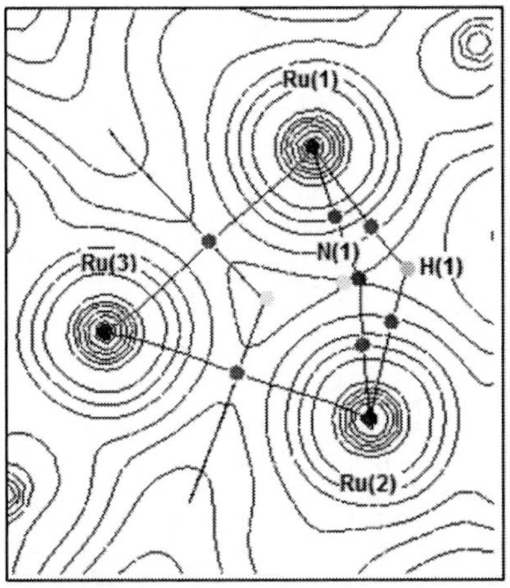

Figure 3: Gradient trajectories mapped on a total electron density plot in the Ru(1)- Ru(2)-Ru(3) plane, showing the atomic basins, bps and bcps.

(Fig. 4) describe a gradient trajectory map of the total electron density in the Ru(1)-H(1)-Ru(2)-N(1) plane. The bps and bcps found between the ligand [H(1) or N(2)] and Ru(2) and Ru(3) metal atoms located in these plane can also be observed.

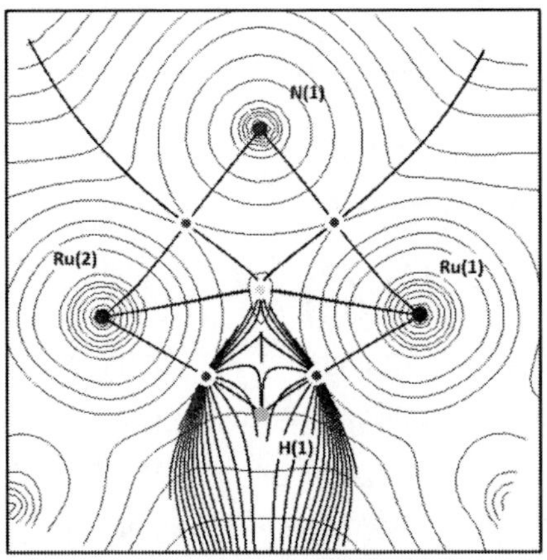

Figure 4: Gradient trajectories mapped on a total electron density plot in the Ru(1)-H(1)-Ru(2)--N(1)plane, showing the atom basins, bps, bcps and rcp.

(Fig. 5) displays a gradient trajectory map of the total electron density in the plane spanned by the (+)-2-amino(4R,5S)-indanyl-2-oxazoline **(Haminox)** ligand. It shows the bps and bcps of the interactions of Ru(3) with the amino oxazoline ligand N(2) atom. On the other hand, both bps and bcps in a ring made between Ru(3)-N(2)-C(1)-N(1)- Ru(1 or 2).

IOP Conf. Series: Journal of Physics: Conf. Series **1032** (2018) 012068 doi:10.1088/1742-6596/1032/1/012068

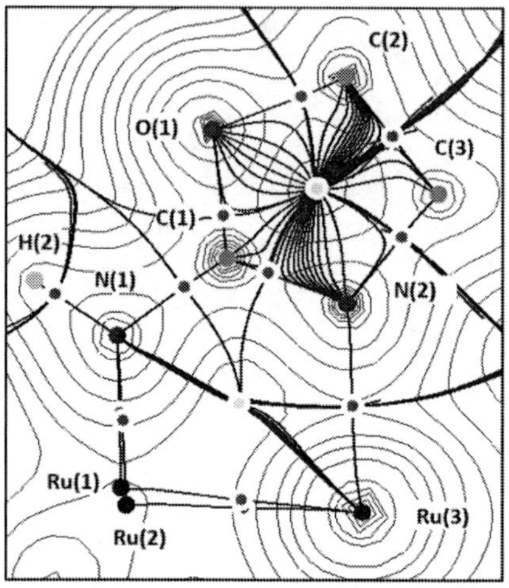

Figure 5: Gradient trajectories mapped on a total electron density plot in the NHC ligand plane, showing the atom basins , bps and bcps .

3.2. Atoms in Molecules Analysis

In the framework of QTAIM, we have calculated the topological properties of the triruthenium cluster complex (Table 1).

Table 1. Selected topological properties at the critical points for triruthenium cluster.

Bond	ρ_b(eÅ^{-3})	$\nabla\rho_b$(eÅ^{-5})	G_b(he^{-1})	H_b(he^{-1})	V(he^{-1})	ε_b
Ru(1)-Ru(3)	0.4430	0.6748	0.2499	-0.0812	0.3311	0.0539
Ru(2)-Ru(3)	0.4370	0.6910	0.2517	-0.0790	0.3307	0.0576
Ru(1,2)-N(1)(ave)	0.7869	-0.0929	1.0250	-0.0961	1.1211	0.0185
Ru(1,2)-H(1)(ave)	0.7665	-0.0450	0.6729	-0.2225	0.8954	0.0723
Ru(3)-N(2)	0.7781	3.8768	1.0460	-0.0768	1.1229	0.0727
C(1)-N(1)	3.3848	-12.6587	1.8135	-4.9781	6.7916	0.1421
C(1)-N(2)	3.8528	-9.9338	4.2871	-6.7705	11.0576	0.3281
C(3)-N(2)	2.6500	-7.4056	1.2541	-3.1055	4.3597	0.0280
N(1)-H(2)	3.4125	-18.0341	0.5481	-5.0566	5.6046	0.0220
C(1)-O(1)	3.1256	0.0860	4.1271	-4.9873	9.1144	0.0394
C(2)-O(1)	2.3873	0.0931	2.3727	-3.3036	5.6763	0.0239
C(2)-C(3)	2.4442	-5.5945	0.5525	-1.9511	2.5035	0.0463
Ru-CO(ave)	1.4142	-0.1289	1.8762	-0.5868	2.4630	0.0357
C-O(ave)*	4.5687	0.1439	10.5306	-7.4839	18.0145	0.0054

*Average values. [a] Bond path length. [b] Electron density at the bcp. [c] Laplacian of the electron density at the bcp. [d] Kinetic energy density ratio at the bcp. [e] Total energy density ratio at the bcp. [f] Potential energy density ratio at the bcp. [g] Ellipticity at the bcp.

3.2.1. Ru- Ru Interactions. In the core the hydride-bridged Ru(1)-Ru(2) interaction exist no bcps or direct bps. the values of ρ(b) for the Ru(1)-Ru(3) and Ru(2)-Ru(3) interactions are [0.4430 and 0.4370 e Å^{-3}], respectively. also it have a small positive values for $\nabla^2\rho(b)$ [0.6748 eÅ^{-5} for Ru(1)-Ru(3) and 0.6910 eÅ^{-5} for Ru(2)-Ru(3)]. The positive value of G(b) were obtained for the both Ru(1)-Ru(3)

[0.2499 he^{-1}] and Ru(2)-Ru(3) [0.2517 he^{-1}]. Above findings suggested that the interactions between Ru(1)-Ru(3) and Ru(2)-Ru(3) are best described as typical open-shell metal-metal interactions.

The small values of ε_b [0.0539 and 0.0576] are comparable to those reported for Os-Os bonds in $[Os_3(CO)_{12}]$, $[Os_3(\mu\text{-}H)(\mu\text{-}Cl)(CO)_{10}]$, $[Os_3(\mu\text{-}H)(\mu\text{-}OH)(CO)_{10}]$, $[Os_3(\mu\text{-}H)_2(CO)_{10}]$[29], and Ru-Ru bonds in $[Ru_3(CO)_{12}]$[13], and $[Ru_3(\mu\text{-}H)_2(\mu^3\text{-}MeImCH)(CO)_9]$[21]. For instance, for $[Os_3(CO)_{12}]$, the electron density ρ_b, Laplacian $\nabla^2\rho_b$, kinetic energy G_b and ellipticity ε_b values for unabridged Os-Os bond are 0.327 eÅ^{-3}, 0.829 eÅ^{-5}, 0.435 he^{-1} and 0.120, respectively. The average value of the electron density of [0.7665 eÅ^{-3} and 0.7869 eÅ^{-3}] and Laplacian of [-0.0450 eÅ^{-5} and -0.0929 eÅ^{-5}] for Ru(1,2)-H(1) and Ru(1,2)-N(1) interactions indicated that the strength of these bonds is comparable to that of pure covalent single bonds between non-metal atoms[15][8]. Additionally, the ellipticity is [0.0723 and 0.0185, and this is comparable with $[Ru_3(\mu\text{-}H)_2(\mu^3\text{-}MeImCH)(CO)_9]$ system (0.086)[21].

The Ru_3(H,N) part of the cluster can be analysed by looking at the Laplacian of the electron density at the Ru(1)-H(1) or N(1)-Ru(2)-Ru(3) plane of the molecule (Fig. 6). Valence shell charge concentration (VSCC) of bridging H(1) and N(1) atoms are polarized toward the midpoint of the Ru(1)-Ru(2) edge. A small polarisation was found, and positive values were obtained for Laplacian of Ru(1,2)-H(1) and Ru(1,2)-N(1) bcps, (Table 1), than those expected in the absence of such polarization. One maximum of Laplacian was found for the hydrogen bridging two ruthenium atoms[21].

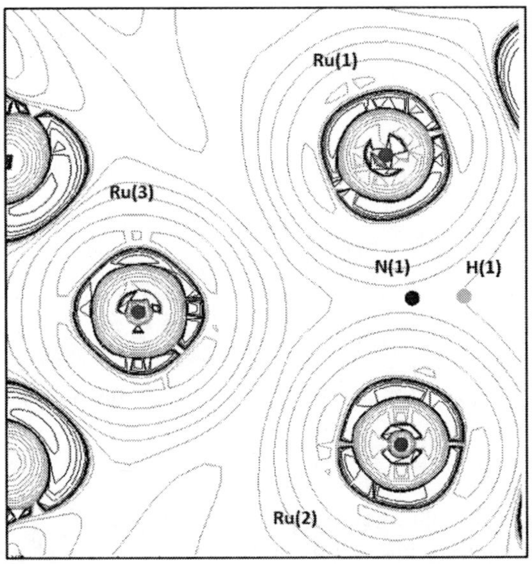

Figure 6: Laplacian map of the electron density in the Ru(1)...Ru(2)-Ru(3) plane of triruthenium cluster.

3.2.2. Delocalization Indices. The delocalization indices for the two unabridged Ru-Ru bonds. For the Ru(1)-Ru(3) and Ru(2)-Ru(3) bonds, calculated values are smaller than 0.474. This is in agreement with the few cases reported in the literature, the Ru-Ru bond showed an average value of 0.458 in $[Ru_3(\mu\text{-}H)_2\text{-}(\mu^3\text{-}MeImCH)(CO)_9]$[21], the delocalization indices of Os-Os bonds in the range of 0.350 - 0.461 for $[Os_3(CO)_{12}]$, $[Os_3(\mu\text{-}H)(\mu\text{-}Cl)(CO)_{10}]$, $[Os_3(\mu\text{-}H)(\mu\text{-}OH)(CO)_{10}]$ and $[Os_3(\mu\text{-}H)_2(CO)_{10}]$[29] and the Fe-Fe bond indicated an average value of 0.398 in $[Fe_3(\mu\text{-}H)(\mu\text{-}COMe)(CO)_{10}]$[30]. Also, similar values were observed for the indexes for other metal-metal interactions[31]. In contrast, Ru(1)...Ru(2) nonbonding interactions of this compound showed a small value for their delocalization indices (0.2007). This is comparable with the reported hydride-bridged M...M nonbonding interactions. For instance, between 0.169 and 0.246 for Ru...Ru interactions of $[Ru_3(\mu\text{-}H)_2\text{-}(\mu^3\text{-}$

MeImCH)(CO)$_9$][21], 0.177 for Os...Os interaction of [Os_3(μ-H)(μ-Cl)(CO)$_{10}$][29], and 0.208 for Fe...Fe interaction of Fe_3(μ-H)(μ-COMe)(CO)$_{10}$][2] and in line with other values found for ligand-bridged M...M interactions in several organometallic compounds[32][8]. On the other hand, the delocalization indices of the nonbonding Ru(1)...Ru(2) interaction is comparable with the previously reported for unbridged metal-metal interaction such as Co-Co interaction of [$Co_2(CO)_8$] (0.460)[10]. As the bridged Ru...Ru interaction of this cluster show a smaller delocalization index than that of unbridged Ru-Ru interaction.

Table 2. The delocalization indices for specific atomic δ(A, B) interactions.

atom pairs (A, B)	δ(A, B)	atom pairs (A, B)	δ(A, B)
Ru(1)-Ru(3)	0.5028	C(2)-O(1)	0.7790
Ru(2)-Ru(3)	0.4849	O(1)-C(1)	0.7800
Ru(1,2)-H(1)	0.4550	C(2)-C(3)	0.8070
Ru(1,2)-N(1)	0.5080	N(2)-C(3)	0.8430
Ru(3)-N(2)	0.5079	Ru(1)...Ru(2)	0.2007
N(1)-C(1)	0.8380	Ru(3)...N(1)	0.0410
N(1)-H(2)	0.7380	Ru(2,3)...C(1)	0.0210
N(2)-C(2)	0.0110	Ru(3)...C(3)	0.0150
		Ru...O_{CO}	0.1907

The delocalization index of the δ(Ru-H) interactions of cluster (Table 2) has an average value of 0.455, which is almost equal to the average value of δ(Ru-H) interactions of [Ru_3(μ-H)$_2$(μ^3-MeImCH)(CO)$_9$] system (0.474) [21]. Also, this is in line with few cases reported to date for the M-H interactions[31]. Interestingly, the calculated delocalization index values for the Ru(1)-H(1) and Ru(2)-H(1) suggested that a half an electron pair is shared between two Ru-H bonds. Additionally, delocalization index values of the Ru-H bonds are very similar to that of the values computed for Ru(1)-Ru(3) and Ru(2)-Ru(3) bonds, although the topology of both kinds of bonds is quite different as reflected by their local topological parameters presented in (Table 1).

It is interesting to note that the sum of all of the delocalization indices values for the bonding and non-bonding interactions in the Ru(1)- H(1),N(1)-Ru(2)-Ru(3) plane is roughly 3 electron pairs. It can be conclude that the interaction in the core part for this compound exists a five centers six electrons (5c-6e).

3.2.3. Ru-CO Interactions. The mean value of the Ru-CO interactions is presented in (Table 1). This value give an agreement with the literature[31]. The $\rho_{(b)}$ value of this bond is closed to 1.4142 Åe^{-3}, which is higher than metal-metal bonds, but lower than the pure covalent single bonds between non-metal atoms.

The value of δ(M...O_{CO}) is an indicator of the presence or absence of π-back-bonding in M-CO bonds, because the π-back-donation involves significant M...O_{CO} interactions[8]. Hence, the delocalization indices for the δ(M...O_{CO}) interactions were calculated (Table 2) to check the π-bonding in the Ru-CO unit. The average values of δ(Ru...O_{CO}) is 0.1907, and this is within the range of δ(M...O_{CO}) for Ru, Os, Co, Fe, and Ni carbonyl complexes (0.15 - 0.25)[29][21][8]. On the other hand, no π-back-donation was observed [$Cu(CO)_2$]$^+$ [δ(Cu...O_{CO}) = 0.09 for and δ(B...O_{CO}) = 0.04 for

H_3BCO][8]. These results are also confirmed by the topological indices of the C-O bonds calculated in (Table 1), which are consistent with a weak C-O bond compared to a H_3BCO.

3.2.4. Ru-N(1)oxazoline ring and Ru-N(2) Interactions. The values of the topological indices for the Ru(3) atom and N(2) atom of oxazoline ring, $\rho_b(eÅ^{-3})$ = 0.7781, $\nabla^2\rho_b(eÅ^{-5})$ = 3.8768, $G_b(he^{-1})$ = 1.0460, $H_b(he^{-1})$ = -0.0768, V =1.1229, ε_b = 0.0727 (Table 1). These values are similar to the bond between Ru(1) and Ru(2) with N(1) atoms [$\rho_b(eÅ^{-3})$ = 0.7869, $\nabla^2\rho_b(eÅ^{-5})$ = -0.0929, $G_b(he^{-1})$ = 1.0250, $H_b(he^{-1})$ = -0.0961, V= 1.1211, ε_b= 0.0185)]. Importantly, the ellipticity (ε_b) are (0.0727 and 0.0185) for the former and latter Ru-N bonds, these results indicate that the Ru(3)-N(2) bond has a slightly greater double bond character. The delocalization indexes computed for the two Ru-N bonds interactions are very similar, 0.510 for Ru(1)-N(1) and 0.507 for Ru(2)-N(2) and Ru(3)-N(2). These numbers indicate that about 0.5 electron pair is shared between bonded atoms. The valence shell charge concentrations (VSCCs) of both these two nitrogen atoms, as shown in (Fig. 7), are distorted toward their bonded ruthenium atoms.

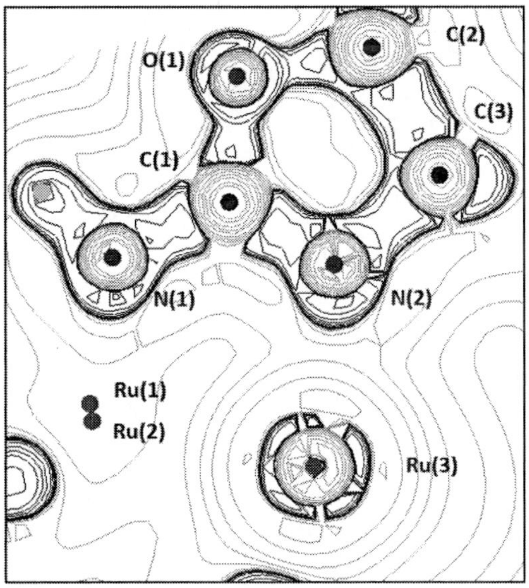

Figure 7: The Laplacian of the electron density in the Ru(1)-N(1), Ru(3)-N(2) and Ru(2)-N(2) atoms.

As discussed above, the value of $\delta(M...O_{CO})$ is a reasonable sign of π-bonding in M-CO bonds. To establish the presence or absence of π-bonding in the Ru(1)-N(1), Ru(2)-N(1) and Ru(3)-N(2) bonds, the delocalization indices for the Ru(1)...C(1,3) and Ru(2,3)...C(1) interactions were calculated (Table 2). These values are very small and nearly equal to the delocalization indices of the Ru(3)...N(1) interactions, where no π-bonding exist. Therefore, the oxazoline ligand bonded to Ru atom behaves as a pure σ-donor.

3.2.5. C-C, C-N and C-O Interactions within oxazoline ring. The gradient trajectory maps as show in (Fig. 8) describe the bcps and bps between the atoms of the oxazoline ligand and their atomic basins. The positions of bcps are not symmetrical, and they are shifted toward the less electronegative atoms C1, C2 and C3. The O(1)-C(1), O(1)-C(2) and N(1)-C(1) bonds nearly equal to the value of topological parameters of N(1)-C(1) bond, the electron density = 3.3848, $\nabla^2\rho_b(eÅ^{-5})$ = -12.6587, $G_b(he^{-1})$ = 1.8135, $H_b(he^{-1})$ = -4.9781, V =6.7916, ε_b = 0.1421. This is comparable with the reported pyridyl ligands. For instance, the N atom and C atom of pyridy ligand, $\rho_b(eÅ^{-3})$ =3.386, $\nabla^2\rho_b(eÅ^{-5})$ = -10.578,

$G_b(he^{-1})$ = 2.203, $H_b(he^{-1})$ = -4.848, V =7.051, ε_b = 0.020 for N-C interactions of $[Os_3(\mu\text{-H})(\mu\text{-}\eta^2\text{-dpa-}N,N)(CO)_{10}]^{**}$. That indicate the O(1)-C(1), O(1)-C(2), N(1)-C(1) and N(2)-C(1) bonds have a relatively smaller double bond character. These eve dances refers to found π-donation from the p-π orbitals of the N and O atoms to the empty p-π orbital of the C of d-π orbitals of metal carbene atom[33]. The topological parameter values of the oxazoline ligand, the highest electron density value, the most negative value of Laplacian, and the highest value of the ellipticity at bond critical points are consistent with the presence of double bond character within the oxazoline ring.

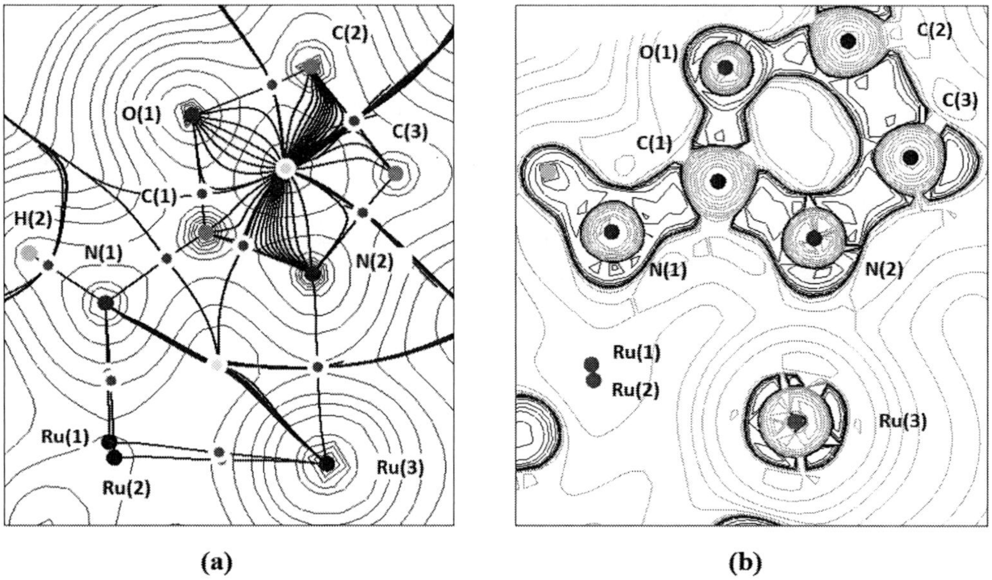

Figure 8. (a) Gradient trajectories mapped (b) Laplacian of the electron density in the oxazoline ring.

4. Conclusions

The chemical bonding in the triruthenium mono-hydride cluster **[Ru$_3$ (μ -H)(μ_3-κ^2-Haminox-*N,N*)(CO)$_9$]** is explored using the Quantum Theory of Atoms-in-Molecules (QTAIM). According to our calculations, we propose a multi-centre (5c–6e) type for the bridged core part (Ru_3H). As expected, the local topological parameters of the unbridged Ru–Ru bonds differ considerably from those of the bridged Ru…Ru bond, for which a bond path was not observed. A non-negligible delocalization indices δ(Ru...Ru) was observed for the non-bonding interactions between Ru(1) and Ru(2). Topological parameters of Ru-N support for the presence of a σ-bond. In the oxazoline ligands, the analysis of the topological parameters of their bonds confirmed the existence of π-electron delocalization within the oxazoline ring.

The metal–metal and metal–ligand bond critical points properties and bond delocalization indices δ(A, B) are aligned with the data from the previous studies on Os, Ru, Co, Fe, and Ni carbonyl complexes[29][21][8].

References

[1] F. a. Cotton and C. B. Harris, 1965 "The Crystal and Molecular Structure of Dipotassium Octachlorodirhenate(III) Dihydrate, $K_2[Re_2Cl_8]2H_2O$," *Inorg. Chem.* **4(3)** 330–333.

[2] M. Niskanen, P. Hirva, and M. Haukka, 2009 "Computational DFT study of ruthenium tetracarbonyl polymer," *J. Chem. Theory Comput.* **5(4)** 1084–1090.

[3] J. F. Van Der Maelen and J. A. Cabeza, 2012 "QTAIM analysis of the bonding in Mo − Mo bonded dimolybdenum complexes," *Inorg. Chem.* **51(13)** 7384–7391.

[4] J. Overgaard, H. F. Clausen, J. A. Platts, and B. B. Iversen, 2008 "Experimental and theoretical charge density study of chemical bonding in a Co dimer complex," *J. Am. Chem. Soc.* **130(12)** 3834–3843.

[5] L. J. Farrugia, C. Evans, and M. Tegel, 2006 "Chemical bonds without 'chemical bonding'? A combined experimental and theoretical charge density study on an iron trimethylenemethane complex.," *J. Phys. Chem. A* **110(25)** 7952–7961.

[6] L. J. Farrugia, C. Evans, D. Lentz, and M. Roemer, 2009 "The QTAIM Approach to Chemical Bonding Between Transition Metals and Carbocyclic Rings: A Combined Experimental and Theoretical The QTAIM approach to chemical bonding between transition metals and carbocyclic rings: a combined experimental and theoretical," *J. Am. Chem. Soc.* **131(3)** 1251–1268.

[7] G. Gervasio, D. Marabello, R. Bianchi, and A. Forni, 2010 "Detection of Weak Intramolecular Interactions in $Ru_3(CO)_{12}$ by Topological Analysis of Charge Density Distributions," *J. Phys. Chem. A* **114(34)** 9368–9373.

[8] P. Macchi and A. Sironi, 2003 "Chemical bonding in transition metal carbonyl clusters: complementary analysis of theoretical and experimental electron densities.," *Coord. Chem. Rev.* **238–239** 383–412.

[9] C. Bo, J. P. Sarasa, and J. M. Poblet, 1993 "Laplacian of charge density for binuclear complexes: terminal vs bridging carbonyls," *J. Phys. Chem.* **97(24)** 6362–6366.

[10] A. A. Low, K. L. Kunze, P. J. MacDougall, and M. B. Hall, 1991 "Nature of metal-metal interactions in systems with bridging ligands. 1. Electronic structure and bonding in octacarbonyldicobalt," *Inorg. Chem.* **30** 1079–1086.

[11] R. Bianchi, G. Gervasio, and D. Marabello, 2000 "Experimental electron density analysis of Mn2(CO)10: Metal-metal and metal-ligand bond characterization," *Inorg. Chem.* **39(11)**, 2360–2366.

[12] C. Gatti and D. Lasi, 2007 "Source function description of metal-metal bonding in d-block organometallic compounds.," *Faraday Discuss.* **135** 55-78-149, 503–506.

[13] G. Gervasio, R. Bianchi, and D. Marabello, 2005 "Unexpected intramolecular interactions in Ru3(CO)12: An experimental charge density study at 120 K," *Chem. Phys. Lett.* **407(1–3)** 18–22.

[14] P. Macchi, L. Garlaschelli, S. Martinengo, and A. Sironi, 1999 "Charge density in transition metal clusters: supported vs unsupported Metal–Metal interactions," *J. Am. Chem. Soc.* **121** 10428–10429.

[15] M. Feliz, R. Llusar, J. Andrés, S. Berski, and B. Silvi, 2002 "Topological analysis of the bonds in incomplete cuboidal [Mo3S4] clusters," *New J. Chem.* **26(7)** 844–850.

[16] I. Bytheway, C. S. Griffith, G. Koutsantonis, B. W. Skelton, and A. H. White, 2007 "A linear Ru-Tl-Ru complex obtained from halide abstraction: An example of metal-dative bonding," *Eur. J. Inorg. Chem.* **2007(20)** 3240–3246.

[17] L. J. Farrugia, C. Evans, D. Lentz, and M. Roemer, 2009 "The QTAIM approach to chemical bonding between transition metals and carbocyclic rings: A Combined experimental and theoretical study of (η5-C5H5)Mn(CO)3, (η6-C6H6)Cr(CO)3, and (E)-{(η5-C5H4)CF=CF(η5- C5H4)}(η5-C," *J. Am. Chem. Soc.* **131(3)** 1251–1258.

[18] J. F. Van der Maelen et al., 2007 "Experimental and theoretical characterization of the Zn-Zn bond in [Zn2(eta5-C5Me5)2].," *Acta Crystallogr. B.* **63(Pt 6)** 862–868.

[19] P. Macchi, D. M. Proserpio, and A. Sironi, 1998 "Experimental electron density in a transition metal dimer: Metal-metal and metal-ligand bonds," *J. Am. Chem. Soc.,* **120(51)** 13429–13435.

[20] P. Macchi, L. Garlaschelli, and A. Sironi, 2002 "Electron density of semi-bridging carbonyls. Metamorphosis of CO ligands observed via experimental and theoretical investigations on [FeCo(CO)8]-," *J. Am. Chem. Soc.* **124(47)** 14173–14184.

[21] J. A. Cabeza, J. F. Van der Maelen, and S. García-Granda, 2009 "Topological Analysis of the

IOP Conf. Series: Journal of Physics: Conf. Series **1032** (2018) 012068 doi:10.1088/1742-6596/1032/1/012068

Electron Density in the N-Heterocyclic Carbene Triruthenium Cluster [Ru 3 (μ-H) 2 (μ 3 - MeImCH)(CO) 9] (Me 2 Im = 1,3-dimethylimidazol-2-ylidene)," *Organometallics*, **28(13)** 3666–3672.

[22] J. A. Cabeza, I. Da Silva, I. Del Río, R. A. Gossage, D. Miguel, and M. Suárez, 2006 "Triruthenium carbonyl clusters derived from chiral aminooxazolines: synthesis and catalytic activity.," *Dalton Trans.* **3(20)** 2450–2455.

[23] M. J. Frisch et al., 2004 "*Gaussian 03*." Wallingford, CT.

[24] M. J. Frisch et al., 2009 "*Gaussian 09*." Wallingford CT.

[25] F. Biegler-König and J. Schönbohm, 2002 "*AIM2000." Büro für Innovative Software.*

[26] W. J. Hehre, R. Ditchfield, and J. A. Pople, 1972 "Self—consistent molecular orbital methods. XII. further extensions of Gaussian—type basis sets for use in molecular orbital studies of organic molecules," *J. Chem. Phys.* **56(5)** 2257–2261.

[27] C. Adamo and V. Barone, 1999 "Toward reliable density functional methods without adjustable parameters: The PBE0 model," *J. Chem. Phys.* **110(13)** 6158.

[28] R. F. W. Bader, 1998 "A Bond Path: A Universal Indicator of Bonded Interactions," *J. Phys. Chem. A* **102** 7314–7323.

[29] J. F. Van Der Maelen, S. García-granda, and J. A. Cabeza, 2011 "Theoretical topological analysis of the electron density in a series of triosmium carbonyl clusters: [Os3(CO)12], [Os3(μ-H)2(CO)10], [Os3(μ-H)(μ-OH)(CO)10] and [Os3(μ-H)(μ-Cl)(CO)10]," *Comput. Theor. Chem.* **968(1–3)** 55–63.

[30] L. J. Farrugia and H. M. Senn, 2010 "Metal-Metal and metal-ligand bonding at a QTAIM catastrophe: A combined experimental and theoretical charge density study on the alkylidyne cluster Fe_3(μ-H)(μ-COMe)$(CO)_{10}$," *J. Phys. Chem. A* **114(51)** 13418–13433.

[31] P. Macchi, D. Donghi, and A. Sironi, 2005 "The electron density of bridging hydrides observed via experimental and theoretical investigations on [Cr2(μ2-H)(Co) 10]-," *J. Am. Chem. Soc.*, **127(47)** 16494–16504, 2005.

[32] C. Gatti, 2005 "Chemical bonding in crystals: new directions," *Zeitschrift für Krist.* **220(5-6)** 399–457.

[33] F. E. Hahn and M. C. Jahnke, 2008 "Heterocyclic carbenes: synthesis and coordination chemistry," *Angew. Chemie - Int. Ed.* **47(17)** 3122–3172.

IOP Conf. Series: Journal of Physics: Conf. Series **1032** (2018) 012069 doi:10.1088/1742-6596/1032/1/012069

Comparison of mixed Ultrasound waves with the extracts of *Nerium oleander* L. on the growth of cyprid barnacles *Balanus amphitrite Amphitrite*

H M Kredy*, M E Aldokheily and I H Khlewy
College of Science - University of Thi-Qar, Thi-Qar, Iraq.
Corresponding author: hmk20001999@gmail.com,

Abstract. This study was conducted to demonstrate the effect of plant extracts from the *Nerium* oleander L. plant and ultrasound waves of different frequencies (10-50) (20-45) (50-60) (30-65) KHz on the barnacle larvae cyprid of barnacle *Balanus amphitrite amphitrite* obtained from Nasiriya Thermal Power Station inlet water. Samples were collected bimonthly three replicates. The results of the Oleandrin 1 and 2.5 mg/500 ml showed a an effect of losing the larvae cyprid, 2.5 mg showed an increasing effect on the seventh day. A maximum losing of larvae cyprid was attained when using ultrasonic with conjunction of 2.5 mg Oleandrin extract reached 100% on the tenth day

1. Introduction

Barnacles are marine dominant fouling organisms which accumulate on ship hulls, piers ,and under water structures. Biofouling by these invertebrates results in an increase in the power and fuel consumption [(1)]. Figure 1 shows a colony of barnacles at water intakes of Nasiriya Province. The growth of barnacles proliferation at port pipes that carry river water to condenser power plants, cause the blockage and closing of those tubes. This will lead to the deterioration in the production of electrical energy in the power plants of Al-Nasiriya power station in the province of Thi-Qar, southern Iraq, leading to losses in the energy production which in most cases exceeds 200MW.

Some technologies were used to deter fouling on stationary objects or niche areas and these include sound[(2)3)(4)],vibration[(5)], Sonication[(6)], electricity[(7)(8)], microwave[(9)] and aeration[(10)(11)]. Moreover, Pulsed electric fields (PEF) on planar-interdigitated electrodes (IDE) have been investigated to deter Pseudomonas aeruginosa biofilms [(7)]. Barnacle cyprids are organisms in aquatic biofouling, they search for a substratum for settlement and metamorphosis [(12)]. As a result of their broad distribution in coastal and estuarine regions [(13)], and their prominence as a fouling species in warm waters worldwide [(14)].

IOP Conf. Series: Journal of Physics: Conf. Series **1032** (2018) 012069 doi:10.1088/1742-6596/1032/1/012069

Figure 1: Growth of Barnacles Colonies at water intakes Nasiriya Province

The aim of the present paper was to study the effects of ultrasound in inhibition of the growth of cyprid larvae *B. a. amphitrite* and comparison of mixed ultrasound effects with those of extracts of *Nerium oleander*.

2. Experimental Design

2.1 Description of the study area

Nasiriya thermal generation power station in southern Iraq is located on the Euphrates river, and just 7.34 km from the city centre, Which is located °31 2'18."58 North and °46 11'25."66 east of the city centre. The station consists four main units of electric power generation per unit capacity of 210 megawatts / hour, and although of the total capacity of 840 megawatts / hour when the build, but the total production at the present time is between (500-550) megawatt / hour due to, among other thing, growth of barnacles.

The samples were collected from the cooling water processing (C.W.Ps). It is an iron platform containing concrete ponds, down the platform includes a set of pumps needed for water to cool the thermal condensate per unit of electrical energy generation as well as a number of filters (rough, soft, swivel) that operate on the incoming water purification through massive pipes to capacitors. Barnacles settle colonies on the concrete surfaces of the platform as well as the previously mentioned rail parts,

IOP Conf. Series: Journal of Physics: Conf. Series **1032** (2018) 012069 doi:10.1088/1742-6596/1032/1/012069

causing clogging of the pipes and block the cooling water taken directly from the river. In addition, some barnacles penetrate through the water pipes to heat exchangers causing the deterioration of their efficiency in thermal exchanger.

Figure 2: Sample collection site

2.2 Extraction of Oleandrin

Leaves of plant *Nerium oleander* were collected, atmospheric dried samples were ground to fine powder with the aid of a mechanical grinder and sieved through coarse 2.5μm mesh. Then 100g of powder was extracted with 150 ml of methanol to water (90:10) using a homogenizer for 5 minutes. The mixture was then filtered through Whitman No.1 filter paper. The sample was re-extracted following the same above procedure for 4 hour many times. The extracts were collected, and evaporated by rotary evaporator to 10 ml, the extract was stored in a refrigerator until analysis [(15)]. Ethyl acetate as an intermediate polar solvent to extract Oleandrin through the alcoholic extract after dissolve in water and mixing it with ethyl acetate. A small amount of ammonium chloride NH_4Cl added to prevent emulsifying phenomenon, then separating funnel left for a period to be separated into two layers. The organic layer was collected, concentrated by a rotary evaporator with controlling the temperature which should not exceed 50 ^{o}C [(16)].

2.3 Treatment with Oleandrin

The concentrations 0, 1, and 2.5 mg /500 ml of oleandrin extract, were used in three replicates. After the collection river water samples, filtered using a sieve of a diameter of 2.5μm to isolate the Cyprid larvae from the rest of the pond. placed in a container with a volume of 1000 ml. 500 ml of river water was added to the container. Cyprid larvae were monitored during the larvae control period (10 days). The percentage of inhibition was observed daily by utilizing a microscope MT4300L.

2.4 Treatment with ultrasound waves and Oleandrin

To achieve this purpose, the following devices were used:(i) New dual speaker Ultrasound pest Repeller CO-350902 (30-65KHz), (ii) Solar Ultrasound Repeller AN-B030 (10-50KHz). Water samples were subjected in outlined addition to the extract in 2.3 and specifically with a concentration of 2.5 mg/500 ml for the ultrasonic waves source of frequency (30-65, 10-50KHz) power (2watt). Ultrasound device were located at a distance of 5cm from the container in conjunction with the oleandrin extract, except for the control sample. Cyprid larvae were monitored within 10 days. The percentage of inhibition was observed daily by utilizing a microscope MT4300L.

2.5 Treatment with ultrasound waves and mixed ultrasound with Oleandrin

To perform this purpose, the following devices were used (i) Solar Ultrasound Repeller AN-B030 (10-50KHz), (ii) LW-11DS automatic sweep Ultrasound Repeller YL-35-090200D (20-45KHz). Ultrasound devices were located at a distance of 5cm from the container except for the control sample were subjected to the mixed effect of the extract and Ultrasound. Cyprid larvae were monitored within 10 days of the experiment. The percentage of in inhibition was observed daily using a microscope.

The following devices were also used to perform the experiment:

(i) New dual speaker Ultrasound pest Repeller CO-350902 (30-65KHz).

(ii) Double wave pest Repeller SS-788 (50-60KHz).

Ultrasound devices were located at a distance of 5 cm from the container in conjunction with the Oleandrin extract (2.5 mg/500 ml) except for the control sample. cyprid larvae were also monitored within 10 days of the experiment.

2.6 Statistical Analysis

Statistical analyses were conducted using social mediated statistical program SPSS (version 19) and using an ANOVA test to find the least significant difference L.S.D below $p \leq 0.05$ probabilities of a comparison between the plant extracts and ultrasound and duration of exposure to the treatment. Death percentage of cyprid is corrected according to the Abbott equation knowledge by Schneider and Orell Formula as following as[17]:

$$\text{death percentage of Cyprid} = \frac{\text{\% death percentage in the treated sample} - \text{\%deathe percentage in control sample}}{100 - \text{\% deathe percentage in control sample}} \times 100\%$$

3. Results and Discussion

The collected water samples had cyprid larvae, these larvae are shown in Figure 3

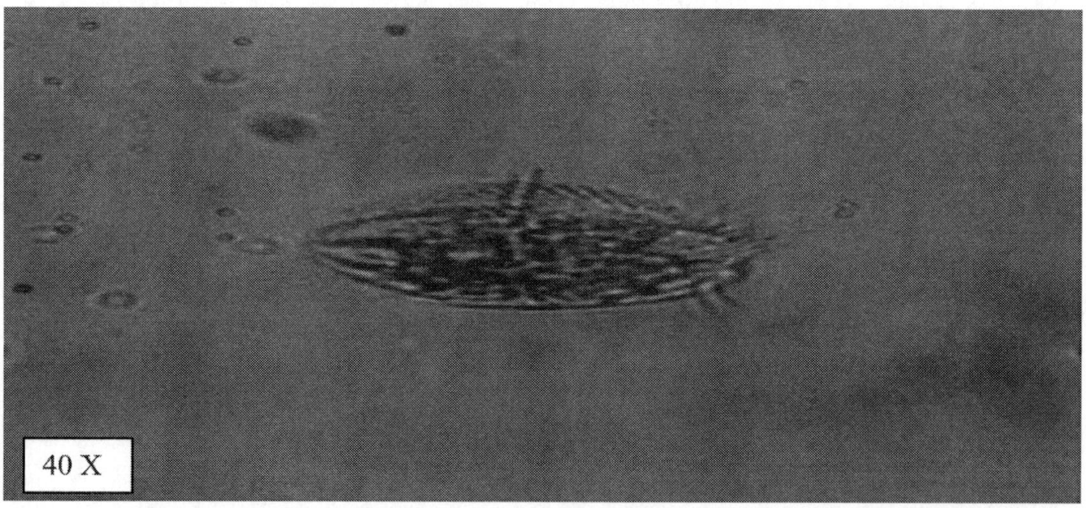

Figure 3: Cyprid larva in the samples collected from the Euphrates river in Nasiriya, south of Iraq.

IOP Conf. Series: Journal of Physics: Conf. Series **1032** (2018) 012069 doi:10.1088/1742-6596/1032/1/012069

3.1 Impact of treatment with Oleandrin

The results listed in Table 1 and Figure 4 illustrate the effective compound extract of *Nerium* leaves had an impact in the destruction of the larvae cyprid of barnacle *B.a.amphitrite*. The results show that there is no significant difference between the working groups appeared in a concentration of 1 mg/500ml compared with the control group during the first day of the study. The highest percentage of the loss of the larvae cyprid was on Day 10. The loss was 12 larva out of 12 larva, and consequently the rate of destruction was 100% compared with the control group in which the loss percentage of 2 larva out of 12 larva and the rate of loss of 16.6% reached.

The results showed that the lowest rate of destruction of the larvae was on first day of the experiment. The loss percentage of cyprid larva was one out of 12 larva and therefore the percentage of loss of 8.3% when using the concentration of 2.5 mg/500ml compared with the control group that did not experience any percentage of the loss of larvae under those circumstances.

The percentage of the loss of larvae cyprid of barnacle *B.a.amphitrite* increased as the concentration of active compound oleandrin extract increases. Table 1 illustrates that the concentration of 2.5 mg/500ml caused the highest percentage of loss of larvae and the differences are significant compared to the concentration of 1 mg/500ml. Moreover, the table shows that the loss percentage of the larvae spread was 8 larvae and the percentage of loss is 59.9% on seventh day.

Table 1: Measurements of Oleandrin impact on the Inhibition on the growth of cyprid barnacles *B.a.amphitrite* (the total number of larvae = 12).

Time (day)	Conc.(mg/500ml)		
	1	2.5	Control
1	0.00±0.00	1.00±0.00	0.00±0.00
2	1.00±0.00	2.00±1.73	0.00±0.00
3	2.00±1.73	2.00±1.73	0.00±0.00
4	4.00±1.73	4.00±3.46	0.00±0.00
5	4.00±1.73	6.00±0.00	0.00±0.00
6	5.00±1.73	8.00±1.73	1.00±0.00
7	5.00±1.73	8.00±1.73	2.00±1.73
8	7.00±1.73	9.00±0.00	2.00±1.73
9	8.00±1.73	10.00±1.73	2.00±1.73
10	9.00±0.00	12.00±0.00	2.00±1.73
LSD	3.51		

* Number of larvae ± Standard deviation

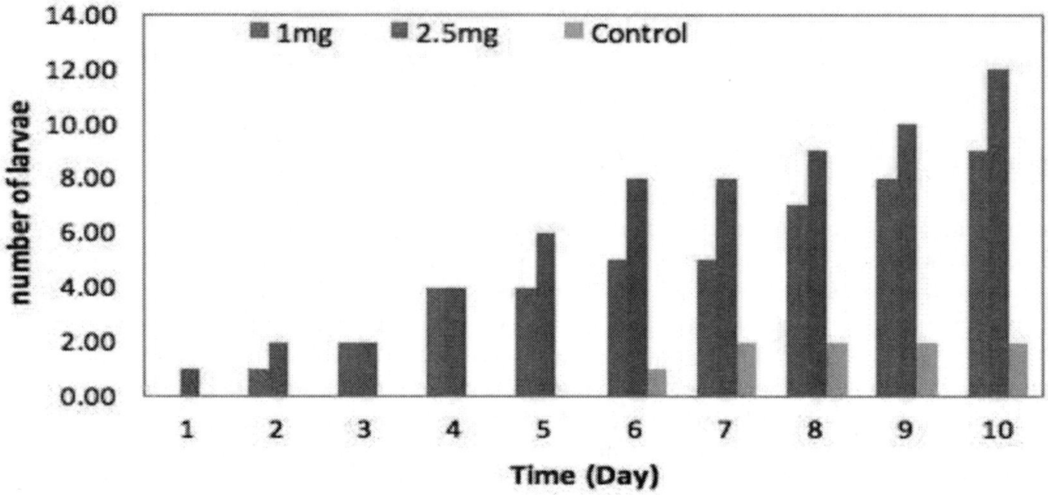

Figure 3.1: The effect of mixed time and concentrations of Oleandrin extract on the growth of cyprid barnacles *B.a.amphitrite*.

3.2 Effect of mixed ultrasound and Oleandrin

The results listed in Table 2and Figure 3 illustrate the ultrasound and the extract effective composite plant oleander had an impact in the destruction of larvae cyprid of barnacle *B.a.amphitrite*, the result show that is a significant difference between the working groups appeared the in a concentration of 2.5 mg /500ml compared with the control group during the first day of the study. The highest a percentage of the loss of larvae cyprid was on tenth day.The loss was 12 larva out of 12 larva , and consequently the rate of destruction was 100% compared with the control group.

The results showed that the lowest rate of destruction of the larvae was on Day one of the experiment. The loss percentage of cyprid larva was one out of 12 larva and therefore the percentage of loss of 8.3% when using the concentration of 2.5 mg/500ml compared with the control group that did not experience any percentage of the loss of larvae under those circumstances.

The results showed that the highest rate of destruction of the larvae was on tenth day of the study. The loss percentage of cyprid larva was 12 out of 12 larva and therefore the percentage of loss of 100% when using the concentration of 2.5 mg/500ml and ultrasound waves compared with the control group that did not experience any percentage of the loss of larvae under those circumstances.

This is a unique study of its kind, is considered as the use of ultrasound and the concentration of 2.5 mg / ml of the active compound (Oleandrin) and were compared between the two separately, and in other studies used different frequencies of ultrasound (70 to 445 Hz), the Water basins contain on the larvae. It was observed to increase the vibration amplitude and a decrease in the number of marine barnacles conjoined at the frequency of 260 Hertz [18] [19].

Table 2: Effect of mixed ultrasound and oleandrin extract on the growth of cyprid barnacles *B.a.amphitrite* (the total number of larvae = 12).

IOP Conf. Series: Journal of Physics: Conf. Series **1032** (2018) 012069 doi:10.1088/1742-6596/1032/1/012069

Time (day)	Treatment		
	Ultrasound waves	Oleandrin extract (2.5mg/500ml)	Control
1	1.00±0.00	0.00±0.00	0.00±0.00
2	2.00±1.73	1.00±0.00	0.00±0.00
3	3.00±0.00	3.00±0.00	0.00±0.00
4	5.00±1.73	5.00±1.73	0.00±0.00
5	6.00±0.00	5.00±1.73	0.00±0.00
6	6.00±0.00	6.00±5.20	0.00±0.00
7	9.00±0.00	8.00±3.46	0.00±0.00
8	11.00±1.73	9.00±3.00	0.00±0.00
9	11.00±1.73	10.00±3.46	0.00±0.00
10	12.00±0.00	12.00±0.00	0.00±0.00
LSD	3.46		

* Number of larvae ± Standard deviation

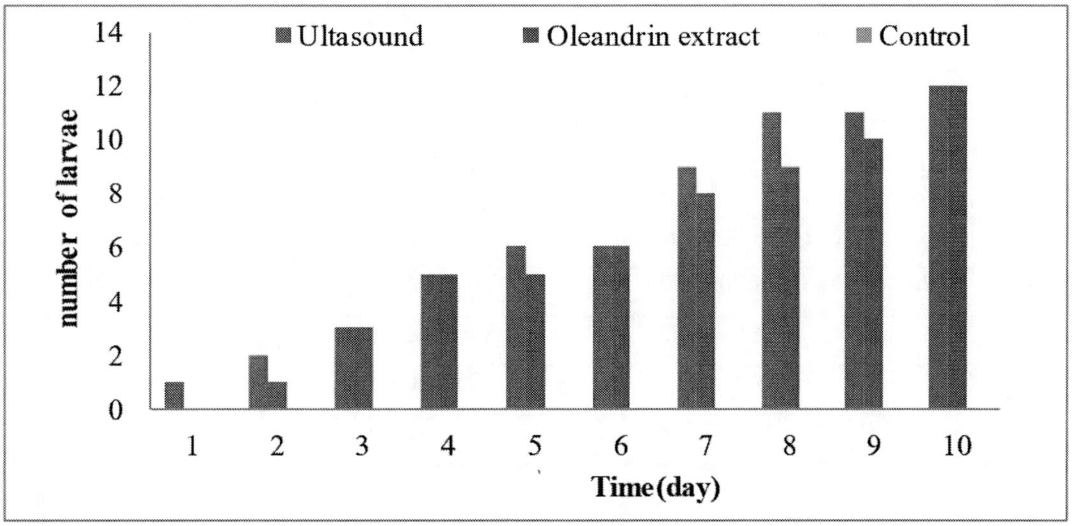

Figure 3.2 : Effect of mixed time , ultrasound and Oleandrin extract on the growth of cyprid larvae of barnacle *B.a.amphitrite*

3.3 Effect of ultrasound waves and mixed ultrasound waves with Oleandrin.

The results listed in Table 3 and Figure 4 illustrate that the ultrasound waves and the mixture (ultrasound + active compound extract) had an impact in the destruction of the larvae cyprid of barnacle *B.a.amphitrite*. The result also showed that there is no significant difference between the working groups compared with the control group during the first day of the study, as well as, on second day of the working group using a mixture of ultrasound and effective compound, there were no significant differences compared with the control group.

The results showed that the highest rate of destruction of the larvae was on tenth day of the study. The loss percentage of cyprid larva was 12 out of 12 larva, and therefore, the percentage of loss was 100% when using the ultrasound waves and mixed compared with the control group that did not experience any percentage of the loss of larvae under those circumstances. Moreover, the loss percentage of cyprid larva was 12 out of the 12 larva at Day nine of the study.

In this study, the ultrasound waves were used with low frequencies and a mixture of these waves with the oleandrian extractor. As shown in Table 3 when only ultrasound waves were used, the 100% of destruction occurred on tenth day while when the mixture of ultrasound waves and the oleandrian extract were used, the 100% destruction was on Day 9. The use of low frequency (30 Hz) was studied before and there was no destruction of the larvae, but when the waves were applied to the basins full with larvae; it was noted that there was a small and insignificant lack in the number of larvae [(9)].

Table 3: Efeect of ultrasound waves and mixed ultrasound waves with Oleandrin extract at the same experience on the growth of cyprid barnacles *B.a.amphitrite* (The total number of larvae = 12).

Time (day)	Treatment		
	Ultrasound waves	Ultrasound waves and Oleandrin extract (2.5mg/500ml)	Control
1	0.00±0.00	0.00±0.00	0.00±0.00
2	1.00±0.00	0.00±0.00	0.00±0.00
3	2.00±1.73	2.00±1.73	0.00±0.00
4	4.00±1.73	5.00±1.73	0.00±0.00
5	5.00±1.73	6.00±0.00	0.00±0.00
6	6.00±0.00	8.00±1.73	0.00±0.00
7	8.00±1.73	10.00±1.73	0.00±0.00
8	8.00±1.73	11.00±1.73	0.00±0.00
9	10.00±1.73	12.00±0.00	0.00±0.00
10	12.00±0.00	12.00±0.00	0.00±0.00
LSD	2.37		

* Number of larvae ± Standard deviation

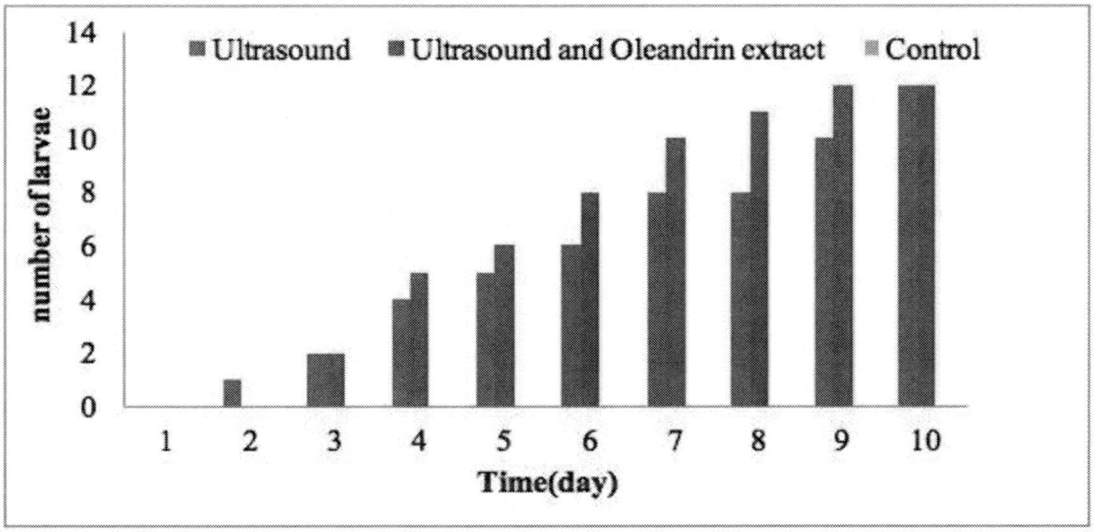

Figure 3.3: The effect of time, ultrasound waves and a combination of ultrasound waves and extract the effective compound of the oleander plant in the inhibition of the barnacle larvae cyprid *B.a.amphitrite*.

4. Conclusion

In this study, the impact of using a mixture of extracts of the oleander plant (Oleandrin) and ultrasound waves of different on the cyprid larvae *Balanus amphitrite amphitrite* which grow on Al-Nasiriya Power Station inlet water pipes was investigated. The rate of losing the larvae cyprid ws 100% when using ultrasound with Oleandrin extract of the oleander plant at different frequencies,and this indicates that the ultrasound is a perfect choice with the plant extracts in the inhibition larvae cyprid.

Reference

[1] Holm E R, 2012 Barnacles and biofouling. *Integr Comp Biol.* **(52)348–355**

[2] Rittschof D, Forward RB, Cannon G, Welch J M, McClary M, Holm E.R, Clare A.S, Cues and context, 1998 larval responses to physical and chemical cues. Biofouling. **(12)31–44** Guo S.F, Lee H.P, Chaw K.C, Miklas J, Teo S.L.M, Dickinson G.H, Birch W.R and Khoo B.C, 2011 Effect of ultrasound on cyprids and juvenile barnacles. *Biofouling*. **(27)185–192**

[3] Kopel M, Degtyar E and Banin E, 2011 Surface acoustic waves increase the susceptibility of Pseudomonas aeruginosa biofilms to antibiotic treatment. *Biofouling*. **(27)701–711** Nakagawa N, Fujihara M, Wu C and Satonobu J, 2006 Removal of pipe fouling inside pipes using ultrasonic waves. *JSME Int J Ser C.* **(49)713–718**

[4] Gavand M.R, Mc Clintock J.B, Amsler C.D., Peters R.W and Angus R.A 2007 Effects of sonication and advanced chemical oxidants on the unicellular green alga Dunaliella tertiolecta

and cysts, larvae and adults of the brine shrim Artemia salina: a prospective treatment to eradicate invasive organisms from ballast water. *Mar Pollut Bull.* **(54)1777–1788**

[7] Perez-Roa R.E, Anderson M.A, Rittschof D, Orihuela B, Wendt D.E, Kowalke G.L and Noguera D.R, 2008 Inhibition of barnacle (*Amphibalanus amphitrite*) cyprid settlement by means of localized, pulsed electric fields.*Biofouling. (*24**) 177–184**

[8] Perez-Roa R.E, Tompkins D.T, Paulose M, Grimes C.A, Anderson M.A and Noguera D.R, 2006 Effects of localised, low-voltage pulsed electric fields on the development and inhibition of Pseudomonas aeruginosa biofilms. *Biofouling*. **(22)383–390**

[9] Boldor D, Balasubramanian S, Purohit S and Rusch K.A, 2008 Design and implementation of a continuous microwave heating system for ballast water treatment. *Environ Sci. Technol.* **(42)4121–4127**

[10] Scardino A.J, Fletcher L.E and Lewis J.A, 2009 Fouling control using air bubble curtains: protection for stationary vessels. *J Mar Eng. Tech*. **(13)3–10**

[11] Bullard S.G, Shumway S.E and Davis C.V, 2010 The use of aeration as a simple and environmentally sound means to prevent biofouling. *Biofouling*.**(26) 587–593**

[12] Aldred, N. and Clare A.S, 2008 The adhesive strategies of cyprids and development of barnacle-resistant marine coatings. *Biofouling.* **(24)351–363**

[13] Mare chal J.P and Hellio C, 2011 Antifouling activity against barnacle cypris larvae: do target species matter (*Amphibalanus amphitrite versus Semibalanus balanoides*) Int. *Biodeterior. Biodegrade*. **(65) 92–101**

[14] Clare A and Aldred N, 2009 Surface colonization by marine organisms and its impact on antifouling research. In Advances in marine antifouling coatings and technologies (eds C Hellio, DMY Yebra) *Cambridge, UK. Woodshead Publishing*, **pp.46–79.**

[15] Mizhir, A. H, 2013 Influence of copper on larval and adult phases densities of the barnacle *Balanus amphitrete amphitrite* on the cooling system water supply of Al-Najibiyah thermal power plant / Iraq. *Journal of Purity, Utility Reaction and Environment*.**2(4) 85-91**

[16] Ghaleb Tayoub, et al, 2014 Analysis of Oleandrin in Oleander Extract *Nerium oleander*. **(7)73-78**

[17] Abbott W. S, 1987 A method of computing the effectiveness of an insecticide. *Journal of the American mosquito control association*.**3(2):302-303**

[18] Choi C.H, Scardino A.J, Dylejko P.G, Fletcher L.E and Juniper R, 2013 The effect of vibration frequency and amplitude on biofouling deterrence, Biofouling: The *Journal of Bio adhesion and Biofilm Research*.**29(2) :195-202**

[19] Shi Feng, G and Jason M, 2011 Effect of ultrasound on cyprids and juvenile barnacle. **27(2):185–192**

IOP Conf. Series: Journal of Physics: Conf. Series **1032** (2018) 012070 doi:10.1088/1742-6596/1032/1/012070

Determination of antimicrobial and antioxidants activity of Salvia przewalskii seed oil against pathogenic bacteria and fungi

M Alsirrag[1] and R Ali[2]
[1] University of Karbala, Karbala, Iraq
[2]University of Kufa, Najaf, Iraq
Dr_manal@yahoo.com
riyadh.alhabeeb@uokufa.edu.iq

Abstract. In this study antimicrobial and antioxidant effect of methanolic and aqueous extracts of Salvia przewalskii seed oil were tested against four strain of bacteria (*Streptococcus bovis, Proteus mirabillis, Enterococcus gallinarum , Escherichia coli*) and two strain of fungi (*Rhizopus oryzae and Rhizopus stolonifer*) . Aqueous and methanolic extracts of Salvia przewalskii seed oil were prepared by solvent extraction, the results showed the hydro distillation of seed parts of *S. przewalskii in vitro* yellow oils in dry weight (0.3% and 0.5 % (v/w; ml 50 g) products, antibacterial data showed higher ability towards E.coli (26.67±0.89) compared with standard streptomycin (22.78±0.49) and highest inhibition zone measured 25 mm . as well as methanol extract at (200mg/ml) recorded highest antioxidant ability (86.87±0.76) and DPPH scavenging activity at 20 μl/ml (89.56±0.47).

Key words: *Salvia przewalskii*, antioxidant, antimicrobial.

1.Introduction

Salvia or sage its large genus includes many type of species about 1000, its has wide uses in Pharmaceutical and many of artificial or commercial industries includes the production of fixed oils and additives substances. *Salvia* oils isolated by vacuum distillation from the plant's dried aerial parts haveFeatured biological ability and medicinal properties. its oils contain different type of vital compounds, mainly include polyphenols hydrocarbons, tanines, penzohydrocarbones, carmophenoles (1).

The fixed oils were is derived from upper part of the plant have a high spectrum of applications: antimicrobal, anti-inflammatory, antioxidant, antiviral and anticoagulant.(2,3). An fixed oils is aconcentrated hydrophobic contain volatile aromatic liquids obtained from plant substances .

Fixed oils are substances extracted by several ways using steam. Its contain fatty acids and its esters manufacteared by solvent extraction, hot and cold pressing. And it is utilize in perfume , cosmetic and several substances, for flavouring edible and softers , and cleaning materials. (4). Sage is perennial plant included in the *Salvia* species, belongs to the Labiatea family. It is known as sage and garden sage. Medical and culinary use of this medicinal herbs belong to ancient times. Sage has antimicrobial, antioxidant, antiviral, and immunosuppressive effects, therefore its had medical and aromatic usage is important In addition to this, sage is used traditionally for different purposes, like as preservation of natural and processed food, as a sweetener, for food coloring some of the aromatic compounds of saliva is derived from this genus hase shown higest antibacterial ability, antiflamntry and anti-mutagenic effect of saliva is often related to the presence of active compounds such as: carnosic acid, carnosol, rosmarinic acid and camphor

IOP Conf. Series: Journal of Physics: Conf. Series **1032** (2018) 012070 doi:10.1088/1742-6596/1032/1/012070

(5). the fixed oils of some *Salvia* species have recently been investigated, showing strong antimicrobial ability. (6) demonstrated that *S. chloroleuca* oil exhibited moderate to high antimicrobial ability especially for *Bacillus subtilis*, *Staphylococcus epidermidis and S. bovis* with inhibition zones of 21, 19, 15 mm Some polyphenolic extracts from *Salvia* were examined by (7) for antioxidant ability in correlation with their polyphenolic content. Haigh active free radicals present in vital systems derived from tissue metabolic activities (8), more researches revealed that the polyphenolic extracts had highest chelating ability towards DPPH_ and ABTS_+ radicals and protected from hydroxyl and peroxyl radical-induced DNA damage.
DPPH method is an antioxidant test deepened on chemical action between ion indicator produces acolorant agent in solvent . This free ion fixed at room conditions, it changed according to an inhibition substances , gave high change colorless in extracts solution. The benefit of this methods provides simple and rapid methods to determinate inhibatory radicals by colorimetry so it can be perfect assay to show different products at experiment time (9). The aime of this study was to evaluate the antioxidant potential and antibacterial capability of the fixed oil extract from *Salvia przewalskii* seed.

2.Material and methods

Collection of *Salvia przewalskii* seed
Salvia przewalskii seed were collected from the local market (herbalists) in 2017 at Kerbala city, it was cleaned of impurities and grind and preserved at sterilized bottle in can sealed and it was stored in the refrigerator.

Preparation of fixed oil:
The air –dried and ground leaves collected and was submitted to water 1:3 distillation for 3hours using a Clevenger - the obtained essential oil was dried over anhydrous barium sulphate and after filtration it was kept in dark -18C° in glass vial until use the yield was 2ml /100g dry leaves .(10).

Preparation extract:
Alcoholic extract
(40 g) of dried powder of saliva extracted by (220) ml 80% methanol overnight and filtered using vacuum pump The residue was re extracted 3-4 times with methanol until it was exhausted . Methanol evaporated using vacuum rotary at 50C .The methanol-free extract was dried using lyophilize and keep it at dark condition until use(11).
Water extract:
100g of salvia ground samples were extracted by soaking in cold distilled water (1:4 w/v) and homogenized for 3 min at 3000 rpm, The mixture was filtered through cheesecloth and centrifuged for 15 min at 2000 rpm ,The supernatant was filtered through Whatman No1 and dried by lyophilized and kept in dark at 4°C.(12)

The antioxidant evaluation tests of Saliva extract
1- DPPH Scavenging experiment:
The free radical scavenging of various extracts was performed according to (13) 1.5 ml of (0.2 mM) solution of DPPH mixed with 3ml of each concentration (5-200 mg/ml)for the each solvent extract and (5-20µl/mol) for the fixed oil .the absorbent of DPPH reagent was read by spectrophotometer at 517nm after 15 min of keeping in incubator at 25C.
Inhibition percent (IP) % = $(A_{blank} - A_{sample} / A_{blank}) \times 100$.

2- Test of ABTS :

Antioxidant by this method measured by using 2ABTS test free radical ABTS dissoleved in distilled cold water reached 7 mM concentrate after this added of 1.45mM sodium persulfate the mixture kept

IOP Conf. Series: Journal of Physics: Conf. Series **1032** (2018) 012070 doi:10.1088/1742-6596/1032/1/012070

at room condition overnight in the away from light before used , the radical cation was added to 0.01M PBS(phosphate buffer saline PH7,untile to gave absorbant 0.70 nm then the test substance was added to 100 withABTS solution to make volume to 1 ml after that absorbance was readed spectrophotometry quickly with 1 min after added of sample this was made in triplicate and controls contain 990 µl of buffer solution to replace ABTS(14) ,The Trolox chemical was added as indicator (2 to10 µM). This test were calculated according to the capability of ion to scavenge hydrogen ion, and the percent of supernatant ABTS measured at study object from four separate testing solvents.(15).

Test microorganisms :
All the microorganisms which includes *Streptococcus bovis, Proteus mirabillis, Enterococcus gallinarum, Escherichia coli* and two strain of fungi *Rhizopus oryzae* and *Rhizopus stolonifer* were isolated from stools of patients from Allhusainii hospitals and purified on selective media as MacConkey Agar, and XLD Agar also bacterial isolate were diagnosed by API 20E of Enterobacteraceae family . It have been obtained from the laboratories of the Department of Health Surveillance at a holy city of Karbala. Strains of bacteria have been grown on nutrient agar (Hi-media) at 37C° used for 24 hour.

Antimicrobial experiment:
The antimicrobial ability was tested by the standard disc seperation way(16). Inocula of *Streptococcus bovis, Proteus mirabillis, Enterococcus gallinarum, Escherichia coli* plates (Petri dish) and each sample had been added to a concentration of tested oil (4% , 2%, v/v in ethylene glycol). Zones of an activation were measured
The standard of the antibiotic streptomycin was 2 mg/ml. samples read were done with triplicate.
Antifungal activity assay (Disc diffusion technique)
Preparation of inocula
7 days old cultures grown on Czapec dox agar media at room temperature were selected. Culture plates were flooded with 0.45% sterile saline solution and inoculum turbidity was adjusted to 0.5 McFarland standards according to CLSI document M2-A8. Antifungal activity assay (Disc diffusion technique) Disc diffusion method was performed according to CLSI M44-A document. 100µl of the inoculum was seeded on the plates containing CDA medium. The plates were allowed to dry for 2-5 minutes. 100µl of the extract was loaded to the sterile discs of 6mm diameter and placed on the test plates. The plates were incubated at 37°C for 72- 96 hrs. The diameter of the inhibition zones were measured in mm ,Discs loaded with solvents without extract served as negative control and standard fungicides served as positive control. All the tests were performed in triplicates (17).

3.Statistical Analysis

The results were expressed as mean values, standard deviations and analyzed for variance (SAS). Significant differences between control and treatments were determined using Duncan's multiple range test at a level of P 0.05. Statistical analysis was performed using the statistical analysis system(18).

4.Results and Decision

Table (1) showed the of antimicrobial ability of the tested oil from seeds of *S. przewalskii*. each concentrations of sample appeared antibacterial action toward *Streptococcus bovis, Proteus mirabillis, Enterococcus gallinarum, Escherichia coli* as well as antifungal ability toward *Rhizopus oryzae and R. stolonifera* the samples 4(%) concentration showed higher antibacterial ability against E.*coli* and comprised with the lower sample concentration (2%) and it corresponded with literature data in (19). The standard antibiotic streptomycin recorded high degree activity against all tested pathogenic microorganisms. This study recorded that tested oil of *S. przewalskii* appeared significant antibacterial and antifungal ability and it can be used as antimicrobial inhibitor, at same study on some of pathogenic microorganism found that inhibitory zone of methanolic extract of *C. molmol* showed highest antibacterial ability towards each of *Streptococcus bovis* and *Escherichia coli* (20)

Table (1) a Zone inhibition of *saliva przewalskii* seeds (water extract)

Microorganism type	Conce 1:50 (2%)	Conce 1%25(4%)	Streptomycin (2 mg/ml)	Zone diameter (mm)
Streptococcus bovis	20.33±0.56	22.67±0.63	22.43± 0.45	22
Proteus mirabillis	19.76±0.67	15.90±0.87	23.76±0.43	19
Enterococcus gallinarum,	18.98±0.54	14.87±0.78	29.90±0.41	21
Escherichia coli	25.98±0.56	26.67±0.89	22.78±0.49	25
Rhizopus oryzae	23.76±0.78	22.99±0.76	23.87±0.45	19
Rhizopus stolonifera	19.76±0.81	18.64±0.76	24.78±0.51	23
Streptomycin				22

Table :(1) b Zone inhibition of *saliva przewalskii* seeds (methanol extract)

Microorganism type	Conce 1:50 (2%)	Conce 1%25(4%)	Streptomycin (2 mg/ml)	Zone diameter (mm)
Streptococcus bovis	22.33±0.53	22.27±0.63	22.43± 0.45	22
Proteus mirabillis	19.76±0.57	13.90±0.89	23.76±0.43	19
Enterococcus gallinarum,	17.98±0.54	14.87±0.58	29.80±0.41	22
Escherichia coli	25.98±0.76	26.55±0.69	22.78±0.39	23
Rhizopus oryzae	20.76±0.79	21.96±0.76	21.85±0.35	20
Rhizopus stolonifera	18.76±0.51	19.44±0.77	23.48±0.51	22
Streptomycin				22

In the DPPH ion test alleyways high effect radical ha been on laboratory scale used in the examination of the antioxidant capability of active substance as biological supernatant by the capability of component to play as free ion and ion inhibitor after that we can estimate the antioxidant ability (21) Table 2 appeared the range data of ion inhibition capacities to each seed supernatants of the scavenging ion.

At this experimental degrees of samples, all tested solvents appeared a minimum - activates and inhibition capability at different added . soluble and alcohol supernatant of saliva recorded high activity as well as its inhibition capacities salive fixed oil showed lowest antioxidant ability compared with BHA the commercial standard . (Table 2) recorded highest results showed that the free ion ability (%) recorded increased significantly with adding of the solvents from 2.5 to 20 µl/mol saliva fixed samples and the testing added about 100 µg/mol, the methanol extract showed the highest (86.87±0.76) scavenging ability compared to other extracts and BHA another study revealed that there are many solution from *Salvia euphratica*,

Saliva sclarea had antioxidant ability *and* the commercial BHA concentration from18.8±1.21 to 23.4±0.97 µg/ml (23), this results agreed with(18) that fixed oil extracted from species *Salvia sclarea* had component of phenolic compound it has ability as antioxidant ability .as well as results agreed with (23) that found (*Salvia officinalis* L.) methanolic solution had higher antioxidant ability measured by DPPH scavenging ability . at same time results in this study showed antifungal ability of extract towered each *Rhizopus stolonifera* and *Rhizopus oryzae* this agreed with (24,25,26) its founded that fixed oil of saliva *sclarea* had antifungal activity . Parameters at table (4) showed the ability of different extract to reduced ABTS and the highest value recorded it 43.56mg/MM for methanol extract with 100mg/ml conc. This result compatible with (27) it found that highest value of antioxidant ability recoded 36.42 at ABTS methods .

Table (2): DPPH ion capacities range of solutions extracts

Conce	Extraction with water solvent	Extraction with alcohol solvent	BHA
5	44±0.24	48±0.29	50.78±0.78
20	44.67±0.65	51.72±0.32	65.45±0.32
25	48.87±0.45	55.76±0.46	69.31±0.41
50	50.76±0.36	64.89±0.61	73.21±0.34
100	55.76±0.78	79.32±0.53	80.32±0.32
125	67.90±0.45	80.21±0.54	85.34±0.43
150	78.90±0.67	84.89±0.61	86.90±0.74
200	81.42±0.51	86.87±0.76	

Table (3) :DPPH ion capacity (%) of saliva sp. testing oil

Conce	DPPH ion capacities (%)
2.5	29.67±0.34
10	43.67±0.56
15	67.89±0.57
20	89.56±0.47

Table (4) Determination of TEAC test

Sample	TEAC value for DPPH
Water extraction 50mg/ml	22.65
Water extraction 100mg/ml	25.74
Mthanol extraction 50mg/ml	33.12
Mthanol extraction 100mg/ml	43.54

References:

[1] Fu Z, Wang H, Hu X, Sun Z and Han C, 2013.The Pharmacological properties of *Salvia* essential oils. *Journal of applied pharmaceutical science* Vol. 3 **(07)**, pp. 122-127.

[2] Peana A.T, Moretti M.D and Juliano C, 1999 Chemical composition and antimicrobial action of the essential oils of *Salvia desoleana* and *S. sclarea*. *Planta Med.*, ***65***, 752-754.

[3] Burt S, 2004 Essential oils: their antibacterial properties and potential applications in foods—a review. *International Journal of Food Microbiology* **94** :223–253.
[4] Yanishlieva-Maslarova, N.N. and M. Heinonen, 2001 Sources of Natural Antioxidants. In J. Pokorny, N. Yanishlieva and M. Gordon (Eds.), Antioxidants in Food. pp: **210-249**. Boca Raton: CRC Press.
[5] Yousefzadi M, Sonboli A, Ebrahimi S. N and Hashemi S 2007 Antimicrobial Activity of Essential Oil and Major Constituents of Salvia chloroleuca. Z. Naturforsch. **63**:337–340.
[6] Stagos D, Portesis N, Spanou C, Mossialos D, Aligiannis N, Chaita E, Panagoulis C, Reri E, Skaltsounis L, Tsatsakis AM and Kouretas D, 2012 Correlation of total polyphenolic content with antioxidant and antibacterial activity of 24 extracts from Greek domestic Lamiaceae species. *Food Chem Toxicol*; 50(**11**):4115-24.
[7] Tepe B.D, Daferera A, Sokmen M, Sokmen and M, Polissiou, 2005 Antimicrobial and antioxidative activities of the essential oils and various extracts of *Salvia tomentosa* Miller (Lamiaceae). *Food Chemistry*, **90: 333-340**.
[8] Boakye A. A, Wireko-Manu F. D, Agbenorhevi J. K, and Oduro I, 2015 Antioxidant activity, total phenols and phytochemical constituents of four un derutilised tropical fruits, IFRJ 22(**1**): 262-268
[9] Garcia E.J, 2012 Antioxidant activity by DPPH assay of potential solutions to be applied on bleached teeth, *Braz Dent J* 23(**1**): 22-27.

[10] Tepe B.D, Daferera A, Sokmen M, Sokmen and M. Polissio 2005Antimicrobial and antioxidative activities of the essential oils and various extracts of *Salvia tomentosa* Miller (Lamiaceae). *Food Chemistry*, **90: 333-340.**

[11] Kansoh A.L, S.T. El-Sayed and E.W. Jwanny, 2001Bioactivity of natural compounds from some Egyptian higher plants: 1-isolation, purification and some properties of antimicrobial compounds from romman (*Punica granatum*) peels. *Arab J. Biotechnol.*, 4(**1**) 1-8.
[12] Kansoh A.L., S.T. El-Sayed and E.W. Jwanny, 2001 Bioactivity of natural compounds from some Egyptian higher plants: 1-isolation, purification and some properties of antimicrobial compounds from romman (*Punica granatum*) peels. *Arab J. Biotechnol.*, 4(**1**) 1-8.

[13] Ingolfsdottir K, Hjalmarsdottir M.A, Sigurdsson A, Gud-jonsdottir G.A, Brynjolfsdottir A and Steingrrimsson O, 1997 Antimicrob. *Agents and Chemother*. **4**, 215-218.
[14] Chang H.Y, Ho Y.L, Sheu M.G, Lin Y.H, Tseng M.U, Wu S.H, Huang G.H and Chang Y.S, 2002 Antioxidant and free radical scavenging activities of Phellinus merrillii extract ., *Botanical studies* **48: 407-41**.
[15] Pedriali C.A, Fernandes A.U, Santos P.A.and Silva M.M, 2010. Antioxidant activity, cito- and phototoxicity of pomegranate Punica granatum L.) seed pulp extract., Ciênc. *Tecnol. Aliment., Campinas*, **30(4)** 1017-1021.
[16] Miladinović D and Miladinović Lj. 2000. Antimicrobial activity of essential oils of sage from Serbia,: Physics, *Chemistry and Technology* Vol. 2, No **2**, pp. 97 - 1.

[17] Akhilesh Dubey , Neeraj Mishra,and Neha Singh. *2010 International Journal of Applied Biology and Pharmaceutical Technology*, I (**3**): 94-99.
[18] SAS 2012 Statistical Analysis System, Users Guide Statistical . *Ver 9.1th ed .Inst.Cary .N.C.USA.*

[19] Adam E.M and Selim S.E. 2013 Antimicrobial activity of essential oil and methanol extract from *Commiphora molmol* (Engl.) resin, *Int.J.Curr.Microbiol.App.Sci* , 2(12): 1-6
[20] Jao, C.L. and W.C. Ko, 2002 1, 1-Diphenyl-2-picrylhydrazyl (DPPH) radical scavenging by protein hydrolyzates from tuna cooking juice. *Fish. Sci.*, **68**: 430-435.

[21] Tepe B. M, Sokmen, H.A, Akpulat and A. Sokmen,2006. Screening of the antioxidant potentials of six Salvia species from Turkey. *Food Chem.*, **95**: 200-204.

[22] Kuźma L, Kalemba D, Różalsk M, Różalska B, Więckowska-Szakiel M, Krajewska U and Wysokińska H, 2009 Chemical composition and biological activities of Essential oilfrom *Salvia sclarea* plants regenerated *in vitro*; *Molecules*, **14**, 1438-1447.

[23] Rasmy N.M, Hassan A.A, Foda M.I and El-Moghazy M.M. 2012. Assessment of the antioxidant activity of sage (*Salvia officinalis* L.) extracts on the shelf life of mayonnaise; *World Journal of Dairy & Food Sciences* **7 (1):** 28-40.

[24] Dzamic A, Sokovic M, Ristic M, Jovanovic S.G, Vukojevic J and MarinP.D. 2008. Chemical composition and antifungal activity of *saliva sclare* (Lamia ceae) essential oil ; *Arch. Biol. Sci.*, Belgrade, **60** (**2**), 233-237.

[25] Abu-Darwish M.S, Cabra C, Ferreira I.V, Gonçalves M.J, Cavaleiro C, Cruz M.T, Al-bdourT.H and Salgueiro L. 2013. Essential oil of common sage (*Salvia officinalis* L.) from Jordan: Assessment of safety in mammalian cells and Its antifungal and anti-Inflammatory potential; *BioMed Research International*, vol **3** :1-9.

[26] Pitarokili D, Couladis M, Petsikos-Panayotarou N, Tzakou O. 2002 Composition and antifungal activity on soil-borne pathogens of the essential oil of *Salvia sclarea* from Greece. *J. Agric. Food Chem* ***50***, 6688-6691.

[27] Rana M.G, Katbamna R.V, padhya A.A, dudhrejiya A.D, jivani N.P and shete N.R. 2010 In vitro antioxidant and free radical scavenger studiesof alcholicextract of medcago sativa L. , ROM. J. *BIOL.PLANT BIO*L., Vol 55, N1, P. **15–22**.

IOP Conf. Series: Journal of Physics: Conf. Series **1032** (2018) 012071 doi:10.1088/1742-6596/1032/1/012071

Antibacterial activity of Pomegranate molasses (alone and in combination with Ampicillin and Ciprofloxacin) on multidrug resistant *Serratia marcescen*

W AL-Wazni [1], R AL- qarraawi [2], and S Jaber[3]

[1,2,3] Department of Biology, College of Science,University of Kerbala, Karbala, Iraq.

alwazniwafaa@gmail.com

Abstract. The antibacterial activity of Pomegranate molasses against *Serratia marcescens* bacteria was determined. The correlation between molasses and antibiotics were evaluated by using fractional inhibitory concentration.

1.Introduction

Serratia marcescens is gram-negative bacteria and one of the most important bacterial species responsible for different infections in human body. It is opportunistic pathogen in immunocompromised patients because its presence in the air, water, and food as well as the spread in the hospital environment [1]. Many virulence factors were secreted by these bacteria to play important role in increasing their ability to causes disease, such as the hemolysin (which lead to lyses the erythrocytes cells by form holes in its wall) and biofilm which consists of complex made up of viable and dead bacteria cells held within an extracellular matrix made up of polysaccharide, protein and extracellular DNA)[2]. In addition, to the ability of bacteria to form adhesion factors that play an important role in its pathogenicity [3].

The antibiotics are one of the most common methods that use to eliminate pathogenic bacteria and of great benefit in maintaining human health, however, the resistance of many pathogenic bacteria to these antibiotics is one of the biggest challenges facing human as well as causing toxic reactions in humans' body [4]. Therefore, there is a need to find alternative method to avoid antibiotic resistance problem, one of this method is use the medical plants extracts. Some researchers highlighted the use of medicinal plant extracts as an alternative to antibiotics because of its own ability to improve the immune system of humans and work as the killer or inhibitor of bacterial growth inside the body [5]. Currently, there is considerable interest toward evaluating plant sources for use against multidrug-resistant pathogenic bacteria. Various components found in many types of medicinal plants, some of which can operate in synergy with drugs to kill pathogenic bacteria but others are able to sensitize the pathogen to antibiotic [6].

Pomegranate is one of the most important medicinal plants, which contains many therapeutic compounds. Also, is an important source of phenolic compounds with effective antioxidants on the one hand and anti-bacterial activity on the other hand [7].

The purpose of this study was to determine the safe way to reduce the virulence of pathogenic bacteria by using a new method. The simultaneous use of antibiotics (which began to lose their ability to kill bacteria) with plant extracts as a natural source of some substances active in increasing the sensitivity of bacteria to antibiotics and inhibit their ability to produce some virulence factors.

IOP Conf. Series: Journal of Physics: Conf. Series **1032** (2018) 012071 doi:10.1088/1742-6596/1032/1/012071

2. Materials and Methods Bacterial isolates

2.1 The microbiology lab in the biology department, College of Science, Kearbala University, supplied 30 *Serratia marcescens* isolates (these isolates were obtained from women with urinary tract infection aged from 26-40 years attending Maternity and Women's Hospital in Kearbala province). The isolated bacteria were already diagnosed biochemically according to methods described by Collee *et al.* [8] and Baron *et al.* [9]. Finally, the diagnosis of isolates bacteria were confirmed by using the APi-20E system applied depending on the instructions manufacturer.

2.2 Ability of isolated bacteria to produce virulence factors

As described by Collee *et al* [8] and Baron *et al* [9] the ability of isolated bacteria to produce hemolysin and protease enzyme were determined by culture isolates on blood and skim milk agar plates respectively. In addition, the ability of isolated bacteria to form biofilm was determined by tube and tissue culture plat methods as described by Mathur *et al* [10] and Maldonado *et al* [11].

2.3 Determination of the minimum inhibitory concentration (MIC) of molasses and antibiotics.

Pomegranate molasses got ready from the local markets of the Kearbala city. The minimum inhibitory concentration (MIC) of the dry pomegranate molasses (the 1g of dry pomegranate molasses were dissolved in 10 ml of distal water to prepare stock solution 100 mg/ml diluted with sterile to prepared gradual concentration ranging from 10-80 mg/ml) and to antibiotic (the 1mg of antibiotic were dissolved in 10 ml of distal water to prepare stock solution 100 μg/ml diluted with sterile to prepared gradual concentration ranging from 10-80 μg/ml) as described by CLSI [13].The MIC was taken as first dilution that inhibited the growth of the tested bacteria or represented by the last concentration with little or no visible growth. The MIC of each antibiotic were mixed with each dilution of molasses in ratio 1:1 to determine the MIC of molasses in the presence of antibiotic then the MIC of molasses were mixed with each dilution of antibiotic (Cipo &Amp) in ratio1:1 to determined the MIC of antibiotic in the presence of molasses. Then by using agar well diffusion methods determined the antibacterial activity of MIC of molasses and antibiotic alone and in compensation through account of diameter of inhibition zone as design in [12].

2.4 Effect of pomegranate molasses on the bacterial virulence factors.

The bacterial suspension was mixed with MIC of molasses, and all tests for determine virulence factors were made as mentioned in step2.

2.5 Combination tests of molasses and antibiotic

The fractional inhibitory concentrations (FIC) was determined as described by Manda *et al* [14] to evaluate the effect of combination between pomegranate molasses and antibiotics on the antibacterial activity of each one, where the value calculated using the following formula.

FIC (antibiotic) = MIC of antibiotic in combination/MIC of antibiotic alone

FIC (extract) = MIC of extract in combination/MIC of extract alone The interactions indices between the antibiotics and the molasses were assessed by using the FIC indices as described by Pankey *et al.*[15] and Kamatou *et al.*[16] which were calculated by using formula:

FIC index = ΣFIC = FIC (antibiotic) + FIC (molasses)

The combinations types were depended on value of FIC indices, so classified on this basis to several types such as synergistic (FIC indices were <1), additive (FIC indices were 1), indifferent (FIC indices were between 1 and 2), and antagonistic (FIC indices were >2).

3.Result and Discussion

3.1 Determine of the minimum inhibition concentration (MIC)

The minimum inhibitory concentration (MIC) of pomegranate molasses was 40 mg/ml, while was 70μg/ml for ciprofloxacin and 110 μg/ml for ampicillin (Table 1). The results show that molasses has a good antibacterial effect compared with antibiotics. The MIC value of molasses decreased to 10 mg /ml when used with the ciprofloxacin antibiotic, while it was 20 mg /ml when used with the ampicillin antibiotic, indicating that the synergism use of the molasses with antibiotics has led to an increase effectiveness of molasses against pathogenic bacteria. The antibacterial activity of pomegranate molasses may be related to different compounds such as hydrolysable tannins, polyphenols and alkaloid, which was known as punicine that play an important ability of the molasses to effect on bacteria by changing the protein nature that causes kills of bacteria [17]. The molasses contain several chemical compounds such as tannins(which display toxic effect on molecular level of the bacteria after cross the cellular wall of bacteria through their ability to precipitate proteins and inhibit the work of many enzymes such as glycosylate transferees) and polyphenol structures (which have directly influence on the bacterial cell wall and inhibit enzymes by oxidized agents in addition their ability to interact with proteins and disturb co-aggregation of microorganisms [18].

Table 1: MIC of pomegranate molasses and antibiotic against *Serratia marcescens*

Type of material	MIC of each material alone	MIC of molasses in combination with antibiotic	MIC of antibiotic in combination with molasses
Pomegranate molasses	40 mg/ml	-	-
Ciprofloxacin	70μg/ml	10 mg/ml	50 μg/ml
Ampicillin	110μg/ml	20 mg/ml	60 μg/ml
D.W	0	0	0

3.2 Antibacterial activity of pomegranate molasses

The results showed clearly that pomegranate molasses were active against *S. marcescens* in comparison to antibiotic (ciprofloxacin and ampicillin) as a positive control and distilled water as a negative control (table 2). The rate diameter of inhibition zone of molasses reach to 1.5cm which regard good in compere to antibiotic that reach to 2.4 and 2cm for ciprofloxacin and ampicillin respectively. The results agree with Rathinamoorthy *et al.*[19] who attributed the antibacterial activity of pomegranate peel molasses to the presence of the broad spectrum antimicrobial compounds that act against both selected isolates. The result also showed that the mixture of antibiotic and molasses was more active against bacteria in comparison with antibiotic alone or molasses alone .The ability of plant to potentiate antibiotic has not been well describes. Where expected that it may be due to the inhibition of drug efflux and to many change in the bacterial cell wall that help the antibiotic in access to the target point [20].

Table 2. Antibacterial activity of Pomegranate molasses against *Serratia marcescens* bacteria

Type of material	Inhibition zone diameter rate (cm)
Pomegranate molasses (g/ml)	1.5cm
Ciprofloxacin MIC (μg/ml)	2.4 cm
Ampicillin MIC (μg/ml)	2 cm
MIC of Pomegranate molasses(gm/ml) and MIC of Ciprofloxacin (μg/ml)	3.8 cm
MIC of Pomegranate molasses(gm/ml) and MIC of Ampicillin (μg/ml)	3.1 cm
DW	0

3.3 Antivirulence activity of pomegranate molasses

The studied bacterial isolates had the ability to produce a number of virulence factors such as hemolysin , protease and biofilm, This ability was influenced by the addition of MIC of molasses or antibiotics to bacterial suspension. The results of the current study showed decline in the secretion of hemolysin toxin and protease enzyme after treatment compared to control.

Figure 1 and 2 illustrates that *S. marcescens* was a high producer for biofilm formation; this was shown through its optical density that reaches to 1.8 when it was measured for bacterial suspension without molasses (regarded as control). However, the optical density showed reduced largely when the MIC of molasses (0.0588) or antibiotic such as ciprofloxacin (0.21) and ampicillin (0.016) was added to *S. marcescens* suspension before reading the optical density. While reached to 0.09 in compensation of molasses and ciprofloxacin and to 0.05 in compensation of molasses and ampicillin . The biofilm formation activity of *S. marcescens* bacteria were decline after treatment with MIC of molasses to transfer from high producer (OD high than 0.5) to producer(OD between 0.1-0.5) but when treated with molasses and MIC of ciprofloxacin antibiotic transfer to poor producer(OD lower than 0.1) in compere to bacteria treated with MIC of ciprofloxacin antibiotic only which transfer high producer to producer. Therefore, some plant extracts (such as pomegranate, which is one of the most important) are a good alternative to antibiotics in order to overcome the accelerating risk in the emergence of antibiotic-resistant pathogenic bacteria [21 & 5].

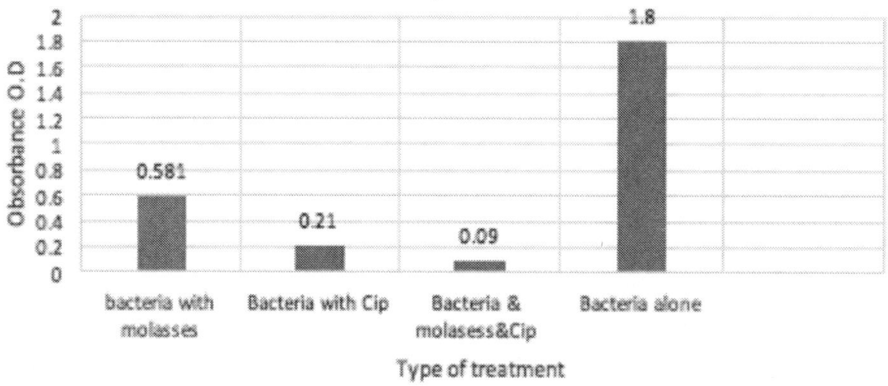

Figure 1: Effect of molasses and ciprofaxicin on biofilm formation of bacteria.

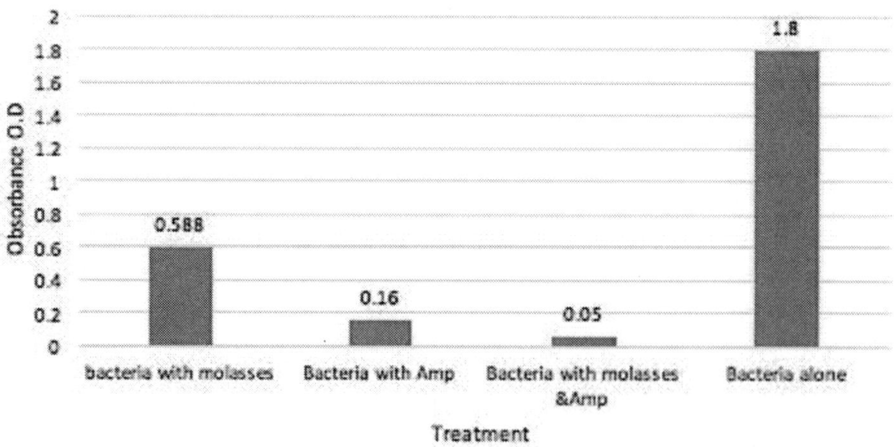

Figure 2: Effect of molasses and ampicillin on biofilm formation of bacteria

3.4 *Interactions between the molasses and antibiotics in vitro*

The results of the current studies showed that the interaction between the studied antibiotics and molasses had positive effects in the elimination of pathogenic bacteria and found that the effectiveness of antimicrobial agents was improved by combining the antibiotics with pomegranate molasses against *S.marcescens* bacteria *in vitro*. As shown in Table 3, the interaction of molasses with ciprofloxacin was synergistic. Nevertheless, with ampicillin was indifferent. In addition, showed that this compensation reduced MICs of antibiotics against resistant bacteria.

Some studies have indicated that there is a synergistic effect of most of the compounds extracted from medicinal plants when used with antibiotics. Some of them may be ineffective when used alone, while enhancing the activity of antibiotic when used against the pathogenic bacteria. In addition, the synergies between antibiotics and plant extracts may regard the first step to solve the toxicity problems of the human body resulting from the use of excess doses of antibiotics for the purpose of elimination of the high resist pathogenic bacteria [22&23].

Table 3. Effect of the combination between pomegranate molasses and antibiotics on *Serratia marcescens* bacteria.

Type of Antibiotic	Mean FIC (Antibiotic)	Mean FIC (Molasses)	FIC index	Interaction
Cip	0.714	0.25	0.964	Synergistic
Amp	0.545	0.5	1.045	Indifferent

Cip: ciprofloxacin - Amp: ampicillin

References

[1] Kida Y, Inoue H, Shimizu T and Kuwano K, 2014 *Serratia marcescens* serralysin Induces inflammatory responses through protease-activated receptor 2. *Infect & Immun*.; 75 **(1):** 164-174.

[2] Chanda S and Rakholiya, K. 2011 Combination therapy: Synergism between neutral plant extract and antibiotics against infectious disease. *Science against microbial pathogens: communicating current research and technology advances* ; p.**520-529.**

[3] Flemming H, Wingender, J, Szewzyk U, Steinberg P, Rice S.A and Kjelleberg S, 2016 Biofilms: an emergent form of bacterial life. *Nat Rev Micro*;14**(9)**:563–575.

[4] Mahlen, S.D, 2011 Serratia infections: from military experiments to current practice. *Clin Microbiol Rev*; 24**(4):**755–791.

[5] AL-wazni W.S and Hadi B.S, 2015 Antivirulence effects of pomegranate peel extracts on most common urinary tract infection pathogens in pregnant women. *J.Contemp.Med .Sci.* 1**(4)**:7-12.

[6] Aiyegoro O, Adewusi A, Oyedemi S, Akinpelu D. and Okoh A,20 11 Interactions of antibiotics and methanolic crude extracts of *Afzelia Africana* (Smith.) against drug resistance bacterial isolates. *Int. J. Mol. Sci.* **12:4477–503**.

[7] Gabriel Betanzos C, Perla Y. M, Héctor E. F, Helen Belefant M, and Juan C, 2015 Antibacterial activity of fresh pomegranate juice against clinical strains of *Staphylococcus epidermidis. Food Nutr Res.*; 59 (**10**):3402.

[8] Collee J.G, Fraser A.G, Marmion B.P, Simmons A. M. and McCartney, 1996 Practical medical microbiology, 14th ed. UK: *The Churchill Livingstone. Inc.* pp. **329–41**.

[9] Baron, E.J, Peterson L.R. and Finegold S.M, 1994 Bailey & Scott's diagnostic microbiology, 9th ed. USA: *Mosby-Year Book*.

[10]Mathur T, Singhal S, Khan S, Upadhyay D.J, Fatma T and Rattan A, 2006 Detection of biofilm formation among the clinical Isolates of Staphylococci: an evaluation of three different screening methods. *Indian J. Med. Microbiol.;*24(**1**):25–30.

[11]Maldonado N.C, Silva de Ruiz C, Cecilia M and Nader-Macias M.E, 2007 A simple technique to detect Klebsiella biofilm-forming-strains. Inhibitory potential of *Lactobacillus fermentum* CRL 1058 whole cells and products. Communicating Current Research and Educational Topics and Tends in Applied Microbiology A. Mendez Vilas (Ed).

[12]Egharevba H.O, Kunle O.F, Iliya I, Orji P.N, Abdullahi M.S, Okwute S.K *et al.* 2010 Phytochemical analysis and antimicrobial activity of *Punica granatum* L. (fruit bark and leaves). *New York Sci J.*;3(**12**)91–8.

[13]CLSI 2012. Methods for dilution antimicrobial susceptibility tests for bacteria that grow aerobically; Approved Standard—Ninth Edition. CLSI document M07-A9. Wayne, PA: Clinical and Laboratory Standards Institute.

[14]Mandal S, Mandal M.D and Pal N, 2004 Evaluation of combination effect of ciprofloxacin and cefazolin against Salmonella enterica serovar typhi isolates by *in vitro* methods. *Calicut Med J.*;2(**2**):e2.

[15]Pankey G, Ashcraft D and Patel N, 2005 *In vitro* synergy of daptomycin plus rifampin against *Enterococcus faecium* resistant to both linezolid and vancomycin. *Antimicrob Agents Chemother. Dec*;49(**12**):5166–8.

[16]Kamatou G.P.P, Viljoen A.M, van Vuuren S.F and van Zyl R.L, 2006 *In vitro* evidence of antimicrobial synergy between *Salvia chamelaeagnea* and *Leonotis leonurus. SA. J. Bot. Nov*;72(**4**)**:**634–6.

[17]Dahham S.S, Ali M.N, Tabassum H, and Khan, M, 2010 Studies on antibacterial and antifungal activity of pomegranate (*Punica granatum* L.). *American- Eurasian J Agric & Environ Sci.*;**9**(**3**) 273–81.

[18]Abdollahzadeh S, Mashouf R, Mortazavi H, Moghaddam M, Roozbahani N and Vahedi M, 2011 Antibacterial and Antifungal Activities of *Punica granatum* Peel Extracts Against Oral Pathogens. *J. of Dentistry* ; 8(**1**):1-6.

[19]Rathinamoorthy R, Udayakumar S, and Thilagavathi G, 2011 Antibacterial efficacy analysis of *Punica granatum* L. leaf, rind and *Terminalia chebula* fruit extract treated cotton fabric against five most common human pathogenic bacteria. *Int. J. Pharm. & Life Sci. Oct.* ; 2(**10**):1147–53.

[20]Adwan G, Abu shanab B, and Adwan K, 2010 Antibacterial activity of some plant extracts alone and in combination with different antimicrobial against multidrug resistant pseudomonas aerugenosa strain. *Asian Pacific J. of Tropical Medicine*;**266-269**.

[21]Janecki A, and Kolodziej H, 2010 Anti-adhesive activities of flavan-3-ols and proanthocyanidins in the interaction of group A-streptococci and human epithelial cells. *Molecules* ;15(**10**)**:**7139–52.

[22]Adwan G, and Mohammad M. M, 2008 Synergistic effects of plant extracts and antibiotics on *Staphylococcus aureus* strains isolated from clinical specimens. *Middle-East J .Scientific Res.*;3(**3**)**:**134–9.

[23]Aiyegoro O.A, and Okoh A.I, 2009 Use of bioactive plant products in combination with standard antibiotics: implications in antimicrobial chemotherapy. *J. Med. Plants Res. Dec*;3(**13**)**:**1147–52.

IOP Conf. Series: Journal of Physics: Conf. Series **1032** (2018) 012072 doi:10.1088/1742-6596/1032/1/012072

Epidemiological study on *Cryptosporidium* among children in Basra Province-Iraq

M Salim

Sothern Technical University, Basra province, Iraq

Epru2007@yahoo.com

Abstract. The study included examination of 134 stool samples from several reigns in Basra province includes AL-Hyania reign and some Health Centers in Basra province from February to June 2015 for detection of *Cryptosporidium spp* oocysts. Each sample was concentrated with sedimentation method and stained with a modified Zheil - Nielsento detect of *Cryptosporidium* Oocytes. The study revealed 23.8% were positive with highest infection rate in young children between days - 1> year and 5 - >15 which reached 28% and 25% respectively.
Keywords: *Cryptosporidium* epidemiology, Basra.

1.Introduction

Cryptosporidiosis is a zoonotic and anthroponotic disease which caused by protozoan parasites genus *Cryptosporidium*, it has a worldwide distribution and in the most surveys it is considered to be among the four major pathogens causing diarrheal diseases in children (Xiao *et al.*, 2000). It has major public health implications because infections can result from exposure to low doses of *Cryptosporidium*Oocysts. The Oocysts are highly resistant to chlorination and common household disinfectants, ability to pass through physical water treatment processes, and survive long periods in the environment. Humans can acquire *Cryptosporidium* infections through several transmission routes, such as direct contact with infected persons or animals, and ingestion of contaminated food (food borne transmission) and water (waterborne transmission) (Xiao andCama , 2006), insects play a significant role in Oocysts transmission. House flies, cockroaches, beetles and synanthropic flies can transmit the Oocysts by their exoskeletons and in their digestive tracts too(Graczyk*et al.*, 2005; Yen *et al.*,2007).

There are more than ten species of *Cryptosporidium*, *Cryptosporidiumparvum* and *Cryptosporidiumhominis*are the two species responsible for the most cases of human Cryptosporidiosis worldwide (Victoria *et al.*,2000 ;Mosallanejad*et al.*,2010). Age group play an important role in *Cryptosporidium*infection, the pathogen can be found infecting all age groups but it is more common among children and elderly as most studies have indicated (Amin, 2007).This study was done to evaluate the correlation between prevalence and some factor e.g. age and sex.

2.Material and Methods

Source and collection of specimens: The study included examination of 134 stool samples from several reigns in Basra province includes AL-Hyania reign and some Health Centers in Basra province from February to June 2015 for detection of *Cryptosporidium spp* oocysts. A total of 134stool samples taken from children age between 1month to 15 years (66 males and 68 females). The stool samples were collected directly from the rectum in plastic containers with a detailed history about age (days - <1),(1-5), (5-15), sex and were transported to the parasitological laboratory.

Oocyst detection: The samples were concentrated by formalin method (Nasser, 2014). A drop was taken from each deposit by Pasture pipette and was smeared on a glass slide; the slides were stained by the modified acid - fast method. In this method Oocysts of *Cryptosporidium* were visualized as red spots, all the smears were examined by light microscope with 40X and 100 X objectives. The Percentage of infected samples was calculated by using the following:

$$\text{Percentage of infected samples} = \frac{\text{Number of infected samples}}{\text{Number of total samples}} \times 100\%$$

3.Statistical Analysis:

The data was put into a Microsoft Excel Spreadsheet, after data cleaning; the data was transported into SPSS (Statistical Package for the Social Sciences version 15.) package software program for statistical analysis. Descriptive statistics (percentage) were calculated for all variables, as well as analytical statistics was done to find the relations between the variables. The relation between variables was calculated by using the appropriate statistical tests by Chi-square. Values were considered to be statistically significant when the p-value obtained was less than 0.05.

4.Results

134 stool samples were examined in a trial for studding *Cryptosporidium* prevalence in Basra city. High prevalence (23.8%) of the parasite was recorded among human population (Fig. 1,Table 1).

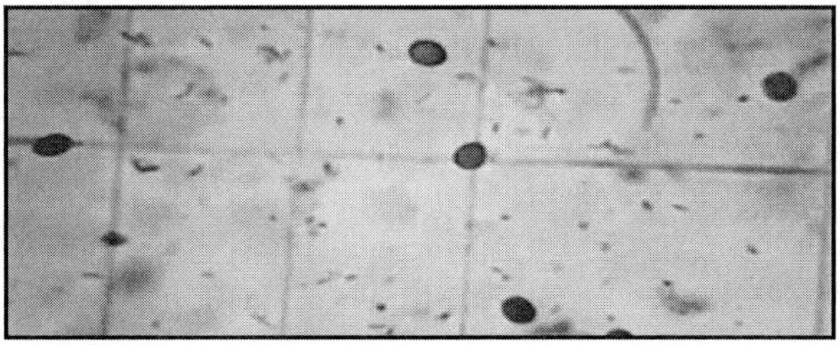

Figure 1. The oocyst of *Cryptosporidium* Staining with modified ZeihlNeelsen100X

Table 1. The Overall Prevalence of *Cryptosporidium*

Name of Parasite	No. of Examined samples	No. of infected samples	Percentage of infected samples
Cryptosporidium	134	32	23.8%

There were no significant differences between the sexes as to the prevalence percentage in males (24.2%) which is nearly the same in females (23.5%).(Table2).

Table 2: Prevalence of *Cryptosporidium* sp. according to sex

Genus	No. of Examined samples	No. of infected samples	Percentage of infected samples
Male	66	16	24.2%
Female	68	16	23.5%
Total	134	32	23.8%

The standard Pearson chi-square statistic is defined as:

$$X^2 = \text{SUM}\frac{(\text{Oij-Eij})^2}{\text{Eij}} = \frac{(66-24.4)^2}{24.4} + \frac{(68-23.5)^2}{23.5} = 155.18$$

Table (3) shows the prevalence of *Cryptosporidium* infection according to the age. The highest infection(28.0%) was recorded in age group 1(< **1**) year, while the lowest (21.6%) was appeared in age group 2(1-5) years old. Statistical analysis showed no significant difference appears between age**s**.

Table 3: Prevalence of *Cryptosporidium* sp. according to ages

Year	No. of Examined samples	No. of infected samples	Percentage of infected samples
Group 1 < 1	50	14	28.0 %
Group 2 (1-5)	74	16	21.6 %
Group 3 (5-15)	8	2	25.0 %
Total	132	32	74.6 %

$$X^2 = \frac{(50-28.0)^2}{28.0} + \frac{(74-21.6)^2}{21.6} + \frac{(8-25.0)^2}{25.0} + = 155.95$$

5.Discussion

From 134 stool samples examined 32 samples were positive for *Cryptosporidium* infection with rate of 23.8%, these results agreed with Nasser, (2014) who showed that the increased rate of cryptosporidiosis infection in Basra province were associated with contaminated drinking water supplied to these population or contaminated food which linked to consuming fresh vegetables, This is result in comparison with other studies is a little agrees with Ali(2013) who recorded 15.2%inSulimani

city, and with worldwide prevalence of parasite which was between 0.3- 32% (Obiad*et al.*, 2012), but disagreement with Mahdi and Ali (2002) in Basra, Al-Alousi (2004) in Tikrit and Othman (2000) in Kirkuk were recorded 5%, 9.42%, 12.62%.
The differences between this results and the other studies because of the technique used for the diagnosing or due that the parasite prevalence is not constant from year to year and from one area to another in the same country (Amin, 2007), or due to sample size differences or because most of studies was concentrated on one age group (infants) or children.

There were no significant differences between the sexes as to the prevalence percentage in males (24.2%) which is nearly the same in females (23.5%) which may indicates that both sexes have equal chance of being infected. These results were in agreement with other studies conducted worldwide (Ali, 2008; Hossein *et al.* 2010; Rahi*et al.* 2013), the data of age groups infection showed there is no significant difference among the three age groups in this study the rate of infection first group (< 1) which higher (28.0 %) compare other groups, that due to their weak immunologic status of lacking protective immunity to prevent themselves from getting infection.

6.Conclusions

1. The results showed that (23.8%) of the fecal samples were positive for *Cryptosporidium* Oocysts.
2. Although both genders were sensitive to infection with *Cryptosporidium* a higher rate of infection was recorded in males rather than females with no significant difference of infection.
3. Children of (< 1) year recorded a higher rate of infection than other groups of ages.

7.Recommendation

1. Routine microscopically diagnostic method is recommended and technical medical staff must be trained to diagnose *Cryptosporidium* in our hospital and primary health care centre in Iraqi country.
2. Further study for the detection of *Cryptosporidium* in food.
3. A more detailed epidemiological study is recommended to contain all society factors.
4. Introduce diagnosis based on molecular technique, genotyping and sub-typing for the *Cryptosporidium* Species.
5. Raising the health awareness about cryptosporidiosis among the population through health education, seminars and leaflets.

References

[1] Al-Alousi, T.I. 2004 Prevalence of *Cryptosporidium* spp. in different resources with a trial treatment by using medical plants extracts. Ph. D. Thesis, Coll. Med. Tikrit Univ.

[2] Ali, F. M 2013 Cryptosporidiosis in Sulaimani pediatric teaching hospital and comparison of different diagnostic methods for its detection.M.Sc. *Thesis,Sulaimani Univ.*,**100** pp.

[3] Ali, M. A 2008 Prevalence of *Cryptosporidium* among children at Ramadi city. M.Sc. thesis. College of medicine, *Al-Anbar Univ.*, **46-48** pp.

[4] AminO.M 2007 Prevalence, distribution and host relationships of *Cryptosporidium*parvum(protozoa) infections in the United States,2003-2005. *Parasitol. Cent. Inc.* (PCI) explores 16(1): **22-28**.

[5] Graczyk T. K Knight R and Tamang L 2005 Mechanical transmission of human protozoan parasites by insects. *Clin. Microbio*. Rev., **18(1):** 128-132.

[6] Hossein S, Omid Y, Amene Y, Mohammad M and Mohsen S 2010 Infection Rate of *Cryptosporidium parvum*among Diarrheic Children in Isfahan.*J. Ped.* **20** (**3**): 343-347.

[7] Mahdi N. K and Ali N. H 2002 Intestinal parasites including *Cryptosporidium* species in Iraq patients with sick cell anemia. Basrah. *East. Med., Health* J. **8: 1-7**.

[8] Mosallanejad B, Hamidinejat H, Avizeh R, Ghorbanpoor N. M and RaziJalali M. H 2010 Antigenic detection of *C. parvum*in urban and rural dogs in Ahvaz district, southwestern Iran.J. Iranian Vet. Res, *Shiraz Univ.*, Vol. **11(3)**: 32.

[9] Nasser S. K 2014 The distribution of *Cryptosporidium* in Some Resources of Water, Vegetables and Animals Feces in Basra Province. M. Sc. Thesis, *College of Education for pure sciences, Basra Univ*. **88** pp.

[10] Obiad H.M, Al-Alousi T. I and Al-Jboori A. H. 2012 An epidemiologic study on *Cryptosporidium* spp. in Kirkuk city with some trials for *in vitro* treating the parasite. Second Scientific Conference – Science College – Tikrit Univ.

[11] Othman, N. F. 2000 Comparison between different laboratory methods for diagnosis of *Cryptosporidium* species. (*PDCLI). Tikrit Univ.*

[12] Rahi, A. A, Ali M. . and Al-Charrakh A. H. 2013 Prevalence of *Cryptosporidium parvum* among children in Iraq. *American J. Life Sci.*, **1(6):** 256-260.

[13] Victoria H. B, Vasquez J, Nelson R.G, Forney J. R, Rosowsky A and Sibley C. H 2000 Identification of *Cryptosporidium parvum*Dihydrofolate Reductase Inhibitors byComplementation in *Saccharomyces cerevisiae.* Ant. *Agents and Chemotherapy*. **44** (**4**): 1019–1028.

[14] Xiao L. and Cama V 2006 *Cryptosporidium* and Cryptosporidiosis in: Ortiga,Y. editor."Food Born Parasitology". *Springer Sci. USA*. **289** pp.

[15] Xiao L, Morgan U.M, Fayer R, Thompson R.C and Lal A.A 2000 *Cryptosporidium* systematics and implications for public health. Parasitol.Today. **16**: 287–292.

[16] Yen M. L, Chien C. C, Chin L. M and others 2007 Multilinegedifferentiationand characterization of the human fetal osteoblastic 1.19. cell linc: a possible in *vitro* model of human mesenchymal progenitors. Stem cell. **25(1**): 125-131.

IOP Conf. Series: Journal of Physics: Conf. Series **1032** (2018) 012073 doi:10.1088/1742-6596/1032/1/012073

Biotechnology and Biochemical Analysis Techniques to isolate Vanillin derivatives for Plum Fruit Tree (I)

*A. J. Mahrath[a] , E. I. Yousif [b], T. H.Al-Mgheer [a]
a) Babylon University College of medicine Biochemistry Department
b) Department of Chemistry, college of Education for Pure Science (Ibn-Al-Haitham), Baghdad University
*Corresponding author e.mail : ajmbioorg@gmail.com , or ajm68@uobabylon.edu.iq

Abstract: This application studied some biochemical and technological techniques to identify ,isolate and purified some of biological molecules from fruits of Plum tree . The study involved drying the sample with freeze dryer for 24 hrs . Then samples were performed with Soxhlet apparatus by protic solvent (PSI) , aprotic solvent (ASI) ,and aprotic–protic solvent interaction(PASI) were used to extracted the molecules . Then recovered the excess solvent vy evaporation and condensation. The sample directly transformed to the flash column chromatography technique (FCC)and monitored by Thin layer chromatography (TLC) to isolate m-methoy-p-hydroxy benzoic acid **(vanillic acid) or (VA)** and m-methoxy-p-hydroxy ethyl Benzoate **(Ethyl Vanillate) or EV** in about 95% pure state. These molecules **AV and EV** elucidated by spectroscopic tools (UV-Vis , ATR-IR ,^{1}H-NMR, ^{13}C-NMR , and Mass spectrometry) .

Key words : plum tree , Vanilinic acid ,flash chromatography , purification .

1.Introduction:

It's well known that Green Chemistry techniques consider one of the best choice in the field of Medical chemistry research .alternative medicine depends mainly on herbs and medical plants in the hall world which in turn have a lot of benefits in the field of medical treatments through what they have different types of cyclic organic molecules(1).The Plant products are consider the main source of pharmaceutical agent since long time (2-3).Tree plum is one of the most important plant that possessing medical applications (4). Some applications of the plant were used it's extract as chest pain , throat and urinary infection(5) . It's also for it's diuretic ,anthelmintic , demulcent ,antidiarrheal , anti-inflammatory ,and anti-arthritic activities (6) ,anti-diabetic(7-8) in addition to antimicrobial activity against various pathogenic microorganism(9,10) ,protective ulcer(11) and wound healing activity(12) .the leaves of the plum tree also have potent antioxidant activity and can show anticancer activity too (13-15). On the other hand medicinal plants consider as the major sources of Modern and traditional medicine in the hall world (16).Recently methanolic extract of cordia leaves tested against human prostate carcinoma cell line by Md.Azizur Rahman and coworkers (17).according to all of these applications the investigation of the fruit plum became part of our researches. Phenolic compounds or Vanillin derivatives have widely used as antioxidant activity in addition ant-cancer and pharmacological properties(18,19) . Vaniliic acid and it's derivatives are one of the most flavoring and scent agent that producing nice oders . They are mainly present in many kind of plants .Therefore the aim was trying to Extraction the derivatives of these molecules , isolation and identification them by using Green chemistry principles as possible.

2.Experimental Section

Chemical Reagents :

All the reagents and chemicals (ethanol ,ethyl acetate , ether, silica gel ,magnesium sulphate and alumina) used were of analytical grade and also double distilled water (DI) was used in extraction process .

IOP Conf. Series: Journal of Physics: Conf. Series **1032** (2018) 012073 doi:10.1088/1742-6596/1032/1/012073

Apparatus:
The following instrumentations were used in this application .
Freeze Dryer (Crust Alpha 1-4 LD with Vacuum Pump RZ6 up to 4-10^{-4}mbar) , Soxhlet extraction , pH meter , UV-Vis spectrometer(Specord,PC 200 plus KARL KOLB) ,Shaker water bath ,Thin layer chromatography(aluminum oxide 60,size 20×20 cm fluorescent indicator) , Flash Column chromatography (class column 30 × 500 nm,packed with silica gel with partical size 230-400 mesh,1 gm/25 gm with vacuum pumb ROCKER 300 at 50 kpa) , E-graph image ATTO with Wise UV M20) Electro thermal melting point apparatus (stuart) uncorrected , ATR mode with FT-IR spectroscopy (Bruker Tensor 27) with KBr Disk and ATR unite , Gas Chromatography with-Mass spectrometry (shimadzu GC Mass QP2010 Ultra with ion source temperature 150 ^{o}C ,Ionization mode SCI , DI Temperature 350°C with detector gain : 0.69 KV) . Proton and carbon magnetic resonance (^{1}H-NMR ,^{13}C-NMR) Bruker spectroscopy 300 MHz with tetramethylsilane (TMS)as standard with dimethyl sulfoxide with $CDCl_3$.

Collection of plant sample
The plum fruits were collected from Plum tree for experimental purpose in the zone of Al Mahaweel , Hilla city –Babylon district during June 2016 and stored immediately after collection at 10 ^{o}C in darkness place .

Preparation of plant sample Extract[20]
Step one : Extraction of the sample
About 250 gm of plum fruits **fig(1a)** were washing and drying under dark and reduce pressure for 3 days by laypholyzer then stored in dark at room temperature .The dry sample crashed by blinder then 8 g powder of fruit sample transferred by thimble into the Soxhlet extraction apparatus for1 to 2 hours according to the kind solvent ,**fig (1b)** .Different solvent were used via this apparatus such as number (1) DI water , (2) Ethanol , (3) ethyl acetate ,and (4) Ether . Different collected product were solvated in each solvent , The crude extract concentrated under rotary evaporating and lympholizer specifically to produce 760 , 656 ,370 and 123 mg respectively **Table (1).**
Step two :Examine fruit Extraction in TLC :
The crude extract solvated again in ethanol with the following ratio (10 ,20 , 30 , 50 , 70 ,90 %) respectively . and tested in thin layer chromatography **(TLC)** with the following description (aluminum silica 60 ,size 20×20 cm with fluorescent indicator) . the best mobile phase result of the target products was the highest ratio **90% ethanol with 10% ethyl acetate** . more than two spots in component number (2). UV light view develop the TLC plate after drying at about 267 nm **figure (2)**
Step three : Isolation of the Target molecules by FFC :
The crude that separated in method 1 directly packed in column chromatography with vacuum to accelerated the separation (reducing time for separation) (FCC). The column description was (30 × 500)mm size ,backed with silica gel (220 -380) mesh . the best mobile phase for elution was mixed (Ethanol : ethyl acetate) with the following ratio 4:1 .Ten fractions were collected from the FCC and Tested again with TLC to collect the similar products . the collected samples concentrated again ,Tested with TLC and drying under laypholyzer to produce while the second one was mp = 208-212 ^{o}C respectively . This methodology used according to reference [18-20] . Further purification of the isolated molecules were done by Gas chromatography with Mass analysis (GC-Mass) in College of science Al-Mustansiriah University .

Step Four : Elucidation of the isolated molecules from step three by Spectroscopic Analysis:
Full characterization by spectroscopic Instrumentations such as UV-vis , FT-IR ,^{1}H-NMR , ^{13}C-NMR ,**figures (3-10)** respectively in addition to melting points measurement in **Table (2).**

IOP Conf. Series: Journal of Physics: Conf. Series **1032** (2018) 012073 doi:10.1088/1742-6596/1032/1/012073

3.Results and Discussion

The results of isolated molecules were represented in both Tables (1, 2) and figures (3-10).
the molecules were tested after separation and purification by spectroscopic techniques as follows : UV-Vis spectra shown the absorption bands at 297 and 312 nm for each **VA (vanilic acid)** and **EV** (ethyl Vanillate) respectively . FT-IR spectroscopy with (KBr-disk) and ATR unit of the two isolated molecules confirmed the functional groups structures and matched with the published data([21]) see figure **(3 ,4)**. For instant Two strong absorption bands for the carbonyl groups at 1727 and 1689 cm^{-1} that belong to (Ester and Carboxylic acid) respectively. Also the two different absorption band at 3484 and 3418 cm^{-1} refer clearly to (OH) phenolic groups in **VA and EV** .while the OH group in vanillic acid appeared as a broad absorption at about (3210-2630) cm^{-1} . the aromatic group also present clearly in the range of absorption at 1552 -1418 cm^{-1} for both compounds **figure (5,6)**.The **^{1}H-NMR** spectroscopy in $CDCl_3$ for **VA** (1) appeared the following peaks : (dd,2H, aromatic) at δ=6.90-6.92 , 7.32-7.36 ppm ,and (s,1H, aromatic) at δ=7.51 ppm , three protons (3H,s, OCH_3) appeared at δ= 3.89 ppm belong to methoxy group , and (b,1H, phenolic OH) at δ=9.65 ppm, **figure (5)** .The second molecule **EV** proved by the following peaks . three aromatic protons (3H) appeared at δ= 7.77 ppm (dd) , at δ= 7.52 ppm (dd) ,and 7.16 ppm (s) protons . the phenolic proton appear shielded at δ= 6.46 ppm compared with phenolic group at **VA** .Sharp signal observed at δ=3.81 ppm which belong to methoxy group **(3H,O-CH_3)** protons. On the other hand the ethyl ester group clearly observed as methylene (2H,-**CH_2** , q) at δ=4.18-4.29 ppm , and the methyl (**3H**,CH_3 , t) at δ=1.37-1.34 ppm respectively **figure(6).**
^{13}C-NMR spectrum of both **VA** and **EV** in $CDCl_3$ proved the that are purified with 95% as shown in the **figure (7 , 8)** respectively. Signal of carbonyl group for **VA**(C=O) appeared at δ = 168.18 and 168.61 ppm , while Carbon of methoxy group (C-O) appeared at δ = 53.09 and 56.51 ppm , plus the rest of spectrum .On the other hand Mass spectroscope give with no doubt by presence of molecular ion peak (m/z)=[168 and 198] for both **VA and EV** respectively **figure (9,10)** .

Figure (1a): Plum Tree with Collected outgrow th in Al Mahaweel sector **.**

IOP Conf. Series: Journal of Physics: Conf. Series **1032** (2018) 012073 doi:10.1088/1742-6596/1032/1/012073

Figure (1b): Extract process for sample **Figure (2):** sample after testing TLC.

Table (1): the optimum condition of sample Extract.

Solvent	**(1) DI H_2O**	**(2) C_2H_5OH**	**(3)$CH_3CO_2C_2H_5$**	**(4) $(C_2H_5)_2O$**
Time	120 mins	65 mins	42 mins	24 mins
Volume solvent	350 ml	350 ml	350 ml	350 ml
Evaporation Solvent by rotary	At 55 °C/ 40 kPa	45 °C/ 70 kPa	50 °C /85 kPa	35 °C/ 90 kPa
Drying by Freeze dryer	24 hrs at -50 /0.04 mbar	10 hrs -30 /1.2 mbar	8 hts -20/40 mbar	4 hrs -10 /80 mbar
Mass of extract	760 mg / 8gm	656 mg/8 gm	370 mg/ 8gm	123 mg/ 8gm

Table (2): Physical properties of Isolated Molecules from Plum Fruit.

<table>
<tr><td>Compound Name</td><td>Ethyl Vanillate / Ethyl-3-methoxy-4-hydroxy Benzoate</td><td>Compound Name</td><td colspan="2">Vanillic acid / 4-hydroxy Anisic Acid</td></tr>
<tr><td>Structure</td><td>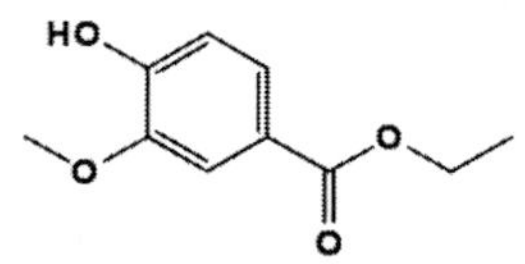</td><td>Structure</td><td colspan="2">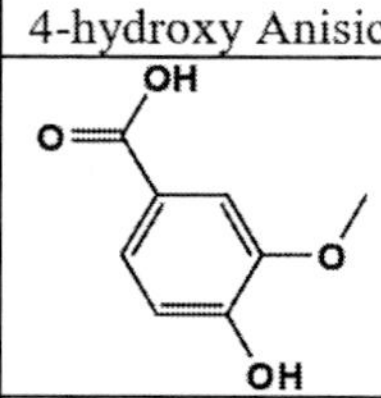</td></tr>
<tr><td>Molecular weight</td><td>196.202 g/mol</td><td>Molecular weight</td><td colspan="2">168.147 g/mol</td></tr>
<tr><td>Solubility</td><td>Methanol</td><td>Solubility</td><td colspan="2">Water</td></tr>
<tr><td>Melting point</td><td>43-45 °C /45-46 °C lit</td><td>Melting point</td><td colspan="2">208-210 °C/ 211-212 °C lit</td></tr>
<tr><td>Boiling Point</td><td>292 /293 °C lit</td><td>Boiling Point</td><td colspan="2">.........</td></tr>
<tr><td>Molecular Formula</td><td>$C_{10}H_{12}O_4$</td><td>Molecular Formula</td><td colspan="2">$C_8H_8O_4$</td></tr>
<tr><td>Color</td><td colspan="2">Light yellow</td><td>Color</td><td>Milk</td></tr>
</table>

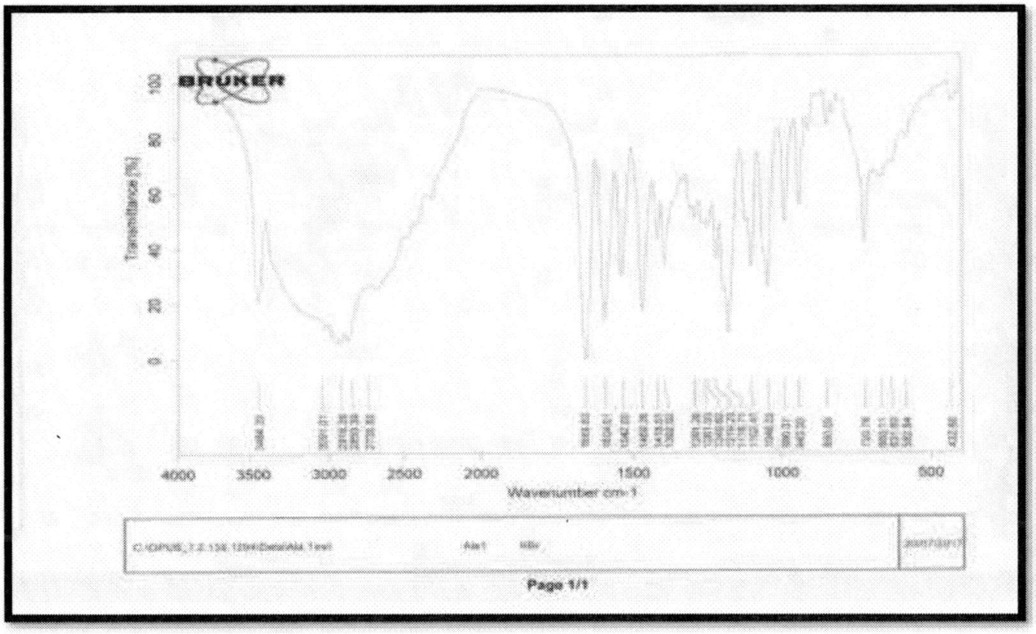

Figure (3) FT-IR spectra of Extract sample (1) Vanillinc Acid

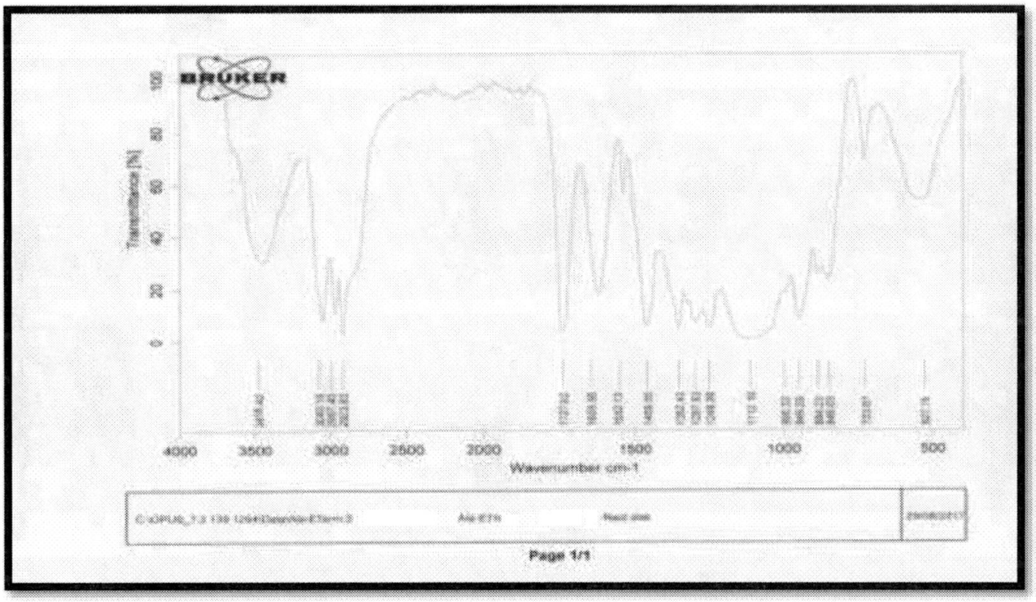

Figure (4): FT-IR spectra of second isolated molecule (2) Ethyl Vanillate

IOP Conf. Series: Journal of Physics: Conf. Series **1032** (2018) 012073 doi:10.1088/1742-6596/1032/1/012073

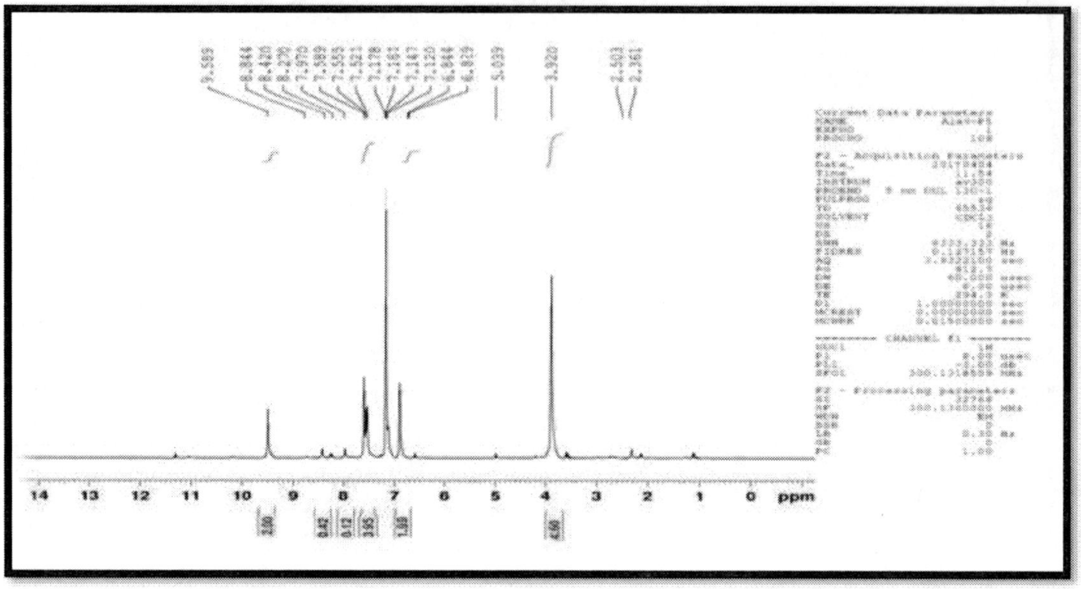

Figure(5): ^{1}H-NMR spectra of Vanillic acid(VA) after purification

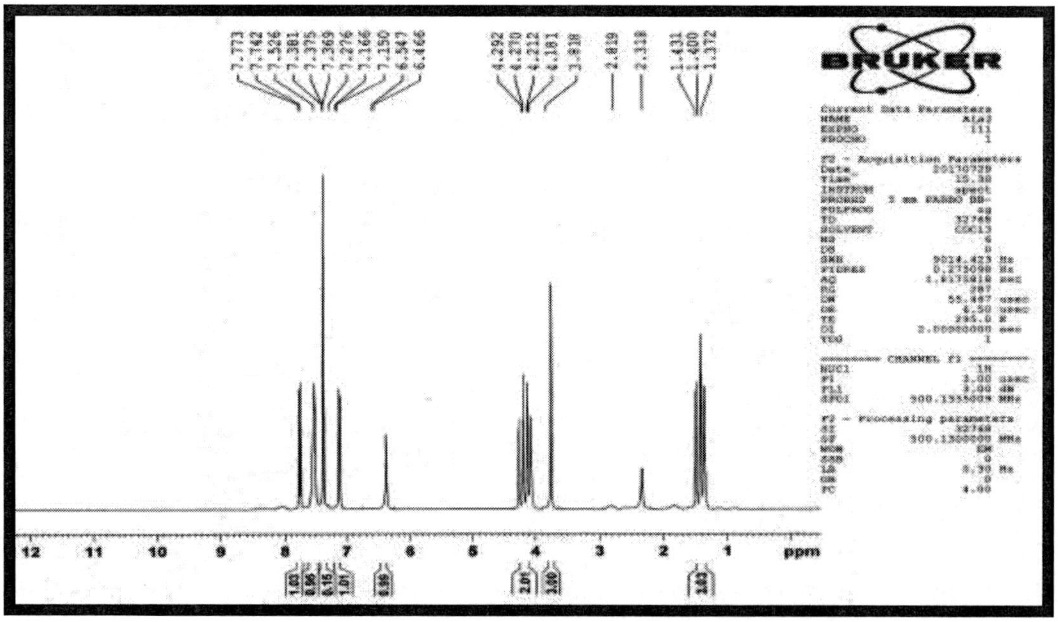

Figure (6): ^{1}H-NMR spectra in $CDCl_3$ of **Ethyl Vanillate** (EV)after purification

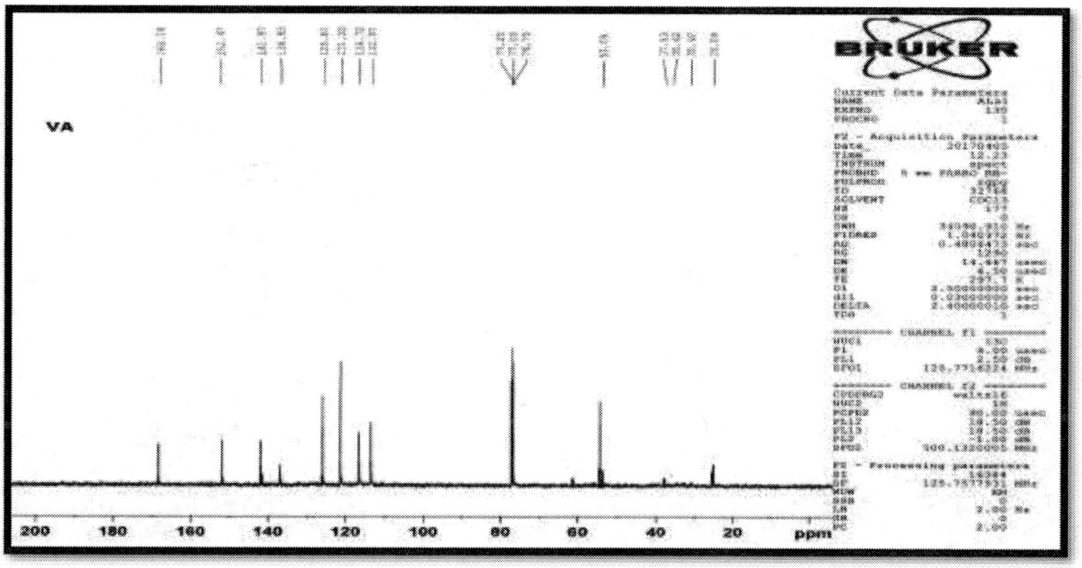

Figure (7): ^{13}C-NMR spectra in $CDCl_3$ for Vanillic acid (VA)after purification.

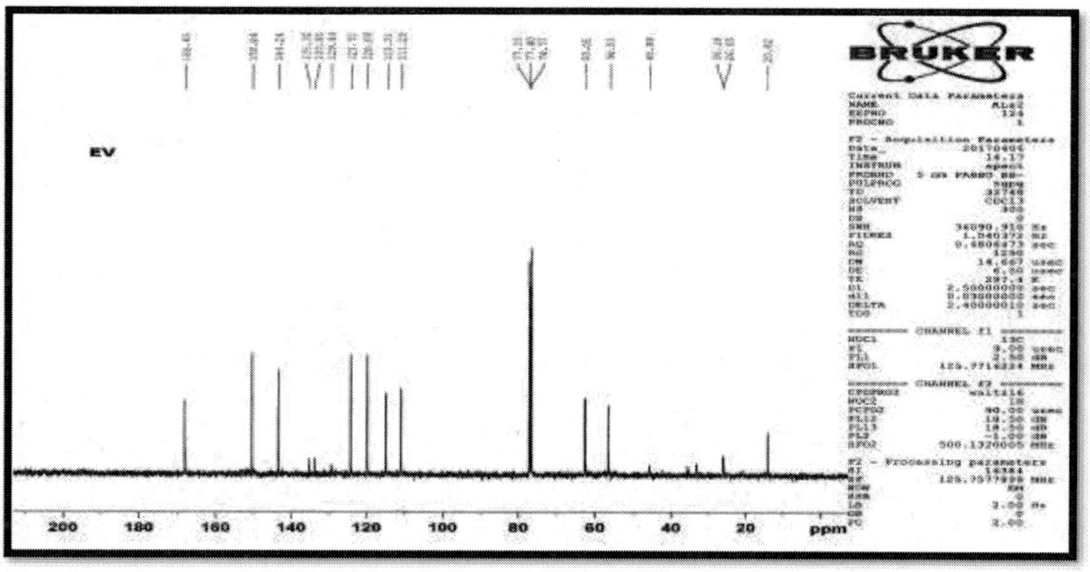

Figure (8): ^{13}C-NMR spectra in $CDCl_3$ for **Ethyl Vanilate** (EV) after purification.

IOP Conf. Series: Journal of Physics: Conf. Series **1032** (2018) 012073 doi:10.1088/1742-6596/1032/1/012073

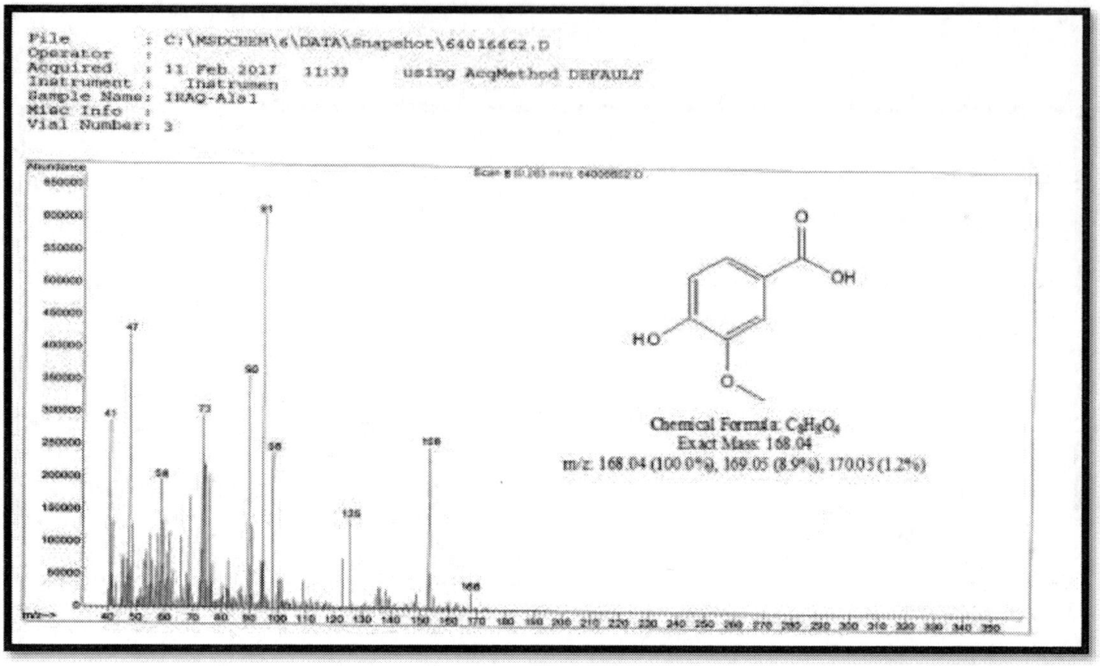

Figure (9) Mass Spectra of **Vanillic acid(VA)** after purification

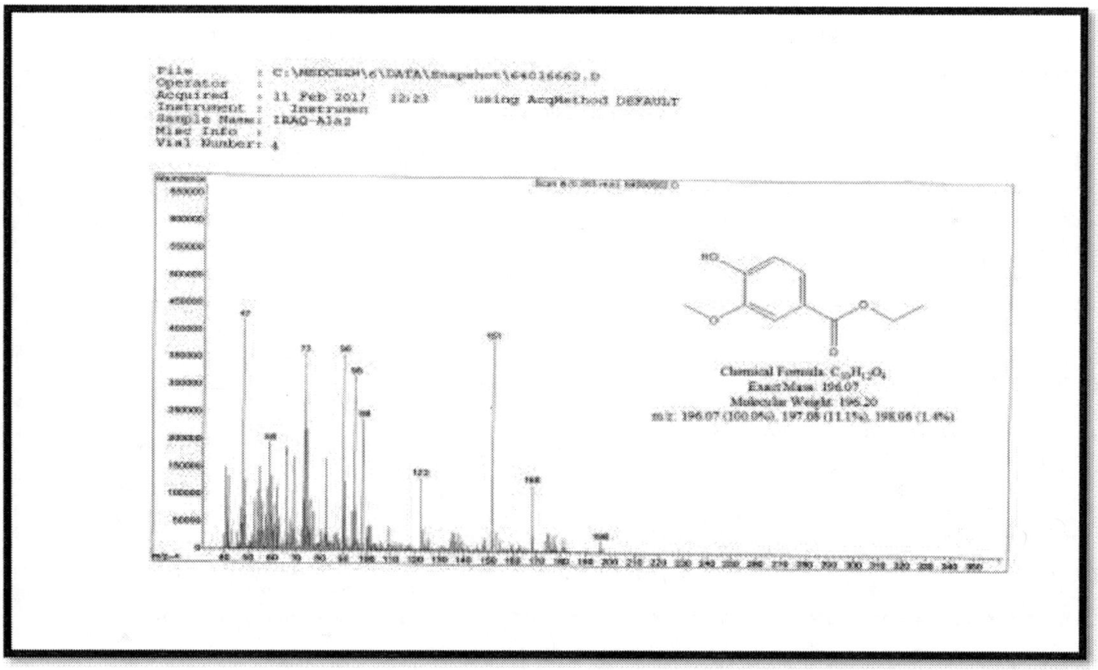

Figure (10): Mass spectra of **Ethyl Vanilate(EV)** after purification

IOP Conf. Series: Journal of Physics: Conf. Series **1032** (2018) 012073 doi:10.1088/1742-6596/1032/1/012073

References:

[1] Julbert C and Baiely G 2007 Isolation of Chlorophylls a and b from spinach by counter-current chromatography. *Journal of chromatography*. A,1104 ,**95** .

[2] Bandow J.E, Brotz H, Leichert L.O, Labischinski H and Hecker 2003 Proteomic approach to Understanding Antibiotic Action. *Antimicrobial agent and Chemotherapy*. **47**: 948.

[3] Yadav P and Singh D 2012 Madhuca longifolia (Sapotaceae) a review of it's traditional uses , phytochemistry and pharmacology. *International Journal of Biomedical Research*. **3**,7:291.

[4] Prasad G.J, amkhande, Sonal R, Barde, Shailesh L, Patwekar 2013 plant profile phytochemistry and farmachology of cordia Asian Pac J.*Trop Biomed*. **3** - **12**, 1009.

[5] Chopra and Doiphode V 2002 Ayurvetic medicine core concept therapeutic principles and current relevance. *Medical Clinics of North America*, **86-75** .

[6] Ficarra R, Ficarra P, Tommasaini S, Barbera and Rapisarda A 1995 Leaf extract of some Cordia species ; analgesic and anti-inflammatory activity as well as their chromatographic analysis , Farmaco : **50**:245 .

[7] Misha A and Garg GS 2011 Ant diabetic activity of fruit pulp of Cordia dichotoma in alloxan induced diabetic rats. *Int J.Pharm.Sci.Res*. **2**:2314 .

[8] Rahman MA and Akhtar J 2016 Phytochemistry and pharmacology of Traditionally used medicinal plant cordia dichotoma ,Curr.*Trends Biotechnol pharm*.**10** :180.

[9] Bhawana P, Bhagyashree D, Sheetal S and Varsha C 2014, Estimation of Elemental Content of Cordia Myxa and it's antimicrobial activity against pathogenic microorganisums .*Indian Journal of science research* **4**, 1 ,39 .

[10] Rawat S, Saini R and Sharma A 2013 Phytochemical Study and antimicrobial activityties of Cordia dichotoma, *International Resaerch of Journal Pharm*. **4** :12;53.

[11] Ganjare AB, Nirmal SA, Rub RA, Patil AN, Pattan SR. Use of Cordia dichotoma bark in the treatment of ulcerative colitis. Pharm Biol 2011; 49(8): 850-855.

[12] Kuppast IJ, Nayak PV. Wound healing activity of Cordia dichotoma forst F. fruits. Indian J Nat Prod Resour 2006; 5(2): 99-102

[13] Rahman MA and Hussain A 2015 Anticancer activity and apoptosis inducing effect of methanolic extract of cordia diachotoma against human cancer cell line. *Bangladesh J.Pharmacol*.**10** :27.

[14] Nariya PB, Bhalodia NR, Shukla Vj, Acharya R and Nariya MB 2013. In vitro evaluation of antioxidant activity of cordia dichotoma Bark Ayu. **34** :124.

[15] Rahman M.A and Hussain A 2015 phytochemical and analytical evaluation of cordia dichotoma Linn , leaves . *Pharmacogn J*. **7**:5 .

[16] Suri R.K , Chaudhari D.C and Jaffer R1992 Commercially important medicinal plant from forest ,*J Ecol Bot Phytochem*. **3**; 129 .

[17] Md.Azizur Rahman , Sahabiada and Juber Akhtar 2017Evaluation of anticancer activity of Cordia leaves.Journal of Traditional and complementary Medicine. **7**:315.

[18] Prasd G, Jamkhande, Sonal R.Brade ,Shailesh L.Patweaker , Priti S.Tidke 2013 Plant profile ,phytochemistry and pharmacology of Cordia dichotoma :A review .*Asian Pascific Tropical Biomedicine*. **3(12)**1009.

[19] Mahrath A.J, Mahmood S, Magtoof 2015 Isolation and identification of green pigment from waste pineapple peels. *Chromatography International Labmate online*.**11**: **12** .

[20] Mahrath A.j ,Ghafil, Hassan S and Ehab K.Obaid 2015 Isolation and characterization of Terpenoid derivatives from medicinal plant.*Int.J. Chem.Sci.***13** :2,983.
[21] Mahrath A.J, Saadon A. Aowda and Sabah N. Kamil 2012 Synthesis and spectroscopic studies of some new oxazepine derivatives throughout [2+5] cycloaddition reactions (III), *Res Chem Intermed*. DOI 10.1007/s11164-012-0877-2.

Effect of Crushing Speed Rate on Crashworthiness Parameters and Energy Absorption Capability of Composite Materials

Hakim S. Sultan Aljibori[1], Firas K. Mohamad Alosfur[2] and Noor J. Ridha[2]

[1]Department of Air-conditioning and Refrigeration Engineering, University of Warith Alanbiyaa, Kerbalaa, Iraq
[2]Department of Physics, College of Science, University of Kerbala, Kerbala, Iraq

Corresponding Author: hakimaljobori@gmail.com

Abstract. Improvement of energy absorber in automotive and aircraft industry is a demanding requirement. Crashworthiness parameters designed to special interesting and could ensure the passengers safety and reduce fuel costs. Thin shells are the main structures of energy absorbing in transportation applications for collisions; it is very important to check their energy absorption and performance. In the present paper, composite materials of glass and epoxy conical shell structures of slipping solid steel cone are studied experimentally. The effect of crush rate and crushing behaviour on failure modes and energy absorption are investigated in details. The specific and volumetric energy absorption capabilities are studied and failure modes for quasi-static analyses were investigated. One type of semi-vertex angle (β) of conical tubes was 10 degrees with stacking sequence of 90/45/-45/90. The composite material specimens were loaded and compressed at various crush speed rates of 5, 10, 15 and 20 mm/min. The results of composite structures are demonstrated and showed that the conical shapes with speed rate of 5 mm/min absorbed high specific energy and displayed more stability in load-stroke curve.

Keywords: Composite, Conical Shells, Energy Absorption, Laminated and Failure Mode

List of Symbols and Abbreviations

FWL	Filament wound laminated	β	Semi-vertex angle of cone
FRP	Fibre-reinforced plastic		
E_s	Energy absorbed/ unit mass	*CFE*	Crush force efficiency
E_v	Energy absorbed/ unit volume	*SE*	Stroke efficiency
P_m	Mean (average) crush load at post crash stage	*D, d*	Maximum and minimum diameter of cone
P_i	Initial crushing load	W_t	Total work done
P_{max}	Maximum peak load	W_p	Work at post crush stage

IOP Conf. Series: Journal of Physics: Conf. Series **1032** (2018) 012074 doi:10.1088/1742-6596/1032/1/012074

L	Maximum stroke	S_i	displacement in initial crash stage
A	Cross sectional area	S_p	Displacement when crushing is completed in post crash stage
M	Mass of the structure	H	Height of the Specimen

1. Introduction

Composite material structures are being used for most applications in automotive and aircraft parts to meet the weight and manufacturing cost constraints. The modes of composite material are significantly different from the metallic materials. Recently, attempts were made to improve energy absorption capability of composite structures [1-3]. These structures have a wide area of applications due to their specific stiffness and strength value. Also, the composite materials have low thermal expansion, good corrosion resistance and non-conductive materials. Understanding of the energy absorption capability of composites has become an area of great academic and practical interest. Many researchers have studied the effect of crushing behavior on energy absorption capability [4-5]. For the metal applications, the trigger put as a chamfer, or crimps nearest to the ends of the structure (Cone, for example), however, the crash effect in these material are still limited because of the complex shape of these materials [6]. Experimental studies are important for providing most the information that concerning materials and structural factors and predicting the crash shape of these materials. For all the crashworthiness effects, thin-shell fiber and resin tubes have been the focus on various investigations, due to cost, easer of manufacturing, fabrication and excellent energy efficiency [7-8]. Experimental findings explained that the composite shells deform in a shape different from components made of conventional materials, such as metals and polymers, since micro-failure modes such as delamination, matrix cracking, fiber breakage and so on. They depend on the geometry of structure, material characteristics and testing procedure and parameters [9-10].
The present paper, reported on an experimental test and analysis of the failure mechanism and crushing behavior to the stable mode of thin-shell fiber glass conical shell under axial slipping loads. The shape of structure was designed by slipping of rigid steel cone to laminated fibre/epoxy cone to predicting the loads and the energy absorbed during the collapse procedure. However, there is a poor of experimental work as well as dynamic pushing of solid structures into hollow structures. The idea comes from: the high energy absorption can be produced if the properties of material (especially, longitudinal properties) are staring utilized and effect in application. The objective of this research paper is to determine with detail the contributions of fiber deformation and cracking to energy absorption in these materials.

2. Design Parameters

2.1 Energy Absorption Capability

Fiber-reinforced materials are low strain to fracture, but it found now has been good characteristics for structures made from these materials and can have energy absorption capability and better than such materials those have other properties and made from some metals. As we kno and that the total work (W_T) of the structure is the area under the curve of load and distance as shown in Eq.1:

$$W_t = \int_{s_i}^{s} F\,\Delta s \;=\; W_t = \int_{s_i}^{s} P\,ds \qquad (1)$$

At the Post and advance stage, the work (W_P) can be calculated as shown in equation 2:

$$W_P = \int_{s_p}^{s} P\,ds \Rightarrow P_m\left(s - s_p\right) \tag{2}$$

and the specific energy absorption per unit mass (kJ/kg) may be representing from the equation 3:

$$E_s = \frac{P_m\left(s - s_i\right)}{\rho \times V} = \frac{W_P}{M} \tag{3}$$

Where,
(P_m): The mean (average) load,
A: Cross-section area of hollow tube,
ρ: the density of the material,
V: volume of structure and M: mass of structure.
So, the energy absorption per unit volume (E_v) calculated by equation 4:

$$E_v = \frac{E_s \times M}{V_s} \tag{4}$$

The volume at current structure occupied by the cone shape before the failure can be calculated from:

$$V_{con} = \frac{\pi H}{12}\left[D^2 + Dd + d^2\right] \tag{5}$$

2.2 *Stroke Efficiency (SE)*

From the experimental tests noticed that the length of crushing (Stroke) depends on the specimen geometry and on the method that structure is dispersed. The stroke-efficiency may give by eq. 6:

$$SE = \frac{S_p - S}{H} \tag{6}$$

2.3 Crush Force Efficiency (CFE)

The evaluation of efficiency and performance of composite structures depend on the loads that support the structure, these loads are average loads and maximum load cause the failure. This loads are demonstrated by evaluating and assessment of this efficiency (CFE), this is calculated by eq. 7:

$$CFE = \frac{P_m}{P_{\max}} \tag{7}$$

3. Fabrication Procedure of Cone

The winding procedure of fibre filament was used to manufacturing the specimens of the glass fibre filament laminated as conical shape as shell. Epoxy resin was mixed with epoxy hardener, and then fibres are pulled to the resin bath across tension devices to wood mandrel for good impregnation. The wet cone is curried out and kept at room temperature for 24 hours for optimum and better combined and then extracted the composite cone from the mandrel. The schematic diagram of the winding procedure is shown in Figure 1. The winding factor and parameters (winding speed, winding tension, resin content and winding timing) were kept constant level during fabrication procedure. BENSO machine was used for cutting and trimming to obtain the desired cone shape. Cutting and finishing processes are shown in Figure 2. The specimens of solid steel cone and glass fibre and epoxy composite conical shell used for slipping test were fabricated and tested as shown in Figure 3.

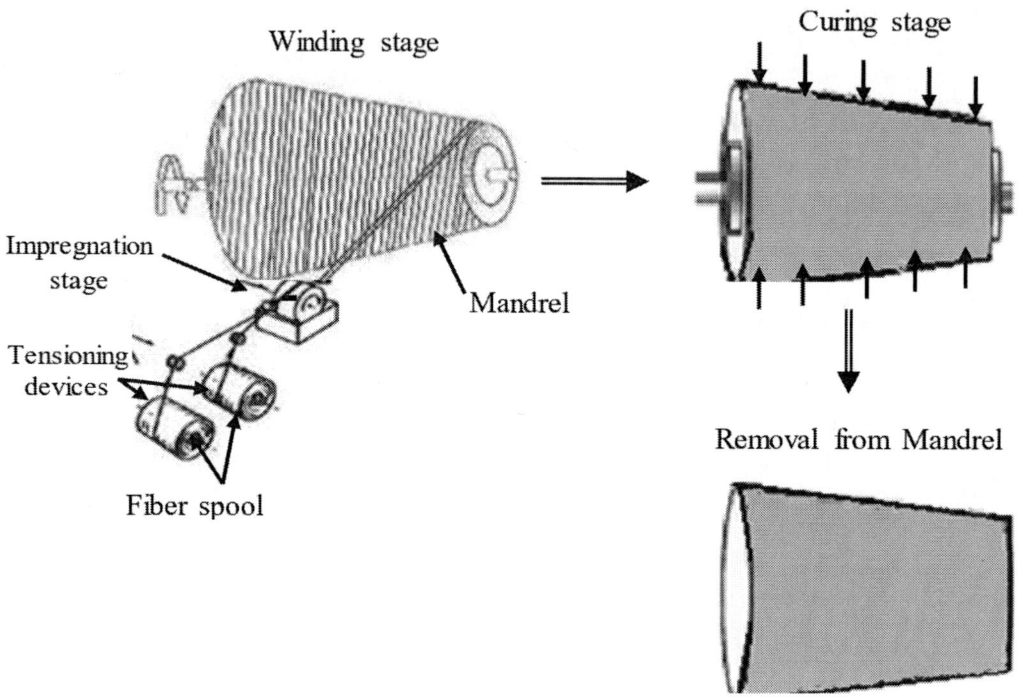

Figure 1. Schematic Representation of the Wet Filament Winding Process

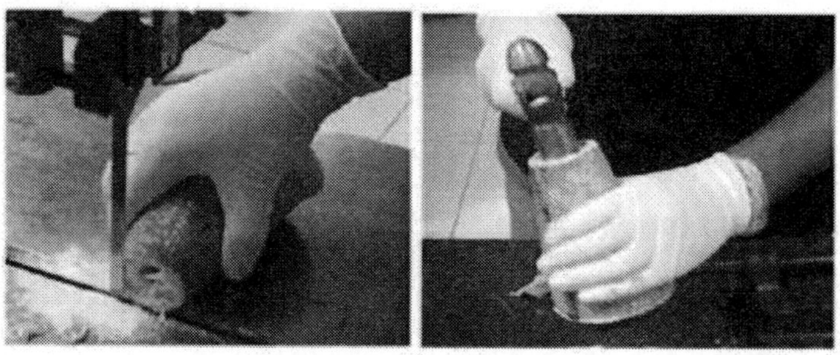

Figure 2. Cutting and finishing processes

Figure3. (a) Photographs of Specimen of Glass Fibre and epoxy
(b) Photograph of Solid Steel Cone
(c) Photograph of Glass Fibre and Epoxy under Slipping Test

3.1 Define a Composite Layers

The sequence of micro-fracture leading to the formation of the stable crush depends on the shape and arrangement of the fibers inside the tube wall and the materials that used in the fabrication. Eeach layer name must be unique and also need to make the thickness of the layer, angle of orientation to be maintained within the plane of the layer, and then it will specify the overall composite orientation. It is to be noted that (0° angle) with the x axis and a (90° angle) with the y axis. This is to be repeated for as many layers as required. The unidirectional laminates in which fibres are positioned parallel to each others in a matrix and the layers oriented at an angle (θ) from reference axis (x and y) to principal material coordinates system 1 and 2 as shown in Figure 4.
On the other hand, one of the most common forms of fibre-reinforced composite structure is the stacking sequence laminate (angle ply-laminates), in which the fabricator lay-up a sequence of fibre reinforced with different fibre orientation θ in alternate layers (/θ/-θ/θ/-θ/) as shown in Figure 5. The conical shell structure has four layers of different fibre orientation. Arrangement of fibers from 90/45/-45/90 within the thickness is shown in Figure 6.

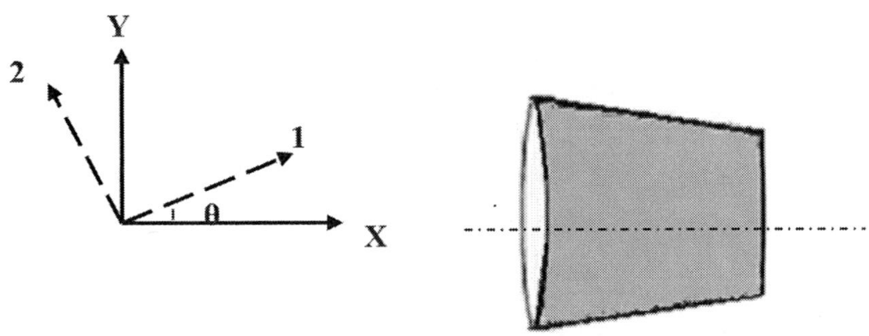

Figure 4. Definition of Principle Material Axes

IOP Conf. Series: Journal of Physics: Conf. Series **1032** (2018) 012074 doi:10.1088/1742-6596/1032/1/012074

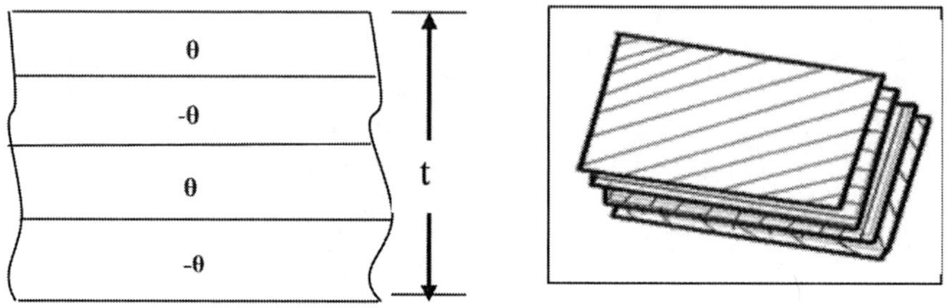

Figure 5: Typical sketch of Stacking Sequence of Laminate

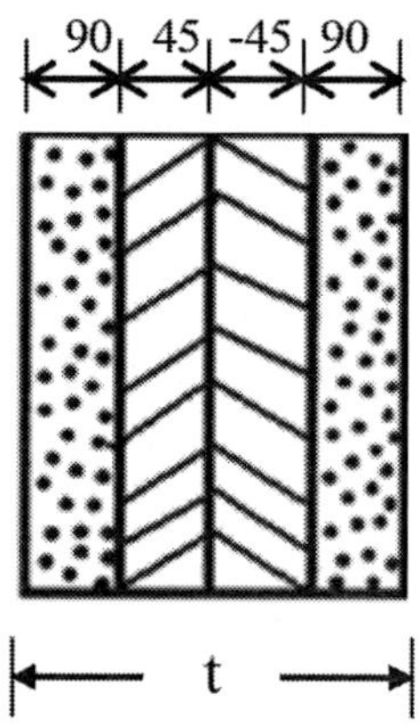

Figure 6. Arrangement of Fibers from 90/45/-45/90 within the Thickness

4. Crushing Process

A total of 16-specimenes of filament laminates were tested under axial crushing test and slipping of rigid steel cone into glass fibre as a composite shell with cone angle (β) of 10^0 and fibre orientations form (90/45/-45/90) as a sequence. Crushing tests were carried out using an digital machine with loading capacity of machine is 250 KN. The entire specimens were compressed and loaded by stroke at various crush rates of 5, 10, 15 and 20mm/min. The load and stroke were recorded by memory data system. The results are represented by the average of the four typical tests under same conditions. All the specimens were fabricated with height of 100mm and outer diameter was 80mm. Some photographs during the test were taken, thus photographs show the specimen at different crushing stroke to provide the history of their crushing process. Test procedure and slipping of rigid cone to composite cone are shown in Figure 7.

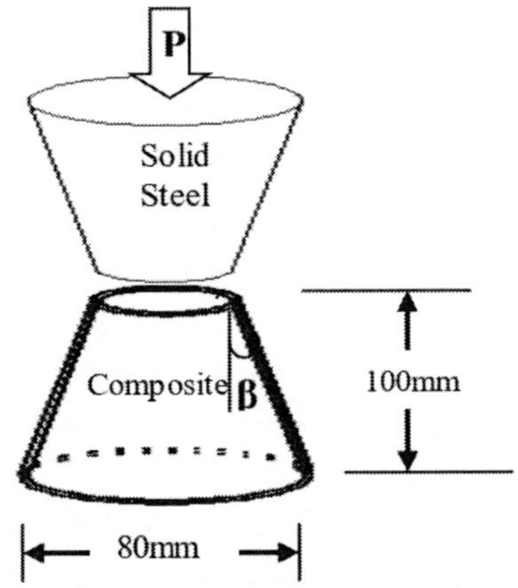

Figure 7. Slipping of Solid Cone into Composite Cone

5. Result and Discussion

The load-stroke curves and energy absorption relations as well as the deformation history and failure modes for composite conical structures are presented in this section. The load-stroke curve was obtained from average of the load-stroke points for four replicated tests with same-crushing speed rate and the same testing conditions. In addition the effect of the crushing speed rate on energy absorption capability is discussed. The energy absorbed by all the cones was evaluated and calculated by numerical methods of the load- stroke curves.

5.1 Loading responses

Figure 8 shows the typical load-Stoke curve for FWL conical shell at first crushing speed (V1) of 5mm/min. To check the strain rate, four variables and parameters was evaluated these are: load at first initial zone P_i, maximum load P_{max}, stroke at initial zone S_i and stroke at complete procedure in post advance stage S_p. From the tests it is noticed that the force-stroke behaviour shows a multi-relationships prior to the initiation, and then a tiny sudden drop is observed as shown in Figure 8.

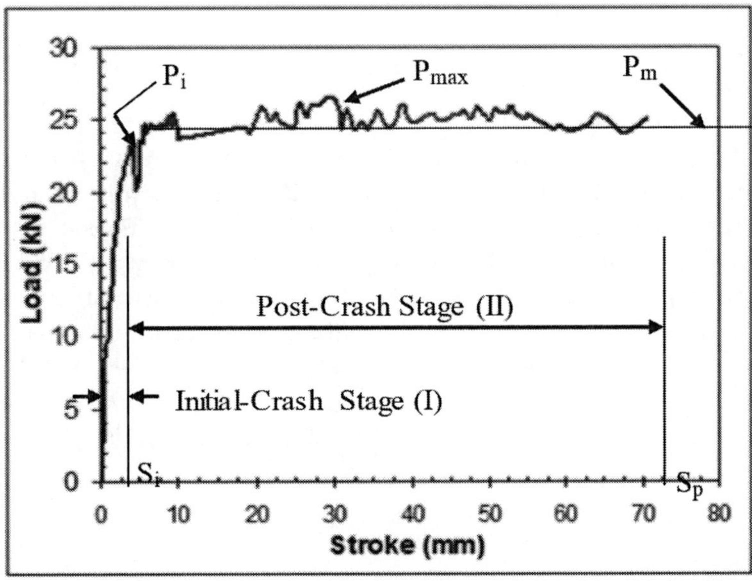

Figure 8. Typical Load – Stroke Curve of V1 Composite Cones Subjected to Slipping Crush Test

IOP Conf. Series: Journal of Physics: Conf. Series **1032** (2018) 012074 doi:10.1088/1742-6596/1032/1/012074

In Figure 9, the load and Stoke curves for all of the tests and crushing speed rate to failure. It is observed that both the loads at first initial failure and at the maximum peak are function of stroke at several speed rates. No clear rate dependence on either of the stroke parameters. This is due to the fact that the epoxy matrix can become brittle material as well as its stiffness decreases when the strain rate increases. As we know that the matrix controls the internal zone growth and if the failure strains of the matrix decrease with increasing crush speed rate, the internal layer crack length increases with increase of crushing speed.

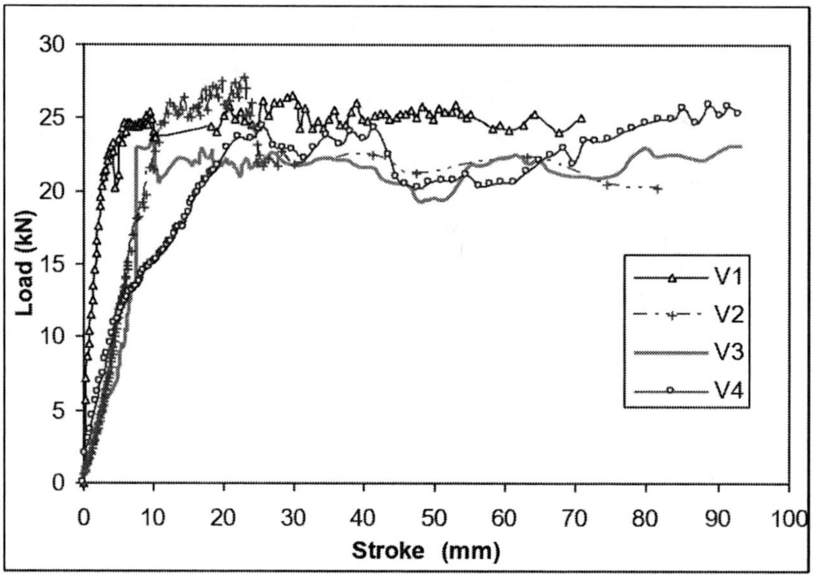

Figure 9. Effect of Crushing Speed Rate on Load – Stroke Curve of Composite Cones Subjected to Slipping Crush Test
Note: V1, V2, V3 and V4 stand for Crushing speeds of 5, 10, 15 and 20mm/min respectively.

5.2 Energy Absorption Capabilities

The main objective of this work is to improve the energy absorption ability of composite conical shell. Energy absorption capability of these devices is the main factor in measuring their performance. It is noted that most of the absorption procedure occurs at post crush zone. Crush failure modes in an initial and propagation were influenced by crushing speed rate. From the previous equations (Eq.3 and Eq.4), there are two kinds of energies were measured, during slipping rigid cone to cone. These are the energy absorption per unit mass (E_s) and energy absorbed per unit volume (E_V). Table1 summarized the results for crush energy absorption capabilities for all the tests. It can be observed that the specific energy absorption at V1 is slightly higher compared to those recorded at others speeds.

IOP Conf. Series: Journal of Physics: Conf. Series **1032** (2018) 012074 doi:10.1088/1742-6596/1032/1/012074

Table 1: Summarised results of the crashworthiness parameters and energy absorption capabilities of composite conical shells.

Crushing Speed	P_i	P_m	P_{max}	CFE	SE	E_s	E_v
		(kN)		kN/kN	mm/mm	kJ/kg	(kJ/m³)
V_1 (5mm/min)	23.30	24.76	26.51	0.93	0.66	5.44	5264.5
V_2 (10mm/min)	24.66	24.82	27.7	0.92	0.64	5.30	5129.1
V_3 (15mm/min)	22.92	22.1	23.7	0.93	0.62	4.7	4548.4
V_4 (20mm/min)	23.6	23.1	24.9	0.92	0.55	4.3	4162.3

5.3 Stroke Efficiency (SE) and Crush Force Efficiency (CFE)

Crashworthiness of a vehicle is measured by its ability to absorb energy and prevent any injury to the occupants. Therefore, experimental results for composite conical shell with several crushing rate subjected to slipping solid steel are presented as dimensionless ratios known as crush force efficiency (CFE) and stroke Efficiency (SE). The values of these crashworthiness parameters for all the tests are tabulated in Table 1.

6. Conclusions

The effect of stacking sequence from 90/45/-45/90 with cone angle of 10 degrees was studied experimentally. All the specimens were tested for compression case and loaded by various speed rates of 5, 10, 15 and 20mm/min. The load-stroke relations and the energy absorption behaviour were discussed. However the following conclusions could be drawn from the investigation:

1. All specimens followed the similar failure mechanism. This shape and mechanism are associated with crushing mechanism that involves matrix cracking, tearing of fibres and delamination between layers.
2. The effects of rate of crushing speed on the energy absorption and failure mechanism
3. All the results and modes of composite structures noticed that the cone shell with stacking sequence from 90/45/-45/90 absorb high specific energy.

Acknowledgement

The authors are grateful to the authorities of the University of Warith Alanbiyaa and University of Kerbalaa for conducting this research.

IOP Conf. Series: Journal of Physics: Conf. Series **1032** (2018) 012074 doi:10.1088/1742-6596/1032/1/012074

7. References

[1] Mahdi, E. Sahari, B. B., Hamouda, A. M. S. and Khalid, Y. A. Effect of hybridisation on crushing behaviour of carbon/glass fibre/epoxy circular cylindrical shells, Int. Journal of Materials and Processing Technology, 2003,132, 49-57.

[2] Mallick, P. K. "Fiber reinforced Composite materials, Manufacturing and Design" New York: Marcel Decker Inc, 1993.

[3] ALkoles, O.M. S., Mahdi , E., Hamouda , A. M. S. , Sahari, B. B. Ellipticity Ratio Effects in the Energy Absorption of Axially crushed composite Tube. Journal of Applied composite material, kluwer academic publishers, 2003, 1-25.

[4] Daining, Tao and Jian, Lu Tiang. Dynamic crashing and impact energy absorption of 3D braided composite tubes, Journal of Materials Letters 2005.V59, 1491–1496.

[5] Russell, A. J. and Street, K. N. Delamination and debonding of materials, ASTM STP 876, 1985, p. 349.

[6] Mamalis, A. G., Manolakos, D. E., Viegelahn, G. L., Min Yap, Sin and Demosthenous G. A. Microscopic failure of thin walled fibre-reinforeed composite frusta under static axial collapse, International Journal of Vehicle Design, 1991, 12, 557.

[7] Zhou and Davies, G. A. Characterization of thick glass woven roving polyester laminates: tension, compression and shear. Int. Journal of composite material, 1995, V26, 579-586

[8] Mamalis, A. G., Manolakos, D. E., Demosthenous, G. A. and Ioannidis, M. B. Axial collapse of thin-walled fiberglass composite tubular components at elevated strain rates, Composites Engineering, 1994, 4, 653.

[9] Price, J. N. and Hull, D. Axial crushing of glass fiber-polyester composite cones, Computer Science Technology, 1987, 28, 211.

[10] Hakim S. Sultan Aljibori, Irfan Anjum Badruddin, A. Badarudin and W.T. Chong. experimental study of composite structures in automotive applications, International journal on Mechanical and Material Engineering *(IJMME)*,V l. 3, 2008, 47-54.

Institute of Physics
Dirac House, Temple Back
Bristol BS1 6BE UK

ISSN: 1742-6588
ISBN 978-1-5108-6407-8
9 781510 864078

CURRAN ASSOCIATES INC.
proceedings.com